CONTENTS

Contributors   ix
Foreword   xi
Preface   xii
SI Units and Conversion Factors   xiii

Section 1   Organization and Management of the Maintenance Function

Chapter 1. Redefining Maintenance—Delivering Reliability   *Scott Franklin*   1.3

Chapter 2. Introduction to the Theory and Practice of Maintenance
R. Keith Mobley   1.9

Chapter 3. Maintenance and Reliability Engineering   *R. Keith Mobley*   1.17

Chapter 4. Cooperative Partnerships   *Jeff Nevenhoven*   1.23

Chapter 5. Effective Maintenance Organizations   *Randy Heisler*   1.31

Chapter 6. Operating Policies of Effective Maintenance   *Tom Dabbs*   1.39

Chapter 7. Six Sigma Safety: Applying Quality Management Principles to Foster a Zero-Injury Safety Culture   *Michael Williamsen*   1.55

Section 2   The Horizons of Maintenance Management

Chapter 1. Corrective Maintenance   *R. Keith Mobley*   2.3

Chapter 2. Reliability-Based Preventive Maintenance   *R. Keith Mobley*   2.7

Chapter 3. Predictive Maintenance   *R. Keith Mobley*   2.19

Chapter 4. Reliability-Centered Maintenance *Darrin Wikoff* 2.35

Chapter 5. Total Productive Maintenance *R. Keith Mobley* 2.41

Chapter 6. Maintenance Repair and Operations—Storeroom Excellence
Wally Wilson 2.59

Chapter 7. Computerized Planning and Scheduling *Thomas A. Gober* 2.79

Chapter 8. Computer-Based Maintenance Management Systems
R. Keith Mobley 2.91

Section 3 Engineering and Analysis Tools

Chapter 1. Economics of Reliability *Robert Fei* 3.3

Chapter 2. Work Measurement *Bruce Wesner* 3.19

Chapter 3. Rating and Evaluating Maintenance Workers *Robert (Bob) Call* 3.65

Chapter 4. Work Simplification in Maintenance *Al Emeneker* 3.89

Chapter 5. Estimating Repair and Maintenance Costs *Tim Kister* 3.107

Chapter 6. Key Performance Indicators *John Cray* 3.121

Chapter 7. Maintenance Engineer's Toolbox *Shon Isenhour* 3.133

Chapter 8. Root Cause Analysis *Darrin Wikoff* 3.153

Section 4 Maintenance of Plant Facilities

Chapter 1. Maintenance of Low-Sloped Membrane Roofs
Donald R. Mapes and Dennis J. McNeil 4.3

Chapter 2. Concrete Industrial Floor Surfaces: Design, Installation,
Repair, and Maintenance *Robert F. Ytterberg* 4.17

Chapter 3. Maintenance and Cleaning of Brick Masonry Structures
Brian E. Trimble — 4.27

Chapter 4. Maintenance of Elevators and Special Lifts *Jerry Robertson* — 4.43

Chapter 5. Air-Conditioning Equipment *Martin A. Scicchitano* — 4.53

Chapter 6. Ventilating Fans and Exhaust Systems *R. Keith Mobley* — 4.87

Chapter 7. Dust-Collecting and Air-Cleaning Equipment
Lee Twombly and Samuel G. Dunkle — 4.111

Section 5 Maintenance of Mechanical Equipment

Chapter 1. Plain Bearings *R. Keith Mobley* — 5.3

Chapter 2. Rolling-Element Bearings *Daniel R. Snyder* — 5.19

Chapter 3. Flexible Couplings for Power Transmission *Terry Hall* — 5.45

Chapter 4. Chains for Power Transmission *Frank B. Kempf* — 5.73

Chapter 5. Cranes: Overhead and Gantry *William S. Chapin* — 5.83

Chapter 6. Chain Hoists *R. C. Dearstyne* — 5.91

Chapter 7. Belt Drives *Dan Parsons and Tim Taylor* — 5.99

Chapter 8. Mechanical Variable-Speed Drives *Carl March* — 5.145

Chapter 9. Gear Drives and Speed Reducers *Robert G. Smith* — 5.161

Chapter 10. Reciprocating Air Compressors *R. Keith Mobley* — 5.185

Chapter 11. Valves *Terry Hall* — 5.197

Chapter 12. Pumps: Centrifugal and Positive Displacement *Carl March* — 5.213

Section 6   Maintenance of Electrical Equipment

Chapter 1. Electric Motors   *Shon Isenhour* — 6.3

Chapter 2. Maintenance of Motor Control Components   *Shon Isenhour* — 6.39

Chapter 3. Maintenance of Industrial Batteries (Lead-Acid, Nickel-Cadmium, Nickel-Iron)   *Terry Hall* — 6.79

Section 7   Instruments and Reliability Tools

Chapter 1. Mechanical Instruments for Measuring Process Variables
R. Keith Mobley — 7.3

Chapter 2. Electrical Instruments for Measuring, Servicing, and Testing
R. Keith Mobley — 7.43

Chapter 3. Vibration: Its Analysis and Correction   *R. Keith Mobley* — 7.69

Chapter 4. An Introduction to Thermography   *R. Keith Mobley* — 7.105

Chapter 5. Tribology   *R. Keith Mobley* — 7.127

Section 8   Lubrication

Chapter 1. The Organization and Management of Lubrication
F. Alverson, T. C. Mead, W. H. Stein, and A. C. Witte — 8.3

Chapter 2. Lubricating Devices and Systems   *Duane C. Allen* — 8.13

Chapter 3. Planning and Implementing a Good Lubrication Program
R. Keith Mobley — 8.27

Section 9   Chemical Corrosion Control and Cleaning

Chapter 1. Corrosion Control   *Denny Bardoliwalla and Klaus Wittel* — 9.3

Chapter 2. Industrial Chemical Cleaning Methods
Robert Haydu, W. Emerson Brantley III, and Jerry Casenhiser — 9.17

Chapter 3. Painting and Protective Coatings *Bryant (Web) Chandler*   9.35

Chapter 4. Piping *Tyler G. Hicks*   9.55

Chapter 5. Scaffolds and Ladders *Colin P. Bennett*   9.91

Section 10 Maintenance Welding

Chapter 1. Arc Welding in Maintenance *J. E. Hinkel*   10.3

Chapter 2. Gas Welding in Maintenance
Engineers of L-TEC Welding and Cutting Systems   10.63

Index   I.1

ABOUT THE EDITORS

R. KEITH MOBLEY is principal of Life Cycle Engineering in Knoxville, Tennessee.

LINDLEY R. HIGGINS was an engineering consultant and senior editor of *Factory* magazine.

DARRIN J. WIKOFF is a senior reliability consultant in Charleston, South Carolina specializing in project management, business process reliability engineering, reliability-centered maintenance, and CMMS/eAM implementations.

CONTRIBUTORS

F. Alverson *Group Leader, Texaco, Inc., Research & Development Department, Port Arthur, Tex.* (SEC. 8, CHAP. 1)

Duane C. Allen *Consultant, LubeCon Systems, Inc., Fremont, Mich.* (SEC. 8, CHAP. 2)

Denny Bardoliwalla *Vice President of Research and Technology, Oakite Products, Inc., Berkeley Heights, N.J.* (SEC. 9, CHAP. 1)

W. Emerson Brantley III *Marketing Director, Bronz-Glow Coatings Corp., Jacksonville, Fla.* (SEC. 9, CHAP. 2)

Colin P. Bennett *Scaffolding Consultant* (SEC. 9, CHAP. 5)

Robert (Bob) Call *Principal, Life Cycle Engineering, Inc., Charleston, S.C.* (SEC. 3, CHAP. 3)

Bryant (Web) Chandler *Cannon Sline, Philadelphia, Pa.* (SEC. 9, CHAP. 3)

William S. Chapin *Director of Engineering, Crane & Hoist Division, Dresser Industries, Inc., Muskegon, Mich.* (SEC. 5, CHAP. 5)

Jerry Casenhiser *Senior Chemist, Bronz-Glow Coatings Corp., Jacksonville, Fla.* (SEC. 9, CHAP. 2)

John Cray *Managing Principal, Life Cycle Engineering Inc., Charleston, S.C.* (SEC. 3, CHAP. 6)

Tom Dabbs *Vice President Life Cycle Engineering, Inc., Charleston, S.C.* (SEC. 1, CHAP. 6)

R. C. Dearstyne *Manager, Product Application, Columbus McKinnon Corporation, Amherst, N.Y.* (SEC. 5, CHAP. 6)

Samuel G. Dunkle *Manager, Electrostatic Precipitators and Fabric Collectors, SnyderGeneral Corporation (American Air Filter), Louisville, Ky.* (SEC. 4, CHAP. 7)

Al Emeneker *Work Control SME, Life Cycle Engineering, Inc., Charleston, S.C.* (SEC. 3, CHAP. 4)

Engineers of L-TEC Welding and Cutting Systems *Florence, S.C.* (SEC. 10, CHAP. 2)

Robert Fei *Managing Principal, Life Cycle Engineering, Inc., Charleston, S.C.* (SEC. 3, CHAP. 1)

Scott Franklin *Senior V.P., Life Cycle Engineering, Inc., Charleston, S.C.* (SEC. 1, CHAP. 1)

Thomas A. Gober *Maintenance Planning Consultant, Life Cycle Engineering, Inc., Charleston, S.C.* (SEC. 2, CHAP. 7)

Tyler G. Hicks *Mechanical Engineer, Rockville Centre, N.Y.* (SEC. 9, CHAP. 4)

Terry Hall *Reliability Engineer, Life Cycle Engineering, Inc., Charleston, S.C.* (SEC. 5, CHAPS. 3 & 11; SEC. 6, CHAP. 3)

Robert Haydu *NACE, ASHRAE President and Chief Chemist, Bronz-Glow Coatings Corp., Jacksonville, Fla.* (SEC. 9, CHAP. 2)

Randy Heisler *Managing Principal, Life Cycle Engineering, Inc., Charleston, S.C.* (SEC. 1, CHAP. 5)

J. E. Hinkel *The Lincoln Electric Company, Cleveland, Ohio* (SEC. 10, CHAP. 1)

Shon Isenhour *Principal, Life Cycle Engineering, Inc., Charleston, S.C.* (SEC. 3, CHAP. 7; SEC. 6, CHAPS. 1 & 2)

Frank B. Kempf *Division Marketing Manager, Drives & Components Division, Morse Industrial Corporation, a subsidiary of Emerson Electric Company, Ithaca, N.Y.* (SEC. 5, CHAP. 4)

Tim Kister Maintenance Planning SME, Life Cycle Engineering, Inc., Charleston, S.C. (SEC. 3, CHAP. 5) Louisville, Ky. (SEC. 4, CHAP. 7)

T. C. Mead Senior Technologist (Ret.), Texaco, Inc., Research & Development Department, Port Arthur, Tex. (SEC. 8, CHAP. 1)

Donald R. Mapes Building Technology Associates, Inc., Glendale, Ariz. (SEC. 4, CHAP. 1)

Dennis J. McNeil Construction Consultants, Inc., Homewood, Ill. (SEC. 4, CHAP. 1)

R. Keith Mobley Principal, Life Cycle Engineering, Inc., Charleston, S.C. (SEC. 1, CHAPS. 2 & 3; SEC. 2, CHAPS. 1, 2, 3, 5 & 8; SEC. 4, CHAP. 6; SEC. 5, CHAP. 1 & 10; SEC. 7, CHAPS. 1, 2, 3, 4 & 5; SEC. 8, CHAP. 3)

Carl March Reliability Engineer Life Cycle Engineering, Inc., Charleston, S.C. (SEC. 5, CHAPS. 8 & 12)

Jeff Nevenhoven Senior Consultant, Life Cycle Engineering, Inc., Charleston, S.C. (SEC. 1, CHAP. 4)

Dan Parsons Application Engineer, Gates Corporation, Denver, Colo. (SEC. 5, CHAP. 7)

Jerry Robertson Maintenance Quality Engineer, Otis Elevator Company, Farmington, Conn. (SEC. 4, CHAP. 4)

Martin A. Scicchitano Carrier Air Conditioning Company, Syracuse, N.Y. (SEC. 4, CHAP. 5)

W. H. Stein Group Leader, Texaco, Inc., Research & Development Department, Port Arthur, Tex. (SEC. 8, CHAP. 1)

Robert G. Smith Director of Engineering, Philadelphia Gear Corporation, King of Prussia, Pa. (SEC. 5, CHAP. 9)

Daniel R. Snyder SKF USA, Inc., King of Prussia, Pa. (SEC. 5, CHAP. 2)

Tim Taylor Application Engineer, Gates Corporation, Denver, Colo. (SEC. 5, CHAP. 7)

Brian E. Trimble E.I.T., Brick Institute of America, Reston, Va. (SEC. 4, CHAP. 3)

Lee Twombly Manager, Scrubber and Mechanical Collectors, SnyderGeneral Corporation (American Air Filter), Louisville, Ky. (SEC. 4, CHAP. 7)

Bruce Wesner Managing Principal, Life Cycle Engineering, Inc., Charleston, S.C. (SEC. 3, CHAP. 2)

Darrin Wikoff Principal, Life Cycle Engineering, Inc., Charleston, S.C. (SEC. 2, CHAP. 4; SEC. 3, CHAP. 8)

Michael Williamsen Senior Consultant, Core Media Training Solutions, Portland, Oreg. (SEC. 1, CHAP. 7)

Wally Wilson Materials SME, Life Cycle Engineering, Inc., Charleston, S.C. (SEC. 2, CHAP. 6)

Klaus Wittel Manager of Technology Transfer, Oakite Products, Inc., Berkeley Heights, N.J. (SEC. 9, CHAP. 1)

A. C. Witte Consultant, Texaco, Inc., Research & Development Department, Port Arthur, Tex. (SEC. 8, CHAP. 1)

Robert F. Ytterberg President, Kalman Floor Company, Evergreen, Colo. (SEC. 4, CHAP. 2)

FOREWORD

Some engineering fields change dramatically from year to year, with radical breakthroughs in technology happening often. These fields may have hundreds or more papers and texts published each year on the latest best practices. Maintenance engineering is a field which, for the most part, hasn't fundamentally changed much over the years. And there aren't many sources for the latest information or best practices.

But in recent years, maintenance engineering has, more and more, put an emphasis on true reliability. A business which is asset-intensive, such as manufacturing, relies on a reliability-centered field of engineering to be successful. In my opinion, reliability engineering itself has become a technology used for the purpose of improving manufacturing capacity, without capital investment.

The *Maintenance Engineering Handbook* has long been regarded as the premier source for expertise on maintenance theory and practices for any industry. This text has been considered invaluable and now, this latest edition defines those practices that are critical to developing an effective *reliability* engineering function within your business.

This text is no longer just about mechanical, electrical, and civil maintenance engineering. Instead, the seventh edition also focuses on recognized and proven best practices in maintenance, repair, and overhaul (MRO) inventory management, root-cause analysis, and performance management. Keith Mobley, the editor in chief of this text, has more than 35 years of direct experience in corporate management, process and equipment design, and reliability-centered maintenance methodologies. For the past 16 years, he has helped hundreds of clients across the globe achieve and sustain world-class performance through the implementation of maintenance and reliability engineering principles.

You may spend your career worrying about excessive downtime and high maintenance costs as a result of repetitive failures. As a fellow veteran maintenance and reliability engineer, I encourage you to recognize that this field is changing and improvements are being made that empower today's business leaders. This text can help you reap the benefits of those changes so that your hard work produces the best possible results.

JAMES R. FEI, PE
CEO, Life Cycle Engineering, Inc.
Charleston, S.C.

PREFACE

This "Maintenance Engineering Handbook" is written, almost exclusively, by those people who have had to face the acute never-ending problems of equipment failures, repairs, and upkeep, day by day, hour by hour, midnight shift by midnight shift. They understand better than most the extraordinary demands that every maintenance manager, planner, and craftsperson must face and overcome to meet the everchanging maintenance requirements of today's plant.

It is the function of "Maintenance Engineering Handbook" to pass along invention, ingenuity, and a large dose of pure basic science to you, the user. This then is your key, your guide, and your chief support in the tempestuous battle of Maintenance in the days and years ahead.

Lindley R. Higgins, as editor-in-chief of the first five editions of this handbook, established a standard for excellence that we have attempted to maintain in this seventh edition. Through the excellent help of maintenance professionals, we have updated those sections that were in the earlier editions and have added new topics that we believe will help you survive in the battle against excessive downtime, high maintenance costs, and the myriad other problems that you as a maintenance professional must face each day.

R. Keith Mobley

SI UNITS AND CONVERSION FACTORS

ACCELERATION

feet per second per second = 30.48 centimeters per second per second
= 0.3048 meters per second per second

free fall (standard) = 9.8067 meters per second per second

kilometers per hour per second = 27.778 centimeters per second per second
= 0.9113 feet per second per second
= 0.27778 meters per second squared

ANGULAR

circumferences = 6.283 radians

degrees = 1.111 grade
= 0.017453 radians

degrees per second = 0.017453 radians per second
= 0.16667 revolutions per minute
= 0.0027778 revolutions per second

minutes = 0.002909 radians

radians = 57.296 degrees (angular)

radians per second = 57.296 degrees per second (angular)
= 9.549 revolutions per minute

revolutions per minute = 6 degrees per second
= 0.01472 radians per second

AREA

acres = 43560 square feet
= 4046.9 square meters
= 0.40469 hectares

circular mils = 0.000007854 square inches

hectares = 2.471 acres
= 107639 square feet
= 10000 square meters

square centimeters = 0.155 square inches

square feet = 0.000022956 acres
= 0.092903 square meters

square inches = 6.4516 square centimeters
square kilometers = 247.1 acres
= 0.3861 square miles
square meters = 0.0002471 acres
= 10.764 square feet
square miles = 640 acres
= 2.59 square kilometers
square yards = 0.00020661 acres
= 0.83613 square meters

CANDLEPOWER

foot candles = 10.764 lumens per square meter

CAPACITY, DISPLACEMENT

cubic inches per revolution = 0.01639 liters per revolution
= 16.39 milliliters per revolution

DENSITY, MASS/VOLUME

grams per cubic centimeter = 0.001 kilograms per cubic meter
= 0.03613 pounds per cubic inch
= 62.427 pounds per cubic foot
pounds-mass per cubic foot = 16.018 kilograms per cubic meter
pounds per cubic foot = 0.016018 grams per cubic centimeter
= 16.018 kilograms per cubic meter
= 0.0005787 pounds per cubic inch
pounds per cubic inch = 27.68 grams per cubic centimeter
= 27.68 kilograms per cubic meter
= 1728 newtons per meter

ENERGY AND WORK

British thermal units = 1055 joules
British thermal units per second = 1.055 watts
British thermal units per minute = 0.02358 horsepower
= 17.58 watts
British thermal units per hour = 0.2931 watts
calories = 0.0039683 British thermal units
= 3.088 foot-pounds
= 4.1868 joules
= 0.4265 kilogram-meters
= 0.001163 watt-hours
ergs = 0.0000001 joules

foot-pounds-force = 0.001285 British thermal units
= 0.3238 calories
= 0.000000505 horsepower hours
= 1.3558 joules
= 0.0003238 kilocalories
= 0.13825 kilogram-force meters
= 3.766E-07 kilowatt hours

horsepower = 42.43 British thermal units per minute
= 33000 foot-pounds-force per minute
= 550 foot-pounds-force per second
= 10.69 kilocalories per minute
= 0.7457 kilowatts
= 1.0139 horsepower (metric)
= 745.7 watts

horsepower-boiler = 33479 British thermal units per horse
= 9.8095 kilowatts
= 34.5 pounds of water evaporated per hour at 212°F

horsepower hours = 2.545 British thermal units
= 1980000 foot-pounds-force
= 2684500 joules
= 641.5 kilocalories
= 273200 kilogram-force meters
= 0.7457 kilowatt hours

joules = 0.0009484 British thermal units
= 0.239 calories
= 0.73756 foot-pounds-force
= 0.00027778 watt-hours

kilowatts = 56.92 British thermal units per minute
= 44254 foot-pounds-force per minute
= 737.6 foot-pounds-force per second
= 1.341 horsepower
= 14.34 kilocalories per minute

kilowatt hours = 3413 British thermal units
= 2655000 foot-pounds-force
= 1.341 horsepower hours
= 3.6 joules
= 860 kilocalories
= 367100 kilograms-force meters

pounds-force = 0.45359 kilograms-force
= 4.4482 newtons

tons of refrigeration = 12000 British thermal units per hour
= 288000 British thermal units per 24 hours

watts = 0.05691 British thermal units per minute
= 0.73756 foot-pounds-force per second
= 44.254 foot-pounds-force per minute
= 0.001341 horsepower
= 1 joules per second
= 0.01434 kilocalories per minute

watt-hours = 3.413 British thermal units
= 2665 foot-pounds-force
= 0.001341 horsepower hours
= 3600 joules
= 0.8604 kilocalories
= 367.1 kilograms-force-meters

There are several definitions of the Btu, and the values of applicable and/or equivalent factors may vary slightly depending on the definition used. For this reason, three or four significant figures are given on this page, and in most cases provide a value near to most definitions of the Btu. However, as always in making calculations of high accuracy, one should reference the appropriate lists and handbooks of standards.

ENERGY/AREA TIME

Btu/cubic feet second = 11348 watts per square meter
Btu/cubic feet hour = 3.1525 watts per square meter

FLOW RATE MASS

pounds per minute = 0.4536 kilograms per minute

FLOW RATE VOLUME

cubic feet per minute = 471.9 cubic centimeters per second
  = 0.0004719 cubic meters per second
  = 1.699 cubic meters per hour
  = 0.4719 liters per second
  = 0.2247 gallons (US) per second
  = 62.32 pounds of water per minute (at 68°F)
cubic feet per second = 0.028317 cubic meters per second
  = 1.699 cubic meters per minute
  = 101.9 cubic meters per hour
  = 448.8 gallons (US) per minute
  = 646315 gallons (imp) per hour
  = 28.32 liters
cubic meters per hour = 0.016667 cubic meters per minute
  = 0.00027778 cubic meters per second
  = 4.4033 gallons (US) per minute
  = 0.27778 liters per second
cubic meters per second = 3600 cubic meters per hour
  = 15850 gallons (US) per minute
gallons (US) per minute = 0.00006309 cubic meters per second
  = 0.0037854 cubic meters per minute
  = 0.2771 cubic meters per hour
  = 0.002228 cubic feet per second
  = 8.021 cubic feet per hour
  = 0.06309 liters per second
liters per minute = 0.0005885 cubic feet per second
  = 0.01667 liters per second
  = 0.004403 gallons (US) per second
  = 0.26418 gallons (US) per minute
  = 0.003666 gallons (imp) per minute
liters per second = 0.001 cubic meters per second
  = 0.06 cubic meters per minute
  = 3.6 cubic meters per hour

= 60 liters per minute
= 15.85 gallons (US) per minute
= 13.2 gallons (imp) per minute

pounds of water per minute at 60°F = 7.5667 cubic centimeters per second
= 0.0002675 cubic feet per second
= 0.00045398 cubic meters per minute
= 0.0075599 kilograms per second

standard cubic feet per minute = 1.6957 cubic meters per hour at STP
= 0.47103 liters per second at STP

stokes = 0.001076 square feet per second
= 0.0001 square meters per second

tons (short) of water per 24 hours at 60°F = 1.338 cubic feet per hour
= 0.03789 cubic meters per hour
= 0.1668 gallons (US) per minute
= 83.333 pounds of water per hour

FORCE

dynes = 0.00001 newtons

grams-force = 0.0098066 newtons

kilograms-force = 9.8066 newtons
= 2.2046 pounds-force

kilopounds = 9.807 newtons
= 1 kilograms-force
= 2.2046 pounds-force
= 70.932 poundals
= 0.002205 kips

kips = 4448 newtons
= 453.6 kilograms-force
= 1000 pounds-force
= 32174 poundal
= 453.6 kilopond

grams-force per centimeter = 98.07 newtons per meter
= 0.0056 pounds-force per inch

kilograms-force per meter = 9.8066 newtons
= 0.6721 pounds-force per foot

newtons = 100000 dynes
= 0.10197 kilograms-force
= 7.233 poundals
= 0.2248 pounds-force

poundals = 0.13826 newtons

pounds-force = 4.448 newtons

LENGTH

centimeters = 0.3937 inches

fathoms = 6 feet
= 1.8288 meters

feet = 30.48 centimeters
 = 12 inches
 = 0.3048 meters
 = 0.3333 yards

inches = 2.54 centimeters
 = 0.0254 meters
 = 25.4 millimeters
 = 25.4 micrometers

kilometers = 3280.8 feet
 = 0.62137 miles

meters = 3.2808 feet
 = 39.37 inches
 = 1.0936 yards

micrometers = 0.000001 meters

millimeters = 0.03937 inches

mills = 0.0254 millimeters

miles = 5280 feet
 = 1.6093 kilometers
 = 1609.3 meters
 = 1760 yards

statute miles = 1.609 kilometers

yards = 0.9144 meters

MASS/WEIGHT

drams (avoir) = 27.344 grains
 = 1.7718 grams
 = 0.0625 ounces

grains = 0.0648 grams
 = 0.0022857 ounces (avoir)

grams = 15.432 grains
 = 0.035274 ounces (avoir)
 = 0.0022046 pounds (avoir)

kilograms = 2.2046 pounds
 = 0.0011023 tons (short)

metric tons (tonnes) = 1000 kilograms
 = 2204.6 pounds

ounces (avoir) = 16 drams (avoir)
 = 437.5 grains
 = 28.3495 grams
 = 0.02835 kilograms
 = 0.0625 pounds-avoir
 = 0.0000279 tons (long)
 = 0.00002835 metric tons

pounds (avoir) = 256 drams (avoir)
 = 7000 grains
 = 453.59 grams
 = 0.45359 kilograms

$\quad$ = 16 ounces (avoir)
$\quad$ = 0.00045359 metric tons
$\quad$ = 0.00044643 tons (long)
$\quad$ = 0.0005 tons (short)

tons (long) = 1016 kilograms
$\quad$ = 1.016 metric tons
$\quad$ = 2240 pounds-avoir
$\quad$ = 1.12 tons (short)

tons (short) = 907.18 kilograms
$\quad$ = 2000 pounds-avoir
$\quad$ = 0.89286 tons (long)
$\quad$ = 0.9072 metric tons

short ton = 0.9072 metric ton/tonne

POWER

Btu/hour = 0.2931 watt
Btu/second = 1055 watt
Horsepower = 0.746 kilowatt

PRESSURE AND STRESS

atmosphere = 1.01325 bars
$\quad$ = 76 centimeters of mercury at 32°F
$\quad$ = 33.96 feet of water at 68°F
$\quad$ = 29.921 inches of mercury at 32°F
$\quad$ = 1.0332 kilograms-force per square centimeter
$\quad$ = 103322 kilograms-force per square meter
$\quad$ = 101.325 kilopascals
$\quad$ = 14.696 pounds-force per square inch
$\quad$ = 1.0581 tons-force (short) per square foot
$\quad$ = 760 torr

bars = 100 kilopascals

centimeters of mercury = 0.013158 atmospheres
$\quad$ = 0.01333 bars
$\quad$ = 0.4468 feet of water at 68°F
$\quad$ = 5.362 inches of water at 68°F
$\quad$ = 0.19337 kilograms-force per square centimeter
$\quad$ = 27.85 pounds-force per square inch
$\quad$ = 10 torr

feet of water (at 68°F) = 0.02945 atmospheres
$\quad$ = 0.02984 bars
$\quad$ = 0.8811 inches of mercury (at 0°C)
$\quad$ = 0.03042 kilograms-force per square centimeter
$\quad$ = 2.984 kilopascals
$\quad$ = 0.4328 pounds-force per square inch
$\quad$ = 62.32 pounds-force per square foot

inches of mercury at 0°C = 0.00342 atmospheres (standard)
 = 0.033864 bars
 = 1.135 feet of water at 68°F
 = 13.62 inches of water at 68°F
 = 0.034532 kilograms-force per square centimeter
 = 345.32 kilograms-force per square meter
 = 3.3864 kilopascals
 = 25.4 millimeters of mercury
 = 70.73 pounds-force per square foot
 = 0.4912 pounds-force per square inch

inches of water at 68°F = 0.002454 atmosphere
 = 0.002487 bars
 = 0.07342 inches of mercury
 = 0.002535 kilograms-force per square centimeter
 = 0.2487 kilopascals
 = 0.577 ounces-force per square inch
 = 5.193 pounds-force per square foot
 = 0.03606 pounds-force per square inch

kilograms-force per square centimeter = 0.9678 atmospheres
 = 0.98066 bars
 = 32.87 feet of water at 68°F
 = 28.96 inches of mercury at 0°C
 = 98.066 kilopascals
 = 2048 pounds-force per square foot
 = 14.223 pounds-force per square inch

kilograms-force per square millimeter = 1000000 kilograms-force per square meter
 = 9.8066 megapascals

kilograms per square meter = 9.807 pascals

kilopascals = 10000 dynes per square centimeter
 = 0.3351 feet of water at 68°F
 = 0.2953 inches of mercury at 32°F
 = 4.021 inches of water at 68°F
 = 0.010197 kilograms-force per square centimeter
 = 1000 pascals
 = 0.145 pounds-force per square inch

kips per square inch = 6894.8 kilopascals
 = 70.307 kilograms-force per square centimeter
 = 68.94 bars
 = 1000 pounds per square inch

megapascals = 0.10197 kilograms-force per square millimeter
 = 10.197 kilograms-force per square centimeter
 = 1000 kilopascals
 = 1000000 pascals
 = 145 pounds-force per square inch

millibars = 100 pascals

millimeters of mercury at 0°C = 0.0013332 bars
 = 0.00468 feet of water at 68°F
 = 0.03937 inches of mercury
 = 0.53616 inches of water at 68°F
 = 0.0013595 kilograms per square centimeter
 = 133.32 pascals
 = 0.0193368 pounds per square inch

ounces-force per square inch = 4.395 grams-force per square centimeter
= 43.1 pascals
= 0.0625 pounds-force per square inch

pascals = 0.00001 bars
= 10 dynes per square centimeter
= 0.010197 grams-force per square centimeter
= 0.000010197 kilograms-force per square centimeter
= 0.001 kilopascals
= 1 newtons per square meter

pascals = 0.000145 pounds-force per square inch

poise = 100 centipoises
= 0.1 pascal-seconds
= 0.0020886 pound-force-seconds per square foot
= 0.06721 pounds per foot-second

pounds-force per square foot = 0.01605 feet of water at 68°F
= 0.0004882 kilograms-force per square centimeter
= 0.004788 kilopascals
= 47.88 pascals
= 0.0069444 pounds-force per square inch

pounds-force per square inch = 0.06805 atmospheres
= 2.311 feet of water at 68°F
= 27.73 inches of water at 68°F
= 2.036 inches of mercury at 0°F
= 0.07031 kilograms force per square centimeter
= 6.8948 kilopascals

pounds per square foot = 4.8824 kilograms per square meter

pounds per square inch = 6895 pascals
= 6.895 kilopascals
= 0.006895 megapascals

THERMAL CONDUCTIVITY

Btu · inch · hour · feet^2 · °F = 0.1442 watt per meter^2 °K

TORQUE: BENDING MOMENT

pound feet = 1.356 newton meters

kilogram meters = 9.807 newton meters

VELOCITY

centimeters per second = 0.03281 feet per second
= 1.9685 feet per minute
= 0.02237 miles per hour
= 0.036 kilometers per hour
= 0.6 meters per minute

feet per minute = 0.508 centimeters per second
= 0.01829 kilometers per hour
= 0.3048 meters per minute
= 0.00508 meters per second
= 0.01136 miles per hour

feet per second = 30.48 centimeters per second
= 1.097 kilometers per hour
= 18.29 meters per minute
= 0.3048 meters per second
= 0.6818 miles per hour

international knots = 0.5144 meters per second
= 1.1516 miles per hour

kilometers per hour = 27.778 centimeters per second
= 0.9113 feet per second
= 54.68 feet per minute
= 0.53996 international knots
= 16.667 meters per minute
= 0.27778 meters per second
= 0.6214 miles per hour

kilometers per second = 37.28 miles per minute

meters per minute = 1.6667 centimeters per second
= 3.2808 feet per minute
= 0.05468 feet per second
= 0.06 kilometers per hour
= 0.03728 miles per hour

meters per second = 196.8 feet per minute
= 3.281 feet per second
= 3.6 kilometers per hour
= 0.06 kilometers per minute
= 2.237 miles per hour
= 0.03728 miles per minute

miles per hour = 44.7 centimeters per second
= 88 feet per minute
= 1.4667 feet per second
= 0.869 international knots
= 1.6093 kilometers per hour
= 26.82 meters per minute

VOLUME

acre-feet = 43.56 cubic feet
= 325851 gallons (US)
= 1233.5 cubic meters

barrels (US liquid) = 31.5 gallons (US)

barrels (oil) = 0.11924 cubic meters
= 42 gallons of oil
= 0.15899 cubic meters

cubic centimeters = 0.06102 cubic inches
= 0.000035315 cubic feet
= 0.000001308 cubic yards
= 0.0002642 gallons (US)
= 0.00022 gallons (imp)
= 0.001 liters

cubic feet = 28317 cubic centimeters
= 0.028317 cubic meters
= 1728 cubic inches
= 0.03704 cubic yards
= 7.4805 gallons (US)
= 6.229 gallons (imp)
= 28.32 liters

cubic inches = 16.387 cubic centimeters
= 0.0005787 cubic feet
= 0.000016387 cubic meters
= 0.00002143 cubic yards
= 0.004329 gallons (US)
= 0.03605 gallons (imp)
= 0.016387 liters

cubic meters = 61024 cubic inches
= 35.315 cubic feet
= 1.308 cubic yards
= 264.17 gallons (US)
= 219.97 gallons (imp)
= 1000 liters

cubic yards = 764550 cubic centimeters
= 27 cubic feet
= 46.656 cubic inches
= 0.76455 cubic meters
= 201.97 gallons (US)
= 168.17 gallons (imp)
= 764.55 liters

fluid ounces (US) = 1.8046 cubic inches
= 0.02957 liters

gallons (imp) = 4546.1 cubic centimeters
= 0.0045461 cubic meters
= 0.16054 cubic feet
= 0.005946 cubic yards
= 1.20094 gallons (US)
= 4.5461 liters
= 10 pounds of water at 62°F

gallons (US) = 3785.4 cubic centimeters
= 0.0037854 cubic meters
= 231 cubic inches
= 0.13368 cubic feet
= 0.0049515 cubic yards
= 8 pints (liquid)
= 4 quarts (liquid)
= 0.8327 gallons (imp)
= 3.7854 liters
= 8.338 pounds of water at 60°F

liters = 1000 cubic centimeters
= 0.035315 cubic feet
= 61.024 cubic inches
= 0.001 cubic meters
= 0.001308 cubic yards
= 0.26418 gallons (US)
= 0.22 gallons (imp)

pounds-mass of water at 60°F = 454 cubic centimeters
    = 0.01603 cubic feet
    = 27.7 cubic inches
    = 0.11993 gallons (US)
    = 0.45398 liters

quarts (dry) = 1101.2 cubic centimeters
    = 67.2 cubic inches
    = 0.0011012 cubic meters

quarts (liquid) = 946.35 cubic centimeters
    = 57.75 cubic inches
    = 0.94635 liters

WATER HARDNESS

grains per gallon (imp) = 14.25 grams per cubic meter
    = 0.01425 kilograms per cubic meter
    = 14.25 parts per million by weight in water

grains per gallon (US) = 17.118 grams per cubic meter
    = 0.017118 kilograms per cubic meter
    = 17.118 parts per million by weight in water
    = 142.9 pounds per million gallons

grains per liter = 58.417 grains per US gallon
    = 1000 parts per million by mass weight in water
    = 0.0622427 pounds per cubic foot
    = 8.3544 pounds per 1000 gallons (US)

milligrams per liter = 1 parts per million

parts per million by mass = 0.0583 grains per gallon (US) at 60°F
    = 0.07 grains per gallon (imp) at 62°F
    = 0.9991 grams per cubic meter at 15°C
    = 1 milligrams per liter
    = 8.328 pounds per million gallons (US) at 60°F

SECTION 1

ORGANIZATION AND MANAGEMENT OF THE MAINTENANCE FUNCTION

CHAPTER 1
REDEFINING MAINTENANCE—DELIVERING RELIABILITY

Scott Franklin
Senior V.P., Life Cycle Engineering, Inc., Charleston, S.C.

In today's competitive environment business sustainability requires manufacturers to capitalize on every possible advantage. Companies often pursue *lean manufacturing* as a means for gaining competitive advantage. Similarly, numerous firms drive initiatives to attain excellence in maintenance and reliability. Unfortunately, few companies address the significant synergies of lean and maintenance excellence that the power of the combination of lean manufacturing and lean maintenance. The concepts presented here are not just theory. They have been proven through more than 300 maintenance step-change efforts in more that 300 Fortune 500 companies and lean implementations in the automotive, consumer products, foods, chemicals, pharmaceuticals, and power generation industries. They represent learnings from more than 25 million hours of experience annually in facility operations, maintenance, and technical support.

The three most common metrics for asset performance are RoNA, RoCE, and EVA:

- *Return on Net Assets* (RoNA) is the earnings before interest and taxes (EBIT) divided by asset book value. RoNA is closely linked to share value for heavy industry corporations
- *Return on Capital Employed* (RoCE) can be calculated in several ways. A good method used by some Fortune 100 companies for new facilities is *Net Present Value* (NPV) divided by the initial facility investment
- Many other successful companies use *Economic Value Added* (EVA), which discounts all related asset cash flow. Figure 1.1 illustrates the primary elements of asset performance. For step changes in asset performance, we need to focus on assets, people, materials, working capital, and capital investment. Too often, companies try to drive these elements by headcount and budget reductions. This rarely generates sustainable asset performance improvements. Rather than demand financial improvements, cut heads, and let those who remain figure out how to proceed, proper step change is about finding work process improvements to drive improved results and financials.

The best approach we have found is to focus on lean, maintenance and reliability improvements simultaneously. Simply put, we must stabilize production processes through equipment and process reliability while we challenge work-in-process, raw materials, and finished goods buffers. By applying lean tools to maintenance, we enhance the synergies achieved by integrating lean.

Every manufacturing facility wants production systems and equipment to operate and be operated in a reliable fashion. When the equipment does what it needs to do when it needs to do it, plant output and profitability is maximized. No organization wants its production systems or processes to break down, to produce poor quality products, or to operate inefficiently. We want

1.4 ORGANIZATION AND MANAGEMENT OF THE MAINTENANCE FUNCTION

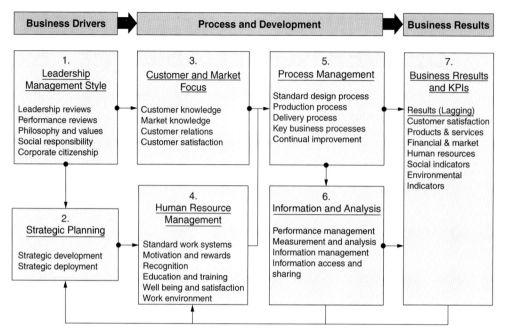

FIGURE 1.1 Primary Elements of Performance.

them to operate perfectly. Unfortunately, we do not live in an ideal world; no physical asset operates flawlessly forever. In most organizations, breakdowns are the norm. Quality and productivity losses are high. Scheduled shipments are missed. Since the majority of these deficiencies are manifest as equipment-related problems, for example, breakdowns or maintenance-related corrective actions, maintenance is too often blamed for all problems that plague most plants, facilities, and corporations. In truth, the reasons for these inherent problems are shared by all functional groups.

The only time anyone pays attention to maintenance is when production demands that they "get it running again, and quickly!" The majority of work is done on a reactive basis. Performing sustaining levels of maintenance is a fundamental requirement of long-term survivability of all plants. Ignoring this requirement is a guarantee that the plant will incur unacceptably, ever-increasing higher operating cost that will assure the loss of the ability to compete in today's world market.

The role of maintenance must change to support the growing worldwide competition. It can no longer limit its role to immediate reaction to emergencies and overpower problems with more bodies and excessive overtime. There is a better way. If the right systems, infrastructure, processes, and procedures are in place and consistently executed well, losses can be minimized; the operation will become stable; production output will be maximized; and consistently high product quality will become the norm. We call this a state of *maintenance excellence*. Maintenance excellence is a subset of *reliability excellence* and redefines the traditional roles and responsibilities, as well as the maintenance processes that are necessary to assure asset reliability, maximum asset useful life and best life cycle asset cost. Under the reliability excellence umbrella the maintenance function becomes an equal partner within the corporation's operation. It is run like any other for-profit business and expected to meet its critical contribution to a fully integrated plant organization.

Achieving high reliability in manufacturing and maintenance operations minimizes waste, maximizes output, as well as minimizes cost. It allows us to get the most out of the assets we have. By redefining the role of maintenance as part of a total plant reliability program provides the infrastructure, processes, and employee involvement that result in improved throughput and lower total cost of goods sold (COG). Specifically, changes such as lower production unit cost, reduced maintenance cost, better process stability, and the like.

LOWER PRODUCTION UNIT COST

Production unit cost is one of the most critical variables impacting an organization's profitability. It is calculated simply as the sum of all manufacturing cost divided by the production volume. Improved asset reliability impacts production unit costs in two ways—by reducing the numerator and by increasing the denominator.

Ensuring that resources such as labor, materials, energy, and fixed costs are used efficiently minimizes expenses. While a major component of these costs is fixed, increasing throughput will decrease the unit cost of production. Base labor cost will remain constant even when production throughput is increased; incremental cost for materials and energy is also reduced as volume increases.

Eliminating losses as described above ensures that production volume is maximized. Even if the additional volume is not needed to support the business, eliminating losses enables an organization to reduce the operating schedule or reduce the production asset base, which further reduces fixed cost.

REDUCED MAINTENANCE COSTS

Improved reliability results in lower maintenance costs. If the assets are not breaking down, a greater percentage of maintenance work can be performed in a planned and scheduled manner, which enables the workforce to be at least twice as efficient. Reducing these losses will also result in requirement of

- Fewer spare parts
- Less overtime
- Fewer contractors

All of these result in significant reductions in maintenance spending. It is not unusual for organizations to experience as much as a 50 percent reduction in maintenance cost as a result of moving from a reactive style of management to a proactive approach.

BETTER PROCESS STABILITY

Equipment breakdowns inevitably result in process upsets. It is difficult to have a stable, optimized process when the production equipment is constantly failing. This inevitably results in problems with final product quality. When reliability is improved, process variability is reduced, and statistical process capability (CpK) is increased. This results in the capability to have a more stable, predictable manufacturing process.

EXTENDED EQUIPMENT LIFE

Many organizations spend an excessive amount of capital funds to replace equipment that failed far earlier than it should have. If routine maintenance is continually deferred due to production demands or resource limitations, the organization is in fact mortgaging the future value of the asset—taking the capital value from the future and spending it today. The end result is a wasted asset that must be replaced. The financial result is excessive write-off expenses and a requirement for a constant infusion of new capital.

Organizations that place a priority on reliability recognize that newer is not necessarily better, and that a small amount of investment in routine care can pay big dividends in extended equipment life. This frees up capital to be used for more productive purposes, such as expansion or to implement new technology.

REDUCED MAINTENANCE SPARE PARTS INVENTORY

All organizations require some level of spare parts inventory to ensure the right parts will be available when needed. Reactive organizations typically find themselves carrying a large quantity of inventory because they cannot predict when the parts will be needed. This ties up working capital and results in excessive carrying costs. Organizations that take a proactive approach to reliability place a high value in knowing the condition of their assets. The need for parts is much more predictable. There are fewer "surprises"; more parts can be purchased on a just-in-time basis. Since the volume of inventory required is based to a large degree on usage, the fewer parts we use, the fewer we need to keep on hand.

REDUCED OVERTIME

Reactive organizations can never predict when a critical equipment failure will occur. Murphy's law typically applies; it will invariably happen at the most inconvenient time and will require craft resources to be called into the facility to correct the problem. To counter this reality, most reactive organizations have a large percentage of the maintenance workforce spread across all operating shifts "just in case" a failure occurs. In this situation, the equipment is in control, not management. Large amounts of overtime are experienced. In organizations that focus on reliability, breakdowns are much less common. A larger percentage of craft resources are on day shift where adequate staff supports is available to increase their productivity. Fewer resources are waiting for breakdowns to occur because equipment condition is known and early warning signs of distress are heeded.

OTHER BENEFITS

In addition to the reduced cost and increased throughput, for example, capacity, *reliability excellence* provides other benefits that improve the overall performance of the plant.

Improved Sense of Employee Ownership

In most reactive organizations, employees don't exhibit a sense of pride in the workplace. The high frequency of equipment failures demands that more attention is paid to making repairs and managing the consequences of equipment failures than to routine preventive maintenance and housekeeping. Dirt and contamination is widespread; little attention is paid to cleanliness. In proactive organizations, however, it is realized that basic equipment care is one of the most critical elements affecting equipment reliability. Emphasis is placed on routine cleaning, inspection for deteriorating conditions, and basic lubrication. In most cases, this is done by the personnel operating the equipment and is a fundamental job expectation. As they take an interest in the condition of equipment, they tend to develop a sense of ownership—in the appearance of the equipment and its operating performance.

Improved Employee Safety

Several studies have indicated that asset reliability and employee safety are closely correlated. When the operations are unstable as in a breakdown environment, employees are often placed in awkward situations. They often take shortcuts in an effort to get the plant back up and running, which increases the likelihood of an injury. In a culture that values reliability, however, these situations are minimized. Additionally, the same behaviors that result in improved reliability—the discipline to follow procedures, attention to detail, and the perseverance needed to find the root causes of problems—result in improved employee safety.

Reduced Risk of Environmental Issues

Equipment failures in many chemical processes can result in releases of hazardous substances to the environment. If we improve equipment reliability, we reduce the risk of environmental releases. In fact, a specific requirement of the OSHA 1901.119 Process Safety statute is that the mechanical integrity of equipment containing hazardous chemical substances must be maintained. Even if the facility is not required to meet the Process Safety statute, there still may be equipment covered by state and local environmental permits. In all cases, the same systems and procedures that protect the reliability of production equipment will protect permitted equipment as well, greatly reducing the risk environmental releases.

CONTINUOUS IMPROVEMENT

No organization can afford to accept its current level of performance or competitive pressures will eventually drive it out of business. An organization must continue to improve. One key element of reliability excellence is an organizational focus on continuous improvement. A great degree of emphasis is placed on systems that provide data on current performance, and the analysis of that data is highly valued.

The bottom line is simply this. Maintenance can no longer be a reactive, fix it when it breaks anchor that prevents plants from achieving their full potential. Instead, maintenance must become an active member of the plant team with its total focus on life cycle asset management and optimum reliability.

SELF-DIRECTED WORK TEAMS: A COMPETITIVE ADVANTAGE

An increasing number of companies are adopting the Toyota Production System (lean manufacturing) and are involving their employees in the daily operations and management through StarPoint teams. These teams are empowered to design how work will be done and to take corrective actions to resolve day-to-day problems. Team members have free, direct access to information that allows them to plan, control, and improve their operations. In short, employees that comprise work teams manage themselves.

Self-directed work teams represent an approach to organizational design that goes beyond quality circles or problem-solving teams. These teams are natural work groups that work together to perform a function or produce a product or service. For example, a team would consist of all operators, maintenance crafts, and support personnel on a given shift that are assigned to a specific production unit or manufacturing job. Each team would have team members, StarPoint, would have the responsibility of coordinating the actions of the team with similar teams or other teams on the same shift and other shifts. These teams not only do the work but also take responsibility for the management of that work—a function that was formerly performed by supervisors and managers.

Why is this concept of self-directed teams growing? The reasons vary from a corporation's misdirected effort to reduce salaried headcount, to genuine efforts to empower the workforce. The real answer is simple, effective use of self-directed work teams have results in

- Improved quality, productivity, and service
- Greater flexibility

1.8 ORGANIZATION AND MANAGEMENT OF THE MAINTENANCE FUNCTION

- Reduced operating costs
- Faster response to technological change
- Fewer, simpler job classifications
- Better response to workers' values
- Increased employee commitment to the organization
- Ability to attract and retain the best people

The employees on the floor, for example, operators, maintainers, and others, understand the problems that limit productivity. Given a chance, they can resolve these issues and radically improve plant performance.

The major challenges organizations face in changing from a traditional environment to a high-involvement environment include developing the teams and fostering a culture of management support. Teams go through several stages of increasing involvement on their way to self-management. This journey can take between 2 and 5 years, and is never-ending from a learning and renewal perspective. Comprehensive training is also critical to developing effective self-directed work teams. The training for these teams must be more comprehensive than for other types of teams. Not only must employees learn to work effectively in teams and develop skills in problem solving and decision-making, they also must learn basic management skills so they can manage their own processes. Additionally, people must be cross-trained in every team member's job. Therefore, it is not uncommon for self-directed work teams to spend 20 percent of their time in ongoing training.

The transition from traditional organizational structures to self-directed work teams is not easy. One of the biggest problems is the reeducation of the front-line supervisors and middle managers. Front-line and middle management can either enable or stifle employee involvement, empowerment and self-directed work teams. Therefore, it is important to elicit management's active support in these efforts. Management also must be involved in the transition. The pragmatic, day-to-day skills in managerial functions that the team will assume currently reside in the supervisors and managers. They need to learn to guide the work group in its transition, development, and empowerment. They need to learn when to hold on and when to let go. This requires planning, training, facilitating, and team-building skills. Supervisors should also learn to provide ongoing coaching support, linking the team's role with the rest of the organization.

Upper management also has a vital role to play in the implementation of self-directed work teams. Senior managers need to strongly champion and sponsor the teams and the process. This commitment must be constantly visible and ongoing. It also should be reinforced with sufficient resources, including time. Last, management must exhibit patience and tolerance because the transition will take time, and delays and mistakes will occur.

The self-directed work team concept is not for everyone. Some corporations simply cannot lose the traditional salaried-hourly mentality that has for so long restricted our ability to compete on the world market. For these corporations, survival may be short-term. For others who are willing to embrace new ideas and new ways of doing business, the future is bright. Try empowering your workforce. I think you'll like the results.

CHAPTER 2
INTRODUCTION TO THE THEORY AND PRACTICE OF MAINTENANCE

R. Keith Mobley
Principal, Life Cycle Engineering, Inc., Charleston, S.C.

As with any discipline built upon the foundations of science and technology, the study of maintenance begins with a definition of maintenance. Because so many misconceptions about this definition exist, a portion of it must be presented in negative terms. So deeply, in fact, are many of these misconceptions rooted in the minds of management and even more so in the minds of many maintenance practitioners that perhaps the negatives should be given first attention.

Maintenance is not merely preventive maintenance, although this aspect is an important ingredient. Maintenance is not lubrication, although lubrication is one of its primary functions. Nor is maintenance simply a frenetic rush to repair a broken machine part or a building segment, although this is more often than not the dominant maintenance activity.

In a more positive vein, maintenance is a science since its execution relies, sooner or later, on most or all of the sciences. It is an art because seemingly identical problems regularly demand and receive varying approaches and actions and because some managers, supervisors, and maintenance technicians display greater aptitude for it than others show or even attain. It is above all a philosophy because it is a discipline that can be applied intensively, modestly, or not at all, depending upon a wide range of variables that frequently transcend more immediate and obvious solutions. Moreover, maintenance is a philosophy because it must be as carefully fitted to the operation or organization it serves as a fine suit of clothes is fitted to its wearer and because the way it is viewed by its executors will shape its effectiveness.

Admitting this to be true, why must this science-art-philosophy be assigned—in manufacturing, power production, or service facilities—to one specific, all-encompassing maintenance department? Why is it essential to organize and administer the maintenance function in the same manner as other areas are handled? This chapter will endeavor to answer these questions. This handbook will develop the general rules and basic philosophies required to establish a sound maintenance engineering organization. And, it will also supply background on the key sciences and technologies that underlie the practice of maintenance.

Let us, however, begin by looking at how the maintenance function is to be transformed into an operation in terms of its scope and organization, bearing in mind its reason for being—solving the day-to-day problems inherent in keeping the physical facility (plant, machinery, buildings, services) in good operating order. In effect, what must the maintenance function do?

SCOPE OF RESPONSIBILITIES

Unique though actual maintenance practice may be to a specific facility, a specific industry, and a specific set of problems and traditions, it is still possible to group activities and responsibilities into two general classifications: primary functions that demand daily work by the maintenance function and secondary ones assigned to the function for reasons of expediency, know-how, or precedent.

Primary Functions

Maintenance of Existing Plant Equipment. This activity represents the physical reason for the existence of the maintenance professional. Responsibility here is simply to make necessary repairs to production machinery quickly and economically and to anticipate these repairs and employ preventive maintenance where possible to prevent them. For this, a staff of skilled engineers, planners, and technicians who are capable of performing the work must be trained, motivated, and constantly retained to assure that adequate skills are available to perform effective maintenance.

Maintenance of Existing Plant Buildings and Grounds. The repairs to buildings and to the external property of any plant—roads, railroad tracks, in-plant sewer systems, and water supply facilities—are among the duties generally assigned to the maintenance engineering group. Additional aspects of buildings and grounds maintenance may be included in this area of responsibility. Janitorial services may be separated and handled by another section. A plant with an extensive office facility and a major building-maintenance program may assign this coverage to a special team. In plants where many of the buildings are dispersed, the care and maintenance of this large amount of land may warrant a special organization.

Repairs and minor alterations to buildings—roofing, painting, glass replacement—or the unique craft skills required to service electrical or plumbing systems or the like are most logically the purview of maintenance engineering personnel. Road repairs and the maintenance of tracks and switches, fences, or outlying structures may also be so assigned.

It is important to isolate cost records for general cleanup from routine maintenance and repair so that management will have a true picture of the true expense required to maintain the plant and its equipment.

Equipment Inspection and Lubrication. Traditionally, all equipment inspections and lubrication has been assigned to the maintenance organization or function. While inspections that require special tools or partial disassembly of equipment must be retained within the maintenance function, the use of trained operators or production personnel in this critical task will provide more effective use of plant personnel. The same is true of lubrication. Because of their proximity to the production systems, operators are ideally suited for routine lubrication tasks.

Utilities Generation and Distribution. In any plant generating its own electricity and providing its own process steam, the powerhouse assumes the functions of a small public utilities company and may justify an operating department of its own. However, this activity logically falls within the realm of maintenance engineering. It can be administered either as a separate function or as part of some other function, depending on management's requirements.

Alterations and New Installations. Three factors generally determine to what extent this area involves the maintenance department: plant size, multiplant company size, company policy.

In a small plant of a one-plant company, this type of work may be handled by outside contractors. But its administration and that of the maintenance force should be under the same management. In a small plant within a multiplant company, the majority of new installations and major alterations may be performed by a company-wide central engineering department. In a large plant a separate organization should handle the major portion of this work.

Where installations and alterations are handled outside the maintenance engineering department, the company must allow flexibility between corporate and plant engineering groups. It would be self-defeating for all new work to be handled by an agency separated from maintenance policies and management.

Secondary Functions

Storeskeeping. In most plants it is essential to differentiate between mechanical stores and general stores. The administration of mechanical stores normally falls within the maintenance engineering group's area because of the close relationship of this activity with other maintenance operations.

Plant Protection. This category usually includes two distinct subgroups: guards or watchmen; fire-control squads. Incorporation of these functions with maintenance engineering is generally common practice. The inclusion of the fire-control group is important since its members are almost always drawn from the craft elements.

Waste Disposal. This function and that of yard maintenance are usually combined as specific assignments of the maintenance department.

Salvage. If a large part of plant activity concerns offgrade products, a special salvage unit should be set up. But if salvage involves mechanical equipment, such as scrap lumber, paper, containers, and so on, it should be assigned to maintenance.

Insurance Administration. This category includes claims, process equipment and pressure-vessel inspection, liaison with underwriters' representatives, and the handling of insurance recommendations. These functions are normally included with maintenance since it is here that most of the information will originate.

Other Services. The maintenance engineering department often seems to be a catchall for many other odd activities that no other single department can or wants to handle. But care must be taken not to dilute the primary responsibilities of maintenance with these secondary services.

Whatever responsibilities are assigned to the maintenance engineering department, it is important that they be clearly defined and that the limits of authority and responsibility be established and agreed upon by all concerned.

ORGANIZATION

Maintenance, as noted, must be carefully tailored to suit existing technical, geographical, and personnel situations. Basic organizational rules do exist, however. Moreover, there are some general rules covering specific conditions that govern how the maintenance engineering department is to be structured. It is essential that this structure does not contain within itself the seeds of bureaucratic restriction nor permit empire building within the plant organization.

It is equally essential that some recognized, formally established relationship exists to lay out firm lines of authority, responsibility, and accountability. Such an organization, laced with universal truths, trimmed to fit local situations, and staffed with people who interact positively and with a strong spirit of cooperation, is the one which is most likely to succeed.

Begin the organizational review by making certain that the following basic concepts of management theory already exist or are implemented at the outset.

1. *Establish reasonably clear division of authority with minimal overlap.* Authority can be divided functionally, geographically, or on the basis of expediency; or it can rest on some combination of all three. But there must always be a clear definition of the line of demarcation to avoid the confusion and conflict that can result from overlapping authority, especially in the case of staff assistants.

2. *Keep vertical lines of authority and responsibility as short as possible.* Stacking layers of intermediate supervision, or the overapplication of specialized functional staff aides, must be minimized. When such practices are felt to be essential, it is imperative that especially clear divisions of duties are established.

3. *Maintain an optimum number of people reporting to one individual.* Good organizations limit the number of people reporting to a single supervisor to between three and six. There are, of course, many factors which can affect this limitation and which depend upon how much actual supervision is required. When a fairly small amount is required, one man can direct the activities of twelve or more individuals.

The foregoing basic concepts apply across the board in any type of organization. Especially in maintenance, local factors can play an important role in the organization and in how it can be expected to function.

1. *Type of operation.* Maintenance may be predominant in a single area—buildings, machine tools, process equipment, piping, or electrical elements—and this will affect the character of the organization and the supervision required.
2. *Continuity of operations.* Whether an operation is a 5-day, single-shift one or, say, a 7-day, three-shift one makes a considerable difference in how the maintenance engineering department is to be structured and in the number of personnel to be included.
3. *Geographical situation.* The maintenance that works in a compact plant will vary from that in one that is dispersed through several buildings and over a large area. The latter often leads to area shops and additional layers of intermediate supervision at local centers.
4. *Size of plant.* As with the geographical considerations above, the actual plant size will dictate the number of maintenance employees needed and the amount of supervision for this number. Many more subdivisions in both line and staff can be justified, since this overhead can be distributed over more departments.
5. *Scope of the plant maintenance department.* This scope is a direct function of management policy. Inclusion of responsibility for a number of secondary functions means additional manpower and supervision.
6. *Workforce level of training and reliability.* This highly variable characteristic has a strong impact on maintenance organization because it dictates how much work can be done and how well it can be performed. In industries where sophisticated equipment predominates, with high wear or failure incidence, more mechanics and more supervisors are going to be required.

These factors are essential in developing a sound maintenance department organization. It is often necessary to compromise in some areas so that the results will yield an orderly operation at the beginning yet retain sufficient flexibility for future modification as need indicates.

Lines of Reporting for Maintenance

Many feel that a maintenance department functions best when it reports directly to top management. This is similar in concept to the philosophy of having departments with umpire-like functions reporting impartially to overall management rather than to the departments being serviced. This independence proves necessary to achieve objectivity in the performance of the maintenance engineering function. However, in many plants the level of reporting for the individual in charge of the maintenance engineering group has little or no bearing on effectiveness.

If maintenance supervision considers itself part of production and its performance is evaluated in this light, it should report to the authority responsible for plant operations. The need for sharply defined authority is often overemphasized for service or staff groups. Performance based on the use of authority alone is not and cannot be as effective as that based on cooperative efforts.

Certainly it is not practical to permit maintenance engineering to report to someone without full authority over most of the operations that must be served by it. The lack of such authority is most troublesome in assigning priorities for work performance.

Maintenance engineering should report to a level that is responsible for the plant groups which it serves—plant manager, production superintendent, or manager of manufacturing—depending on the organization. The need to report to higher management or through a central engineering department should not exist so long as proper intraplant relationships have been established.

Specialized Personnel in the Maintenance Organization

Technically Trained Engineers. Some believe that engineers should be utilized only where the maximum advantage is taken of professional training and experience and that these individuals should not be asked to handle supervisory duties. Others feel that technical personnel must be developed from the line in order to be effective and that the functions of professional engineering and craft supervision must somehow be combined. Both views are valid. The former arrangement favors:

1. Maximum utilization of the engineer's technical background.
2. Maintaining a professional approach to maintenance problems.
3. Greater probability that long-range thinking will be applied, that is, less concern with breakdowns and more with how they can be prevented in the future.
4. Better means of dealing with craftpersons' problems by interposing a level of up-from-the ranks supervision between them and the engineer.
5. The development of nontechnical individuals for positions of higher responsibility.

Combining engineering and supervisory skills assures:

1. Rapid maturing of newly graduated personnel through close association with craftpersons' problems.
2. Increasingly expeditious work performance through shorter lines of communication.
3. Possible reduction in the supervisory organization or an increase in supervision density.
4. An early introduction into the art of handling personnel, making them more adaptable to all levels of plant supervision.
5. Less resistance to new ideas.

Staff Specialists. The use and number of staff specialists—electrical engineers, instrument engineers, metallurgists—depends on availability, need for specialization, and the economics of a consulting service's cost compared to that of employing staff experts.

Clerical Personnel. Here there are the two primary considerations. Paperwork should be minimized consistent with good operations and adequate control; the clerical staff should be designed to relieve supervision of routine paperwork that it can handle.

The number of clerks used varies from 1 per 100 employees to 1 per 20 to 25 employees. These clerks can report at any level of the organization or can be centralized as proves expedient.

MANPOWER REQUIREMENTS

The number of employees—labor and supervision—to assure adequate plant maintenance coverage depends upon many factors. Each plant must be treated as a separate problem with a consideration of all its unique aspects.

Hourly Personnel

Ratio of Maintenance Manpower to Total Operation Personnel. This ratio is too often considered the measure of adequacy and relative efficiency of the department. In practice it will vary with the type of machinery and equipment expressed in terms of an investment figure per operating employee.

To estimate the number of maintenance employees necessary to maintain a plant properly, an approach based on the estimated size of the maintenance bill and the percentage of this bill that will cover labor has proved more realistic. Experience factors, however, can be used in many industries to estimate maintenance cost as a percentage of investment in machinery and equipment. Before building a plant, many companies determine the approximate rate of return on investment that can be expected. One factor to be considered here is maintenance cost. Generally, the annual cost of maintenance should run between 7 and 15 percent of the investment. Building maintenance should run between $1\frac{1}{2}$ and 3 percent, per year. The cost of labor alone, exclusive of overhead, will run between 30 and 50 percent of the total maintenance bill.

In addition, other duties of the maintenance department must be considered and extra manpower allowances made. This supplementary personnel can serve as a cushion for fluctuations in strictly maintenance work loads by adding 10 to 20 percent of the maintenance force estimated to be necessary under normal conditions.

These criteria are only suited to a preliminary study. Actual manpower requirements must be controlled by a continuous review of work to be performed. Backlog-of-work records are a help here; and the trends of the backlog of each craft enable maintenance supervision to increase or reduce the number of employees to maintain the proper individual craft strength and total work force.

Crafts That Should Be Included. The crafts and shops that should exist in any good maintenance operation are set by the nature of the activity and the amount of work involved. This means existence of a close relationship between plant size and the number of separate shops that can be justified.

Another actor is the availability of adequately skilled contractors to perform various types of work. In some plants jacks-of-all-trades can be used with no special problem. Yet, in spite of the difficulties inherent in recognizing craft lines in scheduling, there is a real advantage in larger plants to segregating skills and related equipment into shops. In general, however, it is difficult to justify a separate craft group with its own shop and supervision for less than 10 men.

Supervision

Supervision Density. The number of individuals per supervisor (supervision density) is an accepted measure for determining the number of first-line supervisors needed to handle a maintenance force adequately. Though densities as low as 8 and as high as 25 are sometimes encountered, 12 to 14 seems to be the average. Where a large group of highly skilled men in one craft perform routine work, the ratio will be higher. If the work requires close supervision or is dispersed, a lower ratio becomes necessary. For shops with conventional crafts—millwrights, pipe-fitters, sheet-metal workers, carpenters—one foreman accompanied by some degree of centralized planning can direct the activities of 12 to 15 individuals of average skill. Supervision density should be such that the foreman is not burdened with on-the-job overseeing at the expense of planning, training workers, or maintaining the personal contacts that generate good morale.

Cross-Craft Supervision. The use of first-line supervision to direct more than one craft should be considered carefully. If a small number of people are involved, this arrangement can be economically preferable. But, for the most effective use of specific craft skills, experience indicates that each should have its own supervision.

SELECTION AND TRAINING

Selection—Craft Personnel

Normally, the union contract places sharp restrictions on the means by which applicants for maintenance craft training are selected. If there are no such restrictions, more definitive selection methods can be employed. When this is the case, bases for selection should be education, general intelligence, mechanical aptitude, and past experience. When it is possible, personnel with previous craft experience offer the easiest and most satisfactory method of staffing the maintenance engineering department, particularly when the cost of a formal training program cannot be economically justified. When, however, you must draw on plant personnel, the factors cited above, plus the candidate's age, should be considered. It is an unfortunate fact of life that it is easier to develop a craftsman from someone in the early twenties than someone who is over forty.

Training—Craft Personnel

This activity can be performed in two ways—formal instruction or informal on-the-job instruction.

Formal Instruction. While this subject is covered in some depth later in this handbook, it has a place in this earlier area. Many formalized maintenance training programs are currently available, usually in packaged form. The most common is an apprentice training program that conforms to the National Apprenticeship System of the U.S. Department of Labor's Bureau of Apprenticeships. Moreover it has the added advantage of acceptance by most unions. Graduates are presented with certificates and are considered to be fairly well equipped on a nationwide basis. But the administration of such a system constitutes an expense which must be taken into consideration.

Most other formalized training plans are those developed by major firms for their own use and then made available, at a fee, to others. Often these have an advantage over the federal plan since they can conform to peculiar plant needs. But they are usually even more expensive and lack the universal recognition of the former.

Informal Instruction. This consists primarily of spot exposure of personnel to intensive instruction in some phase of plant activities. It takes the form of lectures, sound-slide films, movies, or trips to suppliers who may, with or without charge, provide instruction on their particular equipment. Usually these are directed more at developing advanced mechanical skills, however.

On-the-Job Training. This is the most prevalent method for training maintenance personnel. Although its short-range effectiveness is difficult to measure, many excellent craftsmen have acquired their skills in this way. Usually a new man is assigned to an experienced craftsman as a helper and learns by exposure to the job and from the instruction he receives from his appointed mentor. The effectiveness is improved if the training is supplemented by routine rotation of the trainee among several knowledgeable craftsmen and is accompanied by personal interviews by the foreman to determine the degree of progress.

Selection—Supervisory Personnel

Only very general rules can be set out for selecting supervisory people for maintenance. For the first and second levels—that is, those directly in charge of craft personnel—prospective candidates should possess better than average mechanical comprehension and be capable of handling a number of diverse problems at one time. And while high craft skill is desirable, it should not be the sole basis of selection. In fact, there is more chance of developing a satisfactory foreman from an individual having all the traits except craft skill than of trying to develop these important abilities in a man with only craft training.

Although there are advantages in selecting people entirely from among maintenance employees, others who display leadership potential should be considered. But too much importance attached to long years of experience and technical skill can result in poor selection of personnel.

Serious consideration should be given to the temperament of technically trained individuals when choosing them for maintenance. For the best results, candidates should be slightly extroverted and prone to take the broader view of their own professional utilization. They should be able to temper professional perfectionism with expediency. Since effectiveness in maintenance depends greatly on the relationships that exist with other plant units, the technical man in this field must have some of the attributes of a salesman.

If possible, aptitude testing and comprehensive interviews should be part of the selection process. The use of comprehensive interviews in the selection of all supervisory employees is of great value. This service can be obtained from professional sources or developed within an organization by training key individuals. These interviews afford good basic information on the inherent personal characteristics of the candidate.

Training—Supervisory Personnel

This training consists of initial orientation; a formalized, sustained program of leadership training; and on-the-job coaching and consultation.

Orientation gives the candidate basic data about the management team he is joining and about company and department policy. Included also are facts about the scope of his personal responsibilities and the limits of the authority delegated to him.

Training ensuring continued effectiveness and improved performance should include such subjects as human relations, conducting of interviews, teaching methods, safety, and many more. Goals of this part of the program should head toward increasing the candidate's effectiveness as supervisor, instilling a feeling of unity with fellow managers, and enhancing personal development.

On-the-job coaching is especially important for the embryo supervisor. There is no substitute for frequent, informal, personal contact with a superior concerning current technical, personnel, or personal problems. This type of development is a force for high morale and job satisfaction. It should include both praise and criticism, the former sincere, the latter constructive. Bear in mind that a cadre of highly capable and motivated supervisors will make overall maintenance management simpler, easier, and, in the long run, far more economical.

CHAPTER 3
MAINTENANCE AND RELIABILITY ENGINEERING

R. Keith Mobley
Principal, Life Cycle Engineering, Inc., Charleston, S.C.

Maintenance engineering is typically defined as a staff function whose prime responsibility is to ensure that maintenance techniques are effective, equipment is designed and modified to improve maintainability, ongoing maintenance technical problems are investigated, and appropriate corrective and improvement actions are taken. Used interchangeably with reliability engineering. In small plants, this definition and the stated interchangeability of titles is generally true. One group of engineers provides the duties of reliability, plant, and maintenance engineering, but these functions have different roles and responsibilities.

The *reliability engineering* function is responsible for risk management and life cycle asset management. It is a strategic resource that has single point accountability for providing the long-term business strategy that ensures production capacity, product quality, and best life cycle cost. Its mission is to provide the proactive leadership, direction, single point accountability, and technical expertise required to achieve and sustain optimum reliability, maintainability, useful life, and life cycle cost for a facility's assets, as well as its processes.

RESPONSIBILITIES OF RELIABILITY AND MAINTENANCE ENGINEERING

Reliability engineering is primarily concerned with application of technical skills and ingenuity to the correction of equipment problems causing excessive production downtime and maintenance work. The position is dedicated to the maintenance function and is focused on the elimination of repetitive failure.

- Ensure maintainability of new installations.
- Identify and correct chronic and costly equipment problems, eliminate repetitive failure.
- Technical advice to maintenance and partners.
- Design and monitor an effective and economically justified preventive or predictive maintenance programs.
- Proper operation and care of equipment.
- Comprehensive lubrication program.
- Inspection, adjustments, parts, replacements, overhauls, and the like, for selected equipment.

- Vibration and other predictive analyses.
- Protection from environment.
- Maintain and analyze equipment data and history records to predict maintenance needs.

Reliability engineering is primarily a strategic activity focused on the future. However, there also exists a need for tactical assistance to the operations and maintenance functions for methods improvement, identification and establishment of best design, procurement, installation, operating, maintenance and repair practices, and developing methods to reduce losses from optimum performance levels. Some of these tasks, for example, failure analyses, sustaining maintenance, primarily deal with maintenance tactics and are usually performed by a *maintenance engineer*. In larger organizations, these roles are separate, but combined in most midrange and smaller organizations. For the purposes of this manual, we will consider the roles combined under the title of reliability engineer.

Reliability engineering is one of the most essential elements of reliability excellence and optimum plant performance. This is the function that is responsible for

- Guiding efforts to ensure reliability and maintainability of equipment, processes, utilities, facilities, control loops, and safety or security systems.
- Reducing and improving maintenance work wherever feasible; assuring efficient and productive operation of facility process and equipment; while protecting and prolonging the economic life of facility assets; all at minimal expense to the facility.
- Serving in a staff capacity, the function relieves maintenance supervisors and planner or schedulers of those responsibilities that are engineering in nature. Reliability engineering is the prime user of equipment history information. This key feature of any maintenance work order system will be ineffective and underutilized without this function and system payback will be significantly reduced.

Reliability is defined as the probability that an item, for example, production or utility asset and work processes, will continue to do what the user needs it to do without failure under specified conditions for a specified period of time. Reliability engineering is the application of engineering knowledge to risk management.

Risk Management. Increase the probability that manufacturing assets and processes will operate or perform when needed by defining strategies to prevent failures, detect the onset of failures in their earliest stages, and minimize all risks associated with the plant or facility.

- Identify potential for cost reduction through extended parts life, reduced labor cost, and other parts-related improvement techniques.
- Participate in review phases of design of capital changes in facility layout to ensure full maintainability of equipment, utilities, and facilities.
- Initiate corrective action using the study of corrosion, fatigue, wear, and erosion rates throughout the facility.
- Alternate solutions to reduce the high costs associated with certain units of equipment.
- Recommended economic studies for equipment retirement, modification, updating, and the like.

Life Cycle Asset Management. Develop, implement, and oversee a viable process or processes that will ensure optimum life cycle asset management.

Configuration Management. Ensure that all physical assets of the plant or facility are designed, installed, operated, and maintained in a manner that will provide maximum useful life and best life cycle cost.

- Develop and standardize a program that influences new construction and equipment purchase(s) including materials, equipment, and spare parts.

- Participate in approval of all new installations, including those done by contractors in order to ensure their maintainability and reliability as influenced by life cycle costing.

Asset Care. Develop, implement, and oversee a viable process or processes that will ensure appropriate levels of sustaining maintenance and operator care will be provided for all physical assets.

- Implement processes to minimize the number and severity of physical asset failures, including operation at non-design performance levels.
- Determine strategies to mitigate the consequences of failures that cannot be prevented.
- Defining, developing, administering, and refining the preventive or predictive maintenance program.
- Specify repair techniques for major repetitive tasks, such as component replacements, develop standards specifying the end product, and the resources needed for these tasks.
- Apply value analysis to make maintenance decisions by repair, replace, or redesign.

Loss Elimination. Develop, implement, and oversee a viable process or processes to continuously eliminate losses and waste from all functional areas, for example, procurement, operations, maintenance, and the like, within the plant or facility.

- Reduce and improve maintenance work wherever feasible; assuring efficient and productive operation of facility process and equipment; while protecting and prolonging the economic life of facility assets; all at minimal expense to the facility.
- Analyze equipment histories to identify specific repetitive failures and effectively address identified areas.
- Review all equipment failures in order to determine what action might have been taken to prevent failure and to protect against reoccurrence. Revise processes and procedures accordingly.

MAINTENANCE ENGINEERING

The role of maintenance engineering function is more tactical in nature, for example, ensuring that the assets within a plant meet the present demands of the company. Where the reliability engineer is looking at the long-term reliability needs, the maintenance engineer is handling the day-to-day reliability responsibilities.

Responsibilities of Position

- Identify, initiate, coordinate, and complete tactical maintenance and process improvement opportunities
- Technical support of operations or maintenance in assigned area, for example, trouble-shooting, turnaround scoping, spares management, and so on
- On new projects, assist project engineering with development and implementation of control plans, that is, criticality, preventive maintenance plan, spares, quality, maintainability, and operability
- On existing assets, perform periodic reviews and upgrade and modify control plans
- Communicates with reliability engineering to ensure long-term operations and maintenance asset problems are appropriately investigated with solutions implemented
- Assist operations or maintenance management with department budgeting and expenditure forecasting
- Continuously track and evaluate operations or maintenance expenditures, for example, cost and labor-hours to ensure effective resource utilization

- Facilitate positive change in the maintenance organization by proactively leading department initiatives. Act as a change agent in implementing cross-functional plant business objectives
- Generate the maintenance reports for area of responsibility
- Completions of environmental health and safety (EHS) tasks, including excavation permits, preliminary hazard analysis (PHA), and the like
- Develop, implement, and survey *best practices* within area of responsibility, that is preventive maintenance practices, operating methods, and the like
- Review major purchases to assure correct specifications and design
- Initiate, develop, and review capital improvement projects
- Generate the monthly maintenance report for area of responsibility
- Backup for maintenance supervisor
- Other duties and projects as assigned

As part of a proactive operations and maintenance philosophy, the maintenance engineering group is responsible for the development, implementation, and periodic evaluation of an effective *asset maintenance* plan (AMP). The objective of this plan is to

- Maintain the function in terms of the required safety.
- Maintain the inherent safety and reliability levels.
- Optimize the availability.
- Obtain the information necessary for design improvement of those items whose inherent reliability proves inadequate.
- Accomplish these goals at a minimum total life cycle cost, including maintenance costs and the costs of residual failures.
- Obtain the information necessary for establishing a dynamic maintenance program that improves upon the initial program, and its revisions, by systematically assessing the effectiveness of previously defined maintenance tasks. Monitoring the condition of specific safety, critical or costly components would play an important role in the development of a dynamic program.

These objectives recognize that maintenance programs, as such, cannot correct deficiencies in the inherent safety and reliability levels of the equipment and structures. The maintenance program can only minimize deterioration and restore the item to its inherent levels. If the inherent levels are found to be unsatisfactory, design modification, operational or procedural changes (such as training programs) may be necessary to obtain improvement.

Maintenance Program Content

The content of the maintenance program itself consists of two groups of tasks.

- A group of preventive maintenance tasks, which include failure-finding tasks, scheduled to be accomplished at specified intervals, or based on condition. The objective of these tasks is to identify and prevent deterioration below inherent safety and reliability levels by one or more of the following means:
  - Lubrication or servicing
  - Operational, visual, or automated check
  - Inspection, functional test, or condition monitoring
  - Restoration
  - Discard or disposal

It is this group of tasks, which is determined by reliability-centered management (RCM) analysis, for example, it comprises the RCM based preventive maintenance program.

A group of nonscheduled maintenance tasks which result from

- Findings from the scheduled tasks accomplished at specified intervals of time or usage
- Reports of malfunctions or indications of impending failure (including automated detection)

The objective of this second group of tasks is to maintain or restore the equipment to an acceptable condition in which it can perform its required function.

An effective program is one that schedules only those tasks necessary to meet the stated objectives. It does not schedule additional tasks that will increase maintenance costs without a corresponding increase in protection of the inherent level of reliability. Experience has clearly demonstrated that reliability decreases when inappropriate or unnecessary maintenance tasks are performed, due to increased incidence of maintainer-induced faults.

CHAPTER 4
COOPERATIVE PARTNERSHIPS

Jeff Nevenhoven
Senior Consultant, Life Cycle Engineering, Inc., Charleston, S.C.

While much is made of the need for a real and active partnership between the various functions within the typical manufacturing organization, the reality is sometimes less than ideal. Nowhere is this partnership more essential than between the operations and maintenance departments. The ideal partnership incorporates some fundamental requirements, and needs much more then passive acknowledgement that there is a partnership in place. The ideal partnership will derive strength from open and frank communications, a common set of beliefs, well-defined expectations on the part of all participants, and a set of common goals that are aligned with business needs. With this in mind, any organization can improve the performance of the manufacturing environment.

While a traditional *maintenance department* can and should improve the internal processes and practices that are used to execute maintenance activities, it is impossible to achieve and sustain world-class reliability levels without the support and cooperation of other, non-maintenance plant functions. Simply stated, maintenance is dependent on these other plant or corporate functions. Conversely, these non-maintenance functions are also dependent on effective maintenance to achieve world-class reliability performance levels.

Most North American plants and facilities are functionally integrated. Each functional group is dependent on inputs from other functions and provides outputs to others. As a result, each function is dependent on other functional groups and cannot operate effectively without proactive cooperation from these groups.

WHO IS RESPONSIBLE FOR RELIABILITY PROBLEMS

Evaluations of the reasons for asset reliability problems confirm the integration of the plant's functional groups. An analysis of reliability related issues with a number of clients through the 1990s indicates that most of the functional groups within the plant have some direct responsibility for reliability-related problems.

Sales

The *sales* function has a proven, direct impact on asset reliability. Historically, 15 percent of all asset reliability problems can be directly attributed to deficiencies within the sales function. The predominant deficiencies that cause reliability problems include improper product mix, unrealistic delivery commitments, and inadequate order or production run size.

Improper Product Mix. Statistically only 3 percent of the sales force that generates plant backlogs are considered professional salespersons. The balance, often referred to as "order takers," tend to accept any combination of standard and nonstandard product orders that their clients are willing to award. As a result, plants are forced to accept production orders for higher volumes of nonstandard and special-order production runs that directly reduce asset utilization and reliability.

Unrealistic Delivery Commitments. Effective asset utilization, for example, production planning and control, is dependent on sufficient lead-time to properly sequence orders through the production process. High volumes of special orders with unrealistically short delivery schedules force production to increase the number of changeovers and reduce the normal longer run campaigns that the production assets are designed to produce.

Inadequate Order or Production Run Size. Asset utilization and reliability are dependent on consistent operation of production and manufacturing assets within their normal design limits. Small production runs and/or orders size force the operations function to increase the number and frequency of changeovers, introduce a higher potential for human errors, and reduces the operating efficiency.

Production

The production function is responsible for 23 percent of all asset reliability problems. These include deficiencies in the production planning and scheduling function, as well as those in the production function. The predominant deficiencies include improper planning and scheduling, poor operating procedures, and operator errors.

Improper Planning and Scheduling. This is primarily a failure to effectively utilize the installed plant capacity. In most cases, the production planning and scheduling function is a clerical activity that simply enters incoming orders from the Sales function into the production schedule without any attempt to optimize the process.

Poor Operating Procedures. In too many plants, the procedures that are used to govern the production process are inadequate, or simply non-existent. Many are outdated and are no longer adequate for proper, effective utilization of the plant's production systems. Also contributing to this problem is a failure of management to enforce universal adherence to those procedures that are appropriate for the production systems.

Operator Errors. While some of these problems are solely the result of operator errors, most are a failure of corporate management to provide adequate training for operating personnel. Most operators have little, if any, real knowledge of the proper operating procedures, or the internal working of the operations equipment. Instead of real production based procedures and knowledge, they are taught the minimal steps that must be taken to operate these critical systems. Perhaps the most critical operations error is to simply ignore the basic fact that reliability is everyone's job, in much the same way that safety is everyone's job. Operations must take critical pride in the appearance and functionality of the production equipment, and this dictates a cooperative working relationship with the maintenance and engineering departments.

Maintenance

While maintenance is not the "bad actor" that many corporate managers believe, this function does contribute 17 percent to asset reliability problems. The dominant deficiencies are similar to those of the production function.

Improper Maintenance. Most maintenance functions permit the crafts to determine how maintenance activities will be executed. As a result, many of these tasks are performed incorrectly and incompletely. The result is chronic reliability problems.

Poor Planning. Too many maintenance functions have eliminated the planning and scheduling function. Instead, work requests are compiled, routed to the supervisors and issued for execution without proper planning. As a result, critical activities are not executed in a timely manner or the procedures used are inadequate.

Failure to Perform Effective Preventive Maintenance Tasks. Preventive maintenance, that is, inspections, lubrication, calibrations, and adjustments must be performed in a timely manner to sustain reliable asset operation. Failure to adhere to these schedules and effective execution of these tasks result in reduced asset reliability.

Procurement

The procurement function contributes 12 percent to poor asset reliability. The predominate deficiencies in this function include substitution of inadequate maintenance, repair, and operating (MRO) parts, late deliveries, and vendor selection.

Substitution of Inadequate MRO Parts. In their zeal to meet their perceived mission-to reduce cost as much as possible, the procurement function substitutes cheaper, often incompatible parts for the repair and maintenance of assets.

Late Deliveries. Too many procurement functions have delegated expediting to the maintenance planner or storeroom clerk. As a result, MRO materials are often late or do not arrive at all.

Vendor Selection. A growing trend is to establish national buy agreements with a select group of vendors. In too many cases, these agreements do not provide adequate, timely support of either the production or maintenance functions.

Plant Engineering

Plant engineering contributes 22 percent to asset reliability problems. Over the past 10 years, the plant engineering function has typically changed from an effective plant support function to a project management group. While this function retains the responsibility for design and acquisition of new production systems, as well as any modification to existing assets, they do not provide active engineering support for these or other critical activities. The dominant factors that cause reliability problems include improper design, inappropriate modifications, and failure to document changes.

Improper Design. Few plants retain the in-house expertise needed for proper design of new, or modification of existing, production systems. Instead, these tasks are contracted out to engineering firms that are assumed to have the required expertise. As a result, many of the new systems and modifications to existing systems are not suitable for long-term reliable operations.

Inappropriate Modifications. The lost of in-house engineering expertise has resulted in modifications to existing production systems that create multiple, long-term reliability problems.

Failure to Document Changes. The addition of new production systems or modifications to existing system must be fully documented and all related issues updated. drawings, bills of materials, operating and maintenance procedures, spare parts inventory, and operator and maintenance crafts training must be upgraded to support these changes. Failure to follow adequate configuration management procedures results in reliability problems.

Management

Eleven percent of recorded asset reliability problems can be directly attributed to deficiencies in the management function. Most of these problems are the result of policies and procedures that are mandated by plant or corporate management that have an adverse impact on asset reliability.

INTERDEPENDENCY

Because of the integrated nature of plant operations and management, maintenance typically cannot control its own destiny. Instead, it is dependent on other plant functions as well as the plant culture. Figure 4.1 illustrates the level of control that maintenance has over the 21 building blocks that are essential for effective maintenance management.

Maintenance has majority control (80 percent) over five of the building blocks:

- Supervision and practices
- Work control
- Planning
- Work measurements
- Maintenance asset history

In these five categories, maintenance management can control, within the constraints of the plant culture, how these activities will be carried out. As a result, the effectiveness of these building blocks lies predominately within the control of maintenance management.

Maintenance management has less control, about 50 percent over the following building blocks:

- Goals and objectives
- Organizational structure
- Training and motivation
- Organizational behavior
- Budgetary control
- Scheduling and coordination
- Master plan

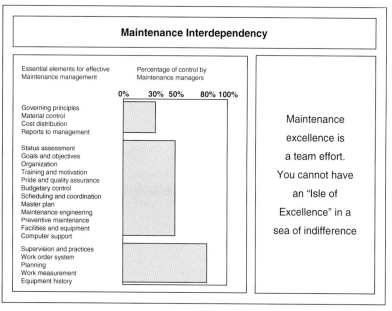

FIGURE 4.1 Maintenance interdependency.

- Maintenance engineering
- Preventive or predictive maintenance
- Facilities and equipment
- Computer support

In each of the blocks, maintenance management has direct input, but may be severely constrained by corporate or plant restrictions that have equal or greater control. In most cases, organizational structure; maintenance engineering; facilities and equipment; training; goals and objectives; and computer support are dictated or controlled by plant or corporate management.

Maintenance has little, less than 30 percent control over the remaining building blocks:

- Governing principles
- Materials management
- Cost distribution
- Reports to management

In almost all cases, control of these building blocks resides with plant or corporate management, and maintenance has little, if any direct control.

As one can see from Fig. 4.1, it is imperative that maintenance develops a cooperative partnership with plant management, as well as with production and other functional groups. Without direct input into the cultural decisions that are made by plant or higher management, maintenance has little chance of optimizing reliability independently.

FUNCTIONAL RESPONSIBILITIES

Each of the functional groups that comprise the plant or corporate team has clearly defined roles and responsibility that must be effectively performed and coordinated with other functions. Each function should sit down with the other functions that they deal with on a regular basis and develop *service-level agreements* that spell out their roles and responsibilities to the other in order to ensure success. Typical roles and responsibilities include operations, maintenance, and engineering.

Responsibilities of Operations

A critical part of the functional partnerships is clearly defined roles and responsibilities. Operations must provide effective coordination and support with other functional groups that directly influence or are influenced by its actions.

Maintenance. The operations or production function has explicit responsibilities that must be provided before maintenance can achieve and sustain world-class performance. These responsibilities include

- Operate machinery and equipment properly.
- Know the conditions and performance of facilities and equipment.
- Maintain surveillance thereof in order to detect unsatisfactory conditions and anticipate essential work.
- Report malfunctions to appropriate engineering or maintenance personnel for diagnosis and action.
- Authorize and describe clearly in writing the repairs, replacements, and alterations. Avoid unnecessary work and fictitious priority. Help to control the volume variance within maintenance budgets. Participate in backlog management using a clearly defined and agreed upon set of priorities, and participate in repetitive failure analysis, using cause codes and effective analysis techniques.
- Accept equipment ownership.

1.28 ORGANIZATION AND MANAGEMENT OF THE MAINTENANCE FUNCTION

- Participate in performance of maintenance work to the degree specified, authorized, and trained for.
- Plan for and provide adequate equipment access for timely performance of programmed and scheduled maintenance.
- Communicate the necessary capacity specifications for systems and equipment.

Responsibilities of Maintenance

The maintenance function must not only perform its duties effectively, but also provide direct coordination and support with other functional groups.

Operations. The maintenance organization must also meet its responsibilities. These include

- Based on authorized requests for maintenance service, define and execute the required work in a timely fashion, with quality workmanship; knowing what is to be done and when and how to do it best. Then do it right, the first time.
- Assist operations in establishing a practical level of maintenance so that long- and short-term operating plans can be met and repairs can be anticipated, planned, and scheduled.
- Maintain facilities at specified levels of operating condition, at lowest possible cost consistent with the goals of producing a quality product as economically as possible.
- Actively participate with production to create and implement a comprehensive preventive maintenance program.
- Convert emergency work to planned work by anticipating it.
- Make repairs and replacements at intervals required for optimal operating efficiency and in a manner creating as little production loss as possible.
- Constantly strive to improve maintenance work methods, completeness, and neatness with the goal of quality work at minimal cost.
- Effectively plan, schedule, and coordinate maintenance work with production, far enough in advance to permit them to plan for out-of-service equipment and to minimize nonproductive time and production shortages.
- Prior to execution, thoroughly review all shutdown work with key production personnel so their intimate knowledge can be fully utilized.
- Prior to start up, review the repairs made and any circumstances of note with the operations personnel in the area.
- Provide regular feedback regarding status of work requests and completion promises.
- Advise production personnel as to the levels of risk and the potential costs related to operating equipment believed to be close to failure.
- Develop techniques for predicting failure of critical facilities with reasonable accuracy.
- Inform operating personnel of facilities requiring excessive maintenance and take appropriate action to reduce it.
- Account for the level of cost incurred in the performance of requested maintenance (standard vs. actual—the performance variance).
- Regard operations as a customer (internal).
- Sponsor and participate in repetitive failure analysis sessions, with the goal of eliminating repetitive failures and isolating the behavioral or mechanical causes of these failures.

Engineering. The maintenance function has several responsibilities to engineering that will enable reliable designs to be implemented:

- Input on plant standard equipment and components
- Realistic assessment of necessary redundancy (not everything requires an installed spare)

- Participation on the design team to identify potential reliability and maintainability issues
- Commissioning assistance to enable thorough check-out of equipment prior to startup
- Realistic maintainability requirements (access platforms and overhead monorails aren't necessary everywhere)
- Resources to conduct or witness performance testing in vendor facilities prior to machine acceptance

Procurement. To enable the procurement function to be effective, the maintenance organization needs to provide:

- Reliability specification information so effective vendor negotiations can be held
- Input on supplier and/or material performance (let us know when it's not right)
- Assistance in minimizing quantities of stocked material
- Compliance with stores procedures to enable that function to work effectively
- Set up *bills of material* so that the right parts can be acquired
- Input on obsolete stocked items
- Realistic delivery requirements
- Proactive work process (enables just-in-time procurement)
- Information to file warranty claims

Responsibilities of Engineering

Maintenance. The plant or project engineering organization has several responsibilities to maintenance to ensure adequate reliability of newly installed assets:

- Design for lowest life-cycle cost instead of lowest installed cost
- Design for reliability and maintainability
- Up-front failure modes and effects analysis to identify potential design changes that may reduce the need for maintenance
- Standardization of components to reduce the need to stock additional parts in the storeroom and to reduce future training requirements
- Vendor input on maintenance strategies, failure modes, mean time between failures (MTBF), and spare parts requirements for all components
- Computerized maintenance management system (CMMS) records updates for all newly installed equipment
- Technical documentation to enable proper preventive or predictive strategy definition as well as to enable planning of future corrective work
- Test and inspection results from performance test at the vendor's facility
- Installation in accordance with good reliability practices, such as good piping alignment, adequate foundation mass, and the like

Responsibilities of Procurement

Maintenance. The procurement function has several responsibilities to the maintenance organization to enable maximum reliability of equipment:

- The right materials in the right place at the right time (and the right price!)
- Commitment to the lowest *total cost of ownership* rather than lowest initial price
- Commitment to standardization of components to reduce training needs

- Hold vendors accountable for performance:
- Compliance to specifications
- On-time delivery
- Effective storeroom layout to enable critical parts to be easily located
- Prekitting and delivery services
- Identification of components under warranty so that claims can be made if necessary

CHAPTER 5
EFFECTIVE MAINTENANCE ORGANIZATIONS

Randy Heisler
Managing Principal, Life Cycle Engineering, Inc., Charleston, S.C.

There are an almost infinite number of organizational structures in use, but most are not configured to provide effective utilization of the workforce and/or have too little or too much indirect—for example, clerical, administrative, and support—personnel. This section provides the practical knowledge needed to evaluate your existing organization to determine whether or not it's effective.

Organization is people with a purpose working together. Good organization is effective people working constructively together toward a common goal. There is a dramatic difference between these two statements. Planning a plant and corporate organization is both a science and an art. One portion of the science of organization lies in the dimensions on which they are designed. The dimensions of structure, culture, systems, and processes are common to all plants. How organizational dimensions are coordinated and governed is more art than science. In no two companies are dimensions combined or managed in the same way. Simply stated, there is neither a clear guideline nor a single, ideal organization structure that is best for all plants or corporations.

A fundamental principle of organization is that the pieces or functions must fit together and there should be effective coordination between functions. As a system, organizations can only be understood or governed as an integrated total. This view argues that the total is something more than merely the sum of its parts and that analysis by differentiation, followed by aggregation does not provide an accurate picture of the total, or of how the total performs. Just as the human body cannot be understood by examining its subcomponents, such as its circulatory, digestive, or respiratory system, a plant organization only has true meaning as a total integrated system. Evaluation of the performance of the engineering, financial, maintenance, or marketing functions of a plant as separate, independent functions cannot provide a characterization of the total plant.

Organizations, like individuals, must preserve their integrity. They are built on unique corporate concepts, with the intention of accomplishing a specific purpose or mission. The integrity of organizations is defined in relation to its congruence, symmetry, or fit. It must have a balance between policy and practice; between philosophy and performance; between decisions and deeds.

One serious problem that limits plant performance is the fact that many organizations have failed to recognize the interrelationship of plant functions. Within these organizations, each function operates as a separate entity, without any coordination or communication between other plant functions.

In 1986, the *Society of Manufacturing Engineers* examined the balance between current levels of manufacturing technology and of company organizations. The conclusion of this study was that American industry, in its drive to become more competitive, is attempting to put *fifth-generation technology into second-generation plant organizations*. The study also concluded that all forms of advanced manufacturing technology require an organizational form, style, and culture that are attuned with the standards and processes of the plant.

The success of the maintenance function depends on the participation, and total commitment of all employees within the company. Based on a 1987 survey conducted by the *U.S. General Accounting Office*, the most common approaches used by domestic industries to gain employee involvement include suggestion systems, information sharing, training, and survey feedback. While these methods had a positive short-term impact on overall plant performance, the probability that they will provide long-term employee commitment is not very great. While the intent of the program is not to increase staff or create an additional management structure, it will require a core group, which has the authority to and responsibility of program implementation. Care should be taken to ensure that this change in organization is not perceived as another quick-fix management solution.

FUNCTIONAL RESPONSIBILITIES

Each function within the organization has specific responsibilities that must be enforced as part of an effective *reliability excellence* change process. These responsibilities include

Responsibilities of Front Line Supervision

Supervision is primarily concerned with *getting today's work done today.* Specific responsibilities include

- Control over quality, duration, cost, and thoroughness of work:
  - Time lost between jobs and at breaks and shift changes.
  - Next job always ready.
  - Follow up on overruns and interruptions.
  - Balance motivation and discipline.
  - Each mechanic visited at least twice a day.
  - Significant jobs visited at least thrice a day.
- Training and motivation:
  - Identify and provide or obtain the skills training required by each crew member.
  - Give adequate time and attention to formal and on-the-job training. Never neglect the development of your people.
  - Act upon requests for support and help. Effective listening reduces grievances.
- Tactical decisions to stay on schedule:
  - Refine and finalize labor, materials, priorities, and methods.
  - Tactics often must be established after job start.
  - Communicate with planners/schedulers, operations/maintenance management, and reliability engineering as frequently as possible.
- Administrative or personnel functions:
  - Control tardiness, absenteeism, and vacations.
  - Assure reasonably accurate distribution of time and materials to specific jobs.
  - Prompt and fair handling of grievances.

Responsibilities of Maintenance Planning and Scheduling

Planning is concerned with preparing work to be done in the future. Specific responsibilities include

- Customer liaison for nonemergency work
- Job plans and estimates
- Full day's work each day for each man

- Work schedules by priority
- Coordinates availability of manpower, parts, materials, equipment in preparation for work execution
- Arranges for delivery of materials to job site
- Ensures even low priority jobs are accomplished
- Maintains records, indexes, charts
- Reports on performance versus goals

Responsibilities of Reliability Engineering

Reliability engineering is primarily concerned with application of technical skills and ingenuity to the correction of equipment problems causing excessive production downtime and maintenance work. The position is dedicated to the maintenance function and is focused on the elimination of repetitive failure.

- Ensure maintainability of new installations.
- Identify and correct chronic and costly equipment problems, eliminate repetitive failure.
- Technical advice to maintenance and partners.
- Design and monitor an effective and economically justified, preventive or predictive maintenance program.
- Proper operation and care of equipment.
- Comprehensive lubrication program.
- Inspection, adjustments, parts, replacements, overhauls, and so on, for selected equipment.
- Vibration and other predictive analyses.
- Protection from environment.
- Maintain and analyze equipment data and history records to predict maintenance needs.

KEY CONSIDERATIONS OF ORGANIZATIONAL STRUCTURE

Following are key considerations for an organizational structure supportive of reliability excellence:

- Set forth organizational principles and ground rules:
- Maintenance management should be structured level with operations management.
- Maintenance is not subordinate to operations.
- Supportive service versus subordinate service.
- Defined roles, responsibilities, and authorities:
  - Operation department
  - Maintenance department

SELF-DIRECTED WORK TEAMS

Lean production is rapidly displacing conventional mass production at manufacturing companies in the United States and throughout the world. Human resource practices play a critical role in any company's program to develop and institutionalize lean methods on the shop floor. One approach that has been successful at many companies involves organizing production workers into self-directed or self-managed work teams.

Teams of between five and fifteen workers take responsibility for an integrated, customer-driven production process. Team members cross-train in many of the tasks within the defined process, and gradually expand their capabilities to include administrative and support roles. As the team matures, it slowly becomes increasingly autonomous, until it functions with minimal supervision.

Self-directed work teams are quite similar to lean production teams. In fact, many students of shop-floor organization have failed to make a distinction between the two. Figure 5.1 clarifies the difference between lean production teams and self-directed work teams by highlighting the similarities and differences. Although these are separate and distinct concepts, many principles of both are common, notably teamwork, empowerment, participation, flexibility, and cooperation. There are two important distinctions, however. First, participation for lean production teams focuses heavily on continuously improving the process and relentlessly eliminating waste. For self-directed work teams, in contrast, participation focuses on allowing shop-floor workers to take on administrative and managerial tasks, in addition to regular production activities.

Second, lean producers expect team members to cross-train in all of the skills within the team's boundaries. As a result, team members cannot ever achieve more than a moderate level of expertise in any one activity, and the team may end up with excessive redundancy. Self-directed teams, on the other hand, recognize that human limitations place a cap on the number of different competencies that any one individual can master. Self-directed teams do not expect every team member to learn every skill within the team's boundaries. Rather, the team leverages the diverse and complementary skills of all of its members to ensure that the team as a whole has all of the needed competencies with just enough redundancy.

Lean production and self-directed work teams developed historically at different times and in different parts of the world. Nevertheless, given that the two concepts share so much common ground, many lean producers have begun to expand the boundaries of their work teams to encompass not only *kaizen* activities, but also administrative and managerial tasks; as a result, they have created truly self-directed work teams under the lean production banner. For the remainder of this paper, we will refer exclusively to self-directed work teams that fit this expanded definition.

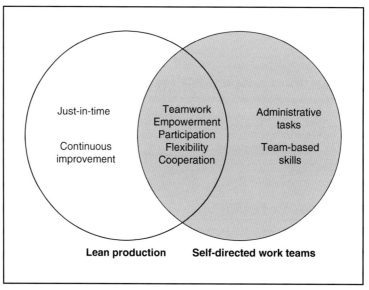

FIGURE 5.1 Differentiation between Lean and Self-directed Teams.

Implementing Self-directed Work Teams

Creating self-directed work teams is an evolutionary process consisting of four major steps: (1) cross-training, (2) enhancing teamwork skills, (3) participating in proactive improvement efforts, and (4) developing administrative skills.

Cross-Training. Workers must learn how to do many but not necessarily all of the production tasks within the mission of their team. Cross-training not only allows team members to substitute for absentee workers on a moment's notice, but it also enables the team to realize job rotation. Rotation has the advantage of adding variety to workers jobs and thus relieving boredom.

Teamwork. Workers must enhance their ability to work with others, by learning or improving such skills as cooperation, conflict resolution, communication, negotiation, and consensus formation.

Proactive Improvement. Management must empower workers to begin diagnosing and analyzing production processes, and to develop and implement ideas for improving quality, increasing productivity, and reducing waste.

Developing Administrative Skills. Team members need to develop production support and administrative skills, including maintenance and repair, quality control, scheduling, purchasing, inventory control, personnel management, performance measurement, and personal computer skills.

In addition, a smooth transition depends heavily on securing management commitment, preferably early in the process. Senior managers can provide the vision and leadership needed to spark change throughout the organization. Equally important, the buy-in of middle managers can help overcome their natural fear of organizational change, and prevent any counter-productive resistance from developing.

Managers and team members alike must remember that this process takes a great deal of time, and teams may experience setbacks en route to full self-direction. Teams that attempt to take on too much responsibility too quickly are destined to fail miserably. Moreover, when things don't work out well for a specific initiative, as is bound to happen occasionally, it is important to pick up the pieces, learn from the mistakes, and move on.

Assigning blame is a counter-productive effort in futility, and managers and workers must overcome this temptation, which is so common under the mass production system. These lessons include continuous improvement, motivated workers, overcoming resistance to change, and appropriate incentives.

Continuous Improvement. Managers and workers must never lose sight of the fundamental purpose of adopting self-directed work teams—to continuously improve the quality of manufacturing and business processes. Eliminating defects, reducing costs, improving on-time delivery performance, promoting safety, and providing stimulating work not only add value for the customer and reward workers with greater job satisfaction, but they also create wealth for the shareholders.

Motivated Workers. Likewise, the active participation of team members demands a lot of hard work and persistence. Periodic setbacks will inevitably hold up the transition, and only highly motivated workers will show the persistence needed to stay on track.

Overcoming Resistance to Change. Organizational change by its very nature threatens the established order, and creates fear and discomfort among workers and managers alike. The natural reaction of entrenched interests and those who fear change is to resist. Proponents of change must keep a constant vigil against subterfuge aimed at derailing the team initiative.

Appropriate Incentives. One of the most effective means of promoting all of these efforts is to develop incentive structures that reward the right behaviors and undermine resistance to change. Pay-for-knowledge, pay-for-performance, excellence awards, business performance scorecards, job performance evaluations—any or all of these programs can provide strong incentives that support the self-directed team concept.

MAINTENANCE ORGANIZATIONAL STRUCTURE

Best practices for a maintenance organizational structure are directly dependent on factors such as operations business plans, maintenance work types, and the like.

Operations Business Plan

The maintenance organization must be established to meet the demands of the operations function. For example, a plant that will be operated 24 hours per day, 7 days per week requires a maintenance organization structure that can support that mode of operation. The maintenance workforce must be distributed to support continuous operation and has an effective planning and scheduling that can effectively take advantage of "windows of opportunity," for example, periods when production demands permit sustaining maintenance activities. On the other hand, when the production cycle is 24 hours per day, 5 days per week, the maintenance organization must be configured to take full advantage of the 2 day window, for example, weekends, to perform sustaining maintenance.

Maintenance Work Types

An effective maintenance organization must be organized to provide different levels of maintenance by work type. As a minimum, the maintenance organization must be configured to provide effective, quality support for three major work classifications or types, namely, emergency maintenance, preventive maintenance, and periodic rebuilds and overhauls.

Emergency. All maintenance organization must provide timely response to emergency work request without adversely affecting its ability to effectively utilize the workforce or negatively impact total maintenance cost. In most cases, this requires an organization structure that dedicates a small percentage of the craft workforce, as well as planning and supervisory support, to emergency response work.

Preventive Maintenance. Preventive maintenance is an absolute requirement of asset reliability and effective management of asset life cycle cost. An effective maintenance organization must dedicate a portion of its craft workforce, as well as planning and supervisory support, to consistent, timely execution of preventive maintenance activities.

Periodic Rebuilds and Overhauls. Without exception production assets require periodic overhauls or rebuilds to replace wear parts, finite-life components and to assure that acceptable levels of reliability are consistently maintained. Because of the liability or risk, as well as higher skill levels associated with major rebuilds or overhauls of capital assets, the organization structure must assure that the best qualified crafts are utilized for this type of work.

Other Considerations

- Work execution.
- Planning and scheduling.
- Maintenance engineering.
- Central and area assignments balanced to extent of economic soundness.
- When one component of any organization maximizes, the organization suboptimizes.
- Planning and scheduling the key.
- Application of technical knowledge.
- Consider the nature of maintenance work and its control.
- Consider the impact of technical advancements on the nature of maintenance and production assignments.

- Organization of future.
- Encompassing job fulfillment.
- Rationalize the maintenance shift schedule.
- Off-shift schedule.
- Primary maintenance shift.
- Split shift needed.

Regardless of which organization is used there should always be a current and complete organizational chart that clearly defines all maintenance department reporting and control relationships, and any relationships to other departments. The organization should clearly show responsibility for the three basic maintenance responses: routine, emergency, and backlog relief.

SUPERVISION AND SUPPORT REQUIREMENTS

Unlike operations, reliance on self-directed or self-managed teams does not work well in maintenance organizations. The nature of maintenance work does not lend itself to natural work teams and must rely on more traditional organizational structures for success. The following information provides guidelines for an effective organizational structure.

Best Maintenance Practice: Span of Control Ratios

To support effective identification, prioritization, planning, and execution of maintenance activities, the organization structure must provide direct and indirect support to the craft workforce. The span of control for these support activities include

Support position	Ratio to technicians
Supervisors	1:10 (8 to 15)
Planner/Schedulers	1:20 (12 to 30)
Reliability/Maintenance engineers	1:40 (40 to 70)
Maintenance clerks	1:40
Training coordinator	1:80
Composite	1:5

These ratios exclude maintenance management above the first-line supervisory level and also the maintenance control manager if staff support exceeds three positions. In very large organizations there may also be provision for *safety inspector* and/or *quality control inspector*. More detailed composition of maintenance staff by departmental size is

Technicians	Manager	Supervisor	Planner/scheduler	Clerk	Reliability engineer	Maintenance control manager	Training coordinator
1–8		1					
9–13		1	1				
14–19		2	1	1			
20–24	1	2	1	1			
25–29	1	3	1	1			
30–34	1	3	2	1			
35–39	1	4	2	1			
40–44	1	4	2	1			
45–49	1	5	2	1	1		

(Continued)

(*Continued*)

Technicians	Manager	Supervisor	Planner/scheduler	Clerk	Reliability engineer	Maintenance control manager	Training coordinator
50–54	1	5	2	1	1		
55–59	1	6	3	1	1		
60–64	1	6	3	1	1	1	
65–69	1	7	3	2	1	1	
70–74	1	7	3	2	1	1	
75–79	1	8	3	2	1	1	1
80–84	1	8	3	2	2	1	1
85–89	1	9	4	2	2	1	1
90–94	1	9	4	2	2	1	1
95–99	1	10	4	2	2	1	1
100–109	1	10	4	2	2	1	1
110–119	1	11	5	2	3	1	1
120–129	1	12	5	2	3	1	1
130–139	1	13	6	2	3	1	1
140–149	1	14	6	2	3	1	1

CHAPTER 6
OPERATING POLICIES OF EFFECTIVE MAINTENANCE

Tom Dabbs
Vice President, Life Cycle Engineering, Inc., Charleston, S.C.

This chapter covers basic policies for the operation of a maintenance engineering department. While many of these policies overlap and are interdependent, they may be grouped in four general categories:

Policies with respect to *work identification*
Policies with respect to *work prioritization*
Policies with respect to *work allocation*
Policies with respect to *work force*
Policies with respect to *intraplant relations*
Policies with respect to *performance management and control*

POLICIES WITH RESPECT TO WORK ALLOCATION

To Schedule or Not to Schedule?

It is generally accepted that, in any maintenance department where there are more than 10 craftspersons and more than two or three crafts, some planning, other than day-to-day allocation of work by supervisor or leadsperson, can result in improved efficiency. As the size of the maintenance organization, for example, scheduling, increases, the extent to which work planning can be formalized and the amount of time that should be spent on this activity are increased. There should be only as much planning as necessary for maximum overall efficiency so long as the system costs less than the cost of operating without it.

How Much Scheduling?

There are practical limitations to any scheduling system. A very detailed schedule that because of emergencies becomes obsolete after the first hour or two of use is of little value. If, however, actual performance indicates from 60 to 80 percent adherence during normal operation, the value of the schedule is real. Justification of any scheduling system requires proof of its effectiveness in dollars saved. Where some form of incentive system or work measurement exists, such proof is readily available. But in most maintenance departments no such definitive method is available and the only criteria of measurement are overall trends in maintenance costs and quality of service.

Some aspects to be considered in arriving at a sound work-scheduling procedure are work unit, size of jobs scheduled, percent of total work load scheduled, and lead time for scheduling.

Work Unit. Most detailed schedules are laid out in terms of labor-hours or, if standard times are used, fractions of hours. Other scheduling systems use a half craft-day as a minimum work unit. Others may use a craft-day or even a craft-week as a basis.

Size of Jobs Scheduled. Some work-scheduling systems handle small jobs as well as large ones. Others schedule only handle major work where the number of craftspersons and the length of time involved are appreciable.

Percent of Total Work Load Scheduled. Although in some cases all work may be scheduled, the most effective systems recognize the inability of any maintenance engineering department to anticipate all jobs, especially those of an emergency nature, and do not attempt scheduling for the entire work force. A portion of the available work force is left free for quick assignment to emergency jobs or other priority work not anticipated at the time of scheduling.

Lead Time for Scheduling. Lead time for scheduling, or the length of time covered by the schedule, is another variable to be considered. Some scheduling systems do not attempt to cover breakdown repairs and are limited to the routine preventive maintenance and to major work that can be anticipated and scheduled well in advance. In these cases a monthly or biweekly allocation of manpower suffices. In most instances, however, a weekly schedule with a 2- or 3-day lead time results in good performance, yet is sufficiently flexible to handle most unexpected work. In extreme situations a daily schedule with a 16- to 18-hr lead time may be necessary to provide the necessary control. A more workable solution for this situation, however, involves use of a master schedule for a minimum of 2 weeks with provision for modifying it daily.

Selection and Implementation of a Scheduling System

Flow-of-Work Requests. Before any formalized scheduling program can be initiated, the method of requesting work from the maintenance department should be formalized. This request may take the form of a work description or job ticket, listing labor hours or equipment requirement, or it can be in the form of a work sheet on which the same type of information is accumulated by either verbal or written communication. Regardless of the form this information takes, it must be routed to one central point if a scheduling system is to be used. In a small plant this can be the supervisor, self-direct team leader, the maintenance superintendent, or the maintenance engineer. In a larger maintenance department it should be through a staff individual or group.

The amount of information on the work request depends upon the type of talent used in the scheduling group. If the individual charged with planning is completely familiar with the job requirements and can determine the craft skills and labor-hours involved, the necessary equipment, and any other information required for scheduling, a summary of the jobs will suffice. On the other hand, where complexity of work is such that it is practically impossible for any individual to have this information, or if the person charged with scheduling does not have the training necessary to analyze the work, then the information on the work request must be presented in more detail. The number of labor-hours required, by craft, the timing, the relation between crafts, the location and availability of parts and equipment, and any special requirements concerning coordination with production schedules or personnel should be included.

In addition to job information required for planning, it is equally important to have a feedback on actual performance in terms of notification of completion and actual time consumed, by craft. This may be incorporated in the work-request system, but provision must be made for channeling this information back to the scheduling center. The scheduling system should also provide for work scheduled but not completed becoming a part of the work backlog. As such, it is considered, along with new work, for new scheduling.

Determination of Priority. In any maintenance organization which is efficiently manned, the work load, in terms of quantity or timing, exceeds the availability of men and/or equipment. For this reason the problem of defining the order in which the work is to be carried out, or establishing priority, exists and is an important factor in scheduling. In a small plant with one operating department and a small maintenance organization, establishment of priorities may amount to casual discussion between maintenance and production. However, as the plant grows and the maintenance department is called upon to provide service to more than one production department, the problem of equitable and efficient priority assignment becomes more involved. One of the most serious problems in maintaining good relations between maintenance and production departments is in this sphere. Too frequently personalities, working conditions, accessibility, or geographic location with respect to central shops influence the order of work assignment. This may decrease the overall efficiency of the plant.

The means for determining work priority figures most importantly in the establishment of a work-scheduling system. On the surface a solution to this problem would reserve decisions concerning priorities to an individual who is in position to judge the effect on overall plant performance. In a plant of any size, it is usually most effective to handle such decisions at a lower level of management, with the plant manager having the final say when no decision as to priority of work can be reached.

A method which has proved satisfactory in many instances has been to assign a rough allocation of craft manpower to each production department, then to establish the priority of work within each department by consultation with its supervision. When it is necessary to vary the allocation of men, this should be done by negotiation between production departments to arrange a mutually agreeable exchange. If such a reallocation cannot be concluded, as a last resort the plant manager must make the decision.

Coordinating and Dispatching. In the execution of an effective scheduling system it is necessary to compromise with the practical considerations of getting the work done, and done economically. If a supervisor or team leader guided his or her craftspersons on the assumption that the job must be completed at the exact time he had estimated and then continued to assign work on the basis of his estimate of the time necessary, it is obvious that confusion, incomplete work, and idle craft time would result. A formal schedule, issued weekly and followed blindly, would have the same effects. Instead, the schedule should be used as a guide, and modifications can be made as needed. Rapid communication of such modifications to the men responsible for carrying them out is essential to the success of a work schedule.

It is also essential that any changes or unexpected work for which provision has not been made in the schedule be funneled through the dispatch center. Usually the dispatch center can incorporate this type of work more efficiently than is possible by random selection of the nearest craftsmen or injection of higher authority into the picture.

Preventive versus Breakdown Maintenance

Preventive maintenance has long been recognized as extremely important in the reduction of maintenance costs and improvement of asset reliability. In practice it takes many forms. Two major factors that should control the extent of a preventive program are first, the cost of the program compared with the carefully measured reduction in total repair costs and improved asset performance; second, the percent utilization of the asset being maintained. If the cost of preparation for a preventive-maintenance inspection is essentially the same as the cost of repair after a failure accompanied by preventive inspections, the justification is small. If, on the other hand, breakdown could result in severe damage to the asset and a far more costly repair, the scheduled inspection time should be considered. Furthermore, in the average plant preventive maintenance should be tailored to fit the function of different items of equipment rather than applied in the same manner to all equipment. Key pieces of equipment in many other integrated manufacturing lines are in the same category. Conversely, periodic inspections of small electric motors and power transmissions can easily exceed the cost of unit replacement at the time of failure.

Indeed, a program of asset or component replacements can result in considerably lower maintenance costs where complete preventive maintenance is impractical. In a plant using many pumps, for instance, a program of standardization, coupled with an inventory of complete units of pumps most widely used, may provide a satisfactory program for this equipment. This spare-tire philosophy can be extended to many other components or subassemblies with gratifying results.

Sometimes, instead of using a centrally administered formal preventive program, qualified mechanics are assigned to individual pieces of equipment, or equipment groups, as mechanical custodians. Operating without clerical assistance and with a minimum of paperwork, these men, because of familiarity with equipment and ability to sense mechanical difficulties in advance, can effectively reduce maintenance costs and breakdowns. These compromise devices can frequently be used to greater advantage, even in plants where equipment is not in continuous operation and a more comprehensive preventive program might be set up.

Periodic shutdown for complete overhaul of a whole production unit, similar to the turnaround period in oil refineries, is another method of minimizing breakdowns and performing maintenance most efficiently. Unfortunately, this is a difficult approach to sell to management of a 7-day, around-the-clock manufacturing plant not accustomed to this method.

One of the most effective methods of tempering ideal preventive maintenance with practical considerations of a continuous operation is that of taking advantage of a breakdown in some component of the line to perform vital inspections and replacements which can be accomplished in about the same time as the primary repair. This requires recording of deficiencies observed during operating inspections and moving in quickly with craftsmen and supervision prepared to work until the job is done. Production supervision usually can be sold the need for a few more hours' time for additional work with repair of a breakdown much more easily than they can be convinced of its necessity when things are apparently running smoothly.

Reliability Engineering

One of the most important tools in minimizing downtime, whether or not a conventional preventive-maintenance program is possible, is called *reliability engineering*. Although this would appear to be the application of common sense and design best practices to asset design and maintenance engineering. It is a field which is often neglected. Too often maintenance engineers are so busy handling emergency repairs or in other day-to-day activities that they find no opportunity to analyze the causes for breakdowns which keep them so fully occupied. While most engineers keep their eyes open to details such as better packings, longer-wearing bearings, and improved lubrication systems, true reliability engineering goes further than this and consists of actually setting aside a specific amount of technical manpower to analyze incidents of breakdown and determine where the real effort is needed; then through redesign, substitution, changes, and specifications, or other similar means, reducing the frequency of failure and the cost of repair.

This can be handled by a special group acting as a cost-reduction unit, or it can be included as one of the functions of the maintenance engineer. Some companies can support groups that actually develop and test equipment to promote more reliability, maintenance-free operation, and optimum life-cycle cost. The aid of equipment suppliers can be solicited in this same effort. It should be emphasized, however, that this type of program requires intelligent direction to ensure that time and money are expended in the areas where the most return is likely. A particular pump, operating under unusual conditions, shows a high incidence of failure but because of the simplicity of repair has a low total maintenance cost, and if it were the only one of its type in the plant, an intensive investigation for maintenance-cost reduction would be difficult to justify. On the other hand, a simple component such as a capstan bearing on a spinning machine, although having a low unit-replacement cost, can fail so often and on so many machines that the total cost per year would run to many thousands of dollars. Here an investigation concentrated on the reason for failure of one unit could be extremely profitable. Effective preventive engineering can result only when it is recognized as an independent activity of a research nature that cannot be effectively sandwiched into the schedule of a man who is occupied with putting out fires.

POLICIES WITH RESPECT TO WORK FORCE

In-House Work Force Or Outside Contractors?

The primary factor in deciding whether to use an outside contractor is cost. Is it cheaper to staff *internally* for the performance of

1. The type of work involved
2. The amount of work involved
3. The expediency with which this work must be accomplished

In studying these relative costs it is not sufficient to consider the maintenance cost alone. The cost to the company, including downtime and quality of performance, must also be considered.

To establish, supervise, and maintain a group of men in any specific craft means a continuing expense over the wages paid to the men. In general, this total cost must be balanced against the estimated cost for the same work performed by an outside contractor who must, in all probability, pay higher wages, carry the overhead of his operations, and realize a profit. By analysis of the work load and evaluation of the relative costs of its performance by plant maintenance or outside contractors, criteria for this division of work can be evolved. This analysis must include other factors such as time required, availability of the proper skills with outside contractors and in some instances, the possibility of process know-how leakage if contractors are employed. In deciding whether to set up your own shop or rely on contractors, the degree of skill required in the particular craft is important. If this requirement is relatively low, and supervision and facilities of some other craft can be expanded to include it, this step can often be profitable. If, on the other hand, the degree of skill is high or the necessary equipment complex or costly, there must be a much greater amount of work for this craft before such a shop can be justified.

Once the basic craft types have been established for the maintenance organization, the question of the personnel strength of these crafts is also a function of the amount of work assigned to outside contractors. In general it is wise to staff the in-plant force so that it can handle a work load slightly above the valleys in anticipation that some of the peak work periods can be deferred. Outside contractors may then be used for the normal peaks and for the unusually high loads resulting from major construction or revision projects.

The preceding discussion has omitted two elements which may have an arbitrary effect on the practical distribution of work between inside and outside labor. If local manpower and contractors are scarce, expeditious work performance often dictates maintaining a much larger and more diversified maintenance group. The use of outside contractors may then be limited to major projects where it is feasible to use imported labor for an extended time, or to highly specialized work performed by a factory representative or a contractor specializing in this work as a job unit.

The other factor which may interfere with the optimum formula is the attitude of the labor organizations involved. This is a problem which varies not only among geographical areas, but frequently among plants in the same area. In some instances the plant union is militant at the prospect of any work being performed within the plant by nonunion workers or by members of another union. In other plants an understanding is reached, generally limiting contractor participation to construction or major revisions. Other plant unions recognize factors, which permit considerable latitude in the use of outside contractors. For instance, the union may recognize that the amount of concrete and masonry work will not justify more than a minimum repair crew and that any new work or major repair in this line would be more economically handled by an outside contractor. Many unions recognize the need for employment of contractors in such fields as refrigeration, window washing, and steeplejack services. An optimum solution for this problem is more likely to result if the union and maintenance supervision arrive at a mutual understanding of the problem in advance. However, this is often difficult, and failure may occasionally necessitate establishment of uneconomical crafts and work allocation for the in-plant work force.

On the other hand, the situation may be reversed, with outside craft unions refusing to perform in-plant work unless granted exclusive rights to all the work or at least to certain clearly defined portions of the work. This presents an entirely different problem and generally favors the expansion of the in-plant group, with respect to both numbers and crafts involved, so as to minimize the use of contract labor, limiting its use to major new construction.

Coverage

In the process industries, where plants frequently operate continuously—three shifts, 7 days a week—some of the maintenance load can be separated and handled simply. Maintenance of buildings and grounds, for instance, is the same for three-shift operations as with one shift. For the rest, however, special consideration is required to provide the service necessary for optimum production. Not only will lubrication and breakdown repairs continue around the clock, but other items such as waste collection, janitor service, elevator maintenance, and fork-truck maintenance must be considered in a different light from the same services in a plant on a one-shift basis. The two extremes in providing maintenance for continuous operation are to provide full coverage during all hours that the plant is in operation or to maintain day coverage only, letting the plant shift for itself during other periods or to accept minimum essential service on call-in, overtime basis. The optimum arrangement is something in between, depending a great deal upon circumstances in an individual plant.

In considering the staffing of a maintenance department to cover more than one-shift operation, many factors are involved.

Efficiency of the Worker. Although exception may be taken to this statement, it is generally conceded that a man who is not paced, by either the equipment he operates or the performance of a large group of individuals, is not so efficient on the off shifts as during the day. This loss of efficiency can be attributed to many causes. First, a man is normally happier living a normal life, which in most communities includes sleeping at night and working days. Most of his out-of-plant relationships are with people living this sort of life. The activities of his wife and children are normally concentrated in the daylight hours. All these factors make for conflict in an attempt to reconcile the schedule of the shift worker with that of his family and friends.

Another cause for a loss in efficiency is the fact that usually the work of a maintenance man must be coordinated with production activity. Even though every attempt is made to plan this activity, unexpected variances occur which call for changes in coordination with production. Since most of these require decisions at supervisory levels normally at work during the day, delays frequently occur, resulting in a loss of efficiency.

While some types of operations may justify both production and maintenance supervision on the scene at full strength around the clock, usually only the supervision necessary to maintain operations on an essentially static basis is available during the off shifts. Around-the-clock maintenance must be weighed against the reduction in efficiency resulting from the absence of adequate authority. Efficiency may also be reduced by the need for unexpected supplies, tools, or equipment which can be procured only from outside suppliers during regular working hours. The alternative may be improper substitutions or costly on-the-spot fabrication, either of which will reduce maintenance efficiency.

There are other factors which argue for around-the-clock maintenance, such as the location of a plant with respect to the homes of the craftsmen, which may make call-in impractical. In other cases a particular production unit may be so critical as to make any maintenance delay intolerable, or a breakdown may create a safety hazard so grave that maintenance coverage must be provided, regardless of its economic justification.

Experience indicates that minimum downtime and lowest maintenance cost result from using the least coverage on the off shifts that can be tolerated from the standpoint of safety and lost production time. Adequate craft supervision should be provided where justified, or this responsibility should be transferred to some other member of supervision. As much work as possible should be handled on the day shift. The cost of call-in overtime should be compared with the cost of scheduled coverage, including the cost of delays resulting from call-in. The cost of finishing jobs of more than 8-hr duration should include comparison of cost of holdover overtime with that of a second or third shift. The amount of routine work that can be assigned to fill out the time of men on off shifts and the amount of application of the men to this work that can be reasonably expected is another factor. Where both centralized and decentralized maintenance groups are available, men on the off shifts, with the possible exception of such specialized crafts as electricians and instrument men, should be from the decentralized group.

It is sometimes possible to use a split-day-shift schedule with one crew working Monday through Friday and another on a Wednesday through Sunday schedule to extend day coverage over a 7-day period. In some instances it may be more economical to have a large day crew, an intermediate afternoon and evening shift, and a skeleton midnight shift.

The best plan for any plant can be determined only after due consideration of all the factors mentioned above and any other special considerations peculiar to the plant. This plan may have seasonal variation or may change with the plant's economic situation.

Centralization versus Decentralization

The subject of centralized versus decentralized maintenance has elicited a great deal of discussion over the past few years, with strong proponents and good arguments on each side. Advantages of a centralized maintenance shop are

1. Easier dispatching from a more diversified craft group
2. The justification of more and higher-quality equipment
3. Better interlocking of craft effort
4. More specialized supervision
5. Improved training facilities

The advantages of decentralized maintenance are

1. Reduced travel time to and from job
2. More intimate equipment knowledge through repeated experience
3. Improved application to job due to closer alliance with the objectives of a smaller unit—"production-mindedness"
4. Better preventive maintenance due to greater interest
5. Improved maintenance-production relationship

In practice, however, it has been found that neither one alone is the panacea for difficulties in work distribution. Often a compromise system in which both centralized and decentralized maintenance coexist has proved most effective. For handling major work requiring a large number of craftsmen, the centralized maintenance group provides a pool which can be deployed where needed. To provide the same availability in a completely decentralized setup would mean staffing at dispersed locations far in excess of optimum needs for each area, plus difficulty in coordinating on the big job. The installation of some of the costly and specialized equipment that is needed for some of the crafts can seldom be justified at other than a central location. On the other hand, a great deal can be accomplished in minimizing downtime by having a decentralized group which can function "Johnny-on-the-spot" and give immediate attention to minor maintenance problems. Familiarity with a smaller sphere of production equipment through experience is almost certain to improve the performance of craftsmen. In general, good overall efficiency will result from the decentralization of a specific number of the less specialized crafts in area shops, augmented by minimum personnel of specialized crafts to provide emergency service in their field. An improvement over this would be the utilization of a general craftsman who can perform the work of many crafts in a decentralized group. This, of course, presents a problem with organized labor and will normally require agreement from the union. It also limits the skill that can be expected of such men, since there are few men who can become experts in all the crafts.

It is suggested that, rather than assign an arbitrary number of people to a decentralized facility, a comprehensive study be made of the type of service required to sustain production in the area under consideration, and that from this service be separated which can be performed by a general area mechanic. The incidence of this type of work and the resulting area mechanic work load should then be determined from some factual records, and a sufficient number of men assigned to handle this work load. The preventive-maintenance program can provide a reservoir of work for the maximum utilization of this decentralized group during periods of low breakdown or modification maintenance activity.

Recruitment

Unfortunately policies for recruitment of personnel for the maintenance department are controlled a great deal more by local conditions and expediency than by the ideal approach. This in itself is a major argument for maintaining as stable a work force as is economically practical.

Where the union contract makes job posting mandatory, the problem of getting men who are or will eventually be satisfactory craftsmen can be difficult. All too often, since this particular problem is only a small part of the overall management-union relationship, little effort is made to arrive at a better method of filling vacancies among the crafts. Many maintenance departments in union plants have become resigned and make the best of the candidates turned up through the bidding procedure, usually selected on a seniority basis. With good union-maintenance supervision relationships supplementary agreements or understandings may be reached which will considerably improve the type of candidate considered. Age, aptitude, past experience, educational background, and general level of intelligence are frequently considered in some mutually acceptable screening technique. An accepted apprentice-training program with recognized entrance qualifications will generally create a source of competent personnel.

In plants where there is no problem of union resistance, local conditions and the makeup of the production work force are the major factors in recruitment of maintenance personnel. Where many of the operations being performed on the production line are of a mechanical nature, the probability of securing competent recruits from production is greater than in the process industries, which do not generally attract the type of individual suited for maintenance. Availability of trained prospects outside the plant is naturally better in highly industrialized communities. The qualifications of candidates for maintenance work and methods for evaluation of applicants is a subject in itself. However, a few generalities can be made concerning the two types which make up the craft groups of a maintenance department. These two types are the untrained candidate, who enters at the bottom of the scale and receives his training while employed in the maintenance department, and the completely trained, skilled mechanic. In evaluating the untrained candidate, primary considerations should be age, mechanical aptitude, manual dexterity, and analytical ability. Some degree of self-assurance and stability of character is important. Also, the candidate's motivation for entering the crafts field should be thoroughly explored during interviews. It is preferable that this motivation be a real liking for the type of work rather than a desire for more money, security, prestige, or some other factor. In selecting trained applicants, age and education should carry less weight. Experience is most important in this case, as well as attitude and type of work he has done, but also regarding the quality of performance, teamwork potential, ability to carry out assignments without constant supervision, and his personal stability. Summarizing, policies with respect to the recruitment of maintenance personnel are controlled largely by the conditions existing at a specific plant. Every device for the best selection of the available personnel should be employed, and the use of advanced techniques in testing, interviewing, and screening is recommended.

Training

There are several methods for training personnel in a maintenance department. The simplest and most effective is an established and recognized apprentice-training program. The details of such a program are available from many sources, but the most widely used is the apprentice-training program sponsored by the U.S. Department of Labor, Bureau of Apprentice Training. Usually the administration of this program is handled by a state organization which will provide all the necessary information, as well as assistance in adapting the program to an individual plant.

Many companies establish their own apprentice-training programs which are similar to this nationally recognized one. This requires considerably more preparatory work by the company but is not so widely recognized and therefore does not have the same appeal to the craftsmen as the national program. Administration of both systems requires about the same attention.

On the other hand, many plants have no formalized training for their craftsmen and depend entirely upon exposure, supervisory job coaching, and association with experienced workmen for their training. In between there is a whole range of possibilities, including such variations as "short-course" on-the-job programs, qualification and skill-development evaluation tests, promotional programs based on either formal or informal evaluations, and less detailed apprentice programs.

The factors that should influence the degree of formality of the training program are similar to those used to determine many other aspects of maintenance operations, that is, size of the plant, attitude of the labor group, availability of skilled craftsmen, and the overall policy of management. A large plant can obviously afford to initiate and maintain a more elaborate training program than can a small plant. The lack of availability of skilled craftsmen increases the need and justification for better training.

Training programs have been installed with and without the support of organized labor, but in general, they are more effective with the wholehearted support of the crafts group, particularly if it is jointly administered by the company and the union.

Above all, the amount of formal training to be used must be based upon the value of the results. It is not good management to have a training program for the sake of having a training program. A training program should result either in improved maintenance performance or in proper staffing of a maintenance department. The availability of some craft skills in certain areas or a change in methods and techniques may be such that the only means of providing the necessary skills is through a training program. Frequently, although a comprehensive program cannot be justified for all crafts, programs for individual skills are a necessity. These can be handled internally or in cooperation with an educational institution or an equipment supplier. Excellent examples of this treatment are the courses run by the suppliers of welding equipment, which make it possible to provide men with up-to-the-minute instructions on developments in welding techniques.

POLICIES WITH RESPECT TO INTRAPLANT RELATIONS

Participation by Maintenance Personnel in Selection of Production Equipment

In some plants one engineering department handles all phases of engineering activity from design through construction and maintenance. In the majority of plants, however, the construction of major facilities or addition of major equipment is engineered by a separate organization, reporting at a higher level, or by outside engineering contractors. The primary mission of these activities is to project pilot-plant operations to production-scale facilities or to expand existing installations to meet increased production goals. Built-in ease of maintenance does not normally receive the same emphasis that would result from the same work done by people who are to be responsible for maintenance. Most progressive companies provide for representation from the maintenance group as well as from the production group in design and selection of new facilities. A trained maintenance engineer can draw upon his experience or that of his department in suggesting modifications or brands of equipment that will result in reduced maintenance cost after it is placed in operation. Good equipment histories on performance of existing facilities are invaluable in assisting this contribution to design and construction.

It is not meant to suggest that the maintenance engineer should attempt to control the design of new equipment. He should, however, be offered the opportunity to review designs and specifications carefully in order to predict maintenance problems and suggest modifications for reduced repair costs. If his recommendations are logical and well presented, they will usually be accepted, particularly when real savings can be demonstrated. All too often the maintenance department is handed a surprise package which can be a nightmare to maintain and quickly requires revision to make maintenance at all practical. This not only results in high maintenance costs but is extremely damaging to the morale of the department. In summary, the maintenance engineer can be of inestimable value to a design group, first, because of the performance records at his disposal and second, because of his ability to suggest changes reducing the maintenance problem.

Standardization of equipment, whether centralized for a multiplant company or delegated to the maintenance department in a single plant, is another factor to be considered in specifying equipment. In this case, also, the maintenance engineering department should play a major part in policy formulation. A considerable reduction in maintenance costs can result from a sound standardization program by

1. Simplifying training of both operating and maintenance personnel
2. Increasing interchangeability of equipment
3. Decreasing capital tied up in spare-parts inventory

As with preventive maintenance, a poorly established or inflexible program of standardization can be an obstacle and can be obstructive and costly. Any program of standardization should provide for transition to improved equipment types as they are available and should take local vendor relationships into account.

Design and construction groups should provide the maintenance department with recommendations concerning spare-parts and preventive-maintenance programs received from equipment suppliers. The former group can transfer their contact with the supplier to the maintenance department with much better effect than is possible when the maintenance department is required to make the contact independent of the work that has gone before.

The use of a group called the *project board* has proved extremely successful in smoothing the way for any new engineering venture. This group functions as a clearinghouse for progress of the work and brings together all the activities that can be expected to have contact with the work during and after installation. The project board consists of a qualified member from each of the departments involved. For example, in an expansion of existing facilities this board might consist of a representative from production, two or more from design engineering, one from maintenance engineering, and a representative from the safety department. If, on the other hand, the project is one involving a new process recently developed by a research group, a member of this organization should be included on the board. In this way the transition to an operating production unit is much easier, since the project board normally is in existence until a successful plant demonstration has been made and tentative operating procedures established. This approach gives the maintenance department, as well as production, the opportunity to grow with the job and to suggest the modifications which familiarity with similar equipment makes possible and which make the final operation so much more satisfactory.

Authority to Shut Down Equipment for Maintenance

The authority of a maintenance department to dictate shutdown of production equipment for needed repairs is controversial and has contributed a good deal to the friction that sometimes exists between maintenance and production departments. In some plants the maintenance department does have this authority and it is generally recognized. In others there is no such prerogative and the decision rests entirely with production. Usually, and preferably, the decision is reached jointly.

Naturally, there are many areas in which the maintenance department has essentially unilateral authority, particularly in building repair, yard maintenance, upkeep of shops, and the like. However, the primary responsibility for total manufacturing costs is usually that of a production department and so, therefore, is the ultimate control of production equipment availability. The maintenance department should have the complete confidence of production so that a recommendation for shutdown results in immediate consideration. A doctor has no authority to order medication or treatment for a patient if the patient refuses. However, the doctor's specialized training and knowledge are generally recognized, and once we retain him, it would be well to follow his advice. The same philosophy is applicable in the maintenance-production relationship.

Responsibility for Safety

Safety is one of the most important aspects of industrial management today. The maintenance department should play a large part in making its plant a safe one in which to work. Although general administration of the safety effort is usually delegated to a specialist group, the maintenance department is often the key to success of the program. Not only is it responsible for the safety of its own personnel, but by definition it also is responsible for providing mechanical safeguards and for maintaining equipment and services in safe operating condition. Because of this collateral responsibility, the safety function is often combined with maintenance in a small plant. In a larger plant there is a definite need for a separate staff group.

The problem of safety of personnel in the maintenance department is somewhat different from that of safety of production personnel. Although mechanical guarding and safe operating conditions can be maintained in the shops, most of the work performed outside the shop is of a nonrepetitive nature, frequently requiring operation of equipment with guards or other safety devices removed. Safety in maintenance department activity, therefore, depends to a much larger extent on the individual safety performance of its men. In a production department where obvious hazards can be kept guarded and personnel instructed in performing a routine operation, programs and specific safety instructions are most effective. In the maintenance department, however, the craftsman must be taught to think safety and translate his thoughts into a multitude of situations without much help from prescribed rules.

Whether the responsibility for the safety program rests with a staff safety department or with the maintenance department, the work to be done must be performed by the maintenance group. Standards for guards, grounding, bumping- and tripping-hazard elimination, warning signals, and safety devices must be closely followed. Installations of this type must be maintained in perfect operating condition. A maintenance department should not presume to ignore requests for work of a safety nature and must find the means for giving top priority to these jobs. Often the actual inspection of safety devices rests with the maintenance department. Where this inspection is a function of the safety department or production department, close liaison must be maintained with maintenance for the immediate correction of deficiencies.

During repair of equipment in production areas maintenance personnel must be continuously alert to the hazards they may be creating for themselves and less experienced personnel in the immediate area. Fire permits, lockout procedures, and warning signs must be used in this connection. Possibility of tools or pieces of equipment falling and injuring others is always present. Protection must be provided from exposure to welding, sandblasting, and oil spillage. Electrical work is always accompanied by potential hazards and deserves special attention.

In conclusion, while the responsibility of staff's safety may be part of the maintenance function in a small plant, usually it is preferable to have an independent safety department either reporting to top management or incorporated in the personnel department. Regardless of its staff responsibility for safety, the maintenance department in any plant has a direct responsibility for implementation of the safety program, and its supervision must recognize this and provide the means for its accomplishment.

Instrumentation

The question of responsibility of the maintenance department for instruments can best be answered by practical consideration of the problems peculiar to the plant involved. Instrument installation and maintenance theoretically should be considered in the same light as the addition of any other equipment. There are, however, several factors which make some other arrangement expedient, such as a separate department or assignment of this responsibility to the production department.

In a plant using relatively simple types of instruments their selection and maintenance is frequently a function of the electrical group. On the other hand, in some industries where instrumentation has been carried much further and includes knowledge of complex electronic components, particularly in fields of automation, instrumentation may be a separate plant department. In some industries instruments are the major tools of production personnel and smooth operation requires their intimate knowledge of the instruments involved. In this situation, except for major changes, the responsibility for instrument care is with the production department.

With the increased use and complexity of instruments the problem of providing trained personnel for selection and maintenance has also increased. Technical men must frequently be used in a maintenance capacity for effective service. Unless there are enough instruments to warrant staffing the maintenance department with this caliber of personnel, the responsibility may be best transferred to those technical personnel operating the plant.

POLICIES WITH RESPECT TO CONTROL

Communications

A starting point in analyzing the problem of communications and the types to be used is a study of the sort of information to be transmitted and the amount of detail involved through these three major channels:

1. Up through the supervisory organization
2. Down through the supervisory organization
3. Laterally across the same level of organization

Generally, all communications should be reduced to a minimum consistent with effective operation. It is also accepted that information should flow upward only as far as is necessary for effective action. Slower response frequently nullifies the value of higher-level judgment that might result from a flow of the information upward beyond this point. In addition, communication upward should be so handled that each level passes on only that information which is of *value* to the next level. Horizontal channels of communication should also be controlled to limit information to that necessary for effective cooperation between various sections of the maintenance group. In a small plant having only two or three levels between first-line supervision and the department head, and where most transactions can be handled by telephone or word of mouth, there is little problem. As the plant gets larger, with more intermediate levels of supervision, more procedural formality and greater specialization of duties develop. This evolution should be accompanied by clearly defined limits of authority for independent action at each level, with *action* communication up from any level limited to the decisions outside its authority. If a foreman has a question concerning his work that can be answered by his supervisor, there is no need to involve the superintendent or plant engineer in the transaction.

Copies of order or performance reports are too often distributed to people who ignore them or at best scan them briefly, with no thought of retention. Detailed information is frequently passed to top levels where it is meaningless unless summarized. It would have been better to transmit only the summary. Indiscriminate requirement for approvals of instruction sheets, order blanks, requisitions, and correspondence can also clutter channels of communication and delay action. This problem is characteristic of fast-growing organizations and should be reviewed periodically. Flow diagrams for all written instructions, reports, and approval systems are helpful in focusing attention on unnecessary steps which increase the work load on the supervisory and clerical organization and delay execution of the work.

This minimum flow should naturally be tempered by recognition of the natural desire of people to know what is going on around them, with respect to their own work, that of other departments, and the company as a whole. For this reason it is important to include provision for passing this type of information to the satisfaction of the personnel involved. In most attitude surveys among first- and second-level supervision more dissatisfaction is evident with the amount of this type of information than with that required for performance of their work. The morale of supervision, which in turn is frequently reflected in the morale of the hourly group, can be considerably improved by the proper dissemination of general information for its own sake, making them feel as though they belong.

There are at least two areas in the activities of a maintenance organization where effective use can be made of special advanced aids to communication. These are the transmission of work requests and job instructions from various sections of the plant to the proper coordinating group or work area, and the quick contact with personnel dispersed throughout a plant.

It is important that requests for maintenance work be promptly and accurately received at the dispatch center or at the individual shops. The advantages of written work requests should be thoroughly explored, since they can be justified at much lower levels of operation than is evident at first glance. When used, these written requests can be transmitted by courier or plant mail service. Telephone orders are frequently used to initiate the work, with a written confirmation following. This introduces considerable chance for error in word-of-mouth instructions, and there is a tendency to neglect confirmation once the work is performed. This in turn results in difficulty in accounting and repair cost analysis.

The use of written job instructions for all work, while it may seem troublesome, is basic to the development of many other control devices, particularly those for accumulation of information used to assist in improving operations and in accurate distribution of the resulting costs. General, or blanket, authorizations are frequently sufficient for repetitive small items such as light-bulb replacement, routine lubrication, and miscellaneous valve packing, but even here, handling of these requests by operating departments will direct attention to trouble spots and allow more equitable distribution of accumulated charges.

A major problem in some organizations is keeping in touch with members of supervision and craftsmen dispersed over a wide area. The perfect solution appears to be a lightweight, pocket-size, continuously operating two-way radio in the hands of every member of the maintenance department. The most common means of locating dispersed supervision is through a plant auto-call system. In a small plant this may be adequate, but as a plant grows and the system becomes more and more complex, with blacked-out areas developing, this method becomes slow and unreliable. If each member of supervision checks out of a central point to a predetermined area, the auto-call system can be augmented by other telephones, but maintenance problems are usually such that a foreman may go from one area to another on a route that is impossible to predetermine. Two-way radio on maintenance vehicles has been very successful for groups working in the open or in areas that are within sound of a parked vehicle. Radio becomes less effective where craftsmen are occupied inside a building, out of hearing from a vehicle-borne receiving unit. An individual paging unit about the size of a package of cigarettes which can be carried on the person at all times is available today. In its present stage of development, however, this device depends upon an induced current, necessitating special installations in each building covered. It has considerable value in compact, multistory plants, but where there are many smaller buildings through which the call system must operate the installation expense is rather high. Any maintenance department should consider the increase in supervisory efficiency that can be realized from quick contact with dispersed personnel and provide the best means that can be justified economically.

USE OF STANDARD-PRACTICE SHEETS AND MANUALS

There are many forms of standard-practice sheets, or standard job-instruction sheets, and instruction manuals used in maintenance departments. They are excellent devices for planning work, ordering materials, improving estimating accuracy, and training crafts personnel. Justification of cost of preparation and their ultimate effectiveness depend entirely on the particular problems of an individual plant.

A plant having a large number of identical machines or of machines having identical components which require a repetitive type of repair can justify more detailed standard-practice sheets than a plant with very little duplication of equipment or maintenance jobs. The need for standard-practice sheets also varies with the complexity of the repair and with the degree of skill and the experience of the men performing the work. Most equipment suppliers will provide excellent manuals which, although do not cover all the detail found in a standard-practice sheet, cost little and provide much assistance for maintenance of the equipment. Every effort should be made to maintain a complete supply of these manuals available to the men directly involved in the maintenance of the equipment. These may be reproduced and divided to provide each craftsman with a copy if this seems advisable. A work measurement or incentive system based on summarized elemental standards makes some sort of standard-practice sheet a must.

Most repetitive repairs can be profitably studied for the best approach, and a standard procedure developed. A typical standard-practice sheet should include specifications for the tools required, the necessary parts and supplies, a sufficiently detailed print of the equipment, indicating the components with sufficient clarity for the craftsman to follow the instructions, a step-by-step procedure with complete notes to cover any unusual or critical steps, and a close approximation of the time required. The development of these sheets is time consuming and expensive, and rapidly changing conditions and equipment may make them obsolete quickly.

Electrical- and piping-layout drawings for the plant should be available to craftsmen and their supervision for quick appraisal in execution of their work. These should be kept up to date, and new work or changes in the field should be recorded by supervision in charge of the change on the master copy of the appropriate print. A great deal of time and expense can be saved by the availability of clear up-to-date drawings for use in planning repair, replacement, or modification of existing installations. This is particularly true of underground systems which cannot be easily traced.

Cost Control

The subject of cost control in the maintenance department is a complex and controversial subject. While it is not intended in this chapter to go into details of cost-control systems or budgets, some generalizations can be made regarding cost-control techniques, cost indexes, and performance checks which may be found useful in establishing the overall cost policy of a maintenance organization.

Any indexes used for internal control should incorporate factors within the control of those people held accountable for performance. For instance, for the lower levels of supervision, man-hours per unit of work, per job, or per department maintained can be directly influenced by the efficiency of the men performing the work and are therefore good measures of performance at this level. At the maintenance superintendent level the overall cost of maintaining the plant in terms of the value of equipment maintained, or in terms of the goods produced or percent of operating time available, can be influenced by good planning, good engineering, and good management, and these broader indexes may be applicable.

Top management is generally interested only in that part of the total cost to manufacture which is chargeable to the maintenance department. It is not interested in high worker efficiency if poor engineering and poor planning result in higher overall cost, nor is it interested in an extremely low maintenance cost with respect to the value of installed equipment if, as a result, the total cost of manufacturing is increased. Good management in a maintenance department should provide such indexes as are needed to permit evaluation of the performance of the department internally and provide top management with the information they need to assess maintenance performance of this function as part of the big picture.

Some of the indexes that are commonly used are maintenance cost as a function of

1. Value of the equipment maintained
2. Pounds produced
3. Total manufacturing cost
4. Total conversion cost
5. Power consumed

Some other useful presentations are

The ratio of labor cost to material cost in maintenance work

The comparison of man-hours used to the level of the activity of the plant

Downtime of equipment expressed as a percent of total scheduled operating time

Sometimes a formula, expressing a combination of two or more of the above relationships, is developed and accepted by a company as a more usable composite index. Some of these formulas take into consideration the value of the equipment, the rate of its utilization, and downtime chargeable to repairs.

A few companies have established work-measurement programs where it is theoretically possible to set a definite standard for maintenance costs at varying levels of activity and then compare actual performance against this standard. This type of comparison can frequently be misleading if carelessly set up. It is possible for it to indicate excellent performance in spite of excessive overall maintenance costs. If properly administered and used for the purpose intended, such a standard can be extremely useful in maintenance management. Unfortunately, the overhead cost of most work-measurement programs detailed enough to be useful for this purpose is high and can be justified only by a large plant or by an industry having many similar plant operations. In this case a study by an independent industrial engineering firm may be practical.

In general, maintenance-cost trends are more important than the cost for any one short length of time. Although high and low peaks may occur monthly, for instance, if the trend of an index is downward, this change is more significant than the month-to-month variation. Most maintenance departments will find that no single index will be sufficient to evaluate their performance completely. A study of the trend of several indexes is more satisfactory.

Cost-Control Systems. In selecting its cost-control system maintenance engineering should conform to the system that is in use in the plant it serves. It may be expedient to adapt this system to the particular needs of the maintenance department, but the plantwide format should be used.

The purposes of a cost-control system include

1. Equitable distribution of repair costs over the departments serviced
2. A source of information necessary for sound administration of the maintenance department
3. Compliance with legal requirements for taxes and earnings
4. A source of information for the plant accounting group in its function of recording and reporting the financial position of the plant

In establishing its cost-control system the maintenance group should keep these purposes in mind and solicit the cooperation of the accounting group in adapting the plantwide program to maintenance needs, eliminating as much detail and duplication of effort as possible. Hand data-processing should be minimized by the use of modern business machines whenever this can be justified.

One of the most effective cost-control systems is the conventional job-cost method which accumulates expense items for labor, supplies, and services on a specific job number which, in turn, is the liability of a specific department. This accumulated charge, together with overhead and fixed charges, then provides the basis for distribution of cost by the accounting department.

In cost control, as with all performance records, it is well to remember that the actual cost of recording information at its source is small compared with the cost of further processing and analysis of the information. For this reason it is well to record information in considerable detail at the source but to scrutinize its further use and analysis carefully to ensure the most economical data-processing system required for the cost-control plan in use. It is then possible to rearrange the processing to fit changes in cost-control systems without affecting the data-accumulation habits at the level of origin.

The subject of cost control is covered in more detail in another section. Regardless of the system selected it should be flexible enough to provide additional information that might be useful in resolving specific cost problems and should operate at minimum overhead cost.

CHAPTER 7
SIX SIGMA SAFETY: APPLYING QUALITY MANAGEMENT PRINCIPLES TO FOSTER A ZERO-INJURY SAFETY CULTURE*

Michael Williamsen
Senior Consultant, Core Media Training Solutions, Portland, Oreg.

Is safety given the same commitment as product quality? Are employees accountable for their own safety? Is safety excellence imbedded into the company psyche? These are the fundamental questions that are driving today's safety revolution.

In much the same way total quality management made significant strides during the 1980s, industrial safety is poised for its own transformation. This article provides an actionable approach to how a zero-injury culture can be driven by adopting the same tools and tactics of product quality's Six Sigma.

Included in this article is a case study that documents the teamwork, methodology, and results of a continuous improvement team led by the author while serving as corporate group manager at Frito-Lay, Inc., overseeing the safety for 40 plants with 10,000 employees.

The Six Sigma tools are nonproprietary, with a growing number of documented references to their statistical origin. The unique aspect of this case is not the Six Sigma tools; rather, it is the documentation of their practical application to safety and their resulting injury breakthroughs shown in the accompanying charts. At the time of the following events (1985–89), the author and his team were unaware of any similar documented continuous improvement (CI) approaches to safety that utilized Six Sigma tools.

SAFETY PERFORMANCE CULTURE

Like all innovations, Six Sigma had the perspective of the great thinkers of manufacturing and production. Although the concept originated with a group of Motorola engineers during the mid-1980s, Six Sigma includes the theory and logic of quality pioneers such as W.E. Deming, Joseph Juran and Philip Crosby to address the age-old question: "Is the effort to achieve quality dependent on detecting and fixing defects? Or can quality be achieved by preventing defects through manufacturing controls and product design?"

---
*Reprinted with permission from Professional Safety, journal of the American Society of Safety Engineers.

At the core of Six Sigma is improvement in effectiveness and efficiency. Its core pursuit is perfection—a never-ending dissatisfaction with current performance. But what separates Six Sigma from conventional quality concepts is its focus on communicating measurable error ratios. By incorporating customer-focused objectives and metrics to drive continuous improvement—and by establishing processes that are so robust that defects rarely occur—Six Sigma quality objectives aspire to reach a three-parts-per-million error ratio at a 99.9996 percent incidence.

Statistically, Six Sigma variations are the standard deviation around the mean, represented by the Greek symbol sigma σ.

Today's Six Sigma quality community includes certification that incorporates formal instruction, performance standards, and applying a wide range of analytical problem-solving tools such as Pareto charts, process maps and fishbone diagrams. Its mastery borrows martial arts vernacular (e.g., black belt, sensei, etc.) to define levels of understanding and performance.

SIX SIGMA CONTROL LEVELS

What Six Sigma did for quality is beginning to be applied to industrial safety. The same desire to eliminate product mistakes is at work to reduce injury rates. In this parallel journey there are six levels, or Six Sigma in safety. Each "sigma control" builds on the previous level until the sixth sigma—a zero-injury culture—is attained.

One Sigma Control: "Reacting"

One sigma is set in the era of the three E's of safety: engineer, educate, and enforce. The tools for these rudimentary safety mechanics include work orders, safety rules, injury investigations, and compliance programs. While barely touching the surface of why injuries occur, one sigma tools nonetheless lay the foundation in establishing a safe workplace. As with one sigma in quality, the performance—conceptually, at least—is 68.5 percent error-free. This first level represents the ability to sustain the essentials in worker safety.

Two Sigma Control: "What we see"

The tools for two sigma control include observation programs, job safety analyses (JSA), and near-miss reporting. At this level, awareness and analysis tools are applied to reach a two sigma level or injury-free rate of about 98.5 percent. Research indicates that a 10 percent error level requires roughly 3000 observations to detect and act on mistakes (Harry 1998, 2000; Walmsley 1997).

As errors decrease, more observations are needed to detect the incorrect activities, which means a 1 percent error level requires about 10,000 observations to be statistically valid (Petersen 1993). It's a benchmark that underscores just how challenging it is for companies to move beyond two sigma control without adding to its traditional safety repertoire of observation programs and "rearview mirror" reporting. Two sigma safety control is focused on "what we see" in the workplace.

Three Sigma Control: "What we do"

Three sigma product quality requires well-defined responsibilities and accountabilities to provide predictable results on a regular basis. The same is true for three sigma safety. Without safety accountability at all levels, the possibility for companies to attain this level is next to impossible. Organizations that have been able to move from two- to three sigma, have generally attributed their success to the introduction of individual accountabilities into their safety programs. Embracing the conventions of accountability and personal responsibility is a critical factor in achieving a workplace that is 99.7 percent injury-free. While three sigma is commendable, companies are still incurring lost-time injuries at a rate of 3 per 1000 employees. Three sigma safety addresses "what we do" in the workplace.

Four Sigma Control: "What we believe"

Beginning in 1979, Dan Petersen teamed up with Charles Bailey to develop a comprehensive and statistically validated safety perception survey on behalf of the U.S. rail industry (Bailey 1993, 1988; Bailey and Petersen 1989). Today, the survey system is used to audit an organization's safety culture and identify perception gaps across 20 categories, cross-tabulated by management, supervisors, and front-line employees. The self-administered questionnaire includes 73 questions and provides companies with a statistically reliable method to answer the questions, "Where do our people believe we are weak?" and "Where do they agree and disagree?" Today's safety perception survey results can be compared with a database that combines more than two million respondents. It's a tool that provides statistically valid data for industry-wide comparison analyses.

The survey system breakthrough added an important dimension to pinpoint opportunities. Not only does it identify safety shortcomings, its implementation is recognized as an invaluable "buy-in" mechanism to set the stage for continuous improvement work teams—a necessary component to reach four sigma control: 99.97 percent injury free. Four sigma control concentrates on the non-observable "What we believe" in workplace safety.

Five and Six Sigma Safety Control: "How we engage and how we lead"

The next challenge is to utilize the data in the previous four levels of safety:

- The fundamentals: injury and work order data
- Observable processes
- Accountabilities of what we do
- Information on what we believe from a safety perception survey

The material from these four safety databases needs to be applied in a rapid, accurate, and functional way. Once a company is nearing four sigma, the major barriers to effective cross-functional continuous improvement are eliminated. A roadmap can be developed to an unprecedented five-sigma (99.997 percent) and six sigma (three injuries per million employees) safety performance. At this point, an organization can approach a virtual zero-injury workplace.

As in a Six Sigma quality programs, all the foundational mechanics—engineer-educate-enforce, observe, investigate, accountability principles, and thought patterns—are necessities to establish an authentic Six Sigma safety culture. The challenge is to create a sustainable safety culture where heightened safety decisions occur without any thought. It's a process that begins by addressing the milestones to continuously improve.

Good data are necessary. However, to achieve four sigma performance and beyond, SH&E professionals need to implement a similar approach to what zero-error quality cultures use in manufacturing. That's why the next two critical success factors to establish a zero-injury safety culture require continuous improvement teams to "own" and implement the following.

- A regular, sanctioned meeting system with actionable rules and mechanisms and trained leaders to manage the continuous improvement process in safety.
- Six Sigma analytical techniques/tools with safety issues and projectible data.

Once these critical success factors are in place, a zero-error safety culture can be a recognized strength alongside the traditional business necessities of customer service, quality assurance, and manufacturing efficiencies. The resulting savings in both cost and hardship can be dramatic.

SIX SIGMA TOOLS IN THE WORKPLACE

Five and Six Sigma injury control requires statistical process control tools, a dedicated continuous improvement (CI) team and active participation from all levels of employees. This latter component emphasizes the importance of effective meetings. Organizing effective "subteams" to execute tasks is essential. Furthermore, because many of the subteams combine cross-functional employees from disparate groups, it is critical to delineate proven principles to create a meeting structure that ensures efficiency, participation, action, and high performance.

EFFECTIVE MEETINGS FOR CONTINUOUS IMPROVEMENT

To achieve results from safety meetings, the person who calls the meeting must focus on its purpose and desired outcomes. By deploying the POP model—purpose, outcomes, process—the group can remain focused and on task.

Purpose

The purpose is a mini-mission statement. Why is the group meeting? If the purpose is unclear, start with an open-ended question, "What is our purpose for this meeting?" If necessary, record responses on a flipchart until agreement is reached. Subsequent meetings of this same group need to restate the purpose and make sure it remains on target. If the meeting starts to wander or branch into a tangent, ask whether the current topic is "on purpose." Atypical safety purpose may resemble a statement such as, "Develop safety accountabilities for all levels of the organization that will help eliminate injuries."

Outcomes

What will be accomplished when the stated purpose is achieved? This is a brainstormed list of the issues that the meeting is designed to address. It is also the metric for whether those tasks have been accomplished. The whole team or group participates in setting these outcomes and, therefore, seeks complete agreement as to definitions of success. Not only will this eliminate future differences, it also helps eliminate discussions that stray from the desired outcome. A typical set of outcomes for a safety team might be: Accountabilities that make a difference in safety for every job in the facility; a tracking system to follow accomplishment of these accountabilities; a reward system that reinforces these activities; reduced injury frequency as a result of doing this work well.

Process

How will the purpose and outcomes be accomplished? What typically follows is a description of how the team will work. Often, it is divided into small problem-solving groups that include volunteers to accomplish small tasks. Why volunteers? When people get to place themselves in performance zones where they are comfortable, they are more likely to succeed. Conversely, quick delegation can lead to having the wrong people assigned to the wrong task. If there are not enough volunteers to perform all the work in the time allotted, time or resources (or both) may need to be increased. One distinction must be remembered throughout: This is not a crisis team; it is an improvement team that fosters the continuous improvement process.

ACTION ITEM MATRIX

In many cases, a significant number of tasks need to be completed by various people in varying time frames. To effectively manage this wide spectrum, it is best to use an action item matrix (AIM), which is a simple five-column spreadsheet (Fig. 7.1). The columns (from left to right) are:

- **Item number**. Each item on the list is numbered. As items are completed, they are moved to the bottom of the list.
- This provides a record of what the team has completed as well as what still needs to be accomplished.

Action Item Matrix: Accountability Team

Date: 6-27-04
Members: Wolf, Lowery, Jennings, Williamsen, Brown, Morrison, Gilbert

Item	Action item	Who	Target date	Comments
1	List all job titles/functions	Morrison	7-2	In database
2	Hand out accountabilities from company "xyz" for examples	Jennings	7-2	Will be e-mailed
3	Each team member to list their own safety accountabilities	Team	8-27	Judy to put in database
4	Critique accountabilities	Team	Until completed	Final copy review by safety council and other potential parties

FIGURE 7.1 Action item matrix: accountability team.

- **Task to be accomplished.** This is a simple, succinct statement of the issue. Each task or action item is a small, manageable portion of the larger project scope.
- **The team.** The list of volunteers who have agreed to accomplish this action item. Each item may have one or more volunteers—or in some cases none, if the assignment is not ready to be worked on.
- **The date.** This indicates the next report date for the task team on this action item. It may be a completion date, a progress report date or other target date.
- **Comments.** This field holds information pertinent to the action item, for example, "awaiting vendor quote."

At this point, the team has its assignments, the POP statement and its progress-tracking mechanism, and the AIM. How often should the teams meet? The whole team meets every 2 weeks, with the task or subteams meeting more frequently as they are problem-solving units. More-frequent whole team meetings do not allow the subteams enough time to complete their tasks and are an inefficient use of time. Less-frequent meetings do not create the needed sense of urgency.

An entire safety program was developed in less than 9 months using this meeting process (Petersen 1988). Hourly and salaried employees applied these guidelines for all 20 safety perception survey categories. Although the impact cannot be entirely attributed to the team initiatives, the number of serious injuries dropped by more than 80 percent over the course of 2 years (Figs. 7.2 to 7.4).

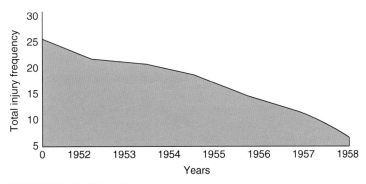

FIGURE 7.2 Total injury frequency.

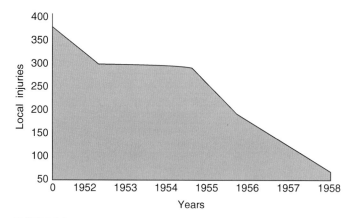

FIGURE 7.3 Lost-time injuries.

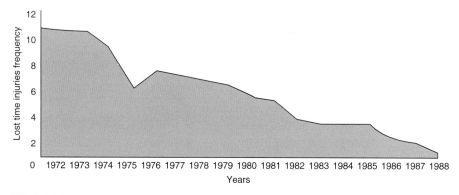

FIGURE 7.4 Lost-time injuries frequency.

EFFECTIVE SAFETY TASK FORCES

How are safety taskforces created? How are tasks priority ranked? The answers are summarized in this process:

- Start with an AIM.
- At Frito-Lay, supervisors trained in CI techniques could generally lead up to two CI teams of 3 to 10 people while still performing their normal work tasks.
- Attempt to enlist only volunteers so people assign them-selves to tasks they want to pursue and are willing to make the time to complete.
- Implement only short-term, 90-day teams that have effective facilitation, leadership and closure.
- If those three characteristics are not achievable, then the teams should not be initiated. The short-sighted approach of trying to "do everything for everybody right now" will only lead to frustration.
- Have teams meet every 2 weeks to reconnect on a regular basis. The time between meetings can be increased to 3 weeks, but the groups should not meet more often than every 2 weeks. Subteams should meet as necessary to test, discuss and resolve problems. The "**TASK: DEFINE MACHINE OPERATOR ROLE**" section provides an example of hourly employee safety accountabilities developed through this process. This process can be used in each of the 20 safety perception survey categories.

CONCLUSION

The case study and figures demonstrate how a CI approach helped to improve safety performance in a manufacturing setting. Injury data were combined with perception survey data to obtain a full spectrum of workplace realities—both observable and hidden. Hourly and salaried employees then team—using Six Sigma tools and effective safety meeting techniques—to develop and implement a zero-injury safety culture, a workplace that neither tolerates, nor experiences, injuries.

TASK: DEFINE MACHINE OPERATOR ROLE

Definition

The key safety accountabilities of the operator are to use safe work practices, use all safety equipment when required and promote safety with coworkers.

Responsibilities

1. Before each shift, inspect/check the work area to identify any unsafe issues and correct or initiate corrective action as needed.
2. Perform daily housekeeping duties to keep/maintain work area in a safe and clutter-free condition.
3. Attend and participate in all shift supervisor safety meetings.
4. Team with the supervisor to present/discuss topics in the supervisor safety meeting (two to four per year).
5. Initiate and follow up on safety work orders.
6. Provide appropriate safety and health training to new/transferred personnel.
7. Review and improve job hazard analyses regularly.
8. Be familiar with all documents in work area.
9. Pay attention to coworkers and outside personnel working in the area. If they are not following proper practices or procedures, talk with them immediately about correcting their activities.
10. Inspect containers to ensure that they are labeled correctly. If not, relabel them immediately.

Measures of Performance

1. Appraisal by supervisor of individual task achievement.
2. Observations by supervisor.

A CASE STUDY: USING SIX SIGMA TOOLS

This case study—first published ASSE's Professional Safety in 2005—illustrates how Six Sigma measurements were applied to a Fortune 500 food product company that was experiencing hundreds of injuries across multiple facilities. The initiative resulted in a rapid improvement in workplace injuries and the start of a zero-injury safety culture.

Pareto Charts

The Pareto chart is one of the most helpful visual tools in the safety Six Sigma tool box. These charts help to pinpoint unacceptable occurrences that warrant high priority. The charts (Figs. 7.5 to 7.11) show the frequency and severity of problems and where they occurred geographically.

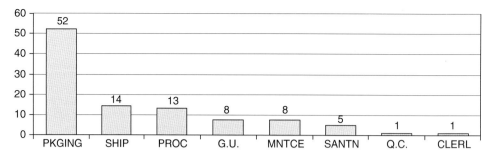

FIGURE 7.5 Department lost time injuries last 10 periods.

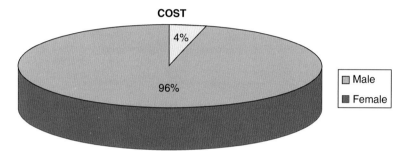

FIGURE 7.6 Injuries by gender.

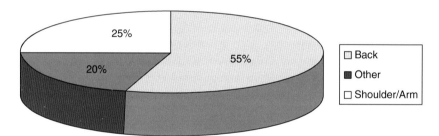

FIGURE 7.7 Lost-time injuries.

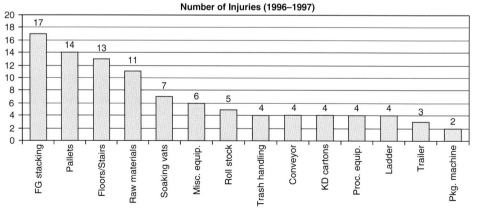

FIGURE 7.8 Injury locations.

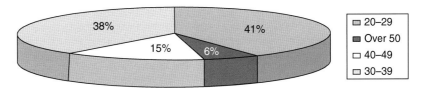
FIGURE 7.9 Back injuries by age.

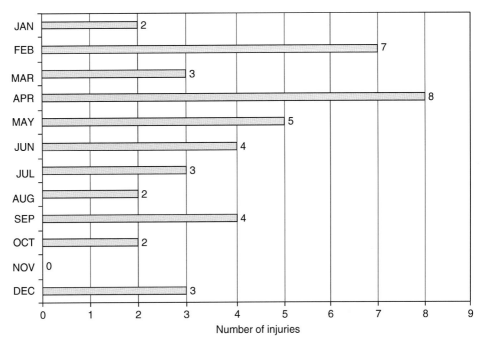
FIGURE 7.10 Injuries vs. month.

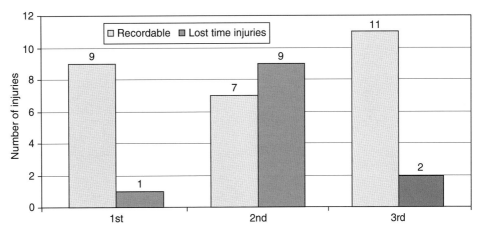
FIGURE 7.11 Injuries per shift.

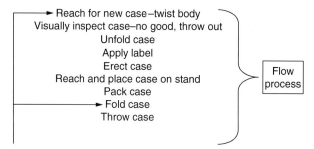

FIGURE 7.12 Case erection process.

Process Maps

Process maps or process flow diagrams graphically illustrate how a task or process can be accomplished effectively within the constraints of time and resources (Figs. 7.12 to 7.15). This tool allows a continuous improvement team to break down a complicated sequence of events into simple metered steps, which result in a "spaghetti diagram." The team then analyzes each step in the process being studied and optimizes each individual task to a point where inefficiencies, errors, complicated "spaghetti" and safety hazards are eliminated.

Cause-and-Effect Diagram

As the CI team continued its efforts to eliminate back and soft tissue injuries, the safety team used another Six Sigma tool, the cause-and-effect diagram (Fig. 7.16, which is also referred to as a fishbone or Ishigawa diagram). Team members were able to refer to the chart to identify multiple potential causes for the problem at hand. The "bones" of the normal potential "cause" categories include people, methods, machinery, and materials. As problem situations vary, this Six Sigma tool has the added benefit of being able to creatively identify different elements to better fit the individual situation. For the food products company, environment and technology were added as potential causes. After listing all potential causes, each team member voted for two or three of the individual fishbone

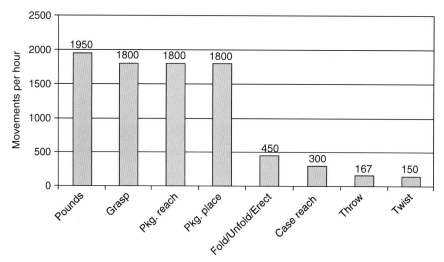

FIGURE 7.13 One-pound package movement per hour.

6:00 AM	7:00 AM	7:45 AM	Start Shift 8:00 AM	9:00 AM	Break 10:00 AM
Get up Eat breakfast Talk with spouse Read the paper Feed the dog Two cups of coffee	Drive 30 minutes thru heavy traffic	Arrive at plane			
	One cup of coffee	One cup of coffee Clock in	Arrive at machine Pack 1 pound bags at 30/min 0 pounds	1950 pounds packed	One cup of coffee Rest, take break 3900 pounds packed

FIGURE 7.14 Packer process flow diagram.

diagram causes deemed most important. This individual voting process is referred to as "Pareto voting" in Six Sigma organizations; other trainers use the term "multivoting" (ReVelle 2004). It is not a rigorous statistical evaluation; rather, it is a method that uses the personal experiences and judgment of the engaged subject-matter experts. It is an efficient way to quickly determine the top "vote-getting" issues believed to warrant more research and detail. These "focus causes" were then placed in an AIM for deeper team analysis and problem resolution.

In the next step, the team began a systematic search for low-cost, highly effective solutions. The cause-and-effect diagram (in group mode) allowed each team member to record what he/she thought was important. In turn, the team began to work on areas of interest believed necessary to be resolved in order to eliminate back and soft-tissue injuries (Fig. 7.17).

From start to finish, the CI team approach to safety-issue resolution worked well for the manufacturing environment. The efforts to apply Six Sigma and other CI tools led to improvements in both total recordable and lost-time injury rates (Figs. 7.2 to 7.4).

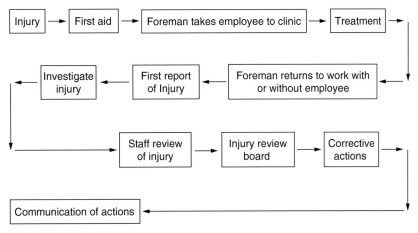

FIGURE 7.15 Injury sequence.

1.66 ORGANIZATION AND MANAGEMENT OF THE MAINTENANCE FUNCTION

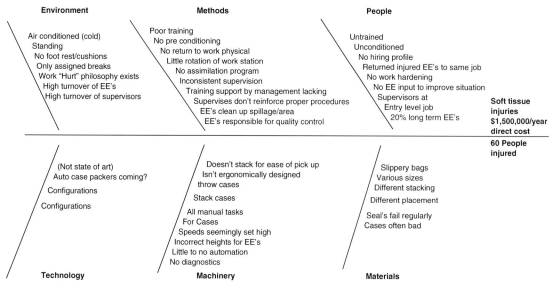

FIGURE 7.16 Cause-and-effect for packers.

TEAM: Packer Safety Improvement
Members: John, Sam, Steve, Dimitri, Sharon, Bob
Date: 11/30

Item	Action item	Who	Date	Comments
1	Hire ergonomist to study and make recommendations	John	12/10	M. Ayub, J.L. Sangre, K State, UT
2	Get EE's to develop training program	Sam	3/3	
3	Retain doctor to develop physical requirements of job	Steve	3/3	
4	Improve case sorting prior to use	John & Packers	12/2	
5	Change supervision rotation program from 6 months to 1 year	Sharon	2/6	
6	Improve mechanical reliability of packaging machines	Bob	12/8	
7	Work with EE's to develop warm up program	Dimitri	1/15	

FIGURE 7.17 Action item matrix.

Although the impact can't be entirely attributed to the team initiatives, the number of serious injuries dropped by more than 80 percent over the course of 2 years.

REFERENCES

Bailey, C.W. 1988. *Using Behavioral Techniques to Improve Safety Program Effectiveness,* Washington, DC, American Assn. of Railroads.

Bailey, C.W. "Improve safety program effectiveness with perception surveys," *Professional Safety,* Oct. 1993: 28–32.

Bailey, C. and D. Petersen. "Using Perception Surveys to Assess Safety System Effectiveness," *Professional Safety,* Feb. 1989: 22–26.

Harry, M.J. "Six Sigma: A Breakthrough Strategy for Profitability." *Quality Progress.* 31(1998): 60–64.

Harry, M.J. "Framework for Business Leadership." *Quality Progress.* April 2000.

Petersen, D. 1988. *Safety Management: A Human Approach.* Goshen, NY: Aloray Inc.

Petersen, D. 1993. *The Challenge of Change: Creating a New Safety Culture.* Portland, OR: CoreMedia Training Solutions.

ReVelle, J.B. "Six Sigma Problem-Solving Techniques Create Safer, Healthier Worksites." *Professional Safety.* Oct. 2004: 38–46.

Walmsley, A. Six Sigma Enigma. *Globe and Mail.* Oct. 1997.

SECTION · 2

THE HORIZONS OF MAINTENANCE MANAGEMENT

CHAPTER 1
CORRECTIVE MAINTENANCE

R. Keith Mobley
Principal, Life Cycle Engineering, Inc., Charleston, S.C.

There are three types of maintenance tasks: (1) breakdown, (2) corrective, and (3) preventive. The principal difference in these occurs at the point when the repair or maintenance task is implemented. In breakdown maintenance, repairs do not occur until the machine fails to function. Preventive maintenance tasks are implemented before a problem is evident and corrective tasks are scheduled to correct specific problems that have been identified in plant systems.

A comprehensive maintenance program should use a combination of all three. However, most domestic plants rely almost exclusively on breakdown maintenance to maintain their critical plant production systems.

BREAKDOWN MAINTENANCE

In these programs, less concern is given to the operating condition of critical plant machinery, equipment, or systems. Since most of the maintenance tasks are reactive to breakdowns or production interruptions, the only focus of these tasks is how quickly the machine or system can be returned to service. As long as the machine will function at a minimum acceptable level, maintenance is judged to be effective. This approach to maintenance management is both ineffective and extremely expensive. Breakdown maintenance has two factors that are the primary contributors to high maintenance costs: (1) poor planning and (2) incomplete repair.

The first limitation of breakdown maintenance is that most repairs are poorly planned because of the time constraints imposed by production and plant management. As a result, manpower utilization and effective use of maintenance resources are minimal. Typically, breakdown or reactive maintenance will cost three to four times more than the same repair when it is well planned.

The second limitation of breakdown maintenance is that it concentrates repair on obvious symptoms of the failure, not the root cause. For example, a bearing failure may cause a critical machine to seize and stop production. In breakdown maintenance, the bearing is replaced as quickly as possible and the machine is returned to service. No attempt is made to determine the root cause of the bearing failure or to prevent a recurrence of the failure. As a result, the reliability of the machine or system is severely reduced. This normal result of breakdown maintenance is an increase in the frequency of repairs and a marked increase in maintenance costs.

PREVENTIVE MAINTENANCE

The concept of preventive maintenance has a multitude of meanings. A literal interpretation of the term is a maintenance program that is committed to the elimination or prevention of corrective and breakdown maintenance tasks. A comprehensive preventive maintenance program will utilize regular evaluation of critical plant equipment, machinery, and systems to detect potential problems and immediately schedule maintenance tasks that will prevent any degradation in operating condition.

In most plants, preventive maintenance is limited to periodic lubrication, adjustments, and other time-driven maintenance tasks. These programs are not true preventive programs. In fact, most continue to rely on breakdowns as the principal motivation for maintenance activities.

A comprehensive preventive maintenance program will include predictive maintenance, time-driven maintenance tasks, and corrective maintenance to provide comprehensive support for all plant production or manufacturing systems.

CORRECTIVE MAINTENANCE

The primary difference between corrective and preventive maintenance is that a problem must exist before corrective actions are taken. Preventive tasks are intended to prevent the occurrence of a problem. Corrective tasks correct existing problems.

Corrective maintenance, unlike breakdown maintenance, is focused on regular, planned tasks that will maintain all critical plant machinery and systems in optimum operating conditions. Maintenance effectiveness is judged on the life-cycle costs of critical plant machinery, equipment, and systems, not on how fast a broken machine can be returned to service.

Corrective maintenance, as a subset of a comprehensive preventive maintenance program, is a proactive approach toward maintenance management. The fundamental objective of this approach is to eliminate breakdowns, deviations from optimum operating condition, and unnecessary repairs and to optimize the effectiveness of all critical plant systems.

The principal concept of corrective maintenance is that proper, complete repairs of all incipient problems are made on an as-needed basis. All repairs are well planned, implemented by properly trained craftsmen, and verified before the machine or system is returned to service. Incipient problems are not restricted to electrical or mechanical problems. Instead, all deviations from optimum operating condition, that is, efficiency, production capacity and product quality, are corrected when detected.

PREREQUISITES OF CORRECTIVE MAINTENANCE

Corrective maintenance cannot exist without specific support efforts. A number of prerequisites must exist before corrective maintenance can be properly implemented.

ACCURATE IDENTIFICATION OF INCIPIENT PROBLEMS

Both preventive and corrective maintenance programs must be able to anticipate maintenance requirements before a breakdown can occur. A comprehensive predictive maintenance program that has the ability to accurately identify the root cause of all incipient problems is the first requirement of corrective maintenance. Without this ability, corrective actions cannot be planned or scheduled.

PLANNING

All corrective repairs or maintenance must be well planned and scheduled to minimize both cost and interruption of the production schedule. Adequate time must be allowed to permit complete repair of the root cause and resultant damage caused by each of the identified incipient problems.

Proper maintenance planning is dependent on well-trained planners, a viable maintenance database, and complete repair procedures for each machine train or system within the plant.

Trained Maintenance Planners

Many plants do not have full-time maintenance planners or their planners lack the knowledge or skills that the job demands. It is therefore imperative that proper training is provided to ensure that each planner has the skills necessary to properly plan repairs and maintenance tasks.

Maintenance History Database

The planner must have accurate maintenance history in order to properly plan repairs. As a minimum, he must know the standard mean-time-to-repair for every recurring repair, rebuild, and maintenance task required to maintain optimum operating condition of critical plant machinery, equipment, and systems. Without this knowledge, he cannot plan an effective repair.

In addition, the planner must know the specific tools, repair parts, auxiliary equipment, and craftsmen skills required to complete each maintenance task. This information, in conjunction with proper repair sequence, is an absolute requirement of a viable repair plan.

This type of information requires a comprehensive maintenance database that compiles actual mean-time-to-repair, standard repair procedures, and the myriad of other information required for proper maintenance planning.

PROPER REPAIR PROCEDURES

Repairs must be complete and properly implemented. In many cases, poor maintenance or repair practices result in more damage to critical plant machinery than the observed failure mode.

A fundamental requirement of corrective maintenance is proper, complete repair of each incipient problem. To meet this requirement, all repairs must be made by craftsmen who have the necessary skills, repair parts, and tools required to return the machine or system to as-new condition.

Craft Skills. A growing number of maintenance craftsperson do not have the minimum skills required to properly maintain or repair plant equipment, machinery, or systems. In many cases, they cannot properly install bearings, align machine trains, or even balance rotating equipment. Few have the knowledge and skills required to properly disassemble, repair, and reassemble the complex machinery or systems that comprise the critical production systems within plants.

A prerequisite of corrective maintenance is skilled craftsperson. Therefore, plants must implement a continuous training program that will provide the minimum craft skills required to support their production or manufacturing systems. The training program should include the means to verify craft skills and periodically refresh these skills.

Standard Maintenance Procedures

All recurring repairs and maintenance tasks should have a standard procedure that will specifically define the correct method required for competition. These procedures should include all of the information, such as tools, safety concerns, and repair parts, required for the task and a step-by-step sequence of tasks required to complete the repair.

Each procedure should be complete and contain all information required to complete the repair or recurring preventive maintenance task. The craftsperson should not be required to find or have supplemental information in order to complete the repair.

ADEQUATE TIME TO REPAIR

One of the fundamental reasons that most plants rely on breakdown maintenance is that tight production schedules and management constraints limit the time available for maintenance. The only way to reduce the number and frequency of breakdown repairs is to allow sufficient time for proper maintenance.

Plant management must permit adequate maintenance time for all critical plant systems before either preventive or corrective maintenance can be effective. In the long term, the radical change in management philosophy will result in a dramatic reduction in the downtime required to maintain critical production and manufacturing equipment. Machinery that is maintained in as-new condition and not permitted to degrade to a point that breakdown or serious problems can occur will require less maintenance than machinery maintained in a breakdown mode.

VERIFICATION OF REPAIR

The final prerequisite of corrective maintenance is that all repairs or rebuilds must be verified before the machine train or system is returned to service. This verification process will ensure that the repair was properly made and that all incipient problems, deviations from optimum operating conditions, or other potential limitations to maximum production capacity and reduced product quality have been corrected.

ROLE OF CORRECTIVE MAINTENANCE

Corrective maintenance will remain a critical part of a comprehensive plant maintenance program. However, the objective of a viable preventive program is to eliminate all breakdown maintenance and severely reduce the number and frequency of corrective maintenance actions.

The ultimate objective of any maintenance program should be the elimination of machine, equipment, and system problems that require corrective actions.

CHAPTER 2
RELIABILITY-BASED PREVENTIVE MAINTENANCE

R. Keith Mobley
Principal, Life Cycle Engineering, Inc., Charleston, S.C.

Preventive maintenance, as its name implies, are specific tasks that are designed to prevent the need for corrective or breakdown maintenance, as well as prolong the useful life of capital assets and auxiliary equipment. Most preventive maintenance programs are a loose conglomeration of inspections, cleaning, adjustment, lubrication, and similar tasks that do little, if anything, to preserve the reliability of critical production assets. Statistically, between 33 percent and 42 percent of so-called preventive maintenance tasks add no value, in terms of reliability or maintenance prevention.

Reliability-based preventive maintenance replaces these no-value tasks with specific maintenance activities that both prevents failures and prolong the useful life of plant assets.

Development of a reliability-based preventive maintenance program follows logic diagrams shown in Fig. 2.1 and the task selection criteria, illustrated in Table 2.1, which are it's principal tools. The logic diagrams are the basis of an evaluation technique applied to each *functionally significant item* (FSI) using all available technical data, as well as the "native knowledge" of plant personnel. Principally, the evaluations are based on the items' functional failures and failure causes. The development of a reliability-based preventive maintenance program is based on the following:

- Identification of FSIs
- Identification of applicable and effective preventive maintenance tasks using the decision tree logic.

A functionally significant item is an item whose failure would affect safety or could have significant operational or economic impact in a particular operating or maintenance context. The process of identification of FSIs is based on the anticipated consequences of failures using an analytical approach and good engineering judgment. The process also use a top-down approach, and is conducted first at the system level, then at the subsystem level and, where appropriate, down to the component level. An iterative process should be followed in identifying FSIs. Systems and subsystem boundaries and functions are first identified. This permits selection of critical systems for further analysis, which involves a more comprehensive and detailed definition of system, system functions, and system's functional failures.

The procedures such as information collection, system analysis, and so on outline a comprehensive set of tasks in the FSI identification process. All these tasks should be applied in the case of complex or new equipment. However, in the case of well-established or simple equipment, where functions and functional degradation/failures are well recognized, tasks listed under the heading of *system analysis* can be covered very quickly. They should, however, be documented to confirm that

2.8 THE HORIZONS OF MAINTENANCE MANAGEMENT

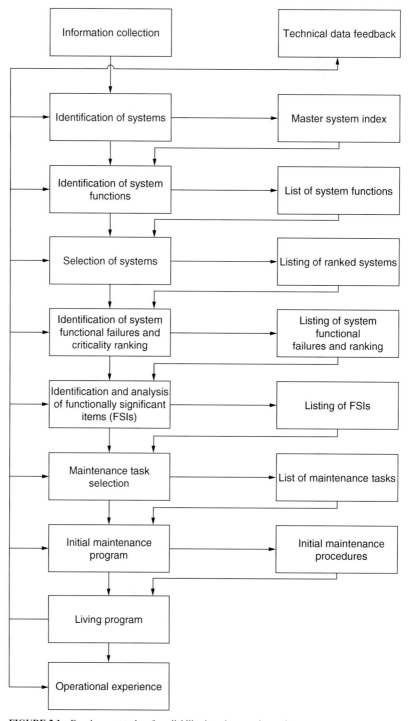

FIGURE 2.1 Development tasks of a reliability-based preventive maintenance program.

TABLE 2.1 Task Selection Criteria

Tasks	Application Criteria	Effectiveness Criteria		
		Safety	Operational	Direct Cost
Lubrication or Servicing	Replenishment of consumables shall reduce the rate of functional deterioration	The task shall reduce the risk of failure	The task shall reduce the risk of failure to an acceptable level	The task shall be cost-effective
Operational, Visual or Automated Check	Identification of the failure shall be possible	The task shall reduce the risk of failure to assure safe operation	Not applicable	The task shall ensure adequate availability of the hidden function in order to avoid economic effect of multiple failures and shall be cost-effective
Inspection, Functional Check or Condition Monitoring	Reduced resistance to failure shall be detectable and rate of reduction in failure resistance shall be predictable	The task shall reduce the risk of failure to assure safe operation	The task shall reduce the risk of failure to an acceptable level	The task shall be cost-effective, i.e. the cost of the task shall be less than the cost of the failure prevented
Restoration	The item shall show functional degradation characteristics at an identifiable age and a large proportion of units shall survive to that age. It shall be possible to restore the item to a specific standard of failure resistance	The task shall reduce the risk of failure to assure safe operation	The task shall reduce the risk of failure to an acceptable level	The task shall be cost-effective, i.e. the cost of the task shall be less than the cost of the failure prevented
Discard	The item shall show functional degradation characteristics at an identifiable age and a large proportion of units shall survive to that age	A safe-life limit shall reduce the risk of failure to assure safe operation	The task shall reduce the risk to an acceptable level	The task shall be cost-effective, i.e. the cost of the task shall be less than the cost of the failure prevented

they were considered. The depth and rigor used in the application of these tasks will also vary with the complexity and newness of the equipment.

INFORMATION COLLECTION

Equipment information provides the basis for the evaluation and should be assembled prior to the start of the analysis and supplemented as the need arises. The following should be included:

- Requirements for equipment and its associated systems, including regulatory requirements
- Design and maintenance documentation
- Performance feedback, including maintenance and failure data

Also, in order to guarantee completeness and avoid duplication, the evaluation should be based on an appropriate and logical breakdown of the equipment.

SYSTEM ANALYSIS

The tasks described in the preceding section (Information collection) define the procedure for the identification of the functionally significant items and the subsequent maintenance task selection and implementation. It should be noted that the tasks can be tailored to meet the requirements of particular industries and the emphasis placed on each task will depend on the nature of that industry.

IDENTIFICATION OF SYSTEMS

The objective of this task is to partition the equipment into systems, grouping the components contributing to achievement of well-identified functions and identifying the system boundaries. Sometimes it is necessary to perform further partitioning into the subsystems, which perform functions critical to system performance. The system boundaries may not be limited by the physical boundaries of the systems, which may overlap.

Frequently, the equipment is already partitioned into systems through industry specific partitioning schemes. This partitioning should be reviewed and adjusted where necessary to ensure that it is functionally oriented. The results of equipment partitioning should be documented in a master system index that identifies systems, components, and boundaries.

IDENTIFICATION OF SYSTEM FUNCTIONS

The objective of this task is to determine the main and auxiliary functions performed by the systems and subsystems. The use of functional block diagrams will assist in the identification of system functions. The function definition describes the actions or requirements which the system or subsystem should accomplish, sometimes in terms of performance capabilities within the specified limits. The functions should be identified for all modes of equipment operation.

Reviewing design specifications, design descriptions, and operating procedures, including safety, abnormal operations, and emergency instructions, may determine the main and auxiliary functions. Functions such as testing or preparations for maintenance, if not considered important, may be omitted. The reason for omissions must be given. The product of this task is a listing of system functions.

SELECTION OF SYSTEMS

The objective of this task is to select and prioritize systems, which will be included in the reliability-centered maintenance (RCM) program because of their significance to equipment safety, availability, or economics. The methods used to select and prioritize the systems can be divided into

- Qualitative methods based on past history and collective engineering judgment.
- Quantitative methods, based on quantitative criteria, such as criticality rating, safety factors, probability of failure, failure rate, life cycle cost, and the like, used to evaluate the importance of system degradation/failure on equipment safety, performance, and costs. Implementation of this approach is facilitated when appropriate models and data banks exist.
- Combination of qualitative and quantitative methods.

The product of this task is a listing of systems ranked by criticality. The systems, together with the methods, the criteria used and the results, should be documented.

SYSTEM FUNCTIONAL FAILURES AND CRITICALITY RANKING

The objective of this task is to identify system functional degradation/failures and prioritize them. The functional degradation/failures of a system for each function should be identified, ranked by criticality and documented.

Since each system functional failure may have different impacts on safety, availability, or maintenance cost, it is necessary to rank and prioritize them. The ranking takes into account probability of occurrence and consequences of failure. Qualitative methods based on collective engineering judgment and based on the analysis of operating experience can be used. Quantitative methods of *simplified failure modes and effects analysis* (SFMEA) or risk analysis can also be used.

The ranking represents one of the most important tasks in RCM analysis. Too conservative a ranking may lead to an excessive preventive maintenance program, and conversely a lower ranking may result in excessive failures and a potential safety impact. In both cases, a nonoptimized maintenance program will result. The outputs of this task are the following:

- Listing of system functional degradation/failures and their characteristics.
- Ranking list of system functional degradation/failures.

IDENTIFICATION OF FUNCTIONALLY SIGNIFICANT ITEMS

Based on the identification of system functions, functional degradation/failures and effects, and collective engineering judgment, it is possible to identify and develop a list of candidate FSIs. As said before, these are items whose failures could affect safety; be undetectable during normal operation; have significant operational impact; have significant economic impact. The output of this task is a list of candidate FSIs.

FUNCTIONALLY SIGNIFICANT ITEM FAILURE ANALYSIS

Once an FSI list has been developed, a method such as failure modes and effects analysis (FMEA) should be used to identify the following information that is necessary for the logic tree evaluation of each FSI. The following examples refer to the failure of a pump providing cooling water flow:

- *Function:* The normal characteristic actions of the item (e.g., to provide cooling water flow at 100 to 240 gpm to the heat exchanger).
- *Functional failure:* How the item fails to perform its function (e.g., pump fails to provide required flow).
- *Failure cause:* Why the functional failure occurs (e.g., bearing failure).
- *Failure effect:* What are the immediate effect and the wider consequence of each functional failure (e.g., inadequate cooling leading to overheating and failure of the system).

The FSI failure analysis is intended to identify functional failures and failure causes. Failures not considered as credible, such as those resulting solely from undetected manufacturing faults, unlikely failure mechanisms or unlikely external occurrences, should be recorded as having been considered and the factors which caused them to be assessed as not credible should be stated.

Prior to applying the decision logic tree analysis to each FSI, preliminary worksheets need to be completed which clearly define the FSI, its functions, functional failures, failure causes, failure effects, and any additional data pertinent to the item (e.g., manufacturer's part number, a brief description of the item, predicted or measured failure rate, hidden functions, redundancy, etc.). These worksheets should be designed to meet the user's requirements.

From this analysis, the critical FSIs can be identified (i.e., those that have both significant functional effects and a high probability of failure, or have a medium probability of failure, but are judged critical or have a significantly poor maintenance record).

MAINTENANCE TASK SELECTION (DECISION LOGIC TREE ANALYSIS)

The approach used for identifying applicable and effective preventive maintenance tasks is one that provides a logic path for addressing each FSI functional failure. The decision logic tree (Fig. 2.2) uses a group of sequential "YES/NO" questions to classify or characterize each functional failure. The answers to the "YES/NO" questions determine the direction of the analysis flow and help to determine the consequences of the FSI functional failure, which may be different for each failure cause. Further progression of the analysis will ascertain if there is an applicable and effective maintenance task that will prevent or mitigate it. The resultant tasks and related intervals will form the initial scheduled maintenance program.

Note: Proceeding with the logic tree analysis with inadequate or incomplete FSI failure information could lead to the occurrence of safety critical failures, due to inappropriate, omitted, or unnecessary maintenance, and to increased costs due to unnecessary scheduled maintenance activity, or both.

Levels of Analysis

Two levels are apparent in the decision logic.

- The first level (questions 1, 2, 3, and 4) requires an evaluation of each functional degradation/failure for determination of the ultimate effect category, that is, evident safety, evident operational, evident direct cost, hidden safety, hidden non-safety or none.
- The second level (questions 5, 6, 7, 8, and 9, A to E, as applicable) takes the failure causes for each functional degradation/failure into account in order to select the specific type of tasks.

First Level Analysis—Determination of Effects. Consequence of failure, which could include degradation, is evaluated at the first level using four basic questions (Fig. 2.2).

Note: The analysis should not proceed through the first level unless there is a full and complete understanding of the particular functional failure.

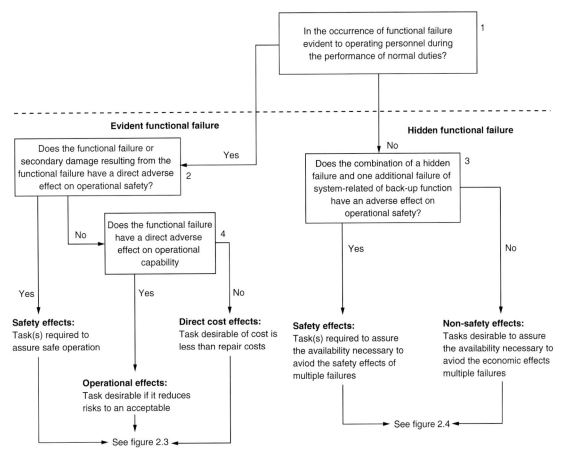

FIGURE 2.2 Reliability decision logic tree (level 1)—effects of functional failures.

Question 1—Evident or hidden functional failure? The purpose of this question is to segregate the evident and hidden functional failures and should be asked for each functional failure.

Question 2—Direct adverse effects on operating safety? To be direct, the functional failure or resulting secondary damage should achieve its effect by itself, not in combination with other functional failures. An adverse effect on operating safety implies that damage or loss of equipment, human injury or death, or some combination of these events is a likely consequence of the failure or resulting secondary damage.

Question 3—Hidden functional failure safety effect? This question takes into account failures in which the loss of a hidden function (whose failure is unknown to the operating personnel). This type of failure does not directly affect safety, but in combination with an additional functional failure, has an adverse effect on operating safety.

Note: The operating personnel consist of all qualified staff who are on duty and who are directly involved in the use of the equipment.

Question 4—Direct adverse effect on operating capability? This question asks if the functional failure could have an adverse effect on operating capability:

- Requiring either the imposition of operating restrictions or correction prior to further operation
- Requiring the operating personnel to use abnormal or emergency procedures

Second Level Analysis—Effects Categories. Applying the decision logic of the first-level questions to each functional failure leads to one of five effect categories, as follows:

Evident safety effects—Questions 5A to 5E. This category should be approached with the understanding that a task (or tasks) is required to ensure safe operation. All questions in this category need to be asked. If no applicable and effective task results from this category analysis, then redesign is mandatory.

Evident operational effects—Questions 6A to 6D. A task is desirable if it reduces the risk of failure to an acceptable level. If all answers are in the logic process, no preventive maintenance task is generated. If operational penalties are severe, a redesign is desirable.

Evident direct cost effects—Questions 7A to 7D. A task is desirable if the cost of the task is less than the cost of repair. If all answers are "NO" in the logic process, no preventive maintenance task is generated. If the cost penalties are severe, a redesign may be desirable.

Hidden-function safety effects—Questions 8A to 8F. The hidden-function safety effect requires a task to ensure the availability necessary to avoid the safety effect of multiple failures. All questions should be asked. If no applicable and effective tasks are found, then redesign is mandatory.

Hidden function non-safety effects—Questions 9A to 9E. This category indicates that a task may be desirable to assure the availability necessary to avoid the direct cost effects of multiple failures. If all answers are "NO" in the logic process, no preventive maintenance task is generated. If economic penalties are severe, a redesign may be desirable.

TASK DETERMINATION

Task determination is handled in a similar manner for each of the five effect categories. For task determination, it is necessary to apply the failure causes for the functional failure to the second level of the logic diagram. Seven possible task resultant questions in the effect categories have been identified, although additional tasks, modified tasks, or modified task definition may be warranted depending on the needs of particular industries.

PARALLELING AND DEFAULT LOGIC

Paralleling and default logic play an essential role at level 2 (Figs. 2.3 and 2.4). Regardless of the answer to the first question regarding "lubrication or servicing," the next task selection question should be asked in all cases. Then following the hidden or evident safety effects path, all subsequent questions should be asked. In the remaining categories, subsequent to the first question, a "YES" answer will allow exiting the logic. (At the user's option, advancement is allowable to subsequent questions after a "YES" answer is derived, but only if the cost of the task is equal to the cost of the failure prevented).

Default Logic. Default logic is reflected in paths outside the safety effect areas by the arrangement of the task selection logic. In the absence of adequate information to answer "YES" or "NO" to questions in the second level, default logic dictates that a "NO" answer be given and the subsequent questions be asked. When "NO" answers are generated, the only choice available is the next question, which in most cases provides a more conservative, stringent and/or costly route.

Redesign. Redesign is mandatory for failures that fall into the safety effects category (evident or hidden) and for which there are no applicable and effective tasks.

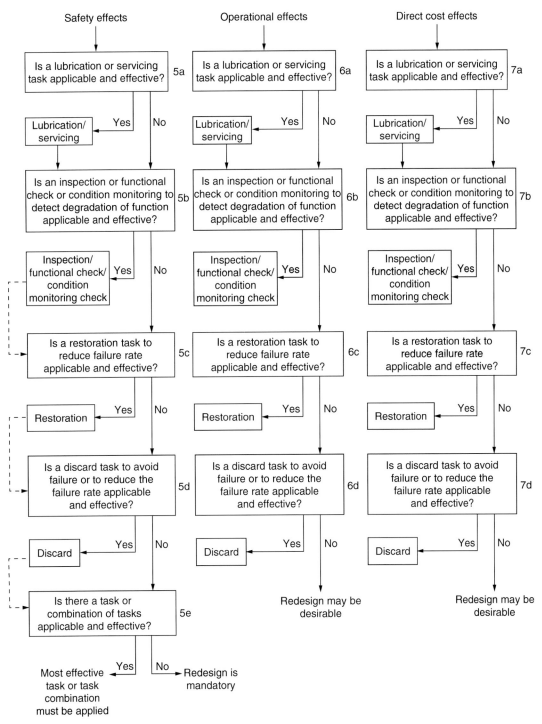

FIGURE 2.3 Reliability decision logic tree (level 2)—effects categories and task determination.

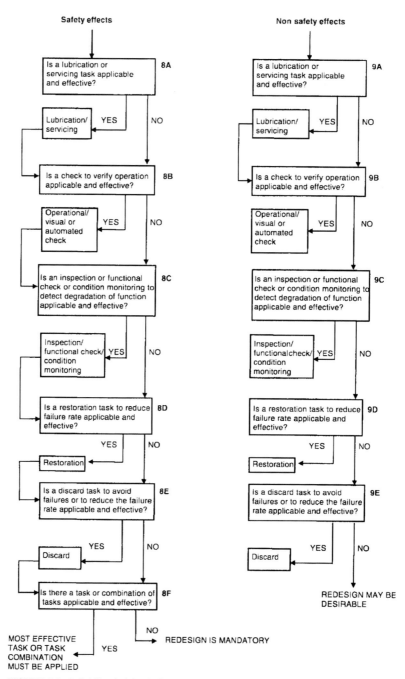

FIGURE 2.4 Reliability decision logic tree (level 2)—effects categories and task determination.

MAINTENANCE TASKS

Explanations of the terms used in the possible tasks are as follows:

- Lubrication/servicing (all categories)—this involve any act of lubricating or servicing for maintaining inherent design capabilities.
- Operational/visual/automated check (hidden functional failure categories only)—an *operational check* is a task to determine that an item is fulfilling its intended purpose. It does not require quantitative checks and is a failure-finding task. A *visual check* is an observation to determine that an item is fulfilling its intended purpose and does not require quantitative tolerances. This, again, is a failure-finding task. The visual check could also involve interrogating electronic units that store failure data.
- Inspection/functional check/condition monitoring (all categories)—an *inspection* is an examination of an item against a specific standard. A *functional check* is a quantitative check to determine if one or more functions of an item perform within specified limits. *Condition monitoring* is a task, which may be continuous or periodic to monitor the condition of an item in operation against preset parameters.
- Restoration (all categories)—restoration is the work necessary to return the item to a specific standard. Since restoration may vary from cleaning or replacement of single parts up to a complete overhaul, the scope of each assigned restoration task has to be specified.
- Discard (all categories)—discard is the removal from service of an item at a specified life limit. Discard tasks are normally applied to so-called single-cell parts such as cartridges, canisters, cylinders, turbine disks, safe-life structural members, and the like.
- Combination (safety categories)—since this is a safety category question and a task is required, all possible avenues should be analyzed. To do this, a review of the tasks, which are applicable, is necessary. From this review, the most effective tasks should be selected.
- No task (all categories)—It may be decided that no task is required in some situations, depending on the effect. Each of the possible tasks defined above is based upon its own applicability and effectiveness criteria. Table 2.1 summarizes these task selection criteria.

TASK FREQUENCIES OR INTERVALS

In order to set a task frequency or interval, it is necessary to determine the existence of applicable operational experience data that suggest an effective interval for task accomplishment. Appropriate information may be obtained from one or more of the following:

- Prior knowledge from other similar equipment, which shows that a scheduled maintenance task has offered substantial evidence of being applicable, effective, and economically worthwhile.
- Manufacturer/supplier test data, which indicate that a scheduled maintenance task will be applicable and effective for the item being evaluated.
- Reliability data and predictions.

Safety and cost considerations need to be addressed in establishing the maintenance intervals. Scheduled inspections and replacement intervals should coincide whenever possible, and tasks should be grouped to reduce the operational impact.

The safety replacement interval can be established from the cumulative failure distribution for the item by choosing a replacement interval that result in an extremely low probability of failure prior to replacement. Where a failure does not cause a safety hazard, but causes loss of availability, the replacement interval is established in a trade-off process involving the cost of replacement components, the cost of failure, and the availability requirement of the equipment.

Mathematical models exist for determining task frequencies and intervals, but these models depend on the availability of the appropriate data. This data will be specific to particular industries and those industry standards and data sheets should be consulted as appropriate.

If there is insufficient reliability data, or no prior knowledge from other similar equipment, or if there is insufficient similarity between the previous and current systems, the task interval frequency can only be established initially by experienced personnel using good judgment and operating experience in concert with the best available operating data and relevant cost data.

CHAPTER 3
PREDICTIVE MAINTENANCE

R. Keith Mobley
Principal, Life Cycle Engineering, Inc. Charleston, S.C.

Predictive maintenance is perhaps the most misunderstood and misused of all the plant improvement programs. Most users define it as a means to prevent catastrophic failure of critical rotating machinery. Others define predictive maintenance as a maintenance scheduling tool that uses vibration and infrared or lubricating oil analysis data to determine the need for corrective maintenance actions. A few share the belief, precipitated by vendors of predictive maintenance systems, that predictive maintenance is the panacea for our critically ill plants. One common theme of these definitions is that it is solely a maintenance management tool.

Because of these misconceptions, the majority of established predictive maintenance programs have not been able to achieve a marked decrease in maintenance costs or a measurable improvement in overall plant performance. In fact, the reverse is too often true. In many cases, the annual costs of repairs, repair parts, product quality, and production have dramatically increased as a direct result of the program.

Predictive maintenance is much more than a maintenance scheduling tool and should not be restricted to maintenance management. As part of an integrated, total plant performance management program, it can provide the means to improve the production capacity, product quality, and overall effectiveness of our manufacturing and production plants.

DEFINITION OF PREDICTIVE MAINTENANCE

Predictive maintenance is not a panacea for all the factors that limit total plant performance. In fact, it cannot directly affect plant performance. Predictive maintenance is a management technique that, simply stated, uses regular evaluation of the actual operating condition of plant equipment, production systems, and plant management functions to optimize total plant operation.

The output of a predictive maintenance program is data. Until action is taken to resolve the deviations or problems revealed by the program, plant performance cannot be improved. Therefore, a management philosophy that is committed to plant improvement must exist before any meaningful benefit can be derived. Without the absolute commitment and support of senior management and the full cooperation of all plant functions, a predictive maintenance program cannot provide the means to resolve poor plant performance.

Predictive technology can be used for much more than just measuring the operating condition of critical plant machinery. The technology permits accurate evaluation of all functional groups, such as maintenance, within the company. Properly used, predictive maintenance can identify most, if not all, factors that limit effectiveness and efficiency of the total plant.

TOTAL PLANT MANAGEMENT

One factor that limits the effective management of plants is the lack of timely, factual data that defines operating condition of critical production systems and the effectiveness of critical plant functions, such as purchasing, engineering, and production. Properly used, predictive maintenance can provide the means to eliminate all factors that limit plant performance. Many of these problems are outside the purview of maintenance and must be corrected by the appropriate plant function.

High maintenance costs are the direct result of inherent problems throughout the plant, not just ineffective maintenance management. Poor design standards and purchasing practices, improper operation, and outdated management methods contribute more to high production and maintenance costs than do delays caused by catastrophic failure of critical plant machinery. Because of the breakdown mentality and myopic view of the root cause of ineffective plant performance, too many plants restrict predictive maintenance to the maintenance function. Expansion of the program to include regular evaluation of all factors that limit plant performance will greatly enhance the benefits that can be derived.

In a total plant performance mode, predictive technology can be used to accurately measure the effectiveness and efficiency of all plant functions, not just machinery. The data generated by regular evaluation can isolate specific limitations in skill levels, inadequate procedures, and poor management methods as well as incipient machine or process system problems.

MAINTENANCE MANAGEMENT

As a maintenance management tool, predictive maintenance can provide the data required to schedule both preventive and corrective maintenance tasks on an as-needed basis. Instead of relying on industrial average-life statistics, such as mean-time-to-failure, to schedule maintenance activities, predictive maintenance uses direct monitoring of the operating condition, system efficiency, and other indicators to determine the actual mean-time-to-failure or loss of efficiency for each machine train and system within the plant.

At best, traditional time-driven methods provide a guideline to normal machine-train life spans. The final decision, in preventive or run-to-failure programs, on when to repair or rebuild a machine must be made on the basis of intuition and the personal experience of the maintenance manager. The addition of a comprehensive predictive maintenance program can and will provide factual data that define the actual mechanical condition of each machine train and operating efficiency of each process system. These data provide the maintenance manager with factual data that can be used to schedule maintenance activities.

A predictive maintenance program can minimize unscheduled breakdowns of all mechanical equipment in the plant and ensure that repaired equipment is in acceptable mechanical condition. The program can also identify machine-train problems before they become serious. Most problems can be minimized if they are detected and repaired early. Normal mechanical failure modes degrade at a speed directly proportional to their severity. If the problem is detected early, major repairs, in most instances, can be prevented.

To achieve these goals, the predictive maintenance program must correctly identify the root cause of incipient problems. Many of the established programs do not meet this fundamental requirement. Precipitated by the claims of predictive maintenance system vendors, many programs are established on simplistic monitoring methods that identify the symptom rather than the real cause of problems. In these instances, the derived benefits that are achieved are greatly diminished. In fact, many of these programs fail because maintenance managers lose confidence in the program's ability to accurately detect incipient problems.

Predictive maintenance cannot function in a void. To be an effective maintenance management tool, it must be combined with a viable maintenance planning function that will use the data to plan and schedule appropriate repairs. In addition, it is dependent on the skill and knowledge of maintenance craftsmen. Unless proper repairs or corrective actions are made, the data provided by the predictive maintenance program cannot be effective. Both ineffective planning and improper repairs will severely restrict the benefits of predictive maintenance.

Predictive maintenance utilizing vibration signature analysis is predicated on two basic facts: (1) all common failure modes have distinct vibration frequency components that can be isolated and identified, and (2) the amplitude of each distinct vibration component will remain constant unless there is a change in the operating dynamics of the machine train. Predictive maintenance utilizing process efficiency, heat loss, or other nondestructive techniques can quantify the operating efficiency of nonmechanical plant equipment or systems. These techniques used in conjunction with vibration analysis can provide the maintenance manager or plant engineer with factual information that will enable them to achieve optimum reliability and availability from their plant.

PRODUCTION MANAGEMENT

Predictive maintenance can be an invaluable production management tool. The data derived from a comprehensive program can provide the information needed to increase production capacity, product quality, and the overall effectiveness of the production function.

Production efficiency is directly dependent on a number of machine-related factors. Predictive maintenance can provide the data needed to achieve optimum, consistent reliability, capacity, and efficiency from critical production systems. While these factors are viewed as maintenance responsibilities, many of the factors that directly affect them are outside of the maintenance function. For example, standard operating procedures or operator errors can directly influence these variables. Unless production management uses regular evaluation methods, that is, predictive maintenance, to determine the effects of these production influences, optimum production performance cannot be achieved.

Product quality and total production costs are other areas where predictive maintenance can benefit production management. Regular evaluation of critical production systems can anticipate potential problems that would result in reduced product quality and an increase in overall production costs. While the only output of the predictive maintenance program is data, this information can be used to correct a myriad of production problems that directly affect the effectiveness and efficiency of the production department.

QUALITY IMPROVEMENT

Most product quality problems are the direct result of (1) production systems with inherent problems, (2) poor operating procedures, (3) improper maintenance, or (4) defective raw materials. Predictive maintenance can isolate this type of problem and provide the data required to correct many of the problems that result in reduced product quality.

A comprehensive program will use a combination of data, such as vibration, thermography, tribology (the science of friction, wear, and lubrication of interacting surfaces), process parameters, and operating dynamics, to anticipate deviations from optimum operating condition of critical plant systems before they can affect product quality, production capacity, or total production costs.

PREDICTIVE MAINTENANCE TECHNIQUES

There are a variety of technologies that can and should be used as part of a comprehensive predictive maintenance program. Since mechanical systems or machines account for the majority of plant equipment, vibration monitoring is generally the key component of most predictive maintenance programs. However, vibration monitoring cannot provide all of the information that will be required for a successful predictive maintenance program. This technique is limited to monitoring the mechanical condition and not other critical parameters required to maintain reliability and efficiency of machinery.

It is a very limited tool for monitoring critical process and machinery efficiencies and other parameters that can severely limit productivity and product quality.

Hence, as previously noted, it must be iterated that a comprehensive predictive maintenance program must include other monitoring and diagnostic techniques. These techniques include (1) vibration monitoring (2) thermography, (3) tribology, (4) process parameters, (5) visual inspection, and (5) other nondestructive testing techniques.

VIBRATION MONITORING

Vibration analysis is the dominant technique used for predictive maintenance management. Since the greatest population of typical plant equipment is mechanical, this technique has the widest application and benefits in a total plant program. This technique uses the noise or vibration created by mechanical equipment and in some cases by plant systems to determine their actual condition. Using vibration analysis to detect machine problems is not new. During the 1960s and 1970s, the U.S. Navy and petrochemical and nuclear electric power generating industries invested heavily in the development of analysis techniques based on noise or vibration that could be used to detect and identify incipient mechanical problems in critical machinery. By the early 1980s, the instrumentation and analytical skills required for noise-based predictive maintenance were fully developed. These techniques and instrumentation had proved to be extremely reliable and accurate in detecting abnormal machine behavior. However, the capital cost of instrumentation and the expertise required to acquire and analyze noise data precluded general application of this type of predictive maintenance. As a result, only the most critical equipment in a few select industries could justify the expense required to implement a noise-based predictive maintenance program.

Recent advancements in microprocessor technology coupled with the expertise of companies that specialize in machinery diagnostics and analysis technology have evolved the means to provide vibration-based predictive maintenance that can be cost-effectively used in most manufacturing and process applications. These microprocessor-based systems have simplified the data acquisition, automated data management, and minimized the need for vibration experts to interpret data.

Commercially available systems are capable of routine monitoring, trending, and evaluation of the mechanical condition of all mechanical equipment in a typical plant. This type of program can be used to schedule maintenance on all rotating and reciprocating and most continuous process mechanical equipment. Monitoring the vibration from plant machinery can provide direct correlation between the mechanical condition and recorded vibration data of each machine in the plant. Any degradation of the mechanical condition within plant machinery can be detected using vibration monitoring techniques. Used properly, vibration analysis can identify specific degrading machine components or the failure mode of plant machinery before serious damage occurs.

Most vibration-based predictive maintenance programs rely on one or more trending and analysis techniques. These techniques include broadband trending, narrowband trending, and signature analysis.

Broadband Trending

This technique acquires overall or broadband vibration readings from select points on a machine train. These data are compared with either a baseline reading taken from a new machine or vibration severity charts to determine the relative condition of the machine. Normally an unfiltered broadband measurement that provides the total vibration energy between 10 and 10,000 Hz is used for this type of analysis. Broadband or overall rms data are strictly a gross value or number that represents the total vibration of the machine at the specific measurement point where the data were acquired. It does not provide any information pertaining to the actual machine problem or failure mode. Ideally broadband trending can be used as a simple indication that there has been a change in either the mechanical condition or the operating dynamics of the machine or system. At best, this technique can be used as a gross scan of the operating condition of critical process machinery. However, broadband values must be adjusted to the actual production parameters, such as load and speed, to be effective even in this reduced role. Changes in both the speed and load of machinery will have a direct effect on the overall vibration levels of the machine.

Narrowband Trending

Narrowband trending, like broadband, monitors the total energy for a specific bandwidth of vibration frequencies. Unlike broadband, narrowband analysis uses vibration frequencies that represent specific machine components or failure modes. This method provides the means to quickly monitor the mechanical condition of critical machine components, not just the overall machine condition. This technique provides the ability to monitor the condition of gear sets, bearings, and other machine components without manual analysis of vibration signatures.

As in the case of broadband trending, changes in speed, load, and other process parameters will have a direct, often dramatic, impact on the vibration energy produced by each machine component or narrowband. To be meaningful, narrowband values must be adjusted to the actual production parameters.

Signature Analysis

Unlike the two trending techniques, signature analysis provides visual representation of each frequency component generated by a machine train. With training, plant staff can use vibration signatures to determine the specific maintenance required by plant machinery. Most vibration-based predictive maintenance programs use some form of signature analysis in their program. However, the majority of these programs rely on comparative analysis rather than full root-cause techniques. This failure limits the benefits that can be derived from this type of program.

THERMOGRAPHY

Thermography is a predictive maintenance technique that can be used to monitor the condition of plant machinery, structures, and systems. It uses instrumentation designed to monitor the emission of infrared energy, that is, temperature, to determine their operating condition. By detecting thermal anomalies, that is, areas that are hotter or colder than they should be, an experienced surveyor can locate and define incipient problems within the plant.

Infrared technology is predicated on the fact that all objects having a temperature above absolute zero emit energy or radiation. Infrared radiation is one form of this emitted energy. Infrared emissions, or below red, are the shortest wavelengths of all radiated energy and are invisible without special instrumentation. The intensity of infrared radiation from an object is a function of its surface temperature. However, temperature measurement using infrared methods is complicated because there are three sources of thermal energy that can be detected from any object: energy emitted from the object itself, energy reflected from the object, and energy transmitted by the object. Only the emitted energy is important in a predictive maintenance program. Reflected and transmitted energies will distort raw infrared data. Therefore, the reflected and transmitted energies must be filtered out of acquired data before a meaningful analysis can be made.

The surface of an object influences the amount of emitted or reflected energy. A perfect emitting surface is called a *blackbody* and has an emissivity equal to 1.0. These surfaces do not reflect. Instead, they absorb all external energy and reemit as infrared energy. Surfaces that reflect infrared energy are called *graybodies* and have an emissivity less than 1.0. Most plant equipment falls into this classification. Careful consideration of the actual emissivity of an object improves the accuracy of temperature measurements used for predictive maintenance. To help users determine emissivity, tables have been developed to serve as guidelines for most common materials. However, these guidelines are not absolute emissivity values for all machines or plant equipment.

Variations in surface condition, paint, or other protective coatings and many other variables can affect the actual emissivity factor for plant equipment. In addition to reflected and transmitted energy, the user of thermographic techniques must also consider the atmosphere between the object and the measurement instrument. Water vapor and other gases absorb infrared radiation. Airborne dust, some lighting, and other variables in the surrounding atmosphere can distort measured infrared radiation. Since the atmospheric environment is constantly changing, using thermographic techniques requires extreme care each time infrared data are acquired.

Most infrared monitoring systems or instruments provide special filters that can be used to avoid the negative effects of atmospheric attenuation of infrared data. However, the plant user must recognize the specific factors that will affect the accuracy of the infrared data and apply the correct filters or other signal conditioning required to negate that specific attenuating factor or factors.

Collecting optics, radiation detectors, and some form of indicator comprise the basic elements of an industrial infrared instrument. The optical system collects radiant energy and focuses it upon a detector, which converts it into an electrical signal. The instrument's electronics amplifies the output signal and processes it into a form which can be displayed. There are three general types of instruments that can be used for predictive maintenance: infrared thermometers or spot radiometers, line scanners, and imaging systems.

Infrared Thermometers

Infrared thermometers or spot radiometers are designed to provide the actual surface temperature at a single, relatively small point on a machine or surface. Within a predictive maintenance program, the point-of-use infrared thermometer can be used in conjunction with many of the microprocessor-based vibration instruments to monitor the temperature at critical points on plant machinery or equipment. This technique is typically used to monitor bearing cap temperatures, motor winding temperatures, spot checks of process piping temperatures, and similar applications. It is limited in that the temperature represents a single point on the machine or structure. However, when used in conjunction with vibration data, point-of-use infrared data can be a valuable tool.

Line Scanners

This type of infrared instrument provides a single-dimensional scan or line of comparative radiation. While this type of instrument provides a somewhat larger field of view, that is, area of machine surface, it is limited in predictive maintenance applications.

Infrared Imaging

Unlike other infrared techniques, thermal or infrared imaging provides the means to scan the infrared emissions of complete machines, process, or equipment in a very short time. Most of the imaging systems function much like a video camera. The user can view the thermal emission profile of a wide area by simply looking through the instrument's optics. A variety of thermal imaging instruments are on the market, ranging from relatively inexpensive black-and-white scanners to full-color microprocessor-based systems. Many of the less expensive units are designed strictly as scanners and do not provide the capability of store-and-recall thermal images. The inability to store and recall previous thermal data will limit a long-term predictive maintenance program.

Inclusion of thermography into a predictive maintenance program will enable you to monitor the thermal efficiency of critical process systems that rely on heat transfer or retention, electrical equipment, and other parameters that will improve both the reliability and efficiency of plant systems. Infrared techniques can be used to detect problems in a variety of plant systems and equipment, including electrical switchgear, gearboxes, electrical substations, transmissions, circuit breaker panels, motors, building envelopes, bearings, steam lines, and process systems that rely on heat retention or transfer.

TRIBOLOGY

Tribology is the general term that refers to design and operating dynamics of the bearing-lubrication-rotor support structure of machinery. Several tribology techniques can be used for predictive maintenance: lubricating oil analysis, spectrographic analysis, ferrography, and wear particle analysis.

Lubricating oil analysis, as the name implies, is an analysis technique that determines the condition of lubricating oils used in mechanical and electrical equipment. It is not a tool for determining the operating condition of machinery. Some forms of lubricating oil analysis will provide an accurate quantitative breakdown of individual chemical elements, both oil additive and contaminates, contained in the oil. A comparison of the amount of trace metals in successive oil samples can indicate wear patterns of oil-wetted parts in plant equipment and will provide an indication of impending machine failure.

Until recently, tribology analysis has been a relatively slow and expensive process. Analyses were conducted using traditional laboratory techniques and required extensive, skilled labor. Microprocessor-based systems are now available which can automate most of the lubricating oil and spectrographic analysis, thus reducing the manual effort and cost of analysis.

The primary applications for spectrographic or lubricating oil are quality control, reduction of lubricating oil inventories, and determination of the most cost-effective interval for oil change. Lubricating, hydraulic, and dielectric oils can be periodically analyzed, using these techniques, to determine their condition. The results of this analysis can be used to determine if the oil meets the lubricating requirements of the machine or application. Based on the results of the analysis, lubricants can be changed or upgraded to meet the specific operating requirements. In addition detailed analysis of the chemical and physical properties of different oils used in the plant can, in some cases, allow consolidation or reduction of the number and types of lubricants required to maintain plant equipment. Elimination of unnecessary duplication can reduce required inventory levels and therefore maintenance costs.

As a predictive maintenance tool, lubricating oil and spectrographic analysis can be used to schedule oil change intervals based on the actual condition of the oil. In middle-sized to large plants, a reduction in the number of oil changes can amount to a considerable annual reduction in maintenance costs. Relatively inexpensive sampling and testing can show when the oil in a machine has reached a point that warrants change. The full benefit of oil analysis can be achieved only by taking frequent samples trending the data for each machine in the plant. It can provide a wealth of information on which to base maintenance decisions. However, major payback is rarely possible without a consistent program of sampling.

Lubricating Oil Analysis

Oil analysis has become an important aid to preventive maintenance. Laboratories recommend that samples of machine lubricant be taken at scheduled intervals to determine the condition of the lubricating film that is critical to machine-train operation. Typically 10 tests are conducted on lube oil samples:

Viscosity. This is one of the most important properties of a lubricating oil. The actual viscosity of oil samples is compared with an unused sample to determine the thinning or thickening of the sample during use. Excessively low viscosity will reduce the oil film strength, weakening its ability to prevent metal-to-metal contact. Excessively high viscosity may impede the flow of oil to vital locations in the bearing support structure, reducing its ability to lubricate.

Contamination. Oil contamination by water or coolant can cause major problems in a lubricating system. Many of the additives now used in formulating lubricants contain the same elements that are used in coolant additives. Therefore, the laboratory must have an accurate analysis of new oil for comparison.

Fuel Dilution. Oil dilution in an engine weakens the oil film strength, sealing ability, and detergency. It may be caused by improper operation, fuel system leaks, ignition problems, improper timing, or other deficiencies. Fuel dilution is considered excessive when it reaches a level of 2.5 to 5 percent.

Solids Content. This is a general test. All solid materials in the oil are measured as a percentage of the sample volume or weight. The presence of solids in a lubricating system can significantly increase the wear on lubricated parts. Any unexpected rise in reported solids is cause for concern.

Fuel Soot. An important indicator for oil used in diesel engines, fuel soot is always present to some extent. A test to measure fuel soot in diesel engine oil is important, since it indicates the fuel-burning efficiency of the engine. Most tests for fuel soot are conducted by infrared analysis.

Oxidation. Lubricating oil oxidation can result in lacquer deposits, metal corrosion, or thickening of the oil. Most lubricants contain oxidation inhibitors. However, when additives are used up, oxidation of the oil itself begins. The quantity of oxidation in an oil sample is measured by differential infrared analysis.

Nitration. Fuel combustion in engines results from nitration. The products formed are highly acidic and may leave deposits in combustion areas. Nitration will accelerate oil oxidation. Infrared analysis is used to detect and measure nitration products.

Total Acid Number. This is a measure of the amount of acid or acidlike material in the oil sample. Because new oils contain additives that affect the total acid number (TAN), it is important to compare used oil samples with new, unused, oil of the same type. Regular analysis at specific intervals is important to this evaluation.

Total Base Number. This number indicates the ability of an oil to neutralize acidity. The higher the total base number (TBN) the greater its ability to neutralize acidity. Typical causes of low TBN include using the improper oil for an application, waiting too long between oil changes, overheating, and using high-sulfur fuel.

Particle Count. Tests of particle count are important to anticipating potential system or machine problems. This is especially true in hydraulic systems. Particle count analysis made a part of a normal lube oil analysis is quite different from wear particle analysis. In this test, high particle counts indicate that machinery may be wearing abnormally or that failures may occur as a result of temporarily or permanently blocked orifices. No attempt is made to determine the wear patterns, size, and other factors that would identify the failure mode within the machine.

Spectrographic Analysis

Spectrographic analysis allows accurate, rapid measurements of many of the elements present in lubricating oil. These elements are generally classified as wear metals, contaminates, or additives. Some elements can be listed in more than one of these classifications. Standard lubricating oil analyses do not attempt to determine the specific failure modes of developing machine-train problems. Therefore, additional techniques must be used as part of a comprehensive predictive maintenance program.

Wear Particle Analysis

Wear particle analysis is related to oil analysis only in that the particles to be studied are collected through drawing a sample of lubricating oil. Where lubricating oil analysis determines the actual condition of the oil sample, wear particle analysis provides direct information about the wearing condition of the machine train. Particles in the lubricant of a machine can provide significant information about the condition of the machine. This information is derived from the study of particle shape, composition, size, and quantity. Wear particle analysis is normally conducted in two stages. The first method used for wear particle analysis is routine monitoring and trending of the solids content of machine lubricant. In simple terms the quantity, composition, and size of particulate matter in the lubricating oil are indicative of the mechanical condition of the machine. A normal machine will contain low levels of solids with a size less than 10 µm. As the machine's condition degrades, the number and size of particulate matter will increase. The second wear particle method involves analysis of the particulate matter in each lubricating oil sample. Five basic types of wear can be identified according to the classification of particles: rubbing wear, cutting wear, rolling fatigue wear, combined rolling and sliding wear, and severe sliding wear (see Table 3.1). Only rubbing wear and early rolling fatigue mechanisms generate particles predominantly less than 15 µm in size.

TABLE 3.1 Five types of wear

Type	Description
Rubbing wear	Result of normal wear in machine
Cutting wear	Caused by one surface penetrating another machine surface
Rolling fatigue	Primary result of rolling contact within bearings
Combined rolling and sliding wear	Results from moving of contact surfaces within a gear system
Severe sliding wear	Caused by excessive loads or heat in a gear system

Rubbing Wear. This is the result of normal sliding wear in a machine. During a normal break-in of a wear surface, a unique layer is formed at the surface. As long as this layer is stable, the surface wears normally. If the layer is removed faster than it is generated, the wear rate increases and the maximum particle size increases.

Excessive quantities of contaminate in a lubrication system can increase rubbing wear by more than an order of magnitude without completely removing the shear mixed layer. Although catastrophic failure is unlikely, these machines can wear out rapidly. Impending trouble is indicated by a dramatic increase in wear particles.

Cutting Wear Particles. These are generated when one surface penetrates another. They are produced when a misaligned or fractured hard surface produces an edge that cuts into a softer surface, or when abrasive contaminate becomes embedded in a soft surface and cuts an opposing surface. Cutting wear particles are abnormal and are always worthy of attention. If they are only a few micrometers long and a fraction of a micrometer wide, the cause is probably a contaminate. Increasing quantities of longer particles signal a potentially imminent component failure.

Rolling Fatigue. This is associated primarily with rolling contact bearings and may produce three distinct particle types: fatigue spall particles, spherical particles, and laminar particles. Fatigue spall particles are the actual material removed when a pit or spall opens up on a bearing surface. An increase in the quantity or size of these particles is the first indication of an abnormality. Rolling fatigue does not always generate spherical particles, and they may be generated by other sources. Their presence is important in that they are detectable before any actual spalling occurs. Laminar particles are very thin and are thought to be formed by the passage of a wear particle through a rolling contact. They frequently have holes in them. Laminar particles may be generated throughout the life of a bearing, but at the onset of fatigue spalling the quantity increases.

Combined Rolling and Sliding Wear. This results from the moving contact of surfaces in gear systems. These larger particles result from tensile stresses on the gear surface, causing the fatigue cracks to spread deeper into the gear tooth before pitting. Gear fatigue cracks do not generate spheres. Scuffing of gears is caused by too high a load or speed. The excessive heat generated by this condition breaks down the lubricating film and causes adhesion of the mating gear teeth. As the wear surfaces become rougher, the wear rate increases. Once started, scuffing usually affects each gear tooth.

Severe Sliding Wear. This is caused by excessive loads or heat in a gear system. Under these conditions, large particles break away from the wear surfaces, causing an increase in the wear rate. If the stresses applied to the surface are increased further, a second transition point is reached. The surface breaks down and catastrophic wear ensues.

Normal spectrographic analysis is limited to particulate contamination with a size of 10 μm or less. Larger contaminants are ignored. This fact can limit the benefits that can be derived from the technique.

Ferrography

This technique is similar to spectrography, but there are two major exceptions. First, ferrography separates particulate contamination by using a magnetic field rather than burning a sample as in spectrographic analysis. Because a magnetic field is used to separate contaminants, this technique is primarily limited to ferrous or magnetic particles. The second difference is that particulate contamination larger than 10 μm can be separated and analyzed. Normal ferrographic analysis will capture particles up to 100 μm and provides a better representation of the total oil contamination than spectrographic techniques.

There are three major limitations with using tribology analysis in a predictive maintenance program: equipment costs, acquiring accurate oil samples, and interpretation of data.

One factor that severely limits the benefits of tribology is the acquisition of accurate samples that represent the true lubricating oil inventory in a machine. Sampling is not a matter of opening a port somewhere in the oil line and catching a pint sample. Extreme care must be taken to acquire samples that truly represent the lubricant that will pass through the machine's bearings. One recent example is an attempt to acquire oil samples from a bullgear compressor. The lubricating oil filter had a sample port on the clean, that is, downstream, side. However, comparison of samples taken at this point and one taken directly from the compressor's oil reservoir indicated that more contaminants existed downstream from the filter than in the reservoir. Which location actually represented the oil's condition? Neither sample was truly representative of the oil condition. The oil filter had removed most of the suspended solids, that is, metals and other insolubles, and was therefore not representative of the actual condition. The reservoir sample was not representative since most of the suspended solids had settled out in the sump.

Proper methods and frequency of sampling lubricating oil are critical to all predictive maintenance techniques that use lubricant samples. Sample points that are consistent with the objective of detecting large particles should be chosen. In a recirculating system, samples should be drawn as the lubricant returns to the reservoir and before any filtration. Do not draw oil from the bottom of a sump where large quantities of material build up over time. Return lines are preferable to the reservoir as the sample source, but good reservoir samples can be obtained if careful, consistent practices are used. Even equipment with high levels of filtration can be effectively monitored as long as samples are drawn before oil enters the filters. Sampling techniques involve taking samples under uniform operating conditions. Samples should not be taken more than 30 min after the equipment has been shut down. Sample frequency is a function of the mean time to failure from the onset of an abnormal wear mode to catastrophic failure. For machines in critical service, sampling every 25 hr of operation is appropriate. However, for most industrial equipment in continuous service, monthly sampling is adequate. The exception to monthly sampling is machines with extreme loads. In this instance, weekly sampling is recommended.

Understanding the meaning of analysis results is perhaps the most serious limiting factor. Most often results are expressed in terms that are totally alien to plant engineers or technicians. Therefore, it is difficult for them to understand the true meaning, in terms of oil or machine condition. A good background in quantitative and qualitative chemistry is beneficial. As a minimum requirement, plant staff will require training in basic chemistry and specific instruction on interpreting tribology results.

PROCESS PARAMETERS

Many plants do not consider machine or systems efficiency to be part of the maintenance responsibility. However, machinery that is not operating within acceptable efficiency parameters severely limits the productivity of many plants. Therefore, a comprehensive predictive maintenance program should include routine monitoring of process parameters. As an example of the importance of process parameters monitoring, consider a process pump that may be critical to plant operation.

Vibration-based predictive maintenance will provide the mechanical condition of the pump and infrared imaging will provide the condition of the electric motor and bearings. Neither provides any indication of the operating efficiency of the pump. Therefore, the pump could be operating at less than 50 percent efficiency and the predictive maintenance program would not detect the problem.

Process inefficiencies, like the example cited, are often the most serious limiting factor in a plant. Their negative impact on plant productivity and profitability is often greater than the total cost of the maintenance operation. However, without regular monitoring of process parameters, many plants do not recognize this unfortunate fact. If your program included monitoring of the suction and discharge pressures and ampere load of the pump, you could determine the operating efficiency. The brake-horsepower (bhp) formula

$$bhp = \frac{gpm \times TDH \times sp.gr}{3960 \times efficiency}$$

could be used to calculate operating efficiency of any pump in the program. By measuring the suction and discharge pressure, the total dynamic head (TDH) can be determined. A flow curve, used in conjunction with the actual total dynamic head, would define the actual flow (gpm) and an ammeter reading would define the horsepower. With these measured data, the efficiency can be calculated.

Process parameters monitoring should include all machinery and systems in the plant process that can affect its production capacity. Typical systems include heat exchangers, pumps, filtration, boilers, fans, blowers, and other critical systems. Inclusion of process parameters in predictive maintenance can be accomplished in two ways: manual or microprocessor-based systems. However, both methods will normally require installing instrumentation to measure the parameters that indicate the actual operating condition of plant systems. Even though most plants have installed pressure gages, thermometers, and other instruments that should provide the information required for this type of program, many of them are no longer functioning. Therefore, including process parameters in your program will require an initial capital cost to install calibrated instrumentation. Data from the installed instrumentation can be periodically recorded using either manual logging or a microprocessor-based data logger. If the latter is selected, many vibration-based microprocessor systems can also provide the means of acquiring process data. This should be considered when selecting the vibration monitoring system that will be used in your program. In addition, some microprocessor-based predictive maintenance systems provide the ability to calculate unknown process variables. For example, they can calculate the pump efficiency used in the example. This ability to calculate unknowns based on measured variables will enhance a total plant predictive maintenance program without increasing the manual effort required. In addition, some of these systems include nonintrusive transducers that can measure temperatures, flows, and other process data without the necessity of installing permanent instrumentation. This further reduces the initial cost of including process parameters in your program.

ELECTRIC MOTOR ANALYSIS

Evaluation of electric motors and other electrical equipment is critical to a total plant predictive maintenance program. To an extent, vibration data isolate some of the mechanical and electrical problems that can develop in critical drive motors. However, vibration cannot provide the comprehensive coverage required to achieve optimum plant performance. Therefore, a total plant predictive maintenance program must include data acquisition and evaluation methods that are specifically designed to identify problems within motors and other electrical equipment.

Insulation Resistance

Insulation resistance tests are important, although they may not be conclusive, in that they can reveal flaws in insulation, poor insulating material, the presence of moisture, and a number of other problems. Such tests can be applied to the insulation of electrical machinery from the windings to the frame, to underground cables, insulators, capacitors, and a number of other auxiliary electrical components. Normally these tests are conducted using (1) megger, (2) Wheatstone bridge, (3) Kelvin double bridge, or (4) a number of other instruments.

A megger provides the means to directly measure the condition of motor insulation. This method uses a device which generates a known output, usually 500 V, and directly measures the resistance of the insulation within the motor. When the insulation resistance falls below the prescribed value, it can be brought to required standards by cleaning and drying the stator and rotor.

The accuracy of meggering and most insulation resistance tests varies widely with temperature, humidity, and cleanliness of the parts. Therefore, they may not be absolutely conclusive.

Other Electrical Testing

A complete predictive maintenance program should include all testing and evaluation methods required to regularly evaluate all critical plant systems. As a minimum, a total plant program should also include (1) dielectric loss analysis, (2) gas-in-oil analysis, (3) stray field monitoring, (4) high-voltage, switchgear discharge testing, (5) resistance measurements, (6) Rogowski coils, and (7) rotor bar current harmonics.

VISUAL INSPECTION

Regular visual inspection of the machinery and systems in a plant is a necessary part of any predictive maintenance program. In many cases, visual inspection will detect potential problems that will be missed using the other predictive maintenance techniques. Even with the predictive techniques discussed, many potentially serious problems can remain undetected. Routine visual inspection of all critical plant systems will augment the other techniques and ensure that potential problems are detected before serious damage can occur. Most of the vibration-based predictive maintenance systems include the capability of recording visual observations as part of the routine data-acquisition process.

Since the incremental costs of these visual observations are small, this technique should be incorporated in all predictive maintenance programs. All equipment and systems in the plant should be visually inspected on a regular basis. The additional information provided by visual inspection will augment the predictive maintenance program regardless of the primary techniques used.

ULTRASONIC MONITORING

This predictive maintenance technique uses principles similar to vibration analysis. Both techniques monitor the noise generated by plant machinery or systems to determine their actual operating condition. Unlike vibration monitoring, ultrasonics monitors the higher frequencies, that is, ultrasound, produced by unique dynamics in process systems or machines. The normal monitoring range for vibration analysis is from less than 1 to 20,000 Hz. Ultrasonics techniques monitor the frequency range between 20,000 and 100 kHz. The principal application for ultrasonic monitoring is in leak detection. The turbulent flow of liquids and gases through a restricted orifice, that is, leak, will produce a high-frequency signature that can easily be identified using ultrasonic techniques. Therefore, this technique is ideal for detecting leaks in valves, steam traps, piping, and other process systems.

Two types of ultrasonic systems are available that can be used for predictive maintenance: structural and airborne. Both provide fast, accurate diagnoses of abnormal operation and leaks. Airborne ultrasonic detectors can be used in either a scanning or a contact mode. As scanners, they are most often used to detect gas pressure leaks. Because these instruments are sensitive only to ultrasound, they are not limited to specific gases as are most other gas leak detectors. In addition, they are often used to locate various forms of vacuum leaks. In the contact mode, a metal rod acts as a waveguide. When it touches a surface, it is stimulated by the high frequencies (ultrasound) on the opposite side of the surface.

This technique is used to locate turbulent flow and/or flow restriction in process piping. Some of the ultrasonic systems include ultrasonic transmitters that can be placed inside plant piping or vessels. In this mode, ultrasonic monitors can be used to detect areas of sonic penetration along the container's surface. This ultrasonic transmission method is useful in quick checks of tank seams, hatches, seals, caulking, gaskets, or building wall joints.

In a typical machine, many other machine dynamics will also generate frequencies within the bandwidth covered by an ultrasonic instrument. Gear meshing frequencies, blade pass, and other machine components will also create energy or noise that cannot be separate from the bearing frequencies monitored by this type of instrument. The only reliable method of determining the condition of specific machine components, including bearings, is vibration analysis. The use of ultrasonics to monitor bearing condition is not recommended.

OPERATING DYNAMICS ANALYSIS

This analysis method is driven by machine or system design and is not limited to traditional analysis techniques. The diagnostic logic is derived from the specific design and operating characteristics of the machine-train or production system. Based on the unique dynamics of each machine train or system, all parameters that define optimum operating condition are routinely measured and evaluated. Using the logic of normal operating condition, operating dynamics can detect, isolate, and provide cost-effective corrective action for any deviation from optimum.

Operating dynamics analysis combines traditional predictive maintenance techniques into a holistic evaluation technique that will isolate any deviation from optimum condition of critical plant systems. This concept uses raw data derived from vibration, infrared, ultrasonics, process parameters, and visual inspection but applies a unique diagnostic logic to evaluate plant systems.

OTHER TECHNIQUES

Numerous other nondestructive techniques can be used to identify incipient problems in plant equipment or systems. However, these techniques either do not provide a broad enough application or are too expensive to support a predictive maintenance program. Therefore, these techniques are used as the means of confirming failure modes identified by the predictive maintenance techniques identified in this chapter. Other techniques that can support predictive maintenance include acoustic emissions, eddy-current, magnetic particle, residual stress, and most of the traditional nondestructive methods.

PROGRAM COSTS

The initial and recurring costs required to establish and maintain a comprehensive predictive maintenance program will vary with the technology and type of system selected for plant use. While the initial or capital cost is the more visible, the real cost of a program is the recurring labor, training, and technical support that is required to maintain a total plant program.

VIBRATION MONITORING

The capital cost for implementing a vibration-based predictive maintenance program will range from about $8000 to more than $50,000. Your costs will depend on the specific techniques desired.

Training is critical for predictive maintenance programs based on vibration monitoring and analysis. Even programs that rely strictly on the simplified trending or comparison techniques require a practical knowledge of vibration theory so that meaningful interpretation of machine condition can be derived. More advanced techniques, that is, signature and root-cause failure analysis, require a working knowledge of machine dynamics and failure modes.

THERMOGRAPHY

Point-of-use infrared thermometers are commercially available and relatively inexpensive. The typical cost for this type of infrared instrument is less than $1000. Infrared imaging systems will have a price range between $8000 for a black-and-white scanner without storage capability and over $60,000 for a microprocessor-based, color-imaging system.

Training is critical with any of the imaging systems. The variables that can destroy the accuracy and repeatability of thermal data must be compensated for each time infrared data are acquired. In addition, interpretation of infrared data requires extensive training and experience.

TRIBOLOGY

The capital cost of spectrographic analysis instrumentation is normally too high to justify in-plant testing. Typical cost for a microprocessor-based spectrographic system is between $30,000 and $60,000. Because of this, most predictive maintenance programs rely on third-party analysis of oil samples. Simple lubricating oil analysis by a testing laboratory will range from about $20 to $50 per sample. Standard analysis will normally include viscosity, flash point, total insolubles, total acid number (TAN), total base number (TBN), fuel content, and water content. More detailed analysis, using spectrographic or ferrographic techniques, including metal scans, particle distribution (size), and other data, range to well over $150 per sample.

ULTRASONICS

Most ultrasonic monitoring systems are strictly scanners that do not provide any long-term trending or storage of data. They are in effect point-of-use instrument that provide an indication of the overall amplitude of noise within the bandwidth of the instrument. Therefore, the cost of this type of instrument is relatively low. Normal cost of ultrasonic instruments ranges from less than $1000 to about $8000. Used strictly for leak detection, ultrasonic techniques require little training to utilize. The combination of low capital cost, minimum training required to use the technique, and potential impact of leaks on plant availability provides a positive cost benefit for including ultrasonic techniques in a total plant predictive maintenance program. However, care should be exercised in applying this technique in your program. Many ultrasonic systems are sold as a bearing condition monitor. Even though the natural frequencies of rolling-element bearings will fall within the bandwidth of ultrasonic instruments, this is not a valid technique for determining the condition of rolling-element bearings.

BENEFITS

Properly implemented predictive maintenance can do much more than just schedule maintenance tasks. Typical results of predictive maintenance, based on operating dynamics, can be substantial. Using the four major loss classifications, first-year results from a maintenance improvement program based on a comprehensive predictive maintenance program include the following.

Breakdown Losses

In the first year at a large, integrated steel mill, plant delays as a result of machine and system breakdowns were reduced by more than 15.4 percent as a direct result of a comprehensive predictive maintenance program. As Fig. 3.1 illustrates, all divisions of the mill reflected a marked reduction in total

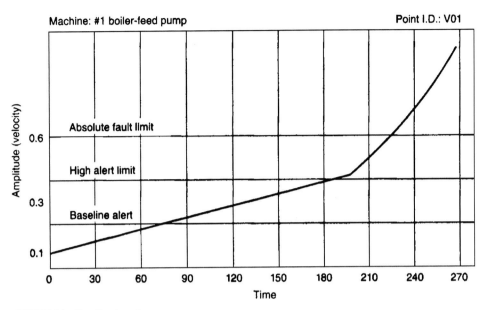

FIGURE 3.1 Broadband trend.

delays. The key to reduction in delays is not limiting the scope to unscheduled delays. A focused effort must be made to reduce scheduled maintenance as well.

Arbitrary acceptance of planned delays for maintenance severely limits available production time. Too many plants accept historical data as the only reason for planned maintenance downtime. A comprehensive predictive maintenance program must include specific methods to evaluate all delays and downtime. The objective of predictive maintenance is to achieve 100 percent availability. The 15.4 percent improvement does not include the added production capacity that resulted from elimination of scheduled downtime. This classification added an additional 5 percent to the availability of the mill.

Quality Defects

Rejects, diversions, and retreats were reduced by more than 1 percent across the integrated mill. This reduction reduced the negative costs of poor quality by more than $5 per ton of product produced, or a reduction of 13 percent. After 2 years, the total costs associated with poor quality had been reduced by more than 24 percent, or $10 per ton.

Capacity Factor

Setup and adjustment, reduced speed, and start-up losses as well as operating efficiency of plant processes directly affect the overall capacity factor of a plant. Reduction of these major losses, in conjunction with the reduction in delays and rejects, resulted in an overall increase of 2.5 percent in net production capacity. The net result, in prime quality product, was an additional 477,000 tons produced by the end of the first year.

Predictive maintenance, when implemented as an integral part of a total plant improvement program, can dramatically improve the net operating profit of the company.

Maintenance Costs

Traditional maintenance costs, that is, labor and material, are not included in the total productive maintenance (TPM) indexes recommended by the Japanese. However, they are a major factor that should be addressed by any plant improvement program. Traditional applications of predictive maintenance will do little to decrease the overall maintenance costs within a plant. In most cases, the only reduction will result from an incremental reduction in overtime costs. Material costs, such as bearings and couplings, will increase and the net overall effect will be a slight increase in the overall costs.

Predictive maintenance, based on operating dynamics, will dramatically reduce both the labor and material, that is, maintenance, costs. After 2 years, the example steel mill reduced its total labor costs by more than 15 percent, or $45,000,000 per year. In addition, their material costs were reduced by more than $6,000,000 per year. One simple example of this reduction in material cost is rolling-element bearings. In the years preceding implementation of the program, the client purchased an average of $9,100,000 of bearings each year. During the first year after the operating dynamics program was implemented, the total expenditure for bearings dropped to $4,000,000 and was further reduced to less than $2,000,000 in the second year. In this one line item, the client was able to eliminate more than $7,000,000 per year in repair parts costs.

CHAPTER 4
RELIABILITY-CENTERED MAINTENANCE

Darrin Wikoff
Principal, Life Cycle Engineering, Inc., Charleston, S.C.

A reliability-centered maintenance (RCM) process systematically identifies all of the functions and functional failures of assets. It also identifies all likely causes for these failures. It then proceeds to identify the effects of these likely failure modes and to identify in what way those effects matter. Once it has gathered this information, the RCM process then selects the most appropriate asset management policy.

RCM considers all asset management options: on-condition task, scheduled restoration task, scheduled discard task, failure-finding task, and one-time change (to hardware design, operating procedures, personnel training, or other aspects of the asset outside the strict world of maintenance). This consideration is unlike other maintenance development processes.

SEVEN QUESTIONS ADDRESSED BY RCM

Fundamentally, the RCM process seeks to answer the following seven questions in sequential order.

Functions

What are the functions and associated desired standards of performance of the asset in its present operating context (functions)? The specific criteria that the process must satisfy are

- The operating context of the asset shall be defined.
- All the functions of the asset or system shall be identified (all primary and secondary functions, including the functions of all protective devices).
- All function statements shall contain a verb, an object, and a performance standard (quantified in every case where this can be done).
- Performance standards incorporated in function statements shall be the level of performance desired by the owner or user of the asset or system in its operating context.

The operating context is the circumstance in which the asset is operated. The same hardware does not always require the same failure management policy in all installations. For example, a single pump in a system will usually need a different failure management policy from a pump that is one of several redundant units in a system. A pump moving corrosive fluids will usually need a different policy from a pump moving benign fluids. Protective devices are often overlooked; an RCM process shall ensure that their functions are identified. Finally, the owner or user shall dictate the level of performance that the maintenance program shall be designed to sustain.

Functional Failures

In what ways can it fail to fulfill its functions (functional failures)? This question has only one specific criterion: All the failed states associated with each function shall be identified. If functions are well defined, listing functional failures is relatively easy. For example, if a function is to keep system temperature between 50°C and 70°C, then functional failures might be

- Unable to raise system temperature above ambient.
- Unable to keep system temperature above 50°C.
- Unable to keep system temperature below 70°C.

Failure Modes

What causes each functional failure (failure modes)? In failure modes effects and criticality analysis (FMECA), the term *failure mode* is used in the way that RCM uses the term functional failure. However, the RCM community uses the term *failure mode* to refer to the event that causes functional failure. The standard's criteria for a process that identifies failure modes are

- All failure modes reasonably probable to cause each functional failure shall be identified.
- The method used to decide what constitutes a *reasonably probable* failure mode shall be acceptable to the owner or user of the asset.
- Failure modes shall be identified at a level of causation that makes it possible to identify an appropriate failure management policy.
- Lists of failure modes shall include failure modes that have happened before, failure modes that are currently being prevented by existing maintenance programs, and failure modes that have not yet happened, however they are thought to be reasonably likely (credible) in the operating context.
- Lists of failure modes should include any event or process that is likely to cause a functional failure, including deterioration, human error whether caused by operators or maintainers, and design defects.

RCM is the most thorough of the analytic processes that develop maintenance programs and manage physical assets. It is therefore appropriate for RCM to identify every reasonably likely failure mode.

Failure Effects

What happens when failures occur (failure effects)? The criteria for identifying failure effects are:

- Failure effects shall describe what would happen if no specific task were done to anticipate, prevent, or detect the failure.
- Failure effects include all the information needed to support the evaluation of the consequences of the failure, such as
  - What is the evidence (if any) that the failure has occurred (in the case of hidden functions, what would happen if multiple failures occurred)?
  - What it does (if anything) to kill or injure someone, or to have an adverse effect on the environment?
  - What it does (if anything) to have an adverse effect on production or operations?
  - What physical damage (if any) is caused by the failure?
  - What (if anything) must be done to restore the function of the system after the failure?

FMECA usually describes failure effects in terms of the effects at the local level, at the subsystem level, and at the system level.

Failure Consequences

In what way does each failure matter (failure consequences)? The standard's criteria for a process that identifies failure consequences are

- The assessment of failure consequences shall be carried out as if no specific task is currently being done to anticipate, prevent, or detect the failure.
- The consequences of every failure mode shall be formally categorized as follows:
  - The consequence categorization process shall separate hidden failure modes from evident failure modes.
  - The consequence categorization process shall clearly distinguish events (failure modes and multiple failures) that have safety and/or environmental consequences from those that only have economic consequences (operational and nonoperational consequences).

RCM assesses failure consequences as if nothing is being done about it. Some people are tempted to say, "Oh, that failure doesn't matter because we always do (something), which protects us from it." However, RCM is thorough, it checks the assumption that this action that "we always do" actually does protect them from it, and it checks the assumption that this action is worth the effort.

RCM assesses failure consequences by formally assigning each failure mode into one of four categories: hidden, evident safety/environmental, evident operational, and evident nonoperational. The explicit distinction between hidden and evident failures, performed at the outset of consequence assessment, is one of the characteristics that most clearly distinguishes RCM, as defined by Stan Nowlan and Howard Heap, from MSG-2 and earlier U.S. civil aviation processes.

Proactive Tasks

What should be done to predict or prevent each failure (proactive tasks and task intervals)? This is a complex topic, and so its criteria are presented in two groups. The first group pertains to the overall topic of selecting failure management policies. The second group of criteria pertains to scheduled tasks and intervals, which comprise proactive tasks as well as one default action (failure-finding tasks).

The criteria for selecting failure management policies are

- The selection of failure management policies shall be carried out as if no specific task is currently being done to anticipate, prevent, or detect the failure.
- The failure management selection process shall take account of the fact that the conditional probability of some failure modes will increase with age (or exposure to stress), that the conditional probability of others will not change with age, and the conditional probability that others will decrease with age.
- All scheduled tasks shall be technically feasible and worth doing (applicable and effective), and the means by which this requirement will be satisfied are set out under scheduled tasks in the failure management section.
- If two or more proposed failure management policies are technically feasible and worth doing (applicable and effective), the policy that is most cost-effective shall be selected.

Scheduled tasks are tasks that are performed at fixed, predetermined intervals, including *continuous monitoring* (where the interval is effectively zero). Scheduled tasks should fit the following criteria:

- In the case of an evident failure mode that has safety or environmental consequences, the task shall reduce the probability of the failure mode to a level that is tolerable to the owner/user of the asset. In the case of a hidden failure mode where the associated multiple failure has safety or environmental consequences, the task shall reduce the probability of the hidden failure mode to an extent which reduces the probability of the associated multiple failure to a level that is tolerable to the owner/user of the asset.
- In the case of an evident failure mode that does not have safety or environmental consequences, the direct and indirect costs of doing the task shall be less than the direct and indirect costs of the

failure mode when measured over comparable periods of time. In the case of a hidden failure mode where the associated multiple failure does not have safety or environmental consequences, the direct and indirect costs of doing the task shall be less than the direct and indirect costs of the multiple failure plus the cost of repairing the hidden failure mode when measured over comparable periods of time.

Categories of Tasks. There are three general categories of tasks that are considered to be proactive in nature, namely, on-condition tasks, scheduled discard task, and scheduled restoration tasks.

On-condition Tasks. An on-condition task is *a scheduled task used to detect a potential failure.* Such a task has many other names in the maintenance community such as:

- *Predictive* tasks (in contrast to *preventive* tasks, a name that these people apply to scheduled discard and scheduled restoration tasks.)
- *Condition-based* tasks, referring to *condition-based maintenance* (CBM) (again, in contrast to *time-based maintenance* or scheduled discard and scheduled restoration tasks)
- *Condition-monitoring* tasks, since the tasks monitor the condition of the asset.

Scheduled Discard Task. The next kind of task is a scheduled discard task, defined as a scheduled task that entails discarding an item at or before a specified age limit regardless of its condition at the time. A scheduled discard task must be subjected to the following criteria before accepting the task:

- There shall be a clearly defined (preferably a demonstrable) age at which there is an increase in the conditional probability of the failure mode under consideration.
- A sufficiently large proportion of the occurrences of this failure mode shall occur after this age to reduce the probability of premature failure to a level that is tolerable to the owner or user of the asset.

Scheduled Restoration Tasks. The next kind of task is a scheduled restoration task, defined as a scheduled task that restores the capability of an item at or before a specified interval (age limit), regardless of its condition at the time, to a level that provides a tolerable probability of survival to the end of another specified interval. The following criteria must be applied to a scheduled restoration task before accepting the task:

- There shall be a clearly defined (preferably a demonstrable) age at which there is an increase in the conditional probability of the failure mode under consideration.
- The task shall restore the resistance to failure (condition) of the component to a level that is acceptable to the owner or user of the asset.
- A sufficiently large proportion of the occurrences of this failure mode shall occur after this age to reduce the probability of premature failure to a level that is tolerable to the owner or user of the asset.

Default Actions

What should be done if a suitable proactive task cannot be found (default actions?) This question pertains to unscheduled failure management policies: the decision to let an asset run to failure, and the decision to change something about the asset's operating context (such as its design or the way it is operated.)

FAILURE-FINDING TASKS

A failure-finding task is defined as a scheduled task used to determine whether a specific hidden failure has occurred. Failure-finding tasks usually apply to protective devices that fail without notice. This task represents a transition from the sixth question (proactive tasks) to the seventh question (default actions, or actions taken in the absence of proactive tasks.) Failure-finding tasks are scheduled tasks like the proactive tasks. However, failure-finding tasks are not proactive. They do not predict or prevent failures. They detect failures that already have happened, in order to reduce the chances of a multiple failure and the failure of a protected function while a protective device is already in a failed state.

RUN TO FAILURE

If a process offers a decision to let an asset run to failure, the following criteria should be applied before accepting the decision:

- In cases where the failure is hidden and there is no appropriate scheduled task, the associated multiple failure shall not have safety or environmental consequences.
- In cases where the failure is evident and there is no appropriate scheduled task, the associated failure mode shall not have safety or environmental consequences. In other words, the process must not allow its users to select "run to failure" if the failure mode, or (in the case of a hidden failure) the associated multiple failure, has safety or environmental consequences.

Each of these continuous improvement programs contains invaluable changes that would improve plant performance. In many cases, these changes are common to all of the approaches. However, some are unique to only one of the approaches. The greatest weaknesses in these approaches include:

- Lack of focus or inclusion of effective culture change, for example, change management process
- Lack of holistic approach. Each is focused on a single function or activity within the plant
- Requires permanent organizational structure to *manage* the effort

CHAPTER 5
TOTAL PRODUCTIVE MAINTENANCE

R. Keith Mobley
Principal, Life Cycle Engineering, Inc.,Charleston, S.C.

Total productive maintenance (TPM) provides a comprehensive, life cycle approach, to equipment management that minimizes equipment failures, production defects, and accidents. It involves everyone in the organization, from top level management to production mechanics, and production support groups to outside suppliers. The objective is to continuously improve the availability and prevent the degradation of equipment to achieve maximum effectiveness. These objectives require strong management support as well as continuous use of work teams and small group activities to achieve incremental improvements. TPM is not a radically new idea; it is simply the next step in the evolution of good production and maintenance practices.

Asset maintenance has matured from its early approach of "breakdown maintenance." In the beginning, the primary function of maintenance was to get the equipment back up and running, after it had broken down, where the attitude of the equipment operators was one of "I run it, you fix it." The next phase of the maintenance history was the implementation of *preventive maintenance*. This approach to maintenance was based on the belief that if you occasionally stopped the equipment and performed regularly scheduled maintenance, the catastrophic breakdowns could be avoided. The next generation of maintenance brings us to TPM. In TPM, maintenance is recognized as a valuable resource. The maintenance organization now has a role in making the business more profitable and the manufacturing system more competitive by continuously improving the capability of the equipment, as well as making the practice of maintenance more efficient. To gain the full benefits of TPM, it must be applied in the proper amounts, in the proper situations, and be integrated with the manufacturing system and other improvement initiatives.

TOTAL PRODUCTIVE MAINTAINANCE CONCEPTS

Different sources provide several different descriptions of what makes up total productive maintenance. Some list five different concepts, others list up to seven different concepts that fall under the umbrella of TPM. Rather than try to decide which the correct quantity is, the concepts of TPM will simply be collected into three different groupings: autonomous maintenance, planned maintenance, and maintenance reduction.

Autonomous Maintenance

The central idea of autonomous maintenance is using the equipment operators to perform some of the routine maintenance tasks. These tasks include the daily cleaning, inspecting, tightening, and lubricating that the equipment requires. Since the operators are more familiar with their equipment than anybody else, they are able to quickly notice any anomalies. The training required to make autonomous maintenance effective comes in several forms: The manufacturing and maintenance staff and their management are educated on the concepts of TPM and the benefits of autonomous maintenance, the maintenance staff trains the operators on how to properly clean and lubricate the equipment, and special safety awareness training is provided to address the new tasks performed by the equipment operators.

Implementing autonomous maintenance often includes the use of *visual controls*. Visual control is an approach used to minimize the training required to learn new tasks, as well as to simplify inspection tasks. The equipment is marked and labeled to make identification of normal versus abnormal conditions easier. For example, the face of a gauge will be colored to show the normal operating range, lubrication points will be color coded to match the container that stores the proper lubricant; bolts will be match-marked with the surrounding structure so any movement is obvious. All of these inspections are also documented on simple check sheets that include a map of the area and the appropriate inspection route.

The equipment operators are also expected to collect daily information on the health of their equipment: downtime (planned and unplanned), product quality (preferably statistical process control [SPC] data, rather than just reject rates), any maintenance that was performed (tightening loose bolts, adding coolant fluid, etc.). This information is useful to both the operator and the maintenance staff to identify any signs that the equipment is beginning to degrade, and may be in need of more significant maintenance. Additional data collection requirements are discussed in the TPM Metrics section, later in this chapter.

Even though autonomous maintenance is supposed to be implemented in a supportive environment, using a cross functional team approach, there are a few common concerns that need to be addressed. First, the equipment operators are now being asked to assume additional responsibilities. These new tasks must be treated as a priority by management, and the operator's performance measures should be modified to include these new activities. Second, the maintenance staff is being asked to give up part of their responsibilities. This can cause the maintenance staff to worry about their job security, especially if the company is currently down-sizing. To address these fears, management must communicate their support for the new maintenance approach, and provide the opportunity for the maintenance staff to assume new responsibilities. Ideally, the maintenance staff will now be free from their daily fire-fighting activities, and can focus on planned maintenance, equipment analysis, and equipment design activities. Third, these changes in roles and responsibilities need to be developed with union representative involvement. In some instances, union contracts may have stipulations that constrain, or encourage these changing roles.

Planned Maintenance

By removing some of the routine maintenance tasks through autonomous maintenance, the maintenance staff can start working on proactive equipment maintenance. Planned maintenance activities (also known as *preventive maintenance*) are scheduled to repair equipment and replace components before they breakdown. This requires the production schedule to accommodate planned downtime to perform equipment repairs, and allowing these repairs to be treated as a priority on par with running the equipment to produce parts. The prevailing theory is that as the planned maintenance goes up, the unplanned maintenance (breakdowns) goes down, and the total maintenance costs go down as a result. The following graph (Fig. 5.1) shows this theoretical trade-off curve.

However, most of the TPM programs that I have witnessed have failed to collect the necessary data to prove this theory. Even if this trade-off is not always achievable, it is easy to see that the equipment will likely receive better care than it was prior to implementing TPM. The manufacturing and maintenance organizations, as a team, should determine the proper amount of planned maintenance, based on the health of the equipment and the type of manufacturing process. Performing excessive amounts of maintenance can be as costly as not performing enough maintenance; their needs to be a balance point determined by careful analysis of the equipment.

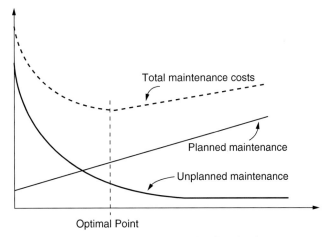

FIGURE 5.1 Trade-off between planned and unplanned maintenance.

Performing planned maintenance, in the proper amounts, requires an in-depth understanding of the production equipment, down to the equipment component level. This understanding needs to start with the products and their critical features, and flow down through the equipment, the equipment's processes, to the process parameters. Figure 5.2 shows a graphic example of how the critical top-level requirements (key characteristics) of a product can be traced down to the manufacturing process parameters.

Once the manufacturing and maintenance team has identified what they believe to be the critical process parameters, they need to validate these, as well as determine the proper parameter settings. The best way to accomplish this is through design of experiments (DOE). These DOE will identify which of these process parameters provide the greatest amount of leverage for improving the equipment performance that is linked to the critical product features.

Planned maintenance uses data from process capability and machine capability studies to determine acceptable performance levels. The process capability studies evaluate the equipment's ability to manufacture consistently high quality parts. The machine capability studies analyze the

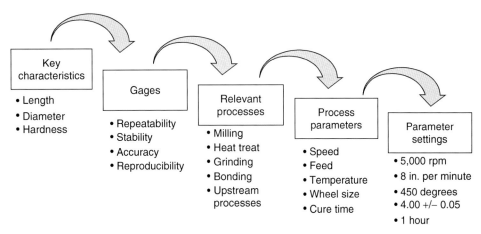

FIGURE 5.2 Key characteristic flowdown to process parameters.

equipment's ability to perform a specific set of operations, and compare the results to industry standards. Both of these studies, when performed on a periodic basis, can provide indicators that the equipment's performance is going downhill, and that it will start producing bad parts, or have a breakdown, in the near future. This data can also be stored in a maintenance database so that similar equipment, or similar equipment components, can be analyzed together to look for chronic problems. Carefully combining the data in this way can help reduce the problems from making decisions with insufficient data.

Maintenance Reduction

The final TPM concept is really made up of two concepts, equipment design and predictive maintenance that are focused on reducing the overall amount of maintenance that is required. By working with equipment suppliers, the knowledge that is gained from maintaining equipment can be incorporated into the next generation of equipment designs. This *design for maintenance* approach results in equipment that is easier to maintain (easy to reach lubrication points, access covers to inspection points, etc.) and can be immediately supported with autonomous maintenance. The equipment manufacturer can even include the visual control markings and labels that the customer currently uses for cleaning, inspecting, and lubricating. This communication between supplier and customer can also be used to establish equipment performance criteria. Both the supplier and the customer should be able to achieve the same results from their machine capability studies, and this can serve as an equipment acceptance test. Also, the equipment supplier may be able to provide data on their components that will help to determine the required frequency of inspections and planned maintenance.

The other method of reducing the amount of required maintenance is to perform special equipment analysis to collect data that can be used to predict equipment failures. This type of analysis includes thermography, ultrasound, and vibration analysis, which allows a technician to gather information on what is happening inside the equipment. Thermography is used to detect equipment "hot spots," where the excessive heat may be related to bearing wear, poor lubrication, or plugged coolant lines. Ultrasound analysis is used to detect minute cracks in the equipment that are invisible to the human eye. If these cracks are detected early enough, repairs can be made before a catastrophic failure occurs. Vibration analysis is used to detect unusual equipment vibration (both in magnitude and frequency). Well-performing equipment will have a certain vibration *signature,* and any changes in this signature can be an indication that internal components are wearing out or coming loose. These types of equipment analysis can be performed on a periodic basis, the frequency of which can be fine tuned as historical data starts to show trends. These studies are also useful for finding the causes of chronic problems that can not be eliminated with the data collected by the operator's inspections and the regular planned maintenance.

Equipment Effectiveness

When people use the term *equipment effectiveness* they are often referring only to the equipment *availability* or up-time, the percentage of time it is up and operating. But the overall or true effectiveness of equipment also depends upon its performance and its rate of quality. One of the primary goals of TPM is to maximize equipment effectiveness by reducing the waste in the manufacturing process. The three factors that determine equipment effectiveness: equipment availability, performance efficiency, and quality rate are also used to calculate the equipment's *overall equipment effectiveness* (OEE) measure which is described later in the TPM Metrics section.

Equipment Availability

A well-functioning manufacturing system will have the production equipment available for use whenever it is needed. This doesn't mean that the equipment must always be available. For example, in a synchronized production system there is little benefit to having equipment up and running when the products aren't necessary. This simply builds up the system's inventory. However, if there is a need to increase the production rate, the equipment must be capable of satisfying the increased demand. The equipment management program must strike a balance between the costs of keeping the potential utilization of the equipment high versus the costs of storing excessive inventory to avoid missing a sales opportunity.

The equipment availability is affected by both scheduled and unscheduled downtime. In a well-functioning system the unplanned downtime is minimized, while the planned downtime is optimized; based on the amount of inventory in the system and the equipment's ability to change production rates. The in-process inventory can often be used to satisfy the downstream demands while the equipment is temporarily shut down to perform maintenance tasks. Determining the proper amount of inventory becomes a function of how often the equipment is down for both scheduled and unscheduled repairs.

The most common cause of lost equipment availability is unexpected breakdowns. These failures affect the maintenance staff (which must scramble to get the equipment running) and the equipment operator (who often has to wait for the equipment to be repaired to continue working). Keeping back-up systems available is one way to minimize the effect of lost equipment availability. However, this is rarely the most cost effective approach since it requires investing in capital equipment that wouldn't be needed if the equipment performed more reliably. Another drain on the equipment availability is the time required to change-over the equipment to run different products. This *setup* time is often overlooked, even though it has the potential to eliminate a significant amount of non–value-added time in the production cycle.

Performance Efficiency

Equipment efficiency is commonly used to metric when evaluating a manufacturing system. The efficiency is typically maximized by running the equipment at its highest speed, for as long as possible, to increase the product throughput. The efficiency is reduced by time spent with the equipment idling (waiting for parts to load), time lost due to minor stops (to make small adjustments to the equipment), and lower throughput from running the equipment at a reduced speed. These efficiency losses can be the result of low operator skill, worn equipment, or poorly designed manufacturing systems.

However, just measuring the equipment's efficiency can lead to poor decision making, because the manufacturing system may not benefit from traditional goal of 100 percent efficiency. The important criteria are how many parts should the equipment be producing, not how many can it produce if run at a breakneck pace. The target efficiency needs to consider how many parts the equipment is designed to produce, and how many parts are needed to satisfy the downstream requirements. When the equipment is up and running, it should be capable of being run at its designed speed. But when the parts aren't needed, shut the equipment down and use this time to perform other tasks, rather than slow the equipment down to reduce the throughput. This occasional downtime can be very useful for performing autonomous maintenance, planned maintenance, and equipment analysis.

Quality Rate

If the equipment is available and operating at its designed speed, but is producing poor quality parts, what has really been accomplished? The purpose of the manufacturing system is not to run equipment just to keep people busy and watch machines operate; the purpose is to make useful products. If the equipment is worn to the point where it can no longer produce acceptable parts, the best thing to do is shut it down to conserve the energy and raw materials, and repair it. Quality losses also include the lost time, effort, and parts that result from long warm up periods or waiting for other process parameters to stabilize. For example, the time lost and parts scrapped while waiting for an injection molding machine to heat up should be considered part of the equipment's quality rate.

The effort to improve the quality rate needs to be linked back to the critical product requirements. There is little benefit from producing parts that are perfect in almost every feature, except for the critical feature that matters most to the customer. Once again, the concept of key characteristics is useful for aligning the critical product features with the responsible equipment parameters. These are the parameters that need to be improved in order to have the maximum benefit to the overall system. These are also the parameters and features that should be measured when determining the quality rate of the equipment.

TPM METRICS

As documented in the previous section, there are data collection requirements that are a prerequisite to even starting a TPM program. Once the organization has decided that TPM is appropriate for their current situation, there are additional data collection requirements inherent in TPM.

Overall Equipment Effectiveness

The concept of overall equipment effectiveness (OEE) is included in nearly all TPM literature. OEE is calculated by multiplying the equipment availability, performance efficiency, and quality rate; which were previously described. The data required to determine these values is scheduled downtime, unscheduled downtime, and throughput (both good and bad parts); which is collected by the equipment operators on a daily basis. Implementing control charts on the equipment availability, performance efficiency, and quality rate provides aggregate data that is useful for tracking any changes in equipment performance. However, these control charts should have predefined thresholds to determine when more detailed data collection is required, so that the necessary changes can be made prior to catastrophic failure. Determining these thresholds requires first collecting a history of the OEE data, along with a history of more detailed data, where undesirable events and their causes can be identified.

OEE gives a useful yardstick for tracking the progress and improvements from the TPM program; but it does not give enough detail to determine why the equipment is better or worse. For example, OEE will reflect a drop in product quality, but it will tell you nothing about why quality is suffering, or what can be done to resolve the problem. To determine the cause of the observed events, supplemental data is required.

Supplemental Data Collection. The data collection methods documented here is not described in the TPM literature, but can be found in process control and capability literature. These supplemental data are intended to be more useful for problem solving and decision making than the aggregate measure of OEE.

Statistical process control data, collected on the critical features of the products, can provide feedback to equipment operators on the repeatability of specific equipment operations. If the process goes out of control, the SPC data should immediately relay this information to the operator.

SPC data collected by monitoring the equipment itself is one step closer to true in-process data collection. For example, a strain gauge may be mounted to a milling machine's spindle. As the cutter wears, more force is required to maintain the established speed and feed rates, which is recorded by this strain gauge. Historical data on strain readings and correlated part quality can be used to determine thresholds that define the allowable wear before the cutter needs to be replaced. This approach differs from the previously described predictive maintenance; this is continuous monitoring and predictive maintenance is only performed periodically.

Collecting SPC data on critical process parameters such as feed, speed, temperature, and time. is another step closer to measuring the underlying process. This approach requires first identifying the critical process parameters (those that affect the critical product features), then determining their optimal settings. This can be accomplished using DOE techniques, and the resulting data is also useful for determining the *planned maintenance* requirements described earlier in this chapter. Once the critical parameters are established, the operators can collect data or use continuous monitoring to track the parameter's performance.

For SPC analysis on equipment and process parameters, special "short run" methods may be required due to limited data quantities. Also, an effective method for monitoring these parameters is the *western electric rules* for control charts. These rules state that a process is to be considered out of control when any of the following are detected in the control charts:

- One point is more than three standard deviations from the process mean
- Two out of three points are at least two standard deviations from the process mean
- Four out of five points are at least one standard deviation from the process mean
- Eight points in a row lie on the same side of the process mean

The on-going measurement of the process must strike a balance between providing data that are too aggregated to be useful, and providing so much data that nobody has time to analyze it. If the data can not be easily monitored, the burden of analysis can outweigh the benefits that come from the analysis.

TPM IMPLEMENTATION

The following description of TPM implementation provides an overview of the critical issues that should be considered. The chart in Fig. 5.3 depicts the principal activities that make up the majority of the TPM implementation effort. This implementation plan uses the TPM Concepts that were described at the beginning of this chapter, and should be given approximately 3 to 5 years to be completed.

The following is a brief description of each of the TPM implementation activities:

Master Plan

The TPM team, along with manufacturing and maintenance management, and union representatives determines the scope/focus of the TPM program. The selected equipments and their implementation sequences are determined at this point. Baseline performance data is collected and the program's goals are established.

Autonomous Maintenance

The TPM team is trained in the methods and tools of TPM and visual controls. The equipment operators assume responsibility for cleaning and inspecting their equipment and performing basic maintenance tasks. The maintenance staff trains the operators on how to perform the routine maintenance, and all are involved in developing safety procedures. The equipment operators start collecting data to determine equipment performance (see TPM Metrics section).

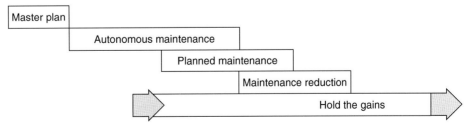

FIGURE 5.3 TPM implementation activities.

Planned Maintenance

The maintenance staff collects and analyzes data to determine usage/need based maintenance requirements. A system for tracking equipment performance metrics and maintenance activities is created (if one is not currently available). Also, the maintenance schedules are integrated into the production schedule to avoid schedule conflicts.

Maintenance Reduction

The data that has been collected and the lessons learned from the TPM implementation are shared with equipment suppliers. This "design for maintenance" knowledge is incorporated into the next generation of equipment designs. The maintenance staff also develops plans and schedules for performing periodic equipment analysis (thermography, oil analysis, etc.). This analysis data is also fed into the maintenance database to develop accurate estimates of equipment performance and repair requirements. These estimates are used to develop spare part inventory policies and proactive replacement schedules.

Holding the Gains

The new TPM practices are incorporated into the organization's standard operating procedures. These new methods and data collection activities should be integrated with the other elements of the production system to avoid redundant or conflicting requirements. The new equipment management methods should also be continuously improved to simplify the tasks and minimize the effort required to sustain the TPM program.

Summary

Total productive maintenance is a world-class approach to equipment management that involves everyone, working to increase equipment effectiveness. Successful implementation requires shared responsibilities, full employee involvement, and natural work groups. A simple analogy to describe the TPM approach is to consider the activities of the average car owner. The owner (equipment operator) performs minor maintenance activities such as checking the oil, checking the air in the tires, perhaps even giving the car a tune up. However, if something major goes wrong, an expert auto mechanic (maintenance technician) is called in to perform the difficult tasks. The important distinction between this car analogy and production equipment, is that most traditional organizations treat their equipment as if it were a rental car.

TPM is often implemented as a stand alone improvement activity. However, it is my belief that it should be done in concert with the other elements of a world class manufacturing system. The synergy of the world class manufacturing concepts such as inventory reduction, hardware variability control, and cycle time reduction, can provide benefits that are greater than the sum of their parts. I would like to close this chapter with an old proverb that was shared with me: "Production equipment is like the goose that lays the golden eggs. If you want to keep getting golden eggs, you need to take care of the goose!"

COMMON BARRIERS ENCOUNTERED

Each of the TPM programs described above have had their share of challenges and setbacks. Although some of these are specific to one particular site, many of the barriers that have been encountered are common to many of the TPM programs. The following information is a brief description of some of the challenges that affected not only some of the four sites described above, but other TPM programs as well.

Strategic Direction

Possibly the most significant challenge to the success of TPM is the lack of strategic direction provided. At this time, each TPM program that is initiated is unique to the organization that decides to implement a TPM program. As a result, each of these organizations must either seek out a previous TPM installation, or start from scratch to create an implementation plan. One of the drawbacks of this situation is that the TPM program tends to be "owned" by just one or two individuals within the organization. If these individuals leave the group, the TPM program typically experiences a gradual decline in direction and support. The end result of these dissimilar programs is that information, tools, and data can not be shared among all of the separate programs. Also, if BCAG should develop an overall TPM strategy and approach, some of these organizations may be forced to backtrack to achieve commonality.

Priority Given to TPM

Most manufacturing cells still view TPM as a maintenance issue, rather than as a manufacturing issue. This observation is backed up by the fact that most of the TPM locals are either maintenance or training personnel. This situation may be magnified by the fact that the maintenance organization is totally separate from the manufacturing organization. These two groups do not report to the same organization until above the vice president level. As a result, the manufacturing personnel are rarely measured on equipment performance, and the maintenance group is rarely measured on production part quality or cycle time. This separation of maintenance and manufacturing essentially eliminates any incentive for managers of both organizations to pool their resources to achieve a successful TPM program. Evidence of this barrier is provided by the noticeable lack of TPM goals and metrics in business plans and performance plans.

Conflicting Processes

At any given time, nearly every manufacturing cell is working on implementing a handful of new manufacturing process, as well as a couple of process improvement initiatives. This usually creates more work than the organization is capable of handling simultaneously. In the end, some of the projects are successful, some get canceled, and others simply get ignored. Unless all of the activities within the organization are documented and prioritized, adding another process such as TPM will simply force some other project to "fall off their plate." No organization has unlimited budget, time, and people to allow them to implement every good idea that comes along. As manufacturing and maintenance managers are expected to implement additional processes, they must develop methods for prioritizing their work load. A major step in this prioritization, that commonly gets overlooked, is an evaluation of the possible process integration that can be achieved. Rather than forcing the processes to compete for resources, several processes can be combined into one cohesive process that is implemented over a longer time frame.

Data Availability

Several of the efforts to implement TPM have been thwarted by the lack of reliable data to utilize for planning purposes. The existing data collection methods do not emphasize the benefits that can be achieved by accurately monitoring equipment performance. Without this data, it is very difficult to determine the relationship between equipment performance, product quality, and manufacturing costs. The data that is being collected is often not used for any decision making so the quality of this data is never verified. The end result is that there is some data available, but it may be of poor quality, and is not collected in a manner that allows easy analysis. Without reliable data, the organization can not develop accurate prioritization plans and they can not quantify any of the benefits received from their TPM program.

SUCCESS FACTORS AND ENABLERS

As can be expected, several of the factors that have led to the successful TPM programs are simply doing the opposite of the barriers. However, there are a few additional activities that seem to be common to the more successful TPM programs. For the purpose of this evaluation, I am defining a successful TPM program as any program that has developed implementation plans, followed those plans through, and realized the expected benefits of their implementation. This does not necessarily mean that these *successful* projects have resulted in major financial savings. Some of these programs had less ambitious goals, yet I perceive them as successful if they met their stated goals.

Management Support

The TPM implementation plans that have been successfully deployed typically had the benefit of an extremely supportive management team. This means that management did more than just *allow* TPM to be implemented, they were actually a part of the driving force behind the implementation. The management activities include rewarding teams for proactive maintenance, revising business plans to include TPM goals, allowing production workers to attend training sessions, and communicating the TPM goals to the entire organization. By having management's full support, the TPM program should not "die on the vine" if the TPM coordinator transfers to another organization.

Focused Approach

Most of these facilities used a different approach for determining where they could achieve the greatest benefits from their TPM program. Although there may not be one best method for prioritizing the TPM implementation sequence, the important decision is to actually take the time to perform this prioritization. No organization has all the necessary resources to simultaneously solve all of their problems, so they must pick and choose their opportunities to utilize their resources. The organizations that developed a clear master plan for how they would transition to TPM practices have made more progress toward their goals than the organizations that have failed to lay out a focused plan of attack.

Operator Ownership

Although management needs to assume a leadership role in TPM implementation, they must also allow the equipment operators to take a prominent role in the development and implementation of TPM. One of the essential concepts of TPM is encouraging the operators to assume more responsibility and authority for decisions affecting their production equipment. If the operator is detached from the TPM program, it is extremely difficult to get proactive equipment inspection and maintenance. The benefit of having the equipment operator deeply involved is that the person that knows the most about the equipment (the individual that runs the equipment day in and day out) is providing input to the plans. Achieving operator ownership requires early involvement by the equipment operators, so that they can feel a sense of belonging to the TPM implementation team. This enables them to have some of their own blood, sweat, and tears invested in the TPM plans.

Just-In-Time Training

Training that is delivered too early is almost as ineffective as training that is delivered too late. If the affected individuals are trained immediately prior to their hands-on use of their new knowledge, they are given the chance to reinforce their classroom learning by getting direct and immediate feedback. This just-in-time (JIT) training approach also reduces the impact of the extensive TPM training required. Since the training is given in small doses over a long period of time, it has less of an impact on the organization's ability to meet their production schedule.

Integrated Processes and Schedules

As mentioned above in the barriers encountered, failing to integrate the various processes levied upon the organization creates many conflicts. However, the element of this integration that often gets overlooked is the scheduling activities. Many organizations use totally isolated scheduling systems for their production schedule and their maintenance schedule. In this situation, every preventive maintenance activity must be scheduled by two organizations reaching a compromise on when to shut the equipment down for maintenance. By integrating all of the entities that require access to the production equipment, the organization can avoid the conflicts that arise over who has the highest priority for the equipment. Groups other than manufacturing and maintenance may also need access to the equipment to perform tests (Quality Assurance), or to run prototype parts (Research and Development).

Union Buy-In

Last, but certainly not least, is the issue of getting the employee's union representatives involved in the TPM implementation planning. The introduction of autonomous maintenance almost always involves the migration of responsibilities from maintenance technicians to equipment operators. It is ridiculous to wait until an employee files a grievance with the union to consult with these organizations. The union need not be treated as an adversary in these workplace transformations. History has proven that the union may very well support the concepts of TPM, since the affected employees are developing additional skills that make them more valuable to the company. TPM can effectively be used to create a more multiskilled workplace, which usually improves employee job security.

RECOMMENDATIONS FOR IMPROVEMENT

The following conclusions have been compiled from interviews and surveys administered during my research as well as personal observations. These conclusions provide common themes that appeared in several different organizations that were working to implement TPM within the Boeing Commercial Airplane Group.

Top Down Direction

Successful implementation of any new process benefits from a coordinated approach using pilot implementation, lessons learned, and then full-scale deployment. Accomplishing this requires clear direction and priorities from upper management. The goal is to avoid a haphazard implementation approach where each organization can decide if they will implement the new process, and then develop their own unique approach. If this occurs, the organizations can not easily share and transfer knowledge, tools, or procedures.

Integrating Processes

New processes must be integrated with the existing system and positioned to allow for future process introductions. Establishing stand-alone processes will create "process silos" where each process has to compete with the others. This process competition creates a great deal of confusion within the organization, and makes resource allocation extremely difficult. Metrics and incentives that support more than one process (e.g., machine capability data evaluates both product quality and equipment performance) make the prioritization of improvement activities more obvious.

Data Driven Decisions

There is a noticeable lack of useful data available to the lower levels of many of the BCAG organizations. Further, the data that is available is all too often not effectively utilized by the intended organization. This makes it virtually impossible to drive decision-making authority down to these lower levels. The lower level organizations need to develop simple data collection methods to gather information that can immediately be used by the factory floor employees. This includes both qualitative and quantitative data. Also, data collection plans should be established at the very beginning of any improvement activities. These improvement activities need to address data collection prior to implementing any changes in the workplace, so that the results of the change can be quantified later.

Impact on Organizations

Most of the organizations that implement new processes fail to fully identify the required changes in the organization's culture. Part of the implementation planning should be dedicated to explicitly identifying the necessary organizational changes. This includes current and future roles and responsibilities for the individuals affected by the proposed process changes. These changes need to be communicated to the organization, and the appropriate training needs to be provided, to make the process change a smooth transition.

Breaking Down the Barriers

In many organizations, TPM is still treated as a facilities maintenance issue rather than as a manufacturing issue. The solution to this dilemma is to tie these two organizations metrics, goals, and rewards together. For example, the maintenance typically has no goals related to product quality, and the manufacturing organization has no goals related to equipment performance. If both organizations were working to maximize the overall equipment effectiveness, they would be encouraged to work together and support each others needs. Another barrier that needs to be addressed is the basic human reaction to change: we usually don't like change. This is often the result of poor communication and training, which leads to a fear of the unknown; not necessarily a resistance to change.

Effect of Equipment Performance

Teams that have started implementing TPM have quickly realized that equipment performance is critical to reducing manufacturing costs. The equipment variation leads directly to hardware variability, which has repeatedly been proven to increase manufacturing costs. This is due to scrapping parts, reworking parts, and adjusting production schedules to accommodate production delays. This poorly performing equipment is typically countered by increasing coordination activities and storing extra inventory. Also, the unreliable equipment prevents the organization from implementing other manufacturing initiatives such as pull production/JIT scheduling. Previous studies have indicated that on average, overall equipment effectiveness is well below *world-class* levels. The benefits of improving the equipment management practices across BCAG have been estimated to be in the hundreds of millions of dollars.

TOTAL PRODUCTIVE MAINTENANCE DETAILED DESCRIPTION

The objectives, components, benefits, and activities of TPM were briefly described in Chap. 3. This chapter also contains an examination of how TPM fits into the overall manufacturing system, and brief description of the activities required to implement TPM. This information is primarily based on the writings of Seiichi Nakajima, Terry Wireman, Charles Robinson, and Kunio Shirose. Additional information has been included from TPM training materials developed by Westcott Communications and BCAG. The intent is to synthesize these references, along with personal observations, to create a complete description of TPM and how it relates to the overall manufacturing system.

Objectives of TPM

The ideal goals of TPM are to achieve zero failures, zero defects, and zero accidents. Although these goals are extremely difficult to achieve, and in many cases may not be economically feasible due to the high cost of eliminating all failures, defects, and accidents; they provide a directional target that the organization can shoot for. To move an organization toward these goals, TPM provides the means to increase the amount of time that a piece of equipment is reliably available for production use. This requires significant effort to reduce the equipment degradation that may lead to equipment failures or production part variation. Additionally, TPM affects the equipment operator by providing an increased awareness, sense of ownership, and responsibility for the production equipment. The TPM approach to equipment management also includes a new set of metrics to measure the overall equipment effectiveness.

The implementation of TPM should not be seen as a short-term fix to production problems. The full deployment of a TPM program typically takes from 3 to 5 years. Following the implementation of TPM through to fruition requires a focus on the costs of equipment over its entire life cycle. The elements of life cycle costs include acquisition costs, operating costs, maintenance costs, and conversion/decommission costs. Each of these costs can be further subdivided. For example, the maintenance costs contain costs of repairs, maintenance labor costs, operator labor costs (during downtime), and equipment component parts costs.

Eliminating Process Losses

To help focus improvement efforts, it is often helpful to categorize, or group the problems that the equipment is experiencing. Thinking in terms of production-system losses and grouping the problems based on the cause of these losses helps point out the common sources of equipment problems. However, to eliminate chronic problems that are reducing the equipment performance requires tracing the problems down to their root cause. The process losses described in this section are broken into groupings that will be used later, when discussing the calculation of overall equipment effectiveness.

Losses Due to Downtime. The production system losses that fall into this grouping are the result of the equipment being temporarily unavailable for production. These losses can be subdivided into two categories: *Breakdowns* and *Setup and Adjustment*. Sporadic breakdowns, that are sudden or dramatic, are usually obvious and easy to correct. However, frequent or chronic minor breakdowns are often ignored or neglected after repeated unsuccessful attempts to cure them. To maximize equipment effectiveness, breakdowns must be reduced to the lowest possible frequency. This goal can be achieved by proactive replacement of worn parts during scheduled maintenance. It is almost always more cost effective to replace a questionable part than to allow its failure to shut down equipment (remember: "a stitch in time saves nine").

Losses during setup and adjustment result from downtime, and producing defective products, when production of one item ends and the equipment is modified to meet the requirements of another item. Setup can be reduced considerably by making a clear distinction between internal setup times (operations that must be performed while the machine is down) and external setup times (operations that can be performed while the machine is still running) and by reducing internal setup time.

Losses Due to Poor Performance. This category focuses on equipment utilization that is lost as a result of the equipment being operated at less than maximum speed. The lost production capability falls into the subcategories of *reduced speed* and *idling and minor stops*. If the maximum operating speed of a piece of equipment falls below the original design speed, a loss is incurred. This may occur due to excessive equipment wear, or due to lack of operator confidence in the manufacturing process. Well maintained equipment and a reliable manufacturing process will help minimize both of these problems.

Idling and minor stops refers to brief interruptions in processing. These stoppages generally stem from the need for some slight adjustment, such as tightening a bolt or holding fixture. Resolving a minor stop may require the correction of a small problem, such as equipment jams. The difference between these minor stops and equipment breakdown is typically a function of time and severity. Idling and minor stops can be rectified quickly, often without completely shutting the equipment down; breakdowns are associated with large or catastrophic failures.

Losses Due to Poor Quality. Just because the equipment is running, and running at full speed, does not guarantee that it is producing a satisfactory product. If the equipment output is not useable, the equipment may as well be shut off to conserve energy. The losses that come from producing poor quality products are separated into two classifications process defects and startup losses.

Process Defects. Defects in output are often generated by defects in the process related to equipment performance. The way to improve quality in these cases is to eliminate the root cause of the loss, by improving the equipment. These process defects include both chronic and sporadic production problems that result in parts that are not acceptable or must be reworked.

Startup losses. Startup losses are created by reduced product yield that occurs during the early stages of production; from machine startup to stabilization. The longer it takes for the equipment to achieve stabilization, the greater the amount of unusable output. Examples of this situation include producing unacceptable products, or having a reduced output, during the time required to reach operating temperature or speed. Reducing the time required for equipment parameters to reach their necessary state will minimize the amount of startup losses.

Major Components of TPM

Total productive maintenance involves improvements to production equipment, as well as making the practice of maintenance more efficient. This requires accurate planning and scheduling, access to reliable equipment information, and a good spare parts inventory system. It can also involve designing or redesigning equipment to make maintenance easier and quicker to perform, or purchasing equipment that requires less maintenance. Total productive maintenance activities can effectively be collected in the following six separate components.

Education and Training. This element supports all the other TPM components by ensuring that employees have the necessary knowledge and skill to do a quality job while performing TPM related tasks. The affected employees include management, maintenance personnel, operators, and other stakeholders in the equipment operation. Education and training also provides a common vocabulary and an accurate understanding of the TPM goals for the manufacturing process.

Autonomous Maintenance. Autonomous maintenance requires the proactive involvement of equipment operators to eliminate accelerated equipment deterioration through cleaning, monitoring, fastener tightening, data collection, and reporting equipment conditions and problems to the maintenance staff. Further, the operators must work to develop a deeper understanding of their equipment which should improve their operating skills. Daily cleaning reduces wear on the machines and provides an opportunity to inspect for excessive wear and minor equipment malfunctions. The appropriate person can be notified or corrective action taken prior to excessive damage. Minor adjustments by operators, where appropriate, help keep overhead costs low by avoiding a special trip to the machine by a maintenance mechanic. This immediate operator response assures adjustments are made before they can contribute to equipment breakdown or variations in production parts. Autonomous maintenance, practiced by an operator, or manufacturing work cell team member, will help to maintain high machine reliability, low operating costs, and high quality of production parts. Information collected by the equipment operators contributes to overall equipment effectiveness measures and to reliability and maintainability improvements for both new and existing machines.

Preventive Maintenance. Preventive maintenance is based on planned servicing of equipment and in-depth inspections to detect and correct conditions that might cause breakdowns, production stoppages, and premature wear. This function consists of periodic inspections, planned restoration of deterioration, and proactive replacement of suspect equipment components. While performing preventive maintenance, data is collected for equipment effectiveness measurements, reliability studies, maintainability metrics, and operating costs. Another target of this element is to reduce the time required for planned maintenance and eventually eliminate the requirement for unplanned repairs of equipment.

Planning and Scheduling. This element coordinates the production schedules, the preventive maintenance schedules and other activities requiring utilization of the equipment. Further, it assures that trained technicians with proper tools, equipment, parts, documents, and safe work instructions are coordinated with the equipment availability. Also included are dispatching activities to conduct planned maintenance or breakdown repairs in such a manner that maximizes overall equipment effectiveness and availability for production, while avoiding schedule disruptions.

Reliability Engineering and Predictive Maintenance. Predictive maintenance provides a process for improving equipment effectiveness by using a dedicated technical core group to identify and focus on chronic equipment problems. In addition, reliability engineering performs analysis for the root cause of problems and identifies actions and resources required to improve machine reliability and maintainability. This function is also responsible for developing predictive maintenance techniques such as vibration analysis, thermal analysis, and lubricant analysis. Further, it is responsible for development and feedback of reliability and maintainability data to equipment engineers and equipment suppliers.

Equipment Design and Start-Up Management. This component of TPM is responsible for incorporating the knowledge gained from maintaining the existing equipment into new equipment designs. This information includes equipment performance, life cycle costs, reliability and maintainability targets, equipment testing plans, and operating documentation and training. This process requires joint planning and coordinating with other stakeholders involved in equipment start-up. The goal is to accomplish rapid and reliable ramp-up to designed production rate performance.

BENEFITS OF TPM

When facing the decision of whether or not to implement a TPM program, a good question to ask is *What's in it for me?* (what are the benefits?). The bottom-line answer to this question is: *TPM helps you reduce your manufacturing costs!* This answer is particularly true for organizations that make extensive use of automation and sophisticated production equipment. Of course, the actual amount of cost savings will depend heavily on the current status of the manufacturing system and the type of production process. If the equipment is already performing well, the organization may be better off focusing on other opportunities to improve the production system (e.g., inventory reduction, employee skill training, cycle time reduction, etc.). Also, if the plant currently has excess capacity, looking for new customers for your products may provide better returns than improving the equipment reliability. Even though TPM has been proven to yield extraordinary benefits for many companies; each facility must carefully evaluate their current situation to determine if their manufacturing system can gain any advantage from these benefits.

Although TPM implementation is not free; it requires training, new data collection procedures, and changing roles and responsibilities; the benefits and paybacks have been documented in many industries. These benefits are achieved through improved equipment reliability and utilization, reduced equipment variation due to wear and tear, and less maintenance "fire fighting." Further, the increased availability of the equipment also allows the organization to defer the purchase of additional equipment to satisfy increases in production demand.

Many companies have demonstrated that increasing scheduled maintenance activities (preventive maintenance) will drastically reduce unscheduled maintenance (breakdown repairs), and that the total maintenance costs decrease as planned maintenance replaces unplanned breakdowns. The incentive to reduce these maintenance costs is motivated by the observation that maintenance costs are typically 15 to 40 percent of the cost of goods sold for a manufacturing firm. These costs can increase drastically when excessive breakdowns occur. However, there is a trade-off that must be made between the costs of performing excessive amounts of preventive maintenance and allowing rare breakdowns. The organization must determine their costs incurred from breakdowns and balance this against the cost of avoiding these breakdowns to identify the optimal amount of preventive maintenance. In an effort to identify the proper maintenance mix, some companies have estimated that the average cost of equipment failure is four times more than the cost of the repair. In this case, it would be cheaper to allow the equipment to fail once than to perform preventive maintenance four times.

Implementing TPM has the additional benefit of improving product quality, which reduces rework costs and increases customer satisfaction (due to consistently superior quality). The following sections provide many examples of the benefits that may be reaped as companies successfully implement TPM programs.

Reduced Variation

Variation in a manufacturing system comes in many forms: hardware variability, throughput variation, inventory variation, and so on. In many cases TPM can effectively decrease the sources of these variations, the frequency of their occurrence, and improve the robustness of the production system.

In the case of hardware variability, TPM can eliminate sources of variation by improving the repeatability of the production equipment. This reduction is achieved by systematically tracing the sources of variation from the critical product features (key characteristics) down to the equipment and process parameters. Tracing the relationships between product features and process parameters is not a simple task. The first step in this activity is to determine which product features are most critical to satisfying customer requirements. From this point, DOE techniques can be employed to identify the process parameters that have the greatest impact on these key characteristics. Additionally, DOE techniques will help identify the parameter values that produce the best products.

Once these parameter settings have been determined, they can be monitored by the operator using SPC control charts to provide in-process data. If an anomaly is observed in the SPC data, the equipment operator is alerted that the resulting production part may not be acceptable and requires additional investigation. Further, these process parameters should be aligned with the maintenance task that has a direct impact on their stability. For example, if a machine cutting tool can not achieve the preferred operating speed, it may be due to excessive bearing wear due to poor lubrication.

Improving the reliability of the equipment, by implementing TPM, will also reduce variability in the manufacturing system throughput. Sporadic equipment breakdowns and unscheduled repairs are major causes of fluctuations in throughput. Implementing TPM will minimize these equipment reliability problems. An effective TPM program permits the equipment to run at full speed when necessary, with the only downtime being for planned maintenance. Also, the planned maintenance can be scheduled with the production scheduling system so that there is no disruption to the forecasted throughput.

Inventory fluctuations can also be attributed to equipment reliability and the resulting throughput fluctuations. Additionally, reliable equipment with high possible utilization rates allows the manufacturing system to react more effectively to changes in customer demand. However, it is important to remember, just because the equipment can run at full speed 24 hours per day, does not mean that it should be run this hard. The higher possible utilization rates means that the organization no longer needs to store excess inventory, or work excessive overtime, to cover customer demand fluctuations. The production equipment can simply ramp-up to a higher throughput when necessary. This extra production capacity comes from having less breakdowns and unscheduled maintenance, and from making the equipment capable of running at its maximum designed speed.

Companies that have successfully implemented TPM have seen reductions in breakdowns of up to 80 to 90 percent, cost of defects drop by 55 percent, product lead times cut by 50 to 75 percent, and on-time deliveries increase by 50 to 95 percent. These levels of improvement can not be expected by every facility implementing TPM. However, they document the large potential gains that some companies may achieve.

Increased Productivity

Eliminating unscheduled downtime and excessive rework allows the organization to spend more of their time on the value-added tasks, such as producing good parts. Implementing TPM establishes processes and metrics that focus attention on minimizing the non–value-added tasks. The resulting increase in productivity applies not only to the equipment, but to the people working in the manufacturing system as well. Production workers are no longer forced to wait around while their equipment is repaired, and the maintenance staff is no longer required to put off planned maintenance and equipment analysis work while they scramble to fix the broken equipment.

An effective TPM program also establishes metrics that focus on reducing equipment setup and change over times. TPM encourages changing equipment setup processes to allow the next product's configuration to be set up while the equipment is still running on the existing products. Setup reduction is one of the major components of implementing a pull production system, such as *just in time*.

The documented gains from TPM implementation include equipment productivity increases from 50 to 80 percent, value-added time per person increasing by 100 to 150 percent, labor productivity increases of up to 150 percent, and setup times dropping by 50 to 70 percent.

Reduced Maintenance Costs

The changing role of maintenance from breakdown repair to proactive improvement enables the organization to reduce its overall maintenance costs. The traditional fire fighting approach to equipment maintenance forces the maintenance organization to carry extra staff members to handle the wildly fluctuating and unpredictable workload. By using scheduled maintenance events, the organization can level-load their work across all staff members. Further, the implementation of TPM's autonomous maintenance removes many of the less technically challenging tasks from the maintenance staff's workload. This frees up the maintenance staff to focus on proactive equipment improvements, equipment performance analysis, and simplification of existing maintenance practices. This transition of responsibility requires an enlightened management team that focuses on the potential gains from improved maintenance, rather than focusing on the cost savings from simply reducing the maintenance staff headcount. There is an additional benefit from running the equipment more efficiently: reduced energy costs. The equipment spends less time idling and operates more effortlessly due to TPM. Although the gains from reduced energy consumption may not be staggering, they are still reductions in the overall manufacturing costs.

The following data provide examples of the benefits that very successful companies have received from their TPM program: maintenance spending reduced by 40 percent, energy conserved by 30 percent, and reduced maintenance labor by 60 percent.

Reduced Inventory

Any manufacturing organization that uses unreliable equipment must maintain an unnecessarily large stock of finished goods to fulfill the customer demand while the equipment is nonoperational. The more unreliable the equipment, the larger the necessary stock of finished goods. If a given production line is composed of unreliable equipment, the work-in-process inventory must be kept higher than desirable to accommodate equipment performance uncertainty. All of this extra inventory can create many problems: changes in customer requirements take too long to incorporate; the new product lead time must allow for using up the finished goods and in-process inventory. Further, any defective parts that are produced can sit in the in-process inventory waiting to be discovered at the next step in the production process. The inventory is effectively hiding these production problems. Implementing a TPM program removes much of the uncertainty in the throughput and cycle time of the production system.

The spare parts for the production equipment are another source of unnecessary inventory holding costs. The spare parts are used to make repairs to the equipment, which could occur at any time on unreliable equipment. Once again, the uncertainty in the equipment performance requires extra inventory. Through reliability engineering, data collection and analysis, the maintenance staff can develop an accurate estimate of the necessary spare parts and the frequency of their usage. Implementing TPM will allow the maintenance technicians to perform the necessary analysis to optimize their spare parts inventory policy.

Companies that have implemented TPM have been able to increase inventory turn rates by as much as 200 percent, slash inventory levels by 35 percent, and reduce spare parts costs by 20 to 30 percent. However, these gains are not likely to be achieved by simply implementing TPM in isolation. Additional effort should be applied to reducing inventory via improved scheduling systems and synchronized production processes.

Improved Safety

The initial steps in implementing the autonomous maintenance activities of TPM create an environment that could easily reduce safety and increase accidents. This is the result of equipment operators taking on additional and unfamiliar maintenance tasks, for which they may not have been effectively trained. Since these tasks are new to the operators and they often involve potentially hazardous activities (removing debris from equipment, inspecting chains and gears, etc.), they pose a new threat to the safety of the operator. Therefore, ensuring the safety of the operators must be a primary function of the TPM implementation plan. This requires extensive training, developing "fool-proof" maintenance tasks, and implementing improved procedures. Also, by performing the routine maintenance tasks on a frequent basis, the operators develop a better understanding of their equipment. This new knowledge helps the operator make more intelligent decisions to reduce the potential hazards that the equipment presents. The safety of all individuals involved with the equipment must be a top priority of any good TPM program.

The benefits of the improved safety within TPM have allowed some companies to *reduce their accidents essentially to zero*. Another side benefit of the TPM program is that pollution is often reduced due to more efficient equipment, which extends the safety improvements to include the surrounding community.

Improved Morale

The final benefit discussed here (although additional benefits certainly exist) is employee morale. As with any change in the workplace, their is bound to be some disruption from implementing TPM. However, this does not necessarily have to be all negative. Since TPM uses employee teams to develop the implementation plans and to deploy these plans, the operators are "in the driver's seat," and can now be empowered by management and given increased levels of control and ownership over the equipment. This ownership allows the operator to take more pride in their equipment and to make informed decisions on how best to run the equipment. Obviously, this requires management support, since the operators are now assuming decision-making authority. If the managers are unwilling to relinquish control of these decisions, morale may end up suffering, rather than improving.

The maintenance technicians now have the time to perform equipment analysis, work with equipment designers, and work on other technically challenging tasks. The maintenance staff will not necessarily see a drop in their workload due to handing routine maintenance tasks to the operators. They may simply see a shift to more proactive maintenance activities such as working to develop preventive maintenance requirements for the equipment. This change also requires management support to allow the maintenance staff to develop their skills in these areas.

CHAPTER 6
MAINTENANCE REPAIR AND OPERATIONS—STOREROOM EXCELLENCE

Wally Wilson
Materials SME, Life Cycle Engineering, Inc., Charleston, S.C.

The MRO Best Practices seminar is a definitive training program that presents current best practices in stores operation and identifies the milestones to achieve a world-class maintenance repair and operations (MRO) storeroom. Knowing the status of your MRO inventory and providing a quality part for the technician to perform scheduled repair work is essential in the overall purpose of the storeroom. Training begins with identifying strengths and weaknesses in current work processes and then defining the foundation of best practices for target storeroom operation. If we don't know we have a problem, how can we fix it?

With the best practices identified and target workflow processes mapped, baseline information for MRO inventory management can be established and key performance indicators selected to track storeroom operation. Since the storeroom does not function in a vacuum, there are several areas of maintenance and operations that have shared work processes that flow into and out of each department. These shared functions require that partnerships be established between maintenance planners, technicians, production operators, and supervisors in each of these areas. As the workflow processes are mapped, it is important that responsibility and accountability is assigned to ensure departmental partnerships are successful and problems can be identified and resolved quickly. Finally, the course defines what is needed to affect a positive change, and gives solid direction and support on the methods to achieve it.

OBJECTIVES OF INVENTORY

The main objectives of inventory management include the following:

- Reducing cycle-time by
  - Lead-time improvement
  - Transportation-time reduction
  - Repair- and return-time reduction
  - Improving the kitting and delivery process
- Reducing inventory and associated carrying costs
- Reducing expedite freight costs
- Improving profitability
- Increasing inventory accuracy

TYPES OF INVENTORY

- Finished goods—end product
- Work in process (WIP)—parts in the manufacturing process
- Raw materials—materials before processing
- MRO supplies—parts supporting maintenance and operations. Operating supplies supporting both maintenance and operations
- Office and facility supplies—any office supplies and equipment, all janitorial and sanitary supplies
- Hardware—small tools, fasteners, free stock, vendor managed inventories, consumables

In MRO materials management the focus of attention is on the last three categories. However, MRO materials management is an integrated part of operations performance and as such is directly linked to finished goods, work in process, and raw materials due to control activities and material management processes.

BEST PRACTICES INVENTORY MANAGEMENT

1. Implementing and sustaining lay-up maintenance for parts in storage. A lay-up program ensures that all rotating stock is maintained under a preventive maintenance (PM) program and other items like "O" rings, belts, gaskets, and so on, that are affected by dust, dirt, and temperature or humidity changes receive special attention for their storage needs.
2. Vendor-managed inventories that are managed effectively and have a good partnership established between the vendor and their customer can be very beneficial.
3. Cycle-counting should be part of a daily routine for the storeroom. The ABC classification or counting by selected areas are both acceptable methods to manage an inventory cycle-counting program.
4. Identifying obsolete parts and removing them based on a monthly budget is the best practice to keep dead inventory at an acceptable level.
5. Effective salvage of obsolete and scrapped materials.
6. Controls over the repair and return process are important for the storeroom to maintain.
7. Storeroom layouts need to stress efficiency and effectiveness.
8. Manning levels should be optimized and inventory levels controlled.
9. Housekeeping practices meet 5S standards.
10. A defined receiving process is in place.
11. Stocks in stores meet the FIFO (first in, first out) guidelines for shelf administration.
12. The workflow process for kitting is mapped and put in place for all planned work.
13. All storerooms are closed and physically secured.
14. A dashboard has been established to measure key performance indicators.
15. There is an approved supplier list.
16. There is a defined locator system for inventory and tools.
17. All processes are mapped and analyzed to streamline the workflow process. RASI's have been identified and assigned to task functions. Step definitions, training plans, and job descriptions have been developed.

BARRIERS TO BEST PRACTICE INVENTORY MANAGEMENT

1. Not utilizing the ABC inventory management classification.
2. No standard operating procedures to follow for storeroom operation, workflows are not defined.
3. Storeroom key performance indicators are not measured to monitor the following areas: inventory accuracy, supplier performance, emergency purchase orders, emergency freight costs, overtime, inventory turns, carrying cost, and quality.

   Note: If you want something to improve, monitor the performance and it will get better.
4. The lack of coordination, communication, and cooperation between departments.
5. Obsolete material is not identified and removed from the inventory as well as the computer-based maintenance management systems (CMMS).
6. Failure to have purchasing practices in place, like economic order quantity (EOQ), lot-for-lot (L4L), and promoting vendor managed inventory (VMI).
7. Lack of an accurate equipment bill of material (BOM).
8. Work orders system is not used by the maintenance department.
9. Kitting and delivery of planned work is not practiced.
10. Critical assets and parts are not identified so these items can be available in the storeroom inventory 100 percent of the time.
11. Hidden storerooms (lockers, tool boxes, and locations not identified in the CMMS) with inventory "not on the books."
12. No rating system for vendors or suppliers.

INVENTORY CONTROL

Minimizing poor use of the company's working assets can be accomplished through improved inventory turn-rate, cost control, efficient purchasing practices, inventory cycle-counting, recorded issuances against actual equipment and work orders, secured access, and staffed coverage. Minimizing stocking or squirreling of parts can go a long way to ensure best use of inventory dollars. It has always been difficult to forecast what inventory you need to stock, when you will need it, and in what quantity. The goal is to stock the lowest level of inventory possible but have the parts readily available when they are needed. Forecasting for many is just a wild guess but the best-in-class plants are reviewing past inventory records to determine item activity, min/max levels, supplier contracts, and options for vendor-managed inventory. There are many software packages available that can greatly assist in forecasting inventory needs. Improving your inventory management practices can convert your storeroom investment from a liability to a highly valued asset.

REPLACEMENT ASSET VALUE

The replacement asset value (RAV) is the dollar amount it would cost to replace the equipment assets if a disaster occurred that destroyed the entire plant. Insurance companies value buildings and equipment for replacement much like the average person secures insurance for their home and automobiles. The replacement value of the plant equipment can be used as a gauge to determine the dollar amount of MRO inventory they needed to maintain in the storeroom. With this in mind, the best-practice inventory value for the average MRO storeroom is estimated to be in the range of 0.50 to 0.75 percent of the RAV. The RAV percentage will vary according to the industry and is usually determined by a corporate directive or expectation communicated to the plant manager.

RISK MANAGEMENT

The risk associated of not stocking some items in the MRO inventory is a decision the maintenance and plant management should make. It is not logical to stock every part for every piece of equipment you have on site, the inventory cost would go through the roof. Using a parts standardization program and assessing the availability of parts from suppliers in a few hours or days if you are a proactive site are good business decisions that must be made to stock the most efficient MRO inventory possible. Equipment parts that have been identified by the reliability engineers as strategic or critical are not candidates for non-stock or offsite supplier stocking. Even though you decide to stock parts on site, the job is not complete. A good lay-up and PM program should be in place or you are placing your organization in a position of high risk of the part not performing as expected when put into service.

The probability of failure is an ongoing analysis since part availability and suppliers can change, equipment can get modified, and who knows, the weather can cause a problem in some instances. A failure modes and effects analysis (FMEA) becomes a basis for establishing a probability for failure evaluation. To conduct a valid FMEA, there has to be a history of equipment repair and some idea of the life expectancy of the parts you are trying to evaluate. Without repair history on equipment and parts life, decisions to stock or not to stock become very subjective and logic takes a back seat in the final decision. Some of the rules of logic applied when conducting a FMEA are listed below and as you can see, it becomes a "what if" scenario and if in doubt, error on the side of safety stock.

- Is there a possibility for the supplier to go out of business?
- Have engineering changes been made that make the part more reliable?
- Are there technological advances that have occurred, are we keeping up with the times?
- Is the equipment in a maintainable state?
- Is there a possibility of a labor strike at the supplier site and how would this affect us?
- Can a natural disaster occur and what are the probabilities?
- What are the safety issues, and what are the risks?

There are many other factors in a FMEA and this list gives a very general look at some of the questions plants should be asking themselves when evaluating their MRO inventory stock or non-stock policy.

LOSS ELIMINATION

Inventory losses occur when items are damaged, lost, stolen, or not issued using the proper procedure. Most of the time, when an item is missing from the storeroom inventory, a well-intentioned employee took the item to do a repair and either forgot or didn't see the need to complete the necessary paper work to complete the job. Following the proper work-process in any area of the plant is essential for efficient operation. There are instances where an employee takes advantage of a situation and decides they have a need for a particular item at home so it goes in the lunch box or backpack as they leave the plant. At any rate, a loss is a loss and the cost has to be absorbed by the company. On top of it all, the inventory is incorrect. Requiring maintenance to have an approved work order to perform work will help get parts assigned to the correct repair asset. *No work order, no work* is a good policy so repair histories can be tracked and maintenance activity is monitored. Maintaining 95 percent overall inventory accuracy in the storeroom will mean catching just about all of the inventory errors before they become problems. Common reasons for inventory losses are

- Wrong part received
- Items received damaged, no claim processed
- Damage to items while in inventory

- Located in the wrong location
- Move to a new location and new location not entered
- Wrong part issued
- Damage from exposure to the environment

    Loss prevention can be done by

- Shipping documents on all outbound materials
- Implementing quality control checks in receiving
- Monitoring shelf life of parts in storage
- PM activities for parts in storage
- Proper packaging
- Accurate warranty claims
- Removing obsolete materials
- Enforcing the security policy in the storeroom
- Requiring a work order when issuing materials from stores
- Proper handling and storage of hazardous materials

INVENTORY CRITICALITY

Knowing what parts are important and what parts are not so important is essential to having the right inventory available when you are called on to issue it. Maintenance and engineering play a key role in determining the criticality of MRO inventory and the priority assigned to each item. Equipment bill of materials is the list of parts that have been evaluated by the reliability engineering team to determine an equipment hierarchy and part criticality for each identified asset group. The part criticality determined by the reliability engineers is used to prioritize the MRO inventory using an ABC or some type of item priority assignment in the CMMS. The priority assignment provides a data source that can be sorted and retrieved to support the inventory management processes that rely on the ABC classification. Inventory flagged as critical or insurance spares by maintenance and engineering are managed with extreme care under a policy that addresses environmental conditions, humidity, dust, dirt, and exposure to high traffic areas.

Accurate equipment BOM are also essential to the planning & scheduling department as they prepare a work schedule for the maintenance technicians. Efforts to implement or manage a kitting program to support planned work is going to prove to be a very frustrating experience as well if the equipment BOM do not contain current and accurate information. Effective planning is essentially impossible if the planners are not able to determine or trust the BOM or the inventory data they get from the CMMS. When plants have no idea what is in their inventory, critical parts are assumed to be available, and if available, they may or may not perform as expected if installed. Equipment modifications or inventory added to the current inventory must be entered into the CMMS using the approved standard part description and the criticality priority assigned for inventory management. The accuracy of the BOM ensures that the critical parts are identified, and then it becomes the responsibility of the storeroom to ensure that these parts are in inventory 100 percent of the time.

Certain types of MRO inventory require special attention and facilities to protect the integrity of the item. Electronic drive boards and drive units are examples of the inventory, which is affected by being exposed to wide ranges of temperature, humidity, or dust and dirt. This high-dollar inventory is sensitive to these extremes and if not housed or protected in a controlled environment can be damaged. The MRO bearing inventory is a prime example of another item that needs to be properly stored to prevent storage damage that will cause the bearing to fail prematurely when placed in service. The supplier is a good resource for proper guidelines to follow to ensure bearings are stored correctly and are not being damaged during storage from vibration or exposure to a harsh environment.

Hazardous materials in the form of flammables, certain chemicals, paints, radioactive materials, and hazardous waste are items that storeroom personnel must be trained to handle and react quickly to accidental spills of these materials to prevent environmental or personal injury. These items must have special storage accommodations like flame-proof cabinets, segregation from the regular storeroom inventory or secure facilities built to house only that classification of hazardous material. Training to understand the materials safety data sheets (MSDS) and to use the personal protective equipment required to safely handle this material must be part of a safety awareness program for your storeroom.

STOCKING LEVELS

When materials are needed, it is essential that they are available either from the MRO inventory or from a supplier in a timely manner. The goal is to operate at full capacity, maximizing the need to have zero downtime, and maximum production output. Maintenance means capacity assurance, which in general terms means that repair work becomes seamless and a goal of attaining 85 percent planned work becomes a way of doing business. The goal of the storeroom is to support maintenance and operations by being able to supply parts by the date required. Cost prohibits large inventories and the facility must learn to run lean and hold inventories at a minimum. Suppliers are accountable for meeting the due date indicated by the planner and or the requestor. If there is a problem and the delivery date is going to be missed, the supplier must contact purchasing to let them about the delay and provide a new date when the material will be arriving. Purchasing and the storeroom must communicate with the planners when parts are going to be delayed so planned jobs can be adjusted. This type of communication is essential between departments and suppliers, surprises are not an option.

Expected service levels of MRO parts:

- *Critical or insurance spares (available 100 percent of the time)* are spares used on critical components of an equipment asset. The MRO parts identified as insurance spares are required to be in stock to receive reduced insurance premiums and hedge against lost production downtime insurance claims.
- *Components to insurance spares (98 percent availability)* are parts used in critical equipment and equipment components. The use is unpredictable since the mean times between failures are unpredictable. The result of not having insurance parts in stock can cause extended downtime and major losses of production.
- *Standard replacement parts (95 percent availability)* parts that can be used on more than one component or piece of equipment. Suppliers for a number of users generally stock these parts. Delivery lead times are predictable so replenishment can be managed.
- *Hardware items (90 percent availability)* Nuts, bolts, washers, cotter pins, and other items those are low in unit costs, carried in ample quantity, and readily available from suppliers. These should be stocked in adequate supply in a convenient location so that daily needs can be met.
- *Small tools (90 percent availability)* Company-owned tools that are controlled by the storeroom tool crib and assigned to craftspeople for use on a specific job then returned to the storeroom.
- *General supplies (90 percent availability)* All other supplies that includes office and sanitary, operating supplies, chemicals, water, uniforms, gloves, and so on, as well as low value parts (LVP).

CARRYING COST

Carrying costs are calculated by applying an annual percentage against the total value of MRO inventory per time period. The yearly percent is prorated, according to the time span being used, (day, week, or month). Many controllers do not use a complicated formula to determine the carrying cost of the storeroom inventory but rather use a percentage plus the prime-lending rate. Using the general calculation like this is not going to give an accurate picture of your cost to maintain the MRO inventory and will without doubt be far down the list of urgent needs. A direct correlation between inventory and carrying cost is noted when you have excessive obsolete or slow moving inventory.

We can use the example of a plant that has a $10 million MRO inventory with $2 million wrapped up in obsolete inventory. Using 20 percent as the carrying cost factor, it becomes clear that the $2 million obsolete inventory is costing us $400,000 annually to keep in on the books. It may seem hard to believe but most businesses we work with have 18 to 20 percent of their current MRO inventory identified as obsolete but have a hard time letting go of the items. I don't know of many plant managers who would have much heartburn about saving $400,000 a year and reducing their inventory value by $2 million. Most cost-cutting efforts are focused on reducing head count and seem to be the first option exercised. As Ron Moore states in his book, *Making Common Sense Common Practice,* "You cannot cost cut your way to profitability." It might work short-term, but long-term, you are making a serious tactical business mistake. The following are some of the areas that contribute to carrying cost, many are considered to be a cost of doing business and a necessary evil. Best-practice inventory management and following defined workflow processes is the first step to controlling carrying costs and must be considered in plant cost-reduction lean efforts. We just need to manage our business much better.

Elements of MRO Carrying Costs

- The cost of money
- Property taxes
- Liability/property insurance
- Obsolete inventory
- Shrinkage and deterioration
- Scrap
- Damage
- Theft
- Facility rent
- Utilities
- Storeroom labor cost
- Storeroom security

Carrying Cost Calculation

$$\text{Carrying cost} = (\text{carrying cost \%}) \times (\text{total inventory value})$$

STOREROOM MANAGEMENT

Role of a Storeroom Supervisor

What is the role of the storeroom supervisor in supporting the maintenance department? The storeroom is where the maintenance technicians will come, expecting to find the parts they need to do the necessary repairs to operational equipment. The storeroom supervisor is running a business within a business and has the ultimate responsibility to provide the correct items, in good condition, of the right quantity, and of good quality. Much like an auto mechanic that goes into a parts store, if they are given the wrong part, it is dirty, corroded, with parts robbed off it, will that auto mechanic return to that part store or find another supplier? The only vendor is the storeroom supervisor and the maintenance department is the only customer. If the maintenance department gets upset with the storeroom, they can always force the planners to become expediters and now we are well on the road to a highly reactive situation that is going to drive the cost of repair through the roof and directly to the bottom line of the balance sheet. Do you see the important role of the storeroom supervisor and their performance in managing the storeroom operation plays in the overall performance of the maintenance department to support the plant overall equipment effectiveness (OEE)?

The following are some key responsibilities of the storeroom supervisor.

- Maintain a clean and orderly storeroom
- Plan the storeroom layout for efficient order picking and inventory care
- Organize and manage staffing levels to ensure adequate support for maintenance
- Work closely with the planning/scheduling department
- Provide inventory reporting to purchasing
- Monitor min/max levels and order point information
- Manage inventories by ABC classification
- Use best practice inventory management practices
- Coordinate activities with other disciplines (purchasing, accounting, engineering, operations, management) to ensure a high support of maintenance activities
- Provide monthly key performance indicators (KPI) reports to the maintenance manager
- Coordinate special parts orders with maintenance
- Provide reports to management such as inventory valuation reports, negative inventory reports, cycle count variances, scrap, and obsolescence
- Attend maintenance and plan meetings
- Become familiar with plant equipment and operational processes
- Maintain open lines of communication

Functions of a Storeroom

- Receive goods
- Store inventory correctly
- Issue items from inventory
- Utilize the kitting process for planned work
- Respond to emergency breakdowns
- Maximize effective use of resources
- Perform PM as required
- The single point for shipments leaving the plant

Who Should the Storeroom Report to?

The organizational structure of every business is different and who has responsibility for each specific function will vary as well. Who the storeroom reports to is a topic of debate and there are strong arguments for the ownership and management in each case. The MRO storeroom is a facility that houses millions of dollars in spare parts with lesser amounts of operating supplies and acts as a funnel for all purchased items received into the organization. The storeroom is responsible for distributing spare parts and supplies upon request to the job site in the form of a kit or assembly to support maintenance. The storeroom orders its own replacement inventory and ensures the items are stored properly to protect them from handling or environmental damage. Storeroom employees perform job tasks much like accountants, cycle counting the inventory, monitoring the inventory turns, and accounting for each dollar of inventory on a monthly basis.

Looking at the activities and function of the storeroom on these factors, it would be a good argument that accounting or purchasing should control the purse strings of such a large investment. Who would know better how much of each item to stock and the best investment of parts to produce the highest return. In the case of the storeroom it is not so much about return on the investment it is more about having the right part ready to go when you need it. Kitting and scheduling maintenance work are two activities that require seamless communication and cooperation every step of the way. Not having the parts available when needed or finding out you have a part that doesn't meet the engineering specifications will discount any cost saving and in fact cost more in operational downtime and rework.

The storeroom inventory will be composed of about 20 percent of items identified as critical or insurance spares. These parts are determined to be critical by maintenance engineering and operations, compose about 80 percent of the total value of storeroom inventory. Many of these items will require long-term storage and regular preventive maintenance in the storeroom to ensure they will perform at 100 percent when they are put into service. Seventy percent of the stores inventory will consist of items that are not identified as critical but have been listed as needing to be on-hand in the store stock inventory. Maintenance and engineering must play an active role in obsolete inventory identification and removal, layup policies, job kitting and planning/scheduling, PM, and PdM programs. Again, the decision of what is critical and what needs to be stocked in the storeroom comes down to a decision that must be made by maintenance, engineering, and operations.

While all departments within the business have some ownership in the storeroom, the main focus is to supply maintenance with the parts and materials to perform their work in the most efficient manner possible. Accounting and purchasing have a definite interest in the dollars invested in the MRO inventory but they lack the expertise to make the decisions necessary to support maintenance and operational excellence. Maintenance engineering and the maintenance department have the expertise and are ultimately on the hook to ensure operations will produce at maximum capacity. There is no doubt that it is essential to have the storeroom report to the maintenance department/manager.

Housekeeping and 5S

Storeroom excellence begins with a clean house. It has been proven time and time again that employees that work in a dark, dirty, and poorly run storeroom have extremely low morale, don't like their jobs, and feel that their contribution to the business is meaningless. Assigning housekeeping responsibilities to the storeroom employees and conducting a weekly walk through to audit their performance is an essential start to a world-class storeroom operation. A good example of how environment affects behavior is a comment a speaker made about the condition of the subways in New York. The waiting areas and subways were covered with graffiti, the subways were filthy, broken windows, and most of the equipment was in a state of disrepair. When Rudy Giuliani, the recent mayor of New York, directed the city manager to clean up the subways, remove the graffiti, repair the broken equipment, and instructed the NYPD to start patrolling the subways, people felt safer and the idea of subway travel became more appealing. The result of a clean and safe work area is swift and the employees start to get excited about the changes taking place. Most obvious is the pride and ownership that results from these minor cosmetic changes. These small changes in business operation are the foundation to storeroom excellence.

We have seen many programs that are designed to facilitate change in business operation. Many of these programs have been adopted and Americanized throughout the years. Not that the American businesses didn't realize that a clean work environment was important, we have just moved it down the priority list of important things. Many of the methods used to organize the housekeeping effort have been adopted from the Japanese manufacturing techniques. The one we use to train and is pretty direct to manage is the "5S" program. There are five basic steps to the 5S process to implement good housekeeping practices at the client site.

- *Sort*—Clutter is sorted into categories and organized
- *Stabilize*—The work place is systemized so that everything has its place
- *Shine or sweep*—Daily maintenance of the work area occurs, so that the floors are swept, tables dusted and polished, and the work place organized
- *Standardize*—The methods learned above are kept and maintained so that the cleanup is not just for visitors but part of the daily routine
- *Sustain*—The 5S method is expanded throughout the facility, and is a part of the business practice

The first two steps of the 5S program are simple and require bodies and muscle to get these done. Deciding what is needed and what is trash is always an interesting project. The question employees ask is "why are we keeping this?" The answer is, if we are asking this question, we probably don't need it or it belongs in another area of the plant. As the storeroom moves into the daily shine and sweep phase of 5S, things become more of a routine and behavior change is starting to take place. We are always amazed at the activity in a plant when a visiting customer or a VIP is coming to tour the plant. All kinds of activity starts to take place for the visit, things getting thrown away, things getting painted, and in a very short time, the plant looks good along the tour route the group will be taking. Why not keep the plant tour ready at all times? It's your house, employees spend a good amount of their life working there and it is good for your business, so keep it clean.

Key Performance Indicators

Key performance indicators are measurements of your storeroom's effectiveness and efficiency and benchmarks of where you are currently compared to where you have been or should be now. KPIs are trend indicators telling you through charts and graphs where you are trending or heading unless you change something to affect the current performance. Employees need to know how they are performing and what is expected from them to perform their jobs effectively. Many times the department workers are showing up to work, doing their jobs, thinking they are doing great and not understanding why the supervisor and manager are always so stressed out. The 80/20 rule applies with your employees as with most situations in society. The majority of your employees want to do a good job. All they just need to know is what level is good. KPIs are a great tool to manage any process and are critical in facilitating a change management program. We will discuss KPIs in more detail later in the storeroom management section.

Material Care and Storage

The environment of the storeroom should be sensitive to the type of parts stored. Computer boards and sensitive electronic equipment need to be in a controlled temperature and humidity atmosphere while in stores inventory. At any time parts are not suitable to install in a piece of equipment, they must be removed from the inventory. Racking is also an essential part of inventory storage. There are several types of racks available and selecting the correct application for your site is essential to proper inventory storage. Rack loading should never exceed the rated capacity and the capacity rating for each rack must be clearly indicated to prevent overloading. If the storeroom has multiple floors or the floor has a rating capacity it may be necessary to calculate total rack capacity and load restrictions to prevent overloading the floor capacity. Many storerooms incorporate standard pallet size racking and purchase pallets to be used to store MRO inventory and remain in or are returned to the storeroom. Some storerooms have problems with lighting and will use a light colored floor epoxy paint that reflects available light causing the entire storeroom to be better lit with a very small investment.

All inventories require some type of special care while stored in the storeroom. Bearings are sensitive to vibration, dirt, humidity, and being dropped. Vibration dampening cabinets or racking in a climate-controlled environment is a good idea for bearings that will be stored for extensive periods of time. Electric motors, gearboxes, and other rotating stock should be under a PM program that rotates the motor shaft or gearbox shaft 450 degrees from the original position. If the motor or gearbox is identified as a critical item, the PM should be conducted every 45 to 60 days. Heating blankets and low-voltage current applied to critical motors and transformers also decreases the possibility of moisture causing damage to windings and internal electrical components.

Some maintenance storerooms have a central storeroom with several smaller satellite stores throughout the plant. The practice of satellite stores allows maintenance employees to have parts and supplies closer to their work areas but many times this is a convenience to compensate for a reactive maintenance program. If the decision to have satellite stores is accepted, the main storeroom must control all inventory that is entering the satellite stores and this inventory is subject to the same guidelines as the main storeroom when it comes to PM's, cycle counting, housekeeping, and inventory levels.

Sometimes the storeroom has been allowed to develop some very poor business practices and the organization needs to revise the current practices. The first step in organizing or revising a storeroom is to develop a plan of action. The plan should include an organization chart, a matrix of resource allocation, and an analysis of the type of material stored, the type of storage required, and the inventory levels that are needed to support maintenance. Who are we and what do we expect to get accomplished? The organization chart describes the positions of the personnel that are needed including the reporting structure required to manage a storeroom. An allocation of resources matrix determines the workforce that will be required to perform the work and provide the expected service in a timely manner. The matrix may include slots from some or all of these positions: engineering, purchasing, planning and scheduling, storeroom management, storeroom personnel, shop repair technician, quality control and assurance, materials management, and possibly a representative from operations. The maintenance storeroom will interact with all of these areas at some point in time and the expected areas of participation must be communicated so employees understand their role in the change process.

Materials stored need to be reviewed for the following considerations:

- Environmental exposure
- High-dollar critical items
- Security requirements
- Bulk storage
- Operating supplies, consumables
- Specialty tools
- Electric motors, gear boxes (PM program)
- Packaging (replace broken packages or aged package)
- VMI and consignment inventory

After the inventory is reviewed, the type of storage racks and cabinets can be planned and the general warehouse layout determined. The inventory cannot be just placed on a rack and forgotten, it has to have an assigned location so it can be retrieved when an issue requisition and work order are received from maintenance. Warehouse rack locations use the row, section, tier, and position to identify the location of items. Cabinets that provide additional protection from dust, dirt, and the environment use much the same principle to identify their locations, row, cabinet, drawer, and drawer position to identify item locations.

It is impossible to maintain every item on each bill of material in the plant. Supplier partnerships and certified suppliers are critical to support the maintenance effort. It is necessary for the purchasing and storeroom managers to communicate lead times from suppliers what parts are available within hours, a day, 2 days, a week, month so the maintenance crews can be confident that we will have the required parts when they need them to do their work. Best practice storeroom inventory value in industry is usually 0.5 to 0.75 percent of the replacement asset value of the plant equipment. Not the site replacement, but the value of the equipment itself. Much of the inventory value is dependent on the number and value of identified critical insurance spares and components that are required to be in inventory 100 percent of the time. Equipment repair history of demand and forecasts in the CMMS are better ways to determine future needs and trends as to inventory levels than a strict percentage due to the nature of most businesses. By planning the organizational needs of your storeroom in regards to materials and labor, the daily operation will be more predictable and efficient from day to day. The goal of the storeroom is to provide outstanding support and service to the maintenance organization to ensure that equipment is repaired providing the lowest mean time to repair possible with the right parts, the right quality of parts, and the right quantity. Planned purchases, detailed forecasts, and a sound min/max program administration will minimize stock outages, and provide a sound basis for parts management.

If overflow storage locations are required, a specific location needs to be designated the quantity in the overflow location entered in the CMMS. Creating an overflow location for the inventory will cause your CMMS to indicate two locations or in some cases, it will only identify one of the two locations causing the storeroom employees to manage a manual location system for overflow inventory. The goal of a locator system in your CMMS is to have each part in a specific location in the computer system with all empty bins reported and marked in the system. The empty locations allow the storeroom manager to plan locations for future receipts. Any relocation of inventory requires the storeroom employees to reassign the inventory to the new location. Bar coding is a great tool to track inventory movement and will help employees to move the inventory easily.

Types of Locator Systems

1. *Fixed Locations* are now considered inefficient due to the fact that inventories fluctuate in size and usage reducing the amount of space needed.
2. *Random Locations* are most efficient and used more frequently due to bar coding and radio frequency identification (RFID) locators. When a location is emptied a new item can be located in that location allowing full utilization of warehouse space.
3. *Floating Slots* works best for capital project inventory management or items that will not be held in storage for more than a few months.
4. *Logical Address System* is most efficient method to group inventory by commodity. The system is easily understood by employees and can reduce search time if the MRO inventory is not cataloged in a CMMS.

The preference or best practice is to have one primary fixed location that is determined using an ABC classification system. Critical spares and low use items are located in low traffic areas and higher demand items are concentrated closer to the counter for easy picking access. If an overstock location is required the second location is listed in the CMMS for the overstock. Inventory will continue to be picked from the primary warehouse location and the inventory in the secondary location moved to the primary location as room is created. Employees conducting cycle counts will need to go to both the primary and secondary locations to count the inventory. When the cycle count list is generated from the CMMS, both locations will be indicated to let the employees know they have two locations for this material.

The kitting area within the storeroom should have locations assigned so the kits can be located quickly and the status of each kit determined at a glance. Each planner will have an assigned set of racks where all kits they have in progress will be located. The kits in each planner's area will have the work order number displayed prominently and any tags or stickers to indicate the status of the kit on the work order sheet for easy viewing.

The CMMS should have the capability to identify and report all empty bin locations available throughout the warehouse. Utilizing this report, stores employees can plan the stock location work for inventory stock received each day. The warehouse manager can also use the empty bin data to calculate and report the warehouse availability from week to week and plan for peak periods of seasonal inventory. Cabinets can be tricky to determine how much space you have available. A good method to ensure space utilization is to use magnetic buttons to indicate open slots in cabinets and closed containers. Commodity grouping of inventory is a very good method to keep the unused space problem in cabinets to a minimum.

Returnable plastic containers that break down when empty are used in place of cardboard boxes that have to be disposed of by the customer. Some customers have agreements with their motor suppliers to crate large critical spare motors in plywood boxes, leaving the drive shaft exposed for easy access to conduct the scheduled PM program. As the new motor is put in service, the motor taken out of service is placed in the crate and sent out for rebuild, reusing the crate over and over. If the racking is constructed to use a standard pallet for item storage, many sites are going to the colored plastic pallet that will last for several years and can easily be tracked if someone takes the pallet out of the storeroom. Using a standard pallet that items can be strapped to will reduce the possibility of an unsecured item weighing 20 lb from falling off the pallet and striking the employees below. Returnable pallets from the supplier promote cost saving by allowing the supplier to reuse the pallets to reduce cost that is normally passed along to the customer. Totes for oils and chemicals are reusable bulk containers and are returned to the supplier eliminating the need for metal or plastic barrels. Totes are much easier to transport than barrels and hook up and disconnect of supply hoses is simple with the chance of spills greatly reduced.

Dust and dirt are constant enemies of the MRO inventory causing damage to unprotected inventory. Exposed "V" belts and seals can be destroyed from exposure to the daily environment of the average storeroom. Belts can be stored in plastic containers with lids rather than hanging them on hooks that exposes them to the environment and causes damage. Manufactures of belts, hoses, seals, "O" rings, and other rubber like material that is affected by dust, dirt, and heat will not guarantee their products past 2 years. We are always concerned when we see 35 to 40 "V" belts hanging on the wall of a storeroom when we walk in. Isn't there a supplier that can provide these belts in less than an hour if needed? More over, if the scheduled PM is conducted, unexpected belt damage is going to be minimal. There are times when motors or pumps seize up causing the belts to burn off, but this should be the exception rather than the rule. If you are going to have rubber products in the MRO inventory, the items need to be stored in vacuum sealed packages or in plastic containers to prevent damage.

Free issue inventory are items that are of low value and high volume. These items are usually stored outside the secured storeroom and contain items like nuts, washers, and bolts. To manage this inventory, we usually suggest using a vendor-managed inventory agreement and using a two-bin Kanban system. The supplier checks the inventory periodically and fills the bins as required. This system is visual and easily managed by the storeroom employees. Once the front bin is empty, the supplier refills it and puts the newly filled bin in the back of the current bin to promote the first in, first out inventory management system.

Bar Coding

Bar coding programs were developed back in 1962 to track railroad cars as they travel around the United States. When the bar codes started appearing rumors abounded and everyone thought it was a plot to take over and control the people. As in most rumors, it turned out to be false. Everyone later discovered that the bar codes were simply a future means the grocery stores planned to use to control their inventory. Imagine that, having a plan to better manage their business. Since the early 1960s bar coding has become a common method to monitor and verify information from the auto industry to the medical fields. Bar coding is a valuable tool in maintenance reliability to have real-time data on inventory control, work order status, production, and distribution activities at multiple sites.

Bar coding is available in over 40 symbolisms which give the customer and industry several options to choose from. There are two major categories of bar codes to choose from, a one dimensional (1D) and a two dimensional (2D). The 1D is the most common and least expensive of the two and serves all the needs of a large MRO inventory. The more sophisticated 2D bar code system is used by large parcel carriers like United Parcel Service and other commercial carriers that ship and track deliveries of millions of parcels each day. With technology advancing as quickly as it does, we would expect to see other businesses like the pharmaceutical and medical industry incorporate the 2D technology into their business operation in the future. The character set and code formats are described in the American National Standard Code for Information Interchange, ANSI X3.4-1977.

Bar code scanners are used to read bar code labels and in some cases, transmit the data in real time using a radio frequency (RF) to the CMMS where the transaction is recorded immediately. If the scanner unit stores the data internally, the data will need to be downloaded to the CMMS at a later time. There are two primary technologies used in bar code scanners. The laser scanner or the charged coupled devices (CCD) are available in today's market. The laser scanner is most functional and can scan a bar code from over 20 ft away making it the choice of most storeroom employees. The CCD is useful at close range and is most applicable for office or scanning bar codes on documents on the shop floor. Hand-held scanners are capable of reading the bar coded information from left to right or right to left so it doesn't matter if you hold the scanner unit upside down or right side up. The scanner can still do its job very well. Laser scanners can also be mounted directly to a vehicle like a lift truck where the driver doesn't have to leave the seat to scan the location and product label. One of the benefits of implementing a bar code program in the MRO storeroom is to increase productivity of the stores employees. The big drawback to the bar code system is that the employees have to scan items to a location, scan inventory moved to another location, and record all issues and receipts for the system to maintain an accurate inventory. The system will only work as well as the people operating it use it.

Why is Bar Coding Attractive in MRO Inventory Control?

Using a bar code system and radio frequency transmitters to record transactions allows inventory management to be recorded in real time. Inventory received is entered into the CMMS inventory management system as soon as it is scanned and removed from the inventory as soon as it is issued to a work order. Bar coding the MRO inventory is rather simple and the small amount of time required to print and apply the bar code labels is time well spent when cycle counting is increased from 10 to 15 line items per shift to 200 line items per shift.

LCE toured an MRO storeroom recently that had a very mature bar code system and the storeroom inventory accuracy was consistently at 98 percent. When we spoke with the storeroom employees about their operation the one thing that came out repeatedly was that they conducted cycle counts religiously and used hand-held radio frequency (RF) bar code scanners to record the data. The employee that was showing us around the storeroom had been with the company for over 25 years, long before bar coding was implemented at this site, he gave all the credit to the storeroom manager and the bar code system they implemented a few years earlier for the high level of inventory accuracy.

Tracking and recording repair histories of operational equipment is simplified using a bar code system to track work orders and part life when put into service. If you are using thermography or vibration analysis as part of the PM program, bar codes can be assigned to motors, gearboxes, and pumps and scanned each time the equipment is checked. Employees that conduct regular lubrication of equipment can use hand-held scanners to record when they add oil or grease to a motor or gearbox. Using a bar code system takes away the temptation to exaggerate equipment downtime and gives an accurate accounting for the time of the maintenance employee conducting the repairs. Training the workforce to use scanners and bar codes is fast and easy. Having employees scan bar codes in place of writing information and numbers on a work order eliminates errors in transposing numbers, forgetting to record an item issue transaction, or just not completing the necessary paperwork. If you are currently using a clerk to enter the data recorded by the storeroom employees, using a bar code system may allow you to reposition that clerk to a position that will be more productive to the business. All in all, the move to a bar coded system improves efficiency and leads to the MRO world-class storeroom activities.

Common mistakes made when implementing a bar code system are not communicating to the employees the benefit the business is expecting to get from purchasing and installing the new technology. Everyone knows a little bit about bar coding and that is just about enough to get you in trouble. Ample time to plan, design, and install the new system is essential. It is usually a good idea to involve some of the shop-floor employees for their input, which in most cases will prevent some unexpected problems later during the implementation phase. Communicate to the people that will be expected to use the new bar code technology each day as part of their job responsibility. We use the word "new" because to many employees the only exposure they have had to a bar code system is at the local grocery store where the scanner is a laser light with many beams to scan the bar code for the price. They probably don't know that when the item is scanned it removed that item from the store's inventory and recorded the transaction in several areas of the business software. Setting unrealistic schedules for the completion of the project is another mistake made when implementing a bar code system or any other change to the activity of shop-floor employees. The employees need to understand what the bar code system will do for them, not to them. It takes a little time to turn a battleship so the captain has to plan and implement his turn long before to prevent colliding with another ship or ending up off course. Implementing new technology is much the same and employees should be coached and informed about the changes on the horizon that will affect their job. The last mistake is not allowing enough time to train the employees on how the system works. If they don't understand who, what, when, and why of the system they will be unsure and you will not receive the full benefit of the system.

Bar code technology is expensive and is a large capital expenditure. It is obvious the corporate leadership will want some kind of cost justification and operational benefit that can be expressed in dollars. The main benefit of a RF bar code system is the real-time control of the MRO inventory which will equate to a reduced total inventory investment if managed properly. Using a RF bar code system reduces the paperwork necessary for stores operation moving you much closer to being a paperless department with the possibility of reducing the clerical workforce. Scanning item transactions in the CMMS inventory management system will aid in the elimination of errors caused by an employee's handwriting being misread resulting in mistakes in inventory or item descriptions in the CMMS.

For the organization wanting to step headlong into the high-tech age, RFID is the solution to your needs. RFID has been available for several years and has recently been targeted by the U.S. Department of Defense and the retail giant Wal-Mart as a tool they want their vendors to use on items supplied so they can better manage their retail inventory. RFID has many applications and can be applied to a large piece of equipment the size of a shipping container or inserted into a bar code label applied to a small package. RFID tags can be active or passive, with an electronic product code or a global tag that can be used for asset tracking internationally. Active RFID tags have their own power supply and passive tags are powered by the signal generated from the reader device. The electronic product code can track individual items as well as cases and pallets of product.

MRO applications for RFID technology is not cost affective for most organizations at this time. Using a bar code system to reduce the time required for receiving and issuing inventory or the capability to complete inventory cycle counts quickly and accurately could outweigh the cost of an RFID system.

Types of Storage Equipment

When implementing a storeroom, it is important to understand the storage techniques and types of equipment available to organize and bin properly.

1. *Bulk storage:* Pallets are stored on the floor in specified areas. These parts do not need or involve storage equipment. These parts are stored in lines, with each line being 48″ wide, and then 12″ separation from the next set of lines. This enables a person to walk down the aisles of bulk material, to count the material, pick it, or check the material for quality.
2. *Demand flow racks:* Material is stored so that the oldest material gravitates to the front on a roller. New material is binned from the back so that the rotation of stock occurs.
3. *Pallet racks:* These are normally used for bulk material when cubic feet are important. The racks allow a better utilization of space since it can go up. It is cheaper to go up than build out.
4. *Vidmar storage cabinets:* Used for a variety of smaller stock. Some of these include fasteners, computer parts, tools, rubber products, and products where dust needs to be minimized.
5. *Cantilever racks:* Used for steel tubing, long shafts, springs, axles, and other long parts.
6. *Drive in/drive through:* Parts are stored in a manner in which the fork truck enters from the rear and pulls from the front. It is most efficient in a finished goods warehouse so that older material is picked first for the shipment outbound.
7. *Flow-through rack:* A flow-through rack utilizes the higher cubing allowed by the pallet rack and incorporates demand flow technology to rotate material. It is loaded from the rear and removed from the front.

8. *Sliding rack/shelving:* The shelves can be moved on the floor to consolidate floor space. When picking the bins are moved apart and when stored moved together. This is a space saving method of cubic feet.
9. *Shelving:* Metal shelving is a must for the bin rooms. This shelving is easy to install and can be adjusted to any height easily.
10. *Rotating shelving:* It is just like a lazy susan concept with parts that can rotate to the part that is to be picked.
11. *Rack entry module system (REM):* It is a method used to assist in rotating material more effectively. Racks are movable and on a track.
12. *Carousel storage:* Parts are stored so that shelves can be rotated automatically. It is designed to store a number of parts in a small amount of cubic feet.

STOREROOM OPERATION

Utilization and Control

Standard operating procedures must be followed for each of the workflow processes from the receipt of the part into the MRO inventory to issuing parts to an approved work order. It is important that the storeroom manager recognize the importance of following the prescribed workflow processes to manage their operation and insist that the storeroom employees also follow each process. A security policy that restricts access to the storeroom during off hours and doesn't let non-storeroom employees enter unless accompanied by a stores employee is essential. Sounds harsh, but letting employees into the storeroom for self service is a huge mistake and will affect your inventory accuracy. Storeroom security and inventory accuracy go hand in hand. There are going to be times when employees need to get into the storeroom for parts during off hours if the storeroom is not on the same schedule as the production and maintenance folks. The maintenance and supervisory employees that have access to the storeroom during off hours must know how to find inventory in the CMMS and how to issue the parts to a work order just like the storeroom employees do when they issue parts out of inventory.

The MRO inventory has several built in controls that continuously monitor and adjust the inventory requirements if all the workflow processes have been implemented properly and employees are following the prescribed workflow process. Since the min/max reorder points automatically create an order to replenish the inventory, the min/max parameters should be under constant review, looking at usage, balance-on-hand inventory levels and forecast needs for parts in the coming months. In a proactive site, these needs and decisions are part of the daily work activity and communication between the storeroom, planning, and purchasing are expected. Reviewing the min/max inventory levels on a regular basis, you are able to control overstock and stock-out occurrences much better which will show in the monthly storeroom KPIs.

Store stock repair parts are most overlooked by many sites because they are removed from service, the new rebuilt part installed and the problem is solved. Not so fast! Parts that are identified as store stock repair items have to be sent out for rebuild or repair or repaired in-house by the maintenance shop. Either way, the job is not complete until this item is repaired and returned to the storeroom inventory. A durable tag should be placed on the repair item after it is repaired that indicates the store stock number, part number, and other information that will identify this item when it is put into and pulled from the storeroom inventory. If the part is going off-site for repair or being repaired by the maintenance shop, the item needs to have documentation that accompanies it that ties it to a work order and purchase order. Shipping documents in the form of some kind of material service order (MSO) should accompany the part with transportation documents for the carrier. Some store stock repair items can only be repaired so many times then it is time to purchase a new replacement to maintain the inventory at the expected level.

Through a quality check during the receiving process, some items are not going to be the correct part, wrong specification, or for some reason, not meet the quality standards required. Parts that fail the quality check must be returned to the supplier for credit. A return authorization number and form are sometimes required by the supplier to receive proper credit for the returned item. The purchasing department or the buyer for the item will usually contact the supplier and make the necessary arrangements for the return authorization and documentation required to accompany the returned part.

The planners and maintenance technicians use the CMMS inventory system everyday and need to be able to trust the inventory levels that are stated when they pull up the parts. Storeroom credibility is on the line each time the inventory is accessed and the part may or may not be available in the inventory as indicated in the CMMS. The storeroom cannot be a reactive organization. The maintenance department can get away with a reactive situation here and there, but the storeroom should always be the safety net that is able to supply the parts needed to get the equipment running. The kitting and delivery processes are probably a new experience for most storerooms and will allow the storeroom to take an active role in contributing to the support of maintenance

Reliability Excellence Project. Planning and scheduling work, kitting planned jobs, and delivery of those kits to the work site so maintenance workers using less time to go to the storeroom to pickup parts are factors that will drive the overall equipment effectiveness of operations and production departments. The refined data in the bill of materials and an accurate MRO inventory are essential to supporting maintenance

Reliability Excellence and Sustaining the Results Long Term. It also ensures the quantities and part numbers accurately reflect what is needed for the repair. The excepted outcomes below lead to the storeroom becoming a profit center for the organization instead of an expense center.

Expected Outcomes of Storeroom Best Practice Implementation

- Reduced inventory levels
- Increased inventory accuracy
- Obsolete inventory identified
- Overstock inventory identified
- Accurate min/max inventory levels
- Work processes identified
- Reduced emergency buying
- Reduced inventory stock out occurrence
- Increased operational efficiency of equipment
- Reduction in production downtime
- Supplier partnerships

COMPUTERIZED MAINTENANCE MANAGEMENT SYSTEMS

Enterprise Asset Management

Successful maintenance practices depend a great deal on a robust information system that can track equipment histories and help manage the MRO inventory levels. An exciting trend in the world of CMMS is the increasing sophistication of enterprise asset management (EAM) features and functions software offers and maintenance professionals can actually use.

EAM is being incorporated into CMMS programs in a number of ways. The simplest packages allow manual input of data such as condition readings for triggering PM routines and inventory management. The more sophisticated CMMS software connects online to PLCs or other shop-floor devices for automated data collection. Linking the CMMS to a radio frequency identification software package for inventory control allows for real-time inventory management and tracking PM activity by scanning the tag. The software analyzes incoming data to ensure that trends are on target and within user-defined control limits. When data strays outside the defined limits, the software can automatically initiate a work order or notifies the appropriate individuals that an action needs to be taken to change the current situation. The software also tracks variance from target as well as the worst and best readings.

EAM and other monitoring programs like it are a form of proactive, preventive, and predictive maintenance that can be defined as maintenance initiated on the basis of an asset's condition. Physical properties or trends are monitored on a periodic or continuous basis for attributes such as vibration, particulates in the oil, wear, and so on. EAM is an alternative to failure-based maintenance initiated when assets break down, and use-based maintenance is triggered by time or meter readings.

The CMMS that can perform all of these necessary functions is user friendly and well supported, and is the heart and sole of your business operation. One of the common issues is that so little resources are applied to the CMMS implementation that it gets off to a slow and inefficient start. Another setback to overcome is the improper training of personnel, which should include basic computer literacy training for those who have little to no experience in computer usage. Investing the time up front to develop failure codes and action codes, as well as developing common conventions as to what things will be called and how parts will be numbered, can improve the quality of the CMMS. An essential element of a CMMS is its reporting capabilities that include reporting tools to analyze and make decisions on facts and data, rather than on opinions and assumptions.

When upgrading the current system or just starting to look for a good, solid CMMS, the main consideration is that the system must support both operations and maintenance reliability. There are a number of reasons for implementing, upgrading, and improving the current CMMS. Technology is changing so quickly and the cost to implement a new CMMS is so high that it is very important to research and select the system that is going to supply your needs for the next 5 to 6 years of operation. Upgrades to the current system is also very expensive and once they are in place, it is probably going to be your system until the choice is made to upgrade again or change systems altogether.

Stock Keeping Units

Stock keeping units (SKU) are defined as a number assigned to a specific inventory item with a detailed description of the part in the inventory database. There can only be one SKU in the inventory records for each inventory item. For example, if warehouse "A" has part number A987xx and warehouse "B" has part number A987xx, the items would be identical, as would the part description and SKU or stock number. Many businesses are realizing how much they have invested in MRO inventory at plants globally and the dollars are staggering to say the least. The task to standardize the SKUs for parts across such a large platform is huge and according to the size of the corporation, it could take 3 to 5 years to complete such an undertaking. The reward is worth the effort and with the EAM (enterprise asset management) systems available today, multisite operations can use the information to their benefit to reduce inventories throughout the corporation.

Part Nomenclature. If you show the exact same item to a group of people and ask each of them what it is, they will almost certainly give you different answers. Some will offer only a brief description, others more detailed. Several may actually give you the exact same information but in a different sequence.

This is often what happens when different individuals are asked to provide descriptions for new storeroom items, or when multiple people are responsible for entering part descriptions into the CMMS. Without providing some guidance, each individual is free to express his or her own style, preferences, and biases. When this happens, the Part Master can quickly be cluttered with disorganized or incorrect data. Establishing and following standard guidelines for item descriptions will help identify similar parts, reduce the likelihood of duplicate CMMS numbers for the same part, and facilitate queries for parts within and across sites.

While there is no single standard or best practice for establishing part descriptions, there are some generally accepted rules that can be applied. Using these guidelines with some common sense and a disciplined approach is generally sufficient to establish adequate part descriptions.

The most commonly used method for establishing part descriptions follows a sequence of

- Noun
- Attribute
- Specification
- Further description

For example

Motor, AC,	3.0 HP, 3500 RPM, FRAME 182T, 115V, 32A, SF1.15,		AO Smith H699
(Noun)	(Attribute)	(Specification)	(Further description)

Generally the noun is one word (or at most two words), which best describes what the item is or what most people would call it, for example, fan, motor, belt, pump, and so on.

Attributes are adjectives that further refine the noun description, and distinguish items within a noun group from one another. For example, within the noun grouping "fan", there may be axial fans and centrifugal fans. "Axial" and "centrifugal" are the attributes. Specifications can be many things, such as material type; size, weight, or other physical dimensions; operating parameters; color; and so on.

Further description is generally used to convey other key information that isn't contained elsewhere in the description. This might include things such as manufacturer, model number, or if relevant, even supplier information.

The key to establishing effective part descriptions is having discipline in gathering the right descriptive and parametric data, formatting it correctly, and entering it accurately into the CMMS. To aid in this process, there are several tools that can be used.

To ensure the integrity of noun descriptions, it is useful to develop a standard list to choose from when a new item is added. This helps minimize interpretation and prevent similar items from being called different things (e.g., "fuse" vs. "breaker" or "breaker" vs. "circuit breaker"). Many CMMS systems support the use of list values or tables that must be predefined, and only values from these lists or tables can be used to populate fields such as description, class, category, and so on. Using these tools helps to minimize the variability in descriptions and also helps to avoid many typographical errors.

Sometimes there is confusion about the difference between the noun part of a description and the attributes. Essentially this boils down to a decision about how specific the noun descriptions should be. The more generic the noun is, the more likely it will be to need one or more attributes to distinguish different parts from each other. For example, if the noun is "fitting," the attributes should distinguish whether the fitting is a tee, an elbow, a union, a coupling, and so on. If instead the noun is tee, elbow, union, and so on then the need for additional attributes is not as great.

If multiple attributes are required, it is important to remember that they should be listed in a consistent manner. If there is a logical sequence or hierarchy, it should be followed. If not, then again the use of value lists or tables can help ensure that this is done. Depending on the type of item, there may be a number of specifications required to uniquely identify one part from another. For these it is helpful to develop and utilize templates that contain the basic specifications suggested or required for each type of item. The template should have the specifications listed in the order they will appear in the CMMS description, and each field should have a predefined list of typical values that can be entered. For example, in developing standard part descriptions for motors, the specifications might be

- Horsepower
- Operating RPM
- Frame type
- Operating voltage
- Service factor

For certain types of fittings, however, the key specifications might be

- Material type
- Diameter
- Length

Developing and using templates for these types of items will facilitate the capture of required part description data, help to ensure that the data is consistent and accurate, and also makes it easier to create, enter, or even automatically import the description into the CMMS.

Obsolete Inventory

The storeroom is expected to identify and remove obsolete materials on a regular basis. This does several things for the storeroom. One, it adds capacity and availability to store more parts correctly and reduces the over-crowding in the storeroom which can lead to damaged and lost inventory. Two, it reduces the chances that obsolete parts are picked and installed during an unplanned work activity causing equipment failure. Three, it helps to rotate inventory properly so older inventory is picked first to avoid exceeding the expected shelf life of the item. Finally, it is a recognized process to continually monitor the parts in the inventory to ensure they will provide the expected service life when installed.

The storeroom should have a monthly budget to write off obsolete items and a disposal process to scrap, recycle, or otherwise dispose of these items. Some items that become obsolete in your plant might be a needed asset at another operation, or maybe an inventory liquidator has a customer that is in immediate need of a motor. Liquidators are usually reluctant to buy inventory if they don't have a customer looking for and ready to buy that particular item. Very few liquidators have a warehouse to hold inventory to keep their overhead down and broker a quick profit. One caution on using liquidators, some of the inventory items have legal liabilities that are attached and if there is a problem that relates to the item, your company could be liable for legal action. For this reason, many businesses do not allow material to leave the plant site unless it is going to a recycle or scrap dealer. As stated earlier, it can be very costly to, in many ways, maintain obsolete items in your MRO inventory.

CHAPTER 7
COMPUTERIZED PLANNING AND SCHEDULING

Thomas A. Gober
Maintenance Planning Consultant, Life Cycle Engineering, Inc., Charleston, S.C.

Computerization of any management function has become possible and in some cases relatively common. Computerization has yielded significant benefits, but it has also yielded disasters. Thus, if we are to be successful in computerizing maintenance management, it is necessary to know the components of the function The components described here are appropriate for the various types or levels of maintenance work including routine, preventive, corrective, shutdown, facility, and the like.

Work Request. A document that instructs the maintenance department that work is required. It identifies the equipment number, a unique job number, the work requested, any approvals required, and the priority of the work

Work Order. A document that instructs the maintenance person in what is to be done. It identifies crafts, if appropriate, materials, special tools, critical times, and provides other necessary information to accomplish the job. The format for the work order must be agreed upon at the beginning of the computerization process. It is the key to definition of the job, planning, scheduling, and control of the work, plus developing histories for future analysis. Although no database is preloaded, one is rapidly generated.

Prioritizing. The act of determining which jobs have precedence. Since the function of maintenance has limited resources available at any given time, this act is always performed in a formal or informal manner.

Work Plan. The asking of why, what, who, where, when, and how the maintenance group will respond to a work request. It provides logical answers to these questions.

Job Sequence. Frequently called scheduling. It recognizes priorities and resource availability and can be done at several levels.

Total Backlog. A listing of all work in the computerized maintenance management system (CMMS) that is yet to be done.

Ready Backlog. A listing of all work in the CMMS that is ready to be scheduled.

Control Reports. An after-the-fact record, or accounting, or what has been done and some form of measurement.

Computerization inherently means organization. In the case of maintenance management it means an organized database. The components of the maintenance management function determine what the elements of the database are and include assignment of costs, equipment identification, employee lists, and so on.

Assignment of Costs. This usually follows the patterns established by accounting procedures used at a particular facility. It generally recognizes cost centers, departments, divisions, and so on, and frequently, but not necessarily, is geographically oriented. Although a variety of approaches can be used to develop sort levels or to accumulate costs, a clear definition of the approach is critical at the start of a program to computerize maintenance management.

Equipment Identification. This normally takes the form of equipment numbering and includes physical assets or functions on which maintenance resources will be expended. Careful consideration should be given here as to how finely equipment should be identified. For example, should each door in a facility be numbered or should all doors of a certain type be grouped together as one equipment? A case can be made for both approaches depending upon the type of facility and its needs.

Employee Lists. Those people who will be charging time to maintenance work have to be identified. In many cases this will involve the development of a trades or crafts list. In the case of multicraft facilities it may include several levels within the maintenance multicraft category. Trades or crafts are then associated with employee lists for time and cost analysis. Employee lists should be reconcilable to payroll but not necessarily generate or drive payroll.

Priorities. The type of priority approach must be agreed upon when building the database for a computerized maintenance management system.

Stores Catalog. In a complete maintenance planning and management control system, materials play an important role. Thus, a parts list or catalog for stocked material is necessary. These should be numbered, categorized, quantified, located, and priced when loading the database. In addition, provisions should be made for purchased parts that are not stocked.

Equipment Bill of Materials. The listing of all parts in stores inventory that are associated with a specific equipment number. The list includes both stocked and non-stocked parts. The list should include the part description, manufacturer, vendor, unit cost, delivery lead time and how many are required on the equipment.

Cause Codes. Standardized identification of the basic causes of the work generated. This may be the basis for preventive maintenance (PM) programs or schedules that are part of the system. Cause codes provide a means for analyzing work and developing a corrective maintenance program.

Action Codes. Standardized identifications of what was done to respond to a work request. This is basically used to identify what level of "fix" was done and again can initiate a corrective maintenance program.

It should be noted here that a computerized maintenance management program does not, by itself, truly plan maintenance work. It cannot

- Determine if the work request provides adequate information so that maintenance people understand what is to be done
- Make a sketch to illustrate what is to be done
- Decide what materials to use
- Ascertain what time constraints exist because of production or other needs

A computerized maintenance management system is a fine and powerful tool for assisting in maintenance planning. A well-conceived computerized maintenance management system should provide manpower backlogs, equipment histories, equipment parts lists; determine material availability; provide preventive maintenance schedules; and track costs.

WHY COMPUTERIZE MAINTENANCE MANAGEMENT?

It has become commonly accepted that maintenance represents a significant portion of the cost of doing business or providing a service. The portion of the cost that maintenance represents will continue to increase as the various forms of automation increase. It therefore behooves us to make optimum use of that resource called maintenance.

Planning and scheduling of maintenance is one of the ways of optimizing the use of this resource. Normally, however, one of the problems is the amount of clerical work, or "paper shuffling," associated with such planning and scheduling. Computerization, if properly conceived, can minimize this problem.

Computerization can provide backlog information for various types of work; availability of materials; costs by job, facility, or type of work, and so on, easily. It can increase effectiveness of planning, scheduling, and cost tracking by as much as 50 percent. In addition, it can frequently provide types of information not normally available, at no additional cost.

ORGANIZING FOR COMPUTERIZATION

This first step in computerization of maintenance management is program definition. This is necessary whether buying a software package in the marketplace or developing programs in-house. The question of what you want done is paramount.

The ability to track costs is an obvious requirement. What costs? The computerized program should give information on material availability. What information? Location? Quantity on hand? Vendor? Should the program generate purchase orders and when? What type of equipment histories should be generated? Is the program to be maintenance management oriented? Should the program provide information to corporate headquarters and between other facilities within the corporate structure? These and many other questions must be addressed in order to define the program.

Normally the definition cannot be achieved by information technology, engineering, management, or any other single individual. In addition, it is not normally practical for a single individual to implement a computerized maintenance management program. An interdisciplinary team provides the most workable approach for defining the program and implementing it. Heading up or coordinating such a group should be a high-level person responsible for the maintenance function. Disciplines represented on the project team include maintenance, data processing, and accounting. Other disciplines that are frequently helpful are industrial engineering, purchasing or material control, payroll, and production. The use of the latter depends upon the personality of a particular organization. The team should be kept to a workable size, three to six people.

After establishing a project team and developing the program definition, technical evaluation takes place. Answers to questions in the technical evaluation may modify program definition, but the definition must be made first. Technical evaluations include hardware evaluation and make-or-buy software considerations.

IMPLEMENTATION OF A COMPUTERIZED MAINTENANCE MANAGEMENT PROGRAM

Organizing for implementation of a computerized maintenance management program is one of the jobs of the project team. At this time of implementation, definitions of the following items should have been made:

- Who is going to perform the function of maintenance planning and/or scheduling?
- Who is going to front-end load semipermanent information such as cost centers, employee lists, parts lists, equipment lists, cause codes, action codes, budgets, or other information that is not changed on a daily or weekly basis? Included in this is the definition of the sequence necessary for the loading of these data.
- Who is going to maintain the files on a daily and weekly basis? Several persons may be involved, such as a planner to load work orders, a time clerk to load employee time sheets, a storeroom clerk to load material requisitions, or any other appropriate combination. It is advisable to have individual information flowcharts for each of the information flows of the system.

- What type of security is going to be part of the system? That is, definitions and procedures should be established to designate who can enter and/or modify information in each of the systems' various segments, who can view information, and who can call for reports.
- Forms for data entry should be designed for data entry. They should be compatible with the computer's entry format and include work orders, time sheets, and material requisitions.

Orientation of the various people concerned with the new computerized maintenance management program is the next step after organization. This may take place as part of organizing. Orientation will be necessary from the highest levels of the facilities organization to the lowest and includes production and staff as well as maintenance people. The details of the orientation appropriate for the various groups are of course different. It is also appropriate to have concise progress review sessions as the implementation takes place.

Orientation and training are similar, but there is a difference. *Orientation* is informing people what is going to be done or what is being done, while *training* is instructing people in how to do something.

Training of people is a critical aspect of the implementation of a computerized maintenance management program. At a minimum, one should train

- All users to write a work request
- All users how to exercise the priority system
- All data-entry people on the necessary procedures for correct data entry
- Appropriate maintenance people (planners, supervisors, etc.) on how to retrieve information
- Maintenance management and supervision, and appropriate production management how to read and interpret reports and other available information

In summary, although computerizing maintenance management is a demanding task, the benefits are normally significant:

Better labor utilization	5 to 25 percent
Equipment utilization	1 to 5 percent
Stores inventory reduction	10 to 20 percent

OPERATING CHARACTERISTICS OF A GOOD SYSTEM

On-Line Inquiry. A typical complaint made about computerized systems is that they generate too much paper. Thus one of the prime requirements of a good computerized maintenance management system is that it have on-line inquiry and provides screen viewing for the areas of work orders, material, and equipment.

Work Orders. The work order is the controlling document in any maintenance management control program. When viewing a work order on a screen in a computerized program, the following information should be readily available:

- Equipment the work is to be done on.
- Description of the work to be done.
- Priority of work.
- Charging centers to assign costs incurred.
- Date of work order, date it was last worked on, and date it was completed.
- Current backlog status of the work.
- Estimates of how much time by craft will be needed to perform the work.
- Costs, both labor and material, charged against the work order.

Other information may be desired but the foregoing are necessary for good maintenance management.

In addition, the program should be able to selectively bring up work orders for observation. The selection criteria should include

- Cost areas
- Equipment numbers
- Dates written, issued, completed, or last worked
- Priority level
- Backlog status
- Cause or action taken
- Trade
- Supervisor
- Planner

Through the use of any one or combination of the above selection criteria, file search time by the viewer is greatly reduced.

Material. A good storeroom is necessary in optimum equipment and manpower utilization. Screen viewing of material in that stockroom should include

- Parts catalog by part number
- Parts catalog by location
- Parts status summary
- Open purchase orders
- Parts issued and returned

Included in this section should be the ability to view

- Vendors
- Maximum and minimum quantities
- Costs
- Reorder quantities
- Usage information on a periodic basis
- Delivery lead time on parts not kept in stock

Equipment. The third leg of a good maintenance management program concerns equipment. Screen viewing for this section should include

- An equipment list by hierarchy
- A work order list for the piece of equipment
- A parts list or bill of materials for the piece of equipment
- Maintenance expenditures (labor and materials) for the piece of equipment

Custom Report Generation. A second characteristic of a good computerized maintenance management system is the ability to generate specific reports on demand. The selection criteria should be similar to those for screen viewing and include the following reports:

Work Orders

- Work order status
- Backlog status

- Closed jobs
- Equipment downtime
- PM schedule
- Time sheet transactions
- PM compliance

Materials

- Parts catalog
- Inventory status
- Inventory usage
- Reorder report
- Physical inventory reports
- Location catalog
- Parts activity list
- Vendor names and addresses
- Purchase order lists

Equipment

- Equipment lists
- Equipment status
- Equipment parts list
- Equipment parts usage

Performance Reports. The third characteristic that a good system should have is measured performance against some type of target or budget. These reports should be brief yet meaningful for maintenance management and should include

- Hours analysis
- Backlog summary both total and ready
- Closed job summary
- Schedule compliance by hours
- PM compliance

The most important characteristic is that the program be user-friendly. This means that users needing information can get it easily, usually by means of a series of menus and questions asked by the computer.

Teaching the user to respond to menus and questions is easier than teaching the user how to query the computer in its own language. The program should be able to efficiently handle huge amounts of information and many kinds of input and output, yet be understandable at its interface by the least skilled operator, clerk, and maintenance worker.

WORK ORDER SYSTEM

Purpose. A formal work order system provides an information network incorporating inputs and outputs for all the various phases of the maintenance program.

Scope. The work order system and its procedures provide a uniform means of information flow for requesting, planning, scheduling, controlling, recording, and analyzing the performance of all the work done by the maintenance department. The work order form serves as the vehicle for communicating information related to specific work requested of maintenance.

More specifically, the work order system provides

- A single common means of transmitting requests for services by the maintenance department that increases the probability that the information needed by the maintenance department to perform the work will be included in the order.
- A means whereby all work requested may be screened and analyzed to ensure that it is needed.
- A means whereby most work can be preplanned and estimated as to time, methods, and materials to further ensure optimum performance.
- A means of controlling the work going to the various maintenance groups through the scheduling procedure, to assure that the most important work is performed first.
- A means whereby management can track performance relative to time, cost, and materials used for specific work that was requested.

All too often verbal work orders result in performance of unimportant, unauthorized, unnecessary, and even unwanted work. To reduce the possibility of this happening, it is essential that all requests for maintenance services be submitted in writing on the work order form. Requests for service may be initiated by maintenance or production personnel desiring to have work performed.

The work order is the basic authorization for performance of maintenance work and, together with the *planning package*, will produce an accurate account of individual maintenance jobs. It may be used as a reference for similar jobs in the future.

PRIORITY SYSTEM

Purpose. In general, it can be said that there is never enough time, money, or manpower to perform all of the maintenance work that is needed and/or desired.

The decision as to what maintenance work is performed and when, if not systematized, will be made according to the subjective judgment of one or several persons.

If the intent is to perform the most needed and important work first (and it generally is), then it is desirable to have a reasonably objective system to identify priorities for maintenance work that can serve as a guide for the maintenance department.

It is important, therefore, to develop a relative priority ranking system for maintenance work based on the collective judgment of those responsible for the operation of the facility.

The best method of achieving this is one which produces a quantitative index of the relative importance of a job at the time that the need for the job occurs. Thus, personal judgments are less likely to influence the resultant ranking of jobs by priority.

Scope. The system for establishing priority values is called Ranking Index for Maintenance Expenditures (RIME). This system, which provides a wide range of priority numbers, will best provide a true ranking of all the varied jobs which maintenance must perform.

To establish a sound priority system, the following three elements are essential. Without them, the system will function improperly.

1. The priority system must encompass everything within the plant.
2. All production and maintenance personnel involved must understand and respect the priority system.
3. The priority system must be based on profit.

The computing of job priority indexes considers the equipment and facilities (equipment criticality) in conjunction with the importance of the work (work class). These two considerations provide multiples which establish the value of the work requested. The higher this value, the more important the request.

PLANNING PROCEDURE DEFINED

Planning is advance preparation of selected jobs so they can be executed in an efficient and effective manner during job execution, which takes place at a future date. It is a process of detailed analysis to determine and describe the work to be performed, task sequence and methodology, plus identification of required resources—including skills, crew size, man-hours, spare parts and materials, special tools and equipment. It also includes an estimate of total cost and encompasses essential preparatory and restart efforts of operations as well as maintenance.

The following outline describes the procedure for work order planning in detail.

1. Decision has been made that the work order requires planning.
2. Analyze the work requested:
   a. Is the information complete and adequate?
   b. Is the work needed? Have the required approvals been made?
   c. Why is the work needed?
   d. Determine the required level of planning.
   e. Visit the job site and analyze the job in the field.
3. Determine the basic approach to be followed:
   a. What is the priority of the work?
   b. What effect will it have on operating? What are the downtime requirements?
   c. What are the future plans for the equipment?
   d. Is the equipment scheduled to be overhauled, replaced, or phased out?
   e. Can repair provide the same reliability as a replacement at a lower cost?
   f. Make sketches as required.
4. Identify special considerations, needs, and conditions: Is production assistance required?
5. Identify need for engineering: If required, notify engineering or change the status of the work order to awaiting engineering.
6. Identify need for contract service: If required, contact contractor or have the appropriate person do so.
7. Identify work for the maintenance department.
8. Identify information needed.
9. Identify instructions required: Provide a job plan detail by task.
10. Identify manpower required.
11. Estimate man-hours required.
12. Identify any special tools or permits required:
    a. Is work overhead?
    b. If welding is required, is a welding permit needed?
    c. List any special tools.
13. Identify material/parts required: Are any special materials/parts needed? If so, list them.
14. Determine if materials/parts are stock items; if so, list location.
15. If materials/parts are non-stock or out-of-stock items:
    a. Order.
    b. Record the work order number on the purchase order and the purchase-order number on the work order.
    c. Change the status of the work order to awaiting parts in the backlog file.
16. Material/parts received:
    a. As the material/parts are received they are charged to the work order.
    b. When all the parts are received, the work order status is changed from awaiting parts in the backlog file.
17. Stage all material/parts and tools according to the schedule.
18. Identify the supervisor that will be responsible for the work.

19. Plan the work order.
20. Backlog file: Planned work order's status is changed to the appropriate ready-to-schedule category, and it is placed into the backlog file.

SCHEDULING PROCEDURE DEFINED

Scheduling is a distinct process from planning but is closely tied to planning. Scheduling is when to do the job: Scheduling is the process by which required resources are allocated to specific jobs at a time the internal partners can make the associated equipment or job site accessible. Accordingly, the preferred reference is *scheduling and coordination*. It is the marketing arm of a successful maintenance management installation. Planned work orders are scheduled: The work order is pulled from the backlog file and given to the supervisor with the weekly schedule. Work scheduling is the vehicle which facilitates the ability of maintenance to meet the challenge of the plants needs. The goal of scheduling is to ensure that resources are available at a specific time when the equipment is available.

The objectives of maintenance are shown in the form of schedules. The schedule should represent

a. The best utlization of craftpersons
b. A statement of priorities acceptable to both maintenance and operations
c. A means of communicating those commitments

Weekly preliminary schedule requirements:

a. Meet with maintenance supervisors to determine labor availability for the coming week. Schedule work for all available labor hours.
b. Review all PM's and ready-to-schedule work orders. Ensure all PM/PDM work is on the schedule.
c. Determine downtime requirements and review priorities.
d. Prepare a preliminary schedule for each supervisor/area listing jobs in descending order of priority for the entire week by day. This preliminary schedule is the maintenance organization's priority list for the week.

To ensure that all departments have input to the schedule and to allow for a statement of priorities that all parties will agree to and take ownership of a final schedule meeting is required. Before the final schedule meeting the preliminary schedule should be distributed to the operations department for review and markup.

Weekly final schedule requirements:

a. Meet with operations representative and maintenance supervision.
b. With the operations representative and the maintenance supervision, determine the final schedule of priorities for the week.
c. Verify all parts and special tools are on hand.
d. Prepare the final schedule for each supervisor in descending order of priority for the entire week by day and distribute to all parties.

PLANNING FOLLOW-UP

Effective planning requires observing preplanned job progress to eliminate potential delays or problems that may arise. The planner should occasionally observe planned work in progress with the intention of improving preplanning expertise.

During the follow-up process, the planner may address problem areas which are common to planning functions. These areas are

- Were the communications clear and adequate for all personnel involved on the job?
- Was the planner's time properly utilized?
- Was translation of the job plan in clear and understandable terms?

When work orders are completed they should be returned to the planner for review. During this part of the process the planner should review any comments listed on the order with the intent of improving the job plan. This feedback is a critical step necessary for planner improvement, along with improvement of the job plans.

Remember, the goal of planning is to eliminate delay in work execution. To repair quickly is not the basic goal. To repair correctly and efficiently without delays is the ultimate goal of planning. Constant work order audit and review is the method used to drive this goal.

ANALYSES AND REPORTS

Work Orders

The work order system contains information pertaining to specific work for a piece of equipment or facility. This work could be corrective or preventive maintenance, emergency, routine, or standing orders. The data collected provide answers to the standard questions of what, when, where, why, who, how much, and how often. Output from the system consists of two types of reporting.

1. Planning process data and specific information about job orders should be provided on request by the following reports:
   - Work order status
   - Work order recording and update
   - Backlog status report
   - Closed job status report
   - PM master schedule
   - Equipment downtime report

   Backlog and completed job reports can be created based on a multitude of selection criteria, including

   - Customer organization
   - Equipment number or range of numbers
   - Work order priority
   - Key dates (written, issued, completed)
   - Component codes/action codes
   - Backlog status
   - Cause codes
   - Craft

2. The second type of output data provides management with performance control data on a weekly and monthly basis through the following listed reports:
   - Hours analysis
   - Backlog job summary
   - Closed job summary
   - Report of schedule compliance
   - PM compliance report

   The control reports should provide summaries by operating organization and maintenance craft.

Work Control

The work order section provides control of the first key element in an effective system: what has to be done. Maintenance is controlled by controlling the backlog of work requested. The limits for ready to schedule backlog fall into the range of 2 to 4 weeks. The limits for total backlog are 4 to 6 weeks. These are the norm based on 80 to 90 percent of the work being planned. If the ready to schedule backlog falls below 2 weeks effective scheduling becomes difficult if not impossible. If total backlog exceeds 8 weeks, necessary work will not be completed before failure occurs. If the total backlog consistently exceeds 8 weeks the emergency rate will tend to control the maintenance process. Maintenance performance metrics are tools that can help management manage the equipment reliability and thus the maintenance process.

Some examples of reports that a computerized system should provide for effective control are

- Work order status report
- Backlog status report, both total and ready
- Closed job status report
- Work order craft list
- Preventive maintenance master schedule
- Preventive maintenance compliance report
- Equipment downtime report
- Work order cost report
- Hours analysis
- Report of scheduled compliance
- Backlog job summary
- Closed job summary
- Time sheet transaction list

Parts Inventory

The parts inventory system provides control over the second key element in an effective maintenance system (the first key element being what has to be done, i.e., work orders). The primary functions of the inventory system are

- To identify and locate spare parts in the storeroom
- To determine the availability, status, and levels of inventory
- To maintain purchase order status and vendor information
- To provide cost and usage history on parts issued from and returned to inventory
- To record cost and usage data against work orders written and equipment repaired

Some examples of reports that a computerized system should provide are

- Inventory status report
- Inventory reorder report
- Vendor name and address list
- Physical inventory listing
- Parts activity transaction list
- Purchase order listing

Equipment

The equipment section contains data used to identify individual pieces of equipment or physical locations and to track history on labor and material costs. The accumulation of data retained in the system provides the means for

- Evaluating changes needed in preventive and corrective maintenance programs
- Analyzing the history of work performed through work orders
- Comparing for trends in parts issued through the inventory
- Assisting in the decision to repair or replace equipment
- Determining what parts are common between different pieces of equipment

Some examples of reports that a computerized system should provide are

- Equipment listing
- Equipment status report
- Equipment parts catalog listing (by equipment no.)
- Equipment parts catalog listing (by part no.)
- Equipment status summary

CHAPTER 8
COMPUTER-BASED MAINTENANCE MANAGEMENT SYSTEMS

R. Keith Mobley
Principal, Life Cycle Engineering, Inc., Charleston, S.C.

A computer-based maintenance management system (CMMS) is an integrated set of computer programs and data files designed to provide its user with a cost-effective means to manage the massive amounts of data that are generated by maintenance and inventory control organizations. In addition, these systems can provide the means to effectively manage both the human and capital resources in a plant. It is imperative to understand that the CMMS is a *tool* used to assist in improving maintenance and related activities. In and of itself, the CMMS only manages data that have been input to it or that it has created as a result of data input. It does not manage the maintenance operation. In this chapter we will discuss typical functions that make up a CMMS, data management files, which use the system, and what it will and will not do.

CMMS FUNCTIONALITY

A computer-based maintenance management system (CMMS), computerized maintenance management system (CMMS), or a computerized asset management system (CAMS) is a set of integrated software programs, data files, and tables that provide functionality for a number of activities. The functionality is normally grouped into subsystems for specific activity sets within the CMMS. These subsystems may include, but are not limited to

- Equipment/asset input and maintenance
- Equipment/asset bills of material input and maintenance
- Equipment/asset and work order history
- Inventory control
- Work order creation, execution, and completion
- Preventive maintenance (PM) plan development, maintenance, and scheduling
- Work order planning; work order scheduling
- Human resources
- Purchasing and receiving
- Invoice matching and accounts payable

There are also programs for creating and maintaining the tables used by the CMMS and for printing reports.

CMMS FILES

The typical CMMS provides storage, manipulation, and retrieval of the following types of information:

Equipment/Asset Identification and Specifications

The equipment/asset file is a primary, and usually a mandatory, file in a CMMS. Most work orders are written against one of the equipment/asset records maintained on this file. Most systems allow codes to be used as input for many of the data fields. These codes are normally defined by the user and are maintained in tables within the CMMS. A code table should also provide space for a description of each code entered. By establishing code tables, and controlling who can update them, you are assured of consistent input to the related fields in the records. Examples of code tables include the following: Equipment type used to group similar equipments such as motors, pumps, cranes, and the like; building, floor, and room where equipment/assets are located; and units of measure. The number and type of code tables will be unique for every CMMS. Specification data includes unique equipment/asset identifiers such as size, weight, color, amps, rpm, flow rate, and so on. These data may be maintained on the equipment/asset file, on a separate specification file, or in tables that are referenced to each equipment/asset record. Specifications may also be associated with an equipment/asset type. For example, equipment/asset type motors will contain specification codes such as horsepower, amps, volts, cycles, and the like. For each code, a value specific to this equipment/asset record can be assigned, that is, amps = 60. Specifications provide two major benefits. First, it is beneficial to have as much information about the equipment/asset on the computer as possible. This eliminates having to search file cabinets or desk drawers for needed information. It will all be together in one easily accessible place. Second, most systems allow sorting equipment/asset records by their specifications. For example, to find all 50 hp, 60 A, two cycles, AC motors, you input those values into the corresponding specification sort fields and the system will find and display all records that match the combined specifications entered.

When a work request is initiated, the requester will normally be required to reference an equipment/asset identifier for which the work is to be performed. The CMMS will first verify that the equipment/asset identifier is valid by checking the equipment/asset file. If the identifier is not valid, the requester is notified and will be required to enter a valid ID. If valid, the CMMS should automatically copy certain information from the equipment/asset file to the work request. This information will normally include equipment/asset description, assigned cost center, physical location, and possibly warranty information. Depending upon the CMMS, other information that is beneficial to the planner and craft personnel may also be retrieved and placed in the work request.

Equipment/Asset Hierarchies

An equipment/asset may be a component of a larger equipment/asset, a process, an area, a department, a plant, a division, and a company. For example, a motor may be part of a drive system that is part of a process that is located in the finishing area of the fine-paper department in the Detroit plant of the newsprint division of ABC Paper Company (Fig. 8.1).

Each of these units may be set up on the equipment/asset file as either an equipment/asset record or a location record.

Most systems allow hierarchies to be built from their lowest level up. They also allow movement of a record from one hierarchy to another or to a different location within the existing hierarchy. Establishing hierarchies is not mandatory, but there are numerous benefits if you do. First, maintenance costs may be automatically rolled up from any level in the hierarchy against which a work order was written to higher levels. Therefore, the costs for repairing the motor in the example are not only maintained for the motor, but are also rolled up and maintained at all or selected levels in the hierarchy above the motor. This feature is invaluable when you need to quickly determine total maintenance cost for a department, area, process, and so on. With one inquiry, you will see the total maintenance cost for all equipment assigned to whatever level of the hierarchy you entered. The inquiry can be made at any level in the hierarchy.

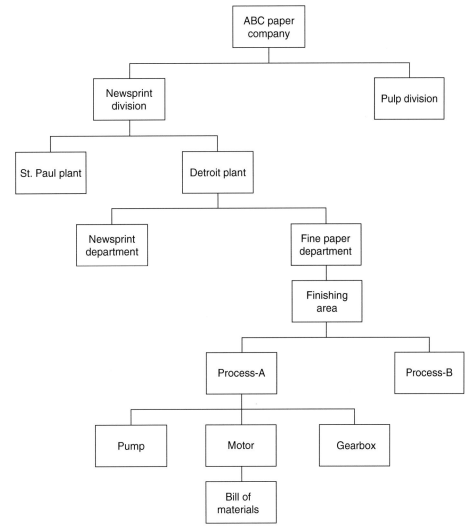

FIGURE 8.1 Equipment/asset hierarchy example.

Second, by inquiring at any level in the hierarchy, all items above and below the level selected can be viewed. This quickly shows what the selected item is made up of as well as what it is part of. From this inquiry, a planner might determine that the work request was initiated against the wrong equipment/asset identifier in the hierarchy and can easily determine which identifier should be assigned to the request.

Third, hierarchies provide a quick method of determining the physical location of an item. This results from including locations as records on the equipment/asset and including them in the hierarchy. By entering an equipment/asset ID, you will immediately see where that item is located. This inquiry can be invaluable if an equipment item in a critical process breaks down and there is no spare available. You may locate an identical item in a noncritical location that can be used until a spare is available.

Fourth, historical work order information should be stored for the equipment/asset item as well as the parent. This information allows you to determine every location the item has been in and the work orders written against it at each location. Conversely, you can inquire on a location and see every equipment/asset item that has been at that location and the work orders completed against it while there. This information is invaluable when trying to determine the cause of certain failures. For example, does a particular type motor always fail at a specific location or does a particular motor fail no matter where it is located and always for the same reason. If all motors fail at the same location and for similar reasons, the problem is likely due to the location.

Equipment/Asset Bills of Material

The equipment/asset bills of material files are usually separate from, but linked to, the equipment/asset file. The bill of materials is the lowest level in the hierarchy for a specific equipment/asset record and contains all, or at least the major parts and components of the equipment/asset. For example, the bill of materials for a pump might contain the housing, shaft, bearings, seals, impeller, and so on. How detailed the bill of materials is for any given item is determined by the user. There are companies, including some of the CMMS vendors, that sell bills of materials for common equipment. Some equipment suppliers may provide them at no cost for equipment purchased from them. The real advantage here is to get these bills of materials on some form of electronic media that can be loaded directly to your CMMS.

Manually creating bills of material can be extremely tedious and time consuming. If you do elect this method, it is recommended that you begin with the most critical parts and components of your critical equipment, then add less significant items as time permits. Some systems will automatically build a bill of materials for you based on the parts issued against work orders. The system should allow you to flag any items on the inventory file that you do not want added to a bill of material. This automatic process can literally require years before a significant bills of material file is developed.

A bill of materials is very beneficial when planning work orders. It shows the planner the exact parts that are to be used on the item to be serviced. The system should also allow the planner to select required parts directly from the bill of materials and add them to the work order plan. Bills of material also provide information when making decisions about which parts to maintain in inventory and in what quantities. If an equipment/asset item is to be permanently removed from service, can all of its associated parts be removed from inventory? Or, do the reorder points and reorder quantities need to be adjusted because these parts are still required for other equipment/asset items? To determine other equipment/asset items that use a part, most systems provide a *parts where used* function that is basically an inverse of the bill of materials. If you enter a part ID into the where used function, the system will display all equipment/asset records that use a particular part and the quantity of the part required for each one.

Spare Parts and Stores Inventory

To fully control and account for maintenance cost and to ensure reliable maintenance practices, control of maintenance inventories must be in place. Most systems provide a maintenance inventory file and all of the programs necessary to create, maintain, and access the file. The inventory file, along with its associated tables, is where all the information about maintenance repair order (MRO) parts is stored. Each inventory record will include descriptive information about the part as well as numerous quantity fields. An MRO inventory file differs from raw materials inventory files in that the MRO inventory file usually contains fields for a cumulative total of an item required by open work orders as well as a cumulative total of an item reserved or committed to open work orders. Other quantitative data either stored on the file or calculated by the inventory programs include reorder points and reorder quantities.

With a fully integrated CMMS, parts can be selected from an equipment/asset bill of material and pulled straight to the work order plan. The CMMS will automatically validate the selected parts against the inventory file and, if valid, add the required quantity to the required quantity on the inventory file. It will also inform the planner if a part is not valid, or not currently available in the required quantity. Quantity available is the balance on hand minus the total quantity reserved or committed. The system should also notify the planner if there is an open purchase requisition or purchase order for a part and the quantity on each. Note that this information will only be available if there is an interface of some type with the purchasing functions. Many systems will automatically generate purchase requisitions for parts when the quantity required or reserved (depending upon the CMMS) will cause the current balance on hand to drop to or below the reorder point.

Parts are issued from inventory to the work order on which they were planned. Most systems also allow unplanned parts to be issued to the work order or to some type of control number such as an account number. Unused parts may be returned to inventory with the quantity and cost of the returned items being automatically credited to the work order and/or the account number associated with the work order.

Some systems provide separate files for stores stock inventory and spare parts inventory based on the premise that spare parts are not the same as stores stock inventory and vice versa. These systems will also provide separate functionality to access and maintain these files. Other systems consider a spare to be an inventory item until it is placed into service. Then, it becomes an equipment/asset item and is maintained on the equipment/asset file until it again becomes a spare.

Some companies either prefer or are required to use their existing inventory system and files instead of those provided with the CMMS. If this happens, an interface between or integration of your existing system and the CMMS should be developed. Without an integrated system, one of two things can happen, neither of which is good. First, you will have two separate, stand-alone systems. Your personnel will have to maneuver between the two systems, often using separate terminals or work stations if the two systems are on different computers. They will be required to manually update the CMMS files with inventory issues and associated cost data.

The second option is to not include inventory on the work orders at all. This is a disaster that results in the loss of one of the major benefits of the CMMS. On average, about 30 percent of maintenance repair costs are for inventory items used. Without including inventory on the work orders, you lose these costs in your accounting, and you lose the history of what parts were used for the repair.

Inventory to Equipment/Asset Where-Used Cross References

This is actually not a file in the CMMS but a process the CMMS provides. The where-used cross reference is normally the inverse of the equipment/asset bill of material. For each inventory item, the where-used will show all equipment/asset items that use the item. If bills of materials have not been created for the equipment/asset records, you probably will not have this cross reference capability. The where-used cross reference can be very beneficial in determining quantities of inventory to maintain as well as determining whether or not an inventory item may be discontinued. As an example, if an equipment item has been removed from service, is it necessary to maintain its parts or components in inventory? By performing the where-used cross reference on each of that equipment/asset part and component, you will see if any are used on other equipment/asset items and how many. If there are no other requirements, you should be able to remove those parts or components from inventory. If there are other requirements, you may still be able to reduce the inventory quantity because demand will be reduced.

Another benefit of the where-used cross reference function is its ability to locate a required part that is not currently in inventory. This part may be on an equipment/asset item that is currently not in service or is not part of a critical process. As a temporary measure, the required part may be "borrowed."

Work Orders

Work orders are the backbone of the CMMS. A work order defines the activities to be performed, the equipment/asset that is to be worked on, the procedures to be followed, the skill/crafts required to perform the activities, an estimate of the time required for each skill/craft, and materials and tool requirements. The work order also provides the means of reporting what was actually done, by whom, how long it took, when it was done, whether or not the work was completed, if there is callback work required, production time lost, and comments about the work. Many systems also allow for input of codes to specify the cause of the problem and the resulting effect, such as motor overheating, resulting in bearing replacement.

Most systems allow multiple types of work orders. Work order types may go by different names, but the basic types are: PM; project; emergency; miscellaneous or unplanned; corrective; and repetitive. Usually, work orders are originated as work requests. The work request may be a paper document that is originated by anyone who wants to request maintenance action or the work request may be input directly into the CMMS by the requester via a computer terminal. With direct input, some control can be built into the process. For example, when the equipment/asset ID is input, the CMMS will validate the equipment/asset ID to the equipment/asset file and if valid, will automatically add the description, cost center, and physical location to the work request enabling the requester to verify that this is the equipment/asset item they want serviced.

The planner has the opportunity to review the work request before converting it into a work order. With most systems, the planner can change the equipment/asset ID if the requester input the wrong ID. The requester may have selected an item in the hierarchy above or below the actual item to be serviced. The planner may also determine that the request needs to be broken down into multiple work orders. Many systems allow a single work order to be broken down into multiple steps or tasks. Each step or task may be for a specific skill or craft or for a specific work activity to be performed. Depending upon the CMMS, steps or tasks may contain multiple skills or crafts and each step or task may be for a different equipment/asset ID. The costs charged to each step or task will roll up to the work order. Normally, the work order cannot be closed until all of its steps or tasks are closed.

The information maintained on the work order file will vary between CMMS systems and, within a CMMS, may vary between the types of work order. At the completion of the work, actual hours worked by craft/skill, actual parts/materials used, and completion comments will have been included.

Most PM work orders are created automatically by the CMMS. When they are created is determined by their execution frequency. This frequency may be time based, cycle based, or conditioned based. This process is performed by the system copying a standard PM plan into a PM work order. Once copied, the work order will be placed automatically in the work order backlog. Repetitive work orders are similar to PM work orders (their plans are often maintained on the same file as PM plans) except repetitive plans are manually copied into work orders on demand. This type of work order is most often used for work such as equipment/asset rebuilds or overhauls.

Preventive Maintenance Plans.   Preventive maintenance plans contain information very similar to a work order. These are work plans that are associated to equipment/asset records and have a defined frequency for when they are to be executed. In most systems, one plan may be associated with multiple, like equipment/asset records with each association having a unique frequency or trigger point. Normally the PM plan is automatically copied into a work order when it is time to execute the PM or at some defined lead time. Once the PM plan is copied into a work order, the PM work order is added to the work order backlog and is tracked and executed like other work orders.

Repetitive Maintenance Plans.   Repetitive maintenance plans are identical to PM plans except there is no associated execution frequency. Often, repetitive maintenance plans are stored on the PM plan file. These plans are manually copied into work orders as required. These plans are used for such maintenance jobs as rebuilds and overhauls or any repetitive work that does not require a fixed maintenance schedule. A repetitive maintenance plan may be used for many similar equipment/asset items. An example of a repetitive maintenance plan would be one to rebuild a certain class of motor. When it is time to rebuild one of the motors, the plan is selected from the file and copied into a work order. The work order may then be modified for this specific rebuild.

Cost Accounting Data

A major advantage of a CMMS is its ability to capture and retain cost accounting data. For example, labor hours and cost, inventory/material quantity used and cost, contract cost, and miscellaneous costs are automatically charged to the cost center, area, and department associated with the equipment/asset for which the work order was written. The system stores these data automatically on the equipment history file and can, at the same time, automatically pass them to your general ledger file. By using the equipment/asset hierarchy for cost roll up, the cost data may be passed to the general ledger at a level that is more useful or meaningful to the accounting organization. For example, accounting may want to receive, as a single entry, total labor or material cost for a department, line, or process. The CMMS eliminates the need for accounting to consolidate the records after they are received.

Work Order History

Work order or equipment/asset history data is the heart of the CMMS when it comes to analyzing how maintenance is meeting their goals. When a work order is closed, it is automatically stored on a history file. Stored work orders can be retrieved by either the work order identification or the equipment/asset identification for which the work order was written. Summary data of skill/craft hours expended as well as labor, material, and other cost should also be provided for each equipment/asset item. A summary of all hours and costs for an area, department, cost center, and so on should also be available. These data should be shown as month-to-date, year-to-date, and life-to-date. With this information, maintenance personnel can determine actual expenditures versus budget for any period in the physical year. This information may also be used as the basis for planning the coming year budget.

Trend analysis can be performed on historical data. If cause and effect codes are used on all work orders, the CMMS should provide the means of locating and reporting all closed work orders in history for an equipment/asset type that has the same cause and/or effect code. From this information, trends by cause can be determined. Repetitive problems for a specific equipment/asset item can also be determined.

From historical data, mean-time-between failure and mean-time-to repair can be determined. It may also be determined when it is more cost effective to rebuild or replace an item than it is to continue to maintain it. By using equipment/asset hierarchies, you can determine where an item was every time maintenance was performed on it as well as what items were at a particular location when maintenance was performed at the location. This is because most systems that allow for hierarchies maintain work order history not only for the equipment/asset item but, also for the parent of the item. This information is invaluable for moveable equipment/assets. Let's use as an example a motor or type of motor experiencing high failure rates. With the CMMS, the first search would be the history of one of the motors to see where it has been and its failure history. If it has a high occurrence of failure at a particular location, look at the history of all motors that have been at that location. If all or most of the motors failed for similar reasons at that location, the problem is likely not with the motors but with the location. It could be an alignment problem causing excessive vibration. If, on the other hand, one motor is failing no matter where it has been located, the problem is with the motor.

Many systems allow you to copy a closed work order from history into a new work order. This is a true time saver for planners. Once copied, the new work order data can be modified to meet the requirements of the specific job to be performed. Locating the work order in history may be done by searching on cause code for an equipment/asset type, searching work order descriptions, or searching history for the equipment/asset item or a similar item for which the new work order is to be written.

Craft/Skill Data

Most systems provide an employee file where data about each employee who can charge to a work order is maintained. Basic data on this file should include the following data: the employee ID (normally badge number); craft/skill code(s); and an hourly rate(s). The system should accommodate multiskilled crafts with multirates. Some systems allow storing additional information such as home address, home telephone, emergency contact, training/educational history, accident history, promotion, pay-raise history; and more. This file is primarily used by the CMMS to obtain the actual hourly rate to be charged to the work order for a specific employee. In many systems, the craft/skill code is used when planning the work order. Actual hours charged are by employee ID with the rate being for that employee.

Purchase Requisitions

The purchase requisitions file is where all requests for the replenishment of maintenance stores and direct-buy items are maintained. There are basically two ways for a purchase requisition to be created in a CMMS. The first is the automatic creation of a requisition by the system to replenish stocked inventory. Each CMMS may perform this function in a somewhat unique way but the basics are that a requisition is created when the balance on hand or the available quantity of an inventory item reaches its reorder point. The information contained on a requisition record may also differ for each CMMS. Examples of what might be on the requisition record include: part number or ID, description; quantity to requisition; a recommended vendor; vendor part number or ID; lead time, and priority. Some systems will have the ability to automatically change the requisition quantity before the requisition record is converted to a purchase order. These updates are based on changes to the inventory record on hand or available quantity. Inventory record changes are the result of returns to inventory, cancellation of a work order, an inventory adjustment, or additional requirements for the item. Once the requisition record is moved to a purchase order, additional updates to requisition are usually not permitted.

The second method of creating a requisition is to manually input it using the CMMS requisition entry function. Manual requisitions are created for the purchase of direct-buy, non-stock materials and services. They may also be created for stock items when it is known that an above normal quantity will be required such as for a project. Some systems allow stock inventory items to be flagged when they are not to be included in the automatic reorder process. For example, these would include seasonal items. Flagged items will require a manually created requisition.

In most systems, requisitions may be reviewed and updated until they are moved or added to a purchase order. It is beneficial that the requisition records remain intact, at least until the required parts or materials are received. The requester, or stores personnel in the case of automatic requisitions, may need to access a record for information. It is also desirable that the CMMS refers the purchase order number to the requisition once the purchase order is created. This allows interested personnel to find critical information about the order without having to contact purchasing personnel.

Purchase Orders

The purchase order file contains information about open purchase orders. This file, or an associated file, will maintain a record of each closed purchase order and each associated line item. The historical records should include receipt information such as a date(s) received, quantities received, and whether or not an overage or short quantity was accepted for closing the record. In many systems, the purchase order cannot be closed until all line items are closed. The purchase order may not be considered complete until after the invoice has been either approved for payment or paid.

Purchase orders are normally created from approved purchase requisitions. How this process is executed varies between systems. Most systems do allow selection and consolidation of requisitions, by vendor, into a single purchase order or into groups of purchase orders by vendor and commodity code or other selection criteria. A cross reference of the purchase order number to the requisition should automatically be created so that the requester can look up the requisition and determine if a purchase order has been created and, if so, what is its number. The creation of purchase orders without a requisition may also be permitted.

Parts and materials are received, on line, against the purchase order. The system should allow partial receipts against a line item on the purchase order. It should also allow for overage receipts, within controlled limits. These limits may be defined as not to exceed a percentage of the quantity ordered or a specific dollar amount. The system should also allow for closing a line item when the amount received is less than the amount ordered. Normally, all line items on a purchase order must be closed before the purchase order may be closed. The system should automatically close the purchase order when all of its line items have been closed.

If the requisition/purchase order was for a direct-buy, non-stock item, and there is an associated work order or project number, that number should automatically be added to the purchase order record. Since the associated number will display the line item on the receiving function, receiving personnel will know that this material likely will not require warehousing and that they should notify the requester of the receipt.

It is common for a company to use a purchase order system other than the one provided by the CMMS. When this happens, it is desirable to have an interface or integration of the two systems. This allows the requisitions created in the CMMS to be passed to the purchasing system and information about the purchase order to be passed back to the requisition on the CMMS. Also, when materials are received, they are normally received via the purchase order system. The CMMS inventory records must be updated with the receipt data. This may be done either manually or with an interface or integration, automatically.

WHO USES A CMMS AND HOW

There may be a misconception that maintenance personnel are the only users of a CMMS. While the maintenance organization is the primary user, many other plant organizations may benefit from access to information that is available within the CMMS. These organizations include engineering, production, inventory control, purchasing, accounting/finance, and executive management. How the personnel in each of these organizations use the CMMS may differ from plant to plant.

Maintenance

As the name implies, a computerized maintenance management system or computer-based managed maintenance system was originally designed and developed for the maintenance organization. Over the years, functionality was added that made the system quite meaningful to organizations other than maintenance, but the basic system is still a maintenance tool. The software programs and associated data bases provide the means to acquire, store, manage, and retrieve the myriad of data needed to effectively utilize all of the maintenance resources. As a minimum, the functions performed by maintenance personnel using the CMMS are work order initiation, PM planning, work order and resource scheduling, and so on.

Work Order Initiation. Work orders may be initiated several different ways depending upon the CMMS and the policies and procedures of the CMMS user. Initiation may begin with a written request, usually on a preprinted form. This method is often a carryover from pre-CMMS days. The request must be input to the CMMS by someone, usually a maintenance clerk or a planner. The advantage of the written request is that it can be reviewed and either approved for entry or rejected before entry to the CMMS. The disadvantage is the request, in effect, has been written twice. First the requester must prepare a paper record and then a clerk must enter the data into the CMMS.

A second method is the telephone call-in request. This method is also a carryover from pre-CMMS days. There are very few advantages to the call-in method except for an emergency request. The disadvantages are many and often costly. The person receiving the call must either write down the information on paper or key it directly into the CMMS as they are receiving it. Untold errors can occur through misinterpretation and misunderstanding. Taking the calls is frustrating to the receiver and the errors that occur are frustrating to the requester. Productive maintenance hours may be lost because craft personnel are sent to the wrong location, have the wrong parts, are prepared to repair the wrong problem, and the like. At least one CMMS vendor is providing the capability of telephone requests using a touch-tone telephone. The original application was for hotel type maintenance where a room number could be keyed in along with a problem code such as three for TV.

A third method is for the requester to input the request directly into the CMMS. The advantages of this method are the request is only written once, the software will assist the requester and it is the least time consuming for the maintenance organization. Software assistance means that the requester normally enters only the equipment/asset ID, a description of the problem, and his or her ID or badge number. The CMMS will validate the equipment/asset ID and, if valid, will display on the request screen such information as equipment/asset description, location, area, department, cost center, and the like. This allows the requester to verify that the request is for the correct equipment/asset. If the equipment/asset ID entered is invalid, the CMMS will immediately notify the requester. Some systems have included problem description tables associated with specific equipment/asset types. The requester selects one of the codes to define the problem. This makes entry easier for the requester and standardizes problem descriptions for maintenance. It also eliminates the too often used description of "broken." Requests entered directly by the requester normally do not become actual work orders until after they are reviewed by maintenance personnel. Requests may be approved, disapproved, or modified by maintenance personnel before becoming a work order. The disadvantages to this method are terminals must be easily accessible to anyone who can input a request and all personnel who can input a request must be trained to do so.

Automatic creation of PM work orders by the CMMS is a fourth method. Preventive maintenance plans that define what is to be done, parts or materials required, craft/skill required and other relevant information, are created and stored on a CMMS file. Each plan is tied to one or more equipment/asset ID for which it is to be performed. A frequency or execution schedule is defined for each PM plan and equipment/asset relationship. When the frequency or execution schedule is triggered, the CMMS will automatically copy the PM plan into a work order. The difference with this method is that a work order is created and not a work request. Repetitive work plans are very similar to PM plans in that the plan has been written and stored in the system. These plans are used for such tasks as rebuilds and overhauls. When a work order is required, the repetitive plan is manually copied, via the CMMS, into a work order.

Work Order Planning. Work order planning is the task of defining what resources are required to perform the job and what instructions or procedures are to be followed when executing the job. Resources include labor, parts, materials, tools, and contracts. The process of planning usually begins with a work request that has been entered into the system. The planner may review the request for accuracy and may make changes as required. It may also be necessary for the planner to physically inspect the equipment/asset to be worked on to determine requirements for the job. The planner determines which craft/skills are required and adds them to the plan. Normally, estimates of the time required for each assigned craft/skill are also included. In many systems, only the craft/skill type (welder) is planned, not the specific individual who will perform the work. Selection of specific individuals is left up to the maintenance supervisor or foreman when the job is assigned. Some systems allow only one craft/skill type per work order. Other systems allow multiple craft/skill types while others allow the planner to break the work order down into steps or tasks. Depending upon the CMMS, each step or task may allow only on craft/skill or each may allow multiples. There are advantages to having multiple steps or tasks. It may be that the job is going to cover a long duration and each step or task can be for a specific segment of the total job. Steps or tasks can usually be sequenced in the order they should be performed even though they may have been planned in a different sequence. If contract labor is to be used along with in-house labor, separate steps or tasks can be used for the contractors.

Planned instructions include specific steps or actions to be performed such as standard operating procedures, safety procedures, lock-out and tag procedures, and possibly sign off requirements. Many systems provide the software to allow retrieval of documents such as procedures and drawings from other systems where these documents are maintained. Required documents may be retrieved and printed with the work order plan.

Work order planning is paramount to controlling maintenance performance and cost. Therefore, it is very important that we define a planner. A planner is a well-trained, intelligent, conscientious, highly motivated individual. Work order plans that are accurate and concise can result in incredible savings to the company. Therefore, planners are not clerks or "gofers" for the maintenance manager or the maintenance department. Their position is key to the success of the maintenance operation and should be equal to or very nearly equal the maintenance manager.

Preventive Maintenance Planning. Preventive maintenance planning is very similar to normal work order planning. The difference is that a PM plan is created once, stored in the CMMS, and is used many times. This plan will have an association link to an equipment/asset record or multiple equipment/asset records for which it is to be used. Each PM plan and equipment/asset record relationship will have an execution frequency or frequencies that govern when the plan becomes a work order. The CMMS will automatically create a PM work order from the PM plan when its frequency or interval is reached.

Another type of plan that is often stored with the PM plans is the repetitive work plan. This plan is very similar to the PM plan and is for jobs that are done repeatedly but not on a predefined frequency. An example is a plan to rebuild a particular type motor. The plan has a list of required parts, materials, tools, crafts/skills, and estimated times for each craft/skill. The plan is copied into a work order, as required, through a CMMS function. Both PM plans and repetitive work plans may be modified as required. Work orders created from these plans should also be modifiable.

Work Order and Resource Scheduling. Once the total planned hours for each craft/skills exceed the number of hours available in one work day, and that should happen the first day of planning, decisions will have to be made about scheduling the backlog. One of the major activities in which a CMMS provides assistance is scheduling work orders and their required resources. All open work orders are maintained on a file that is referred to as the work order backlog. Each work order will have indicators to be used in determining the schedule. The indicators include work order type, status and priority, equipment/asset criticality, and a requested completion date. There may be other indicators depending upon the CMMS but these are the basics. In addition to the indicators, the parts, materials, and tools are required to be available unless the work can begin without all or some of them. Most systems use a work order status code to define where a work order is in its cycle. There should be a status code indicating that all of the parts or materials planned for the work orders are not yet committed or reserved.

The CMMS should allow the scheduler to input criteria for the selection of work orders to be scheduled. These criteria include but may be not limited to, area, department, supervisor, craft/skill, and work order type. The scheduler should also have the flexibility of specifying sort or sequencing criteria that the selected work orders will be scheduled by. These criteria include but may not be limited to work order input date, requested completion date, work order priority, and work order status. From the input provided by the scheduler, the CMMS will select and sequence all work orders that meet the selection criteria and in the sequence specified. Selected work orders will be maintained on a schedule file. Work orders can be removed from the file (unscheduled) and the sequence can be rearranged to meet the specific requirements of maintenance and operations. Many systems allow schedules to be created for periods of up to 12 months. A schedule of longer than 1 week is probably of limited value as too many unknowns can occur to make the schedule invalid.

Once the work orders selected for the schedule are determined, the CMMS should compare the planned labor requirements for each work order in the schedule to the actual labor hours available to produce a workable schedule. The scheduler should be able to adjust the available labor hours and have the system recreate the final schedule. Adjusting labor hours means adding overtime hours, additional personnel (possibly from another group) or contract labor to the available hours.

Requisition of Non-Stock, Direct-Buy Parts and Services. A CMMS should allow for the creation of purchase requisitions for non-stock or direct-buy parts, materials, and services, online. Direct-buy requisitions usually follow the normal (stores stock replenishment) purchase requisition process and may require approval before a purchase order is written. A CMMS provides several advantages over manually requisitioning these items. First, the requisition may reference the work order for which the items are being purchased. This work order reference should carry through to the receipt process so that receiving personnel will know who requested the material. This helps ensure quick notification or delivery of materials to the requester upon receipt. Second, the requester can inquire the CMMS for the status of a requisition. Is it approved, has it been ordered, what is the purchase order number, who is the vendor, what is the expected receipt date? The answers to these questions should be available online. The ability to look up this information eliminates the need for the requester to call the buyer for information.

Analysis of Equipment/Asset Repair History. A key benefit of a CMMS is its ability to serve as a repository for large amounts of data. There are many files containing thousands of records that can be combined, sorted, and displayed or printed in a meaningful format. These displays and reports often provide critical information for analyzing equipment/asset repair history. Historical data includes what the problem was, where it occurred, when it occurred, what caused the problem, the resulting effect, corrective action taken, resources used, how long the repair took, and comments concerning these activities. The CMMS can quickly retrieve information from many different records, combine this information, perform any required calculations, sort the results, and present them in an organized format.

Analyzing history for a single equipment/asset can show breakdown trends, provide information for estimating future breakdowns and provide repair cost data for any period. With repair cost data, a decision can be made about whether to continue repairing the item, to overhaul or rebuild it, or replace it. Failure trends for a specific type of equipment/asset can also be produced. What may be determined is the majority of failures or types of failures are occurring with items from a particular manufacturer. Another trend that might be presented is that most failures are occurring at a specific location (could it be an operator problem?). The types of analysis that can be performed with a CMMS are basically limited only by the imagination.

Craft Utilization. Craft utilization is another major benefit available with a CMMS. This benefit alone may result in enough savings to pay for the implementation of the system. If your CMMS performs automated resource (labor) balancing or leveling, and you use it, you should be able to schedule all personnel full time on "wrench turning" activities. As discussed in the section Work Order and Resource Scheduling, the CMMS will automatically match all work orders on the schedule against available skill/craft hours and finalize the schedule with those work orders for which the required hours are available. If there are sufficient hours planned for each skill/craft, the system will schedule them at 100 percent unless you have specified that certain individuals are to be scheduled at a lesser percent. Most systems allow you to specify what percent to schedule each skill/craft.

If your system does not have automated resource balancing or leveling, you should still be able to accomplish this goal of maximum labor utilization. You accomplish it with excellent job planning and manual scheduling. Excellent job planning will eliminate or greatly reduce delays caused by starting a job only to find that required personnel, parts, tools or permits were not planned. When this happens, the result is usually personnel waiting, unproductively, for the unplanned items. Obviously, the more the jobs are planned, the more productive the personnel. To manually balance the available skill/craft hours to the planned hours on the schedule, requires sorting the schedule by skill/craft type and determining total planned hours for each type. Select as many work orders from the schedule as will equal all available hours for the next day. This becomes the final schedule for that day. Do this each day and you will achieve maximum utilization. When a job requires multiple skill/crafts and at different times, planning the work order in steps, if your CMMS allows this, can make scheduling the personnel easier to manage. You should be able to schedule each step independent of the other steps.

Budget Preparation and Tracking. Because the CMMS maintains all maintenance repair costs for past periods and can present these costs in many sequences or groupings, reliable information is available for preparing budgets. Cost data may be grouped for specific areas or departments and may be separated into labor, material, contract, and miscellaneous cost. These costs may be further separated by type of work performed such as PM, corrective maintenance, and projects. The CMMS and its myriad of data make estimating a future budget fairly routine.

The CMMS should also track actual costs against the budget on a daily, weekly, monthly, or user-defined basis. For PM plans that are executed on a calendar basis, the CMMS should provide a look-ahead capability for determining future labor and material requirements, by week or month, for at least a 1-year period. With this capability, PM schedules can be adjusted to balance labor requirements over specific periods and part and material requirements can be determined by week or month, for up to a year. These factors may not seem important but, the planning capability they provide is extremely cost effective.

Engineering

Engineers may use CMMS to plan projects just as maintenance planners plan work orders. They may also use the CMMS to store and retrieve data on equipment specifications, drawing references, and modifications to equipment/assets. They will be able to quickly and accurately identify and locate identical equipment/assets throughout the facility. This is very important when an engineering change is to affect all similar equipment/asset items.

Project Planning and Tracking. Engineering personnel may use the CMMS for planning any size project. One or more work orders may be required depending upon the size of the project. Separate work orders may be used for each project activity or, if allowed by the CMMS, multiple steps may be planned on a single work order. Project work orders normally have a field for assigning a project number or identification. Using the project identifier field as the sort criteria, all work orders associated with a project may be selected as a group for viewing or printing.

One advantage the CMMS provides through the use of project work orders is it's ability to group and display or print all planned resource requirements and their cost. As work on the project progresses and charges are made to the work orders, comparisons of actual labor hours and costs expended to planned hours and cost will show if the project is within, below, or exceeding the project plan. Contract labor planned on work orders allow tracking both actual and committed hours and cost. If the purchasing module is integrated with the maintenance module, purchase order commitments may also be tracked through the CMMS.

Review Equipment/Asset Specifications. The CMMS should provide for the storage and retrieval of multiple specifications for each equipment/asset record. Many systems allow one set of specifications to be associated with all identical equipment/asset records. Specifications are usually in the form of user-defined codes with descriptive text for each code. Specifications may be maintained on the equipment/asset records, and/or on a separate file or table with a link to the equipment/asset record. Specifications provide several benefits to engineering. If a specification is to be modified for all similar equipment/asset items, one update to the specification table or file will be all that is required. Not only is this quick, it also ensures that no equipment/asset item was overlooked. Engineers may also use these specification data when purchasing a new equipment/asset item that is to be identical to one already owned.

Using specification data provides one of the best ways to locate all identical equipment/asset items on the file. For example, to locate all GE motors, AC, 50 hp, 30 A, specific frame size, cycles, volts, and so on, key these specification values into the CMMS search function and it should return a display or report of all matching items. It should also show where each item is physically located.

Equipment/Asset Modification History. Often equipment/asset items are modified to meet specific needs. If a CMMS work order is used to make these modifications, a history of the modification will be maintained. The history records will include what was modified, when, by whom, and why the modification was required. If the same modification is to be done to all identical or like items, the CMMS provides a very quick, accurate way of locating and grouping these items. This was described under the section *Review Equipment/Asset Specifications*.

Production

In many plants and facilities where total productive maintenance (TPM) or other team concepts are in use, production personnel may use the CMMS as frequently as the maintenance personnel. Where team concepts are not in use, production personnel may still use the CMMS to inquire about the status of a work request without having to contact maintenance. There are several other areas where the CMMS may be beneficial to production personnel.

Downtime Scheduling. Production personnel can use the CMMS to inquire on all open work orders for an equipment/asset item, line, or process that is to be taken out of service. The inquiry will show all work orders, both scheduled and unscheduled, that have been planned for the equipment that is to be taken out of service. Maintenance may then be notified that the equipment will be available for servicing during this production downtime. The benefit to production is that routine work may be scheduled and performed when it is most convenient. When an emergency breakdown occurs, the backlog of work for the line or process can be quickly viewed to see what routine work can be performed during the unplanned stoppage. Production personnel must take the initiative to review the CMMS work order backlog when a process or equipment/asset item is to be out of service but, the benefit will definitely justify the effort.

Repair Request Backlog. The CMMS provides production personnel the means of inquiring on the status of work requests or work orders without having to contact maintenance for the information. Usually, when people have the ability to look for themselves, they have a higher level of satisfaction with the answer. This results in improved communication which leads to improved relations between maintenance and production. Many companies are doing away with paper work requests and call in requests. The standards are now work requests entered directly through a computer terminal by the requester. The benefits of this method can be tremendous. First, paper work is greatly reduced.

Second, the work request will not be lost between the requester and maintenance. Third, when personnel input a request to the computer, they often have a more secure feeling that it is in the computer and will be attended to. Fourth, eliminating or greatly reducing call-in service gives maintenance managers more time on the floor with their personnel

Equipment/Asset Repair History by Cause and Effect. Too often maintenance personnel are labeled with doing poor or incomplete work because of repetitive breakdowns. If a *cause code* for the failure is input to the work order at completion of the work, it will be possible to review the history of why an item is failing. It is possible that some failures were the result of operator error. Cause codes may also show that the equipment in question is not suited for the job it is required to do, that is, a motor that is insufficient to pull the load placed on it. By reviewing the *effect codes* associated with each cause code, they can also see how serious the results of each failure were. It is advisable to provide production personnel the ability to utilize the CMMS for purposes of analyzing failures for themselves. Give them as much opportunity to use the CMMS as they are willing to take and encourage them to use it.

Inventory Control

Inventory control, storeroom, and receiving personnel are normally active users of a CMMS. It is through the CMMS that they receive and issue parts, materials, and tools and adjust inventory balances. Additional uses include parts usage history, parts to equipment/asset cross reference, advance notice of parts requirements for planned work, and storage and retrieval of material safety data sheets.

Parts Usage History. A CMMS should only permit issues of parts and materials to a work order, an account number or some other control number. When this practice is followed, unauthorized and unaccounted issues are eliminated. However, no system will ever totally eliminate the casual issues that occur from time to time. When all parts are issued through the CMMS, a history record of each issue will be created. Each history record should contain the quantity issued as well as the work order and/or account number against which the issue was made. This historical information is used to determine part usage trends including abnormal issues and inactivity for any period. By tracking tool issues, inventory personnel will have a record of whom the tool was issued to, when it was issued and when it is to be returned.

Parts to Equipment/Asset Cross Reference. Often referred to as the where used function, the part to equipment/asset cross reference is the inverse of the equipment/asset bill of material or parts lists. This means that a cross reference inquiry of a part number should result in a list of all equipment/asset items that use the part. The cross reference allows inventory control personnel to make critical, cost-saving decisions. One relates to equipment/assets that are being permanently taken out of service. Inventory control personnel can print the equipment/asset bill of materials. For each item on the bill of materials, they can do a cross reference inquiry to determine other equipment/asset items using those parts. From the results of this inquiry, a decision can be made about eliminating a part from inventory or adjusting the reorder point and reorder quantity. When inventory can be reduced, savings result.

A second use of the cross reference is being able to locate a part that is currently out of stock but, is needed in an emergency. The cross reference may show that the needed part is on an equipment/asset item currently out of service or being used in a noncritical location. The part may be removed from this equipment/asset item for use in an emergency.

Advance Notice of Parts Requirements for Planned Work. In most systems, planning a part on a work order will result in an automatic update to the quantity required field on the corresponding inventory record. This field is the cumulative total quantity of planned work orders. At some point in the work order planning and scheduling process, such as approval of the work order, the quantity required will be added to a field referred to as quantity reserved or quantity committed. Either the required quantity or the reserved/committed quantity, depending upon the CMMS, will update a system calculated value referred to as quantity available. Quantity available is, in effect, the balance on hand minus the required or reserved/committed quantity. If a part is requested by a means other than a work order, such as an over-the-counter unplanned request, stores personnel can immediately determine if the part is available for issue.

The CMMS should print pick lists or pick tickets corresponding to the planned parts on a work order. The pick list will show, for each part planned, the part number, description, quantity planned, and location of the part in the store room. By having the pick list before the job is to be started, parts may be picked and ready for the maintenance personnel to pick up or may be delivered to the job site.

Automatic Requisitioning of Parts to Meet Reorder/Stocking Requirements. The CMMS will automatically create a reorder requisition to replenish inventory. In some systems, the trigger point for creating the requisition is the quantity available. In other systems, it is the balance on hand quantity. When the trigger quantity is equal to or, in some systems, falls below the defined reorder point, the requisition is created. It is important that you know which trigger point your CMMS uses because that will determine where to set the reorder point. If balance on hand is used, the reorder is based on the actual quantity of parts remaining. If quantity available is used, the reorder is based on actual parts remaining fewer parts required or reserved/committed. In the second case, the reorder point will probably be set lower than if the reorder is based on balance on hand.

Many systems allow selected parts to be flagged so an automatic requisition will never occur while the flag is in place. Requisitions for these parts will be created through some other means, probably manually. Seasonal parts such as those for snow removal equipment may fall into this category.

Work Order to Purchase Order Cross Reference for Direct-Buy Items. Often, direct-buy, non-stocked items are requisitioned for a specific work order. They are planned on the work order and the requisition is either created automatically by the CMMS or manually created depending upon the functionality of the system. In either case, the work order number is carried forward to the requisition and subsequently, to the purchase order. When the items are received, the receiving personnel will automatically have the work order number for which the item was purchased. With this information, they can inquire the CMMS to determine the requester so that immediate notification of receipt can be given. In many systems, the direct-buy item for a work order does not have to be stored or issued. It is, in effect, issued to the work order when received, thus saving stores personnel valuable time.

Storage and Retrieval of Material Safety Data Sheets. The CMMS should provide for the storage, maintenance and retrieval of material safety data sheets (MSDS) for parts and materials requiring them. It should also allow for either automatic or manually selected printing of MSDS when a part is issued.

Purchasing

Many companies have a purchasing system in place before they buy a CMMS. Whether to use the purchasing functions provided with the CMMS or another purchasing system must be resolved before the CMMS is implemented. Often, corporate policy mandates that the existing system be retained. The CMMS purchasing module will be fully integrated with the CMMS inventory module.

If you must use another purchasing system, integration with or an interface to the CMMS is highly desirable. Integration eliminates the need for any double entry of data and ensures data integrity throughout the process. Other benefits of an integrated system are automatic requisition of stores stock inventory, consolidation of requisitions for same vendor to single purchase order, and receipts against the purchase order.

Automatic Requisition of Stores Stock Inventory. How the CMMS can automatically create requisitions for replenishment of stores inventory based on reorder points was described in the section titled Inventory Control. These requisitions, as well as manually input requisitions, are stored on a file for purchasing personnel to review, modify, and transfer the purchasing system as required. Purchasing may, for reasons best known to the buyer, change the vendor, order quantity, or unit price before transferring the requisition to a purchase order. The ability to perform these tasks using one terminal and basically one system is possible with a fully integrated system.

Consolidation of Requisitions for Same Vendor to Single Purchase Order. Normally the requisitions are stored as individual records on the requisition file. Each requisition should identify the recommended or preferred vendor from whom the item is to be purchased. Purchasing personnel should be able to request all requisitions for a specific vendor be transferred to a single purchase order. Additional selection criteria, such as commodity code, may also be used to select all requisitions for a specific vendor and specific commodity code or codes for transfer to a purchase order.

Receipts Against the Purchase Order. Items are normally received in the system against their purchase order. The system should allow partial receipts with a back order or partial receipt and closure of an item. Overage receipts within predefined maximum levels should also be allowed. These levels may be either a set quantity or dollar value or a percent of the purchase order value or quantity. When all items on the purchase order have been received and completed, the system should automatically close the purchase order. Some systems allow the purchase order to be reopened if additional items are received after the purchase order was closed.

When stock items are received, the inventory file for that item should automatically be updated with the quantity received and the purchase price of the item. How the purchase price updates the inventory record will depend upon the CMMS in use and your inventory accounting methods. Accounting methods may include average unit cost, first-in first-out (FIFO), last-in first-out (LIFO) or the item may be expensed at receipt. Automatic updates to the inventory records will not occur without an integrated system. Separate systems mean double entries, first for receiving, then for inventory updates.

For direct-buy purchases for a work order, the system may charge the work order directly at receipt and, in effect, automatically issue the item to the work order. This eliminates the need for inventory personnel having to store and issue these items.

Accounting/Finance

The CMMS can provide accounting and finance with accurate maintenance cost data in a consolidated format. It is very important that accounting or finance personnel be included in the implementation planning process so that early decisions can be made as to how costs are to be accumulated within the CMMS. They can help determine the hierarchy structures and cost roll up levels. Their input is also needed in determining cost center and account codes. Benefits of the CMMS for accounting/finance are automatic costs allocation, cost history evaluation, and ISO 9000 compliance.

Automatic Costs Allocation. Cost input to a work order is automatically allocated to the cost center(s) and general ledger account(s) associated with the equipment/asset record the work order was written against. In some systems, work orders may be written directly against a cost center in place of an equipment/asset item. Using the equipment hierarchy and cost roll up features of most systems and the multiple work order types, accounting should be able to receive costs just about any way they require. Examples are by area, department, process, line, project, or any combination of these. The work order also maintains the date that costs were input to the work order and the date the work order was completed. All of the information required by accounting can easily and accurately be transferred to the general ledger system in the format required by accounting. A simple interface program provides the means of transferring these data.

Cost History Evaluation. Accounting personnel should have direct, online access to the CMMS so they can review maintenance cost history data. They can review cost history for labor, material and other costs by area, cost center, department, process, line, project, and individual equipment/asset items. They can also review cost summaries for specific time periods that correspond to their accounting periods.

Executive Management

Executive and upper level managers will be able to retrieve valuable information from the CMMS, when they need it and in summary form. Types of inquiries that upper management would be interested in include budget tracking and ISO 9000 compliance.

Budget Tracking. Included in the maintenance information that should be of interest to executive management is budget tracking. The CMMS can provide budgeted cost versus actual expenditures for any time period quickly and in summary form. For example, they may select the data to review by cost center, area, department, process/line, project or individual equipment/asset item and either view the results online or print them. The data to review can be selected for specific time periods such as year to date or month to date.

Managers may also review, online, areas or equipment/asset items that are experiencing exceptionally high maintenance costs. They can request a listing of all work orders completed during the period in question. A review of the selected work orders will pinpoint reasons for major expenditures or an abnormal number of work orders.

ISO 9000 Compliance. The CMMS should provide for the storage, maintenance, and retrieval of many types of documentation including standard operating procedures, standard maintenance procedures, material safety data sheets, and drawings. The CMMS also maintains a complete history of all work performed. ISO 9000 requires standardization of documentation and compliance and the CMMS can provide it along with complete audit trails. ISO compliance is usually a concern to upper managers.

WHAT A CMMS WILL DO

Maintain, Sort, Summarize, and Display Data

One of the primary functions of a CMMS is to maintain, sort, summarize, and display data for personnel to review and make decisions. The computer and the programs that make up a CMMS can perform these functions faster, more accurately and in greater volumes than humans can even attempt to do manually. Changes made to a data field on one CMMS file can automatically be changed on every other file where that data field is maintained. This ensures that accurate data is always available. Most systems provide many different selection or sort fields for organizing data that is to be displayed or printed. You can tell the system what information you want to see and how you want to see it. The CMMS will also provide input selection fields for producing summary displays or reports. It is very uncommon to lose or misplace data that has been input to the CMMS files. System back up precautions should, of course, be taken for safety's sake.

Automate and Control a Reliable PM Program

A good PM program is a major item for improving maintenance performance, equipment reliability and reducing maintenance cost. Once the PM plans and schedules are developed in the CMMS, the system will automatically create and schedule PM work orders. The system will automatically provide notifications of PM work orders not completed within their scheduled cycle.

Automate and Control a Reliable Inventory Replenishment Program

The CMMS will automatically create purchase requisitions for stock inventory items based on either defined or calculated reorder points. Systems may differ in how they determine when a reorder point is reached, but all are very consistent in creating the purchase requisition once the point is reached. Some systems may base the reorder point on the actual balance-on-hand quantity. Others may base it on total work order requirement quantities for the part while others use total work order commitment quantities. Many systems allow individual parts to be tagged as to which method to use or to never create an automatic requisition for that part. Whatever method is used, there is assurance that it will be consistent and accurate.

Provide Accurate Job Scheduling Based upon Resource Availability

Manually scheduling jobs to be performed, especially when there is a large backlog of work, can be difficult and tedious. It can also be biased and fall under the "squeaky wheel" syndrome. The CMMS scheduling program can schedule work orders based on such criteria as work order priority, equipment/asset criticality, work order type, requested completion date, and origination date of the work request. A good scheduling program will also schedule based on availability of resources. For example, it will have the ability to match the required hours, by craft or skill, of all work orders on the planned schedule against the available (unassigned) hours of each craft or skill for the schedule period. The availability of required parts and materials for each work order is also determined. For the schedule sequence established by priority, criticality, and so on the final schedule will be based on the availability of required labor, parts, and materials.

WHAT A CMMS WILL NOT DO

A CMMS will do many things for many people. It is a tool to be used for improving the way maintenance and other organizations store, manipulate, and retrieve data. It will also generate work orders based on defined trigger points, will schedule work orders, and will automatically reorder replacement parts. In other words, the CMMS can be set up to automatically "remember" and perform activities based on defined parameters. Contrary to what some people believe or want others to believe, the CMMS will not replace good maintenance practices and management. For example, the CMMS will not replace a maintenance manager, replace planners, assign work, and so on.

Replace a Maintenance Manager

Too often the maintenance manager is fighting fires. He can't get control of scheduling the work to be performed and the assignment of personnel to do the work. He is trying to determine what jobs need to be worked first, who is available to do the work, and who is most qualified for each job. A CMMS will maintain, in priority sequence, a backlog of all work to be done and can quickly and efficiently create a schedule for this backlog. The CMMS will take into account the availability of all resources required for a job and schedule based on availability. This activity alone should free up considerable time for the manager to spend on what should be the primary objective, managing personnel. The CMMS can accomplish many other tasks for the manager to free up time to more efficiently manage the organization.

Replace Planners

There have been many instances where the number of maintenance planners has increased as the result of a CMMS implementation. Where the objective is 90 percent planned work and the CMMS can make that possible, more planners may be required to reach the objective. Planning with a

CMMS becomes easier and more efficient because the CMMS contains so much valuable, well-organized information that is available to the planner. For example, the bill of materials for the equipment/asset to be serviced. From that bill of materials, the planner should be able to select, straight to the work order plan, the parts needed for the repair. Another example is the ability of the planner to copy a previously developed plan into a new plan and modify the new plan as required for this job. With a CMMS, the ability to plan quickly and accurately should increase planned work and decrease "fire fighting" to the point it may become necessary to increase the planning staff.

Assign Work

Normally, a CMMS will not assign work to specific individuals or groups, although some now provide the functionality to automatically assign individuals to work orders based on availability. In either case, the CMMS does provide vast amounts of information to enable the maintenance manager and/or scheduler to make decisions about the assignment of work. One example is the ability of the CMMS to display work order backlogs in several different sort sequences. Sort selection usually includes a combination of fields such as area, department, process, equipment/asset type, labor/craft requirements, work request date, requested completion date, work order category/type, and so on. The sorted backlogs should be available for viewing either online or as printed reports. With this capability, the maintenance manager should be able to make the best decision about which individuals are best suited for assignment to a work order. Many systems produce a recommended schedule based on dates, priorities, and skill/craft availability. With this schedule as a guideline, the maintenance manager or scheduler, working with the operations or production managers, should be able to efficiently assign all work to the most qualified personnel and schedule the work in the most beneficial sequence.

Bring Order to Chaos

Great care must be exercised in preparing for the selection and implementation of a CMMS. The very first step in the preparation process is to do a comprehensive evaluation and assessment of your current operation. This evaluation must go beyond the maintenance operation to include production, engineering, inventory control, accounting, purchasing, human resources, and information systems. The objective is to first determine what the current situation is in each of these areas and then determine what changes need to be made in order to reach your objectives. All of the above mentioned areas will be affected by your CMMS decision so they must be included in the evaluation. When you have determined which improvements are required, you may begin developing the specifications for the CMMS. Another reason for the evaluation is to determine what changes should be made in addition to what the CMMS can provide. To carry existing bad practices over to a CMMS will almost always result in compounding the very problems you are trying to solve.

Improve Equipment/Asset Reliability or Product Quality

CMMS will not improve equipment/asset reliability or product quality, nor will it decrease maintenance costs or reduce labor requirements. The CMMS is definitely the tool that will allow you to accomplish these goals, but it is only a tool. Along with a good predictive maintenance program, a CMMS is probably the best, most cost-effective tool that maintenance can use. The CMMS will enable maintenance to accomplish many cost-saving activities, and once implemented, should pay for itself in 18 to 24 months. You must set it up properly, train personnel in its use, and control the data that goes into it. If not properly controlled, the CMMS could actually increase maintenance and inventory costs.

WHY A CMMS FAILS

The term CMMS is usually translated as a *computerized maintenance management system.* Frankly, this is a misnomer. Most CMMS systems, as purchased and installed, are not *management* systems. While it is true that these systems manage vast amounts of data, they are not designed to *manage* the maintenance function. Many of the CMMS systems that are commercially available have limited *management* capability. As software developers, many vendors do not have a viable understanding of the management tasks required to achieve an effective maintenance organization. They are very good at developing software programs that will store massive amounts of data, manipulate the data, automate recurring tasks, and generate standard reports, but they do not provide the real management tools needed to have an effective maintenance organization.

Another restriction on the success of a CMMS is self-imposed. The infrastructure and work culture in many plants and facilities restrict effective management of the maintenance function no matter what improvements are attempted. The addition of a CMMS will have a limited effect on the ability of maintenance managers to improve conditions.

Finally, CMMS systems fail as a management tool because of the way the system is implemented. Poor initial planning results in misdirected resources, increased implementation time, loss of interest on the part of key personnel, and underfunding of the project. Too little training or training at the wrong time results in misunderstanding and confusion for the users. This leads to a lack of confidence in the CMMS with the result being a lack of use. Improper or insufficient data initially loaded to the system results in inadequate information available from the system. This too results in a loss of confidence and again a lack of use. Proper planning and implementation are the keys to a successful CMMS.

The failure rate for CMMS installation is extremely high and we are sure that some of you have already been involved in a failed attempt. For those who have tried, unsuccessfully, to implement a CMMS or are replacing existing systems, this book will provide the knowledge you need to ensure optimum performance from your CMMS.

A survey of failed CMMS implementations disclosed the dominant factors that prevented successful implementation.

Partial Implementation

A large percentage of attempted CMMS implementations fail because the CMMS is not fully implemented. Most companies lack the expertise required to fully implement a CMMS. Since their in-house personnel do not have a working knowledge of these programs or fail to fully understand the capabilities of the system, they fail to recognize all of the tasks that are required to directly or indirectly support the installed system. As a result, the project team cuts corners or only implements the minimum tasks that are absolutely required to install the basic hardware and software required to run the CMMS program. The project team fails to recognize all of the factors, including many non-maintenance issues, that are absolute requirements for successful implementation of a CMMS and effective maintenance management. It is estimated that, on average, only 30 percent of the modules of a CMMS are used and of these, only 30 percent of the functionality is used. The result is a 9 percent overall utilization of the CMMS.

Installation of a few computer terminals and a CMMS software package will not generate any change in the effectiveness of your maintenance organization. You *must* identify and resolve all factors that limit maintenance effectiveness.

Lack of Resources

Limited resources are a major cause of failure for *any* project. This failure is typically the result of either (1) poor planning or (2) lack of management/labor commitment. Generally, the failure results from the former. Most CMMS justification packages and project plans, if developed at all, fail to estimate the level of manpower and financial resources that will be required to fully implement the CMMS and correct all of the limiting factors that preclude effective maintenance management. As a result, the resources required to implement the project will not be available.

The second contributor to this problem is the lack of commitment from both corporate and line management. In many cases, this results from the failure to *sell* the program to all levels of plant personnel. It is imperative that all levels of plant management and hourly personnel buy into the program before implementation is begun. In order to buy in, they must first understand the purpose of the system, all the resource requirements for implementing and maintaining the system, and their part in the project. This requires a very thorough project plan.

Fragmentation of Effort

Simply stated, many plants do not apply effective project management to the implementation of their CMMS. Like all major projects, implementing a CMMS is a complex, long-term project that must have strong management and leadership. It is imperative that an experienced project manager be assigned to the project and be given the authority to accomplish timely completion of the project.

Internal politics, labor relations, and a variety of other factors contribute to the fragmentation of the CMMS implementation effort. The project plan must anticipate these problems and include effective means to limit their impact on the project schedule.

Staff Overload or Not Enough Staff

Most implementations attempt to use in-house personnel for most, if not all, of the tasks required to implement a CMMS. Implementing a CMMS often requires several man-years of effort. In most cases, plants do not have the extra resources required to accomplish a successful implementation. As a result, salaried and hourly personnel are asked to perform the implementation in addition to their regular duties. The resulting conflict between meeting production and maintenance goals and implementing a CMMS creates a total overload on all personnel involved. They become frustrated with the slow progress being made and in the system not meeting expectations. The normal result is that the CMMS implementation becomes the second priority and is never fully implemented. The solution may be to hire outside consultants who specialize in CMMS implementations. They can supplement the in-house team and at the same time, provide the leadership and knowledge that only experience can bring.

Inappropriate Expectations

Too many organizations expect that the implementation of CMMS hardware and software will automatically result in an effective maintenance organization. As we have discussed, this is absolutely not true. CMMS is a tool that will provide the information required to effectively manage the maintenance function, but it cannot overcome the myriad of other factors that preclude effectiveness. If not implemented properly and in a timely way, the CMMS may actually increase the ineffectiveness of a maintenance organization.

Lack of Behavioral Expectations

The inherent expectation that all employees will embrace the new CMMS system and the work culture change that are required to properly utilize this tool is a major contributor to CMMS failure. Without a radical change in the human factor, a CMMS system cannot provide expected benefits. The CMMS will definitely bring change to the organization and the way they conduct business. This is especially true if there was no formal system in place prior to the CMMS. There will now be structure to the way work is assigned and scheduled. Reporting on work completed will be required. Equipment and parts will have to be identified.

Treating Computers as Deliverables

Many organizations have a myopic view of CMMS implementation and never realize that the simple installation of computers, networks, and software is a small part of a CMMS implementation. Failure to fully implement all of the changes in work methods, procedures, organization, employee attitude, skills, and so on will prevent success.

Confrontation Instead of Collaboration

Almost every plant or facility has some level of internal politics that prevent effective coordination and cooperation among and within its functions. In the case of CMMS, the major adversarial relationships will develop between maintenance, information systems, procurement, finance, and production. Each of these organizations will be users of the CMMS system and each has its own agenda of features, implementation schedule, and desired results. During the planning phase and throughout implementation, there should be a team consisting of at least one representative from each affected organization. This team will resolve differences and ensure the implementation stays on track. A senior management person should act as arbitrator to resolve differences the team cannot resolve. This person must have final and absolute authority.

Poor Communications

Project management is the fundamental requirement of successful implementation. Clear, concise communication is an essential part of good project management. Too many projects lack a master project plan and schedule that clearly identify all tasks and the sequence in which they must be performed to meet the implementation schedule and budget. The lack of this master plan leads to poor communication, adversarial relationships, and slippage of both time line and budget.

Lack of Expertise

Many organizations do not have a staff with the experience and expertise to properly implement an effective CMMS. Typically, they will select a project manager from either the maintenance organization or information systems. In the former case, the maintenance manager will normally lack experience in (1) computer-based systems, (2) human behavior and motivation, (3) effective organization requirements and other skills that are fundamental requirements of success. In the latter case, an information systems manager should have the required computer-based systems knowledge, but may lack the other required skills. As stated earlier, implementation requires a team that will bring all the required skills and knowledge together into a cohesive unit. The team should be headed by a member of senior management who will work for the good of the total organization.

Reliance on Consultants

Many plants attempt to resolve the limited in-house knowledge by hiring a CMMS consultant to provide the expertise and experience to properly implement an effective maintenance management program and CMMS. While this approach is valid, extreme caution must be exercised in the selection process. One reason for CMMS failure can be directly attributed to poor leadership provided by an outside expert.

Verify the practical capabilities of the proposed consultant before you hire them. Just because an individual has written books, magazine articles and is on every CMMS conference program does not mean that he or she can provide the practical leadership that you need to implement a system. Talk to previous clients and verify that the consultant has a proven record of actual implementation. Do not use consultants from your CMMS vendor. This is especially true on a time and material contract. No matter how ethical the individual consultant may be, the conflict between his position as an employee of the CMMS vendor and his responsibilities to you as a client will prohibit cost-effective, successful implementation. A consultant must protect your interest and be absolutely committed to the implementation of the most effective system for you. Any conflicts with these goals will seriously limit your program.

Modification of the CMMS

Many organizations elect to modify the CMMS to match their existing business practices even before they have had an opportunity to see if changing some of their practices to meet the CMMS will be acceptable. In most cases, this is counterproductive. While some companies have an effective maintenance organization without a CMMS, the majority do not. If you elect to duplicate your normal business practices in the CMMS, there is a high probability that few benefits will be achieved.

In addition, modification of standard CMMS software can be very expensive and time consuming. Many CMMS vendors will gladly modify their software to meet your unique demands. A major portion of their income is generated by modifications. If you find the system that best meets your needs, there should be no reason for anything other than minor changes such as report formats.

Work Culture Restrictions

Fundamental management, philosophical, or procedural issues, not CMMS system issues, can impede the smooth implementation of, or transition to, a new CMMS. The system may fit your specifications perfectly, but if there is no internal agreement on how the tool can best be used, the system will fail to deliver the desired results. Radical changes are sometimes required to break the habits of the past. If these issues are not dealt with prior to implementing your new system, everyone will blame the CMMS for the continuation of your chronic maintenance problems.

In some cases, work culture issues can impact the specification of the system. For example, one common philosophical issue is what extent the machine operators will be involved in the maintenance of their equipment. This may not impact system requirements, because the system doesn't really care whether a maintenance or production worker does the maintenance, enters the data, or outputs the reports. However, the success or failures of the CMMS can rest with this single issue. Unless operators begin to take the care and maintenance of their equipment seriously, the maintenance workers will feel that it is a waste of time to keep filling out work orders for the same old problems. Operators will complain that the system is not improving the response rate of maintenance to their problems nor the quality of the repairs. Maintenance will insist that nobody looks at the reports from the CMMS to see that it is the same problem caused by operators who are poorly trained and don't care about the equipment. A "catch-22" situation will exist that will eventually cause failure of the CMMS.

In order to realize the potential benefits of implementing a new CMMS, the deep-rooted philosophical, management, and procedural issues must be identified and quickly resolved to the satisfaction of all stakeholders.

SECTION 3

ENGINEERING AND ANALYSIS TOOLS

CHAPTER 1
ECONOMICS OF RELIABILITY

Robert Fei
Managing Principal, Life Cycle Engineering, Inc., Charleston, S.C.

The practice of reliability and maintenance engineering involves selecting alternative designs, procedures, plans, and methods that consider time and economy restrictions in their implementation. The satisfaction of the engineer's sense of perfection does not necessarily assure the best alternative. Preferred rational approaches to the selection of alternatives are based on methods of engineering economy that are concerned only with alternatives that already have been established as technically feasible and providing economic analysis of their prospective differences. The word "prospective" indicates that the analysis looks into the future and cannot be viewed as an absolute prediction because of the always-present element of uncertainty associated with time, uncontrolled variations in the value of parameters considered, and imperfections in cost estimation methods. The need for engineering economy comes from the fact that engineers and managers do not work in an economic vacuum but are under strong impact from the processes that managers use to allocate limited resources to produce and distribute various products. Because engineers are basically trained as physical scientists, it is necessary repetitively to stress that economics, which studies these processes, is an empirical social science that provides us with the conceptual framework known as a *theory of choice*. Positive economics, as opposed to normative economics, which is outside our interest, is concerned with questions of facts to which assumptions relating theoretical constructs to real objects are added. Economic models then represent the purely logical aspects of these theories and serve as tools for decision making, behavior or consequence predictions, and testing or refutation of underlying assumptions and propositions based upon economic rather than physical consequences.

The managerial responsibility for the quality of decisions is usually discharged within a framework of

- Clear definition of alternatives.
- Identification of aspects common to all alternatives, which then become irrelevant.
- Establishing appropriate viewpoints and decision criteria.
- Considering consequences and their measurability.

Given the frequent conflicts of requirements, the different relative position of quality and reliability, as opposed to product performance, project schedule, cost, and their potential impact on business success, the economic aspects of reliability alternatives with their consequences should be studied in detail and well understood. Considering the power and availability of current methods of economic analysis, the principal difficulty in accomplishing this task is in the low availability of valid and detailed reliability cost data and lack of good reliability models. We just do not know well enough the reliability dependencies on design effort, testing, quality control, application environments,

3.4 ENGINEERING AND ANALYSIS TOOLS

and so on. Current trends in all major manufacturing industries, emphasizing the importance of quality and its associated attributes, present a significant driving force to alleviate this problem.

A word of caution is in order. Excessive use of reliability cost models can be a pedantic and fruitless exercise. Modeling works best in steady-state conditions, when the state of the art is not being changed. The less quantifiable aspects of industrial processes, for example, workers' attitudes, levels of standardization, sales practices, are usually left out of the models altogether. And we often fail to recognize that results of our logical arguments, applicable to models, might be irrelevant to the real situation. Experience and common sense help in deciding where to put the effort and how to set realistic expectations.

RELIABILITY AND VALUE

The task of economic science is to devise optimal ways of allocating scarce resources among competing proposals for their use. One of the schemes studied many times in detail, and historically proven as effective in free markets, is based on the price of a given product or service. The concept of price can represent buyers' and sellers' convictions, willingness to pay, personal preferences, efforts to obtain favors, and so on, but fundamentally, it refers to the value of the item concerned and its relationships to market situation (described in terms of supply and demand) and to manufacturing cost.

Relative Importance of Reliability, Price, and Performance

The wealth of data about customer behavior, values, beliefs, and attitudes often confirms reliability as the most important product quality attribute, and by that, its impact on the value in exchange, expressed by price, and value in use, expressed, for example, in terms of users' return on investment. Figure 1.1 illustrates two extreme cases. The decisions in the Apollo Space Program were dominated by reliability considerations (weight 0.7 on scale of 1.0) because of the severity of adverse consequences of a failure. Most of the purchase decisions in the area of consumer electronics consider product price much more important than reliability because of small impact of failures, protection

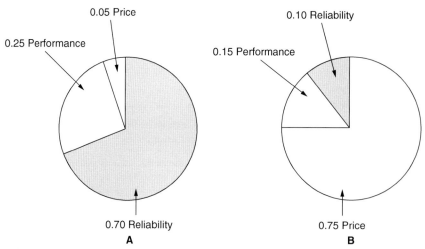

FIGURE 1.1 Examples of the relative importance of price, performance, and reliability: *A*. space program, *B*. consumer electronics.

by a warranty period, seller's goodwill, and so on. The knowledge of relative position of reliability versus other product characteristics, expressed in quantitative weight factors, is important for both formal and practical considerations and can significantly improve the rationality of many decision processes.

Reliability as a Capital Investment

Considering the relationship of reliability to value and its impact on price, product can be analyzed for its attractiveness as a capital investment convenient measure of this attractiveness is *return on investment* (ROI). Other measures of capital investment effectiveness can be easily derived from ROI.

The ROI without consideration of the time value of money is an approximation that assumes indefinite life of the project, and the payback method ignores the desired profit or "interest" on the investment. These approximations can be used for "quick-and-dirty" analysis of proposed investments in reliability programs to prevent or reduce the number of costly failures of the products. It is highly recommended that the more accurate methods involving the use of present worth or equivalent uniform annual payments and returns be used as described in Grant, Ireson, and Leavenworth cited in the bibliography of this section. Where the required rate of return on the investment is high (to compensate for the uncertainties of the future), these correct methods, as illustrated in the example below, can make great differences in the ROI and the payback period.

- *Payback period* is simply the reciprocal of the simple ROI and is a quick test that, however, ignores total cash flow over time and the time value of money.
- *Benefit/cost ratio* is the ROI multiplied by the expected years of useful life.
- The *net return* is calculated by multiplying the benefit/cost ratio by the investment cost, in our case the cost of a reliability improvement program, and subtracting from it again the investment cost.

Benefit/cost ratio and the net return can be discounted to reflect the time value of money, by multiplying the ROI by the present worth factor, which is found in standard interest tables.

Example. A reliability improvement program under consideration has an expected cost C_R, of $50,000 and will avoid annually $N = 250$ failures with average repair cost C_{rep} of $850 each for expected useful life of $L = 8$ years. Let us express the effectiveness of this investment in different measures, considering general and administrative overhead OH = 30%, profit coefficient $P = 10\%$, and 10% discounting schedule defining present worth factor for 8 years of service $F_{8'} = 5.335$.

Return on investment

$$\text{ROI} = \frac{NC_{rep}}{C_R(1+\text{OH})(1+P)} = \frac{250 \times 850}{50{,}000\,(1+0.30)(1+0.10)} = 2.97$$

Payback period

$$\text{PP} = \frac{1}{\text{ROI}} = \frac{1}{2.97} = 0.336 \text{ years}$$

Benefit/cost ratio

$$\text{B/C} = \text{ROI} \times L = 2.97 \times 8 = 23.76$$

Discounted benefit/cost ratio

$$\text{DB/C} = \text{ROI} \times F_8 = 2.97 \times 5.335 = 15.84$$

Net return

$$NR = (B/C \times C_R) - C_R = (23.76 \times \$50,000) - \$50,000 = \$1,138,000$$

Discounted net return

$$DNR = (DB/C \times C_R) - C_R = (15.84 \times \$50,000) - \$50,000 = \$742,000$$

All the terms used can be expanded to account for particulars of a given reliability program. With the aid of a computer, additional investigation is possible to assess sensitivities, impact of uncertainties, or compare alternative programs. Similar studies can be made to assess the economic impact of unreliability on product value via performance and availability degradation or increased cost of ownership.

Reliability Impact on Product Positioning

The communicated and perceived reliability levels have a strong impact on the total perceived value of the product and acceptability of its price. The clarity of the communication about product reliability depends objectively on the published data and subjectively on the position of the product reliability in the mind of a prospective buyer. If the market and competitive analyses confirm the importance of reliability in the customer's set of needs, then reliability can form a base for product positioning and can increase its perceived value by taking advantage of some long-term product-independent issues, such as previous product reliability, company image, competitive product weaknesses, and existing goodwill. Successful positioning allows for higher-profit pricing strategy and for advertising that is more effective.

RELIABILITY AND COST

The study of product reliability impact on cost usually runs into problems of technical nature: cost estimation, lack of data, analysis complexity, optimization methods sensitivity, and so on. A wide variety of economic tools exist to help in solving problems of reliability cost modeling.

Manufacturers' Viewpoint: Cost of Reliability

The concept of *cost of reliability* (COR), the total cost a manufacturer incurs during the design, manufacture, and warranty period of a product of a given reliability, can be developed around the generally accepted notion of *cost of quality* (COQ). The principles of COQ, established in the 1950s, have been verified and found valid in all segments of the manufacturing industry. COQ is applied to measure economic state of quality, to identify opportunities for quality improvement, to verify effectiveness, and to document impact of quality improvement programs. The accounting for COQ tries to identify all the cost items associated with defects in products and processes and then set them in contrast with the cost of doing and staying in business. The classic categorization of COQ applied to COR expresses as unique the costs associated with the following elements:

- *External failure.* Cost of unreliability during the warranty period, cost of spare parts inventories, cost of failure analysis, and so on
- *Internal failure.* Yield losses caused by reliability screens and tests, cost of failure-caused manufacturing equipment downtime, cost of redesign for reliability, and so on
- *Reliability appraisal.* Life testing, environmental ruggedness evaluation, abuse testing, failure data reporting and analysis, reliability modeling, and so on
- *Prevention.* Design for reliability, reliability standards and guidelines development, customer requirements research, product qualification, design reviews, reliability training, fault-tree analysis, failure modes, effects, and criticality analysis, and so on

Understanding these cost categories is a must for rational planning of reliability assurance resources, environmental and life-testing facilities, training programs, warranty policies, and other services needed for successful reliability programs.

For COR management, the cost information must be further restructured by products, process segments, or departments to identify major contributors and by that, opportunities for improvement. The identified cost levels are usually compared with some measure of revenue or value added to form ratios as management indexes. From the viewpoint of cost management, it is also important to recognize that prevention and appraisal costs are controllable by planning and budgetary mechanisms. Both types of failure cost are, on the other hand, expected or actual results of our inabilities to assure defect-free design, manufacturing, and distribution processes, and to control users' application conditions and environment.

There are also some dangers and pitfalls associated with the use of both COQ and COR. Managers often forget that COQ and COR are dependent variables reflecting successes and failures of the quality and reliability programs. By doing so, they run the risk of degrading reliability levels to achieve short-term cost savings. Other possible dangers are caused by the often difficult identification of defect cause, conflicts coming from cost charges transfer rules, preoccupation with reporting systems, tendencies toward perfectionism, and so on. To prevent these and other pitfalls, managerial prudence and knowledge are required in search for facts, understanding, realistic objectives, priorities with rational execution plans, and progress monitors.

The total COR can be developed similarly as a tool for managing cost and resources associated with a design for reliability, reliability manufacture, and warranty cost reflecting residual unreliability. Because of the different slopes of individual cost curves as functions of increased reliability, the applicability of the concept of cost optimum is self-evident.

In some industries, for example, semiconductor industry, enough experience-based information has been accumulated to rationally model dependencies of the final product reliability on the number of redesign cycles, levels of screening and testing or, in general, on the level of reliability assurance. This information confirms the intuitive expectation that increase in planned and controlled reliability assurance cost and activities significantly reduce unplanned failure cost. These models, combined with physical failure rate models and warranty cost calculations, allow optimization of the cost of reliability, and by that, allow rational planning of the project, manufacturing, and support resources very early in the design phase. Development of optimal cost versus reliability strategy starts with a simple formula for total cost of reliability:

$$COR_{total} = CRD + CRM + WC$$

where CRD is the cost of reliability design, modeled by an empirical relationship f_1 among the cost and number of design for reliability cycles, application environment stresses, complexity, and expected general level of quality expressed, for example, via quality coefficient π_Q. This coefficient, regularly used in MIL-HDBK-217, *Reliability Prediction of Electronic Equipment*, which contains reliability models, reflects the relative impact of reliability program actions on the base reliability defined by the physical nature of components used.

$$CRD = f_1(\pi_Q)$$

The *cost of reliability of manufacturing* (CRM) is a function f_2 of the effectiveness of manufacturing screens, expressed again by the values of the coefficient π_Q and associated fixed and variable costs:

$$CRM = f_2(\pi_Q)$$

The *warranty cost* (WC) is a function of the initial failure rate but also takes into account the effect of the bathtub curve, learning curve, and, of course, repair cost. These individual expressions allow us to study the total cost of reliability

$$COR_{total} = f_1(\pi_Q) + f_2(\pi_Q) + WC - F(\pi_Q)$$

3.8 ENGINEERING AND ANALYSIS TOOLS

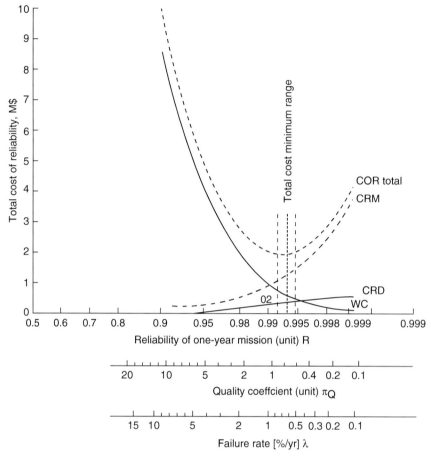

FIGURE 1.2 Optimum set of reliability cost curves.

as a function of the π_Q and allow us to search for an optimum (see Fig. 1.2) under the conditions of different warranty periods, learning factors, inflation rates, shapes of bathtub curves, and so on. The optimum conditions found must be then translated into resources levels and explained in technical terms of reliability engineering. Strategy for reliability, based on minimum total cost and its optimum apportioning among design, manufacturing, and warranty, is generally acceptable to all managers involved.

Users' Viewpoint: Life-Cycle Costing

The birth of the life-cycle costing (LCC) method is traceable to investigations by the Logistics Management Institute for the Assistant Secretary of Defense in the early 1960s. Primary concerns then were the consequences of changing vendors as a result of lower bid prices. Since that time, LCC has evolved into a costing discipline, a procurement technique, an acquisition consideration, and a design trade-off tool. LCC requires the identification of all potential system costs through all the phases of the product life cycle: conceptual, development, manufacturing, installation, operation, support, and retirement, which is obviously a very difficult task. During its history, LCC also often tended to degrade into a method for accumulating and reporting cost, but formalization and computerization of models used are changing LCC into an integrated part of the decision support analysis based on relevant, user-oriented concepts of utility and cost.

In application of the LCC methodology, the design trade-offs usually start with attempts to balance acquisition cost with cost of ownership for maximum system capability and affordable total cost. During the planning phase, these trade-offs will propagate through all system levels down to component selection for hardware implementation. This costing and trade-off process, as shown in Fig. 1.3, which illustrates its generalized structure, requires a very close customer-vendor interface.

The customer has to find balance between the contemplated mission requirements and the budgeted resources, while taking into account the internal constraints in available:

- Internal acquisition logistics, reflecting, for example, status of incoming inspection and testing facilities, installation opportunity, asset management system and procedures
- Support system, which will impact the actual application environment stresses, maintenance strategy, and so on

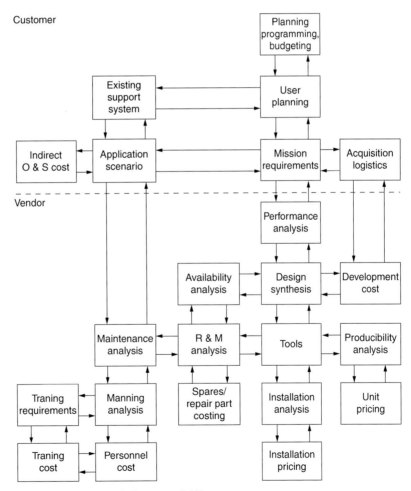

FIGURE 1.3 LCC model of system availability.

The vendor, who wants to satisfy the proposed mission requirements, must perform many analyses to gain insight into available implementation alternatives to find the one that guarantees minimum cost. Major constraints for this activity depend on available:

- Tools, skills, information, and resources for accurate analysis
- Creative design implementation alternatives
- Cost-estimating skills and completeness of historical cost database

This process of finding optimum trade-off for both customer and vendor, internally and between them, must be repeated at different levels of detail throughout the whole acquisition phase. If the vendor is an equipment manufacturer or system integrator, the LCC process is also repeated and refined for each new phase of the product life.

- During the conceptual phase, usually the cost of only one of the proposed alternatives is developed using parametric cost estimating relationships. Cost of other alternatives is calculated from estimated cost differences. When design specifications and implementation strategies are defined, the project enters the validation phase. Here more detailed cost estimates are needed for justification of design-to-cost objectives and need for demonstrable LCC characteristics.
- In the phase of full-scale development, operating and support cost estimates become more accurate because reliability, maintainability, serviceability, and supportability characteristics have been demonstrated. The issues of warranty conditions and pricing are resolved also and form a base for new total LCC reassessment.
- In the life-cycle phases of full-scale manufacturing, delivery, installation, application, and use, the actual cost is measured and stored in cost databases to allow evaluation of accuracy of previous estimates and improvement of future ones.

The magnitude of the LCC formulation problem and cost estimation difficulties is reflected by the large number of cost-influencing variables. The complexity of the LCC process warrants continued monitoring to provide management with the ability to assess progress. Effective local applications of LCC methodology usually start by acceptance and experimentation with existing systems.

Constantly increasing requirements for LCC minimization motivate program managers to use computerized methods to optimize cost by varying cost elements; to consider design alternatives, new concepts, or implementation strategies; and to account for uncertainties.

In profit-oriented environments, the minimization of some cost is not necessarily the best solution available, so LCC models must be developed to assure strategies for net profit maximization. Profit is a random variable that depends on the amount of up-time logged during the period of use. Key to the calculation of the total life-cycle profit is the profitability characteristic, which relates profit to the operating and support cost and is a function of equipment age and often exhibits the same three life stages (early, useful, and wear out periods) as does the failure rate.

In a majority of practical cases, the user is facing decisions in situations much less complex than major military procurement contracts. Because the development and manufacturing costs have been already incurred and are reflected in the equipment purchase price, the user's degrees of freedom in search for LCC minimum are limited only to evaluation of different support and maintenance strategies. The equipment's intrinsic reliability, given the same way as its price, impacts only the cost of maintenance M, which can be estimated assuming (each per some time period, usually a month or year):

- The cost of spare parts inventory C_{SPI}, reflecting the original manufacturing cost of spare parts C_M and inventory cost rate I_{CR} (as a percentage), including depreciation, interest, handling cost, and so on
- Preventive maintenance cost C_{PM}
- Corrective maintenance cost C_{CM}

$$M_C = C_{SPI} + C_{PM} + C_{CM} = C_M I_{CR} + WH \frac{T_R^P + T_T^P}{T_I^P} + WH \frac{MTTR + T_T^C}{MTBF}$$

where W = hourly rate of service engineer, including hourly parts cost
H = equipment usage in hours per time period considered (in-use time)
T_R^P = scheduled time for preventive maintenance
T_T^P = expected travel time for preventive maintenance
T_I^P = scheduled preventive maintenance interval, hours
MTTR = mean time to repair
T_T^C = expected travel time for corrective maintenance
MTBF = mean time to failure, expressed in terms of in-use time, not calendar time

For example, typical values of these factors for minicomputer system maintenance in the early 1980s were as follows:

C_M = $5000 at user's site I_{CR} = 50% per year
W = $250 per hr H = 4000 hr per year
T_R^P = 0.25 hr T_T^P = 0.50 hr
T_I^P = 2 months T_T^C = 1.5 hr
MTTR = 0.5 hr MTBF = 3000 hr

For competitive analysis or comparison and evaluation of alternative strategies, it is customary to express yearly maintenance cost as a percent of the purchase price or to standardize it, for example, per $1000 of price.

PARTICULAR ISSUES

The concepts of total COR and LCC discussed above provide an overall framework for cost optimization and for addressing and studying individual reliability issues, their impact on total cost, revenue, and selection of local implementation alternatives.

Economic Aspects of Product Safety

The close relationship of reliability to issues of product safety is evident from a simple definition of risk:

$$\text{Risk} = \text{probability of a failure} \times \text{exposure} \times \text{consequence}$$

Results obtained by techniques of fault-tree analysis, failure modes and effects analysis, or hazard analysis can be interpreted in terms of negative utility, event probabilities, and severity (criticality), which are standard subjects of cost-benefit studies. The cost factor, strongly dependent on the criticality level, must take into account cost of design for safety, manufacturing for safety, and losses caused by complaints, claims, suits, legal cost, unfavorable publicity, government intervention, and so on. The best investments usually result from the lowest cost-benefit ratio, but implementation alternatives must be selected carefully because cost can significantly increase on both probability extremes:

1. Low probability requirements could result in over-design.
2. Accepted high probability of occurrence could be simply perceived as negligence.

A cost-benefit analysis can be complemented by a comparison of the cost of different alternatives to achieve a defined acceptable level of safety. Analytical approach to decision making about safety issues usually follows this simple process:

1. Identify potential hazards and resulting probable accidents associated with current product design.
2. Obtain credible data on accident rates by product over time, for example, from the National Emergency Injury Surveillance system or insurance companies.

3. Provide a set of alternatives for providing additional increments of safety.
4. Get cost data or cost assessments for these alternatives.
5. Analyze alternatives provided for their effects and cost.

During the analysis, it is necessary to follow some practical principles, such as

- Analysis should be a support for judgment, not a substitute for it.
- Analysis must be open and explicit to be useful as a framework for constructive critique and improvement suggestions.
- There is seldom only one single best solution.
- Conclusions from the analysis should be simple.
- Be realistic about improvement implementation prospects.

Economics of Incoming Inspections

The testing objectives may differ from manufacturer to manufacturer (e.g., improved process control, minimized number of field failures, and so on) but are always combined with the basic tendency to minimize production cost. Four basic incoming inspection (I.I.) techniques are available to implement these objectives:

- *100 percent test-stress-retest.* This comprehensive I.I. technique is the most effective strategy forcing infant failures to appear and failed parts to be removed before they are used in assembly process. The correlation of test results prior to and after burn-in stress often provides important information for vendor's process control and quality improvement. Implementation of this strategy requires investment in both testing and bum-in equipment, indicating its best suitability for high component volumes.
- *100 percent test only.* One hundred percent testing strategy assures removal of defective parts delivered in particular batches and, by that, lower assembly rework cost, improved test effectiveness, and total cost of quality. Functional test is only partially effective as a reliability screen.
- *Buying pre-burned and tested parts.* This variant of a very effective strategy may bring benefits of the economy of scale because of the expected parts volume differences between vendor and user's facilities. Sample testing for vendor's performance audit purposes and risk control will increase the total cost without any additional impact in improved quality or reliability.
- *No I.I.* This is the best strategy in conditions of good vendor's process control and mutual trust in vendor-user relationships. In less favorable conditions the cost of doing no I.I. will be recognized in process disruptions on subassembly or system levels and, potentially, also in the field.

To determine the most suitable alternative, numerous details of all strategies need to be examined and a composite picture formed I.I. is often difficult to justify on purely economic terms, especially for low volumes. For high component volumes, the cost benefit of I.I. is unquestionable in most industrial situations. Also, let us not underestimate the importance of I.I. information feedback to vendors.

A quick estimate of the break-even component volume can be obtained by this simple reasoning. The break-even point is defined by equality of the cost of not testing and the cost of testing per given period of time, usually one year. Cost of not testing is given by

$$C^{NT} = NR\left(P^I C_R^I + P^W C_R^W\right)$$

Cost of testing

$$C^T = P + \frac{NL}{n}$$

where N = number of component units under consideration
R = total fraction defective previously observed or estimated
P^I = fraction defective which fails in-plant
C_R^I = average cost of in-plant repair
P^W = fraction defective which fails during the warranty period
C_R^W = average cost of warranty repair
P = cost of the test equipment
n = number of units tested per hour
L = labor and overhead rate per test hour

So, the break-even volume can be estimated from

$$C^{NT} = C^T$$

$$NR = \left(P^I C_R^I + P^W C_R^W\right) = P + \frac{N}{n} L$$

$$N = \frac{P}{R\left(P^I C_R^I + P^W C_R^W\right) - \frac{L}{n}}$$

Other Topics

For detailed optimization of the complete LCC, the task of reliability management requires a wide variety of economic analyses that address many other diverse topics such as:

- Reliability technology selection
- Manufacturing screening alternatives evaluation
- Optimum maintenance strategies
- ROI in analytical instrumentation

All of them are addressable by the methods of economic evaluation.

REVIEW OF ECONOMIC TOOLS

Methods of Economic Evaluation

In a majority of manufacturing industries, money will not be allocated to a project or spent unless a good argument is presented to the management that this particular investment will assist in reaching company financial goals. To apply this criterion rationally, technical and economic evaluations of proposed projects are performed. Sometimes these judgments are intuitive, based on experience; in other situations, formal detailed analyses are required.

The general concepts forming the framework of sound decisions are quite simple. In the cases of economic decisions, they evolve around notions of profit, growth rate, return on investment, cost-benefit ratio, cash flow, value of money, and so on.

The fundamental step in any economic analysis is the selection of decision criteria or figures of merit, which may significantly vary between the private sector (motivated by profit) and public sector (driven by social benefits or possibly by political motives). The strictly economic-benefit-driven situations are much easier to subject to formalized analyses, and usually define figure of merit in maximal profitability, maximal ROI, or in minimal time required to recover investment. But even in obvious situations of highly favorable cost-benefit ratios, it is necessary to assess possible changes in assumed conditions to assure stability of expected benefits. We need to understand the risks associated with inflation rates, interest rates, business cycles, errors in cost estimates, effects of obsolescence, depreciation, taxation, and sometimes even social cost. The basic methods of economic analysis most frequently used are as follows:

Uniform cash flow method requires conversion of all cash flows to a time adjusted, equivalent, uniform annual amount.

Break-even analysis, suitable for problem solving with incomplete data sets, requires sensitivity analysis to minimize the consequences of errors in estimates.

For technical aspects and details of methods of economic analysis.

Present worth method is based on conversion of all cash flows to an equivalent amount discounted to time zero by application of appropriate interest rate formulas.

Rate of return method computes the discounted cash flow rate of the interest return on invested capital by trial and error.

Benefit-cost method, frequently used by federal and state governmental agencies to estimate the economic attractiveness of an investment, computes ratio of probable annual benefits to the equivalent uniform annual cash flow.

Cost effectiveness analysis allows comparison of alternatives on other than solely monetary measures but requires very careful definition of effectiveness, via, for example, performance, safety, reliability, and relationships to corporate objectives. This presents many difficulties.

Replacement studies, concerned with identification of the most economical time for replacement of existing assets, utilize concept of already incurred (sunk) cost.

Actual implementation of economic analyses and evaluations encounters frequent difficulties in creating consensus on figures of merit, decision criteria and their importance, cost estimation credibility, and unclear power of a wide variety of optimization methods. The major shortcoming of all methods of engineering and managerial economics is their complete disregard of the noncash flow elements involved (which must be balanced with managerial experience and judgment in the final selection of an alternative).

Cost Estimation Methods

Accurate cost estimation of competing engineering design alternatives is essential for project planning and budgeting decisions by both equipment manufacturer and user. Traditionally, systems cost estimates have been prepared using industrial engineering techniques involving detailed studies of necessary operations and materials. These costly estimates, accompanied by volumes of supporting documentation, were subject to frequent extensive revisions in case of even small design changes. Publicized evidence of frequent cost overruns in highly visible government projects indicates their questionable accuracy. These shortcomings resulted in increased interest in statistical and other approaches to cost estimation.

All cost estimating is done by means of analogies and always reflects the future cost of a new system by relating it to some known past experience. Historical cost data incorporate experience with setbacks, design requirements changes, and other difficult to identify and control circumstances in opposition to industrial engineering methods, which tend to be optimistic and not allowing for unforeseen problems. The role of analogy, and the methods of reasoning behind it, is crucial. The art of cost estimation is based on a seven-step process, described, for example, by Barry W. Boehm (see bibliography).

1. Establish objectives for the cost estimating activity to assure support and development of important decision-making information reevaluate and modify these objectives as the process progresses. Objectives should also help to balance expected accuracy, ratio of absolute to relative estimates, and expected level of conservatism.

2. Assure adequate resources for this mini-project and form a simple project plan.
3. Spell out reliability requirements and document all assumptions.
4. Explore as much detail as feasible to assure good understanding of technical aspects of all parts of the product under consideration in a given phase of its life cycle.
5. For actual estimation, use several independent methods and data sources. Most frequently used methods are:
   a. Algorithmic estimating models (see below)
   b. Expert judgment, which could reflect also group consensus derived by formalized methods, such as Delphi technique
   c. Analogy based on similarities with experience, but taking into account impact of inflation, new technologies, productivity growth, and so on
   d. Top-down and bottom-up estimating
6. Compare and iterate estimates to eliminate the optimist-pessimist biases often reflecting a person's roles and incentives. Evaluate the importance of estimates via Pareto analysis, taking into account observed tendencies to overestimate costs related to physically bigger and complex parts of the system.
7. Follow up your estimates with regular comparison with actual cost data collected during the project implementation.

The majority of cost estimating methods is based on the premise that the system cost is in a quantifiable way logically related to some of the system's physical or performance characteristics and, in general, derived from historical cost data by regression analysis. The most common forms of estimating algorithms are:

Analytical models

$$\text{Cost} = f(X_1, ..., X_n)$$

where f is some mathematical function relating cost to cost variables $X_1, ..., X_n$, correlated with some physical or performance characteristics of the system.

Tabular (matrix form) models provide easy-to-understand, -implement, and -modify relationships that are difficult to be expressed by explicit analytical formulas. Analytical cost models are excellent in describing cost-to-cost-estimating relationships. Tabular models are more suitable if cost-to-non-cost functions need to be represented. To date, the analytical models used usually contain a small number of variables and are insensitive to many sometimes-important factors. Some models take a form of a composite function, which can better represent the historical data, but at the expense of increased complexity.

Examples of simple models:

1. *Analytical cost estimate model.* Cost of spare parts procurement support C_{PS}

$$C_{PS} = 0.037 P_{rc}$$

where P_{rc} is parts repair cost.

2. *Matrix model.* Relative cost of software qualification testing C_{sq} as a function of required reliability

C_{sq}	Required reliability
$0.55 C_N$	Very low
$0.75 C_N$	Low
$1.00 C_N$	Normal
$1.25 C_N$	High
$1.75 C_N$	Very high

where C_N is qualification cost of nominal reliability software product.

Note: C_N itself is usually a function of product size and complexity, design group skill and experience, user programming language, and so on, and can be expressed via analytical models.

The strength of algorithmic models is in their objectivity, repeatability, and computational efficiency in support of families of estimates or sensitivity analysis. They do not handle exceptional conditions and do not compensate for erroneous values of cost variables and model coefficients. Actual experience with cost estimation leads to a conclusion that no single method is substantially superior in all aspects and that strengths and weaknesses of many methods are complementary.

Cost Accounting

To assure a reasonable rate of diffusion of methods of economic analysis to support rational decision making about reliability, a system of actual cost data feedback is necessary to allow evaluation of prediction accuracy, cost estimates, and effectiveness of analytical methods used. In the first-time attempt to identify and measure cost of reliability, it is highly probable that our cost data requirements will not match the established cost accounting system. In this situation the decision, to start with a single project study usually prevents deadlock and allows development of information for both reliability improvement program and identification of reliability categories of key importance. Data collection will be manual, necessary forms will be designed separately for this particular study, data compression and interpretation will be done, most probably, on reliability engineers' personal computers.

Only after a demonstrated success of reliability cost management in a single study or project environment can steps be taken to establish a reliability cost accounting system, with the objective of a continuing scoreboard. This continuing scoreboard should be based on a formal reliability cost accounting and reporting system, which parallels the accounting systems required for general financial management and legal purposes. The systematic approach requires at least:

- A list of projects, products, and programs of interest
- A list of departments involved
- A list of accounts where relevant charges are accumulated
- Cost categories (see example in Table 1.1 suitable for LCC model)
- Definitions of data entry requirements and formats
- Definitions of data process flow with control points
- Formats and frequency of reports and summaries
- Rules of data and results interpretation
- An established base for result comparison and evaluation
- A methodology for cost standards creation and improvement

Collected data must be sorted and compressed to accommodate evaluation from many different viewpoints:

- Product, subassembly, part, component, and so on
- Organization responsibility
- Place and time of occurrence or reporting
- Project or program association

In formatting the data and results, the general preference of graphical representations in forms of tables, Pareto-type distributions, pie charts, trend lines, scattergrams, control charts, and so on is well established. Narratives by specialists or representatives from responsible teams can help in interpretation and in assessing the seriousness of data, especially when reports result in transfer of charges between departments and accounts.

TABLE 1.1 Examples of Reliability Cost Categories

Prevention costs
Hourly cost and overhead rates for design engineers, reliability engineers, material engineers, technicians, test and evaluation personnel
Hourly cost and overhead rates for reliability screens
Cost of preventive maintenance program
Cost of annual reliability training per capital

Appraisal costs
Hourly cost and overhead rates for reliability evaluation, reliability qualification, reliability demonstration, environmental testing, life testing
Average cost per part of assembly testing, screening, inspection, auditing, calibration
Vendor assurance cost for new component qualification, new vendor qualification, vendor audit

Internal failure cost
Hourly cost and overhead rate for troubleshooting and repair, retesting, failure analysis
Replacement parts inventory cost
Spare parts cost
Cost of production changes administration

External failure cost
Cost to repair a failure
Service engineering hourly rate and overhead
Replacement parts cost
Cost of service kits
Cost of spare parts inventory
Cost of failure analysis
Warranty administration and reporting cost
Cost of liability insurance

The importance of reliability cost accounting for expense controls of projects or routine activities in every sector is self-evident. But the analysis of data from previous projects, contracts, and economic studies may have longer-term and more fundamental impact on the improvement of managerial decisions, resource allocation, and effectiveness, and by that significantly contribute to the improvement of the business unit's competitive position and probability of success.

Summary

Reliability engineering and management can greatly benefit from prudent application of tools from engineering economics and operations research, especially when facing decisions about costly investments, high risks, or complex situations, such as those occuring in high-technology environments. The power of these tools, with their rigorous methods and computational accuracy, must be understood in the context of their dependency on quality and relevance of underlying assumptions, historical data, cost estimates, and known unequal treatment of results from physical versus human sciences. The growing computerization of described methods and availability of computers for daily work, should allow management to concentrate more on the human and strategic aspects of decision making, while being assured of the rigor and accuracy of their technical aspects.

CHAPTER 2
WORK MEASUREMENT

Bruce Wesner
Managing Principal, Life Cycle Engineering, Inc., Charleston, S.C.

Introduction. The end objective of the various techniques for work measurement is to supply standard data for determining the time required to complete specific segments of work. Whether the resulting standards are used for labor control on a straight time basis or for labor control and wage incentives, the techniques are the same. The principal techniques for establishing standards for work measurement are discussed as to principles, procedures, advantages, and disadvantages; examples of development and application are given.

Participation with Maintenance Engineering in Measuring Work and Establishing Standards. Examine the positions of the foremen and maintenance engineers. These two management groups, one line and the other staff, have the common goal of increasing productivity and decreasing operating costs through various industrial management techniques. However, it is not uncommon for the two groups to be enemies rather than allies. Successful work control depends on the total cooperation and coordination of these two functional groups.

Supervisor's or Team Leader's Role. The supervisor or team leader is the front-line administrator of management methods and management policies. Assume a company is staffed with highly skilled engineers, mechanical, electrical, industrial, and so on. These engineers, through engineering logic, constantly create sound improvements within the framework of their function. The improvements are of no value unless they are placed in operation. The team leader or supervisor or team lead is the only person who can do that. He or she is on the spot where the work is being done, where labor, material, and overhead costs are realities, not just figures on paper.

With the maintenance function, there are many former technicians holding the position of front-line cost administrator. Many of these people carry with them the same resistance to change that they had as technicians. It is management's responsibility to take these courses of action:

1. If a supervisor or team leader or team lead is capable of administering costs, he should be trained in every phase of management theory and management control practices; not with the objective of having him perform the control function but rather to appreciate, understand management controls, accept, support, and enjoy the competitive challenge of management in the cost-reduction field.

2. If the he or she supervisor or team lead is not capable of administering costs, he is not adaptable to progress, which is a necessity for every company's future. Management must find another field for him or her.

Maintenance Engineer's Role. The maintenance engineer is a technical resource to the front-line supervisor or team lead. As such, he is in a position to come in contact with people who work for the supervisor or team leader or team lead. There is often a tendency for the maintenance engineer to talk to the workers or the union steward in the area about work methods, equipment, and standards. This is not a wise course of action, if the engineer is planning to work with the supervisor or team lead. The supervisor or team leader or team lead, in addition to representing management in administrative matters, is also the leader of the workers. Any staff person who bypasses the front-line supervisor or team lead in dealing with the workers concerning their jobs is infringing upon the supervisor's or team lead leadership. A front-line supervisior or team lead can be a leader only when he is not undermined by staff activities. The maintenance engineer's communication line to the workers is the supervisor or team lead.

Engineers have good creative ideas, but they allow these ideas to supersede good management. The development phase of a new project or activity usually proceeds too far before the supervisor or team lead is brought into the picture for his contributions. In many cases, the engineers neglect to ask the team leader or team lead's assistance in the creative stages and present a finished package, expecting immediate acceptance of their work. This method of operation does not allow the supervisor or team lead to contribute his practical experience and know-how in forwarding the project. The best approach for the Reliability engineer is to call the supervisor or team lead in at an early stage, ask for and use his suggestions and those of other staff people, and attempt to get him so interested in the project that it is impossible for him to be anything other than a willing participant. A good method is to form a small committee of the people involved in launching a particular project. Make the supervisor or team leader or team lead chairman, and he immediately leads the activities. There can be no backing out. It is a much more satisfying and easier job for a supervisor or team lead to install his or her project rather than someone else's.

A successful installation of a project usually is followed by a bit of bragging, and that's part of the job. In order to get advanced assignments and freedom to work on future and possibly more difficult projects, the prior requisite is a history of success on the preceding projects. A letter initiated by maintenance engineer to the front-line leader or supervisor, with a copy to his boss, thanking him for his participation and giving him an estimate of the value of the results of the first year's operation under the improvements, will put the supervisor or team lead on a receptive basis for participation in future projects.

A maintenance engineer is successful in his operations only when he has other management people coming to him with basic ideas or telling him of trouble spots and asking his assistance. There is more chance of success when the maintenance engineer is doing work that someone requested rather than digging up projects on his own that have to be sold to others before they are willing to participate.

In comparing the two jobs of supervisor or team lead and Reliability engineer, the supervisor or team lead's is more difficult in respect to completing improvement projects. After the Reliability engineer has the supervisor or team lead participating in the project, he uses all the ideas he can find, his own, the supervisor or team lead's, and the other staff groups', to set the job up. He assembles a mass of data through time study, MTM, estimates, or other means. He outlines the methods procedure, develops standards, and gives the results to the supervisor or team lead. This is where the real work starts. Initially the supervisor or team lead has to orient the workers and the union in regard to this staff activity in his department. He must coordinate and install any changes in work methods. He must train his workers in new methods. He must man his area in balance with work-load requirements. The Reliability engineer has worked with logic and facts. The supervisor or team lead installs the results of the logic and fact procedure. His problem is with people. A successful installation depends upon acceptance by the majority of the workers and this is a far greater problem than getting results from logic and facts.

The supervisor or team lead must deal directly with employees as work-measurement standards are introduced to remove some of the mystery from the activity. Each new standard or standard-data series should be introduced by the supervisor or team lead. He must present the picture of the new standard in detail with a positive approach and not in a hesitant, backward fashion. In the course of this introduction, the supervisor or team lead, by his presentation, must build confidence and create in the worker the desire to give the standard a fair trial. These are some of the impressions that he must develop in his discussion with the worker in the introduction of new standards. That he, as supervisor or team lead:*

---

*W. Colebrook Cooling, "Front Line Cost Administration," p. 121, Chilton Publishing Co., Philadelphia, Pa. 1955.

1. Is positive that the job was time-studied under normal operating conditions.
2. Has determined that the standard was developed around those conditions.
3. Will guarantee that the job will continue to be performed under the same conditions or the standard will be adjusted.
4. Has thoroughly reviewed the standard and firmly believes it to be fair and attainable.
5. Is sure the standard will be given a fair trial.
6. Sincerely hopes the worker will do well on the standard.
7. Expects the worker will contact him, not the union, not another worker, and not the engineer, if there are any questions on any standard.

Without the supervisor or team lead's activity participation no Reliability engineer would be a success. Top management must recognize this fact. They must provide the thinking and direction to impress the supervisor or team lead with his responsibility and the need to operate as a front-line administrator rather than a pusher. Top management must train their line and staff groups in every phase of management control in order to operate in a manner that makes active participation a habit with all improvement projects. With the line groups, the training program should not have the objective of having foremen perform all the control functions, but to have foremen:

1. Understand and appreciate controls.
2. Recognize the need for assistance from staff groups.
3. Immediately call the proper staff function when a course of action is required.
4. Be equipped to use control functions to evaluate their own performance.

With this outlook and method of operating, the most successful work-measurement program is one that calls for the most participation and responsibility on the part of the supervisor or team lead. This thinking is not too common in the field of maintenance-work measurement. Recently, in a comparison of advantages and disadvantages of various methods of work measurement, this sentence was noted under disadvantages: "Requires cooperation of foremen and technicians." There is a lot of revision necessary in the thinking and working of management in that particular company. Costs cannot be controlled without the active participation of the person primarily responsible for material usage and getting the work done, the supervisor or team lead.

Methods. The most important contribution from method studies, with regard to maintenance, is standardization of work. Before work can be measured there must be a standard method by which the work elements are to be completed. The workers must be trained in this method. The data obtained from work measurement are based on the same method, and worker efficiencies against various jobs can be determined. Should there be a change in work methods, the standards must be reviewed to determine the time required to complete the task under the new method.

It is not a requisite to have the best method when establishing standard data. The primary goal is to establish standard methods so that the work can be measured, standard data developed, and maintenance efficiencies determined against the standards. Methods improvements usually require acceptance by management, additional equipment, training, and orientation of personnel, and other problems concerned with the installation of changes in work methods which could slow progress in work measurement. With present methods standardized, work-measurement studies can proceed. Standard data can be changed or added to in the future as improvements are made.

Figure 2.1 shows a print of a typical job for portable flame-cutting machines. Figure 2.2 is an example of standard-data development for that job. The development of the standard shows separate setup and separate machine-burning times for square and bevel cuts. This is the way the job was standardized and time-studied. Note, however, that this method contains extra operations. For square

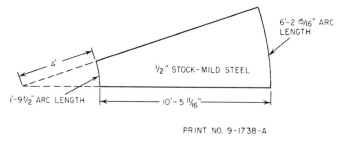

PRINT NO. 9-1738-A

ASPHALT TANK BOTTOMS

FIGURE 2.1 Print of a typical job for flame-cutting machines.

and bevel cutting, standard attachments are available for each cutting machine. The attachments consist of a mounting fixture to provide an extra torch to be mounted on the standard machine, providing the means to square and bevel-cut simultaneously. For work-measurement purposes, the present method of making each cut separately can be studied and data developed. At a later date, with the new method of cutting installed, additional studies can be made and a standard-data table provided for change in work methods. In the meantime the present method is standardized and measured; worker performance can be determined on the old method. As a matter of interest, Fig. 2.3 illustrates the development of the standard for the same job as Fig. 2.2 using the new operation. There is a 21 percent reduction in time. In this particular shop, where a large volume of plate is flame-cut for fabrication, the annual savings would be substantial.

STANDARD DEVELOPMENT SHEET

FLAME CUTTING - PORTABLE FLAME CUTTERS

DESC: ASPHALT TANK BOTTOMS (2)   PRINT NO: 9-1738-A   MATERIAL: MILD STEEL   DATE ISSUED: 3/201

STANDARD DEVELOPED BY: E. Wilmer   DATE: 3/201

PER SEGMENT (10 SEGMENTS/TANK)

ELEMENT	TABLE	UNIT	NO. OF UNITS	STD. MAN HRS/ UNIT	TOTAL STD. MAN HOURS
M/C BURN - SQUARE	18.02	100 IN.	3.49	.150	.523
- BEVEL	18.02	100 IN.	2.52	.180	.454
HANDLE MATL (WT. OF FIN. PC. APPROX. 510#)	18.05	PER PC.	1	.351	.351
SET UP CM - 16 M/C					
SHORT ARC	18.06	PER OCC.	1	.154	.154
LONG ARC	18.06	PER OCC.	1	.061	.061
SET UP CM - 30 M/C					
STRAIGHT LINES 1st OCC.	18.07	PER OCC.	1	.118	.118
3 ADDITIONAL OCC.	18.07	PER OCC.	3	.069	.207
PREHEATS	18.06	PER OCC. 0.8.	6	.014	.084
CHIP EDGE AFTER BEVEL	18.09	100 IN.	252	.040	.101
TOTAL TIME PER PLATE					2.503
PER 20 PLATES					41.060
ADD JOB PREPARATION					.100
					41.160
ADD 5 ADDITIONAL JOB PREP					.500
					41.660

FIGURE 2.2 Standard-data development for flame-cutting operation shown in Figure 2.1. Square and bevel cuts are made independently.

ELEMENT	TABLE	UNIT	NO. OF UNITS	STD. MAN HRS/ UNIT	TOTAL STD. MAN HOURS
M/C BURN - SQUARE	18.02	100 IN.	97	.150	.145
- SQ. & BEVEL (SIMO)	EST.	100 IN.	252	.180	.454
HANDLE MATERIAL (WEIGHT FINISHED PIECE APPROX. 510#)	18.05	PER PC.	1	.351	.351
SET UP CM-16					
M/C SHORT ARC	18.06	PER OCC.	1	.154	.154
M/C LONG ARC	18.06	PER OCC.	1	.061	.061
SET UP CM-30					
SQ. & BEVEL	EST.	PER OCC.	1	.187	.187
ADDITIONAL OCC.	EST.	PER ADDIT.	1	.069	.069
PREHEATS					
O.B. ONE TORCH	18.06	PER OCC.	2	.014	.028
O.B. SQ. & BEVEL	EST.	PER OCC.	2	.030	.060
CHIP EDGE AFTER BEVEL	18.09	100 IN.	252	.040	.101
TOTAL TIME PER PLATE					1.610
PER 20 PLATES					32.200
					.100
ADD JOB PREPARATION					32.300
ADD 5 ADDITIONAL JOB PREP.					.500
					32.800

FIGURE 2.3 Standard-data development for flame-cutting operation shown in Fig. 2.1, with the improved method. Square and bevel cuts are made simultaneously.

Figures 2.13 and 2.15 illustrate the documentation of methods on a job basis. With repetitive jobs it is imperative to be specific in all considerations with the method, that is, layout, equipment, dimensions, and work methods.

On an equipment basis such as Fig. 2.11, the documentation, as far as handling is concerned, is more general. This is due to the fact that every job is different. However, the work area is described in detail, setup methods are described, and the elements of the operation noted.

Figure 2.4 illustrates the procedure for standardizing methods on a craft basis. This particular example is welding pipe. The standard practice prescribes for each size of pipe and the manner in which it is positioned, the size of electrode, and the number of passes required. An illustration (Fig. 2.5) accompanies this to further define the standard method of welding pipes. Similar procedures are outlined for all other work performed by the welding craft.

Methods work cannot be confined to equipment improvements or motion improvements. A company, during the installation of incentive standards on tool- and die-makers, found that interpretation of prints and sketches delayed the start of jobs. In addition, there was quite a bit of stock spoilage. To solve both problems, the company, with consulting engineers, developed a simplified drafting technique. This technique eliminates circles and arrowheads and all dimensions are scaled from one point. This simplification of prints resulted in a reduction of drafting time, a reduction in interpretation time by the mechanic, and less spoilage. Figures 2.6 and 2.7 illustrate the old and new drafting techniques.

In addition to the fact that methods studies are required for standardization, this area also provides immediate savings of maintenance dollars. Maintenance methods should be approached in the same manner as production methods. In most maintenance installations, maintenance methods are still in the hands of superintendents, foremen, and other management without the aid of centralized direction, which is similar to the way in which production methods were handled 50 years ago.

It has been production's experience that a staff engineering function is necessary for methods development. A production supervisor or team lead effectively administering and directing production does not have time to develop methods improvements. This is also true of a maintenance supervisor or team lead who has more mileage to cover to administer and supervise his work than the production supervisor or team lead. To expect him to keep abreast of all technological changes in

3.24 ENGINEERING AND ANALYSIS TOOLS

Size of pipe, in.		Outer circumference, in.	Rolled position				Fixed position			
			No. passes per electrode			Total passes	No. passes per electrode			Total passes
			⅛ in.	⁵⁄₃₂ in.	³⁄₁₆ in.		⅛ in.	⁵⁄₃₂ in.	³⁄₁₆ in.	
1	Std	4.2	2	...	...	2	2	...	...	2
	X Hy	4.2	2	...	...	2	2	...	...	2
	XX Hy	4.2	2	...	...	2	2	...	...	2
1½	Std	6.0	2	...	...	2	2	...	...	2
	X Hy	6.0	2	...	...	2	2	...	...	2
	XX Hy	6.0	2	...	...	2	2	...	...	2
2	Std	7.5	2	...	...	2	2	...	...	2
	X Hy	7.5	2	...	...	2	2	...	...	2
	XX Hy	7.5	2	...	...	2	2	...	...	2
2½	Std	9.0	2	...	...	2	2	...	...	2
	X Hy	9.0	2	...	...	2	2	...	...	2
	XX Hy	9.0	2	...	...	2	2	...	...	2
3	Std	11.0	1	1	...	2	1	1	...	2
	X Hy	11.0	...	2	...	2	...	2	...	2
	XX Hy	11.0	...	4	...	4	...	4	...	4
3½	Std	12.6	...	2	...	2	...	2	...	2
	X Hy	12.6	...	3	...	3	...	3	...	3
	XX Hy	12.6	...	2	2	4	...	4	...	4
4	Std	14.1	...	2	...	2	...	2	...	2
	X Hy	14.1	...	3	...	3	...	3	...	3
	XX Hy	14.1	...	1	3	4	...	4	...	4
4½	Std	15.7	...	3	...	3	...	3	...	3
	X Hy	15.7	...	3	...	3	...	3	...	3
	XX Hy	15.7	...	1	3	4	...	4	...	4
5	Std	17.3	...	3	...	3	...	3	...	3
	X Hy	17.3	...	1	2	3	...	3	...	3
	XX Hy	17.3	...	1	3	4	...	5	...	5
6	Std	20.8	...	3	...	3	...	3	...	3
	X Hy	20.8	...	1	2	3	...	3	...	3
	XX Hy	20.8	...	1	3	4	...	5	...	5
8	Std	27.1	...	1	2	3	...	3	...	3
	X Hy	27.1	...	1	3	4	...	4	...	4
	XX Hy	27.1	...	1	3	4	...	5	...	5
10	Std	33.7	...	1	2	3	...	4	...	4
	X Hy	33.7	...	1	3	4	...	5	...	5
	XX Hy	33.7	...	1	3	4	...	...	...	...
12	Std	40.1	...	1	2	3	...	5	...	5
	X Hy	40.1	...	1	3	4	...	5	...	5

FIGURE 2.4 Standard methods are necessary before work measurement can be attempted. With nonrepetitive work, methods can be standardized as illustrated. On specifications given, the work can be measured and standards applied, based on the standard method for welding specific pipe sizes and the number of welds required.

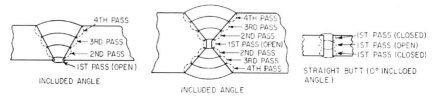

FIGURE 2.5 Further definition of standard method for butt-welding pipe as illustrated by Fig. 2.4.

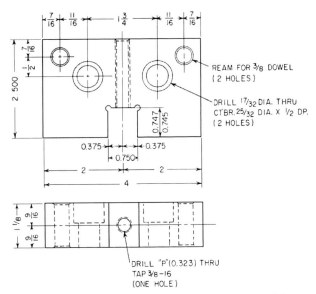

FIGURE 2.6 Conventional method of drafting a part for tool-and-die work. (*Methods Engineering Council.*)

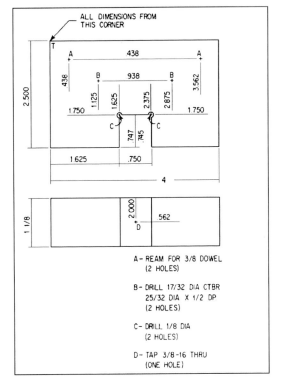

FIGURE 2.7 Improved technique which reduced thinking time of mechanic and material spoilage. (*Methods Engineering Council.*)

maintenance equipment, to develop new methods, and at the same time do a combined supervisory and administrative job is asking too much. Set up a staff function to review all maintenance work methods and plan production. The methods engineer must have a staff relationship with the plant engineer and maintenance supervisor or team lead. With this arrangement, maintenance foremen still are the keymen in charge of setting up work methods. They review proposals with the methods engineer and accept, add to, or reject them. Out of this review will come recommendations for improved methods which will result in more efficient usage of men and equipment.

Providing Handling Equipment. The bulk of the improvements in maintenance work are possible through changes in handling methods. Maintenance operations still contain much of the brute work that has been eliminated from production jobs. Here is a typical example:

In analyzing the operation of straightening paraffin press plates a man-machine chart was developed (Fig. 2.10). This chart demonstrates that the two helpers functioned as holding fixtures to position the plates and to hold them while the operator gaged the plates. It is evident that one man could be eliminated from the job through the substitution of some mechanical device to take his place as a holding fixture. As illustrated in the sketch (Fig. 2.8), the simplest was a short section of roller conveyor. Through rearrangement of operator's duties (as illustrated in Fig. 2.10), he assists the remaining helper in positioning and removing the plate and the conveyor acts as a holding fixture for the gaging operation.

It is quite possible that this method could be further improved. A stop could be placed on the end of the conveyor to catch the plate as it is ejected from the roll, eliminating that function of the helper. An air cylinder could be installed to push the plate into the rolls, thereby eliminating another function of the helper. To position the plates to and from the conveyor, a chain hoist can be used, eliminating the last function of the helper. These three modest improvements could make a one-man job out of this operation.

In addition to providing handling equipment, there is also the possibility of reducing the amount of handling involved. Figure 2.9 is a methods proposal based on eliminating handling operations. As illustrated in the proposal, it was felt that heat exchangers could be cleaned and tested on the pipe-still unit location rather than lowering them to the ground, moving them into position, moving them again, and raising them back on the still. Note that this eliminates from 30 to 40 man-hours per heat exchanger. At this particular location, the savings in time amount to slightly over 1 man-year.

It is difficult to separate material-handling problems from methods. Since a large percentage of maintenance jobs consist of selecting material, getting the material to a location, removing material, installing the new material, and scrap removal, detailed studies must be made to reduce the percentage of handling time in the total working time. An analysis must be made first on repetitive work, as illustrated by the two case histories; then an analysis must be made by craft lines. This must be made with consideration for the size of the plant maintenance function together with the kind of materials used. In a small plant, millwrights provide the material-handling service with an occasional boost by fork-lift trucks. Even though on a small scale, there still are enough hours for a methods study. In a

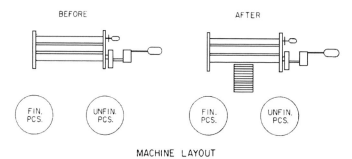

FIGURE 2.8 Before-and-after work-area sketch, showing addition of short roller conveyor that replaced one helper. See Fig. 2.10 for elemental description of the operation.

(This outline contains the results of a savings investigation regarding the discussion on bundle maintenance of all bundles on the old pipe stills. From the discussion, the proposed changes with the savings have been outlined on these operations; lowering bundles to ground, cleaning bundles, testing bundles, and raising bundles back on still.)

Present Method
Description:
1. Lower bundles to ground level and position in alley.
2. Clean and flange test bundles in alley.
3. Remove bundles from alley and raise to still.
Avg No. of riggers used: 5
Elapsed hr for oper.:
Bundle Nos.: (0–4) (5–6) (7) (8–10)
            8   8   8   6
Bundle Nos.: (11–13) (18–22)
            6   6
Act. hr for oper.:
Bundle Nos.: (0–4) (5–6) (7) (8–10)
         40   40 40 30
Bundle Nos.: (11–13) (18–22)
         30   30

Proposed Method
Description
1. _____
2. Clean and flange test bundles on location.
3. _____
Avg No. of riggers used: 0
Operation eliminated

Yearly savings:
Bundle Nos.:   (0–4) (5–6) (7) (8–10) (11–13) (18–22)
No. times pulled per year: 16   6   3   18   12   18
Hours saved per year: 640 240 120 540 360   540
Total rigger hours saved per year: 2440 (a little over 1 man-year)

FIGURE 2.9 A proposal for reducing material-handling requirements.

large maintenance installation, the problem is increased according to the amount of equipment required. This type of installation calls for a central transportation system which might include radio control with a complete dispatch system. A project of this size will justify the assignment of a group of reliability engineers for study. The material-handling problem in the small plant will be a consideration during job, equipment, or craft studies.

Standardization of Crews. One of the common complaints, with regard to the performance efficiency on maintenance jobs, is "too many men on the job." This may be due to the old "craftsman and helper" tradition, or it may be due to poor job planning. In either case, there is a loss in maintenance dollars.

Standardization of materials, equipment, and work methods is a must before maintenance work can be planned, scheduled, or measured. Standardization of crew size enters the problem under the classification of work methods. These are the normal considerations when establishing crew size: (1) physical limitations of the individual with respect to equipment and material handled, (2) individual's safety considerations on particular jobs (a standby may be necessary in areas where a craftsman is not permitted to work alone), and (3) the urgency of the job.

Crew size must be determined by work study based on the considerations listed above. Work studies to determine crew size should be made as follows: first, on a job basis where work is repetitive; second, on handling methods of equipment that is regularly used; and third, on a craft basis where nonrepetitive type does not involve stationary equipment (millwright work, plumbing, field welding, and so on).

Figure 2.10 illustrates an example of establishing a standardization of crew size in repetitive jobs. With the old method, three men were required, one operator and two helpers. One helper would not handle the plates without assistance, and since the operator was required at the machine controls, the crew size was standardized at three men. With small changes two helpers were no longer required. The standard crew size was changed to two men and a new standard time was established. As pointed out previously, further improvements on the job could reduce the manning to the operator alone.

In the cases cited, the work-measurement standards were based on gang size as determined by the methods of completing the work. With the newer method (two-man gang), shop supervision scheduled

3.28 ENGINEERING AND ANALYSIS TOOLS

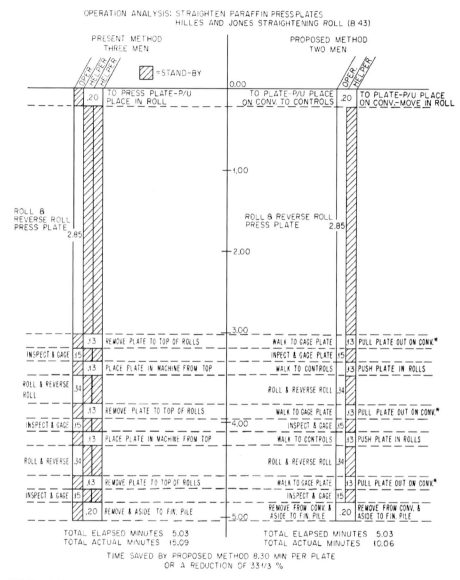

FIGURE 2.10 Man-machine chart illustrating a practical method for determining crew size, analyzing handling equipment, and collecting time data prior to developing a job standard.

the work for two men and planned the efficient disposition of the former helper who was no longer necessary for the job. Figure 2.11 is the work area for two pieces of equipment requiring a gang. The operations, bevel shear and scarp, are performed on light plate stock prior to welding. The weight of plate handled, within equipment capabilities, determines the crew size. The work must be scheduled by crew size and the standard data developed for work measurement must tie in with crew size. With each gang size the machine speed remains the same. However, since work handling is the determining factor in gang size, extra time must be provided during machine operation for the increased gang (see Fig. 2.12). The handling elements vary with the weight of each plate and are completed accordingly

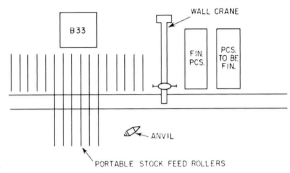

FIGURE 2.11 Bevel-shear-machine work area. Within capabilities of handling equipment, the bevel-shear machine and the weight of the plate handled, standard crew sizes are determined for operations performed on this equipment. At the same time work is scheduled the correct gang is scheduled, as gang size can vary between jobs.

Bevel Shear and Scarp

MACHINE: Hilles and Jones bevel shear (B33)
OPERATIONS: Bevel shear plate stock
   Scarp plate stock corners (by hand)
GANG: The normal gang for jobs consisting of plates weighing from 51 to 100 lb each shall consist of two men.
   The normal gang for jobs consisting of plates weighing over 100 lb each shall consist of three men.

CONDITIONS: The application of these standards is contingent on the following conditions:

   6. The gang size shall be determined by the weight of the pieces to be processed.

Standard-time Table No. 9.2
Bevel Shear—Cutting Time Only

Symbol	Plate thickness, in.	Man-hr per ft of bevel	
		Gang size	
		2 men	3 men
C	$\frac{1}{8}$	0.0048	0.0071
	$\frac{3}{16}$ and $\frac{1}{4}$	0.0059	0.0089
	$\frac{9}{32}$ and $\frac{5}{16}$	0.0062	0.0092
	$\frac{3}{8}$	0.0062	0.0092

FIGURE 2.12 Weight of material handled with equipment available determines gang size. Standard hours for machine operations must be based on gang size.

Standard-time Table No. 9.4
Material-handling Time

Sym-bol	Description of element	Man-hr per plate					
		Weight of plate, lb					
		51–100 lb.	51–100 lb.	101–300 lb.	301–900 lb.	901–1,200 lb.	1,201–1,800 lb.
		(1)	(2)	(3)	(4)	(5)	(6)
V	Initial get and position plate to bevel only, move plate aside to finished pile, no scarping	0.037	0.054	0.221	0.250	0.362	0.362
W	Initial get and position plate to bevel and scarp, position plate for scarping after beveling, move plate to finished pile after scarping	0.064	0.094	0.259	0.297	0.399	0.458
X	Reposition plate to bevel additional edges—once per each additional edge	0.011	0.013	0.045	0.055	0.077	0.122
Y	Reposition plate for additional scarping, once per additional scarp	0.011	0.013	0.038	0.038	0.055	0.055
Z	Initial get and position plate to scarp only, move plate to finished pile after scarping	0.037	0.054	0.183	0.210	0.294	0.294

Material-handling Time—Standard-time 9.4

Column 1. To be used on all jobs consisting only of pieces under 101 lb. The standards in this column are based on a two-man gang.

Column 2. To be used on all jobs consisting of pieces under 101 lb intermingled with pieces over 100 lb. The standards in this column are based on a three-man gang and shall be applied to those pieces 100 lb or under only.

Columns 3, 4, and 5. The Standards in these columns are based on three-man gangs. Use Column 3 for all plates weighing 101 to 300 lb, use Column 4 for all plates weighing 301 to 900 lb, and use Column 5 for all plates weighing 901 to 1,200 lb.

FIGURE 2.13 Crew size is based on weight of material handled, but there are some jobs with plate sizes that can be handled by a two-man gang intermingled with plates requiring a three-man gang. In those jobs, since it is sometimes impractical to split crews, standard times are provided to give fair work standards to the crew. If the job is large enough, the work can be split into two jobs at this operation, one for the two-man and one for the three-man crew. This is a function of planning and scheduling and can result in a 33 percent cost reduction for part of the job.

with the proper gang size built into the standard. Figure 2.13 illustrates the handling elements. Note that there are two columns (Columns 1 and 2) for handling all material at the two-man gang size level. Column 1 applies to a two-man gang, and Column 2 applies to a three-man gang. This is to provide flexibility in those jobs where portions of the job require three men and other portions require two men. Planning and scheduling require some flexibility, and it is not practical to separate jobs into mixed gang sizes. For those instances where various weights of plate would require a combination of two- and three-man gangs on the same job, a standard is provided for the two-man portion that will allow time for the three-man gang, avoiding mixed gang sizes on the same job.

The determination of gang size by craft cannot be defined as clearly as the job or equipment method. However, standard methods of work must be outlined for each craft (see Fig. 2.4). It becomes the planner's or supervisor or team lead's job to review the amount and type of work, consider the standard method of performing the work, and determine the most efficient gang size. The work-measurement standard for the work will be established in accordance with the standard methods of performing the work. After this standard method has been reviewed by the planner or supervisor or team lead and the amount of available work has been taken into consideration together with a completion date, the proper gang size will be scheduled. The work-measurement standards will be adequate since they are also based on the standard method.

Equipment Records. For expediting control functions such as methods, planning and scheduling, and work measurement, equipment records must contain the following information:

1. Machine and parts specification numbers
2. A breakdown history
3. A preventive-maintenance history
4. Maximum and minimum spare-parts record

The machine and parts specification numbers are necessary for the planning function and work-measurement function. With this information readily available, the delays in scheduling repair work and nonproductive time by maintenance crafts will be minimized since the proper parts can be located at the job before the work starts.

The breakdown history must be analyzed to determine repetitive work. The analysis of repetitive work must be made to determine if improper design or improper installation caused the breakdown. If so, this can be corrected. If either of these two is not the cause, each type of breakdown must be analyzed from the methods viewpoint to see if there is a possibility of eliminating or reducing repetitive breakdowns of each type. If this is not possible, it may be necessary to plan preventive-maintenance measures which will allow this type of work to be scheduled and work-measurement standards developed based on the preventive-measurement procedure. Preventive maintenance can be planned and scheduled and job standards for work measurement can be easily applied on this type of work. Equipment records should be a source of information to point out areas for repetitive job standards.

When it is necessary to repair or make parts for equipment, the most economical method is to process this work by job lots instead of one or two parts at a time. It is more efficient to set up a job-lot method and standard quantity rather than process one or two parts. This is illustrated under Job Standards—Time Study. For a part's repair, Straightening of Paraffin Press Plates, the lot size was established at 50 plates. This enables planning to schedule a long run of this work, normally a day's work for two men. With a run of this type, job preparation and setup costs are spread over a large number of pieces, the most efficient crew can be scheduled, and the job becomes a production run without the delays and interference problems encountered in one or two pieces to be processed as a job.

After analyzing parts requirements, provision must be made to have as many lots of parts as necessary to provide for equipment repair, and parts maintenance. For example, one lot of parts must be in use; one lot can be in spare-parts inventory, readily available; and one lot could be in the shop in process or to be scheduled at the shop's convenience. This method of establishing repetitive jobs for parts repair will effectively reduce the cost of maintaining equipment.

Planning and Scheduling. This function is to maintenance as production control is to production. Production labor measurement is impossible without planning how, where, and when the work is to be done. Work measurement of maintenance jobs is not practical if this function is slighted.

Work must be planned ahead of time to eliminate or reduce to a minimum nonproductive hours or inefficient work resulting from waiting for job assignment of instructions, wrong crafts, wrong number of men on the job, inefficient work methods, not enough material, wrong material, and performing shopwork in the field.

Work must be scheduled ahead of time to eliminate or reduce to a minimum nonproductive hours or inefficient work resulting from failure to schedule equipment; excess travel time due to job chasing; failure to schedule material drops; and failure to balance manpower requirements.

There is no method to measure efficiency or pay incentive bonus on nonproductive time. The only choice is to pay base rate, and this penalizes the worker as far as the opportunity to earn incentive pay, and of course the job is credited with excess costs. For that reason, planning and scheduling are interrelated with methods and work measurement. Without a standard method, there is no way to estimate time; without time on a job, there is no basis to schedule jobs, equipment, and men.

Job Standards—Time Study. As previously discussed under Methods, there is some repetitive maintenance work. With this type of work, it is more practical to establish job standards through production-type time studies than by using the standard-data approach. The time study on a specific job supplies the data for developing a job standard based on specific elemental times and allowances for specific conditions for that job.

The method is the same as that used for establishing an ordinary production standard through time study. The processing of several complete units is observed for the purpose of determining the elements of operation. Again, the most important consideration in this mental process of breaking an operation into elements is the separation of constant times from variable times as previously discussed under Standard Data—Time Study.

After the elemental breakdown has been determined, the elements are described on the time-study observation sheet with complete definitions of the movements required for each element. During the time, the breaking points between elements are observed very closely in order to record the correct actual elemental times. Each element must be consistent within itself as far as the starting and stopping points are concerned. If the time breaks between elements are inconsistent, the resulting elemental times will mean nothing. They will have no value for building standard data to be used as a basis for other standards or for use in investigating various improvement proposals involving costs.

The number of cycles to be studied during a time study is normally decided by the time-study observer. There are statistical methods in use to determine if enough time-study readings have been taken on each element and whether they are representative of the time required for that element. However, it is more economical to employ trained time-study engineers who can adequately judge whether sufficient usable data have been obtained than to determine statistically if every elemental time has enough data to support it. The length of study varies with the type of operation, the amount of variables in the operation, the length of the operation cycle, and the performance of the operator. The important consideration is to be sure that sufficient cycles are observed in order to time, record, level the operator's performance, and document all the normal conditions of the operation. If this is an operation that has never been placed on standard, several 8-hr time studies may be taken to make sure all normal conditions have been observed. If this is a variation of a similar job that has already been placed on standard, the 8-hr studies may not be required to determine if all normal conditions have been provided for in the elemental times or allowances. The variation from the existing standard might be studied for elemental times and the conditions found from the previous 8-hr studies used in combination with these new elemental times to establish the production standard.

It is possible to use the ratio-delay technique to determine allowances, but in terms of calendar days this technique takes longer. The results of 8-hr time studies can be developed and in use, with the resulting savings in effect while ratio-delay data are still being collected. The frequency of the occurrence of repetitive work (e.g., one lot every week or one lot every month) will also delay any work or allowances based on ratio delay.

During the time study it is important that a complete record be made of the conditions existing at the time of the study. An observer can hardly document enough of the pertinent facts on a job to answer all the questions that will be asked at a later date. The value of such descriptive and accurate information can best be appreciated by considering that in the past questions or disagreements with the production standard turned into grievances when this documentation was lacking.

Figure 2.10 illustrates the elemental buildup of the standard. This operation, Straighten Paraffin Press Plates, is the same operation that was previously used to illustrate a methods change. When the

STRAIGHTEN PARAFFIN PRESS PLATES

OPERATION: All straightening of paraffin press plates performed on the Hilles and Jones straightening roll (B43).

STANDARD TIMES INCLUDE:

1. Change tickets, to press plate, pick up, place in roll, roll and reverse roll, remove from rolls to gage, inspect and gage, reroll when necessary, straighten with hammer when necessary, remove and aside to finished pile and stack on jig.
2. Conditional allowances: The standard-time table includes a conditional allowance for the noncyclical elements necessary to accomplish the major operation. This allowance was computed to be 3.5% of the constant times and includes the following: starting preparation of getting tools out, oil equipment, clean tools and work area, move out of way of overhead crane, necessary interruptions by foreman, necessary conversations pertaining to work.
3. Personal allowance: A personal allowance of 7% has been included in the standard-time table.

CONDITIONS: The application of these standards is contingent on the following conditions:

1. These standards apply to those operations performed on the below listed machine, in the boiler shop, as specified in the tables.
   Hilles and Jones straightening roll (B43)
   Manufacturer, Hilles and Jones, 1892
   Manufacturer's description, No. 1
      7-in. rolls
      Gear-driven
      No manufacturer's specifications available
      Motor, 7.5 hp, 25 cycle, 440 volts
      Motor No. 382
   Accessories
      Overhead-crane service
      Sledge hammers
      Leveling board
      5 jigs to position plates and transport
      Roller stand
2. These standards are supplementary to the Boiler Shop Standard Plan No. 1-45.
3. These standards shall apply only as long as the tools and equipment are of standard quality and the machine is in good working condition.
4. When changes in tools, equipment, machine, materials, or procedures are effected, the industrial engineering department shall be informed in writing.
5. All press plates with respect to the quality of finished plate shall be subject to the approval of the foreman.
6. The gang size consists of two men.

APPLICATION OF STANDARD: Standard time shall be applied to each job and computed on the job performance biweekly basis.

Standard-time Table: 08.01

Operation	Man-hours
Straighten paraffin press plates (per lot of 50)...........	10.00

EXPLANATION OF TABLE: Allow the standard time once for each lot of 50 plates.

FIGURE 2.14 Data essential to document the conditions for work properly for a job standard.

standard is released to the shop, it is accompanied by data for this job standard that include a definition of the work involved (see Fig. 2.14), a listing of the elements of work that are included in the standard times, and a definition of the allowances to further define the basis of the job standard. The job standard is applied only when certain conditions are met. These conditions specify the equipment and the accessories, which include handling devices and miscellaneous jigs and fixtures. The gang size is also stated (Fig. 2.14) and the responsibility for quality is placed with the supervisor or team lead. In this section it is specified also (items 3 and 4) that the standards are applicable only as long as equipment and working conditions remain the same. In the standard timetable it is noted that the lot size on which the standard is based is 50 plates. Under discussion of repetitive work in the section on Methods, a mention of constant lot size for repetitive jobs was made. To further define the standard, sketches of part and layout are supplied (see Fig. 2.15). For future changes of equipment or method all this documentation is necessary to demonstrate the "before" and "after" elemental times, the basis of changing standards.

Much of the same procedure would be followed if the standard had been developed through MTM data. The elemental times would have been developed through the use of MTM data rather than time study. The conditional allowances would properly be developed through continuous observation of the job or ratio-delay studies. The MTM procedure will be discussed in detail in the following pages.

Standard Data—Time Study. After preliminary work has established standard methods, the next step is to take a few experimental time studies to determine standard elements. These standard elements then become the breaking points between work elements for watch readings in the process of determining elemental times. When building standard data, the watch must be read at the same instant for every like element. After standard elements have been determined, time studies can be started to obtain elemental times for the development of standard data.

The need for elemental standard data can be explained in this manner. While the elements of work on equipment, or as performed by a craft, are the same, they vary in degree from job to job, depending upon the frequency of occurrence or amount of work (i.e., length of cut). Another way of stating this condition is this: While each job (within the same craft) has a similar work cycle, the number of times or degree that each element is performed in the cycle varies between jobs. In order to compensate for this variance, elemental standard-data times are established to measure properly the work on each job. These elemental production standards are issued as time values for performing an element, or a degree of an element, and whenever possible extended into a group of elements (all constant

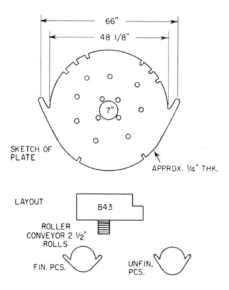

FIGURE 2.15 Parts and layout sketches serve to document the work and work methods in anticipation of future improvements in material or equipment.

times per each occurrence). When establishing a standard time for a job, the worker must be credited with the number of times and the degree to which each element must be used to complete the work. To illustrate, the development of the elemental standard data for the work of Flame Cutting with Portable Cutting Machines will be used.

Since this work involved machines, standard cutting speeds and standard equipment specifications had to be determined. Through a series of experiments prior to study (a portion of the methods phase), the size of the cutting tip, fuel pressure, oxygen pressure, and cutting speeds were established (standard cutting-speed table, Fig. 2.16). The allowed cutting times (standard cutting table, Fig. 2.17) reflect these specifications.

Before the flame cut could commence, the material had to be preheated by the fuel; then the oxygen was turned on and the machine started. In the course of obtaining machine speeds, it was also found that the preheat time was a variable, by stock thickness, and whether the cut started at the edge of the plate stock (O.B. preheat) or in from the edge of the plate (I.B. preheat). The inside preheat took longer because of the greater mass of material that had to be brought up to burning temperature.

Flame Cutting

MACHINE CUTTING

Standard Cutting-speed Table 18.00

Size plate, in.	Size tip (Oxweld)	Fuel pressure	Oxygen pressure	Cutting speed, in. per min
1/16	4	3	30	20.4
1/8	4	3	30	18.5
3/16	4	3	30	17.3
1/4	4	3	30	16.0
5/16	4	3	30	15.0
3/8	4	3	30	14.3
7/16	4	3	30	13.8
1/2	4	3	40	13.0
9/16	4	3	40	12.3
5/8	6	3	40	11.9
11/16	6	3	40	11.3
3/4	6	3	40	10.8
13/16	6	3	40	10.5
7/8	6	3	40	10.1
15/16	6	3	40	9.8
1	6	3	50	9.7
1 1/8	8	3	50	9.0
1 1/4	8	3	50	8.5
1 1/2	8	3	50	8.1
1 3/4	10	3	50	7.4
2	10	3	50	6.8
2 1/2	10	3	50	6.0
3	10	4	50	5.5
4	12	4	60	5.5
4 1/2	12	4	60	5.0

Standard Job-preparation Table 18.01

Job	Man-hr per job
1. Flame cutting, CM-16 and CM-30 portable machines, tray cutting machines..	0.10

FIGURE 2.16 The first step in the development of standard data is to define machine specifications (standard cutting-speed table). Job preparation was determined to be a series of elemental times that were constant for all flame-cutting jobs.

Standard Cutting Table 18.02

Plate* thickness or depth of bevel, in.	Man-hr per 100 in.		Plate* thickness or depth of bevel, in.	Man-hr per 100 in.	
	Square burn	Bevel burn†		Square burn	Bevel burn†
1/16	0.10	0.16	13/16	0.19	0.27
1/8	0.11	0.17	7/8	0.20	0.28
3/16	9.11	0.17	15/16	0.20	0.28
1/4	0.12	0.19	1	0.21	0.30
5/16	0.13	0.20	1 1/8	0.22	0.31
3/8	0.15	0.22	1 1/4	0.24	0.32
7/16	0.15	0.22	1 1/2	0.25	0.34
1/2	0.15	0.22	1 3/4	0.27	0.37
9/16	0.16	0.24	2	0.29	0.39
5/8	0.17	0.25	2 1/2	0.33	0.44
11/16	0.17	0.25	3	0.36	0.48
3/4	0.19	0.27	3 1/2	0.36	0.48

* For any intermediate thickness of plate, use next higher thickness.
† Substitute the depth of bevel for plate thickness, chip edge after bevel included in standard times.

Standard Material-handling Table 18.04
CM-16 Machine

Description	Weight of finished piece, lb	Man-hr per piece
Position material to flame cut, burn scrap, scrap aside, aside finished piece.	0-25	0.02
	26-50	0.03
	51-100	0.08
	101-200	0.16
	201-300	0.23
	301-500	0.29
	501-1,000	0.35
	Over 1,000	0.37

Standard Setup Table, Tray Cutting 18.05

Description	Unit per job	Man-hr per circle
1. Prepare stock, and place machine on stock.	1st circle cut.	0.22
2. Prepare stock and reposition machine.	Each additional circle cut.	0.01

FIGURE 2.17 Standard cutting times were developed from the same data that established the machine specifications. Chipping the edge of plate after beveling, a variable by length of bevel, was added to the bevel-cutting time since measurement and application of these two elements are handled in the same manner. The material-handling table was based on a curve determined by charting weight of finished piece against the time of the material-handling elements for each piece studied. From a number of studies, the time required to set up the tray-cutting machine was determined. An additional setup allowance is necessary to compensate for repositioning the machine for additional cuts in the same job.

Flame Cutting

Standard Preheat Table
Man-hr per Occurrence

Plate thickness, in.	O.B. preheat (1)	I.B. preheat (2)
Up to ¼	0.007	0.010
Over ¼ to ⅜	0.013	0.019
Over ⅜ to ½	0.014	0.024
Over ½ to ⅝	0.015	0.028
Over ⅝ to ¾	0.016	0.031
Over ¾ to 1	0.018	0.034
Over 1 to 1½	0.020	0.039
Over 1½ to 2	0.022	0.044
Over 2 to 2½	0.024	0.048
Over 2½ to 3	0.025	0.051
Over 3 to 3½	0.026	0.053

Standard Table, Miscellaneous Operations 18.09

Description	Man-hr per occ.
1. Change or adjust cascade oxygen system pressure.	0.25
2. Exchange bottles in propane bank system.	0.04

FIGURE 2.18 The total time of flame cutting consists of a preheat time, a constant value by thickness of plate and type of preheat (O.B. or I.B). The above standard times for preheating were determined at the same time machine-cutting specifications were established. There are two miscellaneous operations that can occur in the burner's workday. These operations were measured and the standard established for each type of grouping the elemental times values required to perform the operation.

The same series of experiments that determined cutting speeds were used to establish preheat times (see standard preheat table, Fig. 2.18).

The first elements of work encountered on actual job study were preparatory elements, changing job tickets, and requisitioning material. These elements were determined to be a series of constant times for all flame-cutting jobs.

The next elements of work were those involving the positioning of material. When developing the standard data for material handling, the elements of work required in the removal of finished pieces were added to the positioning elements. These times were charted by weight of finished piece and the material-handling time of each piece. This chart was the basis for the standard material-handling table (Fig. 2.17). This charting procedure is explained in detail in Phil Carroll's book, *How to Chart Time-study Data*.*

The machine setup times (Fig. 2.18A) were established by grouping and totaling the elemental time values required for the various types of setup. From the large number of jobs that were studied, the necessity for the series of setup times was easily recognized.

In addition to the regular work elements, as outlined previously, additional time has been included in the standard timetables as a conditional allowance. This conditional allowance is to compensate for noncyclical elements necessary to supplement the previously detailed work elements. This allowance was computed to be 10 percent of the standard time of the cyclical work elements and included the following: get tools from box or locker; notify crane of lifts to be made; clean

---

*McGraw-Hill Book Company, New York, 1960.

Flame Cutting

Standard Setup Table, CM-16 Machine

Description	Man-hr per occurrence per job	
	First occur.	Each additional occur.
1. Prepare and lay out sheet stock to burn OD of circle—set up machine to burn—for square cut.	0.132	0.073
2. Prepare and lay out sheet stock to bevel OD of circle after square cut—set up machine.	0.059	0.029
3. Measure and mark landing on stock to be beveled—circles only.	0.093	0.093
4. Prepare and lay out sheet stock to burn ID of circle—set up machine to burn—for square cut.	0.067	0.040
5. Locate center of circle (underside of piece only) prior to making second bevel.	0.076	0.076
6. Prepare stock and set up to burn arc—up to 36 in. radius.	0.068	0.038
7. Prepare stock and set up to burn arc—over 36 in. radius.	0.154	0.061

Standard Setup Table, CM-30 Machine 18.07

Description	Man-hr per occurrence per piece	
	First occur. per piece	Each additional occur. per piece
1. Prepare stock and set up track and CM-30 machine to square cut or bevel up to 60 lin in. per cut.	0.044	0.044
2. Prepare stock and set up track and CM-30 machine to square cut or bevel over 60 lin in. per cut.	0.118	0.069

FIGURE 2.18A Separate setup times are necessary for each machine by type of work.

flame-cutting tips when necessary; necessary interruptions by supervisor or team lead, and other necessary conversation pertaining to work. A personal allowance of 7 percent was also included in the standard timetables.

Application of these flame-cutting standards has been illustrated by Figs. 2.1 and 2.2.

Extension of Standard Data into Job or Component Standards. After elemental standard data have been developed from time studies, they must in turn be developed into larger increments of time wherever possible to facilitate the application of standards. An example of this development is the common job of flame-cutting blanks prior to machine-shop operations. Using the elemental standards to determine the job standard for the job, as illustrated by Fig. 2.19, nine elemental-time calculations are necessary to develop the standard. To simplify the application of time, tables were developed combining basic times for flame-cutting circumferences with a portable cutting machine, reducing calculations to five, and supplying a work sheet. The machine setups, preheat times, and burn times have all been grouped in a job table (Fig. 2.20). Since there can be a number of combinations of inside- and outside-diameter sizes, it was not practical to include material-handling time, which is based on weight of finished piece. For ease in calculating material-handling time, a basic

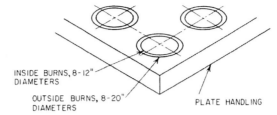

FIGURE 2.19 Flame-cutting blanks—ideal job for standard-data application. Diameter and plate thickness vary; however, any combination of size, within limits of the equipment used, can be worked on incentive through application of the standard data.

weight table with steel circles (Fig. 2.21) is included with the job tables. A preprinted form is designed to reduce rate-setting time further on this operation. Figure 2.22 illustrates this form and the development of a standard for flame-cutting blanks.

Advantages and Disadvantages—Time Study. The development of standard data based on time study has several immediate cost considerations that must be considered as disadvantages:

1. *High initial cost.* A large number of elemental time studies are required. In addition to the actual studies, the development of the data is an expensive project.
2. *Training costs.* Maintenance management must have an application of time-study and work-measurement techniques. The reliability engineers must be trained in craft practice.

STANDARD TIME TABLE 18.11-FLAME CUTTING CIRCUMFERENCE
CM-16 MACHINE-JOB TABLE (CONT'D.)

DIAMETER TO BURN	PLATE THICKNESS TYPE OF CUT	1 3/4"		2"		2 1/2"	
		I.B.	O.B.	I.B.	O.B.	I.B.	O.B.
11"	FIRST OCCURRENCE	0.204	0.247	0.208	0.251	0.229	0.270
	ADDITIONAL OCCURRENCE	0.177	0.188	0.181	0.192	0.202	0.261
12"	FIRST OCCURRENCE	0.213	0.256	0.217	0.260	0.239	0.280
	ADDITIONAL OCCURRENCE	0.186	0.197	0.190	0.201	0.212	0.221
13"	FIRST OCCURRENCE	0.221	0.264	0.225	0.268	0.250	0.291
	ADDITIONAL OCCURRENCE	0.194	0.205	0.198	0.209	0.223	0.232
14"	FIRST OCCURRENCE	0.230	0.273	0.234	0.277	0.260	0.301
	ADDITIONAL OCCURRENCE	0.203	0.214	0.207	0.218	0.233	0.242
15"	FIRST OCCURRENCE	0.238	0.281	0.243	0.286	0.270	0.311
	ADDITIONAL OCCURRENCE	0.211	0.222	0.216	0.227	0.243	0.252
16"	FIRST OCCURRENCE	0.247	0.290	0.252	0.295	0.281	0.322
	ADDITIONAL OCCURRENCE	0.220	0.231	0.225	0.236	0.254	0.263
17"	FIRST OCCURRENCE	0.255	0.298	0.261	0.304	0.291	0.332
	ADDITIONAL OCCURRENCE	0.228	0.239	0.234	0.245	0.264	0.273
18"	FIRST OCCURRENCE	0.264	0.307	0.269	0.312	0.302	0.343
	ADDITIONAL OCCURRENCE	0.237	0.248	0.242	0.253	0.275	0.284
19"	FIRST OCCURRENCE	0.272	0.315	0.278	0.321	0.312	0.353
	ADDITIONAL OCCURRENCE	0.245	0.256	0.251	0.262	0.285	0.294
20"	FIRST OCCURRENCE	0.281	0.324	0.287	0.330	0.322	0.363
	ADDITIONAL OCCURRENCE	0.254	0.265	0.260	0.271	0.295	0.304
21"	FIRST OCCURRENCE	0.289	0.332	0.296	0.339	0.333	0.374
	ADDITIONAL OCCURRENCE	0.262	0.273	0.269	0.280	0.306	0.315

FIGURE 2.20 Portions of job tables for flame-cutting circumferences. Material handling and job preparation are the only additional times to be added to this table. Similar tables were developed from the elemental data tables for flame-cutting arcs and straight lines, bevel and square cuts.

3.40 ENGINEERING AND ANALYSIS TOOLS

STANDARD WEIGHT TABLE 18–10
WEIGHT OF STEEL CIRCLES
POUNDS

DIAMETER	PLATE THICKNESS IN INCHES		
	1 3/4"	2"	2 1/2"
11"	47.20	53.98	67.42
12"	56.18	64.24	80.23
13"	65.93	75.40	94.17
14"	76.46	87.44	109.21
15"	87.78	100.38	125.37
16"	98	112	141
17"	112	128	160
18"	125	142	178
19"	140	160	200
20"	154	176	221
21"	170	194	243

FIGURE 2.21 A table of weights is furnished the rate setter to accompany job tables for determining weight of parts. From the material-handling table, by weight of finished part, material-handling times can be determined.

3. The payoff date is far away from the start of a complete work-measurement program. A large amount of standard coverage should be available before any work-measurement controls or incentive payments are in effect.

The advantages are:

1. Training of supervision in management's methods.
2. Application costs are reasonable once the standard data have been developed.

```
              FLAME CUTTING BLANKS
DATE    5/5                WORK ORDER NO  870-A
DESCRIPTION  Eight (8) Blanks - 20" OD - 12" ID -
2" steel plate stock - no bevel

BURN OB  - 1 ST PC   _____ =  .330
BURN OB  - ADD PCS    7  X  .271   = 1.897
BURN IB  - 1 ST PC   _____ =  .217
BURN IB  - ADD PCS    7  X  .190   = 1.330
MAT HDLG              8  X   16    = 1.280
JOB PREP                           =  .100
                                    ======
                   ALLOWED HRS.      5.154
```

FIGURE 2.22 Rate-setting form to speed development of standard times from job tables. While diameters and plate thicknesses vary among jobs, job tables provide times within the limits of the equipment used for all blank-cutting jobs.

3. Forces tighter control over:
   a. Work methods.
   b. Material specifications.
   c. Maintaining maintenance equipment.
4. Flexibility of the system as shown by the fact that standards can be established for group or individual payment and that elements of work are reproducible.
5. A common unit of measurement is available to evaluate the performance of individual craftsmen and foremen. (Employees can relate earnings to their productivity by direct comparison of time.)
6. Audits are possible to determine if proper labor payments have been made.
7. Information is available to improve management controls in the areas of:
   a. Planning and scheduling work.
   b. Planning and scheduling equipment.
   c. Planning and scheduling manpower standards, providing an accurate means of determining excess maintenance costs in specific areas or on specific equipment or parts which would justify modifications or replacement of equipment.
8. With synthetic standard data, such as MTM, specialists in MTM are required to install and maintain the work-measurement installation. Standard data, developed by time study, require experienced time-study men; thus ease any turnover problem by reducing training costs of replacement Reliability engineers.

MTM Data. Methods-time measurement data are predetermined basic-motion time values described in classifications of Reach, Move, Turn, Grasp, and so on. The unit of time, a TMU, used for measurement, is 0.00001 hr. The tables (Fig. 2.23) give the number of TMU's "required by the operator of average skill working with average effort to make the designated motion under average conditions."*

To illustrate the use of MTM data in developing maintenance standards together with their application, examples will be presented of typical plant work. Maintenance work within plant areas was chosen over the shop maintenance work as the former might be considered by the reader to be the more difficult of the two. An example of a shop job is given in an illustration under the time-study section. The development of standards for this shopwork is quite similar to the development of standards for job-lot production work.

Formulas are developed for each variation of work a craftsman encounters. For example, in the painting craft, some of these variations might be "Preparing and Brush Painting Walls, Preparing and Brush Painting Ceilings, Cleaning and Brush Painting Pipe, Cleaning and Brush Painting Machine Equipment." Figure 2.27 is "Cleaning and Brush Painting Machine Equipment," which is a common machine-maintenance job occurring in production and maintenance areas.

The first step in the development of standards, using MTM, is the collection of complete information as to the work, methods, equipment, and materials used. Every motion required to complete the work must be analyzed, classified, and recorded. The observer must watch the accepted methods in order to determine the sequence in which these motions occur as well as to obtain a correct mental picture of the motions required to do the entire operation.

These motions are then broken down into MTM elements and entered on a "methods-analysis chart," together with the TMU required to complete the motion. In Fig. 2.24, note Element A, "Pick up 1-in., round, all-purpose brush—remove excess solvent." The painter reaches for the brush in a can of solvent (R14B or Reach, 14 in., Case B), and takes hold of the brush (G1A or Grasp, Case 1A). He moves the brush to the lip of the can (M8C or Move 8 in., Case C), and positions the brush to remove excess solvent (PISE, position Class 1 fit, symmetrical, easy to handle). He presses brush slightly to remove excess solvent (AB2 apply pressure, part up to 2 lb) and turns body to machine (TBCL, turn body, Case 1). He moves brush to machine surface (M12B, move brush 12 in., Case B). The total time is 0.053 min for that element. In a similar manner, the same

---

*H. B. Maynard, G. H. Stegemerten, and J. L. Schwab, Methods-Time Measurement, *Factory Management and Maintenance,* New York, 1948.

TABLE I—REACH—R

Distance Moved Inches	Time TMU A	Time TMU B	Time TMU C or D	Time TMU E	Hand In Motion A	Hand In Motion B
¾ or less	2.0	2.0	2.0	2.0	1.6	1.6
1	2.5	2.5	3.6	2.4	2.3	2.3
2	4.0	4.0	5.9	3.8	3.5	2.7
3	5.3	5.3	7.3	5.3	4.5	3.6
4	6.1	6.4	8.4	6.8	4.9	4.3
5	6.5	7.8	9.4	7.4	5.3	5.0
6	7.0	8.6	10.1	8.0	5.7	5.7
7	7.4	9.3	10.8	8.7	6.1	6.5
8	7.9	10.1	11.5	9.3	6.5	7.2
9	8.3	10.8	12.2	9.9	6.9	7.9
10	8.7	11.5	12.9	10.5	7.3	8.6
12	9.6	12.9	14.2	11.8	8.1	10.1
14	10.5	14.4	15.6	13.0	8.9	11.5
16	11.4	15.8	17.0	14.2	9.7	12.9
18	12.3	17.2	18.4	15.5	10.5	14.4
20	13.1	18.6	19.8	16.7	11.3	15.8
22	14.0	20.1	21.2	18.0	12.1	17.3
24	14.9	21.5	22.5	19.2	12.9	18.8
26	15.8	22.9	23.9	20.4	13.7	20.2
28	16.7	24.4	25.3	21.7	14.5	21.7
30	17.5	25.8	26.7	22.9	15.3	23.2

CASE AND DESCRIPTION
A Reach to object in fixed location, or to object in other hand or on which other hand rests.
B Reach to single object in location which may vary slightly from cycle to cycle.
C Reach to object jumbled with other objects in a group so that search and select occur.
D Reach to a very small object or where accurate grasp is required.
E Reach to indefinite location to get hand in position for body balance or next motion or out of way.

TABLE II—MOVE—M

Distance Moved Inches	Time TMU A	Time TMU B	Time TMU C	Hand in Motion B
¾ or less	2.0	2.0	2.0	1.7
1	2.5	2.9	3.4	2.3
2	3.6	4.6	5.2	2.9
3	4.9	5.7	6.7	3.6
4	6.1	6.9	8.0	4.3
5	7.3	8.0	9.2	5.0
6	8.1	8.9	10.3	5.7
7	8.9	9.7	11.1	6.5
8	9.7	10.6	11.8	7.2
9	10.5	11.5	12.7	7.9
10	11.3	12.2	13.5	8.6
12	12.9	13.4	15.2	10.0
14	14.4	14.6	16.9	11.4
16	16.0	15.8	18.7	12.8
18	17.6	17.0	20.4	14.2
20	19.2	18.2	22.1	15.6
22	20.8	19.4	23.8	17.0
24	22.4	20.6	25.5	18.4
26	24.0	21.8	27.3	19.8
28	25.5	23.1	29.0	21.2
30	27.1	24.3	30.7	22.7

Wt. Allowance: Wt. (lb.) Up to	Factor	Constant TMU
2.5	0	0
7.5	1.06	2.2
12.5	1.11	3.9
17.5	1.17	5.6
22.5	1.22	7.4
27.5	1.28	9.1
32.5	1.33	10.8
37.5	1.39	12.5
42.5	1.44	14.3
47.5	1.50	16.0

CASE AND DESCRIPTION
A Move object to other hand or against stop.
B Move object to approximate or indefinite location.
C Move object to exact location.

TABLE III—TURN AND APPLY PRESSURE—T AND AP

Weight	30°	45°	60°	75°	90°	105°	120°	135°	150°	165°	180°
	Time TMU for Degrees Turned										
Small— 0 to 2 Pounds	2.8	3.5	4.1	4.8	5.4	6.1	6.8	7.4	8.1	8.7	9.4
Medium—2.1 to 10 Pounds	4.4	5.5	6.5	7.5	8.5	9.6	10.6	11.6	12.7	13.7	14.8
Large— 10.1 to 35 Pounds	8.4	10.5	12.3	14.4	16.2	18.3	20.4	22.2	24.3	26.1	28.2

APPLY PRESSURE CASE 1—16.2 TMU. APPLY PRESSURE CASE 2—10.6 TMU

TABLE IV—GRASP—G

Case	Time TMU	DESCRIPTION
1A	2.0	Pick Up Grasp—Small, medium or large object by itself, easily grasped.
1B	3.5	Very small object or object lying close against a flat surface.
1C1	7.3	Interference with grasp on bottom and one side of nearly cylindrical object. Diameter larger than ½".
1C2	8.7	Interference with grasp on bottom and one side of nearly cylindrical object. Diameter ¼" to ½".
1C3	10.8	Interference with grasp on bottom and one side of nearly cylindrical object. Diameter less than ¼".
2	5.6	Regrasp.
3	5.6	Transfer Grasp.
4A	7.3	Object jumbled with other objects so search and select occur. Larger than 1" x 1" x 1".
4B	9.1	Object jumbled with other objects so search and select occur. ¼" x ¼" x ⅛" to 1" x 1" x 1".
4C	12.9	Object jumbled with other objects so search and select occur. Smaller than ¼" x ¼" x ⅛".
5	0	Contact, sliding or hook grasp.

TABLE V—POSITION*—P

CLASS OF FIT		Symmetry	Easy To Handle	Difficult To Handle
1—Loose	No pressure required	S	5.6	11.2
		SS	9.1	14.7
		NS	10.4	16.0
2—Close	Light pressure required	S	16.2	21.8
		SS	19.7	25.3
		NS	21.0	26.6
3—Exact	Heavy pressure required.	S	43.0	48.6
		SS	46.5	52.1
		NS	47.8	53.4

*Distance moved to engage—1" or less.

TABLE VI—RELEASE—RL

Case	Time TMU	DESCRIPTION
1	2.0	Normal release performed by opening fingers as independent motion.
2	0	Contact Release.

TABLE VII—DISENGAGE—D

CLASS OF FIT	Easy to Handle	Difficult to Handle
1—Loose—Very slight effort, blends with subsequent move.	4.0	5.7
2—Close — Normal effort, slight recoil	7.5	11.8
3—Tight — Considerable effort, hand recoils markedly	22.9	34.7

TABLE VIII—EYE TRAVEL TIME AND EYE FOCUS—ET AND EF

Eye Travel Time $= 15.2 \times \frac{T}{D}$ TMU, with a maximum value of 20 TMU.

where T = the distance between points from and to which the eye travels.
D = the perpendicular distance from the eye to the line of travel T.

Eye Focus Time = 7.3 TMU.

TABLE IX—BODY, LEG, AND FOOT MOTIONS

DESCRIPTION	SYMBOL	DISTANCE	TIME TMU
Foot Motion—Hinged at Ankle.	FM	Up to 4"	8.5
With heavy pressure.	FMP		19.1
Leg or Foreleg Motion.	LM —	Up to 6"	7.1
		Each add'l. inch	1.2
Sidestep—Case 1—Complete when leading leg contacts floor.	SS-C1	Less than 12"	Use REACH or MOVE Time
		12"	17.0
		Each add'l. inch	.6
Case 2—Lagging leg must contact floor before next motion can be made.	SS-C2	12"	34.1
		Each add'l. inch	1.1
Bend, Stoop, or Kneel on One Knee.	B,S,KOK		29.0
Arise.	AB,AS,AKOK		31.9
Kneel on Floor—Both Knees.	KBK		69.4
Arise.	AKBK		76.7
Sit.	SIT		34.7
Stand from Sitting Position.	STD		43.4
Turn Body 45 to 90 degrees— Case 1—Complete when leading leg contacts floor.	TBC1		18.6
Case 2—Lagging leg must contact floor before next motion can be made.	TBC2		37.2
Walk.	W-FT.	Per Foot	5.3
Walk.	W-P	Per Pace	15.0

TABLE X—SIMULTANEOUS MOTIONS

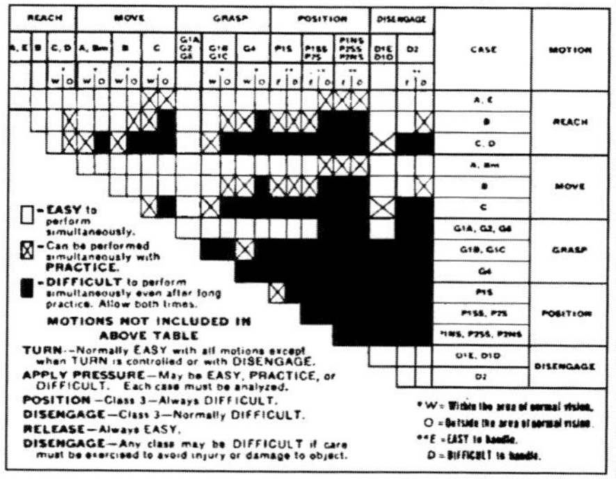

1 TMU = .00001 hour = .0006 minute = .036 seconds

FIGURE 2.23 MTM times tables containing basic motion-time values as classified by the type of motion.

METHODS ANALYSIS CHART I-54-5-6

PART MACHINE EQUIPMENT DEPT. ______ DWG. ______ SH. 1 OF 7

DESCRIPTION—LEFT HAND	NO	L. H. SYM	T. M. U	R. H. SYM.	DESCRIPTION—RIGHT HAND
A Pick up 1" Round All-Purpose Brush - Remove excess solvent					
			14.4	R14B	Reach for Brush
			1.7	G1A	Grasp
			10.6	M8C	Brush to lip
			5.6	P1SE	Align on lip
			10.6	AP2	Press slightly to get rid of excess
			31.9	AB	
				(TBC1)	Turn
			13.4	M12B	Brush to mach. surface
			88.2		
				X .0006 = .053 Minutes	
B Dip 1" Brush into Solvent					
			29.0	B	
				(TBC1)	
			15.2	M12C	Brush into can
			5.6	P1SE	Dip Brush
			10.6	M8C	To lip of gallon can
			5.6	P1SE	Align on lip
			10.6	AP2	Pressure squeeze
			31.9	AB	Arise
			13.4	M12B	Brush to mach. surface
			121.9		
				X .0006 = .073 Minutes	
C Brush Solvent onto smooth SPLASH GUARD - About one square foot area.					
			10.6	AP2	
	12		321.6	2M12B	
			332.2		
				X .0006 = .199 Minutes	
		TOTAL		T. M. U. X .0006 =	MINUTES

ELEMENT Cleaning for Paint SYMBOL A, B & C

DATE ______ BY ______

FIGURE 2.24 The first step in developing standard data with MTM time values. MTM data (Fig. 2.23) were applied to a series of basic motions required to complete elements A, B, and C.

procedure is repeated for all the other elements necessary to "Clean and Brush Paint" machine equipment (B through V). These elements are shown in Fig. 2.25. At this point the MTM basic data have been converted into motions having greater time values to facilitate future formula development.

If these data were developed from stopwatch time studies, the same elements should and probably would be separated for formula application. One advantage of MTM over the stopwatch is the simplicity of work in arriving at these standard times as compared with the work involved in the accumulation of a large amount of elemental time-study data and developing standard data by thorough time-study procedures.

The work observed was then analyzed and classified and the formulas developed based on the work required (Fig. 2.26). With these formulas, time standards, based on a square-foot unit, were established for cleaning and painting. These standard times given in leveled minutes per square foot of area covered were then combined in table form (Fig. 2.27) showing examples of different classes of work in order to better define the application limits. For handling ease, the standard times from Table 1 (Fig. 2.27) were extended into another table (Fig. 2.28), by square foot of area with the various cleaning and painting combinations.

Maintenance—Plant-wide
Title: Cleaning and Painting

Formula #1-54-5-6
Sheet 2 of 15 sheets

ANALYSIS

The tools required for this job are 1- and 2-in. scrapers, a wire brush, round all-purpose paint brushes from ½ to 2 in. in diameter, and an air hose.

The materials used are a carbon tetrachloride solvent, gray paint (Roxalin 13-494), rags, and empty gallon cans for holding the solvent.

The areas cleaned and painted are measured and figured, then classified, and then calculated for their leveled times. See Sheet 8 for a sample calculation.

PROCEDURE

The operator dips a 1-in. brush into a can of solvent, and brushes the solvent onto the machine with a rubbing back and forth motion. When necessary, the operator will also use a scraper or wire brush to remove the grime not removed by the brush and solvent. After the entire area has been cleaned in the above manner, the operator will saturate a rag with solvent and go over the cleaned area.

The operator dips his paint brush into an open can of paint, brushes off the excess paint on the lip of the can, and paints the machine. He will use several sizes of brushes ranging from a ½-in. round to a 2-in. round brush. Where trimming is involved, it may be necessary for the operator to wipe the overlap.

TABLE OF ELEMENTS

A. Pick up 1-in. round all-purpose brush and remove excess solvent = a constant per square foot. — 0.053 min
B. Dip 1-in. brush into solvent = a constant per dip. — 0.073 min
C. Brush solvent onto a smooth surface (Example: Splash guard) about 1 sq ft in area = a constant for smooth surfaces. — 0.199 min
D. Lay brush aside = a constant per square foot. — 0.028 min
E. Pick up scraper and lay aside = a constant per square foot. — 0.059 min
F. Scrape a 1 sq ft area of smooth surface using 2-in. scraper = a constant for smooth surfaces. — 0.199 min

Date: June 30

Maintenance—Plant-wide
Title: Cleaning and Painting

Formula #1-54-5-6
Sheet 3 of 15 sheets

G. Take rag from can of solvent, and return to can = a constant per square foot. — 0.093 min
H. Clean 1 sq ft area of smooth surface with a rag saturated with solvent = a constant per square foot of smooth surface. — 0.193 min
J. Brush solvent on an irregular smooth surface = a constant per square foot of irregular surface. — 0.295 min
L. Brush solvent on very irregular surface = a constant per square foot of irregular surface. — 0.392 min
M. Scrape irregular or curved surfaces of about 1 sq ft = a constant per square foot of irregular surface. — 0.296 min
N. Scrape very irregular surfaces = a constant per square foot of very irregular surface. — 0.579 min
P. Dip brush in can of paint = a constant per dip. — 0.080 min
Q. Paint smooth machine surface of about 1 sq ft—use 2-in. round all-purpose brush = a constant per job. — 0.419 min
R. Additional painting on an irregular surface of about 1 sq ft—use ½- to ¾-in. round all-purpose brush = a constant for irregular surfaces. — 0.418 min
S. Lay aside and pick up next size brush = a constant per brush. — 0.021 min
T. Additional painting on a very irregular surface, trimming around projections, levers, handwheels, etc. = a constant for very irregular surfaces. — 1.056 min
U. Pick up rag—lay aside = a constant for very irregular surfaces. — 0.048 min
V. Wipe paint away with rag = a constant for very irregular surfaces. — 0.109 min

Date: June 30

FIGURE 2.25 Elements as developed from basic MTM values (see Fig. 2.24). These elements are outlined in the "procedure" above and are all the elements necessary to develop formulas for "cleaning and brush painting machine equipment."

Maintenance—Plant-wide
Title: Cleaning and Painting

Formula #I-54-5-6
Sheet 4 of 15 sheets

SYNTHESIS

The areas studied were classified into three distinct groups of work.

Class I. Work is for smooth surfaces that are relatively easy to clean and paint. The surfaces are smooth and flat such as base trays and splash guards.
Class II. Work is for irregular surfaces that are not smooth and flat such as curved chip guards and curved belt guards.
Class III. Work is for very irregular surfaces that have projections, piping, levers, and handwheels.

Class III is very difficult cleaning and painting. Examples of Class III are the complete turret assembly, the saddle and cross slide assembly, the lathe head, and the motor.
Class I is relatively easy work, whereas Class III is very difficult with many more highly controlled motions.
Class II primarily fits the work in between the two extremes.
Cleaning was analyzed as follows:

Leveled minutes per square foot of Class I work equals

$$EL\ A + C + D + E + F + G + H$$
$$= 0.053 + 0.199 + 0.028 + 0.059 + 0.199 + 0.093 + 0.193$$
$$= .82 \text{ min per square ft}$$

Leveled minutes per square foot of Class II work equals

$$EL\ A + J + D + E + M + G + 1\tfrac{1}{2}H$$
$$= 0.053 + 0.295 + 0.028 + 0.059 + 0.296 + 0.093 + 1\tfrac{1}{2}(0.193)$$
$$= 1.11 \text{ min per sq ft}$$

Date: June 30

Maintenance—Plant-wide
Title: Cleaning and Painting

Formula #I-54-5-6
Sheet 5 of 15 sheets

Leveled minutes per square foot of Class III work equals

$$EL\ A + 2B + L + D + E + N + G + 1\tfrac{1}{2}H$$
$$= 0.053 + 2(0.073) + 0.392 + 0.028 + 0.059 + 0.579 + 0.093 + 1\tfrac{1}{2}(0.193)$$
$$= 1.64 \text{ min per sq ft}$$

Painting was analyzed as follows:

Leveled minutes per square foot of Class I work equals

$$EL\ 2(P) + Q$$
$$= 2(0.080) + 0.419$$
$$= 0.58 \text{ min per sq ft}$$

Leveled minutes per square foot of Class II work equals

$$EL\ 4(P) + Q + R + 2(S)$$
$$= 4(0.080) + 0.419 + 0.418 + 2(0.021)$$
$$= 1.20 \text{ min per sq ft}$$

Leveled minutes per square foot of Class III work equals

$$EL\ 8(P) + Q + 3(S) + T + U + V$$
$$= 8(0.080) + 0.419 + 3(0.021) + 1.056 + 0.048 + 0.109$$
$$= 2.34 \text{ min per sq ft}$$

Combining the three classes of work, Table I (Fig. 2.27) was developed.
During time checks on painting equipment, it was found that painting the second coat averaged about 60% of the time required for the first coat. This occurs because of the difference of degree of coverage of the first coat. Thus, fewer brush dips and fewer brush strokes are required. When a second coat of paint is required, apply 60% of the first coat values. This also is incorporated in Table I.

Date: June 30

FIGURE 2.26 The areas studied to develop the table of elements (Fig. 2.25) were separated into three distinct groups of work. Formula was based on the elements required to complete work in each grouping for cleaning and painting.

| Maintenance—Plant-wide | Formula #1-54-5-6 |
| Title: Cleaning and Painting | Sheet 7 of 15 sheets |

Table 1. Leveled Minutes per Square Foot of Area

	Class I	Class II	Class III
Cleaning	0.82	1.11	1.64
Painting (one coat)*	0.58	1.20	2.34
Total	1.40	2.31	3.98

* Use 60% for second coat.

Examples of the Different Classes of Work

Class I (smooth surfaces)	Class II (irregular surfaces)	Class III (very irregular surfaces)
	Turret Lathe	
Base or drip trays	Lathe bed	With levers, windows, handwheels
Flat splash guards	Switch boxes	Turret assembly
Flat drain covers	Curved belt guards	Turret fixtures
	Irr. shaped splash guards	Saddle assembly
	Curved chip guards	Cross slide
	Turret base (on an automatic)	Lathe head
		Motor
	Milling Machine	
	Base up to spindle	Spindle
	Knee support	
	Radial Drill	
	Column or post base	Arm
		Gearbox
	Air Conditioner	
Side panels	Side panel—close confined areas	The area around the piping (gingerbread work)

Date: June 30

FIGURE 2.27 Time values developed from the application of the formulas (Fig. 2.26). These values are expressed in "leveled minutes per square foot of area" for cleaning, brush painting, and combined cleaning and brush painting of machine equipment. Examples of the different classes for work are included to clearly define the differences in the three groups of work.

An example of an application of the "cleaning and brush painting machine equipment" standard is also included in the formula report. This is to further define and illustrate the method of applying times to a typical job in a particular classification of painter's work. Figure 2.29 illustrates the method of measuring the work required to clean and brush-paint machine-equipment items. This is only a portion of the total work required on a particular turret lathe (Fig. 2.30). At this point, a standard for this piece of equipment, and all others like it, can be filed for future job standard reference.

Such items as job preparation, cleanup, travel, personal time, and general allowances to cover specific job conditions (working in restricted areas, working from scaffolds, and so on), together with unavoidable-delay allowances, have not been included in the basic standards for "cleaning and brush painting machine equipment." The times are in leveled decimal man minutes and required additional times (job preparation, cleanup, travel, and so on) or allowances (personal, working in restricted areas, and so on) to complete the job standard.

Figure 2.31 is a sketch showing another common type of maintenance work, electrical installation. An office area must be wired for fluorescent lights and a water cooler. MTM data (Fig. 2.32) have been developed into elements (Fig. 2.33) and then expanded further, to facilitate usage, in the

Craft: Painters
Title: Machine Cleaning and Painting

Values from Formula 6

Class I.   Smooth surfaces such as flat splash guards
Class II. Irregular surfaces such as lathe bed and switch boxes
Class III. Very irregular surfaces such as motor and turret assembly

Sq ft	Class I clean and paint	Class I paint second coat	Class II clean and paint	Class II paint second coat	Class III clean and paint	Class III paint second coat
1	1.4	.5	2.3	1.4	4.0	2.4
2	2.8	1.0	4.6	2.8	8.0	4.8
3	4.2	1.5	6.9	4.2	11.9	7.2
4	5.6	2.0	9.2	5.6	15.9	9.6
5	7.0	2.5	11.6	7.0	19.9	12.0
10	14.0	5.0	23.1	14.0	39.8	24.0
15	21.0	7.5	34.7	21.0	59.7	36.0
20	28.0	10.0	46.2	28.0	79.6	48.0
25	35.0	12.5	57.8	35.0	99.5	60.0
30	42.0	15.0	69.3	42.0	119.4	72.0
35	49.0	17.5	80.9	49.0	139.3	84.0
40	56.0	20.0	92.4	56.0	159.2	96.0
45	63.0	22.5	104.0	63.0	179.1	108.0
50	70.0	25.0	115.6	70.0	199.0	120.0
60	84.0	30.0	138.7	84.0	238.8	144.0
70	98.0	35.0	161.8	98.0	278.6	168.0
80	112.0	40.0	184.9	112.0	318.4	192.0
90	126.0	45.0	208.0	126.0	358.2	216.0
100	140.0	50.0	231.1	140.0	398.0	240.0
110	154.0	55.0	254.2	154.0	437.8	264.0
120	168.0	60.0	277.3	168.0	477.6	288.0
130	182.0	65.0	300.4	182.0	517.4	312.0
140	196.0	70.0	323.5	196.0	557.2	336.0
150	210.0	75.0	346.6	210.0	597.0	360.0

FIGURE 2.28 For ease and efficiency in applying standard times to machine cleaning and brush painting, the times from Table 1, Fig. 2.27, have been extended into square-foot areas by the various combinations for work—Class I. Clean and paint one coat; Class I. Paint 2nd coat; etc.

development of operation standards covering the various combinations of electrical work encountered in installing junction boxes and conduit (Fig. 2.34). Additional tables similar to Table 1 (Fig. 2.35) supply the necessary data to complete the application of standard values of time for the work involved (Fig. 2.36). In this MTM application example, the times for "preparation," "cleanup," "travel," and other "allowances" are included.

Advantages and Disadvantages. There are several points concerning MTM applications in maintenance work that have been regarded as disadvantages, but on analyzing these points with regard to operating costs and improved management control methods, individual plants might tend to make these so-called disadvantages appear as advantages. They are:

1. *High initial cost.* However, the MTM installation is made at such a speed that there are savings in engineering time as well as moving up the payoff period due to early increased maintenance efficiency.

MAINTENANCE – PLANT WIDE
TITLE: CLEANING AND PAINTING

FORMULA #1-54-5-6
SHEET 12 OF 15 SHEETS

11. TURRET (LESS ATTACHMENTS AND BASE)
$$\begin{aligned}
\text{TOP AND BOTTOM AREA} &= \pi 10^2 = 314\ \square" \\
\text{SIDE AREA OF HOLE} &= 10\,(\pi 14) = 440\ \square" \\
& \overline{754\ \square"} \\
&\text{OR 5.24 SQ. FT.}
\end{aligned}$$

12.
$$\begin{aligned}
\text{SIDE } 132 \times 24 \times 2\,(\text{SIDES}) &= 6336 \\
132 \times 18 &= 2376 \\
&\overline{8712\ \square"} \\
&\text{OR 60.6 SQ. FT.}
\end{aligned}$$

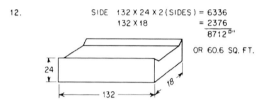

13. SADDLE ASS'Y AND CROSS SLIDE
CROSS SLIDE
$$\begin{aligned}
\text{TOP} = 30 \times 8 &= 240 \\
30 \times 4 \times 2\,(\text{SIDES}) &= 240 \\
8 \times 4 \times 2\,(\text{SIDES}) &= 64 \\
&\overline{544\ \square"}
\end{aligned}$$

SADDLE
$$\begin{aligned}
\text{FRONT} = 22 \times 19 &= 418 \\
\text{SIDE} = 12 \times 22 &= 264 \\
\text{SIDE} = 2(12 \times 19) &= 456 \\
&\overline{1138}
\end{aligned}$$

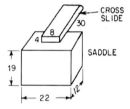

TOTAL AREA = 544 + 1138 = 1682 $\square"$ OR 11.7 SQ. FT.

FIGURE 2.29 Example of the measurement of various machine parts for application of standard times. This is a portion of the total job of cleaning and brush-painting one coat, turret lathe 5319 (Fig. 2.30).

2. Cost of training maintenance management personnel in work methods, equipment care, craft know-how, management control methods, and the MTM system. While this might be regarded as a disadvantage, the cost is more than returned as experienced by progressive companies having competent trained foremen and supervisors who have become cost-conscious controllers.
3. Maintenance engineering personnel must be trained in MTM. This is a disadvantage in small maintenance engineering departments if personnel turnover is a problem. However, the cost must be compared with that involved in the accumulation of the same data through time-study procedures.

The advantages are:

1. The speed of installation which accelerates the payoff date.
2. Training of supervision in management's methods.
3. Application costs are reasonable once the standard data have been developed.

Maintenance—Plant-wide
Title: Cleaning and Painting

Formula #I-54-5-6
Sheet 15 of 15 sheets

Time Summary for Cleaning and Painting One Coat—Turret Lathe #5319

Item	Area, sq ft	Rate	Level. time, decimal min
Class I			
1. Base or drip tray	114.0		
2. Mounted splash guard	14.7		
4. Unmounted splash guard	26.0		
5. Drain cover	8.1		
	162.8	1.40	227.0
Class II			
3. Unmounted splash guard	10.0		
6. Chip guard	9.0		
7. Switch boxes	10.8		
8. Belt guard	15.5		
12. Bed	60.6		
	105.9	2.31	245.0
Class III			
9. Turret fixtures	19.7		
10. Turret base assembly	19.1		
11. Turret	7.3		
13. Saddle assembly and cross slide	11.7		
14. Lathe head	56.0		
	113.8	3.98	453.0
Class III			
15. Motor—painting only	11.2	2.34	26.0
			951.0

Date: June 30

FIGURE 2.30 Time summary for cleaning and brush-painting all surfaces required on turret lathe 5319. Figure 2.29 illustrated the method of measuring various items to obtain the square-foot area. The "rate" was taken from the values in Table 1, Fig. 2.27.

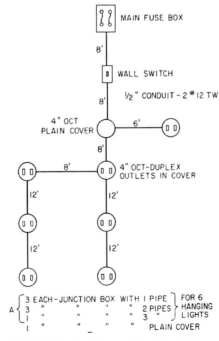

FIGURE 2.31 Sketch of work involved in the installation of fluorescent lights and a water cooler.

3.49

3.50 ENGINEERING AND ANALYSIS TOOLS

METHODS ANALYSIS CHART					FORMULA 49
PART ELECTRICAL CONSTRUCTION	DEPT. _____		DWG. _____		SH. 15 OF 28

DESCRIPTION—LEFT HAND	NO.	L.H.SYM	T.M.U.	R.H.SYM	NO.	DESCRIPTION—RIGHT HAND
K3 Screw nut on pipe end w/fingers						
			3.4	G1A	2	Grasp nut
			8.4	M2B	2	Turn nut down
			3.4	RL1	2	Release nut
			7.4	R2A	2	Reach back
			22.6			
			X.0006 =			.014 mins./rev.
B3 Get tool from belt kit						
			.044 mins. per occ. (El.C1 - formula #44)			
C3 Aside tool to belt kit						
			.053 mins. per occ. (El.G1 - formula #44)			
L3 Tighten nut on pipe end w/hammer and screwdriver						
Reach for screwdriver		R10A	11.3	M10A		Screwdriver to L.H.
Transfer to L.H.		G3	5.6			
Regrasp		G2	5.6			
Screwdriver near nut on pipe						
		M16B	15.8			
Screwdriver to slot on nut edge	4	M2C	16.8			
Position screwdriver	4	PISSD	58.8			
Pressure on screwdriver	4	AP2	42.4			
			18.7	M16C		Hammer to screwdriver
			22.4	PISE	4	Pos. hammer on screwdriver
			55.2	M4B	8	(Hammer stroke - up
			48.8	M4A	8	and down
			301.4			
			X.0006 =			.18 mins./nut
		TOTAL	T.M.U. X .0006 =			MINUTES
ELEMENT _____						SYMBOL _____
				DATE 9-27		BY J.F.

FIGURE 2.32 MTM data used to develop elemental times for electrical work.

4. Forces tighter control over:
   a. Work methods.
   b. Material specifications.
   c. Maintaining maintenance equipment.
5. Flexibility of the system as shown by the fact that standards can be established for group or individual payment and that elements of work are reproducible.

Summary of Suboperation No. 10
Tighten Two Nuts on Mounted Switch Box

Element

(K3) Screw nut on pipe end with fingers (0.014 per rev)

       Run down inside nut   = 0.014 × 2 − 0.028
       Run down outside nut   = 0.014 × 4 − 0.056

(B3) Get tool from belt kit (0.044 per occ.)

       Screwdriver and hammer = 0.044 × 2 − 0.088

(C3) Aside tool in kit (0.053 per occ.)

       Screwdriver and hammer = 0.053 × 2 − 0.106

(L3) Tighten nut on pipe end with hammer
       and screwdriver (0.18 per nut) = 0.18 × 2 − 0.360 *To Operation 4*

                                                    (0.638)

FIGURE 2.33 Basic elemental times of Fig. 2.32 extended into greater portions of work termed suboperations.

Illustration of Summary Value
Junction Box with Outlet—One Pipe Box—Wood Mounting

1. Measure to determine pipe length   3.82
2. Cut off, chamfer, and thread end of pipe   4.51
3. Bend two offsets per pipe   3.64 *See Summary of Operation 4*
4. Assemble junction box and piece of pipe to wood beam   7.39
5. Assemble outlet to junction box—one pipe   5.31 *To Sub-analysis Sheet*

                                            (24.67)

Summary of Operation 4
Assemble Junction Box and Piece of Pipe to Wood Beam

1. Get junction box and four nuts; knock out 2 plugs, avg.   1.106
2. Get pipe and assemble nut to each end   0.717
3. Assembly box to pipe, assemble nut and tighten   0.558
4. Get helper, wait for helper to move ladder, and give helper nut and pipe   1.338
5. Two men ascend ladder with pipe—helper inserts pipe into mounted box   0.502
6. Helper starts nut on pipe (inside box), descends ladder, and returns to his job   0.427
7. Position junction box to wood, drill two holes with ratchet hand drill   0.367
8. Finish hanging junction box   0.983
9. Descend ladder, walk to second ladder, ascend and descend second ladder   0.752
10. Tighten two nuts on mounted junction box   0.638

                                            (7.388)

See Summary of Suboperation No. 10
To Junction Box with Outlet—One Pipe Box Wood Mtg.

FIGURE 2.34 Suboperation (Fig. 2.33) are extended into operations such as "assemble junction box and piece of pipe to wood beam." Operation times are grouped with each other to form combinations for work of larger time increments for ease of application (junction box with outlet, one pipe box, wood mounting).

3.52 ENGINEERING AND ANALYSIS TOOLS

Dept: Plant-wide—Maintenance Department
Title: Electrical Construction ½ and ¾ Conduit
Formula #I-54-5-49
Sheet 16 of 16

Table 1. Electrical Construction ½ and ¾ Electrical Conduit

From Summary Sheet

	Leveled minutes No. of conduits on J. box				Variable
1. Junction box with outlet	1	2	3	4	
Wood mounted or equivalent	24.7	25.6	27.2	27.8	} Per box
Brick mounted or equivalent	33.9	34.9	36.4	37.0	
2. Junction box—cover plate					
Wood mounted or equivalent	—	21.6	24.1	24.7	} Per box
Brick mounted or equivalent	—	30.8	33.3	33.9	
3. Wall switch					
Wood mounted or equivalent				24.5	} Per switch
Brick mounted or equivalent and flush mtg.				33.7	
4. Condulet with cover				11.3	Per condulet
5. Pipe clamp					
Wood mounted or equivalent				1.8	} Per clamp
Brick mounted or equivalent				4.8	
6. Pipe coupling				0.8	Per coupling
7. Conduit bands (exclusive of box offsets)				0.9	Per band
8. Greenlee knockouts				3.3	Per knockout
9. Insert wire				0.44	Per foot of pipe

See Formula #I-54-5-31 for other pipe supports
Date: Oct. 25

FIGURE 2.35 From various summary sheets, times for various combinations of work covering the "electrical construction of ½ and ¾ electrical conduit" are assembled into Table 1.

6. A common unit of measurement is available to evaluate the performance of individual craftsmen and foremen. (Employees can relate earnings to productivity by direct time comparison.)
7. Audits are possible to determine if proper labor payments have been made.
8. Information to increase management controls in the areas of:
   a. Planning and scheduling work.
   b. Planning and scheduling equipment.
   c. Planning and scheduling manpower standards, providing an accurate means of determining excess maintenance costs in specific areas or on specific equipment or parts which would justify modifications or replacement of equipment.

Statistical or Past-Performance Method. A maintenance statistical plan is based on averages of past man-hours expended on jobs. While this type of plan does not give an accurate measurement and in reality is nothing more than an index, management is attracted for two reasons:

1. The minimum cost to collect data for standards
2. The minimum control administration cost or incentive wage administration when the plan is in effect

To obtain standard times for maintenance, it first becomes necessary to make some job classifications into which all hours worked are recorded and charged to separate jobs in the various classifications. The average time clocked in on the jobs under each classification becomes the standard. There are extreme classifications where it is necessary to estimate each job.

Job-analysis Sheet

Job Title: Wire Office—Install ½ in. Conduit System
Location: Dept. 406—General Foreman's Office
Analyzed by: J. Frankhauser
Date: Dec. 13,

Operation	Ref.	Unit	Qty.	Lev. min Unit	Total
1. Junction box with outlet					
a. 1 Pipe box—wood mtd.	F-49	Box	3	24.7	74.1
b. 2 Pipe box—wood mtd.	F-49	Box	3	25.6	76.8
c. 3 Pipe box—wood mtd.	F-49	Box	1	27.2	27.2
2. Junction box—cover plate					
a. 3 pipe box—wood mtd.	F-49	Box	1	24.1	24.1
3. Wall switch (brick mtg.)	F-49	Sw.	1	33.7	33.7
4. Pipe clamps—wood mtg.	F-49	Clamp	4	1.8	7.2
5. Pipe coupling (over 10 ft)	F-49	Coupl.	4	0.8	3.2
6. Insert wire	F-49	Ft.	86	0.44	38.0
7. Wire to main box	F-97	Box	1	26.2	26.2
8. Wire 6 fluorescents	F-62	Fixt.	6	15.3	91.8
9. Hang 6 fluorescents	F-62	Fixt.	6	12.6	75.6
10. Preparation and cleaning					
a. Collect tools	F-28	Man	2	11.0	22.0
b. Buggy loads (load and unload)	F-28	Trip	3	14.0	42.0
					541.9
11. Travel buggy trips	F-1	Trip	2	16.0	32.0
Walk	F-1	Trip	2	8.0	16.0
				Total	589.9
				Allowance 10%	59.0
				Standard min	648.9
				Standard hr	10.8

FIGURE 2.36 The development of the job standard as illustrated by the sketch (Fig. 2.31).

A typical statistical plan for a machine shop might follow on this procedure, commencing with the classification of completed job orders:

1. *Standing orders.* Permanent or perpetuating orders are assigned to highly repetitive tasks. Such work as the recurring repair of paint cups for automatic color-banding equipment or the constant straightening of guide pins for specific assembly equipment.
2. *Repair orders.* Machine-shop orders requiring less than 24-hr labor on items such as repair, adjust, standard part replacement, and so on. This order is not used in making new parts.
3. *Work orders.* Machine-shop orders to cover all types of work other than that covered by standing orders or repair orders, but not exceeding a specific money value, usually $1000. (This amount will vary with each plan.)
4. *Project orders.* Machine-shop orders which apply to jobs where the total estimated cost exceeds the work-order value ($1000).

The next step is to obtain hours worked against individual jobs occurring within the classifications of maintenance orders. A dispatch job-card system requiring clocking in and out is essential for the accumulation of these data.

1. A job number is issued for each job. The job number is then recorded on the paperwork authorizing the job.
2. The workers' time is charged on each job card (by clock rings, verbal reporting, and so on).
3. Check to see that total job times balance the total working time reported daily.
4. All hours against each job must be accumulated and totaled as the job is completed.

At this point a decision must be reached as to the length of the recording period required in order to establish standard man-hours for the various classifications of work. A year is usually considered a representative period of time. The base period will vary dependent upon the potential amount of data available, which is proportionate to the size of the shop (in man-hours) and the pattern of work being processed. The primary goal in gathering those statistical data is to get a good representative picture of the shop operations, or plant operations, as the case may be, and the amount of data to be collected has to be determined for each individual case.

The final step is the development of the standard data from the historical record. These data usually end up in one of four job categories:

1. *Standing orders*. The total man-hours worked during the base period are accumulated and the average number of hours per working day are computed against each standing order. This figure is then the permanent standard for this type of order. Additional allowances must be made on these jobs to compensate for any increased volume of activity due to increased production requirements. This will have to be on a ratio basis as established by production during the base period.
2. *Repetitive jobs*. Certain jobs will be found to be repetitive. This allows individual standards to be calculated for these jobs based on the average man-hours expended for these jobs during the base period. Duties will have to be defined so that the standard will be applied to jobs having the same content as those studied during the base period.
3. *Nonrepetitive jobs*.
   a. *Repair orders*. All hours on this type of order, as previously defined, are accumulated during the base period. This figure divided by the total number of repair-order jobs gives a standard repair-order time.
   b. *Work order*. Orders are accumulated into groups according to the actual hours needed for completion, those requiring 8 hr, 8 to 24, and so on for groups 24 to 48, 48 to 100. In each of these classifications the total hours worked are accumulated against the total number of jobs and the accumulated hours are divided by the total number of jobs, giving the standard under each category. The average job time in each classification becomes the standard time in each case.

   To use these standards, an estimator must judge, in advance, the category into which each nonrepetitive job will fall.
4. *Estimated jobs*. Any job judged as requiring more than 100 hr for completion becomes an "estimated job." All project orders are contained in this category. The allowed standards for these jobs are based on the estimate of required man-hours. In a few instances standard data which have been accumulated within the base period can be used to guide the estimator.

There will be cases where it is impossible to work an accurate estimate of time until the job has started. In such cases the estimate should be made after the job has progressed far enough to foresee all the work required for completion. If the work has commenced and the original estimate is not valid because of unforeseen work, changes in the job, and so on, the estimate must be changed to conform to existing conditions.

Routine checks on all estimated jobs should be made when the actual hours reach approximately one-third of the estimate and again at approximately two-thirds of the estimate in order to establish whether or not conditions are consistent. If not, the estimate must be revised. If these four methods have been used to develop data, a means of measuring maintenance efficiency is available. By comparing actual man-hours worked in any period to corresponding standard hours as applied against the total job orders, a measure of relative performance is available. This is nothing more than a comparison of existing performance with performance during the base period. If incentive payments are made under this type of plan the incentive-pay period must obviously be a long period of time because of the way in which the standards were developed. The period of time for incentive-pay purposes would probably be in the area of 10 weeks. This type of plan is not recommended for wage-incentive purposes (see Fig. 2.37). If serious pay-differential problems exist in a plant (see items 3 and 4 under Advantages), this plan offers a reasonable method of adjustment.

There are many disadvantages to this type of plan because of the method of establishing standards.

1. The standard data developed in this manner exert no control over working methods.
2. There has been little effort to determine a normal working pace; only the pace worked during the base period is reflected.
3. The potential earnings or potential efficiency under this type of plan is unknown.
4. There is no method for adjusting standards for technological changes or change in work pattern.
5. Normally incentives are paid on reproduced work elements. In this instance reproducible work elements are not assured.
6. It is impossible for the employee to relate earnings to productivity.
7. There is no means to check individual employee's output, making it easier for the group to control production during peak and slow periods.
8. Without standard elemental time data upon which to base the application of standards, no audit is possible to see if the correct standard times are allowed for various jobs.
9. Requires no management contribution from supervision.

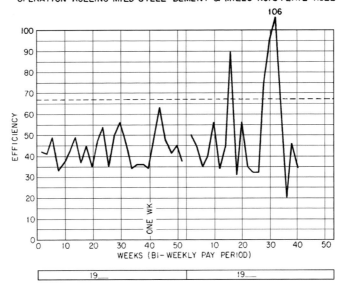

FIGURE 2.37 Past average man-hours is not a sound basis on which to establish maintenance control standards. The standards that have established the above efficiencies were based on past average with separate classifications and standards for weight of stock rolled (a slight improvement over an analysis of man-hours of specific job orders). In the second year of operations with the standard, the chart reflects (1) a change of work pattern, (2) a slight equipment modification. The effect of these two items, in time, could not be calculated to allow a standards adjustment due to the broad, general averages used as a base. (*Factory Management and Maintenance.*)

In addition to the two *advantages* stated earlier under this topic, there are several others:

1. This type of plan adapts itself to group payment which:
   a. Averages difficult with easy work.
   b. Simplifies administration from the maintenance engineering point of view.
   c. Keeps the wage differentials between various skills in line with each other.
   d. Can include everyone in the plan—including workers on standing orders such as tool-crib attendants and stock clerks.
2. This type of plan can be used successfully to eliminate wage-differential problems between production labor and maintenance labor in cases where incentive pay brings the pay of unskilled workers above that of the skilled workers.

Ratio Delay—"Work Sampling." This work-measurement technique is based on the law of probability, the basis for any other sampling procedure. The most familiar use in industry of the law-of-probability theory (as far as work elements are concerned) is quality-control sampling. The probability follows that, by observing a representative number of random occurrences of events from a total universe of events, the percentage distribution of the observed events will be representative of the same distribution in the total universe. The term "ratio delay" comes from a common use of this work-sampling technique, the determination of the ratio of a representative number of random delays to the entire mass of events.

This technique was first used by L. H. C. Tippett in 1934 to determine true operating ratios of machines to properly applied burden and cost rates to spinning and weaving operations in a cotton mill. Since that time there have been many applications of this technique to measure work in production and maintenance areas. The major applications in maintenance are:

1. The evaluation of maintenance efficiency
2. The measurement of man-machine relationships
3. The determination of areas for technical improvements and work simplification
4. The evaluation of the results of improved operations (e.g., incentive installations)

To complete a work-sampling study the following procedure is recommended:

1. Define what is to be measured. The study technique will vary depending upon the end objective.
2. Determine the "elements" to be observed (e.g., adjusting machine, change belt, idle, personal, and so on).
3. Estimate the number of observations required (to determine overall length of study period based on number of observations per day).
4. Outline a random observation plan.
5. Train the observer and other personnel involved.
6. Make the observations.
7. Develop the study data.
8. Determine level of reliability of results.
9. Make recommendations based on study results.

Obtaining a True Random Sample—Ratio Delay. Properly to ensure the absence of bias in a ratio-delay study, a mechanical method must be used to set up a study procedure and observation periods. People develop working habits. For example, the observer making a ratio-delay study might visit an acquaintance in the plant during his rounds. He might fall into the habit of doing this a specific time each day. The observer might take a coffee break with the friend, at the friend's convenience. This could be a certain time each day. The observer might tend to his own personal needs at a certain time each day. It is very easy for these types of working habits to develop which would eliminate observations during these periods and result in biased data. The procedure outlined below will eliminate bias due to habits that an observer might develop. This, or a similar type of procedure, must be used for all ratio-delay observations to ensure the collection of random data.

1. Properly define the problem and establish a list of activities for each problem (Fig. 2.39).
2. Estimate the number of observations required. The accuracy required in the answer depends on the number of observations.
3. Determine how many observations will be made per day.
4. Divide the working day into intervals and number consecutively (Fig. 2.38).
5. Use a table of random numbers and proceed along the table line by line until corresponding numbers in the table appear that are the same as the numbered divisions on the chart. Check off each division as its number appears until the quantity of divisions to be observed equals the number of readings required each day. If duplicate numbers appear, pass over the duplication until the day's total is reached (Fig. 2.38). Make the ratio-delay observation during the period checked off.
6. The activity noted at first glance at the exact instance of observation must be checked. If practical, select a spot in each area where observations of specific jobs must be made. This is not practical for miscellaneous work in the plant, but it is practical for observations on shopwork.

RANDOM OBSERVATION INTERVALS DEPT. 10

WORK SAMPLING STUDY - MACHINE ADJUSTERS

		1ST WEEK					2ND WEEK				
TIME OF DAY	INTERVAL	1	2	3	4	5	1	2	3	4	5
7:45-7:59	1				X	X					
8:00-8:14	2						X		X		
8:15-8:29	3	X	X				X	X			X
8:30-8:44	4				X					X	X
8:45-8:59	5		X					X			
9:00-9:14	6			X	X	X		X		X	
9:15-9:29	7	X	X			X	X	X	X	X	X
9:30-9:44	8						X		X	X	
9:45-9:59	9			X			X				
10:00-10:14	10	X	X	X		X		X	X		
10:15-10:29	11	X						X	X		
10:30-10:44	12		X	X		X				X	X
10:45-10:59	13		X							X	X
11:00-11:14	14	X	X	X					X		
11:15-11:29	15			X		X					
11:30-11:44	16	X	X		X	X	X	X			
11:45-11:59	17				X		X			X	X
12:00-12:14	18			X			X				X
12:15-12:29	19					X	X		X		X
12:30-12:44	20	X			X			X		X	
12:45-12:59	21			X		X					X
LUNCH											
1:30-1:44	22			X	X				X		
1:45-1:59	23				X			X	X		
2:00-2:14	24	X			X			X			
2:15-2:29	25					X					X
2:30-2:44	26	X	X	X	X			X	X	X	
2:45-2:59	27	X	X				X				
3:00-3:14	28						X		X		
3:15-3:29	29					X				X	
3:30-3:44	30				X					X	X
3:45-3:59	31		X	X	X						
4:00-4:14	32	X	X	X			X	X	X		
4:15-4:30	33	X				X				X	X

FIGURE 2.38 To eliminate bias in ratio-delay studies, each day is divided into time intervals and numbered. From a table of random numbers, the 12 observation periods each day are determined.

If the above steps are followed and the number of observations is consistent with the accuracy desired, the ratio-delay study will yield unbiased data.

Assigned Maintenance Manning—Ratio Delay. A common problem in measurement of maintenance hours is the determination of the number of assigned maintenance mechanics required to service a specific area containing a specific amount of production equipment. These mechanics are not concerned with parts, or machine repair, or major parts replacement, but are concerned only with making minor adjustments and keeping equipment operating.

Ratio delay is a practical method for determining the most efficient manning for a given work load. A case history of this type of operation follows.

As outlined previously, the method for obtaining a true sample was used in order to determine the proper observation intervals. Figure 2.38 illustrates the observation intervals in the first 2 weeks of the study.

The next step was to determine an activity list. This was accomplished by continuous observation in the area concerned for a sufficient period to list the activity of the maintenance personnel and the pattern of machine personnel. The activity list for this job is illustrated at the bottom of Fig. 2.39.

RATIO DELAY MACHINE ADJUSTERS

Daily Observation Record Dept. 10

Observations Date: 1/3/55

Machine	1	2	3	4	5	6	7	8	9	10	11	12
1	A	B	A	A	A	A	A	A	A	B	A	A
		1							4			
2	A	A	A	A	B	A	A	A	A	A	B	A
					1						2	
3	A	A	A	A	A	B	A	A	A	A	B	A
						1					2	
4	B	A	B	A	A	A	A	A	A	A	A	A
	1		1									
5	A	A	B	A	A	A	A	A	A	A	A	A
			1									
6	A	A	A	A	A	A	B	B	A	A	A	A
				4			1	1				
7	A	A	B	A	A	A	A	A	A	A	B	A
			1								3	
8	A	A	A	A	C	A	A	A	A	A	C	A
					5						8	
9	A	A	A	C	A	A	B	A	A	A	A	A
		4		5			1	4				
10	B	A	A	A	A	A	A	A	B	A	A	A
	3								2			
11	A	A	B	A	A	A	B	A	A	B	A	A
			1				1			2		
12	A	A	A	A	A	A	8	A	A	A	A	A
Central *	7-7	7-7	-	7-7	7-7	7-7 7	-	7-7	7-7	7-7	-	9-9 6-8

Machine Activity

A-Machine Running.

B-Machine down-adjuster repairs.

C-Machine down-wait for adjuster.

Man Activity

A-Adjust Machine D-Attention G-Idle

B-Change belt F-Walk H-Avoidable delay

C-Clear jam F-Misc. I-Personal

* Central - Location for adjusters when not tending machines.

Each number represents one man

FIGURE 2.39 Daily observation record of machine requirement and human activity.

Observations were made in the department and recorded on a daily observation record (Fig. 2.39). The requirements of each machine were noted and the activity of each man was noted together with the identity of the machine being serviced. Observations were made for a 20-day period and the results were summarized as shown in Fig. 2.40.

Since the actual man loading for 12 machines was 4 adjusters, a total of 960 man observations and 2880 machine observations were recorded. Percentage figures were obtained as shown. Then a synthetic man loading for 3 adjusters tending the same 12 machines was made as follows: Each observation column was scanned with the key spot being "central" (see Fig. 2.39). This indicated whether 0, 1, 2, 3, or 4 of the adjusters were idle or nonproductive when observed. If all were occupied, then with one less adjuster a machine would be down waiting for an adjuster at this time. The daily observation record was then altered so that for each observation only 3 adjusters were considered as being present instead of 4. The new totals were obtained and listed in the column headed "Projected" on summary sheets. From this, the indication was that downtime would increase only 1.5 percent if each man was assigned an additional machine. The cost of this increased downtime and the resulting loss in production were weighed against the potential savings reducing the number of assigned maintenance men. In this particular case, the adjusters were reduced with no appreciable increase in machine downtime.

Work Simplification—Ratio Delay. When used for work-simplification purposes, the study procedure as previously outlined is used. The main use of ratio delay for a work-simplification analysis is to determine the ratio of productive time to nonproductive time or delays. From the data, such items as the amount of time used for job preparation and setup, the amount of time lost by standby time or crew balance, and the amount of time waiting for material or equipment can be determined. When these losses are pinpointed by ratio delay, detailed studies can be made of each lost-time item for reduction or elimination.

Determining Allowances—Ratio Delay. There is always additional time beyond the actual productive work elements that must be included in work standards. These times may include personal time, interference time, crew balance time, and conditional allowances that might apply to the specific

RATIO DELAY — MACHINE ADJUSTERS DEPT. #10

Summary of Observations

Machine condition	Actual 4 adjusters		Projected 3 adjusters	
	Occs.	%	Occs.	%
A. Machine running	2,420	84.0	2,380	82.5
B. Machine down—adjuster at machine	400	14.0	400	14.0
C. Machine down—wait for adjuster	60	2.0	100	3.5
	2,880	100.0	2,880	100.0
Man activity				
1. Adjust machines	240	25.0	240	33.0
2. Change belt	28	3.0	28	4.0
3. Clear jam	48	5.0	48	7.0
4. Attention	58	6.0	58	8.0
5. Walk	38	4.0	38	5.0
6. Miscellaneous	48	5.0	48	7.0
7. Idle	384	37.0	179	25.0
8. Avoidable delay	96	10.0	45	6.0
9. Personal	48	5.0	36	5.0
Total	960	100.0	720	100.0

FIGURE 2.40 Summary of actual observations and projected results of decreasing adjusting crew by one person.

craft, equipment, or jobs. These allowances can be determined by the ratio-delay technique using the same procedure as outlined previously.

This type of study is particularly useful for determining the proper allowances to include in standards based on synthetic data such as MTM. MTM data will supply select elemental times, but not allowances, which must be determined by observation of actual job conditions. Ratio delay is not practical in determining allowance if the basic standard times are based on time studies. When the time-study technique is used to develop standard data, a tremendous number of study hours are required. With continuous 8-hr studies properly documented, and enough hours of study to be representative of the working conditions, standard data and allowances can be developed from the same data, eliminating the need for separate or different types of studies for determining conditional allowances.

Accuracy of Ratio Delay—Random Sampling. With a ratio-delay or random-sampling study, there is a possibility that the results of the study will not be representative of the total universe from which the sample was drawn. There is a mathematical method of determining whether the results of a ratio-delay study are acceptable, provided acceptable tolerances can be defined. The same mathematical technique can be used to estimate the number of observations required to obtain a representative sample with random sampling.

Comments on Ratio-Delay Studies. There have been a number of articles that emphasize the reduced cost and accuracy of ratio-delay studies as compared with time study or other methods. Ratio delay is not a practical method for determining elemental work times. It is a practical method for determining allowances, nonproductive time, and manning requirements. However, the cost of ratio-delay studies can easily exceed the cost of data from continuous time studies because of the following considerations:

1. During the extended period required for ratio-delay studies, no improvements are made, and a potential loss in savings is the result. For example, a job requiring a 3-month ratio-delay analysis could possibly be completed in 2 weeks by continuous time study with two or three reliability engineers on the job. The resulting data, placed in effect, would yield actual savings $2\frac{1}{2}$ months before the same job with ratio-delay data.

2. If an engineer is required to make the ratio-delay observations, using the same illustration as above, it is not likely that he will produce any other useful work during the observation period. This is due to the constant necessity to make ratio-delay observations as scheduled. Therefore, 3 man-months are required to collect the data plus some time to summarize and install the result. With time study, assuming two men for 2 weeks, 4 weeks of engineer's time has been consumed as against 3 months of engineer's time for the ratio-delay method.

3. As for work-simplification improvements with job preparation and setup, interference delays, conditional delays, and other items that can be uncovered through ratio delay, the same items will be uncovered with continuous time studies and corrected as fast as possible at the same time standard data are being collected.

4. For methods improvements in productive work methods (item 3 above considers nonproductive work), there is no substitute for on-the-job detailed analysis. Ratio delay should be considered for this type of study. With assigned maintenance work, as discussed previously, ratio delay can determine high cost adjustments and studies can be initiated to reduce these adjustments through redesign of equipment or parts, or substitution of material.

Estimating Maintenance Efficiency with Ratio Delay. Ratio delay can be used to (1) estimate maintenance efficiency, when work-measurement standards are not available; and (2) audit work-measurement standards after they have been installed. Ratio delay is a practical, accurate, and low-cost method to determine maintenance efficiency.

To make this type of study, devise an observation plan as previously outlined. However, there is a slight difference. In the case-history study of man loading, the work site was in the same area for each observation. With miscellaneous maintenance work in plant areas, there are many jobsites. This requires a different approach in that the observations at a time determined by random numbers necessitate a tour of all maintenance work at the many jobsites. The time necessary to make a complete tour must be estimated and the number of trips per day based on trip time and available observation hours.

For recording observations, use a chart similar to Fig. 2.41. In the first column, enter the total crew size. The number of men working is entered in the next column. There are an additional five columns to give the status of the men who are not working.

With the method outlined above, a determination can be made of the percent of time working, idle, or waiting. However, to obtain a true efficiency, the work pace must be considered. It can be estimated with a technique commonly called "eyeball" speed rating, which is nothing more than snap judgment of working pace. An experienced time-study man can be trained in this technique through the use of leveling films. The technique is necessary so that a speed rating can be made before the subject changes working pace as a result of the presence of the observer and to eliminate bias in the observer's judgment.

At the end of the observation period a summation of the total check marks, divided into the column totals under each category of working, idle, or waiting, will give the percentage of time spent under each of the six categories. If the observations under working time were leveled, this time would have to be adjusted in accordance with the working efficiency.

For a more complete idea of maintenance efficiency in cases where methods, scheduling, and work-measurement techniques have been neglected or an audit is desired, columns can be added to check work methods, whether the crews are properly manned, whether the tools and equipment are proper, whether the best method is being used, and whether or not the work could be scheduled. All these data must be recorded at an extended stay after completing the ratio-delay steps. This type of summary with about 500 jobs should give a good picture of maintenance performance at a low cost.

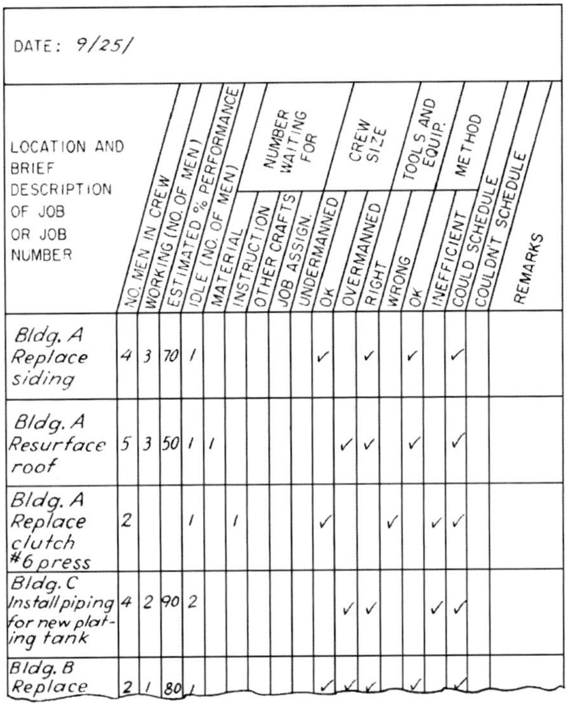

FIGURE 2.41 How to estimate savings potential in applying work measurement to maintenance. Work-sampling technique, illustrated here, provides a simplified method to get data.

In plants with standard practice as far as methods and equipment, where manning and scheduling practices are established, and where a work-measurement system is installed, the same procedure as outlined for estimating maintenance efficiency can be used as an audit of the effectiveness of management cost-control measures.

Another method that has been used to estimate maintenance efficiency is a ratio method consisting of a running index of maintenance labor dollars to direct-labor production dollars. This type of measurement would be effective in a plant where no technological improvements will be made. In a progressive plant, technological improvements tend to reduce direct-labor dollars and cause a slight increase in maintenance dollars. The overall result amounts to cost savings, but a completely different picture will result on the running index; maintenance efficiency would appear to decrease because of the direct-labor reduction. As previously explained, this is far from reality.

Estimating Standards. The degree of accuracy required of maintenance standards depends on the ultimate use of the standard. Estimated standards usually are not accurate enough for incentive-pay purposes, but they do provide an effective base for cost-control measures.

In many plants, it is the responsibility of the supervisor or team lead to review a work order, plan the job, and estimate the labor hours involved. He might also perform the scheduling function. In other plants, a planning department has been established which estimates the job and then performs the scheduling function. The supervisor or team lead's estimate is based on craft know-how and past experience. The estimate from a planner is based on these two items but, in addition, the planning department probably uses a rough form of standard data developed from previous job costs.

The estimated standard usually is in the form of money, rather than standard hours, since the thinking is primarily in terms of job costs. For cost-control purposes, the estimated dollars are evaluated against the actual dollars to measure job performance. If this step is not taken, and if management does not insist on a review of jobs whose completion required money in excess of the original estimate, there is no value to estimating standards.

Advantages and Disadvantages of Estimating Standards. This type of measurement is not practical for incentive purposes, since the standard is determined "by guess and by negotiation." However, it is definitely a step forward in the process of controlling maintenance costs. Many companies have made substantial cost savings through an estimating procedure tied in with the method and planning functions, and their programs may be considered successful. For maintenance departments, where the number of workers and their working efficiency will not support the cost of measured standard data, its application and administration, this type of measurement is most practical.

Advantages:

1. Low cost of determining and applying standards.
2. Can be used effectively for
   a. Methods work.
   b. Planning and scheduling.
3. Can be used as a rough measure of efficiency.

Disadvantages:

1. Accuracy of standard depends upon the educated guess of the estimator.
   a. New methods or new jobs have no past-performance history for estimator to use as a guide.
2. Poor estimates can affect the morale of the workers to the extent that they completely ignore the standard as a goal.
3. Impractical for incentive purposes.
4. *Lack of reproducibility.* Loss of an estimator usually means loss of data, depending on how many data are stored in the head of the estimator.

Conditions Required for the Success of Maintenance-Work Measurement Programs. Certain basic conditions are necessary to ensure success in every management venture. The installation of a work-measurement system for control of maintenance hours is a difficult but rewarding job that must be completed in an environment of acceptance and participation by the entire plant where the installation is to be made.

The following conditions must have recognition and the wholehearted support by the plant to ensure success in work measurement of labor hours.

1. The need for the program must be established, based on a cost picture. Top management, middle management, line supervision, and staff executives must review the maintenance-cost picture both before and as projected after the costs of installation, and the costs of maintenance.

2. Top management should state the objectives of work measurement so completely that those in line and staff functions who are to work with and administer the program can enthusiastically support it.

3. Line supervision must be informed as to their responsibilities in completing the program. Then they must be properly trained to discharge these responsibilities.

4. The hourly worker and the union must be conditioned to the benefits of work measurement. Since this type of work has been traditionally "unratable," there will be a strong feeling of "it can't be done" in many of the workers. The line supervisor or team lead can play a valuable part in orienting these people.

5. The development of standards for maintenance is not a routine time-study job. There will be considerable methods work to be done before and concurrent with the accumulation of standard data. There will be many tricky problems in frequencies of occurrences, in developing standard data, in methods of applying standard data, all of which require a good maintenance engineering background and a high degree of practical judgment.

6. With or without a union, a work-measurement policy must be established. This policy should cover the usual pledges of limitation of earnings, adjustment of standards, payment of earnings, guarantee that workers will receive proper job instruction, and provision of some method of hearing and resolving differences between the workers and management.

7. Work methods and craft-working procedures must be standardized.

8. Adequate administration procedures must be installed for a work-order system, planning and scheduling, and timekeeping.

9. Decisions and their implementation, to be made at the proper level, within boundaries established by the work-measurement policy, must be encouraged.

CHAPTER 3
RATING AND EVALUATING MAINTENANCE WORKERS

Robert (Bob) Call
Principal, Life Cycle Engineering, Inc., Charleston, S.C.

Fair and effective evaluation of maintenance workers is the most challenging work of wage and salary administrators and of responsible managers. Maintenance workers are traditionally craft oriented, have more freedom of action than production workers, and require extensive training and/or experience to be fully qualified. Many times maintenance workers are inclined to compare their wage rates with similarly titled jobs in the construction trades. Maintenance workers usually seek to maintain a margin significantly higher than most production jobs as well. Employer organizations that do not establish and maintain equitable compensation plans are vulnerable to active dissension in production, maintenance, and other specialized units of the plant organization.

BENEFITS OF AN EFFECTIVE COMPENSATION PROGRAM

Long-term economic and intangible benefits are gained by the company; they result from:

- A perception of fair and equitable wages
- Competitive wages within the industry or community
- A logical progression in responsibilities
- A capacity to adjust to a changing organization

To achieve equitable rates a job-evaluation plan that applies to all jobs in the organization is required. Some practices involve separate job-evaluation plans for production and maintenance groups. This approach is complicated and creates confusion and a suspicion of unfair treatment. A broad-based compensation approach and an effective job-evaluation plan ensure consistent, valid, and reliable measurement of the relative value of maintenance, production, and other types of jobs.

A well-designed job-evaluation plan will consistently measure the work content, responsibilities, and special skills required in each job. Maintenance work has distinguishing characteristics that, to some degree, reflect the nature of maintenance employees.

- Maintenance workers usually work without close supervision.
- They take pride in responding well to emergencies.
- They seek to develop corrective modifications to eliminate routine repairs.
- Personal capabilities are dependent upon a strong motivation to learn new skills.

- Written communication abilities may be limited.
- They receive personal satisfaction from solving problems.
- Human relations skills may be important in some maintenance jobs.

The job-evaluation plan will address these and other characteristics. It will provide information which will help to clarify how the demands of the work vary in each of the jobs.

The plan will provide useful information for:

- Establishing performance criteria
- Clarifying specific goals
- Establishing self-development plans
- Preparing future projections of skills needed
- Improving communications with employees and union personnel
- Facilitating changes in job design and changed responsibilities

GENERAL PRINCIPLES OF JOB EVALUATION

Basic factors which are vital to the successful accomplishment of maintenance work must be identified. The relative value of the factors is determined according to their importance to the achievement of the work. The factors must be specific enough to be effective measures of differences but sufficiently broad to cover the spectrum of jobs in the organization and the community.

When valid measures of the relative importance of jobs are established, the compensation can be established. Compensation levels are usually derived from a combination of company policy, negotiations, and individual performance.

The implementation of an appropriate plan for maintenance will require the following:

- Develop and/or modify an existing plan to ensure valid measurement.
- Ensure that the plan is compatible with measurement of other jobs in the organization.
- Make a survey of community and/or industry wage rates for comparable jobs.
- Develop the compensation curves based on company policy, competitive rates, and so on.
- Evaluate all jobs and identify variances from the norm.
- Analyze the variances to determine root causes.
- Make specific plans to address each of the variances.

TRADITIONAL FACTORS

Job-evaluation approaches may include nonquantitative plans such as "rank order" or "comparable" types. The nonquantitative-type plans do not stand the test of time, objective scrutiny, or clarification of the differences between jobs. Quantitative approaches are objective and reduce the judgmental influence to the minimum. Quantitative plans combine point systems and factor comparisons.

The factors used in evaluating jobs can be as many as desired, but all traditional factors can usually be found in one of the following categories:

- Skills
- Responsibilities

- Efforts
- Working conditions

The factors must be complete enough to identify differences over a wide variety of work. An example of a plan in effect over many years is as follows:

- Knowledge
- Experience
- Judgment
- Manual skill
- Responsibility for materials and products
- Responsibility for tools and equipment
- Mental effort
- Physical effort
- Surroundings
- Hazards

This plan was sufficiently broad to cover all the jobs in a large manufacturing plant. Nevertheless, some recently recognized factors or sub-factors which apply specifically to maintenance may be useful in achieving a more accurate evaluation of maintenance jobs. Some examples are:

- Multi-skill capabilities
- Specialized production or processing knowledge
- Troubleshooting knowledge and skill
- Communications with users of maintenance services
- Construction or fabricating skills
- Planning maintenance work
- Purchasing maintenance parts and materials
- Directing or coordinating contractor work
- Preventive maintenance recommendations

RELATIVE WEIGHTS OF FACTORS

The relative weights of these factors should be compatible with the "culture" in which the company operates. In some parts of the world the values given to education as compared with skill and effort are considerably different than the weights that would be assigned in the United States. A recent analysis of the actual points assigned in four well-accepted plans showed the following:

Percent of Total Points

	Plan A	Plan B	Plan C	Plan D
Experience	32.2	52.2	46.3	43.3
Responsibility	14.8	18.1	22.3	23.2
Effort	35.0	15.5	12.4	21.2
Conditions	17.9	14.2	19.0	12.3

The same type of analysis but based only on the maximum point values showed the following:

Maximum Factor Points Percent of Total Points

	Plan A	Plan B	Plan C	Plan D
Experience	36.2	50.0	48.8	46.3
Responsibility	35.6	21.0	24.2	21.5
Effort	18.7	15.0	07.0	17.8
Conditions	09.5	15.0	20.0	14.5

The analyses show that there are differences in each plan, but the weight given to "experience" is predominant in each case and "responsibility" follows. These two factors provide about 70 percent of the weighting.

The plan finally chosen should approximate these guidelines to ensure compatibility with generally accepted practices.

ANALYZING JOBS

The initial work required to design and implement a job evaluation plan should be left to a professional. The work required gathering the data; design the system to meet the specific needs of the organization, and evaluate each job accurately is a full-time project. An outside consultant will provide a more objective view and avoid some of the pitfalls that occur in an internally developed program, even though highly qualified people are on the staff.

When a plan is fully implemented and well accepted, trained administrator within the organization can install the periodic adjustments and changes. Maintenance managers should have a working knowledge of the plan and be prepared to provide appropriate analyses of the nature and character of the jobs of maintenance workers.

Approaches to the accurate analyses of maintenance jobs are as follows:

- Direct personal knowledge and/or observations of the work being performed
- Interviews with incumbents, foremen, managers, and selected staff personnel. (See Exhibit 1 at the end of this chapter.)
- Analysis of the maintenance procedure to identify the degree of art, skill, knowledge, or attention needed to maintain quality and efficiency
- Analysis of the production process controls or maintenance procedures which reduce the chances of costly errors

A brief, thorough, and accurate job description is vital. The description can be a useful tool in administration, manpower planning, organizational development, training, and a host of other human resource functions.

The work is time-consuming and should be accomplished by a trained analyst in the initial implementation. After the plan is accepted, it will be much easier to maintain the job descriptions, following the patterns set in the initial implementation.

The job description should carry a succinct summary of the basic objective and/or the primary responsibility of the job. Each of the specific evaluation factors should be covered with the factual reference in the description. The precise needs of the organization should be clear. Each job should be related to other jobs so that all responsibilities or tasks are covered. Incumbents must understand that they may bring some talents or skills to the organization that is beyond the requirements of the job. They should understand that the job has been carefully defined to reflect the organization's manpower and skill requirements. On the other hand, if the responsible manager can utilize unusual or unrecognized talents in a practical and economical way, it will be useful to reflect this fact in the job-evaluation process and clarify the unique character of the work in the job description.

The job description is the basic document used in evaluating the job. The importance of accurately relating the work content to each evaluation factor cannot be overemphasized. Factual descriptions reduce the time required for evaluation and build the confidence of the organization in the fairness and equity of the plan. (See Exhibit 2 at the end of this chapter.)

EVALUATING JOBS

A common and useful practice is to select a committee of knowledgeable individuals to assign the appropriate degree of value for each factor of each job.

If the committee members are well acquainted with all the jobs under consideration, it is an expeditious and effective technique. Each factor is reviewed for all jobs, and a value is assigned to each job for that factor. Then the next factor is reviewed for all jobs. When all factors values are assigned, the total points for each job are determined. The total points will establish the job value rate.

Under more difficult circumstances, there may be jobs under review by the committee that are not within the prior experience of the committee members. In this situation it will be necessary to bring individuals who are familiar with the work before the committee to answer questions and clarify details which are not fully covered in the description. In some cases a knowledgeable job-evaluation specialist will be asked to evaluate the job for the approval of the committee. The committee will then approve the plan as presented by the specialist or make minor modifications as needed.

The committee will consider one factor at a time. The impact of the factor on each job will be determined before considering the total evaluation for each job. This approach will help to ensure that the relative weighting of the individual factor is consistent.

TYPICAL FACTOR DEFINITION

Definition of requirements and demands of the work provide the basis for determining the relative value for each job. Each factor shows an increasing point value as the defined requirements increase. The rate of increase in point values may be according to an arithmetic or geometric progression. Geometric progressions of point values with each incremental increase in work requirements usually yield a more useful evaluation system. Practical considerations indicate that first-degree equations are desirable in applying individual factors to specific jobs. By ensuring that the points are properly weighted for each factor, it is possible to achieve the final geometric progression of total points. Care should be taken to ensure an evaluation system that is easily understood and contributes to the resolution of conflicting opinions concerning the value of jobs. (See Exhibit 3 at the end of this chapter.)

DETERMINING THE JOB CLASSIFICATIONS

After the point values have been determined for each job, the job classifications are defined. An examination of the array of jobs and point values will usually reveal an appropriate range of points for each classification.

Some job-evaluation plans base wage rates on the point values without defining wage classifications. Limitations on the system of evaluation, the judgments of the evaluation committee, and the difficulties of communicating the basis for minor point differences mitigate against using point values as the basis for determining wage rates.

Wage classifications, based on the designated point range, simplify the administration of the plan, help to accommodate periodic changes in job content and the resulting reevaluation of the point values, and keep the plan easy to communicate to the participants. (See Exhibit 4 at the end of this chapter.)

DETERMINING THE APPROPRIATE WAGE RATE FOR EACH CLASSIFICATION

The procedure for establishing the appropriate wage rates for each wage classification is dependent upon many factors, such as:

- The size of the company and the availability of specialized compensation staff support
- The compensation policy of the company; that is, meet competitive wages or pay less than competitive wages
- The availability of competitive wage rate information

When sufficient data concerning competitive wage rates are available or can be obtained through special surveys, the usual approach is to base wage rates on these data.

The survey data will usually provide accurate wage information for several key or benchmark jobs that are directly comparable with key jobs that have been evaluated through the job-evaluation procedure.

Using the point values at the midpoint of the wage rate classifications and the survey wage rate information, a trend line is calculated. The trend line can be calculated using the least squares method or other appropriate curve-fitting practices. (See Exhibits 5 and 6 at the end of this chapter.)

Once the equation that establishes the point value of the wage class and the initial survey wage rate is determined, it can be used as a basis for making adjustments to meet company or departmental policies. For instance, if the company seeks to pay 5 percent more than the competitive wage rate, then the trend-line equation can be multiplied by 1.05.

Sometimes it is desirable to have some progression within a job classification. If new employees are hired at a beginning rate and give satisfactory performance, progress within a job classification is needed. One convenient way to provide for this progression is to establish a minimum wage rate and steps to higher rates, culminating finally in the fully qualified rate. The wage midpoints provide a way of ensuring that the progressive steps and the entire wage structure are consistent and equitable. Each minimum wage is established as a percentage of the midpoint, for example, 75 to 80 percent.

Periodic adjustments to the wage midpoints can be made in the same way to cover changes in competitive rates, general economic conditions, and company policy. This ensures that the proper relationship between jobs is maintained and that periodic changes in responsibilities and work assignments can be accommodated.

Ideally, the wage rates will be reviewed and compared with competitive wages, company policy, and general economic conditions at least once each year.

EXHIBIT 1

MAINTENANCE JOB SPECIFICATION AND INTERVIEW GUIDE

A. JOB TITLE AND DEPARTMENT
   1. Indicate formal terminology

B. TOOLS
   1. Indicate primary hand and hand-operated tools used; add statement "and similar or equivalent tools."
   2. Indicate measuring tools used, if any.
   3. Describe process equipment tended, operated, or adjusted.

C. SUPERVISED BY OR SUPERVISES
   1. State specifically by title who supervises the job.
   2. State regularly scheduled responsibility for guiding work of others, indicating number of employees, if any, under the guidance of the employee.

D. EDUCATIONAL REQUIREMENTS
   1. State, where applicable, highest degree:
      (a) Required to execute specific verbal orders only.
      (b) Must be able to read and write.
      (c) Must be able to add and subtract whole numbers.
      (d) Must be able to add and subtract decimals and fractions.
      (e) Must be able to multiply and divide decimals and fractions.
      (f) Must be able to convert units of measurement, e.g., heads, linear feet, gallons to pounds, pipe size to G.P.H., etc.
      (g) Must be able to use percentages and simple formulae.
      (h) Must be able to use advanced shop mathematics and handbook formulae.
   2. State, where applicable, highest degree:
      (a) Requires trade knowledge in a specialized field or process equivalent to 2 years high school or trade school.
      (b) Requires trade knowledge in a specialized field or process equivalent to 4 years high school or trade school.
      (c) Basic technical knowledge required to deal with complicated and involved engineering problems.

E. EXPERIENCE REQUIRED
   1. State necessary experience in weeks for an average eligible employee with no experience (but possessing the educational requirements) to reach minimum qualifications of the job.

F. OPERATIONAL COMPLEXITY (INITIATIVE AND INGENUITY)
   1. Indicate the operational complexity as follows:
      (a) Follow well-established routine.
      (b) Perform a variety of simple jobs or follow standard procedures.
      (c) Perform a sequence of semiroutine operations where standard operation methods are available and/or a variety of tasks, some of which are involved.
      (d) Plan and perform unusual and difficult work where only general operation methods are available and/or a wide variety of tasks, some of which are very involved.
      (e) Plan and perform very involved and complex work, and/or requires devising of new or improved methods, and/or many different details or processes requiring considerable versatility.
   2. State, where applicable, the highest degree:
      (a) Few decisions required.
      (b) Minor decisions required involving some judgment.
      (c) General decisions required—judgment required to moderate degree.
      (d) Considerable initiative, ingenuity, and judgment required.
      (e) High degree of initiative, ingenuity, and judgment required.

(Continued)

G. PHYSICAL DEMANDS
  1. Indicate weight, or resultant effort as light, medium or heavy as follows: (divide total weight by number of employees involved)

Unaided (Manually)	*Aided (Mechanically)*
Light effort—approximately 15 lb	Under 100 lb
Medium effort—approximately 35 lb	100 to 300 lb
Heavy effort—approximately 60 lb and over	Over 300 lb

  2. State frequency of such resultant effort as occasional, recurrent, or continuous.
  3. State method of handling, whether performed manually or mechanically.
  4. State working position such as climbing, sitting, standing, walking, stooping, and cramped or awkward position.

H. MENTAL AND/OR VISUAL DEMANDS
  1. Indicate degree of intensity required as either little, moderate, or concentrated.
  2. Indicate degree of duration as either intermittent or semicontinuous to continuous. Omit degree of duration of little intensity.

I. RESPONSIBILITY—MATERIAL OR PRODUCT
  1. Indicate probability of loss through waste or damage as either negligible, slight, moderate, or considerable.
  2. Indicate average monetary loss per occurrence as minor, moderate, or major.
  3. Indicate what primary materials or product are used.

J. RESPONSIBILITY—EQUIPMENT OR PROCESS
  1. Indicate possibility of loss through damage as either:
     (a) Negligible probability.
     (b) Slight probability.
     (c) Moderate probability.
     (d) High probability.
  2. Indicate impact on operations or process flow as:
     (a) Little or no effect on work flow or other operations.
     (b) Some effect on work flow, but adjustments can be made with additional expense in case of mistakes.
     (c) Major effect on work flow, cannot make adjustments in case of mistakes without major expense.

K. RESPONSIBILITY—SAFETY OF OTHERS
  1. Degree of responsibility should be defined as:
     (a) Little or no responsibility—reasonable care in regard to own work will prevent injury to others.
     (b) Reasonable care in regard to own work will prevent lost time and injury.
     (c) Constant attention required to prevent serious injury or fatal accident.

L. WORKING CONDITIONS
  1. State any special conditions such as: heat, cold, excessive noise, fumes, draft, obnoxious gases, smoke, oil, acids, dust, cutting fluids, dirt, vibration, dampness, respirators, etc. Various combinations should be considered.
  2. Indicate frequency of exposure, if present, to elements as being: some, frequent, almost continuous, or continuous.

M. HAZARDS
  1. List major specific causative hazards and probable resultant injuries.

N. CONNECTED EXPENSE
  1. Indicate required personal tool investment.

EXHIBIT 2

JOB SPECIFICATION

CODE NO. _____
DEPARTMENT   Maintenance & Utilities
DATE   October 17, 19

JOB NAME

Maintenance Mechanic

MAJOR PURPOSE

Form, fabricate, and assemble components for experimental projects, repairs and modifications to machines, equipment, and facilities to ensure reliable, consistent operation according to directions and specifications for the approval of supervision.

WORKING PROCEDURE

1. Forms and fabricates parts and components for experimental projects or repairs to machines, equipment, or facilities working from graphic, written, or verbal directions and specifications.
2. Uses machine tools and hand tools working with a variety of metals and other materials within prescribed tolerances as directed.
3. Analyzes malfunctioning machines and systems (electrical, hydraulic, pneumatic, etc.), recommends corrective action, and upon approval, makes repairs or modifications as approved by supervision.
4. Uses past practices, basic data tables, handbooks, and technical references to determine sizes, capacity, dimensions, and other working data for the approval of supervision.
5. Maintains a working understanding of plant production processes equipment, systems, and facilities in order to perform maintenance effectively, under general direction.
6. Works in cooperation with engineering, technical, and production supervisory or outside of the company personnel when directed.
7. Get required tools, equipment, and instruments; clean-up and store as required.
8. Other work as assigned.

MATERIALS

Variety of metals, plastics, and other materials.

TOOLS AND EQUIPMENT

Basic hand tools provided by individual.
Special tools provided by the company.

[This job description reflects the essential information necessary to describe the characteristics of the job and shall not be construed as a detailed list of all job requirements, nor shall it in any way limit the right of management to assign work or direct the work force.]

(Continued)

JOB CLASSIFICATION

DEPT. **Maintenance & Utilities**   TITLE **Maintenance Mechanic**   DATE **Oct. 17, 19**

FACTOR	REASON FOR CLASSIFICATION	EVALUATION DEG.	EVALUATION PTS.
EDUCATION	Ability to use specifications, drawings, shop math, measuring instruments, and general knowledge of mechanical, electrical principles.	4	53
EXPERIENCE	Four years in a wide variety of maintenance construction, machine repair, fabrication, and other similar work.	5−	96
INITIATIVE & INGENUITY	Plan and perform work using recognized methods.	3	40
PHYSICAL DEMANDS	Occasional medium physical effort required.	2	20
MENTAL-VISUAL DEMANDS	Periods of a high degree of concentration are required. May coordinate a high degree of manual dexterity with close visual attention.	4	20
RESPONSIBILITY FOR EQUIPMENT	Using machines of moderate complexity subject to some damage.	3	19
RESPONSIBILITY FOR MATERIAL	Some waste of fabrication or repair materials possible, but not probable with normal care.	1+	10
RESPONSIBILITY FOR SAFETY	Ordinary care and attention to prevent accidents to others.	2	4
RESPONSIBILITY FOR WORK OF OTHERS	Little or no responsibility for the work of others.	1	0
WORKING CONDITIONS	Frequent exposure to dirt, grease, noise, etc.	3	18
HAZARDS	Possibility of lost time accidents; not expected to be incapacitating.	3	15
CONNECTED EXPENSE	Personal tools required. Working clothes provided.	3	6
	JOB CLASSIFICATION	TOTAL	301

(Continued)

JOB SPECIFICATION

CODE NO. _____
DEPARTMENT   Maintenance & Utilities
DATE   December 5, 19

JOB NAME

Electrical Maintenance

MAJOR PURPOSE

Make components for experimental projects, repair and modify machines, equipment, and facilities to ensure reliable, consistent operation according to stated requirements, schedules, approved priorities, subject to the final approval of supervision.

WORKING PROCEDURE

1. Repair production equipment, plant equipment and related systems, particularly when concerned with electrical malfunctions.
2. Uses machine tools and hand tools working with a variety of materials working within prescribed tolerances, as directed.
3. Analyzes malfunctioning machines and systems (electronic, hydraulic, pneumatic, etc.), recommends correction action, and makes electrical or other type repairs, or modifications, as approved by supervision.
4. Maintains a working understanding of plant production processes, equipment, systems, and facilities in order to perform maintenance effectively, under general directions.
5. Works in cooperation with engineering, technical, and production supervisory or outside of the company personnel when directed.
6. Use past practices, basic data tables, hand brakes, and technical reference to determine electrical equipment, materials, sizes, and other working data for the approval by the supervisor.
7. Analyzes and repairs specialized equipment heavily dependent upon electronic controls, electromechanical mechanisms, and related components under general direction.
8. Get required tools, equipment, and instruments; clean up and store as necessary.
9. Other work as assigned.

MATERIALS

Variety of metals, plastics, and other materials.

TOOLS AND EQUIPMENT

Basic hand tools provided by individual.
Special tools provided by company.

[This job description reflects the essential information necessary to describe the characteristics of the job and shall not be construed as a detailed list of all job requirements, nor shall it in any way limit the right of management to assign work to or direct the work force.]

(Continued)

3.76 ENGINEERING AND ANALYSIS TOOLS

JOB CLASSIFICATION

DEPT. **Maintenance & Utilities** TITLE **Electrical Maintenance** DATE **Dec. 5, 19**

FACTOR	REASON FOR CLASSIFICATION	EVALUATION DEG.	EVALUATION PTS.
EDUCATION	Technical knowledge sufficient to perform most skilled jobs involving mechanical, electronic, and other problems.	5	65
EXPERIENCE	Five years in a wide variety of maintenance, construction, machine repair, fabrication, and similar work.	5	103
INITIATIVE & INGENUITY	Plan and perform unusual and difficult work where only general operation methods are available.	4+	57
PHYSICAL DEMANDS	Occasional medium physical effort required.	2	20
MENTAL-VISUAL DEMANDS	Periods of high degree of concentration are required; may coordinate a high degree of manual dexterity with close visual attention.	4	20
RESPONSIBILITY FOR EQUIPMENT	Involved with equipment and systems subject to some damage, some effect on work flow in production areas.	3	19
RESPONSIBILITY FOR MATERIAL	Negligible probability of loss of repair material with normal care.	1+	10
RESPONSIBILITY FOR SAFETY	Ordinary care and attention to prevent accidents to others.	2	4
RESPONSIBILITY FOR WORK OF OTHERS	Responsible for orientation of new personnel on occasion, as directed.	2	6
WORKING CONDITIONS	Frequent exposure to dirt, grease, noise, etc.	3	18
HAZARDS	Possibility of lost time accident, not expected to be incapacitating.	3	15
CONNECTED EXPENSE	Working clothes provided. Personal tools required.	3	6
	JOB CLASSIFICATION	TOTAL	343

(Continued)

	JOB SPECIFICATION
	CODE NO. _____
	DEPARTMENT __Maintenance & Utilities__
	DATE __December 8, 19__

JOB NAME

Painter

MAJOR PURPOSE

Paint equipment and facilities, following established practices to preserve, protect, and improve the appearance of plant and office, for the approval of the supervisor.

WORKING PROCEDURE

1. Prepare surfaces and apply paint and coatings to equipment or building and facilities, according to general directions, following established practices.
2. Repair floors, walls, ceilings using a variety of specialized materials as approved by the supervisor.
3. Make stencils, templates or other auxiliary painting and decorating devices as required.
4. Recommend quantity, type and colors of coating or paint to be used, according to conditions, for the approval of the supervisor.
5. Get required ladders, scaffolding, equipment, and materials; clean up and store, as is appropriate.
6. Other work as assigned.

MATERIALS

Variety of coatings, patching, and repair materials.

TOOLS AND EQUIPMENT

Brushes, scrapers, other coating and patching application devices.

[This job description reflects the essential information necessary to describe the characteristics of the job and shall not be construed as a detailed list of all job requirements, nor shall it in any way limit the right of management to assign work or direct the work force.]

(Continued)

JOB CLASSIFICATION

DEPT. **Maintenance & Utilities**   TITLE **Painter**   DATE **Dec. 8, 19**

FACTOR	REASON FOR CLASSIFICATION	EVALUATION DEG.	EVALUATION PTS.
EDUCATION	Follow verbal and written directions; calculate quantities of coatings required using simple formulas and arithmetic.	3+	44
EXPERIENCE	Two years plant maintenance, construction or other similar work providing a variety of painting and coating experience.	3+	66
INITIATIVE & INGENUITY	Plan and perform unusual jobs where only general methods are known and considerable judgment and initiative are used.	4−	49
PHYSICAL DEMANDS	Recurrent medium physical effort and occasional heavy physical effort is required.	3	30
MENTAL-VISUAL DEMANDS	Required continuous mental and manual attention; coordination of manual dexterity with close attention for recurrent periods.	4	20
RESPONSIBILITY FOR EQUIPMENT	Negligible probability of damage or loss to equipment with normal care.	1	7
RESPONSIBILITY FOR MATERIAL	Limited care required to prevent waste or loss of paint, coatings, and materials.	2	17
RESPONSIBILITY FOR SAFETY	Ordinary care and attention will prevent accidents to others.	2	4
RESPONSIBILITY FOR WORK OF OTHERS	May periodically instruct or direct a full-time helper or small group of part-time helpers.	2	6
WORKING CONDITIONS	Frequent exposure to several disagreeable elements; dirt, paint wetness.	3	18
HAZARDS	Accidents improbable, except for minor injuries.	2	18
CONNECTED EXPENSE	Working clothes provided; paint coverings and tools provided.	1	0
	JOB CLASSIFICATION	TOTAL	271

(Continued)

JOB SPECIFICATION

CODE NO. _____
DEPARTMENT   Maintenance & Utilities
DATE   December 4, 19

JOB NAME
General Maintenance A.

MAJOR PURPOSE
Repair and adjust production and plant equipment, according to instructions.

WORKING PROCEDURE
1. Repair production equipment, plant equipment, and related facilities, as directed.
2. Lubricate and adjust production machines according to prescribed procedures.
3. Under direction, use machine tools or other special equipment to form or fabricate parts or components.
4. Build simple structures from wood, metal, or other materials, as directed, using hand tools, power tools, welding and basic fabricating techniques.
5. Under direction, move machines, install pipe, fixtures, or related equipment and provide support work for plant improvement projects.
6. Get required tools, or equipment, clean up areas, and store tools and equipment.
7. Other work as assigned.

MATERIALS
Wood, metal, plastics, and other common materials, fittings and fasteners.

TOOLS AND EQUIPMENT
Basic hand tools provided by the individual.
Special tools provided by the company.

[This job description reflects the essential information necessary to describe the characteristics of the job and shall not be construed as a detailed list of all job requirements, nor shall it in any way limit the right of management to assign work or direct the work force.]

(*Continued*)

JOB CLASSIFICATION

DEPT. **Maintenance & Utilities** TITLE **General Maintenance A** DATE **Dec. 4, 19**

FACTOR	REASON FOR CLASSIFICATION	EVALUATION DEG.	EVALUATION PTS.
EDUCATION	Follow verbal or written directions, make simple calculations involving decimals, fractions, etc; knowledge of variety of machines and systems.	3+	44
EXPERIENCE	Two years in plant maintenance, construction, machine repair, fabrication, or other similar work.	3+	66
INITIATIVE & INGENUITY	Perform a sequence of operations according to directions and past practices using prescribed methods. Variety of tasks.	3	40
PHYSICAL DEMANDS	Occasional heavy physical effort required.	3	30
MENTAL-VISUAL DEMANDS	Periods of high degree of concentration are required. May coordinate a high degree of manual dexterity with close visual attention.	4	20
RESPONSIBILITY FOR EQUIPMENT	Moderate risk of damage to equipment.	3	19
RESPONSIBILITY FOR MATERIAL	Repair material difficult to damage or waste with normal care.	1+	10
RESPONSIBILITY FOR SAFETY	Ordinary care and attention required.	2	4
RESPONSIBILITY FOR WORK OF OTHERS	None.	1	0
WORKING CONDITIONS	Frequent exposure to dirt, noise, grease.	3	18
HAZARDS	Possibility of lost time accidents, but not expected to be incapacitating.	3	15
CONNECTED EXPENSE	Working clothes provided. Personal tools required, none of which are precision tools.	2	3
	JOB CLASSIFICATION _____	TOTAL	269

(Continued)

JOB SPECIFICATION

CODE NO. _____
DEPARTMENT   Maintenance & Utilities
DATE   October 17, 19

JOB NAME
Machinist

MAJOR PURPOSE
Form, fabricate, and assemble components for experimental projects; make repairs and modifications to machines, equipment, and facilities to ensure reliable, consistent operation according to stated requirements, schedules, priorities, and subject to the final approval of supervision.

WORKING PROCEDURE
1. Forms and fabricates parts and components for experimental projects or repairs to machines, equipment, or facilities working from graphic, written, or verbal requests.
2. Uses machine tools and hand tools working with a variety of metals and other materials, within prescribed tolerances as directed.
3. Analyzes malfunctioning machines and systems (electrical, hydraulic, pneumatic, etc.) and makes repairs or modifications, subject to approval of supervision.
4. Uses basic mathematical formulas, handbooks, and technical references to determine sizes, capacity, dimensions, and other working data subject to approval of supervision.
5. Maintains a working understanding of plant production processes, equipment, systems, and facilities, in order to perform maintenance effectively, under general directs.
6. Works in cooperation with engineering, technical, and production supervisory or outside of the company personnel when directed.
7. Get required tools, equipment, and instruments; clean-up and store as required.
8. Other work as assigned.

MATERIALS
Variety of metals, plastics, and other materials.

TOOLS AND EQUIPMENT
Basic hand tools provided by individual.
Special tools provided by the company.

[This job description reflects the essential information necessary to describe the characteristics of the job and shall not be construed as a detailed list of all job requirements, nor shall it in any way limit the right of management to assign work or direct the work force.]

(Continued)

DEPT. **Maintenance & Utilities** JOB CLASSIFICATION TITLE **Machinist** DATE **Oct. 17, 19**

FACTOR	REASON FOR CLASSIFICATION	EVALUATION DEG.	EVALUATION PTS.
EDUCATION	Technical knowledge sufficient to perform most skilled jobs involving mechanical, electrical, and other technical problems.	5−	65
EXPERIENCE	Five years in wide variety of maintenance, construction, machine repair, fabrication, and other similar work.	5	103
INITIATIVE & INGENUITY	Plan and perform unusual and difficult work where only general operation methods are available.	4+	57
PHYSICAL DEMANDS	Occasional medium physical effort required.	2	20
MENTAL-VISUAL DEMANDS	Periods of high degree of concentration are required. May coordinate a high degree of manual dexterity with close visual attention.	4	20
RESPONSIBILITY FOR EQUIPMENT	Using machines of moderate complexity, subject to some damage.	3	19
RESPONSIBILITY FOR MATERIAL	Some waste of fabrication or repair materials possible, but not probable with normal care.	3	10
RESPONSIBILITY FOR SAFETY	Ordinary care and attention to prevent accidents to others.	2	4
RESPONSIBILITY FOR WORK OF OTHERS	Responsible for introducing new personnel on occasion as directed.	2	6
WORKING CONDITIONS	Frequent exposure to dirt, grease, noise, etc.	3	18
HAZARDS	Possibility of lost time accidents, not expected to be incapacitating.	3	15
CONNECTED EXPENSE	Personal tools required. Working clothes provided.	3	6
	JOB CLASSIFICATION	TOTAL	343

EXHIBIT 3

FACTOR #1 EDUCATION OR TRADE KNOWLEDGE

This factor measures the job requirements in terms of the mental development needed to think in terms of and understand the work being performed. Such mental development or technical knowledge may be acquired by formal schooling or through equivalent experience.

Trade or vocational training is a form of education and should be evaluated under this factor.

Degree		Requirement	Points
1st	1− 1 1+	Ability to recognize material and product variations; to carry out specific verbal orders; ability to read and write; ability to perform simple addition and subtraction; ability to operate a simple machine on simple work.	10(−) 14 18(+)
2nd	2− 2 2+	Ability to identify material and product variations; use simple arithmetic involving addition, subtraction, multiplication, and division and to read simple decimals and fractions. Ability to read simple drawings and use relatively simple tools such as knife, ruler, thread and plug gauges, thickness gauge, and thermometer.	22(−) 26 31(+)
3rd	3− 3 3+	Ability to follow written directions; make calculations involving fractions, decimals, percentages, and simple handbook formulas. Ability to use written descriptions, average drawings, measuring tools such as micrometers. Requires a general knowledge of processes and mechanical or electrical principles.	36(−) 40 44(+)
4th	4− 4 4+	Ability to use specifications, directions for processing, or complicated drawings, advanced shop mathematics and handbook formulas. Ability to use complicated measuring instruments. Requires a thorough knowledge of mechanical and/or electronic principles. [Equivalent to four (4) years of high school or trade school.]	49(−) 53 57(+)
5th	5− 5 5+	Ability to use all types of processing instructions, specifications, or tools. Requires a basic technical knowledge sufficient to perform the most highly skilled jobs involving mechanical and/or electrical or other technical problems. [Equivalent to four (4) years of high school or trade school plus two years of specialized training.]	61(−) 65 70(+)

(Continued)

FACTOR #2 EXPERIENCE

This factor measures the amount of time necessary for the average employee with the required education or equivalent, but without prior experience on the job to obtain the practical knowledge necessary to learn the job duties involved and to be able to perform such duties in the minimum acceptable manner. Part of the experience necessary may have been acquired on related work. Consider only such time as is required to learn to perform the work satisfactorily on the basis of continuous rather than elapsed time. Exclude the theory or general mentality which is measured under "Education." Fundamental knowledge obtained in the form of trades or vocational training should be evaluated under "Education."

Degree	Time Requirement—Working Days		Points
	From	To	
1	0	20	15
1+	21	60	22(+)
2−	61	120	30(−)
2	121	180	37
2+	181	240	44(+)
3−	241	360	52(−)
3	361	480	59
3+	481	600	66(+)
4−	601	720	74(−)
4	721	840	81
4+	841	960	88(+)
5−	961	1120	96(−)
5	1121	1280	103
5+	1281	& above	110(+)

(Continued)

FACTOR #3 INITIATIVE & INGENUITY

This factor measures the job requirements in terms of independent action and exercise of judgment such as devising and developing methods or procedure; analyzing work and adapting methods, equipment, etc., to perform the job; seeing the need for and taking independent action where required. It also includes deciding on matters such as comparison of products with standards. The routine or nonroutine nature of the work; the rules, procedures, or precedents established for performing the work; frequency and significance of changes in the nature of the work; complexity of conditions or facts involved and type of directions provided should be considered.

It is understood that the more routine the operation is or the greater the supervision provided, the less initiative is required.

Degree		Requirements	Points
1st	1− 1 1+	Requires the ability to understand and follow simple instructions, to use simple equipment involving few decisions, or to follow a well-established routine.	10(−) 14 18(+)
2nd	2− 2 2+	Requires the ability to work from detailed instructions and the making of minor decisions, involving the use of some judgment. Perform a variety of simple jobs, usually with easy to understand specifications.	22(−) 26 31(+)
3rd	3− 3 3+	Requires the ability to plan and perform a sequence of operations where standard or recognized operation methods are available and the making of general decisions as to quality, tolerances, operation, and setup sequence. May involve a variety of tasks, some of which are involved. Moderate judgment is required and work to normal tolerances.	36(−) 40 44(+)
4th	4− 4 4+	Requires the ability to plan and perform unusual and difficult work where only general operation methods are available and the making of decisions involving the use of considerable ingenuity, initiative, and judgment. May involve a considerable variety of tasks, some being very involved and or working to close tolerances and specifications.	49(−) 53 57(+)
5th	5− 5 5+	Requires outstanding ability to work independently toward a general result; devise new methods; meet new conditions necessitating a high degree of ingenuity, initiative, and judgment on very involved and complex jobs. May involve many different details or processes requiring considerable versatility.	61(−) 65 70(+)

EXHIBIT 4

Wage Classifications

Classification	Point Range	Base Rate Dollars
1	100–114	4.21
2	115–129	4.38
3	130–144	4.55
4	145–159	4.72
5	160–174	4.89
6	175–189	5.06
7	190–204	5.23
8	205–219	5.40
9	220–234	5.57
10	235–244	5.74
11	250–264	5.91
12	265–279	6.08
13	280–294	6.25
14	295–309	6.42
15	310–324	6.59
16	325–339	6.76
17	340–354	6.93

EXHIBIT 5 Example of Calculation

CONTRACT PAY LINE DATA—"STRAIGHT LINE"
MULTIPLE REGRESSION—GRADES AND WAGE RATES

Twenty subjects:
Constant = 7.0277631E
Coefficient = 1.1109398 − 01
Rate = 0.997528

Grades Predictions	Calculations	Actual	Approximate Difference ($\times 10^{-2}$)
1	7.14	7.215	+7
2	7.25	7.305	+5
3	7.36	7.395	+3
4	7.472	7.490	+2
5	7.583	7.575	+1
6	7.694	7.685	−1
7	7.805	7.785	−2
8	7.916	7.885	−3
9	8.0276	7.985	−3
10	8.138	8.090	−4
11	8.2497	8.180	−7
12	8.3609	8.315	−5
13	8.47198	8.425	−5
14	8.5831	8.560	−2
15	8.6942	8.680	−1
16	8.8052	8.800	0
17	8.9164	8.925	0
18	9.027	9.045	+1
19	9.1385	9.185	+4
20	9.2496	9.360	+11
			+34
			−33
			+ 1

EXHIBIT 6 Example of Calculation

Contract Pay Line Data—"Curvilinear"

Curve Analysis:
- Y = AB^{CX}
- LOGY = LOG A + (LOG 8) C^x
- S.E. EST.: = .009901046
- LOG A = .4895078
- LOG B = .3690950
- C = 1.014032

Ref.	Grade	Actual Wage	Approximate Calculated	Difference
0	1	$ 7.215	$ 7.221	−.006
1	2	7.305	7.308	−.003
2	3	7.395	7.397	−.002
3	4	7.490	7.488	+.002
4	5	7.575	7.582	−.007
5	6	7.685	7.678	+.007
6	7	7.785	7.776	+.009
7	8	7.885	7.878	+.007
8	9	7.985	7.982	+.003
9	10	8.090	8.089	+.001
10	11	8.180	8.199	−.019
11	12	8.315	8.313	+.002
12	13	8.425	8.429	−.004
13	14	8.560	8.549	+.011
14	15	8.680	8.672	+.008
15	16	8.800	8.798	+.002
16	17	8.925	8.929	−.004
17	18	9.045	9.063	−.018
18	19	9.185	9.201	−.016
19	20	9.360	9.343	+.017
20	°21	°9.500	9.489	+.011
				−.079
				+.080
				+.001

°Estimated

CHAPTER 4
WORK SIMPLIFICATION IN MAINTENANCE

Al Emeneker
Work Control SME, Life Cycle Engineering, Inc., Charleston, S.C.

Work simplification can probably be best described as the intelligent employment of well-established human behavioral patterns to encourage and expedite the finding and implementation of more efficient work methods. Over the last 40 years, the work-simplification approach rooted in biomechanics has earned a rapidly expanding popularity. Many industrial firms have sponsored formal work-simplification programs. Most of these have been phenomenally successful in delivering a multitude of cost-reducing and profit-increasing innovations. Many college engineering curricula now include courses in work simplification.

As originally conceived, the application of this approach tended to be concentrated in the area of production methods. But as more experience was gained, its universal applicability became more widely recognized. Work-simplification concepts are now utilized to improve performance in many other activities, including clerical functions, supervisory techniques, research, and *maintenance*. In fact, the term "work simplification" has actually become almost a synonym for "an organized grass-roots methods-improvement technique."

An approach that is particularly well adapted to the study and improvement of maintenance performance. Applications in this area have been exceptionally productive. More frequent and broader applications appear likely to be equally successful.

The traditional approach to methods improvement has been to employ highly trained specialists in industrial engineering techniques to spend full time on this activity. These "experts" are assigned the task of studying one activity after another throughout the entire organization. They are expected to locate opportunities for improved performance, develop ways for these improvements to be achieved, evaluate their desirability, sell their acceptance, and assist in their implementation. A great deal of progress has been achieved in this manner. But the effectiveness of this approach is diluted in two ways:

1. Much time and effort must be expended by the "expert" to become familiar with each new activity studied, in order to be sure that all pertinent aspects and interactions with related activities are uncovered and properly evaluated.

2. The improvements developed and proposed by these "experts" are usually strongly resented by the prospective users. Their implementation is often resisted, even occasionally deliberately sabotaged.

The work-simplification approach is designed to minimize these difficulties. Each employee is assisted to become his own "expert" and is encouraged to study and recommend ways to improve the performance of his own job. Motivation is developed by demonstrating the value to both workers and management of the results they can achieve by working together as a team.

Training in the use of a collection of simple but ingenious techniques provides each employee with know-how adequate to make the required methods-improvement studies.

Work simplification appears to be most productive when there is widespread participation by many individuals from all organizational levels in an "organized" program. Carefully planned indoctrination sessions must be provided to develop effective motivation. All participants should receive training in basic methods-improvement tools and techniques. A means of handling ideas, such as a suggestion system, should be developed (or an existing one adapted) to make a method of communication readily available, to provide a way for obtaining prompt management review of improvement proposals, to facilitate recognition for contributions, and to provide adequate rewards for achievements.

THE LAW OF INTELLIGENT ACTION

When confronted with a problem, the intelligence of an individual's actions is dependent upon his

1. *Desire* to solve the problem.
2. *Ability* to perform the tasks required.
3. *Capacity* to handle the human relations involved.

Desire. Motivations for the actions of human beings can be divided into two basic categories:

1. *To gain* (What's in it for me?)
2. *To avoid loss* (That's mine. Hands off!)

Thus the employee seeks employment as a means of gaining:

1. Security (reasonable control over his own future)
2. Material reward (money to buy things)
3. Opportunity to improve his position (economic and/or social)
4. A sense of participation (belonging to a group and having a say in activities)

Employment is also a means of avoiding the loss of whatever he possesses of these same things.

It can be expected that the attitude of the individual toward an opportunity for personal gain will be almost entirely selfish. His controlling interest will be, "What's in it for me?" But his decisions and actions will tend to be rational, logical, and based upon fact. A direct appeal toward actions that will result in benefit to him and others can be expected to receive objective analysis.

However, the attitude of the individual toward the possible loss of something he already possesses can be expected to be entirely different. Decisions will tend to be based upon emotion rather than fact. Actions taken in connection with a possibility of losing existing possessions may often be devious and will sometimes appear completely illogical.

This difference in attitudes is of great significance when the acceptance of methods improvement is being sought. To an individual not directly involved, the introduction of a cost-saving proposal involving the use of a new piece of equipment or a new method can have the appeal of "intelligent selfishness." But to a person directly involved, *a change from the existing implies the loss of his own know-how applicable to the old procedure or equipment.* The fear generated by the prospect of such a loss can completely cancel out any appeal of mutual benefit. Therefore, to be successful, a work-simplification program must have identified with it specific management policies and practices which will assure the individual that he can gain and will not personally lose as the result of the implementation of work-simplification proposals.

A suggestion system can provide recognition and financial rewards, but an additional guarantee by management that participants will not suffer personal loss through downgrading or layoff is essential. An agreement to achieve force reductions via attrition or transfer of displaced individuals to other expanding activities is often a mutually acceptable approach. With careful planning, this method is usually adequate to absorb force reductions made possible by work-simplification proposals. Reductions via layoffs can eliminate any possibility of a successful program. After all, the cooperation of the individual just cannot be expected if he can see that this cooperation will result in direct losses to himself, his friends, or his associates.

Ability. Until the introduction of participative work-simplification programs, which provided both the receptive climate and the necessary training of the participants, the idea that the average employee could successfully conceive, develop, and implement worthwhile methods improvements was only a hypothesis. Management possessed little evidence and even less faith that such efforts were likely to be really productive of meaningful results. Today, however, the impressive results of many successful industrial work-simplification programs amply document the validity of this hypothesis. It has been unquestionably proved that the latent ability to develop methods improvements exist in the majority of individuals and can be effectively utilized if proper motivation and training are provided. It has been shown that with only minimal training in a few of the simple basic industrial engineering tools, the average individual can develop an amazing ability to recognize opportunities for improvement and to implement workable solutions.

Capacity. The basic pattern of human nature has been fairly well established and demonstrated to be essentially unchangeable. Human behavior, however, can be modified and to a certain extent controlled. In fact, human behavior is relatively predictable and can be measurably influenced by anyone with a thorough understanding of the basic mechanics of human nature plus a willingness to take the prerequisite actions. In respect to influencing attitudes toward prospective methods-improvement installations, it is usually sufficient to learn to recognize and deal with two of the most basic traits of human nature:

1. Resistance to change or to the new
2. Resentment of criticism

The fundamental idea of searching for a better way to perform a task has the built-in assumption that, when it is found, the new way will be substituted for the old. Thus, improvement implies change. From the point of view of the user of the old way, change tends to disrupt complacency and create a fear of possible unfavorable consequences. The firm feeling of "all's well" is replaced with a queasy feeling that perhaps he, the current user, may also become obsolete and have to be retrained, perform a harder task, or perhaps even be replaced. The user can see nothing in the change for him and an excellent chance of insecurity. Naturally, he resists change. It is almost a conditioned reflex. Everyone tends to be critical of, and resistant to, change.

A successful work-simplification program must make provision to assist participants to become familiar with this universal reaction and to learn how to minimize its hampering effect. Participants must learn:

1. To avoid confusing fact and opinion.
   Practice results in habits and can lead to the development of biased opinions that cannot be properly extrapolated.
   Experience increases knowledge of facts which provide a sounder basis for extrapolation.
2. To avoid misunderstandings.
   Failure to ascertain all the facts may result in incorrect conclusions.
   Reliance upon the results of a single nonrepresentative example may lead to erroneous decisions.
3. To avoid snap judgments.
   Time is required for mature judgment.
   Lack of experience must be taken into consideration in making evaluations.

A change for the better implies criticism of the old method and, what is even worse, criticism of the user of the old method. Direct or implied, constructive or destructive, the immediate reaction is fast and always the same. No one likes criticism. It is always taken as a personal affront. It is resented.

To develop a successful work-simplification program, participants must learn to expect this reaction in others and in themselves. They must learn to minimize offending others, to keep criticism from improperly affecting their own judgment, and to help others keep it from confusing their decisions.

However, by far the biggest assist in minimizing both resistance to change and resentment of criticism is the basic premise of the work-simplification approach which substitutes the "participative" development of new methods for the "expert" approach. The participants is most unlikely to develop resistance to, or resentment of, what he believes are largely his own ideas.

THE SCIENTIFIC METHOD

Frederick W. Taylor, the alleged founder of scientific management, once said, "The art of management is knowing exactly what you want done and then seeing that it is done in the best and cheapest manner." Arthur D. Little, the famous research chemist and head of a worldwide management consultant firm, claimed that there were four facets of the scientific approach:

1. The simplicity to wonder
2. The ability to question
3. The power to generalize
4. The capacity to apply

The work-simplification approach applies each of these in a very literal fashion:

Maintaining an Open Mind. (The simplicity to wonder.) The participant with an open mind wonders about everything. He is willing to explore all alternatives. He is not restricted by past practice, precedent, traditions, habits, customs, or fear of the consequences of change.

Observing the Present Way. (The ability to question.) Few people know how to do an adequate job of questioning. Most of them stop asking too soon. Sometimes this is merely to avoid embarrassing the person questioned. To succeed in work simplification, one must "why" the devil out of everything. Work simplification provides an organized plan for questioning. It is called the questioning pattern and is a definite sequence of questions:

What *is done?*	*Is it done at all?*
Where *is it done?*	Is it done here?
When *is it done?*	Why is it done then?
Who *does it?*	Does this person do it?
How *is it done?*	Is it done this way?

This is a training pattern that is to be followed literally at first but which soon becomes simply an organized way of thinking.

Exploring Opportunities for Improvement. (The power to generalize.) From the answers, tentative conclusions (generalizations) can be developed. Possibilities for improvement are investigated:

What?	*Eliminate?*
Where?	Change place?

When?   Why change sequence?
Why?   Change person?
How?   Improve method?

Remember:

- You are searching for possible solutions, not firm conclusions.
- If it has never been done before, it may be a better way.
- Don't admit it can't be done or you are licked before you start.
- Try to find ways to make new ideas work, not to prove them unworkable.
- Don't neglect suggestions by others. Use them, but give credit to the source.

Implementing the New Method.   (The capacity to apply.) It is not enough to wonder; ask why and develop a workable improvement. An idea has no value until it is put to use. The capacity to apply implies two things:

1. The ability to see the application of a general rule to a specific problem
2. The ability to convert an understanding of human nature into an approach to the new method that will gain the cooperation of the people involved

The Five Steps in Methods Improvement

A definite and permanent advance is seldom made until use is made of measurement. This is particularly true where human factors are involved. Human performance tends to vary so much that unless some form of measurement is provided and used as a basis for decisions, there is little possibility of repeating a process accurately or predicting or controlling future conditions sufficiently to allow introduction of improvements.

Mere observation, done objectively, is a form of measurement. It can be used to classify, label, and compare. An interesting demonstration is to pick a task with which you are familiar but not directly involved. Now, subject the performance of this task to your concentrated and undivided attention. Chances are that you will find that you were completely unaware of many important aspects. It can be truthfully said "the commonest article of commerce is misinformation about fundamental things."

An organized pattern of observation is of great assistance. Work simplification suggests a step-by-step program for studying tasks:

1. *Select* the task to be studied. Be careful that only one task is studied at a time. Failure to observe this caution can lead to confusing results or to ineffective efforts. Because time is valuable, doing first things first must make the best possible use of it. Pick the job that needs improvement most. But remember the "human problem." Do not rush in too fast. Start by improving your own job or jobs in your department. Remember, if you work on someone else's problems, they will probably resent your help as implied criticism. Look for:
    a. *Bottlenecks.* Leave the smooth-flowing jobs alone until you crack troublesome ones.
    b. *Time-Consuming Operations.* Lengthy jobs usually offer the greatest opportunity for improvement.
    c. *Chasing Around.* Activities of this type are almost always unproductive and often can be eliminated or drastically reduced.
    d. *Waste.* We become so accustomed to some forms of wasted materials, time, or energy that we have difficulty in recognizing it as such. Increases go unnoticed. Look carefully.
2. *Observe* the present way in which the task is performed. Get all the facts. Be sure to include all the requirements for the performance of the task. Don't forget to determine interactions with related tasks. Make a process or activity chart. Use it to record all details.

3. *Challenge* everything. Question what is done:
   a. Challenge the whole job being investigated. Why is it done? Is it necessary? Can it be done another way or at another time or place?
   b. Next, challenge each "do" operation. This is because if you eliminate the "do," you automatically eliminate the make-ready and put-away that go with it.
   c. Then apply the checklist of questions to every detail:

   What? What is done? What is the purpose of doing it? Why should it be done? Does it do what it is supposed to do?
   Where? Where is the detail being done? Why should it be done there? Could it be done somewhere else?
   When? When is the detail done? Why is it done then? Could it be done at some other time?
   Who? Who does the detail? Why does this person do the detail? Could someone else do it?
   How? How is the detail performed? Why is it done that way? Is there any other way to do it?

   This questioning attitude helps develop a point of view that considers the good of the whole operation rather than that of any one department or individual. It will often bring to light possibilities for eliminating useless or unnecessary work which adds no real value to products. It tends to bring out the type of operation or equipment needed to perform the required work most economically.

   Do not overlook the possibility of obtaining ideas from other people working on the same operation. And do not forget that when you ask for these ideas you have a "human problem." You will get the ideas only if they want to give them to you. They must be convinced that improving performance will help them.

4. *Explore* opportunities for improvement. Consider all possibilities. Examine each in detail. Evaluate, compare, and select the best alternative. Use the flow process chart or multiple-activity chart to pretest and demonstrate the feasibility of new methods.
   a. Can operations be eliminated? What is done? Why is it done? Is it necessary? In far too many instances a good deal of time is spent studying major operations for possibilities of improvement without asking the question, "Why is this operation performed?"

   If it is found that an operation has been in the plant in the same way for a year or longer, it should be questioned. A better way is probably available. If operations cannot be eliminated, perhaps there are unnecessary *transportations* and *storages*. Question every handling. Then, if handling is absolutely necessary, look for:
   (1) Backtracking of work
   (2) Heavy lifting or carrying
   (3) Trucking
   (4) Bottleneck
   (5) Skilled operators doing handling work

   b. Can activities be combined? Can sequence, place, or person be changed? This is an important opportunity for improvements. Whenever two or more operations can be *combined,* they are often performed at a cost approaching or even equal to the cost of one. Likewise, transportations and storages between the operations may be eliminated. If operations cannot be combined, find out if it is possible to combine transportation and an operation. By changing the sequence of an operation, one may eliminate backtracking and duplication of work. The order in which operations are performed may have been derived from the original nature of the process. The process or product design may have been changed since then. But has the order of operations been restudied and changed to regain optimum efficiency?

   Sometimes, just changing the place where the work is done or by whom it is done will help. Better lighting, better ventilation, better tools may be available elsewhere. Perhaps another operator is better equipped to do the operation.
   c. Can the "do" operation be improved? How is it done? Why is it done that way? Is better equipment available? Are other materials available? Can new techniques be applied?

   Unfortunately, it is here that a great deal of work simplification started in the past. We must learn to consider this step the last resort. Major savings can usually be found, but the price of the new equipment, materials, training, and so on, is also usually high, sometimes beyond our reach. Often relatively small rearrangements, method changes, and layout revisions will accomplish almost as much with negligible cost.

5. *Implement* the new method. See that all people involved understand the objective of the task and desirability of the new method. Take care that each person involved knows and understands his or her part in the new method. Be sure that none involved will lose financially or socially as a result of the change. And, even more important, be sure that they know it!

CHARTING TECHNIQUES

There are many charting techniques which have been designed to assist in the development of improved methods. These include:

- Flow process chart
- Multiple-activity process chart
- Gantt chart
- Critical-path network
- PERT (programmed evaluation research technique)

All these charting techniques are similar in principle. They are a means of recording and studying activities required to perform a task. This list is in order of increasing complexity.

The flow process chart is used to record a single sequence of activities. The multiple-activity chart is used when several sequences of activities occur at the same time and their relationships with respect to time are significant. The Gantt chart is used when the number of simultaneously occurring sequences of activities becomes large. The use of the critical-path network is desirable when some of the sequences of activities are time-related and some are not. This approach can become quite complicated, and then computer programs must be used in conjunction with it. PERT is a variation of the critical-path technique into which another variable, probability, has been introduced.

THE FLOW PROCESS CHART

The flow process chart (see Fig. 4.1) was originally developed by Frank B. Gilbreth of "Cheaper by the Dozen" fame. It is a detailed and graphic representation of the sequence of events in a process or procedure and includes measurements considered desirable for analysis, such as distances traveled, time required, delays, and so on, together with reasons for these measurements. It is a widely accepted tool of definitely proven worth.

Two of the most important values of the flow process chart are that the information is presented in condensed form and that, despite this brevity, all the desirable detail is shown. Many a manager, supervisor, foreman, or clerk has been irritated and baffled by an inability to find out or visualize the whole process or procedure under his or her care. Because of this, decisions are often made upon the basis of incomplete knowledge. These may be little more than guesses and have occasionally proved disastrous. A flow process chart can correct this sort of problem.

How to Make a Flow Process Chart

Detailed instructions for making a flow process chart are shown in Fig. 4.1.
Be careful when making up a chart to follow either a product or a person. Stick to one or the other. Do not switch. This is perhaps one of the most common errors made in preparing a chart. If a man is carrying an object, puts it down, and goes to look for a truck and we are following the object, we

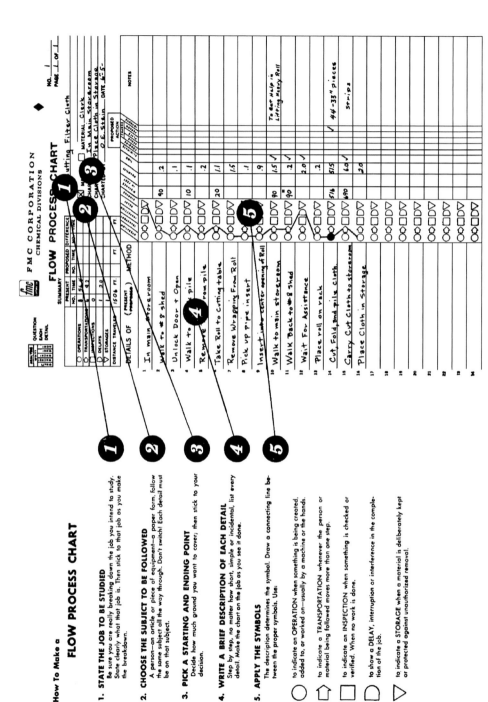

FIGURE 4.1 Flow process chart.

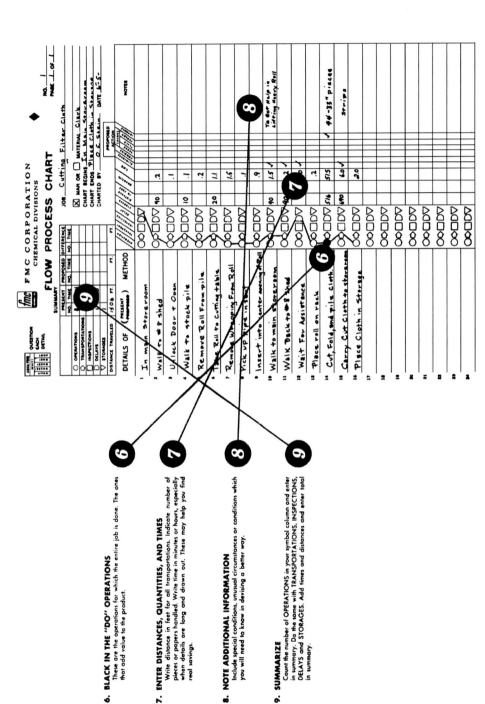

FIGURE 4.1 (Continued)

record a delay. If we are following the man, we do not use the delay symbol but continue to follow the man. The object is delayed, but the man is not. Watch out for this difficulty.

Distinguish carefully between operations and inspections. Use the symbol for inspection only when you are sure no change occurs in the object involved. For instance, when a clerk is looking for a folder in a file, he or she is performing an operation. Looking at a check to see if it is signed is an inspection.

Pointers

1. Do not attempt to cover too much ground in one study.
2. You can make an accurate chart only by making it on the job, actually following the work as it is done.
3. Try not to omit any details.
4. Stick to the subject—one subject per chart.

To summarize, the flow process chart is a valuable tool because:

1. As a still picture it separates the job being studied from a background which may distract attention.
2. It breaks the job down into simple, individual, and reliable details so that one detail can be studied at a time.
3. Its condensed form enables us to visualize the process easily in its entirety.
4. By itemizing, it allows us to express the process or procedure in numerical fashion.
5. The make-ready, "do," and put-away are clearly indicated.
6. Through the mere act of making the close observation necessary to prepare the chart, we often observe ways of improving the process.

Flow Diagram.   It is sometimes helpful to supplement your flow process chart with a flow diagram (Fig. 4.2). A flow diagram is simply a layout of the area involved in the job being studied over which you indicate by a line the path of the object or person followed in the chart. It is often desirable to indicate the action taking place by using the same symbols as those used on the chart. They may, if desired, be keyed to each other by item numbers.

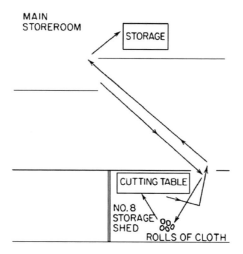

FIGURE 4.2   Flow diagram.

THE MULTIPLE-ACTIVITY PROCESS CHART

There are many situations where two or more interrelated tasks must be studied at the same time. The objective is to reduce the total time, the number of people, or the number of machines required to complete the entire series of tasks. The flow process chart is not adequate for this assignment. A new tool, the multiple-activity process chart, is required. This tool may be described as a graphic method of measuring and demonstrating the time relationships between two or more interrelated operations or procedures. This can be the relationship between a machine and its operator, two or more machines working together on the same job and dependent upon each other, or two or more men working on the same job where each man does part of the work. To illustrate:

Man-machine	Lathe and operator
Multiple machine	Power shovel and trucks
Multiple man	Mechanic and helper

How to Make a Multiple-Activity Process Chart

1. The first step is to prepare a separate flow process chart on each participant in the operation or process being studied. The time required for each detail must be measured and recorded. An ordinary wristwatch is usually satisfactory for timing. In some cases, merely estimating the time will be sufficient.
2. The second step is to choose a time scale large enough to allow plotting of each detail of one complete cycle but not so large as to result in an awkwardly large chart.
3. Then the details of each participant's activities are plotted in separate columns against the single time scale chosen. Be sure that events which occur simultaneously are shown at the same place on the time scale.
4. Next, the activities of each participant are classified into one of the following three categories:
   a. Operating. *Doing work, producing*
   b. *Standby.* Waiting, not producing but presence essential
   c. *Idle.* Neither producing nor standing by
5. Sometimes it is desirable to break operating time down still further into:
   a. Make ready
   b. Do
   c. *Put away*
6. The last step is to prepare the summary. This is especially important because data on percent operating time, percent idle time, and percent waiting time usually give pretty good clues as to where to look for possible improvements.

   The maximum possibilities of interrelated operations are easy to evaluate with this tool. Idle time and the reasons for it are dramatically demonstrated. Existing inefficiencies become less difficult to spot; improvements become easier to develop and demonstrate.

How to Use the Multiple-Activity Process Chart

Whenever machines or equipment are used, there are three possible ways to lose time:

1. Idle operator time
2. Idle machine time
3. A combination of Nos. 1 and 2

Idle operator time is time that is paid for but wasted. Often this idle time is not the operator's fault but inherent in the prescribed method of operation. The multiple-activity process chart will assist in finding such conditions, measuring their significance, and assisting the analysts in exploring ways to eliminate or minimize them. The higher wages go, the more important efforts in this direction become.

Idle machine time can also be critical. As machines become more complex, the cost of having them stand idle part of the time may spell the difference between profit and loss. Often these idle times are individually small and hence not very noticeable, but they may occur so frequently in the cycle of operations that their total effect is considerable. In addition, there often is a tendency to consider the purchase of new facilities without a thorough study of how much the performance of already existing facilities could be improved without major expenditures. This can lead to the purchase of elaborate new facilities which are really not justified. The multiple-activity process chart is an excellent tool for problems in both these areas.

The bar charts shown in Fig. 4.3 represent typical experience in many plants. Looking at bar chart A, it can be seen that idle machine time is 29 percent, and most people would say that the equipment in question is (by subtracting) working 71 percent of the time. But is it productive 71 percent of the time? Obviously, it is not. Actual productive time ("do" time) is only 23 percent, and the true nonproductive time is 77 instead of 29 percent. The difference is handling time (35 percent) and setup time (13 percent) which occupies machine time but is not productive. If the time required for these activities could be cut in half, the productivity of the equipment could be doubled, as is illustrated in bar chart B.

It must be remembered that just because the wheels are turning continuously, the motors humming, or the operators perspiring, it cannot be assumed that maximum possible output is being achieved. A machine is being utilized only when it is actually performing the "do" function for which it was designed.

Sometimes, however, just the opposite situation may exist. When pay rates are high and machines or tools simple and cheap, it may be advisable to reverse our objective and purchase and put into use sufficient machines to obtain maximum effectiveness of labor. We can use the multiple-activity process chart to analyze our problem in either case. We merely change our objective.

Solution of a Typical Problem

Problem. Suppose we have a machine and an operator working together to produce a certain product. It takes the operator 2 min to set up the job in the machine. Then the machine runs for 4 min. The operator completes the job by doing the put-away in 2 min.

A multiple-activity process chart of these two interrelated activities would appear as in Fig. 4.4. Operator activity is shown in the left-hand column, machine activity in the right-hand column. The center is used for the time scale. The summary for our example is:

- Total cycle time   8 min
- Total man time   4 min
- Total machine time   4 min
- Machine utilization   50 percent (machine time divided by cycle time)
- Operator effectiveness   50 percent (man time divided by cycle time)

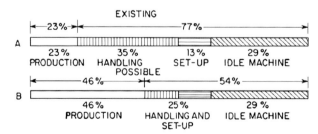

FIGURE 4.3 As many plants do now.

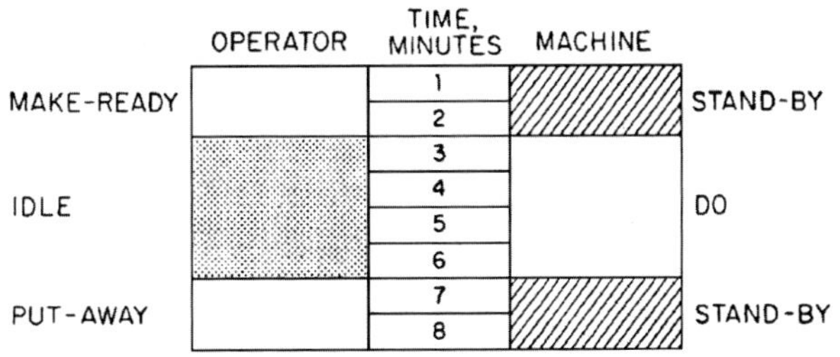

FIGURE 4.4 Multiple-activity process chart.

One solution to this problem is to add another operator. When we do this, our chart would look like Fig. 4.5. Our summary would now appear as follows:

- Total cycle time   6 min
- Total man time   2 min
- Total helper time   2 min
- Machine time   4 min
- Machine utilization   66 percent
- Operator effectiveness 33 percent
- Helper effectiveness   33 percent

We have increased the machine utilization and output at the expense of operator effectiveness.

Now, if we can devise some method so that the make-ready and put-away can be performed while the machine is performing the "do" by the use of some fixture, both machine utilization and operator effectiveness can be improved. The need for the helper can be eliminated. A chart of the new method would look like Fig. 4.6. The summary would be:

- Total cycle time   4 min
- Total man time   4 min
- Machine time   4 min
- Machine utilization   100 percent
- Operator effectiveness 100 percent

This is, of course, a hypothetical and oversimplified situation. In actual practice we would seldom find a problem which has so few factors and in which we can visualize the situation so easily without charting. The 100 percent utilization we have shown is also theoretical, since there would obviously be downtime for lubrication and adjustment of equipment and rest periods for personnel.

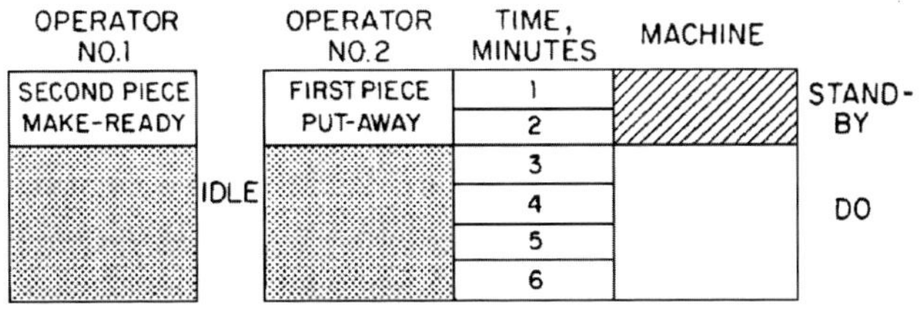

FIGURE 4.5 With an operator added.

OPERATOR	TIME, MINUTES	MACHINE
FIRST PIECE PUT-AWAY	1	DO
	2	
SECOND PIECE MAKE-READY	3	
	4	

FIGURE 4.6 The new method.

THE FOUR BASIC PRINCIPLES OF MOTION ECONOMY

The best use of the human body to yield a given result has received a great deal of study. The results of these investigations may be summed up as the "principles of motion economy." When thoroughly understood and intelligently applied, these principles can have a very favorable effect upon productivity without an increase in work effort. These principles are:

1. *Physical activities or motions should be productive.*
   a. Devise methods and design workplaces to avoid using human hands as holding devices. Put both hands to work.
   b. Keep the workplaces, supplies, and tools in order and in predetermined places. This minimizes searching and fumbling.
   c. Design the workplace to fit human physical limitations. Try to keep activities within normal work areas. Avoid requiring movements outside of the maximum work area.

   Human beings can work better in areas bounded by arcs of circles. This limitation should be considered in laying out workplaces. Note that there are two areas in each plane (Fig. 4.7).

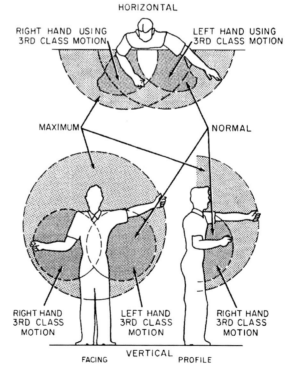

FIGURE 4.7 Normal and maximum working areas for hand motions.

An arc drawn with a sweep of the hand across the work area with the elbow the pivot point determines the normal work area for each hand. Work can be done with a minimum of fatigue within this area. The zone where the swings of the right and left hands overlap is the area where two-handed work can be performed with minimum effort.

An arc drawn with a sweep of the hand and arm across the work area with the shoulder as the pivot point determines the maximum work area for each hand. Work can be performed without experiencing excessive fatigue within this area. The overlap zone limits the area for two-handed work with a reasonable amount of fatigue.

2. *The path of all motions involved should be rhythmic and smooth.*
   a. When both hands are used, the motions should be rhythmic, equal, and in opposite directions. Motions made simultaneously, in opposite directions are natural, and contribute to smooth flow or rhythm. When it is possible for the two hands to move at the same time in opposite directions, by arranging similar work on each side of the workplace, the operator can produce more with considerably less mental and physical effort. Idleness of either hand is not productive. Use both hands for productive work.
   b. Motions should follow curved paths, be continuous, and not require sudden stops. Continuous curved motions give better rhythm than straight-line motions. Sudden stops or changes in direction interrupt smooth flow and spoil rhythm. Motions that use the momentum of the motion itself to aid in performing the work are more effective and more rhythmic than controlled motions. Devices that provide a stop against which momentum may be expended eliminate the need for muscular control of the stop. An example of such a device is the carriage return stop on a typewriter.
   c. Workplace gimmicks, such as prepositioning of tools, magazine-type supply methods, drop discharge of finished work, and the use of foot-operated holding or tripping devices, can do much to keep hand motions rhythmic and smooth. Micro-motion studies have amply demonstrated the value of these procedures. More important than time saved in transporting the material or tool to its destination is the freeing of both hands so that they may proceed simultaneously in unbroken *rhythm—a prime requisite of skill.*

3. *Motions should be as simple as possible and yet be consistent with previous principles.* There are five classes of motions:

Class	
Class I	Fingers only
Class II	Fingers and wrist
Class III	Fingers, wrist, and forearm
Class IV	Fingers, wrist, forearm, and shoulder
Class V	Entire body

   In general, the amount of fatigue resulting from body movements increases with the increasing class number. This relationship is not always strictly true, but it is usually sufficiently accurate to permit meaningful comparison of alternative methods on a numerical basis.

4. *The worker should be at ease.*
   a. Bench or desk tops should be at a height to permit work with minimum fatigue.
   b. Lighting and ventilation should be adequate to minimize discomfort.
   c. Tools and materials should be clearly identified.
   d. Provision should be made for alternate standing and sitting while work is being performed.

A MOTION-ECONOMY CHECKLIST

1. Do motions stay within normal working areas? Could they?
2. Are motions outside maximum working areas required? Could they be avoided?
3. Are hand motions of lowest possible classification?
4. Is the workplace at the right height?

5. Is the workplace the best possible shape or design?
6. Is the workplace orderly?
7. Could tools be prepositioned?
8. Are tools easy to recognize?
9. Could materials or supplies be prepositioned?
10. Are tools and materials arranged in best sequence?
11. Can foot pedals relieve hands?
12. Can handholding be avoided?
13. Does each cycle begin simultaneously with both hands?
14. Does each cycle end simultaneously with both hands?
15. Are arm and hand motions symmetrical and in opposite directions?
16. Can the work be prepositioned for the next operation?
17. Is the workplace designed for sitting/standing operation? Can it be?
18. Can sharp changes in the direction of hand motions be avoided?
19. Are small objects slid rather than picked up and moved?
20. Can your prepositioned tools be grasped quickly and in the position used?

Check each job you study against this list. See if you can make any changes to reduce fatigue and improve the method.

The greatest value of motion economy is the development of the ability to visualize operations in the terms of motions, to recognize good and bad motion practice, and to think in terms of motions as compared with operations. This might be called "motion-mindedness."

APPLICATIONS OF WORK SIMPLIFICATION TO MAINTENANCE

The expanding use of automation is steadily reducing the size of work forces employed in the production area. Electronic data processing continues to whittle at clerical staffs. But these same trends are increasing both the amount and complexity of maintenance requirements. Both opportunities and needs for improving maintenance performance are growing at a rapid pace. Maintenance activities are now and appear likely to continue to be in real need of work-simplification effort.

Maintenance work is different from production work in two basic ways:

1. Most maintenance work input is assigned and controlled on a job-by-job basis rather than unit of time or product output. For this reason, work content is usually nonrepetitive in nature.
2. Direct correlation between work output and product or service output is seldom feasible. This tends to make verification of savings difficult.

These differences do not limit the usefulness of the work-simplification approach. But they do change the emphasis somewhat.

Improving Management Efficiency. The problems encountered in applying intelligent management to maintenance are very complex. Effective management usually requires a great deal of data. A huge volume of paper work and records are often generated. Work simplification can give a big assist to the streamlining of these activities. For instance:

1. *Work-control procedures.* Efficient assignment and control of work on a job-by-job basis requires much planning and a large volume of paperwork. This work is very repetitive in nature and an excellent subject for work simplification. One word of caution: simple elimination of paperwork or arbitrary reduction in the number of work orders is not the answer if it results in loss of control.

Much can be done, however, to reduce the complexity of these documents and decrease the effort and time required to process them without destroying their effectiveness.

2. *History records.* The development of maintenance history records is essential to carrying on a productive preventive-maintenance program. But these records are often quite voluminous and time consuming, in both preparation and use. The methods used for the assembly and retrieval of information from these records represent an excellent area for work simplification.

Improving Technical Decisions. As equipment grows more and more complex, the following are areas of effort which can greatly benefit from the use of the work-simplification approach:

1. *Pre-detection of incipient failures.* Effective preventive maintenance will require improved techniques for predicting when, where, and how failures are likely to be incurred. This will probably involve the development of better inspection techniques, the introduction of the use of more diagnostic instruments, and perhaps the introduction of continuous monitoring techniques.
2. *Post-failure remedial-action decisions.* The determination of the exact nature and extent of equipment malfunctions and remedial action indicated is becoming increasingly difficult as the variety and complexity of facilities increase. The predevelopment of standard diagnostic routines offers an excellent opportunity for the development of better methods.
3. *Repetitive-job standardization.* Use of standardized, preselected procedures for the same or similar jobs will increase the volume of work upon which detailed methods-improvement studies can be justified.

Improving Manpower and Machine Utilization. The multiple-activity process charting technique provides an excellent vehicle for exploring ways to:

1. *Reduce crew sizes.* Use of preplanned, shop make-ready, prefabrication or preassembly, special-handling equipment or tools, and so on, can frequently reduce the amount of work done by field crews.
2. *Reduce out-of-service time.* Careful prescheduling can often appreciably reduce the total time required to complete jobs. The multiple-activity process chart is a good tool for this purpose. When the jobs are large and complicated, it is usually necessary to resort to the more complex critical-path technique.

CHAPTER 5
ESTIMATING REPAIR AND MAINTENANCE COSTS

Tim Kister
Maintenance Planning SME, Life Cycle Engineering, Inc., Charleston, S.C.

Estimating maintenance work is defined as the process of predicting probable costs of any physical change in plant equipment or facilities. A physical change may be the relocation or replacement of machinery or the cleaning, oiling, adjusting, or repairing of machinery, and so on.

The success and effectiveness of a maintenance operation depends to a large degree on the accuracy and timeliness of estimating. Control of labor costs may be accomplished by targeting on estimated standards set by management to limit overtime, regulate crew size, and provide a full work load. Make or buy decisions, methods improvements, and overall management cost controls are necessarily based on some form of estimating. Even projects which do not depend on estimates for their development and execution require estimates for justification and management approval.

Estimating is a matter of judgment, an informed opinion by the estimator. Therefore, the accuracy of an estimate depends not only on the training and experience of the estimator but also on the quality, relevance, and inclusiveness of the data which support it.

PREREQUISITES FOR ESTIMATING MAINTENANCE COST

A maintenance cost estimate is based on two areas of information: the type or classification of the job and the end use to which the estimate will be put.

Classifying the job and obtaining full information about its specifications is the first prerequisite. It is necessary to know the job priority or urgency, work content, and general conditions under which the work will be performed.

How the estimate will be used is the second prerequisite. Together, these major factors will determine who will do the estimating, how the estimate will be made, the amount of detail required, and specific techniques to be followed.

CLASSIFYING THE JOB

What the estimator knows about the job is determined by the degree to which the job is or can be planned before the work is started. Where there is more information, there can be better planning, better estimates, and, usually, better costs. In many cases an important benefit derived from having thorough estimating procedures is more effective management when the work is carried out because the work has to be clearly defined and planned to be accurately estimated.

Maintenance supervisors often feel that all their work is emergency work and that consequently both planning and estimating are impractical. In order to avoid the obvious limitations which result from this position, it is important to have a realistic appraisal of the classifications of work in each individual plant. This means that the real emergencies must be separated from the work which can be planned. Careful consideration of each of the following general classifications will show that at least some of the maintenance work in every plant can be considered as "planned or repetitive." These items can be planned and estimated as accurately as the end use of the estimate requires.

Planned and Repetitive Maintenance

Repetitive repair or replacement of specific items, such as belts, bearings, motors, filters, and screens

Scheduled routine work, such as oiling, cleaning, housekeeping, and inspection

Spare-parts production and overhaul

Planned equipment overhaul

Building and facility repairs

Assigned area service

Planned nonrepetitive replacements and repairs

Relocations

Modifications

Equipment improvements

Repairs on noncritical or lightly loaded equipment that can be economically shut down pending scheduled repair

Emergency Service. While different techniques may be required, estimating procedures may be applied profitably to many emergency service situations as well. Generally the key to accurate estimates here is having repetition of the same or similar problems. In classifying these jobs it is necessary to identify first the highly repetitive items and then the high-cost items which may be expected to repeat after long intervals.

An example of effectively "planning breakdown work" was developed in a foundry for its most critical production unit. Preventive-maintenance inspections generate work orders for specific repetitive parts-replacement jobs to be performed during some shutdown period. These jobs are carefully planned in advance and estimated to permit preparation of a kit including all parts and necessary special tools. The jobs are then classified according to the machine downtime required. When a changeover or breakdown for another reason occurs, the expected downtime is estimated and the previously prepared jobs of that duration are accomplished. This approach permits more major repair jobs to be performed during weekend shutdowns and results in a substantial production increase.

HOW THE ESTIMATES WILL BE USED

The extent of estimating and consequent estimating expense which is justified for a particular situation depends primarily on the end use of the estimate. An easy method for determining relative accuracy requirements is provided by a guide list of uses. The following list is arranged in approximate order of increasing demand for accuracy. It should always be considered along with other criteria for selecting an estimating method.

1. Determination of the extent of approvals required (example: over or under $500?)
2. Evaluation of work-order backlog
3. Long-range forecasting
4. Evaluation of equipment-purchase recommendations
5. Evaluation of method proposals
6. Make-or-buy decisions—limited annual dollar volume
7. Critical-path scheduling
8. Monthly schedules and work-load forecasts
9. Plantwide cost-control reports of work performance
10. Weekly schedules and manpower assignments
11. Departmental cost-control reports
12. Plantwide group incentive
13. Individual cost-control reports
14. Daily manpower assignments and work schedules
15. Make-or-buy decisions—high annual volume
16. Departmental group weekly incentives
17. Small-group daily incentive
18. Individual weekly incentive
19. Individual daily incentive

WHO WILL PREPARE THE ESTIMATES

Estimates may be properly performed by any of four general groups: foremen, engineers, planners, or rate setters. The question of who should do the estimating is really answered when the most appropriate estimating method to fit existing circumstances is established. Each group is better qualified for, or can more conveniently carry out, a particular kind of estimating procedure.

Foremen Estimates. Estimates by the maintenance foreman are generally the quickest and easiest to obtain, can be based on limited advance information, and may be made without formal requests or other controls. In some situations it may be necessary to accept this approach as the only practical answer. Where the end use of the estimate is served as well by an approximation as by a detailed plan or where it appears to be impractical to secure accurate advance information about the job, a foreman estimate may be the best. Furthermore, the foreman must be familiar with the job in order to assign and supervise his men, and for someone else to do the planning and estimating would require some duplication of effort. Why, then, consider having planners or estimators in addition to the foreman? Why not have the foreman do all the estimating?

The answer here is one of practical economics. Few maintenance jobs are so well supervised that they could not be improved to the extent of 5 or 10 percent, or more, by better supervision. If, for example, each of four foremen in four 20-man departments concentrated on direct supervision for 2 hr a day instead of spending that time estimating and a cost reduction of 5 percent was accomplished, the savings would be nearly four times the cost of a full-time estimator to do the 8 hr of estimating work.

Estimating by the foreman should be limited to situations where it does not interfere with needed supervision and where more detailed procedures are not practical.

Engineering Estimates. Again, the source and availability of advance information and the purpose of the estimate indicate the estimating procedure to be used. The procedure determines who should do it.

The design of major construction projects and the selection or design of equipment may require estimates of installation labor costs as well as purchase prices and contractors' quotations. While maintenance foremen or planners may be called in for consultation, the design procedures involved usually require that engineering develop these estimates.

Should engineering estimate ordinary repair projects? Yes, if the required estimating procedure is primarily one of obtaining equipment prices and contractors' quotations and particularly if the estimating information vitally affects design decisions. No, if the required estimating procedure can be done more effectively by the maintenance planner or foreman, because the work to be done by maintenance personnel is the significant part of the job.

Planner Estimates. New production foremen still carry out the wide variety of activities they were expected to handle 30 years ago. Scheduling, timekeeping, wage administration, and methods improvement—these functions are generally developed and carried out by staff people, leaving the foreman free to supervise his people.

The "maintenance-planning" concept is not generally recognized as one of several important steps toward giving the maintenance foreman some of the staff support we have come to expect in production. While the scope of this staff support may vary considerably from plant to plant, it will almost always include estimating. The kind of estimating done by the planner may also vary widely. In fact, flexibility to use various means of estimating to fit different situations is one of the prime advantages of having planners do the estimating. Engineering and foreman estimates can be used ideally for only a limited range of estimating problems; planner procedures can be fitted to almost any requirement.

In most cases, obtaining information about the job is a basic responsibility of the planner. Because he knows the purpose of the estimate, he is in an ideal position to decide what estimating procedure is most appropriate. It is important to assure maximum utilization of this inherent flexibility of method, within a basic framework of company policy. This requires thoughtful supervision of the planning activity.

Rate Setter Estimates. Where detailed standards are applied to maintenance operations for performance measurement or incentives, jobs may be "rated" or "applicated" from basic data during the progress of the work or after it is completed. While these standards cannot be termed estimates by our definition, the people who apply these rates are particularly well qualified to make estimates, using a wide range of estimating procedures. Recommendations regarding flexibility of methods for planners apply in general to this group also.

ESTIMATING TECHNIQUES FOR LABOR COST

The initial task for the estimator, regardless of which of the following techniques will be used, is a thorough analysis of the job. Analysis literally means resolution into elements or constituent parts. This is the most important tool the estimator can use. The most complex major project becomes merely a series of typical jobs when it is divided into its component parts. Without proper analysis, most estimating procedures would be useless.

In a typical application of detailed estimating of machine-shop repair work, breaking the job down into operations requires 90 percent of the estimator's time, whereas actually estimating time values requires only 10 percent. Therefore, it is essential that the degree of analysis for a particular job be in accord with other phases of the estimating method for that job, such as material estimating, and with the end use of the estimate and the scope of available information. As an extreme example to illustrate this point, one would not break down a job into microelements if the time values were to be determined by personal judgment.

Judgment. In many cases, judgment based on personal experience achieves accuracy entirely adequate for a particular situation, with minimum cost for estimating. Clear definition of the scope of the job and analysis in line with the estimator's experience are essential for good results.

The principal objections to estimates based on personal judgment are fundamentally lack of proof of consistency. With a clear-cut job definition, careful analysis, and experienced estimators, the resulting estimates may be well within the accuracy tolerance required yet fail to stimulate confidence because their accuracy cannot be proved, even on a relative basis.

This problem may be intensified by a tendency to issue exact figure estimates where a round figure would be more appropriate (e.g., 267.5 hr instead of 300 hr), with the inevitable result that if the same job is estimated twice, it will have two different estimates.

Slotting. Both the prime objections to estimates based on pure judgment may be partially met by use of another tool called "slotting" in which the job is classified within a cost or time bracket. Classification is usually based on judgment, but judgment may be guided by comparison with benchmarks, typical common jobs for which actual cost is known to fit within the bracket. Issuing the estimate in terms of a slot or cost bracket tends to discourage quibbling about insignificant differences between jobs or between estimates.

A refinement of the slotting technique which offers many advantages over typical slotting procedures requires the use of basic data standards to establish accurate standards for benchmark jobs in each bracket or slot. The benchmark standards are posted on spreadsheets with sufficient material detail given to easily relate to the proposed job. The estimates for labor are in terms of the time which should be required rather than just past average performance. (See Figs. 5.1 and 5.2.)

Standards per Unit. The estimating methods for nonrepetitive jobs described above have one important limitation in common: There is no assurance of consistent application where judgment is the basis for the estimate. For this reason, comparison of the performance results of one department with another or of current results with past performance is meaningless. Improved productivity as measured by judgment estimates may mean only that the estimators are becoming more generous. Unfortunately, this assumption that estimates are generous may be made whenever performance figures based on judgment show improvement, even when the improvement is quite realistic.

The broad category of standards per unit includes a wide variety of estimating procedures, ranging from the builder's estimate of the total cost of construction per square foot of floor space to the application of predetermined elemental time standards to specific craft operations. All have the advantage of being based on fixed values per unit which can be reapplied consistently. It must be admitted, however, that this does not guarantee absolute accuracy of the final estimate, since the standards necessarily are based on some degree of averaging of conditions and requirements. Fundamentally, the probable accuracy of the result is a function of the degree of analysis, or breaking down, of standard application. Estimating standards may be classified on this basis in five basic groups.

1. *Plantwide or industrywide averages or ratios.* This includes such data as total maintenance cost per ton of product per operating hour of primary equipment, basic construction figures per square foot or per cubic foot, and many others. While the limitations of estimates based on these figures are quite obvious, their value for preliminary planning and auditing should not be overlooked. Building-construction data in particular are easily available in handbook form and should be considered as one of the tools of the maintenance estimator. For example, "The Vest Pocket Estimator" by Frank R. Walker includes more than 80 topics, from architectural concrete to weather strips, and covers material and labor costs in a broad range of instances. (For information contact Albert Ramond & Associates, Inc., Hinsdale, Ill.)

2. *Comparative job standards.* This is one of the most effective ways to achieve reasonably good estimating accuracy with minimum cost. Repair jobs on similar equipment covering a large range of total costs per job may be related to one or two simple determinants by comparing standards which are based on detailed analysis. For example, standards for rebuilding and rewinding operations on various sizes of large dc motors were established by detailed application over a 2-year period and then compared graphically on the basis of horsepower and number of coil slots. The resulting formula provides estimates entirely adequate for that particular application and requires only easily available information to determine the standard. Similar comparisons may be made

EXHIBIT 1 Benchmark: Spread Sheet

2.0 Hr		3.0 Hr		4.0 Hr	
1.50	2.50	2.5	3.5	3.5	4.5
Unclog pipe & drain—#2 P.M. Saveall line R & I 2 Fit 4″ Unclog drain line with wire & with high pressure hose 7′–14′ 15# part __1.66__		Replace copper tubing with std. pipe on airline to new case sealer— Hayssen 6 P -33 F ⅜″ R 7 s Cu Tub 0-7′–14′ R & I 4 Cl Bkt. __2.5__		Install temporary hook up to drain over-flow #1 Saveall tank to cellar 0 P —9 F 4″ Thd F1 C 8⅜″ bolts 2 P —0 F 4″ Flg R 8¾″1 Pc Sc Fast Heat part to R 1 8¾″ 2 Pc Sc Fast Clean 16 holes with Rub Hs—Cl both ends H Tap Scrape 2 4″ Flg __3.7__	
Install new float valve assem. on Emulsifier—Beater Rm. 6 P 2″-½″ —12 F 2″ __1.96__		Extend 4″ S.S. line from back water box to tank—#6 P.M. cellar 3 P-4″ S.S. 10 W W bkts. ¼″ 50% 15″ __2.7__		Repipe up #1 P.M. Jordan 2 P-1 F 2½″ Thd R&I Rub HS & F 0 P-3 F 1″ Thd 7 P-0 F 6″ Flg Prep & 1 l gskt (51.0″) (extra) __4.0__	
Disconnect & reconnect sewer pump—#6 P.M. wet end cellar 0 P —30 F ⅜″-½″ R & I gskt (50.2″) R & I 12 Sc Fast Align & level pump __1.98__		Extend 1-½″ sealing water line—#2 P.M. 9 P-31 F 0-7′ __2.8__		Remove sprinklers from skylights in new cutter area (76″-88″) 11 P-22 F 1″ 14′-21′ Drain system __4.1__	

FIGURE 5.1 Benchmark: Spread Sheet.

EXHIBIT 2 Benchmark: Input Sheet

BENCHMARK STANDARDS—INPUT SHEET FOR JOB SLOTTING			CRAFT: PIPEFITTERS DW = DIRECT WORK (CIRCLED)	
ENTER TIME SPAN IN HR	ENTER TIME SPAN IN HR	ENTER TIME SPAN IN HR	ENTER TIME SPAN IN HR	ENTER COMMENTS:
FROM: 4.5 TO: 7.4 DW HR	FROM: 7.5 TO: 10.4 DW HR	FROM: 10.5 TO: 13.4 DW HR	FROM: 13.5 TO: 16.4 DW HR	FIELD BENCHMARKS
6 HOURS DW Field erect 12' × 2" × 80 cs Cut/thread two ends 5.8 hr	**9 HOURS DW** Field erect 12' × 2" × 80 cs Cut/thread 4 lngths-shop 8.5 hr Field erect 20' × 2" × 80 cs Cut/thread two ends 8.2 hr The theory and practice for the development of slotting: For a nine hour day-use: 9 × 15% ± = 1.35 hr Use 1.4 hours ± 7.6* (9.0) 10.4	**12 HOURS DW** Field erect 25' × 2" × 80 cs Cut/thread two ends 11.2 hr Field erect 20' × 2" × 80 cs Cut thread 5 lgths-shop 11.8 hr	**15 HOURS DW** Field erect 25' × 2" × 80 cs Cut/thread 4 lgths-shop 13.9 hr Field erect 35' × 2" × 80 cs Cut/thread four ends 15.3 hr	$\dfrac{39.4}{74 \text{ ft}}$ = .53 hr per/ft $\dfrac{35.3}{75 \text{ ft}}$ = .47 hr per/ft 16.4 hr etc. etc.

*Reduce by .1 hr or 6 min for transition to...... the next slot!
7.5 hr 9.0 hr 10.4 hr 10.5 hr 12.0 hr 13.4 hr 13.5 hr 15 hr 16.4 hr

FIGURE 5.2 Benchmark: Input Sheet.

for projects such as furnace rebuilding by relating the number of bricks to be replaced to the standards for the typical job. For other specific applications, the measure of the relative size of the job may be the number of rolls, tubes, or sections, or the size of the area.

Comparisons based on detailed basic data standards are ideal for this purpose. Where good records of the actual time spent on each job are maintained, these records may be used in the same manner, but with some reservations about the level of accuracy to be attained.

3. *Specific job standards.* Work sampling or time study can provide a measure of the work accomplished per day on a specific assignment such as electrical troubleshooting or maintenance in a production-machine area. Relating the measure of work to determinants, such as the number of trouble calls per day or the operating hours of key equipment per day, results in an estimating figure which can be quite accurate as long as general conditions do not change substantially. Such standards should be periodically audited to provide a measure based on current conditions. Standards for repetitive repair jobs also come under this category, whether established by time study, application of basic data, or adjusted accumulation of actual time.

4. *Operational basic data.* Many maintenance operations are repetitive, although they occur as part of a complete job which may never repeat. For example, in electrical work, installing a complete junction box or mounting a panel box could not be properly called an "element," but the estimating standard for this operation can be added into a job estimate more easily than adding in each element of the operation. A series of these operation standards can provide a relatively quick way of building a complete job standard. Special applications of this technique usually are developed to fit individual plant practices as experience with elemental basic data grows.

5. *Elemental basic data.* The ultimate in estimating repair and maintenance work can be obtained through the application of elemental basic data. The accuracy and consistency attained by this method can fill the most exacting requirements of the end use of the estimate. Where the job information is adequate, the accuracy of job standards set by this method is as good as that normally found in production incentive standards.

Anyone familiar with the process for building up basic standard data for production operations may correctly visualize a tremendous industrial engineering project if the same approach is used to cover the wide variety of maintenance operations. But this has been done over a period of years, and data are now available for most of the common operations in any industry. Several management consultants have developed elemental basic data for repair and maintenance work. Published versions are also available, including a detailed set developed for the United States Navy. Generally, it is far more practical to use existing data than to start from the beginning and develop your own.

When the elemental data are available, there is generally a question about the cost of application. It would be reasonable to assume that this cost would be in proportion to the degree of detail and accuracy involved. Actually, this is not quite the case. Selection of the proper elemental values, factors, and multipliers can be reduced to a simple process by good systems study. The extensions and summarizations are strictly clerical functions, which can be done by a clerk or by data processing. The only really expensive part of the application is defining and describing the job in sufficient detail.

If good job information is easily available, there is usually little question about the practical value of estimating by applying basic elemental data. If such information must be obtained specifically for the purpose of the estimate, then the end use of the estimate must justify the cost of obtaining the detailed job information as well as the minor cost of actually applying the data. In evaluating this approach, the value of consistency in the application should be carefully considered. This is especially important where wage incentives are involved and may be equally valuable where estimated values are used directly for control purposes.

To be effective, the application of basic data must recognize the significant variations in conditions. Consider the relatively simple operation of painting a wall. The standard time per square foot will be determined by the method (roller or brush), the type of paint, and the absorbency, regularity, and smoothness of the surface. In addition, job height, interference and obstructions, extreme heat or cold, humidity, fumes, and unusual factors may have a significant effect. The most practical way to cover these factors is the use of prepared tables relating job conditions and standard allowances, such as those used by consultants experienced in this field.

Quickread Estimating for Labor and Crew Assignments. The effective application of estimates for cost control will become the basis for labor and crew assignments. Work sampling will quickly determine where and when such crew reassignments should be made; however, the use of work sampling percentages permits both the assignment control and cost control when these percentages are combined with basic labor estimating. The estimator converts the work sampling percentages into workable ratio factors, based upon direct craftwork data. Quickread factors can be achieved in four simple steps:

Step 1. Factor Development (Indirect Work)

	Work sample	Conversion	Quickread factor
Direct work	45%	None	QR = 1.00
Makeready indirect work	30%	$\dfrac{0.30}{0.45} = 0.667$	QR = 0.67
Travel and transport	12%	$\dfrac{0.12}{0.45} = 0.027$	QR = 0.27

Lost time = 7%
Personal, etc. = 6%
(Lost time used in step 2)

Step 2. The Rule of Seven (Direct Work). Accumulate previous job history for similar jobs in actual time. Actual records for "Install Faucet."

Date	Reported man-hour	Job no.
07-2	1.1	1326
08-7	2.1	1472
09-9	1.0	1516
10-1	0.9	1540
12-6	0.5	1609

6-month average = 1.06 man-hr
(Extract lost time) 0.93 (93%) (7%)
Direct work = 0.986 hr per job
(Apply quickread factor of 0.7)   0.986 hr × 0.7 QRF
"Rule of Seven" Direct Work (estimated)   0.69 hr

Step 3. Combination Estimate Format. Determine the direct work on site. Use either the (0.7) seven-tenths rule or other methodology. Once *direct work* is determined, utilize your quarterly work sampling factors as shown in step 1:

Typical job,
"Install Faucet"   Direct work hours = 0.69 (0.7 rule)
                                              IDW and travel hours = (×) 1.94
Makeready and travel   Quickread job = 1.3386
0.67 + 0.27 = 0.94 = 1.94 multiplier

Step 4. Determine Which Slot Level of Job Is Used. The estimate alone may become too exact for the estimator. Based upon initial work sampling and actual job history, the estimator determines the level of the job in a typical job slotting technique:

Actual records indicate that faucet   Quickread estimate allows a total of 1.3
installations required a maximum of   man-hr of labor for a typical job
2.1 man-hr (Job 1472)   assignment

Slotted entry level for job would be:

Minimum	Maximum
1.3 man-hr	2.1 man-hr

Job slot average = 1.7 man-hr

$$\text{Calculation:} \quad 1.3 + 2.1 = \frac{3.4}{2} = 1.7 \text{ man-hr slot average}$$

PERT Statistical Approach. In the application of the PERT program of computer-calculated scheduling, the elapsed time to reach each milestone or event is estimated on three bases: pessimistic, expected, and optimistic. The three figures are then used in the program to compute most-probable finished dates. This technique has its greatest value in planning projects which have indefinite scope and which are very difficult to estimate on a sound basis, such as research and development projects. Most maintenance jobs are more tangible and can be more readily estimated by other means.

Estimating the Cost of Deferring Maintenance. The cost of lost production and possible damage to equipment or product because of deferring maintenance generally can be estimated by judgment based on basic guide figures. Out-of-pocket costs will be most realistic and useful, and detailed accuracy is not necessary. It is important to consider carefully any overhead figures used. Common accounting practice of using operating hours on key machines for distribution of costs can produce misleading figures if these rates per hour are applied for evaluation of machine downtime. For example, the production lost during a 2-hr breakdown might be replaced by running 2 hr overtime, with little excess cost except the direct labor hours and overtime premium paid.

Practical and useful approximation of material or product losses can be made by utilizing average ratios of labor to material costs. For example, if direct labor for a product line averages half the material cost and 10 units are produced per day by 4 people at $3 per hr, the approximate labor cost is $10 per unit and the material cost is $20.

HOW TO SELECT APPROPRIATE ESTIMATING TECHNIQUES FOR LABOR COST

From the foregoing discussions on estimating methods for maintenance, it can be seen that a wide variety of techniques may be used for estimating labor cost. It is very important to select the most appropriate method for each situation in order to minimize estimating cost input yet develop the accuracy level required.

As an aid to the development of an estimating program, the guide chart (Fig. 5.3) summarizes the criteria for selecting the best method for estimating labor cost. The major criteria used are the prerequisites for estimating: (1) classification of and knowledge about the job and (2) end use of the estimate.

When using the chart, first consider the purpose for which the estimate is to be used and then select the column whose heading best describes the end use of the estimate. Second, considering the knowledge about the job and the extent of planning done or to be done, select the grouping in the left-hand column which best describes the situation. The most appropriate estimating techniques are indicated for the combination of purpose and job information.

Job description and job information available (planning)	Best Estimating Method When Estimate Is to Be Used for:			
	Approval, backlog, forecasts, equipment purchases	Methods, critical path scheduling, monthly schedules, make-or-buy decisions	Plantwide controls, weekly schedules	Individual controls, daily schedules, large-volume make-or-buy decisions
Group 1: No definite job or plan. Emergencies, such as fires. New equip. run-in. Also repairs without planning.	Judgment	Judgment	Judgment	Not practical
Group 2: Definite end result, with no plan. Breakdowns, troubleshooting. Also poorly planned repairs, repetitive jobs without specific instructions, and construction without planning.	Judgment, past record ratios, construction data	Past record ratios, construction data	Time-study analysis, ratio based on work sampling, construction data	Combination of ratios and average time per call, based on time study and work sampling. Not applicable
Group 3: Definite end result, work description, and methods (planned). Overheads, relocations, building repairs, modifications, spare parts. Also specified routine, preventive, change-over, and repetitive repairs.	Judgment based on details, slotting, past records for repetitive work.	Judgment based on details, slotting, comparative standards, past records modified by work sampling for repetitive work.	Job sampling by elemental basic data, detailed slotting, comparative standards, elemental basic data or time study for repetitive work.	Elemental basic data based on verified methods, detailed comparative standards, elemental basic data or time study for repetitive work.

FIGURE 5.3 Estimating methods at various levels.

ESTIMATING TECHNIQUES FOR MATERIAL COST

Estimating material cost for maintenance and repair work is relatively more precise and straightforward than estimating labor cost. As a result, we frequently find extremely fine details and absolute accuracy in material estimates, even where the vague nature of accompanying labor estimates places the combination of both factors in a rough approximation class. While the savings potential involved in controls which are based on accurate estimates of labor is nearly always substantial, the potential for reducing material cost is usually much less.

So we frequently find an impractical situation, in which the material cost is estimated accurately where there is relatively little need for accurate control, and the cost estimate is combined with rough estimates on labor where the need for accurate control is critical. This results, of course, in an inaccurate total figure although the cost of estimating is relatively high. With this possibility in mind, it is especially important to consider the purpose of the estimate and the available information about the job.

In a typical situation, three well-qualified graduate engineers were occupied in developing material estimates for construction and modification projects which were to be carried out by the maintenance department. Material estimates were developed very carefully, although the primary purpose of these estimates was the obtaining of management approval for the capital expenditure. It was agreed that such approval decisions could be made on the basis of total estimates that were accurate within plus or minus 20 percent. After suggesting shortcut methods for material estimates, it was possible to transfer two of the three engineers to the planning operation, where their skills could make a far greater contribution to the effective management of the maintenance operation.

Purchases of machines, equipment, or construction materials generally should be based on actual bids or quotations from the supplier, even where competitive bids are not officially required. But for repetitive repair and replacement jobs of a smaller scale, it is logical to use the records of machine repair costs which are established for the purpose of determining corrective-maintenance procedures. In large companies, these records will be supplied by data processing, but the same benefits can be obtained in a small company by the simple expedient of establishing a file folder for each piece of equipment and retaining the completed work orders for all the repairs on that piece of equipment within that one file folder.

On nonrepetitive jobs, materials costs can be estimated by comparison with similar jobs which have been completed. The slotting technique used for labor estimates is particularly useful and encourages the shortcut estimating approach which is desirable in most instances. For example, an accuracy level of plus or minus 25 percent from the midpoint of the range can be applied to establish values such as $75 to $125 as one bracket, $125 to $200 as another bracket, $200 to $325 as a third bracket, $325 to $525, and so on.

Another practical method for estimating material is to establish ratios based on labor estimates. Establishing the ratios requires four basic steps:

1. Accumulate actual material and labor cost data for a large number of jobs covering a variety of work.
2. Classify the jobs using natural division which can be easily determined.
3. Calculate the average ratio of material cost to labor cost for each job classification.
4. Analyze the deviation from the average. If the deviation is excessive, the classifications can be divided more finely.

To use this procedure it is necessary only to develop the accurate estimate of labor which is necessary anyway for control of labor cost and scheduling, decide which classification the job fits, and apply the appropriate ratio for that classification.

ESTIMATING TECHNIQUES FOR OVERHEAD COST

Preliminary analysis in estimating overhead cost requires the answer to three questions:

1. What is the purpose of the estimate?
2. How accurate is the information about the job?
3. Is an estimate of overhead cost really necessary?

Careful consideration of overhead cost is necessary when the estimate is to be used for a decision regarding "make or buy." In many cases, the final decision on whether to make a particular part in the maintenance shop or to buy it from an outside supplier will depend greatly on the treatment of overhead. Distribution of cost to various cost centers within a particular facility may be another reason for requiring careful treatment of overhead-cost estimating. In most other situations, the application of an overhead rate to the estimated labor cost for a maintenance job is probably unnecessary and can be very misleading.

An overhead cost is one which is not directly attributable to a specific job or project. In maintenance this means such items as salary supervision, storeroom operation, shop equipment and facilities, and miscellaneous supplies. Fringe benefits such as vacation and holiday pay and group insurance are also overhead costs.

In order to account for all the costs of operating a maintenance department, some means of prorating these general costs to individual jobs is adopted. Usually this is accomplished by dividing the total overhead cost for a given period by the total maintenance labor hours charged to specific jobs and thus establishing an overhead rate per maintenance direct labor hour. As long as this overhead is recognized as a means for arbitrarily dividing up the general cost, it is a useful accounting tool. But when management decisions are based on cost comparison, it is extremely important to know that the overhead estimate is not accepted as anything more than just that—a means of arbitrarily dividing up general cost. For example, suppose it is proposed that a substantial amount of construction work presently done by the maintenance personnel should be contracted to outside firms and that the maintenance force be reduced by 10 employees. In this case, the reduction in direct labor cost can be predicted with acceptable accuracy.

But does this mean that all the overhead cost which has been distributed to this much direct labor will actually be saved? Really the only savings will be in the cost of items which are literally dependent on hours worked, such as the fringe benefits and a portion, a small portion, of the other costs such as supervisory costs. In actual practice, should this change be made, it would become necessary for the accounting department to recalculate the overhead rate per direct labor hour for their cost distribution, since the total overhead cost would not be reduced in proportion to the reduction of direct labor.

TRAINING ESTIMATORS

Application of basic data has some obvious limitations for small maintenance operations and for other situations where the relatively high administration cost cannot easily be justified. We have also pointed out limitations of the pure judgment estimates, the past-experience basis, and slotting which is based on judgment alone. The method which is not universally applicable is the spread sheet, which is a refinement of slotting, using accurately established standards for benchmarks.

These estimates usually are made by the planner, and the planners are usually selected from the ranks of experienced maintenance craftsmen. This raises the next logical question—How does the person who is a craftsman this week become an expert estimator next week?

Probably the first milestone in the training of a new planner is the establishing of an estimating concept where the objective is to determine the time required to perform the operation under conditions as they should be, rather than providing for all the contingencies which could happen.

One way to establish this concept is as follows: First, pick a relatively simple maintenance repair job with which the new planner is thoroughly familiar, and ask him for a rough estimate of the time required. Write this answer in a circle at the top of the page. Almost always these estimates will be in multiples of 8 man-hr. No craftsman is likely to estimate a job to take $3\frac{1}{2}$ hr.

Next, ask the planner to list in detail the steps necessary to perform the job and to estimate the time required for each of these elements in terms of minutes, not hours. You can expect that the total of these minute estimates will be considerably less than the original estimate in hours.

The next step is to take the planner into the shop and observe a typical maintenance job in progress, pointing out specific instances of lost time, wait time, and duplication of effort which could have been avoided by good planning and good supervision, and also pointing out that these elements make the difference between a good estimate and a past record.

Starting the planner's training as outlined above provides a reasonably good basis for developing spread sheets. Where a trained industrial engineer or outside consultant can be utilized, advanced training of the planner/estimator can be carried out by refining his labor estimating technique to include a breakout of the following categories of labor:

Direct productive work (hands-on work discernible in the physical components of the job)

Conditions and manning allowances (modify direct work)

Indirect work (preparation, planning, material, and tools)

Travel time (to and from job based on job length and complexity)

Necessary enforced and controlled delays (to be minimized)

When the planners' training includes the advanced elements, it is frequently recommended that the spread sheets include only the direct work, with the other elements added based on the proposed jobs, work conditions, location, and planning input. In this way, highly accurate standards may be developed where this condition is desired.

In all training situations, self-correcting procedures should be established to permit full development of the individual.

SUMMARY

Estimating maintenance costs involves judgment, forecasting, and predicting. This necessitates the use of past and present data as the base from which to start.

Two areas of information determine which of many techniques is appropriate to a job: (1) the end use of the estimate and (2) the available information about the job.

In estimating labor, the first task is analysis. The estimator will make use of judgment, slotting, PERT, or labor standards per unit. In the last category, the estimator may make use of plant or industrywide averages or ratios, comparative job standards, specific job standards, operational basic data, and elemental basic data. The amount of time to be invested and the techniques to be used are limited by the two factors: end use and information available.

In estimating material, the development of ratios relating material costs and labor cost can greatly simplify the estimating process.

In estimating overhead, careful consideration must be given to: (1) the end use of the estimate, (2) the accuracy of job information, and (3) what portion of overhead cost is truly applicable to the job.

The techniques which the estimator uses must be compatible with job requirements so that the results achieved will warrant the time and cost invested in estimating.

CHAPTER 6
KEY PERFORMANCE INDICATORS

John Cray
Managing Principal, Life Cycle Engineering Inc., Charleston, S.C.

Best practice refers to the doing of something that will achieve superior result or performance. Best practice in key performance indicators (KPI) therefore has to refer to the process of developing useful performance measures for an organization. A process is good in an organization if it is integrated as a system within the total organization. A useful and effective KPI therefore has to come from a well-formalized *performance measurement system* (PMS).

Performance measurement is the process of determining how successful organizations and individuals are in attaining their objectives. It covers all levels, including individuals, teams, processes, departments and the organization as a whole, with the view of continuous improvement of performance against organizational objectives. The performance measurement system is a systematic way of doing this. This is a system that integrates the measurement of performance of all levels within the organization with the view of continuous improvement of performance against organizational objectives. The outcome is the establishment of performance measures—the quantitative indicators that show how well the organization's objectives are being met. Key performance indicators are the performance measures critical to an organization's core business.

These days, everyone likes to talk about the knowledge economy and the knowledge organization. It would be a good idea to relate the use of performance measures to a knowledge-based organization. In a highly competitive business environment, enterprise has to be information-rich and knowledge-orientated in its management. Performance measures become a crucial input for senior managers to perform effectively. At the top of the organization, senior managers have to view so many facets of the business and be on the run always, thus he has time to monitor only the essential indicators on a daily basis. A good set of key performance indicators is necessary.

THE KNOWLEDGE-BASED ORGANIZATION

Figure 6.1 illustrates the key components of a knowledge-based organization. Organization needs to have a model—an organizational excellence model—if we are talking about achieving best managerial practices. Within the model, will be the performance scorecard of key performance indicators. A good set of key performance measures must take into consideration three important aspects of businesses—productivity, total quality, and competitiveness. There should be at least a set of three indicators to monitor these three aspects.

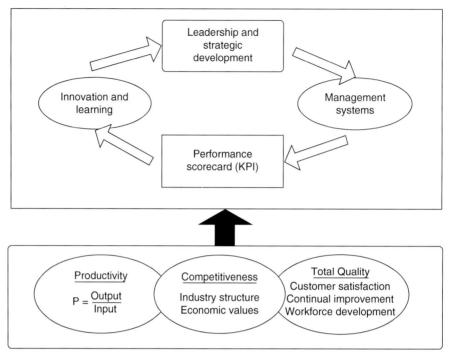

FIGURE 6.1 The organizational model.

The productivity aspect monitors the enterprise's performance in its resource utilization to create value. The competitive environment requires that the enterprise focuses on the efficient use of its input resources and at the same time, continually create new added value for its output. On the quality aspect, the enterprise has to continually improve itself to serve the changing requirements of the customers. Both productivity and quality efforts complement the enterprise ability to compete. Competitiveness is about the enterprise ability to continuously maintain its attractiveness with its customers and its shareholders in the long run.

A highly competitive market means that enterprise has to perform in an excellent manner and it has to keep this up as long as it wants to be in the market. The key indicators to monitor the performance have to build up from the organizational excellence model. Figure 6.2 illustrates the organizational excellence model. If you have a *quality management system* (QMS) like the ISO 9001 model, it will not be sufficient to develop best practice KPI. The QMS is limited to achieving customer satisfaction—only the quality aspect. The organizational excellence model is holistic model to develop the set of best practice KPI, as the model looks at all the three aspects.

- Leadership management style
- Leadership and performance review
- Philosophy and values
- Social responsibility and corporate citizenship
- Business drivers and business results
- Human resource management
- Work systems
- Motivation, reward, and recognitions

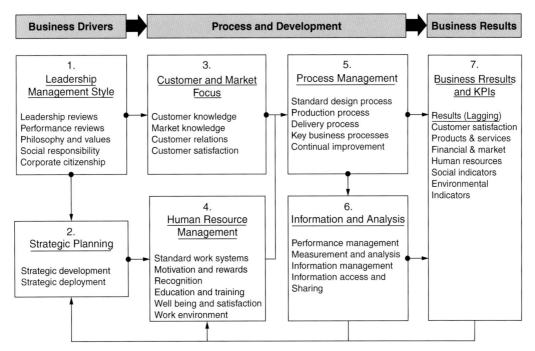

FIGURE 6.2 Organizational excellence framework.

- Education, training, and development
- Well being and satisfaction
- Work environment
- Strategic planning
- Strategy development and deployment
- Information and analysis
- Performance management
- Measurement and analysis
- Information management system and technology
- Information access and sharing
- Process management
- Design process
- Production and delivery process
- Key business processes
- Support processes
- Continual improvement
- Business results and KPIs
- Customer satisfaction
- Products and services results
- Financial and market results

- Human resources results
- Social and environmental indicators
- Customer and marker focus
- Customer and market knowledge
- Customer relationships
- Customer satisfaction
- Process and deployment

MEASURING WITH A PURPOSE

It is not a matter of creating KPI but the need to measure with a purpose. The needs of measurement include:

Planning, Control, and Evaluation. The process of analyzing measurement in orders to make decisions or evaluations, and is central to the operation of an effective and efficient planning, control or evaluation system.

Managing Change. Measures must support management initiatives, and the primary requirement is to integrate measures vertically (across levels) and horizontally (across functions).

Communication. Measurement is required to reduce emotionalism and increase constructive problem solving, increase influence, monitor progress, and give feedback and reinforce behavior.

Measurement and Improvement. Reason for measuring is to support improvement, and at the end provides the scorecard to report on how well improvement efforts are working—if you cannot measure the activity, you cannot improve it.

Resource Allocation. Helps an organization to direct scarce resources to the most attractive improvement activities . . . it is a direct stimulus to action.

Measurement and Motivation. Performance improve if individuals are given achievable and challenging targets.

Long-Term Focus. Appropriate performance measurement can ensure that managers adopt a long-term perspective.

The best practice KPI has come a long way. The focus of performance measurement has changed in recent years. Table 6.1 summarizes the major changes that have taken place in the best practice KPI. These are important shift in managerial thinking in an ever competitive and complex business environment.

TABLE 6.1 Changing Focus of Performance Measurement

Traditional PMS	Innovative PMS
Based on cost/efficiency	Value-based
Trade-off between performance	Performance compatibility
Profit-orientated	Customer-orientated
Short-term orientation	Long-term orientation
Prevalence of individual measures	Prevalence of team measures
Prevalence of functional measures	Prevalence of transversal measures
Comparison with standard	Improvement monitoring
Aim at evaluating	Aim at evaluating and involving

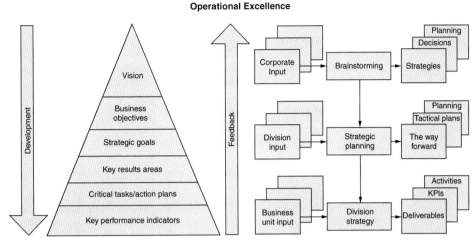

FIGURE 6.3 Hierarchical model.

PMS MODELS

There are three distinct models for the creation of best practice KPI. The first model is strictly hierarchical (vertical), and is characterized by cost and non-cost performances on different levels of aggregation until they ultimately become economic-financial, which connects productivity with the ROI.

The second type is the balanced scorecard, where several separate performances are considered independently. These performances correspond to different perspectives of analyses (financial, internal business processes, customers, and learning and growth). However, the perspectives substantially remain separate and the linkages are defined only in a generally way. The model integrates vertical linkages—from operational measures up to the financial measures. See illustrations in Figs. 6.3 and 6.4.

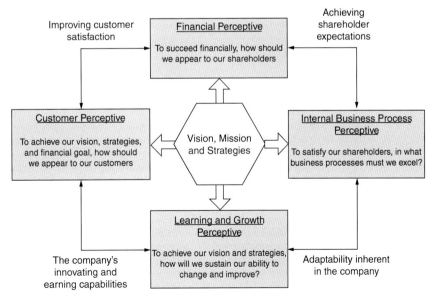

FIGURE 6.4 Balanced scorecard model.

MATCHING MEASURES WITH THE ORGANIZATIONAL CONTEXT

Various outcomes must come out of any business operations. This is the results and performance component in the organizational excellence model. The outcomes must be able to reflect the achievement of the business mission and objectives and to move the organization closer to its vision. In other words, there must be a holistic set of integrated measures that are linked to the mission, objectives, vision, and strategies. A useful framework to do this is the *balanced scorecard* (BSC).

The BSC looks at four perspectives of any business:

- Financial
- Customer
- Business process
- Learning and growth

By integrating these four perspectives with the business mission, vision, objectives, and strategies, the organization will have created the signposts for it to move on its journey toward excellence. This means that the BSC is able to integrate nonfinancial measures with financial measures and to incorporate both lagging and leading measures. BSC also display these characteristics of transparency:

- Simple to understand
- Have visual impact, focusing on improvement
- Visible to all
- Provide timely and accurate feedback
- Relate to specific, stretching but achievable goals
- Based on quantities that can be influenced, or controlled, by the user alone or the user in cooperation with others
- Clearly defined
- Part of a closed management loop
- Have an explicit purpose
- Based on an explicitly defined formula and source of data
- Use data which are automatically collected as part of a process whenever possible

STRATEGY MAPPING FOR A BALANCED SCORECARD

The power of the BSC is not so much the performance indicators but rather the ability to link the performance measures to the strategic intent of the enterprise. The *process of strategy mapping* clarifies the relationships of the four perspectives in a strategic manner. In the strategy map (above), the bubbles represent the various perspectives. The arrows represent the strategic linkages. Each of the bubbles will need at least a key performance measure to monitor and control the business.

The process of strategy mapping when adequately carried out will deliver KPI, which is relevant, purposeful, and useful for strategic monitoring and the management control of the business. The KPI will provide useful and crucial inputs for strategic and management reviews of the business.

The above illustrates a *value-for-money* strategy mapping for an enterprise. The corporate objective is to increase shareholders' value in the company. Revenue growth and productivity will be what shareholders view as important.

From the customer point of view, what is important is getting products/services of highest quality and lowest cost. The ability to satisfy the customers must result in shareholders' value. To achieve

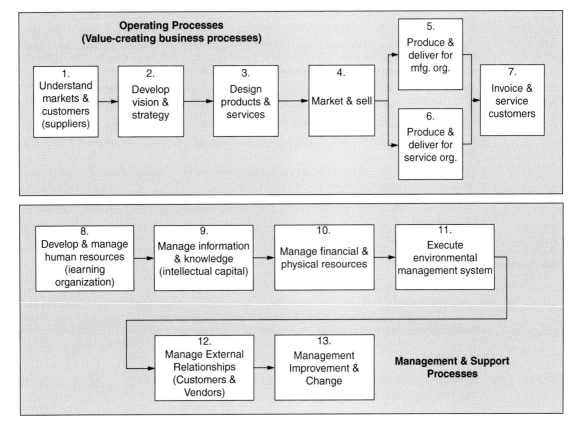

FIGURE 6.5 Value chain model.

outcome that will be perceived as value for money by the customers, the organization must have internal capability to meet customer requirements. The development of internal capability must be able to support the customer-related strategy. As the market is dynamic, the company must be able to maintain its ability to meet future challenges and growth. Strategies to develop new internal competencies are necessary for the continuing survival of the enterprise.

The third model is related to the value chain and also considers the internal and external relationship of customer and supplier. This is as illustrated in Fig. 6.5. Many enterprises surviving in a highly competitive environment, such as those in the consumer businesses has to work on the total value chain to be able to assure that their products are always available to the consumers at the right time, right place, and right price. Quality is a standard and is not a competitive factor anymore.

PERFORMANCE MEASUREMENT SYSTEM

An integrated model of performance measurement includes these levels of implementation:

- Strategy development and goal deployment (Senior Management).
- *Key result areas* (KRA). The limiting number of areas in which results will ensure successful competitive performance for the organization.

- *Key performance indicators* (KPI). The actual measures used to quantitatively assess performance against the critical success factors. There should be at least one KPI for each.
- *Process management and measurement* (PPM) includes measures of inputs, process, and outputs. Outputs from a supplier form the inputs to the next customer. It is only necessary to define the process and output. (Process Management Level)
- *Process measures* monitor the activities of a process and motivate people within a process.
- *Output measures* report the results of a process and are used to control resources.
- *Performance appraisal and management* (PAM) includes two linked elements.
- *Performance management*. Systematic data-oriented approach to managing people at work on an on-going basis.
- *Performance appraisal*. Process by which organizations establish measures and evaluate individual employees' behavior and accomplishments.

The development of best practice KPI is for strategic monitoring at the organizational level by senior managers.

Requirements of KPI

It is assumed that measurement provides a means of capturing performance data, which can be used to inform decision making. Equally important is that people respond to performance measure and they modify their behavior in an attempt to ensure positive performance outcome even if it means pursuing inappropriate courses of action. The key issue in designing measures of performance is then is to match it to the organizational context. The requirements of a good performance measure should consider:

- Its development from strategy and be related to specific goals (targets).
- That they are clearly defined and simple to understand.
- They are timely and accurate feedback.
- They provide fast feedback and are part of a closed management loop.
- They can be influenced or controlled by the user alone or in cooperation with others.
- Have relevant and have an explicit purpose.
- Have explicitly defined formula and source of data.

They provide information and should be precise—be exact about what is being measured:

- *Title* (measure). A good title explains what the measure is and why it is important.
- *Purpose*. Rationale underlying the measure has to be specified.
- *Relates to*. Business objectives to which measure relates to should be specified.
- *Target*. Targets are necessary to evaluate the level of performance.
- *Formula*. The way performance is measured affects how people behave. The right formula ensures the right behavior.
- *Frequency of measurement*. This is a function of the importance of the measure and the volume of data available.
- *Frequency of review*. How often, for example, frequency, of review to evaluate value and benefit.
- *Who measures*. The person who is to collect and report the data should be identified.
- *Source of data*. Consistent source of data is vital if performance is to be tracked (over time).
- *Who owns the measure*. Single-point accountability is essential. One person should have ownership of each of the KPIs.

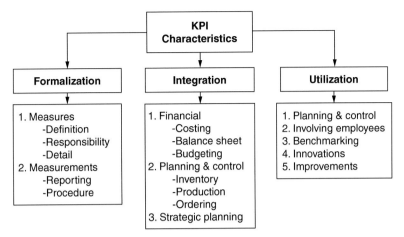

FIGURE 6.6 Formalizing best practice—KPI.

- *Who acts on the data.* The person or function should be identified.
- *What do they do.* Unless the management loop is closed, measurement is pointless.
- *Notes and comments.* Any qualifiers or clarifications, as needed, to ensure clarity.

Formalizing Best Practice KPI

A best practice set of KPI need to be holistically integrated into the organization model. This involves looking at three crucial characteristics—formalization, integration, and utilization (see Fig. 6.6). Formalization deals with the questions:

- What will be measured?
- How will it be measured?

As performance measurement is not an isolated system, best practice KPI must be developed from the integration between various areas of the business and deploying the business objectives throughout the organization (tied inherently into the organization model). The best practice KPI must integrate with the accounting system, the operation (planning and control) system and the strategic planning. Best practice KPI is therefore unique in the context of a particular organization. You cannot copy the best practice KPI but you can learn about the process of developing the KPI.

The third characteristics deal with the purpose of doing the KPI—utilization. There are basically two reasons: to compare with the competitors and to check the accomplishment of the organization's objectives. The idea is to comply, to check, and to challenge in the market.

WORK MEASUREMENT

A necessary ingredient of any operating control system is the generation of quantitative standard information against which to measure and assess performance in the execution of work; to identify areas for corrective action; to measure progress toward goals; and to aid in the decision-making process. The objective of work measurement is to generate this standard information.

Realistic labor estimates are an essential part of a planned maintenance program. There is no effective way of matching workloads against available manpower without measurement, and it is hard to make realistic promises when taking equipment out of production. Work measurement systems support planning. Work measurement and analysis produces maintenance labor and resource estimates that are essential to the entire planning, scheduling, and control process. Realistic labor estimates are essential to planned maintenance in order to:

- Match workloads against labor resources
- Make realistic promises when taking equipment out of production
- Determine correct staffing for each grade of labor
- Determine level of crew performance
- Establish expectations for individual and crew performance

While maintenance work almost always contains unpredictable elements, these elements do not constitute the whole job and are quite often only a minor part of the total workload.

- Estimating is easier and more accurate when the job is broken into separate elements. Long, complex jobs cannot be estimated as a whole.
- The intent is to achieve reasonable accuracy and consistency.

Essentials of Work Measurement

There are several essentials to successful work measurement:

Familiarity with Maintenance Jobs and Plant Equipment. Some of this familiarity the planner brings to the job, some he acquires on the job. There is no substitute for task knowledge when it comes to estimating. A craft background is an ideal starting point; it enables a planner to "visualize" him or herself actually doing the job. The next best thing is actually observing what is involved as jobs are done and, when this is not possible, "talking through" a job with a maintenance craftsperson or supervisor who is familiar with it. Becoming involved in the job and *being seen to be involved* in it has tremendous advantages. It expands the planner's knowledge of the plant and its equipment. More importantly, it builds credibility among the maintenance craftsperson and supervisor for what the planner is doing.

Comparative Time Estimating. Absolute accuracy is neither possible nor necessary for maintaining an acceptable level of efficiency and control in maintenance. What is essential is *consistency*. It is in the nature of human mental processes to estimate by comparison. We compare the unknown with the known and then estimate the degree of similarity (identical, larger, smaller, how much?). We tend to do this automatically and very subjectively. Some people have an almost infallible sense of comparative size, others find it hard to come within a mile. The basis for consistent estimating is for all planners to have access to and use the same library or catalog of job estimates. There are basically four means of structuring comparative maintenance job estimate files:

- Systematic card files
  - By skill
  - By nature of work
  - By crew size
  - By labor hours
- Catalog of standard data for building job estimates
- Catalog of benchmark jobs
- Labor library

Current state-of-the-art rests between the last two structures. Regardless of the comparative approach used, the basic form of work measurement must still be decided. The standards (or estimates) to be loaded into the comparative catalog must first be developed.

The technique of comparative estimating involves the *comparison* of jobs with those in the library, not the *matching* of jobs. This distinction is important because in the former case, a few hundred carefully selected jobs will enable an experienced planner to produce consistent estimates for most maintenance work. In the latter case, many thousands of jobs would be needed to get the same result.

To make a comparative estimate, the planner must first define the scope of the job which is to be done and prescribe the method to be used. He will need a good knowledge of the process and equipment, and will often have to visit the jobsite and talk with the production and maintenance supervisors as well.

The planner's next step is to turn to the appropriate section in the benchmark library to try to find a benchmark with a description similar to the job for which an estimate is required. The planner must then make a judgment based on his own mental comparison of what is involved in doing the benchmark job (for which a time is available) and what will probably be involved in doing the job for which he needs an estimate. Final judgment is based on four basic decisions:

- Is the new job bigger or smaller than the benchmark job it is compared with?
- Is the difference so small that it will remain in the same time interval?
- Will it fall into one of the other intervals, either above or below it?
- Does the new job fall into the next time interval above or below, or two intervals above or below?
- Comparative estimating still involves subjective judgment on the part of planners, but the only choice that they need to make is between one time interval and the next. Comparison with library jobs of known duration greatly reduces guesswork on the part of the planner.
- Maintaining the reference library is a central function requiring the assembly of contributions from all planners to cover the various classes of equipment. In this way, a uniform structure for job estimates can be established and controlled.

Application of Maintenance Work Measurement. Maintenance personnel are fond of the truism: "A given maintenance job never goes the same way twice." While this much of the statement is basically true, they add: "Therefore, maintenance cannot be measured." This is false. The first part of the statement simply influences the precision with which you can estimate, the use to which maintenance estimates should be put, and the measurement period required to level out the fluctuations in individual jobs:

- Industrial engineers attempt to establish production standards with an accuracy of plus or minus 5 percent ($\pm 5\%$). Considering the difference in consistency between production and maintenance work, we cannot expect to get maintenance estimates any closer than plus or minus 15 percent ($\pm 15\%$).
- If the estimates are used for determining percent performance (estimated hours ÷ actual hours) as opposed to schedule compliance (hours of scheduled work completed ÷ hours of scheduled work), the calculation should be made weekly (or even every other week) and should be calculated at the supervisory level—all crews responsible to given supervisor. The 1-week period puts several jobs into the calculation, allowing for an averaging of unusually difficult job occurrences with unusually easy occurrences.
- Accuracy on individual jobs is not reliable, but accuracy over the several jobs completed by a crew on a given week is acceptably accurate, particularly for measuring performance trends.
- Individual (as opposed to crew) performance can be calculated periodically, but only to guide necessary training to the critical needs; never for disciplinary reasons. The use of work measurement for discipline detracts from more important applications; namely backlog control, work scheduling, and group performance trends.

To summarize, there are three considerations, which help planners to be more effective estimators:

- Familiarity with the jobs done and the plant equipment.
- Making sure that everybody compares to the same "known."
- Not trying to estimate with "pinpoint" accuracy.

LEVELS OF MAINTENANCE WORK MEASUREMENT METHODOLOGY

Several forms of work measurement with varying levels of precision can be used in the development of job estimates:

- Supervisor/planner estimates
- Historical averages
- Published job estimating tables (construction trades)
- Adjusted estimates or averages (based on work sampling during a base period)
- Analytical estimating
- Time study
- Standard data (from time study)
- Predetermined times (MTM)
- Predetermined time formulas (UMS)

The form used depends significantly on the focus of the installation, that is, performance measurement or schedule compliance. A performance measurement focus (standard time ÷ actual time = % performance) requires a more precise form of work measurement.

CHAPTER 7
MAINTENANCE ENGINEER'S TOOLBOX

Shon Isenhour
Principal, Life Cycle Engineering Inc., Charleston, S.C.

A primary responsibility of the maintenance engineer is to identify and analyze asset failures and deviations from optimum performance. This responsibility requires tools or methods that the engineer can use to effectively determine the potential failure modes of assets, as well as determine the true root cause of the problem. The maintenance engineer's toolbox includes:

SIMPLIFIED FAILURE MODES AND EFFECTS ANALYSIS

Simplified Failure Modes and Effects Analysis (SFMEA) is a top-down method of analyzing a design, and is widely used in industry. In the United States, automotive companies, such as Chrysler, Ford, and General Motors, require that this type of analysis be carried out. There are many different company and industry standards, but one of the most widely used is the *Automotive Industry Action Group* (AIAG). Using this standard you start by considering each component or functional block in the system and how it can fail, referred to as failure modes. You then determine the effect of each failure mode, and the severity on the function of the system. Then you determine the likelihood of occurrence and of detecting the failure. The procedure is to calculate the risk priority number, or RPN, using the formula:

$$RPN = Severity \times Occurrence \times Detection$$

The second stage is to consider corrective actions, which can reduce the severity or occurrence, or increase detection. Typically, you start with the higher RPN values, which indicate the most severe problems, and work downward. The RPN is then recalculated after the corrective actions have been determined. The intention is to get the RPN to the lowest value.

SFMEA (Fig. 7.1) is a process that permits plant personnel, such as maintenance engineers, craftsperson, and operators, identify the common failure modes of manufacturing and process systems and their associated components.

A SFMEA evaluates three criteria: (1) severity or the impact a failure will have on the plant ability to achieve the capacity or through-put needed to meet delivery requirements; (2) the probability that a specific failure mode will occur, based on in-plant history or industrial statistics; and (3) the probability that current maintenance methods will detect the specific failure mode before it occurs.

For each of the potential failure modes, the investigators identify the anticipated *potential effect(s) of failure*. The team should determine the specific impact that would result from each of the potential failure modes. For each of the identified failure modes and effects, the team must determine

Simplified Failure Modes and Effects Analysis					Severity	Cause of Failure	Probability	Current Control	Detection	RPN	Improvements	New RPN
Process	Asset	Component	Failure Mode	Effect of Failure								
Cigarette Manufacturing	S-3	Packer	Total Failure	Total Loss of Capacity	10	Loss of AC Power	10	None	10	100	None	

FIGURE 7.1 SFMEA template.

the *severity* that the failure would have on the plant's ability to meet its mission. Normally, the severity is based on the failure's impact on production capacity, product quality, or total operating cost; but could also include safety or environmental impacts. The relative severity or impact is ranked on a 1 to 10 scale similar to Table 7.1.

After all of the potential failure modes are identified, each one is evaluated to determine the more probable cause or causes of the specific failure mode. For example, total loss of function might be caused by loss of motive power, mechanical binding or other causes.

The next step is to determine the probability that each of the causes or forcing function could occur. This determination is based on known failures based on industrial or plant-specific histories. A fixed scale, Table 7.2, is used to determine a relative value for each forcing function.

The next step in the SFMEA process is to determine whether or not the current methods used to monitor the system and its components will detect each of the forcing functions or failure modes before failure or serious damage could occur. A column is provided to define the specific preventive maintenance or system monitoring method that is currently used to monitor for the specific failure mode or forcing function. For each of the forcing functions, the specific action item or method should be defined in the appropriate column.

TABLE 7.1 Severity Scale

10	91% to 100% loss of capacity
9	81% to 90% loss of capacity
8	71% to 80% loss of capacity
7	61% to 70% loss of capacity
6	51% to 60% loss of capacity
5	41% to 50% loss of capacity
4	31% to 40% loss of capacity
3	21% to 30% loss of capacity
2	11% to 20% loss of capacity
1	10% of less loss of capacity

TABLE 7.2 Probability of Occurrence

10	91% to 100% probability
9	81% to 90% probability
8	71% to 80% probability
7	61% to 70% probability
6	51% to 60% probability
5	41% to 50% probability
4	31% to 40% probability
3	21% to 30% probability
2	11% to 20% probability
1	10% of less probability

When the preceding task is complete, an inverted fixed scale from one to ten is used to rank the ability to detect the failure mode or forcing functions. The scale is inverted to provide a true representation of the RPN associated with the specific failure mode or forcing function. A low number indicates that the current methods have a higher probability of detecting the problem and a higher number a lower probability of detection. Table 7.3 illustrates the inverted scale used for the probability of detection.

To calculate the RPN, the three values, that is, severity, probability of occurrence, and probability of detection, are added together; the sum is divided by 30 and the resultant multiplied by 100. The resultant number is an approximation of the overall probability of failure for each of the failure modes and forcing functions.

The final step in the SFMEA process is to evaluate the potential for lowering the calculated RPN by improving the preventive maintenance or system monitoring methods. For example, the addition of predictive maintenance technologies could be used to provide early detection of failure modes or forcing functions such as misalignment, abnormal loading, and many others.

In addition to its use as part of root cause analysis, SFMEA is an ideal tool that can be used to establish or enhance preventive and predictive maintenance programs. With the right team, consisting of knowledgeable reliability engineers, operators, and maintenance crafts, the process will identify most, if not all, of the more probable failure modes and forcing functions that would occur. This information can then be used to develop specific preventive or predictive maintenance tasks that will eliminate or substantially reduce the potential for these problems.

Simplified Failure Mode, Effect, and Criticality Analysis

A Simplified Failure Modes, Effects, and Criticality Analysis (SFMECA) is similar to a SFMEA as described above though criticality is usually computed in place of RPN values. This type of analysis is widely used in the military, aerospace, and medical equipment fields for both design and process reliability analyses.

TABLE 7.3 Probability of Detection

10	10% of less probability
9	11% to 20% probability
8	21% to 30% probability
7	31% to 40% probability
6	41% to 50% probability
5	51% to 60% probability
4	61% to 70% probability
3	71% to 80% probability
2	81% to 90% probability
1	91% to 100% probability

Reliability Block Diagram

A reliability block diagram (RBD) allows you to accurately model complex systems. Redundant systems or processes can be easily modeled using an RBD. Individual blocks can be used to represent a single component, a sub-system or any event, which could cause failure. The blocks are joined together graphically to represent the system, and then an analysis is done using special algorithms, which may include system simulations.

Reliability Predictions and MTBF

Reliability predictions form the groundwork for reliability analyses. They are used to compute the predicted failure rate or *mean time between failures* (MTBF) of your system. MTBF is usually expressed in terms of hours. For example, if your system has a predicted MTBF of 1000 hr, this means that, on average, your system experiences one failure in 1000 hr of operation.

By using accepted standards for modeling failure rates of components, you can analyze your system and calculate your system's predicted failure rate or MTBF. The goal is to make sure that your system's predicted MTBF is within acceptable limits. If your prediction analysis shows low MTBFs, this means that you can expect your system to fail more often, and you may need to take steps to improve it. By making changes to your design, for example, maybe lowering temperatures or stress levels, your predicted MTBF may improve and you can expect better product reliability.

Reliability predictions can be done at any level of the design phase. In the early design stages, the reliability prediction may be more of a rough estimate, but then may be refined as the design becomes more stable. Even once the product is in the field, reliability predictions can take into account actual field data for more accurate predictions.

Life Cycle Cost Analysis

Life cycle cost analyses involve evaluating the total cost of a product or system over its entire life span. Life cycle cost will consider the cost of developing or acquiring the asset, the cost of running, operating and maintaining, and the cost of disposal. As life cycle cost is often significantly affected by reliability issues such as frequency of failure and time to repair, it is often included as part of the reliability engineering function.

Reliability, Availability, Maintainability Analysis

This analytical approach is a top-down method that identifies failure modes, predicted failure frequency (MTBF), maintainability (MTTR) from a system or top-level viewpoint.

All electromechanical systems require some maintenance to keep them operating in a satisfactory manner. Preventive maintenance is employed to prevent unscheduled downtime and interruptions in their use; however, it is practically impossible to prevent all interruptions in service. The ease with which the system or product can be restored to operating condition and the time interval required are primary considerations of reliability engineering. Quantitative measures of maintainability are:

MTTR Mean time to repair
MTTRS Mean time to restore system
MTTRF Mean time to restore function
DLH/MA Direct labor hours per maintenance action
TPCR Total parts cost per removal or replacement
PFD Probability of fault detection

Maintainability demonstration is required on complex systems by most customers. This is usually carried out by industrial engineers using regular repair and maintenance personnel with specialized tools, test equipment, and so on, designed for use with the system. Time studies of several trials of each maintenance action are averaged to obtain the expected *mean time*.

FAULT-TREE ANALYSIS

Fault-tree analysis is a method of analyzing system reliability and safety. It provides an objective basis for analyzing system design, justifying system changes, performing trade-off studies, analyzing common failure modes, and demonstrating compliance with safety and environment requirements. It is different from a *Simplified Failure Mode and Effect Analysis* in that it is restricted to identifying system elements and events that lead to one particular undesired event. Figure 7.2 shows the steps involved in performing a fault-tree analysis.

Many reliability techniques are inductive and concerned primarily with ensuring that hardware accomplishes its intended functions. Fault-tree analysis is a detailed *deductive* analysis that usually requires considerable information about the system. It ensures that all critical aspects of a system are identified and controlled. This method represents graphically the Boolean logic associated with a particular system failure, called the *top event*, and basic failures or causes, called *primary events*. Top events can be broad, all-encompassing system failures or they can be specific component failures.

Fault-tree analysis provides options for performing qualitative and quantitative reliability analysis. It helps the analyst understand system failures deductively and points out the aspects of a system that are important with respect to the failure of interest. The analysis provides insight into system behavior.

A fault-tree model graphically and logically presents the various combinations of possible events occurring in a system that lead to the top event. The term "event" denotes a dynamic change of state that occurs in a system element, which includes hardware, software, human, and environmental factors. A *fault event* is an abnormal system state. A *normal event* is expected to occur.

The structure of a tree is shown in Fig. 7.3. The undesired event appears as the "top event" and is linked to more basic fault events by event statements and logic gates.

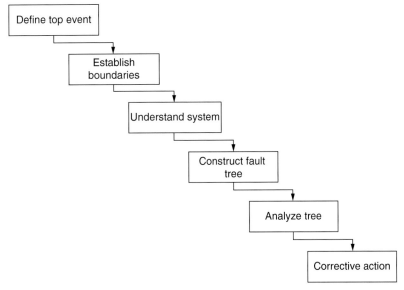

FIGURE 7.2 Typical fault-tree process.

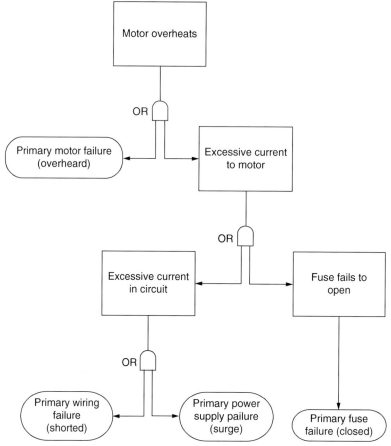

FIGURE 7.3 Example of fault-tree logic diagram.

CAUSE-AND-EFFECT ANALYSIS

Cause-and-effect analysis or Ishakawa diagram is a graphical approach to failure analysis. It is also referred to as fishbone analysis, a name derived from the fish-shaped pattern that is used to plot the relationship between various factors that contribute to a specific event. Typically, a fishbone analysis plots four major classifications of potential causes (i.e., man, machine, material, and methods), but can include any combination of categories. Figure 7.4 illustrates a simple analysis.

Like most of the failure-analysis methods, this approach relies on a logical evaluation of actions or changes that lead to a specific event, such as machine failure. The only difference between this approach and other methods is the use of the fish-shaped graph to plot the cause-effect relationship between specific actions, or changes, and the end result or event.

This approach has one serious limitation. *The fishbone graph does not provide a clear sequence of events that leads to failure.* Instead, it displays all of the possible causes that may have contributed to the event. While this is useful, it does not isolate the specific factors that caused the event. Other approaches provide the means to isolate specific changes, omissions, or actions that caused the failure, release, accident, or other event being investigated.

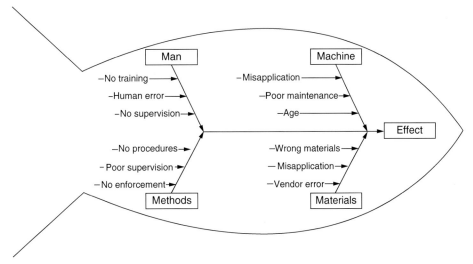

FIGURE 7.4 Typical cause-and-effect diagram.

SEQUENCE-OF-EVENTS ANALYSIS

Sequence-of-events analysis is perhaps the most effective method of evaluating a problem or failure that can be clearly fixed in time. The analysis starts with the exact time of the observed problem or failure and systematically identifies all changes that took place prior to and following that time. Again, the assumption in this analytical approach is that one or more changes caused the observed problem or failure.

The sequence-of-events diagram should be a dynamic document that is generated soon after a problem is reported and continually modified until the event is fully resolved. Figure 7.5 is an example of such a diagram.

Proper use of this graphical tool greatly improves the effectiveness of the problem-solving team and the accuracy of the evaluation. To achieve maximum benefit from this technique, be consistent and thorough when developing the diagram. The following guidelines should be considered when generating a sequence-of-events diagram: use a logical order, describe events in active terms rather than passive, be precise, and define or qualify each event or forcing function.

In the example illustrated in Fig. 7.5, repeated trips of the fluidizer used to transfer flake from the CA department to the preparation area triggered an investigation. The diagram shows each event that led to the initial and second fluidizer trip. The final event, the silo inspection, indicated that the root cause of the problem was failure of the level-monitoring system. Because of this failure, Operator A over-filled the silo. When this happened, the flake compacted in the silo and backed up in the pneumatic-conveyor system. This backup plugged an entire section of the pneumatic-conveyor piping, which resulted in an extended production outage while the plug was removed.

Logical Order

Show events in a logical order from the beginning to the end of the sequence. Initially, the sequence-of-events diagram should include all pertinent events, including those that cannot be confirmed. As the investigation progresses, it should be refined to show only those events that are confirmed to be relevant to the incident.

3.140   ENGINEERING AND ANALYSIS TOOLS

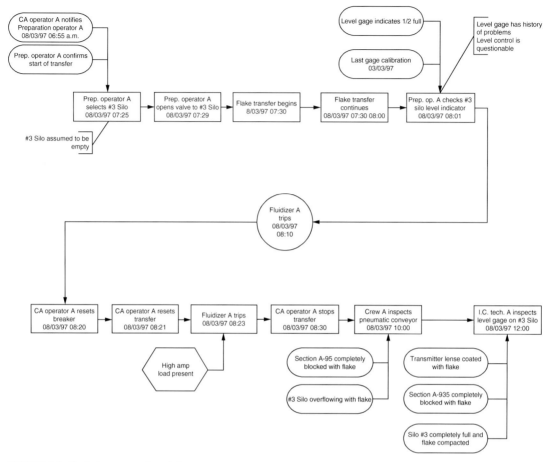

FIGURE 7.5 Typical sequence-of-event diagram.

Active Descriptions

Event boxes in a sequence-of-events diagram should contain action steps rather than passive descriptions of the problem. For example, the event should read: "Operator A pushes pump start button" not "the wrong pump was started." As a general rule, only one subject and one verb should be used in each event box. Rather than, "Operator A pushed the pump stop button and verified the valve line-up" two event boxes should be used. The first box should say "Operator A pushed the pump stop button" and the second should say, "Operator A verified valve line-up."

Do not use people's names on the diagram. Instead use job functions or assign a code designator for each person involved in the event or incident. For example, three operators should be designated Operator A, Operator B, and Operator C.

Be Precise

Precisely and concisely describe each event, forcing function, and qualifier. If a concise description is not possible and assumptions must be provided for clarity, include them as annotations. This is illustrated in Fig. 7.10. As the investigation progresses, each assumption and unconfirmed contributor to the event must either be confirmed or discounted. As a result, each event, function, or qualifier generally will be reduced to a more concise description.

Define Events and Forcing Functions

Qualifiers that provide all confirmed background or support data needed to accurately define the event or forcing function should be included in a sequence-of-events diagram. For example, each event should include date and time qualifiers that fix the time frame of the event.

When confirmed qualifiers are not available, assumptions may be used to define unconfirmed or perceived factors that may have contributed to the event or function. However, every effort should be made during the investigation to eliminate the assumptions associated with the sequence-of-events diagram and to replace them with known facts.

THE FIVE WHYS

The five whys is a technique that doesn't involve data segmentation, hypothesis testing, regression or other advanced statistical tools, and in many cases can be completed without a data collection plan. By repeatedly asking the question "Why" at least five times, you can peel away the layers of symptoms which can lead to the root cause of a problem.

Here is a simple example of applying the five whys to determine the root cause of a problem. Let's suppose that you received a large number of customer returns for a particular product. Let's attack this problem using the five whys:

1. Why are the customers returning the product? Answer: 90 percent of the returns are for dents in the control panel.
2. Why are there dents in the control panel? Answer: The control panels are inspected as part of the shipping process. Thus, they must be damaged during shipping.
3. Why are they damaged in shipment? Answer: Because they are not packed to the packaging specification.
4. Why are they not being packed per the packaging spec? Answer: Because shipping does not have the packaging spec.
5. Why doesn't shipping have the packaging spec? Answer: Because it is not part of the normal product release process to furnish shipping with any specifications.

Using the five whys in this case revealed that a flaw in the product release process resulted in customers' returning of a product.

STATISTICAL ANALYSES TOOLS

Data and the knowledge that it contains is the life blood of the maintenance engineer. He or she relies on a series of statistical analysis tools that provide the ability to convert massive amounts of raw data into knowledge that can be used to improve performance and prevent failures. These tools include, but are not limited to:

Pareto Analysis

All of us are aware that there are always more problems than we have time to work on them, more capital projects than we have available funds, more necessary tasks than time in a week and more demand for solution of problems than resources to address them. Because this is a reality in any operation, the need to look at these challenges utilizing the Pareto principle is all the more crucial. The challenge is not to solve as many problems as we can, but how to take a limited resource (time) and get the maximum return on any time spent with problem solving efforts. Performing a Pareto analysis on a problem gives clear direction as to how to prioritize time and resources to get this maximum return. Pareto can be used for three distinct purposes:

- Prioritize and justify time and resources spent on problems.
- Problem clarification—further defines nature of problem to give direction to problem solving activities.
- Objectively document or test to see if improvement efforts are effective—a benchmark.

80/20 Rule

80/20 Rule: A few causes usually account for a majority of the problem (or 80 percent) while a multitude of other causes accounts for only a very small part of the problem (or 20 percent).

- Eighty percent of rejects are caused by 20 percent of the total potential reasons.
- Eighty percent of instrument downtime is caused by 20 percent of the total potential causes.
- Eighty percent of lab testing time, 20 percent of the total types of tests.
- Eighty percent of time, 20 percent of the total potential tasks.
- Eighty percent of accidents/injuries, 20 percent of the total potential types.

Procedure
- Clearly define problem to be analyzed and its purpose.
- Select a time period ("window") to be analyzed—week, month, year, and so on. Always select a large enough window to allow most sources of variation or problems to occur.
- Select a method to stratify the data (e.g., off-standard occurrences by standard, and off-standard occurrences by parameter or by day of week).
- Collect and organize data by stratified categories. Compute frequency of occurrence for each item (or # of tests, hours, and so on). Try to avoid using percentages. Where possible translate to dollars.
- The totals calculated in step 4 will be equal to the height of the bars on the diagram.
- Along the vertical axis, put units of measurement (e.g., pounds, number of occurrences, hours, dollars).
- Draw a vertical "bar" to the appropriate scale with the biggest category first, the next largest bar to the right of it and continue on to the right. Make the bars equal in width and adjoining each other.
- Create an "all others" category and place it to the far right.

Option
- Construct a vertical scale on the right side from 0 to 100 percent.
- Plot a cumulative percent line graph.

Key Points
- First step in making improvements—justifies spending time.
- Can be applied to variety of problems—process, product, operational, administrative, safety, work-life.
- May be used to tell whether improvement efforts were effective.
- Assists in the efforts to go from effect to cause.
- Procedure is to categorize data, assign a value to each category (preferably in terms of dollar), sort in order from largest to smallest, and construct a bar graph.
- If a number of categories have very small values, they should be lumped together as "other."
- Indicate on Pareto (shaded boxes) those categories that comprise the vital few (80 percent of problem).

- Guidelines for interpretation:
  - Which categories comprise the vital few?
  - If problem solving has taken place, has the category moved or "shifted places"?
  - Is there a second-level Pareto that can be done, on the top category, to shed further light on the problem?

Interpretation
- First few bars will represent the vital few. It is easier to reduce the tallest bar by one-half than to eliminate the shortest bar.
- If improvements take place, order will change over time.
- If no work is done on the problem but the order changes over time, the window is too small or the process is severely out of control.

Application. In the context of problem solving, Pareto charts can be used in one of three ways:
- Determine if a problem is significant—when performing a Pareto analysis, is it part of the vital few?
- Use as a measure of effectiveness for problem solving—construct a Pareto of "before" and compare to a Pareto of "after" and then a change is implemented.
- Further define a problem to reveal possible causes or give better direction to problem solving. This would involve doing a "second-level" Pareto.

Example. We could use Pareto to examine maker performance on relative to machine stops. First, look at number of stops by machine, Fig. 7.6.

Then, looking at Machine #86, by reason, Fig. 7.7.

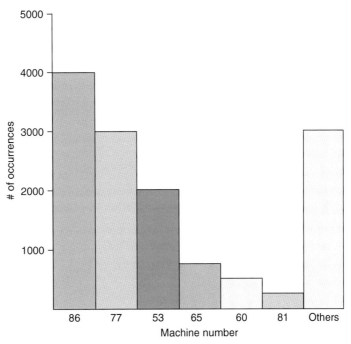

FIGURE 7.6 Failures by machine.

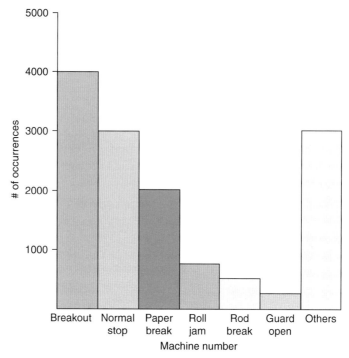

FIGURE 7.7 Failures by type.

Flow Charts

It is vital that we gain a clear understanding of the process that contains the problem to be solved. Constructing a flow chart usually results in clearer thinking, more methodical problem solving and implementation of changes. Flowcharting should be kept simple—following the guidelines below—and it can be used on any problem or process. That process could be filling out a time sheet, performing a calibration, communications from a supervisor, or a simple, daily activity within the department.

A *flow chart* is a pictorial representation showing all of the steps of a process. The picture illustrates all of the changes that occur to the product through each stage. Details of four Ws (Who, What, When, and Where) are helpful if written beneath blocks of the flow chart.

Procedure. The people with the greatest amount of knowledge about the process meet to brainstorm—a flow chart. This might take more than one session to complete.

- List all of the steps in the process. The first few versions may not result in the steps being listed in their actual sequence. It is important at this point to identify all the steps.
- When all steps have been identified, sequence them in proper order.
- Plot flow chart, steps as blocks (equal size) starting at top left of paper and moving step by step from left to right.
- If the process has many steps (operations) and cannot be completed on a single line, drop down one level and continue on, starting at the left side again.
- Only those steps that contribute to the completion of the product/service are included on-line. Non-contributing support type operations (testing, approval, movement between process stages) are plotted off-line as circles.

Key Points
- The easiest way to ensure common understanding of a process is to draw a picture of it.
- Keep it simple but at the same time ensure all steps are shown.
- In using this tool, teams will often realize how "nonstandard" or ill-defined a process might be.
- Before doing a flow chart of "what should be," the team should flow chart "what is."
- Ensure Who, What, When, and Where are answered in looking at the flow chart.
- Keep in mind there are several different ways to use this tool but the procedure remains the same.
- After completing a flow chart of the current process, the team should ask the question: "What should be changed in this process (add a step, delete a step, modify a step) to help eliminate the problem?"

Example. Illustrated below is the proposed process flow chart that a Lab Services Team derived for the Process of Delivering Size Prep PG, Glycerin and Soluble Results.

Control Charts

Control charts (Fig.7.8) are a graphical run chart of a measurement that can, through the use of control limits, distinguish between chance causes of variation due to system elements and special causes of variation due to a significant change in the process.

With the implementation of the Statistical Process Control (SPC) Systems, personnel use this on-line SPC tool to continually monitor processes, take actions when appropriate and document. These activities will help ensure that the demonstrated process capabilities are achieved.

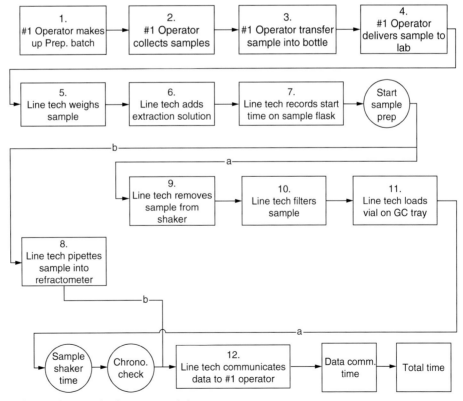

FIGURE 7.8 Example of process control chart.

The focus of this section on control charts is how to use this tool in an off-line SPC activity to go about improving the processes and products.

STRATEGY FOR USE OF CONTROL CHART TOOL

1. Define control limits for a measure (could be experimental variable, TQM, measure of effectiveness, off-line parameter, and so on) based on a given set of current conditions. [Pre-measure]
2. Follow PDCA. Make changes to a process, product, and so on.
3. Test output from the process against previously derived control limits. [Post-measure]
4. Interpret to see if change resulted in significant effect (control chart will shift OOC to "desirable" side).

Key Points

- Always ensure that the conditions under which the first set of data was gathered is as close as possible to when the second set was gathered—except for the deliberate change of the independent variable.
- If other "uncontrolled" variables might be changing, it is recommended that they be plotted as well for both periods to ensure "equivalent" conditions. These are sometimes called "noise" variables—variables that are not part of the experimental variables, but could influence the dependent variable if it is not held constant.
- If there is more than one independent variable of interest and there might be a relationship between them, then do not use this technique.
- If the team has identified three or four different independent variables to be investigated and there is a strong feeling that one "interacts" with another to have an effect on the dependent variable—then experimental design—or Q3 techniques should be used. The TQI specialist can assist you with an experimental design.
- The power/certainty of your conclusion is related to the quality of your data collection (holding noise variables constant) and the quantity. However, collecting data points beyond 20 in number has a point of diminishing return.
- Collecting at least seven data points in the "after" period will allow all OOC rules to be experienced if, indeed, there is a shift.

Application. Shown are two different illustrations of this strategy. The only difference is the type of measure that it was used on. In the first case, it is the team measure of effectiveness and in the second, it is an experimental variable.

Example 1. In the Plan segment, the team selected as its measure of effectiveness the percent OOC (weekly) for an SPC parameter. The team went out and collected existing data and derived control limits to define typical weekly results given the current set of conditions (system). Figure 7.9 illustrates the results of this exercise.

After establishing an objective to reduce this amount of time OOC, the team proceeded through PDCA. As part of the Act segment, the team went back and collected weekly data on this measure of effectiveness and plotted it against the previously derived limits.

Interpretation. You can see that the "post" data has gone OOC on the low side (See Fig. 7.10). If the team objective was to reduce (lower) this measure of effectiveness, this data would indicate that the work of the team produced significant results. (The next question is: Does this data give clear evidence that the change was significant to the point of meeting the team objective? To answer that question, the "t" test or confidence interval technique of Q3 is appropriate.)

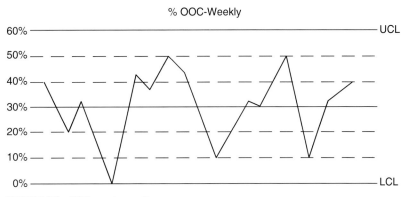

FIGURE 7.9 OOC exercise results.

Example 2. In this case, the team is in the Check segment attempting to sort through the major possible causes identified in the Do segment. The dependent variable—that is the parameter or measurement that the team is trying to affect through elimination of root causes—is machine downtime, Fig. 7.11. Data is gathered and control chart limits derived. Changes are made in the independent variable conditions. (Changes made that are related to the major possible causes.) Continue to plot the maker downtime results on the same chart that was derived earlier.

Interpretation. The control chart is still in control after the condition(s) change. Therefore, this data indicates that the suspected causes—the independent variables—have no significant impact on maker downtime, the dependent variable. *Note:* Be sure to reread some of the cautions listed under key points in this section.

Histograms

Histogram is a graphical means of summarizing information about a set of data: amount of variation, centering point, and shape.

As a team begins to work with data, one of the first tools that is used is the histogram. It is a simple but powerful tool that not only gives the team a good snapshot of current variation and performance level but in many cases can reveal hints of underlying causes of variation (i.e., histograms showing multiple populations, skewed histograms, and so on). To correctly interpret histograms, they must be constructed properly. The following procedure, see Fig. 7.12, guidelines will help ensure correct interpretation of these results.

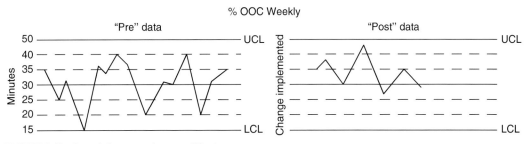

FIGURE 7.10 Control chart pre and post modification.

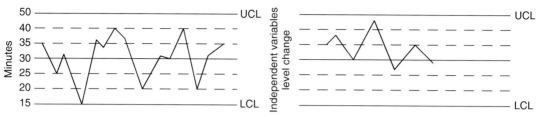

FIGURE 7.11 Machine downtime.

Procedure
- Let the horizontal scale represent observed measurements.
- The vertical scale represents frequency of occurrence.
- Divide the data into "classes," using the following guidelines.

Data points	#Classes
50	5 to 7
50 to 100	6 to 10
100 to 250	7 to 12
250	10 to 20

$$\text{Range} = \text{High} - \text{Low Value}$$

$$K = \text{\# of Classes}$$

$$\text{Class width} = \frac{\text{Range}}{K}$$

- Go through the data, tallying data points into each class.
- Construct bars from frequency tally; bars should be equal in width and adjacent.

Interpretation
- What is the most common value?
- How great is the dispersion?
- Is the distribution symmetrical?
- Is the distribution skewed, cliff-like, comb-like, bimodal, or flat?
- Are there isolated bars? Why?
- Should the data be stratified (separated out by some qualifier—by column, by week, by shift, and so on)?

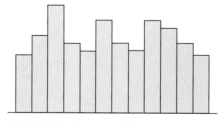

FIGURE 7.12 Example of histogram.

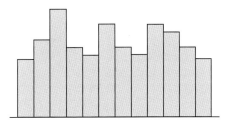

FIGURE 7.13 Dilution.

Key Points

- All distributions have three important characteristics—center, width, shape.
- Within PDCA, histograms are used:
  - To collect and describe "current" conditions. How big is the problem? (Histograms not centered where desired, spread too wide, pattern irregular.)
  - As a method of illustrating the measures of effectiveness: compare the "before" data and the "after" data to illustrate any changes in pattern/shape, center, spread. (*Note*: Use comparable time periods and sample size when doing this.)
  - As a diagnostic tool to further define possible causes.

Example. See Fig. 7.13.

This would indicate that the problem is not one of pure spread but one of multiple populations. Now, the team can investigate what is causing multiple populations.

Histogram Interpretation

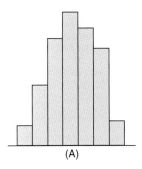

(A)

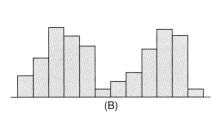

(B)

Normal—A symmetrical shape with a peak in the middle of the range of data. This is the most natural, common shape.

Bi-Modal—A distinct valley in the middle with peaks on either side. This usually signifies a combination of two normal distributions and suggests that *two* distinct processes are at work.

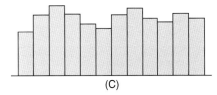

(C)

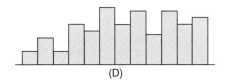

(D)

Plateau—A flat top with no distinct peak and slight tails on either side. This suggests the combination of *many* normal distributions, with centers spread evenly throughout the data range.

Comb—High and low values alternating in regular fashion. This suggests grouping, measurement or rounding errors.

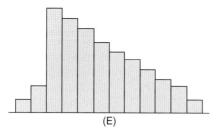

(E)

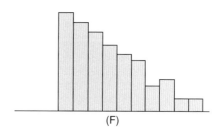

(F)

Skewed—An asymmetrical shape in which the peak is off center and the distribution tails off sharply on one side and gently on the other. This pattern usually occurs when a specification or artificial barrier exists on one side. This pattern is seen often with counting type data.

Truncated—An asymmetrical shape in which the peak is at or near the edge of the data. It is usually the result of the removal of some data by an external force, such as screening.

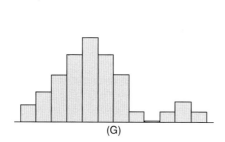

(G)

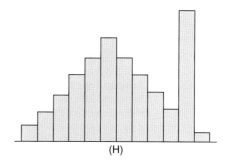

(H)

Isolated Peaked—A small, separate group of data appears in addition to the larger distribution. This is usually the result of an abnormality in processing or measuring.

Edge-Peaked—A large peak is appended to an otherwise normal distribution. This frequently occurs when the extended tail of a normal curve has been cut off and lumped into a single category (i.e., values outside tolerance are reported as being just inside the range.)

Application. Histograms can be used to evaluate a multitude of productions, process and reliability problems. The following examples are but a few of the possibilities:

Example 1. OV – Final weight belt (see Fig. 7.14.)

Example 2. Packer rejects (see Fig. 7.15.)

Scatter Plots

Scatter plot is a plot of pairs of data on an x and y axis that illustrates graphically the *relationship* between two variables. Scatter plots (or scatter diagrams) are a key tool in analyzing the relationship between two variables. Using this graphical technique you can determine the extent to which one variable seems "to follow" another. (If one increases, does the other increase?) This does not imply cause and effect—because two variables are correlated or follow one another does not necessarily mean that one is causing the other to shift. To determine that answer will require the hypothesis testing techniques of Q3, experimental design studies. Additionally in Q3, statistical calculations will be illustrated to determine numerically the amount of correlation between two variables.

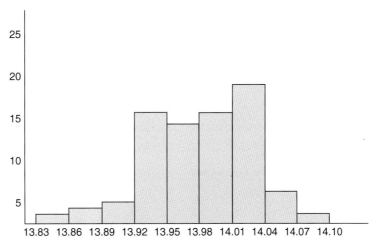

FIGURE 7.14 OV at final weight belt 4–6.

Key Points. In the PDCA cycle, the scatter plot can be used in the Do and Check segments to:
- Check several suspected cause variables to see if they are related or measuring the same thing.
- Check the relationship between a suspected cause variable (independent variable) and the parameter or measure you are trying to impact (dependent variable). If there is not a strong correlation by affecting the independent variable, you will most likely not change or impact the desired variable.
- Know if two variables correlate over the "noise" of other variables.

When viewing a scatter plot, look at the pattern of points to establish which of these three patterns exists.

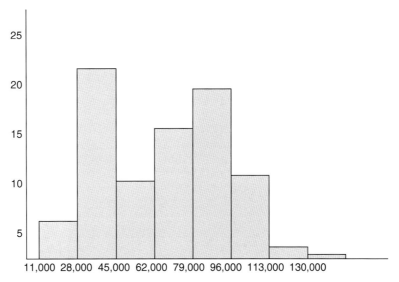

FIGURE 7.15 Packer rejects.

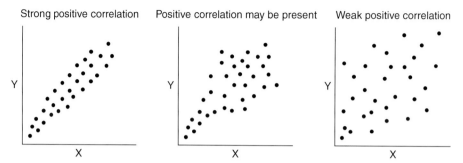

FIGURE 7.16 Positive correlation.

Positive Correlation. If one variable goes up, the other goes up or if one variable goes down, the other goes down (see Fig. 7.16).

Negative (Inverse) Correlation. If one variable goes up, the other goes down or if one variable goes down, the other goes up (see Fig. 7.17).

No Correlation. If one variable moves in a certain direction, it is unpredictable what the other variable will do (see Fig. 7.18).

Application. Interpret the following scatter plot of QB OV values versus oven OV results for matched time periods. What conclusions can you draw from this? (See Fig. 7.19).

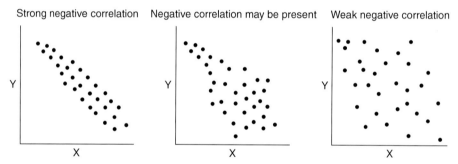

FIGURE 7.17 Negative correlation.

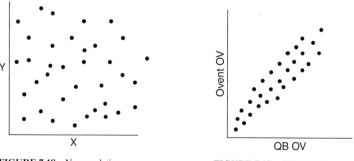

FIGURE 7.18 No correlation.

FIGURE 7.19 OV example.

CHAPTER 8
ROOT CAUSE ANALYSIS

Darrin Wikoff
Principal, Life Cycle Engineering Inc., Charleston, S.C.

America industry has gravitated to quick, simple solutions to symptoms of problems rather than attempt to identify and resolve the real reason or reasons, the root cause that caused the problem. Instead, we ignore these problems or simply swap parts or implement random changes until the symptoms of the problem go away. As a result, these problems tend to recur and the negative impact of these problems continues to plague our business lives. Asset reliability and optimum life cycle cost are essential for world-class performance, as well as survival in today's highly competitive, worldwide economy. Many factors contribute to the loss of asset reliability and inflated life cycle cost. Historically, the myriad of problems can be classified by the primary functional source of the problems.

Based on historical data, 15 percent of all asset reliability and life cycle cost problems can be directly attributed to failures within the sales function of our plants. These deficiencies are driven by the methods that are used to obtain new or continued business from the marketplace. Like many other facets of our business and society, this function has lost sight of proven methods required to properly load our plants with high volume, standard products that are needed to facilitate effective asset utilization and support reliability and optimum life cycle cost. Instead, plants are loaded with small, accelerated delivery, and in many cases nonstandard products. As a result, the plant must violate best practices in production planning and scheduling; maintenance must forego required preventive maintenance; and profits disappear. In effect, the entire plant becomes reactive to the unrealistic demands imposed by an ineffective sales function.

In part because of the failures within the sales function, 22 percent of all reliability and life cycle cost problems are from failures within the production or manufacturing function. In addition to the restrictions imposed by improper loading or backlog, failures in planning and scheduling, as well as the procedures and practices used to product or manufacture product create serious problems. The predominant areas that result in problems that could be resolved using root cause analysis include:

- Failures to properly utilize the installed assets or capacity of the plant
- Outdated or ineffective *standard operating procedures* (SOPs)
- Production practices that are contrary to *best practices*
- Operator errors caused by lack of training, proper supervision, or morale problems
- Improperly adjusted and calibrated production or manufacturing systems

Failures within the maintenance function contribute 17 percent to historic reliability and life cycle cost problems. While some of these deficiencies are caused by the limitations imposed by the preceding functions, failure to universally follow best practices in the planning, management, execution, and evaluation of maintenance activities is responsible for the majority of these problems.

The engineering function, both plant and maintenance is responsible for 27 percent of all reliability and life cycle cost problems. The predominate failures include:

- Improper design, selection, and procurement of new capital assets
- Uncontrolled modifications and changes to existing assets
- Failure to universally apply root cause analysis to recurring problems in capacity, quality, cost, and reliability associated with existing production and manufacturing systems

The procurement or purchasing function contributes and additional 11 percent to reliability and life cycle cost problems. The predominate failures include:

- Procurement of new capital assets and major replacements based solely on "low-bid" and not life cycle cost analysis
- Substitution of improper parts, that is, cheaper, for maintenance, repair and operating (MRO) spares
- Lack of vendor qualification and assessment process
- Improper maintenance on MRO materials

The final 8 percent can be directly attributed to failures within the plant management function. The majority of these problems result from the myopic, short-term management philosophy that dominates corporate and plant management. Everything is driven by monthly and quarterly profitability, not optimum life cycle benefits.

CONCEPT AND METHODOLOGY

The concept and methodology of *root cause analysis* (RCA) are designed to provide a cost-effective means to isolate all factors that directly or indirectly result in the myriad of problems that we face in our plants and facilities. This process is not limited to equipment or system failures; but can be effectively used to resolve any problem that has serious, negative impact on effective management, operation, maintenance, and support of our plants and facilities. It has the capability to identify incipient problems; isolate the actual cause or forcing function that directly resulted in the problem, as well as identifies all factors that directly or indirectly contributed to the problem.

An example of the effective use of RCA is a large, integrated steel mill that exhibited a substantial increase in the annual quantity and cost of rolling-element bearings required to sustain operation of critical production systems. Over a 6-year period, the annual replacement cost rose from $2.7 millions to $14.1 millions. When root cause analysis methodology was used to determine the reason behind this radical increase, the primary reason for it was misapplication of predictive maintenance technologies. When the predictive maintenance program was established, its mission was defined solely as a means to eliminate unscheduled downtime caused by asset failure. Therefore, as soon as a rolling element bearing exhibited any sign of abnormal performance, a work order was written to replace it at the first "schedule" maintenance window. The obvious result was excessive, premature replacement of every bearing within the plant.

The RCA did not stop with the obvious forcing function, that is, predictive maintenance program, but look deeper to determine the underlying reason that these bearing were exhibiting abnormal behavior. The result of the analysis indicated that the real root causes of the bearing problems were:

- Twenty-seven percent of the bearings in a typical year exhibited abnormal wear because of misapplication. The underlying reasons for this misapplication included: failure of maintenance to properly specify replacement bearings and purchasing/electing to substitute less expensive, often light-duty, bearings.
- Improper installation was the cause of 22 percent of the premature bearing failures. The source of these failures was: improper craft skills, lack of clear installation instructions, and the absence of effective supervision.

- Lubrication-related problems contributed 18 percent to the abnormal failure rate. This was a combination of improper lubricants, wrong application methods, and incorrect lubrication frequency. The lack of proper planning and direct supervision, as well as craft skills also contributed.
- Abnormal bearing loading accounted for 17 percent of the premature bearing wear. The contributing factors, or real root causes, of these problems included: improper operating practices and operator errors. Each of these resulted from failures in the planning, training, and supervision within the production departments.
- A variety of other factors, such as electrical arching, caused 11 percent of the premature bearing wear. The actual root cause varied from poor maintenance practices to environmental and corrosive attack. Each application was evaluated and the root cause eliminated.
- The final 5 percent of the annual expenditures for replacement bearing was the result of bearing that had reached the end of their normal (L_{10}) life.

Of the $14.1 millions spent, only $705,000 was for bearings that had reached or exceeded their rated design life. The remaining $13.39 millions were an unnecessary cost. Armed with the knowledge gained through the RCA process, corrective actions were implemented to eliminate each of the identified root causes, as well as the contributing factors. Crafts were provided training in proper installation and lubrication of rolling element bearings; procedures were rewritten to ensure proper operation and maintenance of the bearings; specifications and procurement practices were upgraded to ensure proper replacement bearings were universally used; and first line supervision was increased to ensure universal adherence to best practices. As a result of these changes, the annual replacement cost for rolling element bearings dropped to $750,000 and has remained constant, within $50,000, for more than 10 years.

In this training manual, the terms *root cause analysis* (RCA) and *root cause failure analysis* (RCFA) will be used interchangeably. The methodologies used are identical; the only difference is the trigger or reason for the analysis. Any deviation from an acceptable norm may justify a formal RCA; but only an absolute failure, generally equipment or system related failure, results in a RCFA.

MANAGING PROBLEM-SOLVING PROCESS

Toyota Motor Corporation is famed for its ability to relentlessly improve operational performance. Central to this ability is the training of engineers, supervisors, and managers in a structured problem-solving approach that uses a tool called the *A3 Problem-Solving Report*. We have adapted the approach by articulating 10 steps to proceed from problem identification to resolution in a fashion that fosters learning, collaboration, and personal development. The problem-solver records the results of investigation and planning in a concise, two-page document—the A3 report, also adapted from Toyota—that facilitates knowledge sharing and collaboration.

The term "A3" derives from the paper size used for the report, which is the metric equivalent to $11'' \times 17''$ (or B-sized) paper. Toyota actually uses several styles of A3 reports—for solving problems, for reporting project status, and for proposing policy changes—each having its own "storyline." We have focused on the problem-solving report simply because it is the most basic style, making it the best starting point.

Why Use It?

Most problems that arise in organizations are addressed in superficial ways, what some call "first-order problem-solving." That is, we work around the problem to accomplish our immediate objective, but do not address the root causes of the problem so as to prevent its recurrence. By not addressing the root cause, we encounter the same problem or same type of problem again and again, and operational performance does not improve.

3.156 ENGINEERING AND ANALYSIS TOOLS

The A3 process helps people engage in collaborative, in-depth problem solving. It drives problem-solvers to address the root causes of problems that surface in day-to-day work routines. The A3 process can be used for almost any situation, and our research has found that, when used properly, for example, all of the steps are followed and completed, the chances of success improve dramatically.

Note that the A3 process is rooted in the more basic PDCA (plan, do, check, act) cycle. Steps 1 to 8 are the Plan steps, with step 5 planning and the Do step, and step 6 planning and the Check step. Step 9 is the Do step, and step 10 is the Check step. Based on the evaluation, another problem may be identified and the A3 process starts again (Act).

Figure 8.1 illustrates the Deming PDCA process that is incorporated into the A3 process. Like A3, the PDCA process consists of a series of steps that must be followed as part of the RCA process.

Steps of the A3 Process

The following steps will guide you through the A3 process:

Step 0. Identify a Problem or Need. Whenever the way work happens is not ideal, or when a goal or objective is not being met, you have a problem, or, if you prefer, a need. The preferred source of problem identification is statistical analysis that tracks the actual versus design performance of the plant and all of its functions, for example, sales, production, procurement, maintenance, and so on.

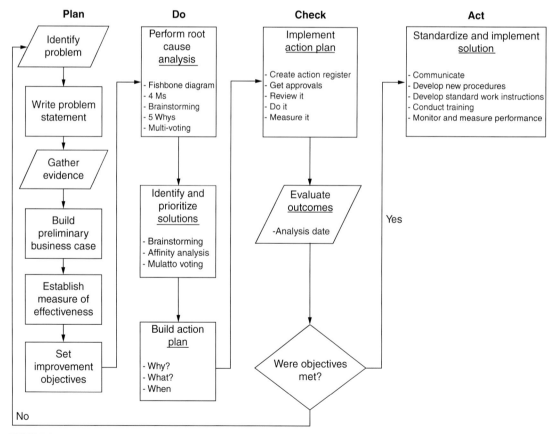

FIGURE 8.1 The PDCA process.

Step 1. Conduct Research to Understand the Current Situation. Before a problem can be properly addressed, one must have a firm grasp of the current situation. To do this, Toyota suggests that problem-solvers:

- Observe the work processes first hand, and document one's observations.
- Create a diagram that shows how the work is currently done. Any number of formal process charting or mapping tools can be used, but often simple stick figures and arrows will do the trick.
- Quantify the magnitude of the problem, for example, percentage of customer deliveries that are late, number of stock outs in a month, number of errors reported per quarter, percentage of work time that is value-added; if possible, represent the data graphically.

Step 2. Conduct Root Cause Analysis. Once you have a good understanding of how the process that needs to be fixed currently works, it is time to figure out what the root causes are to the errors or inefficiency. To accomplish this, first make a list of the main problem(s). Next, ask the appropriate *"why?"* questions until you reach the root cause. A good rule of thumb is that you have not reached the root cause until you have asked *"why?"* at least five times in series. This root cause methodology, called five whys, is proven effective in solving problems with clear, straightforward answers. However, it is also a good starting point on even the most complex problems and should be used by the investigator as part of his or her problem-development or clarification.

There are multiple diagnostic and analysis tools that can be used to conduct the RCA. The guidelines provided will help the investigator select the most effective tool or tools for each classification of problem that will be encountered.

Step 3. Devise Countermeasures to Address Root Causes. Once the current situation is fully understood and the root cause(s) for the main problem(s) has been unveiled, it is time to devise some countermeasures. Countermeasures are the changes to be made to the work processes that will move the organization closer to ideal, or make the process more efficient, by addressing root causes. Generally speaking, we recommend that countermeasures help the process conform to three rules:

- Specify the outcome, content, sequence, and task of work activities.
- Create clear, direct connections between requestors and suppliers of goods and services.
- Eliminate loops, workarounds, and delays.

Step 4. Develop a Target State. The countermeasure(s) addressing the root cause(s) of the problem will lead to new ways of getting the work done, what is called the target condition or target state. It describes how the work will get done with the proposed countermeasures in place.

In the A3 report, the target condition should be a diagram, similar to the current condition that illustrates how the new proposed process will work. The specific countermeasures should be noted or listed, and the expected improvement should be predicted specifically and quantitatively.

Step 5. Create an Implementation Plan. In order to reach the target state, one needs a well thought-out and workable implementation plan. The implementation plan should include a list of the actions that need to be done to get the countermeasures in place and realize the target condition, along with the individual responsible for each task and a due date. Other relevant items, such as cost, may also be added.

Step 6. Develop a Follow-up Plan with Predicted Outcomes. A critical step in the learning process of problem-solvers is to verify whether they truly understood the current condition well enough to improve it. Therefore, a follow-up plan becomes a critical step in process improvement to make sure the implementation plan was executed, the target condition realized, and the expected results achieved. You can state the predicted outcome here rather than in the target condition, if you prefer.

Step 7. Discuss Plans with All Affected Parties. It's vitally important to communicate with all parties affected by the implementation or target condition, and try to build consensus throughout the process. We have included it as a specific step before approval and implementation to make sure it does not get skipped. But the most successful process improvement projects we have witnessed do this step at each critical juncture. Concerns raised should be addressed insomuch as possible, and this may involve studying the problem further or reworking the countermeasures, target condition, or implementation plan. The goal is to have everyone affected by the change aware of it and, ideally, in agreement that the organization is best served by the change.

Step 8. Obtain Approval for Implementation. If the person conducting the A3 process is not a manager, it is imperative to remember the importance of obtaining approval from an authority figure to carry out the proposed plan. The authority figure should verify that the problem has been sufficiently studied and that all affected parties are "on board" with the proposal. The authority figure may then approve the change and allow implementation.

Step 9. Implement Plans. Without implantation, no change occurs. The next step is to execute the implementation plan.

Step 10. Evaluate the Results. Process improvement should not end with implementation. It is very important to measure the actual results and compare to predict. If the actual results differ from the predicted ones, research needs to be conducted to figure out why, modify the process and repeat implementation and follow-up (i.e., repeat the A3 process) until the goal is met.

Using the A3 Form

The A3 Form should be used to manage and track the RCA process, as well as for the final report of the resultants. The form template, as illustrated in Fig. 8.2.

The template should be setup as a PowerPoint slide configured for an 11″ × 17″ landscape printed copy. The resultant of this type of template is that the final document can be directly printed as an 11″ × 17″ hard copy or inserted directly, without any manual manipulation, into a PowerPoint presentation. In addition, this type of template will permit the user to directly insert graphs, plots, flow diagrams, and so on for Microsoft Excel, Visio, Crystal Excelsius, and other accounting or graphics packages.

Steps in Form Use

Use of the A3 format should begin as soon as the potential need for a RCA is identified. This section provides a brief overview of how and when the sections of the form should be completed:

Business Case. Initially, this section should be used to clearly and concisely define the problem that is to be investigated. This definition will define, as well as the business case, for example, cost-benefit analysis will be complete after the investigation is complete.

Background	Target condition
Current condition	
Root cause analysis	Action plan
	Metrics

FIGURE 8.2 A3 template.

Current Conditions. This section includes a concise definition of the current conditions surrounding or as a result of the problem to be investigated. The use of graphs, charts, and other illustrations should be used to clearly convey the message.

Target Conditions. This section defines the resultant of the proposed corrective actions identified by the RCA. Again, the use of graphs, charts, and other illustrative materials will permit the inclusion of much more data as well as provide a more professional report.

Action Plan. This section is your management tool during the RCA and become your "next steps" following management approval to implement the corrective actions. The use of a Gantt, Pert, or other types of project schedules or timelines is ideal for this section. Timelines can be created in MS Visio and inserted directly into this section.

Metrics. This section should be used to define the specific return on investment (ROI) or change that is expected from the recommended changes. It should include both the actual values that represent the change and the source of the data that will be used.

THE RCA PROCESS

The first premise of RCA is that there are always one or more reasons that a deviation from an established or acceptable norm, including equipment or system failure, occurs. Therefore, the first task that is required for effective root cause analysis is the ability to differential between normal and abnormal. Unfortunately, most of the workforce that makes up the management, operations, engineering, maintenance, and support of our plants and facilities lack the knowledge and skills that are needed to accurately recognize, identify, and quantify deviations from acceptable norms. Instead, they react only to obvious symptoms, such as asset or component failures, perceived high costs or other gross changes in plant performance. Because of the lack of practical machine or system dynamics knowledge, no attempt is made to identify the true reason or reasons for the problem. In many cases, only the symptom and not the problem are even recognized. For example, plant personnel will recognize a bearing failure, but no one will question why the bearing failed.

The second premise of RCA is that these problems or deviations from acceptable norm do not just happen, something changed and that change caused the deviation or failure. Therefore, a process must be developed and followed that will permit accurate, effective identification of the change or changes that resulted in the observed deviation, problem, or failure. That process is called RCA.

Root cause analysis takes many forms. It can range from simple, visual inspection of failed parts to a comprehensive process designed to identify; quantify the impact of; develop cost-effective solutions; and implementation of corrective actions for complex capacity, quality, cost, and reliability problems. Regardless of the form, RCA is a systematic process that is based on factual data that is free of prejudice, opinions, or political pressure. It is a logical, practical process that can be used by anyone who is willing to follow it.

Simple Analysis

Many of the equipment-related problems that plague industrial plants and facilities can be resolved by visual inspection of the failed parts. For example, premature failure of rolling element bearings is a common problem in most plants and facilities. Too many plants simply replace the bearing and throw the failed bearing in the nearest trash bin. This approach does little to eliminate the real reason that the bearing failed and there is a high probability that the failure will recur.

Visual Inspection

A simple, visual inspection of the failed bearing, in most cases, will permit plant personnel to identify the underlying reason, that is, root cause, of the premature failure.

FIGURE 8.3 Example of failed bearing.

Figure 8.3 illustrates the inner race of a tapered rolling element bearing removed from a v-belt-driven fan. The bearing was located adjacent to the v-belt drive. Visual inspection clearly shows that the load on this bearing had been concentrated into one quadrant of the inter-race. The deformation, caused by imploding lubricant, caused the premature failure; but what caused the load shift?

Using the information obtained by the visual inspection, the investigator, using root cause logic, can look for changes in the design, installation, mode of operation, and other reasons that could have caused this abnormal loading pattern. In many cases, this can be accomplished by a few simple tests that will isolate the root cause of the load shift and the resultant bearing failure. This type of RCA can be performed by anyone in the plant, including the maintenance crafts and production operators. In this instance, excessive V-belt tension was the actual factor that caused the bearing to fail; but the root causes were:

- *Improper maintenance procedures.* The preventive maintenance task list failed to provide adequate instructions for proper tensioning of the v-belt.
- *Improper craft training.* Crafts were not given proper training thus the correct methods required to properly tension v-belts.
- *Inadequate supervision.* Supervisors, with the proper training, were not available to ensure that maintenance crafts followed best practices.
- Simply correcting the v-belt tension in this example would not correct the true root causes of the problem. Unless the real factors that caused the problem are corrected, the problem will recur in this specific application, as well as other v-belt-driven equipment in the facility.

Visual inspection of other failed components, such as gears, will also provide insight into the potential root cause of premature failures. Figures 8.4 and 8.5 illustrate the more common failure modes of gears.

Figure 8.3 clearly indicates that abrasive contaminates within the lube oil system has damaged the gears. The pattern from root to tip of the gear face could only be caused by abrasives impinged between the mating gears as they roll into and out of mesh.

The severe pitting in Fig. 8.4 is indicative of the implosion of the lubricating film in an overloaded gearbox. The excessive backpressure caused by the overload results in multiple implosions of the oil film and the resultant removal of metal from the gear teeth.

FIGURE 8.4 Abrasive wear caused by contamination.

FIGURE 8.5 Severe pitting caused by overloading gear.

Even simple analysis requires verification of the assumed root cause. This verification may be as simple as using vibration analysis to confirm the forcing function that caused the failure; or more detailed tests designed to eliminate or confirm the assumed root cause. Do not assume anything. Always verify your assumptions.

Five Whys (5W)

As its name implies, this simple root cause tool is an interview process that works best in a cross-functional group of personnel who have direct knowledge of the problem that is being investigated. The process should be repeated as often as necessary to arrive at the true root cause of the problem that is being investigated. For example, a problem exists in one of the primary production modules of a manufacturing plant. The problem statement: "Module 'A' is not meeting its production goals." To resolve this problem, the investigator should arrange a meeting of the operators, maintainers, and other support personnel who have direct knowledge of the "A" Module. After the initial kickoff, the investigator should restate the problem. "Module 'A' is not meeting it production goals" and then ask the first "why." With each answer, the investigator restates the question, but each time it is directed at the previous answer.

- Why is Module "A" failing to meet it production goals? Answer: "We're forced to use relief operators most of the time."
- Why are you forced to use relief operators? Answer: "The regular operators have been in training for the past month."
- Why are the regular operators in training? Answer: "It's mandated training and everyone has to attend before the end of the year."
- Why are all operators being trained at the same time? Answer: "That's how it was scheduled."
- Why was it scheduled that way? Answer: "We were really being pushed earlier in the year and management decided to postpone training until demand dropped."

What is the root cause of the problem? The forcing function was management's decision to wait on mandated training until it was too late to do it efficiently, but is that the real root cause? How would you correct this problem—remember the objective is to prevent a similar situation from reoccurring at some point in the future.

FORMAL PROCESS

More complex problems, such as restricted capacity, product quality, or complex production system failures, require the use of a comprehensive, methodical investigation that evaluates all of the interactions that may have contributed to the deviation or failure. A formal root cause failure analysis will require an investment in both time and manpower. Typically, a formal analysis will require a two to four-person team between 5 and 15 days to complete. If plant personnel cannot resolve the problem within this timeframe, it is unlikely that it will ever be solved without expert assistance.

A classic example is an in-house problem-solving team of six engineers, twelve crafts persons, and three experts from the turbine vendor, worked on a chronic steam turbine-generator problem for more than 10 years without resolution. The turbines exhibited chronic failure of the coupling that connected it to an electrical generator. Over a period of 5 years, each of the six turbines had at least one failure of its coupling. In an attempt to resolve the problem, the team had basically replaced, modified, or changed every part of the turbine-generator drive train without success—in fact the overall reliability of the machine train declined. The incremental cost of this 10-year exercise was well over $1 million with no measurable benefits.

When proper root cause analysis techniques were applied, the problem—lack of sufficient foundation and structural support—was resolved within 7 days. The turbines-generators, mounted on the second floor or the steam plant, were flexing during normal operation, but the most serious damage to the couplings was during startup and coast-down. During these transients, the lack of rigidity in the floor and support structure permitted the entire drive train to move or displace in the horizontal plane. This radical misalignment resulted in premature coupling failure.

Unfortunately, the cost required to undo all of the modifications and changes that had been made during the 10-year troubleshooting exercise were substantial, almost $750,000. The actual cost to correct the real root cause was less than $5000 per turbine-generator or $30,000.

In formal RCA, the investigating team will need input from all plant personnel who may have direct or indirect knowledge of the deviation, event, or problem that is being investigated. This information input activity may be limited to interviews, either individually or in groups; but could entail additional support gathering data, records, and other pertinent information. Obviously, the actual level of effort will depend on the complexity of the problem and the team's ability to determine the root cause or causes.

Purpose of the Analysis

The purpose of RCA is to resolve problems that negatively impact safety, environmental compliance, asset reliability, and plant performance, not to fix blame. Too many companies fixate on finding someone to blame for any failure or deviation from an accepted norm rather than solve the problem. For example, operator error is the most often perceived reason for production system failures. In many cases, the actual reason the accident or failure occurred is because the operator did the wrong thing, but is operator error the real reason for the failure? With the exception of intentional sabotage, the majority of the failures were the result of a failure or failures in the infrastructure not human error.

Little can be gained from the "fixing blame" process. This approach results in lost morale and will condition the workforce to withhold information that is critical to the root cause process and effective plant operation and maintenance. Even in those cases were human error was the sole reason for the problem, reprimanding, punishing, or dismissing the employee will do little to mitigate the impact of the failure or prevent it from recurring.

Everyone involved in the RCA process must clearly understand that resolving the problem is the primary and only objective of the process. Understanding that the investigation is not an attempt to fix blame is important for three reasons. First, the investigator or the investigating team must understand that the real benefit of this analytical methodology is plant improvement through improved reliability, better product quality, and lower life cycle cost.

Second, the investigator or investigating team is dependent on plant personnel to provide factual information and data that is essential to the process. Without a free and open exchange of information from all plant personnel, the RCA process cannot succeed.

Finally, those involved in the failure or deviation that is being investigated will generally adopt a self-preservation attitude and assume that the investigation is intended to find and punish the person or persons responsible. Therefore, it is important for the investigators to dispel this fear and replace it with the positive team effort that is required to resolve the problem.

Effective Use of the Analysis

Root cause analysis cannot be performed sitting in a conference room, office, or in front of a computer. While the RCA process does require working group meetings, as well as individual and group interviews, the heart of the process is gathering factual data that can be used to isolate, identify, and quantify the real reason or reasons that resulted in the abnormal behavior that is being investigated. To do this, the investigator or team must roll up their sleeves and get dirty.

The RCA process requires a hands-on process of interviews, inspections, testing, and evaluations that can only be done in the plant or field. Theoretical evaluations have their place, but to use the RCA process effectively, the investigators must clearly understand the operating dynamics of the investigated system, confirm any and all factors, assumptions, or hypotheses that may be offered by those involved in the event being investigated. Effective use of RCA requires discipline and consistency. Each investigation must be thorough and each of the steps defined in this book must be followed. Perhaps the most difficult part of the analysis is separating fact from fiction. Human nature dictates that everyone involved in an event or incident that requires a RCA is conditioned by his or her experience and their natural tendency is to filter input data based on this conditioning. This includes the investigator! However, this often causes preconceived ideas and perceptions that destroy the effectiveness of the process.

It is important for the investigator or investigating team to put aside their perceptions, base the analysis on pure fact, and not assume anything. Any assumptions that enter the analysis process through interviews and other data-gathering processes should be clearly stated. If the assumptions cannot be confirmed or proven, they must be discarded.

Personnel Requirements

Anyone, from the newest craft employee to specially trained engineers, can perform RCA. The number of people required is dependent on the complexity of the specific event, deviation, or failure that is being investigated. In rare cases, the personnel required to properly perform a RCA can be substantial; but in most cases will require a three to four person, multidisciplined team. For example, failure of a complex production or manufacturing system might require a three person team consisting of: an engineer with the requisite knowledge of the system's design characteristics, an operator who is familiar with the system or similar systems, and a skilled craftsperson who has experience with the system. This combination provides the ability to view the problem from three critical perspectives: engineering, operations, and maintenance. With this three-way view, most problems can be quickly and cost-effectively resolved. Without it, the probability of identifying the true causes of the problem will be elusive and may never be found.

The investigating team will require the involvement of all plant personnel who were directly or indirectly involved in the failure, problem or deviation that is being investigated or have specific knowledge of the boundary conditions when the problem occurred. Regardless of what one may think, problems do not just happen. There is always a reason or reasons for problems that develop in our plants or personal lives. In most cases, the personnel that are involved with the deficient system or process will have the knowledge that will permit the investigator or team to isolate the specific cause or combination of causes that directly resulted in the observed problem.

ROOT CAUSE ANALYSIS METHODOLOGY

Root cause analysis follows a logical sequence or methodology, see Fig. 8.6, which is designed to facilitate the solution of the investigated problem, deviation, or event. The steps shown in Fig. 8.6 should be followed.

IDENTIFYING A POTENTIAL RCA EVENT

There are two primary sources of potential problems. The first and preferred method is through regular analysis of *key performance indicators* (KPI) and asset history to detect deviations from normal conditions. This function is an integral part of the reliability engineering function and should account for 90 percent of the RCA that will be performed in a typical year.

The second source for potential analysis is request from one or more members of the plant's workforce. The concept of *total employee involvement* is a fundamental requirement of *Reliability Excellence*™, as well as any effective continuous improvement program. Therefore, any employee is expected to identify problems or events that may warrant an analysis.

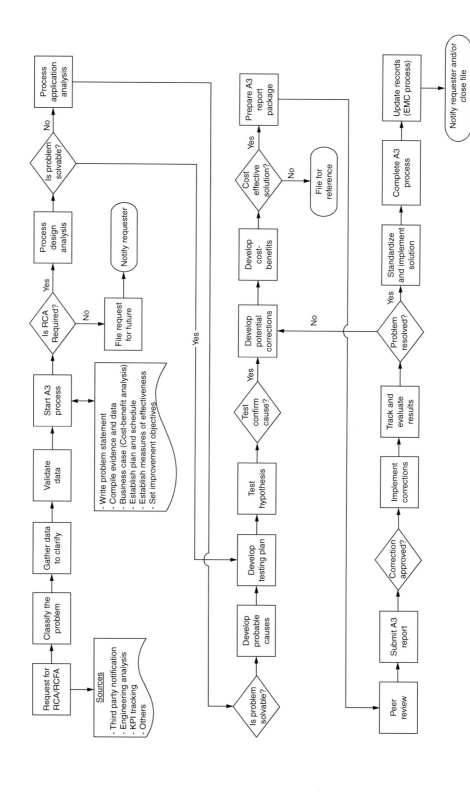

FIGURE 8.6 RCA logic diagram.

Reporting an Incident or Problem

The investigator is seldom present when an incident or problem occurs. Therefore, the first step is the initial notification that an incident or problem has taken place. Typically, this report will be verbal, a brief written note, or a notation in the production logbook. In most cases, the communication will not contain a complete description of the problem. Rather, it will be a very brief description of the perceived symptoms observed by the person reporting the problem.

Symptoms and Boundaries. The most effective means of problem or event definition is to determine its *real* symptoms and establish limits that bound the event. At this stage of the investigation, the task can be accomplished by an interview with the person who first observed the problem.

Perceived Causes of Problem. At this point, each person interviewed will have a definite opinion about the incident, and will have his or her description of the event and an absolute reason for the occurrence. In many cases, these perceptions are totally wrong, but they cannot be discounted. Even though many of the opinions expressed by the people involved with or reporting an event may be invalid, do not discount them without investigation. Each of these opinions should be recorded and used as part of the investigation. In many cases, one or more of the opinions will hold the key to resolution of the event. The following are some examples where the initial perception was incorrect.

One example of this phenomenon is a reported dust collector baghouse problem. The initial report stated that dust-laden air was being vented from the baghouses on a random, yet recurring, basis. The person reporting the problem was convinced that chronic failure of the solenoid-actuated pilot valves that controlled the blowdown of the baghouse was, without a doubt, the cause. However, a quick design review found that the solenoid-controlled valves are *normally closed*. This type of solenoid valve *cannot fail* in the *open* position and, therefore, could not be the source of the reported events.

A conversation with a process engineer identified the diaphragms used to seal the blowdown tubes as a potential problem source. This observation, coupled with inadequate plant air, turned out to be the root cause of the reported problem.

Another example illustrating preconceived opinions is the catastrophic failure of a *Hefler* chain conveyor. In this example, all of the bars on the left side of the chain were severely bent before the system could be shut down. Even though a foreign object such as a bolt was not found, this was assumed to be the cause for failure. From the evidence, it was clear that some obstruction had caused the conveyor damage, but the more important question was why did it happen.

Hefler conveyors are designed with an intentional failure point that should have prevented the extensive damage caused by this event. The main drive-sprocket design includes a *shear pin* that generally prevents this type of catastrophic damage. Why did the conveyor fail? Because the shear pins had been removed and replaced with Grade-5 bolts. In the conveyor example, what is the root cause of the failure? Obviously, someone in the maintenance organization, perhaps a craftsperson or even the maintenance manager, made a decision to replace the shear pin with a Grade-5 bolt, but is this the real root cause? In reality, a failure in the infrastructure is the underlying or root cause of this problem. The absence of a formal, enforced *Engineering Change Management* process that would prevent this type of arbitrary modification is the real problem and unless it is corrected there is a high probability that this and similar problems with recur. Other factors that could have contributed to this problem include: lack of proper supervision, faulty procedures, inadequate training, and absence of a viable maintenance engineering function.

Event-Reporting Format. One factor that severely limits the effectiveness of RCA is the absence of a formal event-reporting format. The use of a format that completely bounds the potential problem or event greatly reduces the level of effort required to complete an analysis. A form similar to the one shown in Fig. 8.7 provides the minimum level of data needed to determine the effort required for problem resolution.

Incident Reporting Form

Date:

Reported by:

Description of incident:

Specific location and equipment/system effected:

When did incident occur:

When was involved:

What is probable cause:

What corrective actions taken:

Was personal injury involed: ☐ Yes ☐ No

Was reportable release involed: ☐ Yes ☐ No

Incident classification: ☐ Equipment failure ☐ Regulatory compliance
☐ Accident/injury ☐ Performance deviation

FIGURE 8.7 Typical incident reporting form.

CLARIFY THE PROBLEM

The logical first step in any problem-solving effort is to fully understand the problem. In too many cases, a RCA is requested without a clear, concise definition of the problem. As a result, much time can be wasted trying to solve a problem that does not really exist or pursuing the wrong problem. This is especially true of product quality, cost and other issues that are nebulous and often difficult to verify or disprove without a substantial investigative effort. Therefore, the investigator or team must first clarify the problem with sufficient definition to: (1) verify that a problem truly exists and (2) that the severity of the problem warrants an analysis.

The first step in the clarification process is an interview with the requestor or initiator of the request. The intent of this interview is to gain a clear understanding of his or her reason for requesting the investigation. The initial interview is relative straightforward. A series of well-designed questions should provide the information needed to confirm the need for a formal RCA.

Questions to Ask

The following questions should be used to clarify the problem:

What Happened? The obvious first question is a clear definition of what triggered the request for analysis. Every effort should be made to gain as much detail as is possible during this part of the interview. Clarifying what actually happened is an essential requirement of RCA. As discussed earlier, the natural tendency is to give perceptions rather than to carefully define the actual event. It is important to include as much detail as the facts and available data permit.

Where Did It Happen? Problems that can be attributed to a specific component, asset, process, or plant area are easier to isolate those that appear to occur at random throughout the plant. Therefore, a clear description of the exact location of the event helps isolate and resolve the problem. In addition to the location, determine if the event also occurred in similar locations or systems. If similar machines or applications are eliminated, the events sometimes can be isolated to one or a series of factors that are totally unique to that location. For example, if Pump A failed and Pumps B, C, and D in the same system did not, this indicates that the reason for failure is probably unique to Pump A. If Pumps B, C, and D exhibit similar symptoms, however, it is highly probable that the cause is systemic and common to all of the pumps.

When Did It Happen? Isolating the specific time that an event occurred greatly improves the investigator's ability to determine its source. When the actual time frame of an event is known, it is much easier to quantify the process, operations, and other variables that may have contributed to the event. However, in some cases, such as product-quality deviations, it is difficult to accurately fix the beginning and duration of the event. Most plant-monitoring and tracking records do not provide the level of detail required to properly fix the time of this type of incident. In these cases, the investigator should evaluate the operating history of the affected process area to determine if a pattern can be found that properly fixes the event's time frame. This type of investigation will, in most cases, isolate the timing to events such as the following:

- Production of a specific product
- Work schedule of a specific operating team
- Changes in ambient environment

What Changed? Regardless of what the bumper stickers say, deviations from the norm, equipment problems, and most other events that warrant a root cause analysis, do not just happen. In every case, there are specific variables that, singly or in combination, caused the deviation or problem to occur. Therefore, it is essential that any changes that occurred in conjunction with the event be defined. No matter what the problem is the evaluation must quantify all of the variables that were associated with it. These data should include the operating setup; product variables, such as viscosity, density, flow rates; operating team; and ambient environment. If available, the data also should include any predictive maintenance data associated with the event.

Who Was Involved? The investigation should identify all personnel involved, directly or indirectly, in the event. Failures and events are often the result of human error and/or inadequate skills. However, remember that the purpose of the investigation is to resolve the problem, not to place blame. All comments or statements derived during this part of the investigation should be impersonal and totally objective. All references to personnel who were directly involved in the incident should be assigned a *code number or other identifier*, such as Operator A or Maintenance Craftsperson B. This approach helps reduce fear of punishment for those directly involved in the incident. In addition, it reduces prejudice or preconceived opinions about individuals within the organization.

Why Did It Happen? If the preceding questions are fully answered, it maybe possible to resolve the incident without further investigation. However, exercise caution to ensure that the real problem has been identified. It is too easy to address the symptoms or perceptions without a full analysis. At this point, generate a list of what may have contributed to the reported problem. The list should include *all* factors, both real and assumed. This step is critical to the process. In many cases, there are a number of factors, many of them trivial, that combine to cause a serious problem. All assumptions included in this list of possible causes should be clearly noted, as should the causes that are proven.

What Is the Impact? The evaluation should quantify the impact of the event before embarking on a full RCA. Again, not all events, even some that are repetitive, warrant a full analysis. This part of the investigation process should be as factual as possible. Even though all the details are not available at this point, attempt to assess the real or potential impact of the event.

Will It Happen Again? If the preliminary interview determines that the event is nonrecurring, the process maybe discontinued at this point. However, a thorough review of the historical records associated with the machine or system involved in the incident should be conducted before making this decision. Make sure that it is truly a nonrecurring event before discontinuing the evaluation. All reported events should be recorded and the files maintained for future reference. For incidents found to be nonrecurring, a file should be established that retains all of the data and information developed in the preceding steps. Should the event or one that is similar occur again, these historical records become an invaluable investigative tool. A full investigation should be conducted on any event that has a history of periodic recurrence, or has a high probability of recurrence, and has a significant impact in terms of injury, reliability, and/or economics. In particular, all incidents that have the potential for personal injury or regulatory violation should be investigated.

How Can Recurrence Be Prevented? Although this is the next logical question to ask, it cannot generally be answered until the entire RCA is completed. Note, however, that if this analysis determines it is not economically feasible to correct the problem, plant personnel may simply have to learn to minimize the impact.

CONFIRM THE FACTS

Root cause analysis should not be based on opinions or assumptions. Before starting an analysis, the investigators must confirm that a problem truly exists and that it warrants a formal investigation. Therefore, all of the information obtained in the initial interview must be confirmed or refuted. If a problem truly exists, there should be data in the *Computer-Based Maintenance Management System* (CMMS) or other records-keeping systems that supports it. The investigators should compile these data to confirm the reported problem.

This step in the process should included sufficient detail to validate the problem, as well as quantify its actual impact on the plant. The impact should include lost capacity, incremental operating and maintenance costs, the potential for accidents or regulatory noncompliance and any other measurable value. It should also include cross-validation of any recorded data. Too often information management systems record erroneous information caused in part by entry errors and fragmentation of data. This erroneous data can distort a potential problem and lead the investigators to the wrong conclusion about whether or not a RCA is necessary. In most cases, data is retained in two or more databases, as well as logbooks and other noncomputerized records. By comparing these various sources, data can be validated so that a true picture of the potential problem can be obtained.

Compile Data and Physical Evidence

The RCA process must be based on factual data. As a result, the first priority when investigating a problem, deviation from acceptable norm or an event involving equipment damage or failure is to preserve physical evidence. If possible, the failed machine and its installed system should be isolated from service until a full investigation can be conducted. Upon removal from service, the failed machine and all of its components should be stored in a secure area until they can be fully inspected and appropriate tests conducted.

If this approach is not practical, the scene of the failure should be fully documented before the machine is removed from its installation. Photographs, sketches, and the instrumentation and control settings should be fully documented to ensure that all data are preserved for the investigating team. All automatic reports, such as those generated by computer-monitoring and control system, should be obtained and preserved.

The legwork required collecting information and physical evidence for the investigation can be quite extensive. The following is a partial list of the information that should be gathered:

- Currently approved Standard Operating (SOP) and Maintenance (SMP) Procedures for the machine or area where the event occurred
- Company policies that govern activities performed during the event
- Operating and process data, such as strip charts, computer output, and data-recorder information
- Appropriate maintenance records for the machinery or area involved in the event
- Copies of logbooks, work packages, work orders, work permits, and maintenance records; equipment-test results, quality-control reports; oil and lubrication analysis results; vibration signatures; and other records
- Diagrams, schematics, drawings, vendor manuals, and technical specifications, including pertinent design data for the system or area involved in the incident
- Training records, copies of training courses, and other information that shows skill levels of personnel involved in the event
- Photographs, videotape, and/or diagram of the incident scene
- Broken hardware, such as ruptured gaskets, burned leads, blown fuses, failed bearings, etc.
- Environmental conditions when the event occurred. These data should be as complete and accurate as possible
- Copies of incident reports for similar prior events and history/trend information for the area involved in the current incident

IS ROOT CAUSE ANALYSIS REQUIRED

Not all problems whether real or perceived justify a formal RCA. Therefore, the clarified and confirmed problem should be evaluated to determine if its impact is sufficient to warrant further investigation. If the initial steps appear to justify a RCA, the next step in the process is to perform a top-level cost-benefit analysis. The intent of this analysis is to verify that the potential benefits generated by resolving the reported problem are greater than the incurred cost associated with the problem.

At this point in the RCA process, the investigators do not know the root cause or causes of the problem or the required corrective actions. As a result, the cost-benefit analysis is limited to the actual cost of the investigation and the verified incurred costs associated with the reported problem. For example, the incremental or elevated cost of repairing a machine with a normal mean time between repair (MTBR) of 12 months but with an actual MTBR of 3 months; the incremental cost is the difference between the rebuild cost. In this case, the pump is being rebuilt three times more often than the norm and the incremental cost is three times higher than norm.

If the cost-benefit analysis indicates that the reported event or problem does not warrant further analysis, the investigator should notify the person or persons who initiated the request. This step is critical and should not be omitted. The source of information that will identify most problems that have a negative impact on plant performance resides with its employees. As long as their efforts are acknowledged and appropriate actions taken, the workforce will continue to report potential or perceived problems. However, when their efforts are ignored, or appear to be ignored, the workforce will quit alerting management when they observe deviations.

DESIGN REVIEW

Most, if not all, problems that warrant a formal RCA involve a manufacturing or production system or some of its components. Therefore, it is essential that the investigators clearly understand the design parameters and specifications of the systems associated with an event or equipment failure.

Unless the investigator understands precisely what the machine or production system was designed to do and its inherent limitations, it is impossible to isolate the root cause of a problem or event. The data obtained from a *design review* provide a baseline or reference, which is needed to fully investigate and resolve plant problems.

The objective of the design review is to establish the specific operating characteristics of the machine or production system involved in the incident. The evaluation should clearly define the specific function or functions that each machine and system was designed to perform. In addition, the review should establish the acceptable operating envelope, or range, that the machine or system can tolerate without a measurable deviation from design performance.

The logic used for a comprehensive review is similar to that of a *Simplified Failure Modes and Effects Analysis* (SFMEA), and *fault-tree analysis* (FTA) in that it is intended to identify the variables or failure modes that could contribute to a problem or failure. Unlike these techniques, which use complex probability tables, theoretical analysis, or break each machine down to the component level, RCA takes a more practical approach. The technique is based on readily available, application-specific data to determine the variables that may cause or contribute to an incident.

While the level of detail required for a design review varies depending on the type of event, this step cannot be omitted from any investigation. In some instances, the process may be limited to a cursory review of the vendor's *Operating and Maintenance (O&M) manual* and performance specifications. In others, a full evaluation that includes all procurement, design, and operations data may be required.

Minimum Design Data

In many cases, the information required can be obtained from four sources: equipment nameplates, procurement specifications, vendor specifications, and the O&M manuals provided by the vendors. If the investigator has a reasonable understanding of machine dynamics, a thorough design review for relatively simple production systems, such as a pump transfer system or compressed air system, can be accomplished with just the data provided in these four documents. Special attention should be given to the vendor's troubleshooting guidelines. These suggestions will provide insight into the more common causes for abnormal behavior and failure modes.

Equipment Nameplate Data. Most of the machinery, equipment, and systems used in process plants have a permanently affixed nameplate that defines their operating envelope. For example, a centrifugal pump's nameplate typically includes flow rate, total discharge pressure, specific gravity, impeller diameter, and other data that define its design operating characteristics. These data can be used to determine if the equipment is suitable for the application and if it is operating within its design envelope.

Procurement Specifications. Procurement specifications are normally prepared for all capital equipment as part of the purchasing process. These documents define the specific characteristics and operating envelope requested by the plant-engineering group. These specifications provide information that is useful for evaluating the equipment or system during an investigation.

When procurement specifications are not available, purchasing records should describe the equipment and provide the system envelope. Although this data may be limited to a specific type or model of machine, it is generally useful information.

Vendor Specifications. For most equipment procured as part of capital projects, a detailed set of vendor specifications should be available. Generally, these specifications were included in the vendor's proposal and confirmed as part of the deliverables for the project. Normally, these records are on file in two different departments: purchasing and plant engineering.

As part of the design review, the vendor and procurement specifications should be carefully compared. *Many of the chronic problems that plague plants are a direct result of vendor deviations from procurement specifications.* Carefully comparing these two documents may uncover the root cause of chronic problems.

Operating and Maintenance Manuals. O&M manuals are one of the best sources of information. In most cases, these documents provide specific recommendations for proper operation and maintenance of the machine, equipment, or system. In addition, most of these manuals provide specific troubleshooting guides that point out many of the common problems that may occur. A thorough review of these documents is essential before beginning the RCFA. The information provided in these manuals is essential to effective resolution of plant problems.

Objectives of the Review

The objective of the design review is to determine design limitations, acceptable operating envelope, probable failure modes, and specific indices that quantify the actual operating condition of the machine, equipment, or process system being investigated. At a minimum, the evaluation should determine design function and specifically what the machine or system was designed to do. The review should clearly define the specific functions of the system and its components.

To fully define machinery, equipment, or system functions, a description should include incoming and output product specifications, work to be performed, and acceptable operating envelopes. For example, a centrifugal pump may be designed to deliver 1000 gal/min of water having a temperature of 100°F and a discharge pressure of 100 lb/in^2.

Incoming-Product Specifications. Machine and system functions depend on the incoming product to be handled. Therefore, the design review must establish the incoming product boundary conditions used in the design process. In most cases, these boundaries include: temperature range, density or specific gravity, volume, pressure, and other measurable parameters. These boundaries determine the amount of work the machine or system must provide.

In some cases, the boundary conditions are absolute. In others, there is an acceptable range for each of the variables. The review should clearly define the allowable boundaries used for the system's design.

Output-Product Specifications. Assuming the incoming product boundary conditions are met, the investigation should determine what output the system was designed to deliver. As with the incoming product, the output from the machine or system can be bound by specific, measurable parameters. Flow, pressure, density, and temperature are the common measures of output product. However, depending on the process, there may be others.

Work Performed. This part of the design review should determine the measurable work to be performed by the machine or system. Efficiency, power usage, product loss, and similar parameters are used to define this part of the review. The actual parameters will vary depending on the machine or system. In most cases, the original design specifications will provide the proper parameters for the system under investigation.

Acceptable Operating Envelope. The final part of the design review is to define the acceptable operating envelope of the machine or system. Each machine or system is designed to operate within a specific range, or operating envelope. This envelope includes the maximum variation in incoming product, startup ramp rates and shutdown speeds, ambient environment, and a variety of other parameters.

CAN PROBLEM BE RESOLVED BASED ON DESIGN REVIEW

Many of the chronic problems that negatively affect critical production systems are caused by inherent design deficiencies. Therefore, the investigator should evaluate the confirmed data develop before and during the design review to determine whether or not the root cause of the problem can be accurately isolated without continuing the RCA process.

In some cases, completion of a thorough design review will identify, or strongly suggest, the probable cause of the problem that is being investigated. For example, cantilevered fans are known to have inherent design weaknesses, such as inadequate bearing support structures, unstable rotating elements and a potential for operating at the first critical speed of the rotating element. If the problem is short bearing life or failure, variable fan output, and fan blade failures, there is a strong probability that these inherent weaknesses are the cause, or a major contributor to the problem.

Extreme caution should be used for this evaluation. The normal tendency is to accept the first or obvious reason for failures and other problems that warrant a root cause analysis. If the cause appears to be an inherent design weakness, the investigator must confirm his or her assumption. Normally, a series of test can be conducted that will confirm or discount inherent design deficiencies as the root cause of a problem. Do not assume that an inherent weakness in the design is the true cause of the observed problem.

APPLICATION REVIEW

The obvious next step in the RCA process is to review the application to ensure that the machine or system is being used in the proper application and that the mode of operation and maintenance are within the operating envelope, as defined in the design review. The data gathered during the design review should be used to verify the application, as well as operating and maintenance records associated with the appropriate system or asset.

In plants where multiple products are produced by the machine or process system being investigated, it is essential that the full application range be evaluated. The evaluation must include all variations in the operating envelope over the full range of products being produced. The reason this is so important is that many of the problems that will be investigated are directly related to one or more process setups that may be unique to that product. Unless the full range of operation is evaluated, there is a potential that the root cause of the problem will be missed.

Factors to evaluate in an application review include: installation, operating envelope, operating procedures and practices, such as standard procedures versus actual practices, maintenance history, and maintenance procedures and practices.

Installation

Each machine and system has specific installation criteria that must be met before acceptable levels of reliability can be achieved and sustained. These criteria vary with the type of machine or system, and should be verified as part of the RCA. Using the information developed as part of the design review, the investigator or other qualified individuals should evaluate the actual installation of the machine or system that is being investigated. As a minimum, a thorough visual inspection of the machine and its related system should be conducted to determine if improper installation is contributing to the problem.

Photographs, sketches, or drawings of the actual installation should be prepared as part of the evaluation. They should point out any deviations from acceptable or recommended installation practices as defined in the reference documents and good engineering practices. This data can be used later in the RCA when potential corrective actions are considered.

Operating Envelope

Evaluating the actual operating envelope of the production system associated with the investigated event is more difficult. The best approach is to determine all variables and limits used in normal production. For example, define the full range of operating speeds, flow rates, incoming product variations, and so on, which are normally associated with the system. In variable-speed applications, determine the minimum and maximum ramp rates used by the operators.

Operating Procedures and Practices

This part of the application review consists of evaluating the standard operating procedures as well as the actual operating practices. Most production areas maintain some historical data that tracks its performance and practices. These records may consist of logbooks, reports, or computer data. These data should be reviewed to determine the actual production practices that are used to operate the machine or system being investigated.

Systems that use a computer-based monitoring and control system will have the best database for this part of the evaluation. Many of these systems automatically store, and in some cases print regular reports, that define the actual process setups for each type of product produced by the system. This is an invaluable source of information that should be carefully evaluated.

Standard Operating Procedures. Evaluate the standard operating procedures (SOPs) for the affected area or system to determine if they are consistent and adequate for the application. Two reference sources, the design review report and vendor's O&M manuals, are required to complete this task.

In addition, evaluate SOPs to determine if they are usable by the operators. Review organization, content, and syntax to determine if the procedure is correct and understandable.

Setup Procedures. Special attention should be given to the setup procedures for each product produced by a machine or process system. Improper or inconsistent system setup is a leading cause of poor product quality, capacity restrictions, and equipment unreliability. The procedures should provide clear, easy to understand instructions that ensure accurate, repeatable setup for each product type. If they do not, the deviations should be noted for further evaluation.

Transient Procedures. Transient procedures, such as start-up, speed change, and shutdown, also should be carefully evaluated. These are the predominant transients that cause deviations in quality and capacity, and that have a direct impact on equipment reliability. These procedures should be evaluated to ensure that they do not violate the operating envelope or vendor's recommendations. All deviations must be clearly defined for further evaluation.

Operating Practices. This part of the evaluation should determine if the SOPs were understood and followed before and during the incident or event. *The normal tendency of operators is to short-cut procedures, which is a common reason for many problems.* In addition, unclear procedures lead to misunderstandings and misuse.

Therefore, the investigation must fully evaluate the actual practices that the production team uses to operate the machine or system. The best way to determine compliance with SOPs is to have the operator(s) list the steps used to run the system or machine being investigated. This task should be performed without referring to the SOP manual. The investigator should lead the operator(s) through the process and use their input to develop a sequence diagram.

After the diagram is complete, compare it to the SOPs. If the operator's actual practices are not the same as those described in the SOPs, the procedures may need to be upgraded or the operators may need to be retrained.

Maintenance History

A thorough review of the maintenance history associated with the machine or system is essential to the RCA process. One of the questions that must be answered is "Will this happen again?" A review of the maintenance history may help answer this question. The level of accurate maintenance data that are available will vary greatly from plant to plant. This may hamper the evaluation, but it is necessary to develop as clear a picture as possible of the system's maintenance history.

A complete history of the scheduled and actual maintenance, including inspections and lubrication, should be developed for the affected machine, system, or area. The primary details that are needed include: frequency of repair and types of repair, frequency and types of preventive maintenance, failure history, and any other facts that will help in the investigation.

Maintenance Procedures and Practices. A complete evaluation of the *Standard Maintenance Procedures* (SMPs) and actual practices should be conducted. The procedures should be compared with maintenance requirements defined by both the design review and the vendor's O&M manuals. Actual maintenance practices can be determined in the same manner as described in earlier or by visual observation of similar repairs. This task should determine if all maintenance personnel assigned to or involved with the area that is being investigated consistently follow the SMPs. Special attention should be given to the routine tasks, such as lubrication, adjustments, and other preventive tasks. Determine if these procedures are being performed in a timely manner and if proper techniques are being used.

CAN PROBLEM BE SOLVED BASED ON APPLICATION REVIEW

More than 27 percent of all reliability problems are caused by misapplication. While the initial design and operations of the system may have been compatible, the myriad of modifications, upgrades, and other changes have historically resulted in operating conditions that are outside the acceptable operating envelope.

In many cases, the combination of an effective design and application review will identify the more likely causes of the observed deficiency or failure. If these can be confirmed as the true root cause or causes of the problem some of the intermediate steps in the RCA process can be eliminated. However, the hypotheses developed by the investigating team must be verified through a series of tests designed to confirm them and to eliminate all other factors that may have contributed to the problem.

ORGANIZE YOUR THOUGHTS

If the preceding steps do not provide a clear understanding of the more probable reasons for the problem, the investigator or team must organize all of the data, assumptions, and hypotheses into a form that can be used for further analysis. The most effective method involves plotting the accumulated facts, assumptions, and hypotheses into a graphical format that facilitates understanding the cause and effect and interactions of all identified variables.

There are a number of techniques that are useful for problem solving. While there are many common, or overlapping, methodologies associated with these techniques, there also are differences. This section provides a brief overview of the more common methods used to perform a RCA.

INCIDENT CLASSIFICATION

There are numerous methods, see Fig. 8.8, that can be used to help the investigator organize his or her thoughts. Each of these methods have their strengths and weaknesses and proper selection is essential for effective, accurate identification of the root cause and all contributing factors that resulted in the problem, event, or failure that is being investigated.

Common problem classifications are equipment damage or failure, operating performance, economic performance, safety, and regulatory compliance. Classifying the event as a particular problem type allows the analyst to determine the best method to resolve the problem. Each of the major classifications requires a slightly different RCA approach.

Note, however, that initial classification of the event or problem is typically the most difficult part of a RCA. Too many plants lack a formal tracking and reporting system that accurately detects and defines deviations from optimum operation condition.

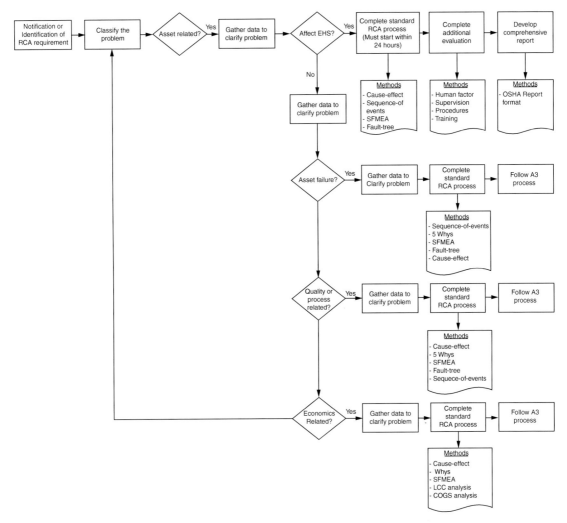

FIGURE 8.8 RCA tools.

System or Equipment Damage or Failure

One of the major classifications of problems that often warrant RCA is an event associated with failure of critical production equipment, machinery, or systems. Typically, any incident that results in partial or complete failure of a machine or process system warrants a RCA. This type of incident can have a severe, negative impact on plant performance. Therefore, it often justifies the effort required to fully evaluate the event and to determine its root cause.

Events that result in physical damage of plant equipment or systems are the easiest to classify. Visual inspection of the failed machine or system component usually provides clear evidence of its failure mode. While this inspection usually will not resolve the reason for failure, the visible symptoms or results will be evident. The events that also meet other criteria, such as safety, regulatory, or financial impacts, should automatically be investigated to determine the actual or potential impact on plant performance, including equipment reliability.

In most cases, the failed machine must be replaced immediately to minimize impact on production. If this is the case, evaluating the system surrounding the incident may be beneficial.

The most effective methods of resolving an equipment or system failure problem are *sequence-of-events analysis* or SFMEA. The methods will be explained in the following sections.

Operating Performance

Deviations in operating performance may occur without the physical failure of critical production equipment or systems. Chronic deviations may justify the use of RCA as a means of resolving the recurring problem. The steps and methods required to resolve this type of problem are discussed earlier. Generally, chronic product quality and capacity problems require a full RCA. However, care must be exercised to ensure that these problems are recurring and have a significant impact on plant performance before using this problem-solving technique.

Product Quality. Deviations in first-time-through product quality are prime candidates for RCA, which can be used to resolve most quality-related problems. However, the analysis should not be used for all quality problems. Nonrecurring deviations or those that do not have a significant impact on capacity or costs are not cost-effective applications.

Capacity Restrictions. Many of the problems or events that occur affect a plant's ability to consistently meet expected production or capacity rates. These problems may be suitable for RCA, but further evaluation is recommended before beginning an analysis. After the initial investigation, if the event can be fully qualified and a cost-effective solution found, then a full analysis should be considered. Note that an analysis is not normally performed on random, nonrecurring events or equipment failures.

The preferred analytical tool for these potentially complex problems is *cause and effects analysis*. In some cases where the exact time the problem first began, *sequence-of-events analysis* can also be used effectively.

Economic Performance

Deviations in economic performance, such as high production or maintenance costs, often warrant the use of RCA. The decision tree and specific steps required to resolve these problems vary depending on the type of problem and its forcing functions or causes. Because of the complexity of economic deviations, the preferred analytical tool is again *cause and effects analysis*.

Safety

Any event that has a potential for causing personal injury should be investigated immediately. While events in this classification may not warrant a full RCFA, they must be resolved as quickly as possible. Isolating the root cause of injury-causing accidents or events is generally more difficult than for equipment failures and requires a different problem-solving approach. The primary reason for this increased difficulty is that the cause is often subjective. In most cases, regulator requirements necessitate using all of the analytical tools, but the primary tool should be *cause and effect*.

Regulatory Compliance

Any regulatory compliance event can potentially impact the safety of workers, the environment, as well as the continued operation of the plant. Therefore, any event that results in a violation of environmental permits or other regulatory-compliance guidelines, such as Occupational Safety and Health Administration, Environmental Protection Agency, and state regulations, must be investigated and resolved as quickly as possible. Since all releases and violations must be reported—and they have a potential for curtailed production and/or fines—this type of problem must receive a high priority.

SECTION 4

MAINTENANCE OF PLANT FACILITIES

CHAPTER 1
MAINTENANCE OF LOW-SLOPED MEMBRANE ROOFS

Donald R. Mapes
Building Technology Associates, Inc., Glendale, Ariz.

Dennis J. McNeil
Construction Consultants, Inc., Homewood, Ill.

This chapter is concerned with the inspection and basic maintenance of low-sloped roofs. Low-sloped or "flat" roof systems are subdivided by membrane type into three major categories: conventional built-up roof systems, modified-bitumen roof systems, and single-ply roof systems.

The roof membrane, regardless of type, is a continuous covering of waterproof material which is typically installed over an insulation system and a structural substrate (nailable or nonnailable roof deck). In some cases, such as on older buildings with roofing installed prior to 1972, on buildings with nailable decks (especially in mild climates), and on most inverted roof system designs, the membrane may be applied directly to the structural deck.

Flashings are continuations of the membrane; they may occur at the edge of a roof, where the roof meets a wall, at expansion joints, and where other appurtenances penetrate the membrane. Flashing problems account for the majority of roof leaks, and if flashings are improperly designed, they can be a continuous source of problems.

The guidelines presented in this chapter will provide those charged with roof maintenance with the information necessary to competently inspect roofs, identify common problems, and correct minor defects.

RECORD KEEPING

Good maintenance begins with good records. An effective roof record-keeping system has never been more important than it is today. Many roof systems are totally incompatible with other roofing systems and products. Maintenance and repairs may not be possible without specific information concerning the particular roofing system on any given building. Also, health and safety precautions dictate that the conscientious facility manager keep good records on exactly what construction materials exist in any building. Additionally, it is very valuable to keep track of all roof activities and expenses.

CONVENTIONAL BUILT-UP ROOFING MEMBRANES

Conventional built-up roofing membranes consist of several layers of roofing felt that are rolled and broomed into hot bitumen (asphalt or coal-tar pitch) or, less frequently, cold-applied asphalt adhesives (usually spray-applied). The bitumen, which is the sole waterproofing agent, serves as an adhesive and provides a protective top coat for the membrane. In some instances, the asphalt may be modified by the addition of a polymer or rubber additives. Felts, which allow for the buildup of succeeding layers of bitumen, provide the necessary tensile strength to support the bitumen and distribute stresses.

There are three basic types of built-up roof membranes:

1. Smooth-surfaced, consisting of asphalt bitumen and organic, asbestos, polyester, or glass-fiber felts, with a top surfacing of asphalt, emulsion, or reflective coating
2. Aggregate-surfaced, consisting of asphalt or coal-tar pitch and organic, asbestos, polyester, or glass-fiber felts with a top surfacing of bitumen and $^3/_8$-in gravel or slag
3. Mineral-surfaced, consisting of asphalt bitumen and organic, asbestos, polyester, or glass-fiber felts, with a "capsheet" covering (heavier-weight felt which is factory-surfaced with small mineral granules)

Coal-tar pitch membranes, primarily made with organic felts (although more recently with glass-fiber felts), *always* have an aggregate over them and are seldom found on roofs with slopes of more than $^1/_4$ in/ft. Asphalt membranes can be made with any of the four types of felts but are now more commonly made of glass-fiber or polyester felts. Asphalt membranes are recommended for roofs with slopes greater than $^1/_4$ in/ft but are all too commonly found on roofs with slopes less than $^1/_4$ in/ft. Although asphalt membranes can be used on very steep slopes, they are seldom found on roofs with slopes over 3 in/ft.

MODIFIED BITUMEN ROOFING MEMBRANES

Modified bitumen roofing membranes are factory-laminated asphaltic membranes consisting of one or more layers of polymer- or rubber-modified asphalt reinforced with polyester fabric, glass-fiber mat, or a combination of both. They are sometimes applied over conventional glass-fiber felts or base plies. There are two basic types of modified bitumen membranes:

1. Torch-applied membranes, which typically use APP (atactic polypropylene) plastic modifiers, are, as the name implies, heated with open-flame torches to achieve adhesion at laps and to the substrate. APP-modified membranes usually do not have factory-applied surfacing and may or may not require additional field-applied surfacing.
2. Mop-applied membranes, which typically use SBS (styrene butadiene styrene) synthetic rubber modifiers, are rolled into hot asphalt or cold-applied asphalt adhesives similar to a conventional built-up membrane. SBS-modified membranes are available with and without factory-applied surfacing; however, those without factory-applied surfacing do require a field-applied surfacing over the final ply for ultraviolet (UV) protection.

Modified bitumen membrane surfacing options are very similar to those used for conventional built-up systems and include uncoated surface, coated surface, mineral granule surface, and aggregate surface options.

Smooth-surfaced membranes are either left uncoated (APP only) or may be field coated with emulsion and/or reflective coatings. Granule surfacing may be factory-applied or field-applied in mastics. Aggregate is always field-applied and may be loose or embedded in hot asphalt or cold-applied adhesives. Particular surface options may be required to meet certain code requirements.

SINGLE-PLY ROOFING MEMBRANES

Single-ply roofing membranes encompass a large range of roofing products and manufacturers which have proliferated in the roofing industry over the past 20 years. Generally, single-ply membranes are large, flexible, factory-manufactured sheets, usually of a homogeneous material, which may or may not include a reinforcing fabric. These large pieces (4 to 50 ft wide and up to 100 ft long) are then spread out and field-spliced together at the job site to form a continuous sheet of material over the entire roof area. Single-ply membranes can be fully adhered to the substrate with adhesives, mechanically attached with various fastening systems, or laid loose and ballasted in place with 10 to 15 lb/ft^2 of concrete pavers or round river rock. Generally, the fully adhered and mechanically attached systems do not require field-applied surface coating.

Single-ply membranes are generally grouped into three categories as defined by their chemical nature:

1. Vulcanized elastomers include EPDM (ethylene propylene diene terpolymer), neoprene (polychloroprene), and epichlorohydrin membranes. They are thermoset materials which cure during the manufacturing process and generally do not include reinforcing fabrics within the membrane. Vulcanized materials cannot be heat- or solvent-welded and therefore use bonding adhesives or tapes to adhere the material to itself.

2. Nonvulcanized elastomers include CSPE (chlorosulfonated polyethylenes, Hypalon, Dupont), CPE (chlorinated polyethylene), PIB (polyisobutylene), and NBP (butadiene-acrylonitrile polymer) membranes. They are uncured elastomers that may cure or vulcanize naturally with exposure to weather. Nonvulcanized membranes may or may not include a reinforcing fabric within the membrane. In their uncured state, most nonvulcanized membranes can be heat-welded together. As the membranes cure, bonding becomes more difficult, and only bonding adhesives will work.

3. Thermoplastics include PVC (polyvinyl chloride), modified PVC, and EIP (ethylene interpolymer) membranes. Thermoplastic membranes can be heat- or solvent-welded together and may or may not include a reinforcing fabric within the membrane.

PROTECTED ROOF MEMBRANE ASSEMBLIES

The protected roof membrane assembly, or "upside-down" roof system, places the waterproofing membrane directly on the structural deck and then insulates above the membrane using rigid extruded polystyrene insulation. The insulation is then ballasted in place using river rock or concrete pavers. The membrane used with these assemblies may be a built-up, modified, or single-ply membrane.

ROOF SURVEY AND DEFECT IDENTIFICATION

Roof inspections should be made at least twice a year—once each during the spring and summer plus additional inspections after major disturbances such as heavy storms, addition of new roof equipment, and repairs or modifications to existing rooftop equipment.

A walk-through examination of the building interior should be made prior to a roof inspection to document leaks or evidence of leakage. Although such an examination may be helpful, it is not always conclusive, since water often travels a distance from where it enters the roof to where it drips on the inside. Also, some interior water stains may be nothing more than condensation from cold pipes. Look for water-damaged ceilings and for stains on walls and the undersides of roof decks. If possible, also check the undersides of roof decks for deteriorated or unsafe areas. Once on the roof, check for the following defects:

Housekeeping

Debris, abandoned equipment, and unused materials should not be allowed to accumulate or be stored on the roof.

Surfacing

Surfacing defects include deteriorated coating, "alligatoring," exposed felts, and eroded granules, aggregate, or ballast.

- Deteriorated coatings are characterized by flaking, chalking or erosion, pinholing, and loss of adhesion.
- "Alligatoring" is surface cracking of the bitumen on a built-up roof. It resembles the pattern of an alligator hide, and in severe cases the cracks may extend through the full thickness of the surfacing bitumen to the felts. Some uncoated modified bitumen membranes exhibit surface crazing similar in appearance to alligatored bitumen.
- Exposed felts are recognized by the absence of any surfacing materials—the membrane felts are simply open to view. The areas affected may be small, due to local conditions such as wind or water erosion, or large, due to overall weathering and aging of the roof.
- Eroded granules, aggregate, or ballast is distinguished by the partial or total loss of granules, aggregate, or ballast, usually occurring in isolated instances.

Membrane

Membrane damage includes physical damage (holes, tears, cuts); fishmouths; loose, dry, or easily delaminated laps; membrane chalking, crazing, or erosion; roofing blisters, cracks, splits, and ridges; and failed, temporary, or incompatible repair materials.

- A fishmouth is a lifting at a lap that results in a short opening between the top and underlying membrane. Loose, dry, or easily delaminated laps are longer sections of overlaps that are not bonded or are inadequately bonded together.
- Membrane chalking, crazing, and erosion are all signs of long-term membrane deterioration and can be recognized by the chalk-like substance which can be easily removed from the membrane, a fine checking of the membrane surface, and a wearing down of the membrane thickness, respectively.
- Blisters are raised areas of the membrane that have a bubble-like appearance. If pressed when warm, they have a spongy feel. Blisters usually form between membrane layers, although they sometimes form between the membrane and the substrate.
- Ridges are long, narrow raised portions of the roof membrane with a maximum height of about 2 in. Ridges usually occur directly above insulation board joints. When they occur over all the joints, the situation is called "picture framing." Sometimes ridges and cracks are also associated with dishing or cupping of the insulation. Ridges are frequently located at drains or other roof projections in loose-laid and partially attached single-ply membranes.
- Splits are very narrow tears that go all the way through the membrane. They vary in length from a few feet to the length of the roof and in width from a hairline to over an inch. When inspected closely, each side of the split has felts that are ragged and look as if they had been pulled apart. Splits generally occur directly above the joints between the long sides of insulation boards. They also can occur at reentrant corners of buildings—for example, at the intersection of the wings in an L- or H-shaped building. Splits occur less frequently with glass-fiber and polyester felts than with other types of felts and are even less common with single-ply membranes.

- Cracks are breaks in the membrane that result when ridges are flexed together and apart until they finally break open. Unlike splits, the felts on each side of the crack do not appear to have been pulled apart.

Membrane Flashings

Flashings are continuations of the membrane that provide watertight tie-ins of the roof membrane to adjacent building components. Flashings occur at the edge of a roof, where the roof meets a wall, at expansion joints, and where other appurtenances penetrate the membrane. Flashing problems account for the majority of roof leaks, and if flashings are improperly designed, they are a constant source of maintenance.

- Check flashing height. Flashings should extend at least 6 in and preferably 8 in above the roof membrane.
- Check for holes, open gaps at the top, open laps or vertical seams, splits or tears, general looseness from substrate, and delamination and slippage.
- Check for surface damage to materials. Deteriorated mineral-surfaced flashings will have lost much of the mineral surfacing. Exposed organic felts will have a dark and fuzzy appearance. Exposed asbestos felts will appear light gray. Exposed glass fiber felt will appear light gray with a noticeable pattern of the glass-fiber strands.
- Check for proper fit and wrinkling. If a flashing has a hollow sound when tapped or moves when pushed, it is loose from the wall and may be delaminated. Such delamination may occur between the entire flashing and the substrate or between plies in the flashing. Diagonal wrinkling of the flashing also may be noted, particularly near building corners or near the ends of walls.
- Check also for sagging of the flashing top edge.
- Check for pitting, chalking, crazing, erosion, and embrittlement of the membrane on single-ply flashings.
- Check membrane flashing and stripping materials at edging and penetration flashings. Inspect for tears, ridges, fishmouths, and other evidence of movement. Also check materials for looseness.

Roof Accessory Components

- Check for watertightness at roof penetration flashings.
- Check drain assemblies for proper drainage, broken or missing components, and dirt and debris.
- Inspect gutters, conductors, and conductor heads for rusted, loose, crushed, or missing components, broken or open joints, and dirt or debris.
- Check the level and condition of fill material in pitch pans (fill material may be dry or cracked and may be separated from the pan side or penetration).
- Inspect expansion joints for crushed or loose components and for holes and breaks.
- Inspect metal edgings for breaks and deterioration of membrane stripping materials and looseness of metal.
- Check metal flashings for attachment, condition of sealant at joints (check for looseness or splits), and deterioration.

Masonry

- Examine parapets, copings, and other walls above roofs, and note any cracks in the masonry or any defective mortar joints.
- Check for spalled or cracked brickwork.

ROOF REPAIR MATERIALS

Roof repairs may consist of hot- or cold-applied materials. Hot repairs may involve the use of heated bitumen, torches, or hot air welding guns and are best left to a professional roofer. Cold repairs may involve the use of roofing or flashing cements; felts, fabrics, and coatings; cleaning fluids; solvents; self-adhering modified bitumen membranes; single-ply membranes; and adhesives.

The repair materials and procedures presented in this chapter are essentially cold-applied. The asphaltic materials listed below are produced by most of the major roofing materials manufacturers to meet certain industry standards. As such, they are widely and readily available. The indicated cements are asphalt-based, and the felts and fabrics are asphalt-saturated and are suitable for cold-applied repairs on either coal tar- or asphalt-based built-up and modified bitumen systems. In sharp contrast, EPDM, neoprene, CPE, CSPE, PIB, NBP, PVC, and EIP single-ply membranes are not compatible with asphaltic materials. Additionally, the various types of single-ply membranes, solvents, and repair materials may not be compatible with each other and are not readily available. Some roofing products may be harmful, dangerous, or toxic. Product information data such as Material Safety Data Sheets (MSDS), precautionary statements, and product use instructions are available from product manufacturers and must be studied carefully and followed with any roofing product. In most cases, nothing more than cold-applied emergency repairs should be attempted.

For repairing built-up or modified bitumen membranes:

Asphalt primer: ASTM D41 (American Society for Testing and Materials)

Aluminum roof coating: ASTM D2824, type III

Flashing cement: ASTM D4586, type I or II, material manufacturer's standard (may be the same as asphalt roof cement)

Asphalt roof cement (plastic cement): ASTM D4586, type I

Glass-Fiber felts: ASTM D2178, type IV

Woven glass fabric: ASTM D1668, type I (asphalt-treated)

Modified bitumen membrane: Material manufacturer's standard

Portland cement (powder)

Galvanized roofing nails (with separate tin caps, if possible)

The following tools also should be made available:

Tape measure or folding rule

Spud bar or straight-clawed hammer

Trowels—pointed and square

Knife and chisel or hatchet

Stiff broom and a square-ended shovel

Wire- and soft-bristle brushes

Rope, gloves, goggles, and rags

For repairing vulcanized elastomers (EPDM and neoprene):

Cleaning solvent and splice primer: Manufacturer's standard

Splice adhesive: Manufacturer's standard

Lap sealant: Manufacturer's standard

Membrane: Manufacturer's standard

Emergency repair materials: Duct tape or silicone sealant reinforced with a nonasphaltic, inorganic fabric

The following tools also should be made available:

Tape measure or folding rule
Scissors
Heavy-duty hand roller
Caulking gun
Solvent-resistant paint brushes
Clean, natural-fiber rags

For repairing nonvulcanized elastomers (CSPE, CPE, PIB, and NBP):

Cleaning solvent: Manufacturer's standard

Lap primer (if applicable): Manufacturer's standard

Uncured (fresh) membrane: Manufacturer's standard (Additional solvents, primers, and bonding adhesives may be required if membrane has cured and is no longer weldable; consult membrane manufacturer.)

Emergency repair materials: Duct tape or silicone sealant reinforced with a nonasphaltic, inorganic fabric

The following tools also should be made available:

Tape measure or folding rule
Hot air gun
Scissors and silicone hand roller
Solvent-resistant paint brushes
Clean, natural-fiber rags

For repairing thermoplastic membranes (PVC, modified PVC, and EIP):

Cleaning solvent: Manufacturer's standard
Lap cleaner (if applicable): Manufacturer's standard
Welding solvent: Manufacturer's standard
Membrane: Manufacturer's standard
Bonding adhesive (if applicable): Manufacturer's standard
Emergency repair materials: Duct tape or silicone sealant reinforced with a nonasphaltic fabric

The following tools also should be made available:

Tape measure or folding rule
Hot air gun
Scissors and silicone hand roller
Solvent-resistant paint brushes
Clean, natural-fiber rags

EMERGENCY ROOF LEAK DETECTION AND REPAIRS

Temporary repairs are made to prevent water damage until permanent repairs can be effected. The leak source must be located as quickly as possible to protect the building and its contents.

In the event of a sudden, serious leak, follow these procedures:

- Determine if all drains and outlets are functioning properly. If the roof is ponded with water due to clogged drains or gutters, clear debris that is stopping water flow.
- Inspect the roof membrane, flashings, and accessory components in the vicinity of leaks for defects, as previously noted. In particular, inspect penetration flashings in the leak area.

On aggregate-surfaced or ballasted roofs, carefully remove loose aggregate or pavers so that the membrane is exposed. Caution is required to avoid overloading the roof structure when temporarily shifting the rock or pavers.

- Check the depth of water against curbs, penetrations, and flashings—water may be splashing up over these areas and leaking into the building rather than leaking through the roof membrane.

If the roof deck is poured concrete or gypsum, or if an underlayment is present above the deck, the only sure way to pinpoint a leak is by trial and error.

- If the deck is of unit construction (panels, planks, or panel slabs), the water will usually leak through the deck joint closest to its membrane entry point. Otherwise, the quickest way to locate roof leaks is from the underside of the roof deck. If ceilings are plastered, look for access openings into the ceiling space.
- If leakage occurs only during rains accompanied by heavy winds from a certain direction, check bituminous and metal flashings and walls. Look for cracks, splits, or openings that could allow water to enter behind and around such defects and into the roof system.

For the repair of visible underwater defects, trowel-apply plastic cement to the crack or hole. (The immediate area of a single-ply membrane contaminated with plastic cement will eventually have to be removed and replaced.)

- Reinforce larger defects with felt or fabric set into and coated with plastic cement.

Light applications of dry Portland cement also may be used as an emergency repair; the cement must later be removed to permit permanent repairs.

- If the defect cannot be seen but is thought to be in an area of ponded water, lightly sprinkle some Portland cement into the water. The cement may be carried by the water to the opening, where it will build up. The cement can then be applied over and around the opening to plug or slow down the leak. Portland cement, as a temporary repair procedure, should only be used when other methods have failed. It should not be used over large areas of a roof.

If water is found to be entering behind a metal counterflashing, use caulking, sealant, or plastic cement, whichever is considered most appropriate, to restore watertight integrity to the flashing and, hopefully, to serve as a permanent repair.

For repair of defects on merely wet or damp surfaces, use felt or fabric plies set into and coated with plastic cement, as previously mentioned.

Temporary repairs must be removed prior to the application of permanent repair materials.

PERMANENT REPAIRS

Unlike emergency or temporary repairs, permanent repairs are made to predetermined and readily recognizable defective areas. Proper preparation of the repair area is vitally important to the long-term effectiveness of the completed repair. As a minimum,

1. Remove surfacing (aggregate, ballast, coatings, and so on) down to the bare membrane. On mineral-surfaced roofs, lightly scrape the surface to remove unembedded granules.

2. Brush away all loose materials, dust, dirt, and debris.
3. Cut out all deteriorated membrane. On built-up, modified, and fully adhered single-ply roofs, also cut away all loose or unbonded membrane. Do not remove any more roofing than necessary to make the repair, and try to keep cuts straight and neat.
4. Sweep the repair area clean.
5. Coat the repair area with appropriate primer or clean with the appropriate solvent, including metal and masonry surfaces, as applicable with the specific membrane product, and allow to dry.

Built-Up and Modified Bitumen Membranes. Most built-up and modified bitumen membrane repair procedures are similar. All roofing defects should be covered with two layers of reinforcing material set into plastic cement—first a fabric, then a felt on top.

1. The fabric should be cut at least 2 in larger than the defect on all sides. The top felt should overlap the fabric by at least 4 in on all sides.
2. Center and embed each fabric and felt into a trowel coating of plastic cement over the defect. Press firmly and smoothly into place until plastic cement is seen at the edges.
3. Trowel coat the top felt with plastic cement, and trowel 4 in past the felt on all sides. Smooth it out to a feathered edge onto the existing membrane. The plastic coating covering the felt should be about $1/8$ in thick.

Coated Smooth-Surfaced Roofs and Mineral Surfaced Roofs. On these surfaces, wait 1 month, and then coat the patched areas with aluminum roof coating. On an aggregate-surfaced roof, wait 1 week, and check the quality of the patch.

1. Touch it up as necessary, and then trowel a $1/4$-in layer of plastic cement over the entire repair area.
2. Embed aggregate into the cement to match the existing material. The aggregate should completely cover the plastic cement.

Vulcanized Elastomers

1. Abrade surface and remove all surface oxidation.
2. Apply primer if specified with particular membrane.
3. Apply lap adhesive to both surfaces to be bonded and allow to dry until tacky but does not transfer to finger.
4. Roll new membrane into place, starting from one side and rolling toward the opposite side so as not to entrap air.
5. Rub into place and roll with a steel roller.
6. Seal edges with appropriate lap sealant.
7. Resurface as appropriate to match existing.

Nonvulcanized Elastomers. If a nonvulcanized elastomer sheet has not cured, follow the same repair procedure as for vulcanized elastomers, except substitute a hot air gun and heat in lieu of the lap adhesive in step 3. Heat temperature should be hot enough to begin softening the membrane after approximately 1 min of heat with the gun 1 in away from the membrane. If the membrane begins to discolor before softening, the temperature is too high or the membrane has cured. If the membrane does not soften after 1 min, the temperature is too low or the membrane has cured. Repair cured membranes by following the procedure for vulcanized materials.

Thermoplastic Membranes. Follow the same procedure used for nonvulcanized elastomeric membranes and heat weld fresh thermoplastic membrane to the existing membrane or use the manufacturer's recommended solvent welding compound in lieu of a heat gun.

Following are repair procedures for commonly encountered defects.

Blister Repair. Repair only blisters that are damaged or deteriorated to the point where water may leak into the roofing.

1. Cut out and remove all deteriorated, loose, or unbonded membrane. Exercise caution not to damage any sound materials which may be located beneath the blister or buckle. Trim away all deteriorated membrane above the blister cavity by cutting horizontally or vertically with a knife or scissors.
2. Look for water that may be inside the blister cavity. Sponge out any water, and wipe the cavity dry with rags.
3. Removal of blistered or buckled membrane generally leaves a depression in the membrane. This depression must be leveled before making the basic repair. Cut membrane filler pieces to fit the depression, and bond each piece into place with appropriate adhesive. Alternate membrane and adhesive until the repair surface is fairly level. Complete basic repair as previously outlined in this section.

Puncture and Hole Repair

1. Avoid making a lot of small patches close together. Use one or more larger patches to simplify your work.
2. Follow the basic surface preparation and repair steps outlined previously in this section.

Split, Crack, or Loose Lap Repair

1. Extend repair membrane at least 4 ft past the ends of the split, crack, or loose lap and at least 6 in on both sides.
2. Follow the basic surface preparation and repair steps outlined previously in this section.

Loose Fastener Repair

1. Remove the loose fastener, and avoid damaging the surrounding membrane.
2. Install new fastener close to the original fastener's position.
3. Cover the new fastener by following the basic surface preparation and repair steps outlined previously in this section.

Basic Repair for Flashing Defects. Flashing repair and membrane repair are basically the same. Follow the basic surface preparation and repair steps outlined previously in this section. If a metal counterflashing must be lifted to complete the repairs, do it carefully, and don't forget to put it back in place afterward. For built-up membranes, be sure to use *flashing* cement, where it is available, instead of plastic cement on any vertical flashing repairs.

Flashing That Is Sagging or Open at Top

1. Wire-brush or scrape off any roughness from the wall.
2. Coat the area to be repaired with primer, and let the primer dry.
3. Apply a coating of flashing cement or appropriate bonding adhesive to the wall behind the loose or sagging flashing and on the back of the flashing membrane.
4. Press the flashing firmly into the coated wall surface, and roll with appropriate roller.
5. Refasten the top of the flashing to the wall to match existing flashing.
6. Reinstall metal counterflashing or termination bar, or seal the top edge with a minimum 4-in-wide strip of fabric set into and coated with a compatible sealant so that the fabric and sealant completely cover and seal all seams, the top edge, and all fasteners.

Flanges of Metal, Gravel Stops, Pitch Pockets, Gutters, and so on. The membrane stripping over metal flanges often has ridges, cracks, splits, or tears. Repair these defects as follows:

1. Remove any surfacing material back about 12 in onto the membrane and 12 in on each side of the defect.
2. Carefully remove defective membrane stripping (it may be cracked, loose, or deteriorated). Take care to avoid damaging or removing any underlaying membrane.
3. If practical, check the flange fastenings; the problem could be that there are not enough fasteners to effectively restrain movement of the metal. Use more nails or wood screws, as necessary, so that the flange is secured about every 4 in. Keep fasteners back at least 4 in from metal overlaps or joints.
4. Strip-in the flange with two layers of reinforcing material set into plastic cement—first a fabric and then a felt on top.
    a. The fabric should be cut at least 2 in larger than the defect on all sides. The top felt should overlap the fabric by at least 4 in on all sides.
    b. Center and embed each fabric and felt into a trowel coating of plastic cement over the defect. Press firmly and smoothly into place until plastic cement is seen at the edges.
    c. Trowel coat the top felt with plastic cement, and trowel 4 in past the felt on all sides. Smooth it out to a feathered edge onto the existing membrane. The plastic coating covering the felt should be about $\frac{1}{8}$-in thick.

On coated smooth-surfaced and mineral-surfaced roofs, wait 1 month and then coat the patched areas with aluminum roof coating. On an aggregate-surfaced roof, wait 1 week and check the quality of the patch.

1. Touch it up as necessary and then trowel a $\frac{1}{4}$-in layer of plastic cement over the entire repair area.
2. Embed aggregate into the cement to match the existing material. The aggregate should completely cover the plastic cement.

Leaking Gutter Joints. To seal open joints or seams inside the gutter,

1. Wire-brush the metal clean on both sides of the joint.
2. Seal with caulking compound.
3. If it is *not* an expansion joint and the ends move freely, tighten them together with pop rivets or sheet-metal screws.
4. Cover the joints with two layers of fabric and plastic cement.

Metal-Base Flashings and Counterflashings. Metal-base flashings are more commonly found on older built-up roofs and PVC single-ply roofs. Defects in metal-base flashings are similar to gravel-stop defects. Another common defect in metal-base flashings is cracked or split soldered or welded joints on PVC-clad metal.

Repair joints at horizontal metal flanges of metal-base flashings in the same manner as gravel stops. To seal cracked and open vertical joints,

1. Wire-brush the vertical joint to clean off dirt and old materials.
2. Prime the metal on each side of the joint, and tape the joint with two layers of fabric, a 4-in width first, followed with a 6-in width, troweled on with flashing cement.

Repair defects on PVC-clad metal used with PVC single-ply membranes by heat welding following the procedures outlined in the basic repair section of this chapter.

Metal counterflashings are sometimes bent out of shape. Straighten these using a hammer and a small wood block to spread the hammer blow.

Pitch-Pan Repair

1. If the pitch pan is deteriorated, it should be replaced. Otherwise, clean out the pan, removing all loose or deteriorated filler material, dirt, rust, and other debris.
2. If the roof-penetrating component has corroded, scrape or wire-brush it clean.
3. Coat the inside of the pan and roof-penetrating component with a brush coat of primer, and let it dry.
4. Fill the pan with plastic cement or other pitch-pan pourable sealant appropriate for use with your particular membrane system. Slope the top to shed water.

Upon completion of repair work, check the entire roof area for debris, equipment repair parts, and any excess materials, and remove such materials from the roof.

MAINTENANCE PREVENTION

Most roof problems stem from either poor design or poor initial installation. A properly designed and installed roof requires little maintenance and should last for 20 or more years. The following guidelines should be considered to minimize the maintenance of a roof system.

Housekeeping

This is the routine removal of debris from the roof, such as trash thrown onto the roof or dropped from windows and the leftovers from mechanics' work on rooftop equipment, such as scraps of wire, rags, pails, paint cans, and broken glass. Such debris can float into drains, causing a stoppage, with subsequent ponding and flooding over roof edges and flashings of roof penetrations. Do not pile up debris for later removal or place debris in barrels on the roof. The debris could be blown off by strong winds, causing damage or creating a hazard to people passing below. Do not place permanent trash barrels on the roof.

Cleaning of Roof Surfaces

This is the cleaning required because of the natural buildup of dust and leaves on the roof. Sometimes it may include heavy, deep snow and ice, which can overload the roof or block the drainage system.

Aggregate-surfaced and ballasted roofs seldom need cleaning, except for housekeeping. Aggregate or ballast that piles up because of wind erosion should be removed if it cannot be reused to resurface eroded areas.

Excess aggregate/ballast piles should be leveled and dressed up to provide a uniform roof surface. Aggregate/ballast should be removed from all roof drain sumps, scuppers, and gutters, as should leaves and other foreign matter.

Any buildup of dust and dirt at the bottom of base flashings should be swept away. These conditions are generally found in wind-sheltered areas and can lead to early flashing deterioration. Check higher roofs, since discharge from some can damage the membrane.

Smooth-surfaced roofs are usually washed clean with each rainfall. However, you can occasionally find areas where leaves, dust, dirt, and other debris are not removed by rain. In this case, brooming or hosing down the area may be necessary. When hosing down the area, direct the stream of water away from membrane laps so that water will not be forced between the laps. Do not attempt to flush such debris through roof drains or gutters—it may clog them. Remove and dispose of loose materials; do not use a shovel, except as a dust pan, because it may cut the roofing membrane.

Snow and ice accumulations heavier and deeper than normal should be investigated. They can be signs of a frozen or otherwise clogged drainage system, or because of the buildup, the thawing water may be forced up over the roof flashings and cause leaks. Never attempt to remove all snow from the roof surface. Leave several inches of snow between your shovel and the roof; your shovel can easily damage the roofing membrane, especially at lower temperatures, since most membranes will be hard and brittle. Ice formations, such as icicles, may be broken with a tool such as a hammer or mallet for easy removal. However, ice on roof surfaces or in roof drains and gutters should not be hammered or chopped. Roofing can be cut or fractured and metal bent or cracked. Use a noncorrosive chemical deicer or hot water to melt the ice or loosen it for easy removal.

Do not use any kind of oil-base cleaning fluids on any roofing surface. They can attack and soften or dissolve the membranes and cements used in all roofing systems.

Roof Drains

Roof drains, gutters, downspouts, scuppers, and the like should be cleaned on a regular basis, since water backup has a potentially catastrophic consequences. All drains should be checked and the strainers cleaned at least twice a year—more often in the fall if trees are growing nearby. Never close or block a roof drain unless authorized to do so for repair purposes.

Corrosion and Rusting of Metals

When galvanized sheet metal loses its protective coating and starts to rust, it is necessary to clean it thoroughly and protect it with paint. Brighten it with coarse steel wool, and then paint it with a yellow chromate primer. After the primer is dry, paint with a red iron-oxide paint; when this is dry, paint with a good grade of aluminum paint.

Storage of Equipment and Materials

A roof should not be used for the storage of equipment and materials. All materials should be stored in properly protected dry-storage facilities.

Physical Protection

When it is necessary to perform any work on the rooftop, protection boards, such as plywood, should be used to prevent accidental damages (e.g., from falling materials or dropping of tools).

Roof Traffic

All unnecessary roof traffic by facility personnel or outside contractors should be avoided. Roof traffic should be restricted to roof walkways or to areas that are easily maintained. Access should be limited to authorized operational or maintenance personnel. Rolling traffic, such as powered carts, wagons, wheelbarrows, and hand trucks, should be controlled and limited to special use by trained operators over designated and protected traffic lanes. Ladders and scaffold legs should be placed on pads or planks so that they do not gouge the roofing. Paint and oil cans should be placed on tarpaulins or plastic sheets to avoid spillage onto roof surfaces.

Cutting Holes Through a Roof

If holes are over 8 in. in diameter, construct a header and stretcher beam system to prevent roof sag.

New Equipment

New equipment mounted on the roof must be properly flashed and mounted at a distance above the roof so that it can be easily serviced and the roofing can still be maintained under the equipment. Where rooftop mechanical units will need regular servicing, install walkways to protect the roofing membrane from damage.

REROOFING

When repair work becomes ineffective or becomes a task other than treatment of merely localized areas, reroofing will need to be considered. Two methods of reroofing are common.

Superimposition

Superimposition is the practice of leaving the existing system in place and installing a new roof system over it. The superimposed system should be a complete roofing system incorporating at least one layer of roof insulation to separate the new system from the existing roof. The condition of the existing membrane surface and its proper preparation prior to overlaying the new system are critical considerations in electing this option.

All defective existing roof systems must be considered as having a potentially high moisture content, which may eventually cause the superimposed system to deteriorate. Therefore, it is always advisable to separate the existing system from the new system with a layer of insulation.

Attachment of the existing roof system is also critical. A poorly bonded underlaying roof will subject the new overlay system to possible wind uplift, as well as unrestricted lateral shrinkage and movement.

Additionally, weight load is of paramount importance, especially for wood-framed or other lightweight construction and in areas subject to heavy snow loading. Local codes may limit the number of separate roofs allowed to be kept in place.

Replacement

If the existing roof system has been recoated, spot-repaired, recovered, and patched for many years and has begun to deteriorate, the safe and economical thing to do is to replace it with a new system. The additional cost of complete removal is of small concern when compared with the possible damage resulting from leaving a deteriorating roof in place.

Replacement typically includes the complete removal and disposal of roofing components down to the structural substrates, including flashings and roof-related metal accessory components. This method of roof replacement allows for examination of the structural substrates, repairs as necessary, and optimal flexibility in the redesign of those elements which may have contributed to the failure of the original system.

Caveat. If the roof is under warranty, contact the installing roofing contractor and/or the materials manufacturer before proceeding with any repair work. Additionally, as recommended by the AIA's professional liability committee and as required by most federal agencies, employ a qualified roofing specialist consultant on any major projects.

CHAPTER 2
CONCRETE INDUSTRIAL FLOOR SURFACES: DESIGN, INSTALLATION, REPAIR, AND MAINTENANCE

Robert F. Ytterberg
President, Kalman Floor Company, Evergreen, Colo.

In the modern industrial building, probably nothing gets tougher use than the floor. The serviceability of the floor surface can affect the entire operation of the plant. Functions such as materials handling and standards such as plant cleanliness are both affected by the quality and performance of the floor.

Ideally, you want to "get it right the first time" by installing the best possible floor at the outset. But in many situations you may not have the opportunity to install a brand new floor or even to resurface an existing floor. For those situations, there are various surface treatments for concrete floors, as well as protective maintenance steps, that can help. This chapter addresses both new floor installation and maintenance of older floors. The chapter is divided as follows:

1. New concrete floors
    a. Cost of floors
    b. Water-cement ratio
    c. Earth subgrade and slab design
    d. Shrinkage and joints
    e. Reinforcing and crack control
    f. Obtaining abrasion resistance
    g. Monolithically finished floors
    h. Separate floor toppings
    i. Superflat floors
2. Resurfacing old floors
3. Maintenance and cleaning floors
4. Repair of dusting floors
5. Floor sealers and finishes
6. Corrosion-resistant floor toppings

NEW CONCRETE FLOORS

Concrete floors can and should be trouble free. When an industrial concrete floor is properly designed and installed, it will last for years, require little maintenance, perform brilliantly, and stay beautiful. It will actually be polished, not destroyed, by use. It can have a nonporous, nondusting surface so that it needs no sealant. It will withstand spills, and oils and chemicals won't soak in but instead will stay on top for easy wipe-up.

But it must be done right. Superior workmanship and highly trained supervisors are necessary for the proper installation of industrial floors. Because a concrete floor is job-built and on-site manufactured, installation must be done by a separate floor subcontractor, skilled in the art and paid sufficient money to provide the workmanship required.

Cost of a New Concrete Floor

As a general estimate, the total cost including all labor, material, and supervision, of a 7-m-thick, lightly reinforced concrete floor slab on ground for an industrial facility runs from \$3.75 to \$5.00 per ft^2 depending on the quality of the finished surface. The cost of a $6^1/_4$-in slab complete with a bonded $^3/_4$-in. Absorption process topping would be at the higher end of the estimate, but this "best floor" usually costs less than 1 percent extra for a \$50-per-ft^2 building and less than 2 percent extra for a \$25-per-ft^2 building than would a medium-duty monolithically finished floor. The initial construction cost for industrial concrete floors is not as telling a factor as their life-cycle cost. Some typical savings and benefits of a low-water deferred floor topping are

- It is longer lasting than other floors.
- It has fewer floor joints, resulting in less maintenance and fewer cracks.
- It has a consistent smoothness and flatness speeding material handling, reducing product damage and cutting forklift truck maintenance.
- The floor remains undamaged by construction because topping installation is deferred until major construction activities are complete.
- It is dust resistant, resulting in cleaner products and lower cleaning costs.
- Floor sealers are unnecessary.
- It is self-polishing, becoming reflective with use.
- It is resistant to pallet nail gouging.
- The surface density is comparable to quarry tile.
- The durability and appearance bring a higher resale value for the building.
- It is cost-efficient because of reduced maintenance.

Water-Cement Ratio

While the principle which governs the strength and durability of concrete—the water-cement ratio law—is widely known and respected, far too often during the installation of concrete floor surfaces its precepts are ignored. The water-cement ratio law states: "The strength of a concrete mixture depends on the quantity of mixing water used in the batch, expressed as the ratio of the volume of cement, so long *as the concrete is workable and the aggregates are clean and structurally sound.* The strength of the concrete decreases as the water-ratio increases."

Three things are involved in the law: (1) workability, (2) a low water-cement ratio to maximize the strength of the cement paste binder, and (3) a good aggregate. It should be noted that achieving both workability and a low water-cement ratio are contradictory goals.

To achieve full hydration, a 94-lb sack of cement needs only slightly more than 3.0 gal (25 lb) of water. This is a water-cement ratio of $25/94 = 0.27$. If, however, a water-cement ratio as low as this were used in concrete floors, the concrete mix would be unworkable and impossible to place, screed, and compact. Because of the lack of workability, the benefits of the low water-cement ratio cannot be realized. The void content of the floor would be so large as to reduce compressive strength and increase permeability. On the other hand, if a sufficiently large quantity of water is used to obtain workability, and if that water is left in the mix, strength will be reduced and the resultant drying shrinkage will cause excessive cracking and curling of the floor.

The primary function of the base slab is to support whatever loads—both static and dynamic—are placed on it. The lowest water-cement ratio consistent with workability will yield the highest compressive strength. Water-reducing admixtures are helpful in achieving high strengths, but they do not reduce concrete shrinkage in proportion to the water reduction because of a chemical effect. For reducing shrinkage to a minimum, the total water per cubic yard must be minimized. This can best be achieved by removing water after the concrete is placed.

Earth Subgrade and Slab Design

Unless a concrete floor is structurally reinforced, its load-bearing capacity is only as good as the load-bearing capacity of the subgrade on which it rests.

When concrete floors are laid directly on an earth subgrade, that subgrade should be uniformly stable. Frequently, the earth is of such composition that changes must be made even in the undisturbed earth cuts to take care of problems such as expandable material and water-bearing soils. It is particularly important to secure such compaction in all newly placed fills, in previously opened trenches, at foundation walls, and around column footings. The services of soil experts are indicated and justified in the design and especially in the construction of any concrete floor slab on ground.

The state of the art in design of floor slabs on ground is covered in an American Concrete Institute document written by ACI Committee 360. The several different slab-on-ground design methods all depend on an assumption about the earth subgrade. One of the design methods covered by the ACI is the Portland Cement Association's method, which is given in their publication 1S195.OID, "Slab Thickness Design for Industrial Concrete Floors on Grade."

It is recommended that floor slabs on ground have a minimum thickness of 6 in. Where the surface is to be a deferred $^3/_4$-in low-water/cement concrete floor topping, a 3000 psi minimum, 5-sack-mix concrete is sufficient for the base slab. Where the surface is to be finished monolithically, a 3500 psi, $5^1/_5$ sack mix is desirable to provide enough cement paste for trowel finishing. The design of each mix must be individually determined according to the availability of fine and coarse aggregates. But the tendency to oversand the mix in order to achieve easy workability and finishing should be avoided, since unnecessary shrinkage can result from the extra water needed in an oversanded mix.

Shrinkage and Joints

Inside a building where temperatures are never high enough to cause expansion, concrete floors always shrink from their original size, and therefore, control joints are required if indiscriminate cracking is to be minimized within the section of concrete floor placed each day.

About 90 percent of shrinkage takes place within 2 years after a concrete floor is placed if it is allowed to dry. Drying shrinkage is a function of the amount of water left in the concrete at the time the concrete hardens, and therefore, anything that can be done to reduce or remove water from the concrete before it hardens also will reduce later drying shrinkage.

Length change per unit due to drying shrinkage of unreinforced concrete is about 600 millionths (0.06 percent) or 0.72 in per 100 ft according to the Portland Cement Association's "Volume Changes of Concrete." Drying shrinkage of normally reinforced concrete is about one half the above amount.

Control joints, one quarter the depth of the floor slab, are sawed into each day's floor slab placement. The concrete cracks irregularly below the saw cut and thus provides load transfer across the joint.

Each day's floor slab placement (usually 5000 to 12,000 ft^2) should be separated into panels, approximately square, and no larger than 50 by 50 ft. The Portland Cement Association's "Concrete

Floors on Ground" recommends joint spacing in feet be no more than two to three times slab thickness in inches, but the initial and maintenance cost of so many joints is outweighed by the presence of some cracking. The most advantageous way of achieving the panel size is to saw in the necessary jointing the day after a pour has been made. Rapid day to night temperature drops, however, will necessitate making the saw cut the same day the slab is placed.

Construction joints are those joints that surround and contain each day's floor slab-placement area. Bulkheads should be straight, rigid, and vertical. Some means of transferring loads from one panel to the next must be provided. This may be done by forming a key in the bulkheads and/or by using steel dowels through the joint fixed on one side and free to move horizontally on the other. Steel dowels are the preferred method, as they provide more positive load transfer across the joint. Also, they tend to resist curling of the slab edges. Keys, on the other hand, are rarely in contact, one side with the other, after shrinkage occurs. Thus they do not provide a positive means of weight transfer, allowing free movement of the slab edge under traffic, which may result in severe cracking behind the joint. Because stripping the bulkhead away from a keyed joint is difficult, sometimes the upper lip of the key is broken.

The exposed edges of joints should be square. They should not be tooled, since this will leave a depression in the surface. The pounding of forklift trucks against such edges will result in rapid deterioration. Joints should be filled with a rigid material to protect the joint edges and also for appearance and sanitary reasons.

Reinforcing and Crack Control

Present design methods referred to above for floor slabs on ground provide only the slab thickness required to resist certain loads on a slab resting on a certain quality of subgrade-subbase. Although the design methods can be used to determine bending moments for the design of structurally reinforced slabs, the methods primarily assume that the floor will be unreinforced.

Reinforcing for floor slabs on ground, therefore, is not needed for structural reasons if the slab thickness is equal to that called for by the above design methods and if the subgrade under the floor slab is everywhere as good as was assumed in the design calculations.

For this condition, the only reason for reinforcing is to resist the shrinkage of the concrete floor and to distribute shrinkage cracks so that the cracks are so small that they are not an eyesore. Since the upper half of floor slabs on ground shrink more than the lower half, shrinkage cracking is always most severe in the slab surface.

Stiff reinforcing located in the upper half of the slab is therefore the best way to distribute shrinkage cracking. Our experience has indicated that $1/2$-in reinforcing bars (#4 rebar) spaced 18 or 20 in on center each way and located in the upper half (upper third if the slab is 6 in or more thick) of the slab is very effective in distributing shrinkage cracking.

The 18- or 20-in spacing of the rebar allows concrete laborers and finishers to walk between the bars, and the $1/2$-in diameter of the bars permits workers to walk on them without bending them. In our opinion, #4 rebar at 18 or 20 in is better temperature steel for shrinkage resistance than light-gauge welded wire mesh with a 4-, 6-, or 8-in spacing between the wires. It is extremely difficult to ensure that welded wire mesh made with light-gauge wires will be located in the upper half of a 6-in floor slab.

Shrinkage cracking also can be controlled by use of a shrinkage compensating admixture cement. If a shrinkage compensating admixture cement is used in the floor slab concrete, then reinforcing *must* be put in the upper half of the slab to restrain the expansion and to put the steel rebar into tension to pull the concrete together. Use of expansive shrinkage compensating cement concrete for floor slabs requires special knowledge and techniques but is very successful in controlling cracking if properly used. Daily expansion testing is a must. See ACI 223 for further information.

If a floor slab on ground is engineered so that reinforcing is needed for structural reasons (to increase the flexural strength of the slab), then rebar, not welded wire mesh, is recommended for this reinforcement. Other alternatives for flexural strength increase are posttensioning the slab or using a wire fiber concrete for the slab.

Obtaining Abrasion Resistance

A durable concrete floor surface cannot be produced merely by troweling it until the surface is smooth. Attention to a number of surface finishing details, however, will produce remarkably durable and cleanable floors. Figure 2.1 shows the relative abrasion resistance of the three basic types of concrete floor finishes.

Several items account for the difference shown in Fig. 2.1. The common monolithic concrete floor finish had no special aggregate embedded in the surface of the ready-mix concrete floor slab, and finishing consisted only of sufficient floating and troweling to make the surface smooth but not necessarily dense. The special monolithic finish had abrasion-resistant aggregates embedded in the floor surface before finishing; also, special finishing techniques were used. For the $3/4$-in-thick deferred topping, the major differences were special coarse and fine aggregates, more cement, and removal of all workability water by the absorption process after the mix was placed.

Less water = better abrasion resistance

Monolithically Finished Floors

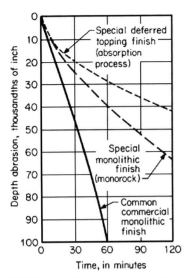

FIGURE 2.1 Relative abrasion resistance of three types of concrete floor finishes. Monolithic finish has hard rock just under the surface. Deferred topping has lower water-cement ratio and a still higher hard-aggregate content near the surface.

Monolithic or single-lift floors are usually given an ordinary, common-commercial trowel finish that provides only minimum wear resistance. Such a monolithic floor is not recommended for industrial buildings. Instead, for medium industrial use, the optimal type is a stone-densified, trowel-polished monolithic concrete floor finish with cement-coated aggregate applied to the ready-mix concrete floor slab surface on the day the concrete is placed.

The best way to densify a monolithic floor surface is to incorporate into the surface of the freshly screeded concrete a large amount of hard, tough stone aggregate (thoroughly mixed with cement with a minimum water-cement ratio) approximately $1\,1/2$ lb/ft^2. When this volume of stone aggregate is embedded into the concrete, the topmost area of the slab is densified in relation to the volume of the material embedded.

The slump of the base slab concrete should be no more than 4 in when it is placed. Otherwise, the cement-coated stone aggregate floor hardener material may sink below the surface where it will be useless.

Other types of stone or mineral aggregate—floor hardeners—can be used for this operation. However, the materials of themselves have no intrinsic hardening quality. They must be incorporated into the surface properly and in sufficient quantity. Finishing techniques for monolithic floors are the same as for deferred toppings and are, therefore, discussed below.

Separate Floor Toppings

The application of a separate $3/4$-in-thick concrete floor topping to a $6\,1/4$- or $7\,1/4$-in base slab that has been previously placed produces a wearing surface that is much stronger than any that can be installed monolithically (Fig. 2.2). The water-cement ratio can be controlled effectively in a separate topping, and specially selected tough, coarse aggregates can be used. The absorption process uses an initial water content sufficient for workability and full hydration of the cement and then removes this water to achieve a theoretically correct water ratio of about 0.30 to 0.35 by weight. By comparison, the average ready-mix concrete has a water-cement ratio of about 0.50 to 0.60 by weight.

The base concrete slab under the topping is struck off $3/4$ in below finish floor grade. When it is partially hardened, the surface of the slab is steel-wire broomed to expose the coarse aggregate. A second brooming is carried out later the same day when the laitance clinging to the coarse aggregate will brush off and show clean stone. If this second brooming does not reveal clean stone, then a third brooming is required. The slab must have a rough texture, but partially exposed coarse aggregate

FIGURE 2.2 The application of a separate $3/4$-in, low water-cement ratio floor topping to a previously placed concrete base slab produces a wearing surface that is denser and more wear resistant than monolithically placed floors.

should not be loose. Since the topping may be installed up to several months after the slab has been poured, this brooming technique is necessary to ensure full bonding of the topping.

The day before the floor topping is applied, the base slab is water-saturated to prevent the slab from drawing too much water from the topping while it is curing. A scrub-in coat of cement grout is then applied. Grout is the adhesive agent that bonds the topping to the base slab.

The aggregate used in a separate floor topping is a basaltic or granitic rock that has been tested for hardness, toughness, and soundness. Each cubic yard of topping mix contains almost a full cubic yard of stone aggregate by loose volume. This volume of stone will have a void content of approximately 40 to 50 percent. Practically all these voids should be filled by fine aggregate, and the voids in the fine aggregate are filled by the cement-water paste. The precise batching will depend on the sieve analysis and specific gravities for sand and stone.

Here is how the absorption technique of installation works: The topping mix is installed relatively wet. This permits maximum workability during placing and straight-edging. After the topping mix is screeded, there is no longer any need for workability. The water needed for this purpose is removed. On large jobs, water removal is accomplished with vacuum pumps and mats except that the floor around the perimeter of the mats must be dried with burlap and cement.

On small jobs, water is removed from the entire floor topping area by placing a burlap blanket on top of the freshly screeded topping mix and then spreading a drier material on the burlap which will remove the sufficient water by capillary action. When the topping mix is dry enough to support the weight of a man without indentation, the drier material is removed. The topping is then returned to a trowelable state by our patented vibratory disc floats.

From this point on, installation steps for both special monolithic and deferred toppings are identical: To make the surface as flat as possible (consistent with maximum production) and to increase abrasion resistance, a cast-on of grit-sized aggregate should now be made. This is then cross-rodded in—straight-edging being performed at a right angle to the original rodding.

The next step, blade floating, is probably the least understood step and perhaps the main reason for many surface failures. Blade floating does two things. First, by going over the floor at right-angle passes, minor elevation differences have a tendency to be corrected. Second, and far more important for durability, blade floating keeps the surface open.

When the concrete begins to lose its plasticity, the floating is stopped and troweling started. Troweling squeezes out more water and tightens the surface structure, increasing abrasion resistance and imperviousness, and should be continued until the surface shines.

It is not enough merely to trowel the floor until it is smooth. This is called "a 4:30 finish," meaning the floor is left in whatever condition exists at the end of an 8-hr work day. The value of extensive and proper troweling (at overtime labor rates except in hot weather) cannot be overemphasized.

Proper curing of the floor after it has been troweled is too frequently neglected. The concrete should be kept moist so that the cement will continue to combine chemically with the water. This curing process should be begun as soon as possible. Delay will cause rapid evaporation in the early stages, with resultant cracking, crazing, or dusting of the surface. Where floors are laid exposed to the sun, it is necessary to start curing the same evening with a moisture spray or vapor. The longer the concrete can be kept wet, the stronger, more impervious, and more wear-resistant the floors will be. The floors should not be used at all for at least 5 days and thereafter only lightly for an additional 10-day period.

If properly installed, this deferred topping should never need to be sealed.

Superflat Floors

The phrase *superflat floors* has come into play in the last 5 to 10 years with the development of the narrow aisle, high-stacking turret truck which automatically stacks pallets into racks on either side of 6-ft-wide aisles as high as 40 ft above the floor. The aisles for turret trucks must be extremely flat for these rigid suspension trucks to operate properly.

We have found through the installation of many floors for turret trucks that a tolerance of $1/8$ in. in 10 ft is necessary for these trucks to operate properly. The Raymond Company, one of the larger manufacturers of these turret trucks in the United States, specifies that the floors for their turret trucks have a tolerance of $1/9$ in. in the 8-ft-long wheelbase of their turret trucks and $1/10$ in. across the 5-ft-wide load wheel dimension of their turret trucks. A $1/8$ in. in 10-ft tolerance will automatically give these other tolerances.

To get a $1/8$ in. in 10-ft superflat floor, the floor must be installed in narrow strips. Floors for turret truck use are installed in 14-ft-wide strips which are equal in length to the length of the narrow aisle—usually 200 to 400 ft long. This width of 14 ft plus a few inches is the usual center-to-center distance between racks in narrow-aisle rack configurations.

The reason for pouring only a 14-ft-wide strip is to put maximum effort into obtaining a flat surface in the 6-ft-wide aisleway in the center of the strip. Pouring only 14-ft-wide strips permits the cement finishers to straight-edge and restraight-edge the floor without walking on the floor.

The temporary bulkheads used to contain the concrete on either side of the narrow strips must be set to exact elevation. If the top edges of the bulkheads have steel strips bolted to them, it is possible to get the tolerance with careful workmanship and supervision. These "bulkhead" or "construction" joints will later be hidden when the steel racks are installed.

Fourteen-foot-wide floor strips with a $1/8$ in. in 10-ft flatness can be either monolithic or a separate deferred topping floor bonded to a previously hardened base slab. If a separate deferred floor topping is used as a superflat floor, then the slab under the topping can be poured the full column bay width and only the topping need later be poured in 14-ft-wide strips. This eliminates the curling problem that occurs at slab construction joints.

Sometimes superflat floors are achieved by minimizing the application of surface aggregate hardeners and by minimizing the amount of trowel finishing. Thus, many superflat floors suffer from a lack of surface abrasion resistance because trowel finishing is minimized so as not to disturb the surface flatness.

Vacuum dewatering of superflat floors, however, will both reduce the shrinkage curling at the construction joint edges of the strips and will also enhance the surface wear resistance of the floor. A vacuum dewatered floor will have twice the wear resistance of a minimally finished nondewatered floor because unlike conventional concrete or even superplasticized concrete, vacuum dewatered concrete has its lowest water-cement ratio rather than its highest water-cement ratio in the upper half of the floor slab.

RESURFACING OLD FLOORS

If you are considering modernization or rehabilitation of an older building or restructuring any building to new uses, it will pay for you to investigate the advantages of installing a new $3/4$-in floor topping over the old floor. Generally, resurfacing economies start in the area of 25,000 ft^2. Resurfacing is recommended prior to occupancy because of unavoidable dusting conditions that result from scarification of the old floor surface.

A leading industrial firm saved almost a year in manufacturing time, and money on new construction as well, by turning a light industrial structure into a heavy manufacturing area with a tough new floor surface. Another firm ripped out looms and converted a textile manufacturing building into a sanitary pharmaceutical plant with a smooth, dense surfaced concrete floor topping.

These are only two examples; there are many more. In all cases, the one, single key element that helped make modernization practical and economical was the ability to get a brand new smooth floor by resurfacing with topping. A good easily cleaned working platform is vital in any industry, and floor resurfacing is the answer to low-cost plant modernization.

The old floor surface is mechanically scarified to expose clean, sound concrete for proper adhesion of the new absorption process topping. On the day the topping is to be placed, a bonding preparation is thoroughly scrubbed into the base slab to provide a bond between the slab and the topping. After the topping mix is placed, it is screeded to the line of finish elevation. It is relatively wet and workable, permitting use of a maximum amount of tough coarse aggregate from specially selected rock quarries.

After the surface is screeded, the absorption process is performed by the use of absorption blankets or special vacuum equipment. The resulting low water-cement ratio provides a high-strength, nondusting, cementitious binder that can hold the tough aggregate in place under a high frequency of traffic.

After completion of the absorption process, low-amplitude, high-frequency vibratory float brings moisture to the surface so that finishing operations can proceed on this low water-cement-ratio, dense topping.

Extensive troweling densifies and tightens the surface until it is burnished. The absorption process floor is an easily cleaned, sanitary surface that is actually self-polishing.

MAINTAINING AND CLEANING FLOORS

Properly constructed concrete floors will require little maintenance other than cleaning. Periodic cleaning is essential to durability, since grit and dirt on floors subjected to considerable traffic will be ground into the finish and accelerate the rate of wear. Since concrete is strongly alkaline and has a high pH, acidic water used to clean the floor will attack and etch the floor surface unless the floor is sealed.

Some cleaners also will actually attack and etch the surface of a concrete floor, thus destroying its dense, shiny, hard trowel finish. Generally, cleaners with ammonium compounds should be avoided because they will attack concrete. See the American Concrete Institute's Committee 515 report for a listing of compounds that attack concrete. Floors should always be rinsed after they are cleaned to avoid leaving aggressive chemicals on the floor that may attack the lime in concrete and leave the floor slippery.

Floors subjected to spilled milk, syrups, fruit juices, brines, fats, oils, and many other industrial products should be thoroughly and frequently scrubbed. Scrubbing with correct detergent solutions removes sticky, adherent soilage, but for regular cleaning, a mild soap does less damage to the floor surface. Automatic scrubbers also allow fast pickup of liquids and similar agents which might stain. Both disk and cylindrical brush scrubbers are available, but usually the latter have greater aggressiveness owing to heavier brush pressure. In many plants, it is necessary to scrub the floors at least once a day. Warm soapy water and stiff brushes should be used, after which the floor should be rinsed clean.

Surfaces subjected to a high frequency of forklift operations should not be allowed to accumulate a crust of dirt, as sometimes happens in warehouses. Trucks ride unevenly over these obstructions, imposing undue impact stresses on the floor finish and increasing the tractive effort of the trucks.

Oil from machining of automotive operations usually has no detrimental effect if the concrete floor surface is dense and the oil is not acidic, but oil detracts from the appearance of the floor and makes the surface slippery. Such floors may be cleaned by scraping off thickened oil crusts, then scrubbing with gasoline, taking due precaution against fire. The floor should then be thoroughly scrubbed with warm soapy water and rinsed. The treatment will not remove stains but will remove the objectionable coating of oil and grease.

The Portland Cement Association publication, "Removing Stains and Cleaning Concrete Surfaces," no. 1S214.01T, offers many solutions to the problems of stain removal.

REPAIR OF DUSTING FLOORS

Exhaust gases from unvented heaters used during winter installation of floors can severely weaken as much as the top $1/8$ in of new concrete surfaces—called *carbonization*—because of reaction between the unvented carbon dioxide and the free lime in cement while the concrete is still in a plastic state. Dusting floors are also created by improper finishing techniques. See ACI Committee 302 report on proper finishing techniques.

Floor finishes that dust under service may usually be improved by one of the chemical hardener treatments discussed in the Portland Cement Association bulletin no. 15147.05T, "Surface Treatments for Concrete Floors."

These chemical hardeners are silicates which react with the free lime in cement to form calcium silicate, thus hardening the surface. Three or more applications of the silicate solution should be made with a warm silicate solution.

Where there is a thin layer of soft, chalky materials at the surface, this may often be removed with 3M plastic buffing pads attached to a scrubbing machine. After the removal of this material, the surface should be thoroughly cleaned, then allowed to dry and one of the hardener treatments applied. In other cases, it is necessary to grind the surface before treatment.

FLOOR SEALERS AND FINISHES

Protective maintenance for existing floors usually involves concrete floor sealers. These products provide a durable layer or coating on top of the floor surface. All such treatments depend on penetration for their effectiveness and have a limited life. Their value is greatest on concrete floor surfaces which are not dense or well-finished.

Successful use of these new materials depends on

1. Soundness of the concrete
2. Correct preparation of the surface
3. Quality of the seal or finish used
4. Correct application methods

Floor sealers are designed to penetrate the concrete surface and fill the pores. Applied by brush, spray, or lamb's-wool applicator, the sealer coats, reinforces, and locks in the brittle, weak cement particles, which can cause recurrent dusting. The result is a concrete surface which allows faster, more effective sweeping. Sticky soilage is also easier to remove. To achieve a greater thickness, a more viscous sealer should be used. Their higher solids content provides a buildup with fewer coats.

For a list of available floor sealers, see the Portland Cement Association publication no. 1S147.05T, "Surface Treatments for Concrete Floors." The most effective of these sealers are the two-package epoxies and the one- or two-package oil-free urethanes. Both have twice the durability of the other types. Opaque pigmented coatings in various colors are available. They can mask discolored concrete and improve plant appearance.

The life of all concrete finishes is heavily dependent on sound concrete of good strength, plus correct, effective precleaning just prior to coating. Many finishes fail because they anchor themselves to an intercoat of soilage residue and degraded cement particles rather than to the sound concrete underneath. Oils, greases, waxes, silicones, and other residues must be completely removed to assure maximum coating life. This stress on preparation and cleanliness is a general rule applying to all concrete floor finishes. The rewards for this extra effort are substantially increased wear life, better gloss retention, and simplified sanitation and maintenance.

In most cases, it is advantageous to acid-etch the concrete surface as part of the preparation cycle. This removes laitance (soft cement crust about $1/20$ in thick).

Thorough rinsing, repeated several times if necessary, is important in obtaining good adhesion. Some detergents have effective cleaning action but often leave a residue which is hard to rinse off.

Since finishing and sealing products usually use solvents, consideration may have to be given to local air-pollution regulations when choosing them.

Even with the best finishes, when used in heavy traffic lanes, it may be necessary to recoat the surface in 1 to 3 years.

It saves money to recoat a finish before it is worn through, because any fracture or scratch in the finish may expose the concrete underneath. Once a stain reaches the abraded porous concrete, it starts to sink in, causing discolorations that are costly and time-consuming to remove.

CORROSION-RESISTANT FLOOR TOPPINGS

Some floors are rough and permanently dirty because they have been attacked by substances inherently harmful to portland cement. Floors in areas subjected to this type of attack should have an aggregate-filled, trowel-applied, corrosion-resistant floor topping to protect the subslab and to keep the surface itself smooth and sound.

A corrosion-resistant floor topping to $3/16$ to $1/4$ in thick can be made using a mixture of aggregate, together with a thermosetting plastic such as epoxy. For new construction, the concrete base slab to which the topping is to be applied should be trowel-finished and water-cured, not sealer-cured. The concrete floor should be carefully pitched to all drains with at least a $1/8$-in/ft slope.

As in portland cement floors, the quality of workmanship in installing epoxy floor topping is extremely important. Preparation of the surface which is to receive the corrosion-resistant topping requires some attention. Clean, sound, and dry concrete is required. It must be free of oil and other contaminants. At severe wear points, such as drains, trench edges, or where epoxy abuts other material, 1-in keys should be formed in the slab to provide extra thickness when the epoxy is placed. Preparation of the base is achieved by acid etching or mechanical scarification.

The thermosetting-resin system which is applied to the bare slab will consist of resin, hardeners, and aggregate. The aggregate is used to increase the abrasion resistance and impact strength of the floor. It also brings the thermal-expansion coefficients of the epoxy topping closer to that of the concrete base slab. This gives the topping an improved tolerance to high temperatures, such as those which occur when cleaning the floor. Nevertheless, epoxy floor toppings should not be steam cleaned because their coefficient of expansion will always be higher than the concrete floor under them and loss of bond can occur.

Corrosion-resistant floor toppings are ideally suited to provide trouble-free surfaces in many areas. They are smooth, virtually joint-free, and impervious. They are easy to clean, can be tinted in a range of colors, have excellent abrasion and impact resistance, and will withstand substantial corrosive attack.

CHAPTER 3
MAINTENANCE AND CLEANING OF BRICK MASONRY STRUCTURES

Brian E. Trimble
Brick Institute of America, Reston, Va.

Good maintenance is important to prolong the life of any building. A good maintenance program begins with routine inspections to pinpoint areas of potential problems. Repairing these problems in a timely manner is essential to preventing continued deterioration. Walls that leak or that are cracked should be repaired. Cleaning of masonry is often desired to remove stains or to upgrade the appearance of a building. Proper cleaning procedures must be followed to ensure a successful job.

ROUTINE INSPECTION

A good, thorough inspection and maintenance program is often inexpensive to initiate and may prove advantageous in extending the life of a building. It is a good idea to become familiar with the materials used in a building and how they perform over a given period of time. Table 3.1 lists various building materials and their estimated life expectancies with normal weathering.

Periodic inspections should be performed to determine the condition of various materials used in a building. These inspections can be set for any given time period: monthly, yearly, and so on. A suggested inspection period is seasonal so that behavior of building materials in various weather conditions can be noted. Inspection records, including conditions and comments, should be kept to determine future "trouble spots" and needed repairs. Minor repairs during the initial stages of deterioration will be less costly than extensive repairs that may be necessary on a major problem.

REPAIRING LEAKY WALLS

Problems resulting from moisture penetration may include efflorescence, spalling, deteriorating mortar joints, and interior moisture damage. Once one or more of these conditions become evident, the direct source of moisture penetration should be determined and action taken to correct both the visible effect and the moisture penetration source. Table 3.2 lists various problems appearing on brickwork due to moisture and the most probable source of moisture penetration. The items checked in the table represent each source that should be considered when such problems occur in brick masonry.

4.28 MAINTENANCE OF PLANT FACILITIES

TABLE 3.1 Estimated Life Expectancy of Selected Materials Exposed to Normal Weathering

Material	Use	Estimated life, years
Brick	Walls	100 or more
Mortar	Walls	25 or more
Metal	Coping, flashing, ties	20–50
Caulking	Sealant	8–10
Plastic	Flashing	5–30
Finishes	Paints, waterproofing	1–10

After investigating all the possible moisture penetration sources, the actual source may be determined through a process of elimination. Many times the source will be self-evident, as in the cases of deteriorated and missing materials. However, the source may be hidden and may need to be determined through some type of building diagnostics. In any case, it is suggested to first visually inspect for self-evident sources. Once the source or sources are determined, measures can then be taken to effectively remedy the moisture penetration source and its effects on the brickwork.

Tuckpointing

Moisture may penetrate mortar which has softened, deteriorated, or developed visible cracks. When this is the case, tuckpointing may be necessary to reduce moisture penetration. *Tuckpointing,* also

TABLE 3.2 Possible Effects and Sources of Moisture Penetration[a]

Effects	Previous acid cleaning[b]	Previous sand-blasting[b]	Plant growth	Deteriorated sealants/caulks	Missing/clogged weepholes[c]	Incompletely filled mortar joints[d]	Capillary rise	Broken/loose units	Differential movement[e]	Missing flashing[f]
Efflorescence (see *Technical Notes* 23 series)	✓		✓	✓	✓	✓	✓	✓		✓
Deteriorated mortar	✓	✓	✓			✓	✓	✓		
Spalled units		✓		✓	✓	✓	✓		✓	✓
Cracked units				✓	✓	✓	✓		✓	✓
Rising moisture					✓		✓			✓
Corrosion of backup materials	✓			✓	✓	✓	✓	✓	✓	✓
Mildew/algae growth	✓			✓	✓	✓	✓	✓	✓	✓
Damaged interior finishes	✓			✓	✓	✓	✓	✓	✓	✓

[a]From Brick Institute of America *Technical Notes* 7F.
[b]See *Technical Notes* 20, revised.
[c]See *Technical Notes* 21B.
[d]See *Technical Notes* 7B revised.
[e]See *Technical Notes* 18 series.
[f]See *Technical Notes* 7 series.

referred to as *repointing,* is the process of placing mortar into cut or raked joints to correct defective mortar joints in masonry.

Prior to undertaking a tuckpointing project, the following should be considered: (1) whether or not to use power tools for cutting out old mortar, since this may damage surrounding masonry and (2) any tuckpointing operation should be done by a qualified and experienced tuckpointing craftsman, since not all bricklayers are good tuckpointers.

In many cases, not all mortar joints in a masonry wall will be defective and need to be replaced. Only those joints which appear to be cracked, deteriorated, have voids, or have separated from the brick should be tuckpointed. Generally, however, it is more economical to tuckpoint all joints in a given area at one time.

Once the determination of defective joints has been made, the mortar in these joints should be cut out by chiseling, grinding, or sawing. The method used will be based on the potential damage to the masonry and the cost. The mortar should be cut to a minimum depth of two times the thickness of the mortar joint, usually $3/4$ in (20 mm), or until sound mortar is reached. Care should be taken not to damage the brick's edges. All debris and dust must be removed from the joint prior to tuckpointing.

Tuckpointing mortar type should be selected based on the purpose for tuckpointing. If the purpose is to preserve the character of the old mortar, the tuckpointing mortar should be mixed with materials and in proportions that match the original mortar. If the purpose is to address specific problems such as moisture penetration, mortar selection should be made with appropriate engineering judgment. Typically, mortar types O and N give best results. Mortars with higher cement contents may be too hard for softer brick and may not bond well to the original mortar.

The actual tuckpointing procedure should be as follows. The tuckpointing mortar should be prehydrated to reduce excessive shrinkage. To prehydrate the mortar, mix all dry ingredients thoroughly with only enough water to produce a damp mix which will retain its shape when pressed into a ball. The mortar should stand in this dampened condition for 1 to $1\,1/2$ hr. The joints to be tuckpointed should be dampened slightly to ensure good bond with the existing masonry but should be surface dry at the time of tuckpointing. Water is added to the prehydrated mix to bring it to a workable consistency, somewhat on the dry side. The mortar is then packed into the joints to ensure a full joint. The joints are then tooled to the desired joint profile.

Face Grouting

If mortar joints develop small hairline cracks, *surface grouting* or *face grouting* may be an effective measure in sealing them. One recommended grout mixture is 1 part Portland cement, $1/3$ part hydrated lime, and $1\,1/3$ parts fine sand (passing a no. 30 sieve). The joints to be grouted should be dampened. To ensure good bond, the brickwork must absorb all surface water. Clean water is added to the dry ingredients to obtain a fluid consistency. The grout mixture should be applied to the joints with a stiff-fiber brush to force the grout into the cracks. Two coats are usually necessary to effectively reduce moisture penetration. Tooling the joints after the grout application may help to compact and force grout into the cracks. The use of a template or masking tape may prove effective in keeping the brick faces clean.

Flashing Replacement

Flashing that has been omitted, damaged, or improperly installed may permit moisture to penetrate a wall and cause damage. If this is the case, a costly procedure of removing brick units, installing flashing, and replacing the units may be required.

To install continuous flashing in existing walls, alternate sections of masonry in 5- to 10-ft (1.5- to 3-m) lengths should be removed. The flashing is installed in these sections, and the masonry is replaced. The replaced masonry should be properly aged (5 to 7 days) before the intermediate masonry sections are removed. The flashing can then be placed in these sections. The lengths of flashing should be lapped a minimum of 6 in (150 mm) and be completely sealed to function properly. The remaining sections of masonry are then replaced.

Water Repellents

The problems described above may cause masonry walls to leak. The specific remedial measures discussed often solve water penetration problems. However, on some specific projects, water-repellent coatings may be recommended in addition to remedial repairs. The use of water repellents may slow down the penetration of water into the masonry but will not stop it and should not be viewed as a catch-all. Water repellents are only effective when all necessary repairs have been made. The indiscriminate use of water-repellent coatings is *not* recommended.

Certain dangers exist in the indiscriminate use of coatings on walls. Among the possibilities are

1. Coatings will not stop moisture penetration through cracks or incompletely filled joints.
2. They will not completely stop staining and efflorescence and may cover it sufficiently to prevent its removal.
3. Some types of coatings may trap moisture behind the coating, which could lead to spalling or disintegration of the masonry.
4. They can make the wall nearly impossible to tuckpoint, if required.

The general recommendation is not to use coatings on brick masonry except under specific controlled circumstances and only after all necessary repairs have been made.

The following checklist is recommended if water repellents are being considered.

1. All obvious cracks or openings in the brickwork should be filled or tuckpointed.
2. All windows, copings, sills, and other joints between brick masonry and other materials that have deteriorated must be cleaned, primed, and caulked. Missing materials should be replaced.
3. There should be minimal efflorescence present before applying any water-repellent material.
4. The wall must be fairly clean and dry at the time of application.
5. The water repellent should be applied to test panels and allowed to weather for several months. The effects of water penetration, color change, and so on should be examined.
6. Depending on the application, the water repellent should permit vapor transmission or penetration through the wall and not form a "nonbreathable" film. Water repellents falling under the generic name of *silane/siloxane blends* appear to permit more vapor transmission than other types.

The cautions discussed above should be studied carefully and applied to the particular project. If the recommendations given above and those of the coating manufacturer are followed and all necessary repairs are made, a successful treatment is possible. Routine inspection is recommended to determine the effects of the water repellent on the masonry and to determine when reapplication is necessary.

REPAIRING CRACKED WALLS

During the lifetime of a building, cracks may occur for a number of reasons. These cracks may be due to movement of the masonry, differential settlement, or excessive deflection of supporting members. Cracks should be repaired to reduce the amount of water penetration and for structural safety reasons. Some cracks may be a sign of serious structural distress. The reasons for cracking should be examined before any repairs are started. For example, a steel shelf angle may need to be stiffened to support the weight of the masonry or a deflection crack may occur again.

Replacing Units

Units that are cracked or spalled should be replaced. The damaged brick is removed by using a toothing chisel to cut out the mortar which surrounds the unit. For ease of removal, the units to be removed can be broken. Once the unit is removed, all the old mortar should be carefully chiseled out, and all dust and debris should be swept out with a brush. If the units are located in a cavity wall, care must be taken to prevent debris from falling into the cavity.

The brick surfaces in the wall should be dampened before the new units are replaced, allowing the surfaces to dry before unit replacement. The appropriate surfaces of the surrounding brickwork and the replacement brick should be buttered with mortar. The replacement brick should be centered in the opening and pressed into position. Any excess mortar should be removed with a trowel. Pointing around the replacement unit will help to ensure full head and bed joints. The joints should be tooled to match the existing profile when the mortar becomes thumbprint hard.

Saw Cutting Vertical Expansion Joints

The source of movement that has cracked masonry must be corrected before repair and replacement can begin. Often a lack of vertical expansion joints has caused the masonry to crack, since thermal and moisture expansion of the brickwork was not accommodated. New vertical expansion joints can be saw cut in the masonry in some circumstances. Non-load-bearing exterior wythes can have new vertical expansion joints cut into them easily. The structural action of load-bearing cavity or multi-wythe walls should be checked to see if cutting a vertical expansion joint would affect the integrity of the wall.

If it is determined that vertical expansion joints are necessary, consideration should be given to where the expansion joints should be placed. The locations of these expansion joints may depend on the locations of structural elements, windows, or doors. Care must be taken to avoid hitting or cutting steel angles, flashing, and so on. An expansion joint must continue to the roof and floor or to intermediate horizontal expansion joints.

CLEANING BRICK MASONRY

The appearance of brick masonry depends primarily on the attention given to masonry surfaces during construction, and proper detailing and maintenance procedures. Although cleaning of masonry walls is often desirable to upgrade the appearance of a building, errors are often made in cleaning such walls. Some walls have been irreparably damaged as a result of a lack of attention to cleaning details and procedures.

Cleaning failures generally fall into one of three categories:

1. *Failure to thoroughly saturate the brick masonry surface with water before and after application of chemical or detergent cleaning solutions.* Dry masonry permits absorption of the cleaning solution and may result in "white scum" or the development of efflorescence or "green stain." Saturation of the surface prior to cleaning reduces the absorption rate, permitting the cleaning solution to stay on the surface rather than be absorbed.

2. *Failure to properly use chemical cleaning solutions.* Improperly mixed or overly concentrated acid solutions can etch or wash out cementitious materials from the mortar joints. These solutions have a tendency to discolor masonry units, particularly lighter shades, producing an appearance frequently termed "acid burn," and also can promote the development of green and brown stains.

3. *Failure to protect windows, doors, and trim.* Many cleaning agents, particularly acid solutions, have a corrosive effect on metal. If permitted to come in contact with metal frames, the solutions may cause pitting of the metal or staining of the masonry surface and trim materials, such as limestone and cast stone.

Before the actual cleaning of a building begins, all cleaning procedures and solutions should be applied to a sample test area of approximately 20 ft^2 (1.9 m^2). The effectiveness of the cleaning agent should be judged by inspection of the sample test area after a period of not less than 1 week after application. The size of the test area may be larger, depending on the cleaning procedure. The indiscriminate use of muriatic acid or the wrong proprietary compounds can cause unsightly, difficult-to-remove stains. Reactions of brick and cleaning solutions are not always predictable, and thus it is safer to use a trial-and-error method on a small test area before committing the entire building to a set procedure. Minute quantities of certain minerals found in some fired clay masonry units and materials, such as manganese, which are used to produce various colors of brick, may react with some acid solutions and cause staining. Sample testing should be performed under conditions of temperature and humidity that will closely approximate the conditions under which the brickwork will be cleaned. Chemical cleaning solutions are generally more effective when the outdoor temperature is 50°F (10°C) or above.

General Cleaning Procedures

Present cleaning methods for new and existing masonry may be classified into three categories: (1) pressurized water cleaning, (2) sandblasting, and (3) hand cleaning. Any method should be assessed as to the impact it might have on the masonry.

These procedures may include the use of chemicals to assist in removing stains. Some chemicals or chemical compounds used to clean brickwork and the resulting fumes may be harmful. Protective clothing and accessories, proper ventilation, and safe handling procedures must be used. The use and disposal of some chemicals or chemical compounds are regulated by federal, state, or local laws and should be researched before use. Manufacturer's material and handling requirements should be strictly observed.

General cleaning procedures recommended for all masonry cleaning projects are as follows:

1. Select the proper solution.
   a. For proprietary compounds, make sure that the one selected is suitable for the brick, and follow the cleaning compound manufacturer's recommended dilution instructions. Many proprietary cleaning solutions perform in a satisfactory manner for their intended cleaning jobs. However, their formulas are not generally disclosed and may be subject to change. It is suggested, therefore, that each product being considered be sample tested on a panel or inconspicuous wall area and judged on a trial basis before being used.
   b. Detergent or soap solutions may be used to remove mud, dirt, and soil. A suggested solution is $1/2$ cup dry measure (0.14 L) of trisodium phosphate and $1/2$ cup dry measure (0.14 L) of laundry detergent dissolved in one gallon (3.9 L) of clean water.
   c. For acid solutions, mix a 10 percent solution of muriatic acid (9 parts clean water to 1 part acid) in a nonmetallic container. Pour the acid into the water. Do not permit metal tools to come into contact with the acid solution. There is the temptation to mix acid solutions stronger than recommended in order to clean stubborn stains. The indiscriminate use of any acid solution may tend to cause further stains.
2. Schedule cleaning at least 7 days after new brick masonry is completed. Prolonged time periods between the completion of the masonry work and the actual cleaning should be avoided when possible. Mortar smears and splatters left over a long period of time (6 months to 1 year) can be very difficult to remove.
3. Protect metal, glass, wood, limestone, and cast stone surfaces. Mask or otherwise protect windows, doors, and ornamental trim from abrasives and cleaning solutions.
4. Presoak or saturate the area to be cleaned. Flush with water from the top down. Saturated brick masonry will not appreciably absorb the cleaning solution. Areas below also should be saturated in order to prevent absorption of the runoff from above.
5. Starting at the top, apply the cleaning solution. Use a long-handled stiff-fiber brush or other type as recommended by the cleaning solution manufacturer. Allow the solution to remain on the brickwork for 5 to 10 min. For proprietary compounds, follow the manufacturer's instructions for application and scrubbing. Wooden paddles or other nonmetallic tools may be used to remove stubborn particles. Do not use metal scrapers or chisels. Metal marks will oxidize and cause staining.

6. Heat, direct sunlight, warm masonry, and drying winds will affect the drying time and reaction rate of cleaning solutions. Ideally, the cleaning crew should be working on shaded areas to avoid rapid evaporation.
7. Rinse thoroughly! Flush walls with large amounts of clean water from top to bottom before they can dry. Failure to completely flush the wall of cleaning solution and dissolved matter from top to bottom may result in the formation of white scum.
8. Work on a small area. The size of the "wash-down" area should be determined after a trial run. This will permit the cleaning crew to examine work for initial results.

Table 3.3 is a guide for various brick units and the recommended cleaning procedures and solutions for each.

High-Pressure Cold Water. This method usually results in a satisfactory job. An ample water supply is necessary. However, disposing of the large volumes of water used is sometimes a problem. On hard-burned brickwork, water at very high pressure can be effective but requires careful application by experienced operators. Pressure should not exceed that which would damage the brickwork being cleaned.

Some pressure systems feature a pressure gun and nozzle equipped with a control switch. This setup permits the operator to apply solutions to a wall over 100 ft (30.5 m) from the base unit. Other systems have two separate hoses—one with plain water and the other with a cleaning solution. Low pressure has been defined as 100 to 300 psi (700 to 2100 kPa), medium pressure as 300 to 700 psi (2100 to 4850 kPa), and high pressure as 700 psi (4850 kPa) or greater. A sand finish or a surface coating may be removed by pressurized water cleaning, resulting in a different appearance. Nozzle pressure in excess of 700 psi (4850 kPa) may damage brick units and erode mortar joints.

Equipment should be as portable as possible. Units may be on wheels, skids, trailers, or pickup truck beds. More elaborate systems include pumps, engines, acid containers, and water storage tanks fixed on truck beds.

Cleaning compounds used with this method should be compatible with the equipment. Some equipment manufacturers are careful to recommend that only specific cleaning compounds be pumped through their equipment. Others build pumps that will resist hydrochloric acid solutions for reasonable lengths of time.

Caution should be taken when using this method, since it is possible for solutions to be driven into the masonry and become the source of future staining. However, if the walls are sufficiently saturated with water before the solutions are applied, the risk of acid penetration is reduced. Experience has shown that this cleaning method has a high probability of changing the appearance of sand-molded brick, sand-faced extruded brick, and brick with glazed coatings or slurries applied to the finished faces. The brick manufacturer should be consulted on the use of high-pressure water cleaning of such brick.

High-Pressure Steam. This method lends itself readily and satisfactorily to various types of masonry and is generally not injurious to most masonry surfaces. Buildings with smooth, hard brick or brick with glazed surfaces should always be cleaned with steam. The more impervious a brick unit, the easier it should clean. Steam cleaning without chemical additives is usually at pressures less than 60 psi (400 kPa).

In most cases, buildings may be cleaned satisfactorily with plain high-pressure steam. For stains, it is sometimes necessary to use a chemical or detergent solution. It should be pointed out that chemicals and high-pressure steam are used primarily to remove applied coatings to masonry, such as paint. This is a highly specialized field, and frequently, the proper cleaning agent can be determined only after an analysis of the various factors involved in a particular project.

Sandblasting. Dry sandblasting has been around for many years and is one method that may be used to clean brick masonry. However, there is also the possibility that, through improper execution, the face of brick units and mortar joints may be scarred. This method is sometimes preferred over

TABLE 3.3 Cleaning Guide for Masonry

Brick category	Cleaning methods	Remarks
Red and red flashed	Pressurized water Bucket and brush hand cleaning Sandblasting	Proprietary compounds, muriatic acid solutions, and emulsifying agents may be used. *Smooth texture:* Stains are generally easier to remove; less surface area exposed; easier to presoak and rinse; unbroken surface, thus more likely to display poor rinsing, acid staining, poor removal of mortar smears. *Rough texture:* Dirt tends to penetrate deep into textures; additional area for water and and acid absorption; essential to use pressurized water during rinsing.
Red body with sand finish or surface coating	Bucket and brush hand cleaning	Clean with plain water and scrub brush using light pressure. Excessive mortar stains may require use of cleaning solutions. Sandblasting is not recommended. Cleaning may affect appearance.
Light-colored units: white, tan, buff, gray, specks, pink, brown, and black	Bucket and brush hand cleaning Pressurized water Sandblasting	Do not use muriatic acid! Clean with plain water, detergents, emulsifying agents, or suitable proprietary compounds. Manganese-colored brick units tend to react to muriatic acid solutions and stain. Light-colored brick is more susceptible to "acid burn" and stains compared with darker units.
Light-colored body with sand finish or surface coating	Bucket and brush hand cleaning	See notes for red body with sand finish or surface coating and light-colored units; sandblasting is not recommended.
Glazed brick	Bucket and brush hand cleaning Pressurized water	Use detergents where necessary and acid solutions only for very difficult mortar stains. Do not use acid on salt-glazed or metallic-glazed brick. Do not use abrasive powders. Do not use metal cleaning tools or brushes.
Colored mortars	Method is generally controlled by the brick unit	Many manufacturers of colored mortars do not recommend chemical cleaning solutions. Most acids tend to bleach colored mortars. Mild detergent solutions are generally recommended.

conventional wet cleaning because it eliminates the problem of chemical reaction with vanadium salts and other materials used in manufacturing brick. Light and heavily sanded, coated, glazed, and slurry-finished bricks should not be cleaned by sandblasting.

Sandblasting by a qualified operator, in conjunction with proper specifications and job inspection, can be satisfactory. Basically, the process involves a portable air compressor, blasting tank, blasting hose, nozzle, and protective clothing and hood for the operator. The air compressor should be capable of producing 60 to 100 psi (400 to 700 kPa) pressure at a minimum airflow capacity of 125 ft^3 (3.5 m^3)/min. The inside orifice or bore of the nozzle may vary from 3/16 to 5/16 in. (4.8 to 7.9 mm) in diameter. The sandblast machine (tank) should be equipped with controls to regulate the flow of abrasive materials to the nozzle at a minimum rate of 300 lb/hr (0.004 kg/s).

There are various degrees of cutting or cleaning desired and, consequently, many types of abrasive materials. They may be of mined silica sand, crushed quartz, granite, white urn sand (round particles), crushed nut shells, and other softer abrasives. Mined silica sands and crushed quartz should have a hardness of approximately 6 on the Moh's scale and be a type A or B gradation (see Table 3.4).

A suggested procedure for sandblasting is as follows:

1. Select sandblast materials that are clean, dust-free, and abrasive.
2. Brick masonry should be dry and well cured.
3. Remove all large mortar particles, as in previous methods.
4. Protect nonmasonry surfaces adjacent to cleaning areas. Use plastic sheeting, duct tape, or other covering materials.
5. Test clean several areas at varying distances from the wall and several angles that afford the best cleaning job without damaging brick and mortar joints. Workers should be instructed to direct abrasives at the units and not on the mortar joints.

Hand Washing. Many buildings of smaller size have been cleaned successfully by hand washing. This is a bit slower method, and it does not have the added advantage of heat, as in high-pressure steam. Usually, this work is done by using soap or detergent with cold water. The method is generally more costly because it is slower and does not lend itself to a job of any great size.

TABLE 3.4 Typical Screen Analysis for Sandblasting Sand Abrasives[*]

Type gradation	U.S. sieve size	Percent passing
Type A, fine texturing[†]	30 mesh	98–100
	40 mesh	75–85
	50 mesh	44–55
	100 mesh	0–15
	200 mesh	0
Type B, medium texturing[‡]	16 mesh	87–100
	18 mesh	75–95
	30 mesh	20–50
	40 mesh	0–15
	50 mesh	0–15

[*]The screen analysis listed above is suggested primarily for mined silica sands and crushed quartz. Reference source: "Good Practice for Cleaning New Brickwork," by the Brick Association of North Carolina, P.O. Box 13290, Greensboro, NC 27415.

[†]Type A gradation is suggested for very lightly soiled brick masonry or where very light, fine texturing of the masonry surface is permitted.

[‡]Type B gradation is suggested for heavy mortar stains or where a medium texturing of the masonry surface is permitted.

Removal of Efflorescence

The removal of efflorescent salts is relatively easy compared with removing some other stains. Efflorescent salts are water soluble and generally will disappear of their own accord with normal weathering. This is particularly true of "new-building bloom." White efflorescent salts can be removed by dry brushing or with water and a stiff brush. Heavy accumulations or stubborn deposits of white efflorescent salts may be removed with a proprietary cleaner. It is imperative that the wall be saturated before and after the cleaning solution is applied. Refer to BIA *Technical Notes* 23 series for more information on the cause and prevention of efflorescence.

Removal of Green Stain (Vanadium Salts)

Brick units can develop yellow or green stains resulting from vanadium salts. These stains can be found on red, buff, or white brick. The vanadium salts responsible for these stains originate in the raw material used to manufacture certain brick units.

As water travels through the brick, it dissolves the vanadium oxide and sulfates. Chloride salts of vanadium require highly acidic leaching solutions and are usually the result of washing brick masonry with acid solutions. Thus the problem incurred with green staining often does not exist until the brickwork is washed down with an acid solution.

To minimize the occurrence of green stain: (1) do not use acid solutions to clean light-colored brick and (2) follow the recommendations of the brick manufacturer for the proper cleaning compounds and procedures.

Should the brickwork have been cleaned with an acid solution and green staining appears, the following procedure may be followed to neutralize the acid:

1. Immediately following the acid wash, flush brickwork with water.
2. Wash or spray the brickwork with a solution of potassium or sodium hydroxide, consisting of $1/2$ lb (0.23 kg) hydroxide to 1 qt (0.95 L) water or 2 lb (0.91 kg)/gal (3.79 L). Allow this to remain for 2 or 3 days.
3. Use a hose to wash off the white salt remaining on the brickwork from the hydroxide.

Various proprietary cleaning compounds have been developed to remove green stain. Their effectiveness on a particular wall can only be determined by test.

Removal of Brown Stain (Manganese Stain)

Under certain special conditions, this stain may occur on mortar joints of brickwork containing manganese-colored units. It appears as a tan, brown, nearly black, or sometimes gray-colored staining. The brown stain has an oily appearance and may streak down over the face of the brick. It appears to be running down from the brick-mortar interface and is the result of manganese used in some brick as a coloring agent. When the solution reaches the mortar joints, the salts are deposited upon neutralization by the cement or lime.

During firing in the manufacturing process of some brick, manganese coloring agents experience several chemical changes. This results in compounds that are not soluble in water but are soluble in weak acid solutions. Since brick can take up acid by absorption, such weak acid solutions can prevail in brick washed with hydrochloric acid. Rainwater also may be acidic in some industrialized areas.

To minimize this problem, do not use any acidic solutions on tan, brown, black, or gray brick. There are special proprietary cleaning compounds available for cleaning bricks containing manganese. These may be tested for effectiveness. The advice of the brick manufacturer should be requested and followed.

The permanent removal of manganese stain may be difficult. After an initial removal, it often returns. The following method has been effective in removing brown stain and preventing its return.

1. Carefully mix a solution of acetic acid (80 percent or stronger), hydrogen peroxide (30 to 35 percent), and water in the following proportions by volume: 1 part acetic acid, 1 part hydrogen peroxide, and

6 parts water. *Caution:* Although this solution is very effective, it is a dangerous solution to mix and use. Acetic acid–hydrogen peroxide also may be available in a premixed form known as *peracetic acid*. This acid, a textile chemical, is also dangerous and may be difficult to purchase.

2. After wetting the brickwork, brush or spray on the solution. Do not scrub. The reaction is usually very rapid, and the stain quickly disappears. After the reaction is complete, rinse the wall thoroughly with water.

An alternate solution suggested for new and light-colored brown stains is oxalic acid crystals and water. Mix 1 lb of crystals (0.45 kg) to 1 gal (3.79 L) of water. There are also proprietary compounds formulated to remove brown stains. Their effectiveness should be judged only after testing.

Proprietary compounds have been used and sometimes found to be effective in keeping the stain from reappearing. Consult the recommendations and directions of the cleaning solution or brick manufacturer when applying proprietary solutions to remove manganese stains.

REMOVAL OF EXTERNALLY CAUSED STAINS

These are stains caused by external materials being spilled on, splattered on, or absorbed by the brick. Each is an individual case and must be treated accordingly.

A large number of external stains can be removed by scrubbing with kitchen cleanser. Others frequently can be removed by bleaching with a household bleach. A combination, such as is found in some kitchen cleansers, may prove most effective.

Poultice

The use of a poultice is included in some of the recommendations that follow. A poultice is a paste made with a solvent or reagent and an inert material. It works by dissolving the stain and leaching or pulling the solution into the poultice. The powdery substance is simply brushed off when dry. Repeated applications may be necessary. Poultices tend to prevent the stain from spreading during treatment and to pull the stain out of the pores in the brick. Poultices are normally used only for small spots.

The inert material may be talc, whiting, Fuller's earth, diatomaceous earth, bentonite, or some other clay. The solution or solvent used will depend on the nature of the stain to be removed. Enough of the solution or solvent is added to a small quantity of the inert material to make a smooth paste. The paste is smeared onto the stained area with a trowel or spatula and allowed to dry. When dried, the remaining powder is scraped, brushed, or washed off.

If the solvent used in preparing a poultice is an acid, do not use whiting as the inert material. Whiting is a carbonate that reacts with acids to give off carbon dioxide. While this is not dangerous, it will make a foamy mess and destroy the power of the acid.

Paint Stains

For fresh paint, apply a commercial paint remover or a solution of trisodium phosphate in water at the rate of 2 lb (0.91 kg) of trisodium phosphate in 1 gal (3.79 L) of water. Allow it to remain and soften the paint. Remove with a scraper and a stiff-bristle brush. Wash with clear water. There are also commercial paint removers in the form of a gel solvent. These should be applied on a small test area on a trial basis. For very old, dried paint, organic solvents similar to the preceding may not be effective, in which case the paint must be removed by sandblasting or scrubbing with steel wool.

Iron Stains

Iron stains are quite common and, in some cases, cover large areas. These stains are easily removed by spraying or brushing with a strong solution of 1 lb (0.45 kg) of oxalic acid crystals per gallon (3.79 L) of water. Ammonium bifluoride added to the solution at $1/2$ lb (0.23 kg)/gal (3.79 L) will speed up the reaction. The ammonium bifluoride generates hydrofluoric acid, a very dangerous material that can etch the brick and glass. Etching will be evident on very smooth brick. Therefore, use this solution only with caution.

An alternate method mixes 7 parts lime-free glycerine with a solution of 1 part sodium citrate in 6 parts lukewarm water, and this is mixed with whiting or diatomaceous earth to make a poultice. Apply a thick paste on the stain with a trowel. Scrape off when dry. Repeat until the stain has disappeared, and wash thoroughly with clear water. A poultice made from a solution of sodium thiosulfate and an inert powder (talc) also has been used for the removal of iron stains.

Copper or Bronze Stains

Mix together in dry form 1 part ammonium chloride (sal ammoniac) and 4 parts powdered talc. Add ammonia water, and stir until a thick paste is obtained. Place this over the stain, and leave until dry. When working on glazed brick, use a wooden paddle to remove the paste. An old stain of this kind may require several applications. Aluminum chloride is sometimes used in this mixture instead of the sal ammoniac.

Welding Splatter

A problem related to iron staining is welding splatter. When metal is welded too close to a pile of brick or completed brickwork, some of the molten metal may splash onto the brick and melt into the surface. The oxalic acid–ammonium bifluoride mixture recommended for iron stains is particularly effective in removing welding splatters.

Scrape as much of the metal as possible from the brick. Apply the solution in a poultice. Remove the poultice when it is dried. If the stain has not disappeared, use sandpaper to remove as much as possible, and apply a fresh poultice. For stubborn stains, several applications may be necessary.

Smoke

Smoke is a difficult stain to remove. Scrub with scouring powder (particularly one containing bleach) and a stiff-bristle brush. Some alkali detergents and commercial emulsifying agents, brush- or spray-applied, given sufficient time to work, also perform well. They should be tested on a small area before use on a large area. They have the added advantage that they can be used in steam cleaners. For small, stubborn stains, a poultice using trichloroethylene will pull the stain from the pores. Exercise caution when using trichloroethylene in confined spaces. Ventilate the fumes.

Oil and Tar Stains

Oil and tar stains may be effectively removed by commercial emulsifying agents. For heavy tar stains, mix the agents with kerosene to remove the tar and then water to remove the kerosene. After application, they can be hosed off. When used in a steam cleaning apparatus, they have been known to remove tar without the use of kerosene.

Where the area to be cleaned is small, or where a mess cannot be tolerated, a poultice using naphtha or trichloroethylene is most effective in removing oil stains.

Also, dry ice or compressed CO_2 may be applied to make tar brittle. Light tapping with a small hammer and prying with a putty knife generally will be enough to remove thick tar splatters.

Dirt

Dirt is sometimes difficult to remove, particularly from a textured brick. Scouring powder and a stiff-bristle brush are effective if the texture is not too rough. Scrubbing with an oxalic acid–ammonium bifluoride solution, recommended for iron stains, has proven effective on some moderately rough textures. For very rough textures, pressurized water cleaning appears to be a most effective method.

Plant Growth

Occasionally, an exterior masonry surface not exposed to sunlight remains in a constantly damp condition, thus exhibiting signs of plant growth, that is, moss. Application of ammonium sulfamate or weed killer, in accordance with directions furnished with the compound, has been used successfully for the removal of such growths.

Ivy

Avoid pulling the vines away from the masonry, since this may damage the brick or mortar. Carefully cut away a few square feet of the vines in an inconspicuous area and examine how much they have rooted into the brickwork. Inspect the exposed area for condition and appearance. There will be some deposits left on the masonry. These are the "suckers" that attached and held the vines. Do not use acids or chemicals to remove the suckers. Leave them in place until they dry and turn dark. Remove with a stiff-bristle brush and detergent.

Egg Splatter

Brick walls vandalized with raw eggs have been cleaned successfully with a saturated solution of oxalic acid crystals dissolved in water. Mix in a nonmetallic container and apply with a brush after saturating the surface with water.

Straw and Paper Stains

Straw and paper stains sometimes result from wet packing materials. Not all packing materials stain brick, but those which do can produce very stubborn stains. Such stains can be removed by applying household bleach. Allow time to dry. Several applications may be required before the stains disappear. The solution of oxalic acid–ammonium bifluoride recommended for iron stain cleans the stain much more rapidly.

White Scum

White scum is a grayish white haze on the face of brick. It is sometimes mistaken for efflorescence but technically is silicic acid scum. This condition results from the failure to saturate the wall before application or thoroughly rinsing acid solutions after cleaning. Generally, it is a film of material that is insoluble in acid solutions except for hydrofluoric acid, which is very dangerous and not generally recommended for this use. Proprietary compounds formulated to remove this scum may be tested and their effectiveness judged.

If removal is too difficult, masking of the haze may be considered. In time, weathering will remove both the masking solution and the white scum.

Masking solutions may consist of paraffin oil and Varsol or linseed oil and Varsol applied by brush to the affected brick units. Linseed oil and Varsol (10 to 25 percent linseed oil) or paraffin oil and Varsol (2 to 50 percent paraffin oil) will darken light-colored brick. Several batches of solutions with various concentrations should be mixed and tested. Generally, solutions of 2 to 25 percent paraffin oil will be satisfactory. Allow 4 to 5 days of warm drying weather to pass, preferably at 70°F (21°C) minimum, before a judgment is made on the effectiveness of the solutions.

Stains of Unknown Origin

Stains of unknown origin can be a real challenge. Laboratory tests of unknown stains may be necessary to determine their composition. Then the appropriate method may be implemented to clean the brickwork. The indiscriminate use of any cleaning agent may aggravate the initial stain and cause further staining. The application of a cleaning agent without identifying the initial stain may result in other stains that are difficult to remove; however, appearance of the stain may be the first clue.

Rust-colored stains may actually be rust. Such stains are quite common and have been known to come from mortar ingredients, wall ties or joint reinforcements with inadequate cover, welding splatter on the brick, or something placed on the pile of brick prior to being laid in the wall.

Green stains may be grass, moss, or vanadium efflorescence. Brown stains also may be vanadium efflorescence or possibly manganese staining.

One test useful in narrowing down the list of possible causes of a stain involves a substance ordinarily not placed on brick masonry. Concentrated sulfuric acid in contact with an organic material will turn it black. This is a quick and easy way to identify stains originating from such a material. Organic stains usually can be removed with household bleach or oxalic acid.

CLEANING HISTORIC STRUCTURES

This type of cleaning endeavor should be referred to a restoration specialist. There are comprehensive papers and publications available on the subject of restoration. Before an old structure is to be cleaned, several questions should be asked before making a final decision, such as (1) Why clean? (2) What is the dirt? and (3) What is the construction of the building?

The query "Why clean?" is addressed in the publication "The Cleaning and Waterproof Coating of Masonry Buildings," Preservation Briefs No. 1, by Robert C. Mack, AIA.

Why Clean?

The reasons for cleaning any building must be considered carefully before arriving at a decision to clean.

Is the cleaning being done to improve the appearance of the building or to make it look new? The so-called dirt actually may be weathered masonry, not accumulated deposits; a portion of the masonry itself thus will be removed if a "clean" appearance is desired.

Is there any evidence that dirt and pollutants are having a harmful effect on the masonry? Improper cleaning can accelerate the deteriorating effect of pollutants.

Is the cleaning an effort to "get your project started" and improve public relations? Cleaning may help local groups with short-term fund-raising yet cause long-term damage to the building.

These concerns may lead to the conclusion that cleaning is not desirable—at least not until further study is made of the building, its environment, and possible cleaning methods.

SUMMARY

Routine inspection is important to prolong the life of all buildings. Elimination of the sources of leaky and cracked brick masonry walls and their repair will avoid more severe future problems.

Cleaning brick masonry still remains, for the most part, a trial-and-error procedure. Therefore, it is strongly suggested that any cleaning procedure and chemical cleaning solution be tested on sample areas first. Such testing should be performed under conditions of temperature and humidity that will closely approximate those conditions under which the brick masonry will be cleaned. Cleaning compounds recommended by the brick or cleaning agent manufacturer also should be trial tested before being committed to the entire project.

BIBLIOGRAPHY

More detailed information on subjects discussed here can be found in the following publications:

"Cleaning Brick Masonry," Technical Notes on Brick Construction 20, Brick Institute of America, Reston, Va., November 1990.

"Colorless Coatings for Brick Masonry," Technical Notes on Brick Construction 7E, Brick Institute of America, Reston, Va., reissued February 1987.

"Efflorescence—Causes and Mechanisms," Part 1, Technical Notes on Brick Construction 23, Brick Institute of America, Reston, Va., May 1985.

Grimm, C. T., *Cleaning Masonry—A Review of the Literature,* Construction Research Center, University of Texas at Arlington, 1988.

Mack, R. C., "The Cleaning and Waterproof Coating of Masonry Buildings," Preservation Briefs No. 1, National Park Service, Washington, D.C., 1975.

"Moisture Resistance of Brick Masonry—Maintenance," Technical Notes on Brick Construction 7F, Brick Institute of America, Reston, Va., reissued January 1987.

CHAPTER 4
MAINTENANCE OF ELEVATORS AND SPECIAL LIFTS

Jerry Robertson
Maintenance Quality Engineer,
Otis Elevator Company, Farmington, Conn.

ELEVATORS AND SPECIAL LIFTS

In any building or plant with two or more floors or levels, vertical transportation is an essential service. People ride from floor to floor in passenger elevators while products move in freight elevators or special lifts.

Materials may be loaded directly into freight elevators or carried by fork trucks or other vehicles which in turn ride the elevators. Increasingly, elevatorlike lifts are being integrated with conveyors, guided vehicles, or other means of horizontal movement into completely automated material-handling systems between various locations and levels.

Operations in a multistory building or production in a multilevel plant depend on the continuous availability of efficient vertical transportation. Elevators and special lifts must be kept operating at peak efficiency, without unscheduled shutdowns, to provide the necessary interlevel movement safely, dependably, and promptly.

Like other essential building and plant systems, vertical transportation consists physically of an installation of elevator equipment integrated with the building structure. The installation is useful only to the extent that it performs a service, in this case moving people and products from level to level. Preventive maintenance becomes the vital link between the vertical transportation system in the building and the service enjoyed by its users.

Maintenance Objectives

Elevators and elevatorlike lifts are built and installed by responsible manufacturers to function efficiently and to keep on doing so as long as they are correctly maintained. With proper preventive maintenance an installation will have a service life as long as that of the building or plant itself. The special nature of elevator equipment and its special place in plant operation determine maintenance objectives.

Among the most complex and specialized in the entire plant or building, the elevator system integrates electrical, mechanical, and in many cases, hydraulic subsystems. Elevator machinery has no close counterpart in other building equipment. Although elevators are designed for ease of maintenance, installation as an integral building system necessitates locating some critical assemblies and wearing parts at various levels from the basement to the roof.

As automation takes over more tasks of elevator operation, equipment becomes more complex and maintenance becomes a weightier responsibility, more difficult to discharge. Each installation requires a preventive maintenance program primarily planned to achieve special objectives: safety, dependability, performance, and economy.

Maintaining Safety. Elisha G. Otis built his first "safety hoister" in 1852, the beginning of the elevator industry. Ever since, quality elevators have been designed and built with attention to the protection of users.

For a modern, well-maintained electric or hydraulic elevator to fall is a virtual impossibility. Preventive maintenance keeps door-protective equipment and leveling controls operating as they should to safeguard people at elevator entrances, where most accidents would otherwise occur. In addition, automatic devices guard against possible hazards at entrances and elsewhere throughout the system. During an elevator's entire lifetime, its safety equipment may never have to operate in an emergency, but safeties must always be maintained in full readiness to function unfailingly, instantly, should the need ever arise.

Maintaining a modern elevator to operate dependably and efficiently also ensures that it will operate safely. Besides preventive care of safety devices, comprehensive maintenance by the manufacturer also includes their periodic inspection and testing to comply with applicable safety codes.

Dependability. Throughout the working day, continuous availability of elevator transportation is essential to profitable operation, making dependability rank high among maintenance objectives. Even an occasional breakdown can prove costly in lost production or delayed delivery.

Like other plant equipment, aging elevators become more vulnerable to possible failure. But a maintenance program engineered to each installation's needs reduces the chance of unscheduled shutdowns to a minimum and protects service continuity even after equipment has been in use for years. If elevators must be taken out of service for essential maintenance work, it can be scheduled to minimize interference with plant or building operations.

Performance. Preventive maintenance can contribute to productivity and profitability by keeping elevators in operation, not only dependably and safely, but also efficiently.

Unless cumulative wear is corrected, power-operated elevator entrances, for example, open and close more slowly and cars take longer to accelerate, decelerate, and level to a stop. Sluggish operation cuts the number of trips each elevator can complete and the total number of loads it can carry. Preventive maintenance, however, keeps elevators operating like new, even after years on the job.

Economy. "Stitch-in-time" economy is inherent in preventive maintenance. Detecting and correcting wear before performance is impaired and failure occurs prevent a chain reaction that could knock out not only one but an entire series of interrelated elevator parts. In this way, maintenance minimizes costly emergency repairs and the even greater cost of interrupted operation.

Among the costliest and potentially longest-lived equipment in a plant or building, elevators represent an appreciable proportion of total investment and deserve the care that will make them last as long as the structure itself. Maintained as the manufacturer recommends, the equipment will serve for the useful life of the plant.

Preventing equipment depreciation proves its value when elevators are modernized and the basic elevator machinery, costliest part of the installation, is found to be so well maintained that it can be retained and reused. Capital conservation, consequently, is a major maintenance objective. Elevator-maintenance economy also demands organizing the program to accomplish the necessary work at the lowest net cost.

Safety Inspections

Cities, states, and other jurisdictions increasingly require annual, semiannual, or quarterly inspections of elevators and certifications of their safety. While specific regulations vary from one locality to another, inspectors generally follow procedures outlined in the "Inspectors' Manual," published by the American Society of Mechanical Engineers. Insurance company inspectors follow similar procedures.

Inspectors from city or state building departments or insurance companies must be fully familiar with operation of the elevator and its safety equipment. In the hoistway, in the pit, or on top of the car, the inspector should conduct his inspection in a safe and orderly manner to prevent injury not only to himself but also to people who might be in immediately adjacent occupied areas.

Inspectors should avoid loose clothing and shoes that have a slick surface or can become slick in contact with oil or grease. They must wear safety glasses or goggles in the hoistway, since airborne particles swept up the hoistway might cause eye injury. Minimum tools include a flashlight, a 6-ft wood rule, a small magnifying glass, a small mirror, and a meter or tester to indicate voltages and grounds.

Hoistway inspection includes:

All mechanical equipment such as sheaves, buffers, door closers, floor selectors, limit switches, hoistway door hangers, closers, and door gibs, to be sure they are intact and securely fastened to their mountings.

Interlocks, to see that they are mechanically fastened to their base or mountings and that the latching head is securely locked when the door is properly closed. Their electrical contacts should not "make" unless the door is in the fully closed and locked position.

Hoist and governor ropes and their fastenings, for wear or rust.

Traveling cables, to make sure that they are properly hung and that the outer wrapping around the electrical wires is not worn so that a short or ground might possibly occur should the traveling cables come in contact with other mechanical equipment in the hoistway.

Rails, for proper alignment and tightness of rail fastenings, brackets, and fish plates.

Steadying plates, to see that the car is securely fastened to the car frame.

Ropes, for vibration and wear.

Guide shoes, for excessive float.

Gibs, for wear.

Door operators, for alignment.

Pit inspection includes:

Oil, level in the buffer.

Rope stretch.

Debris or water leaks.

Safety shoes are checked from the bottom of the car, in the pit.

Machine-room inspection includes:

Motors and generators, to see that commutators are clean, properly undercut, and equipped with the correct brushes. Brush holders must be clean and brushes well seated, with proper spring pressure, the commutator free from flat spots, high bars, and pitting.

Bearing wear, to determine if it is affecting the armature air gap and rotor clearances.

All electrical equipment should be grounded through ground clamps, and machinery should operate without overheating.

Brakes, to make sure that each will hold 125 percent of full load and that lifting application is unimpaired by frozen or worn pins.

Brake linings, to make sure that they are securely fastened to the shoe and in good contact with the brake pulley.

Shafts into the pulley, for proper alignment and secure fastenings to the brake pulley.

Gears and bolts, to make sure they are not loose or broken.

Worm and gear, for backlash and end thrust.

Sheave grooves, for uneven wear or bottomed ropes.

Gland packing, on a geared machine.

Gear-case oil.

Machine fastenings.

Controllers, for correct fuse capacity, broken leads, loose connections in lugs, loose or broken resistances, improper contacts, worn contacts, weak springs, improper contact-spring tension, proper contact wipe, and worn pins or brushes.

Switches, for residual magnetism and gummy cores.

Safety equipment, for blocked or shorted contacts.

Governors, to be sure that the rope is well seated in the sheave, that the operating mechanism is well lubricated and free to move, and that no pins are stiff from rust or paint.

Landing equipment, for broken buttons and lamp fixtures. Emergency key glasses and keys must be in place, and lighting at landings must be bright enough so that people leaving an elevator enter well-lighted areas.

Hydraulic-elevator inspections follow procedures outlined above as they apply. In addition, hydraulic-elevator safety inspection includes:

Cylinder-head packing and piston, to make sure the oil is not leaking excessively or returning to the tank through a bypass.

Hydraulic machines, to be sure that there is sufficient oil so that the car reaches the top landing with oil left in the tank.

Power-medium oil, for cleanliness.

ORGANIZATION FOR MAINTENANCE

Keeping elevators in condition to satisfy safety requirements is but one part of a comprehensive program for preventive maintenance. Such a program must be organized to integrate qualified personnel, effective methods, the necessary tools, records, spare parts, and provisions to handle emergency conditions safely. The entire program must emphasize safety for elevator users as well as for service technicians.

Personnel

Skilled labor, the major element of maintenance expense, is scarce and costly. Expert management is therefore essential to avoid the twin perils of slightly needed attention on the one hand and incurring costly, unnecessary expenditure on the other. An elevator plant incorporates a variety of electrical, mechanical, and hydraulic equipment not common to most other machinery. Mechanics, electricians, and other craftsmen therefore need additional specialized training and experience to maintain elevator equipment.

Qualified elevator mechanics must understand such safety features as governors, contacts, and interlocks, their design and operation, and how wear will affect performance. Familiar with functions of the machine, controller, operating devices, signals, door devices, and their operating sequence and able to read wiring diagrams to localize and correct troubles readily, technicians systematically inspect, clean, lubricate, adjust, and replace defective parts anywhere in the elevator.

Otis Elevator, the industry leader, has developed maintenance methods through experience accumulated maintaining nearly 100,000 elevators in all 50 states and Canada, as well as hundreds of thousands more around the world. Since check charts to guide maintenance mechanics are based on average conditions, inspection frequencies may have to be modified where unusual conditions exist.

Check charts generally use terms such as *check, clean, lubricate,* or *adjust,* while methods charts specify frequencies and lubrications. Bulletins or manuals generally cover adjustments. This was the traditional method for many years, but recently, one major manufacturer has adopted a computerized maintenance management system. This will ensure that all maintenance is performed in a predictive manner based on usage and mean time between failure (MTBF) of components.

Machine-room equipment is cleaned by using approved cleaning compounds and wiping cloths, brushes, vacuum cleaners, or other mechanical means. Cleaning hoistway equipment is especially important. Failure to clean periodically in these locations creates a serious fire hazard, since flammable dust and lint gather on the oil and grease used in lubricating hoistway equipment (Fig. 4.1).

Minor repairs are usually accomplished by replacing a worn or broken part with an identical replacement part. Major repairs may require a planned shutdown of the equipment and replacement of major components or subassemblies of a generator, machine, selector, controller, or other piece of equipment.

Spare parts must be available if equipment is to be repaired properly. Since parts subject to wear are essential to the safety of elevator equipment as well as to its efficient performance, worn parts must be replaced by new parts at least as good as, if not better than, the parts originally installed.

Special tools are also essential to accomplish maintenance procedures. Any mechanic not properly equipped either requires more time or fails to maintain quality when making repairs and adjustments (Fig. 4.2).

Modern automatic elevators or special lifts, which may be controlled in unison with automated conveyor or guided cart systems, require comparably advanced, specialized tools to troubleshoot and adjust. With the advent of microprocessors, maintenance technicians can no longer watch mechanical switches open and close and must therefore use event recorders, oscilloscopes, and recording data meters (Fig. 4.3).

FIGURE 4.1 Machine-room equipment in an elevator installation protected by comprehensive maintenance undergoes inspection and testing at intervals prescribed to detect and correct the effects of wear and tear before they can appreciably affect performance.

FIGURE 4.2 Cathode-ray-tube oscilloscope typifies the more advanced instruments that maintenance technicians use in analyzing the performance of modern elevator equipment. Checking the operation of automated elevators with great accuracy, the technician can promptly make any necessary adjustments to restore "like-new" service.

Emergency Service

Qualified elevator maintenance personnel can assist in emergencies that include fires, earthquakes, and electric power failures. Elevator technicians are trained to protect people first and equipment second and to release trapped riders with minimum damage to a building or to its elevator equipment. With certain remote monitoring equipment and centralized maintenance dispatching, some contractors are immediately aware of emergencies and can dispatch a technician before the customer calls for assistance.

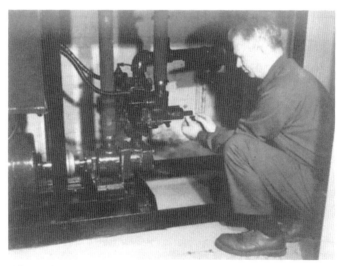

FIGURE 4.3 Oil-hydraulic elevators for freight or passenger service receive the attention of the manufacturer's maintenance specialists skilled in servicing the interrelated electrical, hydraulic, and mechanical subsystems that must function efficiently in unison for the elevator to provide quiet, dependable service. Valve adjustment is one of many conditions that must be checked with a high degree of accuracy.

Ready to restore elevator service promptly in an emergency, trained technicians are familiar with the use of hand and power tools, ladders and stepladders, and proper methods of hand lifting hoisting materials. Wearing personal protective equipment such as proper clothing, shoes, hard hats, and safety glasses, maintenance men have available other safety equipment such as goggles, welder's masks, respirators, hand-protective creams, and leather-palm gloves. Technicians also follow proper safety precautions while working in the machine room, in the hoistway on top of the car, or in the pit.

Do-It-Yourself Maintenance

Plant and building operators have three options for maintaining elevators: (1) having the plant staff perform the necessary maintenance, (2) engaging an independent maintenance contractor, or (3) turning the maintenance responsibility over to the elevator manufacturer.

Do-it-yourself elevator maintenance requires hiring and training specialized craftsmen, supervising them, and supplying replacement parts and servicing equipment.

To compete for qualified elevator mechanics, the building or plant operator must pay at least union scale, including fringe benefits, for routine lubrication and adjustments as well as for emergency repairs. Costly specialized skills must be available in all phases of elevator repair, including roping, major adjustments, bearing replacement, and wiring. For continuity of service in case of illness, retirement, or death, new elevator mechanics must be trained or experienced mechanics hired away from established elevator companies.

Carrying out maintenance in accordance with a properly planned program requires supervising the mechanics actually performing the work. Few plant engineers have either the specialized knowledge of elevator maintenance or the time to supervise the work of the elevator crew properly.

As elevators age, parts are not always immediately available for replacement should the need arise. It is therefore mandatory that the owner stock a full supply of parts ready for use in case of an unexpected breakdown.

CONTRACT MAINTENANCE

Rather than attempting in-house maintenance of plant or building elevators, management can elect to have the system cared for by either an independent maintenance contractor or the service department of the company that manufactured and installed the elevators. Contracts may range in coverage from "oil and grease" to "full maintenance."

Oil and grease contracts, the simplest form of contractual service, cover oiling, greasing, and minor cleaning of elevator equipment. Such contracts generally do not include replacement parts or overtime callbacks. Protective wording in the maintenance agreement limits the contractor's liability. While "hold harmless" clauses protect the elevator contractor, they are potentially costly to the owner. Contract prices are adjusted annually in step with changes in labor cost during the year.

Full maintenance contracts usually cover all parts and labor required to correct normal wear and tear of the elevator system. Overtime emergency callback service may or may not be included, depending on individual needs. Exclusions in full maintenance contracts are normally fewer than in oil and grease contracts and the contractor's liability is greater since he is responsible for preventive maintenance.

Should a "full maintenance" contract exclude mandatory safety tests, the plant or building management must provide trained personnel to shut down the equipment should unsafe conditions develop. Omission of such items by the contractor increases the ultimate cost of maintenance to the owner.

Price adjustments, usually annual, are made in direct ratio to changes in labor rates and the U.S. Department of Commerce metal-products price index.

MAINTENANCE BY THE MANUFACTURER

Maintenance by the elevator manufacturer is organized on a basis of interlocking responsibility to put the necessary people in local facilities, backed up by other, specialized facilities economically serving wider areas. Comprehensive maintenance contracts specify major responsibilities the manufacturer assumes in caring for the physical condition of each installation, keeping its performance at peak efficiency, minimizing the need for emergency repairs, but providing prompt, effective emergency callback service when necessary.

PARTS

While an independent maintenance contractor is unable to guarantee the availability of repair and replacement parts, the elevator manufacturer is obligated to continue providing parts for equipment the company has made and is maintaining.

Each local office keeps stocks of replacement and repair parts controlled with the help of a continuing computer analysis of maintenance customers' usage patterns. "Spare lending" assemblies are strategically located to serve each geographic area so that, in an emergency, an elevator can be restored to service promptly by installing a loan or exchange part.

Replacement parts come from the same manufacturing plants that produce them for new elevators. Complete records, specifications and drawings, and a metal-casting plant that keeps and uses patterns for current or discontinued parts facilitate the maintenance of all protected elevators, whether installed recently or decades ago.

Developments in solid-state circuitry and other fields of technology are continually being incorporated in new and improved equipment for new elevators. Whenever possible, improved parts are manufactured to be compatible with existing installations as well. Computer analysis of parts usage often indicates opportunities for upgrading elevator performance by engineering new replacements with desired improved characteristics.

SAFETY INSPECTIONS BY THE MANUFACTURER

A comprehensive program of maintenance by the elevator manufacturer also may include a thorough full-load safety test of each elevator every 5 years and no-load tests annually. In addition to ensuring compliance with local, state, and national safety regulations, the inspections contribute to life extension of equipment through preventive care and adjustment.

Inspection includes checking the overspeed governor for lubrication, adjustment of jaws, sheave-shaft wear, spindle-bearing condition, and switch operation. For the 5-year test, the governor is calibrated, adjusted, and resealed. The motor overload-protection relay dashpot is cleaned, refilled with oil, and adjusted.

Crosshead and safety switch operating levers are checked to be certain that setscrews or keys are tight and movement is not restricted. The safety releasing carrier is checked for lubrication and spring tension.

As part of the 5-year test, test weights are placed on the elevator to check the correct counterweight overbalance. Tachometer readings are taken to determine that the elevator performs within 10 percent of the rated speed in both the up and down directions.

Guide rails and fastenings are checked before the safety test and after it are again inspected for damage and loose fastenings and to determine that both safeties are set evenly.

With all safety circuits operating, the safeties are set by manually tripping the governor. For the 5-year test, the governor is tripped with the elevator running at full speed, with the full rated load of test weights in the car. To test and oil the buffers, the elevator is run onto the buffers and their plunger operation is observed. Buffer-switch operation, oil level, and plunger condition are checked.

Upon completion of all tests, data are recorded on the elevator maker's safety-test report and in some localities on a city or state test-report form. The manufacturer makes necessary repairs and takes other measures to assure smooth, efficient operation.

INTANGIBLE ELEMENTS

Effective preventive maintenance that can protect the dependability and efficiency of elevator service for years and decades ahead demands more than mere desire for profit. Recognizing that a successfully functioning installation is the best salesman, the responsible elevator manufacturer puts extra effort into preventive maintenance to increase the number of satisfied customers, who constitute the company's best assurance of continuing growth.

Responsible elevator manufacturers are primarily concerned, not so much with selling hardware as with providing effective long-term vertical transportation that will reflect credit on the equipment maker. The manufacturer never loses interest in its installations but backs them up with comprehensive preventive-maintenance service.

INTEGRITY AND EXPERIENCE

The lasting confidence of its customers testifies to the integrity of a company. Going beyond formal contractual agreements between customer and contractor, integrity includes an implied warranty that service by the manufacturer is in the customer's best interest and will protect his substantial investment in elevator equipment. Experience the manufacturer has accumulated in maintaining comparable elevators pays off in fewer shutdowns and faster repair time.

MAINTENANCE COST

For maintenance service of comparable quality, both the independent contractor and the elevator manufacturer must pay the same union rate for elevator mechanics and buy or make replacement parts at about the same cost. Materials usually cost more for the contractor, who must buy parts from the manufacturer.

Since basic costs differ little whether an independent contractor or the elevator manufacturer maintains an elevator, the owner should investigate closely if one source offers to provide ostensibly the same service at a substantially lower price.

An irresponsible contractor may slight the upkeep of out-of-the-way parts of an elevator installation. These parts are often more difficult and costly to maintain. At the same time, failure to service them properly may go unnoticed for years until deferred maintenance accumulates to cause a costly breakdown. The contractor may then exceed his resources. Plant and building managers therefore choose sources on the basis not only of quoted price but also of reputation.

WARRANTY ON NEW EQUIPMENT

After a new elevator installation is completed and turned over to the owner for use, the manufacturer's maintenance mechanics adjust the equipment during a 3- to 12-month new installation service period. A 1-year warranty covering defective parts and workmanship on the part of the manufacturer or his agent protects the new installation, but the warranty does not provide for preventive maintenance.

Wear and tear begin as soon as an elevator is turned over for use. If preventive maintenance is delayed until the end of the first year, wear and tear accumulates. A shorter period remains over which the cost of anticipated repairs can be prorated. The service will consequently cost more per month if maintenance begins at the end of the first year.

For greatest long-run economy, experienced owners start maintenance contracts on their elevator equipment when it is operating at peak efficiency and all its components are new. The elevator manufacturer is then unquestionably responsible for maintaining "like new" the condition of the equipment and the performance it delivers.

AGE ADJUSTMENT AND PREMAINTENANCE REPAIRS

Most manufacturers base maintenance prices over a period of time on equipment serviced. An age-adjustment factor is added to the monthly contract price if maintenance is started on the equipment after the first year of use, to compensate for the reduced period of time to accumulate maintenance funds for necessary repairs.

Age adjustment may vary from 1 to 15 percent, depending upon the equipment and the expected life of its components. To illustrate the effect of age adjustment, assume that an elevator system has had inadequate preventive maintenance or none at all during the first 5 years of its life. Then, owing to increasing shutdowns, the owner decides to place the elevator under manufacturer's maintenance.

The manufacturer's survey team will examine the elevator from top to bottom and list repairs required to assure safety. For instance, brake shoes and hoist ropes may require immediate replacement, electrical grounds in the system must be traced and eliminated, and other parts must be repaired before the manufacturer is willing to assume liability for safe operation of the equipment.

Premaintenance repairs like these can cost thousands of dollars, depending upon the condition of the equipment. This deferred maintenance must be made good before the start of the service contract.

Other parts may have to be repaired in the near future but do not require immediate replacement. These repairs can be prorated in the maintenance agreement, with the owner paying part of the replacement cost.

Prorating is based on the number of years the elevator was not under service contract compared with the number of years the elevator was serviced by the manufacturer. When the exact future replacement date cannot be determined or if a piece of equipment can be used for some months longer, it is economical to prorate repairs rather than perform them immediately.

CHAPTER 5
AIR-CONDITIONING EQUIPMENT

Martin A. Scicchitano
Carrier Air Conditioning Company, Syracuse, N.Y.

Air conditioning is the control of temperature, humidity, purity, and distribution to provide a specified level of human comfort and/or product quality.

AIR-CONDITIONING LOAD

An understanding of the sources of air-conditioning load and how it is handled is critical to the efficient operation and maintenance of air-conditioning equipment.

The load on an air-conditioning system is the amount of heat that must be removed from or added to a space in a given time. This applies to both the load which manifests itself in terms of the actual sensible temperature of the space, and that involved in the control of humidity, since raising or lowering the moisture content requires the transfer of heat.

If heat is to be added, it is a heating load. If it is to be removed, it is a cooling load.

The combined effects of the sun, outside air temperatures, and internal loads often create both a heating and a cooling load, simultaneously, in the same building.

Cooling Loads

To determine the cooling load of a building and thus establish the amount of refrigeration needed, it is first necessary to know what produces the heat and how much is produced. The following sources of heat will be encountered in any building:

Sun Load. The sun shining on a building and through the windows carries heat into the building. Heat coming through the windows is felt much sooner than heat coming through the walls. The rate at which the sun heats up the walls depends upon wall construction. Thick walls increase in temperature much more slowly than thin walls but will hold more heat. This has a definite effect on conditions in the building, since the walls give up their heat slowly and the effects can be felt long after the sun has passed the wall. The sun load varies according to season and exposure. The south exposure has the highest sun load in winter. During both seasons the east-exposure sun load reaches its peak in the morning and the west reaches its peak in the afternoon. The sun load on the north exposure is due mostly to reflection and is therefore low throughout the year.

Transmission Load. The heat that comes into or goes out of the room through the windows and walls (because of the difference between the outdoor and indoor temperatures) is called transmission load. On a warm day, the heat travels into the room. On a cold day it travels from the room.

Transmission load is thought of as separate from the sun load. For example, the sunny side of the building has a sun load and a transmission load while the shady side has only a transmission load.

Outside-Air Load. Outside air is required for ventilation and odor removal. It represents a cooling load that varies based on temperature and moisture content.

Occupancy Load. The human body generates heat through metabolism and releases it by radiation, convection, and evaporation. The amount of heat generated depends on temperature and activity level.

Load from Lights. Electric lights give off heat in proportion to the wattage. This heat is included when the designer calculates the cooling load.

Miscellaneous Loads. The heat from other sources such as electric motors, elevators, computers, and communications equipment is included as part of the cooling load.

Latent Load. Excess moisture in the room is termed latent load. This load appears from several sources: people breathing and perspiring; ventilation air; moist air infiltrating through window cracks or building openings; and many miscellaneous moisture-producing pieces of equipment such as steam tables, showers, and sterilizers.

Load Removal. All of these loads, including latent load, are removed by cooling action. In the case of unitary equipment, the action takes place at the cooling coil in the unit, which is fed liquid refrigerant directly from the refrigerating assembly. As the refrigerant vaporizes it absorbs heat from the air passing over the coil.

Excess moisture is removed by cooling the air, since air will hold less moisture at lower temperatures.

In central-station systems, similar cooling and humidity control takes place at the central air-handling apparatus through the use of cooling coils, a combination of coils and fine water sprays, or sprays alone. In the larger systems the coils usually use chilled water from the refrigeration machine rather than liquid refrigerant.

In order to exercise complete control over both humidity and temperature, it is frequently necessary to add a reheat coil to warm up the air after it has been dehumidified, because the reduced temperature required for dehumidification may be lower than that required to maintain proper temperatures in the conditioned space.

In high-velocity induction systems of the type used in many office buildings, an additional means of regulating temperature is provided by means of a coil in the room, through which chilled or warm water may be passed. Room air is induced over this coil by the effect of high-velocity air from the central unit as it passes over specially designed nozzles. The amount of heating or cooling is controlled by regulating either the water flow through the coil or the airflow across the coil. All dehumidification, however, is done at the central apparatus.

Size of the Cooling Loads

For a typical room, the distribution of cooling loads might be as follows:

Sun	50%
Transmission	20%
Lights	18%
Occupancy	12%
Total	100%

The above distribution shows that the sun is the largest part of the total room load. It also is a changing load, since it moves from one side of the building to the other and, in effect, can turn on or off any time during the day. Therefore, it has considerable effect on the control of the system. The tabulation also shows transmission to be the second largest load. This load can vary as the outdoor air temperature varies. It is zero when outdoor air and indoor air have the same temperature, and negative when the outdoor air is cooler than the indoor air. When the load is negative, heat passes from the room to the outdoors. Transmission is the only load that can reverse. The others can become zero, but none will take heat from the room.

All the loads can vary, within certain limits. The sun load varies with weather conditions and seasons; the load from lights varies according to the number and wattage of units being used at a given time.

Occupancy, transmission, latent, and miscellaneous loads cannot be so readily defined. The engineer must therefore design the system for the maximum practical load. These conditions are called "design conditions." To limit the occupancy load, the engineer decides how many persons will normally occupy the room. For example, a hotel room might be designed for two persons and an office for three or more.

Transmission load varies with the difference between outdoor and indoor temperatures. The maximum transmission load is based on the maximum design outdoor temperature which is determined by studying weather records for the area.

The design engineer bases the system capacity on design conditions only and not on maximum possible conditions. It would be costly and impractical to use a system whose capacity is great enough to meet the demands of an abnormally hot day or an abnormally crowded room. The oversized system would operate at full capacity only a few hours during the season.

Heating Loads

During cold weather heat must be added to the room to offset the heat loss through the walls and windows. Although heat is still added to the space by occupants, lights, and miscellaneous sources, the transmission loss to the outdoors may be great enough to require additional heat. The effect of direct sunlight may offset the outward transmission load so that cooling may be necessary even during the winter months.

Since outdoor air is used for ventilation, it must be warmed in winter. In a typical system the air is preheated to approximately 50°F as it enters. Air then passes to the room after being further heated to produce the room temperature desired by the occupant. In rooms where the heat gain from sun, light, occupancy, and miscellaneous sources is high, supply air may be used without additional heating.

REFRIGERATION

Definition

Refrigeration has been defined as "the transfer of heat from a place where it is not wanted to a place where it is unobjectionable." This transfer usually results in lowering the temperature in the refrigerated space or substance.

Unit of Measure

The unit for measuring refrigeration effect is the "ton." A ton of refrigeration is equivalent to the cooling effect produced by 1 ton of ice melting in a 24-hour period. To make it more useful, this definition is more often expressed in smaller units and related to a change in temperature or a change in state. The smaller unit of heat is the British thermal unit (Btu), which is the amount of heat required to raise 1 lb of water 1°F. Heat can also bring about a change in state. For example, adding heat to ice will change the ice to water. Adding more heat will change the water to steam. This change of state can take place without a change in temperature. For our purposes, the amount of heat required to change 1 lb of ice at 32°F into water at 32°F is 144 Btu. One ton of refrigeration, then, is equivalent to 2000 lb of ice times 144 Btu per lb in 24 hr, or 288,000 Btu.

If a refrigeration plant is cooling water and both the amount and temperature change of the circulating water are known, the maintenance engineer can calculate the tonnage by using the formula, gallons of water per minute (gpm) times the temperature change (degrees F) divided by 24 equals tons of refrigeration, or

$$\frac{\text{gpm} \times (T_1 - T_2)}{24} = \text{tons}$$

where the refrigeration plant is not cooling water, the variations are too great and the computation too complex to be given here. A diagram of a refrigeration cycle using a reciprocating compressor and an air-cooled condenser is shown in Fig. 5.1.

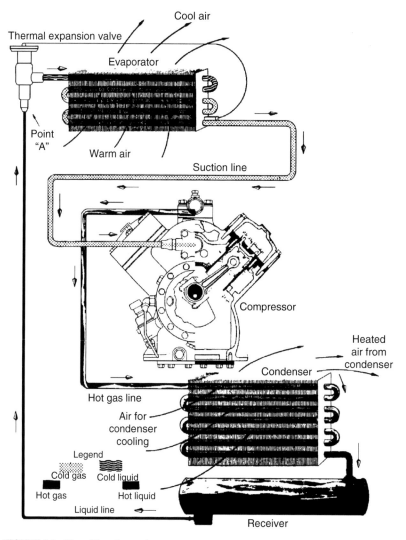

FIGURE 5.1 The refrigeration cycle.

TYPES OF EQUIPMENT

Air conditioning is accomplished by a wide variety of systems and equipment. Some are designed to serve a specific type of space, others to perform a specific function. Generally speaking they may be grouped into two classifications: unitary and central station.

The unitary equipment classification includes those designs such as a room air conditioner, where all of the functional components are included in one or two packages, and installation involves only making the service connections such as electricity, water, and drains.

Central-station systems, often referred to as applied or built-up systems, require the installation of components at different points in a building and their interconnection (see Fig. 5.2).

TYPES OF COMPONENTS

All air-conditioning systems are made up of two major types of components: air handling and refrigeration. The air-handling part of the unit would include the fan, filter, heater, air passages through the unit, and an air outlet. Most of the air outlets have a means of adjusting airflow or direction, or both. The refrigeration part includes the compressor and its drive, the cooling coil, and the condensing coil or condenser.

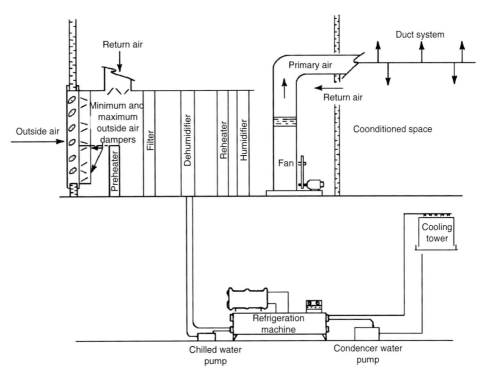

FIGURE 5.2 Central-station system. Air from outdoors enters at the left, is passed over a preheating coil, and then mixed by automatic dampers with the proper proportion of air returning from the conditioned space. This mixed air is then passed through filters, dehumidifiers, reheaters, and humidifiers to get proper temperature and humidity. A fan takes this treated air and distributes it through a duct system to terminals in the conditioned space.

For a central-station system, the air-handling equipment includes the fan, filter, heater, apparatus for treating the air, dampers for control of air volume, and a distribution system of ducts or conduit. In such a system, the refrigeration equipment, consisting of one or several machines, usually is centrally located and can serve many air-handling units located throughout the building.

Control of temperature, humidity, and air quality in the conditioned space is provided by various types of air terminals. These use air and/or water from the centrally located treating and refrigeration plant. Terminals range from simple diffusers to those with self-contained controls capable of varying air volume and temperature to meet individual space requirements. Some terminals provide for induction of room air and controlled passage of this air over a secondary water coil that has either hot or cold water running through it. Another terminal type, classified as a fan coil unit, has its own fan which circulates room air over the coil.

The type of system that is selected will depend on owner needs. Factors to be considered are temperature, humidity and quality requirements, an analysis of first cost versus owning and operating costs, and the flexibility the system offers for meeting possible future needs.

REFRIGERATION EQUIPMENT

Most refrigeration equipment falls into one of four general types. These are reciprocating, centrifugal, screw, and absorption. Reciprocating, centrifugal, and screw refer to the type of compressor used, since all employ mechanical refrigeration cycles. Absorption refers to a fundamentally different cycle, where a cooling effect is produced by the absorption of one fluid by another. Reciprocating-type units usually have capacities up to 150 tons. Centrifugal compressors are available in capacities from 100 to 10,000 tons. Absorption machines are available in sizes up to about 1800 tons.

Refrigerant. Many different chemicals are used as refrigerants in reciprocating, screw, and centrifugal systems. Many others have been tried and discarded. Comparative temperature-pressure relations of the more commonly used refrigerants are shown in Table 5.1.

Most absorption machines use either a water-lithium bromide cycle or an ammonia water cycle. In the first, water is the refrigerant; in the second, ammonia is the refrigerant.

Factors such as safety, cycle efficiency, practical operating pressures, stability, corrosivity, toxicity, flammability, initial cost, and compatibility with other materials must all be considered in selecting a refrigerant. It should be noted that some of the commonly used refrigerants are chlorine-based and apparently contribute to the depletion of the earth's protective ozone layer. As a result, the United States Environmental Protection Agency (EPA) is implementing legislation that will phase out the domestic manufacture of these refrigerants, as well as restrict their emission from existing machines. Similar actions are being taken by most other nations. Operation and servicing of all air-conditioning and refrigeration equipment should be done in accordance with all existing EPA laws, municipal building codes, and other applicable state and local ordinances.

Potential alternate refrigerants have been identified and some are now on the market but each new refrigerant must be evaluated on such properties as toxicity, performance, and compatibility with other system components such as gaskets, O-rings, and oils.

Oil. The oil used in a refrigeration system is highly refined and moisture free. It should be selected based on the equipment manufacturer's specifications. The oil must be kept in sealed cans until ready for use. Under no circumstances should automobile engine grade oil be used. Some of the newer refrigerants that replace the chlorine-based refrigerants require the use of special synthetic oils to ensure compatibility and miscibility. Check with the equipment manufacturer for recommendations on specific oils to use with each refrigerant.

Application. First cost, operating cost, maintenance cost, safety, operating efficiency, space and weight requirements, power availability, and many other factors must be considered in the selection

TABLE 5.1 Comparative Temperature-Pressure Relations of Common Refrigerants
Saturated vapor pressure

Number	11		12		22		113		114		500		502	
Chemical name	Trichloromono-fluoromethane		Dichlorodifluoro-methane		Monochlorodi-fluoromethane		Trichlorotri-fluoroethane		Dichlorotetra-fluoroethane		Azeotrope of dichlorodifluoro-methane and difluoroethane		Azeotrope of monochlorodiflu-oromethane and monochloropenta-fluorethane	
Chemical symbol	$CFCl_3$		CF_2Cl_2		$CHClF_2$		$CCl_2F\text{-}CClF_2$		$C_2Cl_2F_4$		73.8% CCl_2F_2 26.2% CH_3CHF_2		48.8% $CHClF_2$ 51.2% $CClF_2CF_3$	
							Pressure							
Temperature, °F	Gage	Vacuum	Gage	Vacuum	Gage	Vacuum	Gage	Vacuum	Gage	Vacuum	Gage	Vacuum	Gage	Vacuum
−40		27.03		0.58									4.28	
−30		26.01	4.50										9.40	
−20		24.72	9.17		10.31			29.05		22.91	3.14		15.52	
−10		23.95	11.81		16.59			28.69		20.63	7.76		22.76	
0		23.10	14.65		24.09			28.21		17.79	13.26		31.24	
+5		22.73	15.86		28.33			27.92		16.14	16.38		35.99	
10		22.34	17.10		32.93			27.60		14.31	19.75		41.09	
12		21.94	18.38		34.88			27.45		13.52	21.18		43.24	
14		21.52	19.70		36.89			27.30		12.71	22.65		45.45	
16		21.08	21.05		38.96			27.14		11.86	24.16		47.72	
18					41.09			26.97		10.98	25.72		50.05	
20					43.28			26.80		10.07	27.33		52.45	
22		20.62	22.45		45.53			26.61		9.12	28.99		54.91	
24		20.15	23.88		47.85			26.42		8.14	30.70		57.44	
26		19.66	25.37		50.24			26.22		7.12	32.45		60.04	
28		19.14	26.89		52.70			26.01		6.07	34.26		62.70	
30		18.61	28.46		55.23			25.79		4.99	36.12		65.44	
32		18.05	30.07		57.83			25.55		3.85	38.04		68.24	
34		17.47	31.72		60.51			25.31		2.69	40.01		71.12	
36		16.87	33.43		63.27			25.06		1.47	42.02		74.07	
38		16.25	35.18		66.11			24.79		0.22	44.10		77.10	
40		15.61	30.98		69.02			24.52	0.52		46.24		80.20	

(*Continued*)

TABLE 5.1 Comparative Temperature-Pressure Relations of Common Refrigerants
Saturated vapor pressure (*Continued*)

Number	11		12		22		113		114		500		502	
Chemical name	Trichloromono-fluoromethane		Dichlorodifluoro-methane		Monochlorodi-fluoromethane		Trichlorotri-fluoroethane		Dichlorotetra-fluoroethane		Azeotrope of dichlorodifluoro-methane and difluoroethane		Azeotrope of monochlorodiflu-oromethane and monochloropenta-fluorethane	
Chemical symbol	$CFCl_3$		CF_2Cl_2		$CHClF_2$		$CCl_2F-CClF_2$		$C_2Cl_2F_4$		73.8% CCl_2F_2 26.2% CH_3CHF_2		48.8% $CHClF_2$ 51.2% $CClF_2CF_3$	
							Pressure							
Temperature, °F	Gage	Vacuum	Gage	Vacuum	Gage	Vacuum	Gage	Vacuum	Gage	Vacuum	Gage	Vacuum	Gage	Vacuum
42		14.94	38.81		71.99			24.23	1.18		48.44			
44		14.24	40.70		75.04			23.93	1.86		50.69			
46		13.52	42.65		78.18			23.61	2.56		53.01			
48		12.78	44.65		81.4			23.29	3.28		55.39			
50		12.00	46.69		84.7			22.94	4.03		57.82			
60		7.73	57.71		102.5			21.02	8.13		70.96			
70		2.64	70.12		122.5			18.68	12.87		85.81			
80	1.61		84.06		145.0			15.87	18.34		102.5			
90	4.99		99.6		170.1			12.53	24.59		121.2			
100	8.90		116.9		197.9			8.59	31.69		141.9			
102	9.75		120.6		203.8			7.71	33.22		146.3			
104	10.63		124.3		209.9			6.82	34.78		150.9			
106	11.52		128.1		216.0			5.88	36.39		155.4			
108	12.45		132.1		222.3			4.93	38.03		160.1			
110	13.39		136.0		228.7			3.95	39.71		164.9			
112	14.35		140.1		235.2			2.95	41.44		169.8			
114	15.34		144.2		241.9			1.91	43.20		174.8			
116	16.37		148.4		248.7			0.83	45.00		179.9			
118	17.41		152.7		255.6	0.14			46.85		185.0			
120	18.50		157.1		262.6	0.70			48.74		190.3			

of a refrigeration plant. In the smaller sizes, reciprocating equipment is the only generally available equipment. In sizes up to about 150 tons, first cost will normally be in favor of reciprocating equipment. Also, reciprocating equipment adapts more readily than either of the other systems to direct expansion. Direct expansion describes a system in which the refrigerant is expanded in a coil that is located directly in the airstream and thereby directly cools the air. Centrifugal equipment, especially in larger sizes and where a water or brine cooling system is desired, offers advantages in compactness, flexibility, maintenance and operating costs, and adaptability to available power supply. These machines may be driven by slip-ring, synchronous, or squirrel-cage motors, or high- or low-pressure steam turbines.

Absorption refrigeration cycles offer definite advantages over other types for many water cooling applications. The machines are compact, light in weight, may use water as a refrigerant, have only very limited moving parts, and operate on low-pressure steam or high-temperature hot water. Whenever steam or high-temperature hot water is more readily available or cheaper than electric power, the absorption system may offer the greatest advantages. For example, the heating-system distribution system may be used in many cases as the power supply to an absorption machine for summer cooling.

Installation. The factors with which the maintenance engineer should be concerned at the time of installation include selection of the right type of system in the proper size units for flexibility of control; the proper location and/or foundation to avoid transmission of vibration or noise to occupied areas; adequate space for maintaining or repairing the equipment; an absolutely tight system; complete drying of the refrigerant side of the system by mechanical or chemical means before introducing refrigerant; and a detailed check-out of all safety devices at the time of initial start-up.

GENERAL MAINTENANCE

Responsibility. The maintenance engineer is responsible for operating and maintaining air-conditioning equipment so that it provides the intended benefits at the lowest possible cost. Specific operating schedules should be well thought out to meet owner requirements and keep hours of operation to a minimum.

Operating economies can be realized by such things as taking advantage of diversity factors and storage effect, maintaining outside air quantities to just above minimum requirements, proper control of condenser water temperatures, lower throttling ranges on leaving chilled water controls, and operating multiple machine installations at the most efficient operating condition for each individual machine, which usually is at or near full load.

Preventive Maintenance. A thorough preventive maintenance program should be followed, as opposed to a fix-it-when-it-fails approach. This too will lower operating costs by keeping equipment operating efficiency and reliability high. In addition, preventive maintenance will reduce downtime, which in many of today's installations is much more costly to an owner than the cost of making repairs. Preventive maintenance will also extend the useful life of the equipment.

To accomplish thorough preventive maintenance, it is necessary to establish what will be done, make sure it is done, periodically evaluate the effectiveness of the program, and make changes as required to improve it. All these must be done in order to have an effective program.

Checklists should be developed for all mechanical equipment showing each inspection item and the frequency of performance. The manufacturer's operating and maintenance instructions should be used in developing these checklists. A partial checklist for a centrifugal machine is shown in Fig. 5.3.

Report forms should be used by the maintenance personnel to indicate the preventive maintenance checks that are made and also to report on any repairs that should be made.

Some equipment manufacturers and servicing contractors have developed complete preventive maintenance packages and make these available to equipment owners. An evaluation should be made on the merits of using these outside services, especially on complicated and critical system components such as the refrigeration machines and their controls.

INSPECTION CHECK LIST

Machine Model _____ Location _____

Serial Number _____

	Weekly	6 Months	Annual
Review Operating Log	✓		
Check oil level, pressure, & temp.	✓		
Check purge operation	✓		
a. air discharge rate	✓		
b. water accumulation rate	✓		
Adjust refrigerant charge*	✓		
Adjust purge valves*	✓		
Adjust operating controls*	✓		
Check control safeties		✓	
a. chilled water lo-temp. cutout		✓	
b. refrigerant lo-temp. cutout		✓	
c. oil lo-temp. & lo-pressure cutout		✓	
d. chilled water flow switch		✓	
Check purge controls		✓	
Inspect starters		✓	
Tighten flexible couplings		✓	
Inspect compressor			
a. journal bearing **			
b. thrust bearing **			
c. shaft journal **			
d. oil pump **			
e. oil heater **			
Change oil and oil filter			✓
Inspect cooler and condenser			✓
a. tubes and tube sheets			✓
b. division plate and gasket			✓
c. flow switches			✓
d. clean tubes*			✓
Inspect and clean purge unit			✓
Check purge controls			✓
Check electronic controls			✓
a. replace vacuum tubes			✓
Leak test			✓

*As dictated by the Operating Log
**As per Manufacturers' Instructions

FIGURE 5.3 Inspection checklist.

Equipment Operating Logs. It is essential that operating personnel maintain hourly or daily logs on the primary refrigeration and heating equipment. The data would include various operating temperatures and pressures, oil and refrigerant levels, and control motor positions. These data are useful in evaluating machine performance and in troubleshooting. For example, on a centrifugal refrigeration machine, the difference between the condensing and leaving condenser water temperatures at known load conditions would help identify an operating inefficiency caused by air in the machine or scaled or dirty condenser tubes. Figure 5.4 shows a typical operating log for a centrifugal refrigeration machine.

Spare Parts. An inventory of the most frequently used replacement parts should be maintained. It should be controlled by a convenient recording system that will track usage and provide the data required to analyze stock levels and determine proper reorder points and quantities. Equipment manufacturers should be consulted when these inventories are first established.

Training. A thorough training program should be established for all operating and maintenance personnel so that they have a good understanding of the basic cycle and the operating characteristics of the mechanical equipment they are responsible for. Just as important, maintenance personnel should be well indoctrinated on the philosophy, as well as the mechanics, of a thorough preventive maintenance program.

Basic Requirements. Four important maintenance requirements are cleanliness, tightness, adequate lubrication, and operational safety devices. Cleanliness must be maintained throughout the

Operating Log
Hermetic Centrifugal Refrigeration Machine

Plant ——————— Machine serial no. ——————— Machine size ———————

Time	Cooler				Condenser				Compressor						Dehydrator		Operator's initials		
	Gpm				Gpm					Oil									
	Pressure	Shut-down refrigerant level	Water temperature		Pressure		Water temperature		Pressure temperature	Level	Temperature	Pressure	Seal drain reservoir level	Motor amps or vane position	Refrigerant level	Condenser pressure			
			In	Out		In	Out												
1	2	3	4	5	6	7	8	9	10	11	12	13	14	15	16	17	18	19	

FIGURE 5.4 Operating log—hermetic centrifugal refrigeration machine.

air-conditioning system. This applies just as much to the equipment room as it does to the internals of the mechanical equipment.

Dirt and scale in equipment will interfere with heat transfer, fluid flow, and lubrication and will cause cycle inefficiencies and premature failures. Tightness is important to avoid loss of oil and refrigerant and to prevent entrance of water, air, and other noncondensables. Air and water can change the chemical characteristics of some refrigerants, thereby setting up corrosive conditions or lowering the efficiency of the refrigeration cycle. Motors, dampers, and so on, should be lubricated regularly, and lube systems in refrigeration equipment should be kept in good operating condition. Failure to do so will cause premature failures of bearings and bearing surfaces and could lead to major equipment damage.

All safety devices should be kept operative. They prevent minor problems from causing major failures and should never be intentionally bypassed.

EQUIPMENT MAINTENANCE

This section provides data on each general type of equipment used in air-conditioning systems. It includes a brief description of each and gives specific suggestions for operation and maintenance. Because of the many models and variations of each type of equipment, the data will not always apply to all equipment. *It should be understood that the equipment manufacturer's instructions should always take precedence.* The data are presented in two parts; the first covers the equipment, the second covers general subjects that apply to more than one equipment type. For easy reference, the material is presented in alphabetical order as follows:

Equipment:

Absorption machines	Dampers	Humidifiers and
Air washers	Drives	dehumidifiers
Centrifugal compressors	Economizers	Pumps
Coils	Fans	Rooftop units
Condensers	Filters	Room air conditioners
Coolers	Heaters	Self-contained units
Cooling towers	Heat pumps	

General:

Freeze protection

Water conditioning

A discussion of other system components, such as air compressors, controls, electric motors, and circuit breakers, can be found in other sections of this handbook.

Absorption Machines

Some absorption refrigeration machines, especially in the 5- to 25-ton range, use an ammonia-water cycle in which ammonia is the refrigerant and water is the absorbent; however, discussion in this section will be limited to machines with capacities between 100 and 1800 tons which use a lithium bromide-water cycle (Fig. 5.5). In these machines, water acts as the refrigerant and a lithium bromide solution is the absorbent.

The refrigeration effect is produced by maintaining a high vacuum in the evaporator section (between 0.2 and 0.25 in of mercury absolute). At this low pressure, the refrigerant (water) boils at

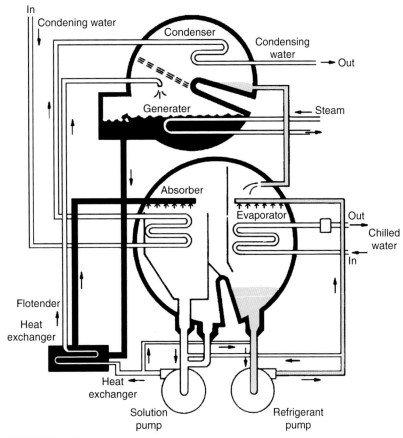

FIGURE 5.5 Absorption refrigeration cycle.

35 to 40°F. The heat required for the boiling action is extracted from the chilled water being cooled. To maintain the high vacuum in the evaporator and to allow the cycle to continue, the water vapor formed by the boiling refrigerant is continuously absorbed by a lithium bromide solution in the absorber section. Since this water-vapor absorption tends to reduce the concentration of the solution, as well as its ability to absorb more water vapor, the weak solution is pumped to a generator where it is reconcentrated by boiling off the absorbed water. The heat required for reconcentration is provided by low-pressure steam, high-temperature hot water, or in the case of direct-fired machines, oil or gas burners. The strong solution is then returned to the absorber, and the water vapor is condensed in the condenser and returned to the evaporator section.

Condenser water circuits in both the absorber and the condenser sections of the machine remove the heat produced by the cycle and the heat extracted from the chilled water.

Leaktightness. Because of the high vacuum that exists in the absorber-evaporator section of the machine, it is very important that a high degree of leaktightness be maintained. Even small leaks will allow sufficient air and other noncondensables to enter the machine and disrupt the water-lithium bromide cycle. The disruptive effect of noncondensables can usually be determined by analysis of the log readings. Further analyses can be made by taking samples of both the lithium bromide solution and the water, measuring the temperature and specific gravity of each, and comparing these values with standard conditions as indicated by various charts and/or equilibrium diagrams.

Purge Units. Purge units are provided to remove air and noncondensables and maintain proper machine operation as long as the leak rate is at a minimum. Purge units and purging techniques vary depending on machine design. Follow the manufacturer's recommendations for operation and maintenance.

Pumps. Pumps are used to circulate the refrigerant and lithium bromide solutions within the machine. On earlier machine models with open-type pumps, mechanical seals are used to prevent leakage of noncondensables, water, and lithium bromide. These seals should be replaced every 2 years. Motors on open pumps should be lubricated every year.

More recent machine models use a hermetic design. The bearings, motors, and other internal components should be inspected approximately every 4 to 7 years, depending on operating conditions.

Service Valve. The diaphragms on service valves located on the machine should be replaced every 2 or 3 years.

Safeties. Various controls such as the refrigerant and chilled water low-temperature cutouts and the chilled water and condenser water flow switches should be checked every 6 months for proper operation.

Leak Testing. Consult the equipment manufacturer for the latest leak testing techniques that are in compliance with EPA regulations regarding refrigerant emissions.

Other Maintenance. Manufacturer's operation and maintenance instructions should be used in establishing a schedule for other preventive maintenance functions. These will depend on machine design and may include such things as reclaiming the lithium bromide solution, adding octyl alcohol, performing running vacuum tests, lubricating capacity control motors and linkages, inspecting and cleaning solution spray headers, periodic analyses of the lithium bromide solution, and making adjustments to the solution, as required, to maintain its effectiveness.

Air Washers

Air washers remove lint, dirt, and other contaminants from incoming air and discharge cleaned air. Air enters the unit, passes through the fan, through the diffuser, and into the spray section, where it is blanketed by water sprays. The water removes the contaminants and the washed air is discharged from the system. The water used in the washing process must be cleaned before recirculating. This is accomplished by passing the water through a strainer, where foreign material is collected and flushed from the system. Maintenance of air washer equipment should include the following procedures:

Cleaning. Remove lint and dirt periodically from damper blades and linkage, if so equipped. Remove all rust appearing on outside damper blades. Check fan and fan motor and remove any accumulation of foreign material. Check the entire spray section at regular intervals. Remove lint and dirt from spray nozzles and piping. Check for any plugged nozzles which could affect pressure, quantity, or distribution of water. Remove lint and dirt from drain bump and from float ball on water level float switch.

Clean unit standpipes. Remove lint from eliminator blades by directing a stream of water between eliminator blades. Use a wire hook to remove any dirt remaining on the blades. Drain water storage tank periodically and remove all foreign matter. If strainer is not self-cleaning, clean it weekly to ensure a constant flow of water. If strainer is nonautomatic, establish the required cleaning cycle in accordance with needs. Check rubber seal and entering end of drum for tightness and remove any lint or dirt. Clean sludge from the collection pan.

Fan. Check fan blade angles and clearance. Adjust where necessary. Remove and dismantle fan motor periodically in accordance with manufacturer's recommendations for inspection.

Lubrication. Check fan bearings and lubricate approximately every 4 months with water-resistant grease. Inspect main shaft bearings of automatic strainer, if so equipped. If bearings are operated continuously under water, lubricate once a year. If operated out of water, lubricate once every 3 months.

Centrifugal Compressors

Centrifugal compressors are used in large-capacity refrigeration systems (see Fig. 5.6). Single compressor capacities range from 100 to 10,000 tons. The refrigerants used include R-11, R-12, R-22, R-113, R-114, and R-500. Depending on the compressor type and capacity, however, some are among the refrigerants for which the EPA has mandated a stop in production. Alternates may be available, such as 134a for R-12. Check with the equipment manufacturer for recommendations on alternate refrigerants for your specific machines.

Centrifugal compressors consist of one or more impellers mounted on a shaft which is rotated at high speeds within a housing. Refrigerant gas entering the eye of the impeller is discharged by centrifugal force (hence the name) to the tip of the impeller at a high velocity.

From here the gas enters a diffuser where velocity pressure is converted to static pressure and then is discharged into the condenser where the refrigerant is condensed back into a liquid and returned to the cooler.

Lubrication. Use only high-grade oil that meets the compressor manufacturer's specifications. Check the oil levels daily and maintain the proper levels in all parts of the lubricating system.

Oil levels should be checked both during operation and when the machine is shut down. Oil level should be marked on the sight glass for reference. Check the operating oil pressure and temperature regularly. Adjust these, if necessary, to meet the compressor manufacturer's specifications. Check the

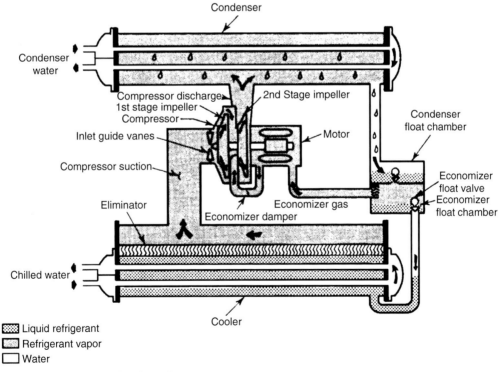

FIGURE 5.6 Centrifugal refrigeration cycle.

oil temperature and pressure cutouts every 6 months for proper operation. During machine shutdown, maintain the reservoir oil temperature recommended by the compressor manufacturer to minimize refrigerant absorption by the oil. Starting the machine with excess refrigerant in the oil will cause excessive foaming, loss of oil, and possible bearing failure. If the oil becomes contaminated during machine repairs, it should be replaced. Refer to the compressor manufacturer's maintenance instructions before changing the oil. Make sure that the replacement oil is compatible with the refrigerant being used.

Bearings. Bearing maintenance consists mainly of maintaining clean oil in the lubricating system. Oil filters should be changed at least once a year. If bearing temperatures rise above normal, check the oil cooling system and oil supply to the bearings. Inspect the bearings once a year and replace when necessary.

Oil Heater. The oil heater should be on during compressor shutdown. Refer to the compressor manufacturer's operating instructions.

Safeties. Safety controls should be checked approximately every 6 months. These include the chilled-water and refrigerant low-temperature cutouts, condenser high-pressure cutout, oil low-pressure cutout, and chilled water and condenser water flow switches.

Leak Testing. Regularly check all compressor joints for refrigerant leaks, and check the purge for air and water leaks.

Purge. The purge unit and its controls should be checked regularly for proper operation. All parts of the purge unit should be checked regularly for corrosion and wear and replaced when necessary. If the purge discharges air frequently, it means air is entering the machine. If it continuously removes water, it indicates the possibility of a water leak. The source of either air or water should be found and corrected as soon as possible to avoid major equipment damage. Keep purge sight glasses clean and check regularly for water accumulation.

Refrigerant. Once every 2 years a refrigerant sample should be removed from the system and analyzed by a competent laboratory. If the refrigerant is contaminated, consult the compressor or refrigerant manufacturer for recommendations.

Extended Shutdown. Refrigerant absorption by the oil can be minimized by proper operation of the oil heaters. If the machine is in an area subjected to freezing temperatures, remove the water from the oil cooler. It may be desirable to remove the oil charge during extended shutdown. However, if the oil is not removed, do not shut off the oil heater.

Coils

Maintenance. Cooling coils and heating coils are made with prime or extended surface tubing. Maintenance in either case is essentially the same and involves two features—tightness and cleanliness.

Tightness. In the case of coils handling refrigerant for direct cooling, a tightness check should include inspection of all joints in the piping connections to the coil, and seasonal inspection for leakage in all return bends and other joints in the makeup of the coil. All leaks should be repaired promptly.

Freeze Protection. Coils handling water should be observed periodically for leakage and must be given careful and thorough attention seasonally to protect against damage due to freeze-up. Positive protection against damage to water coils in the outside airstream can be obtained by removal of all water or by using an antifreeze solution in the coil. Antifreeze acceptable to code and underwriter authorities and in keeping with plant safety rules may be circulated with residual water in the coil and then drained off and reused for the next coil, or next season, until its dilution approaches the minimum concentration for safety. Hand- or power-operated circulating equipment may be used for this antifreeze preparation. Ethylene glycol is a commonly used antifreeze.

Cleaning. Cleaning of coils varies widely as to need and effectiveness. Coils exposed to lint, rug nap, or similar airborne materials will require frequent mechanical cleaning of the entering side. Maintenance personnel should check all coils weekly until a suitable schedule is established. Coils that are several rows of tubes deep may require chemical cleaning.

Coil surfaces from which water is evaporated will sometimes become coated with chemicals left behind by the water. Evaporative condenser coils are particularly subject to this problem. The removal of some of these deposits may be difficult and usually requires chemical cleaning.

The inside of coils handling refrigerants will not need cleaning unless subjected to some abnormal operating condition. The inside of water cooling coils in a closed system will not require cleaning provided suitable water treatment is used.

The cleaning of the inside of any coil or piping system can be accomplished by suitable chemical treatment. Such cleaning will uncover all latent leaks, however, and for that reason, as well as many others, it must be used with judgment.

Air-Cooled Condensers

Air passing over finned tubes of an air-cooled condenser removes heat from a refrigeration system. Condensers may be cooled by gravity or by forced airflow. Forced air-cooled condensers are either draw-through or blow-through. Either a direct- or belt-drive propeller or centrifugal fan may be used to move air through the coil.

Capacity may be controlled to meet load requirements by cycling fan motors if multiple fans are used. Changing the fan motor speed is also a good method, provided that the motor design permits this. Maintenance procedures should include:

Inspection. Condenser must be inspected periodically; how often depends on usage and location. Coil should be inspected for physical damage and airflow restrictions. Examine fan for bent blades and alignment. Belt drive, if used, should be inspected for wear, proper belt tension, and sheave alignment. Motor power and control wiring should be examined for tight connections and deterioration of insulation (and contacts, where possible). Ammeter and ohmmeter tests should also be made. Inspect motor suspension and support for tightness and isolator deterioration.

Cleaning. The condenser should be cleaned at the beginning of each cooling season and at regular intervals during the season, with length of interval depending on usage and location. Airborne dirt may be removed from the air-inlet screen, coil face, and fan by brushing, vacuum cleaning, or spraying with low-pressure water. Preferable cleaning procedure for dry, nongreasy coils is blowing out coil from outlet face with compressed air, then vacuuming inlet face. In areas where oil and grease have accumulated on the coil, a weak solution of detergent in hot water may be brushed onto the dirty coil face. The motor should also be cleaned of any surface dirt, and vent openings should be cleaned by vacuuming or compressed-air blast. If water is used to clean the coil, care must be taken not to spray motor vents. They should be masked shut prior to coil cleaning. Do not use water to clean open-type motors. Finally, clean leaves and debris from the condenser base pan and clean drain holes.

Lubrication. Follow manufacturer's lubrication instructions.

Evaporative Condensers

In an evaporative condenser, heat is absorbed from the coil by the evaporation of water. Water is pumped from a pan in the base of the unit, passes through a series of spray nozzles, and flows over the refrigerant coil. At the same time, air enters through the inlet at the base, passes through the coil and water spray, through eliminators which remove free water, through the fan, and is then discharged from the unit.

The following maintenance procedures can reduce repair costs and improve the efficiency of the condenser.

Cleaning. Clean equipment when unit is shut down for the season. Year-round systems should be cleaned at least once annually. Airborne dirt may be removed from the coil surface by washing down with a high-velocity jet of water or steam. *Caution: If steam is used, coil must first be pumped down.* Dirt may also be removed from the coil surface with a nylon brush. Swab each coil and flush out with clean water.

Evaporative condensers are particularly subject to chemical surface deposits from local water supplies. A water-treatment program should be employed, if possible, to protect the coil from these deposits. Inspect the coil regularly to detect the presence of scaling. If the coil has a deposit of carbonate scale, it may require chemical cleaning. Check the surface of the condenser regularly to detect any tendency to rust. Clean off all rust spots with a wire brush and paint with a rust-resistant paint.

Air inlet screen, water distribution pan, and pump screen should be checked frequently and cleaned if necessary.

Fan Section. Periodically check the current input to the fan motor. Check fan belts at least twice a year. When belts are replaced, replace only in matched sets. Check fan bearings and fan alignment.

Lubrication. If pump motor has grease fittings or cups, check bearings at least twice a year and grease if necessary. Lubricate fan bearings at least twice a year.

Refrigerant Joints. Check regularly the joints of condenser piping and refrigerant piping connections. Repair all leaks promptly.

Freezing Precautions. Evaporative condensers must not be operated with a wet coil when air circulated is below 32°F. If unit is subjected to freezing temperatures, take precautions to prevent damage to pump, drip pan, recirculating-water piping, and supply piping. When unit is shut down, drain all water from the pump. Shut off water supply piping at a point not subject to freezing, and drain supply piping and coil beyond this point. Blow out piping and coil or add antifreeze to ensure complete protection.

Water-Cooled Condensers

Proper maintenance can reduce repair costs and improve the efficiency of operation. Excessively high condensing pressures or temperatures are usually caused by either air in the machine or dirty or scaled tubes.

Cleaning. Local conditions may make it necessary to clean the condenser tubes at more frequent intervals, but under normal conditions cleaning once each year should suffice.

Cleaning of straight tubes can be done either physically or chemically. The simplest process is to swab each tube with a suitable brush, and flush out with clean water. In this method the brush selected should be of a size and bristle stiffness to remove deposited silt and soft scale but must not have bristles stiff enough to score the metal of the tube.

Metallic spinners or other devices that may cut the tube surface should not be used with nonferrous tubes.

Another very effective method is the use of a water gun to drive a nonmetallic plug through the tube. Once set up, the cleaning of an average condenser requires only a few hours by this process.

Chemical cleaning involves the circulation of a suitable inhibited acid solution through the tubes, followed by flushing with clean water, circulation of an inhibitor, and flushing with clean water. Time is an important element and varies with each application. The water-treatment companies are generally able to give detailed assistance on this operation. Chemical cleaning has the distinct advantage of being equally applicable to straight or bowed tubes or to shell and continuous-coil equipment and is the only effective way to remove hard scale deposits.

Water Treatment.   Water treatment may be required to protect the water circuit against scale and corrosion.

Refrigerant Joints.   All piping joints should be checked regularly for leakage, and all leaks should be repaired promptly.

Freeze Protection.   Condensers subject to freezing temperatures must be protected against damage in winter (see Coils—Freeze Protection).

Coolers

Types.   Shell-and-tube and shell-and-coil coolers require essentially the same attention as similar condensers. A cooler may be used in a completely closed circuit where the water introduced to the system is for makeup only. In this case, and with suitable treatment applied to the water, it may be unnecessary to clean cooler tubes with the frequency specified for condensers. An annual inspection of the condition of tubes is nevertheless warranted.

Insulation.   Coolers are insulated, and care should be exercised to keep this covering in good repair.

Refrigerant Joints.   Joints of coolers and the refrigerant piping connections should be checked regularly for leakage and all leaks repaired promptly.

Refrigerant Float Chambers.   Coolers on centrifugal machines usually have float valves to meter the flow of refrigerant from the condenser. The float chambers should be inspected annually and the float valve checked to make sure it operates freely through its full travel.

Control.   Control of direct-expansion coolers used with reciprocating compression is generally by thermal-expansion valve. Once the valves are adjusted to job conditions, they should not be altered or tampered with. Flooded coil coolers are frequently controlled by a high-side float installed in a surge drum. This operates on essentially the same basis as any float and ball cock regulating the level of liquid in an enclosure. Maintenance in this type of equipment is nominal.

Flooded shell-and-tube coolers used with reciprocating compressors are generally controlled by a low-side float. This float with the valve that it regulates is as important to the system as a carburetor to an automobile. Adjustment is critical, and the device should not be tampered with or modified once a correct setting has been made.

Oil Return.   Oil return from the flooded cooler to a reciprocating compressor is most important. Oil return lines are normally provided. The rate of flow through this line may be critical. The hand valve should be opened to the point that oil concentration in the cooler will be kept low (as indicated by the color of liquid flowing through the bull's eye in the oil return line), and not open any wider than necessary to accomplish this. Coolers used with centrifugal machines for air-conditioning duty usually have straight-through tubes with water in the tubes and refrigerant in the shell. Oil lost from the compressor can accumulate in the cooler and reduce heat transfer. It can be removed only by distillation. Keep records on all amounts of oil added to the system.

Cooling Towers

Cooling towers are made in sizes that will handle the heat rejection from 1 ton of refrigeration up to several thousand tons. The larger units are designed for the specific job and usually are made up of several cells, each with its own operating components.

Towers usually fall into one of three groups: atmospheric, forced, or induced-draft. The atmospheric tower has no fan and is commonly used only in smaller sizes. In a forced-draft tower, the fan forces air through the water spray. An induced-draft tower has a fan to draw air through the spray. Tower efficiency is related to the effectiveness of air and water contact, and needs finely divided water or a great amount of wetted surface exposed to airflow. Large air quantities are handled. Noise is inherent. Operating weight is considerable. Tower site selection must take these facts into consideration, as well as the requirement of free movement of large quantities of natural outside air.

A forced or induced-draft cooling tower is an assembly of several functional components. A fan, sprays, motor, drive, and starter are included. Some installations are made of an air washer, housed fan, and apparatus casing similar in all respects to air-conditioning equipment. Maintenance for these parts is given below.

Exposure. Cooling towers use outside air and are usually outdoors. Protection must be provided for motor, starter, disconnect switch, and drive. Also, the structural part must be suitable for such exposure.

Maintenance. Steel casing, basin, and framework should be painted regularly with a good protective paint. In some locations such painting will be necessary annually to prevent rust and deterioration. A regular schedule should be established for such maintenance, and the time between paintings should not exceed 3 years.

Redwood will last without paint. The life of redwood is shortened by painting all surfaces. Painting of a redwood tower, therefore, is for appearance and should be attended as needed. Redwood or cypress fill should not be painted. Bolts in wood towers should be checked annually and tightened while the parts of the tower are dry.

Fans. Fans should turn freely. Propeller-type fans generally can be adjusted for capacity by pitch adjustment of blades. Uniform pitch of all blades is important, and adjustment must keep power requirements within motor capacity.

Alignment and Lubrication. Gear-driven fans require an annual check of alignment, of motor to gear to fan. Alignment must be within limits of couplings, or normally within 0.002 to 0.003 in. Oil level in the gear should be checked weekly. Oil should be replaced annually or after 3000 operating hours, if the tower is used continuously.

Belted drives should be checked monthly, and belt tension and alignment kept in proper adjustment. Belts must not be tight enough to impose undue bearing load but must be tight enough to avoid slipping (see Drives).

Water Distribution. Water distribution must be checked and kept uniform. Gutter distribution must be checked for obstruction. Spray distribution must be checked for clean spray nozzles and design working pressure. Most cooling tower sprays, where fill is included in the tower, operate at approximately 3 psi.

Eliminators. Eliminators must be kept free from algae growth and in good repair. Metal eliminators should be cleaned and repainted annually.

Algae. Algae growth should be cleaned from all parts of the tower, and water treatment put into use to prevent regrowth and to protect pipe and equipment.

Cleaning. The water basin should be drained and hosed out weekly. The strainer should be checked and a regular routine established for cleaning.

Water Level. The float valve should be adjusted to maintain the water level high enough in the basin to prevent serious vortexing at the water outlet and the supply must be adequate to make up for water losses due to evaporation and drift.

Winter Protection. The basin and all piping, valves, and so on, exposed to weather conditions must be drained for shutdown during freezing weather.

Dampers

Automatic Dampers. These dampers are motor operated and under automatic control. Sluggish response to command of the controlling instrument results in poor regulation of conditions.

Maintenance. For good results all operating parts of the control system must move in proper relationship. All automatic dampers should be checked for freedom of movement and lubricated at bearing points. Surplus oil should be wiped off. Blades should be checked in the closed position to be sure that all close tightly, and adjustment should be made to the linkage to close any open blades. Operating motors should be observed through an operating cycle to check defects. The damper-motor anchorage should be checked and verified.

Relief Dampers. A large quantity of outside air is used for conditioning some buildings during certain seasons. This will build up the pressure inside the building to interfere with distribution, and make the opening and closing of doors difficult. Relief dampers are installed to prevent this pressure buildup.

Relief dampers made up of a series of lightweight aluminum blades are provided in some buildings. The purpose of these dampers is to permit passage of excess air from the building.

Maintenance. Trunnion bearings, unless of the oil-impregnated type, should receive a film of light machine oil (with all surplus wiped away) each spring. Blades may become bent or warped and should be checked for complete closing twice annually.

Damaged blades should be repaired or replaced. Dirt, soot, lint, and so on, should not be permitted to accumulate on blades, as this will increase weight and present an unsightly appearance. Calking, where it is used to make damper frames tight to the structure, should be checked and repaired as needed.

Fire Dampers. Fire dampers are installed as a protective feature for life and property. To be sure that the intended safety exists, each damper should be checked annually. The linkage (with fusible link) should be disconnected, to be sure the damper will close under its own weight. Lubrication of fire-damper bearings is usually unnecessary; however, the presence of an oil film will retard oxidation and for that purpose would be helpful.

Occasionally, a damper will be found closed because of a defective fusible link, a source of heat sufficient to fuse the link, or blade weight in excess of the mechanical strength of the fusible link. Corrective measures are, respectively, replacement of the defective link, a nonmetallic shield between the link and the source of heat, or the use of multiple fusible links in parallel (arranged so that fusing of either one will release the damper).

Splitter and Hand Dampers. A multitude of damper arrangements are used to regulate the distribution of air. Under most conditions these dampers are fixed at the time of initial adjustment and will seldom need attention unless the load distribution is altered from original design. In some instances dampers are altered by hand to adjust for seasonal conditions. An example is the hand-operated outside air damper, which is normally closed to minimal position in summer to conserve refrigeration, opened during the intermediate season for the same purpose, and closed to minimal in the winter to save heat.

Drives-Belt and Direct-Connected

Service Life of Belt Drives. The conventional V-belt drive gives exceptionally good service when properly maintained. Two adjustments are paramount—alignment and tension. Misalignment causes excessive belt wear, imposes unnecessary load on the motor, and in many cases accelerates bearing wear. Excess tension imposes overload on the motor and oil bearings, while belts too slack wear out rapidly owing to slippage.

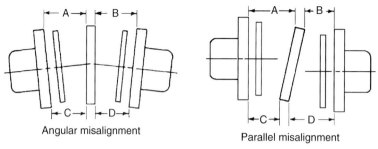

FIGURE 5.7 Belt and coupling alignment.

Alignment.   Realignment should always be carefully done after a pulley is removed and should be checked annually even though no changes have been made. Alignment is made by adjusting the respective pulleys so that a taut chalk line touches both edges of the rim of both pulleys simultaneously. Increasing belt tension causes all motors to assume an out-of-line position (see Fig. 5.7). This must be corrected.

Tension.   A generally applicable rule for correcting tension is difficult to state. The aim should be to run the belts with just enough tension to avoid slippage either at start or during operation. One method is to loosen the belts just to the point where slippage occurs at the start (usually evidenced by belt squeal), then tighten just enough to eliminate the slippage.

Adjustable Pulleys.   Some driver pulleys provide a means of adjusting the width of the slot, which changes the pitch diameter (the diameter of the line of contact between pulley and belt) and changes the speed of the driven pulley. In some multigroove pulleys, it is not possible to keep all faces of the slots in line because of the width changes mentioned. In this case the misalignment should be divided so that the centerline of the two pulleys is in exact alignment.

Replacement.   Minor variation occurs in the manufacture of V belts, which has led to the practice of selection of matched belts for any multibelt drive. As belts are used, they stretch; however, the stretch is about equal for each belt. The result is a pretty well matched set of belts for the life of the drive. Replacement must be made by replacing a complete set of belts.

Couplings in Direct-Connected Drives.   All direct-connected equipment should be carefully checked each year for alignment. Exact requirements will vary from one piece of equipment to another but, generally, all direct-connected equipment should be in exact alignment when operating at normal full-load conditions, and after having operated long enough to rise to full operating temperature. Allowance must be made for load variation, dimensional variation due to expansion from increased temperature, and any other condition that can alter the relationship of driver and driven equipment.

Rigid couplings will need little attention beyond maintaining alignment. Some couplings depend upon the flex of a diaphragm for flexibility and require little attention beyond maintaining alignment and tightness. Flexible couplings must be kept in alignment as indicated and must be lubricated in accordance with the manufacturer's direction. Some of these couplings are oil-lubricated, and others are grease-packed. In the absence of detailed instruction, replace with the same lubricant used in the original application.

Economizers

The concept of utilizing outside air for cooling when conditions are right is not new but, with emphasis now being placed upon energy conservation, the concept is enjoying considerable popularity.

If a building has a cooling requirement regardless of outside temperatures, for example, a center zone surrounded by conditioned space, outside air can be used when its dry bulb temperature and water vapor content are acceptable. Cooling is accomplished with the application of what is commonly known as an economizer section or plenum, usually a part of the fan coil unit or central air apparatus. Outside and return air dampers are linked to a single damper delivery or supply air temperature.

The system must also provide a means of exhausting the quantity of outside air taken in to prevent excessive building pressurization and/or placing a limitation on the quantity of air that can be taken from outside. This is done with gravity relief dampers or power exhaust fans depending upon overall system design. Elements of control are:

1. A device to switch operation from refrigeration cooling to economizer cooling, and back. Early designs used a simple outside air thermostat set to make the change at about 55°F, but a more recent innovation uses an enthalpy controller that will sense both temperature and moisture content of the outside air. This controller will signal the use of outside air when its properties will reduce the load on the refrigeration system and will transfer completely to economizer cooling only when conditions permit.
2. A mixed-air thermostat (MAT) senses supply air temperature and will dictate the position of dampers to maintain its set point. The MAT signal rotates the damper motor shaft in one direction or the other by unbalancing the current flow of a parallel circuit with the coils of a balancing relay. When the new position of dampers satisfies the MAT setting, current flow through the balancing relay coils will equalize, stopping the damper motor.

The economizer control system also makes provision for setting the outside air damper to a position that will take in the required amount of outside air for ventilation when operating in the heating or cooling mode. (*Note*: Some systems may have a requirement for heating.)

When the system shuts down, the outside air damper will close. Performance of an economizer depends upon proper control settings and normal maintenance for the type of devices contained in the system.

Fans

Capacity. Fan capacity is measured in cubic feet per minute (cfm). Fans are selected to deliver the required capacity at the operating pressure required for delivery through the distribution system. Any increase in the operating pressure on a fan reduces the capacity. The fan handles more air as the operating pressure is reduced. The horsepower required increases as the cfm increases.

The frictional resistances to airflow which occur in the outside air intake, coils, filters, washer, duct, and so on, combine to fix the operating pressure of the fan. It will be evident that any increased resistance, such as dirty filters, dirty coils, or restriction of any kind, will reduce the air quantity. Any major reduction in operating pressure may overload the motor by permitting the fan to handle more than rated capacity. Some fans are protected from overloading by features of design, but many are not. Operation of the system with filters out, with access doors standing open, with outlets removed, or under similar conditions may cause overloading of the motor. When it is necessary to operate under any such conditions, the motor should be observed frequently for overheating.

Cleaning. In many locations, fine dust particles find their way into the fan and cling to the blades of the wheel. This should never be allowed to build up, as it will increase operating costs by lowering efficiency and by disturbing the balance of the fan wheel. Fan wheels should be checked as often as operating conditions warrant. Under many conditions the entire inside of the fan should be cleaned and repainted annually.

Bearings. The larger housed fans are normally provided with pillow block, sleeve, or ball bearings. Bearings on these fans are almost all self-aligning. Sleeve bearings are ring-oiled. Ball bearings are generally grease-packed, and provide a means for introducing grease.

The bearing liners of the sleeve bearings are split for ease in assembly and are supported by a ball-and-socket assembly. In the replacement of the bearing assembly, care must be exercised to avoid distorting the bearing shell by overtightening the ball-and-socket adjustment screw.

Locking collars must be properly secured to prevent shaft rotation in the raceway. If this is not done, the shaft may be badly grooved. Should oil be drawn from the bearings, and lubrication instructions have been followed, the felt retainers and slingers should be examined by removing the top housing of the bearing and replacing worn seals.

Direction of Rotation. Most housed fans will deliver some air when running in the wrong direction. The quantity will not be up to requirements and the fan efficiency will be very low. Changes in power source may reverse the motor rotation. The operating staff should observe and check direction of rotation occasionally.

Adjustment of Wheel in Housing. The position of the fan wheel with respect to the inlet is important for efficiency and noise. As a general rule, operation as close as possible without actual striking is the correct position. On unitary equipment with multiple-fan wheels on a common shaft, care must be taken to adjust the position of bearings to avoid striking, as the assembly is designed to operate with close tolerances.

Alignment. Alignment of fan and motor as outlined under Drives is important and should be checked. All bolts in equipment and foundation should be checked annually for tightness.

Lubrication. Babbitt sleeve bearings in normal room temperatures use a high-grade automobile engine oil, SAE 20. Bronze sleeve bearings in normal room temperatures use a high-grade automobile engine oil, SAE 40. Do not use detergent oils. When located in an area of high temperature or when handling air at temperatures over 150°F, special consideration must be given the selection of oil. The local oil companies will select a suitable lubricant for unusual conditions.

Ring-oiled sleeve bearings have a level indicator and oil filter cup located on the side of the shaft where rotation is downward. This cup should be filled to within $1/8$ in of the top (not overflowing) when the fan is shut down. Oil should be drained and replaced with fresh clean oil after 2000 hr of operation. If the oil removed is very dirty, the bearing and well should be flushed out with a light machine oil before new oil is added.

Grease-lubricated bearings should be filled with a good grade of soda-base grease at fixed intervals. Under normal circumstances, grease should be added after 1200 to 1500 operating hours. The bearing should not be overfilled so that the grease retainer felts are forced out, and all excess grease and accumulated dirt should be wiped away.

Vortex Dampers. Some fans are equipped with special dampers, called vortex dampers, at the fan inlet. These dampers are made up of many operating parts, and when used are generally under automatic control. It is most essential that all moving joints of these dampers be carefully and thoroughly lubricated at least once a month. Observations may indicate the need for more frequent maintenance under severe conditions.

Filters

Purpose. A porous material through which a fluid stream passes for removal of solid particles is called a filter. Filters used in the airstream are called "air filters," in the water stream, "water filters," and so on. A filter is used for the very purpose of getting dirty. The more completely the dirt is removed, the more efficient the filter. The more dirt removed by the filter, the greater the need for it. Filters simply must be cared for, and attended to regularly.

Types. There are four general types of air filters:

1. The cell type where the filtering medium is thrown away when dirty, and a clean medium installed
2. The cell type where the medium is cleaned cell by cell and reused

3. The continuous-cleaning type, where some mechanical means is provided to remove dust accumulation from the filtering medium
4. The electrostatic filter, where dust particles are charged electrically and accumulated on a plate of opposite electric charge

Examples of the throw-away cell type are the spun glass mats contained in cardboard frames. Similar cells are made using paper, steel wool, brass wool, or mats of steel wool without a paper frame. There are also cells where a heavy porous paper is stretched over a suitable frame. In the latter, paper replacement is furnished in rolls and used as needed to recover the frames. Cleanable cells are made in many varieties. A common one is the metal mesh included in a metal frame, which is washed when dirty and recharged with a "sticky" surface by dipping in suitable oil. There are also cleanable cells where a water hose is used for washing down.

An example of the continuous-cleaning type is a series of overlapping perforated plates moving as a belt over upper and lower drums. The assembly is moved slowly by a ratchet mechanism. The lower drum is mounted in a tank of special oil. Each plate is released with a snap into the oil bath, where accumulated dust is washed away.

When to Clean. Commercial filters have a great holding capacity and restrict airflow a relatively constant amount for a fairly long period. Beyond this period resistance to airflow builds up rapidly. Observations of operating conditions at different seasons of the year will indicate the rapidity with which the filter collects dust. From those observations, a regular schedule should be established for attention to filters. A draft gage will show when filters need attention. A gage may be permanently installed at each filter bank or a portable unit may be connected to outlets provided for checking.

Cleaning Procedure. Throw-away cell-type filters usually require the use of new cells for replacement when dirty. A supply of spares should always be on hand. Many applications use two layers of throw-away cells in series to airflow. Some economy may be effected by replacement of the cells on the air leaving side with new cells, and reusing the cells from the leaving side on the entering side.

Paper should be replaced on that type of cell when enough dust has accumulated to build up resistance.

Installations using cleanable oiled cells are usually provided with spare cells and dipping tanks. Dirty cells should be thoroughly washed with a strong solution of washing soda, allowed to dry, dipped in an oil bath, and drained for several hours before reuse. The oil used is special and should be obtained from the filter manufacturer. *Do not use lubrication oils for coating filters of any kind.*

Cells designed for hose cleaning should be maintained in full accordance with manufacturer's directions. Satisfactory results depend upon careful adherence to routine as established for each job.

The continuous-cleaning filter requires routine removal of sludge from the oil basin, oiling operating motors annually, and greasing of the chain drive as determined by examination. Oil must be added as needed to maintain the level at the indicated point. Use only an oil recommended by the filter manufacturer. *Do not use lubricating oil for replacement.* The filter must be put into operation at least 12 hr before the fan is started so that the filter grid will have time to complete one circuit through the oil bath. The filter should not be turned off overnight or for week-end shutdown. Electrostatic filters require careful adjustment and maintenance. The manufacturer's written instructions should be followed explicitly. Do not undertake adjustments of the filter without a full understanding of needs. Do not interfere with protective features like the disconnect and manual reset switches, or the time element on the opening devices of access doors. These are provided for protection from the high electrical potentials at which the filters operate. Cleaning is important and is usually done with a water hose.

Oil and Water Filters. Oil filters of the multiplate or felt-on-wire frame type may be used in circuits of oil or water. This multiplate type is equipped with a lever for rotating alternate plates to clean the strainer. Conditions will vary for each job, and only direct observation can determine frequency for routine attention to these filters.

As a starting procedure, the filters should be checked weekly, and the operating lever turned several revolutions to dislodge and collect dirt from the surface of the plates. Operating levers should not be forced, as this will injure the plates and wreck the filter. Cleaning must be done at intervals frequent enough to make forcing unnecessary. Should a filter become so fouled as to need forcing, it should be dismantled and manually cleaned.

The felt-cartridge type of filter, like those commonly used on the oiling system of the automobile, will often be found in liquid circuits. A spare unit should be on hand, and general instructions followed in the establishment of regularly scheduled attention and maintenance.

Many filters are sized to handle only part of the liquid flow through a circuit. The result is the same as using a larger filter, as all the liquid will eventually pass through the filter. It is common practice to install line and bypass valves as part of such an assembly. The operator must avoid undue restriction of flow; however, the valve in the main liquid line must be throttled enough to give the same friction loss through the valve that exists through the filter.

Heaters

Heating Coils. Lightweight coils with external fins are used for almost all heating applications. These coils usually are encased in a galvanized metal casing that can be fitted into the equipment or ductwork. They are made in standard sizes and many different forms. For certain applications, they must be designed so that stratification does not occur. Stratification means that the air passing over one section of a coil may be heated more than that over other sections. Also, coils that are to be used exclusively in air that is above freezing temperatures may be less expensive in construction than coils that are designed against freezing.

Nonfreeze coils in current use are almost always in outside air. The construction of these coils is such as to get the condensate out of the coil before its temperature is reduced to the point of freezing. While it is important to pipe all coils correctly for condensate removal and to avoid the trapping of air in the coils, greater consideration must be given the quick removal of condensate from preheaters.

Where the automatic control of temperature is included, heaters normally work on steam pressures of not more than 10 lb. Lightweight heaters, however, are designed for high-pressure steam, and especially for industrial plants having high-pressure steam available. This provides an economical approach for space heating.

Expansion provisions must be included in the piping to avoid imposing stresses beyond the strength of modern coils, and this becomes of increasing importance as the steam pressure employed goes up. For heaters that are used in outside air, even of the nonfreeze type, it is of great importance that the steam pressure be maintained under all conditions. Even a 5-min period when freezing outside air is passing over the coil is sufficient to condense enough subpressure steam to burst the tubes in the heater.

Coil Maintenance.   See Coils.

Water Heaters.   Two types of steam water heaters are in current use. One is a shell-and-tube unit. Steam is usually supplied to the shell side, while water is circulated through the tubes. Where these heaters are used on closed water circuits and suitable water treatment is employed, the principal maintenance requirement will be observation and repair of leaks. Where the water circulated through the heater is used in open spray, seasonal cleaning of the water circuit through the heater is recommended.

Control of the output from this type of heater is normally accomplished by regulating the amount of steam admitted. However, a heater bypass arrangement for a controlled quantity of water may be used in some applications. Examination of the physical plant will reveal the type of control needed and enable the operator to service it as required.

The other general type of steam water heater is arranged for the admission of low-pressure steam directly into a water stream. Maintenance is normal except that the valves for control of the steam admitted to the unit must be kept in first-class order. The steam line to the heater unit must be closed when water is not flowing through the heater to avoid an open end on the steam line. Electric heaters are designed for use in either air or water. In either case the heater must be loaded before current is applied. In the case of air heaters, it is necessary to establish an airflow over the heater before the current is turned on, which is generally done by electrically interlocking the fan motor and heater circuit. Maintenance includes keeping the contact and working parts of relays in good working order, cleaning heater elements, and replacing burned-out elements.

Capacity of a heater is a function of voltage. High voltage will cause fuse links to fail or the unit to trip on the high-temperature limit switch. Low voltage will cut down on capacity.

Electric water heaters usually depend upon immersion for load. Sometimes the heater control can be electrically interlocked to the water circulating pump; in other cases, it may be necessary to control power input to the unit by a float-operated switch. Maintenance is comparable with the air strip heater. Controls on all heaters should be checked seasonally for operation and for calibration.

Heat Pumps

A heat pump is a reversible heating and cooling system. Through a special four-way valve the normal cycle is reversed so that the condenser becomes the evaporator and the evaporator becomes the condenser. Therefore, the maintenance procedure outlined elsewhere in this chapter for compressor, condenser, and coils also applies to heat pumps. Periodic inspection and maintenance will ensure long, trouble-free operation. The following are a few of the recommended procedures:

1. The most important maintenance action is a regular check of the outdoor coil. Since heat is removed from the low-temperature air passing over the outdoor coil, a large amount of frost will be formed on the coil. This frost is regularly removed by reversing the cycle and circuiting the hot refrigerant gas through the coil. The coil should be checked for rapid and efficient defrost action. Poor defrost action may be traced back to the timer, the defrost thermostat, the reversing valve, the defrost relay, or wind effect.

2. The check valve normally used in a heat pump system forces the refrigerant to go through the expansion device when the flow is in one direction and permits the refrigerant to bypass the expansion device when the flow is in the opposite direction. A slight leak in a check valve will cause a temperature drop across the valve. If the valve is stuck open, refrigerant will flood back to the compressor and high back pressure will result. If the valve is stuck in the closed position, it will tend to back up refrigerant in the condensing coil, thus causing high head pressure and low suction pressure.

3. The reversing valve is a four-way solenoid valve which, when energized, reverses the flow of gas through the indoor and outdoor coils. A defective valve should be replaced. No attempt should be made to repair the valve.

4. Replace air filters with new filters of the same size at least four times each year to ensure maximum efficiency and air circulation. Inspect the filters every 2 months or as often as necessary. If the filters are moderately dirty, they may be cleaned by using a vacuum cleaner or by tapping lightly. Replace the filters with the dirtier side facing into the return airstream. The filters are located in the return airstream.

5. Inspect both the indoor and outdoor coil surfaces. If they are dirty, the surface dirt may be brushed off with a stiff brush and the dirt between the fins cleaned with a good vacuum cleaner. The outdoor coil may be washed out with water, blown out with compressed air, or brushed.

6. Inspect the refrigeration piping for evidence of leaks. Check to see that the piping does not vibrate against any surface that would cause any rattle or abrasion of the piping.

7. Clean and check the condensate drain on the indoor coil.

8. Check all wiring for deterioration and all electrical contacts for tightness and corrosion.

9. Check the fan belt on the indoor unit and adjust to a correct tightness if necessary.

10. Check the mounting arrangement of both indoor and outdoor units.

11. Check defrost thermostat for a good, tight, and clean contact with tubing.
12. Check crankcase heaters by running a continuity check at the terminals when they are energized.

Humidifiers and Dehumidifiers

Humidifiers and dehumidifiers control the moisture content of air. These units are similar in appearance but differ in spray density, air velocity, and other details.

Many types of humidifiers are available to provide humidifying (and evaporative cooling): (1) central station spray, (2) spray heads at outlet of supply air duct, (3) devices in conditioned areas for breaking water into fine particles, (4) steam spray, (5) water pans with steam coil or electric heater, (6) any spray or spray and coil dehumidifier.

Four general types of dehumidifiers are available: (1) spray, (2) coil, (3) coil and spray, (4) chemical. Dehumidification is accomplished by chilling air to a temperature below the existing dew point or by chemical adsorption of air. For effective dehumidification, large wetted areas must be exposed. To reduce repair costs and maintain peak operating efficiency, follow the maintenance procedures listed below.

Cleaning. Remove lint and dirt periodically from air dampers, fan parts, spray chamber and diffuser, controls, strainer, and eliminator. Clean the eliminator wheel by directing a high-pressure stream of water between blades. Use a wire hook to remove any foreign material which remains. Inspect all components for rust and corrosion, and clean if necessary. Inspect and clean surface of reheat coils and supply air ducts.

Lubrication. Lubricate fan and motor bearings according to manufacturer's instructions.

General Maintenance. Remove and dismantle fan motor periodically, in accordance with manufacturer's instructions, to inspect motor bearings. Adjust fan blade clearance. Inspect entire unit for loose connections and tighten if necessary.

Check equipment for carryover. Carryover may be controlled by adjusting eliminator seal gap, altering damper position, or changing air velocity.

Pumps

Centrifugal pumps are used almost exclusively for circulation through sprays and cooling coils, to cooling towers, for the return of condensate to the boiler, and for other air-conditioning services.

Centrifugal pumps are manufactured in sizes that will handle from less than a gallon of water per minute up to several thousand gallons per minute. They are also designed for operating at low heads, such as 2- or 3-ft lift, up to many hundred feet of total operating head. A single-stage pump, that is, with one rotating impeller, will handle almost any requirement in air-conditioning work.

Centrifugal pumps work best when there is suction head. This means that the level of the water from which the pump draws is above the centerline of the impeller. There are applications in air-conditioning work where it is necessary to use a centrifugal pump with a suction lift. In this case, the level of the water from which the pump draws is below the centerline of the pump.

The larger air-conditioning pumps are rated in gallons per minute (gpm). Very small pumps may be rated in gallons per hour (gph).

If the piping system forms a continuous circuit with no open portions, it is referred to as a *closed water system.* There is no static lift involved. If there is a break in the piping system, however, the difference in elevation between the point at which water is discharged to the atmosphere and the point at which the water returns from the atmosphere to the piping system is referred to as *static lift,* or *static head.* This head must be overcome by the pump. In addition, power is necessary for liquid to flow through a piping system. The resistance flow, measured in feet of head, is referred to as *friction head.* The *total head* against which a pump must work is the sum of static head, or lift, and friction head. A pump is rated, therefore, in terms of gallons per minute and *total pumping head.*

The larger centrifugal pumps are usually designed for either belt or direct-connected drive from any standard motor. Many smaller pumps and some in larger sizes are now available with the pump impeller mounted on the end of the motor shaft. This assembly is commonly known as a "close-coupled pump." Pumps that start up with a suction lift will require priming. Once the pump has been primed or has its design flow of liquid, it will continue to operate and pick up the water from a point below its centerline.

Maintenance. Pump bearings must be lubricated, and the information given elsewhere in this handbook is applicable. Where the shaft on which the pump impeller is mounted passes through the casing, a seal must be established between the rotating part and the shell. This is accomplished by suitable packing in a stuffing box, and it is here that the average pump used on air-conditioning equipment needs most attention. It is common practice to permit the passage of a few drops of water each minute through the stuffing box to act as a lubricant between the rotating shaft and the packing. It is of great importance that pump glands be carefully packed with the correct packing and that the gland be pulled up evenly and just to the right tightness to prevent excessive loss of water or inward leakage of air, but not tight enough to bind the shaft.

It is possible to close either the suction line to the pump or the discharge line from the pump on most centrifugal pumps without causing damage. If it is permitted to operate for an extended period of time without flow of water, however, it will overheat. This characteristic of the centrifugal pump makes it easy to adjust water flow without injury to pump.

All the nuts on the packing gland should be tightened uniformly. These should be pulled up only a fraction of a turn at a time, keeping the gland parallel to the shaft at all times.

Most pumps used for air conditioning will be direct-connected through a flexible coupling. It is important that alignment of the motor to the pump be checked regularly and carefully maintained. Misalignment will cause excessive wear on bearings and packing glands, overheating of the pump and motor, and unnecessary noise. Each time piping connections are made to the pump or remade, there is the tendency to pull the pump out of alignment with the motor. Also, large rigid conduit for electrical connections to the motor may cause misalignment. After initial installation or at the time that any connections are remade, alignment must be remade. Alignment should be checked at least semiannually.

Trouble Symptoms and Their Correction. Note: The *symptom* is given; then the *probable cause* (in parentheses); followed by the *correction*.

No water or not enough water delivered. (Needs priming—casing and suction pipe not completely filled with liquid.) Start and stop the pump several times. Vent air from the pump while operating. Check for plugged strainer and adequate water level.

Not enough pressure. (Air in water. Mechanical defects, such as wearing rings worn, impeller clogged, or casing packing defective. Wrong direction of rotation. Leaky suction line. Stuffing box packing worn. Suction lift too high.) Check direction of rotation. Check for an air pocket trapped in the suction line. Replace the stuffing box packing. Unplug the water seal. Repair any air leak in the suction line.

Excessive packing wear. (Misalignment. Shaft scored. Improper packing. Misadjustment.) Realign. Refinish the journal. Pack in accordance with manufacturer's recommendation. Adjust the packing gland with uniform tightness and allow for a slight drip to provide lubrication.

Rooftop Units

Rooftop units are electrically powered cooling units and gas-powered heating units designed primarily for rooftop installation. When available, refer to manufacturer's instructions for inspection and cleaning procedures for heating and cooling sections. Maintenance procedures should include:

Inspection. Heating and cooling sections should be checked periodically, how often depends on usage and location. The thermostat should also be examined periodically.

Condenser, evaporator, and filters must be inspected prior to each cooling season and periodically during the season, how often depends on operating conditions. Coils and filters should be inspected for physical damage and airflow restrictions. Examine condenser and evaporator fans for bent blades and alignment and clearance with venturi orifice or cutoff plate. Belt drive, if used, should be inspected for wear and proper belt tension and sheave alignment. Motor power and control wiring must be examined for loose connections and deterioration of insulation (and contacts, where possible). Ammeter and ohmmeter tests should also be made on motors. Inspect motor suspension and support for tightness and isolator deterioration. Bearing wear on fan and/or motor may be determined by ammeter test and by listening to operating unit or by hand turning when power is off. Inspect panels and ductwork for air leakage. Check condensate pan and drain line for restrictions.

Inspect unit sightglass and moisture indicator. If possible, take suction and discharge readings while unit is operating and check with manufacturer's data. Inspection prior to and during heating season is also essential. In addition to items listed above to be examined for cooling operation, heating components must also be inspected. Combustion air box (and forced-draft blower, if used), orifices, pilot, main burners, heat exchangers, and flues must be checked for dirt, sooting, cracks or distortion, and adjustment. Inspect operation of automatic pilot, gas valves, pressure regulator, and other heating controls (air pressure switch, 100 percent shutoff valves, fan and limit controls, and so on). Burner alignment with heat exchanger flues should also be checked. Check flames for proper primary air.

Cleaning. Condenser and evaporator coils must be cleaned at the beginning of each cooling season and at regular intervals during the season, the length of interval depending on usage and installation location. Evaporator coil should also be cleaned at the beginning of each heating season. Airborne dirt may be removed from the air inlet screen, coil face, and fan wheel and scroll by brushing or vacuum cleaning. Compressed air or refrigerant gas may also be used. The indoor condenser motors should also be cleaned of any surface dirt, and vent openings should be cleaned by vacuuming or compressed air blast. Cleanable-type high-velocity filters should be removed and cleaned at the start of each heating and cooling season, and at least once during the season. Periodic inspection may indicate more frequent cleaning is necessary. Flush dirt from filters with hot water or steam. If dirt is heavily caked, soak filters in mild soap or detergent and water solution. (Refer to filter manufacturer's specific cleaning-agent recommendations.) After drying filters, spray with a water-soluble adhesive recommended by the filter manufacturer. Drain off excessive adhesive and reinstall filters. Throw-away filters should never be cleaned more than once. Clean by tapping gently or vacuuming. Replace with dirty side facing into airstream.

Clear and flush condensate drain lines and condensate pan at the beginning of each cooling season or more frequently if necessary. For winter operation, guard against freezing damage due to residual water in the drain line. Protect with antifreeze or dry completely.

Clean combustion air chamber, pilot, main burners, and flues as required. Refer to manufacturer's maintenance instructions for cleaning of heating section components.

Lubrication. Lubricate motors and bearings periodically, depending on usage. Follow the manufacturer's lubrication instructions.

Room Air Conditioners

The room air conditioner is a piece of unitary equipment designed specifically for a room or similar small space.

Types. The more popular type of room air conditioner is designed with an air-cooled condenser. Since it must have an unrestricted source of outside air, it is most commonly installed on a windowsill. Floor or console models are made by some manufacturers.

Application. All the summer air-conditioning load factors must be considered in selecting a room air conditioner. A unit used in a room which is too large or in which the heat load is far beyond cooling capacity will be completely ineffective.

Units are designed for balance between air side and refrigeration side at nominal comfort conditions. They should not be expected to reduce room temperature below normal comfort conditions even in mild outside weather. Also, since design conditions must be based on the widest general use, maximum outside ambient temperature must be arbitrarily established.

Full voltage at the unit is essential. Almost without exception, existing wiring and power service will be inadequate for the addition of more than one or two room air conditioners.

Leakage around the unit or at any other point into the room seriously impairs effectiveness.

Maintenance. Airflow through the unit must not be restricted. During the operating season, filters and coils must be kept clean. Refrigerant charge in the unit is critical. A seasonal check should include a careful leakage test. The drip pan of the unit, coils, and fan blades should all be cleaned seasonally. Fan motors should be checked for free turning. The service cord and service connections should be examined.

If the unit is equipped with electric strip heaters, voltage must be within manufacturer's specifications to prevent failure or complaints of inadequate capacity.

Detailed Instruction. The manufacturer's product booklet should be followed for detailed instruction.

Self-Contained Units

Self-contained or packaged units are in the unitary equipment class. Some units are made just for summer cooling; others are made for cooling in the summer and for heating in the winter. The latter type has the widest application in residential work. Either may be suitable, however, for use in offices, shops, specialty stores, and other applications. Equipment is available with either air-cooled or water-cooled condensers.

Application. Self-contained units in sizes up to 25 or 30 tons are available in highly styled, finely finished cabinets for a wide variety of application. The larger units are normally intended for industrial use where space saving is given more consideration than styling or finish.

Self-contained equipment has the application advantages of low initial cost, simple installation, and very limited control. Such equipment must be selected to handle the air-conditioning load as outlined elsewhere in this text. It is built with direct-expansion cooling and is designed for balance between air-handling and refrigeration equipment at nominal comfort conditions. Units should be selected to operate at close to full load for best economy.

Maintenance. Full airflow must be maintained. Filters must be cleaned regularly. Drip pans, coils, and other parts must be cleaned seasonally. The refrigerant circuit must be kept tight and the refrigerant charge maintained up to requirements. Control boxes must not be allowed to accumulate dirt. Burners on heating equipment must be kept in order.

Freeze Protection

Damage from Freezing. Freeze-up of water or brines may occur in equipment subject to outside conditions during winter, or in the low side of water cooling refrigeration equipment at any time.

Freeze Prevention. Three common methods are used to avoid freeze-up from winter air:

1. Draining water from vulnerable equipment
2. Lowering the freezing point to a safe limit by adding antifreeze to the circulating water
3. Heating surrounding air to a temperature above freezing

Precautionary measures in the design of refrigeration equipment usually include positive circulation of water before the refrigeration machine can start, automatic control devices to hold liquid temperatures above the freeze point, and temperature-sensitive instruments to stop the refrigeration machine should the other features prove inadequate. Freeze protection devices should be the last line of defense, and equipment *must never be forced to operate when stopped by the protection instrument.* The reason for activating the protective device should be determined and the condition corrected before the machine is put back into service.

Where Attention Is Needed. To prevent freeze-up, particular attention should be given the following:

1. Water coils anywhere in the circuit that may come in contact with air below 32°F
2. Heating coils in outside air
3. Water supply lines and drain lines to evaporative condensers, or other equipment located outdoors
4. Pans of evaporative condensers, cooling towers, and other outside equipment
5. Refrigerant condensers, expansion tanks, and other equipment located in unheated portions of the building, such as penthouses and street vaults
6. Any and all water lines running through unheated spaces or outside the building
7. Well pumps and water lines located outside heated spaces
8. Compressed air lines and air motors of a pneumatic control system in outside air intakes

Protecting Water Coils. Preheating coils cannot provide positive insurance against freezing because they can be effective only with adequate steam supply, and uninterrupted steam supply can seldom be guaranteed. Failure of pressure for only 2 or 3 min in extreme weather, for example, when the boiler or furnace is cleaned, can cause damage.

Positive return air in sufficient quantity to heat outside air above freezing, with a suitable means of mixing before reaching the coil, is reliable. But conditions seldom will permit such a design. The maintenance staff *must* investigate the freeze potential of *every* water coil and exercise the necessary precaution.

Preheating Coils. Design improvements in recent years have provided coils suitable for use in outside air when properly selected and installed. These nonfreeze coils may have the steam supply modulated through a reasonable range to permit better regulation of air conditions, but steam flow must not be reduced below a minimum point under severe outside conditions. Condensate removal must be adequate and effective. Only proven practices of piping installation should be used. Obviously, condensate lines should not pass through the outside air path.

Water-Supply Lines Outside Building. All water lines outside heated space must be completely drained. Water lines passing through an outside air duct or otherwise crossing the path of outside air must be drained.

Pans of Evaporative Condensers. Pans of evaporative condensers, cooling towers, and other equipment located outside heated spaces should be drained when shut down for the season, and at that time cleaning, painting, and general maintenance attention should be provided.

Well Pumps and Water Lines. Well pumps and water lines above ground level and in unheated space must be prepared for winter standby. If the pump is not to be used during freezing weather, it should be properly drained and conditioned for the next season's use.

Compressed-Air Lines and Air Motors. Moisture can enter a pneumatic control system along with air, condense in the system, and seek the point of lowest vapor pressure. The moisture tends to accumulate in any piping, damper, or valve motors located in the outside-air chamber. Continued buildup of moisture at those points leads to the rupture of filled parts when frozen. This problem can be prevented by periodic blow-down of the compressed-air system.

Water Conditioning

The Problem. Water damage in air-conditioning equipment is an ever-present problem which varies widely from area to area, and even from installation to installation. Water, especially when recirculated, can be responsible for scale, which interferes with heat transfer, corrosion, which ravages the equipment, algae or slime, which interferes with performance, and erosion. Each of these actions raises operating and maintenance costs.

Definitions. In general, the term "scale" applies to deposits which result from crystallization or precipitation of salts from the water. The term "corrosion" refers to the decomposition of metal caused by electrolytic or chemical attack. The term "erosion" applies to the impingement of rapidly moving water, particularly when entrained gas bubbles or abrasive solids are present, or where intermittent cavitation occurs. The erosion breaks down protective films and corrosion of metal surfaces results. Slime and algae are microorganisms which are capable of multiplying rapidly and thus produce large masses of plant material.

Treatment or conditioning of the water used in each of the circuits of an air-conditioning system is just as important as the treatment of water used in boilers. Potential problems differ depending on the type of system—closed recirculating, open recirculating, or once-through systems. Unfortunately, because of the many different water conditions encountered in air-conditioning installations, it is difficult to provide guidelines that will apply universally.

Corrective Measures. In cooling towers or evaporative condensers, water is evaporated to produce the cooling effect. This has a tendency to concentrate the chemical impurities in the water which usually raises the scaling or corrosive tendencies of the water. To offset this, a bleed-off should be provided. The amount of water to be bled off will depend on the hardness and other characteristics of the water as well as characteristics of the cooling tower or the evaporative condenser (windage loss, system leaks, and so on). Chemical treatment of the water may also be required in order to provide adequate protection against scale and corrosion, as well as to prevent algae growth. The services of a water-conditioning company may be required in developing a comprehensive treatment program for your installation.

CHAPTER 6
VENTILATING FANS AND EXHAUST SYSTEMS

R. Keith Mobley
Principal, Life Cycle Engineering, Inc., Charleston, S.C.

FAN AND FLOW FUNDAMENTALS

In addition to the task of maintaining existing ventilating systems and the fans which serve them, maintenance personnel are often given the responsibility for adapting them to meet new conditions. Sometimes an entirely new system may have to be designed and installed. Occasionally the existing fan or system can be adapted or modified to meet the new demand. Familiarity with the fundamentals of fan performance, the types of fans available, and their uses as well as their limitations can be very useful where air-moving problems are part of the maintenance function.

A fan or air-moving device is simply the machine which supplies the air with the *energy* required to move it. It overcomes *inertia*, or reluctance to move from rest and *friction*, or resistance to continuous flow, both of which are always present. Mechanical energy supplied to the rotating wheel of the fan is transferred to the air to increase its velocity and raise its pressure enough to overcome the resistance and produce the required flow.

TERMINOLOGY AND DEFINITIONS

Listed below are some of the terms commonly used in the fan industry and their definitions. Abbreviations frequently used are shown in parentheses.

Fan capacity (cfm). Volume of air moved by the fan per unit of time. Stated customarily in cubic feet per minute. Air at a density of 0.075 lb/ft^3 is *always* implied unless specifically stated otherwise.

Static pressure (SP). Pressure produced by the fan which can exist whether the air is in motion or confined in a closed duct. Expressed in inches of water gage (in WG). It can be regarded as simply the *potential* energy produced by the fan and required to overcome resistance to flow offered by the system. Other terms often used synonymously are *fan static, system resistance,* and *system static.*

Velocity pressure (VP). Pressure produced by the fan which can exist *only* when the air is in motion and is *always* exerted in the direction of airflow. Air moving at a velocity of 4000 fpm will exert a pressure of 1 in WG on a stationary object in its path. This is a very useful relationship and should be remembered. VP for any other velocity can be easily calculated, since it varies as the square of the velocity. At 8000 fpm the VP is 4 in WG; at 2000 fpm the VP is 0.25 in WG.

Total pressure (TP). The sum of the static pressure and velocity pressure, in WG.

Outlet velocity (OV). Cfm divided by the *inside* area of the fan outlet (or discharge area) in square feet (ft^2) equals the outlet velocity in feet per minute (fpm).

Brake horsepower (bhp). Power input—the power that must be supplied at the fan shaft to produce the required cfm and pressure.

Static efficiency (SE). Not the true mechanical efficiency but convenient to use for comparing fans, since most published performance data show SP, not TP. Calculated from

$$\text{Static efficiency} = \frac{0.000157 \times \text{cfm} \times \text{SP}}{}$$

Total efficiency (ME). True mechanical efficiency—power output divided by power input. Calculated from

$$\text{Total efficiency} = \frac{0.000157 \times \text{cfm} \times \text{TP}}{}$$

Velocity pressure at the fan outlet must be calculated and added to the SP to arrive at the fan TP. Catalogs usually list outlet velocities, or they can be calculated as shown above.

Fan rating. A statement of fan performance for one condition of operation. Includes fan size, speed, capacity, pressure, and horsepower. Also used merely to specify only the cfm and SP desired. *Always* implies that the fan is handling standard air at 70°F, 29.92 in Hg barometer, and 0.075 lb/ft^3 density unless *specifically* stated otherwise.

Free-delivery capacity. Fan capacity with no resistance or static pressure (a condition rarely, if ever, encountered in practical applications!). It is common practice, however, to publish propeller-fan and occasionally axial-fan capacities at free delivery.

Static no-delivery pressure (SND). Also called shutoff pressure, it means the static pressure produced by the fan at zero cfm.

Tip speed. Peripheral speed of the fan wheel in fpm. Calculated from

$$\text{Tip speed} = \text{wheel diameter} \times \text{pi} \times \text{rpm}$$

(wheel diameter in feet; pi 5 3.14).

USING PUBLISHED PERFORMANCE DATA

Standards. In order to provide uniform standards for evaluating and comparing the performance of air-moving equipment, the Air Moving and Conditioning Association (AMCA) has developed Standard Test Codes which are generally used and accepted throughout the industry. AMCA is a nonprofit association whose membership includes most United States and Canadian manufacturers of air-moving equipment. Under their program, air-moving products are tested in approved laboratories using the appropriate AMCA Standard Test Code and qualified personnel. A list of AMCA technical bulletins available to specifiers and users of air-moving equipment can be obtained by writing to them. Figures 6.1 and 6.2 are AMCA standards.

Most fans are manufactured in a series of sizes in order to cover a wide range of capacities. The fans in the series are geometrically similar. The number of sizes in the series may vary from only a few to as many as 25.

The fan manufacturer publishes performance data based on actual tests of a representative number of fan sizes. Catalogs usually contain other useful and necessary information such as fan weights, dimension prints, optional equipment, and other technical data.

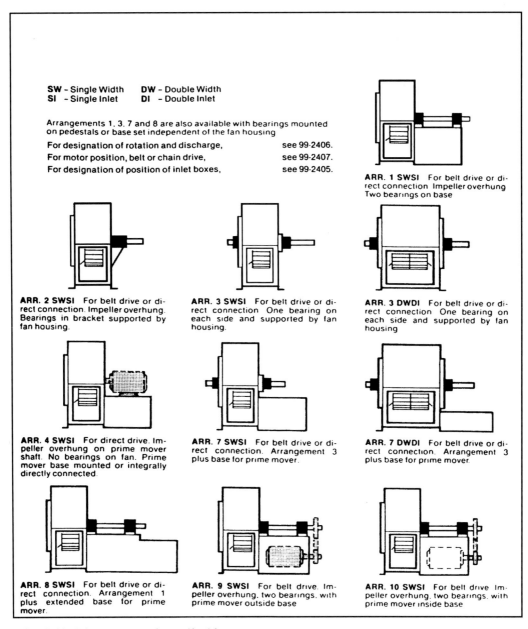

FIGURE 6.1 Drive arrangements for centrifugal fans.

Multirating Tables. Probably the most widely used method of presenting performance data is a series of multirating tables. Each table shows the usable performance range for one size in the series. Figure 6.3 shows a portion of a typical table. Various modifications are often made to aid the user. See Fig. 6.4. The ratings covering the highest range of efficiency are sometimes shaded or printed in boldface. Or a similar device may be used to show the range of maximum safe speeds for the particular size with standard construction.

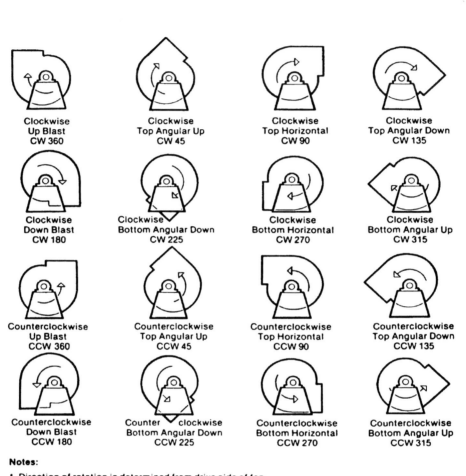

FIGURE 6.2 Designations for rotation and discharge of centrifugal fans.

Wheel Diameter 40 ¼"   Limit Load HP = 24.91 × $\left(\dfrac{\text{RPM}}{1000}\right)^3$ Outlet Area 9.32 sq ft inside.

CFM	OV	¼" SP		⅜" SP		½" SP		⅝" SP		¾" SP		⅞" SP		1" SP		1¼" SP		1½" SP		1¾" SP	
		RPM	BHP	RPM	BHP	RPM	BHP	RPM	BHP	RPM	BHP	RPM	BHP	RPM	BHP	RPM	BHP	RPM	BHP	RPM	BHP
7456	800	262	0.45	289	0.60	314	0.75	337	0.92	360	1.09	382	1.27	403	1.46	444	1.85	483	2.28	520	2.73
8388	900	281	0.55	306	0.72	330	0.89	351	1.06	372	1.25	393	1.44	413	1.63	451	2.05	488	2.49	523	2.96
9320	1000	199	0.66	325	0.85	347	1.04	368	1.23	387	1.43	406	1.63	425	1.83	461	2.27	495	2.72	529	3.21
10252	1100	319	0.79	343	1.00	365	1.21	385	1.42	403	1.63	421	1.84	439	2.06	473	2.51	505	2.99	537	3.50
11184	1200	338	0.93	362	1.17	383	1.40	402	1.63	420	1.85	438	2.08	454	2.31	486	2.79	517	3.29	547	3.81
12116	1300	358	1.10	381	1.35	402	1.61	421	1.85	438	2.10	455	2.34	471	2.59	501	3.09	531	3.62	559	4.16
13048	1400	379	1.29	401	1.56	421	1.83	439	2.10	456	2.37	473	2.63	488	2.90	517	3.43	545	3.98	572	4.54
13980	1500	401	1.50	420	1.78	440	2.08	458	2.37	475	2.66	491	2.94	506	3.23	534	3.79	561	4.37	587	4.96
14912	1600	422	1.74	441	2.03	459	2.35	477	2.67	494	2.98	509	3.28	524	3.58	552	4.19	578	4.79	603	5.41
15844	1700	444	2.01	462	2.32	479	2.65	496	2.98	513	3.32	528	3.64	542	3.97	570	4.61	595	5.25	619	5.90
16776	1800	467	2.31	483	2.63	499	2.97	516	3.33	532	3.68	547	4.03	561	4.38	588	5.06	613	5.74	637	6.42
17708	1900	489	2.65	504	2.98	520	3.33	536	3.70	551	4.07	566	4.45	580	4.81	606	5.54	631	6.26	654	6.98
18640	2000	512	3.02	526	3.36	541	3.72	556	4.10	571	4.49	585	4.89	599	5.28	625	6.05	649	6.81	672	7.57
19572	2100	535	3.43	548	3.77	562	4.15	576	4.53	590	4.95	604	5.36	618	5.78	644	6.59	668	7.40	690	8.19
20504	2200	558	3.87	570	4.23	584	4.61	597	5.02	610	5.43	624	5.87	637	6.30	663	7.16	686	8.01	708	8.85
21436	2300	582	4.36	593	4.72	605	5.12	618	5.54	631	5.95	644	6.41	657	6.86	682	7.77	705	8.66	727	9.54
22368	2400	605	4.89	616	5.26	627	5.67	640	6.10	652	6.54	664	6.99	677	7.46	701	8.41	724	9.35	746	10.27
23300	2500	628	5.46	639	5.85	650	6.26	661	6.70	673	7.16	685	7.63	697	8.10	721	9.09	743	10.07	765	11.04
24232	2600	652	6.09	662	6.48	672	6.90	683	7.34	694	7.81	706	8.30	717	8.77	740	9.80	762	10.83	784	11.84
25164	2700	676	6.75	685	7.15	695	7.58	705	8.04	716	8.52	727	9.01	738	9.53	760	10.56	782	11.63	803	12.69
26096	2800	700	7.47	708	7.88	718	8.32	727	8.78	738	9.27	748	9.78	759	10.30	780	11.35	801	12.46	822	13.57
27028	2900	723	8.24	732	8.66	741	9.11	750	9.58	760	10.08	770	10.60	780	11.13	800	12.20	821	13.35	841	14.49

FIGURE 6.3 Part of a typical rating table. The fan is a backward-curved airfoil-blade type. The column headed OV shows outlet velocities. To use the table, assume the required rating is 12,116 cfm at 1¾ in. SP. Read across to the 1¾ in. SP column. The fan must be driven at 559 rpm and would require 4.16 bhp. The calculated static and total efficiencies are 80 and 84 percent, respectively. The formula for limit load hp gives the maximum hp the fan would demand at any given rpm; in other words, it defines the peak of the bhp curve. In the example used, the curve peaks at 4.35 hp. A 5-hp motor could not be overloaded as long as the speed remained constant and the air density did not increase appreciably above standard.

The user selects from the tables a fan size which will move the desired cfm against the SP of the system. The *ideal* selection when minimum power consumption or lowest sound level are desired is the size requiring the lowest bhp. It will have the highest mechanical efficiency and lowest sound level available for that particular *type* of fan. However, the *optimum* size is more often one or more sizes smaller than the *ideal*. Factors such as initial cost, available space, or weight limitations must also be evaluated.

Interpolation may be used when the desired rating is not one of those tabulated. However, overly tedious calculations are not always necessary. System resistance is rarely predictable with great accuracy, and it may not be possible to select a drive arrangement which will operate the fan at the precise speed required.

Density Changes. When the density of the air differs appreciably from standard, appropriate corrections must be made. Table 6.1 and its appended data describe the most commonly used method of handling this situation.

FAN-PERFORMANCE LAWS

The Basic Laws. Familiarity with the mathematical relationships referred to as *fan laws* can be useful when applied properly. They can be applied *only* to a fan operating in a *fixed system*, to geometrically similar fans, and to the *same point of rating* on the fan-performance curves. The basic fan laws are:

1. Cfm varies *directly* as the rpm.
2. Pressure varies as the rpm *squared*.
3. Hp varies as the rpm *cubed*.

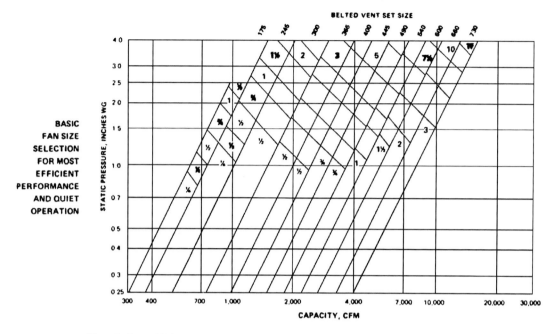

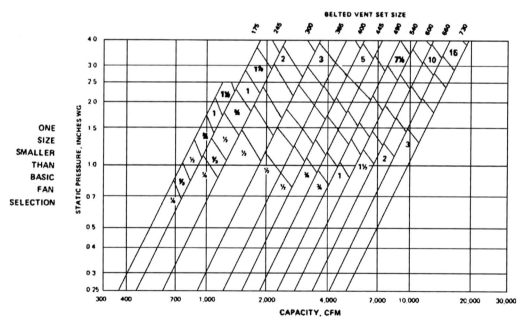

FIGURE 6.4 Examples of rating charts for an entire series of sizes of packaged belt-driven fan and motor units. They are intended to help the user select the optimum size. A cost comparison is included. The series continues and also shows data for three and four sizes smaller than basic. A series of conventional rating tables is included to allow selection of the exact rpm for the rating required. A unit of this type is usually furnished with an adjustable-pitch V-belt drive so that any rating within the motor range can be obtained.

TABLE 6.1 Air-Density Ratios

Air temp., °F	\multicolumn{11}{c}{Altitude, ft above sea level}
	0	1,000	2,000	3,000	4,000	5,000	6,000	7,000	8,000	9,000	10,000	15,000	20,000
	\multicolumn{13}{c}{Barometric pressure, in. mercury}
	29.92	28.86	27.82	26.82	25.84	24.90	23.98	23.09	22.22	21.39	20.58	16.89	13.75
70	1.000	0.964	0.930	0.896	0.864	0.832	0.801	0.772	0.743	0.714	0.688	0.564	0.460
100	0.946	0.912	0.880	0.848	0.818	0.787	0.758	0.730	0.703	0.676	0.651	0.534	0.435
150	0.869	0.838	0.808	0.770	0.751	0.723	0.696	0.671	0.646	0.620	0.598	0.490	0.400
200	0.803	0.774	0.747	0.720	0.694	0.668	0.643	0.620	0.596	0.573	0.552	0.453	0.369
250	0.747	0.720	0.694	0.669	0.645	0.622	0.598	0.576	0.555	0.533	0.514	0.421	0.344
300	0.697	0.672	0.648	0.624	0.604	0.580	0.558	0.538	0.518	0.498	0.480	0.393	0.321
350	0.654	0.631	0.608	0.586	0.565	0.544	0.524	0.505	0.486	0.467	0.450	0.369	0.301
400	0.616	0.594	0.573	0.552	0.532	0.513	0.493	0.476	0.458	0.440	0.424	0.347	0.283
450	0.582	0.561	0.542	0.522	0.503	0.484	0.466	0.449	0.433	0.416	0.401	0.328	0.268
500	0.552	0.532	0.513	0.495	0.477	0.459	0.442	0.426	0.410	0.394	0.380	0.311	0.254
550	0.525	0.506	0.488	0.470	0.454	0.437	0.421	0.405	0.390	0.375	0.361	0.296	0.242
600	0.500	0.482	0.465	0.448	0.432	0.416	0.400	0.386	0.372	0.352	0.344	0.282	0.230
650	0.477	0.460	0.444	0.427	0.412	0.397	0.382	0.368	0.354	0.341	0.328	0.269	0.219
700	0.457	0.441	0.425	0.410	0.395	0.380	0.366	0.353	0.340	0.326	0.315	0.258	0.210

Note: The air density is inversely proportional to the absolute temperature and directly proportional to the absolute pressure. The absolute temperature is 460° plus the ambient temperature in °F. For example, the ratio to standard conditions for air at 200°F and 5000 ft altitude would be

$$\frac{(460 + 70) \times 24.90 \text{ in}}{(460 + 200) \times 29.92 \text{ in}} = 0.668$$

which the table indicates. The actual air *density* would be $0.668 \times 0.075 = 0.05$ lb/ft^3.

A fan is inherently a constant-volume machine. It will handle the same cfm when operating in a fixed system at a constant speed regardless of changes in air density; however, the pressure developed and the horsepower required will vary directly as the air density.

Assume that a fan is required to deliver 16,776 cfm at 1 in SP under the above conditions of 200°F and 5000-ft altitude. Using the rating table in Fig. 6.3, proceed as follows:

1. Density factor is 0.668.
2. Equivalent SP is 1 in/0.668 = 1.5 in
3. Enter table at 16,776 cfm under 1.5 in SP and read 613 rpm and 5.74 bhp.
4. rpm is correct as read.
5. bhp required is $5.74 \times 0.668 = 3.83$.
6. The fan will deliver 16,776 cfm and develop 1 in SP at operating conditions.

When a fan handling hot air must be started under cold conditions, the increased horsepower required must be taken into account. If the motor is sized for the lower bhp required at operating conditions, it may be overloaded at start-up. The possible solutions are providing dampening for the fan at start-up, operating the fan at a lower speed until operating temperature is reached, or sizing the motor for cold conditions.

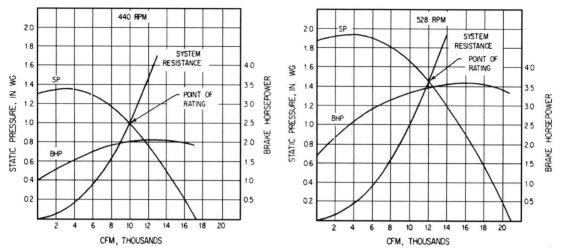

FIGURE 6.5 Typical fan-performance curves. They show the characteristics of the same backward-curved-blade centrifugal fan at two different operating speeds.

To illustrate, see Fig. 6.5 showing two typical fan-performance curves. Note the following:

1. They show the performance of the *same* fan operating in the *same* duct system handling air at the same density.
2. Curve 1 shows a fan that was selected to exhaust (or supply) 10,000 cfm in a duct system whose resistance was calculated at 1 in SP. It will operate at the point where the fan pressure-volume or characteristic curve intersects the curve representing the system resistance. The intersection is called the point of rating. The fan will continue to operate at this point as long as the rpm remains constant and nothing is added to or removed from the system which would change its resistance.
3. The system-resistance curve illustrates that the resistance varies as the square of the cfm.
4. Curve 2 shows the situation which would exist if the user needed a 20 percent increase in capacity, from 10,000 to 12,000 cfm. Applying the fan laws, the calculations are

$$\text{New rpm} = 1.2 \times 440 = 528 \text{ (20 percent increase)}$$
$$\text{New SP} = 1.2 \times 1.2 \times 1 \text{ in WG} = 1.44 \text{ in (44 percent increase)}$$
$$\text{New bhp} = 1.2 \times 1.2 \times 1.2 \times 2 \text{ bhp} = 3.45 \text{ (73 percent increase)}$$

5. The curve representing the system resistance is the same in both cases, since the system has not been altered. The fan will now operate at the same relative point of rating. It will move the increased cfm through the system.
6. The static and mechanical efficiency have *not* changed.
7. The increase in bhp required to drive the fan is an important point. Initially, the fan might have been driven by a 2-hp motor. It would now require a 5-hp motor.

TYPES OF FANS AND RECOMMENDED USES

Fans can be divided into four general categories: propeller, axial, centrifugal, and special-purpose (see Figs. 6.6 and 6.7). Typical performance characteristics are shown, since they will sometimes influence the type of fan selected for a specific task. A brief description of each type including suggested uses and limitations follows.

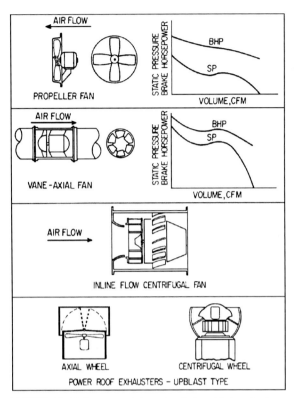

FIGURE 6.6 Various types of air-moving devices. All the units shown are designated as arrangement No. 4, with the wheel mounted on the motor shaft, but they are usually available with V-belt drive as well. A tube-axial fan is similar to a vane-axial fan except that the air-straightening vanes downstream from the wheel are omitted. The performance characteristics shown are typical for propeller and axial fans. The in-line fan is a mixed-flow type utilizing a backward-curved-blade centrifugal wheel in an axial-fan cylindrical housing, with curved air-straightening vanes downstream from the wheel.

Propeller. Consists of a propeller or disk-type wheel within a mounting ring, panel, or cage (see Fig. 6.8). Wheel and/or housing may be sheet metal, cast metal (usually aluminum), fabricated steel (or other metal) plate or various plastics, protective coatings, and combinations of materials. It may be direct-drive with the wheel mounted on the motor shaft or belt-driven with the wheel mounted on its own shaft and bearings.

Axial. A tube-axial fan (see Fig.6.9) is essentially a propeller fan enclosed in a short cylindrical housing. A vane-axial fan (see Fig. 6.10) incorporates specially designed vanes, which may be either upstream or downstream from the fan wheel. The vanes straighten the inherent spiraling flow pattern of the wheel, thereby converting some of the velocity component of the airstream to useful static pressure. Vane-axial fans can develop much higher static pressures than tube-axials. Construction may include fabricated steel, cast metal, plastic, various combinations of materials, and special protective coatings. May be either direct-drive or belt-driven.

They are available with adjustable-pitch blades. This feature permits the use of direct-driven fans to cover the same range of capacities and pressures as belt-driven fans of the same size.

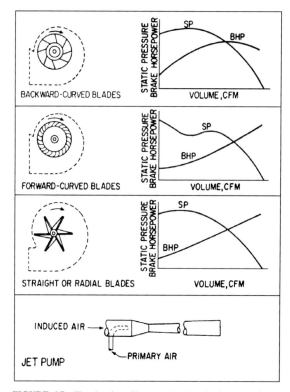

FIGURE 6.7 The drawings illustrate the three basic types of centrifugal-fan wheels with their typical scroll-shaped housing shown in outline. Direction of wheel rotation is indicated by an arrow. The radial-blade wheel shown is a simple paddle type. The large number of blades is typical for the forward-curved type. Backward-curved wheels have fewer blades. Radial types generally have the smallest number of blades. Typical performance characteristics are also shown for each type. The jet pump is an air-moving device sometimes used to avoid passing extremely hot or contaminated air through a fan.

Backward-Curved-Blade Centrifugal. This type of fan is capable of the highest efficiency and lowest sound level of all centrifugal fans, particularly those with airfoil-shaped double-thickness blades. Standard units are fabricated from steel plate or sheet. Also obtainable in a wide variety of special metals and protection coatings.

Wheel blades may be flat, single-thickness, and curved, or double-thickness and curved with an airfoil-shaped cross section.

Forward-Curved-Blade Centrifugal. Characterized by a large number of wide shallow blades, very large inlet area relative to wheel diameter, and low speed compared with other centrifugal fans for equal volume and measure. Sometimes called squirrel-cage fan or multivane fan. Usual construction is fabricated steel but may be other metals when required. Also obtainable with protective coatings or all plastic in the smaller sizes as well as various combinations of materials.

Straight- or Radial-Blade Centrifugal. This group is characterized generally by simple, rugged construction, usually referred to as *industrial* exhaust fans. All do not have flat, strictly radial blades. Many different wheel configurations are available, from simple paddle type to semiopen

FIGURE 6.8 Belt-driven propeller fan for medium duty. Most units can be obtained for airflow in either direction. This fan is for wall mounting and will exhaust air from a space, drawing it over the motor and the drive.

FIGURE 6.9 Inlet end of direct-connected tube-axial fan. Note electrical connections and grease fittings brought outside housing. Arrows are provided near nameplate to indicate direction of wheel rotation and airflow.

FIGURE 6.10 Inlet end of direct-connected vane-axial fan. Motor shaft supports and drives wheel. Vanes behind wheel straighten inherent spiraling motion of air as it leaves the wheel and increase static pressure.

and enclosed types. Blade shapes vary from flat to various bent shapes to increase efficiency slightly or to suit particular applications. Standard construction is usually welded steel plate, but they are available in a variety of special metals, all plastic (glass-fiber-reinforced polyester resin is popular), with a variety of protective coatings, and in a combination of materials. May be belt- or direct-drive. Single-inlet, single-width fans are standard.

Small and medium sizes are available with cast-metal housing and wheels, usually cast iron.

In-Line-Flow Centrifugal. Also referred to as a tubular centrifugal, it is actually a mixed-flow fan. It consists of a centrifugal wheel in an axial housing. Straightening vanes downstream from the wheel convert some of the velocity component of the airstream into useful static pressure.

Power Roof Exhausters. These units usually consist of a package complete with base for roof mounting, backdraft damper, and either propeller, axial, or centrifugal wheels. Available in a variety of styles, direct or belt drive, special metal and plastic construction, or special coatings.

Centrifugal types discharge air around periphery of wheel, deflecting air downward. Others usually discharge air vertically.

Jet Pumps. These devices can be useful where the air or gas is too hot or corrosive to pass through a fan. They are also called injectors or ejectors. See Fig. 6.7. They are available from many manufacturers or may be designed and specially made for a specific application.

Packaged Ventilating Equipment. Many ventilation-equipment manufacturers offer combinations of components or complete systems. They are well worth considering for many applications, since they may save time and money spent in selecting and assembling individual components. A list of those most frequently used includes:

Completely assembled ventilating-fan units with motor, adjustable-pitch V-belt drive, and a variety of optional equipment and accessories

Packaged fan and air-cleaning units including mechanical and fabric-type dust collectors, fume scrubbers, and so on

Tailpipe exhaust systems for automobile and truck service garages

Welding exhaust systems complete with hood and flexible ductwork

SELECTING FANS FOR SEVERE DUTY

Industrial-fan applications often involve troublesome maintenance problems caused by high temperature, corrosion, and abrasion. Some of the problems can be avoided or at least alleviated by using equipment properly designed for the job. An experienced fan supplier can often be a good source of information and assistance. It is vital to give him all the known or expected characteristics of the material which the fan will handle.

A large amount of practical experience has been gathered and studies have been made to find ways of extending the life of fans exposed to these conditions, and great progress has been made. However, since the deteriorating effects are often unpredictable and the operating conditions are beyond the control of the fan supplier, he cannot make any *guarantee* concerning the life of the fan or its degree of resistance to attack.

High Temperature. The fan-maintenance problems associated with high temperatures can sometimes be minimized or eliminated by cooling the airstream before it passes through the fan. The methods worth considering are:

1. *Spray cooling.* Cooling the air with water sprays is very effective. It has the added advantage of sometimes resulting in a considerable reduction in volume and fan size.
2. *Dilution cooling.* Cold air can be bled in on the suction side of the fan to produce a cooler mixture. The cold air must be introduced in a manner that provides enough time and turbulence to produce a mixture with fairly uniform temperature at the fan. It is not as effective as spray cooling but can often be used where only a moderate amount of cooling is needed.
3. *Air-to-air heat exchangers.* The possibilities with this method vary in complexity. They require careful analysis to determine their feasibility. Occasionally, the waste heat can be recovered and used for space heating or in some other process. All these methods increase the system resistance and consequently the power consumed by the fan.

The most elementary method utilizes a long run of ductwork on the suction side of the fan, relying on conduction of heat through the duct walls and natural convection to transfer heat from the airstream to atmosphere. The duct may be serpentined with a series of bends to increase turbulence and length. Further refinements include the addition of cooling fins or a combination of cooling fins and auxiliary air-blast cooling to increase the heat-transfer rate.

Finally, devices are manufactured for this purpose, similar to the regenerative air preheaters used in large steam-generating plants. They are probably the most efficient coolers and the most adaptable for recovery of the waste heat.

4. *Water-cooled ductwork.* This method is sometimes used where exhaust-air temperatures are extremely high, electric-arc steel-melting furnaces, for example. A short section of ductwork, the collecting hood, or both may be water-jacketed and cooled with a recirculating system or with waste water. On very large installations it is feasible to utilize the heated water in a waste-heat boiler or other process.

If possible, do not use a fan with motor or bearings exposed in the airstream if the temperature is appreciably above normal. When this situation is unavoidable, motors are available with special insulation which will operate satisfactorily at maximum total motor-operating temperatures up to 356°F. It is not possible to state a maximum airstream temperature which these motors will withstand, because other variables must be considered. The limit must be obtained from the fan or motor manufacturer.

The maximum airstream temperature to which a particular bearing can be exposed is also impossible to state and must be obtained from the bearing or fan manufacturer.

Axial fans with bearings isolated from the airstream by a cylindrical casing can be modified to handle air up to about 500°F. Special axial fans, such as the bypass or elbow type where the bearings are entirely outside the fan casing, can handle air up to 1000°F.

Both forward- and backward-curved-blade centrifugal fans can be modified to handle air up to about 750°F.

Radial-blade centrifugal fans can be modified to handle air up to 1000°F.

Maximum safe speeds of fans are reduced with increasing temperature. The reduction may be as high as 50 percent at 1000°F. The yield strength of steels decreases, and cooling of bearings is also a problem. Even in normal atmospheres, ordinary steel scales rapidly above 900°F.

Corrosion. All types of fans are available with a variety of special protective paints and coatings, rubber covering, all-plastic construction, special metal construction, and various combinations of materials, all intended to make them more resistant to attack. The problem is so complex and the variety so great that a list of specific recommendations cannot be made. Following are some general recommendations:

1. Avoid, if possible, using a fan with the motor or bearings in a corrosive airstream. Although such fans are satisfactory for some applications, a single-inlet centrifugal fan such as arrangement 1 is the best choice if the problem is severe.
2. Use a shaft seal on centrifugal fans to prevent corrosive material from leaking or being blown out of the housing toward the inboard fan bearing.
3. Keep inlet and outlet connections tight. Flanged and gasketed connections are obviously superior to slip-type in this respect.
4. Use a drain connection at the lowest point in the fan housing if the airstream contains mist or condensable vapor.
5. When it is feasible, install the fan on the clean-air side of the air-cleaning device if the system includes one. There are exceptions, of course; some gases that may be handled easily when dry become corrosive after passing through a wet scrubber or gas absorber.
6. AMCA Publication 99, a standard handbook, includes some very useful information concerning the preparation of centrifugal fans for protective coatings.

Abrasion. The radial-blade centrifugal is by far the best type to consider when the fan must handle air containing abrasive material. Their only disadvantage is their relatively low efficiency compared with the other types of centrifugal fans. When the volume and pressure requirements are such that a very large reduction in power consumption can be realized by using another type of fan, economics may dictate the choice. The options are forward-curved-blade, backward-curved-blade, or modified-radial-blade types. All are available with extra heavy construction and optional features to increase their resistance to abrasive wear. The options may include one or more of the following:

1. Steel wear plates attached to the wheel blades—may be welded on or bolted and replaceable
2. Replaceable wear liners for the fan scroll and sides—may be steel or cast iron
3. Hard-facing the wheel blades with materials such as tungsten carbide

Some manufacturers offer small to medium sizes of radial-blade centrifugals with cast-iron housings and wheels. The thickness of the material is the advantage this type offers.

It is always good practice to install the fan on the clean-air side of the air-cleaning device where the system includes one. When the device is a good fabric-bag-type dust filter, it is fairly common practice to use an efficient backward-curved-blade or airfoil-blade centrifugal fan on the clean-air side. These filters are so highly efficient that the fan can often tolerate the relatively minute amount of material passing through it.

HANDLING FLAMMABLE GASES OR VAPORS

Wherever a flammable or explosive mixture is handled by the fan, either continuously or only infrequently, it is imperative to eliminate all possible sources of ignition. The table below lists the sources and suggested preventive measures.

AMCA Standard 401-66 classifies fans or air-moving devices (AMD) with spark-resistant construction into three types. The standard applies to central-station units, centrifugal fans, axial and propeller fans, and power roof ventilators.

Type A	All parts of AMD in contact with the air or gas being handled should be made of nonferrous material
Type B	The AMD should have an entirely nonferrous wheel or impeller and nonferrous ring about the opening through which the shaft passes
Type C	The AMD should be so constructed that a shift of the wheel or shaft will not permit two ferrous parts of the AMD to rub or strike

1. Bearings should not be placed in the air or gas stream.
2. The user should electrically ground all AMD parts.

It is important that the user check exactly what the state or local codes or ordinances require for fire and explosion prevention. Most of the national standards organizations recommend at least type B construction.

There is one curious contradiction to good practice which some standards organizations and rules, including Occupational Safety and Health Act (OSHA), appear to permit: belt-driven axial fans for paint-spray-booth exhaust systems. It can be argued that the fan bearings are not in the airstream, since most fans used for this duty have the shaft, bearings, and drive enclosed in cylindrical casings. However, the casing is not necessarily vaportight. Also, there must be an opening in it for the fan shaft, and a shaft seal is seldom, if ever, used. The bearings can be regarded as *exposed* to the airstream in most cases, if not actually in it.

Subpart H, section 1910.107 of the OSHA Rules and Regulations dodges the issue by stating, "the shaft preferably to have bearings outside the duct and booth."

Axial fans are almost universally used for paint-spray-booth exhaust. The high volume and relatively low static-pressure requirements as well as the straight-through airflow advantage make them the ideal choice. Centrifugal fans simply cannot compete.

There are some applications so hazardous that both gastight and spark-resistant construction should be considered. When the applicable rules are not specific or stringent, the user must exercise his best judgment to evaluate the hazard and choose the proper fan. A fire or explosion can be the worst kind of industrial disaster.

FAN NOISE

A comprehensive discussion of fan-noise generation, measurement, and control is far beyond the scope of this chapter. In fact, it is still a developing art, although commendable progress has been made by the fan industry and interested engineering societies. Noise is defined simply as *unwanted* sound. The federal government under the OSHA Rules and Regulations now specifies maximum permissible noise exposures to which employees in virtually every industry may be subjected. Unfortunately, fans are all, inherently, noise generators. Consequently, a few recommended references, a brief discussion, and some suggestions about how to avoid or minimize fan-noise problems must be included.

For those interested in more detailed information, the ASHRAE "Handbook of Fundamentals" and "Guide and Data Book" are recommended. AMCA publishes standard sound test codes as well as application data for both ducted and nonducted air-moving devices.

Measurement of the sound level in the vicinity of a fan already installed is a relatively simple matter. There are many sound-level meters on the market which are adequate for this purpose. *Predicting* the sound level which the meter will read *before* the fan is installed is far more difficult even though the fan has been tested according to the accepted code and its sound-producing characteristics are known. The fan manufacturer can furnish sound *power* level ratings which describe the acoustic power level generated by the fan. The meter, however, measures sound *pressure*. The sound level that it will read depends on the acoustical environment in which the fan will operate. The conversion from known sound power levels for the fan to predicted sound pressures *after* it is installed can be made with reasonable accuracy by someone familiar with the application of acoustical data; but he must have *detailed* information about the acoustical characteristics of its environment. Put in simpler terms, the fan will *sound* louder in a small room with hard, reflecting wall and ceiling surfaces than it will in a larger room with walls and ceiling covered with sound-absorbing material. The sound-level meter will say that it is not making as much *noise* in the larger room but is producing the *same* amount of acoustic *power* in either room.

It is obvious that a fan manufacturer *cannot* meet a specification which says only that "the fan must not exceed OSHA noise level requirements."

Sound power levels can be readily used to compare fans of the same type made by different manufacturers if they have been tested and rated in accordance with the same code. Sound measurements cannot be made with great precision, and differences in sound power levels of 2 decibels (dB) or less are not considered significant. When comparing products, differences of less than 4 dB can be disregarded.

Eliminating Fan-Noise Problems. A fan or a fan *system* may generate enough noise to be extremely annoying even though it is well below the legally acceptable limits. If the problem can be foreseen and eliminated by selecting the best type of fan and installing it properly, it is of course much less costly. Unfortunately, this is not always the case. The following general suggestions are included with the hope that they may be helpful with an existing noise problem. One or more of them may apply, depending on the severity of the problem.

1. Inspect the fan for possible *mechanical* sources of noise. Investigate at least these possibilities:
   a. Worn or dry bearings or coupling
   b. Loose set screws on fan wheel or V-belt-drive sheaves
   c. Broken or loose bolts or fasteners
   d. Bent fan shaft
   e. Fan wheel or motor out of balance
   f. Weak or unstable foundation or fan mounting
   g. Loose dampers or inlet vanes
   h. Speed too high
   i. Fan rotating in wrong direction
   j. Foreign material in fan
   k. Fan wheel rubbing on housing
   l. Vibration transmitted to fan from some other source

4.102 MAINTENANCE OF PLANT FACILITIES

2. Locate the fan either in an unoccupied space or as remote as possible from an occupied space. The OSHA requirements apply only to the noise *people* may be subjected to—*not* to the noise generated by the fan.
3. Install a good vibration-isolation base under the fan and motor unit and provide a heavy, rigid foundation for the entire assembly. This will prevent transmitting vibration to the building structure.
4. Install flexible connections between the fan inlet and/or outlet and the connecting ductwork. Do not use the fan housing to support the ductwork.
5. Install sound attenuators on the fan inlet or outlet as necessary. Many commercial units are available for this purpose. They will add resistance to the system which must be taken into account. It may not seem feasible to use them where there is heavy dust loading, for example, but they are worth investigating. They have been designed for some difficult applications.

   It may also be necessary to install a shaft seal. The housing must be tight; calk or gasket as necessary. Noise is readily transmitted through very small openings or through lightweight-fabric flexible connections between the fan and ductwork.
6. The above steps cover all the usual sources of fan noise. The remaining possibility is noise transmitted *through* the fan housing. The possible solutions require careful analysis to determine which is feasible and will achieve the desired reduction to noise at a reasonable cost.

If only a moderate reduction in noise is required, covering the walls and ceiling of the room with sound-absorbing materials may be the answer. It will not reduce the noise transmitted through the fan housing. It will reduce the noise level in the room by absorbing much of the sound energy which would be *reflected* by hard, nonporous wall and ceiling surfaces.

A second possibility is covering the fan housing with material which will act as a sound barrier. It must be a heavy, dense, nonporous material; sheet lead is an excellent example. It will do little good to cover the fan with lightweight, porous, sound-*absorbing* material; it will permit *transmitted* noise to pass through it. A sound *barrier* is required.

The third possibility is to enclose the entire fan assembly in a room or boxlike structure which will serve as a sound barrier. If the noise level is very high, thick concrete or masonry should be considered. The enclosure must be tight with extra heavy doors and all joints or cracks sealed. It will undoubtedly be necessary to ventilate the enclosure to prevent excessive heat buildup. Ventilation might be either by natural convection or mechanical depending on the amount of heat to be dissipated. In either case, the necessary ventilation openings must be treated as separate noise sources and attenuated or baffled.

FAN VOLUME-CONTROL DEVICES

Several options are available where it is necessary to vary the volume handled by a fan. All are readily adapted to automatic control, or they may be regulated manually. The most commonly used devices are outlet dampers, variable-inlet vanes, and variable-speed drivers. The user must decide which is best for a particular application on the basis of adaptability and analysis of the initial and operating costs. Figure 6.11 shows graphically that a damper is least economical in terms of *operating* cost and variable-speed most economical, with variable-inlet vanes somewhere in between. The *initial* cost is reversed; a damper is lowest and variable-speed highest.

Outlet Damper. (see Fig. 6.12). These consist usually of two or more blades or louvers linked together so that they operate in unison and can be regulated by a single lever. Refinements include streamlined blades and additional linkage to rotate adjacent blades in opposite directions. The value of streamlining to reduce resistance is questionable but should produce a stiffer blade; opposite rotation of blades should result in more linear response to control-lever movement.

Dampers perform their function simply by increasing the resistance of the system. Useful static pressure produced by the fan is completely wasted; however, they are simple and reliable.

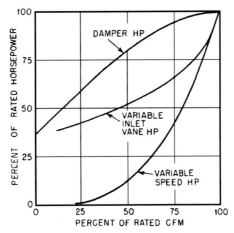

FIGURE 6.11 Comparison of power required with various methods of control.

Variable-Inlet Vanes. (see Fig. 6.13). These consist of a number of radial vanes, positioned as close as possible to the fan inlet. The vanes operate in unison through a linkage arrangement and are controlled by a single lever. As they are moved from full open toward closed position, they cause an increasing whirl or spin of the air as it enters the eye of the wheel. They alter the fan-performance characteristics so that both the pressure and capacity are reduced. Consequently they are not as wasteful as simple damper control.

FIGURE 6.12 Outlet damper with streamlined blades and linkage arranged to move adjacent blades in opposite directions for even throttling.

FIGURE 6.13 Single-inlet arrangement with variable-inlet vanes. Vanes rotate in unison when control lever is moved, varying amount of "spin" imparted to air and modifying fan performance. Access door is equipped with quick-opening latches.

Variable-Speed Control. Varying the fan speed can be accomplished with a number of drivers. They include steam turbines, fluid couplings, magnetic drives, direct-current motors, multispeed ac motors, and adjustable-speed belt drives. Speed control is *usually* although not *always* the most efficient method of control. With multispeed ac motors, the speed reduction is in steps corresponding to the 60-Hz synchronous speeds, and auxiliary damper regulation may be required at intermediate operating points. Even with the best of hydraulic and magnetic drives, which allow stepless speed regulation to almost zero, there are some inherent slip losses at reduced ratings; consequently variable-inlet vanes may be superior in terms of power *input* where only a small reduction in capacity is required.

Steam turbines have an advantage in that they can sometimes be fitted into the plant heat balance at good efficiency. They also will give stepless control.

Varying the fan speed in a fixed system allows the *fan* to operate at constant *efficiency,* and this is the big advantage.

FAN INSTALLATION

Inspecting a New Fan. A thorough inspection of the fan and any auxiliary equipment should be made as soon as it is received, with particular attention to the following:

1. That the shipment is complete and correct in every respect.
2. That the shipment includes *complete* and *explicit* assembly instructions if the fan has been shipped completely or partially disassembled; it is the fan manufacturer's responsibility to furnish this information.
3. That there has been no physical damage either during shipment or due to careless workmanship. This is vital, particularly so if the fan has a special protective coating. Many coatings *cannot* be repaired satisfactorily, and even a slight discontinuity may make the coating worthless.

Handling and Storage. Extra care is required in handling, particularly when a fan is lifted with a crane or hoist. A spreader bar should be used to prevent pressure against the sides of the housing or bearing base. Sheet-metal parts are not designed to withstand these abnormal stresses; misalignment of bearings or poor fit between the wheel and inlet bell can be caused by careless handling. A fan with special protective coating must be given even more than the usual amount of care.

If the fan must be stored, cover the inlet and outlet openings to prevent entry of foreign material. Do not permit the stacking of other stored items on top of the fan. If the fan must be stored outside, cover it completely and coat the exposed shaft with rust-preventive compound or heavy grease. Grease-lubricated ball or roller bearings usually get their initial lubrication at the factory, but oil-lubricated bearings often do not. In *either* case, it is good practice to check before storing, lubricate as necessary, and rotate the wheel by hand to coat the bearing surfaces with a protective film.

Foundations and Mountings. A rigid, heavy, level foundation for the fan and any drive components is a must. Poured concrete is recommended. A generally accepted rule of thumb is that the weight of the foundation should be at least three times the total weight of the fan and drive components. Use L- or T-shaped bolts, surrounded by pipe or sheet-metal sleeves large enough to allow some adjustment. In estimating the length of bolts, allow 1 in for grout and shims plus thickness of fan base, washers, nut, and extra threads for drawdown.

When it is necessary to use a structural-steel foundation, it must be level and sufficiently rigid to ensure permanent alignment. It must be designed to carry with minimum deflection not only the weight of the equipment but the loads imposed by the centrifugal forces set up by the rotating elements. The entire structure should be welded or riveted.

A fan installed above the ground floor should be located above a rigid wall or heavy column if possible. When an overhead platform must be used, it should be strong, level, and rigidly braced in *all* directions.

Prestart Check

1. *Visual.* Make sure that the inside of the housing is free of all foreign objects and installation debris.
2. *Fasteners.* Check and tighten all fasteners, with particular attention to foundation bolts, bearing-mounting bolts, and driver-mounting bolts.
3. *Bearings.* Check alignment and make sure that they have been lubricated properly. Check auxiliaries such as water-cooling, air-cooling, or circulating-oil systems for proper functioning and leaks.
4. *Coupling.* If the coupling is a type which requires lubrication, check to be sure that it has been done. Alignment must be as precise as possible with faces parallel and in angular alignment. Spacing between halves must allow for axial movement of the shaft—particularly if a sleeve-bearing motor is driving the fan. Steam turbines and large motors usually require some allowance for expansion. The turbine half of the coupling is usually set low 0.001 in per inch of turbine height from mounting feet to centerline of shaft. For large electric motors, the rule of thumb is to set the motor half of the coupling low 0.001 in per inch of motor shaft diameter.

Fan Wheel. Make sure that wheel rotation is correct for the housing. The correct rotation is almost always marked on the outside of the housing but not necessarily on the wheel. It is not always obvious visually. If there is the *slightest* doubt, check with the manufacturer. Once it has been established, mark the outside of the fan with the correct rotation where it can be clearly seen and use a marking device that will not be easily obliterated. A metal arrow permanently attached to the housing is suggested.

Tighten the setscrews. If there are two, tighten the one *ahead* of the keyway (relative to the direction of rotation) first; then tighten the one over the keyway.

Rotate the wheel slowly by hand to make sure that it turns freely and will not strike or rub the housing. Set the clearance between the wheel and inlet cone in accordance with the manufacturer's instructions. The clearance may vary greatly depending on the type and size of fan, but generally it is as close as possible without risk of striking. It must not be too large, or fan performance may suffer. On high-temperature applications extra clearance is required cold, as thermal expansion will reduce the space at operating temperature.

V-Belt Drive. Check the sheaves for proper alignment with a straightedge, and tighten setscrews. Check the belts for good tension. It is impossible to be precise about belt tension—judgment and experience must be exercised. The correct tension is just sufficient to prevent excessive squealing at start-up and slipping under full load; more than this will shorten belt life and put unnecessary additional load on fan and motor bearings.

Dampers and Variable-Inlet Vanes. These devices should be checked to be sure that they operate freely and that all blades or vanes operate in unison and close tightly where necessary. The position of the operating lever corresponding to "open" and "closed" should be clearly and permanently marked adjacent to the device; it is not always obvious after the duct connections have been made.

Start-up. Before the fan is brought up to full speed, jog the driver momentarily to make sure the wheel is rotating in the correct direction. If correct, the fan may now be operated. If it is driven by a variable-speed device, it should be started and run at low speed, then gradually brought up to full speed.

It is strongly recommended that tachometer readings should now be made—at full speed and at the low and intermediate points if a multispeed driver is used. Any deviations from design speeds will be obvious and may be corrected. If the fan has an adjustable-pitch V-belt drive, a tachometer reading is essential. If the required operating speed is not correct, stop the fan and adjust the motor sheave as necessary. Recheck the alignment and belt tension.

The fan should be watched carefully for the first few hours of operation. It should be shut down immediately if excessive vibration or any other sign of trouble develops. Watch for leakage of lubricant or cooling water, excessively high bearing or motor temperature, or any unusual noise that may develop. Refer to the section on periodic inspection and the list of possible troubles and causes at the end of the chapter.

After a few days of operation the fan should be stopped and the applicable maintenance operations listed under periodic inspection should be performed.

SAFETY AND PROTECTIVE DEVICES

Guards and screens should be used to protect personnel from contact with rotating elements. They also have the useful function of protecting the equipment from damage.

Protective Screens. Screens should always be installed on a centrifugal or axial fan whenever the inlet or outlet is open while it is in operation. Propeller fans should usually have the entire motor and drive enclosed. Screens are an obstruction in the airstream, and the mesh should not be too fine, generally not finer than 1-in mesh for centrifugal fans and $\frac{1}{2}$-in mesh for propeller fans.

Coupling Guards. The guard must be strong enough to provide complete protection. It must also be easily removable for servicing.

Belt Guards. A V-belt drive is a friction drive and will generate heat. Air must be allowed to circulate freely around all parts of the drive to dissipate the heat; do not enclose it completely. Use an open-mesh expanded metal guard if possible.

Disconnect Switches. Unless there is an established lockout procedure where the fan is remote from its starter, always install a disconnect switch at the fan to prevent accidental starting and injury to maintenance personnel.

Protective Devices. Many devices are available which will automatically detect a fan malfunction. They can be arranged to shut down the fan or send an audible or visible signal to call attention to the problem. Among the many possibilities are sensing excessive vibration, high bearing temperature, loss of coolant or lubricant flow, and loss or reduction in airflow. They are worth considering if the fan is vital to a process or if access to it is difficult.

PERIODIC INSPECTION

Keeping Records. There is probably no need to point out the value of complete and adequate records. How they are kept depends on the user's established practice or preferences. Always keep a permanent record of the equipment manufacturer's order number; he may not be able to furnish the correct spare parts without it. Fan and driver nameplate data should always be recorded; nameplates can be lost or obscured by paint or corrosion. It is good practice to assign a number to each major piece of equipment and attach a permanent tag to the item. Frequency of inspection required will vary depending on the severity of service and type of equipment. The date should be recorded as well as services performed, repairs made, and spare parts used.

General. A good program will include good housekeeping. Keep equipment free of spilled or excess lubricants and debris. Clean and paint when necessary, particularly in outdoor location or corrosive atmospheres.

Bearings. Good bearings of any type, properly selected and installed, will give long satisfactory service when protected from their natural enemies. These include:

1. Too little *or* too much lubrication
2. Wrong *type* of lubricant
3. Dust, moisture, and corrosive atmospheres
4. Heat radiated or conducted from adjacent sources
5. High ambient temperatures
6. Excessive loads imposed by improper V-belt tension, misalignment, fan-wheel unbalance, or vibration transmitted from another source
7. Operating the fan at a speed higher than it was designed for

A bearing that feels hot is not *necessarily* too hot to operate. Bearings on high-speed fans may operate safely as high as 75°F above room temperature. Use the following as an *approximate* guide for maximum total operating temperature:

Ball or roller bearings	165°F
Ring-oiled sleeve bearings	150°F
Water-cooled sleeve bearings	110°F

If there is any doubt, try to get a temperature reading at the hottest spot on the bearing with a contact thermometer and get the opinion of the fan manufacturer or bearing supplier.

Couplings. Misalignment is probably the prime cause of premature coupling failure. Good couplings are capable of giving extremely long satisfactory service if good alignment is maintained and, for those types which require it, proper lubrication is provided regularly. Adhere strictly to the coupling manufacturer's recommendations. It is vital to maintain the best possible alignment with the fan and driver at their *operating* temperatures, particularly on high-temperature fans or where the driver is a steam turbine or large electric motor.

If the motor has sleeve bearings, there may be considerable axial movement of the shaft. The proper clearance between fan and motor shafts must be set with the motor rotor at its magnetic center. It is located by running the motor and noting the normal operating position which the end of the shaft assumes.

V-Belt Drives. Refer to the comments regarding alignment and belt tension under Prestart Check. When it is necessary to replace belts on a multi-V drive, they should be a set matched for length. *Never* use excessive force to pry belts onto the sheaves. Loosen and move the motor enough to allow easy installation. Do not use any belt dressing unless it is recommended by the drive manufacturer.

Fan Balancing. Fan wheels are balanced statically and dynamically by the manufacturer. No further balancing should ever be required for a fan wheel handling relatively clean air as long as it is not damaged and is kept reasonably clean.

Unbalance produces vibration which can be felt by placing the fingertips lightly on a bearing while the fan is running. *Some* vibration will *always* be apparent. Vibration displacement is measured in mils. One mil is 0.001 in. Table 6.2 indicates how much vibration is acceptable.

Wheel unbalance is not the only possible source of excessive vibration; refer to the list of fan troubles at the end of the chapter. The other possibilities should be investigated before attempting to balance the wheel. A good vibration-measuring and -analyzing instrument is extremely useful in locating the source of trouble. It will also make in-place balancing a relatively simple task compared with the cut-and-try method. The procedure is described below. It is done by chalking the motor shaft on arrangement 4 fans and the fan shaft on other arrangements.

1. Bring the fan up to highest operating speed. If the fan and motor and bearing base are on vibration isolators, place blocks under the fan to make the isolators inoperative. *The fan must be rigid while balance is checked.*
2. Clean the shaft by holding a piece of emery cloth against the rotating shaft at the housing on the drive side of single-inlet fans. Double-inlet fans must be chalked on both sides of the fan and balanced as if two separate wheels.
3. Hold a sharpened piece of chalk or soapstone so the point just touches the rotating shaft. The chalk will scribe a line on the shaft, the length of line indicating the amount of unbalance. Make three or four lines so an average can be taken. The chalk must be held firmly so that it will be touched only by the high spot, not allowed to "ride" the shaft. When there is unbalance, the shaft will be forced to deflect outward by the unbalanced weight, producing the high spot.
4. Stop the fan.
5. Turn wheel by hand to see how long and heavy the scribed lines show. Mark the center of the lines and place a trial weight on the heel of the wheel blade opposite the heavy side, 180° away from the center mark.

TABLE 6.2 Acceptable Levels of Vibration

Fan speed, rpm	Vibration, mils			
	Smooth	Fair	Rough	Very rough
600	2	4	8	15–20
900	1.5	2.75	6	8–10
1200	1.0	2	4.5	6–8
1800	0.75	1.5	3.5	5–7
3600	0.4	0.7	2.5	4–5

Notes: *Fair* is not bad but should be improved. *Rough* should be corrected as soon as possible. *Very rough* is too rough to operate.

6. U-shaped or hairpin-type clips can be used as trial weights, being made from sheet metal or bar stock. Weights should be made so they can be forced on the heel (inside) of a wheel blade where they will not fly off. For double-inlet fans, it is a good idea to make clips in pairs.
7. Size of weight will be determined by length of lines on the shaft. Short lines will indicate that the fan is badly out of balance and a larger weight should be used.
8. After the trial weight is placed on the heel of the blade, run the fan again and rechalk the shaft. If the scribed lines are longer and the centers of the lines have not shifted from the ones scribed initially, add more weight on the same blade and run again. Repeat test runs with varying weights until the lines are scribed all around the shaft. A balance will then be obtained.
9. If the center of the lines shifts from that scribed initially, move the weight forward or backward to the next wheel blade, so the center mark (heavy side) will not move from its initial position. Move weights less than 90° if vibration decreases. Move weights 180° if the lines remain in the same original position but vibration increases. When lines return to the original position, increase the weight until balance is obtained.
10. If the weight is too great, the center of the lines will shift 180° from the initial run. Reduce weight until balance is obtained, with the chalk line scribed all around the shaft.
11. When the position and amount of weight are established, the trial weights, or equal weights, may be welded or riveted to the wheel, preferably on the outside of the flange or backplate. The weight of the weld bead or rivets should be taken into account.
12. If chalk marks extend all the way around the shaft but considerable vibration persists, a weak foundation or loose bolts might be the trouble.

FAN TROUBLES

The following list shows the most common fan troubles and the possible causes:

Capacity or Pressure Below Rating

1. Total resistance of system higher than anticipated
2. Speed too low
3. Dampers or variable-inlet vanes not properly adjusted
4. Poor fan inlet or outlet conditions
5. Air leaks in system
6. Damaged wheel
7. Incorrect direction of rotation
8. Wheel mounted backward on shaft

Vibration and Noise

1. Misalignment of bearings, couplings, wheel, or V-belt drive
2. Unstable foundation
3. Foreign material in fan causing unbalance
4. Worn bearings
5. Damaged wheel or motor
6. Broken or loose bolts or setscrews
7. Bent shaft
8. Worn coupling

9. Fan wheel or driver unbalanced
10. 120-cycle magnetic hum due to electrical input (check for high or unbalanced voltage)
11. Fan delivering more than rated capacity
12. Loose dampers or variable-inlet vanes
13. Speed too high or fan rotating in wrong direction
14. Vibration transmitted to fan from some other source

Overheated Bearings

1. Too much grease in ball bearing
2. Poor alignment
3. Damaged wheel or driver
4. Bent shaft
5. Abnormal end thrust
6. Dirt in bearings
7. Excessive belt tension
8. Heat conducted or radiated from another source

Overload on Driver

1. Speed too high
2. Discharging over capacity because existing system's resistance is lower than original rating
3. Specific gravity or density of gas above design value
4. Packing too tight or defective on fans with stuffing box
5. Wrong direction of rotation
6. Shaft bent
7. Poor alignment
8. Wheel wedging or binding on fan housing
9. Bearings improperly lubricated
10. Motor improperly wired

CHAPTER 7
DUST-COLLECTING AND AIR-CLEANING EQUIPMENT

Lee Twombly
Manager, Scrubber and Mechanical Collectors
SnyderGeneral Corporation (American Air Filter)
Louisville, Ky.

Samuel G. Dunkle
Manager, Electrostatic Precipitators and Fabric Collectors
SnyderGeneral Corporation (American Air Filter)
Louisville, Ky.

Increasing demands resulting from environmental regulations such as the Clean Air Act, with Amendments, and the Occupational Safety and Health Act (OSHA) have brought to the forefront the importance of protecting the quality of air in our atmosphere. Contaminants include particulates, vapors, gases and/or acid mists. Sources include any manufacturing process or industrial operation which emits contaminants into the workplace. Maximum allowable airborne concentrations of contaminants are defined in appropriate environmental regulations. Organizations such as the American Conference of Governmental Industrial Hygienists (ACGIH) have established threshold-limit values (TLVs) for airborne substances with health hazard potential.

Maintaining environmentally safe air in the workplace requires a thorough understanding of the regulations, air-cleaning equipment and maintenance requirements for the air-cleaning equipment. The advantages of installing air-cleaning equipment include the following:

1. Prevent contaminants from entering the working environment.
2. Increase quality of products manufactured or processes.
3. Meet the requirements of air pollution regulations.
4. Reduce maintenance downtime of process equipment.
5. Prevent local nuisance or damage to property.
6. Reclaim valuable materials.
7. Recover conditioned air.
8. Reduce or prevent fire or explosion hazards.

Air-cleaning technology can employ one or more of the following basic principles:

- Direct interception
- Centrifugal force

- Impingement
- Diffusion
- Gravity separation and settling
- Humidification
- Condensation
- Electrostatic attraction
- Absorption

TYPES OF EQUIPMENT

Dust-collection air-cleaning equipment can be grouped into the following classes:

Inertial or dry centrifugal
Cyclone
Multiple-tube cycle
Dry dynamic
Wet collectors (scrubbers)
Gas absorbers
Centrifugal
Dynamic
Orifice
Nozzle
Venturi
Fabric filtration
Disposable filters
Cartridge collector
Baghouse
Electrostatic precipitators
High voltage
Low voltage

Inertial or Dry Centrifugal

Simple Cyclone. Relies on centrifugal force to spin the dust particles against the sidewalls of the collector, with the collected dust leaving the airstream through the dust outlet at the apex (Fig. 7.1).

Multiple Centrifugal. Greater centrifugal forces are obtained by using a number of small-diameter tubes with identical elements in parallel (Fig. 7.2).

Dry Dynamic. Dust is precipitated by centrifugal and dynamic forces on numerous specially shaped fan blades. The collected dust travels along the blade surfaces and is discharged into a separate dust circuit within the fan housing. It is then conveyed pneumatically to the dust-storage hopper (Fig. 7.3).

Wet Collectors (Scrubbers)

Wet collectors provide many options for control of emissions. They exhibit a number of relative advantages which include small space requirements; no secondary dust sources; collection of

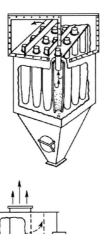

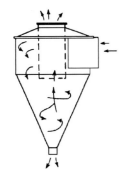

FIGURE 7.1 Simple cyclone.

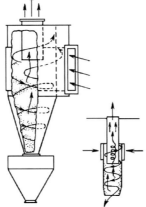

FIGURE 7.2 Multiple centrifugals.

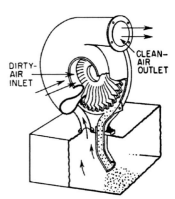

FIGURE 7.3 Dry dynamic.

gaseous and particulate matter; ability to handle high-temperature, high-humidity gas streams; ability to humidify a gas stream; and minimization of fire and explosion hazards. Scrubbers may be used for gaseous and/or particulate contaminant removal. We will break them down into gas absorbers and particulate scrubbers.

Gas Absorbers. Gas absorption is a method of removing soluble and chemically reactive gases from the airstream by contacting them with a suitable liquid. This process occurs when a soluble component of a gas stream transfers to a liquid in contact with the gas. Absorbers provide for intimate contact of the gas and liquid and may vary widely in design and performance. Removal of the contaminant may be by absorption if the gas solubility and vapor pressure promote absorption followed by chemical reaction. Gas absorption is a rate process and therefore does not take place immediately but rather at a rate depending upon several variables. The most important of these is the surface area of contact. The absorption rate varies directly with available surface area.

Contaminant saturation of the scrubbing liquid must be avoided; better yet, the concentration of the contaminant should be kept as low as economically possible. A saturated solution will no longer absorb the contaminant from the gas! Solubility can be improved by introducing chemicals to the liquid which react with the contaminant and/or neutralize it. This may be accomplished by controlling the acidity or alkalinity (so-called pH control) which is very effective for gases such as ammonia and acid gases. No absorption system will make the contaminant disappear. Contaminants are purged from the system by continuous drain of a portion of the liquid which must be made up with fresh absorbing liquid media.

Gas absorption provides an inexpensive means of reducing the concentration of harmful substances in air and rendering them harmless for safe handling and disposal.

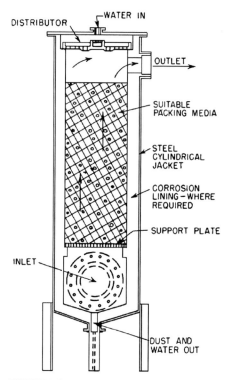

FIGURE 7.4 Packed tower.

Packed scrubbers are gas absorption devices and may be provided in either a horizontal or a vertical configuration (Fig. 7.4). They are essentially contact beds through which gases and liquid pass concurrently, countercurrently, or in crossflow. They are used primarily for applications involving gas, vapor, and mist removal. These collectors can capture solid particulate matter but are not used for that purpose because dust plugs the packing and requires excessive maintenance.

Water rates of 5 to 10 gpm per 1000 scfm are typical for packed towers. Water is distributed over V-notched ceramic or plastic weirs. High-temperature deterioration is avoided by using brick linings, allowing gas temperatures as high as 1600°F to be handled direct from furnace flues.

The airflow pressure loss for a 4-ft bed of packing, such as ceramic saddles, will range from 1.5 to 3.5 in WG. The face velocity (velocity at which the gas enters the bed) will typically be 200 to 600 fpm.

Particulate Scrubbers

Wet Centrifugal. Two types of wet centrifugal collectors are shown in Fig. 7.5. Collection relies on throwing the heavier particles against wetted collector surfaces by centrifugal force. Water distribution can be from nozzles, gravity flow, or induced water pickup by the airstream.

Wet Dynamic. The wet dynamic precipitator (Fig. 7.6) combines the dynamic and centrifugal forces of a rotating fan wheel to cause the contaminant to impinge on the numerous specially shaped blades, on whose surface a film of water is maintained by spray nozzles at the inlet.

Orifice Type. Collection efficiency in orifice-type wet collectors (Fig. 7.7) relies on the pickup or delivery of large water quantities to a collecting zone, where centrifugal forces, impingement, or collision causes capture of contaminant before its removal from the airstream. Water quantities in motion are high, running from 10 to 40 gal per 1000 cfm. Much or all of this water can be recirculated without the use of distribution spray nozzles or recirculation pumps.

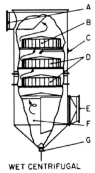

FIGURE 7.5 Wet centrifugal.

SYMBOLS	PARTS
A	CLEAN AIR OUTLET
B	ENTRAINMENT SEPARATOR
C	WATER INLET
D	IMPINGEMENT PLATES
E	DIRTY-AIR INLET
F	WET CYCLONE FOR COLLECTING HEAVY MATERIAL
G	WATER AND SLUDGE DRAIN

FIGURE 7.6 Wet dynamic.

Nozzle Type. High-pressure nozzles normally are installed in tower types of centrifugal collectors (Fig. 7.8). A large number of small nozzles is involved, using water pressure of 250 to 600 psi and water volumes of 5 to 10 gal per 1000 cfm. Distribution of the droplets must be such that dispersion increases the chances of impact between water droplet and contaminant particle, rather than droplet against droplet to form larger, less effective droplets. The use of small nozzle orifices makes recirculation of water impractical for most industrial operations.

Venturi Scrubber. In the venturi design (Fig. 7.9) the shear stresses of the airstream traveling at velocities of 12,000 to 20,000 fpm or higher are used to break up water introduced through open supply pipes in the venturi throat. The turbulence in the venturi occurs within a short time interval.

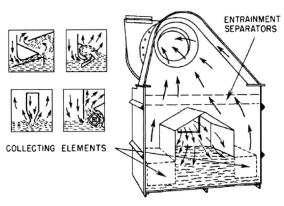

FIGURE 7.7 Orifice type.

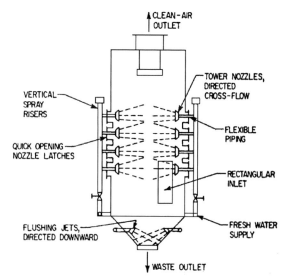

FIGURE 7.8 Nozzle type.   **FIGURE 7.9** Venturi scrubber.

Fabric Filtration

The effectiveness of passing air or gas through a filter medium at low velocity has been used for many years in air-cleaning devices. The collection mechanism is a combination of impaction, direct interception, and diffusion.

The filter media may consist of cellulose (paper), polyester, acrylics, nylon, polypropylene, teflon, glass, cotton, woven metals, and woven ceramics. The filter media can be arranged in many configurations such as a mat, multiplied mats, envelope, cartridge, or a tubular shape (bag).

Disposable filters consist of a filter media which is enclosed in a simple frame and is disposed of when dirty. A typical example would be the filter used in a home heating and air-conditioning system. Cartridge collectors and/or baghouse (Fig. 7.10) use disposable filters; however, the life expectancy

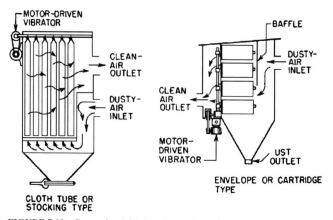

FIGURE 7.10 Conventional fabric collector, intermittent-duty.

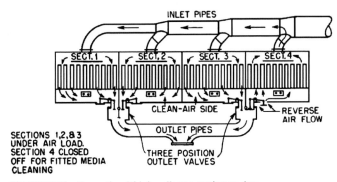

FIGURE 7.11 Conventional fabric collector, continuous-duty.

of the filter media is increased by the type of cleaning and collection mechanism. Filter media cleaning mechanisms consist of shaking (Fig. 7.10), reverse air or gas (Fig. 7.11), or reverse pulse (Fig. 7.12). Shaking cleaning uses direct movement of the filter media to dislodge particulate matter which has been collected. Reverse air or gas cleaning will reverse the direction of air or gas flow through the collector by opening and closing a series of dampers. The reverse pulse cleaning utilizes a short-duration high-/low-pressure pulse of clean, dry air in the opposite direction of air or gas flow. Reverse air or gas and pulse uses the energy in the reverse airflow to dislodge particulate matter from the filter media. Collection of particulate matter takes place in a hopper or similar device located below the collector.

Electrostatic Precipitators

Precipitators are commonly classified in two groups, low-voltage and high-voltage. The low-voltage, small-dust-holding capacity, low-dust-loading designs are used for indoor ambient air filtration. The high-voltage precipitator is used to collect particulate-laden air and gases in high-capacity situations (Fig. 7.13). The principle of collection is defined as using electrical forces to impart a negative charge to either liquid or solid particulate matter in a dirty air or gas stream, collecting the charged

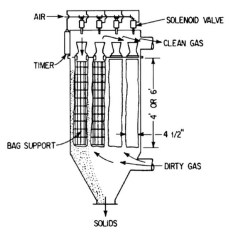

FIGURE 7.12 Reverse-pulse fabric collector.

FIGURE 7.13 Electrostatic precipitator.

particulate matter to a positively charged collecting surface, and finally cleaning the particulate from the collecting surface and collecting the particulate in a hopper or similar storage device.

The operating voltage can vary between 15,000 V for the low-voltage precipitator and 100,000 V for the high-voltage precipitator. Collecting surfaces can be round tubes or flat plates. Precipitators are commonly used to collect acid mist and/or condensed vapors (fog), sometimes referred to as a wet precipitator. Cleaning is accomplished by shaking and/or vibrating the collecting surfaces and usually takes place without stopping airflow through the precipitator. A wet precipitator is usually cleaned by washing with water.

MAINTENANCE OF DUST-CONTROL EQUIPMENT

Adequate and convenient maintenance begins on the drawing board. The eventual removal of parts which wear out or are damaged by other equipment should be taken into consideration by the draftsman and the design engineer. Parts normally exposed to maximum wear or mechanical abuse should be protected with extra heavy construction. Employee safety and convenient maintenance are readily provided by adequate service catwalks, platforms, and ladders. The initial installation and future repair of any fan or motor are always aided by the installation of a simple monorail or bridge crane. Proper upkeep will be ensured by following three general rules:

1. Cleaning the dust-collector storage hoppers at regular, predetermined intervals
2. Definite scheduling of inspection for preventive maintenance
3. Repairing, replacing, and cleaning all parts where indicated by the regular inspection

Inertial or Dry Centrifugal

Simple Cyclone and Multiple Centrifugal. Because these units are frequently used as primary collectors ahead of a more efficient final collector, they are usually subjected to heavy abrasive-dust loadings. Rugged construction is essential, with wear plates or rubber liners sometimes installed in the areas subjected to greatest wear. Installation and repairs are aided by bolted flange construction.

Multiple centrifugals should have the outlet tubes tightly sealed into the header sheet to prevent short circuition. Centrifugals, having multivaned swirl rings to impart a swirling motion to the airstream, should be protected against large or foreign articles by using a $^3/_4$-in mesh wire screen at the hoods. Buildup in the tubes can sometimes be prevented by suspending chains in the center of each cyclone. Typical maintenance on both types includes:

1. Emptying the storage hoppers at regular intervals to avoid reentrainment
2. Avoiding leakage at the dust-discharge point, especially if the unit is under negative pressure
3. Routinely inspecting for buildup and extreme wear
4. Keeping uniform air volume to maintain constant efficiency

Obstruction at the hoods or in the ducts will cause the volume to decrease proportionally.

Dry Dynamic. Routine maintenance includes:

1. Keeping the dust-storage hopper vented back to the collector
2. Inspection of the impeller at scheduled intervals for accumulation and wear on the blade tips
3. Emptying the dust-storage hopper regularly to avoid recirculation of collected dust, which would reduce efficiency and cause extreme wear
4. Because this unit is a combination fan and dust collector, the preventive maintenance outlined below under Fans regarding bearings and V-belt drive should be followed

Wet Collectors (Scrubbers)

As most wet collectors are in the high-efficiency group, they normally provide a nonhazardous dust discharge, but the equipment must be maintained to prevent undue wear, corrosion, and accumulation of sludge. Since corrosion is frequently due to moist air, it is advisable to treat all collector surfaces which come in contact with water or the airstream with a water-resistant or rust-retarding coating. Sheet rubber has been applied to areas subjected to severe abrasion with some success. A crust will sometimes form on the internal surface; it should be removed and the protective coating should be renewed. A periodic flushing will dislodge most accumulations.

Packed Scrubbers. To maintain maximum scrubbing efficiency, make periodic checks of the scrubbing liquor recirculation system, reagent additive system, packing, and demisters. If a pH controller is used, a periodic check should be made to ensure the probe is free of buildup or scale. Consult the specific instructions provided by the probe manufacturer.

Caution: *The reagents used with packed scrubbers are hazardous. As a minimum, eye and skin protection should be used when handling these chemicals. Inhalation and/or ingestion must be avoided. Consult the product safety data sheets provided by the reagent supplier for specific information regarding handling and safety procedures.*

Wet Centrifugal. Maintenance includes:

1. Maintaining an adequate water supply at all times when fan is operating.
2. When water is recirculated, use of adequate settling tanks providing relatively clear water.
3. Continued water flow for $^1/_2$ hr after fan is shut down to flush equipment adequately.

4. Regular inspection of equipment for accumulations, and thorough cleaning when required.
5. Inspection of spray nozzles, when used, at least once a week for plugging.
6. Checking all drains for accumulations. A plugged drain may form a water trap which could reduce the air volume.

Wet Dynamic. The same maintenance requirements as stipulated for wet centrifugals apply to the wet dynamic. Recirculation of the water is not recommended because of the type of nozzles used to provide the proper spray pattern. Since it is a combination fan and dust collector, see the section under Fans for proper maintenance of bearings and V-belts.

Orifice Type

1. The baffles or orifices that force the air to travel through the water should be checked regularly for wear, corrosion, and accumulation.
2. A high-pressure water jet should be used periodically for hosing down the interior.
3. Moisture entrainment baffles should be removed and cleaned. They must be replaced properly, by having the hooked lips opposite the direction of airflow.
4. Water overflow pipe should always be open to prevent reduced air volume resulting from excessive water in the collector.
5. Water level control device should be regularly inspected for proper operation. Improper water level will affect efficiency, air volume, and performance of the unit.
6. Orifice-type collectors incorporating sludge ejectors for removal of the collected dust should be inspected for the same items mentioned below under Sludge Settling Tanks.

Fabric Filtration

Items to be checked in routine maintenance of any fabric filter:

1. Dust leakage through holes in the filter media
2. Unusual wear, or holes in the baffle plates, if used
3. Unusual wear in the spark screens, if used
4. Excessive buildup or accumulations in the dust storage hoppers
5. Leaky or inoperative dust discharge valves
6. Routine inspection of the fabric cleaning mechanism

Dust leakage from holes in the filter media normally will reveal itself on the clean-air side of the unit by staining of the filter media and dust accumulations near the leak. With the fan installed on the clean-air side of the collector, excessive leakage will cause undue wear on the impeller and fan casing. With regular filter media maintenance, a set of filters or bags should last for several years on most applications. Premature wearing or tearing of the filter media can be avoided by not stretching too tightly during installation. The bags should never be under tension when the shaker mechanism is in operation. The pressure differential in fabric filters should not exceed the design pressure drop just before the cleaning cycle for the conventional unit and the design maximum for the reverse pulse unit. Excessive pressure differential can be caused by overloading the filter, plugging of the pores on the filter media by moist or sticky materials, and defective shaker or pulse mechanism.

A permanently installed U-tube to monitor pressure drop will quickly indicate an inoperative condition and provides an easy method to check visually the condition of the fabric filter.

Electrostatic Precipitators

In maintaining electrostatic units, attention must be given to the mechanical system and electrical system.

The mechanical system consists of the cleaning or rapping mechanism. It should receive the same maintenance as given to manufacturing process equipment.

The electrical system is complex, and maintenance should be carried out only by personnel equipped with knowledge of high-voltage electrical equipment. Before opening or entering any part of the unit, it is essential that the current to that portion of the collector be shut off by locking the switch in either the off or the grounded position. A fatal shock from high-voltage current is possible without physical contact because the charge may jump a 6- to 8-in gap.

Caution: Precipitators operate with very high voltages. Do not perform maintenance until power has been disconnected and all approved safety procedures have been followed to ensure the safety of maintenance personnel.

Even with the current turned off, it is not safe to enter the equipment until the static electricity has been dissipated by grounding. Proper grounding procedure consists of connecting a heavy wire to the grounded steel work and then hooking it to the disconnected high-voltage point. In explosive atmospheres, the high-tension points should be grounded to the outside. If this cannot be done, the entire rectifier and transformer must be shut off and the high-voltage line grounded at the rectifier with all high-voltage switches closed.

All grounding procedures should be followed whenever maintenance personnel:

1. Clean the line insulators.
2. Clean precipitator insulators.
3. Contact any high-voltage part.
4. Enter the rectifier screened enclosure.
5. Perform work inside the precipitator chamber.

After all the precautionary measures have been taken, normal maintenance inspection includes:

1. Checking for excessive accumulations on the discharge and electrodes and collecting surfaces.
2. Checking inoperative dust outlets and storage hoppers.
3. Checking performance of the electrical system for good ionization.
4. Maintain electrical clearances inside the precipitator in accordance with manufacturer's tolerances.
5. Inspect and ensure rapping (cleaning) mechanisms are in proper operating condition.

Sludge Settling Tanks. Such equipment may be incorporated with the wet centrifugals and wet dynamic collectors. Routine maintenance includes:

1. Periodic removal of the sludge before the tank becomes completely filled
2. Removing the silt from the bottom of the clean water chamber
3. Checking drag-type chain sludge conveyors, if used, for broken links and bent paddles
4. Checking for accumulations in the hopper bottom to avoid binding of the conveyor
5. Periodically inspecting automatic timers normally used with sludge ejectors to permit operation for 2 hr after the fan has been turned off
6. Routine inspection of chain guides and hopper wear plates for excessive wear
7. Removal of chain links whenever proper tension is lacking in the conveyor
8. Checking pumps handling water from sludge settling tanks for abrasion

Fans. One of the most important items of any exhaust system is the fan. Most fan troubles are caused by:

1. Abrasive cutting of the fan impeller and housing
2. Improper maintenance of V-belt drive and bearings
3. Accumulations causing vibration

Scheduled inspections of fans should include checking:

1. Bearings for proper operating temperature (greasing on an established schedule)
2. Bearings and/or housing for excessive vibration
3. Belt drives for proper tension and minimum wear
4. Correct coupling alignment
5. Fan impeller for proper alignment and rotation
6. Wear or material accumulation on impeller

Heavy-duty industrial fans are recommended for dust-collecting systems. Although paddle wheel fans are not as efficient as the centrifugal type, they give better service and have a more rugged construction. Proper maintenance has to do with:

1. Vibration caused by accumulations or improper mounting on a platform of light construction.
2. Abrasive wear and corrosion of the blades, rivets, and bolt heads. This usually will appear near the impeller disk.
3. Proper rotation. Most fans will discharge some air when running backward; so air movement is not an adequate test for correct fan rotation. Most manufacturers mark direction of rotation on the housing.
4. Proper tension on belts. In the event of a belt break in a multiple belt drive, all the belts should be changed at the shutdown period. Multiple V belts should have matching numbers to avoid having a few of the belts carry all the load.
5. Routine schedule for greasing the bearings. Follow the schedule and methods outlined by the bearing manufacturer. Remember that overgreasing can be as detrimental as lack of greasing.

All fans should be equipped with service doors built into the housing or into the inlet and outlet ducts. These doors should be provided with sturdy clamps and effective seals.

When making repairs to the fan wheels by welding on wear or patch plates, it is essential that the wheel be carefully balanced before reinstalling or operating. All patches should be of the same weight and of a uniform pattern. Balancing is easily done by adding arc-weld beads to the patches. It is possible for the wheel to be in balance in the balancing rolls (static balance) but wobble when the fan is run at operating speed (dynamic unbalance).

MAINTENANCE OF EXHAUST HOODS AND DUCTS

Hoods and ductwork not only must convey the material safely and efficiently but should also be low in first cost, and should require a minimum of maintenance. Factors such as the characteristics of the material to be handled (sticky, abrasive, toxic, oily) determine the type of hood or enclosure.

Future requirements can influence design and maintenance costs. An installation cannot be properly operated and maintained when branches are added indiscriminately. Additions destroy air balances and change velocities, multiplying plugging and other malfunctions. Additions should be estimated when the system is designed. If this has not been done, it may be advisable, from a maintenance point of view, to install a separate main for additional hoods.

Cleanout provisions are necessary. Good design can minimize the number of cleanouts required, but it cannot completely eliminate them. When adequate cleanouts are built into the system, maintenance is simplified. Types of cleanouts in common use are the slide door, hinged access, and dead end cap. They should be placed so that all the header or main can be inspected and cleaned as required. They can be placed in either the side or the bottom of the ductwork. Side cleanout has the advantage of allowing inspection and cleaning without excessive spillage on machinery or personnel.

Preventive Maintenance

The plant maintenance department cannot properly determine the condition of an air-cleaning system unless it knows how the system is to perform. After installation, a check sheet or data sheet should be prepared. This information will be helpful for developing future troubleshooting and maintenance procedures. It should include the following basic information:

1. Location of the system indicated by bay, column, machine, or department numbers. Manufacturer's equipment, operating, and installation instructions recorded or filed for reference in maintenance and ordering spare parts.
2. Operating data. Recorded design volume on each hood, branch, and connection, as well as total volume, pressure, horsepower, and rpm.
3. System characteristics. Pressure readings taken at all checkpoints after the system has been balanced. These figures are used for comparison when checking performance at future scheduled dates.
4. Operating voltages and currents of equipment as compared with design conditions.
5. A daily log of the air-cleaning system operation parameters.

Pressure readings are a guide to system maintenance. When taken at all hood entries, branches, and collection-equipment inlets and outlets, they not only show the requirements for maintenance but indicate plugged collection equipment, slipped belts on exhausters, and other minor ailments.

Weekend work can be largely eliminated when checkpoints are built-in features of the system. The system check sheet and a U gage can discover troublesome conditions without shutdown during regular hours. A comparison of the original readings with those made periodically will point out any need for maintenance. When one hood checkpoint, or the checkpoints on one branch, show lower readings, while other branches give higher readings, an obstruction is indicated. This makes it an easy task to locate and remove the buildup, since it must lie somewhere between the lower reading points and the air mover. The general condition of hoods, pipe hangers, and ductwork can be determined visually at the time of reading. It is important that hoods, access doors, safety shield, and other components be reinstalled after work has been done. (A properly designed hood can be assembled in only one way.)

In cleaning ductwork, the light-gage type can be taken down by removing the draw bands. It can then be emptied into a hooded container if the material is dry. If material containing oil mist or a condensate must be cleaned from ductwork, spray washing may be required. This generally is satisfactory only for short runs of duct.

Condensation in pipes often is a problem, and no certain preventive method has been found. Different plants have tried lower velocities, higher velocities, insulation, heating of the ducts, and built-in scrapers or nozzles. Perhaps the best solution is to keep the ducts as short as possible and mix the high-humidity air with warm, dry air from some other nearby operation. Short runs mean that ductwork can be removed easily and washed out. Strip heaters can be placed on ductwork to warm it before air is pulled through. Auxiliary heated air may be introduced into the duct, but the operation is expensive.

The checkpoint method will more than pay for itself in the long-term operation of the plant.

MAINTENANCE OF DUST COLLECTING SERVICE EQUIPMENT

Don'ts

Don't hope to provide a dust collector by tying a burlap bag over a discharge pipe or directing the discharge duct into a barrel of water or into a settling chamber. Factors involved in the collection of small particles do not lend themselves to such elementary treatment. Hundreds of times each year hopes are blasted in such attempts, at high cost in dollars and man-hours, and loss of face.

Don't put two collectors of the same order of efficiency in series; the second collector will not remove enough material to alter discharge appearance, public nuisance complaints, or settlement in the plant area, unless the initial unit is not functioning satisfactorily. Exceptions could include installations where agglomeration of particles is involved, or where a second collector is used to remove material concentrated in a carrier airstream from the first collector.

Don't visualize discharge appearance based on efficiency data. Fine particles have many, many more light-reflecting surfaces per pound than coarse particles. Appearance of effluent air will be governed by efficiency of the collector, particle size of the contaminant, and concentrations of the solid particles. Often the removal of 85 percent of the solids from a local exhaust system will make no visual change in the discharge stack appearance.

Don't expect dust-collection equipment design to be more advanced than production machine design. Generally, the higher the degree of effectiveness, the more certain the need of periodic servicing, inspection, and part replacement. Such operations will not receive proper attention where collectors are placed in (1) inaccessible locations without ladders, working platforms, and lighting and (2) outdoor locations where workmen are exposed to rain or winter weather.

Don't wait for dust collectors to fail before ordering replacement parts. Periodic inspections will give ample warning. Under present-day working conditions, failure of a dust collector can mean the halting of production lines while long-distance calls and air express shipments attempt to reduce the downtime.

Don't expect collectors employing principles used for lower-efficiency primary collection to be developed that will give efficiencies equal to those of more expensive, high-efficiency units, hopes of the purchaser and enthusiasm of the inventor notwithstanding. The cost to American industry each year is a tremendous figure based on such hopes and enthusiasm.

Don't expect replacement of the exhaust air volume to slip in through cracks in windows and doors without creating cold drafts in winter months, often starving exhaust systems and reducing their effectiveness. Makeup air supply systems can improve control, eliminate drafts, and warm the incoming air efficiently.

Don't attempt to recirculate cleaned air from dust collectors handling toxic material without careful investigation of the policing required to be assured of maintaining effectiveness.

Don't install unit collectors on heavy-duty production operations without careful evaluation of the frequency of servicing required.

Do's

Before making any changes in a process using a dust collector or air pollution device, check the current air pollution regulations. Many newer regulations will require a new permit and may require a more efficient collection device.

When air from a dust-collecting device is being recirculated to the workers' environment, do provide a system or technique to assure the collecting device is operating satisfactorily. Acquaint the operating personnel with the performance characteristics of the equipment and how the equipment is to perform under normal operating conditions.

Install and service the air-cleaning equipment, system, and related accessories in accordance with the manufacturer's recommendations.

BIBLIOGRAPHY

"APTI Course 413 Control of Particulate Emissions," United States Environmental Protection Agency, October 1981.

"Engineering Manual for Control of In-Plant Environment in Foundries," Sec. II, American Foundrymen's Association.

"Industrial Ventilation," 21st ed, American Conference of Governmental Industrial Hygienists, 1992.

Kane, John M.: Operation, Application and Effectiveness of Dust Collection Equipment, *Heating and Ventilation,* August 1952.

SECTION · 5

MAINTENANCE OF MECHANICAL EQUIPMENT

CHAPTER 1
PLAIN BEARINGS

R. Keith Mobley
Principal, Life Cycle Engineering, Inc., Charleston, S.C.

GENERAL

Plain or sleeve bearings are designed to support shafts that rotate, oscillate, or reciprocate. Though seemingly simple, and certainly one of the least expensive of mechanical parts, sleeve bearings are highly engineered components. They range from porous self-lubricated powder metal parts only a fraction of an inch in diameter to stationary power plant bearings, which often exceed 18 in. in diameter.

With few exceptions, sleeve bearing lubrication is hydrodynamic; that is, during operation, the shaft floats on a thin film of the lubricant. Because of this, friction and wear are minimized. However, it is important to realize that this so-called minimum film thickness is *not* the same as the bearing clearance. While the latter may be up to several thousandth of an inch, the minimum film thickness is typically on the order of one ten-thousandth of an inch. Nevertheless, sleeve bearings can have an almost unlimited life, provided proper maintenance practices are followed. When replacement does become necessary, following proper refurbishing and assembly procedures will assure extended life of the replaced parts.

PREVENTIVE MAINTENANCE

Lubricant Supply. Proper bearing design and material are necessary to achieve long service life but are not by themselves sufficient. The lubricant is the key component of the system which determines bearing life. Reduced to simplest terms, if a sleeve bearing is provided with an adequate flow of the proper *clean* lubricant, long life should be realized.

Lubricant flow to the bearings is a function of the equipment design. Oil pressure at specified speeds should be within the limits given by the equipment builder. Lower values suggest worn bearings. In this case, replacement should be made as soon as is feasible. Excessive pressures indicate a blockage or restriction somewhere in the system. This should be investigated immediately. The oil level also should be checked on a routine basis to avoid pump cavitation and subsequent oil starvation. In nonpressurized lube systems, reservoirs should be checked on a regular schedule to ensure that adequate oil is always present. Wick-fed bearings, such as those in fractional horsepower electric motors, should be lubricated periodically according to the schedule called for by the manufacturer.

Cleanliness. Sleeve bearings simply cannot survive without adequate lubrication. Once this is assured, the next most important consideration is the cleanliness of the lubricant. Since minimum film thickness is so small, the presence of oil-borne debris can greatly accelerate the wear process. If foreign materials such as metal chips and abrasives are large and numerous, bearing failure can occur rapidly. It is therefore of the utmost importance to change the lubricant in accordance with the equipment builder's recommendations. The lubricant filter, if one is used, also must be replaced according to schedule. The air filter, if one is present, should be serviced at recommended intervals, for airborne contamination is a primary source of vitreous abrasives that find their way into the oil. If the equipment is operated for extended periods in dirty or dusty environments, more frequent lubricant and filter changes should be adopted. Usually the equipment builder has established recommended change frequencies under these conditions. If not, a good practice is to make changes at intervals from one-third to one-half of that normally recommended.

Lubricant contamination can occur in storage as well. However, simple good housekeeping, such as covering open containers and reservoirs tightly to exclude dirt and water, and keeping anything which contacts the lubricant (oil can, funnels, etc.) as clean as possible will prevent problems.

Lubricant Type. Ensuring an adequate flow of clean lubricant makes long bearing life possible but does not guarantee it. The oil must be the proper one for the application. From a bearing performance viewpoint, lubricant viscosity is the most important parameter. Lower-viscosity (i.e., thinner) oils reduce oil film thickness. This increases the wear rate and can possibly lead to failure. It is critical that the equipment manufacturer's lubricant recommendations be followed.

In addition, the proper combination of oil additives is necessary to prevent rapid breakdown, thickening, foaming, and sludging. All these effects can lead to bearing failure, as well as to the damage of other components. Failure to use the recommended lubricant can have dire consequences and, in most cases, voids the equipment warranty. Extended drain intervals should not be adopted without a strictly monitored oil analysis program.

BEARING MATERIALS

Requirements

Surface Action. Sometimes referred to as *slipperiness* or *compatibility,* surface action is the ability of a material to resist seizure when contacted by the shaft. Contact takes place every time the equipment is started or stopped and can also occur during momentary overloads.

Embeddability. The ability of a material to absorb foreign particles circulating in the oil stream is referred to as *embeddability.* Some particles will go unfiltered, so the material must be soft enough to ingest them.

Conformability. The material also must be soft enough to creep or flow slightly to compensate for the minor geometric irregularities which are present in every assembly. These include misalignment, out-of-round, and taper.

Fatigue Strength. This is the ability of a bearing material to withstand the loads to which it is subjected without cracking. Bearings should not fatigue prior to the normally scheduled overhaul.

Temperature Strength. As operating temperatures increase, bearing materials tend to lose strength. This property indicates how well a material carries a load at elevated temperatures, without breaking up or flowing out of shape.

Thermal Conductivity. Shear of the oil film by the shaft generates significant heat, most of which is carried away by the oil. Nevertheless, it is important for the bearing to transfer heat rapidly from its surface through its back to avoid overheating and resultant reduction in life.

Corrosion Resistance. Oils oxidize with use, and the products of this degradation can be corrosive. Blow by products and fuel or coolant contamination of the oil also promote a corrosive environment. Bearing materials should be resistant to these effects.

Construction

Most hydrodynamic bearings are metallic, primarily for reasons of thermal conductivity. They may consist of one, two, or three layers.

Monometals. Bearings made from a solid bar or tubes of an aluminum or bronze alloy have been available for a number of years. They are generally used where loads are not very high. In order to have the same rigidity as a bearing with steel back (see below) and to avoid yielding at operating temperature, they are made with a comparatively thick wall. As a result, they require a larger housing bore.

Bimetals. A bimetal bearing has a steel back, to which is bonded a liner of Babbitt, copper-lead, or aluminum. *Babbitts* are soft alloys of lead or tin, with additives such as copper, antimony, and arsenic. They have outstanding embeddability, conformability, and surface action but relatively low fatigue strength. Copper-lead and aluminum are harder than Babbitt and have much better strength, but at a sacrifice of the other properties.

Trimetals. In order to achieve the desirable surface properties of a Babbitt bearing and the strengths of harder materials, the trimetal bearing was developed for heavy-duty applications. In this construction, a thin (usually about 0.001-in) layer of a soft material is either electroplated or cast onto the copper-lead or aluminum layer of a bimetal. The surface layer (overlay) imparts the desired "soft" properties to the bearing; however, because it is so thin, it derives improved fatigue strength from the intermediate layer; that is, it is much stronger than a thick layer of the same soft alloy.

Table 1.1 shows the compositions of some of the more popular plain bearing materials in use today.

TABLE 1.1 Bearing Alloys

SAE no.		Nominal composition, %						Bearing construction
Babbits	Pb	Sn	Sb	Cu	As			
12		89	7.5	3.5	—			Bimetal
15	83	1	15	—	1			Bimetal
191	90	10	—	—	—			Plated overlay
192	88	10	—	2	—			Plated overlay
Copper base	Cu	Pb	Sn	Zn				
49	75	24	0.5	—				Trimetal
791	88	4	4	4				Monometal
792	80	10	10	—				Bimetal
793	84	8	4	—				Bimetal
794	72	23	3	—				Bimetal
Aluminum base	Al	Sn	Cu	Ni	Si	Cd	Pb	
770	92	6	1	1	—	—	—	Monometal
780	91	6	1	0.5	1.5	—	—	Bimetal/trimetal
781	95	—	—	—	4	1	—	Trimetal
782	95	—	1	1	—	3	—	Trimetal
787	85	1.5	1	—	4	—	8.5	Bimetal

DESIGN

In order to do a better job of maintaining bearings, it is helpful to have an understanding of certain key bearing design factors. Although this discussion centers on half-shell bearings, much of it applies to full round parts (i.e., bushings) as well.

An assortment of bushings, flanged and straight-shell bearings, and thrust washers is shown in Fig. 1.1. Standard nomenclature for the various aspects of bearing design can be seen in Figs. 1.2, 1.3, and 1.4.

There are two types of insert bearings: precision and resizable. The former is made to precise tolerances and can be installed without modification. A resizable part is manufactured with an extra thick layer of bearing material on the inside diameter (ID). This permits machining to any desired size.

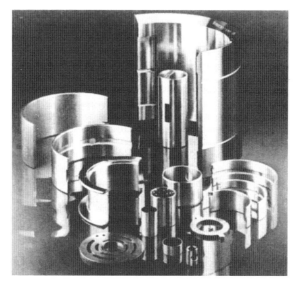

FIGURE 1.1 A sampling of bearings, bushings, and thrust washers.

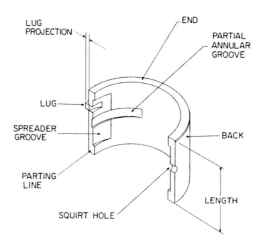

FIGURE 1.2 Bearing nomenclature.

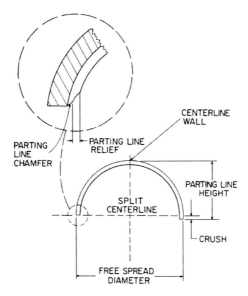

FIGURE 1.3 Bearing nomenclature.

Location and Retention. In order to keep a bearing from shifting sideways during installation and to ensure that its axial position is correct (i.e., so that it doesn't interfere with shaft fillets), a locating lug is formed at the parting line (Fig. 1.2). Another means used less frequently is a dowel in the housing, which protrudes partially into a mating hole in the bearing.

The sole purpose of the design feature, referred to as *free spread* (Fig. 1.5), is to aid in bearing installation. Bearings are manufactured with the distance across the outside parting edges slightly greater than the housing bore diameter. To install the bearing, a light force must be used to snap it into place. Once installed, the bearing will stay in place because of the pressure of the free spread against the housing bore.

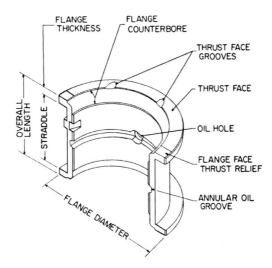

FIGURE 1.4 Flange bearing nomenclature.

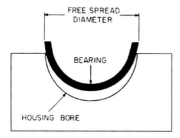

FIGURE 1.5 Free spread ensures bearing retention during installation.

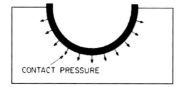

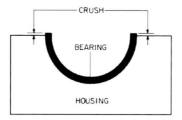

FIGURE 1.6 Crush prevents bearing rotation during operation.

It must be emphasized that neither the lug nor free spread will keep a bearing from spinning during operation. Such an action can cause a catastrophic failure and must be prevented. The means by which rotation is prevented is known as *crush* (Figs. 1.3 and 1.6). Bearings are manufactured so that they are slightly longer circumferentially than their mating housings. Upon installation, this excess length is elastically deformed ("crushed"), which sets up a high radial contact pressure between the bearing and housing. This ensures good back contact for heat conduction and, in combination with the bore-to-bearing friction, prevents spinning. *Under no circumstances* are the bearing parting lines filed or otherwise altered to remove the crush.

Lubrication. As indicated earlier, without adequate lubrication, particularly in the loaded area, a sleeve bearing will not work. Many bearings receive their oil supply through holes drilled in the housing. This necessitates a mating hole in the bearing to introduce oil to the clearance space. Sometimes holes are used to increase oil flow to nearby parts. Since the size and location of oil holes are critical to proper lubrication, replacement bearings must always be examined to ensure that oil holes match original equipment specifications.

Often a hole cannot adequately distribute oil to the loaded area of the bearing because the clearance space offers too much resistance to flow. In these cases, grooves are provided. Examples of typical grooves can be seen in Figs. 1.2 and 1.4. Grooves are used to aid flow both circumferentially (for distribution) and axially (for distribution, flushing debris, or lubricating adjacent parts).

Half-shell bearings are frequently manufactured with a circumferential taper of the wall (shown greatly exaggerated in Fig. 1.7). This *eccentricity* is typically less than 0.001 in and serves two purposes. First, it increases the average clearance in the assembly without raising the noise level while the bearing is in operation. Tight vertical clearances can be run for noise control, while operating temperatures are kept at reasonable levels through the increased horizontal clearance. Second, additional clearance is

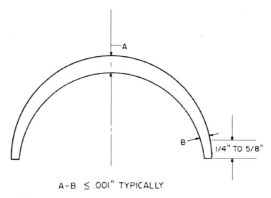

FIGURE 1.7 Bearing eccentricity.

often needed at the split line because of housing deflections that occur during engine operation; for example, four-cycle engine connecting rod bores elongate in the vertical direction due to the tensile load on the exhaust stroke. Without eccentricity, oil flow could be pinched off at the split line.

Another feature used to assure the formation of a good oil film is parting-line relief (see Fig. 1.3). Because of manufacturing tolerances, it is almost impossible to guarantee that the upper and lower bearings will have the same wall thickness at the split line. The thicker bearing can act as a wiper, removing oil from the shaft and hindering formation of a good oil film. By employing parting-line relief this potential problem is avoided.

Parting-time chamfers (Fig. 1.3) are also used to avoid abrupt steps in the bearing ID surface which can disrupt the oil film. They are also used in conjunction with spreader grooves (see Fig. 1.2)—large dirt particles tend to become trapped in the grooves and are then flushed out the bearing ends through the chamfers.

INSPECTION AND RECONDITIONING

Bearings sometimes need to be replaced. Frequently, this is not because they simply wear out, but rather because they have been damaged in use and can no longer perform their intended function. If a damaged part is simply replaced without determining the cause of the problem, it is very likely that the replacement part will become damaged by the same cause. Although the discussion that follows is oriented toward engines, the principles apply equally well to any equipment having sleeve bearings.

Preliminaries

When disassembling an engine to examine the bearings, two things are of utmost importance. First, be sure that the work is being done in a clean area. Dirt is a mortal enemy of an engine's interior, particularly the bearings. The replacement job can be nullified if dirt enters the engine and remains there after the rebuild. Second, lay out all dismantled parts in an orderly fashion. Mark or identify each so that it can be reinstalled in its original position. Lay out the bearings as they were in the engine, that is, with the main bearings in consecutive order, and with the connecting rod bearings placed between the same mains which flanked them in the engine. The following discussion should be helpful in describing various bearing conditions, their possible causes, and corrective actions to be taken to avoid a recurrence.

Analysis of Used Bearings

Normal Appearance and Wear. Most wear occurs during break-in, when minor geometric deviations are being accommodated. Thereafter, in a properly maintained engine, only those dirt particles too small to be filtered will be present to abrade the bearing surface. Two features usually mark normal wear. First, if the bearings are of trimetal construction, some of the overlay will have been removed, exposing the thin barrier layer between the overlay and the intermediate layer and possibly some of the latter as well. If of a bimetal construction, the surface will be noticeably burnished. There also may be minor surface scratches. These are generally not serious unless the intermediate layer has been deeply penetrated.

Dirt. Dirt is responsible for more bearing failures than any other mechanism. When dirt particles are large or numerous, they embed in the bearing lining, deforming the structure beneath and displacing the surrounding metal upward. The resulting high spot may be large enough to contact the journal. (A heavily embedded bearing will have numerous halos from this action.) Rubbing then creates heat which can, in conjunction with the stressed structure beneath, cause a rupture and removal of the bearing lining. If the particles embed only partially, the protruding portions will wear the shaft by a grinding wheel action. Figure 1.8 is an example of severe dirt embedment. Figure 1.9 shows a bearing which was badly scored by circulating particles which were too large to embed.

5.10 MAINTENANCE OF MECHANICAL EQUIPMENT

FIGURE 1.8 Bearing with severe dirt embedment.

Causes of dirt contamination are improper cleaning of the engine and parts prior to assembly, road dirt and sand entering through the air-intake manifold, or wear (including failure) of other engine parts, causing small fragments to enter the oil supply. Poor maintenance practices are generally the root cause of the problem.

Corrective actions: (1) grind and polish the journal surfaces if necessary, (2) install new bearings, paying particular attention to cleaning procedures, and (3) change oil and filters at the intervals recommended by the engine manufacturer.

Fatigue. Generally speaking, bearing fatigue results when either the load or time in service exceed the alloy's capability. There are several possible causes: load concentrations due to dirt, poor shaft or bore geometry, misassembly of the bearing, material weakness caused by high-temperature operation or corrosion, or simply exceeding the bearing's normally expected life span.

Fatigue cracks initiate at the bearing surface and propagate perpendicular to it. Before reaching the steel, the cracks turn, run parallel to the steel, and join. The material can then flake out.

The most common type of fatigue is that of the overlay on trimetal bearings. But since the primary overlay functions are to absorb small dirt particles and provide a slippery surface for starting and stopping conditions, slight overlay fatigue is not regarded as a bearing failure. The load-carrying strength of a bearing is in its intermediate layer. A true fatigue failure involves the intermediate material rather than the sacrificial overlay.

Few bearings escape some degree of fatigue during normal operation. Premature failures occur in stages, beginning with normal "hen track" patterns. As fatigue progresses into the second stage, it takes on the classical "wormhole" appearance, shown in Fig. 1.10 for a lead-base Babbitt bearing.

FIGURE 1.9 Bearing damaged by scoring.

FIGURE 1.10 Fatigue of bearing lining.

Figures 1.11 and 1.12 show fatigue caused by misshapen shafts, while Fig. 1.13 illustrates what happens when the bearing cap shifts. Regrinding the crankshaft will correct the former. Cap shift can be avoided by: (1) alternate torquing from side to side to ensure proper cap seating, (2) using new bolts to ensure against excessive play in bolt holes, (3) making sure the cap isn't reversed when installed, and (4) using the correct size socket to tighten the bolts to avoid interference with the cap.

Excessive Wear.   Some of the same factors that produce fatigue also can cause excessive wear. Generally, what determines the phenomenon that prevails is the load level and the severity of the irregularity which causes the problem. Geometric defects not only concentrate loads but also cause oil films to be thinner than normal. This results in more frequent metal-to-metal contact and wears the lining much faster than normal. Figure 1.14 shows excessive wear caused by a barrel-shaped journal, while Fig. 1.15 shows the skewed-wear pattern from a twisted connecting rod. Regrinding the shaft will correct the former, while replacing the rod is the best course of action in the latter case. Related upper cylinder parts also should be checked and replaced as needed when a bent or twisted rod has been discovered.

Foreign Material on Bearing Back.   Dirt on the bearing back causes high spots on the ID. It also prevents good heat transfer in these areas, which leads to localized overheating. The end result may be either severe local wear or, as in Fig. 1.16, fatigue. Clearly, this type of problem can be prevented through proper cleaning and burr removal prior to assembly.

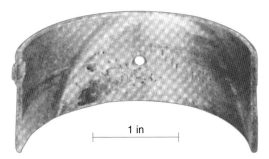

FIGURE 1.11 Fatigue caused by a tapered shaft.

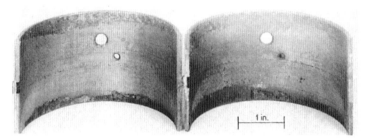

FIGURE 1.12 Fatigue caused by an hourglass-shaped shaft.

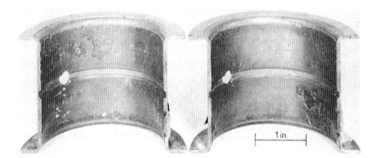

FIGURE 1.13 Fatigue caused by a shifted bearing cap.

FIGURE 1.14 Bearing worn in the center by a barrel-shaped shaft.

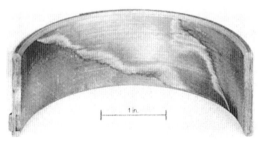

FIGURE 1.15 Skewed wear pattern caused by a bent connecting rod.

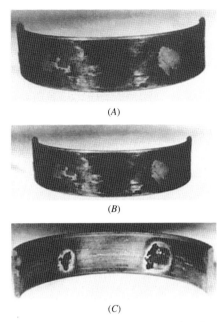

FIGURE 1.16 Foreign particles on bearing O.D. (*top*) can result in I.D. fatigue (*bottom*).

Hot-Short Phenomenon. A bearing suffering this type of failure (see Fig. 1.17) is unmistakable in appearance: large areas of the lining have been cleanly removed from the steel back. The damage occurs when the bearing temperature exceeds the melting point of its lowest-melting-point metal, usually lead or tin. This heat, in conjunction with shear due to shaft-to-bearing contact, leads to the "hot-short" (brittle when hot) condition. Causes of the failure can be insufficient oil flow, excessive dirt in the oil, a rough shaft, or severe misalignment. Historically, dirt has been the most frequent cause of hot-short failures. Correction involves thorough cleaning, regrinding the shaft to fix the damaged journal, checking for blockage of oil passages, the oil-suction screen and oil filter, and making sure the oil pump and pressure relief valves are operating properly.

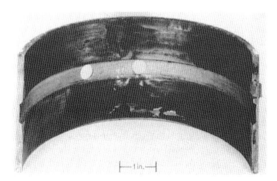

FIGURE 1.17 Hot-short condition (see text for details).

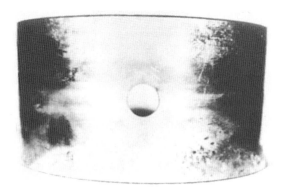

FIGURE 1.18 O.D. fretting.

Crush Problems. If crush is insufficient, relative movement occurs between a bearing and its bore. So-called fretted areas (see Fig. 1.18) will be visible on the bearing back and sometimes on the parting lines. These will appear to be highly polished and/or pitted. Corresponding damage also may be present in the housing bore. Fretted areas are points of stress concentration. If allowed to operate long enough, a fretted bearing may fracture through its steel back, as may the housing. Causes of insufficient crush are (1) filing of the parting lines, (2) dirt or burrs on the contact surfaces of the cap and housing, (3) insufficient bolt torque, and (4) oversize housing bore. Correction of the problem simply involves following good installation practices: bearings should never be altered, mating faces of the assembly should be clean and burr-free, the bore size should be verified as correct, and proper torque should be applied.

Excessive crush can arise if bearing *caps* are filed down in an attempt to reduce oil clearance or if the cap bolts are overtorqued. When too much crush is present, the bearings bulge inward at the parting lines. This can lead to fatigue or metal-to-metal contact and potentially to a hot-short failure if the oil flow becomes pinched off. Correction involves ensuring that proper torque is applied and, if the cap has been filed down, replacing the connecting rod or reworking the bearing bore if a main bearing is involved.

Crankcase and Crankshaft Distortion. These structural defects primarily affect the main bearings. Distortion causes increased loads and lower oil films, with conditions being worst at the point of maximum distortion. The damage varies from bearing to bearing, with the center main usually showing the greatest amount.

If the crankcase is distorted, a wear or fatigue pattern will be present on either the upper or lower bearings (see Fig. 1.19). Alternating periods of engine heating and cooling are the primary cause, though extreme operating conditions and improper torquing of cylinder head bolts also can be responsible. To correct the problem, the crankcase must be line bored.

A bent crankshaft causes a similar failure, but one which involves both the upper and lower bearings. The primary cause is extreme operating conditions. The corrective action is to install a new or reconditioned crankshaft.

Cavitation. This failure is induced by rapid fluctuations in oil film pressure. When the pressure in one area of the film drops below the oil's vapor pressure, a vapor-filled cavity forms. When the pressure increases again, the cavity collapses. This causes the surrounding oil to impinge on the adjacent bearing metal, eventually eroding the surface (see Fig. 1.20). Cavitation appears to be inherent to some applications. Ensuring that there is no air or water entrainment in the oil may help the problem. If possible, changing to higher-viscosity oil and increasing the oil pressure also may help.

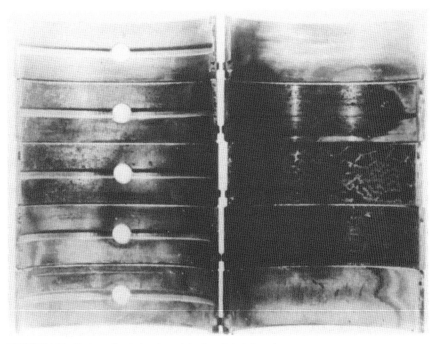

FIGURE 1.19 Fatigue of main bearing set due to a distorted crankcase.

Reconditioning

The preceding section discussed the causes of common bearing failures. It should be obvious that the vast majority of them can be prevented through proper maintenance. When bearings need to be replaced, however, it must again be emphasized that the job is not simply one of removing the old parts and installing new ones. It is mandatory that associated components be inspected and, if required, reconditioned. Every application is unique, and service manuals for the specific equipment should be consulted for detailed instructions. The discussion which follows concerns engines; however, the basic principles apply regardless of equipment type.

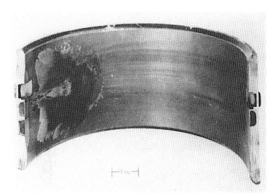

FIGURE 1.20 Cavitation.

Shaft. After cleaning the crankshaft, measure all the main and crankpin journals at several points to determine the out-of-round and taper. If any journal measures 0.001 in less than the manufacturer's specified diameter, has more than 0.001 in taper, or is more than 0.001 in out of round, the crankshaft should be reground before rebuilding the engine. Specifications on tolerances and permissible values for geometric irregularities are given in Table 1.2. Note that the latter are more stringent than the criteria for whether or not to regrind the shaft. This is because most manufacturers permit looser tolerances on rebuilds than on new or remanufactured parts. However, it must be recognized that a penalty is paid for doing this. Namely, one cannot expect to realize the same bearing life from a crankshaft that has not been reground as from one that has been.

In regrinding the shaft, a specific procedure is to be followed. If significant material must be removed from the journals, first turn the shaft (i.e., in a lathe) in the direction of the crankshaft rotation in the engine. Then, with the shaft rotating in the opposite direction, grind toward the high limit. Finally, polish the journals in the direction of the crankshaft rotation in the engine to remove the grinding fuzz. A maximum of 240 grit paper should be used, with a maximum stock removal of 0.0001 in. Oil holes should be blended well into the journal surfaces, with the maximum diameter at the run out of the blend not exceeding twice the hole diameter. The reconditioned shaft and its oil passages should be thoroughly washed in solvent, followed by hot soapy water and a hot water rinse. Use a round, nonmetallic bristle brush to clean the oil passages. Following the final rinse, blow the shaft and oil passages dry with compressed air, and immediately coat all journal surfaces and oil passages with a light film of the oil normally used in the engine.

TABLE 1.2 Shaft Tolerances

	Automotive	Heavy duty
Diameter tolerance:		
Up to 1½-in journal	0.0005 in	0.0005 in
1- up to 10-in journal	0.001 in	0.001 in
Over 10-in journal	0.002 in	0.002 in
Diametral-taper tolerance:		
Up to 1 in of length	0.0002 in	0.0001 in
1 up to 2-in of length	0.0004 in	0.0002 in
Over 2-in of length	0.0005 in	0.0003 in
Out-of-round condition:		
Up to 3-in diameter	0.0005 in	0.0002 in
3-to 5-in diameter	0.0005 in	0.0003 in
Over 5-in diameter	0.001 in	0.0004 in
Maximum misalignment:		
Adjacent main journals	0.001 in	0.0005 in
Crankpin parallel with main journals	0.001 in	0.0005 in
End clearances:		
Shaft diameter		
2–2¾ in	0.003–0.007 in	0.003–0.007 in
2¾–3½ in	0.005–0.009 in	0.005–0.009 in
3½–5 in	0.007–0.011 in	0.007–0.011 in
Over 5 in	0.009–0.013 in	0.009–0.013 in
Shaft hardness:		
Brinell	200 min	300 min
Shaft-journal finish (all applications):		
Microinches 15 max		
Waviness 0.0001 in max		
Lobing 0.0001 in max		
Chatter 0.00005 in max		

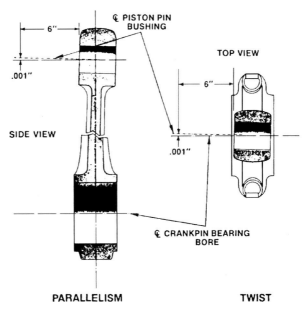

FIGURE 1.21 Connecting-rod parallelism and twist should be checked prior to reassembly with new bearings.

Connecting Rods. Measure the empty rod bores on several diameters. If any bore is more than 0.001 in out-of-round, replace the rod or recondition it. Rods should then be checked for parallelism and twist (Fig. 1.21). A slight bend or twist can be straightened using caution and the proper equipment. Severely bent or twisted rods should be replaced. Table 1.3 shows tolerances and permissible values for geometric deviations for rod bores.

Crankcase. After a thorough cleaning, including all oil holes and passages, measure each empty bore in the same manner as was done for the connecting rods. The standards of Table 1.3 apply to mains as well as rods.

TABLE 1.3 Empty Bore Tolerances

	Automotive	Heavy duty
Diameter:		
Up to 3¼-in diameter	0.0005 in	0.0005 in
3¼ to 10-in diameter	0.001 in	0.001 in
Over 10-in diameter	0.002 in	0.002 in
Taper, hourglass, or barrel shape:		
1-in length	0.0002 in	0.0001 in
1- to 2-in length	0.0004 in	0.0002 in
Over 2-in length	0.0005 in	0.0003 in
Out-of-round:		

0.001-in area if bore is larger horizontally than vertically.
Parallelism and twist between rod bore and wrist-pin bore when measured 6 in from the end of wrist-pin bushing 0.001 in max.

Crankcase alignment must then be checked. The best way of doing this is as follows: Use an arbor ground to 0.001 in less than the low limit of the empty bore specification. The arbor should be slightly longer than the crankcase. Place the arbor into the main bearing saddles (without bearings), position the main caps in place, and torque the bolts to blueprint specification. Rotate the arbor. If it will not turn and earlier measurements did not indicate an out-of-round condition, the crankcase is probably warped and must be line bored.

Reassembly. Prior to reinstallation of the crankshaft, the main bearings should be given a thin coating of oil and assembled clearances verified with a material such as Plastigage. As soon as all the main caps are in place and torqued, the shaft should be rotated. If binding occurs, proceed no further until the cause is found and corrected. The connecting rod bearings also should be prelubricated prior to attaching the rod assemblies to the crankshaft. After doing so, verify clearances and again check crankshaft rotation. Find and correct any binding that occurs. After reassembling the other parts to the engine, it is advisable to use an engine prelubricator to fill all parts of the lubrication system. This prevents the possibility of a dry start.

CHAPTER 2
ROLLING-ELEMENT BEARINGS

Daniel R. Snyder
SKF USA, Inc., King of Prussia, Pa.

GENERAL

Reliable bearing performance is a key factor in reducing maintenance costs and in improving machine availability. When bearings fail, they can bring equipment to an unscheduled halt. Every hour of downtime due to premature bearing failure can result in costly lost product, especially in capital-intensive equipment. To keep machinery in peak operating condition, the bearings should be properly aligned and protected from extreme temperatures, moisture, and contaminants. Knowledge of the proper installation techniques and tools is required to ensure that bearings are not prematurely damaged. Following lubrication and maintenance schedules according to the original equipment builder's operating and maintenaning recommendations and monitoring bearing operating conditions are also important in maximizing bearing service life. For special cases, most bearing manufacturers and equipment builders maintain service departments to render technical assistance. As with any precision mechanical component, bearings should be handled with care and common sense to prevent damage from mechanical abuse and contamination.

WHY BEARINGS FAIL

Only a fraction of all bearings in service actually fail. The vast majority outlive the machinery or equipment in which they are installed. A bearing failure may occur for many reasons, that is, heavier loading than had been originally anticipated, ineffective seals or mountings which are too tight resulting in too little bearing internal clearance, and so on. Each of these factors produces its own particular type of damage and leaves its own imprint on the bearing. Consequently, by examining a damaged bearing, it is possible in a majority of cases to determine the cause and take corrective action to prevent a recurrence.

Of the small fraction of bearings that actually do fail, less than 10 percent die from old age (e.g., fatigue of the bearing surfaces), 24 percent fail because of poor lubrication, 20 percent because of improper installation, 27 percent from misapplication and the rest fail as a result of liquid or solid contaminants entering the bearing or because of handling damage. The pattern of failure can vary according to the specific industry where the bearing is used. For example, in the pulp and paper industry, poor lubrication and contamination are the major causes of failure, not fatigue.

Bearings are selected based on very specific operating conditions. Too often changes involving a different lubricant, higher machine speeds, higher loads, changes in lubrication systems, and so on are made without anticipating the possible negative effects on bearing performance. Therefore, when replacing a bearing, no other changes should be made which would adversely affect bearing operation.

BEARING DESIGNS AND NOMENCLATURE

Rolling bearings include radial and thrust bearings for radial and axial loads, respectively, and some bearing types which are designed for combined radial and axial loads. Generally speaking, ball bearings are recommended for light to moderate loads; roller bearings are recommended for heavy loads. Nine basic types of rolling bearings are shown in Fig. 2.1A to D. Some of these basic types are available in many variations; for instance, cylindrical roller bearings may be obtained with one, two, or four rows of rollers, as shown in Fig. 2.1B. Single-row deep-groove ball bearings are generally available in nine different external configurations, as shown in Fig. 2.2. Taper roller bearings can come in more than 20 different configurations, some of which are shown in Fig. 2.3. The other basic types do not come in large numbers of configurations, but it should be noted that all types of rolling bearings are available in many design variants and thus may vary greatly in internal design, depending on the manufacturer. It is not within the scope of this handbook to describe all the various designs of rolling bearings used in machinery but rather to alert maintenance personnel to their existence. Details are given in manufacturers' catalogs or by contacting the manufacturers directly.

BOUNDARY DIMENSIONS

In general, most ball, spherical roller, and cylindrical roller bearings made to metric boundary dimensions have standardized boundary plans, dimensions, and tolerances according to the International Standards Organization (ISO). Therefore, bearings from all subscribing manufacturers throughout the world are dimensionally interchangeable. Most taper roller bearings are made to inch dimensions and have standardized boundary dimensions and tolerances according to the Anti-Friction Bearing Manufacturers Association (AFBMA), a U.S. standards organization. Metric taper roller bearings utilizing ISO boundary plans are also made. Dimensionally interchangeable taper roller bearing components are thus available from several manufacturers. In most cases, identical basic part numbers are used.

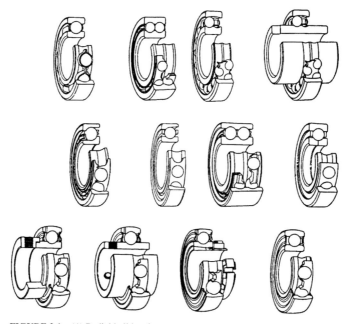

FIGURE 2.1 (*A*) Radial ball bearing types.

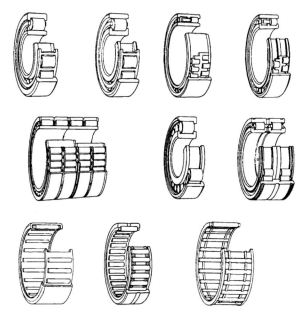

FIGURE 2.1 (*Continued*) (*B*) Radial roller bearing types.

FIGURE 2.1 (*Continued*) (*C*) Roller bearing types for radial and axial loads combined.

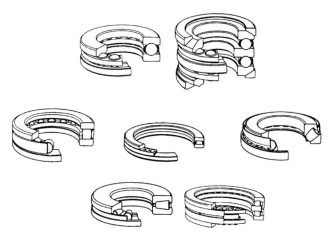

FIGURE 2.1 (*Continued*) (*D*) Thrust bearings.

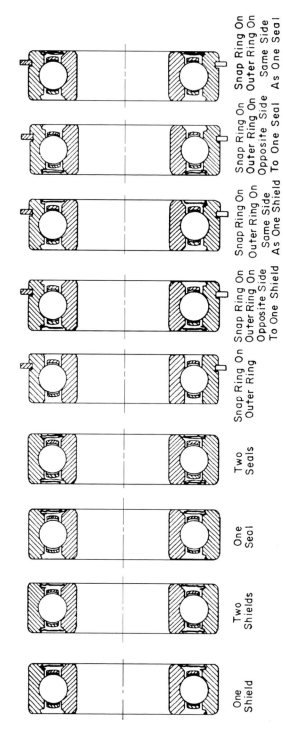

FIGURE 2.2 Single-row deep-groove ball-bearing shields, seals, and snap rings.

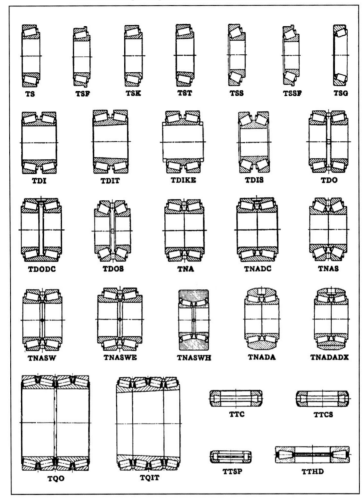

FIGURE 2.3 Tapered roller bearings, single-row, multiple-row, and thrust.

Inch-dimensioned cylindrical roller bearings are not manufactured according to any standard and will vary depending on the manufacturer.

BEARING SERIES

For any given bore size, all types of metric rolling bearings are manufactured in several series each for different severity of service. For instance, most ball bearings are made in three series: light, medium, and heavy duty. These are designated as the 2-, 3-, and 4-diameter series according to the boundary plan shown in Fig. 2.4. Spherical roller bearings are normally available in eight different series, as shown in Fig. 2.5. Taper roller bearings, both inch- and metric-dimensioned, have a larger number of series or duty classifications, but all series are not necessarily available for every bore size (see Fig. 2.6).

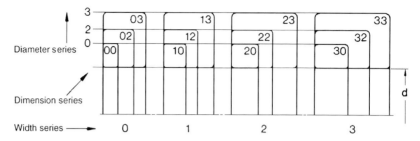

FIGURE 2.4 Metric rolling-bearing boundary dimension plan.

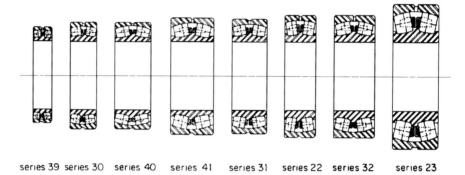

FIGURE 2.5 Spherical roller bearings of different diameter series with common bore size.

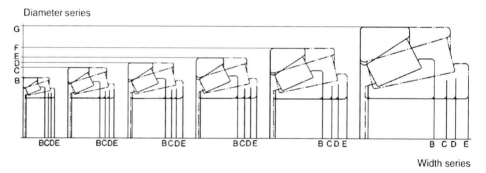

FIGURE 2.6 Metric tapered roller-bearing diameter series with common bore size.

LOAD RATINGS

All manufacturers of rolling bearings establish a dynamic and static load rating for each bearing produced. An ISO method for calculating this method exists, but not all manufacturers adhere to the method. The unfortunate situation therefore exists that two almost identical bearings produced by different manufacturers can have different published load ratings. Ratings are expressed as a load which will provide a basic rating life of a defined number of revolutions. *Basic rating life* is the number of revolutions (or the number of operating hours at a given constant speed) which the bearing is capable of enduring before the first sign of fatigue occurs in one of its rings or rolling elements. The basic rating life in millions of revolutions is the life 90 percent of a sufficiently large group of apparently identical bearings can be expected to obtain or exceed under identical operating conditions. In other words, this is a reliability or statistical rating, the only mechanical component so rated. The ISO definition of the basic rating life is the most common and is at 1 million revolutions. Some taper roller bearings are rated on the basis of 90 million revolutions, or 500 rev/min (rpm) for 3000 hours. Hence it can be easily seen that comparing manufacturers' ratings as published in catalogs can be misleading if appropriate adjustments are not made to published values.

There are several other "bearing lives," including service life and design or specification life. *Service life* is the actual life achieved by a specific bearing before it becomes unserviceable. Failure is not generally due to fatigue, but due to wear, corrosion, contamination, seal failure, etc.

The service life of a bearing depends to a large extent on operating conditions, but the procedures used to mount and maintain it are equally important. Despite all recommended precautions, a bearing can still experience premature failure. In this case it is vital that the bearing be examined carefully to determine a reason for failure so that preventive action can be taken. The service life can either be longer or shorter than the basic rating life.

Specification life is the required life specified by the equipment builder and is based on the hypothetical load and speed data supplied by the builder and to which the bearing was selected. Many times this required life is based on previous field or historical experiences.

SHAFT AND HOUSING FITS

It is a basic rule of design that one ring of a rolling-element bearing must be mounted on its mating shaft or in its housing with an interference fit, since it is virtually impossible to prevent rotation by clamping the ring axially. Generally, it is the rotating ring that is tight, but more correctly stated, it is the ring that rotates relative to the load. In some special cases this is not the rotating ring; for instance, in a vibrating unit where vibration is produced by eccentric weights, the load rotates with the rotating ring, and it is best to have the stationary ring have the tight fit.

Except for special cases as mentioned above, the stationary ring normally can be assembled with the mating shaft or housing with a slip or loose fit.

The magnitude of interference fit will vary with the severity of duty, type of bearing, and different shaft and housing materials. Ball bearings under normal load conditions will have approximately 0.00025 in interference per inch of shaft when the inner ring is the tight fit. Roller bearings will have fits of approximately 0.0005 in per inch of shaft. Fits will be increased for heavy-duty service and decreased for light duty. In general, when the outer ring is the tight fit, the interference is less than a corresponding shaft fit.

All bearing manufacturers show recommended fitting practices for their bearings in their general catalogs. With the exception of inch-taper roller bearings, the recommendations are normally expressed in ISO standards. ISO standards define the fit tolerance between the bearing outside diameter and the housing and utilize a designation system using a capital letter and a number such as H7, J6, P6, and so on. Fit tolerances between the shaft and bore of the bearing are designated by a lowercase letter and number such as g6, m5, r7, and so on. In the ISO system, the letter indicates the class or type of fit, and the number indicates the tolerance range. The diagram in Fig. 2.7 shows the relationship between the nominal diameters and the tolerance grades. The cross-hatched areas indicate the bearing bore diameter variation and the outside diameter variation, respectively. The blackened rectangles show the range of tolerances for shafts (lower half) and housings (upper half).

5.26 MAINTENANCE OF MECHANICAL EQUIPMENT

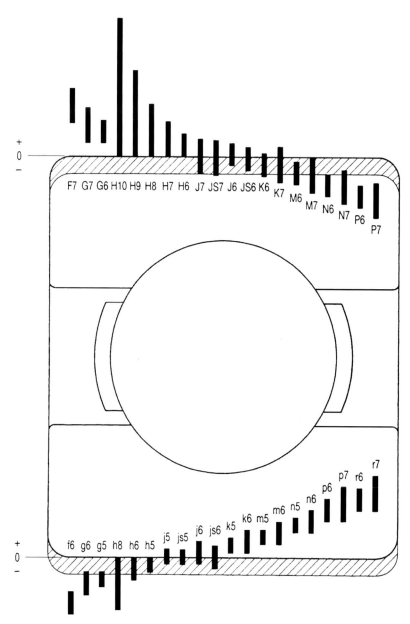

FIGURE 2.7 ISO fit tolerances. Uppercase letters refer to housings; lowercase letters refer to shafts.

BEARING MOUNTINGS

When rolling bearings are mounted on a shaft, some provision must be made for thermal expansion and/or contraction of the shaft. Also, the shaft must be located and held axially so that all machine parts remain in the proper relationship dimensionally. This is normally done by clamping one of the bearings on the shaft. When the inner ring has the tight fit, it is usually locked axially relative to the shaft by locating it between a shaft shoulder and some type of removable locking device. A specially designed nut as shown in Fig. 2.8 is normal for a through shaft. A clamp plate as shown in Fig. 2.9 is normally used when the bearing is mounted on the end of the shaft. For the locating or held bearing of the shaft, the outer ring is clamped axially, usually between housing shoulders or end-cap pilots. This type of mounting restricts axial movement of the shaft to the end movement resulting from the internal clearance of the bearing. If required, this can be 0 if the appropriate bearing type is used. The outer rings on all other bearings on the shaft should not be secured axially, and enough clearance should be provided between the side face of the stationary ring and the nearest housing shoulders to allow for anticipated expansion or contraction. Typical mountings are shown in Fig. 2.10.

General types of cylindrical roller bearings are capable of absorbing shaft expansion internally simply by allowing one ring to move relative to the other, as shown in Fig. 2.10A and C for the non-locating positions. The advantage to this type of mounting is that both inner and outer rings may have a tight fit. This may be desirable or even mandatory if significant vibration and/or unbalance exists in addition to the applied load.

Where bearing center distances are short and minimum thermal expansion is expected, an opposed mounting as shown in Fig. 2.11 may be used. In addition to its simplicity, this mounting has the advantage that thrust in one direction will be taken on one bearing and thrust in the other direction on the other bearing. Obviously, the clearance between the side face of the bearing and the housing shoulder must be controlled carefully or the shaft will shift excessively in an axial direction with the load direction changes.

Single-row tapered roller bearings and angular-contact ball bearings require special consideration. For example, if a radial load is applied to a single-row tapered roller bearing, an axial component of the load is generated by the angle of the roller set. This tends to separate the inner ring from the outer ring unless the "induced thrust" is resisted by another bearing properly mounted to resist

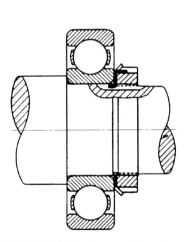

FIGURE 2.8 Bearing mounting with a special nut for a through shaft.

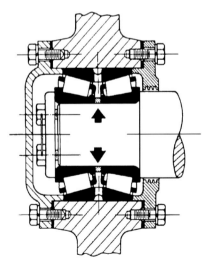

FIGURE 2.9 Bearing mounting with a clamp plate locking device using an inner ring adjustment spacer.

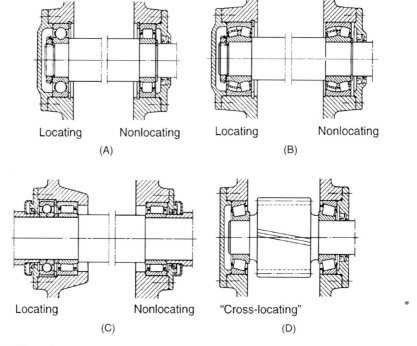

FIGURE 2.10 Typical bearing mountings.

the movement. The other bearing is normally another single-row tapered roller bearing. A mounting of this type may have the bearings arranged in one of two ways, as shown in Fig. 2.12. Figure 2.12*A* shows the included angles of the bearing or the tracks opening away from each other. This is known as an *indirect mounting*. Figure 2.12*B* shows the included angles of the bearings opening toward each other. This is known as *direct mounting*. It should be noted before progressing further with a description of this type of mounting that the point of reaction of the load on the centerline of the shaft, or the *effective center*, is not at the geometric center of a single-row bearing but at some point 0 as determined by the angle of the rolling element centerline relative to the centerline of the shaft.

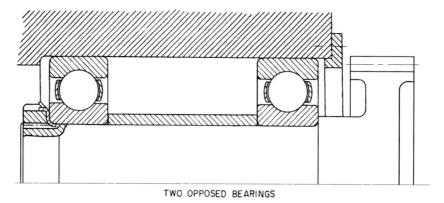

FIGURE 2.11 Example of assembly with opposed mounting of ball bearings.

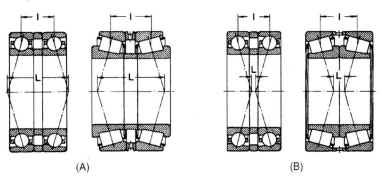

FIGURE 2.12 Direct (*A*) and indirect (*B*) arrangements.

Therefore, if the bearings of two different mountings are physically located the same distance apart with an indirect mounting, the effective centers of the bearings are farther apart than with the direct mounting, a more desirable arrangement when an overturning load exists.

With either a direct or an indirect single-row tapered roller bearing mounting, it is necessary to set the running clearance of the bearings when they are assembled. This is done by adjusting the cones in an indirect mounting and the cups for a direct mounting. Figures 2.13 and 2.14 show two ways of adjusting cones by nuts, and Fig. 2.15 shows a method of shimming for cone adjustment. Figures 2.16 to 2.18 show three ways of shimming cups in a direct mounting. Proper running clearance is controlled by measuring the end movement, or *end lateral*, of the shaft. The machine builder's recommendation for proper end lateral should be strictly followed. It will usually be indicated on the drawing of the particular part or given in the maintenance manual for the equipment.

Obviously, the only provision for thermal expansion in either of these mountings is the end lateral of the assembly. For this reason, they should be used only where bearing centers are relatively short or where little temperature variation is anticipated.

Two-row tapered roller bearings are mounted the same as other types of bearings. Proper end lateral is preadjusted in the factory.

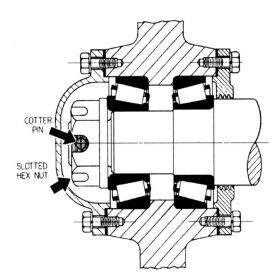

FIGURE 2.13 Slotted-nut adjusting device.

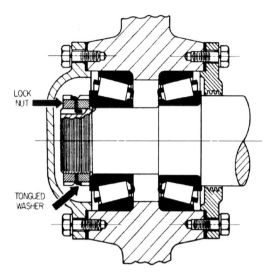

FIGURE 2.14 Double-nut and lock-washer adjusting device.

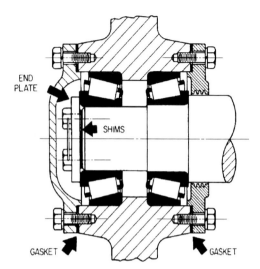

FIGURE 2.15 End-plate and shims adjusting device.

Angular-contact ball bearings are rarely used singly. However, if they are, they must be mounted in a similar manner to single-row tapered roller bearings. The much smaller running clearances used in ball bearings make a mounting of single angular-contact ball bearings very difficult to adjust properly. Angular-contact bearings could be substituted for the tapered roller bearings of Fig. 2.13, and the same comments and nomenclature would apply for single-bearing mountings.

However, angular-contact ball bearings are normally used in pairs. The side faces of these bearings are specially manufactured to permit mounting side by side, as shown in Fig. 2.19. *Tandem* (Fig. 2.19*A*), *back-to-back* (Fig. 2.19*B*), and *face-to-face* (Fig. 2.19*C*) are the common terms for these mountings. When two or more bearings are *stacked* in tandem for high-thrust loads, usually another bearing in the assembly is mounted face-to-face or back-to-back with the tandem *stack*.

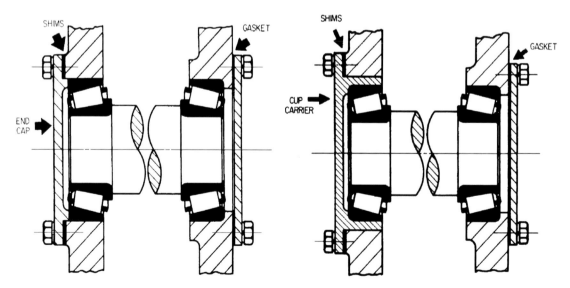

FIGURE 2.16 End-cap and shims adjusting method.

FIGURE 2.17 Cup-carrier and shims adjusting method.

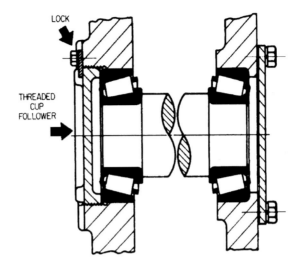

FIGURE 2.18 Threaded-cup-follower adjusting method.

When mounted in any of these arrangements, they may be considered as one multiple-row bearing. Because methods of face modifications may differ from one manufacturer to another, it is advisable not to mix manufacturers in a pair of tandem bearings. The bearing assembly number should indicate in some way that the bearings have been properly manufactured for mounting in pairs. Bearings for single mounting are available and should not be used as part of a pair.

A large percentage of spherical roller bearings are made with tapered bores. Some ball, tapered roller, and cylindrical roller bearings are also available with tapered bores. The bearings may be mounted directly on the shaft, as shown in Fig. 2.20. However, many tapered bore bearings are mounted on one of two types of sleeves, as shown in Figs. 2.21 and 2.22. European machinery builders are particularly partial to the use of sleeve mountings.

The adapter sleeve may be mounted as shown in Fig. 2.21 or with a shaft shoulder ring as shown in Fig. 2.23. With a removable type of sleeve, as shown in Fig. 2.22, the bearing must always be against a shaft shoulder.

The taper is 1 to 12 on diameter in all but the widest series of spherical roller bearings, in which a flatter 1 to 30 taper is used. Some four-row cylindrical roller rolling-mill bearings also will use a 1 to 30 taper in the bore of the inner ring.

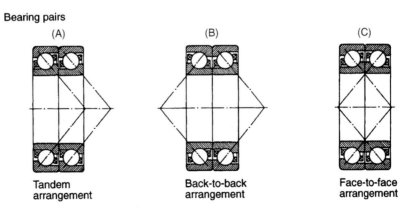

FIGURE 2.19 Angular contact mounting arrangements.

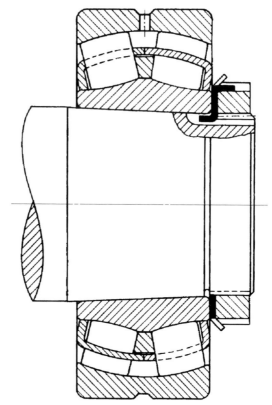

FIGURE 2.20 Direct shaft mounting of a spherical roller bearing.

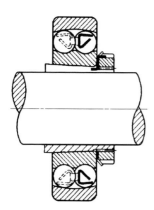

FIGURE 2.21 Mounting of a self-aligning ball bearing with an adapter sleeve.

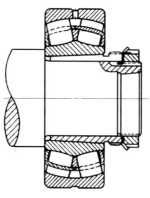

FIGURE 2.22 Mounting of a spherical roller bearing with removable type of sleeve.

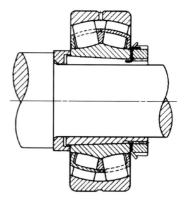

FIGURE 2.23 Mounting with shaft shoulder ring.

MOUNTING AND DISMOUNTING OF ROLLER BEARINGS

The most important thing to remember when mounting or dismounting a roller bearing, of any type, is to apply the mounting or dismounting force to the side face of the ring with the interference fit. Keep this force from passing from one ring to the other through the ball or roller set. This is particularly important during mounting, since damage can easily occur internally to the bearing. Cleanliness is, of course, extremely important. Not only the bearing but also the shaft housing must be free from chips, burrs, dirt, and moisture.

Bearings should be kept wrapped or covered until the last possible moment. Since most modern rust preventives used by bearing manufacturers are compatible with petroleum-based lubricants, the slushing compound is normally not removed. However, there are exceptions to this rule. If oil-mist lubrication is to be used and the slushing compound has hardened in storage or is blocking lubrication holes in the bearing rings, it is best to clean the bearing with kerosene or other appropriate petroleum-based solvent. Obviously, the other exception would be if the slushing compound has been contaminated with dirt or foreign matter before mounting. It is also permissible and sometimes desirable to wipe the rust preventive from the bore or outside diameter of the bearing, depending on which surface will have the tight fit. Before mounting or dismounting a bearing, always take the time to collect the proper tools and accessories. The use of inappropriate tools is a major cause of bearing damage. Also remember, never strike a bearing directly with a hammer, sledge, or mallet.

FIGURE 2.24 Mounting using flat plate.

Cold Mountings. All small bearings (4-in bore and smaller) may and sometimes must be mounted cold by simply forcing them on the shaft or into the housing. However, it is important that this force be applied as uniformly as possible around the side face of the bearing and to the ring to be press-fitted. Mounting fixtures should be used. These can be a simple piece of tubing of appropriate size and a flat plate as shown in Fig. 2.24. Do not try to use a drift and hammer, because the bearing will become cocked. Force may be applied to the simple fixture described above by striking the plate with a hammer or by an arbor press, as shown in Fig. 2.25. It is a good idea to apply a coat of light oil to the bearing seat on the shaft and bore of the bearing itself before forcing on the shaft. It should be noted that all sealed and shielded ball bearings should be mounted cold in this manner.

Temperature Mountings. The simplest way to mount any open straight-bore bearing, no matter what size, is to heat the entire bearing and simply push it on its seat and hold in place until it cools enough to start gripping the shaft. For tight outside-diameter fits, the housing may be heated if practical; if not, the bearing may be cooled by dry ice. However, if the ambient conditions are humid, cooling the bearing introduces the possibility of condensation on the bearing, which will induce corrosion later.

There are several acceptable ways of heating bearings. Some of these are as follows:

1. Hot plate: A bearing is simply laid on an ordinary hot plate until it reaches the approved temperature. The disadvantage of this method is that the temperature is difficult to control. A Tempilstik or pyrometer should be used to make certain the bearing is not overheated.

2. The temperature-controlled oven: This method needs little comment. The bearings should be left in the oven long enough to heat thoroughly. However, never leave bearings in a hot oven overnight or over a holiday or weekend.

3. Induction heaters: These are available and can be used to heat bearings for mounting. One of these is shown in Fig. 2.26. It must be remembered that this is a very quick method of heating and that some method of measuring the ring temperature must be used or the bearing may be damaged. A Tempilstik or pyrometer can serve this purpose. Bearings must be demagnetized after using this method.

FIGURE 2.25 Arbor press.

FIGURE 2.26 Induction heater.

4. A hot-oil bath: This method may also be used to heat the bearing and, in fact, is the most practical means to heat larger bearings. This method has some drawbacks, since the temperature of the oil is difficult to control and may overheat the bearing or even become a fire hazard. A mixture of soluble oil and water can eliminate both these disadvantages. Make the mixture 10 to 15 percent soluble oil. This solution will boil at approximately 210°F, which is hot enough for most bearing fits. The heating solution should be placed in a tank or container which has a grate or screen several inches off the bottom, as shown in Fig. 2.27. This will allow any contaminants to sink to the bottom and keeps the bearings off the bottom of the container.

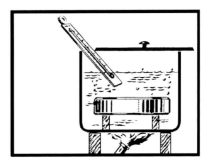

FIGURE 2.27 Hot oil for bearing.

As mentioned above, 210°F is not enough to mount most bearings. If you are using one of the other methods of heating or another solution, 250°F maximum will do the bearings no harm. However, this temperature should not be exceeded for small ball bearings (2-mm bore and smaller). Larger bearings can be heated somewhat higher than this without harm, but metallurgical damage will occur at approximately 300°F.

Mounting Tapered-Bore Bearings. Tapered-bore bearings can be mounted simply by tightening the locknut or clamping plate, which will locate it on the shaft until the bearing has been forced up the taper the proper distance. However, especially for large bearings, this technique will require a good amount of brute force. There are special techniques that may be used to reduce the amount of force required.

Before reviewing the mounting techniques for tapered-bore roller bearings, we will discuss the special case of self-aligning ball bearings. The bearing should be put on its tapered seat and the locknut hand tightened until all looseness is removed between adjacent parts. Then, using a spanner wrench, not a drift and a hammer, tighten the nut one-eighth turn further. Bend a lock-washer tab into the nut slot nearest to a washer tab in a tightened direction. At this point, the outer ring should rotate as well as swivel freely.

Tapered-bore spherical roller bearings can be mounted a bit more scientifically. Since the internal clearance in a roller bearing is significantly larger than in a ball bearing, this clearance can be measured with a thickness feeler gauge. As the bearing inner ring is pushed up the tapered seat, the inner ring expands, thereby reducing the internal clearance. Hence the amount of this reduction is a direct function of the interference fit between the bore of the bearing and the shaft. Therefore, if we measure the internal clearance of the bearing unmounted and control the amount the clearance is reduced during mounting, we control the shaft fit within very close limits. The internal clearance of a spherical roller bearing is measured as follows:

The bearing is unwrapped and placed on a table so that it can be easily handled. With one hand grasping the lower portion of the inner ring, oscillate the inner ring and roller set in a circumferential direction to seat the lower rollers properly in the sphere of the outer ring, on the roller paths of the inner ring, and against the separate guide ring between the two rows of rollers. Select a gauge blade of perhaps 0.003- or 0.004-in thickness or less for small bearings. The usable length of the blade should be somewhat longer than the length of a roller. It should not be equal to or greater than the width of the bearing. While pushing the top roller against its guiding surface, inset the blade between two rollers and the outer ring and slide the blade circumferentially toward the roller at the top of the bearing, as shown in Fig. 2.28. The blade should pass between the uppermost roller and the inside of the outer ring. Do this with successively thicker feeler blades until a blade will not pass. Move it so that it approaches the bite between a roller and the outer ring sphere; then, with one hand grasping the inner ring as described earlier, slowly roll the uppermost roller under the feeler blade. With the blade between the uppermost roller and the sphere, attempt to swivel the blade and withdraw it axially. The swiveling motion helps to center the roller in its proper operating position, and

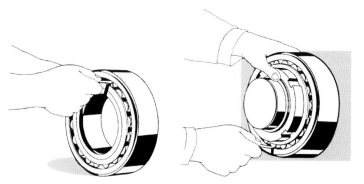

FIGURE 2.28 Determining internal bearing clearance for a spherical roller bearing.

withdrawing it with the characteristic wiping feel of a line-to-line contact will show that thickness to be the looseness over that roller. If the blade becomes looser during the swiveling and withdrawing process, attempt the same procedure with a blade 0.001 in thicker and continue until a blade cannot be swiveled or withdrawn. The internal clearance over that roller will be the blade that can be swiveled and withdrawn after a thicker one has jammed.

Repeat this procedure in two or three other locations by resting the bearing on a different spot on its outside diameter and measuring over different rollers in one row. Either repeat the above procedure for the other row of rollers or measure each row alternatively in the procedure described above. Make a note of this unmounted internal clearance.

After the unmounted radial clearance is measured, the bearing is placed on its tapered seat. If the shaft provides for a locknut, it is then assembled, but the lock washer is left off the shaft at this point. The locknut should then be tightened against the bearing, pushing it up the taper until the internal clearance is reduced by the specified amount, as shown in Table 2.1. An impact-type spanner wrench as shown in Fig. 2.29 is ideal for tightening the nut.

FIGURE 2.29 Use of impact-type spanner wrench.

TABLE 2.1 Recommendation for Driving a Spherical Roller Bearing on a Tapered Seat

Mounting of spherical roller bearings with tapered bore

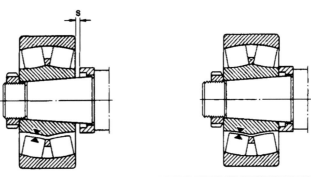

Bearing bore diameter d		Reduction in radial internal clearance		Axial drive-up s[1] Taper 1:12 on diameter		Axial drive-up s[1] Taper 1:30 on diameter		Minimum permissible residual clearance[2] after mounting bearings with initial clearance		
2 220	2 250	0.000	0.000	0.0	0.0	00.0	00.0	0.000	0.000	
over	incl.	min	max	min	max	min	max	Normal	C3	C4
mm		mm		mm				mm		
24	30	0.015	0.020	0.3	0.35	–	–	0.015	0.020	0.035
30	40	0.020	0.025	0.35	0.4	–	–	0.015	0.025	0.040
40	50	0.025	0.030	0.4	0.45	–	–	0.020	0.030	0.050
50	65	0.030	0.040	0.45	0.6	–	–	0.025	0.035	0.055
65	80	0.040	0.050	0.6	0.75	–	–	0.025	0.040	0.070
80	100	0.045	0.060	0.7	0.9	1.7	2.2	0.035	0.050	0.080
100	120	0.050	0.070	0.75	1.1	1.9	2.7	0.050	0.065	0.100
120	140	0.065	0.090	1.1	1.4	2.7	3.5	0.055	0.080	0.110
140	160	0.075	0.100	1.2	1.6	3.0	4.0	0.055	0.090	0.130
160	180	0.080	0.110	1.3	1.7	3.2	4.2	0.060	0.100	0.150
180	200	0.090	0.130	1.4	2.0	3.5	5.0	0.070	0.100	0.160
200	225	0.100	0.140	1.6	2.2	4.0	5.5	0.080	0.120	0.180
225	250	0.110	0.150	1.7	2.4	4.2	6.0	0.090	0.130	0.200
250	280	0.120	0.170	1.9	2.7	4.7	6.7	0.100	0.140	0.220
280	315	0.130	0.190	2.0	3.0	5.0	7.5	0.110	0.150	0.240
315	355	0.150	0.210	2.4	3.3	6.0	8.2	0.120	0.170	0.260
355	400	0.170	0.230	2.6	3.6	6.5	9.0	0.130	0.190	0.290
400	450	0.200	0.260	3.1	4.0	7.7	10	0.130	0.200	0.310
450	500	0.210	0.280	3.3	4.4	8.2	11	0.160	0.230	0.350
500	560	0.240	0.320	3.7	5.0	9.2	12.5	0.170	0.250	0.360
560	630	0.260	0.350	4.0	5.4	10	13.5	0.200	0.290	0.410
630	710	0.300	0.400	4.6	6.2	11.5	15.5	0.210	0.310	0.450
710	800	0.340	0.450	5.3	7.0	13.3	17.5	0.230	0.350	0.510
800	900	0.370	0.500	5.7	7.8	14.3	19.5	0.270	0.390	0.570
900	1 000	0.410	0.550	6.3	8.5	15.8	21	0.300	0.430	0.640
1 000	1 120	0.450	0.600	6.8	9.0	17	23	0.320	0.480	0.700
1 120	1 250	0.490	0.650	7.4	9.8	18.5	25	0.340	0.540	0.770

[1] Valid for solid steel shafts only. Larger axial displacements are necessary for hollow shafts depending on wall thickness.

[2] The residual clearance must be checked in cases where the initial radial internal clearance is in the lower half of the tolerance range and where large temperature differentials between the bearing range can arise in operation. The residual clearance must not be less than the minimum values quoted above.

The amount of force required to drive a tapered-bore bearing can be greatly reduced if the shaft is drilled and grooved as shown in Fig. 2.30. If these fittings are available, attach a hydraulic pump to the connection at the end of the shaft. Drive the bearing on the taper just enough so there is some interference; then build up hydraulic pressure under the bore of the bearing. A pressure of 3000 to 6000 psi will be needed, but with this pressure between the bore of the bearing and the shaft it is possible to float the bearing up the taper with much less torque applied to the locknut or clamp plate than in a dry mounting.

Another convenient way to mount a tapered-bore bearing is to use a hydraulic nut or mounting tool, as shown in Fig. 2.31. This technique also can be adapted to sleeve mountings that are large enough to be drilled and grooved.

Cylindrical and tapered roller bearings with tapered bores are not as common as their spherical counterparts, and the manufacturer will have specific mounting instructions for each application.

Dismounting of Bearings. A wide variety of tools are available commercially which are designed to remove a rolling bearing from its seat without damage. Typical bearing pullers are shown in Fig. 2.32. In removal, we should again keep in mind the basic rule to apply force to the ring with the

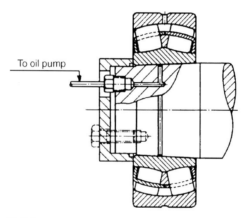

FIGURE 2.30 Drilling and grooving of the shaft to reduce bearing driving force.

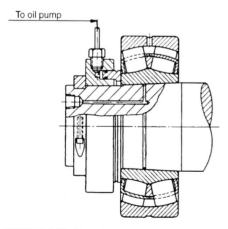

FIGURE 2.31 Use of a hydraulic nut to mount a spherical roller bearing.

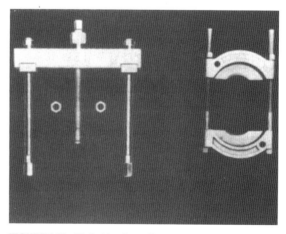

FIGURE 2.32 Typical bearing pullers.

tight fit. Pullers normally can be applied to bearings so that this rule is observed. However, sometimes supplementary plates or fixtures may be required.

For smaller bearings, an arbor press is equally effective at removing as well as mounting bearings. Also, techniques such as the one shown in Fig. 2.33 may be used where size permits.

Hydraulic Removal. Where shafts have been designed to apply hydraulic pressure to the fit between shaft and bearing, removal is quite simple. First, the locking device, whatever it is, should be backed off a distance greater than the axial movement of the mounting; $1/4$ in will be sufficient in virtually every case. Then connect a hydraulic pump to the fitting provided at the end of the shaft, as shown in Fig. 2.34, and start building up pressure. When pressure becomes great enough to break the fit, usually about 3000 to 6000 psi, the bearing will literally jump off the taper with a sharp bang. The retaining device, still being loosely connected, will prevent the bearing from coming off the end of the shaft. Never completely remove the retaining device.

Hydraulic pressure may be used with straight-bore bearings, but a puller must be used in conjunction with the hydraulic pump, since there will be no axial component of the hydraulic pressure to blow the bearing off its seat. See Fig. 2.34.

Larger sleeve mountings also may be designed to utilize hydraulic pressure for dismounting. If this feature is available, follow the same procedure as outlined above. However, if the sleeve mounting does not have this feature, other techniques such as shown in Fig. 2.35, must be used. For withdrawal sleeves, a special nut must be used, as shown in Fig. 2.36. For large sleeves, a hydraulic nut is desirable for dismounting.

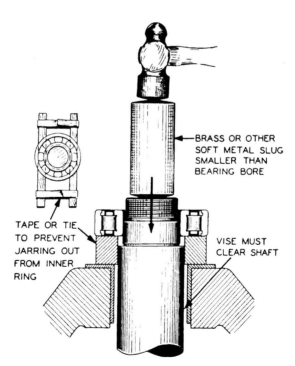

FIGURE 2.33 Method to remove small bearings by driving shaft through supported bearing.

5.40 MAINTENANCE OF MECHANICAL EQUIPMENT

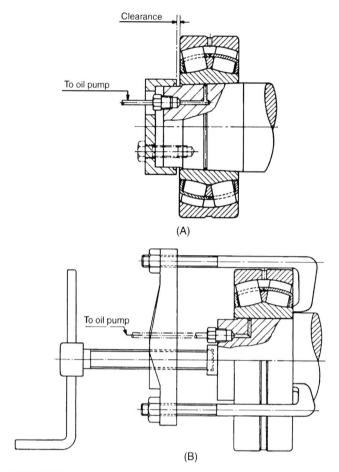

FIGURE 2.34 Hydraulic removal. (*A*) By connection to pump. (*B*) In conjunction with a puller.

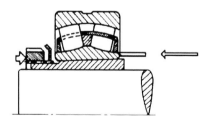

FIGURE 2.35 Bearing removal by driving bearing down tapered sleeve.

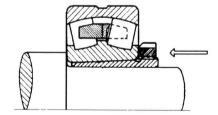

FIGURE 2.36 Bearing removal by pulling sleeve from under bearing.

LUBRICATION

The primary purpose of lubrication in a rolling bearing is to separate the contacting surfaces, both rolling and sliding. This purpose is rarely achieved, and boundary lubrication or partial metal-to-metal contact frequently occurs. By far the most common lubricants are petroleum products in the form of grease or liquid oil. Synthetics are, however, finding more use in high-temperature applications.

Generally, the machine builder decides whether a bearing will be a grease- or oil-lubricated component and normally will recommend the basic specifications of the required lubricant. However, because the machine designer cannot foresee all the variable conditions under which the equipment will operate, some judgment is required on the part of maintenance personnel. Some knowledge of lubricants is therefore useful.

Oil Lubrication. For oil lubrication, the Annular Bearing Engineers Committee (ABEC) has issued the following recommendations:

The friction torque in a ball bearing lubricated with oil consists essentially of two components. One of these is a function of the bearing design and the load imposed on the bearing, and the other is a function of the viscosity and quantity of the oil and the speed of the bearing.

It has been found that the friction torque in a bearing is lowest with a very small quantity of oil, just sufficient to form a thin film over the contacting surfaces, and that the friction will increase with greater quantity and with higher viscosity of the oil. With more oil than just enough to make a film, the friction torque will also increase with the speed.

The energy loss in a bearing is proportional to the product of torque and speed, and this energy loss will be dissipated as heat and cause a rise in the temperature of the bearing and its housing. This temperature rise will be checked by radiation, convection, and conduction of the heat generated to an extent depending upon the construction of the housing and the influence of the surrounding atmosphere. The rise in temperature, due to operation of the bearing, will result in a decrease in viscosity of the oil, and therefore, a decrease in friction torque compared with the friction of starting, but soon a balanced condition will be reached.

With so many factors influencing the friction torque, energy loss, and temperature rise in a bearing lubricated with oil, it is evidently not possible to give definite recommendations for selection of oil for all bearing applications, but two general considerations are dominant:

1. The desire to reduce friction to a minimum, which requires a small quantity of oil of low viscosity.
2. The desire to maintain lubrication safely without much regard for friction losses, which results in using larger quantities of oil and usually of somewhat greater viscosity in order to reduce losses from evaporation or leakage.

This second condition is most frequently met when bearings have to operate in a wide range of temperatures. An oil that has the least changes with respect to changes in temperature, that is, an oil with high viscosity index, should be selected.

In the great majority of applications, pure mineral oils are most satisfactory, but they should, of course, be free from contamination that may cause wear in the bearing, and they should show high resistance to oxidation, gumming, and deterioration by evaporation of light distillates, and they must not cause corrosion of any parts of the bearing during standing or operation.

It is self-evident that for very low starting temperatures an oil must be selected that has sufficiently low pour-point so that the bearing will not be locked by oil frozen solid.

In special applications, various compounded oils may be preferred, and in such cases, the recommendation of the lubricant manufacturer should be obtained.

Grease Lubrication. Where grease lubrication is used, we need to consider a few of the basic physical and chemical characteristics of the lubricant. Greases are a mixture of lubricating oil and usually a soap base. The base merely acts to keep the oil in suspension. When moving parts of a bearing come in contact with the grease, a small quantity of oil will adhere to the bearing surfaces. Oil is therefore removed from the grease near the rotating parts. Bleeding of the oil from the grease obviously cannot go on indefinitely, so new grease must come in contact with the moving part or a lubrication failure will result.

Many maintenance departments want to use one grease to lubricate all bearings in the plant. Some lubricant suppliers even advocate this technique. However, it is a risky procedure at best, since there is no true universal ball and roller bearing grease. A ball bearing is best lubricated with a fairly stiff grease which will channel. On the National Lubricating Grease Institute (NLGI) code, greases of the number 2 consistency, or 265 to 295 worked penetration, are normally recommended. For roller bearings, a grease stiff enough to channel is not desirable, since the full width of the roller track would soon be starved for lubricant if the grease is not soft enough to slump back into the bearing when it is pushed aside. This generally means greases in the number 0 or 1 consistency class with worked-penetration numbers of 355 to 380 for grade 0 and 310 to 340 for a number 1 grease.

Whatever the consistency of the grease, it is still the properties of the oil compounded in the grease that determine if the bearing will be satisfactorily lubricated. All statements and guidelines outlined above in the discussion of oil lubrication also apply to grease-lubricated bearings.

Another characteristic of a grease that must be considered is its *drop point*. This is the temperature at which the grease passes from a semisolid to a liquid. Typical dropping points are as follows:

Calcium	$+14 \pm 140°F$
Sodium	$-22 \pm 176°F$
Lithium	$-22 \pm 230°F$
Bentone	$-22 \pm 266°F$
Silicone	$-22 \pm 266°F$
Calcium complex	$-4 \pm 266°F$
Aluminum complex	$-22 \pm 230°F$

The drop point is the characteristic referred to when a grease is advertised as being good up to 400°F. Whether it will lubricate a bearing or not is still a function of the viscosity of the lubricating oil, not of the drop point of the base. In fact, common industrial bearings made of standard through-hardened or case-hardened materials have temperature limitations of 200 to 300°F depending on the material and how it was heat-treated. The bearing manufacturer should be consulted for specific information.

Never mix greases that are incompatible. If two such greases are mixed, the resulting mixture usually has a softer consistency which will eventually cause failure through leakage. If you don't know what type of grease a bearing was lubricated with originally, do not regrease without first removing the old grease both from the bearing and the surrounding environment.

Generally, bearings are not lubricated until after mounting. The most important reason for this is cleanness. The later grease is applied, the greater are the chances of avoiding contamination. The bearing should be lubricated prior to mounting only when pregreasing is the only way to obtain an even distribution of grease.

The right quantity of grease is as important as the right type of grease. Follow these general rules for quantity:

A bearing should be filled completely with grease, but free space in the housing should only be partially filled (between 30 and 50 percent). However, in nonvibrating applications, many lithium soap greases, also called *total-fill greases,* can fill up to 90 percent of the free space in the housing without any risk of a rise in temperature. Thus impurities can be prevented from entering the bearing, and relubrication intervals can be extended. Bearings that have to operate at high speeds, for example, machine tool spindles, where it is desirable to keep the temperature low, should be lubricated with small quantities of grease.

In vibrating applications, such as wheel hubs, vehicle axle boxes, and vibrators, grease fill should be no more than 60 percent of the housing.

When relubrication intervals are long, then the housings should be easy to open. If more frequent lubrication is required, the housing should be fitted with some kind of grease-filling device, preferably a lubrication duct with a nipple.

In the optimal situation, grease can be injected with a grease gun. Some bearings are provided with grooves and ducts for relubrication; others have to be relubricated from the side.

Only the grease in the bearing should be replaced. The amount of grease, therefore, depends on the bearing size. If relubrication instructions are available from the original manufacturer, follow them. If not, or if you suspect the lubrication amount is inadequate, use the following formula to determine the correct amount.

$$Gq = 0.114 \times D \times B \quad \text{(in ounces)}$$

where Gq = grease quantity in ounces
D = bearing outside diameter in inches
B = total bearing width in inches

Selection of Lubricant. Research in elastohydrodynamics (EHD) has contributed greatly to the knowledge of lubricants, rolling bearings, and how they work. Results of this work have been published in various forms that may be used as a guide in the selection of the correct lubricant. Figure 2.37 is an example of these data in graph form which plots required viscosity of the lubricant in centistrokes at operating temperature as a function of bearing size and speed. The abscissa of the curve

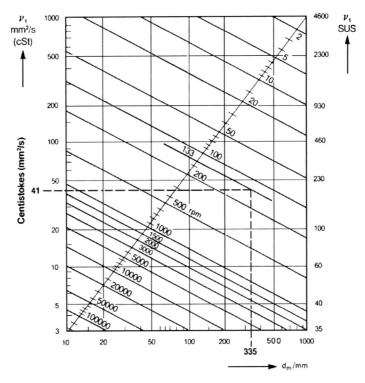

d_m = (bearing bore + bearing O.D.) ÷ 2
ν_1 = required lubricant viscosity for adequate lubrication at the operating temperature

FIGURE 2.37 Minimum required lubricant viscosity for adequate lubricant film.

is the bearing size, expressed as mean diameter in millimeters; the diagonal lines are the speed in rpm; and the ordinate is the required viscosity at the operating temperature of the bearing. It is obvious from an examination of this chart that the larger the bearing and the slower the speed, the higher is the required viscosity. This characteristic of the EHD theory sometimes produces a paradox where the lubricant required may not be realistically used for other reasons. Lubrication experts should be consulted when this situation exists.

When grease lubrication is used, the values obtained from the chart apply to the base oil of the grease. The example shown on the graph indicates that a rolling bearing with a mean diameter of 335 mm, running at 133 rpm, will require a lubricant having 41 centistokes at its operating temperature.

Reasonable estimates of bearing operating temperature can be made ahead of time based on analytical calculations or experience. In most cases, a combination of the two gives the most realistic estimate. The machine builder and/or the bearing supplier can help in this area and will normally make lubrication recommendations for new equipment. Some adjustment may have to be made to the original selections after operating experience is obtained.

In current literature and specification sheets, lubricants are normally grouped into viscosity grades according to standards established by the ISO. Lubricants are rated from ISO 2 to ISO 1500, the numbers indicating the mean value of a specified range of viscosity at a temperature of 104°F. Figure 2.38 plots these lubricants on a temperature-viscosity diagram.

In the selection of a lubricant, it should be kept in mind that the oil temperature in the bearing may vary 5 to 10°F from the temperature of the housing. There are no exact rules for this estimate, but it is wise to add these degrees if housing temperature is used as a criteria for determining operating level of the bearing.

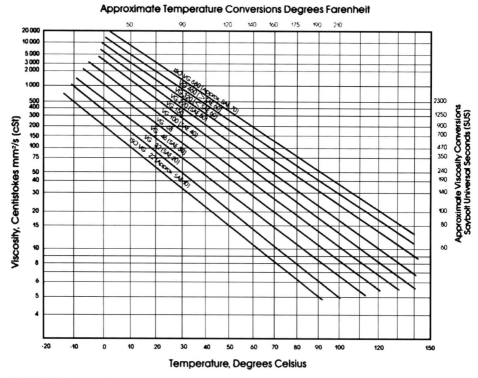

FIGURE 2.38 Approximate temperature conversions. Viscosity classification numbers are according to International Standard ISO 3448-1975 for oils having a viscosity index of 95. Approximate equivalent SAE viscosity grades are shown in parentheses.

CHAPTER 3
FLEXIBLE COUPLINGS FOR POWER TRANSMISSION

Terry Hall
Reliability Engineer, Life Cycle Engineering, Inc., Charleston, S.C.

GENERAL

A *flexible coupling* is a mechanical device used to connect two axially oriented shafts. Its purpose is to transmit torque or rotary motion without slip and at the same time compensate for angular, parallel, and axial misalignment. There are many supplementary functions, which include providing for or restricting axial movement of the connected shafts; minimizing or eliminating the conduction of heat, electricity, or sound; torsional dampening; and torsional tuning of a system. Basically, all flexible couplings can be categorized as either mechanical flexing or material flexing. While most available flexible couplings fall strictly into one or the other of these basic categories, a few combine both principles.

The mechanical-flexing group provides flexibility by allowing the components to slide or move relative to each other. Clearances are provided to permit movements to within specific limits. Lubrication is usually required to reduce wear within the coupling and to minimize the cross-loading in the connected shafts. The most prominent in this category are the chain, gear, grid, and Oldham flexible couplings.

The material-flexing group provides flexibility by having certain parts designed to flex. These flexing elements can be of various materials, such as metal, rubber, plastic, or composite. Couplings of this type generally must be operated within the fatigue limits of the material of the flexing element. Most metals have a predictable fatigue limit and permit the establishment of definite boundaries of operation. Elastomers (rubber, plastic, etc.) usually do not have well-defined fatigue limits, and service life is determined primarily by the operational conditions. The material-flexing group includes laminated-disk, diaphragm, spring, and elastomer flexible couplings.

DESCRIPTION OF TYPES

A wide variety of concepts are available within each of the two basic groups. A detailed account of each would be impractical. Only the most well-established forms will be described and discussed in generalities.

Chain couplings (Fig. 3.1) are compact units capable of transmitting proportionately high torques at low speeds. They consist of two hubs having sprocket teeth which are connected by a strand of single-roller, double-roller, or silent chain. Shaft misalignment is accommodated by clearances

FIGURE 3.1 Roller-chain coupling. (*Rexnord Corporation, Coupling Division.*)

between the chain and the sprocket teeth and/or clearances within the chain itself. A number of special features such as hardened sprocket teeth, special tooth forms, and barrel-shaped rollers are available which are designed to increase flexibility and reduce wear. Nonmetallic chains are used on light-duty drives where the use of a lubricant is prohibited.

Coupling covers are recommended for all drives where the rotating speed is capable of slinging the lubricant or where the atmosphere is wet, corrosive, or abrasive. They protect the coupling and greatly extend its life by retaining the lubricant and preventing dirt or other foreign materials from coming in contact with or between the sliding parts. Most covers rotate with the coupling. Grease holes permit lubrication without disturbing the gaskets or seals. Covers should be half-filled with light grease (number 1 is normally recommended). A heavier grease should be used if the coupling is operated in a high-temperature environment. Routine flushing and relubricating is required. It is generally recommended that a roller-chain coupling be relubricated every 6 months or sooner depending on the conditions of operation. Stationary oil-bath covers are available for large, slow-speed chain couplings.

Where a cover is not required, the chain and sprockets should be coated thoroughly with a good-quality bearing grease. The use of a stiff brush is suggested because it will give better penetration of the grease into critical areas of the chain. This is generally required on a weekly basis.

To obtain maximum service with roller-chain couplings, misalignment of the connected shafts must be restricted to the manufacturer's recommendations. Excessive amounts can cause rapid wear of the chain and sprockets as well as early failure of the cover seals and a resultant loss of lubricant. A straightedge and caliper can usually be used to check angular and parallel misalignment as well as the shaft gap. A properly aligned coupling will allow the chain to be wrapped around the sprockets and the connecting pin inserted without any significant force.

Gear couplings (Figs. 3.2 to 3.5), the most prominent type in the mechanical-flexing group, are available in a wide range of sizes and styles. They are capable of transmitting proportionately high torques at either low or high speeds. In their most common form, they are compact and consist of two identical hubs with external gear teeth and a sleeve or sleeves with matching internal gear teeth. Shaft misalignment is accommodated by clearances between the matching gear teeth. Special tooth forms are available which are designed to reduce wear and increase flexibility without increasing clearances. These include crowned tips, curved flanks, and curved roots. The sleeve may be a single tubular piece, or it may consist of two flanged halves bolted rigidly together.

Floating-shaft gear couplings usually consist of a standard coupling with a two-piece sleeve. The sleeve halves are bolted to rigid flanges to form two single-flexing couplings. These are connected by an intermediate shaft which permits the transmission of power between widely separated machines. On high-speed or short-span drives, spools are used to separate the two half couplings.

Spindle couplings are a modification of the floating-shaft gear coupling. They are used extensively on mill-roll drives and other related equipment which has unavoidable offsetting of the driving and driven shafts. In addition to accepting large angles of misalignment, they must operate with a relatively uniform angular velocity. These couplings are subject to severe operating conditions and are therefore a relatively high maintenance item. Numerous special features are available which are designed to reduce maintenance and downtime.

Some light-duty gear couplings have nonmetallic sleeves such as nylon or urethane, which eliminates the need for lubrication. Generally, all other gear couplings require proper lubrication in order to realize their potential life and are designed to have a static supply of lubricant inside. As the coupling is rotated, this lubricant is thrown outward by centrifugal force to engulf the load-carrying

FLEXIBLE COUPLINGS FOR POWER TRANSMISSION **5.47**

FIGURE 3.2 Gear-tooth coupling, standard double-engagement type. (*The Falk Corporation.*)

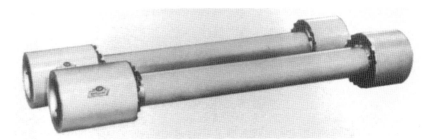

FIGURE 3.3 Gear-tooth coupling, spacer type. (*Zurn Industries, Inc.*)

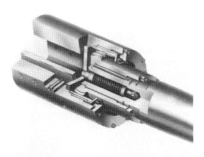

FIGURE 3.4 Gear-tooth coupling, spindle type. (*Zurn Industries, Inc.*)

FIGURE 3.5 Gear-tooth coupling, high-speed spacer type. (*Zurn Industries, Inc.*)

teeth. The relative sliding motion between the hubs and the sleeve permits the establishment of a film between the mating gear teeth.

There are several methods of lubricating gear couplings. These are grease pack, oil fill, oil collect, and continuous oil flow. The vast majority of drives operate at 3600 rpm or less and use grease as the lubricant. Both grease and oil are used at speeds of 3600 to 6000 rpm. Oil is normally used as the lubricant in couplings operation over 6000 rpm. Most high-speed couplings use a continuous oil flow to carry away the heat generated within the coupling. Conditions of operation should be reviewed in selecting the method of lubrication. In addition to the available maintenance program, consideration should be given to torsional loads and their characteristics, high or low temperatures, rotating speeds, and environmental conditions.

The grease-packed and oil-filled units have end rings or seals which are used to retain the lubricant and restrict the entry of dust, grit, moisture, or other contaminants. Sleeves are provided with lubrication holes which permit flushing and relubrication without disturbing the sleeve gasket or seals.

The oil-collector couplings have extended lips on the sleeve through which oil can be induced. The proper oil level can be maintained without shutdown.

The continuous-oil-flow couplings also have extended lips on the sleeve through which oil is induced, while discharge holes in the sleeve permit the excess oil to escape. This arrangement requires that an oil-tight enclosure encompasses the coupling and that an oil-circulating and oil-filtering system be provided.

Lubrication within the coupling is susceptible to heat, centrifugal force due to high speed, water, foreign oil, solids, or other contaminants. In addition to ambient temperatures, heat can be transferred from the connected equipment or generated within the coupling itself. High centrifugal forces can cause a soap separation and oil bleeding. Soap separation results in improper lubrication and sludging, which can cause the coupling to lock up. Oil bleeding can lead to leakage. When low temperatures are encountered, special low-temperature grease may be required.

An adequate supply of the proper type of lubricant is essential to good coupling performance and long life. In view of the many problems that could arise with the use of an improper lubricant, it is necessary that the coupling manufacturer's instructions be adhered to closely.

The retention of the coupling's lubricant is essential. Leakage can result in high costs of lubrication, reduced coupling life, and the creation of hazardous conditions around the connected equipment.

It is equally important to prevent any abrasive materials or contaminants from entering the coupling. Abrasives will cause excessive wear and prevent proper performance. Contaminants such as water, solvents, and foreign oils can cause a deterioration of the coupling's lubricant.

Care should be used during installation of a new coupling or reassembly of an existing unit. Keyways and keys should be coated with a sealing compound that is resistant to the coupling's lubricant. Seals should be checked for pliability and condition. They must be seated properly in the sleeve, and the lip must be in intimate contact with the hub. Sleeve flange gaskets must be whole and in good condition. Sleeve flange faces and rabbet must be clean and free of nicks. This joint should be tightened in accordance with the manufacturer's recommendations. Lubricant plugs must be clean and tight in the sleeve.

Periodic inspection of the coupling is recommended. Whenever possible, inspection and relubrication of the coupling should be done on the same schedule as that established for connected machines. If, during inspection, leakage is evident, plugs, gaskets, seals, and keyways should be checked and the situation corrected. The condition of the lubricant also should be noted. If it is abnormal, the coupling should be flushed and refilled as required.

Metallic-grid couplings (Fig. 3.6) are compact units capable of transmitting proportionately high torques at moderate speeds. They consist of two flanged hubs with special grooves or slots cut axially on the outside. The flanges are joined by interlacing a serpentine metallic grid. Flexibility is achieved by sliding movement of the grid in the slots. Flexure of the grid in the curved slots provides some torsional resilience. The grid may be of one piece or may be provided in two or more sections. Grids with tapered cross sections are available from some manufacturers and are designed to ease installation and removal.

FIGURE 3.6 Metallic-grid coupling. (*The Falk Corporation.*)

Covers are provided to retain the coupling's lubricant and to prevent dust, grit, and other foreign materials from coming in contact with or between the sliding parts. The cover may be split either horizontally or vertically. Grease holes permit lubrication without disturbing the gaskets or seals.

Proper lubrication is essential. The manufacturer's recommendations as to the type of lubricant should be followed. Seals must be in good condition and properly seated. The gasket must be whole and the cover joint tight. Plugs must be tight. Regular intervals of inspection for leakage and condition of lubricant are recommended. If the lubricant is abnormal, the cover should be opened and all parts thoroughly flushed before the new lubricant is added.

The grid members on this type of coupling are replaceable. If they are significantly worn, they should be replaced.

Misalignment of the connected shafts should be kept within the manufacturer's recommendation. Excessive amounts can cause rapid wear of the grid and hub slots as well as early failure of the cover seals. A spacer bar of caliper and straightedge usually can be used to check angular and parallel misalignment as well as shaft gap.

Oldham couplings (Fig 3.7) are also known as *block-and-jaw couplings.* They are compact units normally relegated to light or medium duty and moderate speeds. They consist of two jaw flanges and a floating-block center member. The jaw flanges are positioned at right angles to each other and engage opposite parallel surfaces of the block.

FIGURE 3.7 Oldham coupling. (*Zurn Industries, Inc.*)

FIGURE 3.8 Laminated disc-ring coupling, standard double-engagement type. (*Rexnord Corporation, Coupling Division.*)

Small-sized light-duty couplings usually have flanges of die-cast metal and blocks of nonmetallic material such as laminated phenolic plastic. These units do not require any lubrication. The larger-sized couplings usually have hubs of cast iron and blocks of oil-impregnated sintered metal. On some styles, the bearing surfaces of the block are provided with replaceable nonmetallic strips. The block has a reservoir for lubricant which is fed to the bearing strips through orifices.

Shaft misalignment is provided for by slippage of the block between the jaw flanges. Alignment of the connected shafts should be within the coupling manufacturer's recommendation and usually can be accomplished with the use of a straightedge.

Laminated disk-ring couplings (Figs. 3.8 to 3.10) are the most prominent type in the material-flexing group and are available in a wide range of sizes and styles. They are capable of transmitting proportionately high torques at either low or high speeds. In their most common double-flexing form, two flanged hubs are connected to a floating-center member through laminated disk rings. Each of the disk rings is alternately bolted or riveted to a hub flange and center member. The disk rings in tandem allow the coupling to accommodate angular and parallel misalignment as well as a limited amount of end float. In their single-flexing form, they consist of two flanged hubs and one laminated disk ring. The disk ring is alternatively bolted to the flanged hubs. These single-flexing units are capable of supporting a radial load and provide concentricity of connected three-bearing assemblies. They will accept only angular misalignment and a reduced amount of end float. Shaft misalignment is provided for by flexure of the disk rings. Since these units are normally of all-metal construction, they are free of backlash and are relatively rigid in a torsional plane.

Under normal conditions, the metal parts are not subject to deterioration. Most manufacturers have available couplings that are resistant to corrosion. These units usually have the components plated or are made of a corrosion-resisting material such as stainless steel. Laminated disk-ring couplings have no sliding parts that can wear, so no maintenance is required other than occasionally checking the condition of the laminated disk rings to make sure that all bolts are tight and that the equipment is still in proper alignment. Periodic visual inspection of the condition of the coupling is recommended. This can be done without disassembly or disturbing the connected equipment. When the equipment cannot be shut down conveniently, a stroboscopic light can be used. During inspection, special consideration should be given to the outer sheets of the disk ring. If any deterioration or broken sheets are found, the entire disk ring should be replaced. Significant deterioration and breaking of the sheets are normally indications of excessive flexure due to misalignments beyond the coupling's capacity. Realignment of the equipment must be done immediately. If a coupling has been operating with loose bolts, they should be removed and inspected. If there are significant scour marks or indentations on the body, the bolts should be replaced. Most couplings of this style are completely repaired.

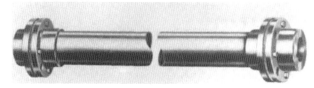

FIGURE 3.9 Laminated disc-ring coupling, spacer type. (*Rexnord Corporation, Coupling Division.*)

FIGURE 3.10 Laminated disc-ring coupling, high-speed spacer type. (*Rexnord Corporation, Coupling Division.*)

FIGURE 3.11 Multielement diaphragm coupling. (*Zurn Industries, Inc.*)

Misalignment of the connected shafts should be restricted to within the manufacturer's recommendations. When accurate measurements are required, dial indicators should be used. Alignment with a caliper and straightedge is usually satisfactory for slow-speed drives.

Diaphragm couplings (Figs. 3.11 and 3.12) are also in the material-flexing group and are used primarily for the high-speed, high-horsepower applications. They are relatively light for the horsepower transmitted. The diaphragm coupling is available in many sizes and styles, including a reduced-moment design. This coupling uses two flexing elements separated by a floating-center member. The diaphragm is normally attached at the outside and inside diameter by bolts welding to connect the hubs to the floating-center member. The torque goes through the diaphragm assembly for the outside to inside diameter, or vice versa. The flexibility of the diaphragm design accommodates angular and parallel shaft misalignment as well as a limited amount of end float. Each flexing

FIGURE 3.12 Single-element diaphragm coupling. (*Lucas Aerospace Corporation.*)

FIGURE 3.13 Elastomeric coupling, shear-type flexing element. (*Rexnord Corporation, Coupling Division.*)

element is made up of one or more diaphragm elements depending on the design. The coupling is radially rigid and maintains its original balance because there are no wearing parts.

Under normal conditions, the metal parts are not subject to deterioration. Most manufacturers have available couplings that are resistant to corrosion. These units usually have the components plated or are made of a corrosion-resisting material.

Misalignment of the connected shafts should be restricted to within the manufacturer's recommendation for long coupling life. If the connected equipment experiences high vibration, the coupling should be inspected for possible damage.

Elastomeric couplings (Figs. 3.13 and 3.14) are available in an almost infinite number of versions. They are generally categorized into two types. There are those in which the elastomer is placed

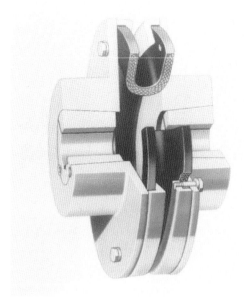

FIGURE 3.14 Elastomeric coupling, shear-type flexing element. (*The Falk Corporation.*)

in shear and those in which it is placed in compression. Their ability to compensate for shaft misalignment is obtained by flexure and/or displacement of the elastomeric element. These couplings are generally relegated to light- or medium-duty service at moderate speeds.

In their basic concept, they consist of two hubs separated and connected by elastomeric element. On shear-type couplings, the elastomer may be bonded, clamped, or fitted to matching sections of the hubs. The compression-type couplings usually utilize projecting pins, bolts, or lugs to connect the components. The elastomeric flexing elements may be polyurethane, rubber neoprene, or impregnated cloths and fibers.

Elastomeric couplings are normally maintenance-free, but it is suggested that occasional checks be made of the elastomer's condition and equipment alignment. If the elastomer shows signs of deterioration or wear, it should be replaced and the equipment aligned to within the coupling manufacturer's recommendations. This usually can be accomplished with the use of a caliper and straightedge.

CAUSES OF COUPLING FAILURE

In the event of a coupling failure, a thorough investigation should be made to determine the cause. Failure may be due to either faults within the coupling itself or external conditions.

Most failures due to internal faults are the result of improper or poor machining. They usually have to do with concentricities, squareness of the mating face, and tolerances on the various piloting or registering diameters. Defective materials and materials with inadequate strength and/or hardness also have contributed to many premature failures. Another major cause of failure due to internal faults is improper product design. On mechanical-flexing couplings, the major problem is to provide adequate lubrication between the sliding contact faces, since lack of a lubricating film between these high-pressure surfaces will result in rapid wear. On material-flexing couplings, improper design of the flexing-element section and method of attachment to the hubs are the main causes of premature fatigue.

Most common causes of failure due to external conditions have to do with improper selection, improper assembly, and excessive misalignment. Consideration of these is given in the following pages.

Coupling Selection

Maintenance personnel are frequently faced with the problem of replacing a worn-out or broken coupling. After the cause of failure has been determined, careful consideration should be given to the type, size, and style of coupling that will be used as a replacement. Whenever possible, it should satisfy all the needs of the drive.

Proper selection as to the type of coupling is the first step of good maintenance. A well-chosen coupling will operate with low cross-loading of the connected shafts, have low power absorption, induce no harmful vibrations or resonances into the system, and have negligible maintenance costs. The primary considerations in selecting the correct type of flexible couplings, as well as its size and style, are

1. Type of driving and driven equipment
2. Torsional characteristics
3. Minimum and maximum torque
4. Normal and maximum rotating speeds
5. Shaft sizes
6. Span or distance between shaft ends
7. Changes in span due to thermal growth, racking of the bases, or axial movement of the connected shafts during operation
8. Equipment position (horizontal, inclined, or vertical)
9. Ambient conditions (dry, wet, corrossion, dust, or grit)
10. Bearing locations
11. Cost (initial coupling price, installation, maintenance, and replacement).

The coupling should be selected conservatively for the torque involved. Consideration must be given to all peak and shock loads encountered in normal service. If the coupling is to operate at high speeds, it should be dynamically balanced. Special coupling modifications dictated by the connected equipment should be made.

If any doubt exists as to the proper type or size of coupling to use, it is recommended that the manufacturer be consulted. Most manufacturers have representatives in all large cities, and they are usually qualified to make recommendations and assist in the coupling procurement.

Installation

On most applications, it is necessary to disassemble the coupling before installation. The arrangement of the components should be noted, since they must be replaced in the same order.

The driving and driven shafts, as well as the bore in the hubs, should be inspected to make sure they are free of burrs, dirt, and grit. Check the keys for proper fit in both shafts and hubs. Clearance over the key is essential. Normal practice uses 0.005 in of clearance.

Next, the hubs should be mounted on their respective shafts. If an interference fit has been specified, it will be necessary to heat the hubs in water, oil, or a furnace and quickly position them on the shafts. Spot heating with a sharp, concentrated flame must be avoided because it will cause distortion and affect the capabilities of the material.

Finally, the equipment should be brought into its approximate operating position and the coupling reassembled. The equipment is now ready for alignment.

Alignment

Remember that misalignment is the major cause of coupling problems. Therefore, machines connected by a flexible coupling should be aligned with the greatest possible accuracy. The better the initial alignment, the more capacity the coupling has to take care of subsequent operational misalignment. Changes from the initial condition can occur through pipe strain, bearing wear, settling of foundations, base distortion due to torque, thermal changes, and vibrations in the connected equipment. To get the potential life of the coupling, the alignment of the connected machines should be checked at regular intervals and corrected as necessary.

The closer you get the hot running alignment, the better the connected equipment will run, giving longer bearing and seal life.

Normally, there are three conditions of misalignment that a flexible coupling must accommodate. These are angular misalignment, parallel misalignment, and axial misalignment (end float). These conditions combine to form the results shown in Fig. 3.15. No specific alignment procedure can be used on all drives. They must be worked out individually to suit the conditions at hand. There are fundamentals that do apply to all, however.

The first and most often overlooked step in equipment alignment is to bring the shafts into their proper axial position (Fig. 3.16). The shaft gap must be in accordance with the coupling manufacturer's recommendations. The mating surfaces or flexing elements of the coupling must be in their normal or relaxed position when the shafts of the connected equipment are in their operating position.

Some equipment has shafts with an inherent freedom to float. A typical example is the frequently encountered sleeve-bearing motor. Such motors may have as much as $1/2$-in free shaft movement. Usually, the shafts are marked to indicate where they should be located relative to the bearing housing. When the shafts are unmarked, it should be assumed that the motor rotor is on magnetic center when it is located halfway between the float limits.

If the exact shaft gap for the coupling is not known, it usually can be obtained from the manufacturer's catalogs, certified prints, or data sheets. Most manufacturers furnish dimensional sheets and installation instructions with all new couplings. In the case of old couplings, this information is usually available upon request. If the gap dimension is not available, it usually can be obtained by measuring the coupling or its components and calculating the normal position. Slight adjustments to the initial position may be required during the later processes of eliminating angular and parallel misalignment. After the equipment has been properly spaced, it is ready for alignment.

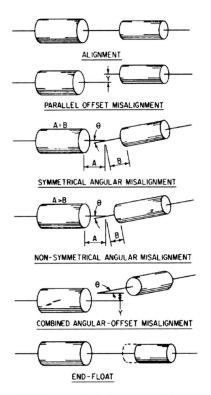

FIGURE 3.15 Shaft-alignment conditions.

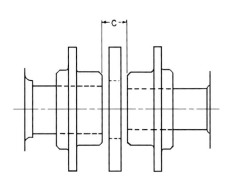

FIGURE 3.16 Axial spacing.

For many years, the accepted method of aligning equipment was with the use of a straightedge and caliper or scale, and this method is still frequently used with some types of couplings on low-speed drives. With the continuing development of small, high-speed equipment, however, the need for accurate measurements of misalignment has increased. On most drives, the use of a straightedge and scale has given way to dial indicators, optics, and lasers. When properly applied, the devices will give precise measurements as to the amount of misalignment as well as give its phase or direction.

No specific method can be established for mounting this equipment. This must be worked out to suit conditions at hand.

The preferred method of aligning equipment is to have the coupling mounted and completely assembled. The dial indicator then rotates with the equipment, but its stem remains in contact with a specific surface with no sliding of the stem over a large surface. In the following comments on alignment, it is assumed that the coupling is completely assembled and that the driving and driven pieces of equipment can be rotated together.

By graphically plotting the shaft misalignment, the solution becomes apparent. A picture is worth a thousand words.

Correct alignment is mandatory for successful operation of rotating equipment. A flexible coupling is no excuse for misalignment.

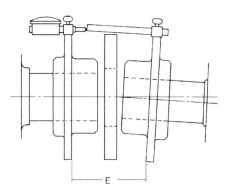 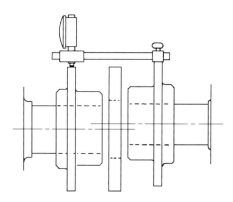

FIGURE 3.17 Using a dial indicator to check angular misalignment.

FIGURE 3.18 Using a dial indicator to check parallel misalignment.

SHAFT CENTER LINE RELATIONSHIP

How do you easily determine the relationship of one shaft center line relative to the other? It is hard sometimes to visualize this.

This is done by drawing a picture on a piece of graph paper to show visually where the equipment is and how far it needs to be shifted to get it into perfect alignment.

There is some confusion between shaft alignment and coupling alignment. We will restrict the comments to couplings using two flexing elements. The coupling flex element sees angular misalignment only. It is possible, as you can see in Fig. 3.17, to have misalignment between two shafts, yet one end of the coupling can still be in perfect alignment with its center member.

This could be the situation if all the problems are at one end of the coupling. In Fig. 3.18, the two ends of the coupling share the misalignment equally.

SHAFT MISALIGNMENT

How do you go about correcting for shaft misalignment?

These instructions will be broken down into four sections:

1. Items that must be considered *before starting* any of the alignment procedures.
2. *Reverse-indicator alignment* graphic analysis.
3. *Face and rim alignment* graphic analysis.
4. *Across-the-flex-element* graphic analysis.

Before Starting

Before we get into the procedure itself, there are several items that must be considered before starting. See Figs. 3.19 and 3.20.

Soft Foot. Soft foot occurs when the equipment, say, a motor, is not sitting flat on its base or it rocks. Now this rocking can be eliminated by tightening all the hold-down bolts. What this does,

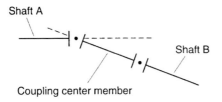
FIGURE 3.19 Coupling center member.

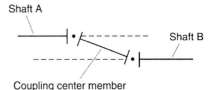
FIGURE 3.20 Coupling center member.

however, is to put the motor bearing under strain. This, in turn, can cause vibration. It also may give erroneous alignment readings. The soft foot must be corrected first. This is easily done by shimming under the motor foot until it no longer rocks.

Indicator Sag. Calibrate the dial indicator setup sag. In other words, determine the difference in the dial indicator reading when it is on top of a shaft as opposed to when it is on the bottom. This is a gravitational effect. It is not necessary to eliminate sag but rather to know the amount of sag. The indicator setup should be as rigid as practical, and then it should be calibrated. It is easily calibrated by mounting the setup on a piece of pipe, allowing the dial indicator to ride on the pipe itself. See Fig. 3.21.

Set the indicator at 0 on top. Now roll the pipe over until the indicator is at the bottom of the pipe. The gravitational effect on the setup can be determined by reading the indicator deference from top to bottom. This delta reading can be subtracted algebraically from the alignment readings obtained at the bottom position.

In the example shown in Fig. 3.21, setup sag checked out to be 20.005 in. This reading will always be negative. The indicator setup sag does not have to be considered for the horizontal or side-to-side reading.

Alignment Readings. It is suggested that the dial indicator be zeroed at the top for convenience. The coupling hub should be marked at 0°, 90°, 180°, and 270° with a reference mark on the equipment so that the units can be turned through 90° increments. This eliminates any runout that might exist between the point at which the indicator rides and the theoretical centerline of the shaft. Now rotate the coupling in 90° increments, recording all readings. It is important to keep the side-to-side readings straight. A suggestion is to refer to the sides of the unit as "near" and "far" ("near" being the side where you are standing). After making the four-position check, return to the top to make sure that the indicator returns to 0. If it does not, disregard the readings and repeat the procedure. It is a

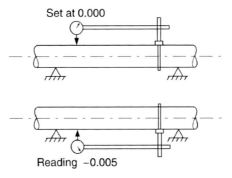
FIGURE 3.21 Indicator sag.

5.58 MAINTENANCE OF MECHANICAL EQUIPMENT

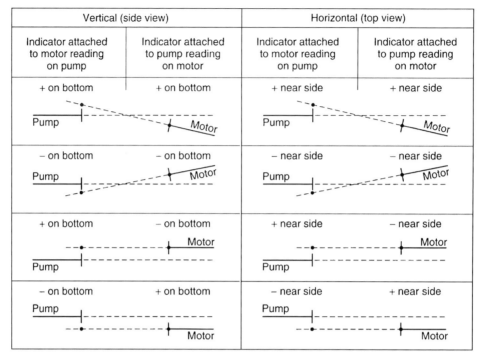

FIGURE 3.22 Pump-to-motor alignment guide.

good practice to take several sets of readings to make sure they are consistent. It is a lot easier to take another set of readings than it is to move the units a second time. See Fig. 3.22.

Thermal Growth. Now consider any thermal growth values for the equipment. For example, if the pump is pumping hot water, it will grow vertically from the ambient to the hot running condition. The whole objective is to have the equipment in good alignment when it is running under normal operating conditions. These predicted thermal movements can be obtained from the equipment manufacturer and should be taken into account before making the alignment changes.

Shaft Relationship. How do you easily determine the relationship of one shaft center line relative to the other? It is hard sometimes to visualize this. To help, when using the reverse indicator method, refer to Fig. 3.22.

Reverse Indicator

To explain the reverse indicator alignment procedure, a motor-to-pump example will be used. First, correct the vertical misalignment by shimming, and then correct the horizontal misalignment by sliding the equipment from side to side. With proficiency, these two steps can be done together.

Before starting the alignment work, determine which piece of equipment is easiest to move. This is not to eliminate the option of moving both units if a problem occurs. The pump, in this example, will be fixed. Therefore, the motor will be moved into alignment with the pump.

Now, on a sheet of graph paper, lay out the equipment being aligned. See Fig. 3.23.

The horizontal scale on the graph used here is one small division equals 1 in The distances needed are

FLEXIBLE COUPLINGS FOR POWER TRANSMISSION **5.59**

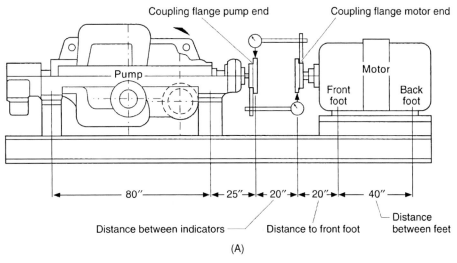

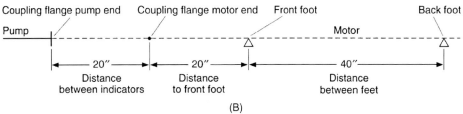

FIGURE 3.23 Typical alignment solution-motor stationary.

1. The distance from the first indicator riding on the pump hub to where the second indicator is riding on the motor hub. In the example, this is 20 in.
2. The axial distance between the motor hub where the second indicator is riding and the center of the motor front foot. In the example, this is 20 in.
3. The distance from the center of the front motor feet to the center of the back motor feet. In this example, this is 40 in.

Vertical Alignment Solution. The alignment can be done either with the coupling totally installed, with the coupling hubs mounted, or with the coupling totally removed from the shafts. Find a spot to mount the dial indicator bracket. The shaft behind the hub or the hub itself is good. A chain-clamp alignment bracket usually fits in well. With the indicator bracket attached to the motor hub and the dial indicator reading off the pump hub, rotate both units in 90° increments and take readings. These are shown in Fig. 3.24.

The bottom reading is then corrected for indicator setup sag. The indicator setup sag in this example was determined to be -0.005 in. Thus -0.005 was subtracted algebraically from the -0.025 indicator reading. The corrected reading is $-0.025 - (-0.005) = -0.020$. The readings taken are total indicator readings (T.I.R.), which are two times the actual shaft-to-shaft relationship. This means that the readings taken must be divided by 2. To solve for the vertical (up and down) part of the problem, take the corrected bottom reading of -0.020 and divide it by 2, that is, $-0.020/2 = 20.010$. With a minus reading on the bottom of the pump flange, the motor center of the extension must be lower than the center line of the pump. For further clarification, see Fig. 3.22.

5.60 MAINTENANCE OF MECHANICAL EQUIPMENT

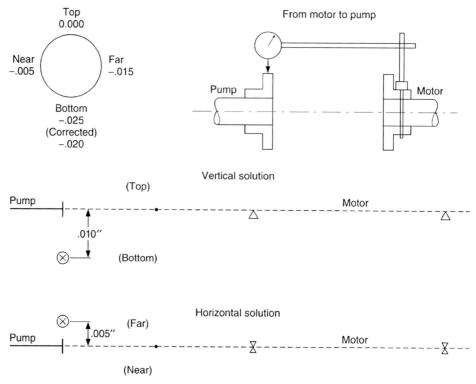

FIGURE 3.24 Typical alignment solution-pump stationary.

Use a convenient scale in the vertical of 0.001 in per small division. Plot this point as shown in Fig. 3.24. Do nothing with the horizontal (near/far) readings at this time.

Reverse the bracket setup by attaching it to the pump hub, as shown in Fig. 3.25, with the dial indicator reading off the motor hub.

Take a set of alignment readings. In this case, the indicator shows a reading on the bottom of +0.005 in. Correct for indicator sag by algebraically subtracting the −0.005 indicator setup from the +0.005, which gives a +0.010 corrected reading. Now divide this number by 2, since it is a total indicator reading. With a positive indicator reading on the bottom, the motor is low relative to the theoretical center line of the pump. To help you see the shaft-to-shaft relationship, refer again to Fig. 3.22. Use the same scale of 0.001 in per small division. Now plot the +0.005 on the graph, as shown in Fig. 3.25.

The solution to the problem now becomes easy. Draw a line through the two points plotted, and extend it beyond the plane of the motor feet, as shown in Fig. 3.26.

Then, by counting the graph graduations in line with each of the motor feet, it can be seen that the motor must be lowered by 0.010 in at the back foot and left alone at the front foot. By making this correction, the motor center line extension falls on top of the pump center line. This would solve the vertical alignment problem if there was no thermal correction to be made for either the pump or the motor.

Note: If the vertical scale chosen is too big, it may realize some minor shimming errors.

Let's say that the pump does grow 0.005 in at both ends because it is handling a hot liquid. In order to compensate for this, add an additional 0.005 in to each motor foot in order to have the motor in line with the pump when the pump is running under normal operating conditions. The total solution now

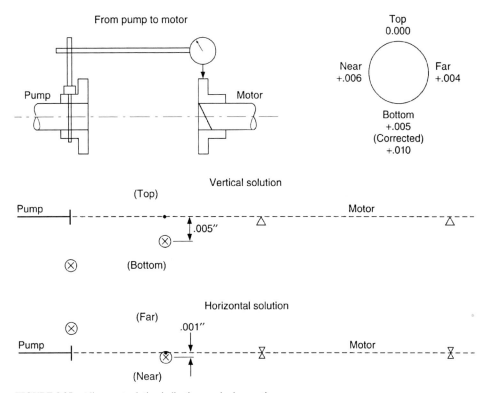

FIGURE 3.25 Alignment solution indicating required corrections.

is to add 0.005 in of shims to the front motor feet and subtract 0.005 in of shims from the back motor feet. The vertical alignment is now solved.

Horizontal Alignment Solution. In a similar manner, the side-to-side movements can be plotted. Refer to Fig. 3.24, and imagine looking straight down on the pump-motor combination. The alignment readings obtained were −0.005 in on the "near" side (this is the side you are standing on) and −0.015 in on the "far" side. By adding a value of 0.015 to each side, the "far" side becomes 0.000 and the "near" side becomes +0.010 T.I.R., or +0.010/2 × +0.005 in actual. With a plus on the "near" side, the motor center line extension falls on the "far" side of the pump center line. See Fig. 3.22.

Now plot this point as shown in Fig. 3.24. Moving on to Fig. 3.25, the side-to-side alignment readings obtained were +0.006 in on the "near" side and +0.004 in on the "far" side. By subtracting 0.004 from each side, the "far" side becomes 0.000 and the "near" side becomes +0.002 T.I.R., or 0.002/2 = 0.001 in actual. With a plus on the "near" side, the motor center line is on the "near" side of the pump center line extension. See Fig. 3.22.

Now plot this point as shown in Fig. 3.25.

Now draw a line through the two plotted points, extending it past the front and back motor feet, as shown in Fig. 3.26.

As the graphic plot shows, the motor must be moved 0.007 in toward the "far" side at the front foot and 0.019 in toward the "far" side at the back foot. This solves the horizontal alignment provided there is no side-to-side movement cause by operating conditions.

Where this graphic procedure really has its greatest value is when there are more than two units in one train to be aligned. As shown in Fig. 3.27, there are three pieces of equipment to be aligned.

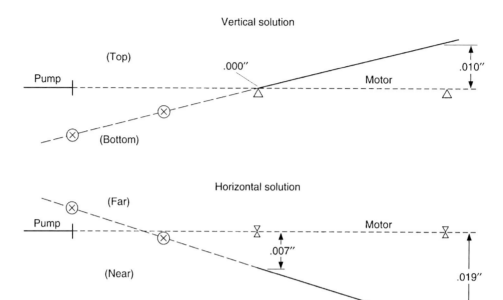

FIGURE 3.26 Alignemnt solution for multiple machine-train components.

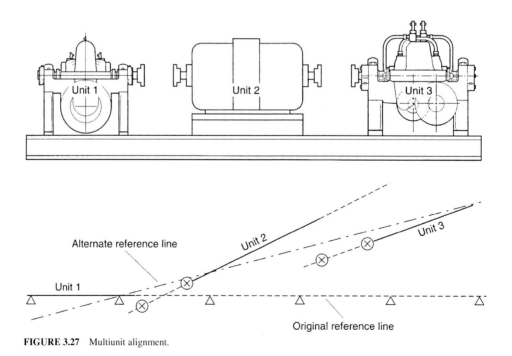

FIGURE 3.27 Multiunit alignment.

If you start by aligning the second unit to the first unit and then the third unit to the second unit, the third unit may have to be moved a considerable distance. In the actual installation, there may not be room to move that third piece of equipment the required amount. For example, there may not be enough shims under the third unit to lower it, or there may not be clearance in the bolt holes to move it sideways. The approach that should be taken is shown in Fig. 3.27. Plot all three units on one piece of graph paper after taking reverse indicator alignment readings across both couplings. As can be seen, the third unit is a long way off the original reference line. By drawing an alternate reference line close to the actual position of the three units, it can be seen that minimal movement is required to get all the units in alignment with respect to the alternate reference line.

Alignment problems can be made easier. There are a lot of tools on the market to make the mathematics simple; however, a graphic picture shows the whole problem at a glance and makes the solution apparent. The alignment chart in Fig. 3.28 may be helpful.

Face/Rim

To explain the face/rim alignment procedure, a motor-to-pump example will be used. First, correct the vertical misalignment by shimming, and then correct the horizontal misalignment by sliding the equipment from side to side. With proficiency, these two steps can be done together.

Before starting the alignment work, determine which piece of equipment is easiest to move. This is not to eliminate the option of moving both units if a problem occurs. The pump, in this example, will be fixed. The motor will be moved into alignment with the pump.

Vertical Alignment. Now, on a sheet of graph paper, lay out the equipment being aligned. See Fig. 3.29.

The horizontal scale on the graph used here is one small division equals 1 in. The distances needed are

1. Distance from where the indicator rides radially on the pump hub to the center of the motor front feet. In the example, this is 15 in.
2. Diameter of the pump hub flange at the location the face indicator rides. In the example, this is 10 in.
3. Distance from the center of the motor front feet to the center of the motor back feet. In the example, this is 25 in.

The alignment can be done either with the coupling totally installed or with just the coupling hubs mounted. Find a spot to mount the dial indicator bracket. The shaft behind the hub or the hub itself is good. A chain-clamp alignment bracket usually fits in well.

Angular (Face) Solution. With the indicator bracket attached to the motor hub, reading off the pump hub face, rotate the shafts in 90° increments and take readings (see Fig. 3.30).

Note: Make sure that neither of the equipment shafts being aligned moves axially, since this will distort the face readings.

Reading on the face at a 10-in diameter and getting +0.005 at the bottom, we know that the motor shaft is off 0.005 in for every 10 in of length. The plus reading at the bottom means that the indicator was compressed. This can only happen when the motor shaft is low compared with the pump center line extension. This can be shown graphically; use the pump flange center as a pivot point. Extend the 10 in (diameter of hub flange) along the graph (pump center line). Using a vertical scale of one small division on the graph equals 0.001 in, plot the 0.005 in as shown in the example. See Fig. 3.30. Draw a line from the center of the pump flange face through the 0.005-in point and extend it past the motor feet.

Note: If the vertical scale chosen is too big, it may realize some minor shimming errors.

5.64 MAINTENANCE OF MECHANICAL EQUIPMENT

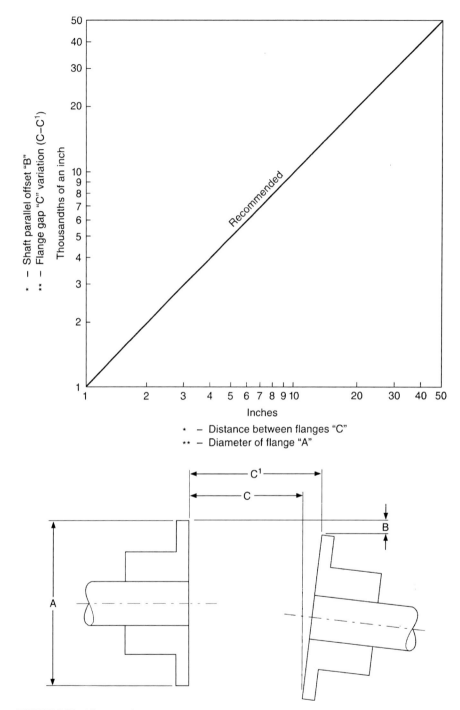

FIGURE 3.28 Alignment chart.

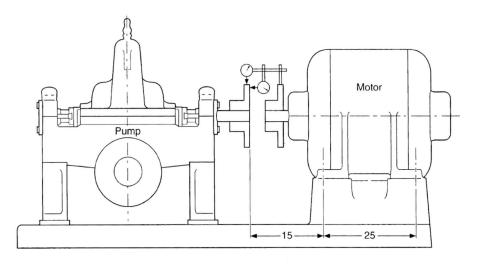

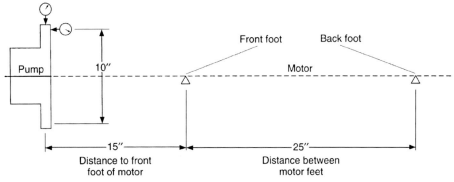

FIGURE 3.29 Example of rim and face vertical alignment.

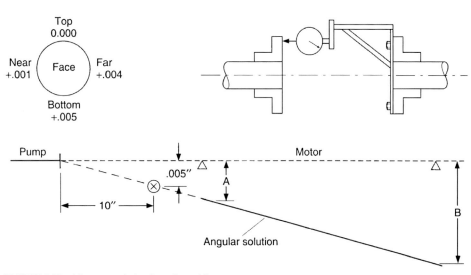

FIGURE 3.30 Alignment solution from rim and face measurements.

The (face) angular vertical misalignment could now be corrected. From the graph, the solution is to add 0.0075-in shims to the front foot (A) and add 0.020-in shims to the back foot (B). Since it is easier to make one shim change instead of two, solve the parallel offset (rim) before shimming. Then add the two results together and make one move.

Parallel-Offset (Rim) Solution. Now, with the indicator bracket attached to the motor hub, reading off the pump hub outside diameter, rotate the unit in 90° increments and take readings. See Fig. 3.31.

Bottom reading is then corrected for indicator sag. Indicator sag in the example was determined to be -0.005. The -0.005 was subtracted from the $+0.010$ indicator reading to give an actual $+0.015$ reading, or $+0.010 - (-0.005) = 10.015$.

Since this is T.I.R., it is two times the actual shaft-to-shaft relation. Thus $+0.015/2$, or $+0.0075$, is used to show where the motor center line extension is relative to the pump shaft center line at the pump hub. With a positive reading at the bottom, it indicates that the motor shaft is high compared with the pump. Using a scale of one small division on the graph equal 0.001 in, plot this point as shown in the example. The parallel-offset (rim) misalignment alone could be corrected by removing 0.0075 in of shims from under both front and back feet.

Total Vertical Alignment Solution. By drawing a line parallel to the angular (face) solution and through the parallel-offset (rim) point, the total solution can be read off the graph at C and D. In this example $C = 0$ and $D = 0.0125$ in. Add 0.0125-in shims to back foot. See Fig. 3.31.

If there are any thermal growth considerations, they should be added or subtracted to the results before shim change. In the example, there are none.

Total Horizontal Alignment Solution. For the horizontal (side-to-side) results, the same procedure is used. Algebraically subtract the side-to-side readings. Indicator sag can be ignored because it cancels out. Plot these readings, and the results can be read off the graph. See Fig. 3.32.

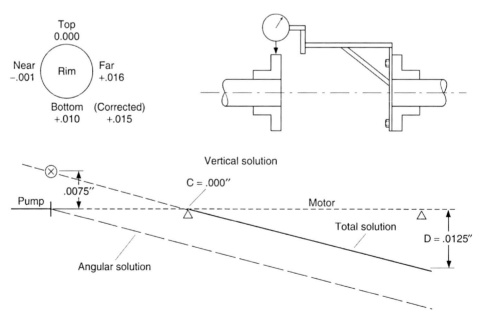

FIGURE 3.31 Parallel-offset (rim) solution.

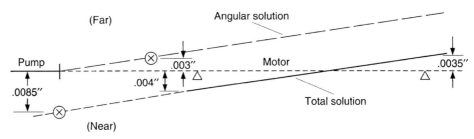

FIGURE 3.32 Total horizonal alignment solution.

The solution for this example is: at the front motor feet, push the motor away from you by 0.004 in, and at the back motor feet, pull the motor toward you by 0.0035 in. The alignment chart in Fig. 3.28 may be helpful.

ACROSS-THE-FLEX ELEMENT

When the distance between the disk packs is long, and where it is not practical to try to span the distance with indicator bracketry, the across-the-flex-element method can be used.

To explain the across-the-flex-element alignment procedure, a motor to a right-angle gear example will be used. First, correct the vertical misalignment by shimming, and then correct the horizontal misalignment by sliding the equipment from side to side. With proficiency, these two steps can be done together.

Before starting the alignment, determine which piece of equipment is easiest to move. This is not to eliminate the option of moving both units if a problem occurs. The gearbox, in this example, will be fixed. The motor will be moved into alignment with the gearbox.

Vertical Alignment Solution. Now, on a sheet of graph paper, lay out the equipment being aligned. See Fig. 3.33.

You should always use a scale that is convenient to the size of the graph paper. The horizontal scale on the graph used here is one small division equals 1 in. The distances that are critical are

1. Distance from the center line of one flex pack to the center line of the other flex pack. In the example, it is 60 in.

2. Distance from the center line of the motor flex pack to the center of front motor foot. In this example, it is 15 in.

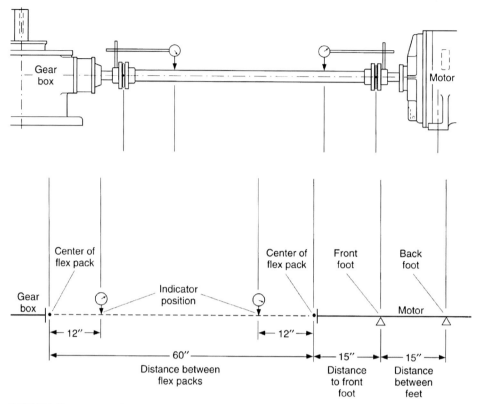

FIGURE 3.33 Across-the-flex alignment.

3. Distance from the center of the motor front feet to the center of the motor back feet. In this example, it is 15 in.
4. Distance from the flex pack to the dial indicator on center member. In this example, the distance is 12 in.

The alignment can be done only with the coupling totally installed. Find a spot to mount the dial indicator bracket. The shaft behind the hub itself is good. A chain-clamp alignment bracket usually fits in well.

With the indicator bracket attached to the gearbox hub, reading out on the center member a convenient distance (in the example 12 in was used), rotate the unit in 90° increments and take readings. See Fig. 3.34.

Bottom reading is then corrected for indicator sag. The indicator sag in the example was determined to be -0.004 in. The -0.004 was subtracted from the $+0.016$ indicator reading to give an actual reading of $+0.020$, or $+0.016 - (-0.004) = +0.020$.

Since this is a T.I.R., it is two times the actual center member center line location relative to the pump shaft extension of $+0.020/2 = +0.010$ (what we are trying to do here is to determine the angle the center member makes with respect of the gearbox shaft).

A plus reading at the bottom indicates that the center member tips down as it extends away from the gearbox. Using a scale of one small division on the graph equals 0.002 in, plot the 0.010 as shown in the example. See Fig. 3.33.

Now with the indicator bracket attached to the motor hub reading out on the center member, rotate the unit in 90° increments and take readings. See Fig. 3.35.

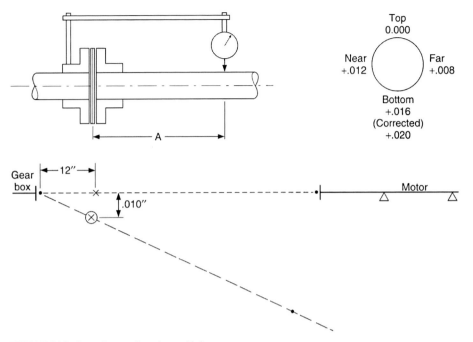

FIGURE 3.34 Secondary reading taken at 90 degrees.

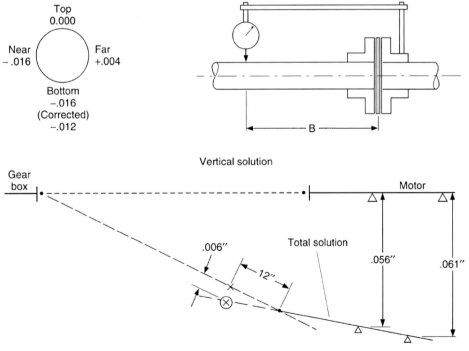

FIGURE 3.35 Vertical alignment solution.

5.70 MAINTENANCE OF MECHANICAL EQUIPMENT

Bottom reading is corrected for indicator sag: $-0.016 - (-0.004) = -0.012$. This is a T.I.R., so the actual figure is $+0.006$ (what we are trying to do here is determine the angle the center member makes with respect to the motor shaft).

The minus reading on the bottom indicates that the center member tips up as it extends away from the motor. Using a scale of one small division on the graph equals 0.002 in, plot the 0.006 in as shown in the example.

The motor shaft can now be drawn in because two points along it have been defined: (1) center of the flex element and (2) the point just plotted 0.006 in below the center member. The shimming requirements can now be read off the plot where the motor shaft intersects the planes of the motor feet.

Note: If the vertical scale chosen is too big, it may realize some minor shimming errors.

In this example, the motor should be shimmed up 0.056 in under the front feet and shimmed up 0.061 in under the back feet.

Horizontal Alignment Solution. For the horizontal (side-to-side) results, the same procedure is used. Algebraically subtract the side-to-side readings. Indicator sag can be ignored because it cancels out. Plot these readings, and the results can be read off the graph. See Fig. 3.36.

The solution for this example is: At the front motor feet, do not move the motor, and at the back motor feet, pull the motor toward you by 0.010 in. See Fig. 3.36. The alignment chart in Fig. 3.28 may be helpful.

The varying coupling configurations and the varying conditions of the drives make it impossible to establish fixed rules of alignment at installation. Normally, the coupling manufacturer's recommendations should be followed.

It must be remembered that all the preceding comments have been relative to cold equipment alignment. Most equipment will shift to some extent as it comes up to operating temperatures. If the movements are known to be slight, no adjustments are necessary. If the movements are unknown

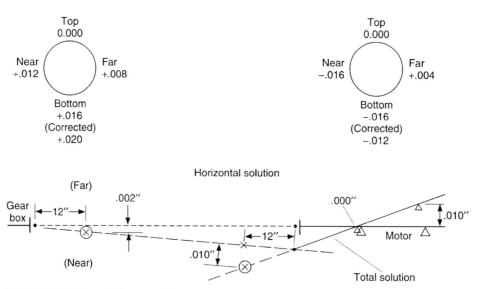

FIGURE 3.36 Horizontal alignment solution.

or suspected of being significant, a hot check of the alignment should be made. If corrections are necessary, the previously outlined procedures should be used.

Long-span couplings are particularly difficult to align. There are known cases where the spans were in excess of 30 ft. The narrow width of the flanged hubs precludes the use of a straightedge. It is again recommended that dial indicators be used and that they be rigidly mounted. The procedure outlined for angular misalignment applies. When measurements are taken and corrections made at both end joints, the system is in alignment.

Three bearing drives require the use of a single engagement coupling. In that they are normally required to support a heavy radial load, these couplings are capable of accepting only angular misalignment. Normal double-flexing couplings cannot be used. Typical examples are engine to single-bearing generators, belt drives, and compressors driven by a single-bearing motor. The procedures outline for angular misalignment under "Face/Rim" apply.

CHAPTER 4
CHAINS FOR POWER TRANSMISSION

Frank B. Kempf
Division Marketing Manager, Drives & Components Division,
Morse Industrial Corporation, a subsidiary of Emerson Electric Company, Ithaca, N.Y.

This section on precision chain drives places before the plant engineer general information and instructions for their installation, lubrication, and maintenance. It will aid him in advising his maintenance departments on how to obtain the best results from chain drives.

Chain drives consist of an endless series of chain links which mesh with toothed wheels, called *sprockets*. The sprockets are keyed to the shafts of the driving and driven mechanisms.

A roller chain has two kinds of links—roller links and pin links—alternately assembled throughout the chain length. A roller link consists of two sets of hollow rollers and bushings, the bushings being press-fitted into the apertures in the roller link plates, the rollers being free to rotate on the outside of the bushing. The pin link has two pins press-fitted into the apertures of the pin-link plates.

When the chain is assembled, the two pins of the pin links fit within the cylindrical bushings of the two adjacent roller links. The pins oscillate inside the bushings, while the rollers turn on the outside of the bushings. This latter action eliminates rubbing of the rollers on the sprocket teeth.

Roller chain is identified by three principal dimensions: pitch, width, and roller diameter (Fig. 4.1).

The term *silent chain* has been generally adopted to describe the inverted-tooth-link type of chain. This type of precision chain is a series of toothed links alternately laced on pins or a combination of joint components in a manner permitting joint articulation between adjoining pitches (Fig. 4.2).

The ends of the toothed links engage the faces of alternate sprocket teeth. The center sections of the links are recessed to provide clearance so that the links straddle one tooth and engage the adjacent teeth (Fig. 4.2).

The chain is retained on the sprockets by means of guide links (not recessed) assembled in the chain. The guide links track in a groove cut in the sprocket teeth (Fig. 4.3).

The use of precision chain drives has steadily increased during the past quarter century through their constantly expanding field of application. This is owing primarily to the following advantages:

1. *Drive efficiency:* This is normally in excess of 98 percent.
2. *Uniform driven speed:* Roller and silent chain drives are positive; the principle of teeth, not tension, results in no loss in rotative speeds through slippage or creep.
3. *Low bearing loads:* Slack side tension is not required.
4. *Larger ratios:* Less wrap on driver sprocket is required, which permits a higher speed ratio in given area than can be obtained from belt drives.
5. *More power per inch of width:* Strength of steel permits greater loads for any given diameter and speed.

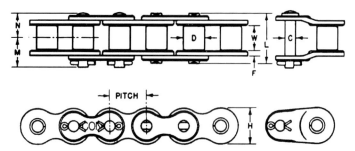

FIGURE 4.1 Dimensions for roller-chain identification.

6. *Relatively unrestricted center distances:* Chain can be made endless to any length within limits.
7. *Ease of installation:* Center distances and alignment do not require close tolerances.
8. *Standardization:* Industry standardization of chain and sprockets means that replacements are available from many sources.
9. *Repair on the job:* Repair links are available for quick emergency replacement of worn or damaged links.
10. *Drive multiple shafts:* Chain is one of the most convenient methods of driving several shafts from one power source.
11. *Long drive life:* Wear is reduced through distribution of load over a number of sprocket teeth. Normal chain wear is a slow process and therefore requires infrequent adjustment.
12. *No deterioration:* Adequately lubricated chains do not deteriorate with age, nor are they adversely affected by sun, oil, and grease.

No one type of chain is ideal for all kinds of service. Certain chain drives are most efficient at very low speeds, others at intermediate speeds, and still others are capable of fairly high-speed operation.

Chain speed, quietness of operation, service life, freedom from maintenance, and the relative first cost will vary within limits according to the combination of chain and sprockets selected. Too much emphasis should not be placed on first cost. Low cost at the expense of other requirements is false economy. Usually it is a matter of compromise to arrive at the best possible combination of specifications to fulfill the requirements of any one drive. Therefore, it is well to evaluate the relative importance of various drive requirements.

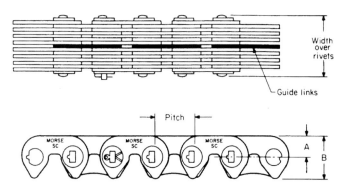

FIGURE 4.2 Silent-chain construction.

FIGURE 4.3 Silent-chain sprocket guide grooves.

Most chain manufacturers' catalogs contain adequate design information on the method of selecting a chain drive. Manufacturers will offer design suggestions for particular applications if they are furnished adequate information.

Basic data needed for the correct design of chain drives:

1. Horsepower and rpm or torque and rpm to be transmitted and whether rpm is exact or approximate.
2. Center distance fixed or adjustable and, if adjustable, the amount of adjustment to be provided.
3. Type of driver (electric motor, gasoline engine, diesel engine, torque convertor, jack shaft, etc.).
4. Type of driven unit (machine tool, hoist, conveyor, agricultural equipment, construction equipment, timing and motion control, lift hoist, tension member, etc.).
5. Service—continuous or intermittent—average number of hours per day.
6. Type of load—smooth and uniform, load reversals, moderate shock or heavy shock.
7. If speed is variable, the maximum, minimum, and usual speeds; the horsepower or torque to be transmitted at each speed; and the approximate percentage of operating time at each speed.
8. Shaft diameters and lengths, keyways and setscrew dimensions.
9. Available space dimensions.
10. Approximate position of the drive (horizontal, vertical, etc.) and the direction of rotation.
11. Operating conditions (wet, dusty, etc.).
12. Lubrication—whether drive will be encased or otherwise adequately lubricated.

Horsepower-rating tables appearing in chain manufacturers' catalogs are based on an average life expectancy of approximately 15,000 service hours under optimal drive conditions and a service factor of 1.

SERVICE FACTORS

Horsepower ratings (see Fig. 4.4) for roller chains must be modified according to the type of load induced by the driver and driven equipment. It is impossible to give service factors by which rated capacities must be multiplied without knowing the type of driver and driven equipment. However, the most prevalent conditions and their accompanying service factors are given in Table 4.1. Conditions which may require a modifying or service factor to be applied to the horsepower ratings are given in the following list.

Favorable service conditions which will contribute to chain life:

1. Drive for intermittent or for standby service
2. Less than maximum service life required
3. Slow speeds, smooth steady load
4. Low ratios permitting larger number of teeth in the sprockets
5. Long centers, on adjustable center distance drives
6. Exceptionally good lubrication

Unfavorable service conditions which restrict chain life:

1. Small sprocket having fewer teeth than recommended by chain manufacturers
2. Unusually large sprockets
3. Inpulse, load reversals, or shock loading
4. Three or more sprockets in the drive
5. Long centers, on fixed center distance drives

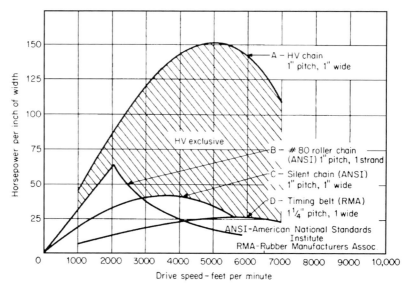

FIGURE 4.4 Horsepower capacity—comparable drives.

TABLE 4.1 Service Factors

Type of load	Internal-combustion-engine hydraulic drive	Electric motor or turbine	Internal-combustion-engine mechanical drive
Smooth	1.0	1.0	1.2
Moderate shock	1.2	1.3	1.4
Heavy shock	1.4	1.5	1.7

6. Poor lubrication
7. Dirty or dusty conditions

Impulse or shock loading should not be confused with high starting or momentary overloads. Because of the high factors of safety with respect to tensile strength, high starting loads or peak loads of short duration do not necessarily require an increase in horsepower rating.

Service factors as given in Table 4.1 should be used in connection with the horsepower ratings, the load being multiplied by the factor to obtain the required chain capacity.

Normal chain wear is caused by the flexing of the chain joints in both roller and silent chains. Wear in the chain joints is usually the limiting factor in the life of a chain. Such wear results in chain elongation, or in other words, the chain pitch is increased. This increase in pitch permits the chain to ride out on the sprocket teeth, which are usually designed to permit moderate pitch elongation. When excessive pitch elongation occurs, the chain must be replaced before it overrides the sprocket teeth.

SPECIAL INVERTED-TOOTH CHAINS

HV, a typical version of an inverted-tooth chain designed expressly for high-horsepower and high-speed application where service is severe, utilizes a unique tooth form on the sprocket and a modified chain-link profile, materially reducing the effect of chordal action and linear pulsation while substantially increasing chain endurance strength.

The chain assembly is similar to silent chain, with inverted tooth links laced in alternate sections across the width of the chain, and assembled with two pins having the same sectional geometry, one called the *pin* and the other the *rocker*. This forms the joint which articulates between the joining links.

Chordal action is a serious limiting factor in roller-chain performance. It may be described as the vibratory motion caused by the rise and fall of the chain as it goes over a small sprocket. Figure 4.5 shows schematically a roller chain entering a sprocket (A); the line of approach is not tangent to the pitch circle. The chain makes contact below the tangency line, is then lifted to the tangent line (B), and then is dropped again (C) as sprocket rotation continues. Because of its fixed-pitch length, the pitch line of the link cuts across the chord between two pitch points on the sprocket and remains in this position relative to the sprocket until the chain disengages. This chordal action seriously detracts from chain performance and life.

1. There is a very definite surge of force in the chain caused by the acceleration and deceleration of the chain as it makes this chordal rise and fall.
2. When the chain enters the sprocket, the tooth gap into which the joint is to fall is rising while the chain strand is falling. Therefore, at contact, there is a definite impact. This impact is very much aggravated by any increase in velocity.

Chordal action, therefore, not only produces pulsations in the chain and generates noise and vibration but also, because of all these things, considerably curtails the power-transmitting capacity and speed range of a roller chain.

The HV drive is designed to minimize chordal action. Smooth engagement with the sprocket minimizes shock loading and stresses in the links as well as noise, vibration, and heating. Figure 4.5D shows how the chain meets the sprocket. It enters approximately tangent to the pitch circle and maintains this position as it travels around the sprocket. This is made possible by two design features: (1) pitch elongation produced by the compensating joint action and (2) mating contours of the sprocket's involute-tooth form and the chain links.

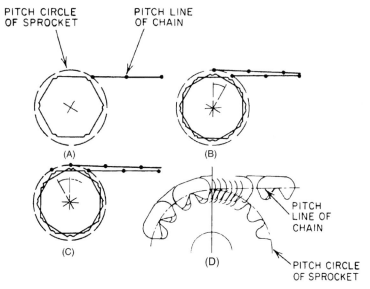

FIGURE 4.5 Chordal action.

The compensating joint is so designed that as flexure takes place the pitch of the chain actually elongates. The joint consists of a pin and rocker of identical cross section—the curved surfaces in contact with each other being tilted in such a manner that the contact point is below the pitch line of the chain. As the joint articulates, the contact point moves upward and the pitch of the chain elongates; the amount of pitch rise is near to that required for the chain to wrap the sprocket along the pitch circle. This is known as *chordal compensation.* The combination of involute-tooth sprocket design and the compensating joint ensures approximate tangential engagement of the chain into the sprocket for smooth and quiet operation.

INSTALLATION OF CHAIN DRIVES

A chain drive is essentially a flexible medium, and its installation is less difficult than many other forms of power transmission. However, care during installation will more than repay the time involved. Improper or careless installation will destroy the precision of any finely designed engineering system.

The shafts must be well supported by suitable and rigidly mounted bearings. Shafting, bearings, and foundations should be suitable to maintain the initial static alignment. Shaft displacement will destroy alignment and so shorten chain life. All shafts should be horizontal and parallel with each other.

Sprockets must be aligned axially on the shafts and secured against axial movement.

Proper chain tension is essential. Too tight a chain will cause excessive bearing loads. Too loose a chain will result in noisy operation and chain pulsations which will cause abnormal chain and sprocket wear.

Contact between the drive and surrounding objects must not be permitted. Ample clearance should be provided to allow for chain pulsations and for possible end float of the shafts. If loose material such as coal, dust, and gravel is present, sufficient clearance is essential to prevent accumulation around the drive.

Installation Procedure

Align each shaft with a machinist's level applied directly to the shafts. (Shafting with silent chain or multiple-width roller chain sprockets may be aligned by applying the level across the sprocket teeth.) Check shafts for parallelism with a feeler bar (Fig. 4.6). After adjusting for parallelism, recheck the shaft levels. Repeat these adjustments until both level and alignment are satisfactory.

Mount sprockets on shafting, and align by checking with a straightedge along the finished sides of the sprockets (Fig. 4.7). A taut wire may be used if the center distance is too long for a straightedge. If a shaft is subject to end float, block it in its running position before aligning the sprockets. Secure the sprockets against axial movement by tightening setscrews.

Before installing the chain, recheck the preceding adjustments and correct any that may have been disturbed.

Wrap the chain around the sprockets, bringing the free ends together on one sprocket. To accomplish this, shorten shaft centers sufficiently. Connect the free ends by use of the connecting link or pins provided.

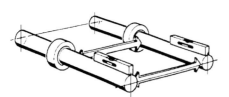

FIGURE 4.6 Shaft alignment.

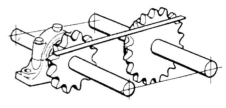

FIGURE 4.7 Sprocket alignment.

Readjust shaft centers to check chain tension. Chains should be installed fairly tight with only a small amount of slack. In the case of vertical drives, the chain should be kept snug and provision for adjustment of chain may be necessary.

New chains will loosen slightly owing to the seating of the joints as the chain is cycled over sprockets under load. After the first several weeks of operation, it is advisable to adjust the centers, if needed, particularly on long center drives. After this initial elongation, with proper care and lubrication, precision chain drives will give long service without undue elongation or wear.

LUBRICATION OF PRECISION CHAIN DRIVES

With the increased speed and horsepower capabilities of modern chain drives, the role of lubrication has increased. The precision roller chain is actually a series of connected journal bearings, and it is essential that lubrication minimize the metal-to-metal contact of the pin-bushing joints of the chain. Many lubrication factors affect chain life.

Heat

Proper chain-drive lubrication will serve to increase the drive life by dissipating frictional heat generated in the joint area. This heat, which varies according to the chain speed, horsepower transmitted, center distance, ratio, drive size, amount and viscosity of lubricant, and others, will range from surrounding temperature of 60 to 70°F above the ambient temperature. Normal chain-drive temperatures should not exceed 180°F.

Improper Lubrication

A lubrication-starved chain drive will show a brownish (rusty) coloration around the joints and in the roller-bushing areas when the link is disassembled and the pin is inspected. The normal highly polished surface of the pin will have deteriorated to a roughened, grooved, or galled surface, which can eventually destroy the hardened surfaces of the chain parts and increase wear until the drive is completely destroyed.

Windage

A chain drive can be running through a sump of good lubricant and still destroy itself from lack of lubrication if the speed exceeds 2500 ft/min (fpm). The chain is actually blowing the lubricant out of its path. In high-horsepower, high-speed drives, it is necessary to use pressurized streams to ensure proper lubrication of the articulating components and to dissipate the heat generated. This lubricant should be sprayed onto the inside of the chain as it enters the sprockets to allow the centrifugal force to carry the lubricant through the joints.

Contamination

Lubricants should be protected from dirt and moisture. A filtering system should be utilized to remove wear particles and abrasive particles to minimize wear on the drive chain.

Oil Viscosity

A good grade of lubricant should be used between the chain parts to maximize the wear life of a chain drive. Lubricants containing antifoam, antirust, or film-strength-enhancing additives may be useful. It is essential that the lubricant reach the sideplate wearing surfaces and pin bushing areas.

TABLE 4.2 Recommended Lubricants and Temperatures for Roller Chain

Temp, °F	Recommended lubricant
0–20	SAE 10
20–40	SAE 20
40–100	SAE 30
100–120	SAE 40
120–140	SAE 50

TABLE 4.3 Maximum Roller-Chain Speeds, ft/min (fpm)

	Chain no.									
Lubrication	35	40	50	60	80	100	120	140	160	200	240
Type A manual and drip	370	300	250	220	170	150	130	115	100	85	75
Type B bath	2800	2300	2000	1800	1500	1300	1200	1100	1000	900	800
Type C pumped	Suitable up to the maximum chain speed										

Therefore, normally heavy oils and greases are not recommended. The lubricant should be free-flowing at the prevailing temperature (Tables 4.2 and 4.3).

The type of lubrication will be dictated by the speed of the chain and the amount of power transmitted. The choice of lubrication method will be determined by the drive itself.

METHODS OF LUBRICATION

Type A: Manual or Drip Lubrication

When manual lubrication is used, oil is applied periodically with a brush or spout can, preferably once every 8 hours of operation. Volume and frequency should be sufficient to prevent discoloration of the lubricant in the chain joints.

When drip lubrication is used, oil drops are directed between the link plate edges by a drip lubricator. Volume and frequency should be sufficient to prevent discoloration of the lubricant in the chain joints. Precaution must be taken against misdirection of the drops by windage. Drops on the center of the chain will not effectively lubricate the joint areas. The lubricant should be directed at the inside of the pin and roller sideplate surfaces.

Type B: Bath or Disk Lubrication

With bath lubrication, the lower strand of the chain runs through a sump of oil in the drive housing. The dynamic oil level should be at the pitch line of the chain at its lowest operating point. With disk lubrication, the chain operates above oil level. The disk picks up oil from the sump and deposits it on the chain, usually by means of a trough. The diameter of the disk should be sufficient to produce rim speeds between 600 and 8000 fpm.

Type C: Oil-Stream Lubrication

The lubrication is usually supplied in a continuous stream to each chain drive. Oil should be applied inside the chain loop, evenly across the chain width, and directed, preferably, at the slack strand.

MAINTENANCE OF PRECISION CHAIN DRIVES

As in the case of any precision-built mechanism, proper maintenance contributes toward long, satisfactory service life.

Before discussing maintenance procedure, it must be assumed that the drive components have been properly selected for the installation, the chain and sprockets have been properly installed, and adequate lubrication has been provided.

1. Every chain drive should be checked periodically for alignment. Misalignment is conclusively indicated when the sides of sprocket teeth or inside surfaces of the chain-link plates show wear. Immediate steps should be taken to realign the drive when these defects are evident.

2. Chain should be checked for excessive slack. If the chain is running close to the tips of the teeth of the larger sprocket, the chain should be replaced. This can be checked visually while the drive is running or by lifting the chain away from the large sprocket, making sure the chain is in mesh with the sprocket teeth, as indicated by arrows in the drawing (Fig. 4.8). Excess clearance is conclusive evidence that the chain has elongated in pitch, and no amount of tension adjustment will keep it properly meshed with the sprocket teeth. Continued operation will cause the chain to jump teeth, destroying the chain and/or sprocket teeth.

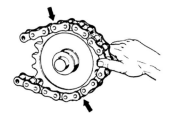

FIGURE 4.8 Elongated chain.

3. Do not install a new chain on sprockets that are badly worn. Worn sprockets should be replaced to ensure proper chain fit on the sprockets, thus eliminating the possibility of premature wear of the replacement chain. The life of a worn sprocket may be extended by reversing it on the shaft to bring a new set of working tooth surfaces into use. If this is done, be careful to check alignment and make sure the sprocket runs true in its new position.

4. New drives should be inspected frequently for any possible interference with the chain. Naturally, if a chain is rubbing or striking against any obstruction, it will necessitate premature replacement.

5. Packing foreign material between the sprocket teeth will occasionally cause the chain to ride high on the sprocket teeth, exert undue stresses and accelerate wear in the chain, and cause abnormal wear of the sprocket teeth.

6. For all types of lubrication, check the quality and grade of the lubricant. For *manual* lubrication, make sure that the lubrication schedule is being followed, and that the oil is being properly applied. For *drip* lubrication, inspect the filling of the oiler cups and the rate of feed; check that the feed pipes are not clogged. Check to ensure that lubricant is applied to link plate edges so that it will penetrate to the pins. For *bath* or *disk* systems, inspect the oil level and check that there is no sludge. Drain, flush, and refill the system at least once a year. For *force-feed* systems, inspect the oil level in the reservoir and check the pump drive and delivery pressure; check that there is no clogging of the piping or nozzles. Drain, flush, and refill the reservoir at least once a year.

7. If roller chains have not been lubricated properly, the joints will have a brownish (rusty) color and the pins of the connecting link of the chain, when removed, will be discolored (light or dark brown). Also, the pins will be roughened, grooved, or galled. Properly lubricated chains will not show the brownish color at the joints, and the connecting link pins will be brightly polished with a very high luster.

8. Even under the best operating conditions, periodic cleaning of the chain is good economy. Gummed lubricant and the products of normal wear cause abnormally rapid pin and bushing wear. A chain exposed to dusty surroundings requires more frequent cleanings.

    Clean a chain as follows:

    a. Remove the chain from the sprockets.
    b. Wash the chain in kerosene. If the chain is badly gummed, soak it for several hours in the cleaning fluid and then rewash it in fresh fluid.

      c. After draining off the cleaning fluid, soak the chain in oil to restore the internal lubrication.
      d. Hang the chain over a rod to drain off the excess lubricant.
      e. Inspect the chain for wear or corrosion. While the chain is off the sprockets, clean the sprockets with kerosene and inspect them for wear or corrosion.

9. Unless properly protected, the components of a chain drive will deteriorate during long periods of idleness. If a chain is to be stored, remove it from the sprockets and coat it with a heavy oil or light grease. Then wrap it in heavy, grease-resistant paper. Store the chain where it will be protected from moisture and mechanical injury. The sprockets may be left in place on the shafts. Cover each with grease, and protect them from mechanical injury. Before placing the drive in service again, thoroughly clean the chain and sprockets to remove the protective grease; then relubricate the chain.

CHAPTER 5
CRANES: OVERHEAD AND GANTRY

William S. Chapin
Director of Engineering, Crane & Hoist Division,
Dresser Industries, Inc., Muskegon, Mich.

GENERAL

Overhead and gantry cranes represent major investment in equipment. Reliable functioning of such equipment is generally vital to operations performed in the areas served by the cranes. Proper installation, operation, inspection, and maintenance of the crane is necessary to ensure performance and to avoid premature breakdowns or accidents which might injure persons working on, under, or near the crane. This chapter is intended to outline recommended procedures. It is by no means all-inclusive.

Manufacturer's instructions should be carefully read, retained, and followed. Attention also must be given to applicable federal standards, OSHA regulations, and state and local codes which include mandatory rules relating to crane inspection and maintenance. (See Fig. 5.1.)

CRANE INSTALLATION

Good maintenance begins with a good installation. Prior to, during, and following erection of the crane, the following precautions should be observed:

1. Crane runway rails should be straight and accurately aligned to the correct span the entire length of the runway.
2. Be sure the crane is assembled in accordance with the match marks and instructions provided by the manufacturer.
3. It is of utmost importance that girders be square with the end trucks, that the trucks be parallel to each other, and that the bolts furnished with the crane be used for making the connection between the girders and end trucks. These bolts are usually of the ground-body type fitted into reamed holes.
4. Check all connecting bolts for tightness, and be sure lock washers or other locking means have been used as furnished.
5. Check for and remove any loose articles, such as bolts, hammers, or wrenches, which may have been left on top of the trolley or girders or on the platform.

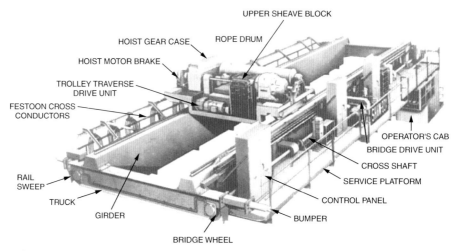

FIGURE 5.1 Typical heavy-duty overhead crane.

6. Grease bearings on the crane as required. Check and add oil to gear housings as required.
7. Grease the hoisting cables.
8. Be sure the hoisting cable is reeved properly.
9. Check for any oil or grease spillage that may have occurred during erection, and wipe all oil spots dry.

Before the crane is placed in service:

1. Check operation of controllers for intended movement of each of the crane motions. In checking the hoisting unit, it is particularly important to note that with a three-phase electrical supply it is possible to have "reverse phasing," causing the hook to lower when the "up" push button, master switch, or controller is actuated. When this condition exists, the automatic limit switch will be inoperative, and hoist operation will be dangerous. To correct this condition, reverse one phase by interchanging any two of the supply wire leads. Do not rewire push buttons or master switches to effect this correction.
2. Check adjustment and operation of all brakes.
3. Check hoist-limit stops, and adjust for proper operation. The trip setting of hoist-limit switches should be determined by tests with the empty hook traveling in increasing speeds up to the maximum speed. The actuating mechanism of the limit switch must be located so that it will trip the switch, under all conditions, in sufficient time to prevent contact of the hook or hook block with any part of the trolley.
4. Check operation of other limit stops or locking or safety devices which may be installed on the crane or runway.
5. Operate the crane slowly through all motions, over the entire length of runway, entire length of bridge, and entire length of lift, checking for proper performance throughout.
6. New cranes and those in which load-sustaining parts have been altered, replaced, or repaired should be subjected to a load test confirming the load rating of the crane. The load rating should be not more than 80 percent of the maximum load sustained during the test, and test loads should not be more than 125 percent of the rated load, unless otherwise recommended by the manufacturer.

CRANE INSPECTION

The frequency of inspection and degree of maintenance required for cranes vary with the service to which the cranes are subjected, heavily used cranes requiring more attention than standby or lightly used cranes. Close attention should be given to the crane in the first few days and weeks of operation, following which routine inspection procedures should be instituted.

Daily to monthly inspections are recommended to include

1. Operation of all limit switches, without load on hook (the crane motion should be inched or run into the limit position at slow speed for these checks)
2. All functional operating mechanisms for misadjustment, damage, or wear
3. Air and hydraulic systems components for deterioration or leakage
4. Hooks for deformations, cracks, and wear
5. Hoisting ropes for broken wires, abrasions, kinks, or evidence of not spooling properly on drum

Monthly to yearly inspections should include checks on

1. Loose connections, bolts, nuts, rivets, keys, etc.
2. Cracked, worn, deformed, or corroded members, including rails or beam flanges on which the crane operates
3. Cracked, worn, or distorted mechanical parts such as shafts, bearings, pins, wheels, rollers, gears, pinions, and locking or clamping devices
4. Excessive wear on brake parts, pawls, pins, levers, ratchets, linings, etc.
5. Rope drums and sheaves for excessive wear or cracks
6. Electric or other types of motor for performance and wear of commutator, slip rings, brushes, etc.
7. Chain and sprockets for excessive wear or stretch
8. Crane hooks for cracks, by magnetic particle, dye penetrant, or other reliable crack-detection method; and hook-attaching means including hook nut, locking pin, etc., for security of hook attachment to lower block
9. Load-limiting or other safety devices installed on the crane
10. Electrical devices, controls, and wiring for signs of deterioration or wear; electrical contactor points for excessive pitting

Standby cranes should be inspected at least semiannually, and more frequently in adverse environment.

Written, dated, and signed inspection reports and records need to be maintained, particularly on critical items such as crane hooks, hoisting ropes, sheaves, drums, and brakes. Figure 5.2 shows the first sheet from the Overhead Crane Inspection and Maintenance Checklist published by and available from the Crane Manufacturers Association of America, Inc., 8720 Red Oak Blvd., Suite 201, Charlotte, NC 28217. Use of this or similar checklist is recommended. Cranes with identified safety hazards should be removed from service until repairs are completed unless other appropriate precautions are taken to eliminate possibility of an accident or injury to personnel.

CRANE MAINTENANCE

A preventive-maintenance program should be established based on the manufacturer's or a qualified person's recommendations. Service schedules and dated detailed records should be maintained.

Since the original equipment manufacturer is usually in the better position to provide replacement parts and ensure their safety, interchangeability, and suitability for the application, it is recommended that such parts be obtained from the original equipment manufacturer.

CRANE MANUFACTURERS ASSOCIATION OF AMERICA
CRANE INSPECTION SCHEDULE AND MAINTENANCE REPORT

Customer: _____ Date: _____

Capacity: _____ Span: _____ Type: _____
Mfr. Ser. No.: _____ Cust. Idnt. No.: _____

Location	Component	Inspection Interval			Condition						Corrective Notes
		Weekly	Monthly	Semi-An'l	OK	Adjust	Repair	Replace	Lubricate	Clean	Describe, Initial, and Date When Corrected
Bridge	Motor			○							
	Brake & Hydraulics	○									
	Control Panels		○								
	Control Operation	○									
	Resistors		○								
	Lights	○									
	Trolley Conductors		○								
	Runway Collectors	○									
	Reducer	○									
	Couplings	○									
	Line Shaft Bearings	○									
	Wheels	○									
	Wheel Gearing	○									
	Wheel Bearings	○									
	Girder Connections		○								
	Align. & Tracking		○								
	Trol. Rails & Stops		○								
	Guards & Covers	○									
	Bumpers		○								
	Rail Sweeps		○								
Cab	Master Switches	○									
	Mainline Disconnect	○									
	Warning Device	○									
	Fire Extinguisher		○								

FIGURE 5.2 This inspection report is being employed by many users of cranes. (Partial checklist shown.)

A good preventive-maintenance program identifies parts requiring replacement sufficiently in advance of actual need to permit ordering of parts after approaching need is identified; however, for cranes in regular service, it is generally advisable to carry on hand a reasonable minimum inventory of repair parts. The needed inventory will vary with type and age of crane, service to which it is subjected, repair history, and general availability of parts. The crane manufacturer can provide lists of recommended spares.

Typical recommended spare-parts lists are apt to include

Brake solenoids, coils, disks, linings
Hoist-limit switches
Contactors
Contact kits
Timing relays
Push-button stations or parts
Crane wheels and guide rollers
Motor couplings and brushes
Current collectors or collector shoes
Bearings
Load hooks, nuts, and thrust bearings
Hoisting ropes
Load brake parts

On-hand availability of parts such as these can often spell the difference between a long, costly wait for repair and efficient replacement at a convenient time in advance of actual breakdown.

When ordering parts for a crane, observation of the following points can save time and expense:

1. Identify the crane by manufacturer's serial number.
2. Refer to the parts manual furnished by the manufacturer, and identify parts by the numbers given.
3. For cranes with auxiliary hoists, specify whether parts are for main or auxiliary hoist; if for a bucket crane, if parts are for the holding-line or closing-line mechanism.
4. If the crane carries more than one trolley or hoist, specify which trolley or hoist the parts relate to.
5. When ordering brake parts, specify which brake, whether hoist (main or auxiliary), trolley, or bridge.
6. If parts are for electrical equipment or other equipment not shown on the parts lists, describe the part and identify the serial number of the unit for which it is required.

Before adjustments or repairs are started, several precautions should be taken:

1. The crane to be repaired should be located where it will cause least interference with other operations in the area.
2. All controllers should be placed in the off position.
3. Main and emergency switches should be locked in the open position.
4. Warning signs should be placed on the crane and on the floor beneath the crane or on the crane hook if near the floor.
5. If other cranes are operating on the same runway, rail stops or other means should be provided to prevent collision with the idle crane, or a signal man may be employed to warn off approaching cranes.

Following completion of adjustments and repairs, all guards and safety devices must be reinstalled and maintenance equipment removed before the crane is returned to service.

Adjustments and repairs to cranes should be done by designated and qualified personnel as soon as possible after identification of need, and before further use of the crane if a safety hazard is involved. Adjustments should be maintained for optimal crane performance and to ensure the safe functioning of all systems and components.

Hook deformation is usually a sign of tip loading of the hook or overloading of the crane. If overloading is suspected, other load-bearing parts of the crane should be checked for possible damage due to overloading. Hooks with cracks, or having a throat opening in excess of normal, or with twist from the plane of the unbent hook should be considered for replacement (see ANSI and other applicable standards). Hooks showing wear in the saddle of the hook, which indicates reduction of strength of the hook, also should be considered for replacement.

Any load-bearing parts which are cracked, bent, or excessively worn must be repaired or replaced.

Pitted or burned electrical contacts should be corrected only by replacement and only in sets.

All control stations should be kept clean with function labels intact. Missing or illegible warning labels must be replaced promptly.

Lubrication should be applied regularly to all moving parts for which lubrication is specified and/or indicated by lubrication fittings. Follow the manufacturer's recommendations as to frequency and types of lubricants to be used. Avoid overgreasing of bearings and overfilling of gear cases. Unless the crane is equipped with automatic lubricators, the same preliminary precautions should be taken for lubricating the crane as for making repairs.

Hoisting ropes require special attention. On cranes in continuous service, ropes should be inspected visually daily, and a thorough inspection should be made at least once a month, with written record made as to rope condition and possible need for replacement. All inspections should be made by an appointed or authorized person. Any form of rope deterioration which could result in appreciable loss of original rope strength should be carefully noted. Conditions such as the following require a determination as to whether continued use constitutes a safety hazard.

1. Kinked, crushed, cut, or unstranded sections
2. Broken, worn, or corroded outside wires
3. Reduction of rope diameter due to loss of core support, corrosion, or wear
4. Damaged end connections or damaged rope wires at the connections

Ropes which have been out of service for long periods of time should be checked carefully before service is resumed.

Replacement rope should be the same size, grade, and construction as the original rope furnished by the crane manufacturer, unless otherwise recommended by the crane manufacturer.

Rules governing replacement of ropes are not precise and hence require judgment by an appointed person as to whether the remaining strength and life in the rope are sufficient to permit continued use. This judgment should take into account the service to which the crane is subjected, as well as the observed condition of the rope. For example, many users consider the following to be conditions which produce serious question as to safety of rope for continued use:

1. Crushed, kinked, birdcaged, or otherwise distorted rope
2. Twelve or more randomly distributed broken wires in one rope lay or four or more broken wires in one strand of one rope lay
3. Wear on outside individual wires exceeding one-third of the original wire diameter
4. Evidence of heat damage
5. Reductions in nominal rope diameters of more than

   $1/16$ in for ropes of $5/16$ in diameter and smaller

   $1/32$ in for ropes of $3/8$ to and including $1/2$ in

   $3/64$ in for ropes to $9/16$ and including $3/4$ in

   $1/16$ in for ropes of to $7/8$ and including $1 1/8$ in

   $3/32$ in for ropes of to $1 1/4$ and including $1 1/2$ in

Wire rope should be stored and handled in a manner which avoids damage or deterioration. Unreeling or uncoiling rope requires care to avoid damaging the rope or introducing twist.

Before a rope is cut, seizings must be placed on each side of the cut location to prevent unraveling of the rope when cut. Apply seizings as follows:

Preformed rope: one seizing each side of cut

Nonpreformed rope $7/8$ in or smaller: two seizings each side of cut

Nonpreformed rope 1 in or larger: three seizings each side of cut

Hoisting ropes should be maintained in well-lubricated condition to reduce internal friction and prevent corrosion. See crane and/or rope manufacturer's recommendations.

GOVERNMENTAL REGULATIONS

The reader is cautioned to recognize that a great many of the recommendations presented in this chapter are now mandatory under OSHA regulations which became effective in 1972 and are also likely to be mandatory under various state safety codes. It should be further noted that OSHA regulations require cranes to conform to the National Electrical Code (ANSI C1). These various regulations are subject to change. OSHA and other regulations applicable to your location should therefore be carefully checked for current requirements. Such codes generally deal with the crane equipment and its operation as well as with maintenance.

For further information on crane maintenance, the reader is referred to the following standards, codes, and regulations:

American National Safety Standards, published by The American Society of Mechanical Engineers, United Engineering Center, 345 East 47th Street, New York, NY 10017:

ANSI B30.2.0: Overhead and Gantry Cranes (Top Running Bridge, Multiple Girder)

ANSI B30.9: Slings

ANSI B30.11: Monorails and Underhung Cranes

ANSI B30.16: Overhead Hoists

ANSI B30.17: Overhead and Gantry Cranes (Top Running Bridge, Single Girder Underhung Hoist)

CMAA Crane Specifications, published by The Crane Manufacturers Association of America, Inc., 8720 Red Oak Blvd., Suite 201, Charlotte, NC 28217

HST Performance Standards for Hoists, published by ASME, 345 East 47th St., New York, NY 10017

OSHA regulations as published in the *Federal Register.* (*Note:* Portions of ANSI standards, such as ANSI B30.2.0, have been adopted as OSHA regulations.)

CHAPTER 6
CHAIN HOISTS

R. C. Dearstyne
Manager, Product Application, Columbus McKinnon Corporation, Amherst, N.Y.

Chain hoists, both manually and power operated, are a widely used type of hoisting equipment. Their simplicity, dependability, and relatively low cost have made them standard equipment in manufacturing plants, foundries, mills, refineries, repair shops, garages, and practically every phase of the construction field.

This chapter describes the various types of chain hoists, explains their relative advantages and usual applications, and provides information on preventive maintenance, inspection, and upkeep.

TYPES OF CHAIN HOISTS

Manually Lever-Operated Chain Hoists (Pullers)

These lightweight, portable tools can be used for pulling horizontally, vertically, or at any angle. A reversible ratchet mechanism, located in the lever, permits short-stroke operation for both tensioning or relaxing. An automatic friction-type load brake, sometimes known as a *releasable ratchet*, is often used to control the load and permit accurate positioning. Other types of lever-operated units utilize a ratchet and pawl-type of brake, which involves ratcheting in both directions, for load control.

Lever-operated hoists are commonly available in both link and roller chain types, in capacities from $\frac{1}{2}$ thru 15 tons. See Fig. 6.1.

Hand-Chain Manually Operated Chain Hoists

These hoists are most frequently used for overhead lifting applications where the use of a powered hoist may not be practical. These would include many maintenance- or construction-type applications where a power source may not be readily available or the portability, load-spotting accuracy, or close-quarter capability of a hand-chain operated hoist may be required.

There are a number of types and classes of hand-chain manually operated chain hoists available, ranging from heavy-duty, high-speed ball bearing spur-geared units, designed for heavy-duty industrial service, to much lighter-weight units with less efficient gearing and power trains designed for more infrequent service.

Heavy-duty, high-speed, spur-geared units, incorporating low-friction bearings and a Weston self-energizing disk-type load brake, typically have very high mechanical efficiency ratings, permitting operation at high speeds in the heavy-duty applications often encountered in industry. Separate hand and load chains operate over pocket wheels which are connected by a gear train and load brake. The brake is disengaged during hoisting by a one-way ratchet mechanism. To lower the load, the hand chain must be pulled continuously in the reverse direction to overcome the holding force of the brake.

FIGURE 6.1 Puller or ratchet-level hoist with link chain.   **FIGURE 6.2** Spur-geared hoist.

Modern hand-chain manually operated chain hoists utilize more compact designs and lightweight alloys to achieve much greater portability through a significant weight reduction as compared with earlier hoist models. See Figs. 6.2 to 6.4. Lower-cost, lighter-duty, spur-geared hand-chain manually operated chain hoists often incorporate gears which are not accurately machined, unlike those employed in high-speed units, and bushings are frequently used in place of low-friction bearings. While these lighter-duty units are somewhat lower in cost, they also operate at mechanical efficiencies 30 to 40 percent lower than those required for high-speed industrial service. Further, these lighter-duty units will typically not withstand the more severe service encountered in rigorous industrial-type applications.

Worm gearing is sometimes used in hoisting applications to provide an inexpensive power train incorporating a high ratio in a relatively small space. These types of power trains are approximately 60 percent less efficient than high-speed spur gearing, however. This is especially true if a high enough gear ratio is utilized to make the power train self-locking. See Fig. 6.5.

A differential hoist is generally the simplest and least expensive chain hoist. It has an efficiency rating approximately 30 percent of that of a high-speed spur-geared unit. A dual-pocketed upper sheave and grooved single lower sheave are connected by an endless reeved chain for lifting and operating. The mechanical advantage is gained by the two upper sheaves differing by one link pocket so that more chain passes over the larger wheel than the smaller one to produce a net raising or lowering of the load. The difference in diameter of these two sheaves results in such a small turning moment on this combination that normal friction holds the load in the suspended position at any point, and this serves as the hoist braking means. An effort must be applied to either raise or lower the load hook. See Fig. 6.6.

There are a number of variations to the hook-suspended hand-chain manually operated chain hoist, which are sometimes encountered, as follows:

Low-headroom army-type trolley hoists are spur-geared units with an integral trolley to provide greatly reduced headroom for close-quarter operation. Figure 6.7 illustrates a low-headroom hoist with plain trolley. Hand-geared trolleys are also available.

FIGURE 6.3 Spur-geared single-reeved hoist with roller chain.

FIGURE 6.4 Multiple-reeved spur-geared hoist with link chain.

Twin-hook hoists are spur-geared hoists with two hooks which are operated simultaneously by one hand chain and are adapted to the handling of bulky objects which require two-point suspension. A twin hoist will lift a total load equal to the capacity marked on the hoist. The load often can be all on one hook or the other or divided any way between the two hooks. Typical extensions range from 3 to 16 ft. See Fig. 6.8.

Extended-handwheel hoists have a handwheel extending from a spur-geared hoist and are designed for service which requires that the hand chain and operator be at a distance from the load being raised. Typical extensions range from 3 to 16 ft. See Fig. 6.9.

Often, chain hoists are used to lift a load, the weight of which may be totally unknown. Because of this fact, many chain hoists are available with overload limiting devices designed to prevent the lifting of dangerous overloads.

Powered Chain Hoists (See Figs. 6.10 to 6.12)

Powered chain hoists are typically used for repetitive higher-speed lifting, as often encountered in production applications. Most powered hoists utilized electrical power, although quite a number of chain hoists are air-powered. Both types of powered units are equipped with either push-button or pendant rope controls. Both link-type chain and roller chain are utilized in powered hoists. Link chain has the advantage of being flexible in all directions, whereas roller chain is flexible in only one plane. Also, powered hoists are normally equipped with travel limit switches to restrict upper and lower extremes of travel, thus preventing the load hook from jamming against the bottom of the hoist or the chain from running out of the hoist.

Electric chain hoists are available for use with most types of current. Many small-capacity models are equipped with single-phase 115-V motors, which can be plugged into a standard receptacle which receives a standard three-prong plug. Some manufacturers offer three-phase dual-voltage single-speed

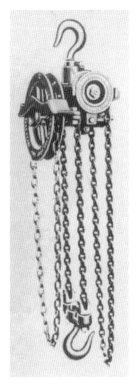

FIGURE 6.5 Screw-geared hoist.

FIGURE 6.6 Differential hoist.

FIGURE 6.7 Low-headroom trolley hoist with link chain.

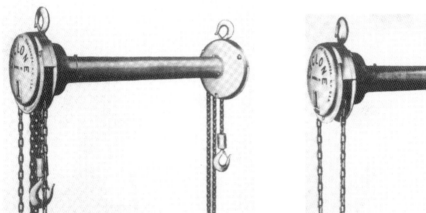

FIGURE 6.8 Twin-hook hoist with link chain.

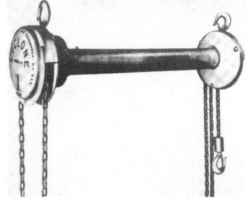

FIGURE 6.9 Extended-handwheel hoist with link chain.

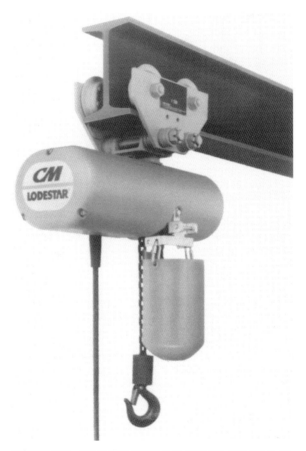

FIGURE 6.10 Lightweight electric chain hoist with push-button control and low headroom trolley.

FIGURE 6.11 Lightweight electric chain hoist with push-button control and hook suspension.

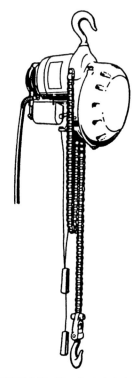

FIGURE 6.12 Electric chain hoist with roller chain and pendant rope control.

models as well as three-phase single-voltage two-speed models. Most single-speed dual-voltage units are reconnectible for operation on either 230 or 460 V, 60-Hz power. Two-speed units are generally built to operate on one specific three-phase voltage (i.e., 230 V, 460 V, etc.). Also, other voltages such as 575 V, three-phase, and 230 V, single-phase, as well as a large number of 50-Hz export voltages, are also readily accommodated.

Powered hoists are generally available with a wide variety of suspensions and accessories. These units may be equipped with swivel or rigid hook, lug, plain trolley, hand-geared trolley, or powered trolley suspensions. Chain containers are often used to collect the unloaded loose end of the load chain at the hoist. A wide variety of power-distribution systems are also available to allow for travel of powered hoists on monorail or bridge crane applications. Often, powered hoist controls are integrated with trolley travel, bridge travel, mainline power disconnect systems, powered accessories, etc.

Powered chain hoists, in capacities from $1/8$ to 15 tons, are widely used throughout industry because of their convenience, relatively low cost, and durability.

SELECTION OF CHAIN HOISTS

In selecting either manually operated or powered hoists, certain considerations are basic and common to both types. Figure 6.13 provides a graphic comparison of the types of manual hoists used today for overhead lifting, each offering a varied range of mechanical efficiency and price and filling the need

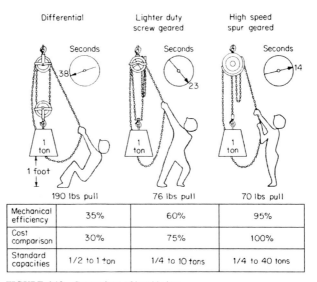

FIGURE 6.13 Comparison of hand hoists.

of a particular condition. Figure 6.14 illustrates the important performance and physical characteristics of both manually operated and powered hoists which must be considered in selecting a unit for a given use and specific installation.

Intended use, safety, labor savings, portability, initial cost, and upkeep are all important factors to be considered in properly selecting a hoist for an application. With high labor costs for both operation and maintenance, low initial cost may not be the true measure of hoist value for a given application.

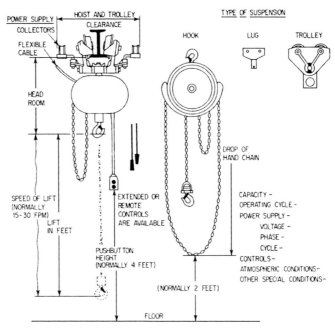

FIGURE 6.14 Hoist installation check diagram.

Hoist capacity, in terms of the heaviest load to be lifted, is of prime importance. Headroom, height of lift, location and height of hand chain or push-button/pendant controls, lifting speed on powered hoists, type of suspension, travel speed on powered trolleys, hoist and trolley clearances, etc. are all factors affecting decisions on specific installations. Figure 6.14 is a self-explanatory diagram and checklist which will be found useful in selecting either a manually operated or powered hoist.

Unusual atmospheric conditions, whether indoors or outdoors, may require sealed enclosures, weatherproof covers, or special protective coatings on the housings, chain, and other fittings. Under normal atmospheric conditions encountered in typical indoor applications, standard hoists are generally satisfactory.

PREVENTIVE MAINTENANCE

Figures 6.15 and 6.16 illustrate typical preventive maintenance tasks that should be performed on chain hoists. For all aspects of chain-hoist preventive maintenance, we suggest that the recommendations contained in ANSI/ASME Standards B30.16 and B30.21 be followed in detail. These ANSI/ASME standards represent an industry consensus published by the American Society of Mechanical Engineers, 22 Law Drive, P.O. Box 2300, Fairfield, NJ 07007-2300, which is nationally recognized and in wide use today.

DESIGN/PERFORMANCE

For additional detailed information on various types of chain hoists, the following standards are referenced:

ANSI/ASME HST 1, Electric Chain Hoists
ANSI/ASME HST 2, Overhead Manual Hoists
ANSI/ASME HST 3, Manual Lever Chain Hoists
ANSI/ASME HST 5, Air Chain Hoists

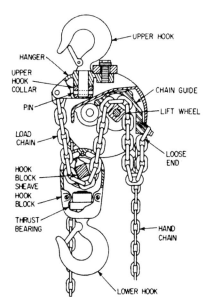

FIGURE 6.15 Hoist parts to be inspected and serviced.

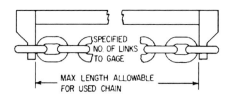

FIGURE 6.16 Load-chain gauging diagram.

CHAPTER 7
BELT DRIVES

Dan Parsons and Tim Taylor
Application Engineers, Gates Corporation, Denver, Colo.

GENERAL

One of the most common elements in power transmission systems, belt drives give dependable and cost effective power transmission with a minimum of maintenance. This chapter contains basic physical dimensions of V-belts and sheaves, synchronous belts, sprockets, plus synchronous application guidelines, installation and maintenance suggestions, and a troubleshooting guide. For actual design of belt drives, manufacturers' design manuals or software should be consulted.

V-BELT TYPES AND NOMINAL DIMENSIONS

Most V-belt drives used in industrial applications fall into two categories: heavy duty (industrial) and light duty (fractional horsepower). There are primarily two types of industrial belts: the *classic* cross sections (A, B, C, and D) (Fig. 7.1), and the *narrow* cross sections (3V, 5V, and 8V) (Fig. 7.2). Most of the sections are available in banded (wrapped) and molded notch (cog) constructions. The banded belt has a fabric cover that completely encloses the exterior of the belt.

The molded notch belt is manufactured with notches on the inside circumference of the belt (Fig. 7.3). The notches allow the belt to be more flexible in bending than the banded belt and may therefore be used on drives where smaller sheaves are required. (Molded notch construction belts are usually designated with an *X* after the section letter. A 3V molded notch belt would be designated 3V*X*.)

Both the classic and narrow section belts are available as joined belts. A joined belt consists of two or more individual belts fastened together with a layer of fabric across the tops of the belts (Fig. 7.4). Joined belts help to prevent belt turnover and maintain belt stability in shock-loaded applications and on drives with long center distances. The horsepower rating of a joined belt is typically equivalent to that of an equal number of individual belts.

Problem Solving Constructions

Belts manufactured with aramid tensile cords are constructed for higher horsepower or extreme shock-loaded applications. These belts come in both single and joined versions and are capable of much higher horespower capacity while elongating less than conventional belts (Fig. 7.5). Care should be taken to ensure that multiple aramid tensile cord belts on the same set of pulleys are matched, or used in sets of belts that are within a tight length tolerance from belt to belt. Also, special belt constructions may be

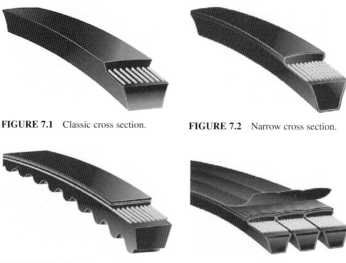

FIGURE 7.1 Classic cross section.

FIGURE 7.2 Narrow cross section.

FIGURE 7.3 Molded notch belt.

FIGURE 7.4 Joined belt cross section.

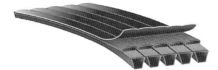

FIGURE 7.5 Heavy problem solving joined construction—gates predator belt.

available to improve load capacity, tolerance for misalignment, high or low temperature capability, or oil resistance. Check with belt manufacturers for construction recommendations and availability.

Fractional-horsepower belts are used most often on drives transmitting less than 1 horsepower. Consequently, they are generally not used in applications requiring multiple belts and are not length-matched out of stock. Length matching will be discussed in the next section. Fractional-horsepower belts are available in the following sections: 2L, 3L, 4L, and 5L.

A V-belt is specified by cross section and length. Nominal dimensions are shown in Figs. 7.6 to 7.8 as an aid in identifying belt cross sections. Other V-belts that are available include V-ribbed belts (for use in high-speed applications and small-diameter sheaves) and variable-speed belts (for use with special sheaves where the belt drive is used for speed variation between the driveR and driveN shafts). Refer to manufacturers' literature for details on these belts.

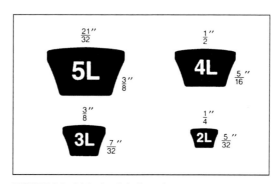

FIGURE 7.6 Light duty belt dimensions.

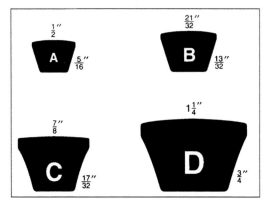

FIGURE 7.7 Classic cross-sectional dimensions.

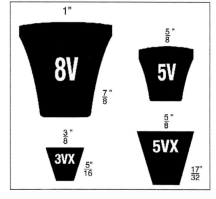

FIGURE 7.8 Narrow cross-sectional dimensions.

V-Belt Length

V-belt length can be measured in three ways: outside circumference (OC), datum length (DL), and effective length (EL). The *outside circumference* is measured by wrapping a tape measure around the outside surface of the belt. This method is useful for obtaining nominal dimensions but does not give a truly accurate belt-length measurement.

Datum Length. *Datum length* is a recent designation adopted by all belt manufacturers in order to retain standard belt and sheave designations while more accurately reflecting the changes that have occurred in belt pitch length and pitch-line location within the belt (*pitch length* is the length of the neutral axis of the belt).

Effective Length. *Effective length* is measured on a length-inspection machine. The machine consists of two parallel shafts on movable centers with a scale to accurately measure the center distance (see Fig. 7.9). Inspection sheaves of equal diameter and grooved in accordance with industry standards are mounted to these shafts. A belt is mounted on these sheaves and tensioned to a specified force. The belt is rotated through at least three complete revolutions to ensure that the tension is equalized around the belt and that the belt is seated in the grooves. The *effective length* is defined as the measured center distance plus the outside circumference of one of the inspection sheaves.

This measurement method accounts for the modulus (stretchability) and dimensional variations among belts with the same cross section. All manufacturers use this designation. To appreciate why this method of measurement is important, consider the following examples:

1. Two belts are both identified as a B105. They both have the same outside circumference; however, one belt has a slightly narrower top width than the other. When the drive is tensioned, the belt with the narrower top width will fit deeper into the sheave groove than the wider top width belt and will therefore have a longer effective length.

2. Two belts are both identified as a B105. They both have the same outside circumference and physical dimensions; however, one is manufactured with a higher-modulus tensile cord than the other. When the drive is tensioned, the belt with the lower-modulus cord will elongate more, resulting in a longer effective length.

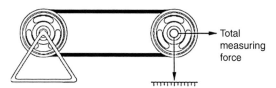

FIGURE 7.9 Schematic of a V-belt measuring fixture.

TABLE 7.1 Standard Matching Tolerances for Industrial V-Belts

A, B, C, D		3V, 5V, 8V	
Standard length designation	Matching tolerance, in	Standard length designation	Matching tolerance, in
26–60	0.15	250–670	0.15
68–144	0.30	710–1500	0.30
158–240	0.45	1600–2500	0.45
270–360	0.60	2650–3750	0.60
390–480	0.75	4000–5000	0.75
540–660	0.90		

The effective length designation ensures that two identically labeled belts will exhibit the same length (within a tolerance) while under tension on a drive regardless of belt modulus and dimensional variations. Accurate belt-length measurements are important in applications which require multiple belts. In these applications, significant differences in belt length will result in unequal load distribution between the belts and subsequent premature belt failure. Belts that are used in these applications must be length-matched. Belt matching is a system in which deviations from ideal effective length are recorded as match numbers. For a given belt-length range, belts to be used on a multiple-belt drive must all fall within a certain match-number range (length tolerance). Most manufacturers have moved away from this system and are able to produce belts within the matching tolerances, thus making belt matching by match numbers unnecessary. Industry standard matching tolerances are listed in Table 7.1.

V-Belt Sheaves

Standard V-belt sheaves are manufactured for industry standard belt sections. V-belt sheaves have exact rather than nominal dimensions. In 1988, the belt-drive industry adopted the datum system as the standard for specifying classic (A, B, C, and D section) V-belt sheaves. Since that time all classical sheaves have been identified by datum diameter rather than by pitch diameter. The physical dimensions of the sheaves have not changed. An old 10.0-in pitch-diameter sheave is directly replaced by a 10.0-in datum-diameter sheave. Datum diameters and datum lengths should be used to calculate center distances and belt-datum length. The often critical speed-ratio and horsepower-rating calculations are now more accurately based on the modern pitch diameter. Deep-groove sheaves are used to increase belt stability in applications where extreme vibration, belt twist, or extreme misalignment are encountered. Specifications for these sheaves are shown in Tables 7.2 and 7.3. (Joined belts cannot be used with deep-groove sheaves.)

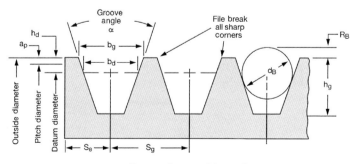

Sheave Groove Dimensions

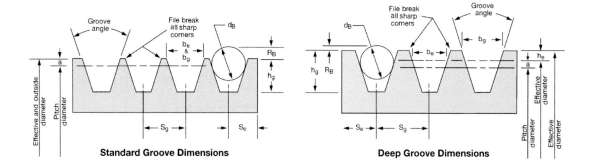

Standard Groove Dimensions **Deep Groove Dimensions**

Minimum Sheave Diameters

When a V-belt bends around a sheave, compressive forces develop in the bottom of the belt, and tension forces develop in the top of the belt. The magnitude of these forces is a function of the diameter of the sheave and the cross section of the belt and has a significant impact on belt life. These forces increase with smaller diameters and larger cross sections. Therefore, minimum recommended diameters were developed for each belt cross section. They are listed in Table 7.4. Using sheaves that are below the recommended minimum will always result in reduced belt life.

Belt drives are often used in applications where the driveR is an electric motor. The motor bearings are specified assuming some maximum overhung load. As the diameter of the sheave used on the motor shaft decreases, the overhung load increases. Therefore, the National Electrical Manufacturers Association (NEMA) developed a standard that specifies minimum recommended sheave diameters for all common motor frame sizes. Using sheaves that are below these minimum diameters could result in reduced motor bearing life. NEMA minimum recommended diameters are listed in Table 7.5.

V-BELT APPLICATION GUIDELINES

The V-belt drives provide a practical and economical means of power transmission. Several unique aspects of V-belt drive application that should be considered, however.

1. V-belt drives can function as a slip clutch within a power transmission system. They may allow slippage during high torque starts, or an overload/jam condition.
2. Joined V-belts are intended to prevent individual belts from interfering with one another when belt span vibrations become excessive due to shock and impulse loading. They also ensure close belt length matching.
3. Notched belts have higher load capacity than banded belts on small sheave diameters.
4. Never mix new and used V-belts on the same drive because uneven load sharing will result. The same is true for mixing belts from different manufacturers.
5. V-belts can safely withstand ambient temperatures up to 165°F (74°C). For every 18°F (8°C) increase in ambient temperature above this point, V-belt service life may be reduced by approximately half.
6. Drive alignment is one of the most common causes of short V-belt life and belt instability.

TABLE 7.2 Standard Groove Specifications for Classic-Section Sheaves

Cross section	Datum diameter range	α Groove angle $\pm 0.33°$	b_d ref.	b_g	h_g min.	$2h_d$ ref.	R_B min.	d_B ± 0.0005	S_g ± 0.025	S_e	Minimum recommended datum diameter	$2a_p$
A, AX	Up through 5.4	34	0.418	0.494 ±0.005	0.460	0.250	0.148	0.4375 (7/16)	0.625	0.375 +0.090 / −0.062	A 3.0	0
	Over 5.4	38		0.504 ±0.005			0.149				AX 2.2	
B, BX	Up through 7.0	34	0.530	0.637 ±0.006	0.550	0.350	0.189	0.5625 (9/16)	0.750	0.500 +0.120 / −0.065	B 5.4	0
	Over 7.0	38		0.650 ±0.006			0.190				BX 4.0	
A-B Combination — A, AX Belt	Up through 7.4(1)	34	(2)	0.612 ±0.006	0.612	0.634 (3)	0.230	0.5625 (9/16)	0.750	0.500 +0.120	A 3.6(1)	0.37
	Over 7.4	38		0.625		0.602	0.226				AX 2.8	
A-B Combination — B, BX Belt	Up through 7.4(1)	34	0.508	0.612 ±0.006	0.612	0.268 (3)	0.230	0.5625 (9/16)	0.750	0.500 −0.065	B 5.7(1)	−0.08
	Over 7.4	38		0.625		0.276	0.226				BX 4.3	
C, CX	Up through 7.99	34	0.757	0.879	0.750	0.400	0.274	0.7812 (25/32)	1.000	0.688 +0.160 / −0.070	C 9.0	0
	Over 7.99 to and including 12.0	36		0.887 ±0.007			0.276				CX 6.8	
	Over 12.0	38		0.895			0.277					
D	Up through 12.99	34	1.076	1.259	1.020	0.600	0.410	1.1250 (1-1/8)	1.438	0.875 +0.220 / −0.080	13.0	0
	Over 12.99 to and including 17.0	36		1.271 ±0.008			0.410					
	Over 17.0	38		1.283			0.411					

(1) Diameters shown for combination grooves are outside diameters. A specific datum diameter does not exist for either A or B belts in combination grooves.
(2) The b_d value shown for combination grooves is the "constant width" point but does not represent a datum width for either A or B belts ($2h_d = 0.340$ reference).
(3) $2h_d$ values for combination groove are calculated based on b_d for A and B grooves.

Machine surface area

Sheave groove sidewalls	125
Sheave O.D.'s and rim edges	250
Rim I.D.'s hub ends, hub O.D.'s	250
Straight bores	125
Taper bores	175
Cast surface area	As cast

Minimum surface roughness height, R_a (Arithmetic avg.) (micro inch)

Face width of standard and deep-groove sheaves

Face width = $S_g (N_g - 1) + 2S_e$

Where: N_g = Number of grooves

| Cross section | Datum[4] diameter range | Deep groove dimensions, in ||||||||| Design factors ||
		α Groove angle ±0.33°	b_d ref.	b_g	h_g min.	$2h_d$ ref.	R_B min.	d_B ±0.0005	S_g ±0.025	S_e	Minimum recommended datum diameter	2ap
B, BX	Up through 7.0 Over 7.0	34 38	0.530	0.747 ±0.006 0.774	0.730	0.710	0.007 0.008	0.5625 (9/16)	0.875	0.562 +0.120 −0.065	B 5.4 BX 4.0	0.36
C, CX	Up through 7.99 Over 7.99 to and including 12.0 Over 12.0	34 36 38	0.757	1.066 1.085 ±0.007 1.105	1.055	1.010	−0.035 −0.032 −0.031	0.7812 (25/32)	1.250	0.812 +0.160 −0.070	C 9.0 CX 6.8	0.61
D	Up through 12.99 Over 12.99 to and including 17.0 Over 17.0	34 36 38	0.076	1.513 1.541 ±0.008 1.569	1.435	1.430	−0.010 −0.009 −0.008	1.1250 (1−1/8)	1.750	1.062 +0.220 −0.080	13.0	0.83

(4) The A/AX, B/BX combination groove should be used when deep grooves are required for A or AX belts.

TABLE 7.3 Standard Groove Specifications for Narrow-Section Sheaves

Cross section	Outside diameter, in	Groove angle ±0.25°	Standard groove dimensions, in							Design factors		
			b_g ±0.005	b_c ref.	h_g min.	R_B min.	d_B ±0.0005	S_g ±0.015	S_c	Minimum recommended outside diameter		2a
3V, 3VX	Up through 3.49	36				0.181						
	Over 3.49 to and including 6.00	38				0.183				3V 2.65		
	Over 6.00 to and including 12.00	40	0.350	0.350	0.340	0.186	0.3438	0.406	0.344 +0.094 −0.031	3VX 2.20		0.050
	Over 12.00	42				0.188						
5V, 5VX	Up through 9.99	38				0.329						
	Over 9.99 to and including 16.00	40	0.600	0.600	0.590	0.332	0.5938	0.688	0.500 +0.125 −0.047	5V 7.10 5VX 4.40		0.100
	Over 16.00	42				0.336						
8V	Up through 15.99	38				0.575						
	Over 15.99 to and including 22.40	40	1.000	1.000	0.990	0.580	1.0000	1.125	0.750	12.50		0.200
	Over 22.40	42				0.585				+0.250 −0.062		

Machine surface area	Minimum surface roughness height, R_a (arithmetic avg.) (micro in)
V-pulley groove sidewalls	125
Rim edges, rim I.D.'s	250
Hub ends, hub O.D.'s Straight bores	125
Taper bore	175

Face width of standard and deep-groove sheaves

Face width = $S_g(N_g - 1) + 2S_e$
Where: N_g = number of grooves

Cross section	Outside diameter, in.	Groove angle ±0.25°	b_g ±0.005	b_e ref.	h_g min.	R_B min.	d ±0.0005	S_g ±0.015	S_e	Minimum recommended effective diameter	2a	2he
3V, 3VX	Up through 3.71	36	0.421	0.350	0.449	0.070	0.3438	0.500	0.375 +0.094 −0.031	3V 2.87 3VX 2.42	0.050	0.218
	Over 3.71 to and including 6.22	38	0.425			0.073						
	Over 6.22 to and including 12.22	40	0.429			0.076						
	Over 12.22	42	0.434			0.078						
5V, 5VX	Up through 10.31	38	0.710	0.600	0.750	0.168	0.5938	0.812	0.562 +0.125 −0.047	5V 7.42 5VX 4.72	0.100	0.320
	Over 10.31 to and including 16.32	40	0.716			0.172						
	Over 16.32	42	0.723			0.175						
8V	Up through 16.51	38	1.180	1.000	1.252	0.312	1.0000	1.312	0.844 +0.250 −0.062	13.02	0.200	0.524
	Over 16.51 to and including 22.92	40	1.191			0.316						
	Over 22.92	42	1.201			0.321						

Deep groove dimensions, in

Design factors

TABLE 7.4 Minimum Recommended Sheave Diameters

Belt cross section	Min. recommended datum diameter (standard groove), in
Classical V-belts	
AX	2.20
A	3.00
BX	4.00
B	5.40
CX	6.80
C	9.00
D	13.00
E	21.00

Belt cross section	Min. recommended outside diameter (standard groove), in
Narrow V-belts	
3VX	2.20
3V	2.65
5VX	4.40
5V	7.10
8V	12.50
Light-duty V-belts	
2L	0.8
3L	1.5
4L	2.5
5L	3.5
Micro-V® belts	
J	0.8
L	3.00
M	7.00
Polyflex® JB® belts	
3M	0.67
5M	1.04
7M	1.67
11M	2.64

7. Applying and maintaining recommended V-belt installation tension levels is essential for good drive performance.
8. Properly maintained V-belt drives will generally provide efficiency in the range of 93 to 97 percent.

The application of these guidelines and characteristics will optimize V-belt drive system performance.

V-BELT SHEAVE INSTALLATION PROCEDURE

V-belt sheaves most commonly used in the belt drive industry generally use either QD (quick disconnect) or TL (taper lock) bushings for secure shaft mounting. Though both shaft mounting systems have been used within the belt drive industry for many years, proper installation procedures

TABLE 7.5 NEMA Minimum Recommended Sheave Diameters

| Frame no. | Shaft dia., in | Horsepower at synchronous speed, rpm | | | | Super HC® V-belts & PowerBand® belts | Hi-Power® II PowerBand® & Tri-Power® V-belts |
		3600 (3450)*	1800 (1750)*	1200 (1160)*	900 (870)*	Min. outside dia., in	Min. datum dia., in
143T	0.875	1½	1	¾	½	2.2	2.2
145T	0.875	2–3	1½–2	1	¾	2.4	2.4
182T	1.125	3	3	1½	1	2.4	2.4
182T		5	—	—	—	2.4	2.6
184T	1.125	—	—	2	1½	2.4	2.4
184T		5	—	—	—	2.4	2.6
184T		7½	5	—	—	3.0	3.0
213T	1.375	7½–10	7½	3	2	3.0	3.0
215T	1.375	10	—	5	3	3.0	3.0
215T		15	10	—	—	3.8	3.8
254T	1.625	15	—	7½	5	3.8	3.8
254T		20	15	—	—	4.4	4.4
256T	1.625	20–25	—	10	7½	4.4	4.4
256T		—	20	—	—	4.4	4.6
284T	1.875	—	—	15	10	4.4	4.6
284T		—	25	—	—	4.4	5.0
286T	1.875	—	30	20	15	5.2	5.4
324T	2.125	—	40	25	20	6.0	6.0
326T	2.125	—	50	30	25	6.8	6.8
364T	2.375	—	—	40	30	6.8	6.8
364T		—	60	—	—	7.4	7.4
365T	2.375	—	—	50	40	8.2	8.2
365T		—	75	—	—	8.6	9.0
404T	2.875	—	—	60	—	8.0	9.0
404T		—	—	—	50	8.4	9.0
404T		—	100	—	—	8.6	10.0
405T	2.875	—	—	75	60	10.0	10.0
405T		—	100	—	—	8.6	10.0
405T		—	125	—	—	10.5	11.5
444T	3.375	—	—	100	—	10.0	11.0
444T		—	—	—	75	9.5	10.5
444T		—	125	—	—	9.5	11.0
444T		—	150	—	—	10.5	—
445T	3.375	—	—	125	—	12.0	12.5
445T		—	—	—	100	12.0	12.5
445T		—	150	—	—	10.5	—
445T		—	200	—	—	13.2	—

*Approximate full load speeds.

may not always be followed. Proper bushing installation is critical for proper and safe V-belt drive operation and are provided below for reference.

QD Type Sheaves

TL Type Sheaves

Sheave/Bushing Installation

1. Preliminarily mount the new bushings and sheaves on the shafts.

   QD-type sheaves

   Determine whether to mount each sheave in a conventional or reverse manner (conventional is easier, but positions the sheave further out on the shaft)

   Conventional mount (shown right)
   a. Insert the key into the keyway on the shaft.
   b. Insert a screw driver blade into the bushing slot to enlarge the bore slightly.
   c. Slide the bushing onto the shaft with the flange side toward the equipment
   d. Place the sheave onto the bushing and insert the cap screws. Align the unthreaded holes in the sheave hub with the threaded holes in the bushing flange. Finger-tighten the screws.

   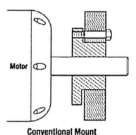
   Conventional Mount

   Reverse mount (shown right)
   a. Insert the key into the keyway on the shaft.
   b. Place the sheave on to the shaft without the bushing.
   c. Insert a screw driver blade into the bushing slot to enlarge the bore slightly.
   d. Slide the bushing onto the shaft with the flange facing outward, away from equipment.
   e. Place the sheave onto the bushing and insert the cap screws. Align the unthreaded holes in the bushing flange with the threaded holes in the sheave hub. Finger-tighten the screws.

   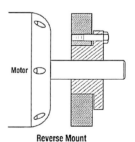
   Reverse Mount

   TL-type sheaves
   a. Insert the key into the keyway on the shaft.
   b. Insert a screw driver blade into the bushing slot to enlarge the bore slightly.
   c. Place bushing in sheave hub and align the holes; all holes should be half threaded.
   d. Insert cap screws loosely into the tightening holes (sheave side of holes threaded).
   e. Slide the assembly onto the shaft with the bushing side facing outward; away from equipment.
   f. Tighten the cap screws finger tight.

2. Slide the sheaves onto the shafts as far as possible (close to the motor/equipment) while remaining within parallel alignment and remaining in full contact with the shaft key.
3. Identify the sheave that first reached a stopping point on the shaft and prepare to secure it by tightening the bushing cap screws.
4. Secure the first sheave to the shaft by tightening the bushing using either of the following two procedures:

   QD bushings
   a. Tighten the set screw down against the shaft key.
   b. Using a torque wrench, tighten the bushing cap screws evenly in an alternating pattern until approximately one half of the recommended torque level has been reached.
   c. Check parallel sheave alignment and sheave lateral runout (wobble), and correct as necessary.
   d. Continue alternate tightening of the cap screws to the recommended torque level. When the recommended torque level is reached, do not tighten the cap screws further. Further tightening may result in over bushing insertion and sheave hub fracture.

   TL bushings
   a. Obtain either hex key sockets or conventional hex head sockets for the bushing cap screws, whichever is appropriate.
   b. Make sure the hex key sockets are not worn excessively, or slippage within the socket head cap screw head may occur. Wrench slippage will result in damage that may prevent proper tightening and/or removal.
   e. Using a torque wrench, tighten the bushing cap screws evenly in an alternating pattern until the recommended torque value has been reached.
   f. After reaching the recommended torque level, further tighten the bushing by driving it further into the sheave hub using a suitable drift or punch.
   g. After tapping the bushing face several times with the drift, retorque the cap screws to their recommended value.

5. Place the second sheave in alignment with respect to the first using a string, straight edge, laser alignment tool, etc.
6. Accounting for axial sheave movement that occurs as the bushing cap screws are tightened, lightly tighten the cap screws to hold the sheave/bushing assembly in place.
7. Using a string, straight edge, laser alignment tool, etc. check both parallel and angular sheave alignment.
8. Make the necessary adjustments in the placement of the second sheave to correct any parallel misalignment.
9. Place the belt on the sheaves and apply light tension by adjusting the motor base, etc.
10. Slowly turn the drive over by hand to allow the belt to seat in the sheave. Several belt revolutions are generally sufficient.
11. Using a string, straight edge, laser alignment tool, etc. check for angular shaft misalignment while adjusting the motor base and applying tension to the belt.
12. Measure the static belt installation tension either directly with the Sonic Tension Meter, or indirectly using the force/deflection method.
13. After reaching the desired static belt installation tension, tighten all mounting fasteners to rigidly secure the motor and related equipment.
14. Turn the drive over by hand again to make sure that the belt is fully seated in the sheaves.
15. Recheck the static belt installation tension along with angular and parallel misalignment, and make any necessary adjustments.
16. Replace the drive safety guard.
17. While watching and listening carefully, start the drive and confirm that all operating characteristics are normal.

5.112 MAINTENANCE OF MECHANICAL EQUIPMENT

MAINTENANCE

A properly designed and maintained V-belt drive will provide long trouble-free service. To maximize V-belt drive performance it is important to follow good maintenance practices. The required maintenance frequency may be influenced by many factors including the operating environment, drive operating speed, load duty cycle, critical nature of the equipment, and belt drive accessibility. Severe operating conditions may require additional inspection and maintenance. The following suggested maintenance intervals should be considered:

1. For critical drives, a quick visual and hearing inspection should be performed every 1 to 2 weeks.
2. For normal drives, a quick visual and hearing inspection should be performed once per month.
3. For normal drives, a shutdown to thoroughly inspect belts, sheaves, and other components should be performed every 3 to 6 months.

Visual Inspection

A visual and hearing inspection consists of the following three items:

1. Look and listen for unusual noise and vibration while observing the drive. A well-designed and well-maintained drive will operate quietly and smoothly.
2. Inspect the guard for looseness and damage. Make sure that it is clean. Any accumulation of foreign material on the guard acts as insulation and could cause excessive heat buildup in the drive.
3. Look for oil and grease contamination that could degrade the material components of a V-belt.

A thorough belt drive inspection requires a complete shut down and should include an inspection of the belt(s), sheaves, belt tension, the belt guard and associated drive components such as bearings, shafts, and takeup rails. Whenever a drive system is inspected thoroughly, use safety precautions to avoid injury. Shut the power off, lock and tag the control box, and place all machine components in a neutral position.

The primary drive troubleshooting tool is belt inspection. Unusual belt wear is a symptom of potential drive problems. Make a mark on the belt, and inspect the entire circumference checking for uneven wear patterns, cracking (back and undercord or notches), frayed covers, burned spots, swelling, and hardening. Belts should be replaced if there is significant cracking or fraying.

A great deal of V-belt damage can be attributed to excessive heat. Properly functioning V-belts operate at approximately 140°F (60°C) and should be able to be held comfortably. Belts that are hot to the touch indicate potential problems.

Pulley Wear

Sheaves should be checked for wear, nicks, sharp edges, as well as for proper mounting and alignment. V-belt sheave wear may be checked with plastic groove gauges, which are available from most manufacturers (see Fig. 7.10). If clearance between the sheave and groove gauge exceeds $1/32$ in or more, the sheave should be replaced.

Misalignment

There are three primary sources of misalignment in belt-drive systems:

1. Driver and driven shafts are not parallel (both horizontal and vertical planes).
2. Sheaves are not located in line axially with respect to one another on the shafts.
3. Sheaves are tilted due to improper mounting (wobble while running).

FIGURE 7.10 Sheave groove inspection.

The effects of these conditions can be illustrated as both parallel and angular misalignment (see Fig. 7.11). Sheaves can be checked for tilting using a bubble level. Axial alignment of sheaves and shaft parallelism can be checked using a straightedge or a string. The total allowable misalignment (angular + parallel) in V-belt drive systems should not exceed $1/2°$ or $1/10$ in per foot of belt span.

Other Drive Components

Check guards for wear, damage, and contamination. Guard wear suggests possible interference with belts or sheaves. Clean the guard to prevent airflow restriction and to allow efficient heat dissipation. Check bearings for correct alignment and lubrication, make sure that drive components are securely tightened down, and ensure that guiderails are free of dirt, obstructions, and corrosion.

Installation and Takeup Allowance

Make certain that adequate shaft movement is provided for belt installation and takeup (see Fig. 7.12). Recommended allowance values are available from belt manufacturers (see Tables 7.6 to 7.8).

Care and Handling

Never pry or roll belts onto sheaves. This will cause invisible damage to the tensile cords and reduce belt life. V-belts should be stored in a cool, dry place. Extended storage of the V-belt in a hot environment will reduce belt life. Belts should be stored on a flat surface where possible. Hanging belts on hooks can cause belt crimping and lead to reduced belt life, especially with larger and heavier belts.

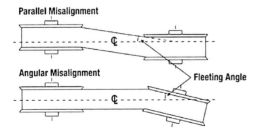

FIGURE 7.11 Types of belt drive misalignment.

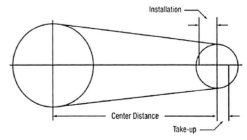

FIGURE 7.12 Installation and takeup allowance.

TABLE 7.6 Center Distance Installation and takeup allowance.

V-belt no.	A		B		C		D		Minimum center distance allowance for initial tensioning and subsequent takeup, in
	Hi-Power II and Tri-Power® molded notch V-belts	Hi-Power® II PowerBand® belt*	Hi-Power® II and Tri-Power® molded notch V-belts	Hi-Power II PowerBand belt*	Hi-Power® II and Tri-Power® molded notch V-belts	Hi-Power® II PowerBand® belt*	Hi-Power® II V-belts	Hi-Power® II PowerBand® belts*	All cross sections — All types
Up through 35	0.75	1.20	1.00	1.50	—	—	—	—	1.00
36 through 55	0.75	1.20	1.00	1.50	1.50	2.00	—	—	1.50
56 through 85	0.75	1.30	1.25	1.60	1.50	2.00	—	—	2.00
86 through 112	1.00	1.30	1.25	1.60	1.50	2.00	—	—	2.50
113 through 144	1.00	1.50	1.25	1.80	1.50	2.10	2.00	2.90	3.00
145 through 180	—	—	1.25	1.80	2.00	2.20	2.00	3.00	3.50
181 through 210	—	—	1.50	1.90	2.00	2.30	2.00	3.20	4.00
211 through 240	—	—	1.50	2.00	2.00	2.50	2.50	3.20	4.50
241 through 300	—	—	1.50	2.20	2.00	2.50	2.50	3.50	5.00
301 through 390	—	—	—	—	2.00	2.70	2.50	3.60	6.00
Over 390	—	—	—	—	2.50	2.90	3.00	4.10	1.5% of belt length

*Also use these figures for individual Hi-Power II and Tri-Power molded notch V-belts in deep groove sheaves.

TABLE 7.7 Center Distance Installation and Takeup Allowance—Narrow-Section Belts

V-belt no.	Minimum center distance allowance for installation, in						Minimum center distance allowance for initial tensioning and subsequent takeup, in
	3V/3VX		5V/5VX		8V		All cross section
	Super HC V-belt	Super HC PowerBand belt	Super HC V-belt	Super HC PowerBand belt	Super HC V-belt	Super HC PowerBand belt	All types
Up to and Incl. 475	0.5	1.2	—	—	—	—	1.0
Over 475 to and Incl. 710	0.8	1.4	1.0	2.1	—	—	1.2
Over 710 to and Incl. 1060	0.8	1.4	1.0	2.1	1.5	3.4	1.5
Over 1060 to and Incl. 1250	0.8	1.4	1.0	2.1	1.5	3.4	1.8
Over 1250 to and Incl. 1700	0.8	1.4	1.0	2.1	1.5	3.4	2.2
Over 1700 to and Incl. 2000	—	—	1.0	2.1	1.8	3.6	2.5
Over 2000 to and Incl. 2360	—	—	1.2	2.4	1.8	3.6	3.0
Over 2360 to and Incl. 2650	—	—	1.2	2.4	1.8	3.6	3.2
Over 2650 to and Incl. 3000	—	—	1.2	2.4	1.8	3.6	3.5
Over 3000 to and Incl. 3550	—	—	1.2	2.4	2.0	4.0	4.0
Over 3550 to and Incl. 3750	—	—	—	—	2.0	4.0	4.5
Over 3750 to and Incl. 5000	—	—	—	—	2.0	4.0	5.5
Over 5000 to and Incl. 6000	—	—	—	—	2.0	4.0	6.0

TABLE 7.8 Center Distance Installation and Takeup Allowance—Poly-V-Belts

V-belt no. Standard effective length, in	Minimum center distance allowance for installation, in			Minimum center distance allowance for initial tensioning and subsequent takeup, in (all cross sections)
	J	L	M	
Up through 20.0	0.4	—	—	0.3
20.1 through 40.0	0.5	—	—	0.5
40.1 through 60.0	0.6	0.9	—	0.7
60.1 through 80.0	0.7	1.0	—	0.9
80.1 through 100.0	0.8	1.2	1.5	1.1
100.1 through 120.0	—	1.2	1.6	1.3
120.1 through 160.0	—	1.4	1.7	1.7
160.1 through 200.0	—	—	1.8	2.2
200.1 through 240.0	—	—	1.9	2.6
240.1 through 300.0	—	—	2.2	3.3
300.1 through 360.0	—	—	2.5	3.9
360.1 through 370.0	—	—	2.7	4.6

Predictive Maintenance

Predictive maintenance will minimize costly, unplanned downtime. Regular replacement intervals should be established so that a belt is replaced near the end of its useful life. Replacement intervals may be established to correspond with a yearly shut down period. The typical life of an industrial V-belt drive ranges between 3 and 5 years depending on the application. In more critical applications, the belt replacement interval should be at least once per year.

BELT TENSION

An overview of the V-belt tensioning process may be summarized with a few simple rules:

1. The ideal installation tension level for V-belts is the lowest tension at which belts will not slip under peak loads. Recommended belt installation levels are available from belt manufacturers.
2. The force-deflection method described below can be used to measure tension levels, a direct force can be applied against the motor or an idler, or the frequency method can be used with an electronic tension gauge.
3. When installing new V-belts, set belt tension to the recommended level, rotate the drive for a few revolutions, and then recheck the tension level. Check the belt tension level once more after running the belts for 48 hours.
4. The V-belt tension level should be checked periodically thereafter.

Force-Deflection Method

The most common method of measuring belt tension levels is the force-deflection method. This method translates the static tension (belt tension level when the drive is at rest) to a specific belt deflection distance with the application of a deflection force. The standard deflection distance is $1/64$ in per inch of belt span length. The force should be applied perpendicular to the belt at midspan (see Fig. 7.13). The

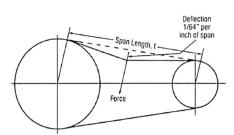

FIGURE 7.13 Force-deflection method of tensioning.

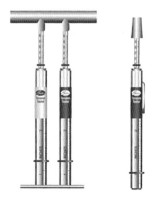

FIGURE 7.14 Pencil-type tension tester and pen method of tensioning.

force required should fall within a calculated range in order for the belts to be properly tensioned. The deflection force is generally applied using a spring type tension gauge, see Fig. 7.13.

When making belt tension measurements if the actual force required to deflect one belt is above the recommended range, the belts are too tight. Likewise, if the actual force is lower than the recommended range, the belts are too loose. Ideally, when tensioning V-belts, the deflection force should be at the high end of the recommended range. When the deflection force drops to the lower end of the recommended range, the drive should be retensioned.

If joined type belts are used, all of the joined belts must be deflected together and the deflection force range must be increased proportionately. For example, if a three-strand joined belt is being deflected, both the upper and lower forces of the recommended range should be multiplied by three.

(*Note:* This has no effect on total drive tension or shaft load.)

Recommended belt deflection force and distance values can be calculated using formulas published in belt manufacturer's design manuals, or can be obtained from manufacturer's drive design software. Required belt tension calculations are based on the drive load requirements, operating speeds, sheave diameters, center distance, and the belt type.

For new belt drive applications, the drive design process can be greatly simplified by using belt drive design software available from many manufacturer's web sites. Older belt drive systems may potentially be overdesigned with the increased load capacity of modern V-belts.

Frequency Method

The frequency method for measuring V-belt tension improves accuracy, saves time, and simplifies the process of checking belt tension. A belt span is "plucked" or impacted so that an electronic tension gauge can measure the frequency of the vibrating span and calculate the belt tension level. Comparing the calculated belt tension level with manufacturer's recommendations will indicate whether belt tension is too loose or too tight. See Fig. 7.15 for an example of the Gates Corporation's Sonic Tension Meter 507C.

FIGURE 7.15 Sonic tension meter tensioning.

TROUBLESHOOTING V-BELTS

Use the following guide as an aid in determining possible causes of belt drive problems.

Step 1: Describe the problem
- What is wrong?
- When did it happen?
- How often does it happen?
- What is the drive application?
- Has the machine operation or output changed?
- What type of belts are being used?
- How is the belt expected to perform in this application?

Step 2: Identify symptoms and record observations of anything unusual

Drive symptoms check list (check those observed)
- **Premature belt failure**
  - Broken belt(s)
  - Belt(s) fail to carry load (slip); no visible reason
  - Edge cord failure
  - Belt delamination or undercord separation
- **Severe or abnormal belt wear**
  - Wear on belt top surface
  - Wear on top corners of belt
  - Wear on belt sidewalls
  - Wear on belt bottom corners
  - Wear on bottom surface of belt
  - Undercord cracking
  - Sidewall burning or hardening
  - Belt surface hard or stiff
  - Belt surface flaking, sticky or swollen
  - Excessive belt stretching
- **Banded (joined) belt problems**
  - Tie-band separation
  - Top of tie-band frayed, worn, or damaged
  - Banded belt comes off sheaves repeatedly
  - One or more belt ribs run out of the sheave
- **V-belt turns over or jumps off sheave**
  - Involves single or multiple belts
- **Belt stretches beyond takeup**
  - Multiple belts stretch unequally
  - Single belt or where all belts stretch evenly
- **Belt noise**
  - Belt squeals or chirps
  - Slapping sound
  - Rubbing sound
  - Grinding sound
  - Unusually loud drive
- **Unusual vibration**
  - Belts flopping
  - Unsusal or excessive vibration
- **Problems with sheaves**
  - Broken or damaged sheave
  - Severe groove wear

- **Problems with other drive components**
  - Bent or broken shafts
- **Hot bearings**
  - Drive requires overtensioning
  - Sheaves too small
  - Poor bearing condition
  - Sheaves mounted too far out on shaft
  - Belt slippage
- **Performance problems**
  - Incorrect driven speed

V-Belt Drive Symptoms

Premature Belt Failure

Symptoms	Probable cause	Corrective action
• Broken belt(s)	1. Under-designed drive 2. Belt rolled or pried onto sheave 3. Object falling into drive 4. Severe shock load	1. Redesign to manufacturers recommendations 2. Use drive takeup when installing 3. Provide adequate guard or drive protection 4. Redesign to accommodate shock load
• Belts fail to carry load, no visible reason	1. Under-designed drive 2. Damaged tensile member 3. Worn sheave grooves 4. Center distance movement	1. Redesign to manufacturers' recommendations 2. Follow correct installation procedure 3. Check for groove wear; replace as needed 4. Check drive for center distance movement during operation
• Edge cord failure	1. Sheave misalignment 2. Damaged tensile member	1. Check alignment and correct 2. Follow correct installation procedure
• Belt delamination or undercord separation	1. Sheaves too small for belt section 2. Use of too small backside idler	1. Check drive design, replace with larger sheaves 2. Increase backside idler to acceptable diameter

Severe or Abnormal Belt Wear

Symptoms	Probable cause	Corrective action
• Wear on top surface of belt	1. Belt rubbing against guard 2. Idler malfunction	1. Repair or replace guard 2. Replace or repair idler
• Wear on top corners of belt	1. Belt-to-sheave fit incorrect (belt too small for groove)	1. Use correct belt/sheave match
• Wear on belt sidewalls	1. Belt slip 2. Sheave Misalignment 3. Worn sheaves 4. Incorrect belt	1. Retension until slipping stops 2. Realign drive 3. Replace sheaves 4. Replace with correct belt size
• Wear on belt bottom corners	1. Belt-to-sheave fit incorrect 2. Worn sheaves	1. Use correct belt/sheave match 2. Replace sheaves
• Wear on bottom surface of belt	1. Belt bottoming against sheave groove bottom 2. Worn sheaves 3. Debris in sheaves	1. Use correct belt/sheave match 2. Replace sheaves 3. Clean sheaves
• Undercord cracking	1. Sheaves too small for belt section 2. Belt slip 3. Backside idler diameter too smalll 4. Improper belt storage	1. Use larger diameter sheaves 2. Retension to manufacturer's recommendations 3. Increase backside idler to acceptable diameter 4. Don't coil belt too tightly, kink, or bend. Avoid heat and direct sunlight

Severe or Abnormal Belt Wear (*Continued*)

Symptoms	Probable cause	Corrective action
• Sidewall burning or hardening	1. Belt slipping 2. Worn sheaves 3. Under designed drive 4. Shaft movement	1. Retension until slipping stops 2. Replace sheaves 3. Redesign to manufacturer's recommendations 4. Check for center distance changes
• Belt surface hard or stiff	1. Hot drive environment	1. Improve ventilation to drive
• Belt surface flaking, sticky, or swollen	1. Oil or chemical contamination	1. Do not use belt dressing; eliminate sources of oil, grease, or chemical contamination
Excessive belt stretching	1. Belt slipping 2. Worn sheaves 3. Underdesigned drive	1. Retension until slipping stops 2. Replace sheaves 3. Redesign to manufacturer's recommendations

Problems with Banded (Joined) Belts

• Tie band separation	1. Worn or incorrect sheaves 2. Improper groove spacing	1. Replace sheaves 2. Use sheaves manufactured to industry specifications

Problems with Banded (Joined) Belts (*Continued*)

Symptoms	Probable cause	Corrective action
• Top of tie band frayed, worn, or damaged	1. Interference with guard 2. Backside idler malfunction or damaged	1. Check and adjust guard 2. Replace or repair backside idler
• Banded belt comes off sheaves repeatedly	1. Debris in sheaves 2. Sheave misalignment	1. Clean grooves and use single belts to prevent debris from being trapped in grooves 2. Realign drive
• One or more belt ribs run out of the sheave	1. Sheave misalignment 2. Belt undertensioned	1. Realign drive 2. Retension belts to manufacturers' recommendations

V-Belt Turns Over or Come Off Sheave

Symptoms	Probable cause	Corrective action
• Involves single or multiple belts	1. Shock loading or vibration 2. Foreign material in grooves 3. Sheave misalignment 4. Worn sheave grooves 5. Damaged tensile member 6. Incorrectly placed flat idler 7. Mismatched belt set 8. Poor equipment structural design	1. Check drive design; use banded (joined) belts 2. Shield grooves and drive 3. Realign drive 4. Replace sheaves 5. Use correct installation tension and storage procedure 6. Place flat idler on slack side of drive close to driveR sheave 7. Replace with new matched set; do not mix old and new belts. 8. Check for center distance stability and rigidity

Belt Stretches Beyond Available Takeup

Symptoms	Probable cause	Corrective action
• Multiple belts stretch unequally	1. Misaligned drive 2. Debris in sheaves 3. Broken tensile member or cord 4. Mismatched belt set 5. Belts from different manufacturers used	1. Realign drive and retension belts 2. Clean sheaves 3. Replace all belts; install properly 4. Install matched belt set 5. Replace all belts with belts made by same manufacturer
• Single belt or where all belts stretch evenly	1. Insufficient takeup allowance 2. Grossly overloaded or under designed drive 3. Broken tensile members	1. Check takeup; use allowance specified by manufacturers 2. Redesign to manufacturers' recommendations 3. Replace belt or entire belt set and install properly

Belt Noise

Symptoms	Probable cause	Corrective action
• Belt squeals or chirps	1. Belt slip 2. Contamination	1. Retension to manufacturers' recommendations 2. Clean belts and sheaves
• Slapping sound	1. Loose belts 2. Mismatched belt set 3. Misalignment	1. Retension to manufacturers' recommendations 2. Install matched belt set 3. Realign drive so all belts share load equally
• Rubbing sound	1. Guard interference	1. Repair, replace, or redesign guard
• Grinding sound	1. Damaged bearings	1. Replace, align, and lubricate
• Unusually loud drive	1. Incorrect belt for sheaves 2. Incorrect tension 3. Worn sheaves 4. Debris in sheaves	1. Use correct belt size and type 2. Check belt tension and adjust 3. Replace sheaves 4. Clean sheaves; improve shielding; remove rust, paint; or remove dirt from grooves

Unusual Vibration

Symptoms	Probable cause	Corrective action
• Belts flopping	1. Loose belts (under tensioned) 2. Mismatched belts 3. Misaligned drive	1. Retension to manufacturers recommendations 2. Install new matched belt set 3. Realign drive
• Unusual or excessive vibration	1. Incorrect belt 2. Poor equipment structural design 3. Excessive sheave eccentricity 4. Loose drive components	1. Use correct belt/sheave match 2. Check structure for adequate strength and rigidity 3. Replace defective sheave 4. Check machine components, guards, motor mounts, motor pads, bushings, brackets, and framework for adequate strength and stability and proper installation

Problems with Sheaves

Symptoms	Probable cause	Corrective action
• Broken or damaged sheaves	1. Incorrect sheave installation 2. Foreign objects falling in drive 3. Incorrect belt installation	1. Do not over tighten bushing bolts 2. Use adequate drive guard 3. Do not pry belts onto the sheaves

Problems with Other Drive Components

Symptoms	Probable cause	Corrective action
• Bent or broken shafts	1. Extreme belt overtension 2. Overdesigned drive 3. Accidental damage 4. Machine design error 5. Sheave mounted too far away from outboard bearing	1. Retension to manufacturers' recommendations 2. Redesign to manufacturers' recommendations 3. Redesign drive guard 4. Check machine design 5. Move sheaves closer to outboard bearing

Hot Bearings

Symptoms	Probable cause	Corrective action
• Drive requires overtensioning	1. Worn sheave grooves—belts bottoming and won't transmit power until overtensioned 2. Improper belt tension	1. Replace sheaves and tension belts properly 2. Retension to manufacturers' recommendations
• Sheaves too small	1. Follow NEMA motor manufacturer's recommendations	1. Redesign drive using proper sheave diameters
• Poor bearing condition	1. Bearings underdesigned 2. Bearings not properly maintained	1. Check bearing selection 2. Align and lubricate bearings
• Sheaves mounted too far out on shaft • Belt slippage	1. Drive installation error	1. Move sheaves as close to outboard bearings as possible
	1. Belts undertensioned	1. Retension to manufacturers' recommendations

Performance Problems

Symptoms	Probable cause	Corrective action
• Incorrect driven speed	1. Drive design error 2. Belt slip	1. Redesign drive using correct sheaves sizes for desired speed ratio 2. Retension to manufacturers' recommendations

SYNCHRONOUS BELTS

Synchronous belts are toothed belts in which power is transmitted through positive engagement between belt teeth and pulley or sprocket grooves rather than by the wedging friction of V-belts. The positive drive characteristics of these belts provide exact synchronization between driveR and driveN shafts and also increase power transmission efficiency. Other advantages of synchronous belts over other modes of power transmission include a wider load/speed range, lower maintenance, increased wear resistance, and a smaller amount of required takeup.

Modified Curvilinear

Curvilinear

Timing

FIGURE 7.16 Synchronous belt profiles.

The three general families of synchronous belts are modified curvilinear, curvilinear, and trapezoidal, See Fig. 7.16.

Modified Curvilinear Belts

The modified curvilinear belt tooth form is a refinement of the curvilinear system. The belt tooth and sprocket groove forms were optimized for smoother belt tooth entry/exit properties and improved belt tooth support in the sprocket grooves. The end results are improved belt load capacity, improved drive registration accuracy, and quieter operation.

Curvilinear Belts

The curvilinear belt tooth form was developed to provide increased load capacity and performance over trapezoidal belts. Innovations in tooth profile design and increased tooth depth resulted in better stress distribution in the belt teeth and increased resistance to tooth ratcheting. The curvilinear belt consequently has a higher horsepower capacity than does a comparably sized trapezoidal belt. They are also readily available by a wide variety of manufacturers.

Timing Belts

Timing belts were the first family of synchronous belts introduced to the market and were designed with trapezoidal teeth. The drive system was developed as a means of synchronizing shafts and reducing belt drive maintenance. Timing belts have been around for many years and are readily available by a wide variety of manufacturers. The belt horsepower ratings are relatively low compared to curvilinear or modified curvilinear belts introduced later, but the synchronization qualities are excellent for accurate positioning or registration sensitive applications.

Tooth Pitch and Profiles

Fifteen standard curvilinear and modified curvilinear belt sections currently exist. They are specified on the basis of tooth pitch and belt tooth profile. Refer to the drawings shown below for each belt type.

SYNCHRONOUS BELT TOOTH PROFILES

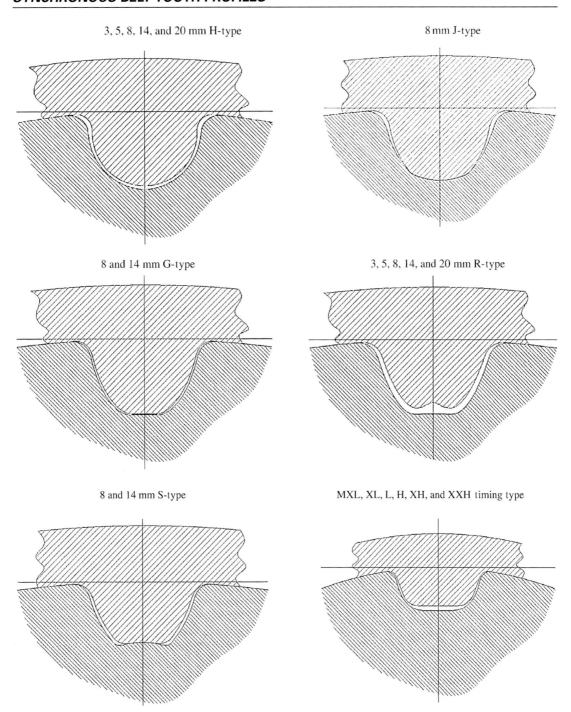

3, 5, 8, 14, and 20 mm H-type

8 mm J-type

8 and 14 mm G-type

3, 5, 8, 14, and 20 mm R-type

8 and 14 mm S-type

MXL, XL, L, H, XH, and XXH timing type

SYNCHRONOUS BELT NOMENCLATURE

Synchronous belts are identified by their pitch length, profile type, tooth pitch (distance between adjacent teeth), and top width.

Belt designations vary by belt manufacturers, but some common general formats are included in Table 7.9 below.

Synchronous belt manufacturers may offer multiple belt types in numerous belt constructions. Belts with the same profile type may perform very differently if manufactured in different belt constructions, or by different manufacturers. When acquiring belt replacements, manufacturers' literature should be consulted to ensure that equivalent products are obtained.

Synchronous belts are most commonly available in conventional single-sided versions. For serpentine applications requiring shaft rotation reversals, double-sided belt constructions are available. Double-sided belt constructions may not be available in all profile types or pitch sizes.

While some manufacturers promote belt interchangeability with other profile types, true interchangeability is quite limited. The best drive system performance can be obtained by matching the correct belt profile with the corresponding sprockets. Each system was designed to perform as such, and all belts perform best when operating in the correct sprockets.

Standard synchronous belt widths and tolerances are shown in Tables 7.10 and 7.11, respectively.

TABLE 7.9 Synchronous Belt Nomenclature

Belt part number	Profile type	Belt pitch	Belt length	Belt width
8MGT-1400-21	G	8 mm	1400 mm	21 mm
1400-8MGT-20	J	8 mm	1400 mm	20 mm
1400-8M-20	H or R	8 mm	1400 mm	20 mm
200-S8M-1400	S	8 mm	1400 mm	20 mm
540H100	Timing	0.500 in	54.0 in	1.0 in

TABLE 7.10 Standard Belt Widths

Belt pitch	Nominal belt widths, mm			
	G-type		H-, J-, and R-type	S-type
	Polyurethane	Rubber		
3M	—	—	6	—
	—	—	9	—
	—	—	15	—
5M	—	—	9	—
	—	—	15	—
	—	—	25	—
8M	12	—	20	15
	21	—	30	25
	36	—	50	40
	62	—	85	60
14M	20	40	40	40
	37	55	55	60
	68	85	85	80
	90	115	115	100
	125	170	170	120
20M	—	—	115	—
	—	—	170	—
	—	—	230	—
	—	—	290	—
	—	—	340	—

TABLE 7.11 Standard Belt Width Tolerances

Belt width, mm	Tolerance on Width for Belt Pitch Lengths		
	Up to and including 840 mm	Over 840 mm up to and including 1680 mm	Over 1680 mm
3–11	+0.4 / −0.8	+0.4 / −0.8	— / —
11.1–38	+0.8 / −0.8	+0.8 / −1.2	+0.8 / −1.2
38.1–50.8	+0.8 / −1.2	+1.2 / −1.2	+1.2 / −1.6
50.9–63.5	+1.2 / −1.2	+1.2 / −1.6	+1.6 / −1.6
63.6–76.5	+1.2 / −1.6	+1.6 / −1.6	+1.6 / −2.0
76.6–101.6	+1.6 / −1.6	+1.6 / −2.0	+2.0 / −2.0
101.7–177.8	+2.4 / −2.4	+2.4 / −2.8	+2.4 / −3.2
Over 177.8	— / —	— / —	+4.8 / −6.4

SYNCHRONOUS SPROCKETS

Synchronous sprockets are identified by the number of grooves, pitch, and the nominal face width. Sprocket designations follow a similar format to belts.

Sprocket designations vary by belt manufacturers, but some common general formats are included in Table 7.12.

Minimum Sprocket Diameters

As synchronous belts bend around sprockets, tensile and compressive forces develop within the belt tensile cords, ultimately impacting belt fatigue life. Belt tensile cord fatigue increases as sprocket diameter decreases. In addition, as sprocket diameter decreases, smooth belt tooth entry and exit properties become more difficult to maintain. Therefore, minimum recommended sprocket diameters were developed for each belt pitch size and are included in Table 7.13. Using sprockets that are below minimum recommended sizes will always reduce belt service.

Synchronous belt drives are often used in applications that are powered by electric motors. Motor bearing sizing is influenced by overhung loads exerted by belt drive systems. Overhung load values are dependent upon sprocket diameters. As motor sprocket diameters decrease, the overhung load increases. Therefore, NEMA publishes a standard that specifies minimum recommended sprocket diameters for all common motor frame sizes. Designing belt drives that use sprockets below these minimums could result in reduced motor bearing life. NEMA recommended minimum sprocket sizes are listed in Table 7.14.

TABLE 7.12 Synchronous Sprocket Nomenclature

Sprocket part number	Profile type	Sprocket pitch	Number of grooves	Face width
8MX-40S-21	G	8 mm	40	21 mm
P40-8MGT-20	J	8 mm	40	20 mm
P40-8M-20	H or R	8 mm	40	20 mm
P40-S8M-200	S	8 mm	40	20 mm
40H100	Timing	0.500 in	40	1.0 in

TABLE 7.13 Minimum Recommended Sprocket Diameter

Belt pitch	Min. recommended sprocket size	
	(no. of teeth)	(in)
Timing belts		
MXL	12	0.31
XL	12	0.76
L	12	1.43
H	14	2.23
XH	18	5.01
XXH	18	7.16
H and R-type pulleys		
3M	12	0.45
5M	14	0.88
8M	22	2.21
14M	28	4.91
20M	34	8.52
S and G-type		
8M	22	2.21
14M	28	4.91
J-type		
8M	22	2.21

TABLE 7.14 Minimum Recommended Sprocket Diameters for General Purpose Electric Motors

Motor horsepower	Motor rpm (60-cycle and 50-cycle electric motors)						Motor horsepower
	575 485*	690 575*	870 725*	1160 950*	1750 1425*	3450 2850*	
1/2	—	—	2.0	—	—	—	1/2
3/4	—	—	2.2	2.0	—	—	3/4
1	2.7	2.3	2.2	2.2	2.0	—	1
1 1/2	2.7	2.7	2.2	2.2	2.2	2.0	1 1/2
2	3.4	2.7	2.7	2.2	2.2	2.2	2
3	4.1	3.4	2.7	2.7	2.2	2.2	3
5	4.1	4.1	3.4	2.7	2.7	2.2	5
7 1/2	4.7	4.1	4.0	3.4	2.7	2.7	7 1/2
10	5.4	4.7	4.0	4.0	3.4	2.7	10
15	6.1	5.4	4.7	4.0	4.0	3.4	15
20	7.4	6.1	5.4	4.7	4.0	4.0	20
25	8.1	7.4	6.1	5.4	4.0	4.0	25
30	9.0	8.1	6.1	6.1	4.7	—	30
40	9.0	9.0	7.4	6.1	5.4	—	40
50	9.9	9.0	7.6	7.4	6.1	—	50
60	10.8	9.9	9.0	7.2	6.7	—	60
75	12.6	11.7	8.6	9.0	7.7	—	75
100	16.2	13.5	10.8	9.0	7.7	—	100
125	18.0	16.2	13.5	10.8	9.5#	—	125
150	19.8	18.0	16.2	11.7	9.5	—	150
200	19.8	19.8	19.8	—	11.9	—	200
250	19.8	19.8	—	—	—	—	250
300	24.3	24.3	—	—	—	—	300

*These RPM are for 50 cycle electric motors.
#Use 8.6 for frame number 444 T only.

SYNCHRONOUS APPLICATION GUIDELINES

Synchronous belt drives are efficient power transmission systems and require minimal maintenance. However, users should consider the unique aspects of synchronous belt drive applications:

1. Synchronous belt horsepower ratings are based on at least six belt teeth in mesh with sprockets or 60° of belt wrap around each loaded sprocket. The belt must be derated when this cannot be achieved.
2. Synchronous belts and sprockets in different profile types should not be mixed. True interchangeability for optimum performance is quite limited.
3. Synchronous belt drives should be considered for increased system efficiency. A well-maintained synchronous belt drive is generally 98 to 99 percent efficient.
4. The precision of shaft synchronization with synchronous belt drives varies with profile type. Contact individual manufacturers for specific values. Interchanging profile types nearly always results in decreased system accuracy.
5. Synchronous belts are free to track across sprocket faces and nearly always track toward one side or the other. Therefore, a synchronous belt must be restrained on both sides by sprocket flanges.
6. Synchronous belts are more sensitive to shaft or sprocket misalignment than V-belts. Synchronous belt drives should not be used in applications with severe inherent misalignment.
7. Installing synchronous belts with the recommended level of belt installation tension is very important in achieving optimum performance. Under tensioning may allow belt ratcheting (tooth jumping). Both over and under tensioning will accelerate belt tooth wear.

The application of these guidelines and characteristics will optimize synchronous belt drive performance.

SYNCHRONOUS BELT INSTALLATION

Sprockets most commonly used in the belt drive industry generally use either QD (quick disconnect) or TL (taper lock) bushings for secure shaft mounting. Though both shaft mounting systems have been used within the belt drive industry for many years, proper installation procedures may not always be followed. Proper bushing installation is critical for proper and safe synchronous belt drive operation and are provided below for reference.

QD Type Sprockets   **TL Type Sprockets**

Sprocket/Bushing Installation

1. Preliminarily mount the new bushings and sprockets on the shafts.

   QD-type sprockets

   Determine whether to mount each sprocket in a conventional or reverse manner. (Conventional is easier, but positions the sprocket further out on the shaft.)

   Conventional mount (shown right)
   a. Insert the key into the keyway on the shaft.
   b. Insert a screw driver blade into the bushing slot to enlarge the bore slightly.
   c. Slide the bushing onto the shaft with the flange side towards the equipment.
   d. Place the sprocket onto the bushing and insert the cap screws. Align the unthreaded holes in the sprocket hub with the threaded holes in the bushing flange. Finger-tighten the screws.

   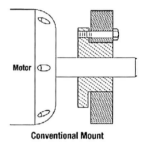
   Conventional Mount

   Reverse mount (shown right)
   a. Insert the key into the keyway on the shaft.
   b. Place the sprocket on to the shaft without the bushing.
   c. Insert a screw driver blade into the bushing slot to enlarge the bore slightly.
   d. Slide the bushing onto the shaft with the flange facing outward, away from equipment.
   e. Place the sprocket onto the bushing and insert the cap screws. Align the unthreaded holes in the bushing flange with the threaded holes in the sprocket hub. Finger-tighten the screws.

   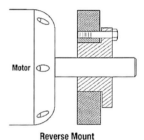
   Reverse Mount

   TL-type sprockets
   a. Insert the key into the keyway on the shaft.
   b. Insert a screw driver blade into the bushing slot to enlarge the bore slightly.
   c. Place bushing in sprocket hub and align the holes; all holes should be half threaded.
   d. Insert cap screws loosely into the tightening holes (sprocket side of holes threaded).
   e. Slide the assembly onto the shaft with the bushing side facing outward; away from equipment.
   f. Tighten the cap screws finger tight.

2. Slide the sprockets onto the shafts as far as possible (close to the motor/equipment) while remaining within parallel alignment and remaining in full contact with the shaft key.
3. Note which of the sprockets reach a stopping point on the shaft first and prepare to secure it by tightening the bushing cap screws.
4. Secure the first sprocket to the shaft by tightening the bushing using either of the following two procedures:

   QD bushings
   a. Tighten the set screw down against the shaft key.
   b. Using a torque wrench, tighten the bushing cap screws evenly in an alternating pattern until approximately one half of the recommended torque level has been reached.
   c. Check parallel sprocket alignment and sprocket lateral runout (wobble), and correct as necessary.

d. Continue alternate tightening of the cap screws to the recommended torque level. When the recommended torque level is reached, do not tighten the cap screws further. Further tightening may result in over bushing insertion and sprocket hub fracture.

   TL bushings
   a. Obtain appropriate hex key sockets or conventional hex head sockets for the bushing cap screws.
   b. Make sure the hex key sockets are not worn excessively, or slippage within the socket head cap screw head may occur. Wrench slippage will result in damage that may prevent proper tightening and/or removal.
   c. Using a torque wrench, tighten the bushing cap screws evenly in an alternating pattern until the recommended torque value has been reached.
   d. After reaching the recommended torque level, further tighten the bushing by driving it further into the sprocket hub using a suitable drift or punch.
   e. After tapping the bushing face several times with the drift, re-torque the cap screws to their recommended value.

5. Place the second sprocket in alignment with respect to the first using a string, straight edge, laser alignment tool, etc.
6. Accounting for axial sprocket movement that occurs as the bushing cap screws are tightened, lightly tighten the cap screws to hold the sprocket/bushing assembly in place.
7. Using a string, straight edge, laser alignment tool, etc. check both parallel and angular sprocket alignment.
8. Make the necessary adjustments in the placement of the second sprocket to correct any parallel misalignment. Remember that synchronous belt drives can tolerate some parallel sprocket misalignment so long as the belt is not pinched between opposite sprocket flanges.
9. Place the belt on the sprockets and apply light tension by adjusting the motor base, etc.
10. Slowly turn the drive over by hand to allow the belt teeth to seat in the sprocket grooves. Several belt revolutions are generally sufficient.
11. Using a string, straight edge, laser alignment tool, etc. check for angular shaft misalignment while adjusting the motor base and applying tension to the belt.
12. Measure the static belt installation tension either directly with an electronic tension gauge, or indirectly using the force/deflection method (belt tension values are available from belt manufacturers).
13. After reaching the desired static belt installation tension, tighten all mounting fasteners to rigidly secure the motor and related equipment.
14. Turn the drive over by hand again to make sure that the belt is fully seated in the sprockets. While the drive is rotating, observe the belt tracking characteristics. Make sure that the belt doesn't track across the sprocket face too fast (should require approximately three belt revolutions or more to track across sprocket face). Rapid belt tracking indicates angular misalignment.
15. Recheck the static belt installation tension along with angular and parallel misalignment, and make any necessary adjustments.
16. Replace the drive safety guard.
17. While watching and listening carefully, start the drive and confirm that all operating characteristics are normal.

MAINTENANCE

A properly designed and maintained synchronous belt drive will provide long trouble-free service. To maximize synchronous belt drive performance, it is important to follow good maintenance practices. The required maintenance frequency may be influenced by many factors including the operating environment, drive operating speed, load duty cycle, critical nature of the equipment, and belt drive accessibility. Severe operating conditions may require additional inspection and maintenance. The following suggested maintenance intervals should be considered:

1. For critical drives, a quick visual and hearing inspection should be performed every 1 to 2 weeks.
2. For normal drives, a quick visual and hearing inspection should be performed once every month.
3. For normal drives, a shutdown to thoroughly inspect belts, sprockets, and other components should be performed every 3 to 6 months.

Visual Inspection

A visual and hearing inspection consists of the following three items:

1. Look and listen for unusual noise and vibration while observing the drive. A well-designed and well-maintained drive will operate quietly and smoothly.
2. Inspect the guard for looseness and damage. Make sure it is clean. Any accumulation of foreign material on the guard acts as insulation and could cause excessive heat buildup in the drive.
3. Look for oil and grease contamination that could degrade the material components of a belt.

A thorough belt drive inspection requires a complete shut down and should include an inspection of the belt, sprockets, belt tension, the belt guard and associated drive components such as bearings, shafts, and takeup rails. Whenever a drive system is inspected thoroughly, use safety precautions to avoid injury. Shut the power off, lock and tag the control box, and place all machine components in a neutral position.

The primary drive troubleshooting tool is belt inspection. Unusual belt wear is a symptom of potential drive problems. Make a mark on the belt, and inspect the entire circumference of the belt, checking for uneven belt wear, cracks, a frayed jacket, edge wear, and swelling, and hardening. Also check for cracked or lost teeth.

Damage to synchronous belts can sometimes be attributed to excessive heat. Properly functioning synchronous belts operate at approximately 140°F (60°C) and should be able to be touched comfortably. Belts that are hot to the touch indicate potential problems.

Sprocket Wear

Sprockets should be checked for wear and for proper mounting and alignment. Inspect sprockets for wear, nicks, and sharp edges. Wear on sprocket running surfaces around the circumference is especially critical, and can be recognized as a ledge between where the belt has been running and the original surface. See Fig. 7.17. Any visible sprocket wear is generally cause for replacement.

Misalignment

Misalignment of synchronous belt drive systems is usually attributed to three sources:

1. DriveR and driveN shafts are not parallel (both horizontal and vertical planes).
2. Sprockets are not properly located in line axially with respect to one another on the shafts.
3. Sprockets are tilted due to improper mounting (wobble while running).

The effects of these conditions can be illustrated as both parallel and angular misalignment (see Fig. 7.18). Sprockets can be checked for tilting using a bubble level. Axial location of pulleys and shaft parallelism can be checked using a straightedge or a string. The total allowable misalignment (angular + parallel) in synchronous belt drive systems should not exceed $1/4°$ or $1/16$ in per foot of belt span.

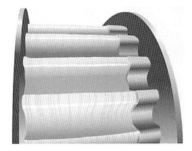

FIGURE 7.17 Sprocket tooth wear picture.

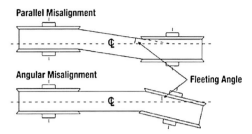

FIGURE 7.18 Types of belt misalignment.

Other Drive Components

Check guards for wear, damage, and contamination. Guard wear suggests possible interference with belts or sprockets. Clean the guard to prevent airflow restriction and to allow efficient heat dissipation. Check bearings for correct alignment and lubrication, make sure that drive components are securely tightened down, and ensure that guiderails are free of dirt, obstructions, and corrosion.

Installation and Takeup Allowance

Make certain that adequate shaft movement is provided for belt installation and takeup (see Fig. 7.19). Recommended allowance values are available from belt manfacturers (see Tables 7.15 and 7.16).

Care and Handling

Never pry or roll synchronous belts onto the sprockets. This will cause invisible damage to the tensile cords and reduce belt life. Careful handling is important with synchronous belts. The higher modulus tensile cords used in most synchronous belts can be damaged by back bending or careless handling. *Do not crimp* a belt by bending it sharply forward or backward (bending the belt to a diameter smaller than the smallest recommended sprocket size) will likely damage the belts tensile members and reduce belt life. Belts should be stored on a flat surface rather than hung on a hook, especially for larger, heavier, belts.

Predictive Maintenance

Predictive maintenance will minimize costly, unplanned downtime. Regular replacement intervals should be established so that a belt is replaced near the end of its useful life. Replacement intervals may be established to correspond with a yearly shut down period. The typical life of an industrial synchronous belt drive ranges from 3 to 5 years depending on the application. In more critical applications, the belt replacement interval should be at least once per year.

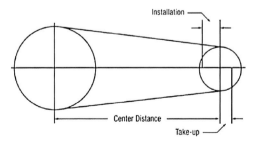

FIGURE 7.19 Installation and takeup allowance.

TABLE 7.15 Installation and Takeup Allowance

Length Belt	(mm) (in)	Standard Installation Allowance (Flanged Sprockets Removed for Installation)	(mm) (in)	Tension Allowance (All Drives)	(mm) (in)
Up to	125 **5**	0.5 **0.02**		0.5 **0.02**	
Over 125 **5** to	250 **10**	0.8 **0.03**		0.8 **0.03**	
Over 250 **10** to	500 **20**	1.0 **0.04**		0.8 **0.03**	
Over 500 **20** to	1000 **40**	1.8 **0.07**		0.8 **0.03**	
Over 1000 **40** to	1780 **70**	2.8 **0.10**		0.8 **0.04**	
Over 1780 **70** to	2540 **100**	3.3 **0.13**		1.0 **0.04**	
Over 2540 **100** to	3300 **130**	4.1 **0.16**		1.3 **0.05**	
Over 3300 **130** to	4600 **180**	4.8 **0.19**		1.3 **0.05**	
Over 4600 **180** to	6900 **270**	5.6 **0.22**		1.3 **0.05**	

TABLE 7.16 Installation and Takeup Allowance for flanges

Pitch	Additional Center Distance Allowance For Installation Over Flanged Sprockets*			
	One sprocket flanged	(mm) (in)	Both sprockets flanged	(mm) (in)
0.080* (MXL)	8.4 **0.33**		12.4 **0.49**	
0.200* (XL)	11.7 **0.46**		18.0 **0.71**	
0.375* (L)	16.3 **0.64**		21.8 **0.85**	
0.500* (H)	16.3 **0.64**		24.4 **0.96**	
5 mm	13.5 **0.53**		19.1 **0.75**	
8 mm	21.8 **0.86**		33.3 **1.31**	
14 mm	31.2 **1.23**		50.0 **1.97**	
20 mm	47.0 **1.85**		77.5 **3.25**	

*For drives that require installation of the belt over one sprocket at a time, use the value for "Both Sprockets Flanged."

BELT TENSION

An overview of the synchronous belt tensioning process may be summarized with a few simple rules:

1. Synchronous belts can be more sensitive to belt installation levels than V-belts. The force-deflection method described below can be used to measure tension levels, a direct force can be applied against the motor or an idler, or the frequency method can be used with an electronic tension gauge. Recommended belt installation levels are available from belt manufacturers.

2. When installing new belts, set belt tension to the recommended level, rotate the drive for a few revolutions, and then recheck the tension level. Check the belt tension level once more after having run the drive for approximately 1 hour.

3. Synchronous belts do not require periodic retensioning like V-belts. After the installation tension is set correctly, no further retensioning procedures should be necessary. A periodic check, however, may be necessary to check for structural shifts or movement, or other component wear.

Force-Deflection Method

The most common method of measuring belt tension levels is the force-deflection method. This method translates the static tension (belt tension level when the drive is at rest) to a specific belt deflection distance with the application of a deflection force. The standard deflection distance is $1/64$ in per inch of belt span length. The force should be applied perpendicular to the belt at midspan (see Fig. 7.20). The force required should fall within a calculated range in order for the belts to be properly tensioned. The deflection force is generally applied using a spring type tension gauge, see Fig. 7.21.

When making belt tension measurements if the actual force required to deflect the belt is above the recommended range, the belt is too tight. Likewise, if the actual force is lower than the recommended range, the belt is too loose. Ideally, when tensioning synchronous belts, the deflection force should be set towards the high end of the recommended range to allow for normal belt tension decay over time.

Recommended belt deflection force and distance values can be calculated using formulas published in belt manfacturers design manuals, or can be obtained from manufacturers drive design software. Required belt tension calculations are based on the drive load requirements, operating speeds, sprocket diameters, center distance, and the belt type.

For new belt drive applications, the drive design process can be greatly simplified by using belt drive design software available from many manufacturer's web sites.

Frequency Method

The frequency method for measuring synchronous belt tension improves accuracy, saves time, and simplifies the process of checking belt tension. A belt span is "plucked" or impacted so that an electronic

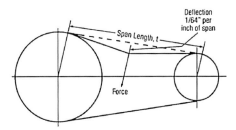

FIGURE 7.20 Force deflection method of tensioning.

5.136   MAINTENANCE OF MECHANICAL EQUIPMENT

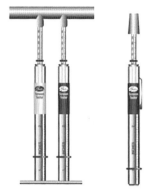

FIGURE 7.21 Pencil-type tension tester.

FIGURE 7.22 Sonic tension meter.

tension gauge can measure the frequency of the vibrating span and calculate the belt tension level. Comparing the calculated belt tension level with manufacturer's recommendations will indicate whether belt tension is too loose or too tight. See Fig. 7.22 for an example of the Gates Corporation's Sonic Tension Meter 507C.

TROUBLESHOOTING SYNCHRONOUS BELTS

Use the following guide as an aid in determining possible causes of drive problems.

Step 1: Describe the problem.
- What is wrong?
- When did it happen?
- How often does it happen?
- What is the drive application?
- Have the machine operations or output changed?
- What kind of belt(s) are you using?
- How is the belt expected to perform in this application?

Step 2: Identify symptoms and record observations of anything unusual.

Drive symptoms check list (check those you observe).

- **Premature belt failure**
  - Broken belt or tensile failure
  - Edge cord pullout
  - Belt delamination or tooth shear
  - Belt ratcheting or jumping teeth
- **Severe or abnormal belt wear**
  - Belt edge wear
  - Premature belt tooth wear
  - Land area worn
  - Belt cracking
  - Belt flaking, sticky or swollen

- **Abnormal belt tracking behavior**
  - Excessive side tracking force
  - Sprocket flange failure
  - Belt running partially off unflanged sprocket
- **Unusual vibration**
  - Excessive belt span vibration
  - Vibration during drive start-up, shut-down, or specific speed range
- **Problems with belt takeup or tensioning**
  - Belt tension loss
  - Belt stretch
  - Center distance collapse
- **Unusual noise**
  - Rubbing sound
  - Grinding sound
  - Unusually loud drive
  - Popping sound
- **Problems with sprockets**
  - Rust or corrosion
  - Severe or rapid tooth wear
- **Problems with drive components**
  - Bent or broken shafts
- **Hot bearings**
  - Drive requires over tensioning
  - Sprockets too small
  - Poor bearing condition
  - Sprockets mounted too far out on shaft
- **Performance problems**
  - Incorrect driven speed

Synchronous Belt Drive Symptoms

Premature Belt Failure

Symptoms	Probable cause	Corrective action
• Broken belt or tensile failure	1. Excessive shock load 2. Sub-minimal diameter 3. Improper belt handling and storage prior to installation 4. Debris or foreign object in drive 5. Extreme sprocket run-out 6. Too low or too high belt tension	1. Redesign to manufacturer's recommendations 2. Redesign drive using larger sprockets 3. Follow proper handling and storage procedures 4. Protect drive 5. Replace sprockets 6. Retension to manufacturer's recommendations

5.138 MAINTENANCE OF MECHANICAL EQUIPMENT

Premature Belt Failure (*Continued*)

Symptoms	Probable cause	Corrective action
• Edge cord pullout	1. Sprocket misalignment 2. Damaged tensile member 3. Debris in sprockets 4. Damaged or bent flange	1. Check alignment and correct 2. Follow correct installation procedures and handle belts properly 3. Eliminate and guard against debris 4. Repair or replace flange
• Belt delamination or tooth shear 	1. Excessive shock load 2. Less than 6 teeth-in-mesh 3. Extreme sprocket run-out 4. Worn sprockets 5. Backside idler 6. Incorrect sprocket groove profile 7. Misaligned drive 8. Belt undertensioned	1. Redesign to manufacturers recommendations 2. Redesign drive 3. Replace sprocket 4. Replace sprocket 5. Use inside idler 6. Use proper belt/sprocket combination 7. Realign drive 8. Retension to manufacturer's recommendations
• Belt ratcheting or jumping teeth	1. Belt undertensioned 2. Excessive shock load 3. Drive framework not rigid	1. Retension to manufacturer's recommendations 2. Redesign drive for increased capacity 3. Reinforce to prevent center distance flex

Severe or Abnormal Belt Wear

Symptoms	Probable cause	Corrective action
• Belt edge wear	1. Misaligned drive 2. Damage due to belt mishandling 3. Flange damage 4. Belt too wide for sprocket 5. Rough flange surface finish 6. Improper belt tracking 7. Belt rubbing against guard or drive structure	1. Realign drive 2. Follow proper handling instructions 3. Repair flange or replace sprocket 4. Use proper belt width for sprocket 5. Replace or repair flange 6. Realign drive 7. Remove obstruction or realign drive
• Premature belt tooth wear 	1. Belt tension too low or too high 2. Belt running partly off unflanged sprocket 3. Misaligned drive 4. Incorrect belt/sprocket match 5. Worn sprocket 6. Rough sprocket teeth 7. Damaged sprocket 8. Sprocket does not meet dimensional specifications 9. Belt rubbing against drive bracketry or other obstruction 10. Excessive load 11. Insufficient sprocket surface hardness 12. Debris in sprockets 13. Cocked bushing/sprocket assembly 14. Rust	1. Retension belt to manufacturer's recommendations 2. Realign drive 3. Realign drive 4. Use proper belt/sprocket combination 5. Replace sprocket 6. Replace sprocket 7. Replace sprocket 8. Replace sprocket 9. Remove obstruction or alter belt path 10. Redesign drive to manufacturer's recommendations 11. Replace sprocket with a more wear resistent type 12. Eliminate and guard against debris 13. Install bushings per manufacturer's recommendations 14. Use stainless steel or nickel plated sprockets

Severe or Abnormal Belt Wear (*Continued*)

Symptoms	Probable cause	Corrective action
• Land area worn	1. Excessive tension 2. Excessive sprocket wear 3. Debris in sprockets	1. Retension to manufacturer's recommendations 2. Check sprocket condition and replace if necessary 3. Eliminate and guard against debris
• Belt cracking	1. Sprockets too small for belt section 2. Idler too small for belt section 3. Extreme low temperature startup 4. Extended exposure to chemicals 5. High temperature environment	1. Redesign drive to manufacturer's recommendations 2. Redesign drive to manufacturer's recommendations 3. Preheat drive environment 4. Guard drive from contamination 5. Improve air flow to drive
• Belt flaking, sticky or swollen	1. Extended exposure to chemicals, fluids, oil, or grease	1. Remove source of oil, grease, or chemical contamination; guard drive from contamination

Abnormal Belt Tracking Behavior

Symptoms	Probable cause	Corrective action
• Excessive side tracking force	1. Misaligned drive 2. Center distance exceeds $8 \times$ small sprocket diameter	1. Realign drive 2. Realign drive paying particular attention to both angular and parallel alignment and belt tracking characteristics
• Sprocket flange failure	1. Belt forcing flange off 2. Belt rides over top of flange	1. Realign drive or properly secure flange to sprocket 2. Retension to manufacturer's recommendations

Abnormal Belt Tracking Behavior (*Continued*)

Symptoms	Probable cause	Corrective action
• Belt running partially off unflanged sprocket	1. Misaligned drive 2. Center distance exceeds 8 × small sprocket diameter	1. Realign drive 2. Realign drive paying particular attention to both angular and parallel alignment and belt tracking characteristics

Unusual Vibration

Symptoms	Probable cause	Corrective action
• Excessive belt span vibration	1. Incorrect belt/sprocket match 2. Belt tension too low 3. Loose drive components 4. Excessive sprocket eccentricity 5. Poor machine design 6. Misaligned drive	1. Use correct belt/sprocket match 2. Retension to manufacturer's recommendations 3. Check machine components, guards, motor mounts, motor pads, bushings, brackets, and framework for proper installation, looseness or adequate strength, and rigidity 4. Replace defective sprocket 5. Check structure for adequate strength and rigidity 6. Realign drive
• Vibration during start-up, shut-down, or specific speed range	1. Drive system hitting belt's natural frequency	1. Increase sprocket diameter; retension to manufacturer's recommendation; use different belt type

Problems with Belt Takeup or Tensioning

Symptoms	Probable cause	Corrective action
• Belt tension loss	1. Poor center distance rigidity 2. Excessive sprocket wear 3. Fixed (nonadjustable) centers 4. Excessive belt loading	1. Reinforce the structure 2. Replace sprockets and consider alternative sprocket materials 3. Use idler to tension belt 4. Redesign drive to manufacturer's recommendations
• Belt stretch	1. Under designed drive 2. Damaged tensile members	1. Redesign drive to manufacturer's recommendations 2. Replace damaged belt; ensure proper installation, handling, and storage
• Center distance collapse	1. Poor center distance rigidity 2. Bent or broken shaft	1. Reinforce the structure 2. Replace damaged components

Unusual Noise

Symptoms	Probable cause	Corrective action
• Rubbing sound	1. Belt rubbing against guard	1. Repair or replace guard
• Grinding sound	1. Damaged bearings	1. Replace, align, and lubricate
• Unusually loud drive	1. Incorrect belt/sprocket match 2. Incorrect tension 3. Worn sprockets 4. Debris in sprockets	1. Use correct belt/sprocket match 2. Retension to manufacturer's recommendation 3. Replace sprockets 4. Eliminate and guard against debris
• Popping sound	1. Belt ratcheting	1. Retension to manufacturer's recommendation

Problems with Sprockets

Symptoms	Probable cause	Corrective action
• Rust or corrosion	1. High humidity environment or exposure to steam or water	1. Use nickel plated or stainless steel sprockets
• Severe or rapid tooth wear 	1. Sprocket has too little wear resistance (i.e., plastic, aluminum, softer metals) 2. Misaligned drive 3. Debris in sprockets 4. Grossly overloaded or under designed drive 5. Belt tension too low or too high 6. Incorrect belt/sprocket match	1. Use alternative sprocket material 2. Realign drive 3. Provide adequate guard or drive protection 4. Redesign drive to manufacturer's recommendations 5. Retension to manufacturer's recommendation 6. Use proper belt/sprocket match

Problems with Other Drive Components

Symptoms	Probable cause	Corrective action
• Bent or broken shafts	1. Extreme belt overtension 2. Overdesigned drive 3. Accidental machine damage 4. Machine design error 5. Sprocket mounted too far away from outboard bearing	1. Retension to manufacturer's recommendations 2. Redesign to manufacturer's recommendations 3. Replace or redesign drive guard 4. Check machine design 5. Move sheaves closer to outboard bearing

Hot Bearings

Symptoms	Probable cause	Corrective action
• Drive requires overtensioning	1. Worn sprocket grooves 2. Underdesigned drive 3. Improper belt tension	1. Replace sprockets and tension belts to manufacturer's recommendations 2. Redesign to manufacturer's recommendations 3. Retension to manufacturers recommendations
• Sprockets too small	1. Follow NEMA motor manufacturer's recommendations	1. Redesign drive using proper sprocket diameters
• Poor bearing condition	1. Bearings underdesigned 2. Bearings not properly maintained	1. Check bearing selection 2. Align and lubricate bearings
• Sprockets mounted too far out on shaft	1. Drive installation error	1. Move sprockets as close to outboard bearings as possible

Performance Problems

Symptoms	Probable cause	Corrective action
• Incorrect driven speed	1. Drive design error	1. Redesign drive using correct sprocket sizes for desired speed ratio

BIBLIOGRAPHY

Figures and tables furnished courtesy of Gates Corporation
Publications of the Rubber Manufacturers Association, Washington, D.C.:

Number	Title
IP-3	*Power Transmission Belt Technical Bulletin.* A complete set of IP-3-1 through 3-13, listed below (also available separately).
IP-3-1	*Heat Resistance of Power Transmission Belts* (1999)
IP-3-2	*Oil & Chemical Resistance of Power Transmission Belts* (1999)
IP-3-3	*Static Conductive Test Method for Power Transmission Belts* (1995)
IP-3-4	*Storage of Power Transmission Belts* (1999)

IP-3-6	*Use of Idlers with Power Transmission Belt Drives* (1999)
IP-3-7	*V-Flat Drives* (1999)
IP-3-8	*High Modulus V-Belts* (1999)
IP-3-9	*Joined V-Belts* (1999)
IP-3-10	*V-Belt Drives with Twist & Non-Alignment Including Quarter Turn* (1999)
IP-3-13	*Mechanical Efficiency of Power Transmission Belt Drives* (1999)
IP-3-14	*Drive Design Procedure for Variable Pitch Drives* (1999)
IP-3-15	*Recommended Fits (Shaft & Bore Tolerances) For Bored-to-Size First & Second Generation Curvilinear Power Transmission Synchronous Pulleys* (2000)
IP-20	*Specification: Joint MPTA/RMA/RAC Classical V-Belts and Sheaves* (1988); A, B, C, and D cross sections
IP-21	*Specifications: Joint RMA/MPTA Double-V Hexagonal Belts* (1997); AA, BB, CC, and DD cross sections
IP-22	*Specification: Joint RMAMPTA/RAC Narrow V-Belts and Sheaves* (1991); 3V, 5V, and 8V cross sections
IP-23	*Specification: Joint RMA/MPTA Single V-Belts* (1997); 2L, 3L, 4L, and 5L cross sections
IP-24	*Specification: Joint RMA/MPTA/RAC Synchronous Belts* (2001); MXL, XL, L, H, XH, and XXH belt sections
IP-25	*Specification: Joint RMA/MPTA/RAC Variable Speed Belts* (1997); Twelve cross sections
IP-26	*Specification: Joint RMA/MPTA/RAC V-Ribbed Belts and Pulleys* (2000); H, J, K, L, and M cross sections
IP-27	*Specification: Joint RMA/MPTA/RAC Curvilinear Toothed Synchronous Belts (2003); H3M, R3M, S3M, H5M, R5M, S5M, H8M, R8M, S8M, H14M, R14M, S14M, H20M cross sections*

CHAPTER 8
MECHANICAL VARIABLE-SPEED DRIVES

Carl March
Reliability Engineer, Life Cycle Engineering, Inc., Charleston, S.C.

Because of their versatility and simplicity, mechanical variable-speed drives have enjoyed wide popularity for almost 100 years. Even today, with the advent of low-cost electronic variable-speed methods, mechanical drives remain in wide industrial usage. In this chapter we will (1) define similar types and principles of operation and (2) offer generalized maintenance comments for similar types. For specific or unusual problems, the best source of information will be obtained from the manufacturers' literature and representatives.

The most popular general type of mechanical adjustable-speed (MAS) drive involves the use of a sliding cone-face pulley with a wide V-belt. The basic mechanism involved entails some method of varying the running pitch diameter of a belt. Two fixed-pitch pulleys, driven at a constant speed, will provide a constant output (driven) speed. This driven speed will produce constant input speed, as determined by the ratio of the pulleys' pitch diameters. If we reduce the pitch diameter of the constant-speed or driving pulley (or increase the diameter of the driven pulley), the output speed (or driven pulley speed) will be reduced, and vice versa. This is the basic principle; the methods of accomplishing it vary widely. Functional problems can frequently be pinpointed by analysis of the problem and principal; that is, if the output speed doesn't change, then for some reason the pulley pitch diameters are not changing.

GENERAL FRICTION-TYPE BELT

Most MAS drives rely on the axial movement of cone-faced disks to vary running pitch diameters. These disks normally have equal and opposite inclined faces with a shaft through their centers. The face angles are designed to mate with the edge angles or tapered sides on the belt and thereby transmit torque through the frictional forces developed between the belt edge and disk face. Using belts with configurations or face angles significantly different from those for which the disks were designed will reduce this frictional force and thus the power-transmitting ability and life of the drive. Since disks and belts are designed as a geometrically integrated package, the use of a belt whose width or thickness is not equal to the recommended belt cross section will modify the position or pitch diameter that the belt attains for a given disk position. The result, if the unit functions at all, will be a reduction in the speed range capability and component life.

With one exception, most MAS belt-type drives rely on dry friction between the belt and disks. Grease, oil, water, or other contaminants on the disk belt interface will reduce the coefficient of friction and hence the torque transmitting capacity of the drive. Belt life also will be affected. Disk faces and belt edges should always be inspected for, and kept as free as possible from, contaminants.

Packaged drives are usually available with sealed enclosures to prevent contamination. Open drives must be protected from these impurities. Wash-down applications are common in the food-processing industry and should be well protected to extend component life. The exception to this occurs with traction and metal-belt drives that run in an oil bath. The lubricant used should be checked periodically for level and contamination and replaced according to manufacturers' recommendations.

One problem in mechanical adjustable-speed drives occurs when the axially movable sliding disks will not slide on their shafts. If movable disks cannot move on command, no speed variation can be obtained. This condition is frequently caused by the products of fretting corrosion lodging between the shaft and the bore of the rotating disk hub, effectively binding the pieces together. Some drives that have run at one setting for a very long time may even be locked by actual wear.

Until a few years ago, most MAS drives used grease lubrication between mating sliding parts. This system, because of its simplicity, proven performance, and long life when properly maintained, is still often used. The present field population of both old and new units demanding periodic relubrication is extensive, exceeding those which do not require such regular service. Therefore, all MAS drives should be inspected to determine whether periodic relubrication is required, and the manufacturers' recommendations for frequency and lube type should be strictly adhered to.

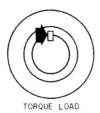

TORQUE LOAD

RADIAL HUB FORCE
(BELT PRESSURE)

FIGURE 8.1 Radial hub force tends to squeeze lubricant from mating surface.

The interface between the axially movable (sliding) disk and its shaft must perform several functions: the fit must be loose enough to allow axial movement yet tight enough to prevent oscillating and impost loading, both of which contribute to the previously mentioned fretting corrosion. See Fig. 8.1. Most designs incorporate single- or multiple-key keyway arrangements in this area to transmit torque from the sliding disk to the shaft. Note that some designs use an external method of transmitting sliding-disk torque, such as a keyed external collar. There are also units employing polygon-shaped or splined shafts for this purpose. Regardless of the specific design, those which have direct metal-to-metal contact must be lubricated to minimize fretting corrosion and to flush out its products. Depending on the load, duty cycle, and environmental considerations, a common recommendation is to use a high-quality grade of NCGI number 1 grease, pumped into the appropriate fittings every 1 to 4 weeks. Enough should be used to ensure flushing of the old grease but not so much as to force large quantities out the ends of the sliding disks. Too much lubricant may lead to disk-face contamination and a general mess. If possible, the drive should be cycled through its speed range both during lubrication and at least once daily (to prevent the start of fretting). This may seem excessive, but experience shows that the results produce many years of reliable service.

A typical example of a grease lubricant design is shown in Fig. 8.2. This is a Reeves unit and incorporates a close grooving hub bore design that effectively distributes lubricant between hub and shaft. This design also prevents the buildup of large fretted areas. If fretting does start, the grooves cause grease to flow into the affected areas, minimizing the corrosive action.

Many new mechanical adjustable-speed devices are designed to eliminate periodic relubrication. These usually incorporate some form of nonmetallic bearing material inside the sliding-disk hubs. In many designs, this bearing material is replaceable. In others, it is not. The replaceable-insert designs have a definite advantage, since periodic inspection and replacement of the insert will return the drive to almost new condition. Periodic inspection is important here. If the insert material is allowed to completely wear through, the hubs and shafts themselves will become worn and not function properly when the inserts are replaced. This will lead to very rapid wear of the new inserts and unacceptable drive performance. The only recourse is to replace the entire shaft and disk-hub assembly.

FIGURE 8.2 "Close grooving" of motor pulley.

Most manufacturers of no-lube pulleys provide instructions for easily gauging wear on the hub inserts. This may be by physical inspection or by rocking the sliding-disk heel and toe or rotating it to detect an increase in internal clearance. If looseness is detected, a more thorough inspection by disassembly must be performed. Any suspect part should be checked immediately; insert kits are much less expensive than disk assemblies. Running a drive to the failure point is false economy.

When replacing no-lube bushings, the manufacturer's recommendations should be followed. These are usually included with the bushing kits. Some designs merely snap into place; others require glues or physical retention methods. For either type, thoroughly clean and inspect all components and remove all traces of old bushings, grease, debris, and the like. Also clean and inspect disk faces, belts, control mechanisms, springs, and the like. A thorough checking and cleaning will help prevent unexpected downtime from undetected sources.

TYPES OF FRICTION DRIVES (BELT AND SHEAVE)

Adjustable While in Motion (Nonenclosed)

These are usually sold as separate, complete pulleys, designed for mounting directly on a motor shaft (see Fig. 8.3). These pulleys are normally spring-loaded and may have either one or both of the disk halves capable of axial movement (see Figs. 8.4 and 8.5). They are designed to drive a fixed-diameter sheave. Output speed variation at the fixed sheave is controlled by varying the center distance between the fixed and spring-loaded sheave. This usually involves a sliding-motor mounting base.

Pulleys with one fixed and one movable disk may be used to drive either a flat-face or a V-sheave. Note that as the pitch diameter traversed by the belt on the spring-loaded sheave changes, the center line of the belt moves axially. If a flat-faced pulley is the driven element, both motor shaft and driven shaft must be parallel, and the motor should move perpendicularly to the driven shaft.

The side movement of the belt results in its tracking across the flat face of the driven units while remaining perpendicular to both shafts. The flat face must be positioned so that the belt does not run off the sides of the sheave at any time. Centering at midrange is best.

If the driven pulley is a V-type, the sliding motor base must be angled so that the motor also moves sideways during forward and backward speed adjustment in such a way as to maintain the position of the belt center line. Note that the motor base must be angled but that the motor must be mounted on the base so that its shaft is parallel to the driven shaft. Manufacturers' instructions as to angle and direction should be followed.

Shortened belt life on sheave arrangements is usually due to misalignment, caused by inadequate sheave alignment or incorrect motor base angle. Extreme wear on the sliding rods or rails of the motor base will allow this angle to change, increasing wear.

The problem of belt-tracing alignment with a V-sheave is solved by using a pulley where both disks are movable axially. In some of these designs, each disk half has its own spring. But all share a common shaft and are sized to accept the motor shaft. With this arrangement, care must be taken to ensure that both disks are easily and equally movable on the common shaft. If one disk begins to bind, the other will take up the slack, resulting in belt misalignment. If this condition occurs, it may be corrected by disassembly, cleaning, and relubrication. Both springs should be inspected carefully for binding or breakage. Required compression force on each spring must be equal by 65 percent.

In order to overcome the problem of unequal spring forces and binding, some double-acting designs use a single spring and a cam or linkage arrangement that mechanically restricts each disk to equal and opposite movement. These linkages must be periodically inspected and lubricated, since most are prone to binding and seizing. The vibrating and oscillating motion that wears disks and hubs can cause rapid linkage wear.

Up to this point, we have considered single open pulleys with movable flanges driving to a fixed-pitch sheave. These devices seldom exceed a 3 to 1 ratio range in speed variation. In order to provide greater speed range at a low cost, the compound pulley arrangement was developed. This system incorporates variable-pitch sheaves on both the driving and driven ends. Such combinations operate on fixed center distances. Speed variation is obtained by adjusting the pitch of the motor pulley mechanically. The driven pulley is spring-loaded. As the pitch diameter of the motor pulley is adjusted, belt slack is taken up by the spring-loaded pulley. Since the pitch diameters of both pulleys vary, speed ranges of a 9 to 1 ratio are common. Maintenance of the hub and

FIGURE 8.3 (*A*) Spring-loaded split pulley, adjustable with drive-in operation.

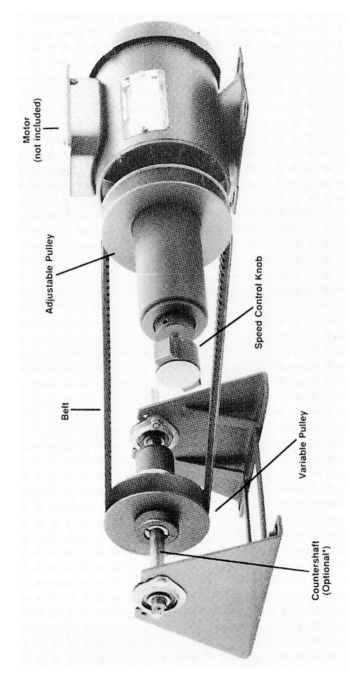

FIGURE 8.3 (*Continued*) (*B*) Fixed-center-distance compound-pulley arrangement. (*Reeves Div.*)

5.150 MAINTENANCE OF MECHANICAL EQUIPMENT

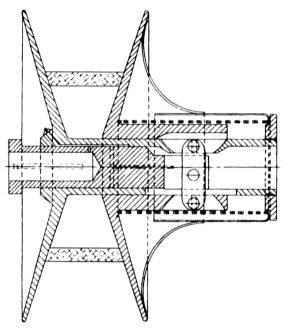

FIGURE 8.4 Pulley halves are given equal lateral movement as speed is adjusted. (*Lewellen Mfg. Co.*)

shaft portion of the motor pulley and the entire driven pulley is similar to that for single pulleys. However, additional attention must be paid to the speed-changing mechanism. This mechanism normally uses either a handwheel and screw with bearings arrangement or a remotely pivoted lever system. The screw and bearing, set within the constant-speed pulley hub, is the most common. This arrangement should be inspected for loose or worn parts or bearing failures. The system incorporates a reaction arm that must be fixed against a stop to prevent rotation while remaining free to move axially with pulley adjustment. Note that since both pulleys are single-acting, with movable flanges on opposite sides, they must be installed with both motor shaft and output shaft parallel. Belt alignment is critical.

One additional type of wide-range assembly uses a compounded variable-pitch sheave assembly mounted on an intermediate countershaft between fixed driving and driven pulleys (see Figs. 8.6 and 8.7). Changing the position of the countershaft, usually on a pivoted arm, causes the compound floating flange to shift in position under belt pressure and thereby vary the operating diameters to achieve the desired speed. The necessarily short hub on the common flange frequently experiences high wear rates.

FIGURE 8.5 Exploded view of motor pulley disks.

FIGURE 8.6 Use of compounded variable-pitch sheave to obtain wide speed range. (*Speed Selector, Inc.*)

Static-Adjustment Types

Adjustable-while-stopped pulleys are designed for the machine which at the outset may require a slight speed adjustment to operate at optimal performance but once set will rarely need adjustment.

The significant makes of stationary-control adjustable pulleys use wide-section V-belts to obtain speed ratios up to 3 to 1 (see Figs. 8.8 to 8.10). They are available in single- and multiple-groove designs from fractional to 100-hp rating.

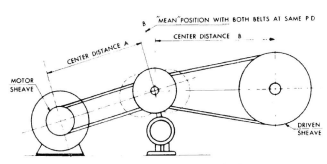

FIGURE 8.7 Diagrammatic scheme of drive shown in Fig. 8.9.

FIGURE 8.8 Adjustable-while-stopped control, speed ratios up to 3 to 1. (*T. B. Wood's Sons Co.*)

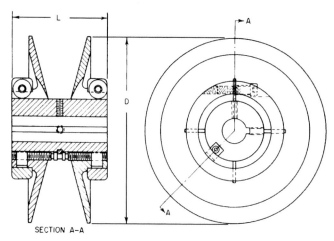

FIGURE 8.9 Section of control shown in Fig. 8.3.

FIGURE 8.10 Adjustable-while-stopped control, speed ratios up to 3 to 1. (*Dodge Manufacturing Corp.*)

Of significance to the maintenance man is the fact that once adjusted, the various parts are designed to be held together in a fixed relation one to another by set screws or holding nuts, thus minimizing the wear and lubrication problems mentioned above. Care must be exercised to ensure that the holding devices are quite secure, for even the slightest oscillatory motion in the absence of lubrication will quickly result in a seizing of the mating surfaces. (See Table 8.1.)

Speed adjustment can be done only while the machine is stopped. The most critical problem with wide-section multiple-belt pulleys is the matching of belt lengths to ensure a sharing of the load. Unfortunately, these must be matched more closely than standard tolerances set up by the belt manufacturers, which call for a trial-and-error matching of belts.

This, together with the definite stretch a belt will take its first several hundred hours of operation, calls for replacement of the complete set of belts when any one particular belt goes bad. Belts can be purchased as matched sets.

TABLE 8.1 Adjustable Pulleys—Stationary Control

Trouble	Cause	Correction	Prevention
I*a*. Accelerated belt wear	Excessive misalignment of driver and driven pulleys	Realign pulleys	
I*b*. Accelerated belt wear	Continuous flexing over small diameters	Select driven pulley of proper diameter to avoid this condition.	
I*c*. Accelerated belt wear	Excessive heat, cold, moisture, acid fumes, abrasives, etc.		
I*d*. Accelerated belt wear	Overloaded, excessive shock loads, excessive belt speeds		Do not exceed ratings set by manufacturers
I*e*. Accelerated belt wear	In multiple-belt pulleys one or a few belts taking entire load	Belt lengths must be matched more closely	
I*f*. Accelerated belt wear	Excessive belt tension	Adjust center distance	
I*g*. Slipping belt	Insufficient tension	Adjust center distance, or use idler roll	
I*h*. Slipping belt	Pulley faces greasy	Clean	
I*i*. Slipping belt	Overloaded		

TABLE 8.2 Adjustable Pulley—Controlled in Motion

Trouble	Cause	Correction	Prevention
II*a*. Accelerated belt wear	See items I*a*, I*b*, I*c*, I*d**		
II*b*. Slipping belt	See items I*h*, I*i**		
II*c*. Slipping belt or belt not running level	Spring-loaded disks do not compensate owing to sticking of disks from improper or insufficient lubrication	Stop at once; disassemble; clean until parts slide freely	Lubricate every 1–4 weeks; shift speed range each day if possible

*Reference is to items in Table 8.1.

This tendency for the belts to stretch during their initial break-in, along with the changing of operating diameters, requires some method of take-up to maintain proper belt tension, which is usually accomplished by adjusting the motor position on slide rails or by some type of idler roll.

Table 8.2 lists troubles, their causes, and cures experienced with the adjustable pulleys, stationary control.

Belt Transmissions

Belt transmissions of the type shown in Fig. 8.11 have, through the years, earned a most enviable reputation among plant maintenance men for ruggedness and reliability with a minimum of attention. They are available in capacities from fractional to 75 hp and up to 16 to 1 speed range. Standard variations include vertical or horizontal mountings, open or enclosed, and with a variety of controls.

The heart of these transmissions is the time-honored block-belt design. Wedge-shaped wooden blocks tipped with leather are bolted to and carried on a wide strip of belting. This design has the advantage of separating the handling of the torque load (belt pull) and the radial wedging forces. Little attention is required to the belt other than an occasional check to ensure that the disk faces are clean and free from grease, acid, or water. If adverse conditions of dust, water, chemical fumes, or live steam are present, an enclosed type of transmission should be used. Table 8.3 lists troubles, causes, and cures experienced with belt transmissions.

Usually, one shaft is driven at constant speed, adjustments in speed is accomplished by a lever arrangement which positively synchronizes the position of all four flanges. Two screw

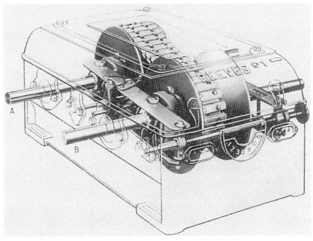

FIGURE 8.11 Belt transmission.

TABLE 8.3 Belt Transmission

Trouble	Cause	Correction	Prevention
IIIa. Accelerated belt wear	Misalignment of disk assembly	Constant-speed and variable-disk assemblies should be parallel at mean speed	
IIIb. Accelerated belt wear	See also items Ib, Ic, and Id*		
IIIc. Slipping belt	Pulley faces greasy, usually from overlubrication of thrust bearings	Clean	Avoid overgreasing thrust bearings
IIId. Slipping belt	Constant-speed shaft too slow	Increase input speed by changing sheaves	
IIIe. Slipping belt	Insufficient belt tension	Adjust tension screw, but only while drive is running	Belt should have a slight sag on the loose side
IIIf. Creaking belt	Excessive belt tension	Adjust tension screw, but only while drive is running	
IIIg. Bearing failures	Belt too tight	Adjust tension	
IIIh. Bearing failures	Excessive overhung load		Do not exceed loads specified by manufacturer
IIIj. Bearing failures	Insufficient or excessive lubrication		See manufacturer's instructions
IIIk. Bearing failure	Atmosphere: abrasive particles, moisture, corrosion	Use enclosures where necessary	
IIIl. Bearing failures	Bent shaft or improperly assembled		
IIIm. Cannot adjust speed	"Sticking disks" due to improper or insufficient lubrication	Stop at once; disassemble; clean disk hub and shaft with solvent	Lubricate every 2–5 weeks; shift through entire speed range each day, if possible

*Reference is to items in Table 8.1.

arrangements are provided: one by adjusting speed by controlling the position of these levers, which, in turn, control the pulley operating diameters, the other, for controlling belt tension and horsepower capacity, by adjusting the center distance of the pivotal points of these synchronizing levers.

Belt tensioning is accomplished in at least two different ways. Older designs rely completely on the natural wedging action of the belt between the disks. The belt-tensioning screw (located between the disk sets and acting on the pivot points of the shifting levers) must be adjusted while the drive is running. The proper adjustment is only to where the belt has some slack or droop on the loose side. Do not pull the belt up tight; if done so it may destroy the belt.

Newer designs tension the belt with short, strong springs acting on the pivot points of the shifting levers. These springs are adjusted with the tensioning screw. These are normally adjusted by tightening the tensioning screw until it stops (tight) while rolling the drive over by hand and then backing the screw off a turn or two.

Note that transmissions all require periodic disk and bearing lubrication. Drives should be inspected carefully for the presence of grease fittings, since some have internal bearing lube points which also require service. As with pulleys, transmissions should be lubricated while running and shifted through the speed range to distribute the lubricant.

All-metal versions of the belt transmissions are also available. These are totally enclosed, oil-filled units that use a belt made up of linked transverse laminations (see Fig. 8.12). These laminations mate with radial grooves on the conical pulley inner flanges. These units offer higher power and positive speed at the expense of greater mechanical complexity and more sensitivity to shock and overload than block-belt types. Adequate lubrication and internal alignment are extremely critical to drive life.

FIGURE 8.12 All-metal belt transmission—P.I.V. (*Link-Belt Div., FMC Corp.*)

Packaged Belt Drives

These drives are very common and find wide use in many industries because of their all-in-one compactness and versatility. Similar in concept and function to the previously described compound-pulley arrangement, they go further by incorporating motor and variable-speed pulleys with control and gear reducer in one package. The available combinations of horsepower, case size, output speeds, controls, and physical arrangements seem limitless—a manufacturer such as Reeves can provide over 10,000 combinations, from $1/4$ to 50 hp at speeds from less than 1 to over 13,000 rpm. Both parallel and right-angle reducers are available. Control arrangements range from manual handwheels through electrical remote control and pneumatic arrangements. These may be tailored to provide output speeds that follow an input signal in any relationship—linear, logarithmic, or exponential.

Maintenance of packaged belt drives is similar to that for open pulleys. Both relubricatable and permanent-lube designs exist, with perm-lube predominating newer drives. However, the population of older drives is very large; therefore, all drives should be inspected for periodic maintenance requirements. Also, some motor bearings have lube fittings. Those units with gear reducers must have oil levels checked and oil replaced periodically. Grade and type of oil should follow manufacturer's specifications. Older units commonly use SAE gear oil or nondetergent crankcase oil. Nondetergent oil is preferred because it allows contaminants to settle out into the bottom of the gearcase rather than being held in suspension in the oil. Oil grade is usually a function on temperature. Newer parallel reducers (and most right-angle worm and worm-helical combinations) may require synthetic or specific lubricants. (See Table 8.4.)

As packaged units, these drives normally require little attention. If disassembly is necessary (internal parts can be easily replaced), all parts should be inspected before reassembly. Most drives require little realignment, the exception being correctly locating the constant-speed disk assembly on the motor shaft before tightening it down if the motor or constant-speed disk assembly must be replaced. Some types locate the constant-speed disk assembly relative to the motor shaft with spacers or adjusting screws in the bottom of the fixed constant-speed disk motor bore.

TABLE 8.4 Packaged Belt Drives

Trouble	Cause	Correction	Prevention
IV*a*. Accelerated belt wear	See items I*b*, I*c*, I*d*, and III*a**		
IV*b*. Slipping belt	Broken spring in variable-speed disk assembly	Replace spring	
IV*c*. Slipping belt	Pulley faces greasy owing to excessive lubrication of thrust bearings	Clean with solvent	
IV*d*. Bearing failures	See Items III*h*, III*j*, III*k*, and III*l**		
IV*e*. Cannot adjust speed; belt not running level	See item III*m**		
IV*f*. Chatter in gearing	Insufficient oil in gear case		Check oil level every 30 days

*Reference is to items in Tables 8.1 and 8.3.

Common practice has been to mount the fixed constant-speed disk directly on the motor shaft (the belt pull becomes an overhung load on the motor bearings) and securing it with either a key and set screws or with a clamp collar or collet arrangement. The latter types may not have a keyway. These types must never be used on a damaged or undersized motor shaft because they may not hold adequately.

Some designs support the constant-speed disk assembly between bearings and connect to the motor with a flexible coupling.

European designs take advantage of the tapped-hole standard available in the end of IEC-type motors. Disk assemblies are secured to the motor shaft with a draw bolt extending completely through the length of the hub of the fixed constant-speed disk and into the tapped motor shaft. Instruction manuals should be consulted for critical measurements. Failure to set these properly will result in premature belt and disk wear.

As with other mechanical devices, a little preventive maintenance and inspection will yield a good return in life, performance, and freedom from downtime.

Friction-Disk-Type Drives

These devices employ an old principle updated. Two disks are used. One is a flat or slightly angled disk coupled directly to and driven at constant speed by the motor. The second disk is usually an annular ring of replaceable friction material attached to a carrier shaft that is supported in bearings at a slight angle from parallel to the first disk. Only a "patch" (theoretically a radial line segment) or friction material actually contacts the driver disk face.

Speed change is affected by altering the contact area radius of the annular ring from the center line of the driver (constant-speed) disk. Single-disk-pair (narrow speed range) drives usually solidly mount the bearings for the output annular friction ring. The motor is usually slide mounted so that it can be moved radially. Speed range is set by stops that limit the radial travel of the motor slide.

Wide-speed-range friction drives may have two disk and ring pairs. See Fig. 8.13 for principles of operation. Drives are usually packaged with the motor and can be obtained with gear reducers.

Maintenance consists of regular bearing lubrication and periodic inspection and cleaning of disk assemblies. The most important maintenance consideration is to ensure that all friction surfaces are absolutely clean, dry, and free from any contaminants. Even fingerprints can cause performance degradation. Some incorporate a torque-sensing (disk-loading) cam arrangement that is subject to wear and also should be checked regularly. Friction disks should be replaced when they become worn, following the manufacturer's recommended minimum thickness.

Premature disk or friction material wear may be due to shock loading that causes skidding, grooving, or flat spots.

FIGURE 8.13 Wide-speed range friction—disk-type drive. (*Reeves Div.*)

Traction-Type Drives

These differ from previously discussed friction types by relying on power transmission through an oil film which microscopically separates extremely hard metal elements rather than through direct contact between parts.

These metal elements may take the form of balls, cones, disks, or rings. Two such designs are shown in Figs. 8.14 and 8.15. Traction drives are built to extreme precision and, as such, are very sensitive to rigid maintenance schedules and correct application. Lubricant levels, operating conditions, and temperature are critical. All use specially developed traction fluids which can serve as lubricants but are not normally interchangeable. Internal repair should only be done by specialists.

These drives must operate free of shocks, overloads, reversals, or excessive temperatures. If the fluid film between mating parts is even momentarily interrupted, scoring of surface may result. Metal particles especially will destroy the drive. (See Table 8.5.)

Geared Differential Drives

Differential gearing can be attached to any parallel-shaft cone-pulley transmission to permit infinite speed variation down to zero speed and, with proper selection of components, on into the reverse direction. Figure 8.16 shows such a differential integrally mounted with an all-metal belt transmission.

The most important consideration with these geared-differential drives is the proper recognition of the internal circulating power, which may reach as high as six times the input power, and if the

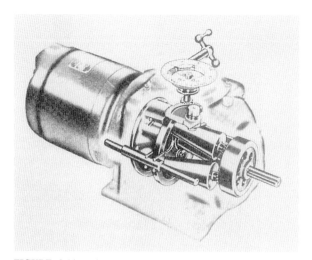

FIGURE 8.14 Drive with metal surfaces in frictional contact. (*Graham Transmission, Inc.*)

FIGURE 8.15 Drive with metal surfaces in frictional contact. (*The Cleveland Worm & Gear Co.*)

TABLE 8.5 All-Metal Traction Systems

Trouble	Cause	Correction	Prevention
V*a*. Cannot shift speed; OK if shifted only slightly; will tend to slide back to original setting	Extended operation at one speed has caused concentrated wear tending to lock the sliding surfaces in a fixed position	Return to factory for replacement of damaged parts	Avoid running at one speed for extended periods
V*b*. Pronounced thumping	Sudden load change reversal or overload has caused scoring of contact surfaces	Return to factory for replacement of damaged parts Replenish to proper level	Avoid use of traction-type drives on this type of application
V*c*. Severe overheating	Insufficient oil	Flush. Replenish with oil of proper viscosity	
V*d*. Excessive slipping	Use of too heavy an oil		Use appropriately thin oil as recommended by manufacturer. Change every 1000 hours of operation

motor transmission and gear components are not properly matched, disastrous internal overloading can result. Since each of the components must be of sufficient size and capacity to handle these maximum conditions, the cost of the total unit approaches that of the more familiar electrical variable-speed drives.

For troubles, their causes, and remedies, refer to the tables which give that information for belt transmissions and gear units.

FIGURE 8.16 Differential gearing integrally mounted with all-metal belt transmission. (*Fairchild Engine and Airplane Corp.*)

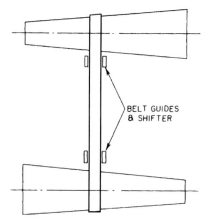

FIGURE 8.17 Flat belt and cone drive.

Flat-Belt Drives

Occasionally, a flat belt driving between two cone-shaped pulleys of the type shown in Fig. 8.17 will be found as a method of obtaining variable speed. There is no particular manufacturer who merchandises a complete line of these pulleys; rather, they are holdovers from the very early days or are specially designed to meet particular requirements.

Line contact is the theoretical ideal and, as such, requires that the belt width be kept quite narrow, which, of course, limits the power-transmitting capacity of the system. This same consideration means that contact surface speed varies across the face of the belt, producing an inherent slip and its consequent effect on belt life. See Table 8.6.

TABLE 8.6 Flat-Belt Drives

Trouble	Cause	Correction	Prevention
Short belt life	Inherent slippage; contact surface speed varies across face of belt	None	
Slipping belt	Overloaded; remember belts are necessarily narrow to approach theoretically ideal line contact	Reduce loading	Do not overload
Slipping belt	Belt has stretched	Extend center distance between pulleys or use idler roll	
Slipping belt	Pulley faces greasy	Clean with solvent	Keep clean

CHAPTER 9
GEAR DRIVES AND SPEED REDUCERS

Robert G. Smith
*Director of Engineering, Philadelphia Gear Corporation,
King of Prussia, Pa.*

Gear drives and speed reducers are widely used where changes of speed, torque, shaft direction, or direction of rotation are required between a prime mover and the driven machinery.

The design approach taken by gear manufacturers today is to consider the gear drive as a component of a mechanical system. Its functional characteristics are engineered to be fully compatible with those of the prime mover and the driven equipment and to take into account such factors as static and dynamic loading, range of torque and operating speed, expected service life, duty cycle, ambient temperature, size and weight restrictions, and total system efficiency.

Gear drives consist of one or more sets of gears mounted on shafts and bearings, a positive method of lubrication, and an enclosed casing with appropriate gaskets, oil seals, and air breathers. They also must be equipped with an integral electric motor, baseplates or other mounting structure, outboard bearings, a device that provides overload protection, a means of preventing reverse rotation, and a variety of other accessory devices.

In many power-transmission applications, the preferred prime mover operates at a relatively high speed because of the superior economy and efficiency of high-speed motors, gas and steam turbines, and the like. The driven equipment, however, often requires a much lower shaft speed and high torque. The gear drive not only reduces shaft speed to the value needed to operate the driven machine but also converts the relatively low torque output of the high-speed prime mover to the high torque needed to drive the low-speed driven device.

Rotary compressors operate more efficiently at high shaft speed and often require a speed-increasing gear drive. In most cases, increasers are not simply speed reducers driven backward but involve design considerations different from those encountered in reducers.

Common Gear Types

Common types of gears used in industrial gear drives include spur, helical, double-helical, bevel, spiral bevel, hypoid, zerol, worm, and internal gears (see Fig. 9.1).

Spur gears transmit power between parallel shafts without end thrust or axial displacement. They are commonly used on drives of moderate speeds such as marine auxiliary equipment, hoisting equipment, mill drives, and kiln drives. Simplicity of manufacture, absence of end thrust, and general economy of maintenance recommend the use of spur gearing wherever practicable.

Helical gear teeth are cut on a helix (oblique) angle across the gear-wheel face. Mating helical gears permit several teeth to be in mesh at the same time. This increases load-carrying capacity,

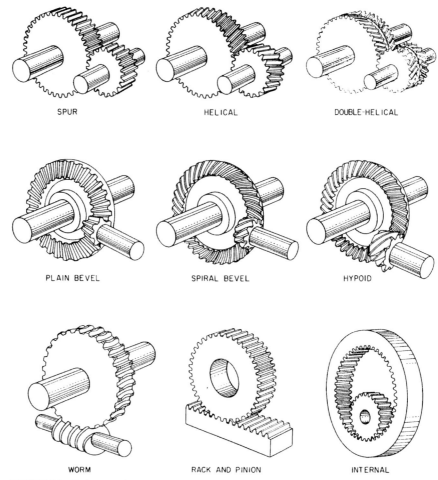

FIGURE 9.1 Basic types of gears.

ensures transmission of constant velocity, and reduces noise and vibration. Helical gears produce end thrust along the axis of rotation, which must be accommodated by thrust bearings.

Helical and double-helical gears are used where loads and speeds may be higher than can be conveniently met by spur gearing. They are also used where shock and vibration are present, or where a high reduction ratio is necessary in a single gear train. Because double-helical gears are actually opposed helical gears, end thrust is practically eliminated. See Fig. 9.2 for examples of double-helical drives.

However, where external thrust loads or thermal expansion of elements in a system are involved, single-helical gears are preferred. External thrust loads may tend to unload one helix of double-helical gears and thus overload the opposite helix.

Where very high speeds (over 20,000 fpm) are involved, critical tolerances of relative tooth positions in each helix may cause the apex of the double-helical tooth arrangement to "run out," which would tend to cyclically unload one helix and set up axial vibrations of the pinion.

Bevel gears transmit power between two shafts, usually at right angles with each other. However, shafts positioned at other than 90° can be used. Straight bevel gears may be used for right-angle power transmission where operating conditions do not warrant the superior characteristics of spiral

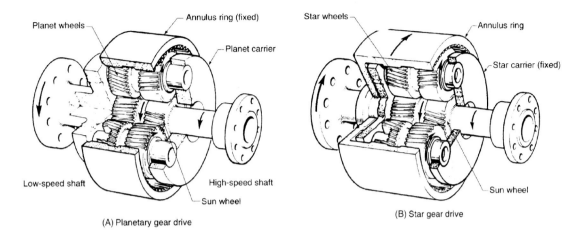

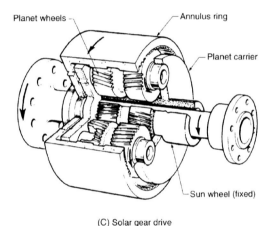

FIGURE 9.2 Double helical drives.

bevel gearing. Since bevel gearing creates thrust loads along the supporting shafts, adequate bearings must be provided.

Spiral bevel, zerol, and hypoid types of gears are generally considered under the heading of spiral-bevel-gear units. Loading of spiral bevel gears is always distributed over two or more teeth. Tooth action is smooth and quiet. Accuracy of tooth contact can be closely controlled through hard cutting, precision grinding, and/or lapping. Axial thrust of spiral bevel gears is slightly higher than for straight bevel gearing and varies with direction of rotation and hand of cut of the gear and pinion. Where possible, gears should be designed so that axial thrust tends to move the pinion out of mesh.

Worm gearing has won wide acceptance for industrial drives because of its many advantages of conjugate tooth action, arrangement, compactness, and load-carrying capacity. Worm-gear drives are quiet and vibration-free and produce a constant output speed. They are well suited to service where heavy shock loading is encountered. The many variable mounting arrangements possible with worm gears allow for compactness of design not otherwise obtainable. Since action between worm thread and the teeth of the driven worm-gear wheel is predominantly sliding rather than rolling, greater heat generation and reduced mechanical efficiencies result at higher speed-reduction ratios.

Internal gears are more compact than external gears of the same ratio. In general, they have greater load-carrying capacity and run more smoothly. Internal gearing usually employs spur, helical, or double-helical teeth. Owing to the nature of their construction, internal gears are limited in speed-reduction ratios obtainable on a given center distance.

Basic Gear Drives

Gear drives are used to transmit power between a prime mover and driven machinery. In addition to the simple transmission of power, gear drives usually change or modify the power being transmitted by (1) reducing speed and increasing output torque, (2) increasing speed, (3) changing the direction of shaft rotation, or (4) changing the angle of shaft operation.

Gear drives are generally considered packaged units, manufactured in accordance with accepted and advertised specifications, to be used for a wide range of power-transmission applications (see Figs. 9.3 through 9.11). Published standards of the American Gear Manufacturers Association (AGMA) are accepted as the basis for design, manufacture, and application of modern gear drives.

In addition to general gear drives and speed reducers, AGMA has established specific standards for certain special types of gear drives which are used explicitly for driving particular types of machinery, such as deep-well pumps, cooling-tower fans, steel-mill pinion stand drives, paper-machine sectional drives, cement mills, and compressors.

Motorized gear drives (commonly called *gearmotors* or *motor reducers*) are used extensively throughout industry. These units differ from conventional gear drives in that the prime mover (usually an electric motor) is designed as an integral component of the assembly. Any of the basic gear drives (see Figs. 9.3 through 9.11) can be manufactured as motorized units.

Epicyclic Gear Drives

In an epicyclic gear drive, power is transmitted between prime mover and driven machinery through multiple paths. The term *epicyclic* designates a family of designs in which one or more gears move around the circumference of meshing, coaxial gears, which may be fixed or rotating about their own axis. Individual gears within an epicyclic drive may be spur, helical, or double-helical.

Because of the multiple power paths, an epicyclic gear drive will normally provide the smallest drive for a given load-carrying capacity. Other advantages include high efficiency, low inertia for a given duty, high stiffness, and a high torque/power capability.

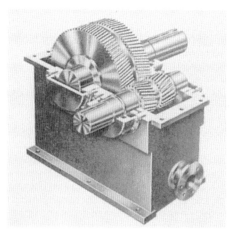

FIGURE 9.3 Single-reduction double-helical drive.

FIGURE 9.4 Double-reduction single-helical drive.

The basic elements of an epicyclic drive are a central sunwheel, an internally toothed annulus ring, a planet or star carrier, and planet or star wheels. Depending on which of the first three elements is fixed, three types of epicyclic drives are possible: a planetary gear drive, a star gear drive, or a solar gear drive.

In a planetary gear drive (Fig. 9.2A), the annulus ring is fixed. In this design, the input and output shafts rotate in the same direction, and ratios are normally in the 3 to 1 to 12 to 1 range. The planetary gear drive is usually the most compact and cost-efficient for a given torque capacity and is therefore the most common. One limitation is that the speed range of the carrier is limited by the centrifugal loading of the planet bearings.

FIGURE 9.5 (A) High-speed gearbox, enclosed. (B) High-speed gearbox, showing internal parts.

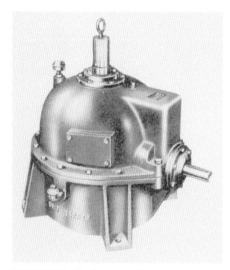

FIGURE 9.6 Cooling-tower drive.

FIGURE 9.7 Vertical pump drive.

FIGURE 9.8 Triple-reduction spiral-bevel helical drive.

To overcome this problem of centrifugal loading in high-speed applications, a star gear drive (Fig. 9.2B) may be used in which the carrier assembly is fixed. In this design, input and output shafts rotate in opposite directions, and ratios in the range of 2 to 1 to 11 to 1 are possible.

The third epicyclic configuration is the solar gear drive (Fig. 9.2C), in which the sunwheel is fixed. With this design, input and output shafts rotate in the same direction, and ratios are on the order of 1.1 to 1 to 1.7 to 1. Because of the limited ratio range, solar gear drives are found only in very special applications.

Select the Proper Drive

Satisfactory performance of a gear drive depends on proper design and manufacture of the drive itself, selection of the proper type and size of unit for a given application, proper installation,

FIGURE 9.9 Single-reduction worm-gear reducer.

5.168 MAINTENANCE OF MECHANICAL EQUIPMENT

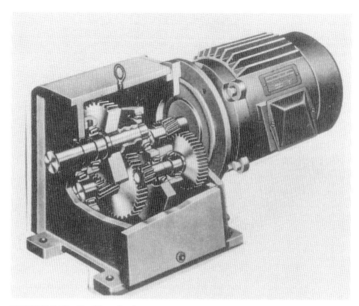

FIGURE 9.10 Double-reduction gearmotor.

proper use of the unit in service, and proper maintenance of the unit throughout its entire service life.

The official AGMA identification plate (see Fig. 9.12) signifies that the manufacturer is a member of AGMA. Do not hesitate to request the assistance of trained engineers from a reputable manufacturer in selecting and properly installing gear-drive units.

In selecting the proper gear drive for any application, you will need to determine the horsepower requirement of the unit and the speed ratio between input and output shafts. You will already know (1) the horsepower and speed required to drive your machinery and (2) the output speed and horsepower of the prime mover.

FIGURE 9.11 Four-speed gearbox.

FIGURE 9.12 AGMA identification plate.

To calculate horsepower required in any gear drive, first adjust for efficiency losses in the unit. Generally speaking, the efficiency of spur, helical, double-helical, straight bevel, spiral bevel, zerol, and hypoid gears is taken as 98 percent per gearset. The efficiency of worm gears in reducers can vary widely (10 to 95 percent) depending on speed, ratio, materials, and the likes. Consult the gear manufacturer for efficiency in specific applications. Where multiple gearsets are used, the overall efficiency is the product of the efficiencies of the gearsets. Most gear-drive manufacturers publish efficiency ratings on their individual units.

Having adjusted required horsepower upward to compensate for efficiency losses, now adjust for the type of service the gear drive must handle. For standardization and convenience, certain commonly used machines have been classified according to the types of service they normally require. Table 9.1 shows this classification, where U = uniform shock load, M = moderate shock load, and H = heavy shock load.

Determine the load characteristics of your application from this table. Then multiply the horsepower (adjusted for efficiency losses) by the appropriate service factor (Tables 9.2 to 9.4 for that type of unit, prime mover, load, and usage.

Next, to calculate the speed ratio of any gear drive, simply divide the rpm of the input shaft by the rpm required on the output shaft. Then, select a gear drive that will meet the speed and increased horsepower requirements.

After a gear-drive unit has been selected for mechanical rating, check the actual horsepower to be transmitted against the manufacturer's thermal rating for that unit. Thermal rating is the maximum average horsepower that can be transmitted continuously without creating a dangerous rise in temperature and without necessitating auxiliary cooling of the unit.

Install Gear Drives Carefully

When installing gear-drive units, be sure that they are well supported, accurately aligned, and securely anchored to prevent misalignment of gears or shafts. Consult the installation and maintenance instructions furnished by the manufacturer.

Good-quality mechanical couplings suitable for the application should be used to couple the shafts of driving and driven units.

Slight angular or linear shaft misalignments may be accommodated by using flexible-type mechanical couplings. In some cases, torsional stresses at starting or during momentary overloads can be compensated for by use of certain types of flexible mechanical couplings.

Proper loading of gear-drive units is essential for a long and trouble-free service life. Assuming that gear drives are properly rated for the particular applications and are properly installed, it is important that they should not be subjected to extreme or sustained overloads.

Torque limit switches are available as optional equipment on the gear drives of some manufacturers. Where the possibility of overloading or machinery jamming (which might produce an overload on the drive unit) is present, it is wise to insist on torque-limiting devices.

TABLE 9.1 Load Characteristics

Application	Uniform load	Moderate shock	Heavy shock	Application	Uniform load	Moderate shock	Heavy shock
Agitators:				Cranes:			
Pure liquids	U			Main hoists	U		
Liquids and solids		M		Bridge travel*			
Liquids—variable density		M		Trolley travel*			
Blowers:				Crusher:			
Centrifugal	U			Ore			H
Lobe		M		Stone			H
Vane	U			Sugar†		M	
Brewing and distilling:				Dredges:			
Bottling machinery	U			Cable reels		M	
Brew kettles, cont. duty	U			Conveyors		M	
Cookers—cont. duty	U			Cutter head drives			H
Mash tubs—cont. duty	U			Jig drives			H
Scale hopper, frequent starts		M		Maneuvering winches		M	
Can filling machines	U			Pumps		M	
Cane knives		M		Screen drive			H
Car dumpers			H	Stackers		M	
Car pullers		M		Utility winches		M	
Clarifiers	U			Dry-dock cranes:			
Classifiers		M		Elevators:			
Clay-working machinery:				Bucket—uniform load	U		
Brick press			H	Bucket—heavy load		M	
Briquette machine			H	Bucket—cont.	U		
Clay-working machinery		M		Centrifugal discharge	U		
Pug mill		M		Escalators	U		
Compressors:				Freight		M	
Centrifugal	U			Gravity discharge	U		
Lobe		M		Man lifts*			
Reciprocating, multicylinder		M		Passenger*			
Reciprocating, single-cylinder			H	Fans:			
Conveyors—uniformly loaded or fed:				Centrifugal	U		
Apron	U			Cooling towers			
Assembly	U			Induced draft*			
Belt	U			Forced draft	U		
Bucket	U			Induced draft		M	
Chain	U			Large (mine, etc.)		M	
Flight	U			Large (industrial)		M	
Oven	U			Light(small diameter)	U		
Screw	U			Feeders:			
Conveyors—heavy duty, not uniformly fed:				Apron		M	
Apron		M		Belt		M	
Assembly		M		Disk	U		
Belt		M		Reciprocating			H
Bucket		M		Screw		M	
Chain		M		Food industry:			
Flight		M		Beet slicer		M	
Live roll*				Cereal cooker	U		
Oven		M		Dough mixer		M	
Reciprocating			H	Meat grinders		M	
Screw		M		Generators (not welding)	U		
Shaker			H	Hammer mills			H

TABLE 9.1 Load Characteristics (*Continued*)

Application	Uniform load	Moderate shock	Heavy shock	Application	Uniform load	Moderate shock	Heavy shock
Hoists:				Notching press—belt driven*			
Heavy duty			H	Plate planers			H
Medium duty		M		Tapping machine			H
Skip hoist		M		Other machine tools:			
Laundry washers:				Main drives		M	
Reversing		M		Auxiliary drives	U		
Laundry tumblers		M		Metal mills:			
Line shafts:				Draw bench carriage and main drive		M	
Driving processing equipment		M		Pinch, drier and scrubber rolls, reversing*			
Light	U			Slitters		M	
Other line shafts	U			Table conveyors			
Lumber industry:				Nonreversing			
Barkers		M		Group drives		M	
Spindle feed		M		Individual drives			H
Main drive			H	Reversing*			
Carriage drive*				Wire drawing and flattening machine		M	
Conveyors:				Wire winding machine		M	
Burner		M		Mills, rotary type:			
Main or heavy duty				Ball†			H
Main log			H	Cement kilns†		M	
Resaw				Dryers and coolers†		M	
Merry-go-round		M		Kilns		M	
Slab			H	Pebble†		M	
Transfer		M		Rod, plain and wedge bar†		M	
Chains:				Tumbling barrels			H
Floor				Mixers:			
Green		M		Concrete mixers, continuous		M	
Cut-off saws:				Concrete mixers, intermittent		M	
Chain		M		Constant density	U		
Drag		M		Variable density		M	
Debarking drums			H	Oil industry:			
Feeds:				Chillers		M	
Edger				Oil-well pumping*			
Gang			H	Paraffin filter press		M	
Trimmer	U	M		Rotary kilns		M	
Log deck			H	Paper mills:			
Log hauls—incline—well type			H	Agitators (mixers)		M	
Log turning devices			H	Barker—mechanical		M	
Planer feed		M		Barking drum			H
Planer tilting hoists				Beater and pulper		M	
Rolls—live-off brg.—roll cases			H	Calendars		M	
Sorting table		M		Calendars—super			H
Tipple hoist		M		Converting machine, except cutters, platers		M	
Transfers:				Conveyors	U		
Chain		M		Couch	U		
Craineway		M		Cutters—platers	U		H
Tray drives		M		Cylinders	U		
Veneer lathe drives*				Dryers			
Machine tools:				Jordans		M	
Bending roll		M		Presses	U		
Punch press—gear driven			H				

(*Continued*)

TABLE 9.1 Load Characteristics (*Continued*)

Application	Uniform load	Moderate shock	Heavy shock	Application	Uniform load	Moderate shock	Heavy shock
Paper mills (*Cont.*):				Collectors, circuline or straight-line	U		
Pulp machine reel	U						
Washers and thickeners		M		Dewatering screws		M	
Winders	U			Scum breakers		M	
Printing presses:*				Slow or rapid mixers		M	
Pullers:				Thickeners		M	
Barge haul			H	Vacuum filters		M	
Pumps:				Screens:			
Centrifugal	U			Air washing	U		
Proportioning		M		Rotary-stone or gravel		M	
Reciprocating				Traveling water intake	U		
Single acting, 3 or more cylinders		M		Slab pushers		M	
				Steering gear*			
Double acting, 2 or more cylinders		M		Stokers	U		
				Sugar industry:			
Single acting, 1 or 2 cylinders*				Cane knives†		M	
Double acting, single cylinder*				Crushers†		M	
Rotary—gear type	U			Mills†			H
Lobe vane	U			Textile industry:			
Rubber industry:				Batchers		M	
Intensive internal mixers				Calendars		M	
Batch mixers				Cards		M	
Continuous mixers		M		Dry cans		M	
Mixing mill—two smooth rolls		M		Driers		M	
Batch drop mill—two smooth rolls		M		Dyeing machinery		M	
Cracker warmer—two roll		M		Knitting machines*			
One corrugated roll			H	Looms		M	
Two corrugated rolls		M		Mangles		M	
Holding feed and bland mill—two roll		M		Nappers		M	
Refiner—two roll		M		Pads		M	
Calendars		M		Range drives*			
Extruders				Slashers		M	
Continuous screw operation		M		Soapers		M	
Intermittent screw operation		M		Spinners		M	
Sand muller		M		Tenter frames		M	
Sewage-disposal equipment:				Washers		M	
Bar screens	U			Winders		M	
Chemical feeders	U			Windlass*		M	

*Refer to gear manufacturer.
†To be selected on basis of 24-hour service only.

If you have any questions about the load capacities of gear drives on your machinery, do not hesitate to consult the manufacturer of the drive units.

Gear-drive housings are usually designed for proper heat dissipation under normal operating conditions. Do not allow units to operate where oil temperatures exceed those recommended by the manufacturer. Where surrounding atmospheric conditions might reduce normal heat dissipation, consult the drive manufacturer for his recommendations.

TABLE 9.2 Service Factors for Double-Helical, Helical, and Spiral Bevel Gear Units

Character of load on driven machine	Electric-motor or steam-turbine drive				Multicylinder internal-combustion engine				Single-cylinder internal-combustion engine			
	8–10 hr per day	24 hr per day	Intermittent 3 hr per day	Occasional ½ hr per day	8–10 hr per day	24 hr per day	Intermittent 3 hr per day	Occasional ½ hr per day	8–10 hr per day	24 hr per day	Intermittent 3 hr per day	Occasional ½ hr per day
Uniform	1.0	1.25	0.8	0.5	1.25	1.05	1.0	0.8	1.5	1.75	1.25	1.0
Moderate shock	1.25	1.5	1.0	0.8	1.5	1.75	1.25	1.0	1.75	2.0	1.5	1.25
Heavy shock	1.75	2.00	1.5	1.25	2.0	2.25	1.75	1.5	2.25	2.5	2.0	1.75

Notes: 1. Ratings shown in horsepower tables are based on a service factor of 1. For service factors other than 1 it is necessary to multiply the actual running horsepower required under normal full load by the service factor. The product of these two, which may be called the *equivalent horsepower*, is to be used when making reducer selection from horsepower table.

2. The horsepower tables permit a maximum momentary or starting load of 200 percent normal (100 percent overload). If peak load on driven machine exceeds twice the normal running horsepower, divide the peak horsepower by 2 and compare with the equivalent obtained by item 1. If larger, use it instead of item 1 in selecting reducer from tables.

3. Extreme repetitive shock and applications where exceedingly high energy loads must be absorbed, as when stalling, require special consideration and are therefore not covered by service factors given in the table.

4. In selecting a unit, the horsepower required should not exceed the thermal rating of the unit. Thermal rating indicates the amount of power that can be delivered through the unit without overheating.

TABLE 9.3 AGMA Standard Practice for Single and Double-Reduction Cylindrical-Worm and Helical-Worm Speed Reducers

Prime mover	Duration of service per day	Driven machine load classifications		
		Uniform	Moderate shock	Heavy shock
Electric motor	Occasional ½ hr	0.80	0.90	1.00
	Intermittent 2 hr	0.90	1.00	1.25
	10 hr	1.00	1.25	1.50
	24 hr	1.25	1.50	1.75
Multicylinder internal combustion engine	Occasional ½ hr	0.90	1.00	1.25
	Intermittent 2 hr	1.00	1.25	1.50
	10 hr	1.25	1.50	1.75
	24 hr	1.50	1.75	2.00
Following service factors apply for applications involving frequent starts and stops				
Single-cylinder internal combustion engine	Occasional ½ hr	1.00	1.25	1.50
	Intermittent 2 hr	1.25	1.50	1.75
	10 hr	1.50	1.75	2.00
	24 hr	1.75	2.00	2.25
Electric motor	Occasional ½ hr	0.90	1.00	1.25
	Intermittent 2 hr	1.00	1.25	1.50
	10 hr	1.25	1.50	1.75
	24 hr	1.50	1.75	2.00

Notes: 1. Time specified for intermittent and occasional service refers to total operating time per day.

2. Term *frequent starts and stops* refers to more than 10 starts per hour.

TABLE 9.4 Service Factors for Gearmotors

The three classes of gearmotors and shaft-mounted reducers as defined by the American Gear Manufacturers Association are as follows:

Class I: For steady loads not exceeding normal rating of motor and 8 hr a day service. Moderate shock loads where service is intermittent. Service factor equals 1.

Class II: For steady loads not exceeding normal rating of motor and 24 hours a day. Moderate shock loads for 8 hours a day. Service factor equals 1.4.

Class III: For moderate shock loads for 24 hours a day. Heavy shock loads for 8 hours a day. Service factor equals 2

Start-Up

Some manufacturers ship new gear-drive units with internal parts protected by a polar-type rust-preventive film. There is no necessity to flush out this film, since it is usually soluble in the lubricant. (Consult the supplier of your particular gear-drive units for confirmation of this fact.) Merely fill the case with the recommended lubricant to the proper oil level. Always check to see if gear-drive units are shipped with or without oil from the factory. Units having bearings requiring grease must be checked and greased as required.

When units furnished with forced-feed lubrication are first put into service, they should be checked to observe that oil is being pumped. When a pressure gauge is furnished with the unit, gauge pressure should be as specified by the manufacturer, or if not specified, the pressure should be approximately 15 to 30 psi with the sump oil temperature at approximately 160°F (71°C). Adjust the relief valve if necessary to obtain the pressure specified in manufacturer's service manual.

Each unit is usually given a short run-in at the factory as part of the inspection procedure. However, for complete run-in under operating conditions, it is recommended that the unit be operated at partial load for 1 or 2 days to allow final wearing in of the gears. After this period, the load should be gradually increased to rated value.

After the unit has been operated under rated load for 2 weeks, it should be shut down in order to drain the oil and flush the housing. If desired, the original oil may be filtered, tested, and replaced. Filters finer than 25 microinches may filter out the additives. After the original oil has been drained, fill the case to the indicated level with SAE 10 straight-run mineral flushing oil containing no additives. The unit should be started, brought up to speed, and shut down immediately as a flushing procedure. Drain off flushing oil, and fill with recommended lubricant to the proper level.

After this initial oil change, an oil change is recommended after every 2500-hour or 6-month period of normal operation, whichever occurs first, unless there are unusually high temperature conditions combined with intermittent high loads where the temperature of the gear case rises rapidly and then cools off quickly. This condition may cause sweating on the inside walls of the unit, thus contaminating the oil and forming sludge. Under these conditions, or if the oil temperature is continuously above 150°F (65.5°C), or if the unit is subjected to an unusually moist atmosphere, oil changes may be necessary at 1- or 2-month intervals, as determined by field inspection of the oil. Synthetic oils, particularly hydrocarbons, may be used to improve oil life. Consult the manufacturer for recommended actions.

Lubrication

Lubricating oils for use with enclosed gears and gear units should be high-grade, high-quality, well-refined, straight mineral petroleum oils, within the recommended viscosity ranges as shown in Table 9.5. They must not be corrosive to gears or ball or roller bearings. They must be neutral in reaction. They should have good defoaming properties. No grit or abrasives should be present.

For high operating temperatures, good resistance to oxidation is needed. For low temperatures, an oil having a low pour point to meet the lowest temperature expected is needed. When the operating temperature varies over a wide range, an oil having a high viscosity index is desirable.

TABLE 9.5 AGMA Lubricant Number Recommendations for Enclosed Helical, Herringbone, Straight Bevel, Spiral Bevel, and Spur Gear Drives

	AGMA lubricant number[b,c]	
	Ambient temperature[d,e]	
	−10 to +10°C	10 to 50°C
Type of unit[a] (low speed center distance)	(15 to 50°F)	(50 to 125°F)
Parallel shaft (single reduction)		
Up to 200 mm (to 8 in)	2–3	3–4
Over 200 mm, to 500 mm (8 to 20 in)	2–3	4–5
Over 500 mm (over 20 in)	3–4	4–5
Parallel shaft (double reduction)		
Up to 200 mm (to 8 in)	2–3	3–4
Over 200 mm (over 8 in)	3–4	4–5
Parallel shaft (triple reduction)		
Up to 200 mm (to 8 in)	2–3	3–4
Over 200 mm, to 500 mm (8 to 20 in)	3–4	4–5
Over 500 mm (over 20 in)	4–5	5–6
Planetary gear units (housing diameter)		
Up to 400 mm (to 16 in) O.D.	2–3	3–4
Over 400 mm (over 16 in) O.D.	3–4	4–5
Straight or spiral bevel gear units		
Cone distance to 300 mm (to 12 in)	2–3	4–5
Cone distance over 300 mm (over 12 in)	3–4	5–6
Gearmotors and shaft-mounted units	2–3	4–5
High-speed units[f]	1	2

[a]Drives incorporating overrunning clutches as backstopping devices should be referred to the gear drive manufacturer as certain types of lubricants may adversely affect clutch performance.

[b]Ranges are provided to allow for variations in operating conditions such as surface finish, temperature rise, loading, speed, etc.

[c]AGMA viscosity number recommendations listed above refer to R&O gear oils. EP gear lubricants in the corresponding viscosity grades may be substituted where deemed necessary by the gear-drive manufacturer.

[d]For ambient temperatures outside the ranges shown, consult the gear manufacturer. Some synthetic oils have been used successfully for high- or low-temperature applications.

[e]Pour point of lubricant selected should be at least 5°C (9°F) lower than the expected minimum ambient starting temperature. If the ambient starting temperature approaches lubricant pour point, oil sump heaters may be required to facilitate starting and ensure proper lubrication.

[f]High-speed units are those operating at speeds above 3600 rpm or pitch line velocities above 25 m/s (5000 fpm) or both. Refer to Standard AGMA 421, "Practice for High Speed Helical and Herringbone Gear Units," for detailed lubrication recommendations.

When the gears are subject to heavy shock or impact loading or when the unit is subject to extremely heavy duty, an extreme-pressure (EP) lubricant should be used.

The EP lubricant must meet the general specifications listed above for straight mineral oil. For severe conditions, synthetic lubricants offer higher viscosity index and extended temperature operating range. All lubricants should meet the recommendations of the gear manufacturer.

The viscosity of the EP lubricant should be approximately the same as that of the recommended AGMA lubricants given in Tables 9.5 and 9.6.

On many types of gear-drive units, pressure fittings are supplied for the application of grease to bearings that are shielded from the oil.

Sufficient grease to form a film over the rollers and races of the bearing is all that is actually required for lubrication of roller bearings; however, ample reservoir space for grease is usually provided.

Gear-drive units are usually shipped from the factory with grease applied. Bearings and seals should be lubricated at definite intervals. Study will be required to determine how frequently this should be done for a particular operation to ensure a proper supply of lubricant to the bearing and seal areas.

Many greases are suitable as lubricants for bearings and seals in gear-drive units. The grease should be high-quality, nonseparating, ball-bearing grade suitable for the operating temperature.

TABLE 9.6 AGMA Lubricant Number Recommendations for Enclosed Cylindrical and Double-Enveloping Worm Gear Drives

Type, worm gear drive	Worm speed‡ up to, rpm	AGMA lubricant numbers* Ambient temperature†		Worm speed‡ above, rpm	AGMA lubricant numbers* Ambient temperature†	
		−10 to +10°C (15 to 50°F)	10 to 50°C (50 to 125°F)		−10 to +10°C (15 to 50°F)	10 to 50°C (50 to 125°F)
Cylindrical worm§						
Up to 150 mm (to 6 in)	700	7 Comp, 7 EP	8 Comp, 8 EP	700	7 Comp, 7 EP	8 Comp, 8 EP
Over 150 mm, to 300 mm (6 to 12 in)	450	7 Comp, 7 EP	8 Comp, 8 EP	450	7 Comp, 7 EP	7 Comp, 7 EP
Over 300 mm, to 450 mm (12 to 18 in)	300	7 Comp, 7 EP	8 Comp, 8 EP	300	7 Comp, 7 EP	7 Comp, 7 EP
Over 450 mm, to 600 mm (18 to 24 in)	250	7 Comp, 7 EP	8 Comp, 8 EP	250	7 Comp, 7 EP	7 Comp, 7 EP
Over 600 mm (over 24 in)	200	7 Comp, 7 EP	8 Comp, 8 EP	200	7 Comp, 7 EP	7 Comp, 7 EP
Double-enveloping worm§						
Up to 150 mm (to 6 in)	700	8 Comp	8A Comp	700	8 Comp	8 Comp
Over 150 mm, to 300 mm (6 to 12 in)	450	8 Comp	8A Comp	450	8 Comp	8 Comp
Over 300 mm, to 450 mm (12 to 18 in)	300	8 Comp	8A Comp	300	8 Comp	8 Comp
Over 450 mm, to 600 mm (18 to 24 in)	250	8 Comp	8A Comp	250	8 Comp	8 Comp
Over 600 mm (over 24 in)	200	8 Comp	8A Comp	200	8 Comp	8 Comp

*Both EP and compounded oils are considered suitable for cylindrical worm gear service. Equivalent grades of both are listed in the table. For double-enveloping worm gearing, EP oils in the corresponding viscosity grades may be substituted only where deemed necessary by the worm gear manufacturer.

†Pour point of the oil used should be less than the minimum ambient temperature expected. Consult gear manufacturer on lube recommendations for ambient temperatures below −10°C (14°F).

‡Worm gears of either type operating at speeds above 2400 rpm or 10 m/s (200 fpm) rubbing speed may require force-feed lubrication. In general, a lubricant of lower viscosity than recommended in the above table shall be used with a force feed system.

§Worm gear drives also may operate satisfactorily using other types of oils. Such oils should be used, however, only upon approval by the manufacturer.

The lubricant should not be corrosive to gears or to ball or roller bearings; must be neutral in reaction; should have no grit, abrasive, or fillers present; should not precipitate sediment; should not separate at temperatures up to 300°F (148.8°C); and should have moisture-resisting characteristics. The lubricant must have good resistance to oxidation.

Every precaution should be taken to prevent any foreign matter from entering the gear case. Sludge is caused by dust, dirt, moisture, and chemical fumes. These are the biggest enemies of proper and adequate lubrication in gear-drive units.

Good Maintenance Practice

During normal periods of operation, gear-drive units should be given daily routine inspection, consisting of visual inspection and observation for oil leaks or unusual noises. If oil leaks are evident, the unit should be shut down, the cause of the leakage corrected, and the oil level checked. If any unusual noises occur, the unit should be shut down until the cause of the noise has been determined and corrected. Check all oil levels at least once a week. The operating temperature of the gear-drive unit is the temperature of the oil inside the case. Under normal conditions, the maximum operating temperature should not exceed 180°F (82°C). Generally, pressure-lubricated units are equipped with a filter which should be cleaned periodically.

Shutdown

If it becomes necessary to shut down the unit for a period longer than 1 week, the unit should be run at least 10 min each week while it is idle. This short operation will keep the gears and bearings coated with oil and help prevent rusting due to condensation of moisture resulting from temperature changes.

Troubleshooting Gears

Someone has observed that "gears wear out until they wear in . . . and then they never wear out." The AGMA describes this phenomenon more precisely as follows:

It is the usual experience with a set of gears in a gear unit . . . assuming proper design, manufacture, application, installation, and operation . . . that there will be an initial "running-in" period during which, if the gears are properly lubricated and not overloaded, the combined action of rolling and sliding of the teeth may smooth out the manufactured surface and give the working surface a high polish. Under continued proper conditions of operation, the gear teeth will then show little or no sign of wear.

Despite the truth of this statement, failure of metallic gear teeth may occur as a result of excessive deterioration of the working surfaces of the teeth or as actual tooth breakage. In many such situations, early recognition of possible trouble may suggest a remedy before extensive damage occurs.

GEAR-TOOTH WEAR AND FAILURE

Experience indicates that the vast majority of gear-tooth wear and failure types may be summed up under nine basic headings in two classifications:

Classification A: Surface deterioration

1. Wear
2. Plastic flow
3. Scoring
4. Surface fatigue
5. Miscellaneous tooth-surface deteriorations

5.178 MAINTENANCE OF MECHANICAL EQUIPMENT

Classification B: Tooth breakage

6. Fatigue
7. Heavy wear
8. Overload
9. Cracking

The following discussion (which conforms to AGMA Standard 110.04 nomenclature) may be used as a guide in identification of gear-tooth trouble. If discovered early enough, many gear-tooth failures can be avoided through proper corrective maintenance as indicated. (The illustrations were prepared by the AGMA, which has given permission for their use.)

Surface Deterioration

Wear. Wear is a general term describing loss of material from the contacting surfaces of gear teeth.

Normal wear is the slow loss of metal from the contacting surfaces at a rate that will not affect satisfactory performance within the expected life of the gears. (See Fig. 9.13.)

Maintenance Procedure. A certain amount of smoothing and polishing is expected during "running-in" of new gearsets. This type of wear is less noticeable where gears have been shaved or finished-ground during manufacture. Before gears are run at all, they should be checked for proper installation and to ensure that loading is controlled within rating limits as set by the manufacturer. The use of recommended lubricant and filters should eliminate excessive gear-tooth wear during the running-in period. Most manufacturers of assembled gear drives recommend flushing the gear case frequently to remove any metallic particles and to eliminate the possibility of foreign objects circulating through the gear mesh.

Abrasive wear is surface injury caused by fine particles passing through the gear mesh. These particles may be dirt not completely removed, sand or scale from castings, impurities in the oil or from the surrounding atmosphere, or metal detached from the tooth surfaces or bearings. (See Fig. 9.14.)

Maintenance Procedure. Whenever abrasive wear is detected, the unit should be stopped immediately. Oil should be drained. The inside of the housing, gear teeth, and oil passages should be thoroughly scraped, flushed, and wiped down. A light flushing oil should be used for a short time and then drained before the oil reservoir is refilled with clean oil of proper grade. In contaminated atmospheres, special air breathers, oil seals, and filters should be considered as means of eliminating the infiltration of foreign particles into the gear case.

Scratching is a severe form of abrasive wear, characterized by short, scratchlike lines or marks on the contracting surfaces in the direction of sliding. It may be caused by burrs, projections on the tooth surface or material embedded in the tooth surface, or hard foreign pieces passing through the gear mesh. Scratching should not be confused with scoring because the two effects differ by definition, and scratching damage usually is light and does not result in progressive destruction, provided that the cause is removed. (See Fig. 9.15.)

Maintenance Procedure. Since scratching is an accentuated type of abrasive wear, with comparatively deep and widespread grooves up and down the tooth profile, the maintenance procedure

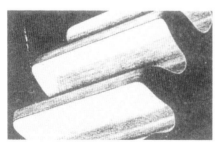

FIGURE 9.13 Normal wear.

FIGURE 9.14 Abrasive wear.

FIGURE 9.15 Scratching.

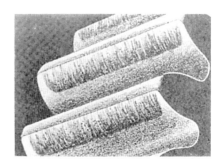

FIGURE 9.16 Overload wear.

is identical with that for simple abrasive wear. Make sure that the case, gears, and lubrication channels are completely free of foreign matter. Protect against recontamination by use of special filters, breathers, and oil seals where conditions indicate.

Overload wear is a form of wear experienced under conditions of heavy load and low speed, in both hardened and unhardened gears. Metal seems to be removed progressively in thin layers or flakes, leaving surfaces that appear somewhat as if etched. (See Fig. 9.16.)

Maintenance Procedure. The only permanent remedy for overload wear is to reduce the rate of wear. Care should be exercised in selecting extreme-pressure lubricants which are free from corrosive substances. Check the manufacturer for recommendations.

Plastic Flow. *Plastic flow* is the surface deterioration resulting from the yielding of the surface metal under heavy loads. It is usually associated with the softer metals but may occur in through-hardened and case-hardened steels.

Ridging is a particular form of plastic flow occurring on the tooth surfaces of case-hardened hypoid pinions and bronze worm gears. It usually appears as diagonal lines or ridges across the tooth surface but may be characterized by a herringbone or fishtail pattern, both occurring in the direction of sliding. Ridging is generally associated with excessive loads or inadequate lubrication, and usually complete failure results unless the material has a great capacity for work hardening. (See Fig. 9.17.)

Maintenance Procedure. Since ridging usually results from localized loading, wherever possible, gears should be adjusted to distribute the load more evenly over the full tooth surface. In the case of bevel gears, backlash should be altered to reduce impact loading. In some cases, the use of extreme-pressure lubricants may help to reduce the rate of tooth-surface deterioration.

Rolling and peening almost always occur together as the result of the sliding action under excessive loads and the impact loading from improper tooth action. They are characterized by fins at the top edges or ends of the teeth (not to be confused with burrs from cutting or shaving), by badly rounded tooth tips, or by a depression in the surface of the driving gear at the start of single-tooth contact, with a raised ridge near the pitch line of the mating or driven gear. The remaining portions of the profiles are usually deformed to a considerable degree long prior to complete destruction. (See Fig. 9.18.)

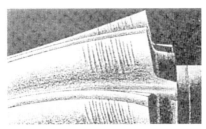

FIGURE 9.17 Ridging.

FIGURE 9.18 Rolling.

Maintenance Procedure. Often, peening can be checked by reducing the backlash of gear teeth. Occasionally, the addition of a flywheel to the gear shaft will serve to smooth out the hammerblow effects of spur gear teeth entering and leaving the mesh. Since peening is "localized" surface deterioration, extreme-pressure lubricant sometimes is effective in reducing this destructive type of plastic flow.

Rippling is a wavelike formation on the surface at right angles to the direction of sliding. It is characterized by a fish-scale appearance, occurs mostly on case-hardened hypoid pinions, and does not constitute failure unless allowed to progress. It may be caused by surface yielding due to "slip-stick" friction resulting from inadequate lubrication, heavy loads, or vibration. (See Fig. 9.19.)

Scoring. The term *scoring* has been selected as preferable to such other terms as *scuffing, seizing,* and *galling.* It is the rapid removal of metal from the tooth surfaces caused by the tearing out of small contacting particles that have welded together as a result of metal-to-metal contact, and the scored surface is characterized by a torn or dragged and furrowed appearance with markings in the direction of sliding. Sometimes surface roughness or foreign matter passing through the mesh will cause localized yielding on the mating profile, without tearing as such, with a similar furrowed appearance as a result of the "plowing" action. It may be localized initially and spread if the causative condition is not corrected. Sometimes, particularly in the case of misalignment, the damage may cease and the surface becomes smoother as the contact area spreads and more load-carrying face is brought into contact. Scoring is usually caused by rupture of the oil film resulting from load concentration at localized contact areas. Excessive unit loading or an unsuitable lubricant has the same effect. Sometimes scoring can be arrested by smoothing up the roughened area by filing or stoning or by use of a different type or grade of lubricant.

Slight scoring is a minor impairment of the gear-tooth surface of a welding nature, showing slight tears and scratches in the direction of sliding. Scoring usually starts at a surface area where there is a combination of high surface stress and sliding velocity—generally occurring at or near the tip of the tooth. (See Fig. 9.20.)

Severe scoring is a more advanced degree of welding, showing deep scratches and adhesions and leading to rapid gear-tooth-surface deterioration. (See Fig. 9.21.)

Maintenance Procedure. Correction of slight surface scoring often can be accomplished through the use of an extreme-pressure lubricant. Consult your lubricant supplier for specific recommendations. In some cases it may be necessary to polish tooth surfaces in addition to using an extreme-pressure lubricant. Where scoring persists, gear teeth may be metallurgically hardened (after polishing damaged areas) to resist further damage.

Surface Fatigue. *Surface fatigue* is the failure of the material as a result of repeated surface or subsurface stresses that are beyond the endurance limit of the material. It is characterized by the removal of metal and the formation of cavities. These cavities may be small and remain quite small; they may be small initially and then combine or increase in size by continued fatigue; or they may be of considerable size at the start.

Initial pitting is the type of surface fatigue which may occur at the beginning of operation and continue only until the overstressed local high areas of the surface have been reduced, thus obtaining

FIGURE 9.19 Rippling.

FIGURE 9.20 Slight scoring.

FIGURE 9.21 Severe scoring.

FIGURE 9.22 Initial pitting.

sufficient area of contact to carry the load without further impairment. It usually occurs in a narrow band just below or at the pitch line. Such pitting is not serious, since it is corrective and nonprogressive. (See Fig. 9.22.)

Maintenance Procedure. Usually, initial pitting is observed as tiny cavities at scattered spots on the surfaces of the gear teeth. In most cases, running-in of gears will tend to polish down surface irregularities, and pitting will cease. Where pitting continues, metallurgical surface hardening of the gear teeth may be necessary. On occasion, grinding and/or polishing of tooth-bearing surfaces will help.

Destructive pitting is a form usually starting below the pitch line, progressively increasing size and number of pits until smoothness of operation is impaired. The remaining surface fails in a similar manner, and finally the tooth shape is destroyed. The pits constitute stress raises which may lead to failure by fatigue breakage. Large pits formed by the joining of smaller adjacent pits are due to failure of the material between them and constitute a form of spalling. (See Fig. 9.23.)

Maintenance Procedure. Destructive pitting may be checked by grinding and polishing gear-tooth surfaces. If polishing fails to retard destruction, metallurgical surface hardening often will eliminate further damage. In some cases, use of extreme-pressure lubricants has met with success.

Spalling of the more usual type is a sporadic fatigue failure, occurring only in fully hardened or, more usually, case-hardened steel, originating with a surface or subsurface defect or from excessive internal stresses due to heat treatment. It is characterized by large particles or chips which spall or flake out the tooth surfaces, usually along the top edges or ends. The cavities are larger, deeper, and of a cleaner break than pits, although the distinction is primarily one of degree. Frequently, it is not a fatigue failure of the usual variety, since it occurs after a relatively few cycles as a result of excessive internal stresses. The joining of several smaller pits by failure of the metal between them is a form of spalling. (See Fig. 9.24.)

Maintenance Procedure. If damage from spalling is not too extensive, use of an extreme-pressure lubricant may retard further damage. In some cases, surface polishing will provide more even distribution of the load across the gear-tooth surface and will relieve excessive pressure at the point where spalling has occurred. If tooth destruction continues, consult the gear manufacturer.

FIGURE 9.23 Destructive pitting.

FIGURE 9.24 Spalling.

Miscellaneous Tooth-Surface Deterioration. Since corrosive wear, burning, interference, and grinding cracks in tooth-surface deterioration are independent sources of trouble, not closely related to one another or to the foregoing groups, they are treated independently.

Corrosive wear is surface deterioration from the chemical action of acid, moisture, or contamination of the lubricant. It may occur under several different circumstances. If the lubricant becomes contaminated with foreign acid, the teeth may become lightly pitted. The wiping action during contact may continually remove all evidence of this, but the rate of wear is excessive. Rusting as a result of contamination with water from condensation, excessive humidity, and so on, will produce similar results. If corrosion or rusting is taking place, evidence also should appear on other surfaces besides the active tooth faces. In addition, corrosive wear can occur as a result of highly active EP ingredients in the lubricant. Under heavy loads, EP oils may react with the metal, permitting operation without scoring but with a uniform and low rate of wear under load conditions that could not otherwise be tolerated. Gear teeth that are wearing as a result of EP activity usually have a smooth appearance. If the oil temperature becomes excessive, more active reaction of the EP materials with the metal can take place, resulting in accelerated high-temperature corrosive wear. (See Fig. 9.25.)

Maintenance Procedure. Drain and flush gear case and gears to remove the source of existing contamination. Be sure that new lubricant is clean, of high quality, and uncontaminated. If corrosive wear persists, consult the manufacturer for recommendations as to special breathers and oil seals.

Burning can result in severe wear and surface deterioration of the previously described types owing to loss of hardness from high temperatures. The fatigue life also may be adversely affected—depending on the degree and location of the burn. It is characterized by temperature discoloration of the contacting and/or adjacent surfaces and is the result of excessive temperature, from either external sources or the excessive friction from overload, overspeed, or inadequate lubrication. On gears that have not been put into service, the same discoloration would indicate improper grinding, but generally, grinding burns can be detected only by etching. (See Fig. 9.26.)

Maintenance Procedure. To reduce friction, look first to the lubricant. In many cases, extreme-pressure lubricants will eliminate gear-tooth burning. Be sure gears are not being run in excess of their rated load and speed capacities. If burning persists, request the gear manufacturer to test for proper backlash and gear-tooth spacing.

Interference wear occurs when improper or premature contact concentrates the entire load at the point of engagement of the driving flank with the mating tip or at the disengagement of the driven flank and mating tip. It may range from a light line of wear or pitting of no serious consequence other than noisy operation to a more severe damage in which the flank is gouged out and the tip of the mate heavily rolled over, usually resulting in complete failure of the pair. (See Fig. 9.27.)

Maintenance Procedure. Since interference usually is the result of improper gear design or manufacture, or deflection, or assembling the gears at too close a center distance for the profile shapes existing on the teeth, remedy for the situation should be left to the gear manufacturer.

Grinding cracks are fine surface cracks developed in grinding, usually in a definite pattern or network, caused by improper grinding technique or heat treatment or both. Sometimes they do not appear until the surface has been subject to load. Such cracks can be originating points for fatigue

FIGURE 9.25 Corrosive wear.

FIGURE 9.26 Burning.

FIGURE 9.27 Interference.

FIGURE 9.28 Grinding checks.

breakage, although sometimes the failure may be of the surface alone with large areas spalling out. (See Figs. 9.28 to 9.31.)

Maintenance Procedure. In some cases, grinding cracks will not cause serious gear-tooth deterioration if gears are properly lubricated. Where overloading, high operating velocities, or high-temperature service cause grinding cracks to enlarge, magnetic inspection and polishing may be useful in overcoming the trouble. If damage continues, consult the gear manufacturer.

Tooth Breakage

Tooth cracking or actual breakage is the end result of gear-tooth deterioration. These conditions are listed for identification purposes only, since their existence indicates a situation already beyond the ability of maintenance procedures to retard.

Fatigue breakage is the most common type of failure by breakage. It results from repeated bending stresses that are above the endurance limit of the material. Such stresses can result from poor design, overload, misalignment, or inadvertent stress raisers such as notches or surface or subsurface defects. It originates as a crack on the loaded side, usually in the fillet at the edge of the face, and progresses to complete failure either along the root or diagonally upward across the tooth. Fatigue fractures are usually characterized by a series of contour lines and a focal point in an area that is smooth by comparison. In the case of a subsurface point of origin, the eye (focal point) at the bottom of the cavity is highly polished. (See Fig. 9.32.)

Breakage from heavy wear is a secondary type, since it is a result of another kind of failure or wear. For instance, severe pitting, spalling, or heavy abrasive wear can remove enough metal to reduce the strength of the tooth below the breaking point.

Overload breakage is a rather common type of failure resulting from sudden shock overload and does not show progression of the crack as in fatigue. The fracture will have a silky appearance in the harder and more brittle materials, and a fibrous and torn appearance without a definite pattern in the more ductile metals. Misalignment which concentrates the load at one end of the face is usually the cause, but overload breakage also may be caused by welding of the teeth due to bearing failure, bent shafts, or large pieces of foreign matter entering the mesh.

FIGURE 9.29 Cracking.

FIGURE 9.30 Quenching cracks.

FIGURE 9.31 Overload breakage.

FIGURE 9.32 Fatigue breakage.

Quenching cracks result from excessive internal stresses developed by heat treatment and can be originating points for failure breakage. Usually they are visible hairline cracks. They may run across the top land or be radial in direction in the fillet region or be at random direction at the ends of the teeth. If large, the cracks may result in a failure similar to overload breakage after relatively few cycles. In either case, the initial portion of a break will be discolored from rusting or oxidation.

CONCLUSION

Maintenance of gear drives involves proper selection, proper installation, proper loading of the unit, proper lubrication, and periodic inspection. Metallic gears have tremendous service life when properly used and cared for.

CHAPTER 10
RECIPROCATING AIR COMPRESSORS

R. Keith Mobley
Principal, Life Cycle Engineering, Inc., Charleston, S.C.

An adequate and dependable supply of air is always necessary for continuous and economical operation of air tools. A specific compressor-maintenance program will go a long way toward obtaining the maximum efficiency from a compressor and eliminating unnecessary shutdown periods. The modern compressor is a precision-built machine, and it should be operated and maintained as such. Too many compressors are installed in out-of-the-way locations and are practically forgotten until trouble develops.

Each major air compressor manufacturer furnishes an installation, operation, and service instruction book with each unit. Many hours of preparation and years of experience are represented in these books. They are included with the compressors so that owners and operators will have sufficient information to install, operate, and maintain the equipment for maximum efficiency. Read the instruction book carefully and become familiar with the compressor construction so that minor adjustments and emergency repairs can be made. Also know who to contact should serious difficulty develop.

Location. For good maintenance a clean, well-lighted location should be selected with enough space allowed to dismantle any parts that may need to be removed for servicing. Too often compressors are located so that it is impossible to remove the pistons, rods, or cylinders without breaking through a wall or moving the compressor. Outline and foundation drawings show the necessary service clearance. Additionally, light overhead lifting capability is highly advantageous when overhaul is required. Near access to clean, cool, outside air will reduce cost of suction air piping. Maintenance and costs are materially reduced where these recommendations are followed.

Foundation. An adequate compressor foundation (Fig. 10.1) is a necessity for satisfactory operation and maintenance of a compressor. A foundation that is designed without sufficient mass and bearing surface will cause vibration of the compressor, resulting in discharge-, suction-, and water-line breakage and excessive wear of compressor parts.

For compressors requiring concrete foundations, the compressor vendor furnishes prints showing the foundation above the floor line plus the weights of the parts to be mounted on the foundation, also the out-of-balance forces that must be absorbed by the foundation. The amount of foundation will depend on the type of soil upon which it is being set. To determine the depth and size of the foundation below the floor line, a competent foundation engineer should be consulted who will take test cores and from these calculate the soil carrying capacity. With this information, along with the weights and out-of-balance forces, a foundation can be designed for satisfactory compressor operation.

FIGURE 10.1 Compressors on proper foundation.

Many small vertical compressors are installed on existing concrete floors and usually operate very well this way, as the large area of the floor forms a more than sufficient mass to offset any out-of-balance forces of the compressor.

At some locations it is impossible to set the compressor on a foundation or concrete floor that is poured on the ground. It must be located on a floor that does not have a solid base under it. For this type of installation, isolation dampers are used under the base supporting the compressor and its driver. Suction, discharge, and water lines should be attached with good flexible connections to prevent vibration and noises from being carried through the building. There are many manufacturers of isolation dampers, and their engineering should be consulted for recommendations for this type of application.

Air Filters and Suction Lines. Every compressor must be equipped with an air filter which should be the most efficient type made for the service it is applied to. The air filter must be located so that an adequate supply of cool, clean, and acid-free air will be had at all times, with explicit instructions for servicing the air filter posted where the maintenance personnel will always be reminded of the regular servicing required for good maintenance (Fig. 10.2).

At some locations it is necessary to place the air filter away from the compressor because of unfavorable surrounding conditions. Care must be used in providing a suction line to a compressor. It must be tight, free of dirt, chips, and scale, corrosion-resistant, and of adequate size for the length necessary to connect the air filter to compressor suction. PVC piping can provide a good solution to most of these requirements. Care should be taken to avoid acoustically resonant pipe lengths, which can result in excessive pulsations within the piping. The compressor manufacturer should provide a list of acoustically resonant pipe lengths to avoid. Normally the shortest possible suction line is preferable.

FIGURE 10.2 Typical reciprocating compressor.

The time interval for cleaning an air filter depends on the type of filter and its location, and is best determined by the differential pressure across the filter. Reputable air filter manufacturers offer differential pressure devices which indicate when the cleaner requires servicing.

Air-Receiver Location and Capacity. Air receivers often are considered accessories to air compressors and, for many applications, are not correctly installed or properly sized. Proper installation and proper sizing are very important for both compressor and air-line systems. An air receiver absorbs pulsations in the discharge line from the compressor and smooths the flow of air to the service lines. It serves as a reservoir for the storage of compressed air to take care of sudden and unusual momentary demands in excess of the capacity of the compressor. Another of its functions is to precipitate moisture that may be condensed in the receiver and prevent it from being carried into the air-distribution system.

The preferable location for an air receiver is as near the compressor as possible so that the discharge line can be of minimum length, eliminating pressure drop between the receiver and compressor. Many receivers are located outside the compressor room and are exposed to the weather, offering difficulties when the temperature drops low enough to cause freezing. An ordinary top-outlet safety valve can be frozen shut, creating a hazard; the valve should be placed with opening down, thereby keeping water out and allowing the valve to function if necessary. Should the compressor be shut down, allowing no air to pass through the receiver, the drain valve or mechanism can freeze, resulting in possible breaking of the parts making up the drain.

The size of the receiver usually is recommended by the compressor vendor, who has charts listing the necessary receiver sizes for various compressor sizes. Start-and-stop compressors require larger receivers than do continuously operated compressors, to keep them from starting too often. Each start requires electrical inrush to the motor, which can cause expense by increasing electrical requirements beyond normal electrical demand.

Air from the compressors should flow into the receiver at the bottom and out at the top. Condensate is a troublesome factor in the system. Use an efficient water-cooled aftercooler and separator between the compressor and receiver. The aftercooler will condense the moisture and collect most of it in the separator, which can be drained manually or automatically. The aftercooler dries and cools air, which promotes efficiency and safety. Most aftercoolers will cool the air within 15°F of the incoming cooling water. Where water supply is short or expensive, air-cooled aftercoolers are available. They are not as efficient as the water-cooled but, if properly sized and of good quality, usually will cool to within 20 to 30°F of the ambient temperature.

Always consult the compressor vendor about receiver problems. Many states are exacting about pressure-vessel requirements; pressure vessel must meet the codes for safety and pass inspection by the insurance companies.

Starting a New Compressor. Before a new or repaired compressor is started, careful check must be made of the lubricating system, making certain all places needing lubrication have been oiled per manufacturer's requirements. On compressors having a forced mechanical lubricator, crank or pump by hand until it is certain the oil is getting to the parts requiring lubrication, as some initial lubrication is required before the unit is started. Tighten all bolts, nuts, and cap screws. Turn the compressor over by hand wherever possible to determine that there is no interference or binding of working parts.

In the case of compressors requiring cooling water from a water main, turn on the water and check for leaks and for circulation through all parts requiring cooling water. For compressors having a self-contained water-cooling system, fill it and check to see that all air is out of the cooling system.

Check the discharge line from the compressor to the receiver, and if there are any globe, gate, or check valves anywhere between the compressor and receiver, be sure the valves are open and that there is a safety valve between the compressor and valves. The safety valve is a necessity, as it is possible that a valve might be left closed when the compressor started, resulting in an explosion should there be sufficient power in the driver, or should the overload protection fail to act.

If all points have been checked, apply driving power momentarily and let the machine coast to rest. Close observation during the coasting period will reveal any excessive tightness and will confirm proper direction of rotation. The time that the unloaded machine continues to roll after driving power has been removed gives a fair indication of no-load friction; if direction of rotation is correct and no other trouble is evident, the unit can be run without load. On units with a pressurized lubrication system for the crankcase, check immediately after start-up for proper oil pressure; if adequate oil pressure is not attained within about 10 sec after start-up, shut the unit down and determine the cause. While running unloaded, check any mechanical forced feed lubricators for proper drops-per-minute feed rates to cylinders and piston packings as specified by the operator's manual.

Operating a water-cooled compressor with too much cooling water through the system will cause excess condensate and cylinder wear because a cold cylinder will not lubricate properly; and because lubrication is affected, excess horsepower is required, adding to both maintenance and operating costs. A good rule is to hold the outlet temperature of the water between 110 and 120°F. This range will allow for good cooling and lubrication and also will keep condensation in the cylinder to a minimum. The introduction of cold water into cylinder water jackets must also be avoided in order to prevent condensation from occurring on interior cylinder walls. Usual practice is to circulate the cold supply water through aftercoolers or other water-cooled heat exchangers nearby and then direct the warmed water to the compressor cylinder jackets. In the case of multistage compressors, the cooling water is initially introduced into the interstage intercoolers and then directed to the cylinder jackets. It is good practice to have the temperature of the cooling water entering the cylinder jackets at least 10°F above the temperature of the air entering the cylinder to preclude condensation.

After running from 1 to 2 hr unloaded, with periodic stops to check for any heating of bearings or other working parts, apply partial load and build up to maximum load and pressure gradually. The entire breaking-in period should consume a minimum of 4 hr.

The importance of a break-in run cannot be stressed too strongly. The time and care spent in giving the running surfaces a polished finish pay dividends by increasing compressor life. After the initial run, compressor operation resolves itself into maintaining a clean air supply, feeding sufficient cooling water, and supplying adequate lubrication.

All the foregoing requirements are necessary to get a compressor ready for efficient operation and to hold maintenance costs to the minimum. Routine maintenance must now be set up and a definite pattern followed.

Lubrication. The most important check for any compressor is the lubrication system (Fig. 10.3). Keep the compressor well lubricated; check the oil level at least once every 8 hr of operation. Use only oil and greases as recommended by the compressor manufacturer. The oil used should have a low carbon-forming tendency and sulfur content and contain an oxidation inhibitor. It is important to use the correct viscosity oil, consistent with existing temperatures. The instruction book lists these conditions.

Because dust, dirt, and atmospheric conditions are different at various locations, it is not practical to state definitely how often the oil should be changed in the crankcase of an air compressor. Oil will become contaminated with foreign materials held in suspension and will also oxidize. The time

FIGURE 10.3 Pressurized lubrication system.

for oil changes is regulated by local conditions and must be determined by the discoloration and physical condition of the oil. Convenient laboratory services are available to analyze mail-in oil samples to assist in determining proper oil change periods and to provide warning of excessive frictional surfaces wear or other unusual foreign material contamination.

When oil changes are made, it will always pay to remove a handhole or cover plate and wipe the inside of the crankcase or power end clean with lint-free rags. If impossible to wipe out, use a good grade of flushing oil to remove any particles that may have settled on the crankcase floor. On units with pressurized crankcase lubrication, clean the oil pump suction strainer located in the lower part of the oil sump. Oil filter elements should also be replaced or cleaned (if applicable). When refilling the compressor oil sump, be certain the filling container is free of all dirt, grit, or dust. This simple point is often overlooked.

Nonlubricated Cylinders. A significant portion of double-acting reciprocating compressors being installed, or currently working, are equipped with nonlubricated cylinders. These cylinders are designed to avoid any metal-to-metal contact between pistons and cylinder bores, and are usually equipped with some type of filled-PTFE (polytetrafluoroethylene) piston rings and rider rings. Piston rod packings are also nonlubricated and usually equipped with either carbon or filled-PTFE packing rings. Nonlubricated, double-acting cylinders are also usually equipped with extended length piston rods and distance pieces so that no portion of the piston rod entering the lubricated crankcase can enter the nonlubricated piston rod packing. Additionally, the piston rod is usually equipped with a baffle ring to prevent creepage of lubricant from the lubricant-wetted portion of the rod to the portion entering the rod packing.

It is critically important to the expected service life and reliability of nonlubricated cylinders that due care be taken to prevent the intake of particulate and liquid contaminants into the cylinder, and the formation of condensate within the cylinder. Start-up procedures are the same as that described for lubricated cylinder units, with the exception of comments related to cylinder forced feed lubricators.

Valves. Reciprocating compressor valves must be kept in first-class operating condition, as leaking or broken valves cause excessive operating temperatures and loss of air delivered. It is therefore important to check the valves periodically and be certain they are always in good operating condition (Fig. 10.4).

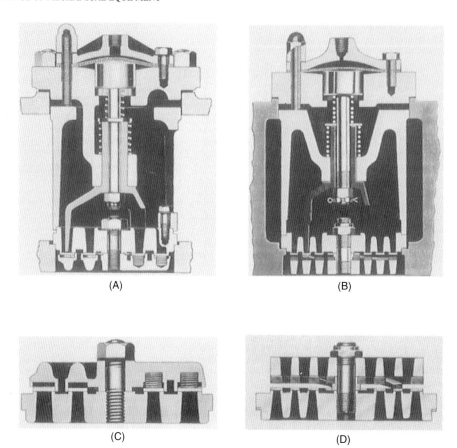

FIGURE 10.4 Compressor valves. *A* and *B*. Different designs of suction unloading valve assemblies. *C*. Compressor valve with individual disks and coil springs. *D*. Compressor valve with plate disk and finger springs.

The checking time for valves depends on several conditions, such as efficiency of the air cleaner, carbon-forming tendency of oil used, and the overall condition of the compressor. If the air cleaner is efficient and regularly serviced, excess dirt will be kept out of the airstream and dirt will not lodge in the valves. By using low-carbon-forming oil, the carbon buildup on the valves is held to a minimum. Synthetic lubricants such as diesters, polyol esters, and polyalphaolefins are commonly available which have characteristics highly desirable for air compressor cylinder lubrication. Although higher in cost than mineral oils, the synthetic oils result in cleaner-running valves and significant reductions in deposits formed in all hot areas of cylinder air passages and air piping. Most currently produced compressors are compatible with the above synthetic oils; however, compatibility with existing and older compressors should be confirmed with the manufacturer. Additionally, prior to conversion of any existing and older compressor and associated air distribution systems from mineral oil to synthetic oil lubrication, due care must be given concerning the solventlike action of most synthetic oils on hydrocarbon deposits within the compressor and air-distribution system. Proper procedures are available from the synthetic oil vendor and compressor manufacturer. For single-acting vertical compressors, the pistons, rings, and cylinder walls should be kept in good condition so that excess oil will not pass these parts. Low oil consumption adds to valve life by eliminating unnecessary carbon deposit. No set checking time can be recommended; it will need to be determined by actual investigation by the maintenance personnel. On a new unit the valves should be checked after 200 hr of operation.

Many compressor owners have found it helpful to have a spare set of valves so that a change of valves can be made immediately, and the replaced set reconditioned when time allows.

When valve troubles occur, there are several means of locating the valve or valves causing the difficulty. The first symptoms usually are low net air delivery and heating around the valve compartments. On a single-stage compressor, the usual method used is to check the temperature of the valve cover plates and examine the valve under the cover plate that is the hottest. If suction valves are leaking, a definite blow-back noise can be heard in the air cleaner when the compressor is operating under load.

On two-stage compressors, the intercooler pressure gage is used as a guide to locate defective valves. When low intercooler pressure occurs, examine the valves on the low-pressure cylinders, and when high intercooler pressure is found, examine those on the high-pressure cylinders. By checking the temperature of the valve cover plates, the defective valve can be located under the cover plate that is the hottest. If high-pressure suction valves are leaking, the intercooler-gage hand will fluctuate above normal intercooler pressure and the intercooler safety valve will pop. If high-pressure discharge valves are leaking, the intercooler-gage hand will rise steadily and pressure will build up in the intercooler until the intercooler safety valve will release it.

When low-pressure suction valves leak, the air will blow back through the suction line and air cleaner if the compressor is operating under load. Leaking low-pressure discharge valves will cause the intercooler pressure gage to fluctuate below normal intercooler pressure.

Since the valves are such an important part of the compressor, the information given in the instruction book must be followed when removing and installing them.

Wear between the valve disks or plate and the valve seat appears as indentations in the valve disks or plate, leaving a shoulder. The valve disks or plate are normally replaced if they show any amount of wear or the presence of any nicks, chips, or cracks.

Most worn valve seats can be resurfaced. On some types of valves, it is necessary to check the lift of the valve after resurfacing the seat, and if found to be more than recommended by the vendor, the bumper will need to be cut down to get the correct lift. Too much lift causes rapid wear and breakage.

Most valves usually have raised valve seats, and when the seat is refinished it is not necessary to do anything to the bumper, as the lift will still be to manufacturer's specifications.

Whenever a valve has been overheated, replace all the valve disks or plates and springs, because excessive temperature resulting from this heat will reduce the life of these parts and may result in breakage, causing damage to the compressor.

Most compressor valves have a gasket under the seat. This gasket must be in first-class condition; should it show any imperfection, replace it, as a leaking valve-seat gasket will eventually blow out.

The cover-plate gaskets are also important, and when installing valve cover plates, be sure the gaskets are in good condition. It is imperative that the valve cover-plate nuts or cap screws be pulled down evenly. Do not completely tighten one side and then the opposite side, as this will cause uneven gasket pressure, resulting in leads or sprung cover plates.

Several types or designs of valves are used by different compressor manufacturers, and in order to get the proper installation in the compressor, refer to the instruction book that was furnished with the compressor. Too much care cannot be used when maintaining and installing the valves and the component parts.

Piston Rings. Valves often are the cause for lost compressor efficiency, but should the valves be known to be satisfactory, the lost efficiency could well be in the piston rings. Piston-ring wear usually is very slow when the rings are properly lubricated, but operating time will eventually wear them so that the gap increases and the piston-ring lands wear to the point where some of the ring valving action is lost, allowing for blowby through the gap and around in back of the ring.

When excess blowby is suspected, remove the pistons and check the piston-to-cylinder wall clearance and the piston rings for the amount of wear, determining the parts that will need to be replaced. Scored cylinders always will allow excess blowby, adding to operating costs due to lost horsepower and fast wear.

Compressors having automotive-type pistons should have the wrist-pin fits checked when new piston rings are installed, and if found loose, the pin bushings should be replaced. Often the added drag on the cylinder walls caused by new piston rings will result in a pin knock when too much clearance is allowed.

Piston Rings. In some cases it is possible to rebore cylinders and obtain pistons and rings for certain oversize conditions, such as 0.005, 0.010, 0.020, or 0.030 in oversize. In still other cases, compressors are provided with replaceable cylinder liners. The compressor operator's maintenance manual should indicate the repair options available.

In the case of nonlubricated cylinders, particular care should be taken to see that piston rider rings do not wear to the point that allows piston-to-bore contact. As might be expected, horizontal and angular-mounted cylinders are subject to higher rates of rider ring wear than vertically mounted cylinders. Each time that valves are removed for inspection or repair, a feeler gage check should be made of the minimum piston-to-bore clearance. If the minimum clearance is 0.015 in or less, it is time to take action. When replacing the rider ring, carefully examine the compression rings for wear and end gap; follow the compressor manufacturer's recommendations as to when rings should be replaced.

Bearings. Crankpin bearings are usually the automotive-insert type (Fig. 10.5). To correct problems with the insert type, the installation of new inserts will serve. Should the crankpin surface be damaged, it can be ground undersize, built up by plating, metallizing, or plasma arc processes, and finish ground to original new condition dimensions. This allows the use of standard dimension replacement inserts. In certain cases, depending upon availability and cost of undersize inserts, it may be more feasible to grind the damaged crankpin undersize to fit available undersize inserts and omit the process of restoring the crankpin to original new dimensions.

Double-acting compressors have crosshead pins which operate in crosshead-pin bushings which have no adjustment. If a failure occurs, the crosshead-pin bushings must be replaced. Because of different fit requirements for the various compressor manufacturers, the running fit must be obtained from the compressor instruction book (Fig. 10 .6).

Many different constructions are used for the main bearings in both single-acting and double-acting compressors, no matter whether they are sleeve or antifriction bearings. Antifriction, single-row, tapered-roller-bearing adjustment is made by removing or adding shims. Double-row tapered roller bearings have an adjusting nut locked on the shaft. Unlock the nut and turn it to move the cone in on the cup. For trial purposes use a feeler gage and get about 0.002 in over the free rolls. Check bearings for heat and noise after starting, as it may be necessary to either tighten or loosen them slightly (Fig. 10.7).

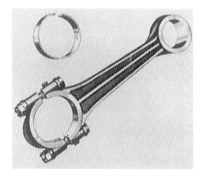

FIGURE 10.5 Automotive insert crankpin bearing with connecting rod.

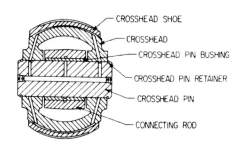

FIGURE 10.6 Crosshead pin bearing and pin.

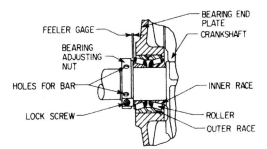

FIGURE 10.7 Crankshaft main-bearing adjustment for double-row tapered roller bearings.

Intercoolers and Aftercoolers. These are important compressor parts that often are neglected to the extent that they become inefficient. The most important maintenance is simple, and that is the proper draining of the moisture traps or compartments. Any type of cooler is a condenser, and the condensate, if not drained regularly, will build up until water is carried over to the high-pressure cylinders in the case of an intercooler, and on into the air receiver and air lines in the case of an aftercooler. Coolers should be drained regularly, according to existing humidity condition. The surest way to ensure draining is the use of automatic or timed drain traps on the intercooler, aftercooler, and air receiver.

Water-cooled intercoolers and aftercoolers are subject to buildup from the mineral content in water, which, if not removed, will eventually affect cooling; therefore, these coolers need inspection for deposit removal.

Air-cooled intercoolers and radiators must have the core sections cleaned on the outside, because dirt will lodge in the core, reducing heat dissipation. For removal of dust, air blown through in a direction opposite the usual flow will do; but in case the dirt is contaminated with oil, a solvent should be applied, allowed to soak for a while, and then blown clean.

Cleaning. An important item for proper compressor maintenance is keeping the compressor clean on the outside surfaces. Dirt and oil will make an insulation which hinders heat dissipation to atmosphere; this is especially true for an air-cooled compressor, which must depend on all heat dissipation through the cylinder and cylinder-head surfaces. When dirt is allowed to accumulate on the surfaces of a compressor, it is certain some will find its way into the working parts. A well-kept clean compressor will pay dividends with a good appearance plus reduced operating and maintenance costs.

Unloading. Practically every compressor manufacturer has his own type of air-unloading and control system; to cover all types would require complete data for each system. Some compressor vendors use several types; so for servicing and unloading system and its control, it is necessary to refer to the instruction book.

Some common unloading systems are suction unloading valves, suction closure device, centrifugal unloaders, and bypass systems. Most of these controls are operated by means of a pressure switch and a three-way valve actuated by a solenoid. Another means of actuating the unloading device is a pneumatic pilot, of which there are several types on the market.

Packing. Double-acting compressors using piston rods have oil-stop-head packing and cylinder-head packing which require periodic checking. The oil-stop-head packing is usually a set of metallic scraper rings. They require very little attention because they are designed to scrape oil off the rod, yet get excellent lubrication. Should the piston rod become damaged, the packing will be ruined and new packing required. Never put new scraper rings on a piston rod that is nicked, scratched, or worn (Fig. 10.8).

5.194 MAINTENANCE OF MECHANICAL EQUIPMENT

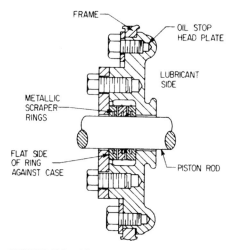

FIGURE 10.8 Oil-stop-head packing.

The cylinder-head packing usually is the full floating design with self-adjusting packing rings. The material, style, and quantity of the packing rings and number of lubrication lines depend on the type of gas being compressed and the discharge pressure. Some applications require special packing, such as vented and/or the elimination of nonferrous packing-ring and gasket materials (Fig. 10.9).

Metallic packing, after it is installed and worn in, requires little attention. However, the piston rod should be checked where it passes through the packing, and if any scratches are present, the packing must be removed and inspected for embedded material causing the scratches. As long as the packing does not leak or show any signs of marking the rod, it should not be disturbed, as the metal rings are self-adjusting for the slight wear that occurs under normal operation. Rod packings for nonlubricated cylinders function basically the same as for lubricated cylinders except that no lubricant is used and materials of construction are different. The packing ring sets are still of the segmental self-adjusting type but are, in most cases, of quite different material. Variations of carbon, filled-PTFE, and resin-bonded composites, among others, are employed for packing rings. Packing cups and glands are either made from corrosion-resistant material (stainless steels, brasses, bronzes, etc.) or are plated to resist corrosion.

The service check chart in Table 10.1 lists the common causes of malfunctions of mechanical parts of compressors.

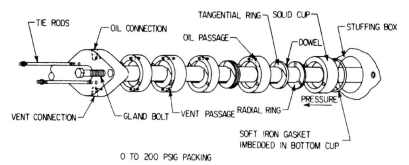

FIGURE 10.9 Cylinder-head packing.

TABLE 10.1 Service Check Chart, Mechanical Parts

1. Low oil pressure
    a. Low oil level
    b. Plugged oil-pump suction strainer
    c. Leaks in suction or pressure lines
    d. Worn-out bearings
    e. Defective oil pump
    f. Dirt in oil-filter check valve
    g. Broken oil-filter-check-valve spring
    h. Oil-pressure-bypass leaks
2. High oil pressure
    a. Plugged oil-pressure lines
    b. Defective oil-filter mechanism
    c. Excessive spring tension on filter check valves
    d. Excessive spring tension on oil-pressure adjusting mechanism
3. Incorrect delivery of mechanical lubricator
    a. Dirty or gummed pumps
    b. Broken spring in check valve at cylinder
    c. Leak in lines or sight feed
    d. Low oil level
    e. Plugged vent in lubricator reservoir
4. Overheated cylinder
    a. Insufficient cooling water
    b. Scored piston or cylinder
    c. Broken valves or valve springs
    d. Excessive carbon deposits
    e. Packing too tight
    f. Insufficient lubrication
    g. Corroded or clogged cylinder water passages
5. Water in cylinders
    a. Leaking head gaskets
    b. Cracked cylinder or head
    c. Condensate caused by too much cooling water or inoperative trap
6. High intercooler pressure
    a. Broken or leaking high-pressure valves
    b. Defective gage
    c. Defective or leaking valve-seat gaskets
7. Low intercooler pressure
    a. Broken or leaking low-pressure valves
    b. Leak in intercooler
    c. Piston-rod-packing leaking
8. Knocks
    a. Excessive carbon deposits
    b. Scored piston or cylinder
    c. Defective lubricator
    d. Foreign material in cylinder
    e. Piston hitting cylinder head
    f. Loose piston or piston pin
    g. Burned-out or worn rod bearings
    h. Loose main bearings
    i. Scored crosshead or crosshead guides
    j. Loose valve set screws

(Continued)

TABLE 10.1 Service Check Chart, Mechanical Parts (*Continued*)

9. Scored cylinder, liner, or piston
    a. Foreign material
    b. Dirty or inefficient air filter
    c. Lack of lubrication
    d. Too much and too cold cooling water causing excess condensate and washing out lubrication
    e. Excessive heat
    f. Plugged water jackets

10. Broken valves and springs
    a. Too much condensation, causing rust
    b. Carbon deposits
    c. Foreign materials not removed by air filter
    d. Incorrect assembly
    e. Acid condition prevailing at location of suction air inlet

11. Control trouble
    a. Suction-valve unloader stuck open or closed
    b. Pressure switch defective
    c. Solenoid burned out
    d. Foreign material in three-way valves
    e. Excessive vibration of control
    f. Voltage drop or loss of power
    g. Plugged air line or strainer
    h. Incorrect voltage or cycle

12. Incorrect operation of suction-valve unloaders
    a. Leaks in unloader line
    b. Foreign material in guides or seats
    c. Worn plungers
    d. Leaking or ruptured diaphragms and O-rings
    e. Broken springs
    f. Manual shutoff partly closed
    g. Wrong pressure-switch settings

Note: Remember to read the instruction book carefully and to keep it and the parts list in an accessible place so that when information to make adjustments and repair is needed, shutdown time can be held to a minimum.

CHAPTER 11
VALVES

Terry Hall
Reliability Engineer, Life Cycle Engineering, Inc., Charleston, S.C.

This chapter discusses the most common designs of industrial valves. The various designs of valves are developed from one idea—to place a disk over a seat opening in such a way that the resulting closure is tight. This one idea branches out into several basic designs of valves, including globe, check, gate, ball, and butterfly valves. Cross sections of globe, check, and gate valves are shown in Fig. 11.1. Cross sections of ball and butterfly valves are shown later in Figs. 11.9 and 11.10, respectively. Each design places the disk over the seat in a different manner.

Valves are usually made of one of three different metals, for the following reasons:

Bronze, for temperatures up to 550°F. Bronze is corrosion-resistant to a large majority of fluids, and it is easy to cast and machine. Bronze valves are usually made in sizes 3 and smaller.

Cast iron, for temperatures up to 450°F. Cast iron is cheaper than bronze, hence the cost of cast-iron valves larger than size 2 is decidedly reduced. Cast-iron valves usually have either a bronze or an all-iron trim. Valves with a bronze trim are called "I.B.B.M." (iron-body, bronze-mounted) valves, while valves with an iron trim are called "A.I." (all-iron) valves. All-iron valves are used for solutions that attack bronze but not iron, such as caustic soda and concentrated sulfuric acid.

Cast steel, for temperatures up to 1100°F. Steel is stronger at high temperatures than bronze or cast iron.

In addition to the different basic designs of valves and the three basic metals, valves are available in different pressure temperature ratings and with different end connections. Pressure ratings will be explained in the next section, and the most common pressure ratings and end connections for each of the basic valve designs and materials will be given in the section on that particular valve.

WHAT IS A VALVE?

In order to understand the operation, application, and maintenance of valves, it might be well if the definition of a valve were considered. For the purposes of this discussion, a valve is a mechanical device usually used in connection with a pressure-containing vessel to completely stop or regulate flow.

As a *mechanical device,* a valve should be selected to do the job expected of it and should be properly installed. It will then give long service before it starts to leak or wear out. After installation, and periodically during service, a valve should be checked to ensure that it has the necessary seat tightness.

When wear or leakage shows up, it will require some maintenance to restore the valve to its original efficiency. General maintenance methods for typical valves are described in this chapter, but the valve manufacturer's literature should be consulted for specific procedures. Another helpful reference

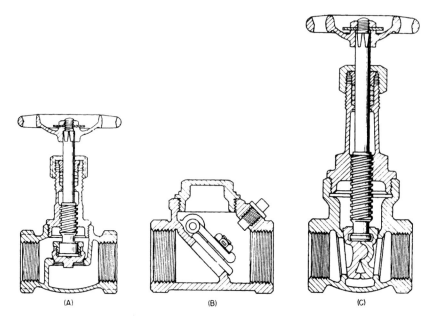

FIGURE 11.1 Three basic designs. *A*. Globe. *B*. Check *C*. Gate.

is MSS-SP-92, "MSS Valve Users Guide." Wear occurs more frequently in globe or check valves, and features are built into these valves to facilitate maintenance or renewal of parts. The seat of all globe valves is directly opposite the top opening of the body, making it easy to get at the seat for inspection or repair. Gate valves are installed where they are not operated very often, and hence do not wear out quickly, and they do not as a rule have the maintenance features of globe and check valves.

Mechanical devices should be operated occasionally. Valves which are placed in lines and then forgotten may become hard to operate. This is especially so in hot-water lines, hard-water lines, or any other lines in which there is a tendency to deposit scale or solids. Valves actually have been known to scale up or coke up so badly over a period of years that they had to be disassembled and cleaned before they were usable.

The statement that a valve is used *in connection with a pressure-containing vessel* deserves consideration because the pressures and temperatures at which a valve of a given size may be used depend upon wall thickness and material of the pressure-containing parts. For the purposes of indicating the pressure temperature ratings of valves, the American National Standards Institute (ANSI) has established pressure class numbers (or classes for short). The pressure class number corresponds to the former steam pressure (SP) or primary pressure rating of the valve. Since the tensile and yield strength of valve materials is higher at room temperature than at the temperature of steam, the rating of a given class of value is higher at room temperature than at the temperature of steam. The pressure rating of the valve at ambient temperature (0 to 150°F for bronze and iron and 0 to 100°F for steel) is called the cold working pressure (CWP) rating of the valve. The CWP rating of the valve corresponds to the former water-oil-gas (WOG) or secondary pressure rating of the valve. The CWP is about two times the pressure class number for Classes 300 and below, and about 2.4 times the pressure class number for Classes 350 and higher, depending upon material and size.

The statement that a valve is used *to completely stop or regulate flow* deserves consideration, as it indicates when a globe valve or when a gate valve is to be used. A globe valve is used to regulate flow, and a gate valve should be used where the service requires the valve to be in full open or closed position. The flow through a throttled globe valve is distributed uniformly around the entire

Globe throttled

Gate wide open

Gate throttled

FIGURE 11.2 Diagrams of flow condition. These diagrams illustrate the answer to the question as to where you use globe valves and where gate valves. Globe valves are used to throttle flow. Note that the flow is around the entire periphery of the disk, giving even wear. Globe valves are easily repaired or reground. Gate valves are used when you want unobstructed flow and little line loss. The illustrations show the uneven wear when gate valves are misused for throttling.

periphery of the disk, giving even and less rapid wear. The flow through a throttled gate valve is concentrated at the bottom of the wedge, giving uneven and more rapid wear. This is illustrated in Fig. 11.2.

Also owing to the construction of the valve, a globe valve is recommended when the valve is to be operated frequently. The disk in a globe valve touches the seat only at the instant of closing. In a gate valve, the wedge travels over the full face of the seat and consequently sliding wear will develop.

When a globe valve in these services finally wears, the globe valve is easier to repair than a gate valve.

Frequently, engineering specifications will state: "Globe valves shall be used on throttling service or where the valve is to be opened and closed frequently. Gate valves shall be used for full-flow conditions or where the valve is normally in an open or closed condition."

When an ordinary globe valve is used in severe throttling service, rapid wear of the seat and the disk can result. For tight closing it is sometimes better to use two valves on the line, one for throttling and one that is either full open or closed.

BRONZE VALVES

Bronze globe, check, and gate valves are available in Classes 125, 150, 200, 250, 300, and 350. They come with integral bronze seats and bronze disks and stems, and with nickel alloy and stainless steel seat rings and disks and bronze stems. Bronze valves are normally made with threaded or solder ends in sizes $1/4$ to 2.

Bronze Globe Valves

Globe valves are of various designs, some of which are illustrated in Fig. 11.3. A discussion of typical globe valves and their maintenance follows.

Threaded-Bonnet Globe Valve (Fig. 11.3*A*).   It is designed for use on inexpensive installations where the valve is not used frequently. Contractors often use it in low-pressure heating systems and on plumbing lines. Maintenance capability is very limited because the threaded body bonnet connection makes it difficult to regrind the valve. This valve can be repacked while under pressure by turning the valve stem full open. Caution should be taken if the valve is to be repacked under pressure as there is a possibility of dirt, scratches, or mars developing on the stem or bonnet back-seat in service, resulting in a poor seating condition. For this reason, repacking under pressure is not recommended unless absolutely necessary, and never for valves in hazardous services.

Union-Bonnet Regrinding Valve.   It is made in two different pressure ratings, 200 and 350 lb SP. It was originally designed for easy maintenance, without removal from the line. It can be reground and repaired. A small metal plate clamped between the end of the stem and the disk is used to prevent the disk from swiveling on the stem during the regrinding operation. The handwheel is used for the tool; the bonnet lip is used for a guide in the body neck. Valve-reseating tools, which can be obtained in sets from any industrial-supply house, can be used to dress up the seat if disk and seat are too worn for regrinding. The hardness of the bronze seat and disk is 85 Brinell, which makes the use of the reseating tool possible. The valve can be repacked under pressure. (See Fig. 11.3*B*.)

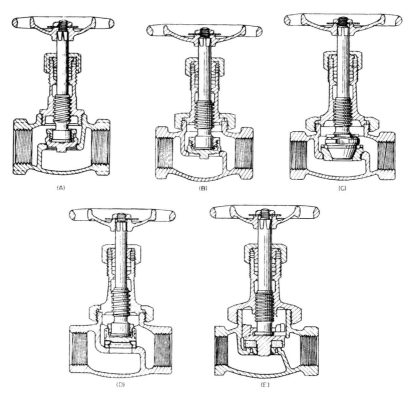

FIGURE 11.3 Bronze globe valves. *A.* Threaded-bonnet globe. *B.* Union-bonnet regrinding globe. *C.* Renewo globe, plug type. *D.* Flat-seat globe, 600 Brinell. *E.* Nonmetallic-disk globe.

Plug-Type Renewo Globe Valve. It is designed for severe throttling, drain, drip, water-column blowdown, and other services demanding high resistance to destructive action on seat bearings. *Maintenance* consists of renewing the seat and disk. Because of their hardness, these valves present quite a regrinding problem by hand. They are cone-shaped and are always installed in pairs. Their hardness is 500 Brinell. The valve can be repacked under pressure. (See Fig. 11.3C.)

600-Brinell Flat-Seat Globe Valve. There is practically no *maintenance* on the valve except for repacking because the flat seats are extremely hard and erosion resistant. (See Fig. 11.3D.) This valve is useful on steam, air, water, oil, gas, or other media, and will be equally tight in all these services.

Nonmetallic-Disk Globe Valve (Fig. 11.3E). It is also known as a composition disk globe valve. This valve is one of the most popular globe valves because of its easy maintenance. *Maintenance* consists principally of renewing the disk as it wears out. It can be easily removed from the disk holder, and a new disk inserted. Two kinds of disks are available: one for steam and hot water and one for cold water, air, gas, oxygen, solvents, acids, and alkalies. If the raised seat becomes worn or grooved, it can be resurfaced with a reseating tool. Brinell hardness is 85. The valve can be repacked under pressure.

Bronze Check Valves

Check valves are the guardians against backflow in a pipeline. They are entirely automatic in action and are of various designs, some of which are illustrated in Fig. 11.4. They fall into two general groups, commonly known as "swing-check" valves and "lift-check" valves. A swing-check valve is usually used where full flow is desired. A lift-check valve is usually used on air or gases or when the operation of the check valve is quite frequent. A discussion of typical check valves and their maintenance follows.

Nonmetallic-Disk Lift-Check Valve (Fig. 11.4A). The seat is rounded to give line contact on the disk, in comparison with the flat seat in a similar globe valve. The line contact is necessary, as there is usually little pressure to hold the disk to its seat. Sometimes a spring is inserted to act on the disk to increase this pressure. *Maintenance* of this valve consists of renewing the disk when necessary, smoothing the upper and lower disk guides when necessary, and removing grooves or worn places from the seat with a reseating tool.

Swing-Check Valve (Fig. 11.4B). Swing-check valves are probably the most popular and the most used of all check valves. They can be installed in horizontal or in vertical lines with flow up. *Maintenance* consists of regrinding the disk to its seat by applying a screwdriver to the slot in the top of the disk and using a grinding compound. If carrier pin, side plugs, or disk carrier become worn, they can be easily and inexpensively replaced with new parts. These maintenance suggestions apply also to I.B.B.M. swing-check valves.

Regrinding Lift-Check Valve. This is a good check valve in that it has both upper and lower guides to guide the disk to its seat. All parts are renewable except the seat, which is integral. *Maintenance* of this valve consists of regrinding the disk to its seat by means of the screwdriver slot in the disk stem. Brinell hardness of seat and disk is 85 Brinell, allowing for the use of a valve-reseating tool. (See Fig. 11.4C.)

Renewo Lift-Check Valve. This check valve has renewable seats and disks of nickel alloy, Brinell hardness 185. The disk has only an upper guide; so it is not as accurate in seating as the regrinding lift-check valve. *Maintenance* consists of regrinding the disk to its seat or replacing it if too badly worn. (See Fig. 11.4D.)

Ball Check Valve. Some people consider the ball check valve the ideal check valve. But in the opinion of valve men, it should be used only on viscous or heavy liquids, such as varnish, molasses, muddy water, or liquids containing solids. Any of these would clog the mechanism of the other check valves. There is little *maintenance* on a ball check valve, as there is no means of holding the ball for

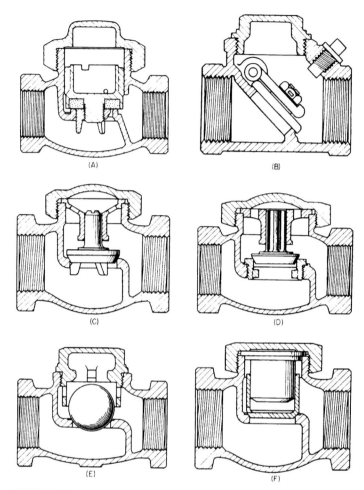

FIGURE 11.4 Bronze check valves. *A.* N-M-D lift-check nonmetallic disk. *B.* Swing-check regrinding seat. *C.* Lift-check regrinding seat. *D.* Lift-check renewable seat. *E.* Ball check. *F.* Air-compressor check.

regrinding the seats. This ball must be as perfect a sphere as possible, and the seat must be perfectly round. (See Fig. 11.4*E.*)

Air-Compressor Check Valve. This check valve (see Fig. 11.4*F*) is especially designed for this service, which is the hardest known for check valves. An ordinary check valve opens and closes once at each revolution of the compressor. Swing check valves have been known to disintegrate in 5 min with this frequency of operation.

The air-compressor check valve shown incorporates a stainless-steel disk operating over a bronze disk guide in such a manner as to provide an air cushion to reduce pounding. This air cushion damps the movement of the disk so the disk opens when the compressor starts, and stays open until the compressor shuts down. The disk then eases itself to its seat and is held tight by back pressure. Carry-over oil in the air line improves the efficiency of the air cushion. *Maintenance* consists of removing the cap and oiling the parts inside of the disk if there is insufficient carry-over oil, replacing renewable parts and reseating the seat. It is best to install the valve as far from the compressor as possible. This will reduce pulsations acting on the disk.

Bronze Gate Valves

Gate valves are by far the most popular and the most frequently used of the three types of valves—globe, check, and gate. As their correct installation calls for usage where they are opened and closed only infrequently, they last a long time and do not require much maintenance. If a gate valve is operated over ten times a day every day, it will quickly wear out and a globe valve should be substituted for it. The wear will be found on the downstream faces of the seat and the wedge because the line pressure forces all the wear on these surfaces. The upstream faces frequently will be found to be in good condition. Very often, worn gate valves can be reversed 180° and they will be as good as new. Gate valves should be installed with the stem vertical if at all possible. Installation with the stem in a horizontal position is permissible, but is not as good.

The three gate valves shown at the top of Fig. 11.5 are kindred valves, in that all are the same body. The first one shown is the most popular valve known—a standard double-wedge bronze gate valve with rising stem.

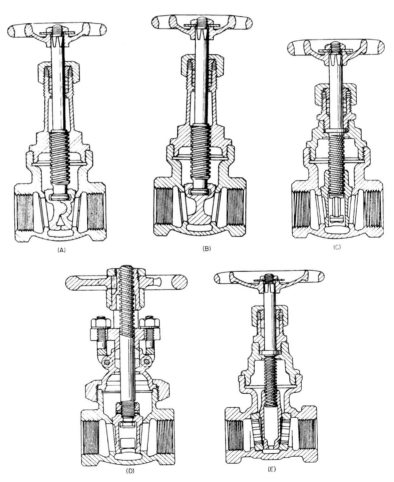

FIGURE 11.5 Bronze gate valves. *A.* Rising-stem, double-wedge disk. *B.* Solid-wedge disk, rising stem. *C.* Non-rising-stem, single-wedge disk. *D.* Outside screw and yoke union bonnet, single-wedge disk. *E.* Nonrising stem, single-wedge disk, renewable seat rings.

Double-Wedge Rising-Stem Gate Valve (Fig. 11.5A). The double wedges are of ball-and-socket construction or of the uniball type where each single wedge has a half socket and half ball. They readily adjust themselves to the taper seats, ensuring a tight valve. It should be evident that line strains could distort the angle of the taper seats in the bronze body. Slight distortion will not affect the tightness of double-wedge valves. The rising stem indicates whether the valve is open or closed. There is little *maintenance* to be done on a gate valve. It should be taken apart occasionally over the years and cleaned, especially valves on hot-water lines. The valve can be repacked under pressure by opening the valve to the limit of the stem travel.

Solid-Wedge Rising-Stem Gate Valve (Fig. 11.5B). This valve is used on heavy liquids such as molasses or varnish or any other liquids that would tend to make ball-and-socket wedges inoperative. The solid-wedge bronze gate valve is not as tight on thin liquids or gases as is the double-wedge valve. *Maintenance* is the same as for the double-wedge rising-stem gate valve.

Single-Wedge Nonrising-Stem Gate Valve (Fig. 11.5C). This valve is popular with contractors and is used extensively on marine service. The nonrising-stem feature allows the valve to be used in cramped spaces where overhead construction interferes with the operating of a rising-stem gate valve.

Small Outside Screw-and-Yoke Rising-Stem Gate Valve (Fig. 11.5D). Note that this valve has a union bonnet and a single wedge. It is required by code on the lines leading to the top and bottom connections on the water column of a boiler. It must be locked in open position by means of a chain and padlock so that there is no wear on it. It is an emergency valve; hence its *maintenance* consists of test operation and inspection to see that it is in working condition. It can become limed up rather quickly, especially the valve on the lower line, as this line contains hot water.

Renewable-Wedge-and-Seat Bronze Gate Valve (Fig. 11.5E). The valve shown has a nonrising stem and renewable nickel-alloy seats and wedge. It is popular in the chemical industry, as the seats and disks can be renewed as they become unsatisfactory in operation. The valve should be removed from the line during this repair. Gate valves are not like globe valves, which can be repaired while on the line.

IRON VALVES

Cast-iron globe, check, and gate valves are available in Classes 125 and 250. They come with bronze seat rings, disk facings, and stem (iron-body bronze-mounted, or I.B.B.M.) and with integral iron seats or steel seat rings, iron disk facings, and steel stems (all-iron, or A.I.). Iron valves are normally made with threaded and grooved ends in sizes 2 to 6 and with flanged ends for sizes 2 to 30.

Iron Globe and Check Valves

Figure 11.6 shows iron-body globe and angle valves, respectively. The application and maintenance of these is similar to bronze globe valves.

Iron-body swing and lift-check valves are also available.

Iron Gate Valves

The size 6 I.B.B.M., outside screw-and-yoke (O.S. & Y.), flanged-end (F.E.) gate valve shown in Fig. 11.7 is the most popular in size and design of all the large valves.

The *maintenance* of the valve follows commonsense lines. The stem threads should be kept lubricated and free from dirt. When the valve is wide open for a long period of time, the exposed stem threads should be protected by a light sheet-iron tube placed over them.

To repack the valve, move the swing gland bolts out of the way. The gland is raised and rests on the ledges provided for that purpose. The stuffing box is then accessible for renewal of the packing.

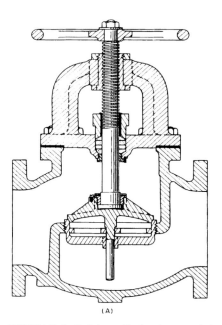
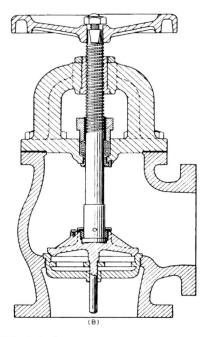

FIGURE 11.6 Variations of basic valves as to design. *A.* Globe. *B.* Angle.

Each ring of new packing should be compressed by the gland before another ring is added. Splits in split ring packing should be staggered. The valve can be repacked under pressure.

Should the downstream seats become scored, the upstream seats will frequently be found to be in good condition. Reverse the valve 180°, and the valve will be as good as new.

Should it become necessary to replace the seat rings, remove the valve from the line and prepare a correct-size pipe with square notches to fit the lugs in the seat rings. As the pipe with lugs is twisted (by means of a bar), tap the body smartly with a hammer to help loosen the ring. Clean all threads and seating surfaces with a wire brush before installing new rings. Graphite or pipe dope can be used. A new disk should be installed with new rings. It may have to be lapped in. Retighten body bonnet bolts uniformly using a crisscross pattern and at least three passes.

STEEL VALVES

Forged-steel globe, check, and gate valves are available in Classes 150, 300, 600, 800, 1500, 2500, and 4500. They are normally made with flanged ends in sizes $\frac{1}{2}$ to 2 for Classes 150, 300, and 600, and threaded and socket weld ends in sizes $\frac{1}{4}$ to 2 for Classes 800, 1500, 2500, and 4500.

Cast-steel bolted-bonnet globe, check, and gate valves are available in Classes 150, 300, and 600. They come with 410 stainless-steel seat and wedge facing and stems for normal service, and hardfaced

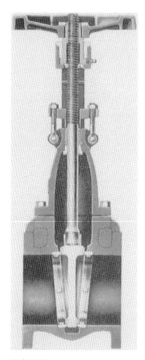

FIGURE 11.7 Iron-body gate valve.

seat and wedge facing and 410 stainless-steel stems for severe service. They are also available in bronze, monel, and 316 trims for special services. They may have screwed-in or welded-in seat rings. They are normally made with flanged or butt-weld ends in sizes 2 to 30.

Cast-steel breech-lock and pressure-seal globe, check and gate valves are made in Classes 600, 900, 1500, 2500, and 4500. They come with hardfaced welded-in seat rings, hardfaced wedges, and 410 stainless-steel stems. They are normally made with butt-weld ends in sizes 3 to 24. These valves are used in high-pressure, high-temperature steam and water service in power plants. Figure 11.8 shows breech-lock gate valves with three different means of actuation.

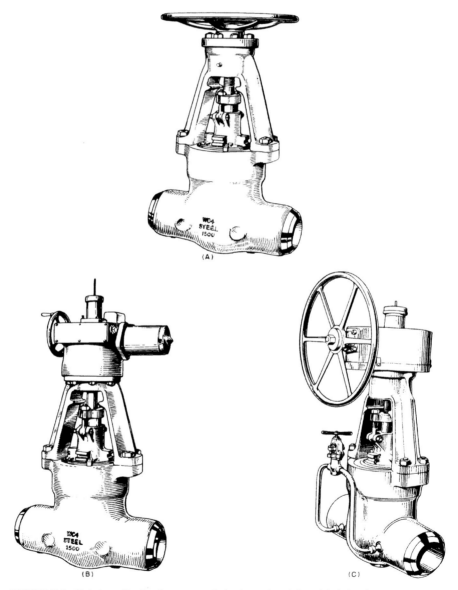

FIGURE 11.8 Variations of basic valves as to methods of operation. *A.* Breech lock, hand-operated. *B.* Breech lock, motor-operated. *C.* Breech lock with bypass, bevel-gear operated.

The *maintenance* of steel valves consists of periodic checks for shell and seat tightness, periodic lubrication and exercising, and repacking when needed. Repair of steel valves is more difficult than for other valves because of their harder materials, integral lay-on or welded-in seat rings, larger size and weight, and use of welded pipe connections. Globe and gate valves must be removed from the line and machine tools used to renew the seats, or else special power equipment can be used to renew the seats in line. Valves with welded-in seat rings must be removed from the line for replacement of seat rings, which is best done by the manufacturer or a valve reconditioning shop. Gate valve wedges can be hand lapped on a surface plate with lapping compound if their condition is not too bad.

BALL VALVES

The ball valve has a spherical closure element (or ball) with a port through it (equivalent to the disk in a conventional valve) mounted on renewable seats of Teflon. To open the valve, the ball is rotated so that the through-port lines up with the seat openings. When the valve is closed, line pressure forces the ball against the downstream seat, in an action similar to that of a gate valve. The ball valve is more compact, tighter sealing, quicker operating, and more easily maintained than conventional gate or globe valves, and has better flow characteristics. It is used in many services where conventional valves would have been used formerly. (See Fig. 11.9.)

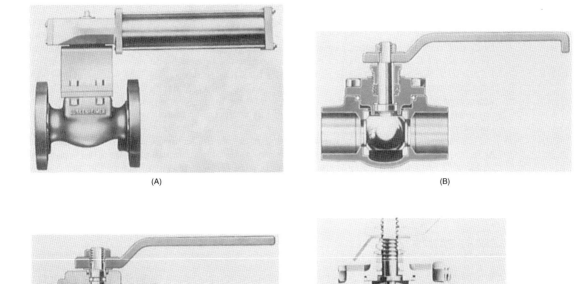

FIGURE 11.9 Variations of ball valves. *A.* Flange-end ball valve with pneumatic actuator. *B.* Top-entry ball valve. *C.* End-entry ball valve. *D.* Three-piece ball valve.

5.208 MAINTENANCE OF MECHANICAL EQUIPMENT

Ball valves come in sizes $1/4$ through 24 and larger. Pressure ratings are 250 to 3000 psig cold working pressure, and the temperature is limited by the sealing elastomer. They are available in bronze, and carbon and alloy steel shell materials with Teflon and other sealing elastomers. They are made with threaded, solder, flanged, socket-welding, and butt-welding end connections, depending upon shell material and size. Design variations include top-entry (Fig. 11.9*B*), end-entry (Fig. 11.9*C*), and three-piece (Fig. 11.9*D*) designs. The top-entry valve can be repaired without removing it from the line. The end-entry design is the most economical and has a simpler or no body joint. The three-piece design offers a built-in union connection.

The maintenance of ball valves is simple. Should the valve show leakage, the seats, seals, and balls can be replaced easily, thus giving a new valve.

BUTTERFLY VALVES

The butterfly valve has a disk-shaped closure element which rotates about a central shaft (or stem). (See Fig. 11.10.) To close the valve, the disk is rotated so that its face is across the pipe, blocking flow. Depending upon the type of valve, the valve seat may consist of a bonded resilient liner, a mechanically fastened resilient liner, an insert-type reinforced resilient liner (Fig. 11.10*C*), a mechanically fastened resilient seal (Fig. 11.10*D*), or an integral metal seat with an O-ring inserted around the edge of the disk. Butterfly valves are much more easily repaired than conventional valves. Butterfly valves find applications in most categories where valves are used. They are used extensively in gathering lines in oil-country installations. They are also used throughout industry for all fluids with which they are compatible.

Butterfly valves are commonly available in sizes $1^1/_2$ through 24. Cold working pressure ratings are 150, 200, 275, and 720 psig, with temperature limited by seat and sealing elastomers. (Butterfly valves with the latter two ratings are often called high-performance butterfly valves.) Body materials are aluminum, cast and ductile iron, and carbon and stainless steel; seat and seal elastomers are Buna N, EPT, viton, and Teflon. There are threaded- and grooved-end body styles for sizes $1^1/_2$ through 6, and wafer and lug styles for sizes $1^1/_2$ through 24. The wafer and lug types have very narrow bodies and are installed by bolting between standard ANSI flanges. On-off detent-type levers; 10-position lift or squeeze-type levers; manual worm gear actuators; and electric, pneumatic, and hydraulic actuators are available. Actuators are recommended for size 8 and larger butterfly valves.

The maintenance of butterfly valves is quite simple. No lubrication is necessary until such time as stem O-rings need replacement. Disks and stems are readily replaceable, as are disk O-rings and replaceable resilient seats. Replacement of the bonded resilient liner is a factory job, but usually when that type of seat deteriorates it is more economical to purchase a complete new valve, since the cost of butterfly valves is low.

ORDERING SPARE PARTS

Whenever it is necessary or advisable to order spare or replacement parts, the first problem that arises is to identify the part, and the second problem is to identify the valve.

Identifying the Parts. It is well to specify the correct name of the part wanted; and for this reason, most valve catalogs carry illustrations naming the parts. This type of information is shown in Fig. 11.11 in the cross sections of the gate and globe valves, with the size and figure number being a necessary part of the information. If you have a valve catalog of one manufacturer and are ordering parts for another make of valve, specify the catalog and page number you are using to identify the parts. The names of each part of a valve are not identical with all the various valve manufacturers.

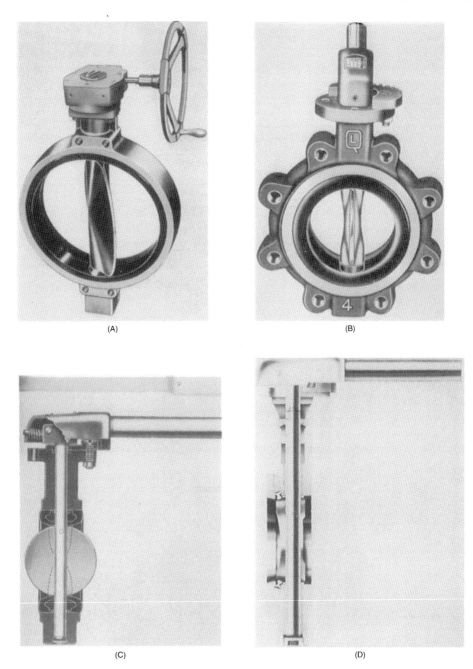

FIGURE 11.10 Variations of butterfly valves. *A.* Wafer type with gear operator. *B.* Lug type with replaceable seat. *C.* Wafer type with replaceable seat and lever operation. *D.* High-performance butterfly valve.

5.210 MAINTENANCE OF MECHANICAL EQUIPMENT

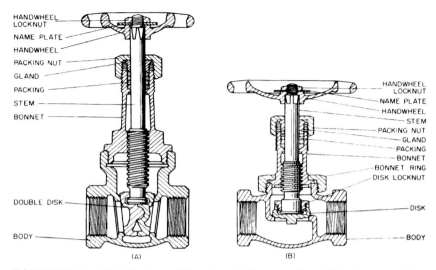

FIGURE 11.11 Parts identification. *A*. Rising-stem, double-wedge valve. *B*. Union-bonnet regrinding globe valve.

Most valve catalogs also show exploded views of their valves with each part illustrated. This makes it a little easier to identify the part desired. This type of information is shown in Fig. 11.12 for a swing-check valve.

Identifying the Valve. Most modern valves carry a nameplate, which makes it easy to identify the manufacturer and figure number. Nameplates originated with steel valves; but after the close of World War II, they were placed on bronze and iron valves, also. Unfortunately, many of the valves without nameplates have not worn out, and we are forced to identify the valve in some other way. If the valve is covered with insulation, it will be necessary to remove this to view the markings on the body.

There you will probably find the name of the manufacturer, the size of the valve, and the steam pressure, and other ratings. This information may be on both sides of the body. The figure number is rarely found on the body casting, as the same casting is often used to make valves carrying various figure numbers.

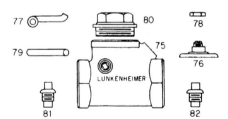

FIGURE 11.12 Ordering parts. Illustrated is bronze regrinding swing-check valve, Figs. 624 and 596 in maker's catalog. Orders should specify size and figure number of valve for which part is intended, quantity and name of repair part, and reference to part key number and catalog page or drawing number, if available.

Copy all this information and include it with the order, which should specify the metal the valve is made of, a brief description together with the type of valve (globe, check, or gate), and the type of ends (screw or flange). In the case of flanged valves, the type of face (whether flat or raised face), the diameter of the flange, the diameter of the bolt circle, the size of the bolts or bolt holes, and finally the face-to-face dimension of the flanges should be given. If possible, give the approximate date of the installation of the valve. Valve designs and details and figure numbers change over the years, and all the above information suggested will be helpful, especially in the case of an old valve. Sometimes it may be necessary to call in a representative of the valve manufacturer to help identify an old valve.

Writing the Order. In writing the order, it is well to specify first the part wanted, and let the other information follow, thus:

Disk only, Size $1^1/_2$ Lunkenheimer
Fig. 554Y, class 200 valve, screw ends.

Frequently repair-parts orders read, "One size $1^1/_2$-in, Fig. 554Y, disk only." This will probably result in a complete valve being shipped, unless some well-informed order checker scans the order.

RECOMMENDED PIPING PRACTICE

Clean the inside of the pipe before installing or repairing a valve. This will remove rust, scale, welding beads, and dirt, which could be carried into a valve and cause trouble. Do not remove flange or thread protectors from the valve until ready for installation. When threading pipe, do not cut the threads too long. Long threads allow the pipe to enter the valve too deeply and distort the seat or will hit the diaphragm. Apply pipe dope or Teflon tape to male threads only when making up a threaded joint. When installing a screw-end valve, do not employ enough force to distort the valve body. Use a crescent wrench or monkey wrench on the valve end that is being made up. Employ a pipe wrench on pipe only. Allow a new valve to warm up gradually. Packing glands are assembled hand-tight at the factory, and on installation should be tightened only enough to prevent leakage.

When installing a globe or angle valve to a pipeline, the direction of flow should be so that the pressure is under the disk except in steam, open-end, and drain lines where the pressure should be on top of the disk.

When installing a check valve in liquid lines, it is suggested that the following guidelines be adhered to:

1. Check valves should be installed as far as possible from the pump discharge.
2. Do not use a swing-check valve on reciprocating liquid pumps; always use a vertical-lift type.
3. In cases of water hammer, noise, or shock during closing of a swing-check valve, it is recommended that the check valve be changed to vertical-lift type, but increase the size by one so that the pressure drop will remain the same for the same flow. If noise is still excessive, a gas-over-liquid damper can be added to the line by inserting a tee and a vertical pipe as high as possible. The pipe should be two or three sizes larger than the supply pipe. An accumulator-type damper can be installed where shock is critical.

Finally, it is extremely important that a valve be closed tightly by hand only. Do not use a wrench or persuader. Dirt under the disk can usually be flushed out by operating the valve a number of times. A valve that is cracked open is subject to the most severe wire-drawing or throttling conditions possible, decreasing valve life and increasing maintenance.

CHAPTER 12
PUMPS: CENTRIFUGAL AND POSITIVE DISPLACEMENT

Carl March
Reliability Engineer Life Cycle Engineering, Inc., Charleston, S.C.

Pumps are designed to transfer a specific volume of liquid at a particular pressure from a fixed source to a final destination in a process system. A pump's operating envelope is defined either by a hydraulic curve for centrifugal pumps or a pressure-volume (PV) diagram for positive-displacement pumps.

CENTRIFUGAL PUMPS

Centrifugal pumps are highly susceptible to variations in process parameters, such as suction pressure, specific gravity of the pumped liquid, backpressure induced by control valves, and changes in demand volume. Therefore, the dominant reasons for centrifugal-pump failures are usually process-related.

Several factors dominate pump performance and reliability: internal configuration, suction condition, total dynamic pressure or head, hydraulic curve, brake horsepower, installation, and operating methods. These factors must be understood and used to evaluate any centrifugal pump-related problem or event.

Configuration

All centrifugal pumps are not alike. Variations in the internal configuration occur in the impeller type and orientation. These variations have a direct impact on a pump's stability, useful life, and performance characteristics.

Impeller Type. There are a variety of impeller types used in centrifugal pumps. They range from simple radial flow, open designs to complex variable-pitch, high-volume enclosed designs. Each of these types is designed to perform a specific function and should be selected with care. In relatively small, general-purpose pumps, the impellers are normally designed to provide radial flow and the choices are limited to either *enclosed* or *open* design.

Enclosed impellers are cast with the vanes fully encased between two disks. This type of impeller is generally used for clean, solid-free liquids. It has a much higher efficiency than the open design.

Open impellers have only one disk and the opposite side of the vanes is open to the liquid. Because of its lower efficiency, this design is limited to applications where slurries or solids are an integral part of the liquid.

Impeller Orientation. In *single-stage* centrifugal pumps, impeller orientation is fixed and is not a factor in pump performance. However, it must be carefully considered in *multistage* pumps, which are available in two configurations: in-line and opposed. These configurations are illustrated in Fig. 12.1.

In-Line. In-line configurations have all impellers facing in the same direction. As a result, the total differential pressure between the discharge and inlet is axially applied to the rotating element toward the outboard bearing. Because of this configuration, in-line pumps are highly susceptible to changes in the operating envelope.

Because of the tremendous axial pressures that are created by the in-line design, these pumps must have a positive means of limiting endplay, or axial movement, of the rotating element. Normally, one of two methods is used to fix or limit axial movement: (1) a large thrust bearing is installed at the outboard end of the pump to restrict movement, or (2) discharge pressure is vented to a piston mounted on the outboard end of the shaft.

The first method relies on the holding strength of the thrust bearing to absorb energy generated by the pump's differential pressure. If the process is reasonably stable, this design approach is valid and should provide relatively trouble-free service life. However, this design cannot tolerate any radical or repeated variation in its operating envelope. Any change in the differential pressure or transient burst of energy generated by flow change will overload the thrust bearing, which may result in instantaneous failure.

The second method uses a bypass stream of pumped fluid at full discharge pressure to compensate for the axial load on the rotating element. While this design is more tolerant of process variations, it cannot compensate for repeated, instantaneous changes in demand, volume, or pressure.

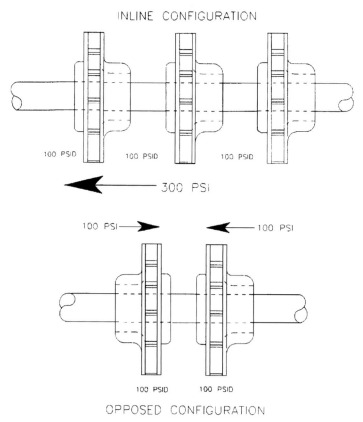

FIGURE 12.1 Impeller orientation of multistage centrifugal pumps.

Opposed. Multistage pumps that use opposed impellers are much more stable and can tolerate a broader range of process variables than those with an in-line configuration. In the opposed-impeller design, sets of impellers are mounted back-to-back on the shaft. As a result, the thrust or axial force generated by one of the pairs is canceled by the other. This design approach virtually eliminates axial forces. As a result, the pump does not require a massive thrust bearing or balancing piston to fix the axial position of the shaft and rotating element.

Since the axial forces are balanced, this type of pump is much more tolerant of changes in flow and differential pressure than the in-line design. However, it is not immune to process instability or to the transient forces caused by frequent radical changes in the operating envelope.

Performance

This section provides the basic knowledge needed to evaluate a centrifugal-pump application to determine its operating dynamics and to identify any forcing function that may contribute to chronic reliability problems, premature failures, or loss of process performance.

Centrifugal pump performance is primarily controlled by two variables: suction conditions and total system pressure or head requirement. Total system pressure is comprised of the total vertical lift or elevation change, friction losses in the piping, and flow restrictions caused by the process. Other variables affecting performance include the pump's hydraulic curve and brake horsepower.

Suction Conditions

Factors affecting suction conditions are the net positive suction head, suction volume, and entrained air or gas.

Net Positive Suction Head. Suction pressure, called net positive suction head or NPSH, is one of the major factors governing pump performance. The variables affecting suction head are shown in Fig. 12.2.

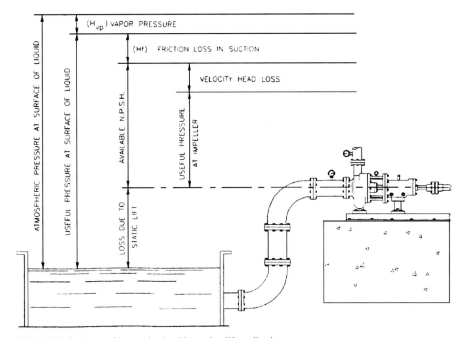

FIGURE 12.2 Net positive suction head in suction lift application.

Centrifugal pumps must have a minimum amount of consistent and constant positive pressure at the eye of its impeller. If this suction pressure is not available, the pump will be unable to transfer liquid. The suction supply can be open and below the pump's centerline, but the atmospheric pressure must be greater than the pressure required to lift the liquid to the impeller eye and to provide the minimum NPSH that is required for proper pump operation.

At sea level, atmospheric pressure generates a pressure of 14.7 pounds per square inch (psi) to the surface of the supply liquid. This pressure minus vapor pressure, friction loss, velocity head, and static lift must be enough to provide the minimum NPSH requirements of the pump. These requirements vary with the volume of liquid transferred by the pump.

Most pump curves provide the minimum NPSH required for various flow conditions. This information, which is generally labeled $NPSH_R$, is generally presented as a rising curve located near the bottom of the hydraulic curve. The data are usually expressed in "feet of head" rather than psi.

To convert from psi to feet of head for water, multiply by 2.31. For example, 14.7 psi is 14.7 times 2.31 or 33.957 feet of head. To convert feet of head to psi, multiply the total feet of head by 0.4331.

Suction Volume. The pump's supply system must provide a consistent volume of single-phase liquid equal to or greater than the volume delivered by the pump. To accomplish this, the suction supply should have relatively constant volume and properties (e.g., pressure, temperature, specific gravity, etc.). Special attention must be paid in applications where the liquid has variable physical properties (e.g., specific gravity, density, viscosity, etc.). As the suction supply's properties vary, effective pump performance and reliability will be adversely affected.

In applications where two or more pumps operate within the same system, special attention must be given to the suction flow requirements.

Generally, these applications can be divided into two classifications: pumps in series and pumps in parallel.

Pumps in Series. The suction conditions of two or more pumps in series are extremely critical (see Fig. 12.3). Since each pump depends on the flow and pressure of the preceding pump, the flow characteristics must match. Both the flow and pressure must be matched to the required suction conditions of the next pump in the series.

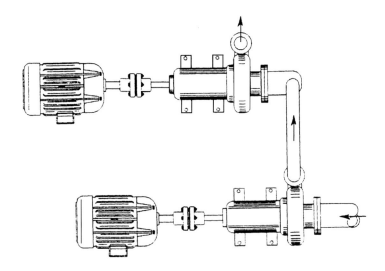

FIGURE 12.3 Pumps in series must be properly matched.

For example, the first pump in the series may delivery 1000 gallons per minute (gpm) and 100 feet of total dynamic head. The next pump in the series will then have an inlet volume of 1000 gpm, but the inlet pressure will be 100 feet minus the pressure losses created by the total vertical lift between the two pumps' centerlines and all friction losses caused by the piping, valves, and the like.

This pressure at the suction of the second pump must be at least equal to its minimum NPSH operating requirements. If too low, the pump will cavitate and will not generate sufficient volume and pressure for the process to operate properly.

Pumps in Parallel. Pumps that operate in parallel normally share a common suction supply or discharge (or both). This is illustrated in Fig. 12.4. Typically, a common manifold (i.e., pipe) or vessel is used to supply suction volume and pressure. The manifold's configuration must be such that all pumps receive adequate volume and net positive suction head. Special consideration must be given to flow patterns, friction losses, and restrictions.

One of the most common problems with pumps in parallel is suction starvation. This is caused by improper inlet piping that permits more flow and pressure to reach one or more pumps, but supplies insufficient quantities to the remaining pumps. In most cases, this is the result of poor piping or manifold design and may be expensive to correct.

Always remember that, when evaluating flow and pressure in pumping systems, they will always take the path of least resistance. For example, given a choice of flowing through a 6-in pipe or a 2-in pipe, most of the flow will go to the 6-in pipe. Why? Simply because there is less resistance.

In parallel pump applications, there are two ways to balance flow and pressure to the suction inlet of each pump. The *first* way is to design the piping so that the friction loss and flow path to each of the pumps are equal. While this is theoretically possible, it is extremely difficult to accomplish. The *second* method is to install a balancing valve in each of the suction lines. Throttling or partially closing these valves can tune the system tuned to ensure proper flow and pressure to each pump.

Entrained Air or Gas. Most pumps are designed to handle single-phase liquids within a limited range of specific gravities or viscosities. Entrainment of gases, such as air or steam, has an adverse effect on both the pump's efficiency and its useful operating life. This is one form of cavitation, which is a

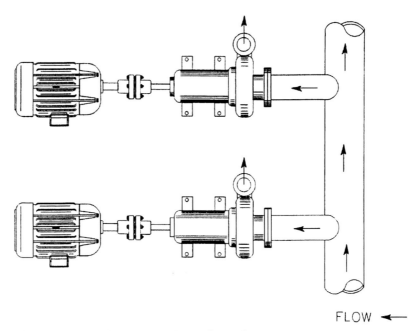

FIGURE 12.4 Pumps in parallel may share suction supply.

common failure mode of centrifugal pumps. The typical causes of cavitation are leaks in suction piping and valves, or a change of phase induced by liquid temperature or suction pressure deviations. As an example, a 1-lb suction pressure change in a boiler-feed application may permit the deaerator-supplied water to flash into steam. The introduction of a two-phase mixture of hot water and steam into the pump causes accelerated wears, instability, loss of pump performance, and chronic failure problems.

Total System Head

Centrifugal pump performance is controlled by the total system head (TSH) requirement, unlike positive-displacement pumps. TSH is defined, as the total pressure required overcoming all resistance at a given flow. This value includes all vertical lift, friction loss, and backpressure generated by the entire system. It determines the efficiency, discharge volume, and stability of the pump.

Total Dynamic Head

Total dynamic head (TDH) is the difference between the discharge and suction pressure of a centrifugal pump. Pump manufacturers to generate hydraulic curves, such as those shown in Figs. 12.5 to 12.7, use this value. These curves represent the performance that can be expected for a particular pump under specific operating conditions. For example, a pump having a discharge pressure of 100 psig and a positive pressure of 10 psig at the suction will have a TDH of 90 psig.

Hydraulic Curve

Most pump hydraulic curves define pressure to be TDH rather than actual discharge pressure. This is an important consideration when evaluating pump problems. For example, a variation in suction pressure has a measurable impact on both discharge pressure and volume. Figure 12.5 is a simplified hydraulic curve for a single-stage, centrifugal pump. The vertical axis is TDH and the horizontal axis is discharge volume or flow.

The best operating point for any centrifugal pump is called the *best efficiency point* (BEP). This is the point on the curve where the pump delivers the best combination of pressure and flow. In addition, the BEP defines the point that provides the most stable pump operation with the lowest power consumption and longest maintenance-free service life.

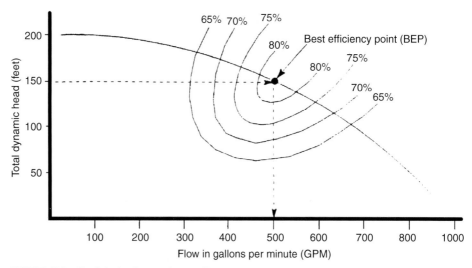

FIGURE 12.5 Simple hydraulic curve for centrifugal pump.

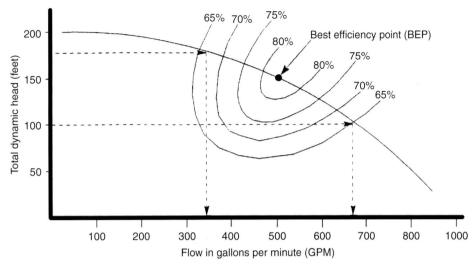

FIGURE 12.6 Actual centrifugal pump performance depends on total system head.

In any installation, the pump will always operate at the point where its TDH equals the TSH. When selecting a pump, it is hoped that the BEP is near the required flow where the TDH equals TSH on the curve. If it is not, there will be some operating-cost penalty because of the pump's inefficiency. This is often unavoidable because pump selection is determined by what is available commercially as opposed to selecting one that would provide the best theoretical performance.

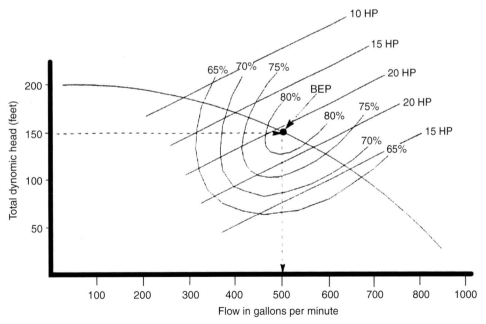

FIGURE 12.7 Brake horsepower needs change with process parameters.

For the centrifugal pump illustrated in Fig. 12.5, the BEP occurs at a flow of 500 gpm with 150 ft TDH. If the TSH were increased to 175 ft, however, the pump's output would decrease to 350 gpm. Conversely, a decrease in TSH would increase the pump's output. For example, a TSH of 100 ft would result in a discharge flow of almost 670 gpm.

From an operating-dynamic standpoint, a centrifugal pump becomes more and more unstable as the hydraulic point moves away from the BEP. As a result, the normal service life decreases and the potential for premature failure of the pump or its component increases. A centrifugal pump should not be operated outside the efficiency range shown by the bands on its hydraulic curve, or 65 percent for the example shown in Fig. 12.5.

If the pump is operated to the left of the minimum recommended efficiency point, it may not discharge enough liquid to dissipate the heat generated by the pumping operation. This can result in a heat build-up within the pump that can result in catastrophic failure. This operating condition, which is called *shut-off*, is a leading cause of premature pump failure.

When the pump operates to the right of the last recommended efficiency point, it tends to overspeed and become extremely unstable. This operating condition, which is called *run-out*, also can result in accelerated wear and premature failure.

Brake Horsepower

Brake horsepower (BHP) refers to the amount of motor horsepower required for proper pump operation. The hydraulic curve for each type of centrifugal pump reflects its performance (i.e., flow and head) at various BHPs. Figure 12.7 is an example of a simplified hydraulic curve that includes the BHP parameter.

Note the diagonal lines that indicate the BHP required for various process conditions. For example, the pump illustrated in Fig. 12.7 requires 22.3 horsepower at its BEP. If the TSH required by the application increases from 150 ft to 175 ft, the horsepower required by the pump increases to 24.6. Conversely, when the TSH decreases, the required horsepower also decreases. The brake horsepower required by a centrifugal pump can be easily calculated by:

$$\text{Brake Horsepower} = \frac{\text{Flow (GPM)} \times \text{Specific Gravity} \times \text{Total Dynamic Head (ft)}}{3960 \times \text{Efficiency}}$$

With two exceptions, the certified hydraulic curve for any centrifugal pump provides the data required to calculate the actual brake horsepower. Those exceptions are specific gravity and TDH.

Specific gravity must be determined for the specific liquid being pumped. For example, water has a specific gravity of 1.0. Most other clear liquids have a specific gravity of less than 1.0. Slurries and other liquids that contain solids or are highly viscous materials generally have a higher specific gravity. Reference books, like Ingersoll Rand's *Cameron's Hydraulic Databook*, provide these values for many liquids.

The TDH can be directly measured for any application using two calibrated pressure gauges. Install one gauge in the suction inlet of the pump and another on the discharge. The difference between these two readings is TDH.

With the actual TDH, flow can be determined directly from the hydraulic curve. Simply locate the measured pressure on the hydraulic curve by drawing a horizontal line from the vertical axis (i.e., TDH) to a point where it intersects the curve. From the insect point, draw a vertical line downward to the horizontal axis (i.e., flow). This provides an accurate flow rate for the pump.

The intersection point also provides the pump's efficiency for that specific point. Since the intersection may not fall exactly on one of the efficiency curves, some approximation may be required.

Installation

Centrifugal pump installation should follow *Hydraulic Institute Standards*, which provide specific guidelines to prevent distortion of the pump and its baseplate. Distortions can result in premature wear, loss of performance, or catastrophic failure. The following should be evaluated as part of a root cause failure analysis: foundation, piping support, and inlet- and discharge-piping configurations.

Foundation. Centrifugal pumps require a rigid foundation that prevents torsional or linear movement of the pump and its baseplate. In most cases, this type of pump is mounted on a concrete pad having enough mass to securely support the baseplate, which has a series of mounting holes. Depending on size, there maybe three to six mounting points on each side.

The baseplate must be securely bolted to the concrete foundation at all of these points. One common installation error is to leave out the center baseplate lag bolts. This permits the baseplate to flex with the torsional load generated by the pump.

Piping Support. Pipe strain causes the pump casing to deform and results in premature wear and/or failure. Therefore, both suction and discharge piping must be adequately supported to prevent strain. In addition, flexible isolator connectors should be used on both suction and discharge pipes to ensure proper operation.

Inlet-Piping Configuration. Centrifugal pumps are highly susceptible to turbulent flow. The *Hydraulic Institute* provides guidelines for piping configurations that are specifically designed to ensure laminar flow of the liquid as it enters the pump. As a rule, the suction pipe should provide a straight, unrestricted run that is six times the inlet diameter of the pump.

Installations that have sharp turns, shut-off or flow-control valves, or undersized pipe on the suction-side of the pump are prone to chronic performance problems. Such deviations from good engineering practices result in turbulent suction flow and cause hydraulic instability that severely restricts pump performance.

Discharge-Piping Configuration. The restrictions on discharge piping are not as critical as for suction piping, but using good engineering practices ensures longer life and trouble-free operation of the pump. The primary considerations that govern discharge-piping design are friction losses and total vertical lift or elevation change. The combination of these two factors is called TSH, which represents the total force that the pump must overcome to perform properly. If the system is designed properly, the TDH of the pump will equal the TSH at the desired flow rate.

In most applications, it is relatively straightforward to confirm the total elevation change of the pumped liquid. Measure all vertical rises and drops in the discharge piping, then calculate the total difference between the pump's centerline and the final delivery point.

Determining the total friction loss, however, is not as simple. Friction loss is caused by a number of factors and all depend on the flow velocity generated by the pump. The major sources of friction loss include:

- Friction between the pumped liquid and the sidewalls of the pipe
- Valves, elbows, and other mechanical flow restrictions
- Other flow restrictions, such as backpressure created by the weight of liquid in the delivery storage tank or resistance within the system component that uses the pumped liquid

There are a number of reference books, like Ingersoll-Rand's *Cameron Hydraulics Databook*, that provide the pipe-friction losses for common pipes under various flow conditions. Generally, data tables define the approximate losses in terms of specific pipe lengths or runs. Friction loss can be approximated by measuring the total run length of each pipe size used in the discharge system, dividing the total by the equivalent length used in the table, and multiplying the result by the friction loss given in the table.

Each time the flow is interrupted by a change of direction, a restriction caused by valving, or a change in pipe diameter there is a substantial increase in the flow resistance of the piping. The actual amount of this increase depends on the nature of the restriction. For example, a short-radius elbow creates much more resistance than a long-radius elbow; a ball valve's resistance is much greater than a gate valves; and the resistance from a pipe-size reduction of 4 in will be greater than for a 1-in reduction. Reference tables are available in hydraulics handbooks that provide the relative values for each of the major sources of friction loss. As in the friction tables mentioned earlier, these tables often provide the friction loss as equivalent runs of straight pipe.

In some cases, friction losses are difficult to quantify. If the pumped liquid is delivered to an intermediate storage tank, the configuration of the tank's inlet determines if it adds to the system pressure. If the inlet is on or near the top, the tank will add no backpressure. However, if the inlet is below the normal liquid level, the total height of liquid above the inlet must be added to the total system head.

In applications where the liquid is used directly by one or more system components, the contribution of these components to the total system head may be difficult to calculate. In some cases, the vendor's manual or the original design documentation will provide this information. If these data are not available, then the friction losses and backpressure need to be measured or an over-capacity pump selected for service based on a conservative estimate.

Operating Methods

Normally, little consideration is given to operating practices for centrifugal pumps. However, some critical practices must be followed, such as using proper startup procedures, using proper bypass operations, and operating under stable conditions.

Startup Procedures. Centrifugal pumps should always be started with the discharge valve closed. As soon as the pump is activated, the valve should be slowly opened to its full-open position.

The only exception to this rule is when there is positive backpressure on the pump at startup. Without adequate backpressure, the pump will absorb a substantial torsional load during the initial startup sequence. The normal tendency is to overspeed because there is no resistance on the impeller.

Bypass Operation. Many pump applications include a bypass loop intended to prevent deadheading (i.e., pumping against a closed discharge). Most bypass loops consist of a metered orifice inserted in the bypass piping to permit a minimal flow of liquid. In many cases, the flow permitted by these metered orifices is not sufficient to dissipate the heat generated by the pump or to permit stable pump operation.

If a bypass loop is used, it must provide sufficient flow to assure reliable pump operation. The bypass should provide sufficient volume to permit the pump to operate within its designed operating envelope. This envelope is bound by the efficiency curves that are included on the pump's hydraulic curve, which provides the minimum flow required to meet this requirement.

Stable Operating Conditions. Centrifugal pumps cannot absorb constant, rapid changes in operating environment. For example, frequent cycling between full-flow and no-flow assures premature failure of any centrifugal pump. The radical surge of backpressure generated by rapidly closing a discharge valve, referred to as *hydraulic hammer*, and generates an instantaneous shock load that can literally tear the pump from its piping and foundation.

In applications where frequent changes in flow demand are required, the pump system must be protected from such transients. Two methods can be used to protect the system.

Slow Down the Transient. Instead of instant valve closing, throttle the system over a longer time interval. This will reduce the potential for hydraulic hammer and prolong pump life.

Install Proportioning Valves. For applications where frequent radical flow swings are necessary, the best protection is to install a pair of proportioning valves that have inverse logic. The primary valve controls flow to the process. The second controls flow to a full-flow bypass. Because of their inverse logic, the second valve will open in direct proportion as the primary valve closes, keeping the flow from the pump nearly constant.

POSITIVE-DISPLACEMENT PUMPS

Centrifugal and positive-displacement pumps share some basic design requirements. Both require an adequate, constant suction volume to deliver designed fluid volumes and liquid pressures to their installed systems. In addition, both are affected by variations in the liquid's physical properties (e.g., specific gravity, viscosity, etc.) and flow characteristics through the pump.

Unlike centrifugal pumps, positive-displacement pumps are designed to displace a specific volume of liquid each time they complete one cycle of operation. As a result, they are not as prone to variations in performance as a direct result of changes in the downstream system. However, there are exceptions to this. Some types of positive-displacement pumps, such are screw-types, are extremely sensitive to variations in system backpressure.

When positive-displacement pumps are used, the system must be protected from excessive pressures. This type of pump will deliver whatever discharge pressure is required to overcome the system's total head. The only restrictions to its maximum pressure are the burst pressure of the system's components and the maximum driver horsepower.

Because of their ability to generate almost unlimited pressure, all positive-displacement pumps systems must be fitted with relief valves on the downstream side of the discharge valve. This is required to protect the pump and its discharge piping from over-pressurization. Some designs include a relief valve that is integral to the pump's housing. Others use a separate valve installed in the discharge piping.

Positive-displacement pumps deliver a definite volume of liquid for each cycle of pump operation. Therefore, the only factor except for pipe blockage that affects flow rate in an ideal positive-displacement pump application is the speed at which it operates. The flow resistance of the system in which the pump is operating does not affect the flow rate through the pump. Figure 12.8 shows the characteristics curve (i.e., flow rate versus head) for a positive-displacement pump.

The dashed line in Fig. 12.8 shows actual positive-displacement pump performance. This line reflects the fact that, as the discharge pressure of the pump increases, liquid leaks from the discharge back to the suction-inlet side of the pump casing. This reduces the pump's effective flow rate. The rate at which liquid leaks from the pump's discharge to its suction side is called *slip*. Slip is the result of two primary factors: (1) design clearance required preventing metal-to-metal contact of moving parts, and (2) internal part wear.

Minimum design clearance is necessary for proper operation, but it should be enough to minimize wear. Proper operation and maintenance of positive-displacement pumps limits the amount of slip caused by wear.

Configuration

Positive-displacement pumps come in a variety of configurations. Each has a specific function and should be selected based on their effectiveness and reliability in a specific application. The major types of positive-displacement pumps are: gear, screw, vane, and lobe.

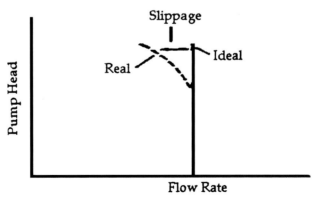

FIGURE 12.8 Positive-displacement pump characteristics curve.

Gear. The most common type of positive-displacement pump uses a combination of gears and configurations to provide the liquid pressure and volume required by the application. Variations of gear pumps are: spur, helical, and herringbone.

Spur. The simple spur-gear pump shown in Fig. 12.9 consists of two spur gears meshing and revolving in opposite directions within a casing. Only a few thousandths-of-an-inch clearance exists between the case, gear faces, and teeth extremities. This design forces any liquid filling the space bounded by two successive gear teeth and the case to move with the teeth as they revolve. When the gear teeth mesh with the teeth of the other gear, the space between them is reduced. This forces the entrapped liquid out through the pump's discharge pipe.

As the gears revolve and the teeth disengage, the space again opens on the suction side of the pump, trapping new quantities of liquid and carrying it around the pump case to the discharge. Lower pressure results as the liquid moves away from the suction side, which draws liquid in through the suction line.

For gears having a large number of teeth, the discharge is relatively smooth and continuous, with small quantities of liquid delivered to the discharge line in rapid succession. For gears having fewer teeth, the space between them is greater and the capacity increases for a given speed. However, this increases the tendency to have a pulsating discharge.

In all simple-gear pumps, power is applied to one of the gear shafts, which transmits power to the driven gear through their meshing teeth. There are no valves in the gear pump to cause friction losses as in the reciprocating pump.

The high impeller velocities required in centrifugal pumps, which result in friction losses, are not needed in gear pumps. This makes gear pumps well suited for viscous fluids, such as fuel and lubricating oils.

Helical. The helical-gear pump is a modification of the spur-gear pump and has certain advantages. With a spur gear, the entire length of the tooth engages at the same time. With a helical gear, the point of engagement moves along the length of the tooth as the gear rotates. This results in a steadier discharge pressure and less pulsation than a spur-gear pump.

Herringbone. The herringbone-gear pump is also a modification of the simple-gear pump. The principal difference in operation from the simple-gear pump is that the pointed center section of the space between two teeth begins discharging fluid before the divergent outer ends of the preceding space complete discharging. This overlapping tends to provide a steadier discharge pressure. The power transmission from the driving gear to the driven gear is also smoother and quieter.

Screw. There are many design variations for screw-type, positive-displacement rotary pumps. The primary variations are the number of intermeshing screws, the screw pitch, and fluid-flow direction.

The most common type of screw pump consists of two screws that are mounted on two parallel shafts and mesh with close clearances. One screw has a right-handed thread, while the other has a left-handed. One shaft drives the other through a set of timing gears, which serve to synchronize the screws and maintain clearance between them.

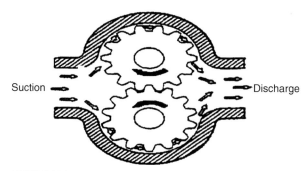

FIGURE 12.9 Simple spur-gear pump.

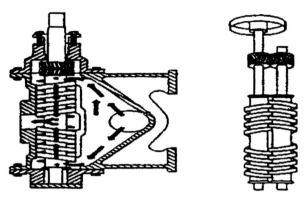

FIGURE 12.10 Two-screw, low-pitch screw pump.

The screws rotate in closely fitting duplex cylinders that have overlapping bores. While all clearances are small, no contact occurs between the two screws or between the screws and the cylinder walls. The complete assembly and the usual flow path for such a pump is shown in Fig. 12.10.

In this type of pump, liquid is trapped at the outer end of each pair of screws. As the first space between the screw threads rotates away from the opposite screw, a spiral-shaped quantity of liquid is enclosed when the end of the screw again meshes with the opposite screw. As the screw continues to rotate, the entrapped spiral of liquid slides along the cylinder toward the center discharge space while the next slug is entrapped. Each screw functions similarly and each pair of screws discharges an equal quantity of liquid in opposed streams toward the center, thus eliminating hydraulic thrust. The removal of liquid from the suction end by the screws produces a reduction in pressure, which draws liquid through the suction line.

Vane. The sliding-vane pump shown in Fig. 12.11, another type of positive-displacement pump, is used with viscous fluids. It consists of a cylindrical bored housing with a suction inlet on one side and a discharge outlet on the other. A cylindrical-shaped rotor having a diameter smaller than the cylinder is driven about an axis position above the cylinder's centerline. The clearance between the rotor and the top of the cylinder is small, but it increases toward the bottom.

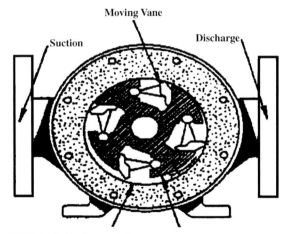

FIGURE 12.11 Rotary sliding-vane pump.

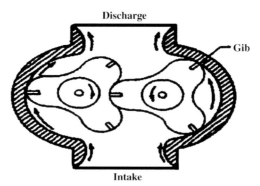

FIGURE 12.12 Lobe-type pump.

The rotor has vanes that move in and out as it rotates, maintaining sealed space between the rotor and the cylinder wall. The vanes trap liquid on the suction side and carry it to the discharge side where contraction of the space expels it through the discharge line. The vanes may swing on pivots or they may slide in slots in the rotor.

Lobe. The lobe-type pump shown in Fig. 12.12 is another variation of the simple-gear pump. It can be considered to be a simple-gear pump having only two or three lobes per rotor. Other than this difference, its operation and the function of its parts are no different. Some designs of lobe pumps are fitted with replaceable gibes, or thin plates, carried in grooves at the extremity of each lobe where they make contact with the casing. Gibes promote tightness and absorb radial wear.

Performance

Positive-displacement pump performance is determined by three primary factors: liquid viscosity, rotating speed, and suction supply.

Viscosity. Positive-displacement pumps are designed to handle viscous liquids such as oil, grease, and polymers. However, a change in viscosity has a direct effect on its performance. As the viscosity increases, the pump must work harder to deliver a constant volume of fluid to the discharge. As a result, the brake horsepower needed to drive the pump increases to keep the rotating speed constant and prevent a marked reduction in the volume of liquid delivered to the discharge. If the viscosity change is great enough, the brake horsepower requirements may exceed the capabilities of the motor.

Temperature variation is the major contributor to viscosity change. The design specifications should define an acceptable range of both viscosity and temperature for each application. These two variables are closely linked and should be clearly understood.

Rotating Speed. With positive-displacement pumps, output is directly proportional to the rotating speed. If the speed changes, from its normal design point, the volume of liquid delivered also will change.

Suction Supply. To a degree, positive-displacement pumps are self-priming. In other words, they have the ability to draw liquid into their suction ports. However, they must have a constant volume of liquid available. Therefore, the suction-supply system should be designed to ensure that a constant volume of nonturbulent liquid is available to each pump in the system.

Pump performance and its useful operating life is enhanced if the suction-supply system provides a consistent positive pressure. When the pumps are required to overcome suction lift, they must work harder to deliver product to the discharge.

Installation

Installation requirements for positive-displacement pumps are the same as those for centrifugal pumps.

Special attention should be given to the suction-piping configuration. Poor piping practices in hydraulic-system application, are primary sources of positive-displacement pump problems, particularly in parallel-pump applications. Often the suction piping does not provide adequate volume to each pump in parallel configurations.

Operating Methods

If a positive-displacement pump is properly installed, there are few restrictions on operating methods. The primary operating concerns are: bypass operation and speed-change rates.

SECTION 6

MAINTENANCE OF ELECTRICAL EQUIPMENT

CHAPTER 1
ELECTRIC MOTORS

Shon Isenhour
Principal, Life Cycle Engineering, Inc., Charleston, S.C.

INTRODUCTION

Electric motors are a vital element in any industrial electric drive process. While all components of a system are important, an electric motor that has been properly applied and maintained and which receives immediate and logical troubleshooting attention in the case of a failure will contribute greatly to the continuous operation and overall success of that system.

As the costs of unscheduled downtime rise, the importance of continuous and regular maintenance of electric motors becomes even more pronounced. With today's increasingly automated systems, loss of a single electric motor because it has failed could mean shutting down an entire line or plant.

Proper application to the driven load is the key step in specifying a motor. Specifying the wrong motor not only can lead to repeated failures but can be hazardous to personnel as well. It is important that the person responsible for specifying a motor understand the motor's role in the system, its operating environment, and its limitations.

CHARACTERISTICS COMMON TO INDUSTRIAL AC AND DC MOTORS

Motor Ratings. Nominal rating data are given on the motor nameplate with allowed tolerances and variations given in the National Electrical Manufacturers Association's (NEMA's) publication MG-1. Nameplate data will include rated horsepower, speed, voltage, and service factor if other than 1.0.

Service factor is a multiplier that may be applied to horsepower for motors that are designed for periodic overloading.

References made to NEMA MG-1 in this chapter refer to NEMA MG-1-1987, Revision 2, with revisions through March 1991.

Enclosures and Cooling Methods. NEMA classifies motors according to Environmental Protection and Methods of Cooling as follows to aid in proper application of motors to governing conditions. Available categories include:

Open Dripproof. Motor has ventilating openings constructed so that operation will not be affected by solid or liquid particles that strike or enter the enclosure at any angle up to 15° downward from the vertical.

Dripproof Guarded. These self-ventilating motors feature louvered covers on the sides with grilles on bottom openings to prevent accidental exposure to live metal or rotating parts.

Dripproof Forced Ventilated. Dripproof-guarded motors with forced ventilation provided by a motor-mounted blower driven by a three-phase ac motor.

Dripproof Separately Ventilated. These motors provide high horsepower ratings with separate ventilation supplied by the customer.

Totally Enclosed. Motors that operate in severe environments require a totally enclosed frame. These motors are either nonventilated, fan cooled, or dual cooled, depending on horsepower.

Totally Enclosed, Nonventilated. Motors not equipped for cooling by means external to the enclosing parts, generally limited to low horsepower ratings or short-time rated machines.

Totally Enclosed, Fan Cooled. Motors with exterior surfaces cooled by an external fan on the motor shaft. Cooling is dependent upon motor speed.

Totally Enclosed Air-Over In-Line. Motors with an external fan driven by a constant-speed ac motor flange mounted to the motor fan shroud. These provide cooling independent of motor speeds. Brakes and tachometers cannot be mounted on the motor end bracket, except for specific small tachometers which can be nested between motor bracket and fan.

Totally Enclosed Air-Over Piggyback. Motors with a top-mounted, ac-motor-driven blower with shroud to direct ventilating air over motor frame.

Totally Enclosed Dual-Cooled Air-to-Air Heat Exchanger. Motors cooled by circulating internal air through the heat exchanger by an ac-motor-driven blower. External air circulated through the heat exchanger by another ac-motor-driven blower removes heat from the circulating internal air. No free exchange of air occurs between the inside and outside of the motor.

Totally Enclosed Dual-Cooled with Air-to-Water Heat Exchanger. Similar to the previous description except that external circulating airflow is replaced by user-supplied water to remove heat from heat exchanger.

Totally Enclosed Pipe-Ventilated or Tally Enclosed Separately Ventilated. Motors which are cooled by user-supplied air which is piped into the machines and ducted out of the machines by user-supplied ducts.

Explosion Proof. Motors whose enclosures are designed and constructed to withstand an explosion of a specified gas or vapor which may occur within them. Prevents the ignition of the specified gas or vapor surrounding the machines by sparks, flashes, or explosions of the specified gas or vapor which may occur within the machine casing.

Dust Ignition Proof. Motors whose enclosures are designed and constructed so as to exclude ignitable amounts of dust or amounts which might affect performance or rating. Will not permit arcs, sparks, or heat otherwise generated or liberated inside the enclosure to cause ignition of exterior accumulations or atmosphere-suspended dust on or in vicinity of enclosure.

Weather Protected. Divided into two types, I and II. A type I motor is open, with ventilating passages constructed to minimize the ability of rain, snow, and airborne particles to come in contact with the electric parts. A type II motor includes ventilating passages at both intake and discharge in an arrangement so that high-velocity air and particles blown into the machine by storms or winds can be discharged without entering the internal ventilating passages leading directly to the motor's electric parts.

Totally Enclosed Pipe Ventilated. Totally enclosed except for openings so arranged that inlet and outlet pipes are connected to them for the admission and discharge of ventilating air. The air may be circulated by means integral with the motor or external to and not a part of the motor, or separately forced ventilation.

Totally Enclosed Water-Air Cooled. A totally enclosed motor cooled by circulating air, which, in turn, is cooled by circulating water. The motor is provided with a water-cooled heat exchanger for cooling ventilating air and with a fan or fans, integral with or separate from the rotor shaft, for circulating ventilating air.

Totally Enclosed Air-to-Air Cooled. Totally enclosed motor cooled by circulating internal air through a heat exchanger, which is cooled by circulating external air.

Application data. Proper application of motors with respect to their service conditions, designated usual or unusual, is the first step in obtaining and maintaining proper motor performance.

Service conditions designated as unusual may include some degree of hazard as opposed to usual. Usual service conditions, as designated by NEMA, include:

Environment

1. Exposure to an ambient temperature in the range of 0 to 40°C (32 to 104°F) or, when water cooling is used, from 10 to 40°C (50 to 104°F).
2. Exposure to an altitude which does not exceed 3300 ft.
3. Installation on a rigid mounting surface.
4. Installation in areas or supplementary enclosures which do not seriously interfere with the motor's ventilation.

Operating

1. V-belt drive in accordance with NEMA standard MG1-14.41 for ac motors, or with NEMA Standard MG-1-14.67 for industrial dc motors.
2. Flat belt, chain, and gear drives in accordance with NEMA Standard MG-1-14.07.

Unusual service conditions, as designated by NEMA, include:

1. Exposure to:
   a. Combustible, explosive, abrasive, or conducting dusts.
   b. Lint or very dirty operating conditions.
   c. Chemical fumes or flammable or explosive gases.
   d. Nuclear radiation.
   e. Salty air, steam, or oil vapor.
   f. Damp or very dry locations, radiant heat, vermin infestation, or atmosphere conducive to the growth of fungi.
   g. Abnormal shock, vibration, or mechanical loading from external sources.
   h. Abnormal axial or side thrust imposed on the motor shaft.
2. Operating where:
   a. There is excessive departure from rated voltage and/or frequency.
   b. The deviation factor of the ac supply voltage exceeds 10 percent.
   c. The ac supply voltage is unbalanced by more than 1 percent.
   d. The rectifier output supplying a dc motor has current peaks unbalanced by more than 10 percent.
   e. Low noise levels are required.
3. Operations at speeds above the highest rated speed.
4. Operation in a poorly ventilated room or in an inclined position.
5. Operation where the motor is subjected to torsional impact loads, repetitive abnormal overloads, or reversing or electric braking.
6. Operation of the motor at standstill with any winding continuously energized.

Power Supply. In order to properly select a motor for an application, the supply voltage must be known. For ac motors, this voltage should normally exceed the nameplate voltage by a slight amount, as illustrated in Table 1.1.

Power Supply Variations. Ideally, power supplies would provide constant voltage, frequency, and phasing. In reality, however, voltages will usually range from 10 percent above to 10 percent below the nominal values, and sometimes more. Frequency is usually closely controlled, but it too can vary.

Phasing, which is balanced when the voltage in each phase of a polyphase system is equal, can be unbalanced by one, two, or more percentage points.

Each type of power variation has a different effect on the operation of an ac induction motor, but the greatest of these is voltage variation.

TABLE 1.1 AC Motor Nameplate Voltages for Corresponding Distribution System Voltages (60 Hz)*

Three-phase motors		Single-phase motors	
System voltage	Motor voltage	System voltage	Motor voltage
208	200	120	115
240	230	240	230
480	460		
600	575		
2,400	2,300		
4,160	4,000		
4,800	4,600		
6,900	6,600		
13,800	13,200		

*From ANSI C84.1-1970, American National Standards Institute, Inc., 1430 Broadway, New York, NY 10018.

AC INDUCTION MOTORS

Theory and Construction. Induction motors are common in most industrial applications. They all have primaries, or stator windings, connected to a power source, with a secondary winding, or rotor, which drives the mechanical load.

The most common type of induction motor is the squirrel-cage motor, so named because its rotor construction entails bars of copper or aluminum conductors, resembling an animal's exercise wheel.

Wound-rotor induction motors entail rotors constructed of polyphase windings, which are connected through slip rings to a controlled external resistance. These are sometimes used when high breakdown torque or variable speed is required.

Induction motors can be either single speed, where the motor operates at a single, fairly constant speed, or multispeed, where the motor can be operated at two or more definite speeds.

Induction motors are likewise classified as either general purpose or special purpose. General-purpose motors carry the NEMA's design B ratings, feature standard operating characteristics and mechanical construction, and are used for applications such as pumps and fans, under normal service conditions. Special-purpose motors are designed for specific torque characteristics, such as design C or design D motors.

Speed-Torque Characteristics. Motor torque, the turning force delivered by the motor shaft, is defined at four points and is usually expressed as a percentage of running torques:

1. Breakaway, starting, or locked-rotor torque.
2. Minimum, or pull-up torque.
3. Breakdown, pull-out, or peak torque.
4. Full load, or running torque.

Application consideration must be given to the selection of a motor to assure that it has the necessary torque characteristics to successfully operate the load.

Breakaway, or starting, torque is that required to start a shaft turning. Proper bearing lubrication has a pronounced effect on this torque requirement.

Minimum, or pull-up, torque is the minimum torque the motor produces during acceleration of the load from standstill to full speed.

Peak torque is the maximum momentary torque a motor can produce during operation. High peak torque requirements for brief periods are required to sustain momentary overloads without stalling the motor.

Full load, or running, torque is that necessary for a motor to produce its rated horsepower at full-load speed.

Torque requirements for an application should be known when specifying a motor. NEMA has designated four design classes—A, B, C, and D—to define categories of torque.

Design A motors are not regularly offered but are usually built to order for special applications. Design A covers a wide variety of motors similar to design B (standard general-purpose), but maximum torque and starting current are usually higher.

Design B, standard general-purpose, motors have normal starting current, normal torque, and normal slip. Their field of application is very broad, including fans, blowers, pumps, and machine tools.

Design C motors have high breakaway torque, normal starting current, and normal slip. This motor is used mainly in hard-to-start applications, such as plunger pumps, conveyors, and compressors.

Design D motors have high breakaway torque combined with high slip. The design is subdivided into three slip design groups, 5 to 8 percent, 8 to 13 percent, used on high-inertia machinery, and 13 percent or greater, limited to short-time duty.

Design Characteristics

Voltage Variations from Nominal Values. NEMA standard MG1-12.44.1 states that motors shall operate successfully under running conditions at rated load with a variation in the voltage up to 610 percent. This does not mean, however, that the motor will operate at rated performance. For example, the motor may not be able to start and accelerate a driven load under a voltage variation condition since the speed-torque curve will change. The major effects on motor operation due to a voltage variation are:

Reduced Voltage

1. Increased temperature rises.
2. Reduction in starting torque.
3. Reduction in maximum torque.
4. Decreased starting current.
5. Increased acceleration time.

Increased Voltage

1. Increased starting and maximum torques.
2. Higher inrush current.
3. Decreased power factor.

Any of these conditions can result in shortening the effective service life of a motor, which necessitates greater maintenance procedures.

50-Hz Operation of 60-Hz Motors. A 60-Hz motor will usually operate, if necessary, at 50 Hz on selected voltages at 80 or 85 percent of the 60-Hz rated horsepower. Table 1.2 illustrates such operation.

AC Motor Insulation.

The overall service life and performance of an ac induction motor are directly related to the quality and maintenance of its insulation system. Insulation systems are designed using materials that have high electrical resistance, confining the path of the current to within the stator winding.

The primary factors in prolonging insulation life and integrity are keeping it clean, dry, and cool. Insulation systems in ac motors include the following:

1. *Turn insulation.* Basic wire coating which is an enamel, resin, film, or film-fiber combination used to electrically insulate adjacent wire turns from one another within a coil.
2. *Phase insulation.* Sheet material used to insulate between phases at the coil end turns.
3. *Ground insulation (slot liners).* Sheet material used to line the stator slots and insulate the stator winding from stator iron or other structural parts.

TABLE 1.2 Comparative Voltage Ratings for 60- and 50-Hz Operation*

Nameplate hp of 60-Hz motor, volts	50-Hz optional voltage rating ±5%	
	80% of 60-Hz motor nameplate hp, volts	85% of 60-Hz motor nameplate hp, volts
230	190	200
460	380	400
575	475	500

*For 48 through 440T frames polyphase only. The motors may operate at less than NEMA torques. Care must be taken in using for hard-to-start and hard-to-accelerate loads. It is advisable to consult the motor manufacturer when operating at frequencies other than that stated on the nameplate.

4. *Midstick (center wedge).* Insulation used in stator slots to separate and insulate coils from one another within the slot.
5. *Topstick (top wedge).* Used to compact and contain the coil wires within the stator slots.
6. *Lead insulation.* Insulation materials surrounding lead wires.
7. *Lacing and tape.* Used to tie lead wires in place on the stator end coils. Also used to tie end coils together adding mechanical strength to the end coils and restricting their movement.
8. *Varnish.* Varnish treatment is employed to increase the resistance of the completely wound stator to environmental attack of chemicals, moisture, and so forth. The varnish treatment bonds the coils, connections, wedges, and stator iron into an integrated structure and improves the components' resistance to electrical and mechanical damage.

Vacuum Pressure Impregnation (VPI). This system is generally found in large (generally above 500 hp) ac form wound motors and is a relatively recent development in methods of motor insulation.

The features and benefits of VPI are a void-free sealed insulation system with great mechanical strength and high thermal conductivity. The magnet wire used is normal film or dacron glass coated. Ground wall insulation varies depending on the system design but always contains some form of mica tape. These tapes can be of mica splittings, mica mat, or nomex mica. Solventless 100 percent-solid resins are used instead of the varnishes used in normal atmospheric dipping. Resins are impregnated into the stator and windings by placing the wound stator in a large vacuum pressure tank and alternately applying vacuum and pressure while the stator is fully immersed in the resin. These may be either polyester or epoxy; however, epoxy gives the greatest strength, abrasion resistance, and chemical resistance to acids and alkalies.

VPI being sealed assures more successful operation in wet and corrosive atmospheres, and the absence of voids greatly reduces electrical insulation losses due to corona at higher voltages.

Electrical properties which should be considered in evaluating a VPI system are dissipation factor, voltage breakdown, voltage endurance, and thermal aging.

B-Stage Insulation. B-stage coils are insulated with resin-treated tapes and are hot and cold pressed to exact dimensions so that they fit very tightly in the stator slots. Pressing cures the resin into a void-free enclosure of accurate dimension. Coil head and connections are insulated with the same B-stage tape and are all final cured as the stator assembly is baked in the oven.

This system is used on very large motors of high voltage when they are too large to fit in a vacuum tank or when a tank is not available. Usually this system is much more expensive than a VPI system and lacks some of the features of VPI, such as the strong mechanical bond of coil to stator.

Motor Temperature Rise. The temperature rise of an ac motor is the difference between the ambient (room) temperature and the average winding temperature. The difference between the average winding temperature and the hottest spot in the motor windings is the hot spot allowance. Note that the motor skin (external surface) temperature is not the temperature rise. NEMA defines the allowed temperature rise for each of the four popular insulation classes.

Insulation Class. Years ago, the popular insulations were either class A or class B, with the former being "organic" materials, and the latter "inorganic." Today, classes F and H have joined A and B as popular systems. And the method of classifying the systems has changed to a "proven by test" basis. Insulation systems are considered to be suitable for operation at hottest spot temperatures of 105, 130, 155, and 180°C for class A, B, F, and H, respectively. Today, the insulation class is given on the motor nameplate instead of a temperature rise. The temperature rise could be determined by subtracting the ambient (also on the nameplate) and the hot spot allowance (defined in NEMA MG-1) from the allowed hottest spot temperature. All components of the insulation system must be capable of withstanding the temperatures that *they* will encounter when the motor is operating at its maximum rated temperature.

DIRECT-CURRENT MOTORS

Theory and Design. A dc motor's speed can be regulated by changing the field strength or armature voltage. Thus, speed regulation is one of the primary advantages of dc motors. Decreasing field strength will increase the speed of the motor. Increasing the field strength decreases motor speed, and increasing armature voltage increases the speed.

DC motors consist of two basic parts, the stationary part (field) and the rotating part (armature). The field is made up of a frame, field poles, and interpoles, which are fastened to the inner circumference of the frame. The poles are usually laminated steel with field windings or permanent magnets and are mounted on these motors to provide excitation for the motor. The armature consists of four parts, the shaft, the core (composed of steel laminations), the windings, and the commutator. Current flows through the armature via carbon brushes which ride on the surface of the commutator. The performance of a dc motor depends upon proper application, brush selection, and maintenance practices.

Types of DC Motors. Four types of dc motors are available. They are as follows:

Shunt-Wound Motors. These motors are used where primary load requirements are for minimum speed variation from full load to no load and/or constant horsepower over an adjustable speed range at constant potential. These motors are suitable for average starting torque loads.

Series-Wound Motors. These are used in applications where high starting torques are required. The load must be solidly connected to the motor and never decreased to zero to prevent excessive motor speeds. The load must tolerate wide speed variations from full load to light load.

Compound-Wound Motors. These are designed with both a series and shunt field winding and are used where the primary load requirement is a heavy starting torque and adjustable speed is not required. They are also used for parallel operation. The load must tolerate a speed variation from full load to no load.

Permanent Magnet Motors. These are motors that use field magnets to provide higher reliability than wound-field motors, as field failure is impossible. No power is needed for field excitation, thus providing higher efficiencies. However, field strength cannot be varied by field voltage changes, and speed is controlled by varying armature voltage.

Application Data. Three basic machine limitations must be considered in the application of dc motors. These are mechanical, thermal, and commutation limits. The selected motor must have enough capacity in each of these categories to satisfy the demands of the application.

NEMA standards specify the minimum thermal and commutation performance of standard ratings, leaving it up to the designer to provide strength and ruggedness. Standard motor ratings are such that performance characteristics are available to successfully handle the majority of applications at a reasonable cost with long life, as long as proper maintenance is provided. Unusual application can be solved by motors with extra mechanical, thermal, or commutation features matching the unusual requirements.

Mechanical Factors

1. *Vibration.* DC motors must have strength in shafts, bearing mounts, frames, and feet to handle full-load torque plus anticipated overload torque. They must also withstand vibration normally found in industrial applications.
2. *Shock.* Motors are subject to shock loading from the driven machine and must be designed to withstand referred shock.
3. *Belted drives.* A longer and larger diameter shaft may be required to handle the higher radial load of belted drives. Roller bearings may be provided at the drive end.
4. *End thrust.* DC motors with antifriction bearings are suitable for normal amounts of end thrust. Special bearings may be necessary for very high end-thrust loads.
5. *Vertical mounting.* Standard design dc motors are suitable for mounting in a vertical position with satisfactory bearing life, provided that no extra thrust is imposed beyond the weight of its armature plus a pulley or half coupling.

Thermal Factors

1. *Insulation life.* Insulation life is chiefly a function of time versus temperature. Standard dc motors are designed for long life when operated within their rating under usual service conditions; but unusual conditions of dirt, dust, oil, chemicals, or vibration can cause a marked decrease in insulation life.
2. *Ambient temperature.* The NEMA standard ambient temperature is 40°C (104°F) and is considered to be the maximum temperature encountered in industrial applications. Other ambient temperatures may be involved for special applications. The ambient temperature is added to the temperature rise of the motor to determine the total temperature imposed upon the insulation.
3. *Altitude.* Standard dc motors are designed for operation at up to 3300 ft. As altitude increases, the air becomes less dense and cooling effectiveness decreases. For this reason, motors are derated at altitudes over 3300 ft. If standard nameplate rating must be provided above this altitude, a special motor must be supplied.
4. *Service factor.* When indicated on a motor nameplate, this numerical value indicates how much above the nameplate rating a motor can be loaded without causing serious degradation. Standard-duty dc motors are rated with a 1.0 service factor.
5. *Losses and cooling.* Losses can be divided into two groups. Current losses include I^2R, eddy current, core and brush contact losses. Mechanical losses consist of brush and bearing friction and windage losses. These losses result in heat and a temperature rise.

    Convection and radiation are the principal means of cooling a motor. In most cases, armature heating is the limiting thermal factor for a dc motor. The most important losses contributing to this are I^2R loss and core loss. Since the I^2R loss is proportional to the square of the load current, heating increases very rapidly with load.
6. *Enclosures.* Dripproof enclosures protect motors from dripping liquids and falling objects and are suitable for many industrial applications. Splashproof covers can be added where additional protection is required.

    Totally enclosed motors are used where ambient air is excessively contaminated with dirt, oil, chemicals, corrosive vapors, or hazardous dust and vapor. Nonventilated, fan-cooled, dual-cooled, and air-over enclosures are generally available in standard or explosion-proof construction.
7. *Duty cycle operation.* If conditions of load, speed, and time are known and the total cycle time is short, motor selection can be made through an rms (root-mean-square) horsepower calculation. In other cases, the motor rating is determined by experience or test.

Commutation

1. NEMA standard. The NEMA standard for industrial dc motors requires that the motor be capable of carrying successfully 150 percent of full-load current for 1 min at any speed within the

standard field-weakened speed range. This is a recognition of the practical necessity of providing for peak overload for starting, accelerating, and decelerating.

There is no sharp definition of successful commutation. Commutation can be considered successful even if sparking occurs, provided excessive maintenance is not required.

The ultimate is black commutation, in which no sparking is apparent. It is common to refer to degrees of commutation, such as number 1, $1\frac{1}{4}$, $1\frac{1}{2}$, 2, and so on, based upon the number, size, and color of sparks. This is a qualitative method, however, since it depends upon the experience and opinion of the viewer.

The best method of establishing commutating ability is the black band test. The test results are plotted on a graph. The black band is established by conducting a "buck-boost" test. As the name implies, the interpoles are first weakened by bucking interpole excitation using a separate isolated voltage source. This is connected in opposition to interpole excitation. Next, the interpoles are strengthened, or boosted, by connecting the external voltage to the poles in a cumulative manner. Poles are bucked and boosted for various $\frac{1}{4}$-load points from no load to 150 percent load. The amperes required to cause the machine to spark, both buck and boost, are plotted against load. All points within the limits of this band represent black commutation, hence the name black band. Separate black bands are measured for full field speed and for field weakened speed.

2. *Peak overloads.* Peak overloads which occur frequently may be cumulatively destructive, while the same peaks can be handled on a less frequent basis. Peak currents of long duration are more apt to be destructive. More current can be commutated without sparking at full field than at field weakened speeds.

3. *Starting peak overloads.* Since at the instant of starting, motor speed is zero, higher peak currents may be tolerated on an occasionally repeated basis.

DC Motor Insulation. DC motors use the same classes of insulation as listed in the ac insulation section (classes A, B, F, and H).

Materials for the various classes of insulation are similar to those for ac motors except that class H systems generally do not use silicone materials because of their harmful effect on commutators, resulting in significantly shortened brush life.

SCHEDULED ROUTINE MOTOR CARE MAINTENANCE PROGRAM

Maintenance Plan. A specific maintenance plan that is carried out on a regular basis is a primary factor in preventing motor problems and failures. And, when problems do arise and system shutdown is inevitable, a logical, step-by-step method of troubleshooting can save time and money.

At the very least, the motor manufacturer's maintenance schedule should be adhered to on all electric motors. The amount of additional maintenance attention required and the stocking of spare parts, or even spare motors, depends to a large degree on the importance, cost, and complexity of the motor. For example, a small, normally stocked, low-horsepower, low-cost standard motor that is not necessarily vital to the continuous operation of a system may not demand such a rigid maintenance schedule. Spare parts may be readily available and stocked. In some cases, it may be less costly to replace the entire motor rather than to proceed with intense repair.

On the other hand, a high-cost specialty motor on which an entire manufacturing process depends will demand more painstaking care and maintenance. Spares may not be readily available, or cost-effective.

General Information. Scheduled routine inspection and service will minimize motor problems. Frequency of routine service depends upon the application. It is usually sufficient to include motors in the maintenance schedule for a driven machine or general plant equipment. If a breakdown could cause health or safety problems, severe loss of production, damage to equipment, or other serious losses, a more frequent maintenance schedule should be adopted.

It is important to plan and document your maintenance program. This includes prepared forms for recording such data as the date of inspection, items inspected, service performed, and general motor condition. These records can help identify specific problems in an application and help avoid breakdowns and production losses.

Routine inspection and service can usually be done without disconnecting or disassembling the motors. The inspection should consider each of the following categories.

Warning: Internal parts of a motor may be at line potential even when it is not rotating. Before performing any maintenance which could result in contacting any internal part, be sure to disconnect all power from the motor.
- *Motors using rectified power units, disconnect all ac line connections.*
- *Motors using rotating power units, disconnect all dc line and field connections.*

Grease-Lubricated Antifriction Bearings. Lubricate bearings only when scheduled. *Do not* overlubricate. Excessive grease creates heat due to churning and can damage bearings. Follow the manufacturer's recommendation for lubrication frequency and amounts.

Before lubricating, thoroughly clean the lubrication equipment and fittings. Dirt introduced into bearings during lubrication probably causes more bearing failure than lack of lubrication. Use only clean, fresh grease from clean containers, and handle so as to avoid contamination.

Overgreasing bearings can cause mechanical churning of the grease and will result in elevated bearing temperature, leaching of the oil from the grease, and shortened bearing life. Excessive lubricant can also find its way inside the motor where it collects dirt and causes insulation deterioration.

The frequency of routine greasing of ball and roller bearings increases with motor size and the application's severity. Actual schedules should be established by the user for the specific conditions.

Before scheduled greasing, both the inlet and drain plugs should be removed. Pump grease into the housing using a standard grease gun, suitable grease fitting, and light pressure.

Should bearings remain hot or noisy even after correction of bearing overloads, remove the motor from service. Wash the housing with a good solvent. Replace the bearings. Repack the bearings, assemble the motor, and fill the grease cavity.

Whenever motors are disassembled for service, check the bearing housing. Wipe away any old grease. If there are any signs of grease contamination or breakdown, clean and repack the bearing system as described in the preceding paragraph. Again, make certain that the inlet and outlet passages are clean and free of dirt or chips.

Oil-Lubricated Sleeve Bearings. As a rule of thumb, fractional horsepower motors with a wick lubrication system should be oiled every 2000 hr of operation, or at least annually. Dirty, damp, or corrosive locations or heavy loading may require oiling at 3-month intervals or more often. Follow the manufacturer's recommendations for quantity, frequency, and type of oil.

Many larger motors are equipped with oil reservoirs and usually a sight gage to check proper level. As long as the oil is clean and light in color, the only requirement is to fill the cavity to the proper level with the oil recommended by the manufacturer. Do not overfill the cavity.

If the oil is discolored, dirty, or contains water, remove the drain plug. Flush the bearing with clean solvent until it comes out clear. Coat the plug threads with a sealing compound, replace the plug, and fill the cavity to the proper level.

If the oil discolors frequently or becomes unusually dark, it is advisable to disassemble and inspect the bearing parts. Possible causes of dark oil are:

1. Wiping or melting of babbit bearing face, caused by improper scrapping, dirt, heat, or vibration.
2. Thrusting the shaft shoulder into the end face of the bearing, caused by improper alignment of the motor to driven equipment, improper internal motor axial clearances, and end float.
3. Rubbing of oil rings on reservoir housing.
4. Improper or insufficient sealing of bearing housing bolted surfaces. This can allow contaminated air to be drawn into the bearing reservoir.

When motors are disassembled, wash the housing with a solvent. Discard any used packing. Replace badly worn bearings. Coat the shaft and bearing surfaces with oil and reassemble. Be certain that new bearings are properly scraped and fitted.

Heat, Noise, and Vibration. Check the motor frame and bearings for excessive heat or vibration with suitable test equipment. Listen for abnormal noise. All indicate possible system failure. Identify and eliminate the source of heat, noise, or vibration. A continuous log comparing vibration levels should be used to predict problems and bearing life.

Heat. Excessive heat is both a cause of motor failure and a sign of other motor problems. Primary damage caused by excess heat is the increase of the insulation's aging rate. Heat in excess of the insulation's rating will shorten winding life.

An approximate rule of thumb is that insulation life is reduced to half its normal life for each 10°C (50°F) increase above the designed temperature. Overheating results from a variety of different problems, grouped as follows:

1. *Improper application.* The motor may be too small or have the wrong starting torque characteristics for the load. This may be the result of poor selection or changes in the load requirements.

2. *Poor cooling.* Accumulated dirt or poor motor location may prevent the free flow of cooling air around the motor. This may also result from the motor drawing heated air from another source. Internal dirt or damage can prevent proper airflow through all sections of the motor. Dirt on the frame may prevent transfer of internal heat to the cooler ambient air.

   a. Wipe, brush, vacuum, or blow accumulated dirt from the motor's frame winding and air passages. Do not use high-pressure sprayers. Thick dirt insulating the frame's outer surface and clogging passages reduces cooling airflow, causing motors to run hot. Heat reduces insulation life and eventually causes motor failure. On open machines, measure the temperature of air in and out using a thermocouple. The difference between these readings in the temperature indicates the rise of the air. Comparison of these readings for a constant or nonvarying load after a sequence of periodic inspections is meaningful. Gradual increase over a period of time indicates a need for cleaning.

   b. Feel for air being discharged from the cooling air ports. If the flow is weak or unsteady, internal air passages may be clogged. The motor should be removed from service and cleaned. Also, for fan-cooled enclosures, check for large ac typically unidirectional fans installed in the reverse direction.

3. *Overloaded driven machine.* Excess loads or jams in the driven machine force the motor to supply higher torque, draw more current, and overheat.

4. *Excessive friction.* Misalignment, poor bearings, and other problems in the driven machine, power transmission system, or motor increase the torque required to drive the load, and raise the motor operating temperature.

5. *Electrical overloads.* An electrical failure of a winding or connection in the motor can cause other windings or the entire motor to overheat.

6. *Incorrect line voltage.* Excessive voltage can cause increased magnetic losses in the motor. Insufficient voltage will cause the motor to draw excessive current under load, thus increasing the resistance losses.

Other possible causes of low voltage at motor terminals include excessively long supply leads, insufficiently sized leads, poor electrical connections, and bad starter contacts.

Noise and Vibration. Noise usually indicates motor problems but usually does not cause damage. It is usually accompanied by vibration, however, which can cause damage in several ways.

Vibration tends to shake windings loose and mechanically damages insulation by cracking, flaking, or abrading the material. Embrittlement, or flaking, of lead wires from excessive movement and brush sparking at commutators also results from vibration. Finally, vibration can bring about bearing failure by causing bearings to Brinell (ball indentations in race), sleeve bearings to be pounded out of shape, or the housings to loosen in the shells.

Whenever noise or vibration occur in an operating motor, the source should be quickly isolated and corrected. What may seem to be an obvious source of the noise or vibration may be a symptom of a hidden problem. Therefore, a thorough investigation is often required.

Noise and vibration can be caused by a misaligned motor shaft or can be transmitted to the motor from the driven machine or power transmission system. They can also be the result of either electrical or mechanical unbalance in the motor.

After checking motor shaft alignment, disconnect the motor from the driven load. If the motor operates smoothly, with both power on and power off, check the source of noise or vibration in the driven equipment.

If the disconnected motor still vibrates, remove power from the motor. If the vibration stops, look for an electrical unbalance. If it continues as the motor coasts without power, look for a mechanical unbalance.

Electrical unbalance occurs when the magnetic attraction between stator and rotor is uneven around the periphery of the motor. This causes the shaft to deflect as it rotates, creating a mechanical unbalance. Electrical unbalance usually indicates an electrical failure, such as an open stator or rotor winding, an open bar or ring in squirrel-cage motors, or shorted field coils in synchronous motors. An uneven air gap, usually a result of badly worn sleeve bearings, also produces electrical unbalance.

The main causes of mechanical unbalance include a distorted mounting, bent shaft, poorly balanced rotor, loose parts on the rotor, or bad bearings. Noise can also come from the fan hitting the frame, fan shroud, or foreign objects inside the shroud. If the bearings are bad, as indicated by excessive bearing noise, determine why the bearings failed.

Brush chatter is a motor noise that can be caused by vibration or other problems unrelated to vibration.

In recent years, the science of vibration analysis has been greatly developed. There are many manufacturers of quality analyzing equipment, varying from the simple handheld meter to the "real-time analyzers," XY plotters for frequency scans, and the like.

Vibration is measured in miles (peak to peak), inches per second, and "G's." Use of this equipment requires some training and skill. Learning to interpret the meaning of the data is all important. In any event, record keeping of data on given motors over a period of time is most meaningful.

Changes in the magnitude of vibration or the frequency signature will predict failure before it occurs. It is advisable to consult with the motor manufacturer and reputable manufacturers of vibration and analysis equipment for detailed guidance on this subject.

Corrosion. Check for signs of corrosion, including corroded terminals and connections inside the conduit box. Severe corrosion may indicate internal deterioration and/or a need for external repainting. Schedule the removal of the motor from service for complete inspection and possible rebuilding.

Brushes and Commutators (DC Motors)

1. Observe the brushes while the motor is running. The brushes should ride on the commutator smoothly with little or no sparking and no brush noise (chatter).
2. Stop the motor. Be certain that:

Warning: Internal parts of a motor may be at line potential even when it is not rotating. Before performing any maintenance which could result in contacting any internal part, be sure to disconnect all power from the motor.
- *Motors using rectified power units, disconnect all ac line connections.*
- *Motors using rotating power units, disconnect all dc line and field connections.*

---

    a. The brushes move freely in the holder, and the spring tension is equal on each brush.
    b. Each brush has a polished surface over its entire working face, indicating good seating.
    c. The commutator is clean, smooth, and has a polished brown surface where the brushes ride.
       Note: Put each brush back into its original holder. Interchanging brushes decreases commutation ability.

d. There is no grooving of the commutator (small grooves around the circumference of the commutator). If grooving is present, schedule the motor for corrective action. This can lead to a serious problem.
3. If there is any chance or indication that the brushes will not last until the next inspection date, replace them. Fit the face of new brushes to the radius of the commutator with sandpaper only; do not use emery abrasive. Final seating of the brushes to the commutator surface may be accomplished with the use of a fine commutator seating stone.

Generally, brushes should be replaced in full sets. However, some motors may have specific brushes which wear at significantly different rates. Selective replacement might be advisable where this pattern is established. Brushes which might wear at a more rapid rate include:

1. The positive brush or brushes on some unidirectional motors.
2. The brushes in the outer track or tracks, where the brushes on each stud are not in a straight line across the commutator.
3. Clean any accumulated foreign material from the grooves between the commutator bars and from the brush holders and posts. Use only clean, dry compressed air.
4. Brush sparking, chatter, excessive wear, or chipping and a dirty or rough commutator are indications that the motor requires prompt service.

Purposely misaligning brushes is known as "stepping" or "staggering" brushes, and is used to improve the electrical characteristics of the motor. It also lengthens the life of all brushes except the last brushes to maintain contact with a given commutator bar.

It is best to have all brushes in the motor be of the same grade of carbon from the same manufacturer and from the same manufacturing lot. Mixing brush manufacturers or carbon grades should be avoided.

Brush and Commutator Care. Many factors are involved in brush and commutator problems. All generally involve brush sparking, usually accompanied by chatter and often excessive wear or chipping. Sparking may result from poor commutator conditions or it may cause them.

The degree of sparking should be determined by careful visual inspection. It is imperative that a remedy be determined as quickly as possible. Sparking usually feeds upon itself and becomes worse with time until serious damage results.

Eliminating sparking requires a thorough review of the motor and operating conditions. Always recheck for sparking after correcting one problem to see that it solved the total problem. Also remember that after grinding the commutator and reseating the brushes, sparking will occur until the polished, brown film reforms on the commutator. In eliminating sparking, consider external conditions that affect commutation. Frequent motor overloads, vibration and high humidity cause sparking. Extremely low humidity allows brushes to wear through the needed polished, brown commutator surface film. Oil, paint, acid, and other chemical vapors in the atmosphere contaminate brushes and the commutator surface.

Look for obvious brush and brush holder deficiencies:

1. Make certain that the brushes are properly seated, move freely in the holders, and are not too short.
2. The brush spring pressure must be equal on all brushes.
3. Be certain that spring pressure is not too light or too high. Large motors with adjustable springs should be set at about 3 to 4 lb/in^3 of brush surface in contact with the commutators. If a choice has to be made between setting pressure higher or lower than recommended, choose the higher setting. Arcing due to low pressure is much more detrimental than the frictional increase of high pressure.
4. Remove dust that can cause a short between brush holders and frame.
5. Check lead connections to the brush holders. Loose connections cause overheating.

Look for any obvious commutator problems:

1. Conditions other than a polished, brown surface under the brushes can indicate a problem. Severe sparking causes a rough, blackened surface. An oil film, paint spray, chemical contamination, and other abnormal conditions can cause a blackened or discolored surface and sparking. Streaking or grooving under only some brushes or flat and burned spots can result from a load mismatch. Grooved commutators should be scheduled for removal from service. A brassy appearance shows low film buildup on the surface and may result from low humidity, wrong brush grade, or prolonged light loading.
2. High mica or high or low commutator bars make the brushes bounce, causing sparking.
3. Carbon dust, copper slivers, or other conductive dust in the slots between commutator bars causes shorting and sometimes sparking between bars.

If correcting the obvious deficiencies does not eliminate sparking or noise, look at the less obvious possibilities:

1. If the brushes were changed before the problem was apparent, check the grade of brushes. Weak brushes may chip. Soft brushes may allow a thick film to form. High-friction or high-abrasion brushes wear away the brown film, producing a brassy surface. If the problem appears only under one or more of the brushes, two different grades of brushes may have been installed. Generally, use only the brushes recommended by the motor manufacturer or a qualified brush expert.
2. The brush holder may have been reset improperly. If the boxes are more than $1/8$ in from the commutator, the brushes can chatter and chip. Setting the brush holder off neutral causes sparking. Normally, the brushes must be equally spaced around the commutator and must be parallel to the bars so all make contact with each bar at the same time.
3. An eccentric commutator can cause sparking and may cause vibration. Normally, concentricity should be within 0.001 in on high-speed, 0.002 in on medium-speed, and 0.004 in on slow-speed motors.
4. Various electrical failures in the motor windings or connections manifest themselves in sparking and poor commutation. Look for shorts or opens in the armature circuit and for grounds, shorts, or opens in the field winding circuits. A weak interpole circuit or large air gap also generate brush sparking.

Brushes and Collector Rings (Synchronous Motors)

1. Remove any black spots on the collector rings by rubbing lightly with fine sandpaper. These spots, if not removed, will cause pitting that requires regrinding the rings.
2. Check for an imprint of the brush, signs of arcing, or uneven wear. Should any of these be present, remove the motor from service and repair.
3. Check the collector ring brushes under the same circumstances as described previously under Brushes and Commutators. Note that these brushes will probably not wear as rapidly as dc commutator brushes.

Windings

Care of Windings and insulation. Routine inspections generally do not involve opening the motor to inspect the windings, unless it is an expensive, high-horsepower motor. Therefore, long motor life requires selection of the proper enclosure to protect the windings from excessive dirt, abrasives, moisture, oil, and chemicals. Routine testing can identify deteriorating insulation and is particularly helpful in severe operating conditions or in an application that shows a history of winding failures. Such motors should be removed from service and repaired before unexpected failures stop production.

Insulation resistance of a winding will vary, depending upon the temperature, moisture in or on the winding, cleanliness, age, test voltage value, and duration of the test voltage. The best method of making these tests is IEEE 43, "Recommended Guide for Testing Insulation Resistance of Rotating Machinery," published by the Institute of Electrical and Electronics Engineers, 345 East 47th Street, New York, NY 10017.

The standard value of insulation resistance is the least value which a winding should have after cleaning and drying.

The ratio of insulation resistance obtained by a 10-min and 1-min voltage application is the polarization index (PI). This is used to determine when drying is complete or to determine winding conditions when no other information is available. The PI of ac stator windings which are clean and dry is usually 1.5 or more for class A windings and 2.5 or more for class B windings.

Keeping windings dry will improve the performance of the insulation. If the motor has been exposed to dampness, it should be thoroughly dried, by either the dc method or the external heat method. Regardless of the method used, the winding should not be heated over 90°C (194°F) measured by resistance or 75°C (167°F) measured by thermometer.

The dc method is done with the motor stationary. A low-voltage, high-current, dc power source is used to dry out armature windings. The armature terminals may be connected either in series or multiple.

The external heat method uses an oven, preferably one with circulating air, at a temperature not exceeding 85°C (185°F). This would be continued until the insulation resistance becomes practically constant.

Whenever a motor is opened for repair, winding care should be as follows:

1. Accumulated dirt prevents proper cooling and may absorb moisture and other contaminants that may damage insulation. Vacuum the dirt from the windings and other internal air passages. Do not use high-pressure air, as this can damage windings by driving the dirt into the insulation.
2. If a motor repair facility is available, windings can be cleaned by steam or by washing in proper detergent solvents. Again, care must be taken to prevent forcing contamination deeper into any existing breaks or cracks in the insulation. Windings cleaned by this method should be dried in a forced draft oven for at least 8 hr at 121°C (250°F). Revarnishing with a polyester or epoxy varnish can be done after the windings are thoroughly cleaned and dried. These varnishes should be baked for 6 to 8 hr at about 148 to 157°C (300 to 315°F).
3. Abrasive dust drawn through the motor can abrade coil noses, removing insulation. If such abrasion is found, the winding should be revarnished or replaced.
4. Moisture reduces the dielectric strength of insulation which results in shorts. If the inside of the motor is damp, dry the motor per information found below under Cleaning and Drying Windings.
5. Wash any oil and grease from inside the motor. Use care with solvents that can attack the insulation.
6. If the insulation appears brittle, overheated, or cracked, the motor should be revarnished or, with severe conditions, rewound.
7. Loose coils and leads can move with changing magnetic fields or vibration, causing the insulation to wear, crack, or fray. Revarnishing and retying leads may correct minor problems. If the loose coil situation is severe, the motor must be rewound.
8. Check the lead-to-coil connections for signs of overheating or corrosion. These connections are often exposed on large motors, but taped on small motors. Repair as needed.
9. Check wound rotor windings as described for stator windings. Because rotor windings must withstand centrifugal forces, tightness is even more important. In addition, check for loose pole pieces or other loose parts that create unbalance problems.
10. The cast rotor bars and ring ends of squirrel-cage motors rarely need attention. However, if they are open or broken, they create an electrical unbalance that increases with the number of bars broken. An open end ring causes severe vibration and noise.

Problems due to open circuits (bars) or high-resistance points between the rotor and end rings are identified by reduced torques, higher slip under load, increased heat, and noise.

These open bars associated with rotors are rarely visible, but can be checked by applying 10 to 25 percent single-phase voltage to the stator with an ammeter in one line, slowly rotating the shaft by hand, and observing the ammeter for significant current changes. A significant current change with rotor position indicates a defect in the winding cage. A pulsing sound related to slip under load is usually the first observation and indicates that the rotor should be replaced.

Rotor construction typical of many large induction and many high-resistance motors consists of a fabricated cage with bars brazed to the end rings. A cracked rotor bar or brazed cage with bar extensions between the iron core and end ring can usually be seen. Brazed joint problems will be reflected by generally localized heating, causing discoloration. Cracked bars should be replaced by a competent repair shop.

Should cracked rotor bars on die-cast rotors be visible where bars and end rings meet, replace the rotor.

Replacing or brazing broken bars on squirrel-cage rotors should be done only by qualified competent personnel.

Testing Windings. Routine field testing of windings can identify deteriorating insulation, permitting scheduled repair or replacement of the motor before its failure disrupts operations. This is good practice especially for applications with severe operating conditions or a history of winding failures and for expensive high-horsepower motors and locations which can cause health and safety problems or high economic loss.

The easiest field test that prevents the most failures is the ground-insulation, or megger test. It applies dc voltage, usually 500, 1000, or 2500 V, to the motor and measures the resistance of the insulation.

NEMA standards require a minimum resistance to ground at 40°C (104°F) ambient of 1 mΩ per kilovolt of rating plus 1 mΩ. Medium-sized motors in good condition will generally have megohmmeter readings in excess of 50 mΩ. Low readings may indicate a seriously reduced insulation condition caused by contamination from moisture, oil, or conductive dirt or by deterioration from age or excessive heat.

One megger reading for a motor is not adequate. A curve of several readings recording resistance, with the motor cold and hot and the date, indicates the rate of deterioration. This curve provides the information needed to decide if the motor can safely remain in service until the next scheduled inspection.

The megger test indicates ground insulation condition. It does not, however, measure turn-to-turn insulation condition and may not pick up localized weaknesses. Moreover, operating voltage peaks may stress the insulation more severely than megger voltage. For example, the dc output of a 500-V megger is below the normal 625-V peak of each half cycle imposed on an ac motor when operating on a 460-V system. Experience and conditions may indicate the need for additional routine testing. A test used to prove existence of a safety margin above operating voltage is the ac high-potential ground test. This test should never be applied until a sufficiently high megger reading has been obtained (preferably infinity). For new motors, IEEE 95 specifies a high-pot voltage of:

$$\text{Test volts} = 2 \times \text{rated voltage} + 1000$$

For motors in service, it is advisable to test at 65 percent of the above value. This is a destructive test, and although this test does detect poor insulation condition, the high voltage can arc to ground, burning insulation and frame. It can also actually cause failure during the test.

Repetitions of the test should be kept to a minimum.

DC high-potential rather than ac high-potential tests are becoming popular because the test equipment is smaller and the low test current is less dangerous to personnel and does not create as a great a degree of damage as ac.

Caution: High-potential testing, whether dc or ac, is destructive testing, and is not generally recommended as a maintenance-type test.

Polarization index is another method of measuring the ground resistance of the insulation. Normally, all three phases are measured together at one time. The test consists of applying 500 V dc and measuring the ground resistance of the insulation for a time period of 10 min. The quotient of the 10-min resistance reading divided by the 1-min reading must give an index number of 2.0 or greater to be acceptable (IEEE 43).

The surge test is a method for evaluating turn-to-turn insulation within the coils. This is a nondestructive test which indicates shorts between turns and dissymmetries such as incorrect number of turns in a given coil. The test is done by comparing the unknown coil or winding to one of known quality. Comparison is made by viewing a decaying sine wave trace of a voltage applied to each coil on a dual beam scope.

Cleaning and Drying Windings. There are basically two ways to clean electric motor windings. For in-plant maintenance it is generally recommended to physically remove dirt and contamination by wiping, brushing, and very carefully blowing with compressed air. A light cleaning with compressed air and ground-up corn cobs is sometimes very effective. The danger, particularly with compressed air, is that of forcing conductive contaminants into cracks or breaks in the insulation and possibly making the problem worse than it was. For this reason, air should only be used as a final touch-up after the majority of dirt has been physically removed.

The second method involves removal of dirt and grease by liquid solution. In the field, this would most likely involve the use of an electrical nontracking solvent; follow the manufacturer's direction for use of such solutions. A most effective method of applying solvents is with a suction-type atomizing spray nozzle. The light air pressure with the atomized solvent does an effective cleaning job. The hazards, however, are that of forcing and flushing conductive contaminants deeper into the winding. Also, some insulation materials, particularly adhesive tapes, may be attacked or softened by solvents. In the repair shop, steam cleaning and flushing, degreasing tanks, and rinsing troughs are common. Slowly rotating a dc armature in a warm detergent solution can be a very effective method for removing carbon dust from deep down in the winding. Motors that have been subjected to immersion because of floods are best cleaned by a low-pressure stream of water or a water detergent solution. This will remove encrusted mud and debris.

In all liquid pressure cleaning, avoid aiming the stream directly at the winding, but rather wash the dirt away by aiming at an angle to the winding surface. This should minimize the chances of forcing dirt into the winding.

In any washing or cleaning involving water, it is necessary to dry out the winding. It is recommended to dry the winding in a forced draft oven for at least 8 hr at about 121°C (250°F).

Motor Testing. Dynamometer load testing will identify performance problems including loss of torque and winding or bearing temperature problems. Load testing is particularly recommended for rebuilt dc motors, where the test can verify the neutral setting, the intercoil strength, and the relative polarities of the various windings, in addition to the load-carrying capability of the motor.

PREDICTIVE MAINTENANCE PROGRAM

Predictive maintenance requires scheduled monitoring of vibration, temperatures, and/or oil contamination (wear particles), with trend analysis to spot the potential problems. A discussion of the techniques used is beyond the scope of this chapter. However, the philosophy offers promise for improved performance, particularly in applications with high failure rates or costly downtime.

Predictive maintenance is currently evolving as a way to anticipate impending equipment failures through the use of modern technologies, and to eliminate the root cause before it can cause catastrophic failure.

The goal of predictive maintenance is to extend the life of machinery by eliminating failures. Some of the benefits of predictive maintenance are:

1. Repetitive problems are identified and eliminated.
2. Equipment installation is performed to precise standards.
3. Performance verification assures that new and rebuilt equipment is free of defects.

Predictive maintenance technologies include:

1. Vibration spectrum analysis for rotating equipment.
2. Oil and wear analysis.
3. Thermography for all electrical equipment.

MOTOR TROUBLESHOOTING

The following troubleshooting flowcharts (Tables 1.3 and 1.4) should provide logical, step-by-step methods for determining solutions to motor problems. They should be used as a guideline by qualified competent personnel. See also Tables 1.5, 1.6, and 1.7.

Warning: Internal parts of a motor may be at line potential even when it is not rotating. Before performing any maintenance which could result in contacting any internal part, be sure to disconnect all power from the motor.
- *Motors using rectified power units, disconnect all ac line connections.*
- *Motors using rotating power units, disconnect all dc line and field connections.*

Index to Table 1.3 Troubleshooting ac Motors

A. Motor won't start or accelerates too slowly.

B. Motor runs noisy.

C. Motor overheats.

D. Motor bearings run hot or noisy.

Index to Table 1.4 Troubleshooting dc Motors

E. Motor won't start.

F. Motor starts but stops and reverses direction.

G. Motor runs but overload protective device trips too often.

H. Motor overheats.

I. Motor runs too slowly.

J. Motor runs too fast.

K. Motor runs noisy.

L. Motor bearings run hot or noisy.

M. Brushes sparking excessively; may be accompanied by brush chatter and/or excessive wear and chipping.

TABLE 1.3 Troubleshooting AC Motors

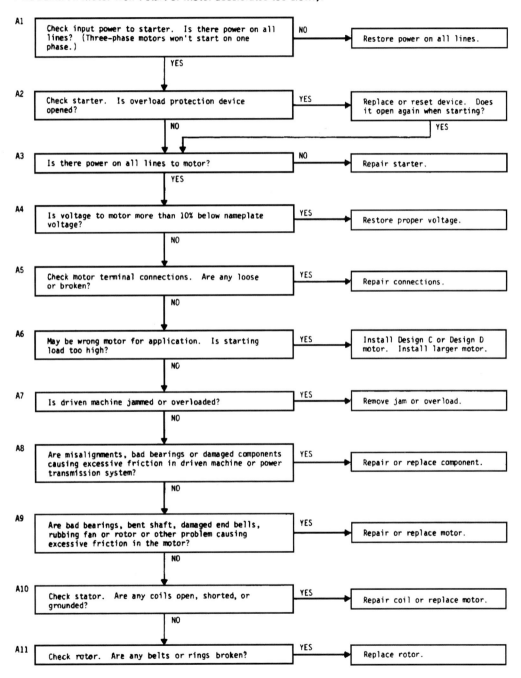

TABLE 1.3 Troubleshooting AC Motors (*Continued*)

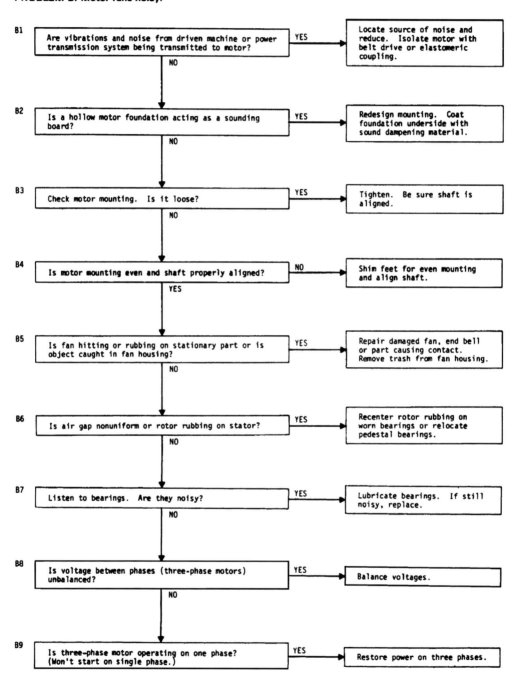

TABLE 1.3 Troubleshooting AC Motors (*Continued*)

PROBLEM C: Motor overheats.

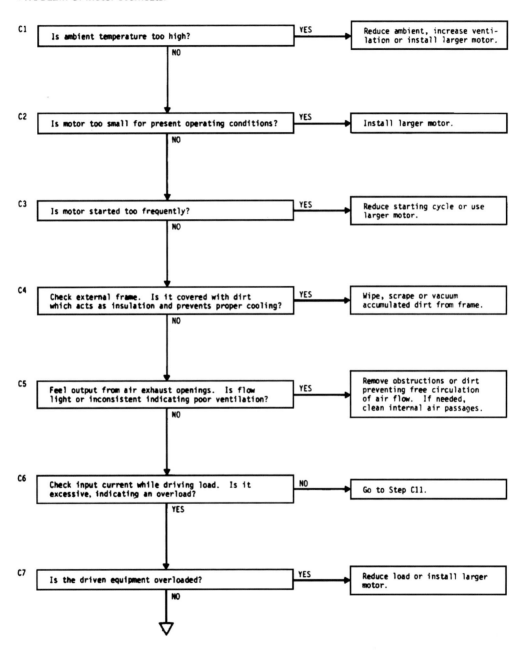

TABLE 1.3 Troubleshooting AC Motors (*Continued*)

PROBLEM C: Motor overheats (continued).

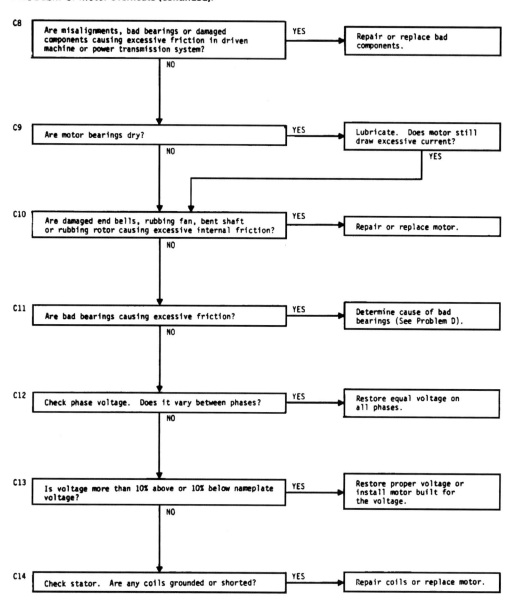

TABLE 1.3 Troubleshooting AC Motors (*Continued*)

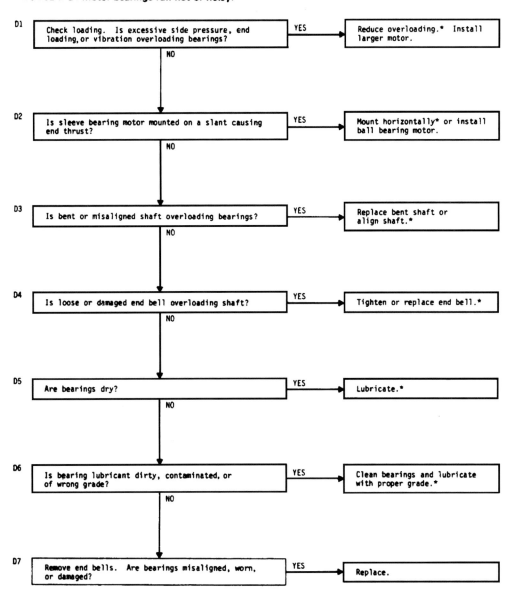

*BEARINGS MAY HAVE BEEN DAMAGED. IF MOTOR STILL RUNS NOISY OR HOT, REPLACE BEARINGS.

TABLE 1.4 Troubleshooting DC Motors

PROBLEM E: Motor won't start.

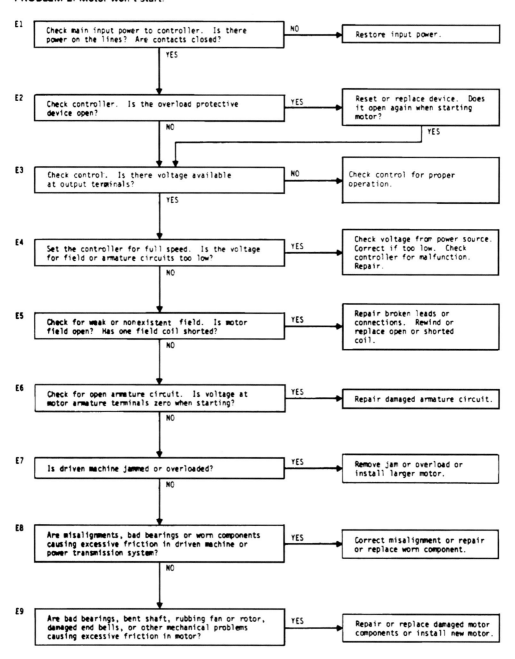

TABLE 1.4 Troubleshooting DC Motors (*Continued*)

PROBLEM F: Motor starts but stops and reverses direction.

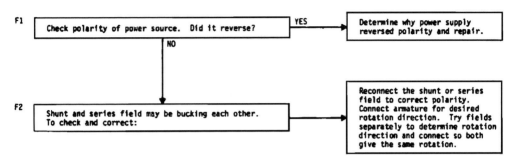

PROBLEM G: Motor runs but overload protective device trips too often.

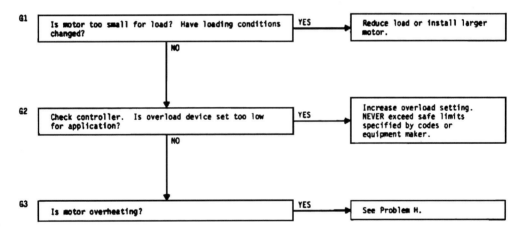

TABLE 1.4 Troubleshooting DC Motors (*Continued*)

PROBLEM H: Motor overheats.

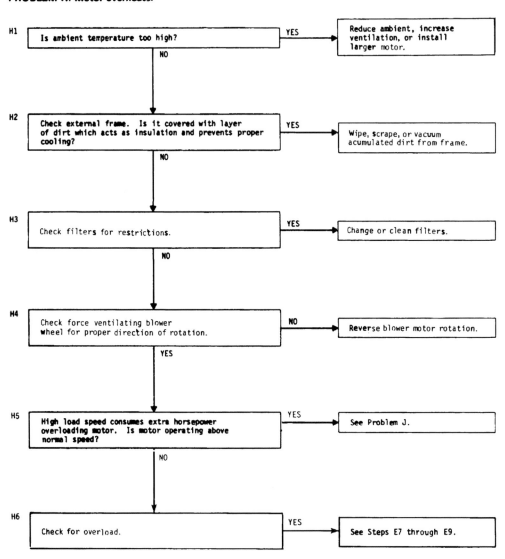

ELECTRIC MOTORS 6.29

TABLE 1.4 Troubleshooting DC Motors (*Continued*)

PROBLEM I: Motor runs too slowly.

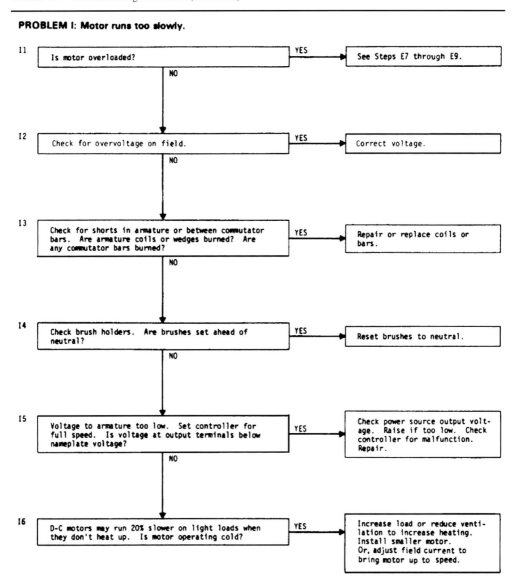

TABLE 1.4 Troubleshooting DC Motors (*Continued*)

PROBLEM J: Motor runs too fast.

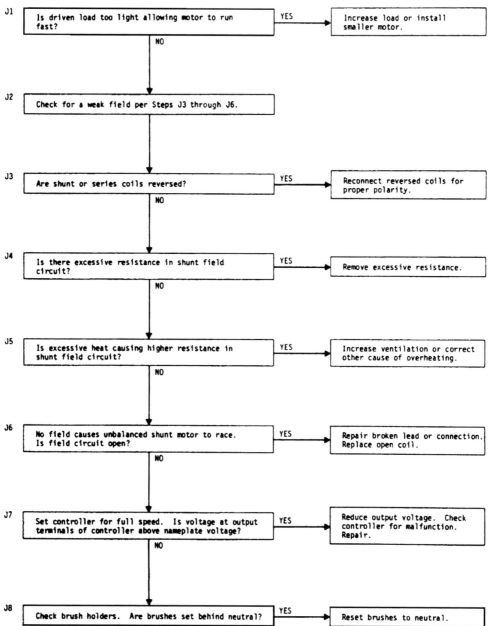

TABLE 1.4 Troubleshooting DC Motors (*Continued*)

PROBLEM K: Motor runs noisy.

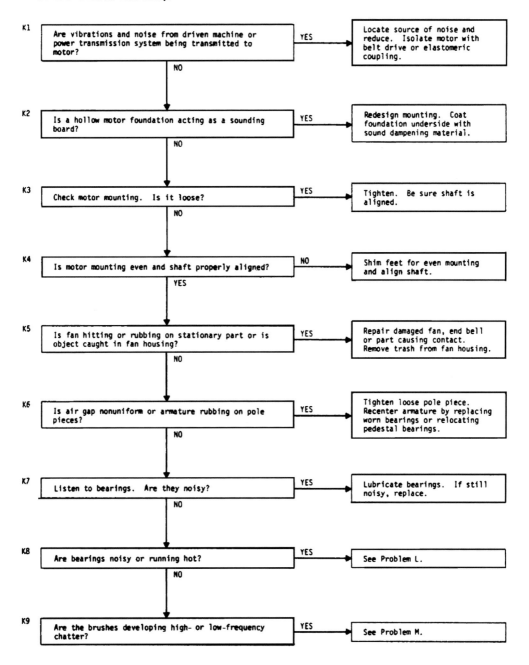

TABLE 1.4 Troubleshooting DC Motors (*Continued*)

PROBLEM L: Motor bearings run hot or noisy.

L1. Check loading. Is excessive side pressure, end loading, or vibration overloading bearings?
 - YES → Reduce overloading.* Install larger motor.
 - NO ↓

L2. Is sleeve bearing motor mounted on a slant, causing end thrust?
 - YES → Mount horizontally* or install ball bearing motor.
 - NO ↓

L3. Is bent or misaligned shaft overloading bearings?
 - YES → Replace bent shaft or align shaft.*
 - NO ↓

L4. Is loose or damaged end bell overloading shaft?
 - YES → Tighten or replace end bell.*
 - NO ↓

L5. Are bearings dry?
 - YES → Lubricate.*
 - NO ↓

L6. Is bearing lubricant dirty, contaminated, or of wrong grade?
 - YES → Clean bearings and lubricate with proper grade.*
 - NO ↓

L7. Remove end bells. Are bearings misaligned, worn, or damaged?
 - YES → Replace.

*BEARINGS MAY HAVE BEEN DAMAGED. IF MOTOR STILL RUNS NOISY OR HOT, REPLACE BEARINGS.

ELECTRIC MOTORS 6.33

TABLE 1.4 Troubleshooting DC Motors (*Continued*)

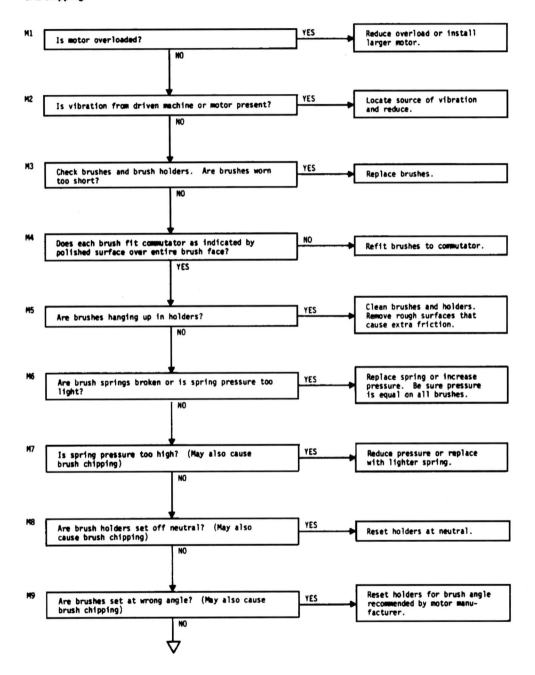

TABLE 1.4 Troubleshooting DC Motors (*Continued*)

PROBLEM M: Brushes sparking excessively; may be accompanied by brush chatter and/or excessive wear and chipping (continued).

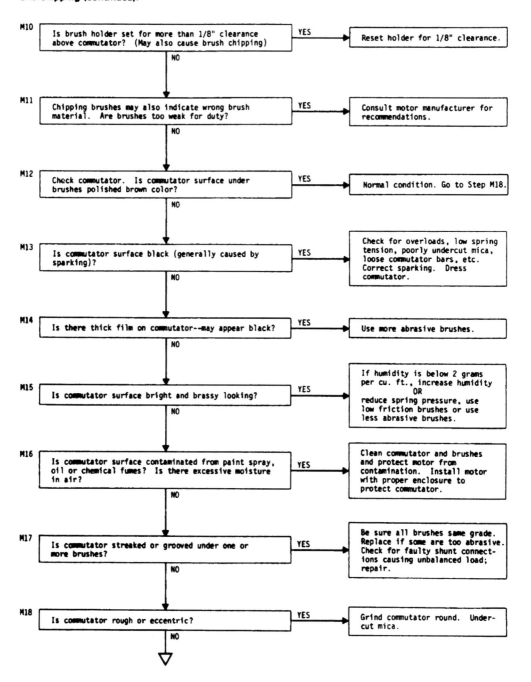

TABLE 1.4 Troubleshooting DC Motors (*Continued*)

PROBLEM M: Brushes sparking excessively; may be accompanied by brush chatter and/or excessive wear and chipping (continued).

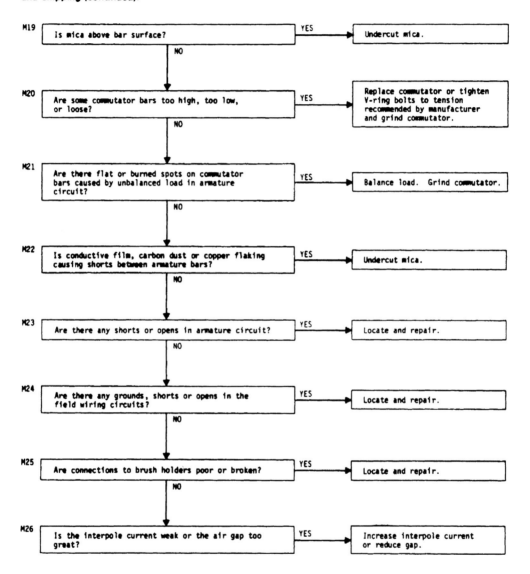

TABLE 1.5 Polyphase AC Motor Nameplate Coding

Code letter*	Locked kVA per hp at full voltage†	Locked amps per hp at 208 volts	Locked amps per hp at 230 volts
A	0–3.14	0–8.73	0–7.88
B	3.15–3.54	8.75–9.83	7.91–8.87
C	3.55–3.99	9.85–11.08	8.91–10.02
D	4.00–4.49	11.10–12.48	10.05–11.28
E	4.50–4.99	12.50–13.88	11.30–12.53
F	5.00–5.59	13.90–15.88	12.57–14.04
G	5.60–6.29	15.60–17.48	14.07–15.68
H	6.30–7.09	17.50–19.68	15.82–17.81
J	7.10–7.99	19.70–22.18	17.83–20.07
L	8.00–8.99	22.20–24.98	20.10–22.58
M	9.00–9.99	25.00–27.78	22.60–25.10
N	10.00–11.19	27.80–31.08	25.12–28.11
P	11.20–12.49	31.10–34.68	28.14–31.38
R	12.50–13.99	34.70–39.88	31.40–35.15
S	14.00–15.99	38.90–44.41	35.17–40.17
T	16.00–17.99	44.44–49.97	40.20–45.20
U	18.00–19.99	50.00–55.52	45.22–50.22
V	20.22–22.39	55.55–62.16	50.25–56.23
	22.40–and up	62.22–and up	56.28–and up

*Nameplates of polyphase AC motors rated $1/2$ hp and larger are marked with a letter code to indicate the locked rotor kVA per horsepower.
†Locked amperes per horsepower for three-phase motors only.

TABLE 1.6 Motor Synchronous Speeds (RPM)

Frequency	Number of poles								
	2	4	6	8	10	12	14	16	18
60 cycles	3600	1800	1200	900	720	600	514	450	400
50 cycles	3000	1500	1000	750	600	500	428	375	334

TABLE 1.7 AC Motor Calculations*

To find:	Two-phase four-wire	Three-phase
Amperes when horsepower is known	$\dfrac{hp \times 746}{2 \times E \times \% \text{ Eff.} \times PF}$	$\dfrac{hp \times 746}{1.73 \times E \times \% \text{ Eff.} \times PF}$
Amperes when kilowatts is known	$\dfrac{kW \times 1000}{2 \times E \times PF}$	$\dfrac{kW \times 1000}{1.73 \times E \times PF}$
Amperes when kVA is known	$\dfrac{kVA \times 1000}{2 \times E}$	$\dfrac{kVA \times 1000}{1.73 \times E}$
Kilowatts	$\dfrac{I \times E \times 2 \times PF}{1000}$	$\dfrac{I \times E \times 1.73 \times PF}{1000}$
kVA	$\dfrac{I \times E \times 2}{1000}$	$\dfrac{I \times E \times 1.73}{1000}$

TABLE 1.7 AC Motor Calculations* (*Continued*)

To find:	Two-phase four-wire	Three-phase
Horsepower—(output)	$\dfrac{I \times E \times 2 \times \% \text{ Eff.} \times PF}{746}$	$\dfrac{I \times E \times 1.73 \times \% \text{ Eff.} \times PF}{746}$

*PF = power factor. E = electromotive force (voltage).
Performance characteristics:

$$hp = \frac{\text{torque (ft-lb)} \times \text{rpm}}{5250}$$

$$kW = hp \times 0.746$$

$$\text{Torque in ft-lb} = \frac{hp \times 5250}{\text{rpm}}$$

$$\text{Motor synchronous speed in rpm} = \frac{120 \times Hz}{\text{number of poles}}$$

$$\text{Three-phase full-load amp} = \frac{hp \times 0.746}{1.73 \times kV \times \text{efficiency} \times \text{power factor}}$$

$$\text{Rated motor kVA} = \frac{hp \times 0.746}{\text{efficiency} \times \text{power factor}}$$

$$\text{kW loss} = \frac{hp \times 0.7463(1.0 - \text{efficiency})}{\text{efficiency}}$$

$$\text{Motor accelerating time in seconds} = \frac{\Sigma WK^2 \times \Delta \text{rpm}}{308 \times T_{avg}}$$

where ΣWK^2 = the sum of all moments of inertia driven by the motor, reflected to the motor, including the driven load, the motor rotor, and any significant gear or pulley inertias, in pound-feet
Δ rpm = the change (increase) in motor speed
T_{avg} = the average motor output torque over the speed change range, in pound-feet

To find the moment of inertia (Wk^2) of a load driven through a belt or gearcase, etc., as reflected to the motor:

$$\text{Load } Wk^2 \text{ at motor shaft speed} = \text{load } Wk^2 \times \left(\frac{\text{load rpm}}{\text{motor rpm}}\right)$$

kVA inrush = percent inrush $\times$ rated kVA

CHAPTER 2
MAINTENANCE OF MOTOR CONTROL COMPONENTS

Shon Isenhour
Principal, Life Cycle Engineering, Inc., Charleston, S.C.

CONTROL COMPONENTS

Control components are those ON-OFF devices that start and stop the flow of electricity to equipment that consumes (utilizes) the electricity. Examples of utilization equipment are motors, ovens, electric furnaces, heaters, lights, welders, electric brakes, and lifting magnets. Control components are located in the branch circuits of an electrical distribution system. They are called power circuit components if their contacts operate in the power circuit and control circuit components if their contacts operate in a logic or control circuit. See Fig. 2.1.

Power circuit components are rated by industry standards in a voltage range of 12 to 7200 V and a current range of 9 to 4000 A. Examples of power circuit components are molded case circuit breakers, disconnect switches, contactors, motor starters, capacitor starting relays, and rheostats.

Control circuit components are rated by industry standards in a voltage range of 12 to 600 V with a maximum current rating typically being 10 A, and occasionally up to 25 A. Examples of control circuit components are pushbuttons, selector switches, pressure switches, temperature switches, foot switches, limit switches, proximity switches, photo switches, control relays, overload relays, time-delay relays (timers), voltage-sensing relays, phase-failure (loss) relays, phase-reversal relays, plugging and antiplugging relays, synchronizing relays, underload relays, jam-detection relays, and auxiliary contacts (electrical interlocks).

In addition to control components with contacts, there are devices without contacts that are typically considered to be part of electrical control schemes to control and monitor the direction of rotation and speed of motors, dimming of lights, and so on. These are solid-state controllers, power circuit resistors, and indicating (pilot) lights.

QUALIFIED PERSONNEL

Control components exist in such a variety of sizes, shapes, ratings, and configurations that their maintenance should be performed only by qualified personnel who know how the equipment is intended to be used, and who are capable of understanding the manufacturer's instructions.

FIGURE 2.1 Typical control components.

To be considered qualified, these individuals should be capable of reading and drawing electrical schematic (sometimes called elementary or ladder) diagrams, as well as connection (wiring) diagrams. They should know how to install and read meters to measure:

Voltage	(voltmeter)
Current	(ammeter)
Power	(wattmeter)
Resistance	(ohmmeter, Wheatstone, and Kelvin bridges)
Dielectric withstand	(high-potential tester)
Continuity	(ohmmeter and voltmeter or equivalent)

They should know how to measure magnet and contact gaps, measure spring force, and apply feeler gages and torque wrenches. They must be able to recognize evidence of overheating and of tracking. They must know the meaning and significance of such terms as available fault current, interrupting rating, polarity, plugging, inching, jogging, wire temperature rating, capacitor charge, contact gap, motor locked-rotor indicating code letters (Article 430 of the National Electrical Code), normally open and normally closed contacts, and the like.

CONTROL COMPONENT ENCLOSURES

Since all control components include exposed terminals, they must be enclosed in some manner to prevent accidental contact. These enclosures vary from simple sheet-metal or cast-metal boxes only large enough to house a single component, to motor control centers and control rooms where the room itself is the enclosure. Another purpose of enclosures is to provide protection against environmental conditions that may have an adverse effect on the equipment. If the control components are contaminated with dust, dirt, or moisture that originates outside of the enclosure, either the enclosure is not the proper type for the environment, conduit entries have not been properly made, or the door of the enclosure has been left open. An enclosure with an open door is a sign of trouble that should be diagnosed and corrected. Maintenance effort will be less with an installation that is suitable for the

TABLE 2.1 Enclosure Types for Nonhazardous Locations

Provides a degree of protection against the following environmental conditions	For outdoor use						
	Enclosures type number						
	3	3R	3S	4	4X	6	6P
Incidental contact with the enclosed equipment	X	X	X	X	X	X	X
Rain, snow, and sleet	X	X	X	X	X	X	X
Sleet	—	—	X	—	—	—	—
Windblown dust	X	—	X	X	X	X	X
Hosedown	—	—	—	X	X	X	X
Corrosive agents	—	—	—	—	X	—	X
Occasional temporary submersion	—	—	—	—	—	X	X
Occasional prolonged submersion	—	—	—	—	—	—	X

Provides a degree of protection against the following environmental conditions	For indoor use									
	Enclosure type number									
	1	2	4	4X	5	6	6P	12	12K	13
Incidental contact with the enclosed equipment	X	X	X	X	X	X	X	X	X	X
Falling dirt	X	X	X	X	X	X	X	X	X	X
Falling liquids and light splashing	—	X	X	X	—	X	X	X	X	X
Dust, lint, fibers, and flyings	—	—	X	X	X	X	X	X	X	X
Hosedown and splashing water	—	—	X	X	—	X	X	—	—	—
Oil and coolant seepage	—	—	—	—	—	—	—	X	X	X
Oil or coolant spraying and splashing	—	—	—	—	—	—	—	—	—	X
Corrosive agents	—	—	—	X	—	—	X	—	—	—
Occasional temporary submersion	—	—	—	—	—	X	X	—	—	—
Occasional prolonged submersion	—	—	—	—	—	—	X	—	—	—

Source: This table is reprinted with permission of the National Fire Protection Association from NFPA 70-1993, *National Electrical Code,* copyright 1993. This reprinted material is not the complete and official position of the NFPA on the referenced subject, which is represented only by the standard in its entirety.

Note: Type 12K enclosures become type 12 after conduit has been installed at knockout locations.

environment and is kept sealed. See Table 2.1 for the types of enclosures available for control components installed in nonhazardous locations. Enclosures other than type 1 should be marked with their type number if they are manufactured after 1984.

Drains and Breathers. Drains and breathers are available for installation in enclosures subjected to wide variations in temperature and humidity. Such drains should be installed at the lowest point in the enclosure and operated periodically to ensure that they do not become plugged with debris washed down by the condensation.

PREVENTIVE MAINTENANCE

Preventive maintenance should be a program, a scheduled periodic action that begins with the installation of the equipment. At that time, specific manufacturer's instruction literature should be consulted, then stored for future reference. Follow-up maintenance should be done at regular intervals, as frequently as the severity of duty justifies. Time intervals of 1 week or 1 month or 1 year may be appropriate, depending on the duty. It is also desirable to establish specific checklists for each control, as well as a logbook to record the history of incidents. A supply of renewal parts should be obtained and stored.

General Guidelines. The whole purpose of maintaining electrical equipment can be summarized in two rules:

1. Keep those portions conducting that are intended to be conducting.
2. Keep those portions insulating that are intended to be insulating.

Good conduction requires clean, tight joints free of contaminants such as dirt and oxides.

Good insulation requires the absence of carbon tracking and the absence of contaminants such as salt and dust that become hydroscopic and provide an unintended circuit between points of opposite polarity.

Maintenance of control components requires that all power to these components be turned off by opening and locking open the branch circuit disconnect device, usually a switch or circuit breaker located in the same enclosure as the control components or in a panel board or switchboard feeding the control enclosure. Separate control sources of power must also be disconnected. If control power is used during maintenance, caution should be used to prevent feedback of a hazardous voltage through a control transformer. Be alert to power factor correction capacitors that may be charged. Discharge them before working on any part of the associated power circuit.

Cleaning. Soot, smoke, stained areas (other than inside arc chutes), or other unusual deposits should be investigated and the source determined before cleaning is undertaken. Vacuum or wipe clean all exposed surfaces of the control component and the inside of its enclosure. Equipment may be blown clean with compressed air that is dry and free from oil. (Be alert to built-in oilers in factory compressed air lines!) If air-blowing techniques are used, remove arc covers from contactors and seal openings to control circuit contacts that are present. It is essential that the foreign debris be removed from the control enclosure, not merely rearranged. Control equipment should be clean and dry. Remove dust and dirt inside and outside the cabinet without using liquid cleaner. Remove foreign material from the outside top and inside bottom of the enclosure, including hardware and debris, so that future examination will reveal any parts that have fallen off or dropped onto the equipment. If there are liquids spread inside, determine the source and correct by sealing conduit, adding space heaters (check manufacturer), or performing other actions as applicable.

Mechanical Checks. Tighten all electrical connections. Look for signs of overheated joints, charred insulation, discolored terminals, and so on. Mechanically clean to a bright finish or replace those terminations that have become discolored. Determine the cause of the loose joint and correct. Be particularly careful with aluminum wire connections. Aluminum wire is best terminated with a crimp-type lug that is attached to the control component. When screw-type lugs (marked CU/AL) are used with aluminum wire, joints should be checked for tightness every 200 operations of the device.

Wires and cables should be examined to eliminate any chafing against metal edges caused by vibration that could progress to an insulation failure. Any temporary wiring should be removed or permanently secured, and diagrams should be marked accordingly.

The intended movement of mechanical parts, such as armature and contacts of electromechanical contactors and mechanical interlocks, should be checked for freedom of motion and functional operation. For example, does a mechanical interlock actually work as intended? Sensitive electromechanical relays should be checked carefully and only in accordance with manufacturer's recommendations.

Lubrication. Lubricate any parts specifically required by manufacturer's instructions. Do not apply lubricants or coatings where such lubrication is not recommended by the manufacturer. Lubrication should be applied sparingly. Wipe off any surplus. It must not be allowed to get onto insulation where it will collect dust and cause a buildup of debris.

Wrap-up. Check all indicating lamps, mechanical flags, doors, latches, and similar auxiliaries, and repair, if required.

Log changes and observations into record book before returning equipment into service. Do not remove any labels or nameplates. Restore any that are damaged.

POWER CIRCUIT COMPONENTS

Power circuit components with butt contacts include manual and magnetic motor starters (a motor starter consists of a contactor combined with an overload relay), magnetic contactors, and drum controllers. See Fig. 2.1.

MOTOR CONTROLLERS

Motor controllers are either magnetically or manually operated control components designed specifically for starting and stopping electric motors. Motor controllers are rated in horsepower and assigned standard sizes by the National Electrical Manufacturers Association (NEMA). Motor controllers are applied by selecting a size having a horsepower rating equal to or greater than the horsepower rating of the motor it is to control at the single-phase or three-phase voltage involved. Manual motor controllers are used where the controller can be mounted close to the operator of the equipment being driven by the motor involved. Most motors used in industry are controlled by magnetic motor controllers.

Magnetic Motor Controllers. Magnetic motor controllers consist of a magnet, an operating coil, an armature which is attracted to the magnet when the coil is energized, a contact carrier (sometimes called a crossbar), a contact system, and terminals. If the device has only these parts suitably mounted on insulating material and baseplate, it is called a contactor. When an overload relay is added to a contactor, the device is called a motor starter.

Magnet Construction. Figure 2.2 illustrates the most frequently used types of ac magnetic operators. The E-shaped magnet with an I-shaped armature uses a single coil positioned around the center pole. This magnet has an air gap formed by grinding the center pole shorter than the two outside poles. This air gap tends to close as pole faces wear. When the air gap closes to zero, the magnet will be very noisy and eventually will stick because of residual magnetism. If this occurs, the magnet and armature must be replaced. The horizontal I-shaped piece atop the magnets in Fig. 2.2 is the armature in both cases.

The U-shaped magnet with an I-shaped armature is assembled so that it has an air gap in the magnet path at a location away from the pole faces. In this design, the air gap cannot be changed by mechanical wear. This construction normally uses two coils, one positioned on each leg, which may be connected in series or in parallel for dual-voltage convenience. Both coils are usually contained in one encapsulated unit.

In either design, a broken or missing shading coil will cause a magnet to be noisy, in which case the magnet and armature should be replaced.

Magnet noise can also be caused by loose laminations when rivets are not properly headed, by poor grinding of pole faces, or by laminations which have shifted after grinding is complete. This last situation can occur if the finished magnet assembly has been dropped.

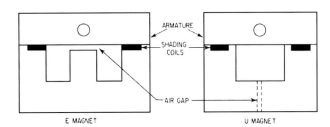

FIGURE 2.2 Shading coils.

When the current in the coil of an ac contactor or relay passes through zero, its magnetic pull or holding power is zero. The device will have a tendency to open unless shading coils (single loops of copper or aluminum) are incorporated in its pole faces. The current, however, is soon effective in the opposite direction, and the device is again pulled closed. The nature of the operation causes a humming noise in any ac operated device and a decided chattering noise in a defective unit. Chattering is normally eliminated by the use of a shading coil embedded in the laminated magnetic circuit of the device. The shading coil produces enough out-of-phase flux to provide holding power to maintain the device closed during the short period when the main flux is zero. Even with shading coils the air-gap surfaces must be free from dirt and well fitted to avoid objectionable noise. Broken shading coils are ineffective and cause noisy operation. Figure 2.3 shows two typical types of magnetic pole pieces (magnet and armature) used in contactors; the outline drawing in Fig. 2.2 shows the location of the shading coils on these E and U types, and they can be seen in Fig. 2.3.

For quiet operation of ac contactors it is necessary to provide well-fitted pole faces. Any dirt in this area introduces a greater air gap when the unit is closed; it increases the load imposed upon the magnet coil and results in noisier operation.

DC coils are not subject to a zero current condition during normal operation; therefore, dc-operated devices are quiet. For this reason, ac-carrying contactors equipped with dc operating coils will operate quietly.

Operating (Magnet) Coils. The operating coils for both ac and dc magnetic devices consist of many turns of insulated magnet wire wound on a bobbin and connected to terminals. The cross section of the wire (wire size) and the number of turns determine the voltage rating of the coil. Coils should be operated at rated voltage. Both overvoltage and undervoltage conditions are undesirable. Coils provide the electromagnetic pull that causes the contacts of relays and contactors to open or close. The coils may be varnish impregnated in a vacuum, molded under pressure with insulating compounds, or merely covered with insulating tape. Many manufacturers use coils encapsulated in epoxy or polyester molding compounds. The coils must withstand temperature and humidity cycling without cracking and reasonable mechanical abuse without damage.

Even at operating temperatures the coils for ac devices must close them at 85 percent of the rated voltage. Coils for dc devices should close them at 80 percent of the rated voltage. Any coil is expected to withstand 110 percent of rated voltage without damage. Measurement of operating voltages should be made at the coil terminals and not at the source of the supply voltages.

A coil with an open circuit will not operate the contactor or relay. Any questionable coil should be immediately replaced by one that is good. Inoperable coils should be checked for open circuits, grounds, or bad lead connections before they are thrown away. Replacement coils must have the proper rating for the application.

If any turns of a dc coil become short-circuited, the resistance of the coil will be reduced and more current will pass through the coil. The increased current may or may not cause higher operating

FIGURE 2.3 Magnets and armatures.

temperatures and coil burnout. Short-circuited turns in an ac coil result in a transformer action which generates a high current at the point of the short circuit. Immediate burnout usually follows the short-circuiting of an ac coil. However, a short-circuited dc coil may continue to operate for an indefinite time or until total insulation failure results.

Coils should be operated at their rated voltage. Overvoltage on coils causes them to operate at a higher temperature which shortens coil life. Overvoltage also operates the contactor or relay with unnecessary force, and this contributes to mechanical wear and bounce when closing. Undervoltage on coils can cause contactors and relays to operate sluggishly. The contact tips may touch, but the magnet may be unable to close the contacts completely against the spring force. Under these conditions, contact pressure is below normal, and the contacts may overheat and weld together. At the same time, it is possible that the coil will overheat and burn out because the magnetic actuator is still in an open-gap condition.

The impedance of an ac coil controlling a magnetic circuit with a large air gap is much lower than the impedance of that same coil when the magnetic circuit has little or no air gap. The current drawn by the coil of an ac magnet is therefore much greater at "open gap" than at "closed gap." Since the closing time is relatively short, ac coils are only designed to withstand full voltage under closed conditions. They will soon overheat if the magnet is blocked open or if the voltage is too low to close the magnetic circuit and the coil remains energized with a large air gap in the magnetic circuit. DC coils are not subject to these conditions because the coil currents do not vary with the air gap. For certain applications, some dc coils are momentarily energized at much higher than rated current. The momentary high energy requirement to close an electrical contactor is not damaging unless the high current is maintained. For this reason, contactors should not be blocked open when testing, since cut-off devices may be made inoperable.

Arc Chambers. When contactors are expected to open circuits carrying currents that are difficult to interrupt, they are designed with arc chambers having arc-interrupting features. The arc chambers (arc boxes) are shown in position over the main contacts in Fig. 2.4 and do not allow a view of the contacts or how they are designed to provide arc interruption. The arc boxes can be raised or removed to inspect the contacts and determine the condition of the ceramic parts and metal-grid plates. These are the principal components in the arc chamber, which is commonly referred to as an arc chute. When arc chutes have been removed for inspection or renewal of contacts, they must be returned to proper position to make arc interruption effective. Make certain that the snap fasteners or holddown bolts have been replaced.

FIGURE 2.4 NEMA size 7 contactor.

6.46 MAINTENANCE OF ELECTRICAL EQUIPMENT

FIGURE 2.5 Direct-current contractor with arc chute removed.

On dc contactors, the arc-quenching structures generally include arc shields of molded ceramic or heat-resistant material. See Fig. 2.5. These arc shields should always be "down," to receive the arc as it is distorted and directed by the magnetic field of the blowout coil; otherwise the shield will not give satisfactory results.

In large ac contactors, magnetic-arc-splitting action confines, divides, and extinguishes the arc without the benefit of a blowout coil. As contacts separate, a magnetic reaction occurs that forms and stretches the arc. It rises into or against metal grids, where it is sliced into a series of arcs. At the next zero point on the current cycle, the air adjacent to each grid is deionized instantly; the voltage per grid is insufficient to reestablish the current, and the arc is out. This type of ac arc quencher, when used, is assembled over the contacts much like an arc shield on a dc contactor; it is easily removed for inspection of the contacts. Small ac contactors have arc chambers with little or no metal in them because the contact gap and arc chamber material are sufficient to interrupt the lower currents.

Braided Shunts. Many power circuit control devices have a single-break construction and use flexible shunts to carry current. See Fig. 2.4 or 2.5. These shunts are made with many strands of fine copper wire. Strands sometimes break where the shunt bends, and the unbroken strands, which must then carry the entire current load, overheat and eventually burn out. Frayed shunts should be replaced if a significant number of strands are found broken.

Manual Motor Controllers. See Fig. 2.6. The majority of manual starters are used to control small motors driving equipment that can be operated independently. An operator must be present at the controller to turn the motor on and off and to reset the starter after an overload. Maintenance of manual motor starter contacts and the overload relay is no different than that of a magnetic starter. Some manual motor controllers include a magnet and coil wired to cause the starter to open in the event of low voltage or a loss of voltage. Such starters require the operator to restart the motor when voltage returns to normal. Manual starters without this feature can cause a motor to restart when conditions

FIGURE 2.6 Manual motor controllers.

return to normal after a power loss. *When checking a manual starter installation, be alert to the status of the starter. The starter can be on when the power is off.*

Contacts.   All of these devices have contacts that are made of either copper or a silver alloy arranged in pairs. One contact of each pair is stationary and the other is movable, controlled by a mechanism that is operated either manually or magnetically. Contact pressure is created by the force of a spring acting on the movable contact. Contact pairs are used in either single-break or double-break per pole configurations. See Fig. 2.7.

Contact Bounce.   Contacts may close one against the other as shown in Fig. 2.7 or may come together in a wiping action as in a rheostat. Contacts of magnetic controllers impact against each other at high speed, driven by the movement of an armature attracted to a magnet energized by a coil. This high-speed impact causes the contacts to rebound or bounce open one or more times for a total period of from 2 to 30 ms while conducting. While the contacts are open, arcing between contact pairs occurs. The higher the current and the longer the arcing time, the more heat is produced. If sufficient heat is produced over an area of contact surface, the contact material melts in this area, and then when the contacts close for the final time, the molten metal reforms to weld the contact pair together. For any given design the probability of welding is highest (1) when the contacts are operated for the first time and the mating surfaces are smooth and silver-rich and (2) when the contacts are badly worn reducing the contact wear allowance (overtravel) to almost zero with the resulting loss of contact pressure in the closed position. Contact pressure can also be lost if the contact overtravel springs become annealed by heat or are broken.

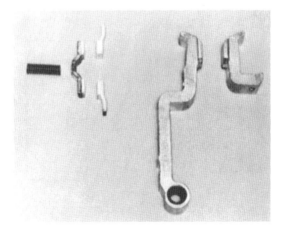

FIGURE 2.7 Single-break and double-break contacts.

Contact Pressure. Weak contact–overtravel springs can cause low contact pressure and can cause contacts to bounce excessively. Low contact pressure can cause the contacts to overheat. Bouncing contacts often cause the contact surfaces to weld. Therefore, inspect the springs for signs of deterioration.

Comparison of a used spring with a new spring as to size, shape, color, and force will indicate roughly whether the used spring has lost its strength. If there is any doubt about the condition of the spring, measure the spring initial and final pressures (forces) and compare them to the recommended values supplied by the manufacturer.

Contact force is measured by using a spring scale as shown in Fig. 2.8. Turn the power *off* and remove the arc box from the contactor.

Around each movable contact tie a loop of strong string or lightweight cord sufficiently large to extend on both sides of the crossbar, and engage the hook on the spring scale. With the contacts in the open position, pull on the spring scale and determine the force required to just begin to move the contact (i.e., to compress the overtravel spring). This measurement is the initial contact force. Measure each contact spring.

Then for a contact with normally open poles, with the contactor energized or the armature held sealed against the magnet by some convenient means, determine the force required to just begin to move the bridging (movable) contact away from the stationary contacts. This measurement is the final contact force.

All force measurements must be taken with a smooth steady pull in a straight line perpendicular to the contact surface. It is not necessary to remove contactors from the enclosures to check contact pressure, but all power to the contacts must be turned *off*.

The moving armature must be held in a closed position on normally open contactors; spring-loaded, normally closed contactors do not present this problem. Check each pole separately; determine and record forces. Always compare results with the manufacturer's recommended or published contact values.

Unless there is a technique recommended by the manufacturer for altering contact pressures, *do not* resort to bending any part (moving or stationary) to increase or decrease spring pressures. The most common causes for loss of spring pressure are overheating or erosion of contact material. When contacts carry too much current or when there are poor connections, contact temperatures can rise and contact springs may become annealed and lose strength. On occasion a spring may break, causing a

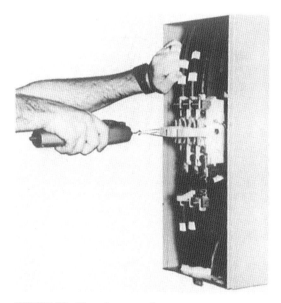

FIGURE 2.8 Measuring contact force.

loss of pressure and a high-resistance contact joint which will heat more than adjacent contacts. If one contact set appears to be burning more than companion contacts, check for the cause. It may be an overloaded phase, insufficient contact pressure, a weak or broken spring, misalignment, or a mechanical-interference problem.

Contact Wear Allowance. Contact wear allowance is a linear dimension that expresses the thickness of contact material that the control component manufacturer expects to be consumed during the life of the contact pair. Contact wear allowance is also called overtravel. Manufacturers sometimes state in the instruction material furnished with their devices the amount of contact wear allowance provided in the design. When the contact wear allowance is not stated for a new magnetically operated control device with normally open contacts, it can be calculated as follows:

$$\text{Wear allowance (overtravel)} = \text{magnet gap} - \text{contact gap}$$

where magnet gap = distance between magnet and armature in the fully deenergized position
      contact gap = distance between contact pairs in the open position

This formula applies for any time in the life of a device. Use it as shown for direct-drive magnet systems. For clapper or bell-crank magnet systems, the formula is modified by the ratio of the levers involved (distance from pivot point to points of application).

Contact Wear and Replacement. Contactors are subject to both mechanical and electrical wear during their operation. In most cases mechanical wear is insignificant. The erosion of the contacts is due to electrical wear. During arcing, material from each contact is vaporized and blown away from the useful contacting surface.

A critical examination of the appearance of the contact surfaces and a measurement of the remaining contact overtravel will give the user the information required to get maximum contact life.

Overtravel Measurement. Contact life has ended when the overtravel of the contacts has been reduced to 0.020 in or less.

Overtravel of the contact assembly is that part of the stroke which the moving contacts would travel to after touching the stationary contacts if they were not blocked from movement by the stationary contacts.

A method of measuring overtravel for direct-drive magnetic systems is as follows:

1. Place a 0.020-in feeler gage between the armature and magnet, with the armature held tightly against the magnet.
2. Check continuity in each phase, that is, determine if the circuit from terminal to terminal for each pole is open under these conditions.
3. If there is continuity through all phases, the remaining overtravel is sufficient. If there is not continuity through all phases, replace all stationary and movable contacts plus movable contact overtravel springs. After replacing parts, manually operate contactor to be sure binding does not occur.

Replace all springs and contacts as a group when inspection or measurement indicates that one or more component part is no longer usable. New contacts have a uniform silver (or copper) color. At the start of service the contact surface will have a blue coloring. The geometric form of the contact is unchanged. The sharp outer corners will be rounded with tiny silver (or copper) beads. See Fig. 2.9. At the end of service life under normal conditions the coloring changes to brown or black distributed with small silvery white or bright copper areas. The surface has a finely chiseled appearance. Material transfer causes small peaks and valleys in the contact button surface. See Fig. 2.10. Abnormal conditions create a contact appearance as described in Table 2.2.

Contact Dressing. Contacts that have some useful life remaining but have been subjected to a heavy current may have large beads of copper or silver as described in Table 2.2. These beads may be removed by "dressing" the contacts with a fine file. Do not change the contact radius or shape.

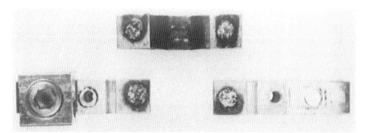

FIGURE 2.9 Contacts with normal service life.

FIGURE 2.10 Contacts at end of service life.

TABLE 2.2 Contact Appearance under Abnormal Conditions

Appearance	Cause
Curling and separation of corner of contact button from carrier	Curing is usually a result of service the produces very high heat, as under jogging or inching duty
Irregular contour or slantwise wear	One corner of a contact may wear more quickly than the other three corners. This wear is normally due to misalignment of the moving and stationary contacts. Contacts should be replaced if it is apparent that one contact is nearly making direct contact with the base metal contact carrier
Large beads of copper or silver on edges of contacts	Breaking an excessive current
Welded spot (core of smooth, shining silver, or copper surrounded by a roughened halo)	Making an excessive current. High frequency of operation, e.g., jogging

There is no need to remove the black coating from silver or silver alloy coating. The silver oxide that forms on such contacts breaks down in the presence of an arc. If all the contacts on a device cannot be salvaged by dressing, replace the entire set. Replacing contacts piecemeal only leads to further deterioration of the device. If a contactor is repeatedly burning up contacts on one particular pole, find out why. There are either electrical load or mechanical problems. Perhaps the arc box is damaged or improperly positioned.

Vacuum Contactors. The contacts of a vacuum contactor are sealed in a ceramic case as a contact pair, one stationary, the other movable. The movable contact is attached to a bellows which together with the case contains the vacuum. This bellows permits the magnet system to drive the movable contact to and from the stationary contact without alerting the vacuum. Each contact pair is assembled into a ceramic and metal container, called a bottle. See Fig. 2.11. The assembly must be replaced when either the vacuum is lost or the contact overtravel has been reduced to the minimum specified by the manufacturer. All poles of a vacuum contactor should be replaced at the same time. When vacuum contacts (bottles) are not mounted in contactors, they appear to be normally closed contacts because atmospheric pressure, working on the bellows and against the vacuum chamber, drives the movable contact against the stationary contact. The spring system of the vacuum contactor accommodates this additional source of contact pressure, and a vacuum bottle with what appears to be a set of normally closed contacts suddenly becomes a normally open pole after it has been properly installed in a vacuum contactor.

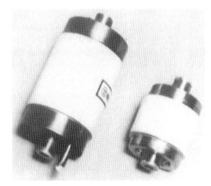

FIGURE 2.11 Vacuum contact assemblies, "Bottles."

Vacuum contactors are so called because their contacts operate in a vacuum rather than in air or oil. They are similar to air-break or oil-immersed contactors in all other respects. They often include electrical and mechanical interlocks that require careful installation or adjustment because the contact gap, and hence the magnet gap, is very small compared to air-break devices having the same rating. Vacuum contactors are usually equipped with a dc coil, often supplied from rectified alternating current. Maintenance of the noncontact portion of a vacuum contactor is the same as for an air-break contactor.

CONTACTOR INSPECTION

After a contactor has been inspected a number of times at weekly or monthly intervals, the frequency of inspection can be increased or decreased to suit conditions, depending upon the severity of the application.

Power should be removed from the starter, and inspection should proceed as follows:

1. General inspection
   - Check for loose, missing, broken, or corroded hardware; inspect pivot points, cotter pins, springs, and other mechanical parts.
   - *Do not oil moving parts unless specifically instructed to by the manufacturer; otherwise oiling will only accelerate the collection of dust and dirt.*
2. Arc boxes
   - Check arc boxes for broken or eroded ceramic insulating parts and steel-grid plates; also check for excessive collections of contact material, carbon deposits, and foreign conducting material on the surface of the insulating parts.

- Replace broken or badly eroded parts.
- Clean or replace all parts that have an excessive buildup of conducting material.

3. Contacts
   - Remove excessive oxide and large beads of contact material from the faces of copper contacts using a fine file.
   - *Do not* file silver alloy contact faces unless large beads have collected on the edges of the contacts.
   - When replacing contacts or other current-carrying parts, clean the surfaces which are to be bolted together.
   - Realign contacts and set overtravel if necessary when adjustments are provided.
   - Never use substitute contacts of different alloy material without factory approval.

4. Shunts
   - Replace shunts having broken or frayed strands.
   - Clean shunt connection points if current-carrying parts are discolored. Whenever silver- or tin-plated terminals are provided, brighten with steel wool before installing. Do not use acids or corrosive metal polishes.

5. Blowout coils
   - Inspect blowout coils for discolored connections, shorted turns, and loose hardware. Inspect turn insulation and coil spacers; replace if burned, cracked, or brittle.

6. Terminals, contact blocks, bus bars, and connectors
   - Discolored connections usually indicate that overheating has occurred, probably caused by loose connections.
   - Clean connection points which are discolored, and tighten all hardware. Observe precautions in item 4 above. Do not use corrosive cleaners or acids. Steel-wool particles *must* be removed if steel wool is used. Fine sandpaper or neutral compounds containing pumice are the preferred abrasives.

7. Insulators
   - Remove dust and dirt from insulating parts; use clean rags or wiping cloths. Use neutral spirits if moisture is required.
   - If carbonized tracks or cracked or broken insulators are found, replace the defective part.
   - As a last resort carbonized tracks may be scraped clean and painted with insulating varnish.
   - Before any repaired insulating part is put into service, it should be given a dielectric test. Set the high-potential tester for a test voltage not less than twice the rated voltage of the equipment or 1000 V, whichever is greater.

8. Magnet assembly
   - Check magnet for dirty or corroded pole faces, pivot points, and other moving parts.
   - Check for loose iron, broken or damaged shading coils, and missing residual shims and hardware.
   - Inspect the operating coil for evidence of electrical or mechanical damage; if in questionable condition, replace it.
   - Operate armature by hand and check for mechanical interference and friction. It should operate freely and not bind in any position.

9. Electrical operation
   - Operate the contactor electrically (without load) and observe magnet operation. Be sure it opens and closes properly and that the armature is fully sealed in the closed position.
   - If it is an ac magnet, check for abnormal magnet noise.
   - Test a sample of the oil used with oil-immersed contactors. Keep the oil tank filled to the proper level with good clean oil (28-kV dielectric-test value or higher).
   - Check insulation between phases, to ground, and to the control circuit.

Troubleshooting Guide for Contactors. Table 2.3 provides troubleshooting information for contactors. Tables 2.4 and 2.5 show the horsepower ratings of single-phase full-voltage magnetic motor controllers for limited jogging and severe jogging duty, respectively. Tables 2.6 and 2.7 show the horsepower ratings of three-phase, full-voltage magnetic motor controllers for limited jogging and severe jogging duty, respectively.

TABLE 2.3 Contactor Troubleshooting Chart

Defect	Cause	Remedy
Overheating	Load current too high	Reduce load. Use large contactor
	Loose connections	Clean discolored or dirty connections and retighten. Replace poorly crimped lugs
	Overtravel and/or contact force too low	Adjust overtravel, replace contacts, and replace contact springs as required to correct defect
	Collection of copper oxide or foreign matter on copper contact faces	Clean with fine file. Use type 12 enclosure for dusty atmosphere
	Load is on in excess of 8 hr on copper contacts	Change operating procedure. Check factory for more suitable contacts
	Ambient temperature is too high	Provide better ventilation. Relocate starter. Use larger contactor
	Line and/or load cables are too small	Install terminal block and run larger conductors between contactor and terminal block
Welding of contacts	Overtravel and/or contact force is too low	Adjust overtravel, replace contacts, and replace contact springs as required to correct defect
	Magnet armature stalls or hesitates at contact touch point	Correct low voltage at coil terminals as coils draws inrush current
	Contactor drops open to contact-touch position because of voltage dip	Maintain voltage at coil terminals. Install low-voltage protective device, sometimes called "brownout protector"
	Excessive contact bounce on closing	Correct coil overvoltage condition
	Contacts rebound to contact-touch position when opening	Correct mechanical defect in stop assembly. Correct mechanical defect in latch if one is used
	Poor contact alignment	Adjust contacts to touch simultaneously within $\frac{1}{32}$ in or to manufacturer's tolerance
	Jogging duty is too severe	Reduce jogging cycle. Check factory for more weld-resistant contact material. User larger contactor
	Excessive inrush current	Motor has locked rotor code letter greater than G. Most contactors are designed for motors with code letters A through G. Therefore, use larger contactor. Check factory for more weld-resistant contact material
	Vibration in starter mounting	Move starter to location having less shock and vibration. Insulate starter from shock and vibration. Provide more rigid support for starter

(Continued)

TABLE 2.3 Contactor Troubleshooting Chart (*Continued*)

Defect	Cause	Remedy
Short contact life	Low contact force	Adjust overtravel, replace contacts, and replace contact springs as required to correct contact force
	Contact bounce on opening or closing	Correct improper voltage applied to coil. Correct any mechanical defects or misalignments
	Abrasive dust on contacts	Use type 12 enclosure. Do not use emery cloth to dress contacts
	Load current is too high	Reduce load. Use larger contactor
	Jogging cycle is too severe	Reduce jogging cycle. Check factory for more durable contact material. User larger contactor
Poor arc interruption	Arc box not in place	Install arc box
	Arc box damaged	Replace broken or eroded insulating parts, arc horns, and grid plates. Clean or replace insulating parts having a heavy coating of foreign conducting material
	Dirt or paint on arc horns or steel-grid plates	Remove contaminating materials which may have accumulated on arc horns and steel-grid plates
	Magnetic hardware substituted for nonmagnetic hardware in arc box and blowout assembles	Replace with correct hardware; brass or stainless steel as available
	Blowout coil reversed or shorted	Replace with new blowout coil or correct defect by reversing coil
	Oil level is low or oil is contaminated (in oil-immersed contactor)	Fill tank to proper level with fresh oil. Test at 28 kV

TABLE 2.4 Single-Phase Motor Controller Ratings for Single-Phase Full-Voltage Magnetic Controllers for Limited Plugging and Jogging Duty

Industry designated size of controller	Continuous current rating, amperes	Single-phase horsepower at 60 Hz	
		115 V	230 V
00	9	$\frac{1}{3}$	1
0	18	1	2
1	27	2	3
1P	36	3	5
2	45	3	$7\frac{1}{2}$
3	90	$7\frac{1}{2}$	15
4	135	10	25

TABLE 2.5 Single-Phase Motor Controller Ratings for Single-Phase Full-Voltage Magnetic Controllers for Jogging Duty

Industry designated size of controller	Continuous current rating, amperes	Single-phase horsepower at 60 Hz	
		115 V	230 V
0	18	½	1
1	27	1	2
1P	36	1½	3
2	45	2	5
3	90	5	10
4	135	7½	15

TABLE 2.6 Three-Phase Motor Controller Normal Service Ratings for Three-Phase Full-Voltage Magnetic Controllers for Limited Plugging and Jogging Duty

Industry designated size of controller	Continuous current rating, amperes	Three-phase horsepower at 60 Hz		
		200 V	230 V	460/575 V
00	9	1½	1½	2
0	18	3	3	5
1	27	7½	7½	10
2	45	10	15	25
3	90	25	30	50
4	135	40	50	100
5	270	75	100	200
6	540	150	200	400
7	810	—	300	600
8	1215	—	450	900
9	2250	—	800	1600

TABLE 2.7 Three-Phase Motor Controller Jogging Duty Ratings for Three-Phase Full-Voltage Magnetic Controllers for Severe Plug-Stop, Plug-Reverse, or Jogging Duty

Size of controller	Continuous current rating, amperes	Three-phase horsepower at 60 Hz		
		200 V	230 V	460/575 V
0	18	1½	1½	2
1	27	3	3	5
2	45	7½	10	15
3	90	15	20	30
4	135	25	30	60
5	270	60	75	150
6	540	125	150	300

DISCONNECT DEVICES

Manually operated power circuit components that are used to isolate a power branch circuit are called disconnect devices. Fusible safety switches, separately mounted molded-case circuit breakers (sometimes mounted in a panelboard), and similar devices mounted in enclosures along with motor starters and contactors serve this purpose. When the disconnect device and motor starter for a low-voltage application (600 V or less) are combined into a single enclosure, the result is a combination starter. Combination starters may have a fusible or nonfusible disconnect switch, a molded-case switch (a circuit breaker without an internal tripping means), a motor circuit protector (a magnetic-trip-only circuit breaker), or an inverse-time (thermal-magnetic trip) circuit breaker as the disconnect means.

Disconnect Switches. The contact mechanism of a disconnect switch consists of one or more blades per pole that mate with a corresponding set of jaws. The blades and jaws should be inspected for signs of abrasion or galling. Any beads that have formed should be removed by filing. Apply a thin coat of petroleum jelly to the blades, jaws, and hinge points. Avoid overlubrication since it will collect dirt and grit that will ultimately do more harm than good. Check terminations for loose connections and signs of overheating. Correct as described earlier in this chapter under Mechanical Checks. Disconnect switches are used in both low- and medium-voltage applications. Figure 2.12 shows a medium-voltage device for application in the range of 2000 to 7200 V.

Circuit Interrupters. The contacts of these devices (also called molded-case switches) are totally enclosed within their molded case and are not accessible for maintenance. Check terminations as described above for disconnect switches.

FIGURE 2.12 Medium-voltage disconnect switch.

TABLE 2.8 Values for Circuit Breaker Overcurrent Trip Test: Tripping Time, in Seconds, at 300 Percent of Rated Continuous Current of Breaker

Voltage, volts (1)	Range of rated continuous current, amperes (2)	Minimum time		Maximum time	
		Electronic sensing breakers (3)	Thermal magnetic breakers (4)	For normal protection (5)	For cable protection* (6)
240	15–45	3	—	50	100
240	50–100	5	—	70	200
600	15–45	5	5	80	100
600	50–100	5	5	150	200
240	110–225	10	5	200	300
600	110–225	10	—	200	300
600	250–450	25	—	250	300
600	500–600	25	10	250	350
600	700–1200	25	10	450	600
600	1400–2500	25	10	600	750
600	2501–4000	25	10	600	750

*These values are based on heat tests conducted by circuit breaker manufacturers on conductors in conduit.

Circuit Breakers and Motor Circuit Protectors. These devices all have molded cases with contacts not accessible for inspection or maintenance. Motor circuit protectors and adjustable magnetic-trip-only circuit breakers have no thermal or electronic sensing elements to cause them to trip at small multiples of their current rating. They are limited to use in combination starters that contain motor starters with overload relays. These overload relays protect the branch circuit against low-level overcurrents and faults. Inverse-time circuit breakers, however, contain an overcurrent-sensing element and a time-delay mechanism. The time delay is inversely proportional to the magnitude of the current. This function can be checked by the following procedure:

1. Remove the breaker from the circuit. Make sure that the disconnecting device ahead of the breaker is *off* and that no power is on the breaker to be tested.
2. Inspect the breaker visually for physical damage or evidence of overheating at terminals.
3. Perform several mechanical ON-OFF operations.
4. Apply 300 percent of the breaker's rated continuous current to each pole to determine that the circuit breaker will trip on an overload. See Table 2.8.
5. If a millivolt drop test is desired, consult the manufacturer for voltage values and detailed test procedures. If millivolt drops are above the manufacturer's limits, check for loose connections at the terminals and, if the breaker is of the interchangeable trip-unit type, at trip-unit connections.
6. Make an insulation resistance test. Resistance should be not less than 1 mΩ.
7. When reinstalling the breaker, make sure all connections are tightened in accordance with manufacturer's recommendations.

STARTING AND SPEED-REGULATING RHEOSTATS FOR MOTORS

Rheostats are used for starting and speed regulation of series-, shunt-, and compound-wound dc motors in nonreversing service. Specific examples are fans, blowers, pumps, machine tools, and similar dc motor applications, ranging from $1/4$ to 150 hp. A typical motor-operated rheostat is shown in Fig. 2.13.

FIGURE 2.13 Motor-operated rheostat with resistor bank.

To keep rheostats in good operating condition, periodic inspection, cleaning, and dressing of contacts with a file is usually all that is needed. The low maintenance factor on rheostats is due to such features as magnetic blowout devices, high contact pressure, contacts raised above the surface of the faceplate, rugged moving parts, and easy accessibility of serviceable components. However, some arcing and burning of the contact-making parts is unavoidable, and servicing of these parts will be necessary. *Contacts should always be smooth.* After each dressing with a file, contacting surfaces should be thoroughly cleaned, as well as the areas between contacts. The contacts may be lightly greased with petroleum jelly. *Do not overgrease.* Sometimes contact surfaces are damaged by burrs or sharp edges on the moving contacts. Look for score marks. On occasion, airborne abrasive particles are responsible for scoring marks on the contacts. If this is the case, refrain from using any lubricant, as the condition will be worsened.

Servicing Rheostats. Many rheostats are built with reversible contacts. The movable and stationary contacts can be turned over and used on the reverse side when necessary. This, in effect, gives the contacts a double life. Reversing of these contacts should be resorted to when abnormal burning and subsequent dressing with a file have made adjacent contact surfaces uneven. In operation, the moving contact should make the transition from one stationary contact to the next with a firm, even pressure if arcing is to be prevented. If this is not accomplished, erratic changes in speeds or voltages can result from poor regulation of resistance. On the larger types of rheostats, or those with movable contacts of the compensated type, a slight variation between contact surfaces will not impair operation. It is generally advisable when turning over one or more contact buttons to turn over all others at the same time.

Oxides will form on the unused copper surfaces of a rheostat. These oxides should not be allowed to accumulate. Remove this oxide film at regular maintenance periods. Thoroughly clean and protect any unused contacts of the rheostat with petroleum jelly to prevent reoxidation.

When load application requires insertion of previously unused portions of a rheostat, it is very important to measure the added resistance. A check should be made to ensure that additional travel of the moving contact is not restricted by the height of the previously unused contacts. Make resistance checks to ensure that these contacts are properly connected to obtain the desired added resistance. Always clean the contact surfaces of a rheostat before placing it in service. A regulation check should be conducted, measuring the amount of variable resistance, preferably at the motor. This will take into consideration the resistance value of motor leads and should reveal any poorly made terminal connections and any irregularity or error in resistance stepping.

Rheostat Contact Pressure Settings. The necessary pressure between moving and stationary contacts of a rheostat must be maintained by proper adjustment of spring tension to minimize pitting, heating, and oxidizing of contacts. Follow the manufacturer's recommendations for maintaining, setting, and adjusting contact force. The recommended moving-contact spring force is usually indicated in pounds or ounces when checked with a spring scale. The force required to separate the contacts should be determined and compared with manufacturer's data. The general practice is to establish a pressure that will provide good conductivity, without damaging the surfaces of either the moving or stationary contacts. At the same time, care must be exercised not to create high frictional forces that would prevent resetting of the arm on rheostats with low-voltage release or automatic resetting capabilities. The turns of the pressure-setting spring should *never* touch. This would indicate a weak or improperly applied spring which must be replaced.

Rheostat Low-Voltage Release Devices. Practically all starting rheostats of a low horsepower rating have a magnet coil as part of the low-voltage release mechanism. This coil is connected directly across the line. The magnet controls a release that is adjusted to hold the operating handle in the last running position as long as the line voltage remains normal.

This release may or may not be adjustable. If it is adjustable, the manufacturer will have made provisions for changing its holding characteristics over a prescribed range of undervoltage. Adjustments may be accomplished by following the manufacturer's instructions for changing the setting to a lower or higher value. If it is not adjustable, the device may be rendered inoperable by filing, bending, or shimming the mechanical parts. The operating arm should *never* be held in the running position by force. A rheostat that does not operate as intended should be thoroughly checked for mechanical defects and the possibility of misapplication. It is more economical to replace the operating mechanism than to risk harming the motor by excessive starting or running current. The control rheostat should *not* be used to stop the motor; a motor starter should be provided for this purpose. The maintenance requirements indicated by the manufacturer should be observed when servicing, and regular inspections should be scheduled to ensure proper maintenance and good operation.

Resistor Replacement. Occasionally, abnormal starting or operating conditions may burn out a section of resistance. Plate resistors must be replaced with units of equal value. If the resistance is a wire-wound bobbin type covered by a ceramic glaze, these units must also be replaced with components of equal or adjustable value. Edge-wound resistors, supported in a ceramic core, lend themselves to temporary repair. Weld or clamp the burned-out sections together until a new resistor can be obtained. As a general rule, heavy-duty resistors using cast grids or strip resistors *cannot* be repaired. Burned-out sections *must* be replaced. To replace burned-out resistors, depending on rheostat design, partial or complete disassembly of the rheostat may be required. In most types of industrial rheostats, the field resistors are removable from the resistor mounting without disconnecting the wiring at the faceplate or disassembling the operating mechanism. Large-motor applications will frequently use separate frames of resistance, remotely located. This resistance is then connected to and varied by a rheostat faceplate and mechanism located elsewhere.

Locating Rheostats and Resistor Banks. Rheostats should always be mounted so that adequate ventilation can be provided. Ventilating openings should be at the top. If resistors are stacked in racks, one on top of the other to conserve space, the accepted practice is to limit stacked height to four frames with ample separation distance between frames. Provide at least 1 ft of space between the floor and the first frame to permit unrestricted airflow through the units. Clearance between the ceiling and the top unit will be determined by the total height of the stacked arrangement; make sure there is adequate top ventilation. In most instances, manufacturers specify the necessary requirements for remote mounting of field resistors and the volume of cooling air to be provided if forced-air ventilation is necessary. Resistance banks and rheostats, like other electrical equipment, must be kept clean and dry. Do *not* store combustible materials on or near the resistor racks. Cooling air should be free of oil mist, lint, and other contaminating particles, which if deposited on the hot grids or resistors, could cause fires. Use dry compressed air occasionally to blow out accumulated dust, dirt, or foreign materials. Since the capacity and the life of resistors depend largely upon heat dissipation, never locate resistors in poorly ventilated areas. Undesirable variations in resistance can result from heating and may necessitate forced ventilation if overheating becomes a problem. Some resistors may operate at temperatures sufficiently high to cause them to glow. However, most applications are kept well below maximum allowable operating temperatures.

Checking Rheostats for Loose Connections. When placing a rheostat or field resistance in service for the first time, it is important to check the resistance value before attempting operation. After placing it in service and during the initial observation period, terminal connections and all interconnections should be checked at least twice to ensure that they have not loosened as a result of heating. Loose connections can cause many annoying delays, are difficult to locate, and create operating conditions that could be hazardous. They are very difficult to find unless arcing or burning occurs. To guard against loose connections developing from cyclic heating and cooling, check frequently for tightness.

It is good maintenance practice to check resistance values occasionally against manufacturer's data. Incremental values and total values should be measured to guard against open and shorted individual resistors, which affect total resistance value. One of the best devices for checking continuity and regulation of a rheostat is an ohmmeter. A Wheatstone or Kelvin bridge or similar precision measuring instrument should be used for low values of resistance.

Most resistors are manufactured with a tolerance factor of 610 percent. For this reason, it is important to know the resistance required by the circuit design when checking or replacing resistors. Unless there is some method of refining the accuracy by tapping or adjusting the "fixed" resistance values in a rheostat, considerable change in motor performance will occur if a portion of a rheostat or resistance bank has open-circuited, burned out, or been replaced. Always consult the motor and application data when restacking or changing out resistors used in rheostats or banks of field resistance.

Arcing at the contacts or burning of contacts on the faceplate of a rheostat is an indication of trouble. There could be shorts or open resistance at these points in the resistance circuit. The cause could also be lack of sufficient resistance at that step.

Continuously Wound Rheostats and Potentiometers. Continuously wound rheostats and potentiometers are fairly common and have many uses in control schemes. Their function is one of regulation by introducing resistance into a circuit smoothly, to maintain steady-state operation, or to provide a means of speed regulation. Figure 2.14 shows three types of such rheostats: a single unit and a double- and triple-stacked array. Each plate has a continuously wound element on a circular ceramic core. The continuous winding is connected to a terminal on each end of the rheostat or potentiometer, and a moving contact is centered in the core. The device can furnish from zero to full resistance without passing over contact buttons as in the case of button rheostats; it relies strictly on contact with the wound-wire resistance for the variation of resistance. These devices usually have much lower ohmic values than button rheostats. They are used principally for low-resistance vernier control and do not handle high currents. They are usually low-wattage/low-voltage components.

FIGURE 2.14 Hand-operated rheostats.

Connections to one end and to the moving contact provide a variable resistance, and the device is called a rheostat. Connecting both ends to a source of voltage and using the movable contact to select a desired portion of the voltage makes the device a potentiometer. Construction is identical. The application is what makes the device either a rheostat or a potentiometer.

Maintenance of continuously wound rheostats and potentiometers closely parallels that outlined for button rheostats. They can become defective from excessive use, lack of maintenance, or by passing more current through them than their resistance windings can carry. Heating is not necessarily a sign of distress, but excessive temperatures should be investigated. Heat losses are natural, and ample ventilation should be provided. If it is suspected that the unit is overloaded or has a burned-out section, check it for total resistance and also for good contact throughout the full range of resistance. Excessive wear on the shaft bearings and bushings and loss of moving-contact pressure can cause arcing. The moving contact must make positive connection and can be affected by corrosive atmospheres or dirt, gritty particles, and oxides which accumulate on unused portions of the wire-wound resistance. Most continuously wound rheostats have a tightly fitting rear cover designed to keep out dirt and contaminates, but this enclosure does restrict air circulation and natural cooling.

Secondary Circuit of Wound-Rotor Motors. Special attention should be given to the maintenance of the secondary control of wound-rotor motors, particularly those used in speed-regulated service. In a large percentage of applications, it is possible for speed-regulating problems to develop without causing either immediate shutdown or failure to start. Since the motor will continue to start and run, even though an open circuit or serious unbalance of resistance may exist in the secondary circuit or at certain points on the controller, it is not always recognized that this condition may have serious consequences. Roasting out of the motor windings, burning and severe arcing at the bushes, damage to collector rings, and overheating or burnout of resistors can result. In addition, undue stress on the equipment may be produced when the smooth steps of acceleration are lost by poor contact or no contact at certain points on the controller. Trouble may develop without the machine operator's noting or reporting any unusual operating conditions until a serious breakdown occurs. Then it may be recalled that "they did have to notch the controller up a step" or that "the machine jumped a bit at that point." A definite and regular inspection schedule is essential, since this class of apparatus typically has such a rugged construction that it normally requires minimum attention and therefore might be neglected.

To ensure against service interruptions, keep all types of secondary control in good operating condition. This requires regular inspection and cleaning and maintenance of contacts. Some arcing and burning of contact-making parts are unavoidable. Keep them smooth to prevent unnecessary burning and to ensure a positive low-resistance contact at all times. Occasional dressing with a file may be necessary. Conducting lubricants should not be used unless endorsed or recommended by the manufacturer.

Rheostat maintenance has been adequately covered in the preceding text. The continuity of resistance related to the operation of controllers should be checked occasionally with an ohmmeter. The continuity check may be made across the contacts of the controller, but the general practice is to raise the collector ring brushes, insert a test lamp, ohmmeter, or other suitable test instrument across the brush holders or outgoing leads, and move the secondary controller through its full sequence, step by step. If this check is repeated across each phase, it not only will verify the continuity of the resistors and tapped connections but also will indicate any open contacts in the controller.

In general, the values of secondary resistance are relatively low and a continuity check will determine only continuity. Many plants lack the resistance bridges needed to measure or set low-resistance values accurately. In these circumstances, it is important to make frequent visual inspections to locate and correct loose connections and low-pressure contact points. Noticeably loose connections on grid-type resistors and resistor connections showing evidence of heating should be removed and the connectors cleaned or replaced. All pressure-type connections should be checked for tightness. If they are found to be loose or poorly conducting, the pressure should be increased by tightening the pressure nuts at the ends of the grid assembly. On ribbon or edge-wound resistors, the clamp connections must be kept tight. Remember that on all new installations at least two checks should be made after initial start-up and energizing of equipment to ensure that resistance connections and all controller contacts are tight.

If they are properly applied and their connections are maintained, resistor banks require no further service, but they should be checked occasionally. Investigate any excessive heating of the resistors to determine if heating is caused by open circuits or unbalanced load connections in the resistance circuit.

DRUM CONTROLLERS

Drum controllers are manually operated power circuit devices that introduce and remove resistance from electric motor circuits. They are used in the armature circuit of dc motors for starting, reversing, and speed regulation. They are also used in the secondary circuit of wound-rotor ac motors. Typical drum controllers are shown in Fig. 2.15.

FIGURE 2.15 Typical drum controllers.

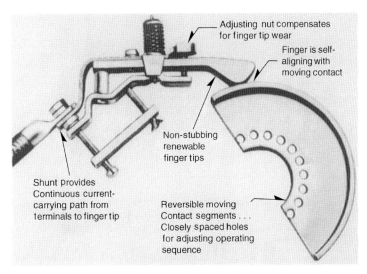

FIGURE 2.16 Drum-controller contact system.

Low-current drum controllers have contact-supporting disks typically assembled on an insulated steel shaft between insulating collars, with reversible contacts bolted directly to the supporting disks. See Fig. 2.16. On larger devices, 150 to 300 A sizes, the heavy curved copper contacts called segment plates are typically mounted on molded insulating supports bolted to a steel shaft. The upper and lower segments usually have the same shape to permit interchange and reversal. As segments wear, they may be reversed and the old trailing edge becomes the leading edge when flipped 180°.

Contact fingers, when open, should be adjusted to drop no more than $1/8$ in below the surface of the drum contacts. The controller bearings, star wheel, and pawl should be regularly cleaned and lubricated. Other points of wear should be routinely checked. It is important to keep all wiring connections tight and to check contact pressures and springs periodically.

EQUIPMENT GROUNDING

Required connections to earth are useful and necessary for safety reasons. They are easily maintained because they need only good electrical connections. Working on an ungrounded system can be a serious personnel hazard that can result in death or injury when the individual or a portion of his body becomes an unintentional path to ground.

Constant vigilance is required to prevent and eliminate undesirable grounds. These grounds can cause operating troubles and erratic and dangerous circuits. Improper wiring practices and grounds may cause motors to start unexpectedly or fail to stop when they should. Overload and other protective schemes may be made ineffective by an unintentional ground.

Unintentional grounding often occurs in pushbutton stations or similar confined spaces where stray strands of wire may make contact at the wrong places. Improper grounds may also occur when insulation on wires is chafed because of vibration, such as at conduit entrances.

Moisture is also a possible cause of ground faults. Conduits should always be installed so that moisture will drain away from equipment and terminations. If enough water collects in a conduit as a result of flooding or condensation, it may become necessary to remove the wires, clean the conduit, provide drain holes, and install new wiring.

Warning: Always make sure circuits are deenergized before working on any conduit or part of a device suspected of containing water, as dangerous ground-fault currents may be present. Failure to heed this warning can cause severe personal injury or death, and damage to property.

6.64 MAINTENANCE OF ELECTRICAL EQUIPMENT

FIGURE 2.17 Typical grounding resistor.

Many medium-voltage systems use a grounding resistor similar to the one shown in Fig. 2.17 to control the potential to ground and the ground-fault current of an electrical system. During a ground fault the grounding resistor bank absorbs a large amount of electrical energy and dissipates this energy as heat. The location of the resistor should provide sufficient ventilation to dissipate the heat generated without adversely affecting any other equipment in the immediate area.

CONTROL CIRCUIT COMPONENTS

Since the contacts of control circuit devices handle much smaller currents than do power circuit devices, contacts of pushbuttons, relays, timers, pressure switches, limit switches, auxiliary contact assemblies for contactors, and the like, seldom need to be replaced because of wear. Typically, contacts for control circuit components (see Fig. 2.18) need to be replaced only because they have been damaged as a result of a short circuit beyond their ability to withstand. Contacts for these devices are either not replaceable (the component must then be discarded) or are replaced as complete assemblies in the form of cartridges, contact blocks, or snap switches, complete with their own housings. Examples are shown in Fig. 2.19.

Manual or Machine-Operated Control Circuit Devices. These devices include pushbuttons, selector switches, limit switches, proximity switches, pressure and temperature switches, and so on. Where any of these components serve as a safety device (e.g., an "emergency stop" pushbutton, an end-of-travel limit switch, an overload relay, or an overpressure relief control), the contacts involved are usually normally closed and hence subject to welding closed in the event of a short circuit. Therefore, every maintenance program should include checking the operation of these normally closed and seldom-used contacts. When welded contacts are found, they should be replaced. Such contacts once welded are prone to welding again.

FIGURE 2.18 Control circuit components.

Contactor-Type Control Relays.   Many magnetic control relays, both ac and dc, are constructed and operate like magnetic contactors. Hence their maintenance is similar, except that contact replacement as a unit is more usual than contact dressing. These control relays are used to perform logic functions as in an automated system where a particular sequence of operation is established or as interposing relays that control a high current or high voltage while permitting an operator to be exposed to only low voltage and low current. The relays shown in Fig. 2.18 are contactor-type relays.

Thermal Overload Relays.   This type of overload relay (the center device in Fig. 2.1) uses the current drawn by the utilization equipment, most often an electric motor, to produce heat within itself and use the magnitude and duration of this heat to cause some mechanical action that results in control circuit contacts opening. These devices protect motor circuit conductors as well as the motors against overcurrents caused by overloads. They also can provide protection in the case of low-level faults where the short-circuit or ground-fault current is too low to cause the branch circuit's short-circuit protective device to operate. Maintenance is limited to cleaning, tightening terminations, and checking for proper selection of the heater element.

FIGURE 2.19 Control circuit contact assemblies.

Overload relays may be of several types:

1. Single-pole bimetallic with removal heaters, either ambient compensating or nonambient compensating. Operation may be either automatic reset, manual reset, or convertible from one to the other. Current rating is sometimes adjustable 615 percent.
2. Three-pole block-type bimetallic with removable heaters. The variations for number 1 above may also be available on these relays.
3. Both single-pole and block-type may be manufactured with directly heated bimetal operating elements, in which case a current range is indicated on the adjustment mechanism. The variations indicated in numbers 1 and 2 above are often available on these devices.
4. Both single-pole and three-pole block-type melting alloy that use a eutectic solder to release a spring-loaded ratchet when the alloy in the solder pot melts. Eutectic alloy overload relays are manual reset only and are not available with ambient compensation, although they can be designed with a high-melting-temperature alloy to make them relatively ambient insensitive.

Magnetic Overload Relays. These overload relays consist of a coil made of wire sufficiently large in cross section to carry motor current without overheating. Such coils are often called series coils because they are wired in series with the motor. These coils create a magnetic pull on a movable plunger which acts as the armature in the magnetic circuit of these relays. Magnetic overload relays may be of the instantaneous type, used to detect a sudden locking or jamming of the equipment (hence the term "jam relay") or they may have an inverse-time function because the movement of the armature is restrained by a dashpot. Maintenance is limited to contact replacement, cleaning, tightening terminations, and adjustment or conditioning of the dashpot.

Dashpots. Keep all dashpots clean. Be sure oil dashpots have the correct type of oil in them. Air or fluid dashpots are used to retard motion. They are machined to close clearances and must be kept clean and free to move. The proper amount and type of fluid is important in dashpots. Since the viscosity of oil changes with temperature, substitute oils should not be used without the manufacturer's approval. If there are approved substitutes, the manufacturer will usually specify what type and brand of dashpot oil can be substituted without altering the timing characteristics of the dashpot or deteriorating gaskets or seals.

Thermal Overload Relays. The primary output contacts of a thermal overload relay are normally closed contacts that open when the overload relay trips. Some thermal overload relays also include an auxiliary set of normally open contacts that close when the overload relay trips. The latter are often called "alarm contacts" since they can be wired to an auxiliary circuit that rings a bell or lights an alarm light to notify an operator that the overload relay has tripped.

Routine maintenance should always include a check to show that the normally closed contacts have not welded. Many thermal overload relays include a "weld check" switch that when depressed will open the normally closed contacts. Where a weld check switch is not provided, devise some means to isolate and operate these output contacts to demonstrate that a fault somewhere in the control circuit has not welded these contacts. Automatically check these contacts and all other normally closed contacts (e.g., EMERGENCY STOP) in the control circuit after any indication that a ground fault or short circuit has occurred in the control circuit.

Solid-State Overload Relays. All solid-state overload relays incorporate some form of current transformer, current sensor, or voltage transducer to measure the current in the power circuit between a motor-rated contactor and a motor. Some solid-state overload relays operate with power taken from the circuits being monitored. Others require a separate source of control power, usually 110 to 120 V ac. All solid-state overload relays are insensitive to the ambient temperature surrounding them, provided the surrounding air temperature does not exceed the temperature limits stated by the manufacturer.

One type of solid-state overload relay functions in a manner similar to a magnetic overload relay and responds only to the magnitude of the motor current. This type will respond instantaneously (as a jam relay) to a preselected overcurrent corresponding to the motor locked-rotor current or can be set to operate only after a selected time delay has expired with the motor drawing locked-rotor current. The latter version is equivalent to a magnetic overload relay with a dashpot.

Another type of solid-state overload relay measures the heating effect of the current drawn by the motor and performs as a thermal overload relay with some additional benefits. This type of solid-state overload relay can respond to overload conditions other than locked-rotor and often offers auxiliary functions such as phase-loss sensing, running overcurrent protection after a long acceleration period without nuisance tripping, and ground-fault sensing. See Fig. 2.20.

With the availability of customized microprocessors and solid-state miniaturization, solid-state overload relays have been built into the frames of motor-rated contactors so that such a starter with built-in overload and related protection is no larger than a contactor alone for any given horsepower rating. See Fig. 2.21.

All separate solid-state overload relays have some type of output, in the form of either a solid-state switch or a small relay with conventional contacts that are wired into the coil circuit of the contactor controlling the motor. These contacts will be closed when the overload relay is energized and appear to be "normally closed" when the overload relay has not tripped. When overcurrent conditions cause the overload relay to trip, the internal relay contacts open to cause the contactor to drop out. Routine maintenance should include checking the output contacts of a solid-state overload relay to be sure that the internal relay contacts are not welded. With control power removed from the overload relay, verify that the output contacts are open. Where the output contacts are a solid-state switch, follow the manufacturer's instructions for verifying the conditions of the switch. Where the overload

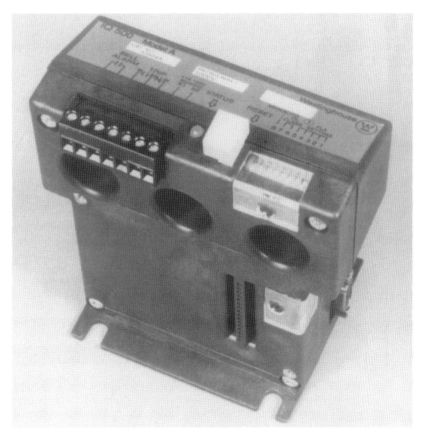

FIGURE 2.20 Stand-alone solid-state overload relay.

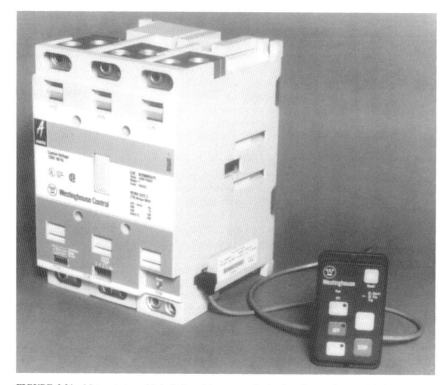

FIGURE 2.21 Motor starter with built-in solid-state overload, phase-loss and ground-fault protection, along with membrane switch-type control station.

protection function is built into the starter as illustrated in Fig. 2.21, there is no overload relay output contact to check.

Note the six membrane switches and the light-emitting diodes (LEDs) marked "Run," "Off," "OL Alarm," and "Trip" mounted in the pushbutton station shown in Fig. 2.21. These sealed low-voltage components are very compatible with microprocessor-based industrial controls when reliable connections can be made using multiconductor cables, sockets, and plugs. However, the sealed components in such control stations generally are not repairable and the entire pushbutton station must be replaced when a malfunction occurs.

Adjustable-Type Relays. A wide variety of adjustable relays have been designed for specific functions where the adjustment of springs, stops, air gaps, and so on, permits the user to obtain the particular operating characteristic desired. Examples of adjustable relays are voltage-sensing (voltage-regulating) relays where the pickup and dropout voltages are set by the user, typically by adjusting spring pressures, plugging and anti-plugging relays that are set to limit plugging to certain voltage or time constraints, and timing relays where the time delay is controlled by adjusting the dashpot openings, a pneumatic valve position, or a potentiometer setting.

Time-Tactors. One of the most complex of the adjustable relay family is a product known as a Time-Tactor, a device that combines the function of a time-limit accelerating relay (time delay, plugging, or antiplugging) and a normally closed (spring-closed) contactor. They are used in dc motor control and in the secondary circuit of ac wound-rotor motors. By explaining the construction and operation of these devices we hope to cover the principles that apply in all adjustable relays.

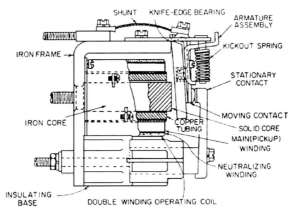

FIGURE 2.22 Adjustable relay with solid core.

Some Time-Tactors are operated with a double-winding coil that is wound on a section of copper tubing. See Fig. 2.22. The two coil windings are known as the main or pickup winding and the neutralizing winding. These Time-Tactors operate on an inductive time-delay principle and the copper tubing serves as a short-circuited, low-resistance winding.

The magnetic circuit which is composed of the armature and frame does not have any substantial air gaps in it because Time-Tactors require that the reluctance of the magnetic circuit be as low as possible.

Figure 2.23 shows a typical hysteresis curve for this family of relays. Flux in the magnetic circuit is plotted versus magnetizing force expressed in percent of rated ampere-turns for the iron circuit. The curve is drawn to illustrate the flux through the various operational sequences from energization to deenergization on the relay.

How the Relay Works.   Referring to the curve in Fig. 2.23, when a dc voltage is applied to the operating coil, the product of the current and the number of turns in the coil winding is the magnetizing force. With the armature in the open position (point 0), the rise of the flux is directly proportional to the increase in magnetizing force. When the ampere-turn value has reached approximately 35 percent of the rated ampere-turns, the armature is picked up (point 1) and rapidly sealed against the core (point 2). Since the reluctance has simultaneously greatly decreased, the flux will increase to approximately

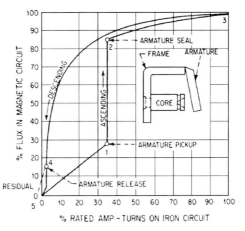

FIGURE 2.23 Flux generated by magnetizing force in d-c coil.

85 percent of the available total with no increase in magnetizing force. As the coil current increases, the iron reaches saturation (point 3) and the increase in flux no longer continues. When the coil is deenergized, the flux decay does not retrace curve 3, 2, 1, 0, but generates a descending curve along points 3, 4, 0 as shown. Unless some decaying means is deliberately provided, the flux decreases immediately from a maximum value (point 3) to the armature release point (point 4). The increase in reluctance due to the opening of the armature drives the flux to zero (point 0). If the armature were held in, then the descending portion of the curve would go from point 4 to 5 as shown by the dotted line. The intersection of line 4-5 and the vertical axis is the measure of the residual magnetism. It would be necessary to reverse the polarity of the current and increase its value so as to produce an equal magnetic field in the reverse direction in order to demagnetize the circuit to zero. The amount of residual magnetism can also be decreased by providing an additional air gap in the magnetic circuit. However, it should be noted that the air gap does not affect the coercive force requirements (0 to 5) to buck down the residual flux in the iron to a zero value.

As stated above, when the operating coil is deenergized, the flux decays rapidly to some predetermined armature-release value resulting from the air gap and kickout spring combination selected. The time period that the armature stays in after deenergization is almost negligible. Therefore, in order to provide a time delay from the moment of deenergization to armature dropout, the rate of flux decay must be retarded. This can be achieved by surrounding the magnetic circuit by a short-circuited turn of very low resistance. Any change in the magnetic field will induce a voltage in the short-circuited turn, causing a large current to flow in it. This induced current will try to maintain the flux at its original value. However, energy dissipation in the form of heat causes the flux to decay slowly to the armature release point, thereby initiating a time delay.

Magnetic Hysteresis. The lagging change of flux in a magnetic substance behind the change of magnetizing force is known as magnetic hysteresis. It appears as the separation of difference between the ascending and descending portions of the curve in Fig. 2.23.

Saturation. The curve in Fig. 2.23 also indicates that the iron in the circuit starts to become saturated at approximately 40 percent of the rated magnetizing ampere-turns. This is desirable from a relay standpoint in order to obtain repeat accuracy in operation. Iron is said to be saturated if there is no increase in flux with an increase in magnetizing force. At this condition the iron is carrying all the flux it can handle and has reached its maximum magnetic capability.

Magnetic Circuit Modification. The magnetic circuit of the relay may be modified by controlling the thickness of the nonmagnetic spacer in the armature end of the core, which changes the dropout characteristics. Pickup characteristics are controlled by the air gap between the core and the armature in the open position plus the force exerted by the kickout spring in the open position. Dropout characteristics are controlled by the air gap between the core and the armature in the closed position, plus the contact force and the kickout spring force. For the effect of the nonmagnetic shim thickness plus the adjustment of the kickout spring, see Fig. 2.24. Figure 2.25 shows a relay with a single-winding coil and a nonmagnetic shim in the magnetic circuit. Some time-delay relays and synchronous relays used for synchronous motor starting use this construction. A nonmagnetic shim of sufficient thickness to provide the required magnetic circuit reluctance is permanently assembled onto the core near the pole face. The built-in shim is analogous to an orifice incorporated into the magnetic circuit whereby the amount of residual flux can be controlled. The resulting iron cross section at this portion of the core is small, and the residual flux is therefore at a minimum. This type of core construction will have approximately one-eighth of the residual magnetism available in a solid-core construction of the same physical size after saturation.

Retarding the decay of flux, for time-delay purposes, is accomplished by nesting a short-circuited copper turn between the coil and the core to provide time-delay adjustment. The short-circuited turn has been divided into approximately five equal volumes, a case and four rings. Additional timing adjustments are possible by substituting a brass case for the copper case, and open- or short-circuiting the coil during deenergization. The time-delay relay is equipped with an adjustable kickout spring and a nonmagnetic shim, which can be applied to the armature for further time-delay adjustments.

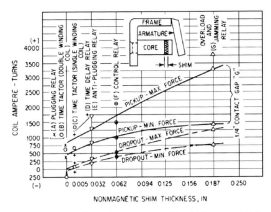

FIGURE 2.24 Effect of nonmagnetic shim thickness and kickout spring adjustment.

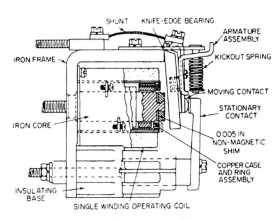

FIGURE 2.25 Relay with nonmagnetic shim.

Time-Tactors are equipped with a nonadjustable kickout spring and are dependent for time-delay adjustment on the case-and-ring combination and the method of interrupting the coil circuit.

The synchronous relay, specifically designed for synchronous motor starting, is frequency-sensitive. It is provided with a double-winding coil, comprised of a pickup and a holding section. The case and rings are omitted, and an adjustable kickout spring is provided for slip-frequency adjustment.

Some Time-Tactors are built with substantially no air gaps or nonmagnetic spacers in the magnetic circuit (see Fig. 2.22) so that after the armature has been closed, the time-delay period can be initiated by opening the main coil circuit to deenergize that winding. When the magnetic field of the main coil collapses, a current will be induced into the relatively large cross-sectional copper tube upon which the coil is wound. The induced current flows in such a direction that it sustains the magnetic field and holds the armature in the closed position for a time period after the main coil winding has been deenergized. The flux in the magnetic circuit must decay and follow the descending portion of the hysteresis loop similar to that of Fig. 2.23. A time delay occurs as the magnetizing force produced by the short-circuited copper tube decays slowly and prevents the magnetizing force imposed on the iron circuit from immediately dropping to zero. The armature will not be released even when the magnetizing force produced by the copper tube drops to zero. The magnetic holding force produced by the residual flux exceeds the armature kickout spring force. In order to release the armature, a magnetizing force in opposition to that of the main coil winding and copper tube must be imposed upon the magnetic circuit to neutralize the residual magnetism. The neutralizing winding is used for this purpose. As the demagnetizing force is increased, the flux will decrease to a value where the magnetic force is less than the armature kickout spring force. Then the armature will be released to end the time-delay period and close the main contacts.

The time-delay period or duration of time from the instant that the main coil circuit was deenergized until the armature is released can be controlled or adjusted very effectively by varying the neutralizing coil current.

DC Relay Maintenance. The relays and Time-Tactors employing this type of frame construction require periodic inspection of the air gap. Rust, dirt, burrs, or foreign particles introduced into the air gap increase the reluctance and cause a decrease in the time delay. Figure 2.26 has been included to illustrate the effects the introduction of particles of various sizes into the air gap have on the time delay.

If the relay stays in for a prolonged period, an increase of the residual magnetism to an undesirable level has occurred. This could be attributed to a reduced air gap or magnetic aging. Where an adjustable kickout spring is provided but increasing the kickout force does not correct this condition, the addition of a nonmagnetic shim between the frame and core of the armature and the core will provide the necessary reluctance to make the relay operative.

Table 2.9 provides troubleshooting data for dc Time-Tactors, time-delay relays, and synchronous relays.

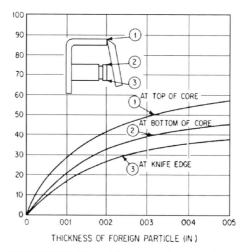

FIGURE 2.26 Effect of foreign particles in the air gap on the time delay.

TABLE 2.9 DC Relay Troubleshooting Chart

If the relay fails to pick up on energization, check for:
1. Damage to coil.
2. Improper contact-gap adjustment.
3. Excessive friction in the moving assembly.
4. Excessive kickout force.
5. Improper coil voltage.
6. Improper wiring connection.
7. Failure to control circuitry to provide ample pickup signal.

If the relay fails to repeat a time-delay setting within permissible variations after deenergization, check for:
1. Excessive friction of the moving assembly.
2. Erosion of the air gap provided by plating.
3. Increase of air gap resulting from nicks, burns, dirt, rust, or other foreign matter.
4. Incorrect circuitry to provide ample response time.
5. Variable kickout force caused by a faulty spring seat.
6. Contact welding.

MAINTENANCE OF MOTOR CONTROLLERS AFTER A FAULT*

In a motor branch circuit which has been properly installed, coordinated, and in service prior to the fault, opening of the branch-circuit short-circuit protective device (fuse, circuit breaker, motor short-circuit protector, and so on) indicates a fault condition in excess of operating overload. This fault condition must be corrected and the necessary repairs or replacements made before reenergizing the branch circuit.

*This section is based on the essence of Part ICS 2-302 as published by the National Electrical Manufacturers Association, November 1989.

It is recommended that the following general procedures be observed by qualified personnel in the inspection and repair of the motor controller involved in the fault. Manufacturers' service instructions should also be consulted for additional details.

Caution: All inspections and tests are to be made on controllers and equipment which are deenergized, disconnected, and isolated so that accidental contact cannot be made with live parts and so that all plant safety procedures will be observed.

Enclosure. Substantial damage to the enclosure such as deformation, displacement of parts, or burning requires replacement of the entire controller.

Circuit Breaker. Examine the enclosure interior and the circuit breaker for evidence of possible damage. If evidence of damage is not apparent, the breaker may be reset and turned on. If it is suspected that the circuit breaker has opened several short-circuit faults or if signs of possible deterioration appear within the enclosure, the test described in the preceding text (preceding Table 2.8) should be performed before restoring the breaker to service.

Disconnect Switch. Where the disconnecting means is a disconnect switch its external handle must be capable of opening the switch. If the handle fails to open the switch or if visual inspection after opening it indicates deterioration beyond normal wear and tear, such as overheating, contact blade or jaw pitting, insulation breakage or charring, the switch must be replaced.

Fuse Holders. Deterioration of fuse holders or their insulating mounts requires their replacement.

Terminals and Internal Conductors. Indications of arcing damage or overheating such as discoloration and melting of insulation require the replacement of damaged parts.

Contactor. Contacts showing displacement of metal, or loss of adequate wear allowance require replacement of the contacts and the contact springs. If deterioration extends beyond the contacts, such as binding in the guides or evidence of insulation damage, the damaged parts or the entire contactor must be replaced.

Overload Relays. If burnout of the current element of an overload relay has occurred, the complete overload relay should be replaced. Any indication that an arc has struck or any indication of burning of the insulation of the overload relay requires replacement of the overload relay.

If there is no visual indication of damage that would require replacement of the overload relay, the relay must be electrically or mechanically tripped to verify the proper functioning of the overload relay contact(s).

Return to Service. Before returning the controller to service, checks must be made for the tightness of electrical connections and for the absence of short circuits, grounds, and leakage.

All equipment enclosures must be closed and secured before the branch circuit is energized.

MAINTENANCE OF MOTOR CONTROL CENTERS*

There are many varieties of motor control centers and molded-case circuit breaker power panels. (See Fig. 2.27.) These maintenance recommendations are general in nature and can be adapted to a wide variety of product types.

---

*This section is reproduced with permission of the National Fire Protection Association from NFPA 70B-1990, *Recommended Practice for Electrical Equipment Maintenance,* copyright 1990 by NFPA. This reprinted material is not the complete and official position of the NFPA on the referenced subject, which is represented only by the standard in its entirety.

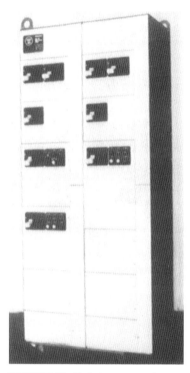

FIGURE 2.27 Typical motor control center.

Control Center Enclosures. Enclosures do not normally require any maintenance in a clean, dry, and noncorrosive atmosphere. Enclosures in a marginal atmosphere should be inspected periodically for excessive dust and dirt accumulation as well as corrosive conditions. The more contaminated the atmosphere, the more frequently the inspections should be conducted. Any accumulation should be removed with a vacuum cleaner or manually cleaned during maintenance shutdown periods for the equipment. Do not use compressed air to clean motor control center compartments. The confined areas and proximity of bus bars and other control gear generally restrict the dispersal of airborne debris so that it is not really removed but merely rearranged. Badly corroded enclosures should be properly cleaned and refinished as required for extended service.

Bus Bars and Terminal Connections. Any loose bus bar or terminal connection will cause overheating that will lead to equipment malfunction or failure. Overheating in a bus or terminal connection will cause a discoloration in the bus bar which can easily be spotted where connections are visible, oftentimes too late to avoid replacement. An overheating bus bar condition will feed on itself and eventually lead to deterioration of the bus system as well as to the equipment connected to the bus such as protective devices, bus stabs, insulated leads, and the like. Aluminum lug connectors are usually plated and should not be cleaned with abrasives.

Bus bar and terminal connections should be inspected periodically to ensure that all joints are properly tightened. Proper torque tightness is a factor of bolt size, bolt type, and type of bus bar and terminal material. Proper bolt tightness torque values for all types of joints involved should be available in manufacturer's maintenance and instruction literature. *Do not assume that bus bar and terminal hardware once tightened to proper torque values will remain tight indefinitely.*

Special attention should be given to bus bars and terminal connections in equipment rooms where excessive vibration or heating and cooling cycles may cause more than normal loosening of bolted bus and terminal connections.

Bus Bar Support Insulators. Bus bar support insulators and barriers should be inspected to ensure that they are free of contamination. Insulators should be periodically checked for cracks and signs of arc tracking. Defective units should be replaced. Loose mounting hardware should be tightened.

Disconnect Devices. Disconnects should be examined on both the line and load side for proper maintenance evaluation. Prior to initiating such an evaluation, the source side disconnect device should be opened and padlocked and tagged to avoid accidental energization by other personnel during maintenance operations. Switches used in drawout units normally supplied in motor control centers can be opened and safely withdrawn and examined on a workbench, thus avoiding this potential hazard.

Never assume that a disconnect is in the open position because the handle mechanism is in the open position. Always double check for safety.

Disconnect switches generally have visible blade contacts and open-type mechanisms which can be susceptible to contamination when not contained in a proper enclosure. Therefore, routine maintenance should include a procedure for inspecting and removing excessive dust accumulations. Nonautomatic molded-case circuit interrupters are often used in motor control circuits in lieu of an open-type disconnect. For this type of application, the internal mechanism will be better protected against contamination. The exterior should be examined and cleaned.

Excessive heat in a disconnect switch can lead to deterioration of the insulation and eventual failure of the device. Loose connections are the major source of excessive heat. Terminal and bus bar connections as well as cable connections should be examined and tightened as required using the manufacturer's torque recommendations. Any device having evidence of overheated conductors and carbonized insulation which could also be caused by arcing should be replaced. Contacts should be examined for evidence of welding or excessive pitting. Damaged disconnects with any evidence of these failure signs should be replaced.

Mechanisms should be manually operated to ensure proper working condition. Factory lubricated mechanisms will sometimes dry out after a period of time in dry, heated atmospheres such as motor control center enclosures. Manufacturer's maintenance literature should be followed for proper lubrication instructions.

Fuses. Fuses are normally used in conjunction with disconnect switches. In no case should a dummy fuse, copper slug, or length of wire ever be used as a proper fuse substitute even on a temporary basis.

Contactors. Since contactors are the working portion of a motor controller, normal wear can be expected. Periodic inspection should be made to ensure that all moving parts are functioning properly. Badly worn or pitted contacts should be replaced in sets to avoid possible misalignment. Dressing contacts should not be performed simply as a cleaning operation; rather, it should be done only to the degree necessary to restore proper contour. Silver alloy or other noble metal contacts should be dressed with a crocus cloth or other suitable nonconductive abrasive. Materials such as emery cloth and steel wool should not be used for this purpose. Routine inspection should always include checks for tightness of terminal and cable connections as well as for signs of overheating. Replacements should be made as conditions dictate. Contactors installed in corrosive or lint-filled atmospheres require more frequent inspection even when enclosed in gasketed enclosures. Many manufacturers of starters, contactors, and relays using silver alloy contacts specifically warn users against filing silver alloy contacts. Manufacturers' recommendations should be followed closely for maintenance and replacement of parts.

Thermal Overload Relays. Motor overload relays perform the vital supervisory function of monitoring the overload current conditions of the associated motor. Heater elements are usually replaceable; however, if a trip or burnout of the element occurs, the cause of this trip or burnout should be identified and corrected. Replacement of the heater element with one of a higher rating should not be done without full consideration of the ambient temperature in which the motor operates. Such overload relays employ a thermal element designed to interpret the overheating condition in the motor windings by converting the current in the motor leads to heat in the overload relay element.

As the heat in the thermal element reaches a predetermined amount, the control circuit to the magnetic contactor holding coil is interrupted and the motor branch circuit is opened. The two most common types of thermal elements in overload relays employ either a bimetal or a melting alloy joint to initiate the opening action of the contactor.

Overload thermal elements are applied on the basis of motor full-load current and locked-rotor data found on the motor rating nameplate. Complete records on all motors including motor full-load amperes together with proper manufacturer's heater selection and application charts should be included as a part of any maintenance file on motor starters. General heater application charts for the size of contactor involved are usually secured inside the starter enclosure.

Routine maintenance should include a check for loose terminal and/or heater connections and signs of overheating. Overheating can cause carbonization of the molding material creating potential dielectric breakdowns as well as possibly altering the calibration of the overload relay. Relays showing signs of excessive heating should be replaced.

Control Devices. Pilot and other control devices consist of control accessories normally employed in motor starters and include: pushbuttons, selector switches, indicating lights, timers, auxiliary relays, and so on. Routine maintenance checks on these types of devices generally include the following:

1. Check for loose wiring.
2. Check for proper mechanical operation of pushbuttons and contact blocks.
3. Inspection of contacts (when exposed).
4. Check for signs of overheating.
5. Replacement of pilot lamps.

Electrical Interlocks. A contactor or starter may be provided with auxiliary contacts which permit interlocking with other devices as well as serve other position-indicating functions. Proper maintenance of these electrical auxiliary contacts includes the following:

1. Check for loose wiring.
2. Check for proper mechanical operation and alignment with the contactor.
3. Inspection of contacts (when exposed).

Mechanical Interlocks. Mechanical interlocks can be classified in two categories according to their application, safety access, and functional performance. Safety access interlocks are designed to protect operating personnel by preventing accidental contact with energized conductors and the hazards of electrical shock. Functional interlocks, such as found on reversing contactors, are designed to prevent the inadvertent closing of parallel contactors wired to provide alternate motor operating conditions. Motor control centers are provided with plug-in starters for ease of inspection and interchangeability. Mechanical interlocks should be examined to ensure that the interlock is free to operate and that bearing surfaces are free to perform their intended function. Interlocks showing signs of excessive wear and deformation should be replaced. Several types of locking and/or interlocking features are used including the following:

Primary disconnect mechanisms are usually mounted directly on the disconnect device in the plug-in unit. They are mechanically interlocked with the door to ensure that the door is held closed with the primary disconnect in the ON position. A maintenance check should be made to ensure that the adjustment is correct and that the interlock is providing proper safety.

Disconnect operating mechanisms are usually provided with padlocking means whereby the mechanism can be padlocked OFF with multiple padlocks while the door is closed or after it is opened. During maintenance checks in the unit as well as downstream at the motor, these mechanisms should be padlocked in the OFF position for personnel safety.

Most starter units are equipped with defeater mechanisms that can be operated to release door interlock mechanisms with the disconnect device in the ON position. The use of this release mechanism should be limited to qualified maintenance and operating personnel.

Plug-in motor starter units are normally locked in their connected cell positions by a unit latch assembly. Maintenance on this assembly is not normally required but should be understood by maintenance personnel.

BIBLIOGRAPHY

National Electrical Code, *NFPA 70-1993, National Fire Protection Association, Quincy, Mass., 1992.*

National Electrical Code and *NEC* are registered trademarks of the National Fire Protection Association, Inc.

Recommended Practice for Electrical Equipment Maintenance, NFPA 70B-1990, National Fire Protection Association, Quincy, Mass., 1990.

CHAPTER 3
MAINTENANCE OF INDUSTRIAL BATTERIES (LEAD-ACID, NICKEL-CADMIUM, NICKEL-IRON)

Terry Hall
Reliability Engineer, Life Cycle Engineering, Inc., Charleston, S.C.

The installation and proper maintenance of three types of storage (secondary) batteries used in industrial applications are discussed: lead-acid, nickel-cadmium, and nickel-iron. Although a detailed discussion on the various considerations that enter into selecting the correct battery type for particular applications will not be presented, appropriate differences in the characteristics of these battery types will be outlined. The selection process is necessarily complex, involving a close examination of the proposed application, together with a thorough familiarity with the operating and other characteristics of the batteries themselves. In most cases it is advisable that the selection of a battery for an application (particularly if it is comparatively novel) be made with the assistance of a knowledgeable representative of one of the major battery manufacturers.

All three of the battery types discussed in this chapter are common in one respect: they are vented to expel safely gases evolved during charge.

A few special terms that the user will encounter in battery literature should be defined. Cell and battery sizes are specified in terms of a nominal or rated capacity. This is the amount of capacity, usually expressed in ampere-hours, that the cell or battery would be expected to deliver under normal conditions through most of its life. It is important to know the discharge rate under which the nominal capacity was established, since the capacity delivered does depend on discharge rate. Hourly rates are convenient means for expressing the charge and discharge rates at which a cell is operated. Hourly rates are established by the manufacturer for each cell type and are reported in his technical literature. The hourly rate is that discharge amperage which will exhaust the cell in the stated number of hours. For example, when a cell is discharged at the amperage given at the 3-hr rate, the cell would be approaching exhaustion and the voltage would start to decline rapidly, at the end of the third hour of discharge. When reading hourly rates, it is important to note the cutoff voltage at which the hourly rate was established; the higher the required voltage at the end of discharge, the shorter the discharge period will be at that given rate. The *cutoff voltage* is that discharge voltage at which a discharge should be stopped; repeated discharging beyond this point may damage the cells. *Efficiency* is that percentage of charge input which can be withdrawn from the cell on the following discharge. *Energy density* is the power delivered per unit weight or per unit volume of the battery; it is expressed as watt-hours per pound or watt-hours per cubic inch.

LEAD-ACID BATTERIES

Two types of lead-acid batteries are manufactured for industrial applications. These are generally referred to as motive-power and stationary batteries. Typical applications for motive-power batteries include material-handling trucks, mine locomotives, mine tractors, mine shuttle cars, floor sweepers and scrubbers, heavy-duty personnel carriers, transport vehicles, golf carts, and lawnmowers. Typical applications for stationary batteries are switchgear and emergency power for electric-utility substations, switchgear and emergency power for generating plants, computer and other no-fail systems, telephone-company equipment for a variety of operations, emergency lighting, and railway signal service. Depending upon the application, three types of stationary batteries are available through most manufacturers, those made with lead-calcium-alloy grids and those made with lead-antimony grids.

The active material of the positive electrode of a lead-acid battery is lead dioxide, and the negative is highly reactive spongy lead. The electrodes are electrically insulated from each other by separators. Many different types of separators are used, such as resin-impregnated cellulose materials, microporous rubber, microporous plastic, and fiberglass-mat separators with micro-porous backing. Most separators are fabricated with vertical ridges or ribs on the surface that face the positive electrode, and the other surface is flat with no ribs.

The electrolyte in fully charged batteries is a solution of sulfuric acid with a specific gravity ranging from 1.215 to 1.300, depending on intended service. The positive and negative active materials are supported on lead-grid structures in all types except the planté. (The positive electrode in a planté battery is a solid piece of pure lead that is scored with evenly spaced ridges to create a large surface area, then electrochemically converted to lead dioxide by an electrolytic forming process.) For most applications, however, the grid is cast from an alloy of lead with 4.5 to 7 percent antimony and small amounts of arsenic and tin. The function of the antimony is to harden the lead and facilitate casting. For telephone standby power where low self-discharge rates and float currents are required, a small amount of calcium (less than 0.10 percent), instead of antimony, is used in the grid alloy. Generally no arsenic or tin is used in the calcium alloy.

In the fully charged state, the negative active material exists as lead, the positive as lead dioxide, and the concentration of sulfuric acid is at its maximum level. As the cell is discharged, the positive electrode is converted to lead sulfate as follows:

$$PbO_2 + 4H^+ + SO_4 = + 2e \rightarrow 2H_2O + PbSO_4$$

and the negative is also converted to lead sulfate:

$$Pb + SO_4 = \rightarrow PbSO_4 + 2e$$

Combining the positive and negative reactions yields

$$Pb + PbO_2 + 2H_2SO_4 \rightarrow 2PbSO_4 + 2H_2O$$

The overall reaction results in the consumption of sulfuric acid and the equivalent production of water. The consumption of sulfuric acid during the discharge of a lead-acid battery provides a convenient method by which the state of charge can be measured.

Installation and Operation. Upon receiving a new battery, it is extremely important to examine the exterior packing case. An examination should be made for wet spots on the sides and bottom of the case. Wet spots may indicate leaking jars, broken in shipment because of rough handling by the carrier. If any damage has occurred, take immediate and proper claim measures. If any jars are damaged and the electrolyte has leaked out, make immediate repairs and replace broken jars at once. If replacement jars are not immediately available, withdraw the elements from the damaged jar (see under "Repairs" below) and place the elements in a glass, porcelain, rubber, or other nonmetallic vessel containing water suitable for battery use. Sufficient water must be added to cover the plates and separators completely. Damage or complete destruction of the cell may result if these procedures are not followed. Note: Use distilled water or tapwater that has been analyzed and approved for battery use.

Special attention should be given to cells that have been put into new jars. The cells should be filled with electrolyte of the same specific gravity as the balance of the cells at the time of initial filling, and a charge should be applied at a low finishing rate until the specific gravity of the electrolyte ceases to rise. If the specific gravity after charging is lower than that of a normal, fully charged cell, a small amount of electrolyte should be withdrawn and replaced with electrolyte of 1.400 specific gravity. The battery should then be given an additional charge for $1/2$ to 1 hr to mix the liquid thoroughly. Another reading for specific gravity should be taken which should indicate full charge. If not, repeat the latter process until the normal specific gravity is obtained.

All specific-gravity readings of electrolyte must be corrected to 80°F (27°C) to compensate for different densities at different temperatures and obtain a constant basis for comparison.

Placing the Battery in Service. Upon receipt of the battery, a useful practice is to give the battery a freshening charge of from 3 to 6 hr or until the specific gravity indicates no further rise. The charge should be given only with a direct-current charger and with the terminals properly connected.

Cell temperature during the charge should not exceed 115°F (46°C). All points of contact between the charger and the battery should be clean to ensure good conductivity to terminal connections. If terminal connections are copper, apply a coat of petroleum jelly or no-oxide grease to prevent corrosion.

If the battery is installed in a vehicle, properly fasten it in place by holddown lugs on the battery or jar, or bars of the vehicle, to reduce vibration and jarring. If the battery is being installed in a metal compartment, make sure the compartment is thoroughly dry and free of moisture prior to installation. If the battery is to be installed in a locomotive, block the battery into position, allowing a $1/8$-in space between the block and battery tray. Do not wedge the battery into position. All connections between the battery and the vehicle must be flexible. All vent caps must be in place while the battery is in service. Failure to keep caps in position will result in loss of electrolyte (and therefore loss of capacity) and will cause corrosion outside the battery.

Charging the Battery. The batteries used in most industrial situations (other than standby emergency use) are used in what is called cyclic operation. That is, the battery is either being charged or being used (discharged). In most such applications batteries are charged about 1500 to 2000 times during their lives. Incorrect charging for a few cycles will do little harm, but incorrect charging day after day will shorten the life of the battery.

Correct charging means charging the battery sufficiently, without overcharging, overheating, or excessive gassing. To accomplish this, the charging of batteries is usually started at a high rate of amperage known as the starting rate. Later in the charge, this rate of current flow is reduced to the finishing rate. Manufacturers generally suggest as a rule of thumb that the finishing rate should not exceed 5 A per 100 A-hr of rated battery capacity. The starting rate may be four or four and one-half times the finishing rate.

Lead-acid batteries should be charged for a sufficient length of time at a rate which will introduce into the battery the same number of ampere-hours removed on discharge, plus a 5 to 15 percent overcharge. The specific value of the overcharge depends almost entirely upon the charging temperature, and the age and history of the battery. In general, it is more harmful to overcharge excessively an older battery at a high rate or a battery operating at high temperature than a freshly manufactured unit or one being charged at room temperature or lower. Any charge rate is permissible which does not produce excessive gassing or cell temperature greater than 115°F (46°C).

Four methods of charging are discussed below; they are:

1. Modified constant-voltage
2. Taper
3. Two-rate
4. Constant-current

The selection of the appropriate method will be governed by considerations such as the type of battery, service conditions, time available for charging, and the number of batteries to be charged at one time. It should be noted that in charging motive-power batteries, the end-of-charge rate (finishing rate) is extremely important and should not be exceeded. Normally, batteries can be charged in 8 hr, assuming a normal-duty discharge; however, if time permits, a longer period can be used.

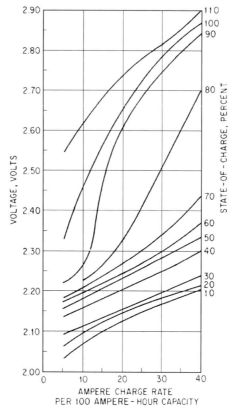

FIGURE 3.1 Lead-acid charging. Voltage per cell vs. charging rate at various states of charge.

Figure 3.1 shows that a discharged battery can absorb high currents at relatively low battery voltages. For example, after the introduction of about 20 percent capacity (20 A-hr at 40 A), a 100-A-hr battery is at a voltage of about 2.22 V per cell. The curves also show that as the charge progresses at a given rate, the voltage increases, the higher charge rates yielding higher voltages. For example, at 110 percent charge (10 percent overcharge), the voltage at 5 A is 2.55 V, and at 20 A, the voltage is 2.74 V.

The generally used finishing rate for lead-acid batteries is the 20-hr rate. With most charging schemes, the normal start-of-charge rate is about the 5-hr rate, or 20 A per 100 A-hr of rated capacity.

Modified Constant-Voltage Method. In the modified constant-voltage method, a fixed resistor is in series with the charger and battery. A 2.63-V-per-cell bus is used for an 8-hr charge. Table 3.1 shows the relationship between volts per cell and time available for charge. In order to use a single fixed resistor and achieve proper start and finish rate, the voltages indicated in the table are required. For an 8-hr charge, the initial current is 22.5 A per 100 A-hr, and for a 16-hr charge, 8.5 A. It should be noted that the charging resistor should be of sufficient current-carrying capacity. The normal "tap" value of the resistor determines the finishing rate. For example, at the 8-hr rate, the "tap" value is 0.022 Ω. This number is calculated as follows: the terminal voltage E_t of the battery at the end of charge, at a finishing rate of 5 A, is 2.52 V (see Fig. 3.2). Therefore, with a bus voltage $E_B = 2.63$ V, the tap resistance must be

$$R = \frac{E_B - E_1}{I_f} = \frac{2.63 - 2.52}{5} = 0.022 \ \Omega$$

TABLE 3.1 Modified Constant-Voltage Charging Design Constants*

Hours available for charge	Bus volts per cell	Resistance values per cell		Rates, amp per 100-A-hr cell	
		Normal "tap"	Max to provide	Start of charge	Resistor capacity
7.0	2.60	0.016	0.027	27.5	32.5
7.5	2.61	0.018	0.029	25.5	30.0
8.0	2.63	0.022	0.031	22.5	26.0
8.5	2.65	0.026	0.035	20.0	23.0
9.0	2.67	0.030	0.039	18.5	21.0
9.5	2.69	0.034	0.043	17.0	19.5
10.0	2.72	0.040	0.049	15.5	17.5
12.0	2.84	0.064	0.073	12.0	13.5
14.0	3.00	0.096	0.015	10.0	11.0
16.0	3.27	0.150	0.160	8.5	9.0

*Design constants based on 100-A-hr cell capacity. For cells of other capacity, external resistance per cell will be inversely proportional and ampere values directly proportional to the capacity. Cell resistance values correspond to electrolyte temperature of 77°F.

When charging several batteries at once, from either a constant-voltage source derived from a motor generator, or from a rectifier, the modified constant-voltage method of charge is preferred because the current tapers during charge, reducing the possibility of excessive charge currents.

The following formula should be used to calculate the kilowatt requirements when using motor generators to charge several batteries from a fixed-voltage bus in 8 hr:

$$kW = \frac{\text{A-hr} \times 0.225\,\text{A} \times \text{No. of cells} \times 2.63\,\text{V} \times 0.8 \times \text{No. of circuits}}{1000}$$

Example. Let four banks of 18 series-connected cells be charged in 8 hr, each cell having a capacity of 500 A-hr.

$$kW = \frac{500 \times 0.225 \times 18 \times 2.63 \times 0.8 \times 4}{1000} = 17\,kW$$

The bus voltage would be $18 \times 2.63 = 47.3$ V, the initial current $0.225 \times 500 = 112.5$ A, and the total initial current from the generator $4 \times 112.5 = 450$ A.

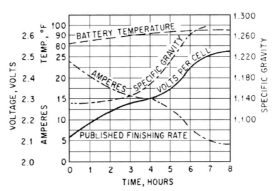

FIGURE 3.2 Lead-acid charging. Typical modified-potential charge profile. Bus voltage 2.63 V per cell; 100-A-hr cell.

6.84 MAINTENANCE OF ELECTRICAL EQUIPMENT

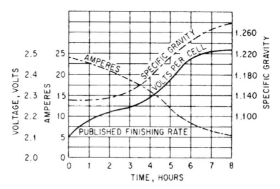

FIGURE 3.3 Lead-acid charging. Typical taper-charge profile for 100-A-hr cell.

Taper Method. This method can be used with either a generator or rectifier equipment, and can be considered a variation of the modified constant-voltage charge method. It is employed when only one size of battery is to be charged. Shunt-wound motor generators, and rectifier chargers can be designed so that their voltage versus current characteristics correspond closely to the modified constant-voltage-type charger. No ballast resistor is required. The circuitry of the charger is such that the initial and finish charge rates are matched to the battery. As was previously mentioned, the finish rate is generally the 20-hr rate, and the initial rate about the 5-hr rate. Figure 3.3 shows the typical voltage, current, and specific-gravity profile of a cell being charged by the taper method. The charge characteristics are nearly the same as those shown in Fig. 3.2 for the modified constant-voltage method. In order to meet the requirements for charging a single battery from a motor generator, the following design parameters must be met.

- The nominal voltage of the generator must be 2.25 V per cell.
- The initial load voltage of the generator should be about 2.135 V per cell.
- At the end of charge, the charging current should be less than 5 A per 100-A-hr battery capacity, and the corresponding voltage 2.52 V per cell.

To calculate the kilowatt requirements for a single motor-generator set, the following formula should be used:

$$kW = \frac{\text{A-hr} \times 0.225 \text{ A} \times \text{No. of cells} \times 2.25 \text{ V} \times 0.8}{1000}$$

Example. Let the battery have 12 series-connected cells, each cell with a capacity of 250 A-hr.

$$kW = \frac{250 \times 0.225 \times 12 \times 2.25 \times 0.8}{1000} = 1.22 \text{ kW}$$

Two-Rate Method. The principle of this method is to begin charging at the recommended start-of-charge rate, then switch to a lower rate when gassing occurs (at about 2.37 V per cell), the proper finishing rate is produced toward the end of the 8-hr period. Figure 3.4 illustrates the charge curve for the two-rate method. When the second resistor is brought into the circuit, a sharp drop in current occurs.

Constant-Current Method. Constant-current charging is seldom used for 8-hr or shift charging of motive power batteries, because this would require manual control during and at the end of charge. If, however, the charge time available were about 12 to 16 hr, constant-current charging could be used. Strictly speaking, the charge period is longer, the initial current is lower than for the 8-hr charge rate, and the taper much shallower. As shown in Fig. 3.5, the initial rate for the 16-hr charge is about 8.5 A, and the finish rate 5 A, yielding a taper ratio (the ratio of the initial to finish current) of 1.7 to 1, and for the 8-hr charge, the ratio is approximately 4.0 to 1.

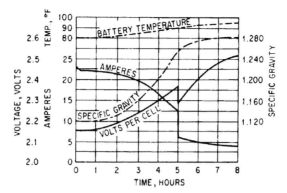

FIGURE 3.4 Lead-cell charging. Charge curves for two-rate method for 100-A-hr cell.

Maintenance. Inspect the battery once every week to make certain all connections are tight. Remove dust or dirt accumulations from the battery top and then wash the battery clean with water and dry with compressed air. At least once each month, neutralize the acid on the battery covers and terminals with either 1 lb of ammonia or sodium bicarbonate solution per 1 gal of water prior to water rinse. Keep terminals and metal parts free of corrosion.

Check the electrolyte level daily and replace any water lost by evaporation. And never allow the electrolyte level to drop below the top of the battery plates. Caution: Never overfill the cells. When replacing water that has evaporated, fill the cells only to the underside of the vent well. Overfilling causes loss of acid, thus reducing battery capacity.

To ensure that water is thoroughly mixed with the electrolyte and to prevent overfilling, additions should only be made while the battery is on charge and gassing at its finish rate. The only exception to this is when the electrolyte level is below the separator protector and not discernible. In this case, add just enough water to bring the level up even with the separator protector prior to charging. Then, make the final adjustment toward the end of the charge period. It is advisable to keep accurate records of the amount of water used and the date of each filling, since the water requirements are an indication of battery overcharging. Battery water should be stored in a covered glass, plastic, earthenware, or other nonmetallic container. Only suitable water should be used for batteries, as certain impurities are harmful and will reduce battery life. Water sources in certain geographic areas are not suitable at any time and in other areas are only satisfactory during certain seasons of the year. If the quality of the local water supply is unknown, arrangements can be made with your battery manufacturer to have an analysis made on a sample at a nominal cost.

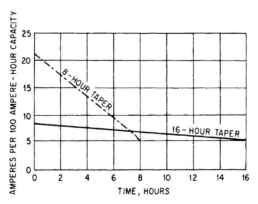

FIGURE 3.5 Lead-acid charging. Constant current at two rates.

Make sure that vent plugs are always kept tightly in place and see that the small gas escape holes do not become clogged. If plugs need cleaning, let them stand in clear water for 30 min or so.

How to Prevent Overdischarging. This is one of the most common causes of battery problems.

Past Experience. This is an obvious but common method. Batteries should be suited to the job for which they are being used. A well-suited battery is a fully charged battery capable of doing the desired amount of work or lasting the desired length of time in a specific service. As long as the job is reasonably standardized (i.e., the equipment powered by the battery is not called on to do extra work during the cycle time), a schedule can be made for battery recharging with very few production failures.

Operator's Experience. An experienced operator can tell from the action of the equipment when the batteries are reaching a point at which they should be charged.

Discharge Indicator. State-of-charge meters which are permanently mounted on material-handling equipment are commercially available. They monitor voltage and rate of discharge and, if properly calibrated and adjusted, will light a warning light on the fuel gage dial just prior to the battery being 80 percent discharged. At 80 percent, a relay is activated which cuts off power to the lifting devices, but allows power to the drive motors.

Ampere-Hour Meter. With this type of meter, the number of ampere-hours removed from the battery is recorded. (Some scales are calibrated in ampere-hours remaining.) Thus, the operator knows how much power is left in the battery.

How to Determine Battery Condition

Records. The purpose of records is to provide a day-to-day case history so that any variations from normal can be detected quickly and acted upon. Daily records should show battery number, identification of the truck the battery was taken from, specific gravity of battery when put on charge (pilot-cell reading), temperature of pilot cell, time put on charge, time taken off charge, and specific gravity when taken off charge. These are enough facts to keep a good case history on the battery.

If specific gravity (corrected for temperature) and time-on-charge data are compared with the previous day's reading, any abnormal battery use or abuse would be indicated and can be acted upon. As a long-run check, most battery manufacturers recommend that special specific-gravity and voltage readings be taken of each cell of the battery every 6 months, after an equalizing charge has been applied. Comparisons of these readings with the readings of the last such test will show any long-term changes in battery condition as well as differences between cells.

Test Discharge. Such a test should be made at any time there is a question as to whether or not the battery is delivering its rated capacity. The procedure is as follows.

The battery is given an equalizing charge and the fully charged specific gravity of each cell is adjusted to normal. Starting time is noted, and the battery is discharged at the standard 6-hr rate given in the operating data supplied by the battery manufacturer. Individual cell voltages and the overall battery voltage should be recorded 15 min after the test is started, and then hourly until the voltage of any cell reaches 1.8 V; thereafter, voltage measurements should be made at 15-min intervals. Record the time when each cell voltage reaches 1.75 V. When the majority of cells reach 1.75 V, record the time and terminate the test. Measure the specific gravity of each cell immediately.

Record all cell voltages and stop the test discharge when the battery voltage reaches the termination voltage of 1.70 times the number of cells in the battery. Record the specific gravity of each cell immediately after terminating the test discharge. The readings will help determine whether the battery is uniform or if any one or more cells are low in capacity. If the battery is uniform and delivers 80 percent or more of its rated capacity, it can be returned to service.

Internal Inspection. If the test discharge indicates that the battery is not capable of delivering at least 80 percent of rated capacity and all cells are uniform, an internal inspection of one of the cells is indicated. Failure to meet capacity ratings may be caused by an internal shunt which can be repaired. The positive plates, which wear out first, should be examined. If they are falling apart or the grids have many frame fractures, a new battery is needed. If the positive plates are in good condition and the cells contain little sediment, the battery may be sulfated. The negative plates of a sulfated battery will have a slatelike feeling, being hard and gritty and having a sandy feeling when rubbed between the fingers. (A good negative plate, when fully charged, is spongy to the touch and gives a metallic sheen when stroked with the fingernail.) Sulfation is such a common condition that a special discussion on its causes and treatment follows.

A sulfated battery is one in which abnormal lead sulfate is formed in the plates. This affects the normal chemical reactions within the battery, causing loss of capacity. The most common causes of sulfation are undercharging, repeated partial charges, neglect of equalizing charge, standing in a partially or completely discharged condition, low electrolyte, specific gravity more than 0.015 above normal, and high temperature. The following steps will usually restore a sulfated battery:

1. Clean battery.
2. If no electrolyte is visible, add water to bring level up to the separator protector.
3. Put battery on charge at the prescribed finishing rate until full ampere-hour capacity has been supplied the battery. If during the charge the temperature of the battery exceeds 115°F (46°C), reduce the charge rate. If any cells give test-voltage readings 0.20 V below the average cell voltage, pull and repair the cell before continuing the charge.
4. Continue the charge at the finishing rate until the specific gravity shows no change for a 4-hr period.
5. Give the battery a test discharge.
6. (*a*) If the battery gives rated capacity, no further special treatment is needed except that the battery should be immediately recharged before being returned to service. (*b*) If the battery does not deliver at least 80 percent of rated capacity, continue the discharge until one or more cells reach 1.0 V. Repeat steps 3, 4, 5, 6*a*, and then go to step 7 if step 6*a* is not met.
7. If the battery does not deliver at least 80 percent of capacity, repeat steps 2, 3, 4, and 5 again. If the battery does not now deliver 80 percent of capacity, assume that it should be replaced.

Causes and Remedies of Common Battery Troubles. In the listing that follows it is impossible to consider all sources of battery trouble. The ones listed are common troubles and will serve as a starting point for investigating the cause of unsatisfactory performance. Eight symptoms are listed; after each are listed possible causes. Where the remedy for the cause is perfectly obvious, it is omitted. Where, however, there might be some doubt as to the correct remedy, it is indicated along with the cause, is marked with the symbol R, and is enclosed in parentheses.

Symptom: Battery Will Not Take a Charge. Possible causes: (1) Direct-current charging-circuit fuse blown or missing. (2) Circuit in charging receptacle or plug open, or connection of cable to stud loose. (3) Alternating-current line fuses blown or missing. (4) Alternating-current line switch open. (5) Circuit in control lead or circuit open, preventing contactor from pulling in. (6) Charging plug not pushed all the way into receptacle. (7) Charging rate too low. (R: Check ammeter for accuracy. See below, under "Symptom: Battery Takes Too Long to Charge.") (8) No voltage output from generator. (R: Check field circuit; if open, correct. Check brush contact to armature; correct by replacing brushes or adjusting so they don't stick.) (9) Bus voltage too low, caused by incorrect tap setting in rectifier or too low voltage from generator. (10) With initial equipment, connections to charging receptacle reversed.

Symptom: Battery Takes Too Long to Charge. Possible causes: (1) Connection poor in charging circuit. (R: Check lugs, bolted connections, charging leads, plugs, and receptacles for high-resistance joints, and correct.) (2) Battery overdischarged. (3) With two-rate charging, charging equipment does not provide high starting rate. (R: Check for open in control circuit to provide high rate. Determine cause and correct.) (4) With two-rate rectifier charging equipment: (*a*) voltage relay connected for smaller number of cells than in battery; (*b*) applied ac voltage too low under load conditions (R: Install greater-capacity line to rectifier to reduce voltage drop, or relocate rectifier nearer to incoming ac source); (*c*) primary transformer taps not set for voltage applied; (*d*) voltage relay operating below standard voltage (i.e., 2.37 V per cell); (*e*) start-of-charging rate too low; (*f*) end-of-charging rate too low; (*g*) start-of-charging rate too high. (5) Where voltage relay is used in control circuit for two-rate charge, temperature of charging control equipment may be materially higher than battery operating temperature. (R: Provide better ventilation for charging equipment, or relocate it to an area where atmospheric temperature is the same as temperature in area where battery operates.) (6) With modified constant-voltage charging: (*a*) bus voltage too low; (*b*) bus voltage decreases as load decreases (R: Adjust generator for flat characteristic); (*c*) ballast resistance too great. (7) Charging leads reversed, or charging-equipment polarity reversed. (8) Battery not placed on proper charging circuit when installation has various battery sizes. (9) Charge not terminated when battery is fully charged.

Symptom: Battery Will Not Work Full Shift. Possible causes: (1) Cell voltages and specific gravity uneven. (R: Give an equalizing charge.) (2) Electrolyte level low. (3) Battery not charged before going into service. (4) Two or more cell leakers in steel tray. (R: Replace broken jars.) (5) One or more cells cut out of battery. (6) Battery with incorrect number of cells assigned to equipment. (7) Specific gravity below normal. (8) Impurities in electrolyte. (9) Operator riding brakes. (10) Operator using reverse instead of brakes. (11) Load too great. (12) Wheels, axles, and bearings need grease. (13) Tires under-inflated. (14) Brakes dragging. (15) Wheels deeply grooved. (16) Ruts in roadbed deep. (17) Series field in motor shorted or grounded. (R: Clear grounds and insulate wiring.) (18) Armature needs repairs. (19) Grounds on equipment. (20) Excessive grades along route traveled. (21) Service required exceeds capacity of equipment. (22) When batteries are in two halves, discharged half has been paired with a charged half. (23) Uneven number of cells in two halves, where split batteries are used in parallel-start, series-run control circuits.

Symptom: Battery Overheats on Charge. Possible causes: (1) Finish rate too high. (2) High-charge rate on too long. (R: Reduce voltage-operating point of voltage relay.) (3) Timer not set correctly. (4) Ampere-hour meter not set correctly. (5) Percent overcharge setting of ampere-hour meter set above correct level of 12 percent. (6) Timer set for too many hours. (7) Two-rate charge did not change over to low rate. (R: check operation of voltage relay. Check for open voltage-relay circuit. Check for open in charge-rate control lead. See if voltage relay is connected for same number of cells as in battery.) (8) Bus voltage too high. (9) Charge rate too high. (10) Charge not stopped—automatic mechanism does not terminate charge. (R: See that voltage relay is connected for same number of cells as in battery. Check timing mechanism. Check for open in control leads. Check operation of voltage relay. Check ampere-hour meter for accuracy and operation at low rates; clean and calibrate it.) (11) Ventilation poor. (12) Separators worn through. (13) Sediment space filled. (14) Internal shunt. (15) Fully charged specific gravity is below normal, and attendant continues charge to increase specific gravity. (R: Adjust specific gravity with acid.)

Symptom: Battery Overheats on Discharge. Possible causes: (1) Overdischarge (beyond allowable limit of 1.130). (2) Battery too small. (3) Ventilation poor. (4) Burn of connectors to cell terminals poor. (5) Load excessive. (6) Battery worn out. (7) Separators worn through. (8) Internal shunt. (9) Battery capacity temporarily reduced because of low fully charged specific gravity. (10) Battery not fully charged before being put in service, resulting in overdischarging. (11) Electrolyte level low. (12) Battery not heat-insulated from resistor in charging equipment. (13) Atmospheric temperature too high.

Symptom: Electrolyte Level Low. Possible causes: (1) Jar broken or cracked. (2) Water additions neglected. (3) Cell overlooked when adding water. (4) Too much overcharging. (R: If automatically controlled, check voltage relay, timer, and charge-rate relay. If manually controlled, terminate charge when specific gravity is 10 points below last equalizing charge value. Change from high rate to low rate when specific gravity reaches 1.200.)

Symptom: Cell Voltages Unequal. Possible causes: (1) Overdischarge. (R: Give an equalizing charge.) (2) Equalizing charges lacking. (3) Internal shunt. (4) Top of battery very dirty. (5) Cells operated with low electrolyte level. (6) Fully charged specific gravity of cell low. (7) Sediment space filled. (R: Replace battery.) (8) Positive plates worn out. (R: Replace battery.) (9) Half tap on cells for lower voltage circuit. (R: Remove tap, and connect load to battery terminals through resistance.) (10) External source (such as charging resistance on locomotive) heating certain cells. (11) Contact poor in controller on split-circuit batteries (parallel and series on discharge, all series on charge). (12) Impurities in cell. (13) Charging rate varies. (14) See also symptom below.

Symptom: Unequal Specific Gravity between Cells. Possible causes: (1) All items under Cell Voltages Unequal above. (2) Overfilled with water. (3) Cell operated with cracked jar. (4) Acid not adjusted properly when jar was changed. (5) Battery operated with vent caps out of place. (6) Sealing compound leaks. (7) Battery operated with broken cover. (8) Neutralizing material in cell.

Repairs. Most of the repairs to storage batteries consist of removing a part of the battery and replacing it. In this section, we shall therefore outline the procedure for disassembling and reassembling a typical battery. The manufacturer's instructions that were received with the battery will undoubtedly outline any specific procedures in handling their units.

Drilling Intercell Connectors. In most batteries, the lead insert of the cover, the cell post, and the intercell connectors are all welded together. To remove a cell from the circuit or an element from a jar, it is therefore necessary to remove the connector or cut it in two.

There are two methods of removing a connector. One is to use a special drill that allows the cell post to remain but cuts the bond to the lead insert of the cover. The other method is to drill through the center of each post, using a $15/16$-in drill, to a depth of $3/8$ in. After the intercell connectors are drilled, they can be lifted off. On some batteries it is possible to saw a connector. It should be cut above the space where the two jars meet. Then the cell can be pulled out of the tray.

Removing Cell from Tray. After the connector is removed, use a warm compound knife and cut the compound from between the jar of the cell and the adjacent cells or tray. Penetrating oil or kerosene mixed with regular oil should be run into the space between cells to act as a lubricant. Work the cell back and forth to see that it is loose; then lift straight up. Small cells can be lifted manually. To lift heavy cells, attach a cell puller (a self-tapping nut with loop attached). Always attach the puller on the negative post. If the cell has two negative posts, use two pullers with a piece of wood through the loops. Lift slowly and carefully, vibrating the lifting rope after a strain is put on to loosen the cell.

Replacing a Jar. Have the new jar ready. Remove the jar to be replaced from the tray (as outlined above). If the cell is to remain out of the battery for a day or two, the space in the tray should be blocked to prevent jars in the tray from bowing out into the space from which the cell was removed.

Cut the compound from around the top of the jar with a warm compound knife, keeping it very close to the inside of the jar cell. Heat the outside of the jar on all four sides with a blowtorch. Place the jar hold-down clips and chains on the jar, and use the cell puller to lift the element halfway out of the jar. Allow the element to remain in this position a minute or so to drain. Then remove the element and lay it down on a wood board surface with the flat side of the negative plate down. The element should not be exposed to air any longer than necessary. (If the element starts to heat, sprinkle it with water and place in a jar.)

After warming the clean jar so that it is pliable, slide it over the bottom of the element carefully, using the compound knife as a guide. Then lift up the jar and lower the element into the jar slowly and carefully so that the separators are not broken or damaged. (If the jar is square, be sure that the element is placed in the jar so that the ribs of the jar are in the correct direction—at right angles to the plate.)

Clean, neutralize, and dry the surface of the jar, and cover. Reseal between the repaired cell and its adjacent cells with compound. (When pouring a seal, use a compound knife in one hand and hold the saucepan of compound in the other. The knife is used to cut off the pour and catch excess compound.) Remove any excess compound that may have run down the outside of the jar. Fill the cell with correct electrolyte (the same as in adjacent cells or higher) to the top of the splash plate. The cell should then receive an equalizing charge, and the acid should be adjusted.

Place the cell in the tray, being sure polarity is correct. The final step is to reburn the intercell connector. This is covered separately below. But first it is important to note that the post and connectors have been cleaned in preparation for burning.

Reconnecting Sawed Connector. As mentioned above, sometimes connectors are sawed through when a cell is removed. The connector can be burned together by using a connector mold, which is simply a shallow trough that fits under the break. It is blocked into place with small wedges. Place the tip of the carbon on the piece of the connector that is in the electric circuit for the carbon burner, and hold the carbon there until it is white hot. Add new lead, and move the tip of the carbon through the molten lead to ensure that the new lead is fused with the lead of each half of the connector.

VENTED NICKEL-CADMIUM BATTERIES

Nickel-cadmium batteries are alkaline batteries, having a solution of potassium hydroxide as the electrolyte. These batteries are very rugged physically and will sometimes withstand more shock and vibration than the equipment they are powering. They are also capable of sustaining considerable electrical abuse (overcharging, standing in the discharged state, and occasional overdischarging). Characteristically they have low internal resistance, and consequently have good charge acceptance and perform well at high rates. Compared with nickel-iron storage batteries, they have a relatively low self-discharge rate. Their performance at low temperatures is excellent; many designs will deliver 80 percent of their rated capacity at temperatures as low as 240°F (115°C). As a rule they are not intended for cyclic applications, being used rather in engine starting, railroad signaling, emergency lighting, communications, alarm, switchgear, marine, and standby applications.

Specific points of comparison with the lead-acid storage battery that may be mentioned are as follows: Nickel-cadmium batteries can be left standing for long periods of time in the discharged condition without fear of deterioration. During charge, no corrosive fumes are released. The nickel-cadmium battery with sintered plates (described below) can, if required, be recharged quite rapidly—in 1 or 2 hr. They can be overcharged with little damage, provided that the temperature is controlled. Further, nickel-cadmium batteries are not damaged by freezing.

A few of the disadvantages of nickel-cadmium cells are as follows: they have a considerably lower voltage than the lead-acid type (both operating and open-circuit); the average discharge voltage is between 1.2 and 1.25 V at ordinary discharge rates. They are not capable of extended deep-cycle service. With the pocket plates (discussed below), the ratio of energy per unit volume and unit weight is no greater, in some cases less, than that of lead-acid storage batteries. The state of charge cannot be readily determined, as can be done with a hydrometer with lead-acid batteries. And, because of the considerably greater cost of the prime metals, nickel and cadmium, on an energy basis their cost is substantially greater than that of lead-acid batteries.

The electrochemical characteristics of the nickel-cadmium system are similar to those of other alkaline batteries such as the silver-cadmium and the nickel-iron but differ greatly from the electrochemical reactions of the lead-acid battery. As was mentioned under "Lead-Acid Batteries," sulfuric acid electrolyte is consumed in reacting with the positive and negative; consequently the state of charge can be determined by measuring the specific gravity of the electrolyte. However, in all alkaline batteries the electrolyte serves only as a carrier of charge. The potassium hydroxide is not consumed but serves only to shuttle electrons back and forth between the positive and negative plates as the battery is being charged or discharged. Consequently, the electrolyte remains relatively constant during both charge and discharge. The discharge-voltage curve of alkaline batteries is relatively flat and the batteries are not as vulnerable to freezing as other storage batteries. To emphasize again: The state of charge of an alkaline battery cannot be measured by the specific gravity of its electrolyte.

Plate Processing and Battery Construction. Two basic types of vented nickel-cadmium cells are available, those with pocket plates and those with sintered plates. Pocket plates are most generally used in vented nickel-cadmium cells and batteries. These plates are extremely rugged and are used in applications requiring maximum life, great resistance to shock or vibration, and maximum cell size. Nickel-cadmium batteries using sintered plates have much lower cell internal resistance and are therefore used in applications requiring very high discharge rates, such as engine-starting and switchgear applications.

Pocket Plates. Production of pocket plates begins with strips of thin steel ribbons, perforated with roughly 2000 holes per square inch, and then nickel-plated. The edges of this ribbon are turned up into a troughlike configuration. The active materials—nickel hydroxide plus graphite for the positive, and cadmium hydroxide (or cadmium oxide) plus iron powder for the negative—are pressed into this trough. A second piece of perforated steel ribbon is applied over this, and the edges of the two ribbons are crimped together, forming a very long flat pocket of perforated steel containing the active material. The material can be cut to pieces of any length to form plates of any desired width. These pockets are then laid horizontally into plate frames stamped from steel sheet; these frames have the length and width of the desired plate and are open in the center to receive the pockets. The pockets are crimped into this frame in such a way that the joints formed along the sides of the plate frame serve also to seal off the cut end of the individual pockets. The plates are then assembled into positive and negative groups, bolted together to the proper terminal post by means of a threaded connector rod passing through the base of the post, or in some cases to comblike teeth extending from the terminal post. The positive and negative groups are then interleaved. The separators—either plastic rods placed vertically between the plates of corrugated or perforated plastic sheet—are inserted. The assembled element is then placed in the cell case, and the cover, with its insulating and sealing washers and nuts, is placed over the terminals and welded (or cemented) in place.

Smaller pocket-plate cells are available either in plastic or in steel cases. Plastic cases have many advantages: They are transparent or translucent. In applications involving large numbers of cells, this means a considerable saving of maintenance time because electrolyte levels can easily be checked visually and it is easy to fill to the proper level when watering. Since the cases are nonconductive, they can be touching, thus saving installation space. There is less likelihood of accidental grounding with these cases, and it is a little safer to work around them with metal tools. Plastic cases are resistant to electrolyte corrosion. For some applications requiring great physical ruggedness, they are assembled into steel battery trays.

Steel cell cases are formed of welded steel sheets, are nickel-plated, and can be produced in a variety of sizes without a large tooling expense. Most importantly, they offer the advantage of great strength. It is for this reason that large cells are built only in steel cases.

Sintered Plates. Sintered plates involve first a sintered nickel plaque, which serves as the plate grid. This plaque is made by sintering fine nickel powder (made by the carbonyl process) to a piece of nickel screen or perforated nickel sheets. The resulting plaque material is a very porous, tough, flexible sheet of pure nickel, usually between 0.025 and 0.08 in thick. Though this material appears solid to the eye, roughly 80 percent of the volume is open space. The positive and negative active materials are then deposited into these pores by any of several different methods of impregnation. The resulting cells have very low internal resistance and consequently perform well at very high discharge rates; this is the prime advantage of the sintered nickel-cadmium cell.

Voltage. The open-circuit voltage of nickel-cadmium cells is about 1.35 V. The average discharge voltage, which can be used for calculating the number of cells used for particular applications, is generally stated as 1.2 V. At lower discharge rates (5 to 8 hr and lower), the average voltage would be about 1.25 V; at these rates, the voltage would drop 0.15 to 0.2 V from the beginning to the end of the discharge. Pocket cells can be discharged at the 1-hr rate and sintered cells at two or three times the 1-hr rate before the average discharge voltage falls below 1.2 V. Sintered cells will deliver 12 to 16 times the 1-hr-rate discharge at voltages no less than 1.0 V. At low-temperature operation, the voltage naturally is reduced to a lower level, but not significantly until the temperature gets down to the range of 220°F (104°C).

For lower-voltage applications (i.e., 6 V, 12 V), the number of cells used is the exact (or the nearest) equivalent to the quotient obtained by dividing the application voltage by 1.2. For example, five cells are used for a 6-V application and 10 cells for 12 V. With higher-voltage applications this factor may not be strictly adhered to. For example, 18 to 20 cells may be used for 24-V circuit-breaker application, and 92 to 95 cells for a 125-V control application. The exact number of cells selected depends on the discharge rate, line loss to be counteracted, float voltage available, and other factors. As a general practice, it is recommended that the manufacturer's service engineer be consulted in selecting the proper number of cells and electrical operating characteristics to ensure a proper interface between the battery system and the proposed application.

Performance Characteristics. Most pocket-plate designs will deliver their normal capacity, though to a lower end voltage, when discharging at rates as high as the 1-hr rate. Some high-rate sintered plate cells will deliver nominal capacity at discharge rates several times the 1-hr rate. When discharging into loads which offer very low resistance (switchgear, engine-starting applications), sintered cells will deliver some 15 to 18 times the 1-hr rate for about 10 sec. Some manufacturers, for sintered-type plates, publish the 5- and 1-sec capacity ratings. Most nickel-cadmium cells will deliver about 80 percent of nominal capacity at 0°F (−17.7°C) at normal discharge rates.

It is suggested that when discharging nickel-cadmium batteries at the 3- to 4-hr rates or lower, care should be exercised to avoid repeatedly discharging below 1.0 V. At very high rates, such as are involved in engine-starting applications, the discharge may, however, be carried down to 0.65 V. Repeated discharging below these limits will lead to a declining capacity. Sintered plates, however, can tolerate overcharging somewhat better than pocket plates.

Life. The life expectancy provided by most pocket plates in float operation is about 15 years, depending on the severity of service conditions. Occasional standby applications have been reported in which a 25-year life was attained. Sintered-plate batteries have a much shorter life; in severe service, such as vehicle-starting applications, lives of 5 to 7 years have been reported. Batteries in emergency lighting, alarms, and communications can be expected to survive for 8 to 15 years. Some manufacturers provide types that are satisfactory for cycling under carefully controlled conditions and will deliver up to 10,000 cycles, depending upon the depth of discharge.

Selection. Having determined the general battery design that should be considered for your application, the approximate number of cells that will be needed, and the ampere-hours to be supplied between chargings, contact the firm manufacturing that cell type and supply them with the detailed information needed to recommend the proper battery configuration for your application. This information should include the following.

Voltage Required. The allowable maximum and minimum values, the degree of voltage regulation preferred.

Capacity and Rate Capability Required. The currents the battery will be expected to deliver and the length of time over which these stated currents will flow. If the current is unknown, state as explicitly as possible the work that is to be done by the equipment which the battery will power, in terms of torque or horsepower delivered, transmitting and receiving power, and so on.

Charging Conditions. Type of charge routine to be used—constant current, constant voltage, float, trickle, type of charge equipment you plan to use, rates and voltages it can deliver, degree of control it exercises, length of time allowable for each charge, and frequency at which charging can be done.

Shock and Vibration Resistance Required.

Angular Inclination. Slant or tilt to which the battery will be subjected.

Installation Conditions. Available space, ambient temperature, ventilation available, proximity of lead-acid batteries, or other contaminating conditions.

Any Special Maintenance Conditions. Desire to minimize frequency of watering, need to avoid special tools, and type of personnel who will care for the battery.

Installation. In the discussion that follows, the assumption is made that the purchaser of a new battery system has received detailed instructions from the manufacturer on installation. In general, nickel-cadmium batteries are shipped charged and filled with electrolyte (except for those exported, which are shipped discharged and dry). Throughout the unpacking and installation procedure, nickel-cadmium batteries should be handled with caution, since they are charged. For example, chains or metal hooks must not be used to hoist cells from the packing crates; rope slings, passed under the intercell connectors, may be used with caution. After having checked for shipping damage, remove whatever shipping plugs have been put into the cell vents and replace with the vent plugs provided. Then check the electrolyte level in each cell. With most medium and large cell sizes, the electrolyte should be at least $1/2$ in above the plates. The maximum level should be obtained from the manufacturer's literature; it will often be about half the distance from the top of the plates to the underside of the cell cover. If electrolyte has been spilled during shipment and the level is below the top of the plates, add refill electrolyte to bring the level to its stipulated level. In general the battery manufacturer is the best source for details on electrolyte replacement, and his recommended procedure should be followed.

Cells in steel cases and battery trays may be permanently installed. Nickel-cadmium batteries do not release corrosive fumes and may be installed next to machinery or instruments; however, this equipment should not be subject to direct spray from the cells. Appropriate battery racks may be purchased from the manufacturer or, if convenient, may be built by the user to the manufacturer's specifications. A small amount of space should be left between trays for circulation of air. Smaller batteries should be placed on shelves. Batteries in trays with special extended sides may be set directly on the floor. Plastic-case cells in plastic or steel battery trays require no special installation. Vehicular or marine batteries must be securely held down. In all cases, if the batteries are to be serviced from the side of the compartment, a minimum of 8 in, preferably 12 in, should be provided between the cell tops and the compartment roof. Batteries should not be installed in areas where the temperature will rise frequently to above 100°F (38°C).

After positioning the batteries and installing the intercell connectors, check the polarity of all cells, following the connectors from one battery terminal throughout the battery to the other terminal to make sure that the cells are correctly connected in series. Any cells connected into the circuit with reverse polarity will be damaged. Look for a "plus" sign marked on the cell terminal or on the cover next to the terminal, a red mark on the side of the vent wall toward the positive terminal, or (in some larger cell types) a red-rubber insulating band on the positive terminal. Then make sure that all terminal-post nuts are tight. Check that the main battery cables to the battery are heavy enough to carry the maximum current that will be required without excessive voltage drop. Battery cables should be fitted with nickel-plated lugs; bare copper lugs are likely to corrode. All cables should be kept off the cell tops. Wipe off any electrolyte that may have splashed out onto the cell tops during installation. When the cell tops of steel-cased cells are perfectly clean and dry, completely cover them with a thin coating of petroleum jelly. This will prevent electrolyte spray from gradually building up into a hard-to-remove crust on the cell tops.

If the battery is not to be placed in service within 90 days, it should either be put on continuous trickle charge or given a charge at the 5-hr rate for 3 hr when it is put back into service.

Charging. It is important to establish a reliable regular charge procedure with nickel-cadmium batteries and to adhere to it, since it is not possible to check the state of charge quickly as can be done with lead-acid batteries. However, if there is a strong need to determine the state of charge, it is possible in many applications to design a method of simultaneously reading voltage and current during a brief high-rate discharge, or reading the amount of current drawn when the battery is placed on a brief constant-voltage charge. The manufacturer's service engineer should be contacted for establishing the most satisfactory method for the particular application.

Nickel-cadmium batteries tolerate overcharging fairly well. In any questionable case, it is always better to overcharge than undercharge. The cautions against excessive overcharge are stated, not because of any direct effect on the plates, but rather so that maximum permissible temperatures are not exceeded and that the electrolyte loss and buildup of conductive film on the cell tops is minimized. On any charge routine the battery temperature should not exceed 115°F (46°C). Occasional temperatures of 125°F (52°C) can be tolerated, but repeated charging at these high temperatures is likely to result in reduced capacity and shortened battery life. When checking temperature, always take it from the cells in the middle of the battery, as these are likely to be the warmest cells.

Nickel-cadmium batteries can be recharged with either of two basic types of charge routine: constant voltage (constant potential) and constant current. They are also maintained in the fully charged state by trickle- or float-charge routines. The equalizing charge discussed below is a variation of the constant-current routine. Constant-voltage charging involves supplying charge current at a fixed regulated voltage. The voltage level is selected so that the current, high at first, tapers off to a very low level as the battery nears full charge, and the countervoltage of the battery rises. This is one of the two most commonly used methods. It readily lends itself to automatic control and can be performed rapidly. The constant-current method offers the advantage of easy calculation of the ampere-hours of charge put into the battery. However, if it is to be performed manually, it calls for frequent adjustment of the rate. With the trickle-charge method, the battery is left permanently connected to a source delivering very small amounts of charge current; an example is the charger/battery combinations included in many emergency-lighting or alarm units. Trickle charging can be done with either constant current or constant voltage.

Float charging, the second very common charge method used for nickel-cadmium batteries, differs from trickle charging in that the battery is permanently connected in parallel across the line between the power source and the equipment to be powered. The power source normally supplies both the equipment and the charge current to the battery. The battery is discharged in the event of failure or inactivity of the power source. This is the typical standby application.

Constant Voltage. In most cases a modified rather than a true constant-voltage method is used so as to limit the high initial surge of current that would otherwise be absorbed by a discharged battery. With this scheme the voltage is automatically reduced below the preselected value until the current taken by the battery at that voltage drops to a value that can be supplied by the charger. This reduces the size and hence the cost of the charging equipment. This method has been found particularly suitable for sintered-plate batteries. For information on the voltage values that should be used for either constant-potential or modified constant-potential charging, the manufacturer's engineers should be consulted.

Constant Current. If it is necessary to charge a battery fully within 7 hr using the constant-current method, the charger must be capable of delivering current at the 5-hr rate, at a voltage of approximately 1.8 to 1.85 V per cell. In many installations smaller chargers are used, delivering a lower rate (8- to 10-hr rate), thus requiring only 1.55 to 1.65 V per cell. Longer charge times are therefore necessary. Water consumption will be lower with these lower charge rates. A variable resistance must be placed in series with any battery to be charged at constant current; make sure that this variable resistor and the ammeter and shunts used in the charging equipment are capable of handling the currents involved. The resistance should be adjusted at least every $1/2$ hr to hold the charge rate steady. Batteries to be series-connected and charged on one charger must be of similar design type and in a similar state of charge. If the state of charge is unknown, charge each battery on a separate charger or at a separate time.

A charge-back factor of about 140 percent is recommended for all nickel-cadmium batteries; that is, the battery is charged until 140 percent of the amperes taken out on the previous charge is returned. If the amount of capacity previously withdrawn is unknown, simply start the charge at a convenient rate, preferably at or near the 5-hr rate. Observe the on-charge battery voltage. Using approximately the 5-hr rate, the initial on-charge voltage of a fully discharged battery will be about 1.35 V per cell. During the charge the voltage will gradually increase to about 1.45 V per cell. At this point a major portion of the normal capacity will have been returned and gassing will begin.

As the cell approaches, and reaches, full charge, the cell voltage will rise quite rapidly to about 1.5 to 1.8 V per cell (depending on the cell design and actual charge rate); charging should be continued until the on-charge voltage has remained steady at this level for 60 min (as indicated by three identical readings taken 30 min apart). When using this method, it is particularly important to watch the end-of-charge voltage point. If the charge is not terminated, the battery will continue to accept current, which will go entirely into the formation of hydrogen and oxygen; water loss will therefore be very rapid, and the battery temperature may rise above the maximum permissible level.

Because of the sharp voltage rise at the end of charge, charging can conveniently be terminated by a voltage-sensing relay or other similar device. In almost all cases nickel-cadmium batteries can be automatically charged on modified constant-current charge equipment designed for use with lead-acid batteries of comparable size. However, note that the end-of-charge voltage of a nickel-cadmium battery differs substantially from that of a lead-acid battery.

Trickle Charge. Trickle charging should be used only to keep a charged battery in the fully charged condition. It is not an acceptable method to charge a completely discharged battery.

Pocket-plate batteries may be maintained on trickle charge at voltages between 1.40 and 1.45 V per cell; for sintered-plate batteries an acceptable voltage level is 1.36 to 1.38 V per cell. However, the manufacturer's instruction should be followed as to the exact values. Self-discharge losses will be replaced when operating at the lower end of the voltage range. (Operating at the higher voltage will ensure return of capacity taken out in a partial discharge.) At all events, stay below the gassing potential of approximately 1.45 to 1.47 V. If water consumption is observed to be excessive, decrease the on-charge voltage. If the battery is cold (32°F [0°C]or colder), raise the voltage by about 0.05 V per cell. These voltages are critical. If charge voltage fluctuates because of changes in line voltage, it may be necessary to monitor the voltage for the initial period of operation and then choose the average value for routine operation.

Trickle charging can also be done by the constant-current method. Set the charger to supply a few milliamperes of current for each ampere-hour rated capacity of the battery. The exact value that will provide a balance between minimizing water consumption and maintaining full charge can be determined through trial and error or by the technical data of the battery manufacturer.

Float Charge. Pocket-plate batteries are maintained on float at between 1.40 and 1.45 V per cell, and sintered-plate types at 1.36 to 1.38 V. As for trickle charging, the lower values cited are adequate to replace self-discharge losses and will ensure minimum water loss but will not replace any significant amount of discharge current withdrawn from the battery. The voltage must be held below 1.45 V to avoid gassing and excessive water loss. Operating at these voltages, the battery will draw current at approximately the 35- to 50-hr rate.

Equalizing Charge. Batteries operating on float- or trickle-charge routines should occasionally be given an equalizing charge to keep the cells in balance. Cells are said to be out of balance when, because of small unavoidable differences in chemical or physical condition, they begin to differ in their state of charge. When this happens, some of the cells in the battery will reach full charge before the others and will exhibit an early increase in cell voltage. In float operation, where the charge voltage is not too far above that voltage at which the cells will accept no charge current, this early rise in the voltage of some cells will result in decreased current delivered to the battery as a whole, before the other cells have reached full charge.

Some commercially available chargers have two charge positions, one for normal charging and one for equalizing charge. In the equalizing position these chargers usually deliver current equivalent to the 15- to 20-hr rate for the battery. This is barely adequate. To ensure complete equalization, charging should be done at the 5- to 10-hr rate if possible. On float or trickle applications, once a year when the battery is observed to have lost capacity or to have gone out of balance, charge it at the equalizing rate until the voltage of each cell, measured individually, has reached a plateau (at about 1.65 V per cell) and has ceased to rise.

Maintenance. Once nickel-cadmium batteries are properly installed and are being operated correctly, the major maintenance effort involved is maintaining the electrolyte level and the specific gravity and keeping the battery exterior clean.

Typical instruments and materials needed for maintaining and overhauling nickel-cadmium batteries are as follows:

- Refill electrolyte (potassium hydroxide of 1.220 specific gravity), or as specified
- Renewal electrolyte (1.240 specific gravity), or as specified
- Adjustment electrolyte (1.300 specific gravity), or as specified
- Petroleum jelly
- Pure mineral oil, acid-free, nonsaponifying
- Asphalt-base paint, caustic- and corrosion-resistant
- Hydrometer (reading 1.150 to 1.300 specific gravity)
- Spirit thermometer (reading 0 to 160°F); special types with scale indicating gravity-correction factors are available
- Electrolyte-level test tube
- Filling squeeze bottle or bulb
- Equalizing bottle or bulb (see below)
- Special post nut and vent tools (as recommended by the manufacturer)

The principles to be observed in maintaining nickel-cadmium batteries are as follows.

Follow carefully the prescribed charge procedures as described previously.

Maintain proper electrolyte level, gravity, and purity. Having ensured that the electrolyte level of new cells as received is correct, set up a schedule for the regular checking of level, to be followed as long as the cells are in operation. Cells in standby applications, which may have to be recharged fairly frequently, should be checked once a month for the first 6 months; by this time the user will see the pattern of electrolyte loss and may find it possible to reduce the frequency of checking. In float- and trickle-charge applications, the electrolyte level should be checked every 3 to 6 months. In the infrequent cases where nickel-cadmium batteries may be used in cycle applications, the level should be checked every other cycle until the pattern of electrolyte loss becomes clear.

Water should be added to maintain solution levels in accordance with the manufacturer's instructions. Where these are not available, the following instructions will give satisfactory results: electrolyte must never be allowed to fall below the plate tops; if the tops of the plates are exposed to the air, serious damage will usually result. The maximum level in most cells is one-half to two-thirds of the distance between the tops of the plates and the underside of the cell cover. If this maximum is exceeded, there is danger that during a heavy-charge routine, electrolyte will overflow, causing leakage currents and loss of potassium hydroxide.

Many manufacturers of cells with transparent plastic cases put two marks on the side of the cell case, the lower one corresponding to the plate tops (the minimum level) and the upper mark indicating the maximum level. In small sintered-plate cells there may be only one mark, indicating maximum level; this mark will be slightly above the tops of the plates. In cells with opaque plastic or metal cases, the electrolyte level may be checked with a level test tube. These tubes may be obtained from the battery manufacturer, but any clean, uncontaminated, clear plastic tube of convenient length (8 to 12 in), having a bore of roughly $3/16$ in, will do. The tube is held vertically and placed into the cell until it comes to rest on the plate tops. The forefinger is then placed tightly over the end, and the tube is withdrawn, permitting one to view the height of the electrolyte above the plate tops. Of course, the electrolyte contained in the tube must be returned to the cell. Wash out the tube in water after each use.

Normal charging procedures do not cause any significant loss of potassium hydroxide. Only water is lost, through the formation of hydrogen and oxygen that is characteristic of any storage cell being charged. Water alone should be added to correct the level drop due to charging and evaporation; potassium hydroxide electrolyte is added to a cell only in the case of spillage. As a general rule, use only distilled or deionized water. In some parts of the country tapwater has the necessary purity, but this can be decided only by chemical analysis; some manufacturers will perform this service if requested.

If it seems that the frequency at which water must be added is excessive, or that spray is building up on the cell tops at an extraordinary rate, check the charging operation. It may be necessary to decrease the charge voltage, or to decrease current and use longer charge periods. If constant-current charging is being done, make sure also that the charge is being terminated at the proper point.

Water may conveniently be added with a squeeze bottle or bulb, sometimes furnished with the battery. When watering cells of larger batteries, establish a regular orderly pattern of working through the cells, and use this pattern consistently. This will decrease the likelihood of missing a cell. Most plastic-cased cells have removable, screw-type vent plugs. When watering these batteries, it is good practice to remove the plugs and soak them in warm water for several minutes to remove crystallized deposits from the vent passages. When replacing these plugs, screw them in with only moderate force; otherwise undue pressure will be exerted on the O ring or washer generally used to provide a seal between the plug and the cover.

If there are many cells to maintain, it may be convenient to provide a second "equalizing" bottle or bulb to withdraw excess electrolyte. This may be acquired from the manufacturer, or it may be prepared by the user. To prepare it, first determine the exact distance above the plate tops at which the maximum electrolyte level falls. Measuring this same distance from the end of the spout of a squeeze bottle or bulb, drill a small hole through the side of the spout. The end of this spout is then seated on the plate tops, and excess electrolyte is drawn into the bulb; the electrolyte level will fall until it reaches the proper height, at which point the hole in the spout will draw air. If there are a great number of cells and watering is fairly frequent, an automatic filler may be justified (check with manufacturer).

The concentration (specific gravity) of the electrolyte is important. Most pocket-plate cells as manufactured are filled with electrolyte of 1.190 to 1.210 specific gravity; the exact value will be specified by the manufacturer for the particular cell-design type. More concentrated electrolyte (1.280) may be used for cells intended for low-temperature operation; this concentration would damage cells operated at room temperature, however. Refill electrolyte also usually has a concentration of 1.190 to 1.210, and is used to replace electrolyte lost by spillage. Renewal electrolyte is generally about 1.240 specific gravity, and is used to replace electrolyte in cells in which plates have been covered by distilled water after shipping or installation accidents, or to replace electrolyte which has become excessively contaminated or diluted through use.

When ordering or preparing refill or renewal electrolyte, gage the quantity needed by the rule of thumb that in most pocket-type cells 1 qt of electrolyte will be needed for each 70 to 90 A-hr of rated capacity. Considerably less electrolyte is needed for sintered cells. For more accurate values check with the battery manufacturer.

When filling cells in which plates have been covered with distilled water following accidental spillage, it may happen that the renewal electrolyte will be diluted by the water in the plates, so that the resulting concentration is below the recommended value. In this case it will be necessary to adjust the gravity, using 1.300 specific gravity electrolyte. This adjustment must be done while the cell is being overcharged, so that the gassing will mix the electrolyte as readings are taken. When the battery has been charging at a steady voltage of 1.6 to 1.7 V per cell for 30 min, check the specific gravity.

Then estimate the total amount of electrolyte contained in the cell. For each quart of electrolyte, a difference of 20 gravity points (0.020) below the necessary value calls for the addition of roughly 60 mL (or 2 fl oz) of 1.300 electrolyte. Add this amount of 1.300 electrolyte, let the cell charge for another 30 min, and check the gravity again. Repeat this procedure until the gravity is correct. If gravity is too high, it may be corrected by withdrawing a portion of the electrolyte from the cell and replacing it with distilled water.

All gravity readings, taken in the course of any maintenance procedure, must be corrected for temperature. This is particularly important when adjusting electrolyte gravity, since the battery is on charge and is likely to be warm. For each 4°F that electrolyte temperature is about 72°F, add 0.001 to the observed gravity reading; for each 4°F below 72°F, subtract 0.001. Similarly, electrolyte must be at the proper level over the plates whenever gravity readings are taken. Always place the hydrometer all the way into the cell, so that its tip rests on the tops of the plates. This will prevent mineral oil from being drawn up into the hydrometer.

With pocket batteries operating on uninterrupted float routines, check the gravity once a year. When operating on any routine that involves recharging, gravity should be checked once every 6 months. The concentration of the electrolyte will decline slowly as small quantities of potassium hydroxide are thrown out along with the gases and spray is released during the charging. When the gravity has dropped to the minimum value specified by the manufacturer (usually in the neighborhood of 1.160), the electrolyte must be renewed. Continued operation beyond this point will result in a fairly rapid decrease in cell life.

The procedure for renewing electrolyte is as follows: first prepare or acquire the necessary amount of renewal electrolyte. Then discharge the battery at the 7-hr rate to a voltage of 0.5 to 0.8 V per cell. This will minimize the danger of shocks or damage through shorting. With cells maintained in wooden trays, disconnect the intercell connectors, incline the tray to one side, and remove the slats from the side. Take out and invert each cell individually, emptying out all electrolyte. Do not allow the cell to touch any conductive material, causing short circuiting. Batteries assembled in steel or plastic trays are simply inverted so that all cells are emptied simultaneously. The electrolyte is injurious to aluminum, copper, zinc, or tin. Do not rinse cells with water or electrolyte. Do not allow any cell to stand empty for more than 30 min, or the plates will be damaged through exposure to air. Fill each cell as it is emptied with renewal electrolyte to the maximum permissible level (halfway between plate tops and cell cover in most cell designs). Wash out the vent cap and replace it immediately.

Clean each cell, preferably by a blast of low-pressure steam followed by compressed-air drying. It is good practice at this time to repaint the cell cases with corrosion-resistant paint. Reassemble the cells into the trays, coat the covers with petroleum jelly, make sure the intercell connectors are tightened securely, and charge the battery at the 7-hr rate for 14 hr. The battery is now ready to be returned to service.

With sintered cells, there is usually not enough free electrolyte in the cell to obtain a gravity reading. Judgment as to when to renew electrolyte is therefore based on electrical performance. If the cell has been cycled considerably or overcharged, or has been used for a period of a few years and is beginning to decline in capacity in spite of good maintenance and proper charging, the electrolyte probably needs renewing. Dump the electrolyte, following the same general procedure as described above. Replace the electrolyte with 1.300 specific gravity solution. The cell will be discharged at this point. Therefore, fill it only to the tops of the plates, and then charge it. Following the charge, the electrolyte level may be brought up the rest of the way to the maximum mark.

Potassium hydroxide reacts with carbon dioxide in the air to form potassium carbonate, which will decrease capacity when its concentration in the electrolyte exceeds a few percent. Formation of carbonate can be minimized by several means: (1) Open cell vents no more frequently, and for no longer, than is absolutely necessary. (2) Make sure that vent components and the glands or washers around the terminals make a good seal against the cell cover. (3) Maintain a layer of oil on the surface of the electrolyte. (4) Minimize overcharging, particularly overcharging at high rates. This condition causes agitation of the electrolyte and formation of crusts of carbonate on the underside of the cell cover, which then fall back into the electrolyte. (5) Control electrical operation and scheduling of maintenance so that frequency of adjustment of electrolyte level is minimized. (6) Store the electrolyte stock in tightly sealed containers only.

Carbonate concentration can be determined by chemical analysis. This is a service that most manufacturers will provide. This service is also available from many commercial testing laboratories. The decision as to when to have an analysis performed can be based on the performance of the battery. It should not be necessary to have this analysis done more frequently than every 2 years. If the battery is not yet approaching end of life but exhibits marginal performance in spite of proper charging and correct electrolyte level and concentration, carbonate contamination should be suspected, and an analysis should be done. On the other hand, if, at the end of a 2-year period, performance is good, an analysis can be deferred. When carbonate concentration in the electrolyte reaches 10 percent by weight, the electrolyte should be renewed.

Electrolyte can be procured from most manufacturers either as dry crystals, to be mixed to the proper concentration by the user, or as solution mixed to a specified concentration. Using dry crystals can save considerable shipping cost and avoids having to order several different concentrations but does involve handling and mixing. Solutions should be mixed in a large glass, porcelain, or plastic vessel that is perfectly clean and free from contaminants. Electrolyte crystals should be ordered from the battery manufacturer; the container will usually include mixing instructions. As a general rule, preparing 1.240 specific gravity solution calls for about 2.56 lb of pure potassium hydroxide per gallon of water—2.33 lb per gal will produce 1.220 solution. Some users may prefer to mix just one solution strength, the strongest needed, for storing, and then dilute this to the other strengths required as they are needed. Starting with 1.300 solution and mixing 7 parts of this with 2 parts water will produce 1.240 specific gravity electrolyte; $1/2$ parts of 1.300 and 4 parts water will yield 1.220 specific gravity. Also, 10 parts of 1.240 solution and 1 part water can be mixed to yield 1.200 specific gravity.

When handling potassium hydroxide in any of these procedures, it should be remembered that it is a corrosive chemical, injurious to skin and eyes. The standard goggles, face mask, and rubber garments should be considered for use. If electrolyte is spilled or splashed on skin or clothes, wash immediately with liberal quantities of water. It is wise to have on hand a stock of boric acid solution to neutralize spilled electrolyte. Diluted pharmaceutical-grade boric acid can be used to rinse the eyes.

Guard against stray currents and shorts by the following means:

Under no circumstances allow metal cell cases to touch each other. Even though both terminals are insulated from the steel cases and covers by rubber glands, current will be conducted by the electrolyte from the plates to the cases, and thence via touching cell cases to plates of the opposite polarity, thus shorting out the battery.

Keep the cases and the covers clean. Films and paths of dirt, moisture, and electrolyte spray not only will conduct current between points of opposite polarity and self-discharge the battery, but also will lead to electrolytic corrosion of the steel cases and covers. Wipe off moisture and carbonate that build up on the cover; keep the cover coated with petroleum jelly. Prevent debris from building up between the cell cases, or under the cells so as to bridge to ground. It is good practice to go over the entire battery periodically with a low-pressure blast of steam, followed by an air blast to dry the cells thoroughly.

Never stack cells or trays on top of one another.

Do not overfill cells with electrolyte and risk overflow during charging.

Dress all cables up and away from cell tops. Never allow cables to lie on cell tops or on intercell connectors.

After installing or doing maintenance on the battery, make sure that no tools, screws, or other metal parts are left in the battery compartment.

Use only spirit thermometers. Mercury is an electrical conductor. If a mercury thermometer should break, allowing mercury to run down into the cell interior between the plates, serious shorting would be likely.

When taking battery voltage readings, check also for possible voltages between each of the battery terminals and ground. Such a voltage is an indication of a ground somewhere in the system.

Instruments or devices which would cause a constant drain of current must not be left connected across the battery permanently. As an example, if the user wants to have a voltmeter connected in readiness, it should be wired through a normally open push-button switch so that it is connected to the battery only when the switch is depressed.

Make sure connections are tight and making good electrical contact and that post and vent seals are maintained.

Good electrical contact at terminal connections will prevent wasteful voltage drop. This can be checked by putting the battery on a high-rate discharge for 15 to 20 min. Defective connections will have resistance and will feel warm to the touch. Take these connections apart and clean the contact areas of the terminal posts, connectors, and nuts with solvent or cleanser and fine emery cloth or steel wool.

If the seal around posts and vents does not remain tight, air and impurities may be admitted to the cell interior, and there will be excessive buildup of carbonate and electrolyte film on the covers. Leaks will be indicated by encrusted carbonate developing at the seal area. In these cases, tighten the lower terminal-post nut or the vent plug. If the rubber sealing glands on the terminal posts or the seal components on the vents have become brittle or deformed, they should be replaced. If special tools are necessary to turn terminal or vent components, they will be available from the manufacturer.

NICKEL-IRON BATTERIES

Nickel-iron alkaline storage batteries are traditionally used in cyclic service, although they have also been used successfully in standby and emergency-power applications.

The number of cells needed in a battery is determined by the voltage requirements of the equipment it is to operate in relation to the average operating voltage per cell. When discharging at their normal rates, all types and sizes of nickel-iron cells have an average discharge voltage of 1.2 V per cell. In most cases the number of cells required for a particular application can be calculated on the basis of 1.2 V per cell. For example, an electrical industrial truck having a 36-V motor should have a 30-cell battery.

The ampere-hour capacity of the cells required in the battery is determined by the rate of current consumed by the equipment and the length of time it is to be operated on a single charge of the battery. This time period, in the principal application in which the battery is the normal power supply for the equipment (cycle service), is usually the regular daily working period (in the majority of cases, one 8-hr shift). In standby applications, it is usually the maximum expected outage of the normal power supply.

After the required ampere-hour capacity has been determined, a cell with 20 percent additional capacity should be selected. This safety factor should be considered adequate for contingencies, and assures that the battery will have ample capacity up to the end of its normal service life.

In standby installations, it is important not only that the battery be of suitable voltage and capacity to carry the load satisfactorily during outages in the primary power supply but also that the power available for charging be ample to recharge the battery without undue delay following the intervals of discharge, and to maintain it in a satisfactory charged condition. How much power is required depends mainly upon how often, how long, and at what rates the battery is on discharge.

If the discharge is infrequent, short, and at low rates, power sufficient only for continuous trickle charging would be sufficient. Emergency power-supply systems for call-bell signals and other equipment having small and infrequent current amounts are examples.

On the other hand, when the discharge is frequent or prolonged, especially at relatively high rates, sufficient power may be needed to charge the battery at an average of its full normal rate if it is to be maintained at a satisfactorily high state of charge.

Operation. The required charging voltage varies according to the method of charging employed and ranges from approximately 1.50 to 1.55 V per cell for trickle charging to 1.84 V per cell or more for charging at normal to high rates. The number of ampere-hours required to charge the battery fully is equal to the number of ampere-hours previously discharged, plus an overcharge factor which averages approximately 25 percent. Charging at an average of the normal rate of the battery usually gives the best overall results and is generally recommended.

For rating purposes, a discharged battery is defined as one that has been discharged to the equivalent of 1 V per cell at the normal rate. This usually represents the lower limit of the range in voltage needed for fully satisfactory operation of the equipment for which the battery supplies power. It is not necessary, however, that the discharge be stopped at this or any other prescribed limit if further output at a low voltage can be utilized. This will not harm the battery. The temperature rise is the principal limitation on charge rates. Any rate is safe as long as it does not result in raising the electrolyte temperature above 115°F (46°C).

Boost charging or supplementary charging at high rates during brief periods of idleness is sometimes useful as an emergency measure only in order to obtain more than the usual amount of work from a battery that is regularly cycled. Regular or frequent boosting is an indication the battery is of inadequate capacity for the work and is not recommended. It is not a substitute for a correctly applied battery. The following information is useful as a guide to determine the amount of current that should be employed in a boosting operation:

- Five times the normal rate for 5 min
- Four times the normal rate for 15 min
- Three times the normal rate for 30 min
- Two times the normal rate for 60 min

When a battery is being boosted, it is useful to take temperature readings of the electrolyte in the cells nearest the center, or warmest part, of the battery and to stop the charge if the temperature rises to 115°F (46°C). Any frothing at the filler openings is also an indication that boosting has gone too far and should be discontinued immediately. A battery that has been discharged need not be immediately recharged. No injurious reactions will take place if charging is delayed.

Charging Batteries That Are Cycled.   Sources of dc power for charging batteries that are cycled may be:

- Direct-current power lines
- Motor generators which accept either dc or ac primary power
- Rectifiers which accept ac primary power

To ensure maximum cooling, be sure that the battery is exposed to free-air circulation while it is on charge. If it is charged in an enclosure of any kind, such as a battery box of an industrial truck or locomotive, open the cover.

Charging Standby-Power Batteries.   At normal temperatures, trickle charging voltage is usually between 1.50 and 1.55 V per cell, and 1.70 to 1.72 V per cell is usual for constant-potential charging at an average of approximately the normal rate.

But these values are not exact; they vary with the age of the battery, the specific gravity of the electrolyte, the temperature, and other conditions. With this in mind it is necessary, therefore, to adjust the voltage on the basis of ammeter readings, not voltmeter readings. Voltmeter readings are useful, however, in determining when a battery is fully charged. Stabilization of the voltage at the battery terminals for about $1/2$ hr while current is flowing through the battery at a constant rate is a trustworthy indication that the battery is fully charged.

For any given charge rate the voltage necessary at the battery terminals varies with the electrolyte temperature. Therefore, for batteries exposed to seasonal changes, a higher setting would be needed in the winter than in the summer.

It is important that the rates employed for trickle charging result in overcharging rather than undercharging. In practice it is virtually impossible to arrive at a charge rate which will result in precisely the amount of input required, especially since the output usually tends to vary from day to day. In the interest of consistently high dependability of operation, the best practice is to use rates ample for the maximum rather than the average or minimum requirements, especially since any overcharging that may result at low trickle rates is not harmful. If a case should arise in which a battery on trickle charge should undergo a prolonged discharge, set the voltage for a higher rate until voltage stabilization occurs, indicating that the battery is again fully charged.

For batteries furnishing large amounts of power each day and requiring correspondingly more input, the best voltage setting is one that results in the highest average rate during the charging interval consistent with rates that are not excessive after the batteries are fully charged, that is, rates that will not result in raising the electrolyte temperature above 115°F (46°C). As long as these rates are not excessive, it is desirable to adopt settings which on an average will tend to result in a slight amount of overcharging rather than undercharging.

Watering. During the operation of the battery, water is dissipated from the electrolyte chiefly as the result of gassing during charge. This loss must be made up by adding distilled or approved water, using as a guide the recommended and minimum levels suggested by the manufacturer. (Caution: Do not add electrolyte, as this will raise the specific gravity of the solution; if the specific gravity is allowed to exceed 1.230 in standard cells or 1.215 in high, wide cells, the battery may be damaged.)

The best time to add water in batteries that are cycled is just before charging; then the gassing during charge will mix the solution. Never add water during or immediately after charging. This avoids getting a false solution-level reading, caused by gassing during and immediately after charging, which makes it virtually impossible to add the correct amount of water.

Maintenance

Putting New Batteries into Service. Always unpack and inspect batteries immediately on arrival so that in case of damage, a claim may be filed promptly with the transportation company.

Test the height of the electrolyte in a few cells to see if any has been spilled. If the electrolyte is below the recommended level but is above the plate tops or can be seen with a flashlight, raise it to the recommended level with distilled water. If it is so low that it cannot be seen, raise to the recommended level by adding refill solution.

Batteries are shipped in a charged condition unless otherwise ordered; so they may be put into service immediately on arrival. In case a charged battery stands idle for a period from a week to a month, charge it at an average of its normal rate for 2 or 3 hr before putting it into service. If you expect to hold a battery idle for more than a month, order it shipped discharged and store it in that condition. Then when you are ready to put it into service, give it a 15-hr charge at its normal rate. A charged or partially charged battery left standing idle for more than a month is likely to become sluggish. Before placing such a battery in service, charge it 15 hr at normal rate; then discharge it at normal rate to an average of 1 V per cell. If it does not deliver normal rate for at least 5 hr before 1 V per cell is reached, it may need further cycling.

Batteries assembled in cradles or demountable boxes have their cell-to-cell connectors in place so that all that is necessary to complete the assembly is to apply the tray-to-tray jumpers. If a battery is assembled in trays only and consists of more than one tray, first arrange the trays so as to ensure correct polarity. Then apply the jumpers. The necessary jumpers and tools (pole-nut wrench and lug-disconnecting jack) are usually included with each shipment.

The lugs on the ends of the jumpers are provided with an inside taper that corresponds to the taper on the poles of the cells. Be sure both these contact surfaces are clean. Remove any oil, grease, or dirt that may stick to them, using a clean cloth. If an abrasive is necessary, use 00 sandpaper or 00 emery cloth; never use a file or other cutting tool that might score or abrade the contact surfaces. Then slip the jumpers into place. If the lugs do not fit exactly on the poles, bend the jumpers until they do; never hammer or force them on. After the lugs are in place, grease the pole threads slightly. Then apply the hexagonal pole nuts.

After completing the connections, you can check their tightness by putting the battery on charge or discharge at its normal rate for 15 or 20 min. Any loose or dirty connections will cause excessive heating of jumper lugs, which will be readily perceptible to the touch. (Caution: Disconnect battery from charging circuit before touching jumper or connector lugs.) Remove any such jumpers, clean the contact surfaces of the lugs and poles, and reapply. Check the tightness of the connector lugs in the same manner. Connectors are removed and applied in the same manner as jumpers. By having all connections clean and tight, you will avoid unnecessary voltage drop in the battery circuit.

Cleaning. Keeping a battery clean is not merely a matter of good housekeeping but is also an assurance of good performance and life. By keeping the cell tops and connectors clean, you lessen the risk of getting impurities into the cells when you open the filler caps to add water. By keeping dirt from accumulating below or between the cells, you reduce the possibility of ground, especially if the battery is exposed to dampness.

Batteries assembled in cradles or demountable boxes are best cleaned when supported so that dirt can be blown out through the bottom. Use a wet steam jet followed by an air blast to blow off any accumulated moisture. Clean the cell tops and connectors first; then blow out any dirt that may become lodged between cells. Be sure all filler caps are closed so that no dirt can get into the cells. Wear goggles when using the steam jet and air blast.

Batteries assembled in trays only can be cleaned by wiping cell tops, connectors, and jumpers occasionally with a wet cloth. In this way you can avoid letting dirt fall down into the spaces between the cells, but if you see dirt beginning to accumulate there, remove the trays to a floor drain or other suitable place and clean them by wet steam or warm water followed by an air blast as already described. Be sure cells and trays are dry before reassembling, also that the contact surfaces of the cell terminals and jumper lugs are clean and that all connectors are tight and of correct polarity.

Inspect the cells for any necessary attention. Make sure the filler caps, hinge bands, and lid springs are in proper alignment to ensure free operation and correct seating of the valves. To prevent contamination of the electrolyte, it is just as important to maintain the valves so that they seat properly as it is to keep the filler caps normally closed. Screw down the gland caps of any cells showing evidence of leakage around the stuffing-box assembly. Use the special wrench available for the purpose, and be careful not to damage the gland caps.

Cycling. A battery that is not kept in regular use or is used only intermittently may become sluggish and deliver less than the capacity of which it is capable. This can be corrected by cycling the battery as follows:

1. Charge the battery if it is not already charged.
2. Discharge through a resistance that can be varied to keep the rate at normal until the potential of the battery falls to the equivalent of 0.5 V per cell (15 V for a 30-cell battery, etc.).
3. Short-circuit each tray, and let stand until the resulting heat is dissipated and the electrolyte cools to not more than 5°F above room temperature.
4. Water as necessary to bring the electrolyte to the recommended level, and charge at normal rate for 15 hr.
5. Discharge at normal rate, and keep a record of the time until the potential of the battery falls to the equivalent of 1 V per cell.

Except while the battery is short-circuited, keep the electrolyte temperature below 115°F (46°C). Take the voltage readings only while current at normal rate is flowing. Usually one such cycle is sufficient, although if the battery still appears sluggish, another cycle or two may bring further improvement.

A discharge at normal rate for 5 hr before the equivalent of 1.00 V per cell is reached indicates full rated capacity. If less capacity is indicated, continue the discharge as in step 1 and repeat steps 2, 3, and 4.

Laying Up. In case a battery is to be laid up for a month or more, discharge and short-circuit as described in operations 1 and 2 under Cycling. Check height of electrolyte solution, and add water if necessary to raise to correct level. Then store in a clean, dry place. Batteries may be left standing idle in this condition indefinitely without injury. When the battery is to be returned to use, charge it at normal rate for 15 hr. If it was laid up for a year or more, follow this charge by a discharge at normal rate to an average rate of 1 V per cell; then follow with operations 1, 2, 3, and 4 under Cycling. Also inspect the cells for any necessary attention as described under "Cleaning."

Renewing Electrolyte. When a battery is new and fully charged, the electrolyte has a specific gravity of approximately 1.200 at 60°F (15.5°C) if thoroughly mixed and at the recommended level. During use of the battery, the electrolyte tends to gradually weaken and must be renewed if its specific gravity falls to between 1.160 and 1.170. Do not operate a battery with an electrolyte of a gravity below 1.160. (Caution: Do not attempt to raise the specific gravity of weakened electrolyte by adding solution.)

To test the electrolyte for specific gravity, use a hydrometer. Take readings only when the electrolyte has been thoroughly mixed by charging, and wait $1/2$ hr or more after the charge has been completed to allow for dissipation of gas. Using a thermometer and a test tube, check the temperature and the height, and correct for any variation from 60°F (15.5°C) and the recommended level.

To renew the solution, proceed as follows:

1. Discharge, short-circuit, and cool the battery as described in operations 2 and 3 under "Cycling."
2. Pour out the old solution.
3. Fill immediately with standard renewal solution.
4. Charge at normal rate for 15 hr.

For ease in pouring out the old solution, disconnect the jumpers so you can do it one tray at a time. Avoid splashing. Do not shake or rinse; just tip the trays so that the old solution will run out. The electrolyte is injurious to wood, brass, copper, lead, aluminum, and zinc. Short lengths of scrap 2 by 4s or similar timbers can be used to support the trays while they are tipped over.

Always keep in mind that the solution is injurious to the skin and clothing. Wear rubber gloves, goggles, and preferably also a rubber apron. If, in spite of these precautions, any solution should be splashed or spilled on the skin or clothing, wash it away immediately with plenty of water. As a further precaution, it may be well to keep available a supply of 4 percent sterile boric acid solution and an eyecup for additional treatment of the skin and eyes. Meanwhile, arrange the containers of standard renewal solution so that you can refill immediately. Do not let the cells stand without solution. The containers may be elevated, and the solution poured into the cells through a hose, or for small cells and small containers, the solution may be poured in directly from the container through a funnel.

Replacing Spilled Electrolyte. An accident which overturns a battery rarely causes damage because of the steel-cell construction but may spill electrolyte solution from the cells. To replace spillage use standard refill solution. Standard renewal solution may also be used in an emergency if diluted with pure distilled water to a gravity of 1.215 at 60°F (15.5°C); an easy way to do this is to mix 1 part of water by volume with 5 parts of renewal solution by volume.

If you have no electrolyte solution on hand, the best thing to do depends on how much solution was spilled. If the solution left in the cells is still above the plate tops or can be seen by a flashlight after the battery has been turned right side up, merely add water and continue the battery in service. If so little solution remains in the cells that it cannot be seen with a flashlight, take the battery out of service, make sure all filler caps are closed in order to keep out impurities, and wait until you can obtain a supply of refill solution.

SECTION · 7

INSTRUMENTS AND RELIABILITY TOOLS

CHAPTER 1
MECHANICAL INSTRUMENTS FOR MEASURING PROCESS VARIABLES

R. Keith Mobley
Principal, Life Cycle Engineering, Inc., Charleston, S.C.

Mechanical instruments provide an accurate and economical means of measuring the status of nearly all process systems. Inclusion of mechanical instrumentation that measures all critical parameters of process systems is essential for all predictive and plant-improvement programs. The information derived from process instrumentation, used in conjunction with vibration, infrared, and other predictive maintenance data, provides the basis for maintenance planning and maintenance prevention.

Typically, this type of instrumentation can be divided into two classifications: (1) monitoring and (2) control.

PROCESS MONITORING

Mechanical instruments can be used to measure, record, and trend variations in process parameters, such as pressure, flow, or temperature. In most cases, the instrument will convert mechanical displacement or movement of its sensor into a corresponding proportional movement of a pen or gauge. These instruments generally transfer the actual value of the measured parameter, by means of a linkage mechanism, to a graph or display that is a part of the unit.

PROCESS CONTROL

The operating principles of the instrument are identical to those used for process monitoring. In this instance, the instrument will translate the actual mechanical displacement into a proportional output, either pneumatic or electrical, that is used by the system's control logic to adjust the process parameters.

Most process variables can be measured with mechanical instruments and include flow, pressure, temperature, level, and positions. The limitations are few, for example, pressures in the micron range, temperatures above 1000°F, and analysis of fluids.

New designs and materials for actuating elements provide acceptable accuracy and sensitivity that are suitable for most applications. These factors, in addition to ruggedness, low cost, and ease of maintenance, have resulted in an increasing use of mechanical instruments for both monitoring and control of critical process systems.

TYPES OF INSTRUMENTS

FIGURE 1.1 Single-pen recording thermometer.

Industrial instruments are classified by case size, type (recorders or indicators), and form (chart or scale). In addition to the record or indication of the process variable, instruments can provide automatic control, integration of flow, and adjustable alarm signals for abnormal operating conditions.

A number of case designs and sizes are available. Designs generally have a rectangular or square front. They can be mounted flush in a panel or on a wall. Mounting accessories are supplied with cases. Rectangular cases are approximately 14-in wide by 18-in high and 4- to 6-in deep. A commonly used smaller size is 14×12 in. Figure 1.1 illustrates a single-pen recorder. In addition, 12×12 in square cases are also available. Since the smaller instruments perform the same functions as those in a rectangular case, the depth also runs from 4 to 6 in. Instruments with smaller case sizes, classified as miniature instruments, are used extensively. They are available in a variety of sizes with width and height of 3×6, 4×6, and 6×6 in. Because of their small cross-sectional area, the depth generally is about 20 in.

Instruments of the sizes listed above are all available as recorders or as indicators. When the pneumatic control function is included, the case assembly includes two 2-in pressure gages, which continuously indicate the air supply to the control unit and the output pressure to the final control element.

Charts are either round or strip in form. Round charts are available in diameters of 3, 4, 6, 8, 10, and 12 in. The smaller sizes are used when accuracy of recording is of secondary importance. The larger sizes provide greater accuracy in reading the value of the process variable and adjusting the set point for automatic control. Most mechanical instruments using strip charts are of the miniature type. Strip charts come in widths of 3 and 4 in.

Indicators are available with concentric and eccentric scales. Concentric scales normally are calibrated for a pointer movement of 270°. Indicators are also available with eccentric scales in the upper or lower portion of the case. A large number of concentric-scale indicators provide only one function—indicating the value of the process variable. Conversely, practically all eccentric-scale indicators provide two functions—indication of the process variable and automatic control. Small-case concentric-scale indicators are also available for automatic control.

Indicating instruments are used when it is not necessary to keep a record of the trend of the process variable for operating use or for future study. Typical applications are to provide operating information of such general-purpose utilities as plant steam, water, and air. Single-point concentric-scale indicators are used for this application.

Where process upsets are not likely to occur, or no record of process trends is required, indicators provide an economical means for automatic control. Not only is the initial cost less, but also there is no continuing expense for charts.

In some applications of indicating controllers, process variables may be recorded separately. This is usually done on three-pen rectangular-, round-, or miniature-case strip-chart recorders. Variables may be recorded continuously or on a selective basis, using plug-in connections when a process upset occurs.

Recording Instruments

A relatively small number of recorders are used to record only the value of the process variable. A large number of these record flow, with the total amount indicated by a totalizing mechanism. The majority of recording instruments are used for automatic-control applications. For these applications,

10- or 12-in charts are the most desirable because of greater accuracy in reading and positioning the set point and a better opportunity of observing process trends.

Recorders are available with one, two, three, or four pens. Inks of different colors can be used. By using an eccentric scale which follows the time line of the chart, and a pointer on the pen arm, it is possible to have a combined recording-indicating instrument.

Printed charts have been in use for so long that they are now available with literally thousands of calibrations, including those for a single variable and others having multiple segments for two or three variables.

The standard chart speed is the 24-hr revolution, but many other speeds and the required drive mechanism are available. The drive mechanism may be electric, hand-wound spring drive, or pneumatically operated.

Miniature Mechanical Instruments

A separate classification of these instruments is warranted by their frequent use. Only a few are used to measure the value of the process variable by a direct connection. The vast majority, known as *receivers,* are actuated by a 3- to 15-psig pneumatic signal from transmitters which measure the value of the process variable. These instruments are available as recorders and indicators. Recorders are supplied with 3- and 4-in wide strip charts, which move either vertically downward or horizontally. Indicators have either concentric or narrow vertical scales. In addition to recording and indication, automatic control of the pneumatic type is also available in all the forms supplied in the larger instruments.

MECHANICAL FLOWMETERS

Mechanical flowmeters are classified as rate meters because they measure and record or indicate the rate of flow. When measuring liquids or gases, the fluids are in direct contact with the actuating elements. When steam is measured, it condenses to water in the connecting piping at atmospheric temperatures and is in contact with the actuating element.

The rate of flow is determined by measuring the drop in pressure of the fluid flowing through a restriction in the flow line. This restriction is known as the *primary element.* The flowmeter is known as the *secondary element,* and the two complete the system.

Several designs of secondary actuating elements are available. An early design is a mercury manometer, with float resting on the surface of the mercury in one leg of the manometer. The float position changes as the flow varies, and by a linkage mechanism, the pen or pointer is moved to correspond to the new rate of flow. For several reasons, such as lower cost and easier installation and maintenance, the use of mercuryless meters is increasing rapidly. These are two general types. One has a differential bellows for measuring the pressure drop, and its motion mechanically positions the pen or pointer. The other type measures the pressure drop with a capsular assembly and transmits a standard 3- to 15-psig pneumatic signal.

Most manufacturers publish an accuracy of 60.5 percent full scale for bellows meters. When an installation requires better accuracy, it can be obtained by recalibration. The general published accuracy for pneumatic transmitters is 60.5 percent full scale. Both types can be operated up to temperatures of 250°F of the fluid in the meter body.

Design of Bellows-Actuated Flowmeters

The general design is shown in Fig. 1.2. It consists of high-pressure and low-pressure chambers that are separated by a liquid-filled bellows assembly. The left-hand assembly includes a calibrated range spring whose primary function, in conjunction with bellows movement, is to measure the pressure drop across the primary element. A connecting rod is connected to the left-hand, or high-pressure, bellows and extends over into the right-hand, or low-pressure, bellows, at which point there is an internal coil spring. This connecting rod has the upper end of the triangular-shaped torque arm attached to it by a cable assembly which is always in tension. The lower end of the torque arm is connected to a torque tube.

7.6 INSTRUMENTS AND RELIABILITY TOOLS

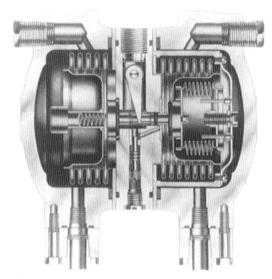

FIGURE 1.2 Bellows-actuated flowmeter.

Any horizontal movement of the connecting rod is therefore converted to a rotary motion of the torque tube. This motion is brought from the meter body into the instrument case as a shaft rotation and then used to actuate the pen linkage. Its angular rotation is approximately 8° between no flow and maximum flow.

Most bellows flowmeters are of the same general design as that shown. The bellows are hydraulically formed; those of another make are of a welded construction. The internal parts of all are available, including the bellows, made of type 316 stainless steel to provide maximum resistance to corrosion.

Various types of liquid fill are used by different manufacturers for covering operating temperatures from 240 to 250°F. The manufacturer should be consulted regarding the special fills available for oxygen measurement and for lower and higher temperatures.

A wide range of springs is available for measuring the pressure drop across the primary element for flow measurement. In addition, ranges in terms of pounds per square inch are available for measuring pressure drop across process vessels. Meters are available for operating pressures to a maximum of 6000 psig.

Operation of Bellows-Actuated Flowmeters

1. Under no flow, there is no pressure drop across the primary element, and the high and low pressures are equal and applied to each of the bellows.
2. When flow starts, the high pressure remains fixed and the low pressure decreases to a value determined by the rate of flow.
3. When this difference in pressure occurs, the forces on each bellows are no longer equal because of the reduction of pressure applied to the low-pressure bellows.
4. The force of the pressure on the high-pressure bellows compresses it and forces liquid to move past the pulsation damper into the low-pressure bellows.
5. When the high-pressure bellows is compressed, the range spring, of which one end is anchored and the opposite end is free to move and is fastened to the bellows, is subjected to increasing tension as it is elongated. This elongation continues until the difference in forces across the bellows, created by the pressure drop across the primary element, is balanced by an equivalent force in the spring.
6. When the bellows and the bellows rod move to the right under the above conditions, the torque arm is rotated in a clockwise direction. Since the bellows assembly is located on the back in both

the mechanical flowmeter and the pneumatic transmitter, the torque-tube rotation will be counterclockwise for increase in flow when viewed from the front of these instruments.

7. The rotation of the torque tube, by means of a suitable linkage mechanism, positions a pen or pointer of a mechanical flowmeter or actuates a pneumatic or electrical transmitter for remote transmission of measurement.

Bellows flowmeters of today's design incorporate two very practical features. They can withstand overrange differential pressures, even up to full-rated line pressure, without damage. Integral compensation is incorporated for ambient temperature variations. The maximum zero shift is 0.5 percent of full scale for a temperature change of 650°F from 77°F.

MEASUREMENT OF TOTAL FLOW

In addition to recording or indicating the rate, it is often necessary to measure the total flow over a period of time. An example is an accounting of steam used in various departments of a plant.

The total flow is determined by means of an integrator. It observes the pen or pointer position or measures a pneumatic signal and changes numbers on a counter in direct proportion to the rate of flow in relation to elapsed time. The operating principle may be either mechanical, pneumatic, or electronic. Integrators that observe the pen or pointer position are within the instrument case. Those operating on a pneumatic signal are separate from the instrument.

Figure 1.3 illustrates an electronic integrator in the recorder case of a bellows flowmeter. This integrator consists of three major assemblies: (1) a scanning unit, located at the left, checks the flow rate, as shown by the pen position, once every 5 s without interfering with pen motion, (2) an electronic detector relay, located at the lower right, is actuated by the scanning unit, which operates the line-voltage motor-driven counting mechanisms, and (3) the motor-driven six-digit counter mechanism, located at the lower left-hand corner of the case, totalizes the successive output signals from the detector relay, thus integrating the flow measurements.

FIGURE 1.3 Electronic integrator assembly in a bellows flowmeter.

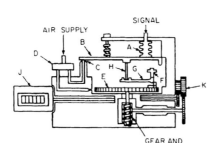

FIGURE 1.4 Pneumatic integrator.

Figure 1.4 shows an integrator which receives a signal from a pneumatic-flow transmitter. It operates independently of a recorder, indicator, or controller and can be installed at any desired location.

Principle of Operation

1. A 3- to 15-psig pneumatic signal from the flow transmitter is applied to the integrator receiver bellows A.
2. The force exerted by the bellows positions a force bar B in relation to nozzle C.
3. With an increase in flow, the force bar approaches the nozzle, and the resulting back pressure at the relay D regulates the flow of air to drive the turbine rotor E.
4. As the rotor revolves, the weight F, which is mounted on a flexure-pivoted bell crank G on top of the rotor, develops a centrifugal force. This force feeds back through the thrust pin H to balance the force exerted on the force bar by the bellows.
5. The turbine rotor is geared directly to the counter J through gearing K. Changes in flow continuously produce changes in turbine speed to maintain a continuous balance of forces.
6. The centrifugal force is proportional to the square of the turbine speed. This force balances the signal pressure, which is proportional to the square of the flow. Therefore, turbine speed is directly proportional to flow; and integrator count, which is a totalization of the number of revolutions of the turbine rotor, is directly proportional to the total flow.

LIQUID-LEVEL MEASUREMENT

Level is measured under a variety of conditions. The liquid may be under atmospheric pressure, as in an open tank or reservoir, or in a process vessel under pressure.

One of the most flexible and convenient ways to measure liquid level is the static-pressure method. It is based on the fact that the static pressure exerted by a liquid is directly proportional to the height of the liquid above the point of measurement, regardless of the volume. A pressure gage, then, can be calibrated in terms of height of a given liquid and used to measure level under atmospheric pressure. When liquids are under pressure in a closed vessel, a differential-pressure-measuring gage must be used.

Measurement under Atmospheric Pressure

Figure 1.5 shows the use of a pressure gage to measure level in a tank under atmospheric pressure. The pressure tap is located at the approximate minimum level. If the liquid contains entrained solids, this location decreases the possibility of plugging the connecting lines. If the liquid is corrosive, a seal should be used, as shown, and the sealing liquid should have a higher specific gravity than that of the liquid being measured.

If the gage must be located below the tap, the difference in elevation, which is an additional head on the gage, can be compensated for by a zero adjustment to a maximum value of about 10 to 20 percent of the instrument range. Gages are also available for head compensation several times the value of the range being measured. For example, the head compensated for may be 90 ft, and the level measured only 20 ft.

For some applications, such as measuring the level in a reservoir, a connection cannot be made at the minimum level. On these applications, level can be measured by means of a diaphragm box

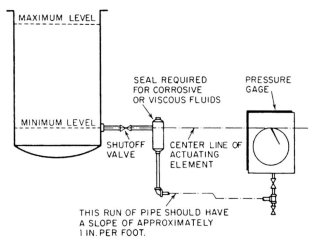

FIGURE 1.5 Liquid-level measurement in an open tank.

shown by the installation on the left in Fig. 1.6. As shown on the right, it also can be connected at the minimum level in order to isolate the gage from the liquid whose level is being measured.

Within the box is a flexible diaphragm made of rubber, neoprene, or other corrosion-resistant material. The box is connected to the gage by small-bore tubing. The volume in the box is large compared with the volume in the tubing and actuating element.

In operation, the pressure of the liquid head is exerted against the underside of the diaphragm, resulting in an upward movement of the diaphragm until the pressure within the closed system is equal to the head of liquid. The gage measures the air pressure but is calibrated in terms of liquid level.

The submerged box shown can be used only with clear liquids. If liquids contain suspended matter, it may collect around or on the diaphragm and cause incorrect readings. The external box should be used for these applications, and when considerable suspended matter is present, a periodic or continuous water backpurge may be necessary. Neither type should be located more than 90 ft from the gage.

Figure 1.7 illustrates an air-purge system. With this system, the location of the gage in relation to the point of measurement is not limited, and the level of corrosive liquids or liquids with large amounts of suspended matter can be measured. A probe is immersed in the liquid to the minimum level, with the pressure and volume controlled by a differential regulator to ensure a slow bubbling of air through the liquid. The air pressure is then equal to the backpressure exerted by the head of liquid. The gage measures the air pressure but is calibrated in terms of liquid level.

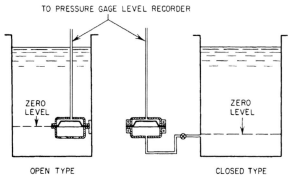

FIGURE 1.6 Liquid-level measurement with diaphragm boxes.

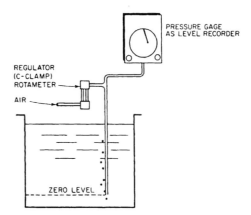

FIGURE 1.7 Bubbler-type liquid-level measurement.

This method makes it possible to measure level where the probe cannot be inserted in the tank because of agitator blades or other reasons. A connection is made in the side of the tank at the minimum level, as shown for the diaphragm-box connection in Fig. 1.6.

The gage can be located up to 100 ft or more from the point of measurement. The longer the distance, however, the slower the response to measuring changes in level. The air purge should always be at or near the point of measurement and not at the gage. Location at the gage causes measurement error due to the friction loss of the airflow through the connecting tubing.

The use of a differential regulator plus rotameter indicator is recommended. The differential regulator results in the same rate of airflow regardless of the level in the tank. If it is not used, larger rates of flow at low level can cause errors in measurement due to friction losses. Its use also decreases the amount of air required. The rotameter provides an indication of whether or not air is flowing and is convenient in adjusting the rate of flow.

Measurement in Closed Tanks under Pressure

Liquid level also can be measured in closed tanks under pressure using the static-pressure method. The static pressure above the level is added to the liquid head and must be compensated for. If the pressure changes, the measurement is in error, and differential-pressure measurement is therefore required.

Bellows-type differential flowmeters can be used for measuring the level in open tanks for applications shown by Figs. 1.5 to 1.7.

Figure 1.8 shows the installation of a bellows meter. The meter body is mounted on the rear of the instrument case. This arrangement provides indication or recording of the level, with or without pneumatic control. In addition, pneumatic transmission can be supplied with the above functions.

The constant-reference head $H1$ is applied to the high-pressure side of the meter body and the variable-level head $H2$ to the low-pressure side. The meter body simply measures the differential between $H1$ and $H2$. The meter is calibrated to read in terms of level, which can be indicated, recorded, or transmitted.

The piping is arranged so that the pressure above the liquid is applied to both the high- and low-pressure sides of the meter body and all effects are canceled. Therefore, regardless of variations in pressure, only changes in level are measured.

For accurate measurement, the constant-reference head must remain fixed. Some bellows meters are designed with an internal volumetric change when measuring changes in differential. For example, for a differential from minimum to maximum, the volume in the high-pressure chamber will increase and that in the low-pressure chamber will decrease by the same amount.

With the installation shown by Fig. 1.8, a decrease in level will increase the volume in the high-pressure side. This will cause a decrease in the constant-reference head $H1$. The volume in the

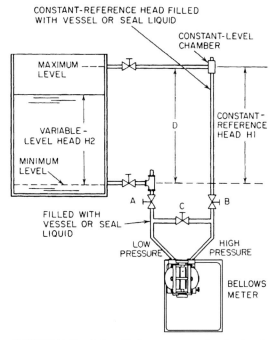

FIGURE 1.8 Installation of bellows meter on a closed tank under pressure.

low-pressure side will increase, but this volume is insignificant with respect to the volume in the tank and will not have a measurable effect on the variable head $H2$.

In order to eliminate the error, a constant-level chamber is installed at the top of the constant-reference head. It should have sufficient area so that the head change will be small for full-range measurement. For example, in a chamber with 3-in diameter and 100 in of water differential, the error will be about $\frac{1}{4}$ percent full scale.

Because of the low volumetric displacement of diaphragm-type transmitters, constant-level chambers generally would not be required for a constant-reference head made of 1-in pipe. As shown by Fig. 1.9, the constant-level chamber is replaced by a tee. This reduces the material and installation costs.

Figure 1.9 shows the installation of a diaphragm-type transmitter on a closed tank under pressure. The transmitter case, shown by dotted lines, has been displaced so that the meter body is not obscured; it actually is in a vertical position.

Note that the piping for both meters is the same except for the constant-level chamber shown by Fig. 1.8. The valves and tees shown are required for filling, placing in service, and checking at zero.

Sealing Liquids

Measuring the level of volatile liquids such as butane, propane, and other hydrocarbons is complicated by the fact that these liquids may be unstable in the constant-reference leg with changes of ambient temperature. Boiling of the liquid may occur in the outer leg, and it will be only partially filled with liquid. When this occurs, the measurement will be in error.

The error can be eliminated by the use of a stable liquid, such as water, in the outer leg. If there is a difference, however, in the specific gravity in the vessel and the leg, it must be compensated for. Almost without exception the stable liquid will have a greater specific gravity than that in the tank. As a result, with the differential designed for the tank liquid for a distance D in inches of water, head $H2$ will not balance $H1$, and with a maximum level the pen or pointer will not reach full scale or the maximum transmission signal will not be generated.

7.12 INSTRUMENTS AND RELIABILITY TOOLS

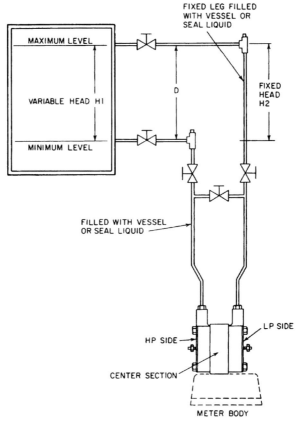

FIGURE 1.9 Installation of a diaphragm-type transmitter on a closed tank under pressure.

It is necessary to compensate for the difference in specific gravities by what is known as *suppression*. The means for doing this either are inherent in the design of level meters or transmitters or can be added in the field.

Cryogenic-Liquid-Level Measurement

Of increasing importance is the measurement of the levels of cryogenic liquids such as oxygen, nitrogen, and helium. A typical installation for a transmitter is shown by Fig. 1.10. On most applications, the connections on the vessel are at the top and bottom, as shown by Fig. 1.9. Fortunately, under most ambient-temperature conditions, the connecting piping is filled with gas. It is important, however, that the lower connection be of sufficient length that PC, the phase change between liquid and gas, be in this run and not in a vertical connection, so that as the position of PC varies with changes of ambient temperature, it will not affect the accuracy of level measurement.

The elevation $H1$ in the vertical line at valve D must be compensated for because this head is also to be measured in addition to head H. Compensation for this elevation is integral in the transmitter design. While Fig. 1.9 shows the transmitter below the vessel, it also will measure the level if located above. The bottom connection also should be such that PC occurs in a horizontal run.

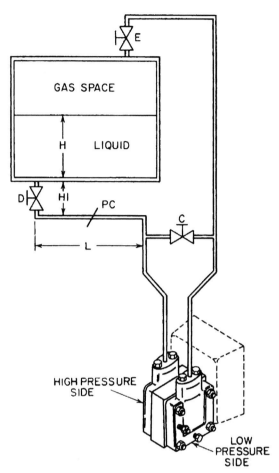

FIGURE 1.10 Cryogenic-liquid-level measurement.

Purge-Type Liquid-Level Measurement

It is often necessary to prevent the liquid whose level is being measured from entering the connecting piping or meter body. Otherwise the liquid might solidify in the piping at ambient temperatures, suspended matter might plug the vessel connections, or corrosive fluids might enter the meter body. Under these conditions, the system can be purged with water, gas, or a light oil at a higher pressure than that of the vessel. The purge connections, as shown by Fig. 1.9, are made at the meter body. Purged systems operate satisfactorily provided the purge is not interrupted.

Figure 1.11 shows a self-purged liquid-level system. It differs from the system shown in Fig. 1.10 for cryogenic liquids in that the connecting piping is heated to higher temperatures than the ambient temperatures to ensure that it is filled with gas. The distance L must be sufficient to permit the location of PC in it and not in the vertical connection.

This system has advantages over those using an external purge. The adjustment at zero is more easily made because it is not necessary first to adjust the purge flow to ensure equal friction loss in each connecting pipe. Another advantage is that if heating is interrupted, piping will again become gas-filled when heating is resumed.

7.14 INSTRUMENTS AND RELIABILITY TOOLS

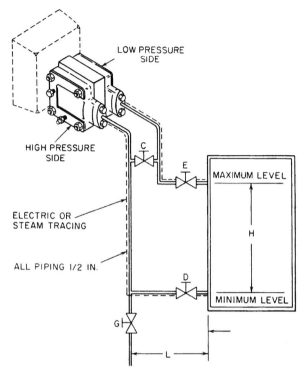

FIGURE 1.11 Self-purged liquid-level system.

MECHANICAL-PRESSURE INSTRUMENTS

Measurement of pressure is carried on over an extremely wide range. The ranges generally encountered in industrial processes go from about 0 to 6 mmHg absolute pressure to 0 to 100,000 psig.

A mechanical-pressure instrument is generally classified by the type of pressure it measures: gage pressures (those above atmospheric pressure), compound range (pressures above and below atmospheric pressure), or absolute pressures (those above absolute zero pressure).

TYPES OF INSTRUMENTS

Figure 1.12 shows a concentric-scale indicator that is the most widely used of all pressure gages. The one shown normally is installed on the process unit and directly supported by the pressure connection. It is also available in cases for flush mounting on a panel or with a back flange for wall mounting. Standard dial sizes are $4^1/_2$, 6, 8, and 12 in with a 270° scale. Ranges are from 0 to 15 to 0 to 100,000 psig. The most commonly used actuating element is a Bourdon tube made of brass, alloy, stainless steel, or Monel. Diaphragm seals are available when process variables must not be allowed to enter the Bourdon tube. These gages are suitable for most applications in the process industries, including pressure measurement of air, steam, water, and hydrocarbons.

Figure 1.13 shows an eccentric-scale indicating pressure gage with pneumatic control. This type is available for direct measurement of nearly every pressure used in processing and also as a pneumatic receiver for a 3- to 15-psig pneumatic signal from a transmitter.

FIGURE 1.12 Concentric scale indicator.

FIGURE 1.13 Eccentric-scale indicating pressure gage.

ACTUATING ELEMENTS FOR PRESSURE GAGES

The actuating elements of pressure gages perform two main functions: measuring the pressure of the process variable and providing sufficient force to position a pen or pointer or to initiate the generation of a signal by a pneumatic transmitter.

Inherent in the design of the actuating elements must be the ability to withstand the pressures being measured. In addition, they must provide the necessary accuracy, sensitivity, and repeatability required for satisfactory recording, indication, and control. To meet these requirements, a variety of actuating elements are available.

Industrial pressure gages have a standard accuracy of 61 percent of full-scale rated pressure. By more careful calibration, accuracy can be increased to 60.5 percent. Special test gages, at a higher cost, provide greater accuracies. The sensitivity and repeatability of commercial gages is 60.25 percent or better.

Bourdon Tubes

The Bourdon tube (Fig. 1.14) is used for measuring the widest range of pressures of any, actuating from 15 psig to as high as 100,000 psig. Its greatest use is in concentric-scale indicators. It is seldom used to actuate recording gages. When correctly designed and made, it meets all the requirements for accuracy, sensitivity, and repeatability.

For high pressures, Bourdon tubes are made of steel, stainless steel, and other materials to withstand the severe service of fluctuating pressures. They are designed to withstand an overload of twice their rated pressures without permanent damage. When subjected to overload pressures, only a zero adjustment is required.

FIGURE 1.14 Bourdon tube.

For lower pressures, Bourdon tubes are generally made of phosphor bronze or brass. In order to provide the operating forces, the width may be at least twice that of tubes for higher pressures.

When pressure is applied to a Bourdon tube, it tends to straighten out. This movement, by means of a link connection, plus a pinion and gear assembly, rotates the pointer of the indicating gage. The usual pointer movement is about 270°.

Helix

This unit is another form of Bourdon tube but is several times as long and wound in a compact helical form. It is made in this form in order to occupy a minimum amount of space in the instrument case, thus providing room for several actuating elements, controllers, chart-drive mechanism, and other devices.

The element is formed by flattening a round tube to an elliptical shape, heat-treating to provide the spring characteristics, and then winding in a helical form. It is available in phosphor bronze, brass, steel, stainless steel, and beryllium copper for a wide variety of applications.

Helix tubes are available for ranges from as low as 0 to 30 psi up to 30,000 psi. When pressure is applied, the helix unwinds, and this movement actuates a pen or pointer or initiates the generation of a signal by a pneumatic transmitter.

Spirals

These actuating elements are similar to helices, except that they are flat, as shown by Fig. 1.15. The principal design feature of spirals is that they can be installed in an instrument case of considerably less depth than required for helices.

FIGURE 1.15 Steel spiral.

Spirals permit the use of suppressed-range, narrow-span measurements where lower pressures are of no value and it is desired to measure higher pressures with greater accuracy, for example, spans of 50 to 100 or 400 to 600 psi.

Spiral actuating elements are available in the same materials as helices. The maximum pressures that can be measured are less, with a limit of about 5000 psig. The range of phosphor bronze or brass spirals generally is limited from 0 to 30 to about 500 psig.

Spring and Bellows

For pressure ranges from about 0 to 5 to 0 to 40 psig, an actuating element of an entirely different design is used. In order to obtain the required actuating forces, the measuring element must have a larger area against which the lower pressures are applied.

Several designs, providing larger effective areas, are illustrated by Figs. 1.16 and 1.17. The design shown in Fig. 1.16 consists of an enclosure around bellows. The source of pressure being measured is connected to the enclosure. Within the bellows, and touching its bottom, is a calibrated steel spring. The spring is supported at its top, and the calibration is such that for any desired pressure range the bottom of the bellows moves the same distance. A commonly used distance is 0.375 in. By means of a linkage connection to the bottom of the bellows the desired actuation is obtained. This design is suitable for measuring pressures from about 0 to 100 in H_2O to 0 to 40 psig.

Figure 1.17 is a similar design except that the pressure-sensing chamber of the unit is the inside of the bellows. This design also makes use of a calibrated spring.

These two designs have several important features. One is that the calibrated spring is never in contact with the fluid whose pressure is being measured. It therefore can be made from spring steel to obtain the best calibration, since it will not be affected by corrosive fluids. The second feature is that the bellows and other portions of the enclosure can be made of different materials as required by the application. For example, to measure the pressure of a noncorrosive fluid a lower-priced gage can be supplied by making the enclosure of brass and the bellows of phosphor bronze. For corrosive applications, the enclosure can be of stainless steel at a higher price.

The use of a spring and bellows makes for better calibration than use of a bellows alone. The spring gradient of a bellows not only is poor but also varies from one to another. Conversely, the calibration

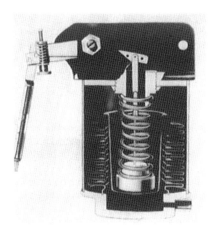

FIGURE 1.16 Intermediate-range spring and bellows.

FIGURE 1.17 Low-range spring and bellows.

of springs is good from one to another. The combined gradient of the spring and bellows enters the calibration. To obtain good calibration, a gradient ratio of 85 percent for the spring and 15 percent for the bellows is used.

For very low pressure ranges from 0 to 1 in to about 10 psig, diaphragm capsule elements (Fig. 1.18) are used. Diaphragms, made of a spring material, are joined together at their circumference to form a compartment. When subjected to pressure, each compartment expands a slight amount. Depending on the full-scale range, a number of compartments are stacked together to provide the proper movement for actuation. In order to obtain the required forces, diaphragms having different effective areas are used. The lower the pressure, the greater the area must be and consequently the larger the diameter. Diaphragms are available in a number of different materials such as brass, phosphor bronze, and stainless steel.

For the measurement of pressures from 0 to 0.2 to 10 in H_2O, the inverted-bell type of measuring element (Fig. 1.19) is used. The lower, or open, end of the bell is sealed by a light oil, and the pressure to be measured is connected to its underside. In most instruments of this type, no springs are used, and the

FIGURE 1.18 Diaphragm measuring element.

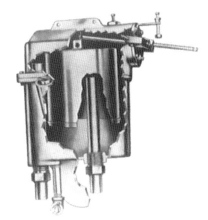

FIGURE 1.19 Inverted-bell measuring element.

movement of the bell with changes in pressure is counterbalanced by weights. In one design the bell, having thick steel walls, is submerged in mercury. With an increase in pressure, the bell moves upward, and the change in pressure is balanced by the loss in buoyancy of the walls emerging from the mercury.

To measure differential pressure two bells are used, one suspended from each end of an arm that is center-pivoted.

Absolute-Pressure Elements

The use of absolute gages is generally confined to pressure measurements, which are of so low a range that they would be seriously affected by variations in barometric pressure if an absolute-pressure sensing element were not used. Measurement of pressures in distillation columns is an example of such a process.

The absolute-pressure sensing element illustrated in Fig. 1.20 is one type used for applications of this nature. It is suitable for measuring pressures as low as 0 to 5 mmHg absolute.

Principle of Operation

The upper bellows is evacuated and sealed at a nearly perfect vacuum, and the inside of the lower bellows is connected to the pressure to be measured. The adjacent ends of both bellows are attached to a movable plate (at the center) which transmits the bellows movement to the recording pen.

The method of operation is identical to that of a conventional spring-opposed bellows element insofar as the actuating bellows is concerned. However, a difference in the resultant operation of the actuating bellows is obtained from the action of the evacuated bellows which expands and contracts in response to barometric-pressure changes. This action prevents any movement of the pen or pointer by the application of an equal and opposite force to the actuating bellows. The evacuated bellows functions as an aneroid barometer in its response to changes in atmospheric pressure.

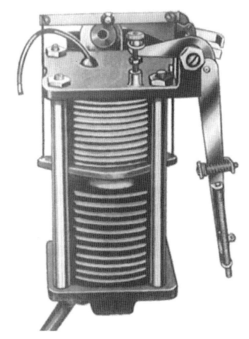

FIGURE 1.20 Absolute-pressure measuring element.

FILLED-SYSTEM THERMOMETERS

A filled-system thermometer consists of a recording or indicating instrument, with its pen or pointer actuated by a Bourdon tube, spiral, or helix connected by a small-bore capillary tube to a bulb, as shown in Fig. 1.21. The combination of actuating element, capillary tubing, and bulb is called a *thermal system*.

The thermal system may be filled with a liquid, a volatile liquid, or a gas under pressure. Since the system is closed, the internal pressure will change when the bulb is subjected to variations in temperature. These changes in pressure result in the movement of the actuating element, which, by means of a connecting link, positions a pen or pointer or initiates a signal in a pneumatic transmitter.

The minimum temperature indicated on the scale shown in Fig. 1.21 is at the right and the maximum temperature at the left. With an increase in temperature of the bulb, the increasing pressure of the fill will cause the end of the spiral to move to the left and the pointer to move upscale.

These thermometers are supplied for temperature measurement from 2350 to 11200°F. Their accuracy, sensitivity, and other functional factors make them suitable for many industrial applications, for indication, recording, controlling, and pneumatic transmission of temperature. They are reasonable in cost and require only nominal maintenance.

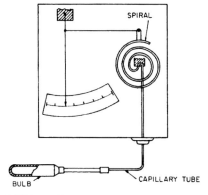

FIGURE 1.21 Filled-system thermometer.

Classification and Descriptions of Filled-System Thermometers

A number of manufacturers of filled-system thermometers have adopted a classification based on the fills. The classifications and figures that follow are taken from this standard.

Liquid-Filled (Class IB). Figure 1.22 illustrates a class IB thermal system, which is completely filled with a liquid (other than a metal such as mercury) and operates on the principle of liquid expansion with changes in temperature. Various liquids may be used, such as distilled water, alcohol, or xylene. The fill used depends on the temperature range to be measured. The temperature limits are from 2300 to 1600°F.

Class IB thermometers are particularly adaptable for applications where narrow spans and small bulbs are required and where conditions prohibit the use of mercury-filled thermal systems.

The accuracy is affected by ambient-temperature changes of the spiral and the connecting tubing. The effects on the spiral can be eliminated by use of a bimetallic compensator attached to the end of the spiral, as shown in Fig. 1.22. This is known as *case compensation*. With an increase in ambient temperature, the internal-system pressure will increase, and the end of the spiral will move to the right, causing the pointer to move upscale, and the indicated temperature will be in error. The compensator is designed to move to the left a distance equal to the movement of the spiral tip; therefore, the temperature indicated remains that measured by the bulb. Because of effects of ambient-temperature change on the connecting tubing, the maximum length is limited to about 15 ft.

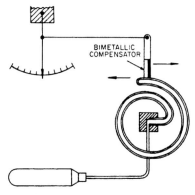

FIGURE 1.22 Case-compensated, liquid-filled thermal system, class IB.

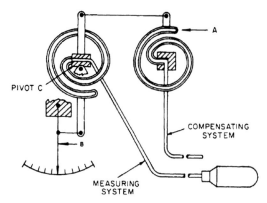

FIGURE 1.23 Fully compensated, liquid-filled system, class IA.

Fully Compensated Liquid-Filled (Class IA). Figure 1.23 shows a class IA, fully compensated liquid-filled thermal system. It has two thermal systems. The one on the left is the measuring system and is complete with a bulb. The compensating system on the right is similar except that it does not have a bulb. This system fully compensates for ambient-temperature changes of the measuring spiral and connecting tubing to the bulb.

The measuring spiral is pivoted at point A and is held in position by the link to the tip of the spiral of the compensating system. The connection from the tip of measuring spiral to the pointer is shown at point B. With changes in ambient temperature the tips of both spirals will move. They are selected, therefore, on a basis of reasonably close matching of movement of each over the range of temperature compensation.

Since changes in the ambient temperature of the tubing also will change the internal-system pressure, in order to obtain good compensation, the internal volume of the two tubes should be as nearly equal as possible. For this reason, tubing is used from the same draw for both systems. Since spiral movement and tubing volumes cannot be matched exactly, calibration adjustments are also provided to obtain compensation.

With the bulb at a constant temperature and a change in ambient temperature of the spirals, point B would move to the left, and indication would be in error. It is prevented from moving because the movement of A rotates the measuring spiral about pivot C and point B remains fixed. With an increase in the ambient temperature of the connecting tubing the internal-system pressure will increase. Tip movement of each spiral will occur, and compensation will be made as described above. Tubing lengths up to 200 ft are supplied.

Vapor-Pressure Thermal Systems (Class II). The system is partially filled with a volatile liquid and operates on the principle of vapor pressure. With changes in temperature of the bulb, variations of vapor pressure within the system occur. Depending on the application, four types are used.

The accuracy of class II thermometers is unaffected by changes of ambient temperature of the spiral or connecting tubing, because the internal-system pressure is that of the boiling point of the volatile liquid in the bulb.

Class II thermometers have expanded scale or chart graduation with an increase in temperature, as shown by Fig. 1.24. This expansion permits more accurate reading of pen or pointer and adjusting the set-point position of controllers.

A class IIA system (Fig. 1.24) is used on applications where the bulb will always be at a higher temperature than the rest of the system. The spiral and connecting tubing are completely filled with liquid. A class IIB system (Fig. 1.25) is used on applications where the bulb will always be at a lower temperature than the rest of the system. All the liquid is in the bulb, and the spiral and tubing are filled with vapor. A class IIC system (Fig. 1.26) is used on applications where the bulb temperature

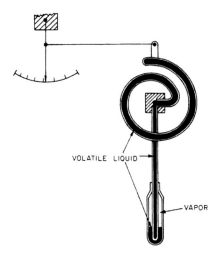

FIGURE 1.24 Vapor-pressure thermal system, class IIA.

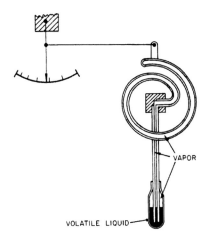

FIGURE 1.25 Vapor-pressure thermal system, class IIB.

may be either above or below that of the rest of the system. Note that as the bulb temperature changes, the vapor-volatile liquid positions become the same as for classes IIA and IIB. A class IID system (Fig. 1.27) is used on applications when it is necessary to measure the temperatures below, at, or above that of the rest of the system. In this system, the volatile liquid and vapor and a second relatively nonvolatile liquid are used to transmit the vapor pressure to the spiral.

In general, the bulb sizes for class II systems are larger than for class I. They can be supplied with tubing lengths up to 200 ft.

Class IIA thermometers must be compensated for the difference in elevation between the bulb and instrument because of the head effect of the liquid in the connecting tubing. If the bulb

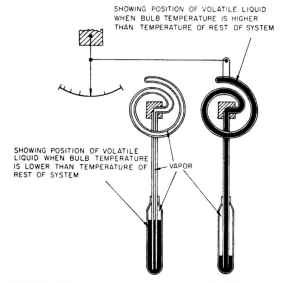

FIGURE 1.26 Vapor-pressure thermal system, class IIC.

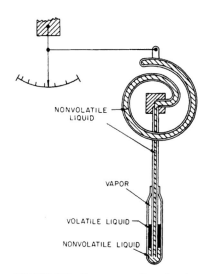

FIGURE 1.27 Vapor-pressure thermal system, class IID.

is below, the instrument will read low. It will read high if the bulb is above. If the difference in elevation is specified when ordering, compensation can be included in the factory calibration. Compensation in the field requires placing the bulb in a bath at a known temperature and making a zero adjustment.

Gas-Filled Thermal Systems (Class III). Class III thermal systems are gas-filled and are suitable for temperature ranges from 2350 to 1000°F. While practices among manufacturers vary, nitrogen is used for temperatures from 2100 to 1000°F. For temperatures down to 2350°F, the gas may be helium or nitrogen.

Figure 1.22 also illustrates a class IIIB thermal system with case compensation that has been previously described. It is used on all class III thermometers except those that are fully compensated. Partial compensation for effects of ambient-temperature changes on the connecting tubing is obtained by a large ratio of bulb to tubing volume, generally about 10 to 1.

Class IIIA is a fully compensated system and uses the same method as shown by Fig. 1.23. Compensation is the same as previously described.

Class III thermometers require larger bulbs than any other thermal system. One reason is the large ratio of bulb to tubing volume required with case-compensated systems. A second reason is the need to have sufficient gas volume in the bulb at high temperatures to produce the required system pressures.

Mercury-Filled Thermal Systems (Class V). A mercury-filled, class VB, case-compensated thermal system is also illustrated by Fig. 1.22. This system is completely filled with mercury or a mercury-thallium eutectic amalgam and operates on the principle of liquid expansion. Partial compensation for ambient-temperature changes of the connecting tubing is accomplished by the ratio of bulb to tubing volume, but tubing length is limited to about 25 ft.

Figure 1.28 shows two fully compensated mercury-filled thermal systems. The double thermal system on the left operates on the same principle as that previously described for class I and III fully compensated thermometers. The method shown on the right uses only one capillary-tubing connection from spiral to the bulb. A metal wire having an extremely small coefficient of expansion with changes of ambient temperature is inserted in the tubing, which is filled with mercury. The wire keeps the volume of the tubing and the volume of the mercury properly related so they change an equal amount and keep the internal pressure constant regardless of ambient-temperature changes along the capillary. Therefore, the pressure in the bulb due to expansion of the mercury is transmitted to the spiral without change, regardless of the ambient temperature of the connecting tubing.

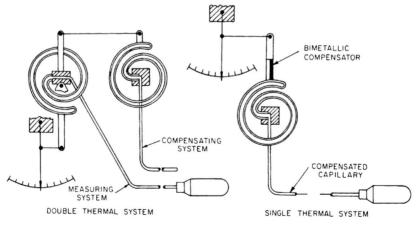

FIGURE 1.28 Fully compensated mercury-filled thermal system, class VA.

Mercury-actuated thermometers are used to measure temperatures from 240 to 1000°F. The lower limit is established by the freezing of mercury. The higher limit is based on the expansion of bulb materials, which increases their internal volume and lowers the required pressure in relation to temperature. The minimum span is 25°F over the temperature range from 240 to 1000°F.

Scales and charts used with class V thermometers are evenly graduated. For most spans, the same ones can be used as for class II and III thermometers.

Bulb sizes are small but do not approach the size of class I. Their size, however, does not increase with the length of connecting tubing, as it does for class II and III thermometers.

The system pressures are high, varying from about 500 psig at minimum to 2000 psig at maximum temperature, regardless of the span. As a result of these high pressures, the calibration and sensitivity of class V thermometers are excellent.

Thermal-System Bulb Designs

Bulbs are available in a wide variety of designs and materials to meet the requirements of most applications. The principal designs are plain, union-connected, coiled, and averaging type. They are available in steel, various types of stainless steel, copper, brass, and Monel. Most of these materials are also available with Teflon, Kel-F, Tygon, and other plastic coatings. Coatings serve two functions: They provide additional corrosion resistance, or they prevent sticking of materials, thus eliminating an increase of heat transfer through the bulb wall.

The bulb is joined to the connecting tubing by means of an extension neck, usually of much smaller diameter but made of the same material as the bulb. The connecting tubing is of two types. The more common type is a capillary tube of small diameter, about 1/16 in, that is protected by a flexible armor of copper or stainless steel. The armor can be provided with the same coatings as the bulbs. The second type, known as *plain* or *smooth tubing,* is made of the same material as the bulb and extension neck. It is generally about twice the diameter of capillary tubing and does not have the flexibility of armored tubing.

Flexibility of connecting tubing is important when making the installation, especially where long lengths are concerned. When a thermometer is installed, the instrument and thermal system cannot be separated. The instrument must be installed first and the bulb and tubing "threaded" to the point of temperature measurement. For this reason flexible armor may be preferred over smooth tubing.

Bulb designs for different applications are described below.

Plain Bulb and Extension Neck. These are used where the bulb can be installed from the top of an open tank. The extension neck may be bent over the top edge of the tank.

Union-Connected Bulbs. These have a union connection at the top of the bulb or on the extension neck. They are used on tanks under pressure. If pressure cannot be removed when the bulb is to be withdrawn, the bulb must be installed in a socket. The socket is screwed into the tank wall, using a pipe thread, or is welded in position.

Coiled Bulbs. Coiled bulbs are made of small-diameter tubing and formed into a coil. Each coil is about 4-in long and have a diameter of 2 in. They are used principally to present more surface for the heat transmitter and obtain faster response. They are used for duct-temperature measurement.

Averaging Bulbs. Averaging bulbs are also made of small-diameter tubing and, because of their long length, present a large surface area and are characterized by a very fast response. They are especially effective for measuring the temperature of air or gas, as well as the temperature of slowly moving liquids where heat-transfer conditions are not favorable. The term averaging is used because the bulbs measure the average temperature over their active portions. They are widely used in oven or furnace chambers.

7.24 INSTRUMENTS AND RELIABILITY TOOLS

HUMIDITY MEASUREMENT

The moisture content of air is a significant variable not only in air conditioning for human comfort but also in industrial processing and manufacturing plants, where it can influence the quality of the product. In the spinning of textiles, for example, better results are obtained when humidity is controlled at optimal values. In drying operations, also, the moisture-laden air from the drier reflects the amount of moisture removal from the solid, and its moisture content can be used as the basis for operation of the drier. Similar needs for moisture measurements are found in other fields, such as in metal heat treating under protective atmosphere, where the moisture content can be quite critical, and in the manufacture of various industrial gases, where absence of moisture in the gas is important.

The three principal methods of measuring relative humidity (RH) use nylon hygrometers, wet- and dry-bulb thermometers, and dew-point measurement.

Nylon-Ribbon Hygrometer Method

The instrument shown in Fig. 1.29 is designed for monitoring the relative humidity and temperature of the ambient atmosphere. The hygroscopic element is a preconditioned nylon unit which responds to a change in relative humidity by a change in length. A suitable linkage connects the elements to a recording pen. The instrument is calibrated for direct reading of percent RH. No conversion charts are necessary to interpret the RH record. A bimetallic spiral is supplied for temperature measurement, if desired.

The practical measuring range is from about 30 to 90 percent RH to an accuracy of 62 percent provided the range of ambient temperature does not exceed 60 to 90°F. The accuracy of temperature recording is 61 percent of span.

The primary application of these recorders is to check on the operation of humidity control of rooms, chambers, and so on, where the actual control is by other apparatus.

Wet- and Dry-Bulb Thermometers. Relative humidity also can be measured by means of a two-pen thermometer. One bulb, known as the *wet bulb,* is kept wet at all times, either by covering it with

FIGURE 1.29 Interior view of two-pen portable temperature-humidity recorder.

a wick, the lower end of which is immersed in water, or by inserting the bulb in a porous tube into which a regulated supply of water is introduced internally. The passage of air over the wet bulb causes evaporation, which results in a lowering of temperature. The second, or *dry, bulb* is adjacent to the wet bulb, and both temperatures are recorded by the thermometer. At every temperature there is a definite relationship between relative humidity and the temperature difference between wet and dry bulbs. Tables giving these relative-humidity values are available. Standard psychrometric charts, which make it possible to determine relative humidity, dew point, vapor pressure, and other quantities, are also available.

These two-pen thermometers are used for humidity control. One controller is actuated by the dry bulb and maintains a constant temperature. The second controller is actuated by the wet bulb and introduces steam or water into the chamber where the atmosphere is being controlled. The individual set points of the two controllers are positioned to obtain the required relative humidity.

Dew-Point Measurement

In some applications, measurement and control of the dew point is more advantageous to use than relative humidity. *Dew point* is the temperature at which water vapor contained in a gas will condense to a liquid. It is related to the amount of moisture in the gas and is not a function of its ambient temperature. A perfectly dry gas would have a dew point of absolute zero while a 100 percent saturated gas is already at its dew point.

Dew point may be measured directly by cooling the temperature-sensing element until moisture condenses on a recognizable surface such as a mirror. The temperature is controlled at this point and the dew point read on a conventional temperature-measuring instrument.

The dew-point sensor shown in Fig. 1.30 uses an ingenious construction which does not require artificial cooling of the sensing element. The sensor consists of two wire electrodes wound side by side on a cloth sleeve which covers a hollow tube. The cloth sleeve is impregnated with lithium chloride which absorbs water from the air and becomes conductive. This allows current to flow from one electrode through the cloth sleeve to the other electrode and produces heat at the bobbin. Moisture is thereby evaporated from the lithium chloride until a heat-moisture equilibrium is established.

FIGURE 1.30 Dew probe dew-point sensor.

The equilibrium bobbin temperature is related to the dew point of the gas sample. By measuring the temperature in the cavity of the bobbin the dew point can be determined.

PNEUMATIC TRANSMISSION SYSTEMS

Central control-room operation of large plants is widespread practice and results in a better-quality product. Further, in plants where corrosive or flammable liquids or vapors are used frequently under dangerously high pressures, remote transmission eliminates control-room hazards by allowing the process to be controlled from a safe location.

Transmitters, which make these advantages possible, are instruments which measure process variables at the tanks and pipes and put out a corresponding proportional signal. The signal may be either electric or pneumatic. The pneumatic type is covered in this chapter.

Pneumatic transmitters are operated with a clean, dry, 20-psig air supply. Their output signal range is 3 to 15 psig. It is 3 psig when the process variable is at the minimum value of its range and 15 psig when it is at maximum value of its range, and a proportional intermediate pressure when the measured variable is at in-between values.

Most are available with or without an indicating dial and pointer. Those which have no indication are known as "blind" transmitters. Frequently the latter instruments are installed with a dial pressure gage connected to a tee in their output tubing, for the purpose of locally indicating the pressure of the transmitted signal.

7.26 INSTRUMENTS AND RELIABILITY TOOLS

Pneumatic flow transmitters have two components: a meter body and a pneumatic assembly. The meter body senses changes in differential pressure as the flow rate varies, and produces corresponding movements in a mechanical linkage. These movements, coming out of the meter body, are used to actuate the pneumatic assembly, which in turn produces the output signal. The meter body and pneumatic assembly of all flow transmitters are close-coupled together.

Diaphragm-Actuated Flowmeter Transmitter

Flow-transmitter meter bodies are sometimes differential bellows assemblies like the one described and illustrated earlier in this chapter. Diaphragm-actuated meter bodies, such as the one illustrated in Fig. 1.31, are much more frequently encountered, however. The diaphragm type has a feature of considerable interest from the viewpoints of installation, maintenance, and operation. This feature is *low volumetric displacement*. It is in the order of 0.1 in^3 in the meter body shown in Fig. 1.32. This means that as the flow rate varies, even from no flow to maximum flow, only a very small change in volume takes place inside the meter body. Therefore, in steam-flow applications, installation can be made without the necessity of using a pair of condensers. In addition, since there is practically no movement of liquid in the connecting piping with variations in flow, the need for periodic cleaning of the piping, by blowing down the lines, is greatly reduced. Further, low volumetric displacement provides inherently high sensitivity.

Flow-transmitter meter bodies are available, as standard, in carbon steel, 316 stainless steel, and Monel. As a special product, other materials are available for unusually corrosive applications. Installation of flow transmitters is generally straightforward and easy, either with a pipe bracket (shown in Fig. 1.31) or else with support from the orifice pipe connections.

In the type of flow transmitter shown, the measured fluid wets only the inside of the measuring chambers. Maintenance on the meter body is thus reduced to a minimum, and consists of very carefully wiping off the two barrier diaphragms.

Compensation for the effects of ambient temperature changes on the accuracy of calibration is integrally included. The maximum error generally will not exceed 1 percent for a 100°F change.

This type of transmitter will withstand, without damage, overrange differential pressure of the meter body.

FIGURE 1.31 Diaphragm-actuated flow transmitter.

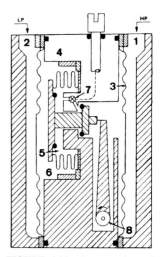

FIGURE 1.32 Cross section of meter body, diaphragm type.

Meter-Body Operation

The operation of the meter body in sensing differential pressure is shown in Fig. 1.32. High (1) and low (2) flow-line pressures are admitted to the two end chambers.

The barrier diaphragms (3 and 4) transfer these pressures into the silicone fill fluid in the center section of the meter body on either side of the measuring element. The higher pressure acts on the inside (5) and the lower pressure acts on the outside (6) of the measuring element.

If there is no flow, these two pressures are equal. With flow, the pressure in the left chamber and the low-pressure side of the fill decreases, the element moves to the left, and high-pressure fill flows through the damping restriction (7). As the measuring element moves, it exerts a proportional torque through a connecting linkage on the force shaft (8). The force shaft extends to the outside of the pressure-tight meter body through a seal tube. The transmitting unit fastens to the outer end of the force shaft.

Pneumatic-Transmitter Operation

Operation of the transmitter is illustrated in Fig. 1.33. The primary beam (1) is fastened to the end of the meter-body force shaft, and differential pressure across the measuring element in the meter body results in a torque being applied to the shaft. The torque attempts to rotate the primary beam in a counterclockwise direction, and as it does so, the baffle (2) is carried closer to the nozzle (3). This causes an increase in nozzle backpressure, which is fed to the pilot (4) and results in an increased output. The output is fed back into the rebalancing capsule (5), which exerts a force on the secondary beam (6); the torque generated by this force is transmitted to the primary beam through the adjustable span rider (7) and rebalances the system. When a suppression or elevation spring (8) is used, it is adjusted to apply the required amount of suppressing force to the primary beam.

The range or span may be changed by repositioning the contact point (7) on the secondary beam (6). The thumbscrew to the right of the primary beam is the coarse-span adjustment. The hex micrometer screw is the fine-span adjustment. Any required span between 0 and 20 in H_2O and 0 and 250 in H_2O may be achieved on this instrument by the use of these two adjustments. Lower- and higher-range meter bodies are also available (0 to 5 in H_2O to 0 to 25 in H_2O, and 0 to 100 in H_2O to 0 to 1000 in H_2O).

The spring and knurled thumbwheel at the lower left is the pneumatic-zero adjustment.

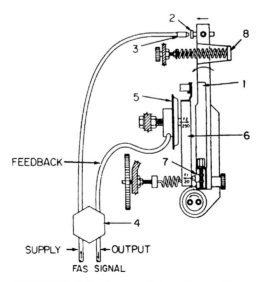

FIGURE 1.33 Schematic, pneumatic-transmitter assembly.

7.28 INSTRUMENTS AND RELIABILITY TOOLS

The spring assembly (8) at the upper end of the primary beam is the suppression adjustment. By turning the screw, a force is applied to the primary beam to prevent movement until a desired range is measured. For example, on a 0- to 100-in range, the adjustment can be such that primary-beam movement will not occur until a range of 50 in is reached.

Suppression is normally used only on flow transmitters for reverse-flow applications. That is, to measure flow in either direction, a center-zero chart is used, and when there is no flow, the transmitted signal is 9 psig. With flow in one direction, the transmitted pressure will increase; when the flow reverses, the pressure decreases. The main use of suppression is when differential-pressure transmitters are used to measure liquid level. When the density of the liquid in the outer fixed leg is greater than that of the liquid in the tank, the difference must be suppressed.

PNEUMATIC PRESSURE AND TEMPERATURE TRANSMITTERS

Pneumatic transmitters have certain components common to all makes and types. The general principle of operation is similar for all.

Each has an integral element to sense changes in pressure, temperature, differential pressure, and so on. These produce a mechanical motion which causes a flapper to approach or leave a nozzle. Resulting changes in nozzle line pressure (generally in the order of 3 to 6 psig) actuate a pilot relay, whose function is to provide an amplified output pressure (between 3 and 15 psig). The pilot-relay output is the transmitter output signal.

In addition, the relay output is branched back into a feedback bellows to give a negative or opposite motion to the flapper. Thus the net movement of the flapper is not as much as it would have been without this motion. Consequently, the transmitter output change will not be as great as it otherwise would have been. Negative feedback is essential to produce a proportional change in output signal. Without negative feedback, the signal would go from maximum to minimum, or vice versa. Figure 1.34

FIGURE 1.34 Indicating type of pneumatic-pressure transmitter.

FIGURE 1.35 Nonindicating force-balance type of pressure transmitter.

illustrates an indicating, motion-balance type of pressure transmitter. It is installed at the point of measurement and supported by a bracket attached to the pipe or process vessel. Its design is such that it will give good performance in locations subject to considerable vibration. The case is designed so that protection is provided for the internal mechanism for installations in corrosive atmospheres. The range of pressures is from 0 to 1 psig up to 0 to 10,000 psig. This same type of transmitter also comes as a temperature transmitter, the only difference being that it has a filled thermal system instead of a pressure-sensing element. Ranges are available from 0 to 100°F to 0 to 1000°F. Figure 1.35 illustrates a force-balance type of pressure transmitter. It is available only as a nonindicating type. If local indication is required, a pressure gage is connected into the transmitted signal line.

MINIATURE MECHANICAL INSTRUMENTS

Recording Type

Many instruments used in the process industries are the miniature type. They generally require only a 4 × 6 in panelboard cutout, with the instrument front having only slightly larger dimensions. They are supplied by almost all the instrument manufacturers.

Figure 1.36 illustrates a miniature recording-control station whose recording pen is actuated by a 3- to 15-psig signal from a process-variable transmitter. This instrument has all the functions of the larger-sized instruments but requires only about 15 to 20 percent of the panelboard area.

7.30 INSTRUMENTS AND RELIABILITY TOOLS

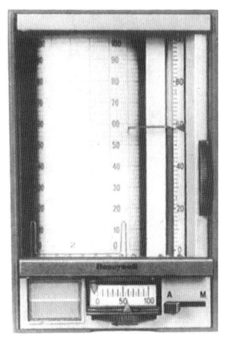

FIGURE 1.36 Miniature recording-control station.

The use of these instruments not only reduces the overall costs but also permits the operator to run the process better. The reduction in size results in the use of shorter panelboards, which in turn reduces the size of the control house and results in considerable savings in costs, and further savings in the heating and air-conditioning loads.

Use of miniature instruments means that more of them are within the direct observation of the operator, permitting closer supervision of the process and generally resulting in improved operation of the plant. They reduce operator fatigue, since the amount of walking required by the operator is materially reduced. In some cases, further savings result from a reduction in the number of operators.

The instrument shown is recording and controlling one process variable. It is also available with two pens, one recording and controlling and the other simply recording. The second-pen record as a control station is generally that of a related variable. For example, the variable controlled may be flow of gas to a burner, with the second variable being gas pressure. Another frequently used application is controlling the flow from a tank, with the second pen recording the level in the tank.

The following items are available for the operator's use:

- A record of the value of the process variable being controlled, plus a red bar-graph indication of the same for better visibility at a distance
- A set-point index and its adjustment (vertical knurled wheel at right)
- An indicating scale and pointer (at bottom) showing "air to valve," a measure of the valve opening, whether the process is on either automatic or manual control
- An "automatic-manual" selector switch to put the process-control valve on either automatic or manual control

When on "manual," the process operator manually controls the pressure to the valve by the horizontal knurled wheel at the bottom. Pressure balancing is not required for bumpless transfer, since the controller output and the manual output are continuously and automatically matched.

The calibrated width of the chart is 4 in, which is approximately the same as the pen travel on the large instruments using 12-in circular charts. Charts last for 30 days at the normal speed of 3/4 in/hr. The capillary-type pen will give a 6-month record, under normal operating conditions, without replenishing its reservoir. Each pen has its own adjustable pulsation damper. The chart drive may be either electric or pneumatic. Chart reroll and daily tearoff are available on all recorders.

To provide an instrument meeting the requirement for small panel area, the depth was increased. The distance behind the panel is about 24 in. The upper chassis can be pulled out to the "service" position, where it is locked by a latch, for the purpose of setting the controller adjustments or performing certain servicing operations.

The controllers are available as proportional-plus-manual reset, proportional-plus-automatic reset, or proportional-plus-automatic reset plus rate.

The process-variable transmitter actuates both the pen-actuating element and the controller. The error signal which initiates the controller action is the difference between the pressure coming in from the transmitter and the pressure of the set point. Any difference in these pressures initiates control action. The sensitivity of these miniature controllers is exceptionally high, which results in outstanding performance.

A relatively recent trend in process instrumentation is an increase in the use of indicating control stations and a decrease in the use of recorders. Frequently, when new process-control panels are designed, permanent recording is not assigned to any controller variable. Instead, quick-connect plug-ins are provided at the back of the indicating instruments so that the operator can select any process variable he wishes and plug it into an adjacent "trend recorder" for either long-term or short-term recording. The trend recorder is somewhat like the recorder control station previously described except that it has no set-point mechanism, no controller, no air-to-valve indication, and no automatic-manual switch. The trend recorder can have as many as three pens, and has a nine-point quick-connect jack board in the lower part of its case so that any three of nine inputs can be recorded at one time.

FIGURE 1.37 Fixed-scale indicating-control station.

The instrument shown in Fig. 1.37 is a "fixed-scale" indicating control station. Here the process variable and set-point indicators operate over 100 percent of the instrument scale. It may have an integrally mounted controller or one which is remotely mounted. It has an automatic-manual selector switch (near the bottom) and a manual adjustment (horizontal knurled wheel, at bottom). The air-to-valve scale and pointer (bottom) indicate the output pressure going to the valve whether the process is on either automatic or manual control. Manual balancing of pressures is not required for bumpless transfer. There are three types of set-point adjustments which can be used with this instrument: (1) Local: the set point may be adjusted from the front of the instrument. In the one shown in Fig. 1.37 the adjustment is the vertical knurled wheel to the right (2) Local-cascade: here the set-point signal may be selected from either an internal or a remote source by a two-position switch. (3) Remote set-point signal only.

The instrument shown in Fig. 1.38 is a "deviation" type of indicating control station. It provides a visual indication of the state of the process variable, set point, and air-to-valve pressure on either automatic or manual control. In addition, it has an automatic-manual selector switch.

FIGURE 1.38 Deviation type of indicating-control station.

The set-point signal to the controller is a pneumatic pressure which is changed by the knurled wheel at the right. This also repositions the movable-tape scale.

Its principal point of interest is the red deviation pointer, which moves from behind the green band on the lens if the process variable departs from its set point. When the process is at its set point, the red pointer remains behind the green band.

The deviation pointer is actuated by a differential–pressure-sensing unit which compares the set-point pressure with the incoming process-variable transmitter pressure. Below the scale is a translucent nameplate covering two optimal signal lights. The operating mechanism is mounted on a chassis which can be withdrawn from the front of the instrument panel for making adjustments or changing the controller settings.

MAINTENANCE OF MECHANICAL INSTRUMENTS

Mechanical instruments are designed so that maintenance has been reduced to a minimum. An inspection program will determine the amount of maintenance required and when it should be done. Protective maintenance, where components are replaced before failure occurs, should be included in this program.

Too often instruments are blamed when the causes of trouble lie elsewhere and are not something the instrument can correct. For example, flow cannot be controlled at the proper rate if a sufficient supply of fluid is not available. If the calibration is found to be correct, or has been corrected, and trouble remains, look elsewhere in the process for the source.

An important factor in maintenance is well-trained instrument service people. Most manufacturers offer training classes at their plants and district offices, and many provide training at customer plants. Taking advantage of this training will generally result in considerable reduction in maintenance costs.

To reduce maintenance costs, it is essential to follow installation instructions as closely as possible. They are based on experienced gained on actual installations of thousands of instruments in many industries. If they cannot be followed, it is important to discuss the installation with the manufacturer; he will be pleased to be of assistance.

Flowmeters

The manufacturer's instructions for the installation of flowmeters should be carefully followed. This is very important in the installation of the primary element with respect to the upstream and downstream piping conditions. Instruction books carefully specify the minimum length of straight pipe required in relation to the ratio of the diameter of the restriction to the internal pipe diameter. Should either the upstream or downstream distance to elbows, bends, valves, etc. be less than specified, the manufacturer should be consulted.

All primary elements are designed for specific operating conditions, which may include pressure, temperature, specific gravity, and other factors. If these conditions change, measurement will be in error, and correction factors must be applied to either the rate or the totalized flow. Most instruction books include tables of correction factors or formulas for calculating them.

If the rate of flow must be frequently used by the operator, as in changing the set point of a controller, applying a correction factor is inconvenient, and the meter should be recalibrated. For example, assume that the correction factor is 1.02 for a meter calibrated for a differential pressure of 100 in H_2O. Recalibrating the meter to a differential of 104 in H_2O will eliminate the need to use a correction factor for the new operating conditions.

Installation Recommendations. A number of general, but important, instructions that are frequently violated are listed below.

1. Follow instructions as closely as possible. If there is a wide discrepancy, be sure to consult the manufacturer.
2. When measuring steam or water, protect the meter body and connecting piping from freezing. If freezing occurs, the pressures developed may blow out gaskets or damage the meter body.

3. Install a pneumatic transmitter as close as possible to the primary-element connections. This eliminates any decrease in speed of response, or failure to measure small changes in differential due to pressure loss in the connecting piping.
4. When measuring fluids that can become viscous at low temperatures, heat the connecting piping and meter body in order to eliminate sluggish operation.
5. When measuring gas containing entrained vapors, install the primary-element connections on the top of the pipe and install the transmitter above them. Condensed liquids will drain back into the pipe. If the meter body is installed below the flow line, unequal heads of liquid in the vertical connecting lines will cause measurement errors.
6. If at all possible, avoid excessive vibration of the flow line. If this cannot be done, use flexible connections in the orifice piping at the meter body.
7. Leaks in connecting piping will result in errors.
8. When measuring liquids, the maximum distance from primary element to meter body should not exceed 100 ft. When measuring gas, it should not exceed 50 ft. Longer distances may result in sluggish action.
9. When measuring liquids, slope the piping at least 1 in/ft from the primary-element connections to the meter body. A lesser slope may result in accumulation of air or vapor that will cause errors in measurement.

Field Maintenance. The need for field maintenance and when it should be done depend solely on operating conditions. The time intervals between checkings can be determined only after the meter has been installed. When clean liquids are being measured and flow is relatively steady, quarterly or semiannual inspections will probably be sufficient. When fluids are viscous, and possibly carry foreign matter, and flow fluctuates over 20 to 30 percent of full scale, more frequent inspections may be required.

After the meter has been installed, a timed program of inspection, as outlined below, should be followed.

Checking at Zero. Stability of calibration is inherent in present-day flowmeters and flow transmitters. After the meter has been in service for 1 week, it should be equalized (using the method given in the instruction book), and the zero reading observed. If not at zero, allow some time for pressures to equalize in the high-pressure and low-pressure chambers. If zero is not correct, check connecting piping for leaks and stop them. Only then make the zero adjustment as directed by the instructions.

If the zero adjustment is within 1 percent, it is safe to assume that calibration is satisfactory and no further adjustments are necessary. This can be verified in other ways; for example, the total flow appears to be correct, or the controlled rate of flow is that expected for a particular process. If a considerable zero adjustment is required, the calibration should be checked; this procedure is discussed below.

After the second week of operation recheck zero. If there is little or no change in zero, this checking procedure can be safely extended for several weeks. By following this procedure, it is possible to establish a plan of maintenance and eliminate the cost of unnecessary steps.

Checking Calibration. The calibration can be checked either in the instrument shop or at the installation. Bellows meters, because of their size and weight, generally are better calibrated at the point of installation. Since a complete overhaul includes disassembly of the meter body and cleaning, it should be done in the instrument shop.

Flow transmitters also can be calibrated in the field. Because of their light weight and ease of installation, it is recommended that a shop-calibrated replacement unit be installed. This procedure eliminates taking checking equipment to the field, where setting up and operation sometimes are difficult. Under some weather conditions, it is often difficult to duplicate shop calibrations in the field.

7.34 INSTRUMENTS AND RELIABILITY TOOLS

Calibration Equipment and Use. To obtain the accuracy of calibration inherent in the design of present-day flowmeters and flow transmitters, standards for pressure measurement must be of high quality.

1. The input differential pressures should be measured with a water column for differential pressures up to 50 in H_2O. For higher pressures, a mercury column may be used, but it should be equipped to read pressures with the same accuracy as an equivalent water column. The mercury column (Fig. 1.39) is designed and built for accuracy and ease of observation. It is designed to be used as a manometer, a differential-pressure indicator, or a differential-vacuum gage.

2. The transmitter output pressures should be measured on a mercury column that can be read to an accuracy of 60.10 psig. This is 62.8 in H_2O, or about 60.5 percent full scale of the 12-psig span. An equivalent indicating gage (Fig. 1.34) is excellent for shop calibrations. This gage is available in ranges from 0 to 120 in H_2O to 0 to 300 in Hg. It can be used for the measurement of pressure, differential pressure, or vacuum. An important feature is the elimination of backlash.

3. For accurate reading, glass tubes of columns should be kept clean. Mercury, red oil, or water must also be clean.

4. To avoid errors, read columns at the same level as the height of liquid.

5. After setting pressures on the column, allow pressure in the meter body to stabilize before reading the corresponding pen position on the chart, or the transmitted pressures.

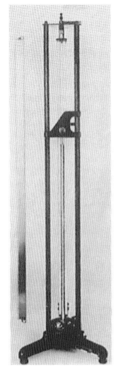

FIGURE 1.39 Precision mercury manometer.

General (Dry-Water Column) Procedure for Calibrating Flowmeters or Flow Transmitters

1. Connect the output of a precision pressure regulator to the high-pressure connection of the meter body. Also tie it into the water-column manometer. Leave the low-pressure connection of the meter body open to atmosphere.

2. With zero differential pressure, the pen should show a 0 percent reading. If the instrument is a pneumatic transmitter, the output signal should be 3 psi. If necessary, use the "zero" adjustment. Its location will be shown in the manufacturer's instructions.

3. Adjust the pressure regulator to the differential pressure which corresponds to a 90 percent pen reading on the square-root chart. Determine this value of water column from "water-column tables" in the flowmeter manual. The pen should read 90 percent, or the transmitter signal should be 12.72 psi. If not, use the "span" adjustment.

4. If a change is made on the span adjustment, repeat step 2.

5. After obtaining correct readings at 0 and 90 percent pen readings (or 3 and 12.72 psi, for flow transmitters), apply a water column corresponding to 70 percent pen reading. If the pen does not read 70 percent (or if the output signal is not 8.88 psi), refer to the manufacturer's manual and make a "linearity" adjustment according to its directions.

6. If a change is made in the linearity adjustment, it will be necessary to rezero and respan. Repeat steps 2, 3, and 4.

Liquid-Level Instruments

In the previous section on liquid-level measurement by the differential-pressure principle, much of the instrument equipment used is similar to instruments used in flow installations. The installation recommendations given for flowmeters therefore also apply, for the most part, to level installations.

To calibrate a liquid-level instrument, the way that the water column is connected up, or even the way that it is used, will depend on quite a few factors:

- Whether the tank is an open (or atmospheric-pressure) tank, or a closed tank under pressure
- Whether there is a constant-reference head as in Figs. 1.8 and 1.9 or no constant-reference head as in Figs. 1.10 and 1.11
- Whether or not the instrument or transmitter is on an application where it requires a suppression or elevation head adjustment

For these reasons, no specific calibration procedure will be outlined here. Instead, it is recommended that reference be made to the manufacturer's manual which describes the sequence of steps for the particular type of instrument involved and for the specific way that it is installed on its application.

THERMOMETERS

Filled-thermal-system thermometers differ from flowmeters and pressure gages in that the process fluids are not in direct contact with the actuating element. The only contact is external, and then when a plain bulb is immersed in the processed fluid. If a protecting socket is used, there is, of course, no direct contact.

Since the tubing lengths generally are short, the instrument often is located in the process area and not in a control house, which means that corrosive vapors may be present. While every effort is made by manufacturers to use corrosion-resistant materials for components, together with the best possible door gaskets, considerable maintenance may be required for some applications.

The connecting tubing is particularly susceptible to corrosive action and mechanical abuse. The most prevalent cause is the manual changing of bulbs from one location to another in order to measure several temperatures. Since the tubing cannot be run through conduit because of the bulb size, it is subject to all manner of damage due to plant operation and maintenance.

Installation Recommendations

Manufacturers' instruction books provide installation information in considerable detail. A number of general, but very important, instructions that are frequently violated are given below.

1. Install the bulb where it will quickly sense changes in temperature. Avoid a location in stagnant areas because a lag in sensing temperature changes will generally result in poor control.
2. Do not place tubing of uncompensated liquid-, gas-, and mercury-filled thermal systems near hot pipes or process vessels because radiant heat will result in higher-than-true readings. Conversely, supporting the tubing on building walls that become cold at low atmospheric temperatures will cause lower-than-true readings.
3. Protect the capillary against breakage. If the capillary tubing is broken, the thermal system will have to be replaced, and the thermometer will be out of service. Armor protection will help avoid damage from general operation or plant maintenance that can lead to time out of service and cost of system replacement.
4. Select internal socket diameters to have a minimum air gap with respect to outside bulb diameter. If bulbs are used in sockets with a greater gap, sluggish operation, and possibly incorrect temperatures as well, may occur.

5. Sockets are supplied with extensions of various lengths between the top of the thread and the socket end for installation on insulated pipes and process vessels. If the socket extends any distance beyond the insulation, it must be insulated in order to avoid a low reading.

Checking of Calibration

The checking of calibration is more difficult than for flowmeters or pressure gages, because an external signal cannot be applied, and the bulb must be placed in baths at several temperatures. This presents two requirements: a well-agitated liquid bath and correct measurement of the bath temperature. Unless the system has been overranged or temperatures shown are obviously incorrect, calibration normally need not be checked more than twice a year.

If the temperature shown appears to be in error, it can often be determined and corrected by recalibration at a single point, assuming that the same error exists throughout the temperature range. Correction is made by checking the calibration at one-third to one-half the maximum temperature and using the zero adjustment to make a correction at that point.

For a thermometer with a range of 0 to 600°F, the bulb would be immersed in boiling water. The water temperature, as indicated by a glass-stem thermometer, would be compared with that shown on the instrument. The glass-stem thermometer should be adjacent to the thermometer bulb or fastened to it, so that both will measure the same temperature.

After this procedure has been followed, the thermometer should be returned to service. If the operating temperature shown appears to be only slightly in error, further correction can be made using the zero adjustment. If the error is considerable, that is, more than 5 percent, complete recalibration is recommended.

For complete calibration, baths at a low checking temperature (10 or 20 percent of full scale) and a high checking temperature (80 or 90 percent of full scale) should be available. They should be well agitated, and an accurate means of reading their temperature should be provided. Glass-stem thermometers are available for high temperatures, although they sometimes are difficult to read above the surface of a hot bath. The use of a thermocouple, wired to the bulb and connected by extension leads to a portable potentiometer, provides a reliable and accurate method.

1. Place the bulb in the low-temperature bath and observe both the pen reading and the reading of the glass-stem thermometer or thermocouple. (Do not make any adjustment.)
2. Place the bulb in the high-temperature bath and observe the pen reading and glass-stem-thermometer reading. (Do not make any adjustment.)
3. If the pen reads an equal amount higher or lower than true temperature in steps 1 and 2, a zero adjustment (only) is required.
4. If the pen reads correct at the low temperature but high or low at the high temperature, a span adjustment (only) is required.
5. If the pen reading is not correct at either low or high readings, both zero and span adjustments are required.
6. Use the zero adjustment when the instrument is at the low reading and the span adjustment when the instrument is at the high reading.
7. Making a change on either zero or span will affect the other adjustment. It may therefore be necessary to go back and forth between the two adjustments several times until the pen reading is correct at both high and low temperatures.

Instrument Maintenance

If the instrument is in a process area, corrosive vapors may enter the instrument case when the door is opened for changing the chart or the set point. Snap-on links, pivot holes, and other connection points should be periodically examined for corrosion or deposits that may cause friction. Instruction books give directions for cleaning.

PRESSURE GAGES

Most pressure gages, like flowmeters, have the pressure-actuating element in direct contact with the process fluid. Exceptions are those equipped with a diaphragm seal located at the pressure tap and connected to the actuating element by capillary tubing. This closed system is liquid-filled, the same as a class IB thermometer. Because of the cost of the sealed type, they are used only where the process fluid cannot be allowed to enter the actuating element.

Because the process fluid is in contact with the actuating element, pressure gages frequently are subjected to severe service. These include corrosive action, the possibility of solidification of fluids under certain ambient-temperature conditions, high overloads, rapid pressure fluctuations, and the high temperature of the process fluid on the actuating element.

Installation Recommendations

The instruction books supplied by manufacturers provide installation information in considerable detail. A number of general, but important, instructions that are frequently violated are given below.

1. When measuring steam, always install a siphon, as shown by Fig. 1.40. It serves two purposes: to prevent high-temperature steam from damaging the actuating element when the gage is first placed in service and following start-up after the main has been shut down. Prior to start-up, and with the main shutoff valve closed, the siphon is filled with water and the gage is connected. To place the gage in service, the main shutoff valve is opened. Line pressure then forces water ahead of the steam into the actuating element and protects it from the steam temperature. When the main is shut down, and not under pressure, the siphon acts as a trap and remains partially filled with water. When the main is pressurized, the actuating element will again be protected.

2. When measuring steam or water, both gage and connecting piping must be protected from freezing. The pressures developed may rupture both actuating element and piping.

3. Select a range so that the maximum operating point is approximately 75 percent of the full-scale value. Longer life can be expected because of lower material stresses of the actuating element, and the extent of overload protection will be increased.

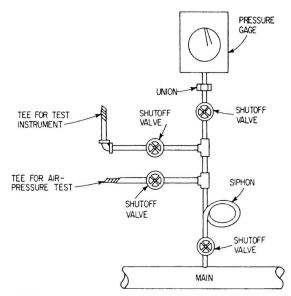

FIGURE 1.40 Indicating pressure gage.

7.38 INSTRUMENTS AND RELIABILITY TOOLS

4. Where pressures being measured fluctuate rapidly and over a considerable portion of the scale, pressure snubbers should be installed. They are available from several manufacturers.
5. When fluids in the actuating element and piping become viscous at low ambient temperatures, the response will be sluggish. It will be necessary to insulate and possibly heat the connecting piping and provide a heated enclosure for the gage.
6. Whenever possible install the gage where it is not subjected to vibration from compressors, pumps, or pipes. It is desirable, under such conditions, to use a flexible connection in the piping to the gage.
7. When measuring liquids, if the gage is installed below the point of pressure measurement, slope the connecting piping at least 1 in/ft to avoid formation of gas pockets. Their formation may cause errors in measurement.

Figure 1.40 represents a method of installation that has many desirable features. It may be too costly, however, for most applications, and can be modified to a considerable extent. For example, when measuring the pressure of low-temperature water with a concentric-scale indicator, the gage connection would be screwed directly into the shutoff valve, thus eliminating the rest of the piping shown. When measuring steam pressure, a gage with a female connection would be screwed to the upper end of the siphon. The system shown is very convenient for field calibration, and its use is explained below.

Checking of Calibration

Two methods are used: one eliminates the need to remove the gage from service, and the second involves checking the gage in the instrument shop. There are also two ways of using standards for checking calibration. One uses a primary standard, such as a water or mercury column or a dead-weight tester. The other uses test gages, which, in turn, are periodically calibrated against a primary standard. These gages have a very high accuracy, and one type is illustrated in Fig. 1.41.

When a considerable amount of checking and recalibration is done, a number of gages may be used to provide a series of increasing ranges. Thus several gages can be used at the most accurate

FIGURE 1.41 Pressure-gage connections.

portion of their range. For example, a 0- to 400-psig gage may be checked against test gages having ranges of 0 to 100, 0 to 300, and 0 to 600 psig, respectively.

Field Calibration

The piping arrangement shown by Fig. 1.40, as previously stated, is very convenient for field calibration. The pressure source for checking can be that in the main, or if the fluid should not be permitted to come in contact with the actuating element of the test gage, an external source must be provided, such as plant air, water pressure, or a nitrogen bottle. The most common method, using pressure from the main, is described below.

1. Close the shutoff valve at the main and open the upper left shutoff valve. If the connecting line to the gage is filled with a liquid, close the valve immediately when the pressure drops to zero so as not to drain the line. If gas pressure is being measured, the upper shutoff valve need not be closed.
2. Do not make any zero adjustments at this time, because with the gage located above the main it may have been adjusted for the head effect of the liquid in the connecting piping.
3. Connect the test gage at the point shown.
4. Apply increasing pressure in successive steps by alternately cracking and closing the main shutoff, and record the readings of the two gages. Decrease pressure in successive steps by alternately cracking and closing the lower left shutoff valve, and record the readings of the two gages.
5. Separately average the up- and downscale readings of each gage. Compare the values of the two gages. This will determine the accuracy of the operating gage compared with that of the test gage.

   Note: If the connecting piping is filled with liquid, the readings of the test gage must be corrected by the pressure equivalent to the elevation between the main shutoff valve and the center of the test gage. The correction factor for water is 1 lb for each 2.31 ft of elevation, and this is to be added to the test-gage reading. The data obtained are used to determine whether recalibration is necessary.
6. For low-pressure ranges, especially where gas pressures are being measured, a water or mercury column is substituted for the test gage. No corrections are required for elevation.
7. If the fluid whose pressure is being measured cannot be allowed to enter the test gage, the system must be completely drained by first closing the main shutoff valve and then opening the lower left shutoff valve. An external source of pressure must be provided at the lower left-hand shutoff valve. The method of checking calibration is the same as that previously described.

Shop Calibration

Checking and recalibration in the shop is better than that done in the field. Not only can it be done more accurately and in less time, but it also provides an opportunity for careful inspection of components, and their replacement if required, as well as general cleaning.

Water and mercury columns should be available and can be considered as primary standards for checking low-range gages. They should be designed so that they can be used for measuring positive pressure and vacuum. A panel of test gages, as previously described, also should be provided if a sufficient number of gages will be checked to warrant the cost.

When test gages are used, they are connected in parallel with the one being checked. Plant air is used for actuation within its available pressure. For higher pressures, a light oil is used, and pressures are generated by a manually operated hydraulic jack. A suitable valve must be installed between the jack and the gage to hold pressures.

If the use of test gages is warranted, a dead-weight pressure-gage tester which is a primary standard should be provided. It will also be necessary for periodic checking of test gages. The tester shown is a conventional type and can be furnished for pressures as high as 25,000 psig. A dead-weight tester consists of a vertical cylinder and piston. On the upper end of the piston is a tray, on which one or more standard test weights are placed. A plunger, operated by a hand screw, applies pressure to the liquid inside the cylinder. This liquid is oil or glycerin, stored in a reservoir connected to the cylinder through a two-way valve.

The gage to be tested is mounted on the dead-weight tester and is connected to the cylinder through a three-way cock, so that it is subject to the same pressure as the liquid in the cylinder. While measurements are being made, the piston carrying the weights is rotated to reduce friction. The pressure in pounds per square inch applied by the dead-weight tester is the sum of the weights and plunger in pounds divided by the cross-sectional area of the plunger in square inches.

The calibration is essentially performed as follows:

1. Apply pressure equal to approximately 10 percent of full scale and observe readings of pen and of test gage, or manometer, or dead-weight tester. (Make no adjustment.)
2. Apply pressure equal to approximately 90 percent of full scale and observe readings of pen, and of test gage. (Make no adjustment.)
3. If pen reads an equal amount higher or lower than the true pressure in steps 1 and 2, a zero adjustment (only) is required.
4. If pen reads correct at the low pressure but high or low at the high pressure, a span adjustment (only) is required.
5. If the pen reading is not correct at either low or high readings, both zero and span adjustments are required.
6. Use the zero adjustment when the instrument is at the low reading and the span adjustment when the instrument is at the high reading.
7. Changing either zero or span will affect the other adjustment. It may be necessary to go back and forth between the two adjustments several times until the pen is correct at both high and low pressures.

RELATIVE–HUMIDITY-MEASURING INSTRUMENTS

As previously stated, two mechanical instruments are used in making measurements of relative humidity: the nylon hygrometer and filled-system thermometers. Since the installation and maintenance of filled-system thermometers has already been discussed, this section will cover only the nylon hygrometer.

Checking of Calibration

Calibration can be checked by means of a sling psychrometer. It consists of two glass-stem thermometers, mounted on a board, with a suitable handle that permits the thermometers to be whirled rapidly. One thermometer bulb is covered with a wick saturated with water. During the whirling the water evaporates from the wick, and this thermometer will indicate a lower temperature than the other. This is known as the wet-bulb temperature. The relative humidity is related to the difference between the two temperatures, and the values can be obtained from tables. The value obtained is then compared with the instrument.

This method of checking must be used with caution. Because of the stratification of air, the relative humidity may differ at the instrument site and at the level where the sling is operated. Such a check is useful, however; for if several values show wide discrepancies, recalibration of the instrument is indicated.

Field calibration is not recommended, because calibration should be done in a cabinet where several values of humidity can be controlled and measured, and will be sensed by the instrument. All manufacturers have means for doing this.

Field Maintenance

Generally little maintenance will be required. The amount necessary will depend on the ambient conditions of the installation. The major effect on measurement is coating of the element with oil, dust, and so on. The nylon element should be cleaned periodically using a clean, dry brush. The elements should never be touched with the fingers.

Installation Recommendations

The installation should be made where the instrument will be in the path of air currents. Avoid installation in a stagnant pocket. Forced circulation by means of a small fan is helpful in obtaining increased speed of response.

CHAPTER 2
ELECTRICAL INSTRUMENTS FOR MEASURING, SERVICING, AND TESTING

R. Keith Mobley
Principal, Life Cycle Engineering, Inc., Charleston, S.C.

With the advent of automation and the extensive use of electronics and machine automation, the maintenance of electrical and electronic equipment has become critical to plant reliability and availability. These changes require the inclusion of new types of test equipment and instrumentation that will facilitate troubleshooting and periodic preventive maintenance.

As in the case of mechanical instrumentation, instruments that can accurately measure, record, and trend the operating condition of electrical or electronic equipment are essential for all predictive and maintenance improvement programs. Without the ability to anticipate incipient problems in this critical classification of plant systems, the reliability and availability of the plant cannot be maintained.

SELECT THE PROPER INSTRUMENT

When selecting the best instrument for the job, you must know something about the electric circuit under test. Once again, seek the advice of the manufacturer on the type of instrument to be used. It is very easy to be fooled when using improper test equipment. Some common examples are

1. Measuring circuits with a low-impedance meter (i.e., 20,000 Ω/V) when the service manual is written around the use of a high-impedance meter (i.e., 10-MΩ input).
2. Using a rectifier-type ac meter to measure waveforms that are not true sinusoidal. This could cause an error of several percent. A true rms responding meter should be used.
3. Improper termination of an oscilloscope or frequency counter to the circuit under test.
4. Attempting to measure only the alternating current in a circuit that also has a dc component value. An ac-coupled instrument (has a blocking input capacitor which excludes the direct current from being measured) must be used.
5. Making ac measurements at a frequency beyond the specification of the test instrument. If this must be done, you can often obtain a typical response curve from the manufacturer for frequencies beyond the standard rating.

This should give some idea of the troubles that can be encountered if special care is not exercised in selecting the proper tool for the job. To cover every electrical instrument made today would require a lengthy textbook. The intent of this text is to cover those electrical instruments most often used in day-to-day maintenance.

MULTIMETERS

In recent years the trend has been to place several functions (volts, amperes, ohms, and the like) within one instrument. This provides a very versatile and portable testing device suitable for varied needs. These multimeters are available with either analog or digital readouts.

Analog Multimeter

The basic analog multimeter shown in Fig. 2.1 is commonly referred to as a VOM (volt-ohm-milliammeter). The VOM shown measures ac and dc volts from 250 mV full scale to 1000 V, dc from 50 μA full scale to 10 A, ohms up to 20 MΩ, and decibels (dB) with a total of 29 ranges. The accuracy on dc is 62 percent of full scale and 63 percent of full scale on ac. The lower-priced and most popular type of VOM has input sensitivities of 20,000 Ω/V on dc and 5000 Ω/V on ac. Typical circuits for the various functions are shown in Fig. 2.2a, b, c, and d. Note that the ac circuits are rectifier-type (average sensing). Maximum frequency coverage is about 100 kHz. VOMs of this type are also available with mirror scales providing dc accuracies of 61 percent of full scale and ac accuracy at 62 percent of full scale.

For those applications requiring higher input impedance, such as 10 MΩ, an field-effect transistor (FET) VOM is available, as shown in Fig. 2.3. The basic dc voltage circuit is shown in Fig. 2.4. In addition to the high input impedance, the FET VOM offers other advantages significant for troubleshooting electronic circuits. It typically has a 100-mV full-scale ac and dc range, a 1-μA dc full-scale current range, and "low-power" ohms selection. Conventional VOM ohms circuits will place

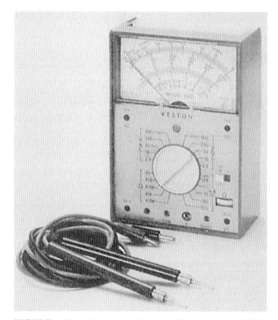

FIGURE 2.1 Typical volt-ohm-milliammeter (VOM). (*Weston Instruments, Inc.*)

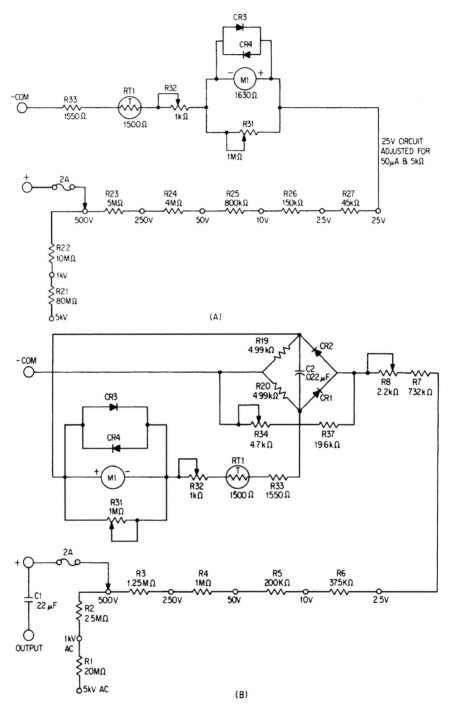

FIGURE 2.2 VOM circuits. (*A*) dc voltage. (*B*) ac voltage.

7.46 INSTRUMENTS AND RELIABILITY TOOLS

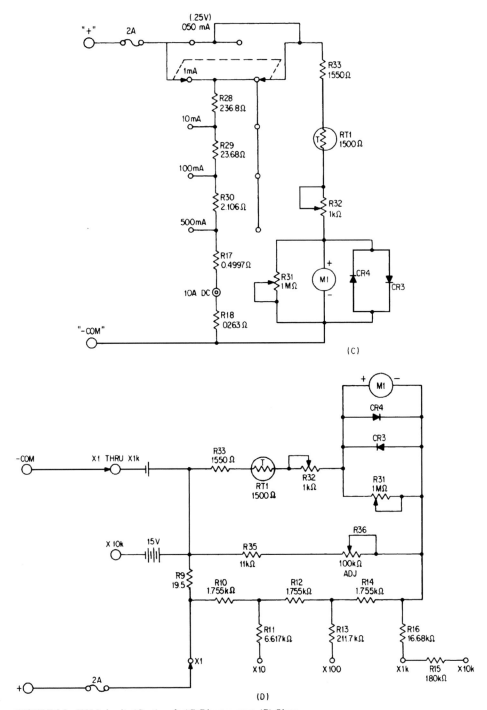

FIGURE 2.2 VOM circuits (*Continued*). (*C*) Direct current. (*D*) Ohms.

FIGURE 2.3 Typical FET VOM. (*Weston Instruments, Inc.*)

from 1.5 to 15 V across the device whose resistance is being measured, and this is sufficient to damage many solid-state electronic components. The FET VOM allows selection of a low-power ohms mode where the voltage placed across the unknown resistance never exceeds 85 mV—safe for most electronic components.

Both types of VOMs, standard and FET, are usually very rugged and well protected against accidental overloads. There are several other versions of analog multimeters designed with specific applications in mind. Figure 2.5 shows an FET VOM which uses a unique set of probes and four terminal inputs, allowing actual measurement of current flow on a printed-circuit board without cutting or disturbing the circuit in any way. Other testers are made primarily for testing transistors and other solid-state devices.

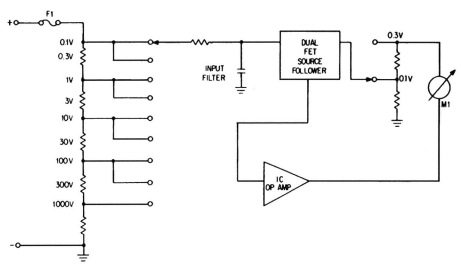

FIGURE 2.4 FET dc volts circuit.

FIGURE 2.5 Special type of "in-circuit-measurement" FET VOM. (*Weston Instruments, Inc.*)

Digital Multimeters

Digital multimeters are now available in fully portable configurations. They are small, accurate, rugged, battery-operated, and light (< $2^1/_2$ lb). Specifications vary from $2^1/_2$-digit (199) units at 1 percent accuracy to $4^1/_2$-digit (19999) instruments with 0.01 percent accuracy, so it is important to know what level of accuracy is actually required. Figure 2.6 shows a four-digit autoranging instrument wit 0.02 percent accuracy, and Fig. 2.7 illustrates a $3^1/_2$-digit (1999) unit with 0.3 percent accuracy. Generally speaking, digital multimeters (DMMs) measure dc and ac volts (rectifier type), dc and ac, and ohms. Ranges begin at 100 mV full scale and go to 1000 V self-contained. Current ranges start at 100 μA up to 2 A. Ohms usually covers from 100Ω to 10 MΩ. In most cases, the input impedance for voltage ranges is high (10 MΩ and up).

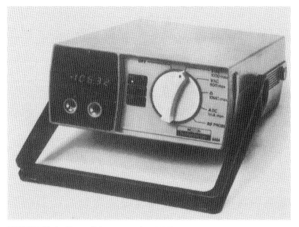

FIGURE 2.6 Four-digit autoranging DMM. (*Weston Instruments, Inc.*)

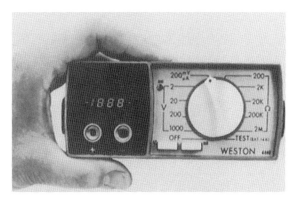

FIGURE 2.7 3½-digit battery-operated DMM. (*Weston Instruments, Inc.*)

Digital instruments can be of great benefit, but on certain applications they encounter problems. The most common one is interference due to electrical noise on the input to the DMM. DMMs are designed with input filtering, but there are cases where the input noise exceeds the filter's rejection limits, and unstable digital readings usually result. This "noise" does not normally affect analog-type meters, since its instrument mechanism inherently rejects the effects of noise—its ballistics are sufficiently slow and damped that it will not respond. What this means is that it is advisable to test a DMM on the actual application prior to purchase.

The language relating to digital instrumentation is relatively new and can be confusing. The following definitions should help in better understanding specifications for this type of equipment.

Number of Digits. A 2½-digit machine gives a maximum reading of 199. A 3-digit machine gives a maximum reading of 999. A 3½-digit machine gives a maximum reading of 1999. A 4-digit machine gives a maximum reading of 9999. A 4½-digit machine gives a maximum reading of 19999.

Long-Term Stability. An accuracy specification for which the instrument is guaranteed over a prolonged period of operation, usually 3 to 6 months.

Monopolar Input. A single-polarity meter—input leads must be reversed to read opposite polarities.

Bipolar Input. A bipolar instrument indicates either positive or negative inputs with appropriate sign.

Common-Mode Rejection. The ability of a meter to be unaffected by noise or interference which is in parallel (common) with the input-signal leads. This is specified in decibel levels.

Normal-Mode Rejection. The ability of a meter to be unaffected by noise or interference which is in series with the input-signal leads. This is specified as a decibel rating.

Overrange. The percentage above the basic range where an instrument will operate and provide useful readings. An instrument with a basic range of 1000 V with 20 percent overrange will read to 1200 V.

Response Time. The time required for the instrument to respond to a step input change and settle down within the stated accuracy.

Resolution. The smallest increment into which the instrument will divide a given input. An instrument that reads 199.99 mV full scale has a maximum resolution of 10 μV (0.01 mV).

FIGURE 2.8 Power analyzer. (*Weston Instruments, Inc.*)

Temperature Coefficient. The effect that varying ambient temperature will have on instrument performance, usually specified per °C from the reference calibration temperature.

Autoranging. The ability of a meter to change ranges automatically within a given function. This provides maximum resolution of the input without any manual switching.

Input Impedance. The effective resistance which the meter presents at its input terminals to the external input circuit.

Where precision measurement is a must (better than 0.5 percent class), one must usually go to an electronic instrument—usually one with digital readout. The variation of digital instruments is rapidly increasing, and it will not be long before there is a digital counterpart for each type of analog meter.

Special-Purpose Industrial Multimeters

As distinct from multimeters, this class of instrument is primarily concerned with 60-Hz measurements of voltage, current, power, and power factor in single- and three-phase measurements. Such a power analyzer has wide acceptance among plant maintenance personnel when these many features are combined in a single package.

The mechanisms of four instruments are contained in one case as seen in Fig. 2.8. The voltmeter, ammeter, wattmeter, and power-factor meter allow true rms analysis of performance of equipment at potential ranges of 150/300/600 V, current ranges of 5/25/125 A, and their corresponding wattmeter ranges.

SINGLE-FUNCTION INSTRUMENTS

Many of the basic test instruments are still of the single-function type, such as tube testers, oscilloscopes, clamp ammeters, Wheatstone bridges, wattmeters, and meggers. This type of instrument is the subject of the following topics.

FIGURE 2.9 Clamp-on ammeter. (*Weston Instruments, Inc.*)

Clamp-On Instruments

Conventional instruments for measuring current are connected in series with the measured circuit. There are, however, many occasions when it is not convenient or desirable to open the circuit so that an ammeter can be inserted. For such applications, an ac instrument has been developed which is inductively coupled to the circuit to be measured.

Clamp-On Ammeter

This type of ammeter utilizes a rectifier-type instrument connected to the secondary of a split-core current transformer so constructed that the core can be conveniently closed around a current-carrying conductor which then forms the primary winding of the current transformer. The method of using a clamp-on instrument to measure current is shown in Fig. 2.9. The electric circuit of the instrument is shown in Fig. 2.10.

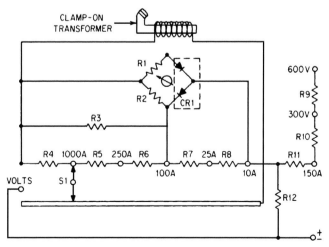

FIGURE 2.10 Schematic diagram of clamp volt-ammeter.

7.52 INSTRUMENTS AND RELIABILITY TOOLS

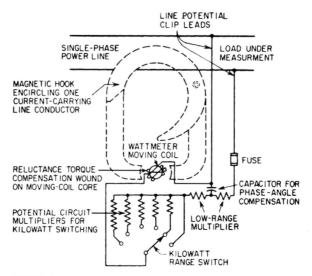

FIGURE 2.11 Schematic diagram for clamp wattmeter.

Current below the maximum calibrated range of the instrument can be measured by looping the line conductor through the transformer opening two or more times; this increases the transformer ratio so that the output indication is increased by a factor equal to the number of primary turns. To increase the utility of the clamp-on instrument, it is sometimes provided with built-in resistors which can be connected in the circuit by means of the range switch, thus making available a rectifier-type ac voltmeter as shown in Fig. 2.10.

Although the rectifier instrument can ordinarily be used in the audiofrequency band when it is incorporated in a clamp-on instrument, the characteristics of the transformer influence frequency response. Consequently, these instruments normally are rated for use between 50 and 70 Hz, although they can be used up to 400 Hz with some decrease in accuracy.

Clamp-On Wattmeter

Another instrument with a built-in split-core transformer is the clamp-on wattmeter for measuring watts in both single-phase and polyphase circuits. Similar in appearance to the clamp-on volt-ammeter, its construction consists of a ferrodynamic wattmeter, the field of which is energized by the conductor carrying the current through the clamp-on transformer. By use of a spring-controlled moving system, the resulting scale capacity is directly proportional to the potential circuit resistance. With this design, range selections can be made by simple one-hand switching of the potential circuit. Figure 2.11 shows a schematic connection diagram of this instrument.

Clamp-On Power-Factor Meter

Another instrument in this family is the clamp-on power-factor meter. It measures the leading or lagging power factor on any balanced three-phase circuit of 100 to 600 V and 15 to 600 A.

The basic instrument is a ferrodynamic galvanometer whose coil moves in the magnetic field produced in the clamp-on transformer by the circuit in a one line conductor. The coil is energized from the potential circuit. Phase relation of the coil voltage is controlled by a potentiometer connected cross-phase. Rotation of the potentiometer dial, until the galvanometer is balanced (zero center), produces a 90° displacement of the coil voltage with respect to the current. Under this condition, power factor can be read directly on the calibrated potentiometer dial. See Fig. 2.12 for a schematic connection diagram of this instrument.

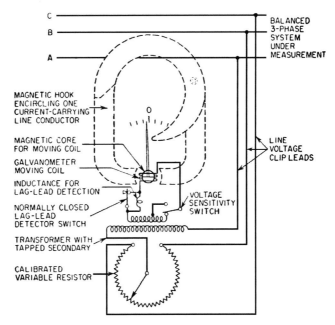

FIGURE 2.12 Schematic diagram for clamp power-factor meter.

AC and DC Volts, Amperes, and Watts

Single-function analog instruments for the measurement of volts, amperes, or watts continue to be very popular for industrial servicing. Some of the reasons are a reputation for years of reliable performance, no requirement for auxiliary power to operate the instrument, usable in almost any environment, and inherent immunity to electrical noise that bothers electronic instruments. Also, some functions, such as wattmeters, are not yet available in electronic counterparts except for laboratory-type equipment.

The main limitations are that the instruments require more power from the circuit under test and accuracies usually run from $1/2$ to $3/4$ percent of full scale with $1/4$ percent full scale being the best available.

DC Volts

Analog dc voltmeters are universally of the permanent-magnet moving-coil type (D'Arsonval). A self-shielding core-magnet version is shown in Fig. 2.13. A typical three-range voltmeter with an accuracy rating of $6 1/2$ percent full scale is shown in Fig. 2.14. These instruments have a standard sensitivity of $1000 \Omega/V$ but can be supplied at $5000 \Omega/V$. Higher sensitivities than this would result in a very delicate meter not suitable for rough industrial use.

Analog instruments with electronic circuitry are made offering input impedances in the 10-$M\Omega$ area. These instruments require either ac power or an optional battery pack to energize the electronics. One such instrument is shown in Fig. 2.15. The typical accuracy is $6 1/2$ percent of full scale.

Direct Current

This instrument appears much the same as the voltmeter in Fig. 2.14 with the basic difference in its internal circuit. Since the moving coil has limited current capacity, it is "shunted" by an internal current shunt which creates a millivolt drop of 50 or 100 mV. It is this millivolt drop that the meter measures, and its value is directly related to the current flow through the shunt.

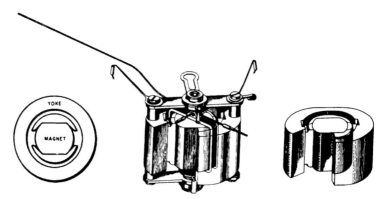

FIGURE 2.13 Core-magnet dc mechanism.

For good portability a minimum range of 20 μA is generally accepted with a maximum self-contained range limit of 50 A. For higher ranges external shunts must be used.

Current ranges lower than 20 μA require an electronically aided instrument such as that shown in Fig. 2.15.

Altenating Current Volts

Instruments in the $1/2$ percent class could be made with permanent-magnet, moving-coil mechanisms through the use of a rectifier bridge network, but this is not the practice. The primary reason is that instruments utilizing iron-vane mechanisms can be manufactured at a similar cost level, and they have one significant advantage: They respond to the true rms value of the input, whereas the rectifier type responds to average values and reads rms correctly only for sine-wave inputs. Since waveforms in industry are seldom perfect sine waves, the iron-vane mechanism is preferred. A form of this mechanism is shown in Fig. 2.16. A typical meter with $3/4$ percent full-scale accuracy is shown in Fig. 2.17.

FIGURE 2.14 Portable three-range dc voltmeter. (*Weston Instruments, Inc.*)

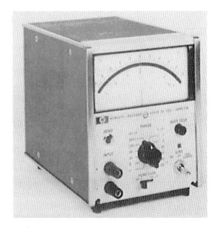

FIGURE 2.15 Electronic-type dc meter with 10-megohm input impedance. (*Hewlett-Packard.*)

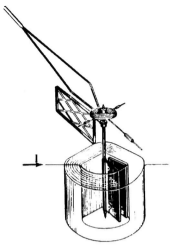

FIGURE 2.16 Iron-vane ac mechanism.

FIGURE 2.17 Portable iron-vane ac voltmeter. (*Weston Instruments, Inc.*)

The iron-vane mechanism is the most rugged and reliable rms unit, but it has some limitations. The major ones are the power taken from the circuit and its frequency limitations. The power required is about $2\frac{1}{2}$ W at 60 Hz, and frequency is limited from 25 to 2500 Hz. This is generally adequate for industrial troubleshooting but may not meet the demands of testing solid-state electronics.

When higher sensitivities and broader frequency coverage are needed, an electronic-type meter with an rms converter (usually a thermocouple type) should be used. Figure 2.18 shows such an instrument. It is limited to ac line operation, however, this type of operation reduces portability.

There is one other class of ac analog portable which utilizes the dynamometer mechanism as shown in Fig. 2.19. This is a true rms sensing mechanism, and it has a significant ability to measure dc inputs with the same accuracy as ac. For this reason it is called a *transfer standard,* meaning that it can be calibrated on dc and then used to measure ac. A portable standard of this type is shown in Fig. 2.20. Rated accuracies are generally $\frac{1}{4}$ percent of full scale on dc or ac with frequency coverage from 15 to 2500 Hz. As with the iron-vane type, power consumption is high. An instrument with a range of 150 V requires about 4 W at 60 Hz.

FIGURE 2.18 Electronic-type true rms meter. (*Hewlett-Packard.*)

7.56 INSTRUMENTS AND RELIABILITY TOOLS

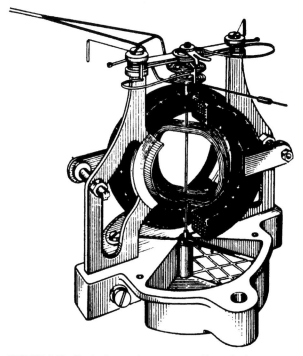

FIGURE 2.19 Single-element dynamometer ac/dc mechanism.

Alternating Current

The comments for ac volts above apply to ac current meters. Current ranges for the iron-vane type start at 15 mA. Approximately $1/2$ W is required from the input at 60 Hz. Electronically aided instruments are available with lower ranges and higher sensitivity. Also electrodynamometer current-transfer standards are used the same as for ac volts. All these instruments are similar in appearance to their ac-volt counterparts.

FIGURE 2.20 Portable dynamometer transfer standard. (*Weston Instruments, Inc.*)

Power Measurements

The field of power measurements in industrial plants is concerned mainly with the electrical loading of machines and equipment supplied from the local power source. In virtually all cases this power source is a 60-Hz single- or three-phase supply, and measurement demands the determination of true rms quantities. For these reasons, the dynamometer instrument virtually dominates the field in view of its versatility as a voltmeter, ammeter, wattmeter, varmeter, and so on. In a wattmeter, irrespective of the number of phases, the stationary coils are supplied with the current to the load and the moving coil with the potential across the load

Single-Phase Connections. Figure 2.21a and b shows the two possible ways of connecting the wattmeter in circuit. Each connection, however, includes the loss of the coil next to the load in the power indicated on the meter. Normally it is easier to correct for the relatively constant load in the potential circuit; hence the usual connection in this form is per Fig. 2.21a. A special type of wattmeter is available that automatically compensates for this error, but in the measurement of power in industrial plants, the above errors are usually insignificant.

To provide instruments having maximum utility in power measurements, it is customary to build portable wattmeters with multirange ratings for the current coil, the potential circuit, or both. The scales of multirange wattmeters normally are graduated for the lowest full-scale rating, and the readings for the higher ranges are determined by multiplying the reading on the low scale by a factor called the *scale constant.*

Instruments having multiple ranges for both current and voltage have scale constants for each range, and the scale reading must be multiplied by both applicable scale constants for the final determination of the power measured. When it is desired to measure power in circuits in which either the current or the voltage (or both) exceeds the ratings of self-contained instruments, the range of a wattmeter can be extended by the use of current or potential transformers. When using external transformers, phase relationships must be carefully observed if correct readings are to be obtained and if the instruments are to be protected against excessive voltage.

Use of Current Transformer. Although wattmeters having current ratings up to 600 A are manufactured, most power measurements involving high current are made by using a suitable current transformer in conjunction with a wattmeter having a current rating of 5 A. The transformer used must have a turn ratio such that its secondary current is within the ampere rating of the wattmeter current coil, and it must be connected in the correct polarity to ensure maintenance of the original phase relationship between the stationary and moving coils of the instrument. The terminals of current transformers normally are marked to indicate direction of current.

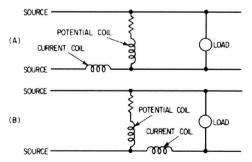

FIGURE 2.21 Two possible connections of a simple single-phase wattmeter. (*A*) Instrument measures load plus power consumed in potential circuit. (*B*) Instrument measures load plus power consumed in current coil.

Use of Potential Transformer. For potentials up to 700 V, series resistors are used in the potential circuit to increase the range of a wattmeter; for higher voltages, potential transformers are employed. As with current transformers, it is essential that the correct polarity be observed in connecting the instrument, the transformers being carefully marked to indicate polarity.

When a wattmeter is used with a current or a potential transformer, or both, the operator must be careful to include all the applicable multiplying factors in the final evaluation of the wattmeter reading. It is necessary to take into consideration the ratio of each transformer and to include in the calculation of indicated watts a multiplying factor corresponding to the ratios.

Precautions in Use

Wattmeters are rated not only in watts full-scale deflection but also in amperes and volts. The ampere ratings indicate the maximum current that should be passed through the field coil, and the voltage ratings indicate the range of potential-circuit voltage for which the instrument is designed. The ampere rating is established by the number of turns and the cross-sectional area of the conductor in the field coil, and the voltage rating is established by the size and heat-dissipating capacity of the moving coil and its series resistors. A wattmeter having a 5-A current rating would have a full-scale rating of 250 W at 50 V, 500 W at 100 V, and 1000 W at 200 V. It is easily seen that if the 200-V instrument were operated at 50 V, a reading of only 250 W would be obtained with the maximum rated current of 5 A in the field coil. To reach the full-scale reading of 1000 W, a current of 20 A would be required. Thus the instrument might be so severely overloaded that it would be seriously damaged, even though its deflection did not exceed full scale.

Three-Phase Wattmeter. If it can be assumed that the load is evenly balanced with respect to all phases, a single-phase wattmeter may be used in one phase as before and the result multiplied by 3. This is not, however, usually to be relied on, and for most circuits where the condition of balanced loads is uncertain, a double-element dynamometer is used. This is accomplished by mounting two sets of rotating potential coils on a single shaft and mounting two sets of coils on the instrument frame in correct relation to the armature coils. Thus each mechanism develops torque proportional to the power in the circuit to which it is connected, and the torques of both mechanisms add to indicate the total power in the three-wire circuit. An instrument constructed in this manner (Fig. 2.22) is called a *two-element wattmeter,* because it consists of two electrically independent dynamometer mechanisms.

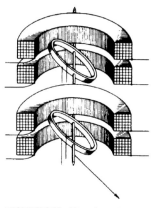

FIGURE 2.22 Two-element dynamometer.

This double mechanism is applicable to the measurement of power in a three-phase, three-wire circuit with either balanced or unbalanced load. Its connection for such measurement is shown in Fig. 2.23. Note that mechanism A produces a torque proportional to the line potential (1–2) and the vector sum of the current from phases a and c. Mechanism B produces a torque proportional to line potential (3–2) and the vector sum of the current from phases b and c. The sum of the torques of the two mechanisms is thus proportional to the total power in the circuit. As with the single-element instrument, the phase differences of voltage and current (power-factor angle) produce corresponding differences in torque of each mechanism, with the result that an indication of true power is obtained.

Blondel's theorem states that true power can be measured by one less wattmeter element than the number of wires of the system, providing one wire can be made common to all element potential circuits. A two-element meter connected as shown in Fig. 2.23 meets this theorem for three-phase, three-wire systems.

Resistance Measurement

Just as in measuring amperes and volts, no one instrument covers the broad range of resistance values that must be measured in an industrial plant. This discussion will cover the major types of resistance-measuring instruments.

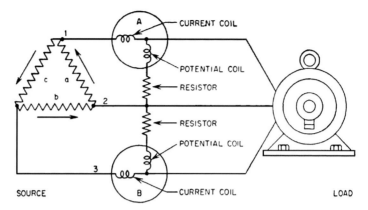

FIGURE 2.23 Simple three-phase, three-wire system with two-element wattmeter measuring load.

Simple Ohmmeter

Figure 2.24 gives the circuit diagram of the conventional ohmmeter consisting of a milliammeter or microammeter, a series resistor, a small battery, and means for setting the current so that the instrument reads 0 Ω, that is, full-scale deflection, with a short circuit across the ohmmeter terminals. This is the basic circuit used in industrial VOMs discussed under "Multimeters" earlier in this chapter.

In this circuit, the deflection of the instrument pointer varies inversely with the value of the unknown resistance and for a given voltage, the scale can be graduated in terms of the series resistance added to the circuit by the test specimen. Full-scale deflection is 0 Ω, half-scale deflection shows a resistance equal to the instrument-circuit resistance, and zero deflection indicates infinite resistance. Hence the scale distribution and the range through which satisfactory accuracy is obtainable depend on the resistance connected in series with the indicating instrument. A high-range ohmmeter would therefore employ a relatively high value of resistance in series with the indicator. The voltage of the battery is determined by the indicator current sensitivity.

Megger

This is another portable piece of equipment widely used for detecting insulation weaknesses. Shown in diagram form in Fig. 2.25, the hand generator serves as a source of emf. The normal maximum

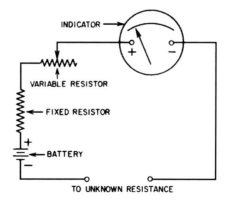

FIGURE 2.24 Diagram of simple ohmmeter.

7.60 INSTRUMENTS AND RELIABILITY TOOLS

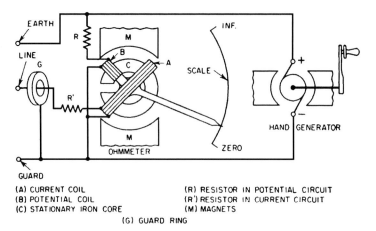

(A) CURRENT COIL
(B) POTENTIAL COIL
(C) STATIONARY IRON CORE
(G) GUARD RING
(R) RESISTOR IN POTENTIAL CIRCUIT
(R') RESISTOR IN CURRENT CIRCUIT
(M) MAGNETS

FIGURE 2.25 Simplified circuit diagram of megger insulation tester. (*James G. Biddle Company.*)

value of this emf is 1000 V, although it may go as high as 2500 for special models. Figure 2.26 illustrates a typical hand-cranked instrument which is supplied with ranges up to 20,000 MΩ. Line-operated units are available.

This instrument is of the crossed-coil-ratio type of design which has two crossed coils fixed on the same shaft. Coil *A* in series with resistance *R* is called the *current coil* and is connected from one side of the generator to the line terminals. The operating torques tend to turn these coils in opposite directions until a position of equilibrium is reached. When the instrument is operated, that is, when the crank is turned, and with an infinite resistance connected between the earth and line terminals, no current will flow in the current coil. The *potential coil* alone, then, will control the motion of the crossed-coil assembly in the gap in the C-shaped iron core until the pointer indicates infinity.

However, when a resistance is connected across the terminals, a current will flow in the current coil, and the corresponding developed torque in the current coil will draw the potential coil away

FIGURE 2.26 Portable megger. (*James G. Biddle Company.*)

from the infinity position into a field of gradually increasing magnetic strength until a balance is obtained between the forces acting on the respective coils. In this manner, the potential coil acts like a restraining spring. When resistances of different known values are introduced across the terminals of the instrument and the corresponding position of the pointer is marked in each case, a scale calibrated in resistance can be obtained. If a very low resistance is connected between the earth and line terminals or should the latter become short-circuited, the pointer simply moves to zero, or off the lower end of the scale of the high-range megger instrument, the ballast resistance R' offering ample protection against excessive current in the current coil. Because the two elements of the true ohmmeter receive current from the same source, namely, the generator, any change in generator voltage will affect both coils in the same proportion, and therefore, the pointer will move to the same position for a given resistance under test. Consequently, the calibration of the instrument is unaffected by the speed at which the crank is turned or by the strength of the permanent magnets. This is one of the outstanding advantages of the megger type of ohmmeter.

Precautions in Use. The moving system is mounted in the usual spring-supported jewel bearings; however, no controlling springs as such are used. Fine filaments are necessary to bring current in and out of the moving coil, but these are made as fine as possible to allow the moving system to reach the final equilibrium position without any spring influence. When the generator is being turned, therefore, there is no controlling torque to return the pointer off scale. Readings thus must be recognized only when the generator is actually turning.

Wheatstone Bridge

A more accurate, though a somewhat less convenient, means of measuring most commonly encountered values of resistance is the Wheatstone bridge. The apparatus, available in portable form (Fig. 2.27) comprises a battery, a galvanometer, and three resistance elements, any or all of which may be adjustable in known steps, as shown schematically in Fig. 2.28. With the unknown resistance connected at X, one or more of the other resistances is adjusted until zero reading on the galvanometer shows that g_1 and g_2 are at the same potential with respect to b_1 and b_2. Since the voltages in a series circuit are proportional to the resistance of the various elements, the value of X is obtained from the formula

FIGURE 2.27 Portable Wheatstone bridge. (*Leeds and Northrup.*)

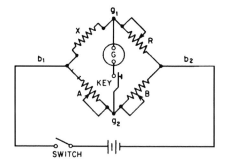

FIGURE 2.28 Wheatstone-bridge circuit.

$X = (A/B)R$. For convenience in solving this equation, two of the resistance arms usually are made adjustable in ratios of 1 to 10, 1 to 100, 1 to 1000, and so on, and the third is made adjustable in unit values.

Precautions in Use. The battery switch should always be closed before the galvanometer key in order to prevent any momentary rush of current through the galvanometer should the unknown resistance be inductive.

Low-Resistance Determination

When extremely low values of resistance are to be measured, the ohmmeter is seldom accurate enough, and the Wheatstone bridge is not satisfactory because of the error introduced by contact resistances and the connecting leads. The Kelvin double bridge (Fig. 2.29) measures resistance down to 0.0001 Ω accurately and conveniently and is available in portable form.

In the double bridge (Fig. 2.30), the unknown resistance X is connected in series with a known resistance K and a battery such that a relatively heavy current flows, producing a voltage drop across the known and the unknown resistances. A galvanometer is connected across the two voltages by two ratio

FIGURE 2.29 Portable Kelvin bridge. (*Leeds and Northrup.*)

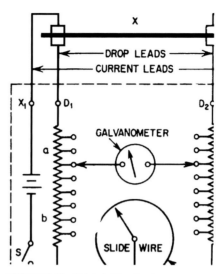

FIGURE 2.30 Kelvin-bridge circuit.

arms so arranged that the ratio a/b always equals c/d. The effect of the voltage across X is to produce a galvanometer deflection proportional to $1X/(a + c)$, and the effect of the voltage across K is to produce an opposite deflection equal to $1D/(b + d)$. For the galvanometer to give zero deflection, these two factors must be equal, and since 1 is the same in both equations, the balance condition can be stated as $X = K(a + c)/(b + d)$.

The ratio arms normally are made with a plugging or switching arrangement for obtaining the various ratios (multiplying factors), and the known resistance K is arranged so that any desired part of its total voltage drop can be applied to the galvanometer circuit. For maximum sensitivity, the current in the measuring circuit is made relatively large, but its exact value does not influence the actual measurement.

The double bridge is provided with four terminals for connection to the unknown resistance.

High-Resistance Determination

In this realm of measurement of high resistance, insulation measurements predominate. For this reason, the voltage applied to the insulator is of the utmost importance, being in almost all cases the determining factor in the test equipment used.

Simple Ohmmeter. Using a higher-voltage battery, say 15 V, the circuit of Fig. 2.24 can indicate up to 100 MΩ.

Megger. This general-purpose instrument described previously is available with generator voltages up to 2500 V and is capable of measurement up to 50,000 MΩ. Voltages up to 10,000 V can be supplied for standard rectifier-operated units with ranges exceeding 100,000 MΩ.

Frequency Counters—Timers

In the past, frequency counters were generally associated with communication equipment, but today they can be a handy service instrument for many types of electronic circuits used in industry. This class of instrumentation has experienced drastic changes, mainly because of new solid-state circuit packages. Small portable units with autoranging up to 110 MHz are common at prices many times lower than a few years ago. Instruments for frequency and time measurement can be divided into several groups.

Counters and counter-timers rely on some source of frequency as their reference, and the quality of this reference directly relates to the accuracy of the instrument. Some of the lowest-priced units use the 60-Hz frequency from standard power lines, but most have a self-contained crystal for the frequency source. These crystals range from the inexpensive to quite high-priced units housed in temperature-controlled ovens for increased stability. All crystals have an aging rate (drift), and it is primarily this factor that affects the accuracy of the instrument. In selecting a counter, make sure the aging rate of the crystal will ensure the instrument remains within the required accuracy.

Basic Counters

Most of these counters are of the five- to nine-digit type designed to count frequency only; quite often they will measure frequency ratio also. A typical unit shown in Fig. 2.31 offers a range of 5 Hz to 110 MHz with a sensitivity of 15 mV and standard 1-MΩ input shunted by 15 power factor. It also has autoranging with a seven-digit display. It uses a 1-MHz crystal for a reference with an aging rate of 7.5×10^{-6}/year. There are many varieties of basic counters available—some with frequency ranges into the gigahertz range and stabilities better than 1.5×10^{-7}/year.

Counter-Timers

The counter-timer does more than just measure frequency. As its name indicates, it is also a timer—something like an electronic stopwatch. These units will generally totalize input pulses; they will measure the time "period" of a single cycle wave, the pulse width of a single pulse, the average period of multiple waveforms over fixed period of time, the ratio of two frequencies, and time intervals between two single pulses that may be related or not, such as two photocells—one at the start of a conveyor with the other at the other end.

It is evident that the counter-timer is a very flexible tool, but it is larger than the simple frequency counter and usually more expensive for a related frequency capability.

Oscilloscopes

The oscilloscope is the TV set of electronics. It allows you to "see" electricity and make measurements which are impractical by any other means. By observing waveshapes, pulse trains, trigger pulses, and so on, you can often troubleshoot and check electronic-circuit operation faster than with any other means.

FIGURE 2.31 Portable electronic frequency counter. (*Weston Instruments, Inc.*)

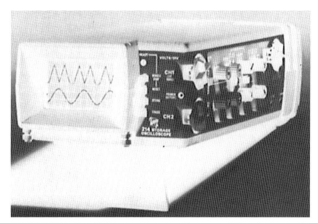

FIGURE 2.32 Portable battery-operated oscilloscope. (*Tektronix Company.*)

Again, recent developments in technology have made it difficult to select the instrument best suited for your needs. Laboratory models offer real-time bandwidths up to 1 GHz and as high as 14 GHz on a sampling basis. Battery-operated portables offer up to 350-MHz bandwidths. Figure 2.32 shows a small compact portable oscilloscope suitable for many industrial applications.

It is not practical to cover all types of oscilloscopes in a text such as this. So many options are offered that a manufacturer's catalog will often run well over 100 pages. Just one example is the option of different phosphors for the cathode-ray tube—some photograph better, while others are better for use in high ambient-light areas.

Solid-State Circuit and Tube Checkers

Tube Checkers. Even though the use of solid-state components increases every year, it will be some time before they will completely replace the electron tube; so the need for tube checkers will continue. Tube checkers come in two basic types: The emission-type and those which are based upon the value of mutual conductance g_m.

The *emission type* measures the electron emission of the heated cathode and is the simplest and least expensive. It also has provisions for checking shorted elements, leakage, presence of gas, and so on. This type of tester is useful for ordinary maintenance, but it does not provide complete data as to the condition of the tube when used in a typical circuit.

The *mutual-conductance type* more closely allows you to monitor tube performance under operating conditions. The reading takes into account not only cathode emission but also the change in plate current caused by a signal introduced into the grid circuit. Usually, the scale of the meter is marked so that a reading may be made directly in micromho of mutual conductance. These testers also allow for noise, gas, shorts, and leakage measurements. It is suggested that if only a limited number of tube checkers are used for service work, they be of the mutual-conductance type.

Another version of the mutual-conductance type is often referred to as a *dynamic tube checker.* In performance, it is between the emission type and the mutual-conductance unit.

The tube testers offered for servicing consumer products such as TVs are quite often this type.

Solid-State Circuit Testers. Testing solid-state devices presents a problem similar to trying to obtain a tube tester in the late 1920s and early 1930s. Solid-state circuit designs are changing so fast that if a portable tester were designed today, it would be obsolete in less than 2 years. The one item that has had some design stabilization is discrete components, such as transistors, diodes, and FETs. Small portable testers do exist for these components.

Troubleshooting other solid-state circuits can be done with other instruments already discussed, such as the digital multimeter, counters, and oscilloscopes. This is not to say that testing devices have not been made for these circuits—only that they are either quite limited or else are of the production or laboratory type not easily transported for field use. There are a number of "logic" probes which are useful in determining the high and low condition of digital-logic integrated circuits but nothing comparable with the portable tube checkers.

Miscellaneous Instruments. As mentioned earlier, it is not practical to cover all forms of electrical instruments in a text such as this. There are many specialized instruments for which you may have need, and if so, it is suggested you contact another text more directly related to that type of instrument or contact some known manufacturer for data. These instruments include

- Signal generators
- Nuclear-radiation detection instruments
- pH/conductivity instrumentation
- Noise-measuring instruments
- Phase-angle meters
- Synchroscopes
- Illumination (light) measurements
- Capacitance bridges
- Temperature instruments
- Pollution-measuring instruments
- Speed-measuring instruments

GLOSSARY*

Accelerating voltage The cathode-to-viewing-area voltage applied to a cathode-ray tube for the purpose of accelerating the electron beam.

Alternate display A means of displaying output signals of two or more channels by switching the channels in sequence.

Astigmatism In the viewing plane of the cathode-ray tube, any deviation of the indicating spot from a circular shape.

Attenuator A device for reducing the amplitude of a signal without deliberately introducing distortion.

Automatic triggering A mode of triggering in which one or more of the triggering-circuit controls are preset to conditions suitable for automatically displaying repetitive waveforms. The automatic mode also may provide a recurrent trigger or recurrent sweep in the absence of triggering signals.

Bandwidth Of an oscilloscope, the difference between the upper and lower frequency at which the voltage or current response is 0.707 (23 dB) of the response at the reference frequency. Usually both upper and lower limit frequencies are specified rather than the difference between them. When only one number appears, it is taken as the upper limit.

Note 1: The reference frequency shall be (1) for the lower bandwidth limit, 20 times the limit frequency, and (2) for the upper bandwidth limit, $\frac{1}{20}$ the limit frequency. The upper and lower reference frequencies are not required to be the same.

―――――――――――――
*Reprinted by permission of Tektronix, Inc.

Note 2: This definition assumes the amplitude response to be essentially free of departures from a smooth roll-off characteristic.

Note 3: If the lower bandwidth limit extends to dc, the response at dc shall be equal to the reference frequency, not 23 dB from it.

Beam finder   A provision for locating the spot when it is not visible.

Blanking   Extinguishing of the spot. Retrace blanking is the extinction of the spot during the retrace portion of the sweep waveform. The term does not necessarily imply blanking during the holdoff interval or while waiting for a trigger in a triggered-sweep system.

Brightness   The attribute of visual perception in accordance with which an area appears to emit more or less light.

Chopped display   A time-sharing method of displaying output signals of two or more channels with a single cathode-ray-tube gun, at a rate which is higher than, and not referenced to, the sweep rate.

Chopping transient blanking   The process of blanking the indicating spot during the switching periods in chopped-display operation.

Common-mode signal   The instantaneous algebraic average of two signals applied to a balanced circuit, both signals referred to a common reference.

Conventional mode   That mode of operating a storage tube where the display does not store but performs with the usual phosphor luminance and decay.

Deflection blanking   Blanking by means of a deflection structure in the cathode-ray-tube electron gun which traps the electron beam inside the gun to extinguish the spot, permitting blanking during retrace and between sweeps regardless of intensity setting.

Deflection factor   The ratio of the input-signal amplitude to the resultant displacement of the indicating spot (e.g., volts/division).

Delay pickoff   A means of providing an output signal when a ramp has reached an amplitude corresponding to a certain length of time (delay interval) since the start of the ramp. The output signal may be in the form of a pulse, a gate, or simply amplification of that part of the ramp following the pickoff time.

Delayed sweep   A sweep that has been delayed either by a predetermined period or by a period determined by an additional independent variable.

Dual beam   A multitrace cathode-ray tube which produces two separate electron beams that may be individually or jointly controlled. In contrast to dual-trace operation, waveforms are not chopped or switched; they are generally brighter and have minimum phase-relationship error.

Dual trace   A multitrace operation in which a single beam in a cathode-ray tube is shared by two signal channels. See alternate display, chopped display, and multitrace.

Focus   Maximum convergence of the electron beam manifested by minimum spot size on the phosphor screen.

Gaussian response   A particular frequency-response characteristic following the curve $y(f) = e^{-af^2}$. Typically, the frequency response approached by an amplifier having good transient-response characteristics.

Geometry   The degree to which a cathode-ray tube can accurately display a rectilinear pattern. Generally associated with properties of a cathode-ray tube; the name may be given to a cathode-ray tube electrode or its associated control.

Graticule   A scale for measurement of quantities displayed on the cathode-ray tube of an oscilloscope.

Input RC characteristics   The dc resistance and parallel capacitance to ground present at the input of an oscilloscope.

Intensity modulation   The process and (or) effect of varying the electron-beam current in a cathode-ray tube resulting in varying brightness or luminance of the trace.

Internal graticule   A graticule whose rulings are a permanent part of the inner surface of the cathode-ray-tube faceplate.

Jitter   An aberration of a repetitive display indicating instability of the signal or of the oscilloscope. May be random or periodic and is usually associated with the time axis.

Magnified sweep   A sweep whose time per division has been decreased by amplification of the sweep waveform rather than by changing the time constants used to generate it.

Mixed sweep   In a system having both a delaying sweep and a delayed sweep, a means of displaying the delaying sweep to the point of delay pickoff and displaying the delayed sweep beyond that point.

Multitrace   A mode of operation in which a single beam in a cathode-ray tube is shared by two or more signal channels. See dual-trace, alternate display, and chopped display.

Resolution   A measure of the total number of trace lines discernible along the coordinate axes, bounded by the extremities of the graticule or other specific limits.

Risetime   In the display of a step function, the interval between the time at which the amplitude first reaches specified lower and upper limits. These limits are 10 and 90 percent of the nominal or final amplitude of the step, unless otherwise stated.

Roll-off   A gradually increasing loss or attenuation with increase or decrease of frequency beyond the substantially flat portion of the amplitude-frequency response characteristic of a system or transducer.

Signal delay   In an oscilloscope, the time required for a signal to be transmitted through a channel or portion of a channel. The time is always finite, may be undesired, or may be purposely introduced as in a delay line.

Tangential noise measurement   A procedure to determine displayed noise wherein a flat-top pulse or square-wave input signal is adjusted in amplitude until the two traces (or portions of two traces) thus produced appear to be immediately adjacent or contiguous. Measurement of the resulting signal amplitude determines a noise value which correlates closely with the value interpreted by the eye for a sampling display and is called the *tangential noise value*.

Trigger   A pulse used to initiate some function (for example, a triggered sweep or delay ramp).

Unblanking   Turning on of the cathode-ray-tube beam.

CHAPTER 3
VIBRATION: ITS ANALYSIS AND CORRECTION

R. Keith Mobley
Principal, Life Cycle Engineering, Inc., Charleston, S.C.

VIBRATION ANALYSIS FOR PREDICTIVE MAINTENANCE

Vibration monitoring and analysis are two of the most useful tools for predicting incipient mechanical, electrical, and process-related problems within plant equipment, machinery, and continuous-process systems. Therefore, they are the most often used predictive maintenance technologies. Used in conjunction with other process-related measurements, such as flow, pressure, and temperature measurements, vibration analysis can provide the means to first schedule maintenance and ultimately to eliminate the need for corrective maintenance tasks.

Vibration monitoring and analysis can be used to evaluate all mechanical and most continuous-process equipment within a manufacturing or production plant. They are not limited to simple rotating machines. Until recently, slow-speed machinery, especially complex, continuous-process lines, were excluded from the useful range of vibration analysis. Recent technology developments have removed this limitation and now permit the use of vibration analysis techniques for machinery with primary speeds as low as 6 rpm.

OTHER USES OF VIBRATION ANALYSIS

The use of vibration analysis should not be restricted to predictive maintenance. The diagnostic capability of this analysis technique has an abundance of useful applications. Some of its other areas of use are as follows.

Acceptance Testing

Vibration analysis is a proven means of verifying the actual performance versus design parameters of new mechanical, process, and manufacturing equipment. Tests performed at the factory and immediately following installation will ensure that new equipment will perform at optimal efficiency and achieve anticipated life-cycle costs before final acceptance. Design problems, as well as possible damage during shipment or installation, can be corrected before long-term damage can occur.

Quality Control

Vibration checks on a production line are an effective method of ensuring product quality. Where machine tools are involved, vibration checks can provide advanced warning that the surface finish on parts is nearing the reject level. On continuous-process lines, such as paper machines, steel finishing lines, and rolling mills, vibration analysis can prevent abnormal oscillation of line components that would result in loss of product quality.

Loose Parts Detection

Vibration analysis can be used as a diagnostic tool that will locate loose or foreign objects within critical plant process lines or vessels. This technique has been used with great success within the nuclear power industry and has similar industrial applications.

Noise Control

Federal, state, and local noise-control regulations require serious attention to noise levels within plants. Vibration analysis can be used to isolate the source of both airborne and machine-generated noise. The ability to isolate the source of abnormal noise will permit cost-effective corrective action.

Leak Detection

Leaks in process vessels, such as valves, are a serious problem in many integrated-process industries. A variation of vibration monitoring and analysis can be used to detect leakage and isolate its source. Leak-detection systems, such as the valve-flow system illustrated in Fig. 3.1, use an accelerometer that is attached to the exterior of a process pipe. The noise or vibration pattern is then monitored to detect the unique frequencies that are generated by flow or leakage.

FIGURE 3.1 Valve-flow monitoring system.

Aircraft Engine Analyzers

Vibration analysis techniques have been adapted for a variety of specialty instruments. For example, vibration monitoring and analysis techniques have been used to develop portable analyzers for turboprop and jet engines. These instruments are normally programmed with logic modules that use traditional vibration data to automatically evaluate the engine's condition and report any deviation from optimal operating condition (Fig. 3.2).

Machine Design and Engineering

Vibration data have become a critical part of the design and engineering of new machines and process systems. Data derived from similar or existing machinery is extrapolated to form the basis of the new design. Prototype testing of new machinery and systems also provides invaluable design data.

Production Optimization

The operating dynamics data derived from vibration analysis can provide the means to increase machine capacity without increasing maintenance or production costs or reducing product quality. Such data can be used to isolate operator-related problems as well as design or operating procedure restrictions that limit capacity. This is especially beneficial for integrated- or continuous-process systems that rely on the effective interaction of multiple mechanical and electrical components.

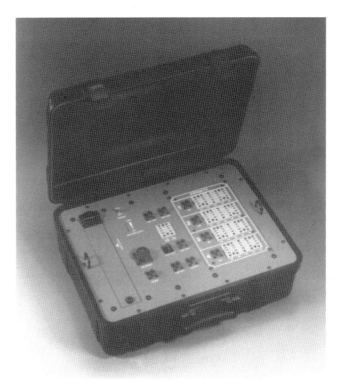

FIGURE 3.2 Jeda jet engine data-acquisition module.

WHAT IS VIBRATION?

While everyone has an intuitive knowledge of what vibration is, when it is to be measured accurately and used as an indicator of machinery mechanical condition, it should be explicitly understood.

Vibration, which is technically defined as the oscillation of an object about its position of rest, can best be described by the simple vibrating system in Fig. 3.3. If the mass in Fig. 3.3 is set in motion, it will move back and forth between some upper and lower limits. This movement of the mass through all its positions and back to the point where it is ready to repeat the motion is defined as *one cycle of vibration.* The time it takes to complete this cycle is the *period of vibration.* The number of these cycles in a given length of time (e.g., 1 minute) is the *frequency of vibration.* Frequency is usually stated in cycles per minute (cpm) or cycles per second (cps). These are called *hertz* (Hz) in honor of the German physicist. For example, a machine may vibrate at 3600 cpm, which is the same as 60 Hz.

Frequency is one of the basic characteristics used to measure and describe vibration. Others include *displacement, velocity,* and *acceleration.* Each of these characteristics describes the severity of vibration; they are illustrated in Fig. 3.4.

It should be noted that the relative amplitudes shown in the figure are for a frequency of 8400 cpm. At lower frequencies, the displacement amplitude would be relatively higher and the acceleration amplitude lower, while the converse would be true at higher frequencies.

Displacement indicates "how much" the object is vibrating. It is usually defined as the maximum distance the object moves (i.e., its lowest to its highest point) and is labeled "peak-to-peak" displacement. It is measured in units of mils (0.001 in), peak-to-peak. In metric units it is measured as microns peak-to-peak (1 μm = 0.001 mm).

Velocity indicates "how fast" the object is vibrating. For the simple harmonic motion in Fig. 3.4, it can be seen that the velocity is highest where the object is passing through its position of rest (i.e., zero displacement) and zero at the upper and lower maximum-displacement limits. When measured, its maximum value, that is, its peak velocity, is usually the value that is recorded. It is measured in units of inches per second peak. In metric units it is measured as millimeters per second peak.

The *acceleration* of the object which is vibrating is related to the forces which are causing the vibration. Again, it can be seen in Fig. 3.4 that acceleration reaches a maximum value as the object reaches its maximum limits of displacement, that is, as it slows down, stops, and starts moving in the opposite direction. The maximum, or peak, acceleration measured is usually the value recorded. It is measured in units of g peak (1 g = 5386 in/s^2 or 1 g = 980 cm/s^2).

One other characteristic of vibration is important for diagnosis and correction of machinery problems. This is *phase,* or *phase angle.* It is used when comparing the motion of a vibrating part with a fixed reference, or comparing two parts of a machine structure vibrating at the same frequency.

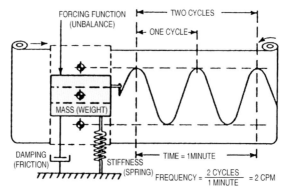

FIGURE 3.3 Simple vibrating system.

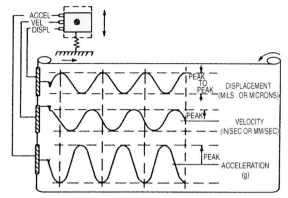

FIGURE 3.4 Relationships among displacement, velocity, and acceleration.

It can be defined as the angular difference at any given instant between two parts with respect to a complete vibration cycle and is usually expressed in degrees. This can perhaps be better seen in Fig. 3.5, which gives a visual representation of phase measured with respect to a fixed position, namely, the position of rest of the simple vibrating-mass system.

Another way of looking at phase is shown in the example in Fig. 3.6. Here, two vibrating parts are compared. In Fig. 3.6a the parts are vibrating in phase (i.e., they are moving together) and are said to have 0° phase angle. In Fig. 3.6b the parts are vibrating out of phase (i.e., they are moving opposite to each other) and have a 180° phase angle.

In actual practice, phase is usually measured with a strobe light whose flash is directed at a reference mark (e.g., a white line painted on a shaft) on the machine's shaft. The flash is triggered by the machine vibration so that when the vibration is at the machine's rotational frequency, the reference mark will appear to stand still. The angle at which the reference mark stops with respect to a fixed reference point is the phase angle for that vibration.

When measuring machinery vibration, it is common practice to measure frequency, phase, and displacement or velocity or acceleration. Since each of the latter three is a measure of vibration severity, in most cases the measurement of one of them is sufficient to describe the vibration. Which one is selected depends on a number of factors which will be discussed later in this chapter. It can be said, however, that for most vibration preventive maintenance programs, velocity is a good choice.

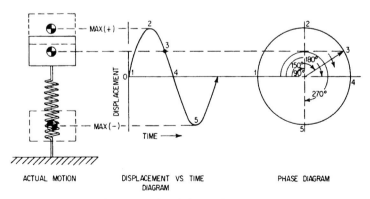

FIGURE 3.5 Schematic representation of phase angle.

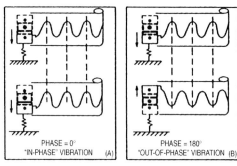

FIGURE 3.6 Phase angle between two vibrating parts.

CAUSES OF MACHINE VIBRATION

All machines vibrate. These vibrations are caused by the tolerances which the machine designer has allowed so that the machine can be built, since some dimensional variations are inherent in any machine's manufacture. These tolerances give a new machine a characteristic vibration "signature" and provide a base line against which future measurements can be compared. Similar machines in good operating condition will have similar vibration signatures which differ from each other only by their manufacturing and installation tolerances.

A change from the base line of the vibration of a machine, assuming it is operating under normal conditions, indicates that an incipient defect is starting to change the mechanical condition of the machine. Different defects cause the vibration signature to change in different ways, thus providing a means of determining the source of the problem as well as warning of the problem itself.

CHARACTERISTICS OF VIBRATION

Vibration is a natural product of the mechanical and dynamic forces within machinery, plant equipment, and process systems. All mechanical equipment and most dynamic plant systems, such as heat exchangers, filters, mixing tanks, and chemical reaction vessels, will generate some level of vibration as part of their normal operation. A clear understanding of machine dynamics and how these forces create unique vibrational frequency components is the key to using vibration data as a diagnostic tool.

Machinery vibration is the result of a series of individual vibration components that are generated by the movement or generated forces of mechanical or process components within the machine or its corresponding system. Each of these individual vibration components has a well-defined periodic motion. That is, the motion will repeat itself in all its particulars after a specific interval of time. The interval or time period T in which the vibration repeats itself is usually measured in seconds. Its reciprocal is the *frequency* of the vibration and is normally measured in terms of *cycles per second* (cps), or *hertz* (Hz).

Each frequency component within a machine train can be calculated by

$$f = \frac{1}{T}$$

For example, a vibration component that repeats itself once each second would have a frequency of

$$f = \frac{1}{1 \text{ s}} = 1 \text{ cps} = 1 \text{ hertz}$$

The simplest kind of periodic motion is a *harmonic motion*. In harmonic motion, the relationship between the maximum displacement and time may be expressed by

$$x = x_o \sin \omega_t$$

The maximum value of the displacement is x_o, and is called the *amplitude of the vibration*. The circular frequency ω is measured in radians per second. From Fig. 3.5 it should be clear that a full cycle of vibration takes place when ω_t has passed through 360°, or 2π radians; then the sine function resumes its previous values.

The amplitude of vibration may be expressed in three basic terms or units: *displacement, velocity, or acceleration*. While all three of these terms are expressions of the vibration amplitude, each of these units expresses totally different values.

Displacement. Displacement, expressed in terms of mils, is a measurement of the actual distance traveled by the machine component that generated the vibration component.

Velocity. In a harmonic motion for which the displacement is given by $x = x_o \sin \omega_t$, the velocity is found by differentiating its displacement with respect to time, or

$$\frac{dx}{dt} = \dot{x}(x_o\omega)(\cos \omega_t)$$

Therefore, velocity is also an expression of harmonic vibration with a maximum value of ωx_o. The resultant of the differentiation of displacement with respect to time is expressed in terms of inches per second (in/s) and is therefore equal to the distance traveled per unit of time, not the total distance of movement.

Acceleration. Harmonic vibration also can be expressed in terms of how fast the vibrating component is moving, or ωx_o. Acceleration or g is expressed in terms of inches per second per second and can be calculated as

$$\frac{d^2}{dt^2} = \ddot{x} = (-x_o\omega^2)(\sin \omega_t)$$

Phase. Phase or phase angle is a means used to define the relationship of harmonic vibration components. Consider the two vibration components expressed as

$$x_1 = a \sin \omega_t \quad \text{and} \quad x_2 = b \sin \omega_t + \phi$$

When these two vibration frequency components are plotted against ω_t as the abscissa, the phase relationship can be visualized. Because of the difference in phase (ϕ), the two vibration components do not attain their maximum displacements at the same time. Instead, the vibration component represented by x_2 achieves its maximum displacement ϕ/ω seconds later than component x_1. The quantity ϕ is known as the *phase angle*, or time difference between two vibration components with identical frequencies. The term has meaning only for two harmonic motions that have the exact same frequency.

One must also consider the phase relationships between displacement, velocity, and acceleration. The nature of these terms creates a radical change in the phase angle. For example, conversion of displacement values into velocity will result in a 90° change in phase; conversion of displacement to acceleration will result in a phase change of 180°. This change in phase, caused by the differentiation or integration of one unit of measure to another, can result in confusion. It is especially troublesome when using balancing systems that acquire data using one unit of measure, such as an accelerometer, and then display the results in velocity. Unless the balancing system adjusts the phase of the resultant velocity values, the reported point of imbalance will have an error of 90° or 180°.

All harmonic motions are periodic, but not every periodic motion is harmonic. Figure 3.6 represents nonharmonic motion of two sine waves where

$$x = a \sin \omega_t + \frac{a}{2} \sin (2\omega_t)$$

The superposition of two sine waves of different frequencies is periodic but is not a harmonic of either frequency. Simply stated, vibration components with different frequencies may combine and generate a composite frequency that is not identical to either component.

Time Domain. To analyze the condition of these machine trains, the analyst must isolate each of the unique vibration components that make up the machine's overall vibration levels. Until a relatively few years ago, this task was extremely difficult. The only means to analyze vibration data was in the *time domain*.

The overall vibration data could be displayed on an oscilloscope or strip chart. These methods displayed the relative vibration levels in relation to elapsed time. By adjusting the time period, an analyst could meticulously separate each of the unique components that made up the overall vibration level. The time, effort, and skill required to isolate each of the unique vibration components generated by a complex machine train or system limited the usefulness of vibration analysis as a plant evaluation tool. In most instances, it was relegated to the role of failure analysis or where a known problem existed.

Frequency Domain. Machine trains and dynamic systems create a complex vibrational energy that consists of a series of unique vibration components. Therefore, the vibration generated by a machine can be separated into these unique frequency components (ω, 2ω, 3ω), or

$$f(t) = A_0 + A_1 \sin (\omega_t + \phi_1) + A_2 \sin (\omega_t + \phi_1) + A_3 \sin (3\omega_t + \phi_3) + \ldots$$

provided that the frequency $f(t)$ repeats itself after each interval

$$T = \frac{2\pi}{\omega}$$

The amplitudes of the various frequency components A_1, A_2, A_3, and so on and their phase angles ϕ_1, ϕ_2, ϕ_3, and so on can be determined analytically when $f(t)$ is given. This calculation is known as a *Fourier series* or *Fourier transform*. The second terms ($2\omega_t$, $3\omega_t$, and so on) are the harmonics of the primary frequency component ω. In most vibration signatures, the primary frequency component will be at the true rotating speed of the machine train (1× or 1ω) and also will have one or more harmonics at two times (2×), three times (3×), and other harmonics of the fundamental frequency.

Frequency

The cyclic movement that occurs in a given unit of time is *frequency*. It is expressed as (1) cycles per minute (cpm) or (2) cycles per second (Hz). The term represents the number of times a specific vibration component will repeat within the descriptive unit of time. For example, a fan with seven blades will generate a blade-pass vibration component equivalent to seven times the shaft speed. The vibration component is the result of a slight imbalance created by each blade as it passes the vibration transducer. If the fan is running at 1800 rpm, the blade-pass frequency will be 12,600 cpm, or 210 Hz, or

$$\text{Blade-pass frequency} = \frac{12{,}600 \, \text{cpm}}{60 \, \text{s/min}} = 210 \text{ hertz}$$

Order

Most vibration components in mechanical equipment can be directly related to the running speed of the machine. The term *order* or *harmonic* is used to equate a unique frequency component to the true speed of rotation or movement of its parent machine. The blade-pass frequency used above also can be expressed as the seventh order or seventh harmonic of the true running speed of the fan. Normally, the order or harmonic frequency is expressed as 7$\times$ to simplify the descriptor.

Amplitude

The magnitude or severity of the dynamic forces or vibration in a machine is expressed in terms of (1) rms, (2) peak, or (3) peak-to-peak. The amplitude of vibration is expressed in specific units of measurement. The most common units of measurement include (1) displacement, (2) velocity, or (3) acceleration.

Displacement. The actual change in distance or in the position of an object relative to a fixed reference point is expressed as displacement. It is expressed in units of mils or microns and should use peak-to-peak values.

Velocity. The time rate of change of displacement is velocity, a relative measurement of how fast the machine or machine component is moving, not how far it is being displaced from its true centerline. It is the unit of measure that best approximates the *energy* generated by the vibration component. Velocity is expressed in terms of inches per second or millimeters per second (mm/s).

Acceleration. The time rate of change of velocity is acceleration. Expressed in terms of g, acceleration is the best approximation of the *force* generated by a vibration component. Vibration frequencies above 1000 Hz (60,000 cpm) should be expressed in terms of acceleration.

MONITORING PARAMETERS FOR PLANT MACHINERY

Mechanical equipment, including all rotating, reciprocating, and other plant equipment or systems that are comprised of moving components or process-related dynamic actions, can be evaluated using vibration monitoring techniques. Three types of vibration pickups are shown in Fig. 3.7.

Rotating Machinery

Many vibration programs are limited to simple machine trains, such as pumps and fans. While appropriate for these simple rotating machines, vibration monitoring and analysis also can be used on complex rotating equipment and a variety of continuous-process systems. This classification should include pumps, fans, compressors, motor-generators, conveyors, paper machines, and many other continuous-process machines.

By definition, a machine train consists of a driver or power source, such as an electric motor or steam turbine; all intermediate drives, couplings, belts, and gearbox; and all driven machine components, that is, fans, pumps, continuous-process components, and so on. Since all the components within a

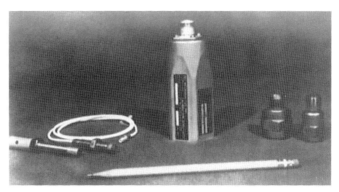

FIGURE 3.7 Three types of vibration pickup.

machine train are mechanically linked, the mechanical forces of one component will be transmitted throughout the entire machine train. Therefore, a problem in any one component will adversely affect other machine-train components.

As illustrated in Fig. 3.8, each of these forces will, in turn, generate specific vibration frequencies. Each of these forces will, in turn, generate specific vibration frequencies that uniquely identify the machine component. For example, a gearset will generate a unique set of vibration frequencies that identifies the actual, normal meshing of the gears. Any degradation of the gearset will change the amplitude and spacing of the unique vibration frequencies generated by the gearset. Since the individual components of a machine train are mechanically linked, the vibration frequencies generated by each individual machine component will be transmitted throughout the machine train. Monitoring the vibration frequencies at specific points throughout the machine train can therefore isolate and identify the specific machine component that is degrading.

To achieve maximum benefit and diagnostic power from your vibration monitoring program, you must monitor and evaluate the total machine train. Many programs have been severely limited by monitoring each component of a machine train separately. This approach limits the ability for early detection of incipient problems. For example, how do you determine that the shafts between a driver and driven machine component are misaligned if you cannot compare the unique vibration frequency components on both sides of the coupling? Initially, this knowledge of the operating dynamics of plant machinery is mandatory to successfully establish a vibration monitoring database, monitoring frequency, alert/alarm limits, and analysis parameters. Later, it will provide the basis for analyzing the vibration data to determine the degrading machine component, severity, and root cause of incipient machine problems.

Reciprocating Machinery

Vibration analysis is directly applicable to reciprocating machinery; however, modified diagnostic logic must be used to evaluate this type of machine train. Unlike pure rotating machines, the vibration patterns generated by reciprocating machines may not be simple harmonics of a rotating shaft. For example, two-cycle reciprocating engines will complete all actions as the crankshaft completes one full complete revolution. This type of action will generate simple harmonics, that is, all forces and their corresponding vibration components, of the crankshaft rotating speed. However, the extra forces created by combustion during the firing cycles or as machine components reverse direction will generate much stronger vibration components at the second harmonic ($2X$) of the crankshaft speed.

In four-cycle reciprocating machines, the crankshaft must complete two revolutions, or 720°, before all actions are complete. Therefore, few of the vibration components or frequencies can be direct harmonics of the crankshaft speed. In four-cycle internal combustion engines, the power stroke will generate a marked increase in the vibration signature. Normally, this increase will occur at a frequency equal to the fundamental rotating speed of the crankshaft.

Complete analysis of reciprocating machinery will require the addition of time-domain analysis. Both time and frequency-domain data should be evaluated in relation to the phase angle of the crankshaft. By evaluating exactly when specific vibration components occur in relation to the crankshaft phase angle, the source of each vibration component can be readily identified.

Linear-Motion Machinery

Linear-motion machinery, such as indexing machines, also have a repeatable pattern of motion and forces. Vibration analysis also can be used to evaluate these machines and systems. Timing of vibration patterns in relation to the stroke or linear movement is key to the evaluation of linear-motion equipment. All data should be recorded and evaluated with an accurate reference to time sequence of the repetitive pattern of movement.

Time-domain vibration data are usually more suitable for this type of analysis. Since the forces and vibration patterns generated by most linear motion do not recur with every rotation of a shaft, frequency-domain data are not necessary for accurate evaluation.

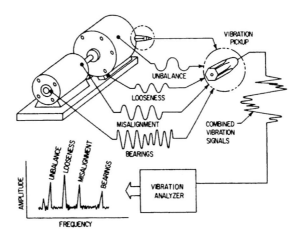

FIGURE 3.8 Pictorial representation of process of vibration analysis.

Continuous-Process Systems

Vibration-based predictive maintenance can be applied to an almost infinite variety of process and manufacturing machinery. However, the section "MACHINERY DYNAMICS" will be limited to machinery that is common to most processes. It is intended to provide guidelines for the information required to implement and utilize this type of monitoring on all mechanical plant equipment.

MACHINERY DYNAMICS

All mechanical plant equipment can be broken down into four classifications: (1) constant-speed, constant-load, (2) constant-speed, variable-load, (3) variable-speed, constant-load, and (4) variable-speed, variable-load. The vibration monitoring system that you select must be able to effectively handle all these combinations of machine operation. Why is this important? Both speed and load will affect the location and amplitude of the unique vibration components generated by the mechanical forces or motion within the machine train.

The location or frequency of individual vibration components will maintain a fixed relationship to the actual running speed of the specific shaft that generated the force. As the shaft speed changes, so will the location or frequency of the individual vibration components generated by that shaft. For example, the gear-meshing-frequency component of a gear with 10 teeth mounted on a shaft turning at 20 Hz will be located at 200 Hz. If the shaft speed changes to 40 Hz, the gear meshing frequency also will move to 400 Hz.

Load changes will not cause the location of individual vibration frequency components to change but will affect the amplitude or energy of each component. Change in machine load will either amplify or dampen the energy of individual vibration components. The variation in vibration energy at 100 percent load cannot be compared directly with that of the same machine operating at 50 percent load. Therefore, your vibration-based predictive maintenance program must compensate for load variations.

To establish and utilize vibration-based predictive maintenance, a complete knowledge of each machine component and how they interact within the machine train is absolutely necessary. Every phase, from implementation through root-cause failure analysis, of a predictive maintenance program is driven by the dynamics and resulting vibration characteristics of each machine train. All rotating, reciprocating, and continuous-process machines have common components, characteristics, and failure modes. Each machine also has totally unique operating dynamics and failure modes. This section will discuss both the common and unique characteristics of typical machine trains found in most manufacturing and process plants. All plant machines have the following common components and characteristics that affect their vibration signature.

Bearings

In modern machinery operating at relatively high speeds and loads, the proper selection and design of the bearings have been the primary limiting factors that dominate the operating life of the machine train. Since all mechanical systems must have some type of bearing or bearings, the first indication of machinery problems will develop in the vibration signature of the machine's bearings. However, bearings are not typically the cause of machine-train problems. They are, by design, the weakest link in most machinery. Therefore, the bearings are usually the first point of machine failure. As a result, it is beneficial to have a good understanding of bearing design and operating dynamics.

Bearings can be divided into two classifications: rolling-element and sleeve. The two classifications have unique operating characteristics and failure modes that can be monitored using vibration-analysis techniques.

Rolling-Element Bearings. Over the past several decades, rolling-element (i.e., ball or roller) bearings have been used for most high-speed applications and in most smaller process machinery. The primary components of a rolling-element bearing include outer race, inner race, cage, and rolling elements.

The general bearing behavior is determined by the interaction between these various elements. The rolling-element to race contacts are the heaviest loaded, and hence all fatigue failures are primarily dominated by this interaction. The rolling-element to cage and race to cage contacts are generally dynamic in nature, since they constitute a number of very high speed, short time-frame collisions.

There are many factors that affect normal bearing life, but we will limit this discussion to the mechanical factors that can be used to predict bearing failure. We also should consider lubrication, design loads, bearing design, and other parameters to achieve maximum bearing life. However, this is normally outside the scope of a predictive maintenance program.

Defective rolling-element bearings will generate vibration frequencies at the rotational speeds of each bearing component. Each of the frequencies can be calculated and routinely monitored using vibration-analysis techniques. Rotational frequencies are related to the motion of the rolling elements, cage, and races. They include ball or roller spin, cage rotation, and ball or roller passing frequencies. The ball or roller spin (BSF) frequency is generated by the rotation of each ball or roller around its own centerline and can be calculated as

$$\text{BSF} = 0.5N \times \left(\frac{D}{d}\right) \times \left[1 - \left(\frac{d}{D}\right)^2\right]$$

Since the ball or roller defect or defects will contact both the inner and outer races each time it completes one full revolution, the ball-spin defect frequency will be at two times the BSF or rotational frequency.

The cage rotation frequency, or fundamental train frequency (FTF) can be calculated as

$$\text{FTF} = 0.5N \times \left[1 - \left(\frac{d}{D}\right)\right]$$

A defect in the outer race of the bearing can be calculated using the ball pass–outer race (BPFO) formula:

$$\text{BPFO} = 0.5Nn \times \left[1 - \left(\frac{d}{D}\right)\right]$$

Inner race defect frequency, or ball pass–inner race (BPFI), can be calculated as

$$\text{BPFI} = 0.5Nn \times \left[1 - \left(\frac{d}{D}\right)\right]$$

where N = shaft speed, in hertz or revolutions per second
D = pitch diameter of bearing, in inches
d = diameter of balls or roller, in inches
n = number of balls or rollers

Many bearing manufacturers have simplified the calculation of bearing defect frequencies by providing a bearing reference guide. This guide provides a constant (i.e., value) for each of the defect frequencies for each bearing manufactured by the vendor. This constant is multiplied by the actual running speed of the machine's shaft to obtain the unique defect frequencies.

Bearing rotational and defect frequencies may be generated as the result of actual bearing defects or by machine- or process-induced loads. Imbalance, misalignment, and abnormal loads will amplify the specific bearing frequencies that must absorb the load. For example, excessive bearing side load created by too much belt tension will amplify the ball spin and both ball pass frequencies. This is the direct result of the abnormal load created by the belt tension. Misalignment of the same belt drive also will amplify the cage frequency, or FTF.

The unique vibration frequencies that will define incipient bearing problems can be identified easily using narrowband monitoring techniques. The FTF, or cage, defect will always occur at about 40 percent of running speed. A narrowband established to monitor the energy in a frequency band from 30 to 50 percent of running speed will automatically detect any abnormal change in the condition of the bearing's cage. Of the remaining three bearing defect frequencies BSF is always the lowest frequency. The ball pass–inner race frequency is always the highest. A single narrowband can be established to monitor these bearing defect frequencies. The narrowband should be established with the lower limit set about 10 percent below the actual BSF to allow for slight variations in running speed. The upper limit should be set about 10 percent higher than the actual inner race defect or BPFI frequency. By using these narrowband monitoring techniques, the microprocessor-based data-acquisition unit can automatically detect any abnormal change in the bearing's condition. The narrowband monitoring of the three higher defect frequencies will not be able to report the specific defect, that is, inner race, outer race, or ball, but it will report the operating condition of the bearing. If you want to know which bearing component is degrading, the full signature must be evaluated manually to detect the specific defect frequency that is in alarm.

Sleeve Bearings. Sleeve or fluid-film bearings can be divided into several subclasses: plain, grooved, partial-arc, and tilting-pad. With the exception of the tilting-pad bearing, they do not generate a unique rotational frequency that would identify normal operation.

Since the tilting-pad bearing has moving parts, that is, the pads, it will generate low-level vibration components at a pad-passing frequency that is equal to the number of pads multiplied by the shaft running speed.

This type of bearing is designed to form a uniform, thin film of lubricant between the bearing's babbitt surface and the rotating shaft. In normal operation, the shaft is centered in this thin lubricating film and will not create a dynamic force or vibration frequency component that uniquely identifies the bearing. However, abnormal behavior of the lubricating film can be clearly identified using vibration-analysis techniques.

If the lubricating film becomes eccentric, the vibration signature will show a marked increase in low-frequency (i.e., less than shaft running speed) energy. Initial breakdown of the uniform oil film will be indicated by an increase in frequency components at even fractions (i.e., $\frac{1}{4}$, $\frac{3}{8}$, $\frac{1}{2}$, and so on) of running speed. These vibration components are created by the eccentric rotation of the machine's shaft. As the condition degrades, the fractional vibration components will consolidate between about 40 and 48 percent of the shaft's true running speed.

If the shaft breaks through the lubricating film, a mechanical rub will become evident in the vibration signature. This rub will show as a very low frequency, that is, between 1.0 and 2.0 Hz, and also will have low-amplitude components at about 25 and 40 percent of true running speed. The low-frequency rub will have the appearance of a "ski" jump and in most cases will generate high amplitudes.

We should note that a limited number of the microprocessor-based vibration-monitoring systems are capable of detecting vibration frequencies below 10 Hz or 600 rpm and therefore cannot be used to detect the mechanical rubs that often occur in machinery.

Monitoring the mechanical condition of bearings should rely primarily on vibration-analysis techniques. However, periodic lubricating oil analysis using either spectrographic or wear-particle methods can provide additional data on their actual operating conditions. The added costs associated with these techniques do not justify their inclusion unless a specific problem has been identified by the vibration-monitoring program.

Bearing temperatures can be added to the monitoring program using point-of-use infrared sensors that can input temperature data directly into the vibration-monitoring instrumentation. The incremental cost is very low, and the added temperature data will assist in early identification of incipient problems. The analysis of vibration requires an understanding of the terminology used to describe the components of vibration. The following subsections define the terms that are normally used in vibration analysis.

Gears

Many machine trains use gear-drive assemblies to connect the driver to various driven machine components. Gears and gearboxes have unique vibration signatures that identify both normal and abnormal operation. Characterization of vibration signatures of a gearbox is difficult to acquire but is an invaluable tool for diagnosing machine-train problems. The difficulty is due to two factors: (1) it is difficult to mount vibration transducers close to the individual gears within a gearbox, and (2) the number of vibration sources in a multigear drive results in a complex assortment of gear mesh, modulation, and running-speed frequencies.

Severe gearbox vibration is usually due to resonance between the system's natural frequency and the shaft speeds. This resonant excitation arises from and is proportional to gear inaccuracies that cause small periodic fluctuations in the pitch-line velocity. Complex machines usually have many resonance frequencies within their operating-speed range. At resonance, these cyclic excitations may cause large vibration amplitudes and stresses. Basically, the forcing torques arising from gear inaccuracies are small. However, under resonant conditions, torsional amplitude growth is restrained only by damping in the mode of vibration. In typical gearboxes, this damping is often small and permits the gear-excited torque to generate large vibration amplitudes under resonant conditions.

One other important fact about gearsets is that they all have a designed preload and create an induced load in normal operation. The direction and force of these loads will vary depending on the type of gears and gearbox design.

The following descriptions of typical gears will provide some insight into the normal operating dynamics of each gear type. To implement a vibration-based predictive maintenance program for gears and gearboxes, a basic knowledge of the dynamic forces that they generate is very important. At a minimum, the following forces and their corresponding vibration components should be identified.

Gear Mesh. This is the frequency most commonly associated with gears and is equal to the number of teeth on the gear multiplied by the actual running speed of its shaft. A typical gearbox will have multiple gears and therefore multiple gear meshing frequencies. A normal gear mesh signature will have a low-amplitude gear mesh frequency with a series of symmetrical sidebands, spaced at the exact running speed of the shaft, on each side of the mesh components. The spacing and amplitude of these side bands will be exactly symmetrical if the gearbox is operating normally. Any deviation in the symmetry of the gear mesh signature is an indication of incipient gear problems.

Gear Excitation. Gears can be manufactured to such a high degree of precision that very slight imperfections can create abnormal vibration components. These imperfections may arise either during manufacturing or from installation or operation. Mounting inaccuracies may cause otherwise perfect gears to run roughly. Measurements of gear error usually reveal a fairly complex pattern of geometric inaccuracies that result in abnormal vibration frequencies.

For vibration analysis of gearbox condition, the lower-frequency harmonics are of greatest interest because these components excite the most destructive drive-train natural frequencies. Higher harmonics, such as tooth-to-tooth gear errors and fluctuations in shaft displacement due to gear tooth flexibility, generate noise rather than vibration in gearboxes.

Backlash. Backlash is an important factor in proper gear operation. All gears must have a certain amount of backlash to allow for the tolerances in concentricity and tooth form. Not enough backlash will cause early failure due to overloading. Too much backlash will increase the contact force and also reduce the operating life of the gearset.

Abnormal backlash will alter the spacing of the sidebands that surround the gear meshing frequency. Rather than maintaining uniform spacing at the shaft running speed, the spacing will be erratic.

Monitoring the mechanical condition of gears and gearboxes using vibration-analysis techniques also must consider the unique forces that specific gears generate. For example, a helical gear will, by design, generate a high thrust load created by the meshing of the mating gears. Degradation of the helical gear's condition will increase the axial force and its corresponding vibration amplitude.

Narrowband monitoring techniques are ideal for detecting incipient gear problems. A narrowband should be established that includes at least five sidebands on either side of the gear meshing frequency. For example, a gearset with a shaft running speed of 20 Hz and a meshing frequency at 200 Hz would have a narrowband with a lower limit set at 100 Hz, that is, 20 Hz $\times$ 5, and an upper limit at 300 Hz. This type of narrowband will allow the microprocessor-based data-acquisition unit to automatically report any abnormal increase in the energy generated by the gearset and therefore any change in the mechanical condition. This automatic function will not provide the root-cause or failure mode. To determine the actual failure mode will require full, manual signature analysis.

Lubricating oil analysis using spectrographic and wear-particle techniques will add useful information about the operating condition of gearboxes and can augment a vibration-monitoring program. However, the added cost will not normally justify the inclusion of these techniques unless a serious problem has been identified by the vibration-monitoring program.

Blades and Vanes

Machines that use blades or vanes have an additional frequency that should be monitored routinely. This frequency, called *blade* or *vane pass,* represents the frequency created by the blades or vanes passing a reference point, that is, the vibration transducer. The passing frequency can be calculated by multiplying the number of blades or vanes by the true running speed of the machine's shaft.

The amplitude and profile of the passing frequency will vary with load. Therefore, it is important to record the actual operating load as part of the data-acquisition process. In a normal machine, the passing frequency should be a low-level, distinct peak at the calculated frequency.

If process-induced instability is present, the passing frequency will increase in amplitude, and modulations or sidebands around the passing frequency will develop.

A narrowband should be established to automatically monitor the blade or vane passing frequency. The lower limit should be set about 10 percent below and the upper limit about 10 percent above the calculated passing frequency to compensate for variations in speed and to capture the sidebands that will be created by instability.

Belt Drives

Machine trains that use belt-drive assemblies have an additional set of frequencies that should be monitored. All belt drives will have a belt pass frequency that will identify the operating condition of the drive system. This unique frequency is generated by the true running speed of the belt and can be calculated for any drive assembly.

Calculate the belt pass frequency by

$$\text{Belt pass} = \frac{d_1 \times 3.1416 \times N}{\text{belt length}}$$

where d_1 = drive sheave diameter, in inches
d_2 = driven sheave diameter, in inches
L = center-to-center distance, in inches
Belt length = $2L + d_1 \times 33.1416 / 2 + d_2 \times 33.1416 / 2$
N = true rotating speed of drive shaft, in hertz

Belt pass frequency is a good means for identifying misalignment, excessive induced load, and other failure modes that are associated with the drive assembly.

A narrowband should be established to automatically monitor the belt pass frequency. The lower limit should be set at 10 percent below the calculated pass frequency. The upper limit should be set at a multiple of the calculated frequency that is equal to the number of belts; that is, a drive with 10 belts would set the upper limit at 10 times the calculated pass frequency.

Many of the common defects or failure modes in mechanical equipment can be identified by understanding their relationship to the true running speed of a shaft within the machine train. The section "Running Speeds" will discuss the general definitions of the most common machine-train failure modes. These definitions are guidelines and should not be accepted as being true in all cases.

Running Speeds

Every machine train will have at least one true running speed. This is the true rotational speed of the shaft or shafts within the machine. In most cases, there will be more than one shaft, and each will have its own unique running speed.

Since most vibration frequencies are related to the running speed of a shaft within the machine, it is important that every running speed and its unique rotational frequencies be identified. The fundamental vibration frequency or running speed will be a primary indicator of many machine-train problems and should be monitored closely.

A narrowband should be established to automatically monitor and trend each true running speed within the machine train. It should be noted that the running speeds within a machine will not remain constant. Even constant-speed machines will have some variation in the true running speed. This variation is primarily a function of the loading factor. As the load increases on most machines, the running speed will decrease. To compensate for this type of speed variation, the narrowband should be established with the lower limits set about 10 percent below and the upper limit about 10 percent above the calculated or normal running speed. This should be sufficient to compensate for slight variations in the true running speed.

Variable-speed machines must be handled in a slightly different manner. Many of the microprocessor-based systems provide an alternate method of automatically setting narrowband filters during data acquisition. This method, called *order analysis,* uses a tachometer input to determine the true running speed and then moves the narrowband filters to the correct setting based on the measured running speed. This approach simplifies automatic monitoring of variable-speed machines.

Critical Speeds

All shafts in rotating mechanical systems exhibit potentially damaging radial vibration when operated at certain speeds. These critical speeds are determined by rotational frequencies that coincide with one or more of the shaft-system natural frequencies. Shafts will vibrate at critical speeds even in precisely balanced machines.

Balancing will narrow but not eliminate the range of speeds in which vibrations build to a peak. Consequently, a well-balanced machine can operate somewhat closer to the system's critical speeds without being damaged.

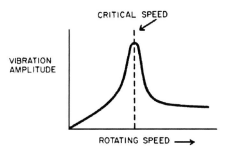

FIGURE 3.9 Critical-speed curve.

Sometimes, the lowest critical speed, or fundamental, is of the most concern because it is the source of the greatest vibration amplitude (Fig. 3.9). Shafts often are selected for their high stiffness and low mass, which helps place the first critical speed above the normal operating speed range. Unfortunately, rigid, light shafts alone do not solve all critical-speed problems. For example, most fans and blowers are designed to operate just below the fundamental, or first critical speed. As long as the fan or blower is operated at its designed speed, it should not experience a critical-speed problem. In actual operation, fans and blowers are often damaged or destroyed as the direct result of a critical-speed problem. The factor that caused the fan to operate at the first critical speed was plate-out. In normal operation, fans and blowers are subject to a buildup of dirt and other solid contamination on the blades. This increase in weight or mass will lower the first critical speed of the fan. As a result, the fan operating at design speed is now within the first critical speed. It is not widely realized that critical speed is a rotating machine phenomenon and not just a dynamic characteristic of rotating shafts. Any component in a rotating machine that reduces stiffness or increases rotating mass also will shift critical speeds closer to the operating speeds of the machine. This can become a real problem when components are designed or selected without regard to potential process influence on the machine's critical speeds. Excessive vibration created by operating at critical speeds will immediately decrease as the machine's speed is changed. Therefore, one test method to determine if a critical-speed problem exists is to change the operating speed. An increase or decrease in machine speed will drastically reduce the overall vibration and specifically the true running-speed component. In addition to the fundamental or first critical speed caused by the centrifugal forces of an unbalanced mass, some vibration components have been observed at one-half the first critical. This effect is typically limited to horizontal shafts, indicating that gravity must be one of the causes. There are two primary causes for this secondary critical speed: gravity in combination with imbalance, and gravity in combination with nonuniform bending stiffness of the shaft.

The theory of actual motion and vibration is very complicated, and a comprehensive technical understanding is not necessary for predictive maintenance applications. Critical speeds should be considered on all rotating machines that are included in the program. Special attention to critical speeds or natural frequencies should be paid on any machine train that has overhung or cantilevered components, such as fans or machines with high rotating mass and low foundation mass. Paper machines are classic examples of the high rotating mass to low support structure mass situation. The majority of the machine's weight or mass is in the rotating components, such as rolls. This mass is supported by minimum structure and is therefore highly susceptible to critical-speed problems.

Mode Shape

All rotating shafts have multiple mode shapes created by either the critical speed or process-induced forcing functions. Since most process machines use relatively flexible shafts, the shafts tend to deflect and operate in a shape or mode rather than rotate on their true centerline. The first mode of operation is a *radial offset* from the shaft's true centerline. In this mode, the shaft actually rotates cylindrically around the normal or static centerline and does not have a true null or zero point between the bearing supports. In this mode, the shaft remains relatively straight but rotates offset or displaced around the true centerline. Single-plane imbalance and other forcing functions will create this mode of operation. The true running speed vibration component will be excited by this mode of operation. The second mode shape taken by a rotating shaft is *conical rather than cylindrical.* In this mode, the shaft is deformed into a mode that resembles an S shape with one null or zero point between the bearing supports. Therefore, for every revolution of the shaft, two high spots will be observed in the vibration signature. This mode, normally associated with misalignment, will create a vibration component at two times running speed, or 2×. Other forcing functions within the machine train or process system also can create this mode shape. For example, out-of-phase imbalance, bent shafts, aerodynamic instability, and many other failure modes can create the second mode shape. The third mode shape of a rotating shaft is a compound *deformation of the shaft.* In this mode, the shaft has two null or zero points between the bearing supports. This mode will create a vibration frequency component at three times (3×) running speed. Many forcing functions can create the third mode shape. The most common are multiplane imbalance and a wobbling motion of the rotating element. A clear understanding of the mode shapes and how they are created and appear in the vibration signature will greatly improve your ability to understand the operating condition of the machine trains included in the program. Understanding mode shapes will make analysis and problem identification easier and much more accurate. Many of the more common failure modes will excite one or more of these mode shapes. Visualizing the forces that these failure modes create and the resulting shaft mode shape will provide a great deal of assistance in identifying the specific problem within a machine train.

Resonance

All machines have a natural frequency, that is, the combination of all machines and structural frequency components. If this natural frequency is excited by one or more running speeds or a defect in the machine train, the machine's support structure will magnify the energy and serious damage may occur. Abnormal resonance of a machine's natural frequency is one of the most destructive vibration forces that you will encounter. It is capable of creating catastrophic failure of machine housings and support structures.

Normally, the frequencies associated with resonance are low and in some cases are below the monitoring capabilities of your predictive maintenance instrumentation. To monitor machine resonance frequencies, your instrumentation must be able to separate frequencies in the 1- to 10-Hz range from the dc noise that usually limits monitoring these bands.

In some cases, this resonance can be transmitted to adjacent machines or plant equipment. This resonance also can mask the vibration signature of other critical machine components. If resonance is suspected, it should be verified and eliminated as quickly as possible.

Preloads and Induced Loads

Design and dynamically induced loads on rotating shafts constitute the most common and yet misunderstood of all machine behaviors. They also are a major contributor to machine failure.

Preload is defined as a directional force that is applied to a rotating shaft by design. An example of preload is the sideload created by a belt-drive assembly. Most machines have at least one designed preload that creates a directional force that is not compensated for with an equal and opposite force. Gravity is one form of preload. All machines have this unbalanced force to overcome during normal operation.

Induced load is also an unbalance directional force within a machine. However, in this case the force is created by the dynamic operation of the machine or system. An example of induced load is aerodynamic instability created by restricting airflow through a fan or blower. All bladed or vaned machines (i.e., pumps, compressors, fans, etc.) are susceptible to this type of abnormal loading.

The result of both preloads and induced loads is the deflection of the rotating shaft into one quadrant of the bearings or into one of the mode shapes. This results in a nonlinear resistance, in that the spring constant of the bearings is much higher opposing the force than it is perpendicular to the load. This will cause premature bearing failure and can cause other serious damage to the machine train.

Understanding these loads and how they affect the machine train is important for two reasons: First, it will enable you to locate the primary data point in the plane opposing the potential induced load. This will ensure early detection of an incipient problem. Second, it will provide assistance in diagnosing developing machine problems.

Preloads and induced loads do not necessarily cause machine malfunctions. In some cases they tend to stabilize the shaft, bearings, and rotating elements. However, if excessive loads are applied to machinery, serious problems can, and in most cases will, develop very quickly. In extreme cases, bent shafts, cracked rotating elements, cracked couplings, and other catastrophic machine problems have developed.

Process Variables

Most mechanical equipments are designed to perform a function within a process system. Therefore, a predictive maintenance program cannot rely strictly on monitoring the vibration data from these mechanical systems. Variations in the process envelope have a direct effect on the operating condition of most mechanical equipments. Pumps, compressors, fans, and other mechanical equipment rely on minimum suction pressures to operate and are limited to the maximum discharge pressure (TDH) that they can generate. Variations in suction pressure, that is, net positive suction head (NPSH), and the discharge pressure demands of process systems can prevent the mechanical equipment from operating within an acceptable environment and will cause catastrophic failure of the mechanical system. Many of the problems that cause premature failure of mechanical systems are the direct result of process-induced loads. A large number of machinery imbalance problems are in truth caused by hydraulic or aerodynamic instability created by process restrictions. In these instances, the system is demanding that the mechanical equipment operate outside its capabilities. A pump, for example, cannot deliver product if the discharge valve is closed. Nor can the pump continue to operate in this condition. Even though vibration monitoring and analysis will detect the abnormal vibration caused by process-induced instability, the addition of recorded process data will greatly enhance a predictive maintenance program. All the process parameters that directly affect the operation of mechanical equipment should be acquired and recorded as part of the routine data-acquisition process. In addition to supporting the vibration-analysis function, long-term trends of these process variables will often identify a potentially serious system problem. Most of the microprocessor-based systems will support at least one method of recording process variables as part of the routine data-acquisition process. Some will allow direct acquisition of process variables from installed plant instrumentation. Others will allow manual entry of process data.

COMMON FAILURE MODES

Imbalance

Imbalance is probably the most common failure mode in mechanical equipment. The assumption that actual mechanical imbalance must exist to create an imbalanced condition within the machine is incorrect. Aerodynamic or hydraulic instability also can create massive imbalance in a machine. In fact, all failure modes will create some form of imbalance in a machine. When all failures are considered, the number of machine problems that are the result of actual mechanical imbalance of the rotating element is relatively small.

Imbalance will take many forms in the vibration signature. In almost every case the fundamental or running-speed component will be excited and is the dominant amplitude. However, this condition also can excite multiple harmonics or multiples of running speed. The number of harmonics and their amplitude have a direct correlation with the number of planes of mechanical imbalance and their phase relationship.

As illustrated in Figs. 3.10–3.13, a single-element rotating machine, a narrowband should be established to monitor the fundamental or true running-speed frequency component. For multiple-element rotating machines, the narrowband should monitor the true running speed and the number of harmonics equal to the number of rotating elements. As in earlier examples, the narrowband limits should allow for slight variations in running speed.

7.88 INSTRUMENTS AND RELIABILITY TOOLS

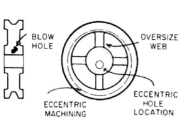

FIGURE 3.10 Sources of unbalance.

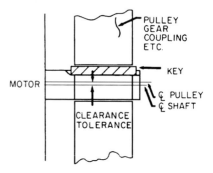

FIGURE 3.11 Stack-up of assembly tolerances.

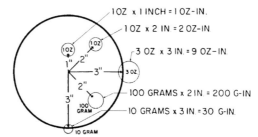

FIGURE 3.12 Unbalance units of measure.

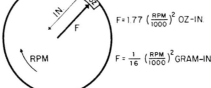

FIGURE 3.13 Force caused by unbalance.

Misalignment

This condition is virtually always present in machine trains. Generally, we assume that misalignment exists between two shafts connected by a coupling, V-belts, or other intermediate drives. Misalignment also can exist between the bearings of a solid shaft and at other points within the machine.

The presentation of misalignment in the vibration signature will depend on the type of misalignment. There are two major classifications of misalignment: parallel and angular.

Parallel misalignment is present when two shafts are parallel to each other but are not in the same plane. This type of misalignment will generate a radial vibration and will duplicate the second mode shape. In other words, it will generate a radial vibration at two times (2×) the true running speed of the shafts.

Angular misalignment exists when the shafts are not parallel to each other. This type of misalignment will generate axial vibration (i.e., parallel to the shaft). Since this form of misalignment can duplicate any of the mode shapes, the resultant vibration frequency can be at the true running speed of the shaft, two times (2×) running speed, or three times (3×) running speed. The key indicator will be an increase in axial vibration.

Bent Shaft

A bent shaft creates an imbalance or misaligned condition within the machine train and should be treated in the same manner as the common failure modes.

Mechanical Looseness

Looseness can create a variety of patterns in the vibration signature. It will most often create a primary frequency component at 50 percent of running speed and will generate multiple harmonics of this primary component. In other words, there will be a frequency component at 50, 150, 250 percent, and so on. In other cases, the fundamental or true-running-speed component (1×) will be excited. In almost all cases, there will be multiple harmonics with almost identical amplitudes.

Mechanical Rub

Many machine trains are susceptible to mechanical rub. This failure mode may be a shaft grinding against the babbitt of a sleeve bearing, the rollers in a rolling-element bearing grinding against the races, or some part of the rotor grinding against the machine housing. In each case, the vibration signature will display a low-amplitude peak, normally between 1.0 and 10.0 Hz. This extremely low frequency peak will have high amplitude and will be accompanied by a lesser set of peaks at about 25 and 40 percent of the shaft's running speed. Machine failure, when the defect is present, is highly probable. Note that not all vibration-monitoring systems can detect this defect. Many have a low-frequency cutoff at 10.0 Hz and will not capture any frequencies below this level.

Failure Modes of Common Machines

In addition to the common characteristics and failure modes, each machine type will have unique requirements for monitoring its operating condition. Now that we understand the common characteristics and failure modes of mechanical equipment, the next step is to determine where and how to monitor specific plant machinery. To maintain continuity and simplify monitoring and analysis, it is suggested that you set up each machine train using a consistent, common-shaft approach to locating measurement points and establishing analysis parameter sets.

Measurement points should be numbered sequentially starting with the outboard driver bearing and ending with the final outboard bearing of the driven machine component. In addition, a consistent numbering sequence that identifies the orientation (i.e., vertical, horizontal, axial, and so on) also should be used.

The term *common shaft* refers to each continuous shaft in the machine train. For example, in an electric motor-driven single-reduction gearbox, the motor and gearbox input shaft is considered a common shaft. Even though the shaft is coupled, all forces acting on the extended shaft, as well as any resulting vibration, will be transmitted throughout the shaft. This approach to setting up a database, monitoring operating conditions, and analyzing incipient problems will provide two benefits: (1) immediate identification of the location of a particular data point during acquisition and analysis of a machine train and (2) the ability to evaluate all parameters that affect each component of the machine.

Understanding the specific location and orientation of each measurement point is critical to the diagnosis of incipient machine problems. The vibration signature is a graphic representation of the actual dynamic forces within the machine. Without knowing the location and orientation, it will be difficult, if not impossible, to identify the incipient problem correctly. The orientation of each measurement point also should be considered carefully during the database setup. There is an optimal orientation for each measurement point on every machine train in your program. In almost all cases, each bearing cap will require two radial measurements, that is, perpendicular to the shaft, to properly monitor the machine's operating condition. There also should be at least one axial measurement, that is, parallel to the shaft, on every common shaft.

These measurement points should be oriented to monitor the worst possible dynamic force and vibration. For example, a belt-driven fan will have a dominant force created by the belt tension. Therefore, the worst vibration should be in the radial direction of the belt drive. To monitor the worst vibration and gain the earliest possible detection of an incipient problem, the primary radial measurement point should be between the shaft and the opposite side of the belt drive. A secondary radial measurement should be taken at 90° to the primary. The secondary measurement will provide comparative energy and will help the analyst determine the actual force vector within the machine. In this example, axial readings on both the motor and fan shafts are also very important. A common failure mode of belt-driven units is misalignment. If the sheaves are not in the same plane, the belt tension will attempt to self-align. This will create axial movement in the two shafts. The axial measurement will detect these abnormal forces and identify the alignment problem.

Electric Motors

Electric motors are often used as the prime mover for process and manufacturing machinery. Depending on size and manufacturer, they may use either sleeve or rolling-element bearings. However, they will rarely have thrust bearings and are susceptible to abnormal axial movement when coupled to process equipment that can create thrust loads.

In addition to the common failure modes, electric motors are prone to several unique problems. These include loss on insulation, loose or broken rotor bars, loose poles, and electrical shorts. These failure modes can be monitored by including narrowbands that monitor line frequency, 60 Hz, and its harmonics, 120, 180. If electrical problems exist, the line frequency and harmonics will clearly indicate a problem. Loose rotors or broken rotor bars can be detected by monitoring the current to the motor. If the condition exists, the motor's slip frequency will be clearly displayed as sidebands on either side of the line frequency. Loose poles will show as a pole pass frequency or at a frequency equal to the number of poles multiplied by the true running speed.

Normally, motors will be used in either a horizontal or vertical position. Horizontal motors should be monitored using two radial measurements points on the inboard and outboard bearings. An axial reading is not necessary unless the motor is driving a machine component that can create an axial or thrust load. Vertical motors will be monitored in the same manner but will require an axial measurement on the lower bearing in the upward direction.

Totally enclosed fan-cooled (TEFC) and some explosion-proof motors are difficult to monitor. The fan housing on these motors will enclose the outboard bearing cap. The best method of acquiring outboard bearing data is to permanently mount transducers on the bearing caps and wire them to a convenient location. If this is not possible, acquire the measurement at the closest point that will provide a direct mechanical link to the bearing.

Narrowbands should be established to monitor (1) imbalance or running speed, (2) misalignment or 23 running speed, (3) electrical problems or 60, 120, and 180 Hz, (4) bearing defects (see "Bearings"), (5) loose poles or pole passing frequency, and (6) mechanical rub or subharmonic. A current loop reading also should be taken to watch for loose or broken rotor bars.

Infrared scanning will provide early detection of both electrical and mechanical problems that may not be detected by vibration analysis. At a minimum, spot motor winding temperatures should be acquired as part of the routine data-acquisition process. Infrared thermometers can be used in conjunction with the vibration-data logger for these data.

Gearboxes

Gearboxes are widely used as intermediate drives to either increase or decrease driver speed. Depending on the application, they may use a variety of gear types and bearings. Therefore, the failure modes, other than the common modes discussed earlier, will vary accordingly.

Regardless of the gearbox internals, two radial measurement points on each bearing cap should be used to monitor operating conditions. If helical gears are used, an axial reading on each shaft is also required. The measurement points should be oriented in the direction opposite the worst anticipated dynamic force and resulting vibration. In most cases, this will be opposing the preload and thrust loads generated by the gearset.

Multiple reduction or increase gearboxes will have idler shafts that in many cases are not accessible from outside the gearbox. In these cases, axial readings at the point closest to the idler shafts should provide reliable data. Make sure that the selected point provides a direct mechanical link to the shaft or bearing housing.

Narrowbands should be established for each shaft that will monitor (1) imbalance, (2) misalignment, (3) gear meshing frequency, (4) bearing defects, and (5) mechanical rub. Bearing cap temperature and motor amp load also should be acquired as part of the measurement set.

Periodic lubricating oil analysis, using spectrographic and wear-particle techniques, will improve the program's ability to detect incipient problems.

Fans and Blowers

The variety of designs of fans and blowers is almost infinite. However, they generally fall into two classes: centerline and cantilevered, or overhung. Both classifications are generally designed to operate just below their first critical speed and are therefore prone to severe imbalance created by the critical speed.

The cantilever design is also susceptible to aerodynamic instability and induced loads. This is primarily the result of the high mass of the rotating element and the overhung configuration of the bearing support structure. All fans and blowers should be monitored for the common failure modes and for process-induced instability. Blade pass is a primary indicator of condition.

Belt-driven units are prone to misalignment and should be watched closely. Two radial measurement points on each bearing cap oriented to oppose the worst dynamic force and at least one axial measurement are required to monitor fan operating condition.

Narrowbands should be established to monitor (1) imbalance, (2) misalignment, (3) bearing defects, (4) blade pass, (5) aerodynamic instability, and (6) mechanical rub.

Attention should be paid to the suction and discharge locations. Restrictions in either of these will try to force the rotating element in the opposite direction. In most cases, this will be the correct monitoring direction.

Process data that include suction pressure, discharge pressure, and motor amp load also should be included in the measurement set. Bearing cap temperatures also will assist in early detection of bearing problems.

Compressors

Like fans, the variety of compressor designs is almost infinite. The major classifications include single-stage centrifugal, multistage, screw, and reciprocating. The multistage centrifugal compressor class can be divided into two subclasses: inline and bullgear. Rolling-element, sleeve, and tilting-pad bearings are used in industrial compressors. Monitoring parameters should be established according to bearing type.

Single-stage centrifugal compressors are similar to blowers and can be monitored in the same manner. Vane pass should be a primary indicator of condition. Narrowbands should be established to monitor (1) imbalance, (2) misalignment, (3) vane pass, (4) bearing defects, (5) aerodynamic instability, and (6) mechanical rub.

Multistage inline compressors are similar to multistage centrifugal pumps and can be monitored in the same manner. Narrowbands should monitor (1) imbalance, (2) misalignment, (3) each vane pass frequency, (4) aerodynamic instability, (5) bearing defects, and (6) mechanical rub.

Multistage bullgear compressors should be treated as a combination of a gearbox and pump. By design, these compressors have a large, helical gear (i.e., bullgear) that drives several smaller gears mounted on impeller pinion shafts. The pinion shaft speeds on this type of compressor are typically in the 30,000 to 50,000 rpm range and should be monitored closely. The pinion shafts normally have tilting-pad bearings that will generate a passing frequency if abnormal clearance or alignment exists. Most bullgear compressors will have displacement transducers permanently mounted to monitor the pinion shafts. Data should be acquired from these sensors in addition to casing sensors established as part of this program. Two radial measurement points on each bearing cap of the bullgear and each pinion shaft should be used to monitor the compressor's operating condition. Because the compressor uses helical gears, axial measurements on each bearing cap are also required. Narrowbands should be established for each shaft that will monitor (1) imbalance, (2) misalignment, (3) gear mesh, (4) vane pass, (5) bearing defects, (6) aerodynamic instability, and (7) mechanical rub.

Screw compressors may use either rolling-element or sleeve bearings and are susceptible to aerodynamic instability. They normally have extremely close axial tolerances that will allow no more than 0.5 mils of axial movement before the rotors contact. Axial measurements on both rotor and screws are absolutely critical on these machines. Two radial and one axial measurement on each bearing cap should be used on screw compressors. The rotor meshing and axial readings are the primary indicators of abnormal operation. Narrowbands should be established for each rotor that will monitor (1) imbalance, (2) misalignment, (3) rotor meshing, (4) bearing defects, (5) aerodynamic instability, and (6) mechanical rub.

Reciprocating compressors will generate forces and resultant vibration at frequencies different from rotating machines. Normally, the second harmonic (2×) will be dominant rather than the true crankshaft speed. In addition to the two radial measurement points on each bearing cap, measurements should be acquired on the cylinder wall to detect rubbing and near the suction and discharge valves to detect valve problems. A full set of process parameters should be acquired on all compressors. Narrowbands should monitor (1) imbalance of crankshaft, (2) misalignment of crankshaft, (3) mechanical rub in cylinder walls, (4) valve defects, (5) bearing defects, and (6) looseness.

Regardless of design, compressors are prone to process-induced problems, and this information is necessary to successfully determine their operating condition. Process parameters provide valuable information about the operating condition of all compressors. All envelope parameters, that is, pressures, temperatures, amp load on motor, bearing cap temperatures, and so on, should be included in the data set. Interstage data on multistage compressors, especially reciprocating units, are critical to analysis.

Periodic lubricating oil analysis, using spectrographic or wear-particle technique, will provide early warning of incipient problems.

Generators

Generators are usually supplied with fluid-film or sleeve bearings and should be monitored with two radial measurement points on each bearing cap. They are also prone to end play or axial movement of the entire rotor assembly. Therefore, at least one axial measurement is required. Narrowbands should monitor (1) imbalance, (2) misalignment, (3) bearing defects, (4) rotor instability (normally 3 running speed), (5) electrical defects (see "Electric Motors"), and (6) mechanical rub. Infrared scanning of the generator will provide early warning of incipient problems that may not be detected by vibration analysis.

Pumps

The variety of pumps is also almost infinite. However, they should be monitored in the same manner as fans and blowers. The vane pass frequency will be the primary indicator of process problems and should be monitored closely. Radial and axial measurement points should be oriented to monitor process-induced loads. The worst radial force should be opposing the discharge on end-suction pumps and should be inline with the suction and discharge on split-case pumps.

End-suction pumps also should be monitored using axial measurement points to detect suction problems.

Multiple-stage centrifugal pumps can, depending on design, create high thrust or axial loads. Multistage inline pumps, all impellers facing in the same direction, must be monitored closely for any increase in axial movement or load. Opposed impeller designs normally balance the axial load and need not be monitored for axial loads.

Narrowbands should monitor (1) imbalance, (2) misalignment, (3) each vane pass, (4) bearing defects, (5) hydraulic instability, and (6) mechanical rub.

A full set of process parameters is required on pumps. Like compressors, they are highly susceptible to process-induced problems, and these data are required to determine their operating condition. These measurements should include pressures, temperatures, flow, motor amp load, and bearing temperatures.

Continuous-Process Lines

Most manufacturing and process plants use a variety of complex, continuous-process mechanical systems that should be included in the program. Included in this classification are paper machines, rolling mills, can lines, printing presses, dyeing lines, and many, many more. These systems can be set up, monitored, and analyzed in the same manner as simple (i.e., pumps, fans, and so on) machine trains. The initial database setup will require more effort, but the same principles apply. Each system should be evaluated to determine the common shafts that make up the total machine train. Using the common-shaft data, evaluate each shaft to determine the unique mechanical motions and dynamic forces that each one generates, the direction of each force, and the anticipated failure modes. This information can then be used to determine measurement point location and the narrowbands that are required to routinely monitor the machine's operating condition.

Narrowband selection will depend on the operating dynamics of each machine. The same methods used for simple machines should be used. Remember to treat each shaft as the basic unit for establishing the narrowbands.

DATABASE DEVELOPMENT

The next step required to establish a predictive maintenance program is the creation of a comprehensive database.

Data-Acquisition Frequency

During the implementation stage of a predictive maintenance program, all classes of machinery should be monitored to establish a valid baseline data set. Full vibration signatures should be acquired to verify the accuracy of the database setup and determine the initial operating condition of the machinery.

Since a comprehensive program will include trending and projected time to failure, multiple readings are required on all machines to provide sufficient data for the microprocessor to develop trend statistics. Normally, during this phase, measurements are acquired every 2 weeks.

After the initial or baseline analysis of the machinery, the frequency of data collection will vary depending on the classification of the machine trains. Class I machines should be monitored on a 2- to 3-week cycle; class II, on a 3- to 4-week cycle; class III, on a 4- to 6-week cycle; and class IV, on a 6- to 10-week cycle.

This frequency can and should be adjusted for the actual condition of specific machine trains. If the rate of change of a specific machine indicates rapid degradation, you should either repair it or at least increase the monitoring frequency to prevent catastrophic failure.

The recommended data-acquisition frequencies are the maximum that will ensure prevention of most catastrophic failures. Less frequency monitoring will limit the ability of the program to detect and prevent unscheduled machine outages.

To augment the vibration-based program, you also should schedule the nonvibration tasks. Bearing cap measurements, point-of-use infrared measurements, visual inspections, and process parameters monitoring should be conducted in conjunction with the vibration data acquisition.

Full infrared imaging or scanning on the equipment included in the vibration-monitoring program should be conducted on a quarterly basis. In addition, full thermal scanning of critical electrical equipment (i.e., switchgear, circuit breakers, and so on) and all heat transfer systems (i.e., heat exchangers, condensers, process piping, and so on) that are not in the vibration program should be conducted quarterly.

Lubricating oil samples from all equipment included in the program should be taken on a monthly basis. At a minimum, a full spectrographic analysis should be conducted on these samples. Wear-particle or other analysis techniques should be made on an "as needed" basis.

Analysis Parameters

The next step in establishing the program's database is to set up the analysis parameters that will be used to routinely monitor plant equipment. Each of these parameters will be based on the specific machine-train requirements that we have just developed.

Normally, for nonmechanical equipment, the analysis parameter set will consist of the calculated values derived from measuring the thermal profile or process parameters. Each classification of equipment or system will have its own unique analysis parameter set.

Signature Analysis Boundaries

All vibration-monitoring systems have finite limits on resolution, or the ability to graphically display the unique frequency components that make up a machine's vibration signature. The upper limit (F_{max}) for signature analysis should be set high enough to capture and display enough data that the analyst can determine the operating condition of the machine train but no higher. Since most microprocessor-based systems are limited to 400 lines of resolution, selection of excessively high frequencies can severely limit the diagnostic capabilities of the program.

To determine the impact of resolution, calculate the display capabilities of your system. For example, a vibration signature with a maximum frequency (F_{max}) of 1000 Hz taken with an instrument capable of 400 lines of resolution would result in a display in which each displayed line will be equal to 2.5 Hz, or 150 rpm. Any frequencies that fall between 2.5 and 5.0, that is, the next displayed line, would be lost.

Alert and Alarm Limits

The method of establishing and using alert/alarm limits varies depending on the particular vibration-monitoring system that you select. Normally, these systems use either static or dynamic limits to monitor, trend, and alarm measured vibration. We will not attempt to define the different dynamic methods of monitoring vibration severity in this book. We will, however, provide a guideline for the maximum limits that should be considered acceptable for most plant mechanical equipment.

The systems that use dynamic alert/alarm limits base their logic on the fact that the rate of change of vibration amplitude is more important than the actual level. Any change in the vibration amplitude is a direct indication that there is a corresponding change in the machine's mechanical condition. However, there should be a maximum acceptable limit, that is, absolute fault.

The accepted severity limit for casing vibration is 0.628 ips peak (velocity). Figure 3.14 shows the machinery-vibration severity chart. This is an unfiltered broadband value and normally represents a bandwidth between 10 and 10,000 hertz. This value can be used to establish the absolute fault or maximum vibration amplitude for broadband measurement on most plant machinery. The exception would be machines with running speeds below 1200 or above 3600 rpm.

Narrowband limits, that is, discrete bandwidths within the broadband, can be established using the following guideline. Normally, 60 to 70 percent of the total vibration energy will occur at the true running speed of the machine. Therefore, the absolute fault limit for a narrowband established to monitor the true running speed would be 0.42 ips peak. This value also can be used for any narrowbands established to monitor frequencies below the true running speed.

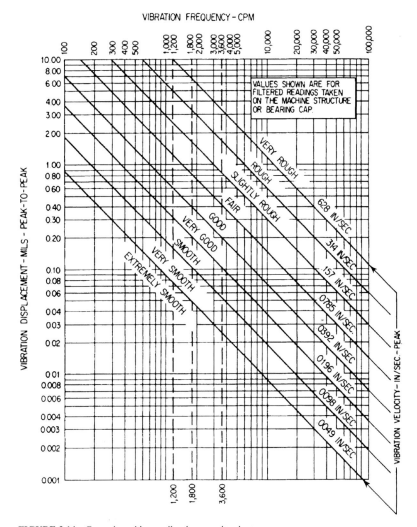

FIGURE 3.14 General machinery-vibration severity chart.

Absolute fault limits for narrowbands established to monitor frequencies above running speed can be developed by using the 0.42 ips peak limit established for the true running speed. For example, the absolute fault limit for a narrowband created to monitor the blade pass frequency of a fan with 10 blades would be set at 0.042, or 0.42/10.

Narrowbands designed to monitor high-speed components, that is, above 1000 Hz, should have an absolute fault of 3.0 g peak (acceleration).

Rolling-element bearings, based on factor recommendations, have an absolute fault limit of 0.01 ips peak. Sleeve or fluid-film bearings should be watched closely. If the fractional components that identify oil whip or whirl are present at any level, the bearing is subject to damage, and the problem should be corrected.

Nonmechanical equipment and systems will normally have an absolute fault limit that specifies the maximum recommended level for continued operation. Equipment or system vendors will, in most cases, be able to provide this information.

INSTRUMENTS FOR VIBRATION MEASUREMENT

The type of transducers and data-acquisition techniques that you will use for the program is the final critical factor that can determine the success or failure of your program. Their accuracy, proper application, and appropriate mounting will determine whether or not valid data will be collected.

The optimal predictive maintenance program is predicated on vibration analysis as the principal technique for the program. It is also the most sensitive to problems created by the use of the wrong transducer or mounting technique.

There are three basic types of vibration transducers that can be used for monitoring the mechanical condition of plant machinery: displacement probe, velocity transducer, and accelerometers. Each has specific applications within the plant. Each also has limitations.

Displacement Probes

Displacement or eddy-current probes are designed to measure the actual movement (i.e., displacement) of a machine's shaft relative to the probe. Therefore, the displacement probe must be rigidly mounted to a stationary structure to gain accurate, repeatable data.

Permanently mounted displacement probes will provide the most accurate data on machines with a low (relative to the casing and support structure) rotor weight. Turbines, large process compressors, and other plant equipment should have displacement transducers permanently mounted at key measurement locations to acquire data for the program (Figs. 3.15 and 3.16).

The useful frequency range for displacement probes is from 10 to 1000 Hz or 600 to 60,000 rpm. Frequency components below or above this range will be distorted and therefore unreliable for determining machine condition.

The major limitation with displacement or proximity probes is cost. The typical cost for installing a single probe, including a power supply, signal conditioning, and so on, will average $1000. If each

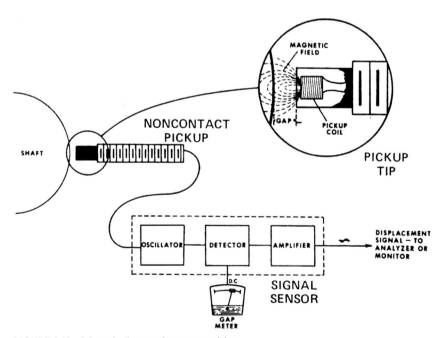

FIGURE 3.15 Schematic diagram of noncontact pickup.

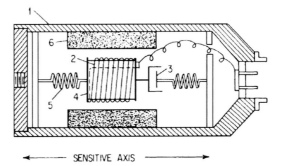

FIGURE 3.16 Schematic diagram of velocity pickup. (1) Pickup case. (2) Wire coil. (3) Damper. (4) Mass. (5) spring. (6) Magnet.

machine in your program requires 10 measurements, the cost per machine will be about $10,000. Using displacement transducers for all plant machinery will dramatically increase the initial cost of the program.

Displacement data are normally recorded in terms of mils peak-to-peak. This value expresses the maximum deflection or displacement off the true centerline of a machine's shaft.

Velocity Transducers

Velocity transducers are electromechanical sensors designed to monitor casing or relative vibration. Unlike the displacement probe, velocity transducers measure the rate of displacement, not actual movement. Velocity data are normally expressed in terms of ips peak and are perhaps the best method of expressing the energy created by machine vibration.

Velocity transducers, like displacement probes, have an effective frequency range of about 10 to 1000 Hz. They should not be used to monitor frequencies below or above this range.

The major limitation of velocity transducers is their sensitivity to mechanical and thermal damage. Normal plant use can cause a loss of calibration, and therefore, a strict recalibration program must be used to prevent distortion of data. At a minimum, velocity transducers should be recalibrated at least every 6 months. Even with periodic recalibration, programs using velocity transducers are prone to bad or distorted data that result from loss of calibration (Figs. 3.15 and 3.17).

Accelerometers

Accelerometers use a piezoelectric crystal to convert mechanical energy into electrical signals. Data acquired with this type of transducer are relative vibration, not actual displacement, and are expressed in terms of g. Acceleration is perhaps the best method of determining the force created by machine vibration.

Accelerometers are susceptible to thermal damage. If sufficient heat is allowed to radiate into the crystal, it can be damaged or destroyed. However, since the data-acquisition time, using temporary mounting techniques, is relatively short, that is, less than 30s, thermal damage is rare. Accelerometers do not require a calibration program to ensure accuracy.

The effective range of general-purpose accelerometers is from about 1 to 10,000 Hz. Ultrasonic accelerometers are available for frequencies up to 1 MHz.

Machine data above 1000 Hz or 60,000 cpm should be taken and analyzed in acceleration or g.

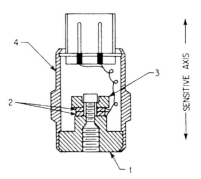

FIGURE 3.17 Schematic diagram of accelerometer. (1) Base. (2) Piezoelectric crystals. (3) Mass. (4) Case.

Mounting Techniques

Predictive maintenance programs using vibration analysis must have accurate, repeatable data to determine the operating condition of plant machinery. In addition to the transducer, three factors will affect data quality: measurement point, orientation, and compressive load.

Key measurement point locations and orientation to the machine's shaft were selected as part of the database setup to provide the best possible detection of incipient machine-train problems. Deviation from the exact point or orientation will affect the accuracy of acquired data. Therefore, it is important that every measurement throughout the life of the program be acquired at exactly the same point and orientation. In addition, the compressive load or downward force applied to the transducer also should be exactly the same for each measurement. For accuracy of data, a direct mechanical link to the machine's casing or bearing cap is absolutely necessary. Slight deviations in this load will induce errors in the amplitude of vibration and also may create false frequency components that have nothing to do with the machine.

The best method of ensuring that these three factors are exactly the same each time is to hard-mount vibration transducers to the selected measurement points. This will guarantee accuracy and repeatability of the acquired data. It also will increase the initial cost of the program. The average cost of installing a general-purpose accelerometer will be about $300 per measurement point, or $3000 for a typical machine train.

To eliminate the capital cost associated with permanently mounting transducers, a well-designed quick-disconnect mounting can be used. This technique permanently mounts a quick-disconnect stud, with an average cost of less than $5, at each measurement point location. A mating sleeve, built into a general-purpose accelerometer, is then used to acquire accurate, repeatable data. A well-designed quick-disconnect mounting technique will provide the same accuracy and repeatability as the permanent mounting technique but at a much lower cost.

The third mounting technique that can be used is a magnetic mount. For general-purpose use below 1000 Hz, a transducer can be used in conjunction with a magnetic base. Even though the transducer/magnet assembly will have a resonant frequency that may provide some distortion to acquired data, this technique can be used with marginal success. Since the magnet can be placed anywhere on the machine, it will not guarantee that the exact location and orientation are maintained on each measurement.

The final method used by some plants to acquire vibration data is handheld transducers. This approach is not recommended if any other method can be used. Handheld transducers will not provide the accuracy and repeatability required to gain maximum benefit from a predictive maintenance program. If this technique must be used, extreme care should be exercised to ensure that the exact point, orientation, and compressive load are used for every measurement point.

Vibration Meters

A number of different portable instruments (Figs. 3.18 and 3.19) are available for making vibration measurements. The basic instrument for a vibration-based predictive maintenance program is a small microprocessor that is specifically designed to acquire, condition, and store both time- and frequency-domain vibration data. This unit is used for checking the mechanical condition of machines at periodic intervals and includes a microprocessor with solid-state memory for recording the overall vibration level of selected plant machinery. This type of meter is limited in that it is restricted to overall vibration levels and cannot provide either time traces or frequency-domain signatures that are an absolute requirement for root-cause or operating dynamics analyses. Preprogrammed prompting messages appear on an LCD display to guide the operator to the correct measurement points. Additional information can be entered using the front panel keyboard. Measurements can be made easily and quickly; it is only necessary for the operator to place the pickup against the point to be measured and press the "store" key to record the overall vibration level.

Vibration Analyzer

As was indicated, the function of a vibration meter is to determine the mechanical condition of critical plant machinery. When a mechanical defect is detected, the meter is not capable of pinpointing the specific problem or its root cause. This is the purpose of the vibration analyzer.

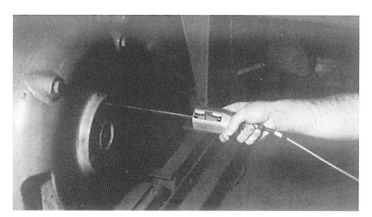

FIGURE 3.18 Common pickup-attachment techniques.

There are a variety of vibration analyzers commercially available. Most are microprocessor-based and combine the capabilities of a vibration meter and vibration analyzer into a single, light-weight unit. The major difference between a meter and analyzer is the ability to acquire, store, and trend complete time-domain, frequency-domain, and synchronous vibration data as well as process variables such as pressure, flow, and temperature. This ability provides the analyst with all the data required to resolve incipient machine or process-system problems. Figure 3.20 illustrates a typical predictive maintenance vibration analyzer. This type of instrument can provide all the capabilities required for a comprehensive predictive maintenance program.

FIGURE 3.19 Hand-held vibration meter used for periodic vibration checks.

Installed Systems

A total plant program may require the inclusion of installed vibration-monitoring systems to achieve maximum benefits. There are two principal reasons for this need: (1) recurring costs of data acquisition using portable instruments and (2) financial impact of failures on some critical plant equipment.

Using portable data-acquisition instruments for predictive maintenance necessitates a recurring labor cost that may become prohibitive in large plants. For example, a typical integrated steel mill may require 20 or more full-time technicians to acquire the monthly vibration data required to maintain a total plant program. At an average burdened cost of $70,000 per technician, a recurring labor

FIGURE 3.20 Microprocessor-based vibration analyzer.

FIGURE 3.21 Typical hard-wired systems.

cost in excess of $1.4 million annually can be directly attributed to the program. These recurring costs can be eliminated by the use of hard-wired, permanent systems.

Some of the critical equipment, machinery, and systems within the plant must be monitored on a continuous basis. Periodic, normally monthly, monitoring cannot provide the level of protection required by the financial impact that premature failure of these machines would generate. Therefore, this classification of plant machinery must be monitored using a real-time or continuous hard-wired system.

Figure 3.21 illustrates typical permanent or hard-wired vibration monitoring systems. These systems use state-of-the-art microprocessor technology and have become extremely cost-effective. Where similar systems 10 years ago have cost several millions of dollars, today's systems can be installed for a fraction of that price. Typical system costs range from less than $20,000 to about $100,000 and can be justified on the basis of a reduction in data-acquisition costs alone.

GETTING STARTED

The steps defined in this chapter will provide the guidelines for establishing a predictive maintenance database. The only steps remaining to get the program started are to establish measurement routes and take the initial or baseline measurements. Remember, the predictive maintenance system will need multiple data sets to develop trends on each machine. With this database, you will be able to monitor the critical machinery in your plant for degradation and begin to achieve the benefits that predictive maintenance can provide. The actual steps required to implement a database will depend on the specific predictive maintenance system selected for your program. The system vendor should provide the training and technical support required to properly develop the database.

Program Maintenance

The labor-intensive part of predictive maintenance management is complete. A viable program has been established, the database is complete, and you have begun to monitor the operating conditions of your critical plant equipment. Now what?

Most programs stop right here. The predictive maintenance team does not continue their efforts to get the maximum benefits that predictive maintenance can provide. Instead, they rely on trending, comparative analysis, or, in the case of vibration-based programs, simplified signature analysis to maintain the operating condition of their plant. This is not enough to gain the maximum benefits from a predictive maintenance program.

The methods that can be used to ensure that you gain the maximum benefits from your program and at the same time improve the probability that the program will continue include trending and analysis techniques.

Trending Techniques

The database established for a vibration-monitoring program should include broadband, narrow-band, and full signature vibration data. It also includes process parameters, bearing cap temperatures, lubricating oil analysis, thermal imaging, and other critical monitoring parameters. What do we do with these data?

The first method required to monitor the operating condition of plant equipment is to trend the relative condition over time. Most of the microprocessor-based systems will provide the means for automatically storing and recalling vibration and process parameters trend data for analysis or hard copies for reports. They also will automatically prepare and print numerous reports that quantify the operating condition at a specific point in time. A few will automatically print trend reports that quantify the change over a selected time frame. All this is great, but what does it mean?

Monitoring the trends of a machine train or process system will provide the ability to prevent most catastrophic failures. The trend is similar to the bathtub curve used to schedule preventive maintenance. The difference between the preventive and predictive bathtub curve is that the latter is based on the actual condition of the equipment, not on a statistical average.

The disadvantage of relying on trending as the only means of maintaining a predictive maintenance program is that it will not tell you the reason a machine is degrading. One good example of this weakness is an aluminum foundry that relied strictly on trending to maintain its predictive maintenance program. In the foundry are 36 cantilevered fans that are critical to plant operation. The rolling-element bearings in each of these fans is changed on an average of every 6 months. By monitoring the trends provided by their predictive maintenance program, management can adjust the bearing change-out schedule based on the actual condition of the bearings in a specific fan. Over a 2-year period, there were no catastrophic failures or loss of production that resulted from the fans being out of service. Did the predictive maintenance program work? In the company's terms, the program was a total success. However, the normal bearing life should have been much greater than 6 months. Something in the fan or process created the reduction in average bearing life. Limiting the program to trending only, the foundry was unable to identify the root cause of the premature bearing failure. Properly used, your predictive maintenance program can identify the specific or root cause of chronic maintenance problems. In the example, a full analysis provided the answer. Plate-out, or material buildup on the fan blades, constantly increased the rotor mass and therefore forced the fans to operate at critical speed. The imbalance created by operation at critical speed was the forcing function that destroyed the bearings. After taking corrective actions, the plant now gets an average of 3 years from the fan bearings.

VIBRATION IDENTIFICATION

CAUSE	AMPLITUDE	FREQUENCY	PHASE	REMARKS
Unbalance	Proportional to unbalance. Largest in radial direction.	1 x RPM	Single reference mark.	Most common cause of vibration.
Misalignment couplings or bearings and bent shaft	Large in axial direction 50% or more of radial vibration	1 x RPM usual 2 & 3 x RPM, sometimes	Single double or triple	Best found by appearance of large axial vibration. Use dial indicators or other method for positive diagnosis. If sleeve bearing machine and no coupling misalignment balance the rotor.
Bad bearings anti-friction type	Unsteady - use velocity measurement if possible	Very high several times RPM	Erratic	Bearing responsible most likely the one nearest point of largest high-frequency vibration.
Eccentric journals	Usually not large	1 x RPM	Single mark	If on gears largest vibration in line with gear centers. If on motor or generator vibration disappears when power is turned off. If on pump or blower attempt to balance.
Bad gears or gear noise	Low - use velocity measure if possible	Very high gear teeth times RPM	Erratic	
Mechanical looseness		2 x RPM	Two reference marks. Slightly erratic.	Usually accompanied by unbalance and/or misalignment.
Bad drive belts	Erratic or pulsing	1, 2, 3 & 4 x RPM of belts	One or two depending on frequency. Usually unsteady.	Strob light best tool to freeze faulty belt.
Electrical	Disappears when power is turned off.	1 x RPM or 1 or 2 x synchronous frequency.	Single or rotating double mark.	If vibration amplitude drops off instantly when power is turned off cause is electrical.
Aerodynamic hydraulic forces		1 x RPM or number of blades on fan or impeller x RPM		Rare as a cause of trouble except in cases of resonance.
Reciprocating forces		1, 2 & higher orders x RPM		Inherent in reciprocating machines can only be reduced by design changes or isolation.

FIGURE 3.22 Vibration-identification chart.

Analysis Techniques

All machines have a finite number of failure modes, see Fig. 3.22. If you have a thorough understanding of these failure modes and the dynamics of the specific machine, you can learn the vibration-analysis techniques that will isolate the specific failure mode or root cause of each machine-train problem. The following example will provide a comparison of various trending and analysis techniques.

Broadband Data. The data acquired using a broadband are limited to a value that represents the total energy being generated by the machine train at the measurement point location and in the direction opposite the transducer. Most programs trend and compare the recorded value at a single point and disregard the other measurement points on the common shaft.

Rather than evaluate each measurement point separately, plot the energy of each measurement point on a common shaft. First, the vertical measurements were plotted to determine the mode shape of the machine's shaft. This plot indicates that the outboard end of the motor shaft is displaced much more than the remaining shaft. This limits the machine problem to the rear of the motor. Based strictly on the overall value, the probable cause is loose motor mounts on the rear motor feet. The second step was plotting the horizontal mode shape. This plot indicates that the shaft is deflected between the pillow-block bearings. Without additional information, the mode shaft suggests a bent shaft between the bearings. Even though we cannot identify the absolute failure mode, we can isolate the trouble to the section of the machine train between the pillow-block bearings.

Narrowband Data. The addition of unique narrowbands that monitor specific machine components or failure modes produces more diagnostic information. If we add the narrowband information acquired from the Hoffman blower, we find that the vertical data are primarily at the true running speed of the common shaft. This confirms that a deflection of the shaft exists. No other machine component or failure mode is contributing to the problem. The horizontal measurements indicate that the blade pass, bearing defect, and misalignment narrowbands are the major contributors.

As discussed earlier, fans and blowers are prone to aerodynamic instability. The indication of abnormal vane pass suggests that this may be contributing to the problem. The additional data provided by the narrowband readings help to eliminate many of the possible failure modes that could affect the blower. However, we still cannot confirm the specific problem.

Root-Cause Failure Analysis. A visual inspection of the blower indicated that the discharge is horizontal and opposite the measurement point location. By checking the process parameters recorded concurrent with the vibration measurements, we found that the motor was in a no-load or run-out condition and that the discharge pressure was abnormally low. In addition, the visual inspection showed that the blower sits on a cork pad and is not bolted to the floor. The discharge piping, 24-in diameter schedule 40 pipe, was not isolated from the blower, nor did it have any pipe supports for the first 30 ft of horizontal run. With all these clues in hand, we concluded that the blower was operating in a "run-out" condition; that is, it was not generating any pressure and was therefore unstable. This part of the machine problem was corrected by reducing (i.e., partially closing) the damper setting and forcing the blower to operate within acceptable aerodynamic limits. After correcting the damper setting, all the abnormal horizontal readings were within acceptable limits. The vertical problem with the motor was isolated to improper installation. The weight of approximately 30 ft of discharge piping compressed the cork pad under the blower and forced the outboard end of the motor to elevate above the normal centerline. In this position, the motor became an unsupported beam and resonated in the same manner as a tuning fork. After isolating the discharge piping from the blower and providing support, the vertical problem was eliminated.

CHAPTER 4
AN INTRODUCTION TO THERMOGRAPHY

R. Keith Mobley
Principal, Life Cycle Engineering, Inc., Charleston, S.C.

Thermography is a predictive maintenance technique that can be used to monitor the condition of plant machinery, structures and systems. It uses instrumentation designed to monitor the emission of infrared energy, that is, temperature, to determine their operating condition. By detecting thermal anomalies, that is, areas that are hotter or colder than they should be, an experienced surveyor can locate and define incipient problems within the plant.

INFRARED TECHNOLOGY

Infrared technology is predicated on the fact that all objects having a temperature above absolute zero emit energy or radiation. Infrared radiation is one form of this emitted energy. Infrared emissions, or below red, are the shortest wavelengths of all radiated energy and are invisible without special instrumentation. The intensity of infrared radiation from an object is a function of its surface temperature. However, temperature measurement using infrared methods is complicated because there are three sources of thermal energy that can be detected from any object: energy emitted from the object itself, energy reflected from the object, and energy transmitted by the object (Fig. 4.1).Only the emitted energy is important in a predictive maintenance program. Reflected and transmitted energies will distort raw infrared data. Therefore, the reflected and transmitted energies must be filtered out of acquired data before a meaningful analysis can be completed.

The surface of an object influences the amount of emitted or reflected energy. A perfect emitting surface, Fig. 4.2, is called a *blackbody* and has an emissivity equal to 1.0. These surfaces do not reflect. Instead, they absorb all external energy and reemit as infrared energy.

Surfaces that reflect infrared energy are called *graybodies* and have an emissivity less than 1.0 (Fig. 4.3). Most plant equipment falls into this classification. Careful considerations of the actual emissivity of an object improve the accuracy of temperature measurements used for predictive maintenance. To help users determine emissivity, tables have been developed to serve as guidelines for most common materials. However, these guidelines are not absolute emissivity values for all machines or plant equipment.

Variations in surface condition, paint or other protective coatings and many other variables can affect the actual emissivity factor for plant equipment. In addition to reflected and transmitted energy, the user of thermographic techniques must also consider the atmosphere between the object and the measurement instrument. Water vapor and other gases absorb infrared radiation. Airborne dust, some lighting and other variables in the surrounding atmosphere can distort measured infrared

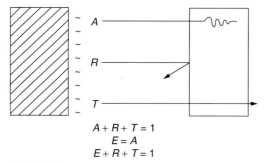

FIGURE 4.1 Energy emissions. All bodies emit energy within the infrared band. This provides the basis for infrared imaging or thermography. A = absorbed energy. R = reflected energy. T = transmitted energy. E = emitted energy.

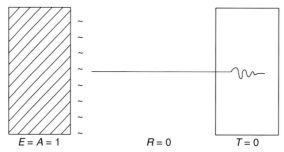

FIGURE 4.2 Blackbody emissions. A perfect or blackbody absorbs all infrared energy. A = absorbed energy. R = reflected energy. T = transmitted energy. E = emitted energy.

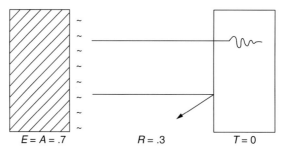

FIGURE 4.3 Graybody emissions. All bodies that are not blackbodies will emit some amount of infrared energy. The emissivity of each machine must be known before implementing a thermographic program. A = absorbed energy. R = reflected energy. T = transmitted energy. E = emitted energy.

radiation. Since the atmospheric environment is constantly changing, using thermographic techniques requires extreme care each time infrared data is acquired.

INFRARED EQUIPMENT

Most infrared monitoring systems or instruments provide special filters that can be used to avoid the negative effects of atmospheric attenuation of infrared data. However the plant user must recognize the specific factors that will affect the accuracy of the infrared data and apply the correct filters or other signal conditioning required to negate that specific attenuating factor or factors.

Collecting optics, radiation detectors, and some form of indicator are the basic elements of an industrial infrared instrument. The optical system collects radiant energy and focuses it upon a detector, which converts it into an electrical signal. The instrument's electronics amplifies the output signal and process it into a form, which can be displayed. There are three general types of instruments that can be used for predictive maintenance: infrared thermometers or spot radiometers, line scanners, and imaging systems.

Infrared Thermometers. Infrared thermometers or spot radiometers are designed to provide the actual surface temperature at a single, relatively small point on a machine or surface. Within a predictive maintenance program, the point-of-use infrared thermometer can be used in conjunction with many of the microprocessor-based vibration instruments to monitor the temperature at critical points on plant machinery or equipment. This technique is typically used to monitor bearing cap temperatures, motor winding temperatures, spot checks of process piping temperatures, and similar applications. It is limited in that the temperature represents a single point on the machine or structure. However when used in conjunction with vibration data, point-of-use infrared data can be a valuable tool.

Line Scanners. This type of infrared instrument provides a single-dimensional scan or line of comparative radiation. While this type of instrument provides a somewhat larger field of view, that is, area of machine surface, it is limited in predictive maintenance applications.

Infrared Imaging Unlike other infrared techniques, thermal or infrared imaging provides the means to scan the infrared emissions of complete machines, process, or equipment in a very short time. Most of the imaging systems function much like a video camera. The user can view the thermal emission profile of a wide area by simply looking through the instrument's optics. There are a variety of thermal imaging instruments on the market ranging from relatively inexpensive, black and white scanners to full color, microprocessor-based systems. Many of the less expensive units are designed strictly as scanners and do not provide the capability of store and recall thermal images. The inability of store and recall previous thermal data will limit a long-term predictive maintenance program.

Point-of-use infrared thermometers are commercially available and relatively inexpensive. The typical cost for this type of infrared instrument is less than $1000. Infrared imaging systems will have a price range between $8000 for a black and white scanner without storage capability to over $60,000 for a microprocessor-based, color imaging system.

Training is critical with any of the imaging systems. The variables that can destroy the accuracy and repeatability of thermal data must be compensated for each time infrared data is acquired. In addition, interpretation of infrared data requires extensive training and experience.

Inclusion of thermography into a predictive maintenance program will enable you to monitor the thermal efficiency of critical process systems that rely on heat transfer or retention; electrical equipment; and other parameters that will improve both the reliability and efficiency of plant systems. Infrared techniques can be used to detect problems in a variety of plant systems and equipment, including electrical switchgear, gearboxes, electrical substations, transmissions, circuit-breaker panels, motors, building envelopes, bearings, steam lines, and process systems that rely on heat retention or transfer.

7.108 INSTRUMENTS AND RELIABILITY TOOLS

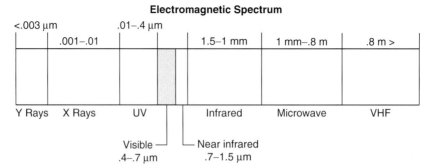

FIGURE 4.4 Electromagnetic spectrum.

Basic Infrared Theory

Infrared energy is light that functions outside the dynamic range of the human eye. Infrared imagers were developed to see and measure this heat. This data is transformed into digital data and processed into video images that are called thermograms. Each pixel of a thermogram has a temperature value and the image's contrast is derived from the differences in surface temperature. An infrared inspection is a nondestructive technique for detecting thermal differences that indicate problems with equipment. Infrared surveys are conducted with the plant equipment in operation, so production need not be interrupted. The comprehensive information can then be used to prepare repair time/cost estimates; evaluate the scope of the problem; plan to have repair materials available, and perform repairs effectively.

Electromagnetic Spectrum

All objects when heated emit electromagnetic energy. The amount of energy is related to the temperature. The higher the temperature, the more electromagnetic energy it emits. The electromagnetic spectrum (Fig. 4.4) contains various forms of radiated energy including x-ray, ultraviolet, infrared, and radio. Infrared energy covers the spectrum of 0.7 to 100 μm.

The electromagnetic spectrum is a continuum of all electromagnetic waves arranged according to frequency and wavelength. A wave has several characteristics (Fig. 4.5). The highest point in the wave is called the *crest*. The lowest point in the wave is referred to as the *trough*. The distance from wavecrest to wavecrest is called a *wavelength*. *Frequency* is the number of wavecrests passing a

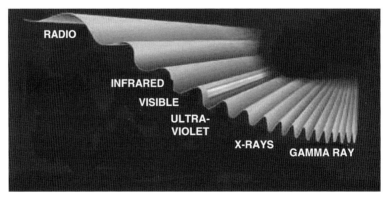

FIGURE 4.5 Wavelengths.

given point per second. As the wave frequency increases, the wavelength decreases. The shorter the wavelength the more energy contained; the longer the wavelength, the less energy. For example: a steel slab exiting the furnace at the hot strip will have short wavelengths. You can feel the heat and see the red glow of the slab. The wavelengths have become shorter crest to crest and the energy being emitted has increased, entering the visible band on the spectrum. By contrast, (infrared energy) when the coil comes off of the coilers it has been cooled. There is a loss of energy. The wavelength have increased crest to crest and decreased in frequency.

Heat Transfer Concepts

Heat is a form of thermal energy. The first law of thermodynamics is that heat given up by one object must equal that taken up by another. The second law is that the transfer of heat takes place from the hotter system to the colder system. If the object is cold, it absorbs rather than emits energy. All objects emit thermal energy or infrared energy through three different types or modes. The three modes are conduction, convection, and radiation. It is important to understand the difference of these three forms.

Conduction. Conduction is the transfer of energy through or between solid objects. A metal bar heated at one end will, in time, become hot at the other end. When a motor bearing is defective, the heat generated by the bearing is transferred to the motor casing. This is a form of conduction.

Convection. Convection is the transfer of energy through or between fluids or gases. If you took the same motor mentioned above and placed a fan blowing directly on the hot bearing, the surface temperature would be different. This is convection cooling. It occurs on the surface of an object. An operator must be careful to identify the true cause and effect. In this case, the difference between good and bad source heating and the surface cooling due to convection.

Radiation. Radiation is the transfer heat by wavelengths of electromagnetic energy. The most common cause of radiation is solar energy. Only radiated energy is detected by an infrared imager. If our motor were sitting outside in the slab storage yard with slabs stacked around it, the electromagnetic energy from the sun and from the slabs would increase the temperature.

The purpose of this exercise was to make the thermography aware that there could be other causes of the thermal energy found or not found. In this case, was the motor hot because of a bad bearing or because of solar radiation? Was the motor missed and failed later because of the fan blowing on it and causing convection cooling? Conduction is the only mode that transfers thermal energy from location to location within a solid, however, at the surface of a solid or liquid, and in a gas, it is normal for all three modes to be operating simultaneously.

Emissivity. Emissivity is the percentage of energy that is emitted by an object. Infrared energy hits an object; the energy is then transmitted, reflected, or absorbed. A common term used in infrared thermography is blackbody.

A blackbody is a perfect thermal emitter. Its emissivity is 100 percent. It has no reflection or transmittance. The objects you will be scanning will each have a different emissivity value. A percentage of the total energy will be due to reflection and transmittance. However, since most of your infrared inspection will be quantitative thermography, the emissivity value will not be as important now.

EVALUATION OF INFRARED EQUIPMENT

Listed below is the criteria used to evaluate infrared equipment. It is important to determine which model fits your needs best before a purchase is made. Some of these points will be important to you and others will not. You will know more about yours needs by the end of this course.

Portability. How much portability does your application require? Does weight and size of the instrument affect your data collection? What kind of equipment will you be scanning?

Ease of Use. How much training is required to use the imager? Can it be used easily in your environment?

Qualitative or Quantitative. Does it measure temperatures? If yes, what temperature range will be measured? Will you need more than one range?

Ambient or Quantitative Measurements. What are the maximum upper and minimum lower ambient temperatures in which you will be scanning?

Short or Long Wave Lengths. Long wavelength systems offer less solar reflection and operate in the 8 to 14 μm bandwidth. Short-wave systems offer smaller temperature errors when an incorrect emissivity value is entered. The operating bandwidth for a short-wave unit is 2 to 5.6 μm.

Batteries. What is the weight and size of the batteries? How long will they last? Will you need additional batteries? How long does it take to charge?

Interchangeable Lenses. Do the ones available fit your application? What are their costs?

Monitor, Eyepiece, or Both. Will you need to show a live image to others while performing an inspection?

Analog or Digital. How will you process the images? Does the imager have analog, digital, or both capabilities?

Software. Can the software package produce quality reports, store and retrieve images? Do you require colonization and temperature editing?

INFRARED THERMOGRAPHY SAFETY

Equipment included in an infrared thermography inspection is almost always energized. For this reason, a lot of attention must be given to safety. The following are basic rules for safety while performing an infrared inspection:

- Plant safety rules must be followed at all time.
- Notify area personnel before entering the area for scanning.
- Qualified electrician from the area should be assigned to open and close all panels.
- Where safe and possible, all equipment to be scanned will be on line and under normal load with a clear line of sight to the item.
- Equipment whose covers are interlocked without an interlock defect mechanism should be shut down when allowable. If safe, their control covers opened and equipment restarted.

INFRARED SCANNING PROCEDURES

The purpose of an infrared inspection is to identify and document problems in an electrical or mechanical system. The information provided by an inspection is presented in an easily and understandable form. A high percentage of problems occur in termination and connections, and especially in copper to aluminum connection. A splice or a lug connector should not look warmer than its conductors if it has been sized properly. All problem connection should be dismantled, cleaned, reassembled, or replaced as necessary.

Type of Infrared Problems

When viewing thermal problems there are three basic types of problems.

- Mechanical looseness
- Load problems
- Component failure

Mechanical Looseness. Mechanical looseness occurs most frequently. A loose connection will result in thermal stress fatigue from over use. Fuse clips are a good example, the constant heat up and cool down creates a poor connection. An accurate temperature measurement, or use of an isotherm will identify a loose condition. When the isotherm is brought down to a single pixel, or temperature, it will identify the source of the loose condition.

Component Failure. Understanding the nomenclature of the problem can identify component failure. Specifically, the actual component will be the heat source. For example, a heat-stressed fuse in a three-phase assembly will appear hotter than the other two fusses.

Common Problems Found and What to Scan

Listed below are the items to scan while performing an infrared survey and some of common problems?

Motor Control and Distribution Centers. Have the switchgear panel covers opened or removed by qualified personnel prior to inspection. Scan cable, cable connections, fuse holders, fuse circuit breakers, and bus.

Main Secondary Switchgear. Have the switchgear panel covers opened or removed by qualified personnel prior to inspection. Scan cables, cables connections, circuit breakers (front and back), and bus.

Circuit Breaker Distribution Panels. Covers on small circuit breaker panels do not have to be removed for scanning. Circuit breakers and conductors are very close to the metal covers. Defective components are usually detectable by the heating of the cover in the area of the problem. If a problem exits, remove the panel cover to locate the problem. Only remove panel covers that can safely be removed.

Bus Duct. Electrical conductors are very close to the metal "skin" of the duct and defective joints are usually detectable by the heating of the cover in the vicinity of the problem.

Motors. Do not scan motor less than 25 hp unless they are critical to production. On motors greater than 25 hp scan the "T" boxes, visible conductors connections and rotors. Bearing problems can be found by comparing the surface temperature of like motors. Overheating conditions are documented as hot spots on the CRT and are usually found in comparing equipment to equipment, end bell to end bell (same type bearings), and stator to end bell.

Transformer—Oil Filled. Scan transformer, transformer fins, cable connections, bushings, and tap changer. On all transformers, the oil level should be inspected during the survey. During the infrared survey, if a transformer appears exceptionally warm, the cooling radiators are near ambient temperature and the transformer is above 50 percent of full load, the oil level is too low to circulate the oil and cooling is not taking place. Oil in the transformers is cooled by convection; as the load increases, the oil expands and the level will increase until then it circulates in the cooling radiators. As a result of repeated oil samples and oil leaks, the reduced volume of oil causes the winding to overheat, thus reducing the life of the transformer. Plugged cooling headers, isolated radiators, and plugged individual cooling fins can also be detected.

Transformers—Dry Type. Scan transfers, cable connections, bushings, and tap changer. Enclosure covers on dry type transformers should only be removed if there is safe clearance between the transformer connections and the enclosure panels. Some models, especially the newer ones, have screened openings for ventilation. Use these openings for your scanning survey.

The iron in these transformers is hot. It will heat the bus work and cause substantial infrared reflection. By increasing the temperature scale and adjusting the level control on the imager, you will be able to get uniform images, which will show hot spots in the secondary bus or the iron. A hot spot in the iron usually indicates a short. Make certain reflection is not a factor.

Compare all windings. If temperatures are over a winding, but there is a difference in temperature of two windings, there may be an unbalance load. A hot spot on a winding may point to shorted turns.

Transformers Bushings. As a scanner moves upward on the transformers main tank and tap changer compartment, the bushings, lighting arresters, and their bus connections should be observed. This is also a critical area because the integrity of the transformer, substation, or the complete system is dependent on proper installation and maintenance of each component. A survey of the transformer bushings, comparing one to the other, will reveal any loose connections or bushing problems. With the scanner, you can determine if the connection is loose internally or externally.

Capacitors. A capacitor has two conductive surfaces, which are separated by an dielectric barrier. Capacitors usually function as power-factor correctors. When energized, all units should have the same temperature if the K_{VAR} size is the same. A high uniform temperature is normal. A cold capacitor usually indicates a blown fuse or bad cell. Isolated spots showing a high temperature on a surface of the capacitor may indicate a bad capacitor.

High Voltage Switchgear. Scan lighting arresters, insulators, cables, cables connections, bushing, circuit breakers, and disconnect switches.

Load Break Switches. In the switch, two metal surfaces act as conductors when they are brought into contact. Usually, problems are restricted to the contact surface. Poor contacts usually show up as hot spots.

Fuses. A fuse is a metal conductor, which is deliberately made to melt when an overload of current is forced upon it. Major problems affected as loose mechanical stab clips cause hot spots, corroded or oxidized external contact surface, and/or poor internal connection, which are bolted or soldered.

Circuit Breakers. Circuit breakers serve the same function as a fuse. It is a switching device that breaks an electrical circuit automatically. Problem areas are caused by corroded or oxidized contact surfaces, poor internal connections, poor control circuitry, and/or defective bushings.

Conductors. The melting points and current-carrying capacity of conductors are determined by the size and base material of the conductors. During a survey, make comparisons between phases and between conductors and connections. An unbalance load will account for some difference between conductors. Use metering devices already installed to check the differences.

The type of load will have an effect on whether the load is balanced. Three-phase motor loads should be balanced; lighting and single-phase load may be unbalanced.

Other Problems
  Broken Strands. These hot spots are found at the support and at the cables termination.

Spiral Heating. This is found on stranded wire, which is heavily oxidized. The problem will show up as a hot spiral from one connection to another connection. There is a load imbalance between the strands, which results in a poor connection.

Ground Conductor. Usually there are no hot spots on a ground conductor. They do show up, however, as hot spots where there is abnormal leakage current to the ground. Be suspicious about such spots. Always point them out in the inspection report.

Parallel Feeders. A cold cable indicates a problem when parallel conductors are feeding the same load.

GLOSSARY

Infrared Thermography Terminology

ΔT (delta temperature)   The delta notation represents the difference in two temperatures.

μ   Symbol for unit ohm/s. Also used to describe microns in the infrared electromagnetic scale.

°C   Celsius degree.

°F   Fahrenheit degree.

A/D conversion   The conversion of continuous type electrical signals varying in amplitude, frequency, or phase into proportional discrete digital signals by means of an analogue-digital converter.

Absorptivity   The ratio of the absorbed to incident electromagnetic radiation on a surface.

Analogue data   Data represented in continuous form, as contrasted with digital data having discrete values.

Atmospheric absorption   The process whereby some or all of the energy of sound waves or electromagnetic waves is transferred to the constituents of the atmosphere.

Atmospheric attenuation   The process whereby some or all of the energy of the sound waves or electromagnetic radiation is absorbed and/or scattered when traversing the atmosphere.

Atmospheric emission   Electromagnetic radiation emitted by the atmosphere.

Atmospheric radiance   The radiant flux per unit solid angle per unit of projected area of the source in the atmosphere.

Atmospheric reflectance   Ratio of reflected radiation from the atmosphere to incident radiation.

Ambient temperature   Ambient temperature is the temperature of the air in the immediate neighborhood of the equipment

Band   A specification of a spectral range (say, from 0.4 to 0.5 μm) that is used for radiate measurements. The term *channel* is also in common use with the same meaning as band. In the electromagnetic spectrum, the term *band* refers to a specific frequency range, designated as L-Band, S-Band, X-Band, and so on.

Bandwidth   A certain range of frequencies within a band.

Conduction   The transfer of heat through\between solids.

Convection   The transfer of heat through\between fluids.

Corona   The glow or brush discharge around conductors when air is stressed beyond its ionization point without developing flashover.

Electromagnetic spectrum   Electromagnetic radiation is energy propagated through space between electric and magnetic fields. The electromagnetic spectrum is the extent of that energy ranging from cosmic rays, gamma rays, x-rays to ultraviolet, visible and infrared radiation including microwave energy.

Emissivity Consideration of the characteristics of materials, particularly with respect to the ability to absorb, transmit, or reflect infrared energy.

Emittance Power radiated per unit area of a radiating surface.

Far infrared Infrared radiation extending approximately from 15 to 100 μm.

Gamma ray A high-energy photon, especially as emitted by a nucleus in a transition between two energy levels.

Hot spot An area of a negative or print revealing excessive light on that part of the subject.

Infrared band The band of electromagnetic wavelengths lying between the extreme of the visible (approximately 0.70 μm) and the shortest microwaves (approximately 100 μm).

Infrared radiation Electromagnetic radiation lying in the wavelength interval from 0.7 to 1000 μm (or roughly between 1 μm and 1 mm wavelength). Its lower limit is bounded by visible radiation, and its upper limit by microwave radiation.

Isothermal mapping Mapping of all regions having the same temperature.

Microwave band The portion of the electromagnetic spectrum lying between the far infrared and the conventional radio frequency portion. While not bounded by definition, it is commonly regarded as extending from 0.1 cm (100 μm) to 30 cm in wavelength (1 to 100 GHz frequency).

Mid-infrared Infrared radiation extending approximately from 1.3 to 3.0 μm and being part of the reflective infrared. Often referred to short wavelength infrared radiation (SWIR).

Near-infrared. Infrared radiation extending approximately from 0.7 to 1.3 μm and being part of the radiative infrared.

Qualitative infrared thermography The practice of gathering information about a system or process by observing images of infrared radiation, and recording and presenting that information.

Quantitative infrared thermography The practice of measuring temperatures of the observed patterns of infrared radiation.

Radar band Frequency and designation with wavelengths within the range of approximately 100 μm to 2 m.

Radiation The emission and propagation of waves transmitting energy through space or through some medium.

Radio band The range of wavelengths or frequencies of electromagnetic radiation designated as radio waves; approximately 4 to 9 Hz in frequency.

Reflectivity The fraction of the incident radiant energy reflected by a surface that is exposed to uniform radiation from a source that fills its field of view.

Spectral band An interval in the electromagnetic spectrum defined by two wavelengths, two frequencies, or two wave numbers.

Temperature gradient Rate of change of temperature with distance.

Thermal emittance Emittance of radiation by a body not at absolute zero due to the thermal agitation of its molecules.

Thermography The recording of the thermal qualities of objects and surfaces by means of scanning equipment in which the infrared or microwave radiation recorded can be converted into a thermal image.

Transmittance The ratio of energy transmitted by a body to that incident on it.

Ultraviolet band That portion of the electromagnetic spectrum ranging from just above the visible (about 4000 Å) to below 400 Å, on the border of the x-ray region.

Visible band The band of the electromagnetic spectrum, which can be perceived by the naked eye. This band ranges from 7500 Å to 4000 Å Being bordered by the infrared and ultraviolet bands.

X-ray Electromagnetic waves of short wavelength from .00001 Å to 3000 Å.

Electrical Terminology

Alternator An ac generator that produces alternating current, which is internally rectified to dc current before being released.

Alternating current (ac) Electrical current which reverses direction periodically, expressed in hertz or cycles per second. Abbreviated ac.

Ampere The quantitative unit measurement of electrical current.

Ampacity A term used to describe the current-handling capacity of an electrical device.

Amperage A term synonymous with current; used in describing electrical current. The total amount of current (amperes) flowing in a circuit.

Armature The main power winding in a motor in which electromotive force is produced, usually the rotor of a dc motor or the stator of an ac motor.

Arrester A device placed from phase to ground whose nonlinear impedance characteristics provide a path for high-amplitude transients.

Attenuator A passive device used to reduce signal strength.

Brush A piece of conducting material, which, when bearing against a commutator, slip ring, and so on will provide a passage for electric current.

Capacitor A discrete electrical device that has two electrodes and an intervening insulator, which is called the dielectric. A device used to store an electrical charge.

Circuit (parallel) An electrical system in which all positive terminals are joined through one wire, and all negative terminals through another wire.

Circuit (series) An electrical system in which separate parts are connected end to end, to form a single path for current to flow through.

Circuit (closed) An electrical circuit in which there is no interruption of current flow.

Circuit (open) Any break or lack of contact in an electrical circuit either intentional (switch) or unintentional (bad connection).

Circuit breaker A resettable device that responds to a preset level of excess current flow by opening the circuit, thereby preventing damage to circuit elements.

Circuit protector A circuit protector is a device that will open the circuit if it becomes overheated because of too much electricity flowing through it. Thus it protects other components from damage if the circuit is accidentally grounded or overloaded. Fuses, fusible links, and circuit breakers are circuit protectors.

Coil A continuous winding arrangement of a conductor, which combines the separate magnetic fields of all the winding, loops to produce a single, stronger field.

Current The flow of electricity in a circuit as expressed in amperes. Current refers to the quantity or intensity of electrical flow. Voltage, on the other hand, refers to the pressure or force causing the electrical flow.

Diode A device that permits current to flow in one direction only. Used to change ac to dc. A rectifier.

Direct current (dc) Electrical current that flows consistently in one direction.

Distribution The way in which power is routed to various current-using sites or devices. Outside the building, distribution refers to the process of routing power from the power plant to the users. Inside the building, distribution is the process of using feeders and circuits to provide power to devices.

Electromagnetic interference (EMI) A term that describes electrically induced noise or transients.

Filter An electronic device, which opposes the passage of a certain frequency band while allowing other frequencies to pass. Filters are designed to produce four different results. A high-pass filter allows all signals above a given frequency to pass. A low-pass filter allows only frequencies below a given frequency to pass. A band-pass filter allows a given band of frequencies to pass while attenuating all others. A trap filter allows all frequencies to pass but acts as a high-impedance device to the tuned frequency of the filter.

Flashover   Arcing that is caused by the breakdown of insulation between two conductors where a high current flow exists, with a high potential difference between the conductors.

Fuse   A device that automatically self-destructs when the current passing through it exceeds the rated value of the fuse. A plug-in protector with a filament that melts or burns out when overloaded.

Ground   A general term that refers to the point at which other portions of a circuit are referenced when making measurements. Power systems grounding is that point to which the neutral conductor, safety ground, and building ground are connected. This grounding electrode may be a water pipe, driven ground rod, or the steel frame of the building.

Harmonic   A frequency that is a multiple of the fundamental frequency. For example, 120 Hz is the second harmonic of 60 Hz, 180 Hz is the third harmonic, and so forth.

Harmonic distortion   Excessive harmonic content that distorts the normal sinusoidal waveform is harmonic distortion. This can cause overheating of circuit elements and might appear to a device as data-corrupting noise.

Hertz (Hz)   A term describing the frequency of ac. The term hertz is synonymous with cycles per second.

Impedance (Z)   Measured in ohms, impedance is the total opposition to current flow in a circuit where ac is flowing. This includes inductive reactance, capacitive reactance, and resistance.

Inductance   This term describes the electrical properties of a coil of wire and its resultant magnetic field when an ac is passed through it. This interaction offers impedance to current flow, thereby causing the current waveform to lag behind the voltage waveform. This results in what's known as a *lagging power factor*.

Inductor   A discrete circuit element, which has the property of inductance. It should be noted that at very high radio frequencies, a straight wire or a path on a printed-circuit board could act as an inductor.

Insulator   A nonconducting substance or body, such as porcelain, glass, or Bakelite, used for insulating wires in electrical circuits to prevent the undesired flow of electricity.

Inverter   An inverter takes dc power and converts it into ac power.

Isolation   The degree to which a device can separate the electrical environment of its input from its output, while allowing the desired transmission to pass across the separation.

Kilovolt-ampere (kVA)   An electrical unit related to the power rating of a piece of equipment. It is calculated by multiplying the rated voltage of equipment by the current required (or produced). For resistive loads 1 kVA equals 1 kW.

Kilohertz (kHz)   A term meaning 1000 cycles per second.

Lightning arrester   A device used to pass large impulses to ground.

Mean time between failure (MTBF)   A statistical estimate of the time a component, subassembly, or operating unit will operate before failure will occur.

Megahertz (MHz)   A term for 1×10^6 Hz (cycles per second).

Motor alternator   A device that consists of an ac generator mechanically linked to an electric motor, which is driven by utility power or by batteries. An alternator is an ac generator.

Motor generator   A motor generator consists of an ac motor coupled to a generator. The utility power energizes the motor to drive the generator, which powers the critical load. Motor generators provide protection against noise and spikes, and, if equipped with a heavy flywheel, they may also protect against sags and swells.

Neutral   One of the conductors of a three-phase wye system is the neutral conductor. Sometimes called the return conductor, it carries the entire current of a single-phase circuit and the resultant current in a three-phase system that is unbalanced. The neutral is bonded to ground on the output of a three-phase delta-wye transformer.

Ohm (Ω)   The unit of measurement for electrical resistance.

Ohm's law   A law of electricity, which states the relationship between voltage, amperes, and resistance. It takes a pressure of 1 V to force 1 A of current through 1 Ω of resistance. Equation: volts = amperes $\times$ ohms ($E = I \times R$).

Radiation   RF energy, which is emitted, or leaks from a distribution system and travels through space. These signals often cause interference with other communication services.

Rectifier   An electrical device containing diodes, used to convert ac to dc.

Relay   An electromagnetic switching device using low current to open or close a high-current circuit.

Resistance (R)   A term describing the opposition of elements of a circuit to ac or dc.

Resistor   A device installed in an electrical circuit to permit a predetermined current to flow with a given voltage applied.

Rheostat   A device for regulating a current by means of a variable resistance.

Rotor   The part of the alternator that rotates inside the stator and produces an electrical current from induction by the electromagnetic fields of the stator windings.

Semiconductor, or silicon-controlled rectifier (SCR)   An electronic dc switch, which can be triggered into conduction by a pulse to a gate electrode, but can only be cut off by reducing the main current below a predetermined level (usually zero).

Shielding   Protective coating that helps eliminate electromagnetic and radio frequency interference.

Shunt   A conductor joining two points in a circuit to form a parallel circuit, through which a portion of the current may pass, in order to regulate the amount of current flowing in the main circuit.

Sine wave   A fundamental waveform produced by periodic oscillation that expresses the sine or cosine of a linear function of time or space, or both.

Single phase   That portion of a power source, which represents only a single phase of the three phases that are available.

Solenoid   A tubular coil containing a movable magnetic core, which moves when the coil is energized.

Stator   The stationary winding of an alternator (the armature in a dc generator).

Switch   A device used to open, close, or redirect current in an electrical circuit.

Single phase   That portion of a power source, which represents only a single phase of the three phases that are available.

Three phase   An electrical system with three different voltage lines or legs, which carry sine-wave waveforms that are 120° out of phases from one another.

Transformer   A device used for changing the voltage of an ac circuit and/or isolating a circuit from its power source.

Volt (V)   Electrical unit of measure (current $\times$ resistance).

Watt (W)   The unit for measuring electrical power or work. A watt is the mathematical product of amperes and volts (W=A $\times$ V).

APPENDIX : MATERIAL LIST

Material Metals		Temperature, °F	Temperature, °C	Emissivity
Alloys	20-Ni, 24-CR, 55-FE, Oxid.	392	200	0.9
	20-Ni, 24-CR, 55-FE, Oxid.	932	500	0.97
	60-Ni, 12-CR, 28-FE, Oxid.	518	270	0.89
	60-Ni, 12-CR, 28-FE, Oxid.	1040	560	0.82
	80-Ni, 20-CR, oxidized	212	100	0.87
	80-Ni, 20-CR, oxidized	1112	600	0.87
	80-Ni, 20-CR, oxidized	2372	1300	0.89
Aluminum	Unoxidized	77	25	0.02
	Unoxidized	212	100	0.03
	Unoxidized	932	500	0.06
	Oxidized	390	199	0.11
	Oxidized	1110	599	0.19
	Oxidized at 599°C (1110°F)	390	199	0.11
	Oxidized at 599°C (1110°F)	1110	599	0.19
	Heavily oxidized	200	93	0.2
	Heavily oxidized	940	504	0.31
	Highly polished	212	100	0.09
	Roughly polished	212	100	0.18
	Commercial sheet	212	100	0.09
	Highly polished plate	440	227	0.04
	Highly polished plate	1070	577	0.06
	Bright rolled plate	338	170	0.04
	Bright rolled plate	932	500	0.05
	Alloy A3003, oxidized	600	316	0.4
	Alloy A3003, oxidized	900	482	0.4
	Alloy 1100-0	200–800	93–427	0.05
	Alloy 24ST	75	24	0.09
	Alloy 24ST, polished	75	24	0.09
	Alloy 75ST	75	24	0.11
	Alloy 75ST, polished	75	24	0.08
Bismuth	Bright	176	80	0.34
	Unoxidized	77	25	0.05
	Unoxidised	212	100	0.06
Brass	73% Cu, 27% Zn, polished	476	247	0.03
	73% Cu, 27% Zn, polished	674	357	0.03
	62% Cu, 37% Zn, polished	494	257	0.03
	62% Cu, 37% Zn, polished	710	377	0.04
	83% Cu, 17% Zn, polished	530	277	0.03
	Matte	68	20	0.07
	Burnished to brown color	68	20	0.4
	Cu-Zn, brass oxidized	392	200	0.61
	Cu-Zn, brass oxidized	752	400	0.6
	Cu-Zn, brass oxidized	1112	600	0.61
	Unoxidised	77	25	0.04
	Unoxidised	212	100	0.04
Cadmium		77	25	0.02
Carbon	Lampblack	77	25	0.95
	Unoxidised	77	25	0.81
	Unoxidised	212	100	0.81

(Continued)

(*Continued*)

Material Metals		Temperature, °F	Temperature, °C	Emissivity
	Unoxidized	932	500	0.79
	Candle soot	250	121	0.95
	Filament	500	260	0.95
	Graphitized	212	100	0.76
	Graphitized	572	300	0.75
	Graphitized	932	500	0.71
Chromium		100	38	0.08
		1000	538	0.26
	Polished	302	150	0.06
Cobalt	Unoxidized	932	500	0.13
	Unoxidized	1832	1000	0.23
Columbium	Unoxidized	1500	816	0.19
	Unoxidized	2000	1093	0.24
Copper	Cuprous oxide	100	38	0.87
	Cuprous oxide	500	260	0.83
	Cuprous oxide	1000	538	0.77
	Black, oxidized	100	38	0.78
	Etched	100	38	0.09
	Matte	100	38	0.22
	Roughly polished	100	38	0.07
	Polished	100	38	0.03
	Highly polished	100	38	0.02
	Rolled	100	38	0.64
	Rough	100	38	0.74
	Molten	1000	538	0.15
	Molten	1970	1077	0.16
	Molten	2230	1221	0.13
	Nickel plated	100–500	38–260	0.37
Dow metal		0.4–600	18–316	0.15
Gold	Enamel	212	100	0.37
	Plate		0.0001	
	Plate on 0.0005 silver	200–750	93–399	0.11–0.14
	Plate on .0005 nickel	200–750	93–399	0.07–0.09
	Polished	100–500	38–260	0.02
	Polished	1000–2000	538–1093	0.03
Haynes Alloy C,	Oxidized	600–2000	316–1093	0.90–0.96
Haynes Alloy 25,	Oxidized	600–2000	316–1093	0.86–0.89
Haynes Alloy X,	Oxidized	600–2000	316–1093	0.85–0.88
Inconel sheet	1000 (538)	1000	538	0.28
	1200 (649)	1200	649	0.42
	1400 (760)	1400	760	0.58
Inconel X, polished	75 (24)	75	24	0.19
Inconel B, polished	75 (24)	75	24	0.21
Iron	Oxidized	212	100	0.74
	Oxidized	930	499	0.84
	Oxidized	2190	1199	0.89
	Unoxidized	212	100	0.05
	Red rust	77	25	0.7

(*Continued*)

(*Continued*)

Material Metals		Temperature, °F	Temperature, °C	Emissivity
	Rusted	77	25	0.65
	Liquid	2760–3220	1516–1771	0.42–0.45
Cast iron	Oxidized	390	199	0.64
	Oxidized	1110	599	0.78
	Unoxidized	212	100	0.21
	Strong oxidation	40	104	0.95
	Strong oxidation	482	250	0.95
	Liquid	2795	1535	0.29
Wrought iron				
	Dull	77	25	0.94
	Dull	660	349	0.94
	Smooth	100	38	0.35
	Polished	100	38	0.28
Lead	Polished	100–500	38–260	0.06–0.08
	Rough	100	38	0.43
	Oxidized	100	38	0.43
	Oxidized at 1100°F	100	38	0.63
	Gray oxidized	100	38	0.28
Magnesium		100–500	38–260	0.07–0.13
Magnesium oxide		1880–3140	1027–1727	0.16–0.20
Mercury		32	0	0.09
		77	25	0.1
		100	38	0.1
		212	100	0.12
Molybdenum		100	38	0.06
		500	260	0.08
		1000	538	0.11
		2000	1093	0.18
	Oxidized at 1000°F	600	316	0.8
	Oxidized at 1000°F	700	371	0.84
	Oxidized at 1000°F	800	427	0.84
	Oxidized at 1000°F	900	482	0.83
	Oxidized at 1000°F	1000	538	0.82
Monel, Ni-Cu		392	200	0.41
		752	400	0.44
		1112	600	0.46
Monel, Ni-Cu oxidized		68	20	0.43
Monel, Ni-Cu Oxidized. at 1110°F	1110 (599)	1110	599	0.46
Nickel	Polished	100	38	0.05
	Oxidized	100–500	38–260	0.31–0.46
	Unoxidized	77	25	0.05
	Unoxidized	212	100	0.06
	Unoxidized	932	500	0.12
	Unoxidized	1832	1000	0.19
	Electrolytic	100	38	0.04
	Electrolytic	500	260	0.06
	Electrolytic	1000	538	0.1
	Electrolytic	2000	1093	0.16

(*Continued*)

(Continued)

Material Metals		Temperature, °F	Temperature, °C	Emissivity
Nickel oxide		1000–2000	538–1093	0.59–0.86
Palladium plate (0.00005	on 0.0005 silver)	200–750	93–399	0.16–0.17
Platinum		100	38	0.05
		500	260	0.5
		1000	538	0.1
Platinum, black		100	38	0.93
		500	260	0.96
		2000	1093	0.97
	Oxidized at 1100°F	500	260	0.07
		1000	538	0.11
Rhodium flash (0.0002	on 0.0005 Ni)	200–700	93–371	0.10–0.18
Silver	Plate (0.0005 on Ni)	200–700	93–371	0.06–0.07
	Polished	100	38	0.01
		500	260	0.02
		1000	538	0.03
		2000	1093	0.03
Steel	Cold rolled	200	93	0.75–0.85
	Ground sheet	1720–2010	938–1099	0.55–0.61
	Polished sheet	100	38	0.07
		500	260	0.1
		1000	538	0.14
	Mild steel, polished	75	24	0.1
	Mild steel, smooth	75	24	0.12
	Mild steel, ÊLiquid	2910–3270	1599–1793	0.28
	Steel, unoxidized	212	100	0.08
	Steel, oxidized	77	25	0.8
Steel alloys	Type 301, polished	75	24	0.27
	Type 301, polished	450	232	0.57
	Type 301, polished	1740	949	0.55
	Type 303, oxidized	600–2000	316–1093	0.74–0.87
	Type 310, rolled	1500–2100	816–1149	0.56–0.81
	Type 316, polished	75	24	0.28
	Type 316, polished	450	232	0.57
	Type 316, polished	1740	949	0.66
	Type 321	200–800	93–427	0.27–0.32
	Type 321 polished	300–1500	149–815	0.18–0.49
	Type 321 w/BK oxide	200–800	93–427	0.66–0.76
	Type 347, oxidized	600–2000	316–1093	0.87–0.91
	Type 350	200–800	93–427	0.18–0.27
	Type 350 polished	300–1800	149–982	0.11–0.35
	Type 446, polished	300–1500	149–815	0.15–0.37
	Type 17-7 PH	200–600	93–316	0.44–0.51
	Type 17-7 PH ÊPolished	300–1500	149–815	0.09–0.16
	Type C1020, Oxidized	600–2000	316–1093	0.87–0.91
	Type PH-15-7 MO	300–1200	149–649	0.07–0.19
Stellite	Polished	68	20	0.18

(Continued)

(Continued)

Material Metals		Temperature, °F	Temperature, °C	Emissivity
Tantalum	Unoxidized	1340	727	0.14
		2000	1093	0.19
		3600	1982	0.26
		5306	2930	0.3
Tin, unoxidised		77	25	0.04
		212	100	0.05
Tinned iron, bright		76	24	0.05
		212	100	0.08
Titanium, Alloy C110M	Polished	300–1200	149–649	0.08–0.19
Oxidized at	538°C (1000°F)	200–800	93–427	0.51–0.61
Alloy Ti-95A,	Oxid. at, 238°C (1000°F)	200–800	93–427	0.35–0.48
	Anodized onto SS	200–600	93–316	0.96–0.82
Tungsten	Unoxidized	77	25	0.02
	Unoxidized	212	100	0.03
	Unoxidized	932	500	0.07
	Unoxidized	1832	1000	0.15
	Unoxidized	2732	1500	0.23
	Unoxidized	3632	2000	0.28
	Filament (aged)	100	38	0.03
	Filament (aged)	1000	538	0.11
	Filament (aged)	5000	2760	0.35
Uranium oxide		1880	1027	0.79
Zinc	Bright, galvanized	100	38	0.23
	Commercial 99.1%	500	260	0.05
	Galvanized	100	38	0.28
	Oxidized	500–1000	260–538	0.11
	Polished	100	38	0.02
	Polished	500	260	0.03
	Polished	1000	538	0.04
	Polished	2000	1093	0.06
Non-metals				
Adobe	68 (20)			0.9
Asbestos	Board	100	38	0.96
	Cement	32–392	0–200	0.96
	Cement, red	2500	1371	0.67
	Cement, white	2500	1371	0.65
	Cloth	199	93	0.9
	Paper	100–700	38–371	0.93
	Slate	68	20	0.97
	Asphalt, pavement	100	38	0.93
	Asphalt, tar paper	68	20	0.93
Basalt		68	20	0.72
Brick	Red, rough	70	21	0.93
	Gault cream	2500–5000	1371–2760	0.26–0.30
	Fire clay	2500	1371	0.75
	Light buff	1000	538	0.8
	Lime clay	2500	1371	0.43
	Fire brick	1832	1000	0.75–0.80

(Continued)

(*Continued*)

Material Metals		Temperature, °F	Temperature, °C	Emissivity
	Magnesite, refractory	1832	1000	0.38
	Grey brick	2012	1100	0.75
	Silica, glazed	2000	1093	0.88
	Silica, unglazed	2000	1093	0.8
	Sandlime	2500–5000	1371–2760	0.59–0.63
Carborundum		1850	1010	0.92
Ceramic	Alumina on Inconel	800–2000	427–1093	0.69–0.45
	Earthenware, glazed	70	21	0.9
	Earthenware, matte	70	21	0.93
	Greens No. 5210-2C	200–750	93–399	0.89–0.82
	Coating No. C20A	200–750	93–399	0.73–0.67
	Porcelain	72	22	0.92
	White Al2O3	200	93	0.9
	Zirconia on Inconel	800–2000	427–1093	0.62–0.45
Clay	68 (20)	0.39	0.39	
	Fired at	158	70	0.91
	Shale at	68	20	0.69
	Tiles, light red	2500–5000	1371–2760	0.32–0.34
	Tiles, red	2500–5000	1371–2760	0.40–0.51
	Tiles,			
	Dark purple	2500–5000	1371–2760	0.78
Concrete	Rough	32–2000	0–1093	0.94
	Tiles, natural	2500–5000	1371–2760	0.63–0.62
	"Brown	2500–5000	1371–2760	0.87–0.83
	"Black	2500–5000	1371–2760	0.94–0.91
Cotton cloth	68 (20)			0.77
Dolomite lime	68 (20)			0.41
Emery corundum	176 (80)			0.86
Glass	Convex D	212	100	0.8
	Convex D	600	316	0.8
	Convex D	932	500	0.76
	Nonex	212	100	0.82
	Nonex	600	316	0.82
	Nonex	932	500	0.78
	Smooth	32–200	0–93	0.92–0.94
Granite		70	21	0.45
Gravel		100	38	0.28
Gypsum		68	20	0.80–0.90
Ice, smooth		32	0	0.97
Ice, rough		32	0	0.98
Lacquer	Black	200	93	0.96
	Blue, on Al foil	100	38	0.78
	Clear, on Al foil (2 coats)	200	93	0.08 (0.09)
	Clear, on bright Cu	200	93	0.66
	Clear, on tarnished Cu	200	93	0.64
	Red, on Al foil (2 coats)	100	38	0.61 (0.74)
	White	200	93	0.95

(*Continued*)

(Continued)

Material Metals		Temperature, °F	Temperature, °C	Emissivity
	White, on Al foil (2 coats)	100	38	0.69 (0.88)
	Yellow, on Al foil (2 coats)	100	38	0.57 (0.79)
Lime mortar		100–500	38–260	0.90–0.92
Limestone		100	38	0.95
Marble, white		100	38	0.95
	Smooth, white	100	38	0.56
	Polished grey	100	38	0.75
Mica		100	38	0.75
Oil on nickel	0.001 film	72	22	0.27
	0.002 "	72	22	0.46
	0.005 "	72	22	0.72
	Thick "	72	22	0.82
Oil, linseed	On Al foil, uncoated	250	121	0.09
	On Al foil, 1 coat	250	121	0.56
	On Al foil, 2 coats	250	121	0.51
	On polished iron, .001 Film	100	38	0.22
	On polished iron, .002 Film	100	38	0.45
	On polished iron, .004 film	100	38	0.65
	On polished iron, thick film	100	38	0.83
Paints	Blue, Cu_2O_3	75	24	0.94
	Black, CuO	75	24	0.96
	Green, Cu_2O_3	75	24	0.92
	Red, Fe_2O_3	75	24	0.91
	White, Al_2O_3	75	24	0.94
	White, Y_2O_3	75	24	0.9
	White, ZnO	75	24	0.95
	White, $MgCO_3$	75	24	0.91
	White, ZrO_2	75	24	0.95
	White, ThO_2	75	24	0.9
	White, MgO	75	24	0.91
	White, $PbCO_3$	75	24	0.93
	Yellow, PbO	75	42	0.9
	Yellow, $PbCrO_4$	75	24	0.93
Paints, aluminum	100 (38)	100	38	0.27–0.67
	10% Al	100	38	0.52
	26% Al	100	38	0.3
	Dow XP-310	200	93	0.22
Paints, bronze	Low			0.34–0.80
	Gum varnish (2 coats)	70	21	0.53
	Gum varnish (3 coats)	70	21	0.5
	Cellulose binder (2 coats)	70	21	0.34
Paints, oil				
	All colors	200	93	0.92–0.96
	Black	200	93	0.92
	Black gloss	70	21	0.9
	Camouflage green	125	52	0.85
	Flat black	80	27	0.88
	Flat white	80	27	0.91

(Continued)

(Continued)

Material Metals		Temperature, °F	Temperature, °C	Emissivity
	Grey-green	70	21	0.95
	Green	200	93	0.95
	Lamp black	209	98	0.96
	Red	200	93	0.95
	White	200	93	0.94
Quartz, rough, fused	Glass, 1.98 mm	540	282	0.9
	Glass, 1.98 mm	1540	838	0.41
	Glass, 6.88 mm	540	282	0.93
	Glass, 6.88 mm	1540	838	0.47
	Opaque	570	299	0.92
	Opaque	1540	838	0.68
Red lead		212	100	0.93
Rubber, hard		74	23	0.94
Rubber, soft, grey		76	24	0.86
Sand		68	20	0.76
Sandstone		100	38	0.67
Sandstone, red		100	38	0.60–0.83
Sawdust		68	20	0.75
Shale		68	20	0.69
Silica, glazed		1832	1000	0.85
Silica, unglazed		2012	1100	0.75
Silicon carbide		300–1200	149–649	0.83–0.96
Silk cloth		68	20	0.78
Slate		100	38	0.67–0.80
Snow, fine particles	20 (_7)			0.82
Snow, granular	18 (_8)			0.89
Soil	Surface	100	38	0.38
	Black loam	68	20	0.66
	Plowed field	68	20	0.38
Soot	Acetylene	75	24	0.97
	Camphor	75	24	0.94
	Candle	250	121	0.95
	Coal	68	20	0.95
Stonework		100	38	0.93
Water	100 (38)	100	38	0.67
Waterglass	68 (20)	68	20	0.96
Wood	Low			0.80–0.90
	Beech planed	158	70	0.94
	Oak, Planed	100	38	0.91
	Spruce, sanded	100	38	0.89

CHAPTER 5
TRIBOLOGY

R. Keith Mobley
Principal, Life Cycle Engineering, Inc., Charleston, S.C.

Tribology is the general term that refers to design and operating dynamics of the bearing-lubrication-rotor support structure of machinery. Several tribology techniques can be used for predictive maintenance: lubricating oil analysis, spectrographic analysis, ferrography, and wear particle analysis.

Lubricating oil analysis, as the name implies, is an analysis technique that determines the condition of lubricating oils used in mechanical and electrical equipment. It is not a tool for determining the operating condition of machinery. Some forms of lubricating oil analysis will provide an accurate quantitative breakdown of individual chemical elements, both oil additive and contaminates, contained in the oil. A comparison of the amount of trace metals in successive oil samples can indicate wear patterns of oil wetted parts in plant equipment and will provide an indication of impending machine failure.

Until recently, tribology analysis has been a relatively slow and expensive process. Analyses were conducted using traditional laboratory techniques and required extensive, skilled labor. Microprocessor-based systems are now available which can automate most of the lubricating oil and spectrographic analysis; thus reducing the manual effort and cost of analysis.

The primary applications for spectrographic or lubricating oil are: quality control, reduction of lubricating oil inventories, and determination of the most cost-effective interval for oil change. Lubricating, hydraulic, and dielectric oils can be periodically analyzed, using these techniques, to determine their condition. The results of this analysis can be used to determine if the oil meets the lubricating requirements of the machine or application. Based on the results of the analysis, lubricants can be changed or upgraded to meet the specific operating requirements.

In addition, detailed analysis of the chemical and physical properties of different oils used in the plant can, in some cases, allow consolidation or reduction of the number and types of lubricates required to maintain plant equipment. Elimination of unnecessary duplication can reduce required inventory levels and therefore maintenance costs.

As a predictive maintenance tool, lubricating oil and spectrographic analysis can be used to schedule oil change intervals based on the actual condition of the oil. In mid to large plants, a reduction in the number of oil changes can amount to a considerable annual reduction in maintenance costs. Relatively inexpensive sampling and testing can show when the oil in a machine has reached to point that warrants change.

The full benefit of oil analysis can only be achieved by taking frequent samples trending the data for each machine in the plant. It can provide a wealth of information on which to base maintenance decisions. However, major payback is rarely possible without a consistent program of sampling.

LUBRICATING OIL ANALYSIS

Oil analysis has become an important aid to preventive maintenance. However, it is limited to maintaining optimum condition of the lubricating oil and is not a means of detecting or anticipating the need for preventive maintenance of critical plant equipment. Oil analysis is overused as a predictive maintenance tool. The misconception that this method can replace vibration analysis and other predictive techniques is the predominant reason for this overuse.

It should be limited to applications where replacement or premature failure of large quantities of lube oil represents a substantial loss. This methodology can be used to detect degradation of the lubricating properties of oil and may permit the user to extend its useful life by filtering, water removal or the replacement of additives that will result in longer life.

With few exceptions, lube oil analysis is conducted by outside laboratories. While there are commercially available oil analysis systems, the cost and training required to perform in-house analysis is generally prohibitive. Laboratories recommend that samples of machine lubricant be taken at scheduled intervals to determine the condition of the lubricating film that is critical to machine-train operation.

Tests Conducted on Lube Oil Samples

Representative samples of lube oil are collected, bottled, and sent to outside laboratories for routine analysis. Generally, these labs will conduct the following tests and provide a written report for each sample submitted.

Viscosity. *Viscosity* is one of the most important properties of lubricating oil. The actual viscosity of oil samples is compared to an unused sample to determine the thinning or thickening of the sample during use. Excessively low viscosity will reduce the oil film's strength, weakening its ability to prevent metal-to-metal contact.

Excessively high viscosity may impede the flow of oil to vital locations in the bearing support structure, reducing its ability to lubricate. Generally, higher viscosity is attributed to oxidation and/or heavy particulate contamination.

Contamination. *Contamination* of oil by water or coolant can cause major problems in a lubricating system. Many of the additives now used in formulating lubricants contain the same elements that are used in coolant additives. Therefore, the laboratory must have an accurate analysis of new oil for comparison.

Fuel Dilution. *Fuel dilution* of oil in an engine weakens the oil film's strength, sealing ability, and detergency. Improper operation, fuel-system leaks, ignition problems, improper timing, or other deficiencies may cause it. Fuel dilution is considered excessive when it reaches a level of 2.5 to 5 percent.

Solids Content. *Solids content* is a general test. All solid materials in the oil are measured as a percentage of the sample volume or weight. The presence of solids in a lubricating system can significantly increase the wear on lubricated parts. Any unexpected rise in reported solids is cause for concern.

Fuel Soot. *Fuel soot* is an important indicator for oil used in diesel engines and is always present to some extent. A test to measure fuel soot in diesel engine oil is important since it indicates the fuel-burning efficiency of the engine. Most tests for fuel soot are conducted by infrared analysis.

Oxidation. *Oxidation* of lubricating oil can result in lacquer deposits, metal corrosion, or thickening of the oil. Most lubricants contain oxidation inhibitors. However, when additives are used up, oxidation of the oil itself begins. The quantity of oxidation in an oil sample is measured by differential infrared analysis.

Nitration. *Nitration* results from fuel combustion in engines. The products formed are highly acidic and they may leave deposits in combustion areas. Nitration will accelerate oil oxidation. Infrared analysis is used to detect and measure nitration products.

Total Acid Number. *Total acid number* (TAN) is a measure of the amount of acid or acid-like material in the oil sample. Because new oils contain additives that affect the TAN number, it is important to compare used-oil samples with new, unused oil of the same type. Regular analysis at specific intervals is important to this evaluation.

It is also an indication of the level of organic acidity in the oil and is normally attributed to oxidation level. For hydraulic oils, the value should not exceed twice the level of new oil.

Total Base Number. *Total base number* (TBN) indicates the ability of oil to neutralize acidity. The higher the TBN the greater is its ability to neutralize acidity. Typical causes of low TBN include using the improper oil for an application, waiting too long between oil changes, overheating, and using high sulfur fuel.

Particle Count. *Particle-count* tests are important to anticipating potential system or machine problems. This is especially true in hydraulic systems. The particle-count analysis made a part of a normal lube-oil analysis is quite different from wear particle analysis. In this test, high particle counts indicate that machinery may be wearing abnormally or that failures may occur because of temporarily or permanently blocked orifices. No attempt is made to determine the wear patterns, size, and other factors that would identify the failure mode within the machine.

Standard lubricating oil analysis does not attempt to determine the specific failure modes of developing machine-train problems. Therefore, additional techniques must be used as part of a comprehensive predictive maintenance program.

SPECTROGRAPHIC ANALYSIS

Spectrographic analysis allows accurate, rapid measurements of many of the elements present in lubricating oil. These elements are generally classified as wear metals, contaminates, or additives. Some elements can be listed in more than one of these classifications. The technique measures the molecular compounds, such as water, glycol, refrigerants, blow-by gases, liquid fuels, and so on, in the oil.

FERROGRAPHY

This technique is similar to spectrography but there are two major exceptions. First, ferrography separates particulate contamination by using a magnetic field rather than burning a sample as in spectrographic analysis. Because a magnetic field is used to separate contaminants, this technique is primarily limited to ferrous or magnetic particles.

The second difference is that particulate contamination larger than 10 μm can be separated and analyzed. Normal ferrographic analysis will capture particles up the 100 μm and provides a better representation of the total oil contamination than spectrographic techniques.

WEAR PARTICLE ANALYSIS

Wear particle analysis is related to oil analysis only in that the particles to be studied are collected through drawing a sample of lubricating oil. Where lubricating oil analysis determines the actual condition of the oil sample, wear particle analysis provides direct information about the wearing condition of the machine-train. Particles in the lubricant of a machine can provide significant information about the condition of the machine. This information is derived from the study of particle shape, composition, size, and quantity. Wear particle analysis is normally conducted in two stages.

Routine Monitoring and Trending

The first method used for wear particle analysis is routine monitoring and trending of the solids content of machine lubricant. In simple terms the quantity, composition, and size of particulate matter in the lubricating oil is indicative of the mechanical condition of the machine. A normal machine will contain low levels of solids with a size less than 10 μm. As the machine's condition degrades, the number and size of particulate matter will increase. See Fig. 5.1

Three Monitoring Techniques. Generally, a combination of three techniques is used to collect and analyze wear particles generated by a machine-train. These include:

In-Line Monitoring. Devices, such as magnets, inductive coils or differential pressure gauges, are installed in the oil circulating system to monitor the main flow through the machine. Magnets and inductive coil may be used to extract ferro-magnetic particles from the circulating oil stream, but are not effective on nonmagnetic particles. Differential pressure gauges, located on in-line filters, can be used to monitor the amount of solids that are being removed from the oil stream. The in-line filter will remove all particles, that is, both magnetic and nonmagnetic, from the oil stream. However, the size of particles extracted from the system is dependent on the absolute rating of the filter media.

Generally, these devices do not provide on-line analysis of wear particles. Instead, they are used as collection devices.

On-Line Monitoring. This method uses a by-passed portion of the main oil flow combined with optical devices that measure turbidity as an indicator of particle concentration. As in the case of in-line monitoring, this method cannot provide definitive root-cause analysis. Rather it is limited to an approximation of the total solids concentration in the sampled volume.

Off-Line. The only method that is capable of definitive analysis of wear particles found in lubricating oil must extract representative samples from the oil volume for laboratory evaluation.

Analysis of Particulate Matter

The second wear particle method involves analysis of the particulate matter in each lubricating oil sample. Five basic types of wear can be identified according to the classification of particles: rubbing

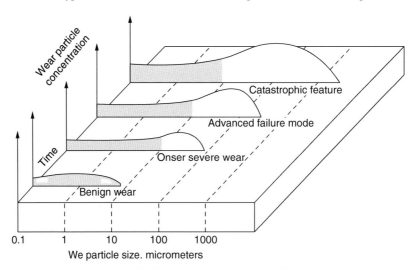

FIGURE 5.1 Particle size distribution equal machine condition.

wear, cutting wear, rolling fatigue wear, combined rolling and sliding wear, and severe sliding wear. Only rubbing wear and early rolling fatigue mechanisms generate particles predominantly less that 15-μm in size.

Normal spectrographic analysis is limited to particulate contamination with a size of 10 μm or less. Larger contaminants are ignored. This fact can limit the benefits that can be derived from the technique.

Rubbing Wear. *Rubbing wear* is the result of normal sliding wear in a machine. During a normal break-in of a wear surface, a unique layer is formed at the surface. As long as this layer is stable, the surface wears normally. If the layer is removed faster than it is generated, the wear rate increases and the maximum particle size increases. Excessive quantities of contaminate in a lubrication system can increase rubbing wear by more than an order of magnitude without completely removing the shear mixed layer. Although catastrophic failure is unlikely, these machines can wear out rapidly. Impending trouble is indicated by a dramatic increase in wear particles. Figure 5.2 illustrates typical rubbing wear. Rubbing wear particles are platelets from the shear mixed layer, which exhibits super-ductility. Opposing surfaces are roughly of the same hardness. Generally, the maximum size of normal rubbing wear is 15 μm.

Cutting Wear. *Cutting wear particles* are generated when one surface penetrates another. These particles are produced when a misaligned or fractured hard surface produces an edge that cuts into a softer surface, or when abrasive contaminate become embedded in a soft surface and cut an opposing surface. Cutting wear particles are abnormal and are always worthy of attention. If they are only a few microns long and a fraction of a micron wide, the cause is probably contamination. Increasing quantities of longer particles signal a potentially imminent component failure. Figure 5.3 illustrates typical cutting wear particles. These particles are generated because of one surface penetrating another. The effect is to generate particles much as a lathe tool creates machining swarf. Abrasive particles, which have become embedded in a soft surface, penetrate the opposing surface causing cutting wear particles. Particles may range in size from 2- to 5-μm wide and 25- to 100-μm long.

Rolling fatigue Wear. *Rolling fatigue* is associated primarily with rolling contact bearings and may produce three distinct particle types: fatigue spall particles, spherical particles, and laminar particles. Particles are released from the stressed surface as a pit is formed. Particles have a maximum size of 100 μm during the initial micro-spalling process. These flat platelets, Fig. 5.4, have a major dimension to thickness ratio greater than 10:1.

Fatigue Spall Wear. *Fatigue spall particles* are the actual material removed when a pit or spall opens up on a bearing surface. An increase in the quantity or size of these particles is the first indication of an abnormality. Rolling fatigue does not always generate spherical particles and they may

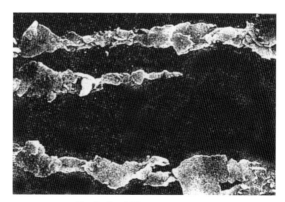

FIGURE 5.2 Typical rubbing wear.

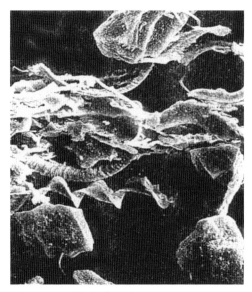

FIGURE 5.3 Typical cutting wear patterns.

be generated by other sources. Their presence is important in that they are detectable before any actual spalling occurs. Laminar particles are very thin and are formed by the passage of a wear particle through a rolling contact. They frequently have holes in them. Laminar particles may be generated throughout the life of a bearing, but at the onset of fatigue spalling the quantity increases.

Combined Rolling and Sliding Wear. *Combined rolling and sliding wear* results from the moving contact of surfaces in gear systems. These larger particles result from tensile stresses on the gear surface, causing the fatigue cracks to spread deeper into the gear tooth before pitting. Gear fatigue cracks do not generate spheres. Scuffing of gears is caused by too high a load or speed. The excessive heat generated by this condition breaks down the lubricating film and causes adhesion of the mating gear teeth. As the wear surfaces become rougher, the wear rate increases. Once started, scuffing usually affects each gear tooth.

There is a large variation in both sliding and rolling wear velocities at the wear contacts; there are corresponding variations in the characteristics of the particles generated. Fatigue particles from the gear pitch line have similar characteristics to rolling bearing fatigue particles. The particles may have a major dimension to thickness ratio between 4:1 and 10:1. The chunkier particles result from tensile

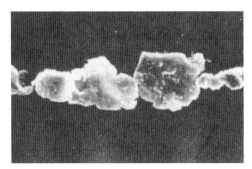

FIGURE 5.4 Rolling fatigue wear.

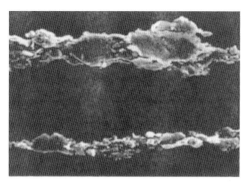

FIGURE 5.5 Combined rolling and sliding wear.

stresses on the gear surface causing fatigue cracks to propagate deeper into the gear tooth prior to pitting. A high ratio of large (20 μm) particles to small (2 μm) particles is usually evident. See Fig. 5.5.

Severe Sliding Wear. Excessive loads or heat causes this type of wear in a gear system. Under these conditions, large particles break away from the wear surfaces, causing an increase in the wear rate. If the stresses applied to the surface are increased further, a second transition point is reached. The surface breaks down and catastrophic wear ensues. Severe sliding wear particles, Fig. 5.6, range in size from 20 μm and larger. Some of these particles have surface striations because of sliding. They frequently have straight edges and their major dimension to thickness ratio is approximately 10:1.

Table 5.1 lists the common materials that are found by wear particle analysis. Many of the metallic contaminates found in the oil can be traced to one or more potential sources. Unfortunately, this technique is not capable of absolutely identifying either the source of contamination or the location of damaged components. For example, an oil sample may contain copper that suggest that rolling element bearing damage is the most likely source of contamination. The analysis cannot confirm that the copper is actually from a bearing nor can it isolate which bearing within the common lube oil system is damaged.

Because of these limitations, tribology has limited value as a predictive maintenance tool. The one exception is as part of a total plant lubrication improvement program that is designed to optimize the useful life of oils and grease that are used within the plant.

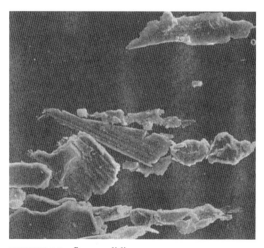

FIGURE 5.6 Severe sliding wear.

TABLE 5.1 Sources of Materials Found in Wear Particle Analysis

Material	Likely source	Material	Likely source
Aluminum	Light alloy pistons Crankshaft bearings Castings	Lead	Plain bearings
Antimony	White metal plain bearings	Magnesium	Wear of plastic components with talc fillers Seawater intrusion
Boron	Coolant leaks Oil additives	Nickel	Valve seats Alloy steels
Chromium	Piston rings Cylinder liners Valve seats	Silicon	Mineral dust intrusion
Cobalt	Valve seats Hard coatings	Silver	Silver-plated bearing surfaces Fretting of silver solder joints
Copper	Copper-lead or bronze bearings Rolling element bearing cages	Sodium	Coolant leakage Seawater intrusion
Indium	Crankshaft bearings	Tin	Plain bearings
Iron	Gears Shafts Cast iron cylinder bores	Zinc	Common oil additive

Wear particle analysis is an excellent failure analysis tool and can be used to understand the root-cause of catastrophic failures. The unique wear patterns observed on failed parts, as well as those contained in the oil reservoir provide a positive means of isolating the failure mode.

LIMITATIONS OF TRIBOLOGY

There are three major limitations with using tribology analysis in a predictive maintenance program: equipment costs, acquiring accurate oil samples, and interpretation of data.

Capital Cost

The capital cost of spectrographic analysis instrumentation is normally too high to justify in-plant testing. Typical cost for a microprocessor-based spectrographic system is between $30,000 and $60,000. Because of this, most predictive maintenance programs rely on third-party analysis of oil samples.

Recurring Cost

In addition to the labor cost associated with regular gathering of oil and grease samples, simple lubricating oil analysis by a testing laboratory will range from about $20.00 to $50.00 per sample. Standard analysis will normally include: viscosity, flash point, total insolubles, total acid number (TAN), total base number (TBN), fuel content, and water content.

More detailed analysis, using spectrographic, ferrographic, or wear particle techniques, that include metal scans, particle distribution (size), and other data can range to well over $150 per sample.

Accurate Samples

A more severe limiting factor with any method of oil analysis is acquiring accurate samples of the true lubricating oil inventory in a machine. Sampling is not a matter of opening a port some where in the oil line and catching a pint sample. Extreme care must be taken to acquire samples that truly represent the lubricant that will pass through the machine's bearings. One recent example is an attempt to acquire oil samples from a bullgear compressor. The lubricating oil filter had a sample port on the clean, that is, downstream, side. However, comparison of samples taken at this point and one taken directly from the compressors oil reservoir indicated that more contaminates existed downstream from the filter than in the reservoir. Which location actually represented the oil's condition? Neither sample was truly representative of the oil condition. The oil filter had removed most of the suspended solids, that is, metals and other insolubles, and was therefore not representative of the actual condition. The reservoir sample was not representative since most of the suspended solids had settled out in the sump.

Proper methods and frequency of sampling lubricating oil is critical to all predictive maintenance techniques that use lubricant samples. Sample points that are consistent with the objective of detecting large particles should be chosen. In a recirculating system, samples should be drawn as the lubricant returns to the reservoir and before any filtration. Do not draw oil from the bottom of a sump where large quantities of material build up over time. Return lines are preferable to reservoir as the sample source, but good reservoir samples can be obtained if careful, consistent practices are used. Even equipment with high levels of filtration can be effectively monitored as long as samples are drawn before oil enters the filters. Sampling techniques involve taking samples under uniform operating conditions. Samples should not be taken more than 30 minutes after the equipment has been shut down.

Sample frequency is a function of the mean time to failure from the onset of an abnormal wear mode to catastrophic failure. For machines in critical service, sampling every 25 hours of operation is appropriate. However, for most industrial equipment in continuous service, monthly sampling is adequate. The exception to monthly sampling is machines with extreme loads. In this instance, weekly sampling is recommended.

Understanding Results

Understanding the meaning of analysis results is perhaps the most serious limiting factor. Most often results are expressed in terms that are totally alien to plant engineers or technician. Therefore, it is difficult for them to understand the true meaning, in terms of oil or machine condition. A good background in quantitative and qualitative chemistry is beneficial. As a minimum, plant staff will require training in basic chemistry and specific instruction on interpreting tribology results.

Cost-benefit

The cost-benefit of tribology is questionable. In too many cases, it is over-used and does not provide enough benefit to justify the cost of regular sampling and analyses.

Lubricating Oil Analysis. As part of a comprehensive lubricant management program, lubricating oil analysis can be a cost-effective tool. However, it must be limited to large volumes of oils, not every oil reservoir in the plant. Also, the frequency of sampling and analysis should be limited to reasonable intervals. Typically, sampling once per quarter is more than adequate for this type of application. Severe applications may warrant more frequent analysis, but extreme care should be exercised to assure that the interval provide meaningful information. Table 5.2 lists the analysis techniques for oil from various types of machines.

Wear Particle Analysis. This technique should be limited solely to a failure modes or root-cause analysis. With few exceptions, wear particle analysis does not add any value to a predictive maintenance program. Other techniques, such as vibration, are better equipped to detect, isolate, and quantify wear within critical plant system.

As a predictive maintenance tool, the potential benefits of this technique will not offset the recurring cost required to acquire and evaluate oil samples.

TABLE 5.2 Analysis Techniques for the Oil from Various Types of Machines

**Essential *Useful	Diesel engine	Gasoline engine	Gears	Hydraulic systems	Air Compressors	Refrigeration compressors	Gas turbines	Steam turbines	Transformers	Heat transfer
Spectro-chemical	**	**	**	**	**	**	**	**		
Infrared analysis	**	**	*	*	*	**	*	*	*	*
Wear debris quantifier	**	**	**	*	*	*	*	*		
Viscosity at 40°C	**	**	**	**	**	**	**	**		*
Viscosity at 100°C	*	*								
Total base number	**	**								
Total acid number			*	**	*	*	**	*	*	**
Water %	**	**	*	**	*		**	**	**	*
Total solids	*	*						*	*	
Fuel dilution	**	**								
Particle counting				**			*			

Note: * = useful, ** = essential.

S·E·C·T·I·O·N · 8

LUBRICATION

CHAPTER 1
THE ORGANIZATION AND MANAGEMENT OF LUBRICATION

F. Alverson, *Group Leader*

T. C. Mead, *Senior Technologist (Ret.)*

W. H. Stein, *Group Leader*

A. C. Witte, *Consultant*
Texaco, Inc., Research & Development Department, Port Arthur, Texas

ORGANIZED PLANT LUBRICATION

An organized lubrication program should be an important component of preventive maintenance. Machinery is costly, and newer models designed for greater precision and faster production certainly require proper lubrication. An organized lubrication program will reduce the possibility of breakdowns and save on repairs, downtime, and lost production. Successful lubrication programs involve both management and plant personnel.

Management Responsibility

In order to establish a plantwide lubrication program, management should arrange to have a survey conducted on each piece of equipment, noting the manufacturer's recommendations and warranty provisions for lubrication along with machine condition, operating speeds and loads, operating conditions (such as contaminants and temperatures), and machine history. This information can be computerized to establish lubrication/maintenance schedules, oil change intervals, and routes to perform the actual lubrication tasks.

A general inspection of plant equipment should be made once a month. At every third month, the plant superintendent or chief engineer should invite an engineer from the lubricant supplier to accompany maintenance personnel on their inspection tour. By working together, they can determine where the use of an improved type of bearing or lubricating device would make a machine run more smoothly, longer, and more economically, where a more suitable grade of lubricant would reduce the frequency of lubrication, or where the housing of overexposed gears would improve safety and reduce fire hazards. Management can promote preventive maintenance by praising operators personally for keeping their machines and working areas clean of dirt and oil. The spoken words "well done" are often much more appreciated than an impersonal memo from the front office with the same message.

8.4 LUBRICATION

TABLE 1.1 Duties of Lubrication Personnel

1. Use correct lubricants in every case and as few types as possible for the plant as a whole.
2. Apply lubricants properly.
3. Apply the correct amount of lubricant.
4. Apply lubricants at proper intervals.
5. Develop schedules for items 1 to 4 for each machine, distribute or post them, and see that they are followed.
6. Train and instruct the oilers, and arrange for lubrication clinics if the number of oilers warrants. Suppliers' sales and engineering representatives frequently can render valuable assistance in the preparation and execution of such programs.
7. Install and use lubricating devices correctly.
8. Keep lubricants clean by keeping the oil room clean and keeping lubricant containers covered.
9. Dispense lubricants through clean, properly identified equipment.
10. Practice preventive maintenance.
11. Cooperate with the maintenance and production departments on lubrication problems.
12. Collect used oils for purification for resale or reclamation if quantity warrants.
13. Keep complete consumption records.
14. Record and analyze all lubrication-connected failures and breakdowns.
15. Eliminate all accident hazards connected with lubrication.
16. Keep abreast of new developments and practices in the lubricating field by periodic consultation with a qualified lubrication engineer—staff, consultant, or supplier's representative.
17. Minimize the total cost of lubrication, remembering that the price of an improper lubricant is a small fraction of its final cost in terms of poor service.

Employee Responsibility

Lubrication supervisors, oilers, and operators have the greatest responsibilities involving lubrication. Lubrication personnel have to be properly trained on storage, handling, application, and use of lubricants. They must then perform the lubrication tasks in accordance with the recommended lubrication schedules. McGraw-Hill's *Factory* magazine provides a list of lubrication duties that is reproduced in Table 1.1.

Safety Considerations

Safety of operating personnel also must be considered. Usually it is practical to plan for inclusion and installation of modern systems of lubrication so that operators will not be exposed to hazards when adjusting fittings or going through normal relubrication procedures. Intricate parts that require hand lubrication can involve personal hazard, especially if this is done while the machine is in operation. The designer should work to prevent such hazards by putting intricate parts on centralized systems, even though it may require external service lines from the main oil or grease supply. Caged ladders and walkways with railings should be installed to safely permit lubrication of equipment where heights are involved.

FLUID MANAGEMENT

Fluid management programs are being used in many plants to extend lubricant life and reduce disposal costs. There are four essential components in a fluid management program: selection and purchase of the lubricant, lubricant monitoring during use, lubricant maintenance using processing and refortification techniques, and finally, disposal of the spent lubricant.

Lubricant Selection and Purchase

Fluid management begins with purchasing the correct lubricant for the application. For most equipment, premium long-lasting lubricants meeting equipment manufacturers' recommendations and specifications should be purchased. During the competitive bidding process, purchasing personnel should carefully consider the supplier, products, and services. A supplier should be chosen on the basis of the quality of lubricants and services (engineering lubrication surveys, troubleshooting, used oil analyses, and so on) offered rather than on price alone.

The overall cost of lubrication compared with the total cost of plant equipment is relatively insignificant. Purchase of lubricants on the basis of price alone is not justified when considering the cost of downtime for repair and lost productivity if attributed to the use of an inferior lubricant. On the other hand, purchase of premium-grade lubricants will not improve or correct lubrication problems if mechanical factors such as misalignment or severe environments (high levels of dirt and water contaminants) are involved.

Lubricant Monitoring Programs

Monitoring programs may be used to determine the condition of the lubricant and to detect early signs of equipment failure. Used oil analyses also can be used to extend lubricant life and establish oil changeout intervals. The properties that should be monitored are dependent on the application and environment. Table 1.2 lists the properties and condemning limits for most large-volume applications of industrial lubricants, namely, turbine/circulating, hydraulic, compressors, and gear oils. Other lubricant applications, such as slideways, rock drills, and so on, which involve small volumes and/or once-through applications, need no monitoring.

The results of monitoring tests can be used in some cases to correct conditions that are contributing to degradation of the lubricant. For example, if the lubricant in a circulating system shows that water is present, it may be possible to locate and eliminate the source of the water. If the viscosity is dropping, it may be determined that an incorrect oil is being used for makeup, or there may be leakage of a different lubricant into the system. The condemning limits shown in the table are intended to serve as general guidelines. The lubricant supplier should provide actual limits for the products being used and interpretation of used oil test results.

Lubricant Maintenance

Lubricant maintenance is closely associated with the monitoring program. When used oil test results exceed the condemning limits, corrective action needs to be taken. Such action could include filtration to remove particulate matter and in some cases oxidation products and/or dehydration. This processing can be done either on site or at a recycle station. Additive replenishment for depleted

TABLE 1.2 Condemning Limits for Used Oils

	Turbine/circulating	Hydraulic	Compressor	Gears
Appearance	Hazy or cloudy indicates moisture or fine particulate matter present. Recondition by filtration, centrifugation, and/or dehydration.			
Water content	If moisture content exceeds 0.1 percent, treat as above or change out. Locate and eliminate the water source.			
Neutralization no.	Change of +0.2 above original is cause for concern. Change of +0.5 requires recycle or replacement.			
Viscosity	Change of ± 10 percent indicates action required. If change is due to oxidation, change or recondition oil; if due to addition of incorrect viscosity grade, correct by addition of appropriate grade.			Change of ±25 percent indicates that corrective action is needed, as indicated for turbine oils.
Flash point	Not usually run unless contamination is suspected (low viscosity or smell). Above 375°F, no action is required. If below, contamination is indicated. Action indicated above should be initiated.			
Particle count	SAE class 6 max	SAE class 3 max	SAE class 6 max	Generally not applicable
Rust test (D665A)	If failing, change oil or reinhibit. ⟶			Generally not applicable
Sediment	If not visually clear, recondition by filtration/centrifugation.			
Oxidation stability RBOT, minutes	If less than 50, change oil or reinhibit	Generally not applicable	Same as turbine oil	Generally not applicable

inhibitors may be feasible for some products in some applications. Since additive replenishment requires a considerable amount of technical expertise, the lubricant supplier should be contacted to provide information and service to reclaim and refortify used lubricants.

Disposal

Disposal is the last step that must be addressed in fluid management when the monitoring results indicate that the oil is severely degraded and/or depleted of additives that cannot be restored. Various options to consider include recycling, burning, land-filling, and re-refining. The most appropriate method of disposal will depend on local, state, and federal regulations. These will clearly be affected by the location, which makes the best method of disposal site-specific. Lubricant disposal needs to be considered carefully on a case-by-case basis.

LUBRICANT PROTECTION

Proper handling and storage of lubricants and greases are important to ensure longevity and satisfactory performance. Premium-grade products should be stored inside to prevent contamination with dirt and water and to protect against temperature extremes. If drums are stored outside, they should be stored on their sides, tilted, or upside down. Drums will expand and contract as the temperature changes, and any water on top of a drum may be drawn through the bung as the drum expands and contracts. Ester- and polyglycol-based lubricants need especially to be protected from atmospheric humidity.

Location and Personnel

A clean, well-lighted room or building is advisable, with provisions for heating in cold weather. It should be specifically kept for lubricant storage and reserve lubricating equipment. In most plants, one or two individuals are assigned the responsibility for inventory and dispensing of lubricants. These individuals should be trained on the importance of protecting lubricants from contamination and commingling with other lubricants. Drums should be labeled clearly to ensure application/use of the correct lubricant.

Facilities for Handling Containers

One-level handling is an important item wherever possible in planning for lubricant storage. If practical, the floor level should be the same as the delivery-truck floor. This facilitates rolling of drums into the storeroom, where racks can be arranged along one or more walls so that oil drums can be raised by a forklift truck and spotted in order to draw the contents off with the least effort into distribution containers. Each drum should have its own spigot to avoid commingling of products. Grease drums are normally stored on end because the contents are removed by paddle, scoop, or pressure pump, according to the consistency of the grease. Paddles, scoops, and other devices must be kept clean to protect against abrasive particles and dirt.

In large plants, where a considerable volume of lubricants must be stored, a set of parallel rails (see Fig. 1.1) is useful for handling full drums to service racks as well as empties for return.

Lighting

This relates to good records. The lubrication and maintenance departments can function most effectively when they have complete records as to lubricant consumption per machine per area. This requires careful inventory (monthly) and recording of amounts of oil and grease issued. Lighting plays an important part. If the storeroom is painted gloss white, if light outlets are well located to obviate glare, and if a comfortable record desk is installed, personnel will keep more careful records.

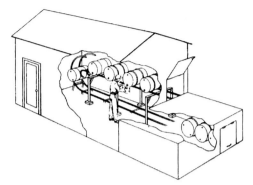

FIGURE 1.1 Parallel rail for handling both full drums and empties.

Bulk Storage

Bulk storage can be an investment that provides benefits in improved efficiency, reduced handling costs, reduced risk of contamination, and simplified inventory. Each product requires its own dedicated bulk storage system, including tank, pump, and receiving line. The tank should be equipped with a water draw-off line, sampling line, and entry to permit periodic tank cleaning. If tanks are equipped with electric heating coils or steam lines, precautions must be taken to prevent overheating and thermal degradation of the lubricant.

Bulk shipments may be supplied in tank cars, tank trucks, or tote bins. Upon arrival of bulk shipments, each product should be inspected visually for clarity and cleanliness and checked for viscosity with a handheld viscometer. Prior to unloading, each tank should be gauged to ensure sufficient room. Tank lines and valves should be checked to ensure that the product is being unloaded into the correct tank. If dedicated lines and pumps are not being used, the system should be flushed with one to three times the volume of the lines to prevent cross-contamination of products. Samples should be obtained from the tank after unloading and labeled with product name, date, invoice number, and batch number. The samples should be stored for at least 6 months.

Fire Protection

The possibility of fire in a well-planned lubricant storage area is remote, assuming that no-smoking rules are observed, that casual visits from other plant personnel are prohibited, that oil drip is prevented or cleaned up promptly, that waste or wiping rags are stored in metal containers and in minimum quantity, and that sparking or arcing tools are used only under conditions of good ventilation. Even so, insurance regulations will require installation of suitable fire-extinguishing equipment and possibly a sprinkler system. The accepted foam-type device for smothering is best. In a small storeroom, one or two hand units may suffice. In a larger area, a multiple-gallon foam cart with adequate hose may be required.

GREASES

A lubricating grease can be defined as a solid to semisolid material consisting of a thickening agent dispersed in a liquid lubricant. Because greases are semisolid materials, they can be used in applications where leakage of an oil will occur. Also, greases can provide a natural sealing action, as in a bearing application where the grease film tends to keep contaminants out and the oil film in. The thickener system used to form a grease also can provide an extra film thickness over that provided by a lubricating oil.

Components of Greases

Greases consist of three components: (1) a fluid lubricant, which usually makes up 70 to 95 wt% of the finished grease, (2) a thickener, which provides the gel-like consistency that holds the liquid lubricant in place and usually makes up from 3 to 15 wt% of the finished grease, and (3) additives, which are added to enhance certain properties of the finished grease and usually make up from 0.5 to 10 wt% of the finished grease. Each of these components must be selected and formulated in the proper proportion to obtain a grease with the desired performance properties. The fluid lubricant most often used is an oil derived from petroleum. Both paraffinic and naphthenic oils are used, and these can be either a single component or a blend. Synthetic-base oils also can be used, but owing to their higher cost, synthetic-base oils are usually used in specialty greases where extreme low-temperature or high-temperature performance characteristics are required. The thickener systems most often used to form greases are soaps such as those formed by the saponification of a fatty material with an alkali metal. These metallic soaps include sodium, calcium, lithium, and aluminum. In addition to soap thickeners, nonsoap chemical compounds such as those based on a urea derivative are also used, as well as inorganic materials such as those made from certain clays (i.e., montmorillonite). Additives, the third component of a grease, are used to modify and improve certain properties of the grease. The principal types of additives used are antioxidants, rust and corrosion inhibitors, extreme pressure/antiwear agents, tackifiers, and solid fillers. Antioxidants are used to protect the grease during extended storage and to extend the service life at elevated temperatures. Rust and corrosion inhibitors provide protection for rusting in the presence of water and inhibit the grease from attacking certain metals such as copper and bronze. Antiwear agents function to extend the loads that can be carried by the lubricant film and reduce wear under boundary lubrication. Tackifiers are used to improve the resistance of the grease to water and to improve the adherence of the grease to a metal surface. Solid fillers are used to improve grease performance under conditions of extremely high loads, where metal-to-metal contact is highly likely.

Grease Properties and Tests

The single most distinguishing property of a grease is its consistency, which is related to the hardness or softness of the grease. The consistency is related to the penetration number obtained on the grease and is defined as the depth, in tenths of a millimeter, that a standard cone penetrates a sample of grease under prescribed conditions of weight, time, and temperature. To ensure a uniform sample, the grease is "worked" 60 strokes in a prescribed manner before running the penetration test (ASTM D217). Based on the worked penetration value, the National Lubricating Grease Institute (NLGI) has devised a classification system using defined consistency grades ranging from 000 (very soft) to 6 (very hard). Each consistency grade has a range of 30 penetration units, with a 15 penetration unit range between each grade. This classification system is shown in Table 1.3.

As indicated in Table 1.3, a no. 2 grade grease will always have a worked penetration in the range of 265 to 295, as determined by the penetration test. Of the grades available, grades 0, 1, and 2 are the most widely used in industry. The more fluid grades, 00 and 000, are used when a thickened oil

TABLE 1.3 NLGI Classification of Greases

NLGI grade	Worked penetration range (tenths of a millimeter)
000	445–475
00	400–430
0	355–585
1	310–340
2	265–295
3	220–250
4	175–205
5	130–160
6	85–115

is desired, such as in the lubrication of gearboxes, where high leakage may occur when using a conventional oil lubricant.

Another important property of a grease is its dropping point. Since greases are semisolid materials, they exhibit a characteristic temperature range wherein they change from a semisolid to a fluid. Greases do not have a sharp melting point but, upon heating, become softer until at some point they become essentially fluid and no longer function as a thickened lubricant. Dropping points are useful in characterizing greases. Each type of soap thickener exhibits a particular dropping point range. Table 1.4 shows the typical dropping point ranges for greases containing different thickeners.

Greases cannot be used at temperatures above their dropping points. However, the dropping point by itself does not establish the maximum usable temperature. The maximum usable temperature is considered to be the temperature limit where the grease should not be used without frequent relubrication. As noted in Table 1.4, the maximum usable temperature is well below the dropping point of the grease. As a general rule, the maximum usable temperature of a grease should be at least 25 to 50°F below its dropping point. With frequent relubrication or continuous lubrication, the maximum usable temperature can be raised. Extended use at temperatures above 350°F usually will result in severe oxidation of the grease. For example, for an organo-clay grease, even though the thickener can withstand very high temperatures (above 500°F), the fluid lubricant component of the grease will be severely oxidized, leaving behind only the solid thickener, which has very poor lubrication properties. Once the fluid lubricant component is removed from the grease, the remaining thickener component becomes dry and abrasive. This is why frequent relubrication is required when operating a grease at high temperatures.

Other grease properties are also important when considering a particular grease for a specific application. These include resistance to softening (shear stability), oxidation resistance, water resistance, antiwear protection, corrosion and rust resistance, and pumpability. Table 1.5 lists tests that are used to characterize these properties.

Grease Thickeners

As noted earlier, greases can be classified by the type of thickener used, such as soap-thickened, complex soap–thickened, or non–soap-thickened greases. The characteristics and principal applications of the various types of greases are discussed in the following paragraphs.

Aluminum Soaps. Aluminum soaps are made by reacting aluminum hydroxide with a fatty acid, such as stearic acid. These finished greases are characterized by having a smooth, gel-like appearance, a low dropping point, good water resistance, and thixotropic behavior (softening or hardening dependent on shearing rate). Aluminum greases were used on slow-speed bearings under wet conditions; however, use of this type of grease has greatly decreased in favor of the higher–dropping-point complex analogues.

Sodium Soaps. Sodium soaps are made by reacting sodium hydroxide with a tallow-derived triglyceride or fatty acid. The finished grease is characterized by having a rough, fibrous appearance, a moderately

TABLE 1.4 Dropping Point and Maximum Usable Temperature of Greases

Thickener type	Dropping point range, °F	Maximum usable temperature, °F
Calcium, water-stabilized	210–220	175
Aluminum	220–230	175
Calcium, anhydrous	275–285	250
Sodium	340–350	250
Lithium	385–395	325
Complex soap	Above 450	350–400
Nonsoap polyurea	Above 450	350–400
Nonsoap organoclay	Above 450	350–500

TABLE 1.5 Tests for Characterizing Greases

Property	Test
Shear stability	Multistroke penetration Shell roll Wheel bearing leakage
Oxidation resistance	Bomb oxidation Bearing life tests Pressure difference scanning calorimetry
Water resistance	Water washout Water spray-off Water absorption
Oil bleed resistance	Oil bleed tests Pressure oil separation
EP/antiwear	Four-ball wear and EP Timken SRV wear test Fafnir
Corrosion/rust	Rust test Emcor rust test Copper corrosion
Pumpability	USS mobility Apparent viscosity Low-temperature torque tests

high dropping point, good adhesive (cohesive) properties, but very poor water resistance. Sodium greases are used widely in certain plain, slow-speed bearings and in gearboxes where water contact is low. These greases should never be used in applications with any appreciable exposure to water.

Calcium Soaps. Calcium soaps are made by reacting calcium hydroxide (lime) with either 12-hydroxystearic acid (derived from castor oil) or a tallow-derived triglyceride. The grease made using the castor oil derivative is usually referred to as an *anhydrous calcium grease*, whereas the grease made from the tallow derivative is referred to as a *water-stabilized calcium grease*. A water-stabilized calcium grease contains up to about 2 wt% water, which aids in the formation of the soap fibers. Both types of calcium greases are characterized by having a smooth, buttery appearance and good water resistance. The major difference between the two types is the dropping points. Calcium greases are used for plain and rolling bearing lubrication, especially in the presence of water.

Lithium Soaps. Lithium soaps are made by reacting lithium hydroxide with a castor oil derivative, such as 12-hydroxystearic acid. The finished grease is characterized by having a smooth, slightly stringy appearance, a high dropping point, good resistance to softening, low leakage, and moderate water resistance. Lithium greases are the most commonly used greases in the United States and are used throughout the automotive and industrial market as chassis lubricants, wheel bearing greases, and ball and roller bearing greases.

Aluminum Complex Soaps. Aluminum complex soaps are made by reacting an organoaluminum compound, such as aluminum isopropoxide, with a fatty acid, such as stearic acid, and a cyclic acid, such as benzoic acid. The finished grease is characterized by having a smooth, slightly gel-like appearance, excellent resistance to softening, good water resistance, and a dropping point above 450°F. Aluminum complex greases are widely used in the steel industry because of their good water resistance properties and good pumpability.

Calcium Complex Soaps. Calcium complex soaps are made by reacting calcium hydroxide with a fatty acid, such as 12-hydroxystearic acid, and acetic acid. Calcium complex greases have a smooth, buttery appearance, good water resistance, inherent EP or load-carrying capabilities, and dropping points above 450°F. These greases are generally used in industrial applications where both high-temperature and water-resistance properties are required.

Lithium Complex Soaps. Lithium complex soaps are made by reacting lithium hydroxide with a fatty acid, such as 12-hydroxystearic acid, and a dicarboxylic acid, such as azelaic acid. The finished grease has properties very similar to those of a regular lithium grease except for its improved high-temperature properties related to the increased dropping point. Lithium complex greases are finding wide applications in the automotive and industrial markets because of their excellent overall performance as wheel and roller bearing greases.

Nonsoap Polyurea. Polyurea thickeners are made by reacting a diisocyanate with a diamine and a monoamine. In addition to the polyurea thickeners, diurea and urea/urethane derivatives also can be used. Diurea thickeners are made by reacting the diisocyanate with a monoamine. The urea/urethane thickeners are made by reacting the diisocyanate with an amine and an alcohol. All these types of greases are generally smooth (but somewhat opaque), exhibit dropping points above 450°F, and have very good resistance to oxidation. They are widely used to lubricate ball and roller bearings, such as those used in electric motors, and to lubricate automotive constant-velocity joints.

Nonsoap Organoclay. The organoclay thickeners used to make greases are referred to as *organophilic bentonites*, such as montmorillonite and hectorite. To make a grease, the clay is mixed with a lubricating oil, after which a polar dispersant, such as water or acetone, is added. As the clay platelets separate, the oil coats each platelet, forming a stable gel structure. The finished grease has a smooth, buttery appearance and good heat resistance, since the clay is nonmelting. Clay greases are often used in high-temperature applications where frequent relubrication can be done, such as on furnace door bearings, kiln car bearings, and bearings on shafts extending through furnaces.

Compatibility of Greases

Because there are so many different types of thickeners used to manufacture greases, the mixing of various types of greases occurs quite often in the field. Grease compatibility is a question that is often raised in the field when one grease is being replaced with another. If two greases are incompatible, serious problems could occur. A definition of *incompatibility* is as follows: "Two lubricating greases show incompatibility when a mixture of the products shows physical properties and service performance which are markedly inferior to those of either of the greases before mixing."

Generally, laboratory testing is required to determine if two greases are compatible. If laboratory tests indicate a compatibility problem, it is usually recommended that the greases not be mixed in service. When changing from one type of grease to another, the piece of equipment should be cleaned of the old grease, or if that is not possible, then the old grease should be flushed from the system by the new grease. Frequent relubrication with the new grease should be followed for a period of time to be reasonably sure that all the old grease has been replaced.

Manufacture of Greases

Modern greases are manufactured using very specific manufacturing procedures that are based on three basic steps. These steps are reacting a material with an alkali metal to form a soap, dehydrating and conditioning the soap, and incorporating the required oil and additives to form a finished grease. Each of these steps must be done under specific conditions of temperature, concentration, and mixing in order to obtain a finished grease with the desired performance properties.

Traditionally, the manufacture of greases has been done as a batch process in large kettles. Such kettles range in capacity from as small as 10,000 lb to as large as 50,000 lb. Most kettles are equipped with a heating jacket, a stirrer to aid mixing and heat transfer, and some type of circulation system for pumping the grease during processing and packaging steps. Successful kettle manufacture is often based on the skill and experience of the greasemaker. Since most grease manufacturing procedures require 16 to 24 hours to complete, the greasemaker must carefully observe each step of the process and sometimes make adjustments as necessary. Batch-to-batch variations are usually greater when using the batch-processing procedure.

Improved efficiency of the batch-processing procedure can be realized by using a pressure reactor, called a *contactor*, to reduce the time for conducting the saponification step. A contactor is a large vessel equipped with internal heating coils and a bottom impeller that provides increased mixing. Use of a contactor can shorten the manufacturing procedure from a typical time of 24 hours to about 8 hours. The contactor allows for pressurized saponification, increased mixing, and improved heat transfer, which accounts for the reduced processing times required. Usually the contactor is used along with one or more kettles such that the soap can be processed in the contactor and then transferred to a finishing kettle for the addition of the final oils and additives.

During the mid-1960s, the first fully continuous process was commercialized by Texaco for the production of greases. This process transformed the conventional batch process to a continuous process. The process consists of three sections: the reactor section, where saponification of the fat and alkali occurs; the dehydration section, where any water is removed and the soap fibers are formed; and the finishing section, where the final oil and additives are added and the grease is fully homogenized to give a finished grease. The continuous process allows for the production of large volumes of grease (throughput rates of 5000 lb/h) under very precise control of operating variables, which greatly reduces the batch-to-batch variations seen in batch processing.

All three of the processes are used commercially to manufacture greases. The production of greases in the United States is about 355×10^6 lb/year. It is projected that the use of greases will decrease slightly over the next few years. This reduction will result from the use of better-quality greases that will allow for longer service before relubrication. Also, grease use should decrease because of an increased focus on the environmental impact of greases, the use of improved seals, and the use of more sophisticated recovery systems. The market will require greases that are highly formulated to give improved performance. It is likely that the older-technology greases, such as the sodium and calcium grease, will gradually disappear and be replaced by the newer complex greases. It is also likely that greases based on synthetic oils will continue to increase in use and that biodegradable oils will become commonplace in applications where environmental exposure is greatest, such as in the timber, mining, and railroad industries. All these changes will challenge grease suppliers to provide the new products required by the market.

CHAPTER 2
LUBRICATING DEVICES AND SYSTEMS

Duane C. Allen
Consultant, LubeCon Systems, Inc., Fremont, Mich.

Good lubrication is simply placing the *right amount* of the *right lubricant* in the *right place* at the *right time*. There are many devices and systems available today to achieve this in a variety of environmental conditions and operating situations.

Manually and electrically powered grease guns are used in some remote locations. Canisters containing compressed air and thin-film/dry-film lubricants and solvents are available to manually apply the lubricant to slides and guides, corroded mechanisms, and frozen bearings to break them loose. Regardless of what system or device is selected to provide lubrication, it is important that a good maintenance and follow-up program be implemented to keep the equipment operating properly.

Manual techniques used to apply lubricants have been improved from the old lever-handled grease gun to portable compressed air or electrically operated pail pumps with hoses, reels, applicator nozzles, and related equipment. Many plants that have adequate aisle space and ready access to equipment provide golf cart type vehicles for their lubricators with these devices and the required lubricants on-board.

AUTOMATIC LUBRICATION

Almost any well-planned and implemented lubrication program can be made more efficient with automatic lubrication equipment. Properly selected, installed, and operated automatic lubrication systems can reduce lubricating man-hours, precisely control lubricant application, and reduce material consumption.

The simplest automatic lubricators are the single-point, small-reservoir units that mount on the bearing in place of the grease fitting and dispense the correct amount of grease to bearings as needed. There are three types of these units available. One is spring-activated, with orifice flow plugs selected for each application. The second is a chemically generated gas-activated unit with fixed flow rates. The type of unit is selected on the basis of feed rate required for the given application. The third unit is an electrochemical gas-activated unit. The grease feed rate is controlled by setting dip switches in the battery-powered lubricator.

The *spring-activated lubricators* (Fig. 2.1) adjust the flow of grease to a bearing automatically by the use of a new, more effective metering control principle. This is accomplished by a special piston O-ring seal that creates a changing level of friction as it moves along the tapered wall of the transparent reservoir dome. The changing resistance is designed to counterbalance the changing force of the compression spring as it gradually expands.

8.14 LUBRICATION

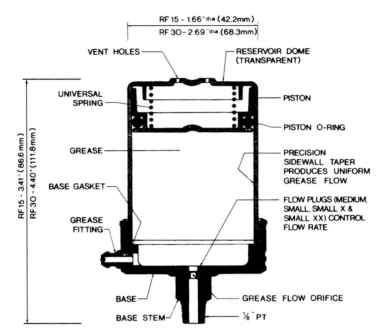

MODELS GP20 AND CM40 HAVE SAME OVERALL DIMENSIONS AS MODEL RF30 ALTHOUGH BASE CONTOUR IS SLIGHTLY DIFFERENT.

FIGURE 2.1 Spring-activated lubricator.

This balancing principle operates entirely above the grease in the reservoir, never allowing the spring pressure variation to be transmitted through the grease. The result is 50 percent less pressure on the grease and less tendency for oil separation. Because these lubricators operate with a single universal spring at the lowest reliable pressure (under 2 psi), no grease is moved into the bearing until it is needed.

The *chemically generated gas-activated lubricators* (Fig. 2.2) discharge lubricant at a constant rate 24 hours a day. These units generate their own power, increasing output pressure up to 136 psi and will not blow the seal on any bearing. This lubricator holds approximately 4.8 oz of lubricant, equivalent to 7.3 in^3. The type of unit selected is based on ambient temperature and desired discharge rate for each application.

In the *electrochemical gas-activated lubricator* (Fig. 2.3), when one of the selector switches is closed, an electrochemical reaction takes place by which electrical energy is converted into nitrogen gas. The gas is trapped in a hermetically sealed bellows-type gas chamber. As the gas is produced, an internal pressure builds up, which is applied against a piston. The piston then forces the lubricant out of the cylinder and into the lube point. The strength of the electric current determines the amount of gas produced, which, in turn, controls the rate of lubricant flow and the length of time the unit will operate. Lubricant discharge rate for a specific application is achieved by setting a selected combination of switches.

Centralized Automatic Lubrication Systems

Centralized automatic lubrication systems can be readily justified for the following reasons.

Safety

1. No climbing around running machinery.
2. Safe lubrication of bearings usually inaccessible because of location, gas, fumes, or height (blast furnaces, overhead cranes, and so on).
3. No excess spillage around machines to cause slippery conditions.

Specifications

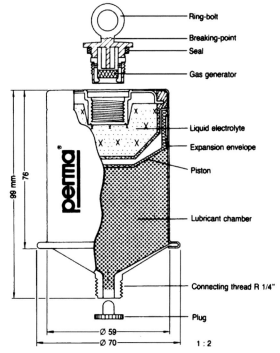

FIGURE 2.2 Chemically generated gas-activated lubricator.

More Efficient Lubrication

1. Lubricant is applied in small, carefully controlled, correct amounts more frequently.
2. No skipping or underlubrication of any bearing or surface.
3. No wasteful overlubrication, as with the old hand-applied methods.

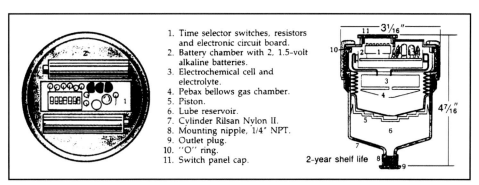

FIGURE 2.3 Electrochemical gas-activated lubricator.

Increased Productivity

1. Increased machine life.
2. Lubrication is done automatically, while machine is running.
3. Increased operating time through reduced downtime.

Reduced Operating Costs

1. Reduced maintenance labor costs.
2. Lower power costs—less friction.
3. Fewer man-hours required for lubrication.

Better Housekeeping. Lubrication points, machines, and surrounding area remain clean of excess lubricants.

Types of Automatic Lubrication Systems

Types of lubrication systems have changed dramatically in the last decade. With the advent of programmable microprocessor controllers, proximity switches, and temperature monitoring devices, it is now possible to precisely control and monitor lubrication systems.

Automatic lubrication systems generally fall into the following simply defined categories:

Oil mist system

Orifice-control system

Injector system

Series-progressive system

Twin-line system

Duoline system

Pump-to-point system

Zone-control system

Ejection system

Injection system

Each of these lubrication systems has its place in industry. They range in applications from machine tool, primary metals, automotive, agriculture, mining, textile, and food packaging and processing to the military and aerospace industries. Each system, however, has its own unique characteristics and specific requirements.

Oil Mist System. Oil mist or fog lubrication from a central system has become widely accepted, especially for high-speed precision service, as in grinding machinery, woodworking, and aircraft compressors. The principle involves injecting oil drop by drop into a stream of low-pressure air to circulate just enough oil to wet the operating surfaces.

Orifice-Control System. This system (Fig. 2.4) limits itself to using only oil as a lubricant, and the viscosity of the oil is usually limited to 300 SUS or less. The system relies on an orifice to control the amount of lubrication going to a bearing. This can become a problem with low-viscosity index oils if the system is in an area of changing temperatures. As temperatures vary, this can dramatically change the lubricant viscosity. This system is also limited in terms of size, since it usually operates in a pressure range of less than 125 lb/in^2 (psi). A single broken line will disable the complete lubrication system, and a low-pressure or high-flow fault switch is recommended. Filters are also recommended, since they should be on all lubrication systems, but particularly on an orifice system, because orifices are prone to plugging due to contaminants. This system lends itself best to

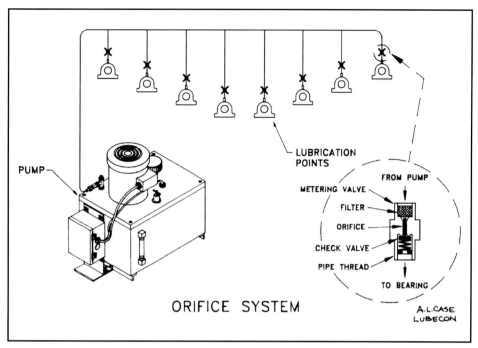

FIGURE 2.4 Orifice system.

small, single-purpose machinery in a stable environment. The systems are generally inexpensive and simple to design and install. Additional lubrication points also can be added without disrupting the original system.

Injector System.   The injector system can be used with oil and light greases, usually limited to NLGI 1 (Fig. 2.5). This system operates by means of a quickly pressurized main system line so that the lubricant on the discharge end of a spool can be injected into a bearing. The pump capacity is usually sized four times higher than the system's requirements. This ensures that proper amounts of lubricant have been discharged to cycle the system.

The system also requires a vent or a dump valve that will allow springs in the injector to reset and prime themselves for the next lubrication cycle. Depending on the viscosity of the lubricant, this can be timely. A broken line before any injector will disable the complete system, and a pressure switch is recommended to signal this loss of pressure. Some manufacturers make an adjustable injector that can be adjusted after installation, making this an easy system to design. However, such a setup provides an opportunity for tampering after installation. Additional lubrication points can easily be added to this system without disrupting the original design.

Series-Progressive System.   The series-progressive system can operate with either oil or grease (suggested minimum viscosity of 100 Saybolt unit seconds (SUS), suggested maximum viscosity NLGI 2). The system (Fig. 2.6) provides a positive displacement of lubricant by supplying each bearing with a predetermined amount of lubricant before lubricating the next bearing. The system requires a fair amount of engineering before installation and is more difficult to comprehend than other systems. However, the system's main characteristic is that when a line becomes plugged and will not receive lubricant, the system is disabled. This can be used to the machine's advantage, especially if lubrication is required on a timely basis. The blocked line can be detected rapidly with a pressure switch or a system cycle switch. However, this characteristic can pose problems to a

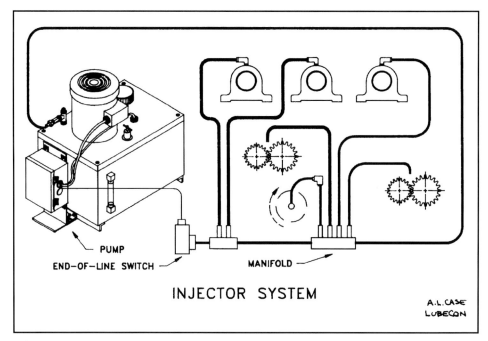

FIGURE 2.5 Injector system.

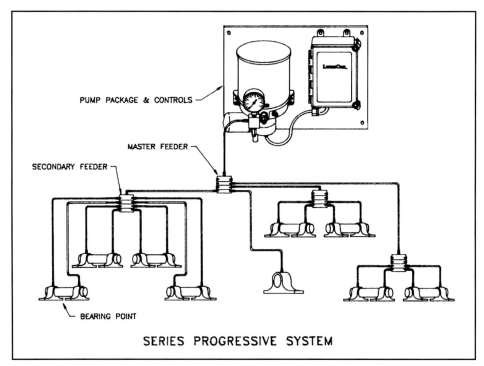

FIGURE 2.6 Series-progressive system.

maintenance person who is not properly trained in troubleshooting the series-progressive system. The system is basically tamper-resistant. Nothing is adjustable except the pump, and this will usually cause the system to fault if tampered with. Additional lube points will require system engineering and cannot be added without disrupting the entire system. The cost of the system is generally higher than most, but positive characteristics generally outweigh this difference.

The divider valves in a series-progressive system are shown in Fig. 2.7.

Twin-Line System. The twin-line system was considered at first to be a competitive system to series progressive, but it did not lend itself well to detection of blocked lines. However, this system (Fig. 2.8) is much more tolerant of contaminants in the system, and the feeders usually do not become blocked as readily. Therefore, the system is used in mining and steel mill applications, where it has been proven very successful in hostile environments. The system utilizes two lines running in parallel to deliver lubricant to the bearing points. The operation of the system is such that the pump supplies flow to one line at a time. The injectors respond by forcing lubricant out to the bearing points. The flow is then redirected to the second line. This resets the injectors and completes the lubrication cycle. Any injectors that were tied to the second line respond at this time, resetting themselves when the first line is fired again. The system can be adjusted at each lubrication point and is fairly simple to design. The use of two main lines to distribute and rest the injectors

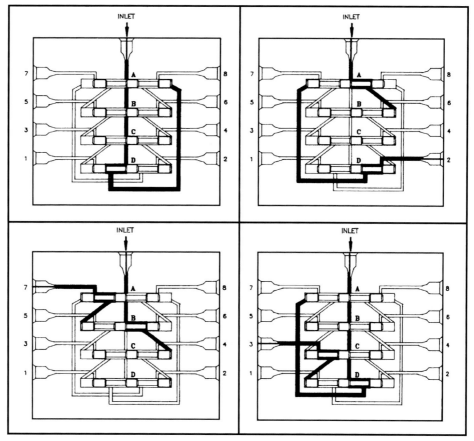

FIGURE 2.7 Divider valve operation.

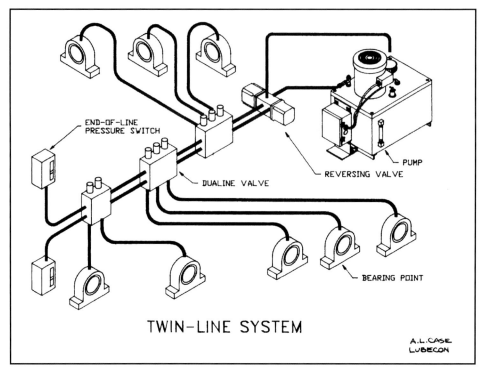

FIGURE 2.8 Twin-line system.

makes it more costly to install. However, this dual-line concept is advantageous in extremely long runs over several hundred feet in length. A suggested minimum oil viscosity is 100 SUS. The maximum grease viscosity of NLGI 2 should not be exceeded in long runs. Consider the twin line to be the best choice to handle heavier lubricants for short runs.

Duoline System. This system is the result of innovative thinking and the combination of two previously designed systems, one being series-progressive and the other twin-line (Fig. 2.9). The combination of the two systems, along with modern technologies in electronics, allows for the best of two worlds. We now have the ability to extend for hundreds of feet the benefits of the series-progressive system along with the benefits of two lines. The concept was first used by Ford Motor Company in Europe in the late 1970s and became widely accepted in the United States a short time later.

By using a series of limit switches, we can now detect if a twin-line feeder actually fed a series-progressive feeder and report the outcome back to the programmable controller. If by chance a limit cycles too slowly or not at all, the controller can disable the machine until the fault is corrected. The system can be used with oil or grease and is very reliable. The minimum oil viscosity to be used in this system is 100 SUS, and the maximum grease viscosity is NLGI 2.

Pump-to-Point System. The pump-to-point system is a simple, easy to design lubrication method (Fig. 2.10). The system operates with a series of pumps that usually are driven off an eccentric cam that is powered by an electric motor. The number of lubrication points the system will service is limited by the number of pumps that can fit onto the eccentric cams of the drive shaft. If the points are numerous and far from the pump location, this can be a very expensive system to install, but if it is a small, simple machine with under 12 lubrication points, it would lend itself well to this type of system. The system contains a series of adjustable pumps that are

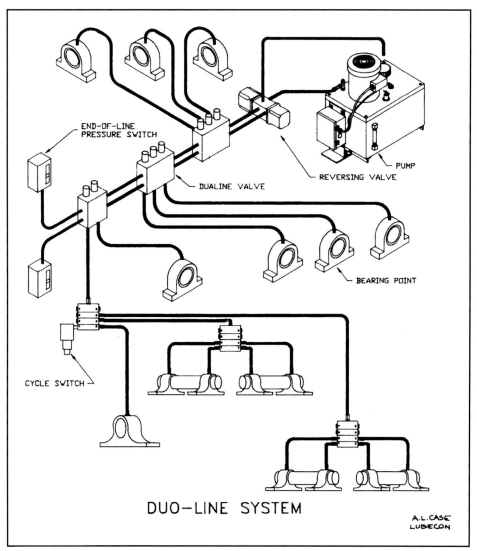

FIGURE 2.9 Duoline system.

positive-displacement, and most models are restricted to oil as a lubricant. This system does not provide a central fault signal.

Zone-Control System. The zone-control system is a relatively new design that was developed primarily for today's flexible machining centers (Fig. 2.11). Basically, it is a series progressive system that utilizes a series of zero-leak solenoid valves that can be turned on to lubricate a particular station or cell of the machine when it is required.

The idea is that multiple parts can be manufactured from a single machine by changing programs and tooling. The program change would add or eliminate the station of operation that would be required for the manufacturing process, thus recognizing whether the station needed to be lubricated or not. This system also can change the frequency of lubrication, if so needed. The system can be used for oil or grease. Minimum oil viscosity is 100 SUS, and maximum grease viscosity is NLGI 1.

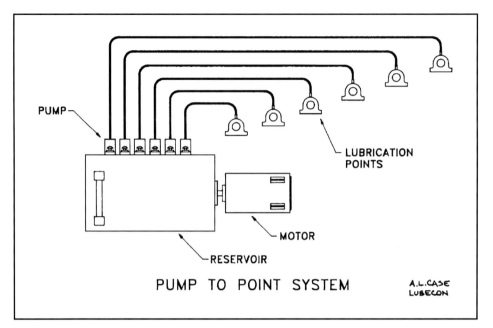

FIGURE 2.10 Pump-to-point system.

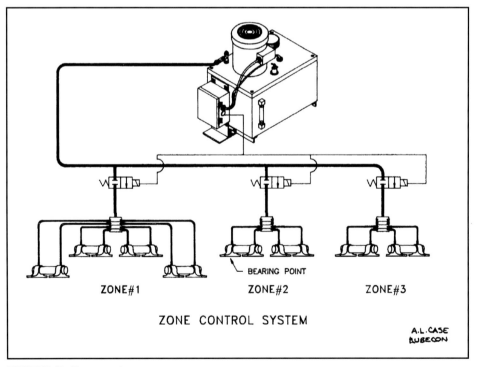

FIGURE 2.11 Zone-control system.

Ejection Systems. The single- and multipoint ejection system ejects lubricant into the wear point or onto the wear surface, such as chain conveyor links and bearings, roller chains, slides and guides, mold car conveyors and drives, shot blast pins and links, crane cables, and open gears (Fig. 2.12). The automatic ejection system is controlled by an electronic head control that counts signals received from proximity switches that read chain links or equipment iterations. Typical lube heads are shown in the Fig 2.12.

Manually adjusted count settings establish when the lubricator goes into lube cycle and how many times it will eject lubricant during that cycle. Ejection lubrication of high-speed chains, cables, or gears can be controlled by a programmable controller timer board that can be programmed to lubricate equipment on the basis of accumulated machine-operating time and controlled valve-open time.

When the control head goes into lube cycle, it turns on a central reservoir pump that supplies lubricant under pressure to the solenoid control valves. A signal is sent from the proximity switches through adjustable timers to the respective ejection valves. These valves open for a preset time and eject lubricant through small tubes to the precise point to be lubricated with minimum waste and no drippage.

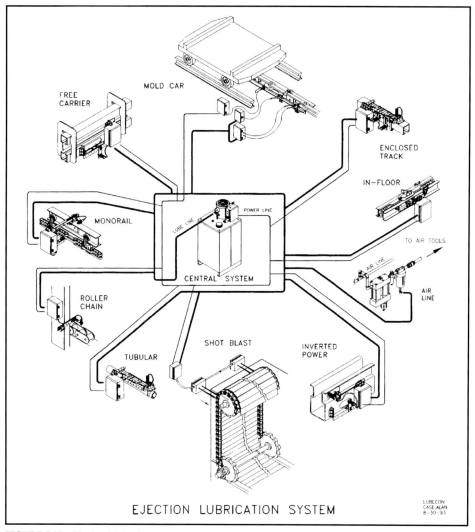

FIGURE 2.12 Ejection lubrication system.

8.24 LUBRICATION

The quantity of lubricant supplied can be controlled by varying the central system pump pressure, the number of times an ejection valve opens in a given cycle, and the time period that the valve is open. A number of lubrication heads installed on a variety of equipment can be serviced by the single central reservoir system, motor control, and power supply, as shown in the Fig 2.12.

The central reservoir system also can be used to supply lubricant to air-line lubricators that are fitted with remote-fill containers. This approach eliminates those empty air-line bottles around the plant that never seem to get filled until an air tool or piece of equipment goes down due to lack of lubrication.

This system is simple to understand and very reliable, if properly maintained. It is best suited for thin-film/dry-film lubricants. The thin film/dry film will not attract dust, nor will it drip onto manufactured product, because the carrier will evaporate, leaving the lubricating film on the bearing surface.

Injection System. The injection system senses the moving lubrication point with a proximity switch, and the lubrication device moves with the point, attaches itself with the help of a pneumatic cylinder, and then injects a predetermined amount of lubricant directly into the moving point (Fig. 2.13). The cylinder then detaches the device from the moving bearing and resets itself properly to lubricate the next moving point.

Automatic injection lubricators with microprocessor controls have been developed to eject a selected amount of grease or thin-film/dry-film lubricant into conveyor trolley wheel bearings on individual conveyors and free carriers. These units with EEPROM memories are designed to count trolleys and a flag trolley and can be programmed to lubricate periodically based on conveyor operation. This equipment will remember if it missed a trolley due to high speed and will lubricate it during the next chain cycle.

The lubricators are supplied with lubricants from a pail grease pump or a central lubricant reservoir.

In order to more effectively implement your lubrication program, contact lubricant and lubrication system suppliers to utilize the talent available from these knowledgeable sources.

Table 2.1 is a checklist to assist in selecting and specifying automatic lubrication systems.

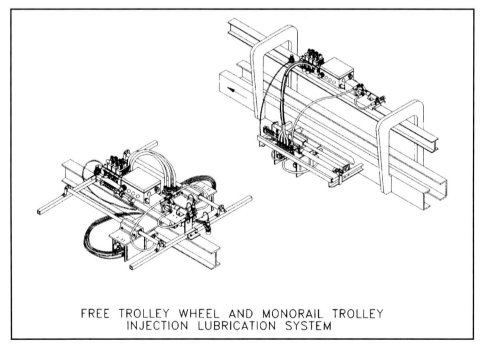

FIGURE 2.13 Free trolley wheel and monorail trolley injection lubrication system.

TABLE 2.1 Lubrication System Selection

The following is a checklist of factors to be considered when selecting and designing an automatic lubrication system:

1. *Scope of project*
   Type of lube points
   Number of lube points
   Size of system requirements
   Available funds

2. *Available power supply and reliability*
   Electric
   Pneumatic
   Hydraulic
   Mechanical

3. *Type of lubricant*
   Thin-film/dry-film
   Oil
   Grease

4. *Type of system*
   Single point
   Oil mist
   Orifice control
   Injector
   Series progressive
   Twin line
   Duoline
   Pump-to-point
   Ejection
   Injection

5. *Lubricant requirements*
   Temperatures at point of lubrication
   Temperature exposure of lubricated equipment
   Bearing volume
   Attrition rate

6. *Type of lube points*
   Stationary
   Rotating
   Moving

7. *System layout*
   Components required and pressure ratings
   Tubing type (rigid, flexible, combination)
   Fitting required
   Central system location

8. *Type of pump*
   Pressure capability
   Volume capacity
   Fluid compatibility

9. *Type of controller*
   Local
   Remote
   Combination
   Timer
   Counter
   Monitor

(Continued)

TABLE 2.1 Lubrication System Selection (*Continued*)

10. *Fault and monitor accessories*
    Low-level switch
    High-pressure switch
    Low-pressure switch
    High-flow switch
    Cycle indicator
    Cycle counter
    Status readout

Proper lubrication system selection, maintenance of lubrication systems, and proper training are a must in today's world of complex machines. A good plantwide lubrication program can be the difference between a company's success or failure.

CHAPTER 3
PLANNING AND IMPLEMENTING A GOOD LUBRICATION PROGRAM

R. Keith Mobley
Principal, Life Cycle Engineering, Inc., Charleston, S.C.

Nearly all moving mechanical components of operating equipment roll or slide against other surfaces. If not properly lubricated, these surfaces can wear rapidly and require excessive amounts of energy. Many equipment malfunctions, repair costs, and resulting downtime can be traced—either directly or indirectly—to inadequate or improper lubrication.

The typical industrial environment contains silica dust, oxides, metal filings, and other abrasive materials. When these materials are mixed with some lubricants, they create a lapping compound that greatly accelerates wear. Unless proper lubricants and lubricating systems are utilized and proper procedures are followed to prevent contamination, premature equipment failure results.

A well-planned and properly implemented lubrication program, designed to place the *right amount* of the *right material* in the *right place* at the *right time,* will more than pay for itself in reduced downtime, lower maintenance costs in both parts and labor, and reduced energy costs.

Selling Management

Because plant management often balks at what it perceives to be an increase in indirect costs for maintenance, it may be necessary to sell management on the cost savings possible through an effective lubrication program. In order to sell management on such a system, it will usually require collecting historical data on the cost of equipment malfunctions, including parts, labor, and downtime. If such records do not exist, this information will need to be collected. If equipment maintenance records have been kept on a computer, it may be necessary to add a few files to collect the data needed to justify the additional expenditure to implement an effective lubrication program.

In addition to historical data, recording of amp meter readings may be used to show the energy savings possible when superior lubricants replace inferior products for given applications. This approach has proved particularly effective in chain and conveyor applications. Some plants are also using vibration analysis equipment with recorded vibration patterns to indicate bearing conditions and to predict failures and lubricant effectiveness.

Selecting Lubricants

The multitude of lubricants recommended by equipment manufacturers can be simplified by selecting good multipurpose lubricants to reduce inventory requirements and the possibility of misapplication.

Thin-film/dry-film molybdenum disulfide lubricants and synthetic lubricants can play an important role in reducing contamination and energy consumption.

Dollar savings from longer lubricant life, reduced equipment maintenance, lower power consumption, and less downtime can be several times greater than the higher cost of these lubricants. On the other hand, premium-grade lubricants will not improve or correct lubrication problems if mechanical factors such as misalignment or severe environments (high levels of dirt and water contaminants) are involved.

These products should be purchased on the basis of in-service results rather than on the price per pound or gallon. It is also necessary to determine the compatibility of the selected lubricants with seals and hose linings before putting them into service.

Lubrication Training

Increasingly, pressure is being placed on maintenance management to hold down or even reduce the number of people performing maintenance functions. During times of personnel cutbacks, the oilers are often the first to go, usually replaced by untrained personnel.

Depending on plant size and contractual obligations, the employee responsible for lubrication should be a machine operator, skilled tradesman, or trained designated oiler. Selecting an employee who knows the equipment will greatly improve the results of any lubrication program. Some plants are using highly skilled preventative maintenance inspectors who also lubricate. These people see every piece of equipment in the operation on a scheduled basis.

In any case, it is important to provide the individual with proper instruction in application methods, types of lubricants, handling methods, and safety procedures. Engineering personnel should be trained on proper design procedures of lubrication systems and be up-to-date on the latest technological advances in the lubrication industry. They also need to be aware of the problems in the plant that maintenance personnel have in troubleshooting lubrication system malfunctions.

Lubrication system specifications should be written to ensure that every piece of equipment entering the plant has a properly designed lubrication system. The lubrication equipment specifications should inform the equipment supplier exactly what is expected of the lubrication system. This would include such items as low-level switches, high/low-pressure sensors, flow switches, or metering devices. All sensors should inform the operator of lubrication system malfunction and, in some cases, provide for machine shutdown if a fault occurs.

Nearly all poor lubrication practices are traced to a lack of training. An employee must know the correct amount of a prescribed lubricant for a specific application and how to measure it. It also is imperative to follow up on any new procedure to ensure that the prescribed practice is followed and the desired results are being achieved.

The level of performance of equipment in an operation is directly proportional to the quality of the lubrication program in that operation and the support provided to the program by management and engineering personnel.

PLANNING AND IMPLEMENTING THE LUBRICATION PROGRAM

The activities to achieve and carry out an effective lubrication program are outlined in this segment. They consist of the plant lubrication survey, establishment of lubrication schedules and improvement in the selection and applications of lubricants, lubricant analysis, fluids management and quality assurance, activities required to implement the programs, and factors to consider if a single supplier source is desired for all plant lubricants.

The program implementor should work closely with plant personnel to determine information now available and programs and procedures presently being used.

The Plant Lubrication Survey

1. Identify equipment and component parts requiring lubrication, the specific location of each machine, and the model, serial number, function, manufacturer, operating instructions, and limitations.
2. Obtain similar information for each subcomponent of the machine, such as drive motors, gears, couplings, and bearings.
3. Examine the lubricant recommendations made by the machine or parts manufacturer and supporting documentation for these selections.
4. Determine the lubricants currently used, including quantity, cost, and supply source.
5. List the schedules in effect for each lubrication point, including frequency, quantity applied, and sampling schedules. Provide similar information for all machine components.
6. Identify the nature of each lubrication point and whether circulating systems are fed from central storage tanks, individual machine sumps, or grease fittings and whether manual, semiautomatic, or automatic equipment is now being used. Operating characteristics, condition, and effectiveness of the lubrication systems encountered should be determined.
7. Make a detailed visual inspection of each machine and its components for indications of problems, such as leakage; excessive noise; high temperature; vibration; and loose, damaged, or missing parts.
8. Record information relating to the adequacy of the machine to perform its intended functions.

Note: An effective approach for conducting the initial lubrication survey is to start with the units of equipment that are critical to maintaining continuous production and work toward the less critical units. This approach will achieve the greatest results in the shortest time period. When surveying an individual machine, start at the power source and follow through each power train, identifying couplings, reducers, bearings, and wear surfaces.

Establishment of Lubrication Schedules and Improvements in Selection and Application of Lubricants

1. Review current lubrication schedules, including type and amount of lubricant used and frequency of application.
2. Determine if it is the best lubricant for the specific application commensurate with the proposed lubricant product reduction program and improved performance requirements.
3. Analyze each piece of equipment to determine if the present lubrication system is adequate and if the lubrication points or central reservoirs are readily accessible.
4. Investigate opportunities to replace inadequate systems, manual systems, and malfunctioning automatic systems with state-of-the-art automatic systems that can be justified through reduced labor, increased equipment reliability, and/or reduced energy costs.
5. Analyze operating records such as frequency of scheduled and unscheduled downtime and reason for each shutdown when preparing the new lubrication schedule.
6. Establish lubrication schedules and routings to minimize travel time and interference with production operations. Determine time required to perform specific lubrication functions and number of workers required to perform the job.
7. Establish a check-off or feedback procedure to indicate that the scheduled lubrication was accomplished with the proper lubricant.
8. Record and report the amount and type of lubricant consumed in each area and on major pieces of equipment.

9. New equipment lubrication specifications are to be determined prior to installation of the equipment.
10. Place tags at each fill point that calls out lubricant to be used, amount of lubricant, and lubrication schedule.

Lubricant Analysis

1. Establish the objectives of the analysis program, that is, monitor and track wear and lubricant quality to detect problems caused by adhesion, friction, and corrosion before there is major component damage and to determine when lubricant should be filtered, replaced, and/or fortified with additives.
2. Select the plant equipment to be included in the analysis program. Equipment selection is usually based on the importance of the equipment to continuity of plant operations.
3. Determine the sampling frequencies for each component.
4. Design the testing packages to meet the selected objectives. *Typical tests for gear reducer lubricants include*

    Wear particle analysis—wear metals, contaminate metals, and additive metals

    Total solids percentage volume—contamination leaks or environmental conditions

    Viscosity—fluidity of the lubricant

    Infrared analysis—oxidation/nitration (general lube degradation)

    Neutralization number—reserve alkalinity (Total base number [TBN]) or total acidity (Total acid number [TAN])

    A glossary of tests for lubrication and fuel analysis is shown in Table 3.1.

5. Select a lubricant testing laboratory that can accurately test the parameters chosen and report the results in a comprehensive manner on a timely basis.
6. Determine the cost of the analysis program.
7. Develop the sampling procedures and modify equipment as necessary to extract representative samples while the equipment is in operation.
8. Establish sampling, testing, and reporting schedules.
9. Develop procedures and lines of communication to report results and to initiate actions dictated by the test results.
10. Establish a program review schedule.

Note: A close liaison should be maintained between the lubricant analysis program and other predictive maintenance activities.

Fluids Management and Quality Assurance

1. Establish handling, storing, dispensing, application, housekeeping, and safety practices. Table 3.2 is a lubricant handling checklist.
2. Obtain *material safety data sheets* on all lubricants in the plant.
3. Periodically inspect and test lubricant shipments to determine if they meet established quality standards. Table 3.3 shows typical tests for oil quality.
4. Prevent comingling of lubricants.
5. Discard contaminated and obsolete lubricants in an environmentally acceptable manner.
6. Record ammeter readings on drive motors to monitor the effectiveness of lubrication.

TABLE 3.1 Glossary of Tests for Lube and Fuel Analysis

Acid number (ASTM D664)	Determines total acidic and acid-acting materials in the lube (*Note:* Some new lubes exhibit an acid number normally from additives.) *Application: Engines burning high-sulfur fuel; most plant equipment types*
Ash (ASTM D482)	Determines tendency of lube or fuel to leave residual deposits upon combustion. *Application: Fuels (spec test); lubes containing no metallic additives*
Ash, sulfated (ASTM D874)	Determines residual ashing tendency of a lube or fuel containing metallic additives. *Application: Primarily motor oils, some fuels*
Base number (ASTM D664, D2896)	Determines alkaline reserve of a lube (i.e., its ability to neutralize corrosive acids). *Application: Most internal combustion engines, separate cylinder lube systems*
Cetane number, calc (ASTM D976)	Estimates ignition tendency of a diesel fuel; normally a higher number is indicative of greater ease of combustion. *Application: Diesel fuel (spec test)*
Chloride ion concentration (Lubricon)	Determines amount of chlorine as chlorides in the lube. *Application: Saltwater contamination, refrigeration systems using Freon*
Cloud point (ASTM D2500)	Determines temperature point at which fuel begins to appear cloudy; forming wax crystals which could plug filters. *Application: Diesel fuels (spec test), particularly for cold climates*
Color test (ASTM D1500)	Determines a numeric color rating of a fluid, which can sometimes be revealing as to quality or contamination (also cosmetic). *Application: Situations where appearance is important, product consistency*
Copper corrosion (ASTM D130)	Determines if a lube is inherently corrosive to copper and its alloys (e.g., bearings). *Application: Mainly in qualifying new lubes, occasionally applied to used lubes*
Direct reading ferrography	Determines ratio of large (>15 μm) and small (<15 μm) iron-based particles in lubricant; amounts and ratio suggest wear severity. *Application: Reciprocating equipment; systems having coarse or no filtration*
Distillation (ASTM D86)	Determines boiling range of a fuel as to combustion characteristics; a gradual, smooth increase is usually best. *Application: Fuels (spec test)*
Ferrography, analytical	Derives microscope slide of (mostly) iron-based particles as well as other particles; ability to view particles greatly aids evaluation. *Application: All circulating lube systems; very useful for inspection decisions*
Fire point (ASTM D92)	Determines minimum temperature at which a small flame passed over the lube will cause sustained ignition. *Application: Safety in transporting lubes (spec test)*
Flash point, COC (ASTM D92)	Determines minimum temperature at which a small flame passed over the lube will cause a small, temporary flare or flash. *Application: Safety in transporting lubes; product verification*
Flash point, PM closed cup (ASTM D93)	Determines fuel's flash point in a semisealed environment (simulating a combustion chamber). *Application: Diesel fuels (spec test)*
Foam test (ASTM D892)	Determines a lube's tendency to foam in service. *Application: Industrial plants, various other systems*
Fuel contamination (Lubricon)	Derives total amount of fuel and "light" flammable materials in a lubricant, generally as a contaminant. *Application: Liquid-fueled recip engines or if solvent contamination suspected*

(Continued)

8.31

TABLE 3.1 Glossary of Tests for Lube and Fuel Analysis (*Continued*)

Test	Description
Glycol (Lubricon)	Determines if any significant amount of ethylene glycol is present in the lube, which could promote wear in critical locations. *Application: Units utilizing glycol-based coolant*
Gravity, API (ASTM D287)	Determines density of a fuel or lube. *Application: Fuels (spec test), crude oil, and new finished lubes, product verification*
Infrared analysis (several formats)	Determines deterioration of lube (e.g., oxidation, nitration, and/or structural changes in the base stock), occasionally, product ID. *Application: All lube systems*
Insolubles, pentane (ASTM D893)	Determines by weight percent of the amount of solid material in a lubricant. *Application: Primarily used reciprocating engine lubes*
Insolubles, toluene (ASTM D893)	Determines by volume percent of the nonresinous solids in a lube (in conjunction with pentane insolubles). *Application: Primarily used reciprocating engine lubes*
Microorganisms (presence of)	Determines if any microbes exist in a fuel or lube, such contamination can render a product useless or harmful to equipment if used. *Application: Primarily fuel tanks, tankers*
Micropatch study (Lubricon)	Determines (using a microscope) particulate nature after a quantity of sample has been drawn through a filter. *Application: All fuels and lubes, except those which are opaque or nearly opaque. Note: Can also be performed gravimetrically to yield percent solids by weight.*
Particle count (ISO, OSU Methods)	Counts particulates in specific size ranges to derive indication of contamination level of fluid (range >5, >10, >15, >20, >30, >40 microns). *Application: Hydraulic systems, two-cycle gas engines, turbines, filtered systems, QC*
Pour point (ASTM D97)	Determines temperature at which fluid becomes immobile (i.e., it freezes or gels). *Application: Fuels (spec test), new lubes*
Solids contamination (ASTM D91)	Determines solids volume percent of contamination in a lube sample (occasionally heavy fuels). *Application: Diesel engines, particularly recip compressors, gearboxes*
Solids, gravimetric (ASTM D4055)	Determines weight percent of solids contamination by solvent-drawing sample through a filter patch. *Application: Most lube systems*
Spectrometals, solid (deposit)	Derives a coarse indication of metals which may exist in a solid material for clues as to the source of the solid (relative only). *Application: Filter, piston deposits, sediment in sample, other solid materials*
Spectrometric metals	Determines type and quantity (ppm) of wear, contaminant, and additive metals in lubricant; can also be applied to fuels and coolants. *Application: Virtually all systems*
Sugar (Lubricon)	Determines presence of sugar for verification of vandalism. *Application: Failure analysis*
Sulfur (ASTM D1552)	Determines amount of sulfur in a fuel or lube. *Application: Primarily fuels (spec test); some new lubes*
Viscosity (40 or 100°C) (ASTM D445)	Determines a fluid's flow rate with respect to a given temperature, a standard inspection for any fluid. *Application: Virtually any lube; occasionally diesel fuel (spec test)*
Viscosity index (ASTM D2270)	Determines lube's resistance to thinning as temperature increases. (*Note:* Requires viscosity testing at two temperatures.) *Application: New lubricants, particularly multiviscosity lubes*
Water by distillation (ASTM D95)	Detects water at levels of 0.1% or greater with reasonable precision. *Application: Lube or diesel fuel inspections*
Water, Karl Fischer (ASTM D1744)	Detection of minute quantities of water contamination (as low as 10 ppm in some applications). *Application: Critical lube systems (e.g., refrigeration, some hydraulics)*
Water and sediment (ASTM D1796)	Determines amount (vol %) of solids and water (combined answer) in a fuel or lube. *Application: Primarily performed on diesel fuels (spec test)*

TABLE 3.2 Lubricant Handling Checklist

	Yes	No

I. Storage
- A. Lubricant containers, in use, are stored in a heated, power-ventilated room. Drums are kept off the floor and are supported by a rack, platform, or blocks at least several inches high.
- B. Inventory records include such details as quantity of each lubricant in stock, its location, and minimum order quantities.
- C. Drums stored outside (even for brief time) are under shelter or covered with tarpaulin or drum covers.
- D. Drums are stored on their sides with the two bungs aligned horizontally (rather than up and down) so they are backed by fluid that prevents air from entering through bung threads.
- E. Lubricant storage room is separated from areas of contamination sources, i.e., metal particles, dust, and chemical fumes.
- F. Solvent drums are grounded to avoid fires caused by electrical charges.
- G. Storage area is accessible both to delivery vehicles and machines to be serviced.
- H. Provision is made for clean and orderly storage of rags, swabs, paddles, cleaning supplies, sample cans, and other lubrication accessories.
- I. A specific area is provided for each of the following:
  - (1) Unopened containers and bulk tanks.
  - (2) Opened containers from which lubricants are being drawn.
  - (3) Lubrication accessories and spare parts.
  - (4) Oil filtering equipment and supplies.
  - (5) Cleaning and storing of dispensing equipment.
  - (6) Recordkeeping.
  - (7) Empty returnable containers.
  - (8) Expansion (if expected).
- J. Emulsifiable oils are protected at all times from extremes of temperature, both high and low.
- K. Bulk storage tanks are used when possible to reduce cost, save space, and avoid contamination.
- L. Storeroom personnel are properly trained in storage techniques, inventory control, and preventing contamination.
- M. Containers and hoses are clearly marked to avoid misapplication of lubricants.

II. Handling
- A. Drums are not dropped or bounced off freight cars, trucks, and racks.
- B. Metal drum slings and overhead drum hoists are used when drums are tiered.
- C. Usually, fork trucks and wheeled hoists are used for drum-handling. Drums manually handled are rolled rather than dragged to avoid damage.
- D. All containers are filled under clean conditions.
- E. Grease drums are emptied completely before discarding.

III. Dispensing
- A. Oldest lubricant is used first.
- B. Lubricant which is suspect because of long storage is checked for condition before use.
- C. Drum spigots are used for fast dispensing, quick cutoff, and to prevent dripping waste.
- D. Different brands, grades, and types of oil are not mixed in dispensing containers.
- E. Dispensing cans are inspected regularly for proper functioning of plungers, spring-closed lids, etc.
- F. Dispensing cans are checked regularly for cleanliness and freedom from contamination.
- G. Containers are kept tightly closed when not in use.
- H. When using open containers, oil is drawn from drums and bulk tanks only when it is ready to be used.
- I. Use of galvanized containers is avoided to prevent reaction with the zinc.

IV. Safety
- A. Dispensing equipment is not left unattended. This prevents personnel tripping and falling over it.
- B. Spilled or leaking lubricants are removed from floors to prevent slipping.
- C. Drum slings are used instead of rope slings for lifting.
- D. Lubricant and solvent containers are not left in direct sunlight, in very hot areas, or where there are sparks because of possible spontaneous combustion.
- E. High-pressure grease guns are used with care to prevent accidents.
- F. Oily rags are put in tightly closed safety containers and disposed of regularly.
- G. Storage tank vents are kept open.
- H. Smoking in lubricant and solvent storage areas is prohibited.
- I. Where necessary, machines are properly shut down before lubricating.
- J. Secure ladders are used to reach high lubrication points.
- K. Light oils and solvents are properly vented to prevent excessive inhaling of fumes by personnel.

TABLE 3.3 Typical Tests for Oil Quality

Viscosity—Reflects the flow characteristics of an oil. The right viscosity is essential for good penetrability and protection. Viscosity is measured in Saybolt unit seconds (SUS) or centistokes (cSt). For example, a synthetic heavyweight oil for high-temperature applications up to 500°F has a viscosity of 335 cSt at 40°C or 250 SUS at 210°F—about the weight of SAE 140 gear oil.

Viscosity index (VI)—Indicates the effect temperature change has on the viscosity of an oil. A high VI, typically over 100, is generally recommended for high-temperature applications. Some synthetics offer a VI above 200.

Carbon residue test—Measures the amount of carbonaceous material left after evaporation and pyrolysis. The more residue, the more likely the oil will "coke" and leave deposits on the chains. Look for numbers less than 1 percent.

Four-ball wear test—Uses a steel ball under load. It is rotated within well formed by three identical stationary balls. Wear is reported as the average scar diameter formed on the stationary balls. A small scar diameter (less than 0.50 mm) indicates excellent antiwear properties.

Falex wear test—Also known as the pin and vee block test. The lower the number, the better the wear protection. The number indicates the number of teeth on a ratchet that advances on the test apparatus as wear reduces the dimensions of a test pin rotating between the jaws of the two vee blocks.

Flash point—Determines the lowest temperature at which an oil gives off sufficient vapors to ignite but not continue burning when a flame is passed over the sample.

Pour point—The lowest temperature where the oil flows under test conditions. The rule of thumb is to select a pour point at least 20°F below the lowest operating or starting temperature.

Rust (ASTM D655) *rating*—Indicates the relative rust protection of an oil. It is a pass/fail rating.

Activities Required to Implement the Lubrication Management Program

1. Establish lines of communication with plant personnel, and determine responsibilities for the following:
    a. *Ordering lubricants*
    b. Receiving, handling, storing, and disposing of lubricants
    c. Dispensing and applying lubricants
    d. Ordering equipment and aides to implement the program
    e. Drawing lubricant samples for test
    f. Processing test samples
    g. Reviewing test results
    h. Taking corrective action as dictated by test results
    i. Housekeeping and safety practices related to the program
    j. Reporting the status and results of the lubrication management program to plant management
2. Establish the cost of the program.
3. Assign a trained lubrication specialist to the plant full time to carry out the program as outlined.
4. Train plant personnel, as necessary, to participate in the program.
5. Maintain records, and provide reports as required by the program and plant management.

Note: This activity will require a terminal on the plant or maintenance department's computer or a separate computer that is compatible with the plant's operating system.

6. Contact plant management to receive approval for recommendations to eliminate or replace presently used lubricants and equipment or to introduce new lubricants and equipment.

Factors to Consider if a Single Source for All Plant Lubricants Is Desired

1. The proprietary recommendations by the supplier must be thoroughly scrutinized and evaluated by plant management.
2. Can the supplier provide a total systems approach, that is, lubricant, equipment, and service?
3. Many of the presently used lubricants have been selected on the basis of plant experience. Changes should be made carefully on the basis of total maintenance savings rather than just lubricant price and should be approved by plant management.

S·E·C·T·I·O·N·9

CHEMICAL CORROSION CONTROL AND CLEANING

CHAPTER 1
CORROSION CONTROL

Denny Bardoliwalla
Vice President of Research and Technology, Oakite Products, Inc., Berkeley Heights, N.J.

Klaus Wittel
Manager of Technology Transfer, Oakite Products, Inc., Berkeley Heights, N.J.

INTRODUCTION

Corrosion is the reaction of materials with their environment. It is not restricted to metals. Rusting of iron and steel, pitting of turbine blades, chalking of paint are all examples of corrosion.

Corrosion damage describes all damage induced by corrosion; its cost runs to many billions of dollars a year. Corrosion damage impairs the function of equipment and structures. Corrosion does not always lead to corrosion damage. A common example of corrosion with hardly any damage is provided by railway rails: They are all rusting without any appreciable damage.

Corrosion control includes all aspects and measures undertaken to reduce corrosion damage. These include:

- Monitoring corrosion
- Selecting proper materials of construction
- Designing to minimize corrosion
- Controlling corrosion by regular maintenance practices
- Taking specific countermeasures, for example, cathodic protection, protective coatings, use of inhibitors

The maintenance engineer is definitely confronted with corrosion, corrosion damage, and corrosion control. To be more specific, he has to deal with

- Monitoring corrosion progress
- Recognizing corrosion defects, damages, and failures
- Maintenance which will prevent corrosion
- Application and also selection of process chemicals and materials which will not induce corrosion damage

A basic description of corrosion, its type and its mechanism, is given here to help the maintenance engineer perform his task. Corrosion control is discussed in some detail, such as questions of material selection, design principles, cleaning practices, and corrosion preventive measures, for example, electroplating, conversion coating, painting, cathodic protection, and inhibitors.

9.4 CHEMICAL CORROSION CONTROL AND CLEANING

The description given here, however, is not intended to replace more elaborate treatments. The reader is referred to the ASTM standards on corrosion, the standards on materials, data sheets of suppliers and producers of materials regarding chemical compatibility, and References 1 to 3. ASTM has published numerous standards and recommended practices regarding corrosion, corrosion testing, and corrosion resistance, of which we list just three (References 4 to 6). Specific reference is made to the publications and conferences of NACE, the National Association of Corrosion Engineers.

CORROSION: TYPES AND MECHANISMS

All materials used for building and construction (metals; inorganics like glass, ceramics, or concrete; organics like rubber, paints, and plastics) corrode. Major factors determining type and extent of corrosion include:

- Temperature
- Medium which is in contact with the corroding material, for example, air, gasoline, oil, water, engine coolant
- Mechanical stresses
- Flow rates
- Electrical potential differences

Metals still represent the most important class of construction materials. Their corrosion is best understood and well described by the electrochemical theory of corrosion: Corrosion proceeds in two simultaneous but locally separate (on a microscopic scale) reactions. In the anodic reaction, the metal loses electrons and dissolves; it might then form insoluble reaction products. In the cathodic reaction, an oxidizing agent is reduced by accepting electrons. Both reactions are coupled and proceed with the same speed. As these reactions involve electrical charges, they are heavily influenced by electric potentials. Anything which hinders one half of the reaction hinders the other reaction to the same extent. Table 1.1 gives a schematic summary. Metals dissolution (reactions 1 and 2, Table 1.1) is rather straightforward; the cathodic reactions listed describe

- The dissolution of metals in acids (reaction 4), table
- The oxidation of metals in moist air (reaction 5), table
- Corrosion of metals by water (reaction 6), table

Metals are grouped into the so-called electrochemical series according to their ease of oxidation. All but the noble metals would thus be expected to oxidize and corrode easily. The first certainly holds true. The less noble metals oxidize rapidly. This oxidation, however, is often slowed down, if not stopped completely, by the formation of oxide layers on the metal surface, which may hinder or prevent further oxidation. It is therefore the properties of these protective oxide layers, or the lack of such layers due to instabilities, that determine the corrosion properties of the base metal. Iron oxide, for example, dissolves easily in acids but is stable in alkaline media. Zinc oxide dissolves in both acid and strong bases. Consequently, iron may well be used and often is the material of choice for alkaline media, whereas galvanized surfaces must not be used for either acidic or strongly alkaline media. The interplay between electrochemical potential and pH value, the major measure of acidity, is described by Pourbaix diagrams, for which the reader is referred to textbooks on corrosion.

Corrosion of plastics, paints, or rubber, on the other hand, usually proceeds via attack by free radicals resulting from ultraviolet radiation, chemical reactions, or strong oxidizing agents. Plastics being nonconducting, an electrochemical description of such processes would offer only few insights. In addition, such materials might degrade—corrode—owing to the loss of additives or absorption of solvents. More detail is presented below under Corrosion of Polymeric Materials.

TABLE 1.1 Corrosion of Metals

ANODIC REACTION

$$Me \rightarrow Me^{n+} + ne^- \quad (1)$$

e.g.

$$Fe \rightarrow Fe^{++} + 2e^- \quad (2)$$

CATHODIC REACTION

$$Ox + nH^+ + n \cdot e^- \rightarrow H_nOx \quad (3)$$

e.g.

$$2H^+ + 2e^- \rightarrow H_2 \quad (4)$$

$$O_2 + 4H^+ + 4e^- \rightarrow 2H_2O \quad (5)$$

$$H_2O + 2e^- \rightarrow H_2 + 2OH^- \quad (6)$$

Me: Metal
Ox: Oxidizing Agent

Corrosion at high temperature, as in exhaust streams, and corrosion due to contact with liquid metals and molten salts is not discussed here.

Corrosion may be uniform or localized. Uniform corrosion can be characterized by a loss of material per time per surface area ($g/m^2/hr$) or by the loss of thickness per time (mm/year). If corrosion is localized, such data will carry no meaning. Corrosion control is easy under conditions where uniform corrosion prevails; monitoring of wall thickness and provisions for the loss of material during the working life can be made without problem. Localized corrosion, on the other hand, is difficult to predict, and also hard to make provisions for. It should be avoided by all means. Localized corrosion, for example, pitting, occurs when the anodic reaction, the metal dissolution, proceeds in a small area only, whereas the accompanying cathodic reaction, for example, oxygen reduction, is spread out over a large surface area. Inhomogeneities, resulting from the following factors, drive localized corrosion:

- Inhomogeneous materials
- Local precipitations or local bacterial growth
- Poor design
- Improper pairing of metals
- Local turbulences

A special kind of corrosion damage is due to interfilm corrosion such as filiform corrosion or delamination of paints.

CORROSION: FORMS AND DEFECTS

Uniform corrosion is characterized by a laterally constant speed of corrosion. Typical examples are given by the atmospheric corrosion of galvanized steels—the thicker the zinc coating, the longer the service life—or the chalking of paints, which are degraded uniformly by radiative oxidation in air.

Erosion Corrosion. Uniform corrosion may proceed faster when there is a flow. The flow rate velocity will locally reach high values if turbulent situations occur, which will prevent the formation of protective layers on the metals. Abrasive particles, in a stream of gas or liquid, may mechanically disrupt the protective surface film and thus enhance corrosion, often in a nonuniform, localized way.

Cavitation Corrosion. Formation and collapse of tiny gas bubbles in a liquid stream, called cavitation, may mechanically destroy any protective layer, causing localized corrosion.

Erosion and cavitation corrosion are controlled by proper design and selection of materials. Inhibitors may help in special cases.

Galvanic Corrosion. Galvanic corrosion often occurs when two different metals are in contact in the presence of a conductive solution. What results is similar to the reaction of a car battery where an anode and cathode arrangement occurs. A potential difference may develop between areas on the metal surface if foreign metals, for example, copper on iron, precipitate. The potential differences of metals are given by the galvanic series. The more active metal will oxidize more quickly in the galvanic arrangement, while the inert or noble metal will essentially be unaffected. The greater the potential difference or distance in the series between two metals, the greater the tendency for galvanic corrosion (pits) to result. The relative sizes of the anode (active metal) or cathode affect the relationship between the active and inert metals significantly. A small anode in the presence of a large cathode will corrode more quickly than the opposite. A practical working illustration of this is cast iron, where graphite is more inert than iron. As a result, cast-iron pipe fittings eventually lose their strength because the iron is lost to solution, leaving the weaker graphite behind.

The most practical way of avoiding the galvanic action is to keep dissimilar metals apart. If it is unavoidable, then the metals which are similar or close together in the galvanic series should be utilized.

An alternate method is to interrupt the current flow. This may be done by providing an insulating material between the two metals.

Pitting Corrosion. This is the result of galvanic action where the metal surface appears to have pinholes. The pit is the anode with surrounding surface as cathode. Pitting may occur as a result of one of the following: The first is a change in the acidity of the pit area. The pH of the immediate environment favors the dissolution of ions. In most instances, the lower pH causes the surface to corrode more easily than a neutral or alkaline pH.

A second contributing factor to the increase in pits is differential aeration. This situation readily occurs since most solutions are in contact with air, and, because of convection or diffusion, the transport of oxygen through the solution leads to areas of high or low oxygen concentration. Therefore, where the metal surface contacts the solution, the variation may cause the area with the higher oxygen concentration to become a cathode while an area of lower oxygen concentration becomes the anode, resulting in localized attack.

Depletion of an inhibitor will also cause pits to form. This inhibitor does not have to be a commercially available inhibitor; naturally formed oxides on the metal surface act as inhibitors and protect the substrate from pits. When this oxide film is removed, it leaves the surface exposed and prone to corrosion.

Pitting corrosion is a common occurrence with aluminum and stainless steels in aqueous environments containing metallic chloride salts, whereas copper and brass are less likely to suffer from pitting corrosion. Both aluminum and stainless steel are highly reactive, were it not for inert layers of alumina (Al_2O_3) and chrome-nickel oxides, respectively. Copper, and also brass, on the other hand, react more sluggishly, and defects in their protective oxide films have a greater chance to heal. Pitting corrosion on stainless steel can be inhibited to some extent by providing a strongly oxidizing environment, such as chromic or nitric acid.

Crevice Corrosion. Crevices are present in all equipment. They occur naturally around bolts, rivets, and so on. They are also created by scratches on the metal surface. Crevice corrosion may also occur where gaskets are used, because the gasket material absorbs and draws solution toward the reactive area.

Crevice corrosion is influenced by the same factors as pitting corrosion and is, indeed, a specific form of pitting corrosion.

In order to avoid crevice corrosion, materials should be chosen that are corrosion-resistant. Stainless steels are particularly prone to crevice corrosion and are thus not recommended for use where this condition can occur.

Improper geometric design is the most common factor resulting in damage from crevice corrosion.

Most instances of crevice corrosion also occur in neutral to near-neutral solutions by setting up a reaction with the dissolved oxygen (see reaction 5, Table 1.1). The exception to this is copper, and the exact mechanism for this metal has not been thoroughly determined.

Microbiological Corrosion. Bacteria and especially fungi may grow in certain areas on the walls of water tanks, pipes, etc. A corrosion may develop which often is a combination of crevice corrosion due to reduced availability of air and chemical action from the metabolism products, which are mostly acids.

Intergranular Corrosion. Metals usually are not homogeneous. Impurities or alloying elements may segregate in grain boundaries. Heat treatment or localized heating by welding may provoke changes in composition, localized in or near grain boundaries. These inhomogeneities may drive galvanic corrosion along grain boundaries, loosening up the metals. A most common example of intergranular corrosion is the formation of chromium carbides in grain boundaries during welding of stainless steels, and subsequently poor corrosion resistance along the weld. As a countermeasure, very low carbon steels or steels stabilized by columbium or titanium should be used.

Stress-Cracking Corrosion. This type of corrosion is the result of stress to the metal due to contraction after heating or during cold working. Most metals and alloys exhibit this type of corrosion.

Cyclic stresses, both of high frequencies (<1000 sec^{-1}) and of very low frequency (>10^2 sec), have caused considerable corrosion damage in the past. Careful testing, therefore, is necessary during the design and material selection process.

Selective Leaching or Dissolution. This occurs when one of the components of a metal is more active than the others. The result is the selective removal of a particular component. The best-known example of this is the selective removal of zinc from brass, otherwise called dezincification. This results in significant weakening of the metal. Dezincification is obvious because the color of the brass turns from yellow to the pink-orange of copper.

Dezincification may occur in two ways, by layer or plug. The plug dezincification is a superficial patch on the surface which is readily removed by abrasion. The layer type occurs over the whole surface and is generalized. One factor which affects the type of dezincification is pH. The layer type dominates in acid pH ranges while the plug form is prevalent in slightly acidic, neutral, or alkaline ranges.

The method used to retard the dezincification reaction is to use alloys of brass which contain Sn, As, Sb, or P. There has not, to date, been any truly effective method for completely stopping dezincification from occurring.

Interfilm Corrosion. Coatings, such as paints, conversion coating, or metallic coatings may lose the adhesion with their substrate. This may be due to diffusion through the actual coating or to reaction starting from defects like pinholes or scratches. Residues of soluble salts, acids, or bases will attract water through a paint film because of the osmotic effect, and blisters filled with water will form. Filiform corrosion is a wormlike delamination of a paint film, driven by salt residues and high (80 to 90 percent relative) humidity. Scab corrosion develops from paint defects. Corrosion proceeds underneath the paint film; iron oxides, by their volume, build up scabs. Adhesion may be completely lost on large areas, for example, after water immersion of painted galvanized steel.

CORROSION OF POLYMERIC MATERIALS

The development and use of plastics have been rapid in recent decades. Plastics are finding use in more and more areas, particularly those which were previously thought to be strictly for metal. The reason for this is that metal corrodes and loses its properties or wears away, necessitating replacement. Plastic, on the other hand, can be compounded in a multitude of ways so that it will meet the requirements of the jobs.

Plastic is a broad term covering a series of synthesized chemicals. Plastics are made from what may be thought of as chemical units, or building blocks, called monomers. These chemical units (each plastic contains only a few different types of units), when subjected to certain conditions, join together to form a polymer which we commonly call plastic.

Plastics corrode by reaction with their environment. Polymers are degraded by photochemical oxidation which, for example, produces the chalking of paint films. Plastics may absorb solvents and thus change their properties. And then, they may lose additives because of leaching or by migration to the surface.

Plastics are furthermore prone to three types of cracking (cracking is the term used for describing plastic corrosion): solvent, environmental stress, and thermal cracking. *Solvent cracking* is caused by materials which are borderline soluble in the plastic and which cause cracks by partially dissolving the plastic. *Environmental stress cracking* is the most common form of corrosion of plastics. This occurs frequently on the plastic bottles which are moved by chain conveyors. These conveyors use lubricants which contain soaps, surfactants, and other chemicals which are absorbed by the polymer, causing a change in the properties and stress cracking to occur. The least common is *thermal cracking.* Once formed, the plastic can crack by being exposed to air temperatures of 316 to 352°F (158 to 176°C).

CORROSION CONTROL

Corrosion control is needed to prevent or at least minimize corrosion damage.

Material Selection

The selection of proper construction materials is the basis for successful corrosion control. Mechanical stability, chemical stability, operating temperature, formability, and, of course, cost are of major concern. Any mistake in material selection is hard to overcome. Coating, cladding, and sometimes cathodic protection may be used to limit the damage. For the proper selection of materials, it is strongly recommended that the reader consult the data sheet of the materials suppliers, standards, and common reference books like References 1 to 3.

In order to prevent corrosion damage from galvanic corrosion, great care has to be exercised when joining two different metals.

A few short comments regarding major materials are in order:

Steel. This is the most common of all materials. It rusts easily; corrosion often is uniform. In aqueous alkaline environment, unless strong alkalines plus complexants are present, steel is stable.

Stainless Steels. These come in various grades, generally are stable in alkaline, neutral, and acidic environment. This stability is due to a protective layer of chromium and nickel oxides on the surface which forms readily. However, this protective layer may be locally destroyed by chloride, fluoride, or ferric ions at acidic pH values, resulting in most dangerous pitting corrosion. Sulfuric acid also attacks the standard stainless steels. If stainless steels are to be used in such environments, highly alloyed grades, up to nickel-base alloys (Inconel, Hastelloy, Incoloy) can be used. Especially in maintenance, welding may induce a problem: The welding electrodes must be selected so that they would not constitute an anodic material with respect to the stainless steel being welded. Dissolution of the weld due to galvanic corrosion might result otherwise, and the heat generated during welding might, by weakening the area adjacent to the weld, lead to intercrystalline corrosion cracking, especially with unstabilized, "higher" carbon grades. Simply because stainless steels are viewed as corrosion-resistant, any corrosion of these will be the more surprising and damaging.

Aluminum. This is well suited for a variety of organic chemicals, but it reacts rapidly with water. Perforated aluminum parts often demonstrate the result of improper materials selection.

Zinc and Galvanized Steel. These provide good protection against atmospheric corrosion unless conditions of high acidity and temperature, for example, near exhaust fumes, prevail.

Graphite. This material is stable against most nonoxidizing liquids and chemicals. Careful joining and sealing are, of course, required.

Ceramics. Including glass, ceramics can also generally be used. They are attacked by strong alkalies and by hydrofluoric acid.

Polymers. These include:

- Natural rubber
- Butyl rubber
- Chloroprene rubber
- Fluorinated rubber
- Polyethylene (PE)
- Polypropylene (PP)
- Polyvinylchloride (PVC)
- Polyvinylidenefluoride (PVDF)
- Glass-fiber-reinforced polyester (GF-UP)

They have very specific properties with respect to organic chemicals. Uptake of chemicals, swelling, loss of additives, and chemical degradation are the major concerns. These polymeric construction materials are available in a variety of grades, and the suppliers' data sheets should be consulted. A discussion of some of the plastics of industrial importance follows.

Polyvinylchloride (*PVC*) is a very common plastic and in terms of volume produced is one of the two most widely used plastics. It is a rigid plastic. It does not burn easily and can be mixed with rubbers and other polymers to improve rigidity. Flooring and surface coatings may be made of PVC, as well as pipe fittings. PVC is also used for electrical application such as wire and cable insulation.

Polyurethane is also a widely used material classified as a resin because it belongs to a group of materials that exhibits better heat resistance than thermoplastics such as PVC. These are used in making flexible rubbers with excellent abrasion resistance for surface coatings. These materials do, however, lack a resistance to acids and alkalies. Also, prolonged contact with water and steam should be avoided. Industrially, these materials may be used for roofing over metal.

Polyethylene (*PE*) is very resistant to swelling by chemicals but is susceptible to environmental stress cracking in the presence of some detergents and alcohols. Certain containers, such as beverage bottles, may be made of PE. It is replacing metal pipe because of its light weight, low cost, corrosion resistance, and flexibility. Polyethylene is used for wire and cable for power and communications insulation applications. It may also be used to insulate high-voltage wiring because of its ability to withstand heat, and when applied as a very thin film, it is used as a coating over metal to provide a barrier against moisture.

Polystyrene is a plastic with good heat resistance, oil resistance, and impact strength. Most often this material is blended with styrene-butadiene rubber to provide these properties. By itself, polystyrene has found applications in packaging food. Meat and poultry trays as well as insulated containers for hot food are made of polystyrene.

ABS is a plastic formed by the polymerization of acrylonitrile, butadiene, and styrene. Since ABS is made from three different monomers, varying the levels alone can provide a wide variety of properties. This material has low temperature retention properties and flame retardancy, and it can be chrome-plated. It is used for pipe fittings, particularly in drains. The more flame-resistant varieties are being used in building components.

Acetal homopolymers have been finding increasing use in areas which were traditionally metal castings of zinc, brass, aluminum, and steel. The plumbing industry is using acetal homopolymers for shower heads, ballcocks, flushometer components, and faucet cartridges. Machinery is being equipped with acetal homopolymer conveyor plates, pump impellers, and couplings.

Polyethersulfone is strong and rigid and possesses excellent heat and electrical properties. It is also solvent-resistant. Electrical applications include coil formers and edge and round multipin connectors. The good mechanical strength exhibited by polyethersulfone accounts for its use as pump housings, bearing cages, and power-saw manifolds.

Rubber and Synthetic Elastomers. Natural rubber is a material which has been used for years as lining material over metal. Its usage and versatility increased during the late 1800s when it was discovered that the addition of sulfur under heat improved the toughness and resiliency of the products. This process is known as vulcanization.

Natural rubber is resistant to a good variety of chemicals. However, it is attacked by mineral oils and chlorinated solvents and by strong oxidizing agents. The temperature limitation for continuous exposure for most soft rubber compounds is about 140°F (60°C) and for harder rubbers it is about 180°F (82°C). Heat-resisting compounds are available which may be used at higher temperatures. Soft rubber specially compounded for maximum temperature resistance may be used for continuous exposures under some chemical conditions up to 200°F (93°C), and hard rubbers may be compounded for service temperatures as high as 230°F (110°C). Synthetic rubbers and elastomeric materials have been developed which override the chemical resistance drawbacks to the natural product. Synthetic rubbers can be made with any accumulation of physical properties at the will of the formulating chemist.

Many years back when the flexibility of natural rubber was first discovered, it was also determined that a rubber lining over a steel tank would prohibit the attack of strong acids. It was not until after World War I, though, that the proper adhesives were developed for direct bonding of the rubber to the steel. It is this unique bonding that enables the metal to be corrosion-resistant. The lining uses of this synthetic or natural rubber are several. Included are linings for storage tanks, plating baths, pipelines, fans, and other chemical equipment.

Some synthetic rubbers which are of particular interest include chloroprene, nitrile (used for gasket and base on farm equipment and automobiles because of its resistance to the elements), and butyl rubber which is used in areas where low gas permeability is needed.

Table 1.2 attempts to give some preliminary ideas regarding suitability and stability of materials. This table is not intended to replace laboratory testing or careful studies of actual properties, as material may vary widely even within one group.

TABLE 1.2 Stability of Materials

+ stable
o stable under certain conditions
- unstable

Construction Material	strong acid, dilute aqueous solutions	strong alkali, dilute aqueous solutions	acid oxidizing solutions	alkaline oxidizing solutions	chloride containing aqueous solutions, acidic	fluoride or hydrofluoric acid containing aqueous solutions	short chain alcohols	aliphatic hydrocarbons	aromatic hydrocarbons	mineral oil, white spirit, gasoline	chlorinated hydrocarbons	aliphatic amines	short chain ketones	short chain esters	fatty acids	fats, fatty oils, waxes	soap, in aqueous solution
steel	-	+	-	+	-	-	+	+	+	+	+	+	+	+	o	+	+
stainless 304L/321	o	+	o	+	-	o	+	+	+	+	+	+	+	+	+	+	+
steels 904L	o	+	+	+	o	o	+	+	+	+	+	+	+	+	+	+	+
aluminum 1050	-	-	-	-	-	-	+	+	+	+	+	+	+	+	+	+	o
zinc	-	-	-	-	-	-	+	o	o	o	o	o	+	o	-	o	o
tin	-	-	-	-	-	-	+	+	+	+	o	o	o	o	o	+	o
graphite	+	+	o	o	+	+	+	+	+	+	+	+	+	+	+	+	+
acid resistant ceramics	+	o	+	o	+	-	+	+	+	+	+	+	+	+	+	+	+
natural rubber	+	+	-	-	+	o	o	-	-	-	-	o	o	-	o	-	o
Butyl rubber	+	+	o	o	+	+	+	-	-	-	-	o	o	o	o	-	+
Chloroprene rubber	+	+	o	o	+	+	+	-	o	-	o	o	o	-	o	-	o
Fluorinated rubber	+	+	o	o	+	+	o	o	o	+	-	-	-	-	+	o	+
Polyethylene/polypropylene	+	+	o	o	+	+	+	-	-	-	-	o	o	o	o	o	+
PVC (hard)	+	+	o	o	+	+	+	+	-	+	-	-	-	-	+	+	+
PVDF	+	+	+	+	+	+	+	+	+	+	o	o	o	o	+	+	+
perfluorinated plastics	+	+	+	+	+	+	+	+	+	+	+	+	+	+	+	+	+
glass fiber reinforced unsaturated polyester	+	+	o	+	+	o	+	+	o	+	-	-	-	-	+	+	+

inorganic | organic

Design

Design should take into account incompatibilities between different materials which may lead to galvanic corrosion. Design must also be such that maintenance and maintenance cleaning is feasible. Then, design must prevent corrosion due to differential aeration, which will induce crevice corrosion. An example of this is given in Fig. 1.1. A material that is perfectly stable when immersed in a solution still may experience severe crevice corrosion. Washers and O-rings therefore need some consideration.

9.12 CHEMICAL CORROSION CONTROL AND CLEANING

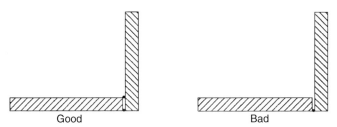

FIGURE 1.1 Example of design avoiding and favoring crevice corrosion, respectively.

Cleaning Practices

Maintenance cleaning is often done to remove corrosion products. A scaled heat exchanger or a clogged nozzle are common examples of corrosion products affecting performance. Deposits may also be the source of corrosion, be it crevice corrosion due to differential aeration or bacterial fouling, where metabolites may attack the substrate underneath the bacterial slime. It must be remembered that scaling, if it is not excessive, *may* considerably slow down corrosion by providing a protective layer.

Removal of dirt and scale is often done by using chemicals such as acids. Table 1.2 should be consulted before starting the work. Inhibited acids (see below under Inhibitors) may be used in some instances. After cleaning, rinsing off the cleaning chemicals is a good practice. This will prevent any contamination, and it will also prevent damages due to prolonged exposure. After cleaning and descaling, some temporary protection may be needed. Normally, corrosion protective oils would be used on steel. Detailed procedures for pipework are described in a NACE publication (Reference 7), stainless-steel cleaning is extensively covered in an ASTM Recommended Practice (Reference 8), to which the reader is referred.

Special Methods

Often recourse is made to some special method, which substantially changes the corroding system.

New Surfaces. Coating the surface of the construction material definitely changes the corrosion performance. One may think of oils and greases, waxes, inorganic conversion coatings, metallic layers produced by electroplating, painting, lining with plastic or rubber, or some combination thereof.

Plating. Plating is done to combine the properties of the base material, often mechanical stability and low cost, with those of the coating metal, often corrosion resistance or abrasion stability. Plating with zinc, chromium, or nickel is a common method of corrosion control. An area of concern is again galvanic corrosion, which might develop around and in pores of the plated metal: If pores should persist, the plating metal should be anodic with respect to the base metal.

Conversion Coatings. There are basically four types of conversion coatings: iron phosphate, zinc or manganese phosphate, and chromate. The first three are used primarily on steel while the last one is for aluminum, aluminum alloys, magnesium, and zinc. These coatings provide a good surface for paint adhesion, and the coating and paint combination provides corrosion resistance.

The method used to produce a phosphate coating involves exposing the metallic surface to the acidic phosphatizing solution. Attack by the acid—mostly phosphoric—dissolves part of the metal and the pH value rises considerably in the boundary layer between metal and phosphatizing solution. As a consequence, insoluble phosphates of iron, zinc, or manganese are deposited on the metallic substrate. Eventually the entire surface is covered uniformly, and the attack stops since the chemical products formed are protecting the surface by being chemically bonded to that surface.

Chromate-type conversion coatings result from a similar process; the deposited coating is caused by the chemicals in the bath reacting with the metallic substrate. This causes a drop in the pH at the immediate interface of bath and metal, which finally results in a lowering of the solubility of the chemical reaction products and the deposition onto the substrate.

Conversion coatings as such are used for corrosion protection, for example, chromate coatings on aluminum, zinc, and magnesium. They are used in combination with oils and waxes on iron and steel, also on galvanized steel, where the combination actually boosts corrosion protection. All types of conversion coatings are used as a paint base. They increase adhesion and prevent corrosion—delamination of the paint—if the paint film gets damaged. Conversion coatings are produced using proprietary chemicals. More details on their advantages and properties can be found in the literature (References 9 and 10).

Closely related to conversion coatings are the protective oxide coatings on aluminum, which mostly are generated by anodic oxidation. Aluminum oxide coatings provide a good paint base, they prevent atmospheric corrosion of aluminum, they may be colored and thus be optically pleasing, and some variants are abrasion-resistant.

One of the shortcomings of protective oxide coatings results from their lack of alkaline stability: Cleaning with alkaline cleaners or household detergent will invariably lead to complete loss of the corrosion resistance these anodic coatings provide.

Organic Coatings. The film-forming organic compounds such as alkyds, vinyls, acrylics, epoxies, phenolics, polyurethanes, and perfluoropolymers are the types of materials that are used as organic coatings. These polymers set and adhere to the metallic substrate by chemical bonding (chemisorption). This method of corrosion protection is very widely used because it provides a heavy barrier which keeps oxygen and moisture away from the metal and also prohibits local galvanic currents from beginning the corrosion process. These organic coatings are used on everything from bridges to marine equipment, to protect the metal from particularly aggressive environments.

One can also mention other types of organic coatings: linings with rubber or other plastic material, and also the polyethylene and polypropylene coated steels, produced by melting the polymer onto the surface.

Paints are applied on top of a conversion coating whenever this seems feasible.

Cathodic Protection. The effect of galvanic corrosion may be used with advantage by electrically connecting a less noble, easily corroding metal with the substrate to be protected. Anodic dissolution will then proceed at the less noble metal; the other will only act as cathode, hence the name "cathodic protection." It is a common practice to protect pipes, storage tanks, ship propellers, and so on, by connecting them with a sacrificial electrode made of zinc or magnesium alloys. Painting the metal to be protected usually increases the lifetime of the anodes. As they are consumed, they must be replaced as part of maintenance.

A similar effect is achieved by using insoluble anodes and protecting the metals by impressing a current (Fig. 1.2).

Inhibitors. The use of inhibitors is the most common method of preventing corrosion (Reference 12). Inhibitors are chemicals which are put into contact with (but do not react with) either the metal substrate or the solution (usually corrosive in nature) which contacts the metal surface. The type of inhibitor will vary depending on the metal and the function of the solution or environment which contacts it. Inhibitors may act by inhibiting the anodic reaction—metal dissolution—or the cathodic reaction—oxygen reduction. There are usually mixed types, with anodic inhibition as the major mechanism. Inhibitors are therefore classified as anodic, cathodic, or mixed type.

Inhibitors act by adsorption, and also by reaction with the metal surface. If the concentration of inhibitors in a solution is insufficient to provide complete coverage of a metallic surface, *and* if they are anodic by type, localized corrosion may set in. Another course for localized corrosion due to insufficient inhibition may occur in crevices, where inhibitors may not diffuse into these areas. Consequently, anodic inhibitors are sometimes called "dangerous" inhibitors, while cathodic inhibitors are called "safe." As a consequence, maintenance involves refilling inhibited process fluids by adding water or oil to make up for liquid losses or evaporation. Hence a sufficient amount of inhibitor must be added.

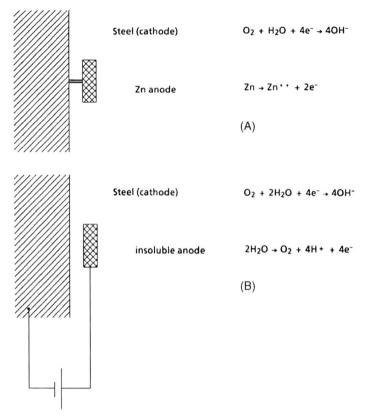

FIGURE 1.2 Cathodic protection with (A) sacrificial anode, (B) impressed current.

Rust-Preventive Oils. The effect of inhibition is not limited to water-based systems. Rust-preventive oils contain inhibitors. Sulfonates are a most common type.

Inhibitors for Aqueous Solutions. In discussing this type of inhibition, it is easier to think in terms of its application. The area of boiler water and cooling tower treatment is one where corrosion inhibition plays a very important role in the efficiency and working of the equipment. Boilers are prone to several different types of attack, with pitting of boiler tubes being the most common. Pitting in boilers is predominantly caused by dissolved gases, oxygen and carbon dioxide. The first step taken to control these offenders is to neutralize them. By adding alkali, such as soluble phosphate, to the boiler water, the carbon dioxide which accelerates the pitting reaction is neutralized. Chromate chemicals, such as sodium chromate, were popular many years ago, but, with the advent of an environmentally concerned society, these quickly became unpopular. Sodium silicates were also used during the same period as chromates and became unpopular for similar disposal reasons. Sodium sulfite and hydrazine are today the main chemicals added to boiler water for the removal of oxygen. Both of these are called oxygen scavengers, very appropriately, because that is exactly what they do. These materials remain in solution until they can react with the available oxygen in the system. Once this reaction occurs, it does not reverse.

In the condensate return lines of boilers, film-forming amines such as cyclohexylamine or morpholine have been used. These are volatilized by the heat in these systems and condense on the metal pipes forming a monomolecular film which keeps any acidic gases from attacking the metal.

Where corrosion of cooling towers is of concern, sodium molybdate, polyacrylates, phosphonates, triazoles, thiazole, and soluble zinc salts have been widely used. Zinc salts are being used with decreasing frequency because of the problem of heavy metal in the effluents. Molybdates, on the other hand, are gaining acceptance because of their low toxicity and flexibility. Molybdate maintains its performance despite high levels of contaminants such as chlorides and fluctuation of pH. This inhibitor functions on ferrous metals by forming a transparent passivating film with the iron corrosion products.

Nonferrous metals such as brass and copper are widely used in cooling towers and must also be protected from corrosion. The chemicals used in this application are tolyltriazole; 1,2,3-benzotriazole; and 2-mercaptobenzothiazole (MBT). MBT is a very effective inhibitor for copper at very low concentrations (2 ppm) and at wide pH ranges as well. The only drawback with this material is that it deactivates polyphosphates for inhibiting the attack on steel or other ferrous metals. Tolyl- and benzotriazole have many of the same characteristics as MBT. MBT is unfortunately deactivated by chlorine, which may be added in the system for slime control. However, once the chlorine is dissipated, the reaction is reversible.

Phosphonates are used in cooling water for general corrosion control. These materials have rather lengthy names and are often referred to by shorter terminology, such as HEDP and EDP. These are usually amine phosphonates. They inhibit the formation of calcium carbonate scale which lies on the surface of the tower and causes the metal to begin corroding. The latest technology in the treatment of this type of corrosion is the use of polyacrylic acid salts. These keep corrosion products from depositing onto the surface.

A second area which uses aqueous corrosion inhibitors is machining fluids. It is important in these systems to inhibit against attack on the ferrous parts of the machine as well as the attack on the ferrous and nonferrous metals which may be machined. The first line of protection in these systems comes from the greases and oils used to lubricate slides, ways, and other moving pieces of the equipment.

The amines supplied in the lubricant are second most important in protecting the surfaces. These are usually alkanolamines, and they function by altering the pH of the metal surface to inhibit corrosion. Amine soaps also have been proved to be very good because they provide a thin film on the metal substrate.

Some inhibitors which have been traditionally used in machining lubricants are *p-t*-butyl benzoic acid and sodium nitrite. The former has been banned for use in this application because of its ability to be absorbed through the skin and cause cancer in the reproductive system of male rats. Sodium nitrite, although still in use, is also suspected of causing cancer. The amines which are basic to lubricant formulating may combine with the nitrites to form nitrosoamines. The nitrosoamines are suspected carcinogens. Molybdates, triazoles, and phosphorus-containing materials, as previously discussed, may also be used.

The third application of corrosion inhibitors for aqueous systems is rather unique unto itself. Pickling inhibitors are used to inhibit the attack of the acid on the base metal. They must not, however, impede the attack on rust or scale. Typical inhibitors come from such chemical classes as thioureas, quaternary ammonium compounds, or acetylenic derivatives.

REFERENCES

1. "Corrosion," in "Metals Handbook," vol. 13, ASM, Metals Park, Ohio.
2. Rabald, E., ed., "Dechema—Materials Tables," Deutsche Gesellschaft für Chemisches Apparatowesen, Frankfurt (Main), 1948 ff.
3. deRenzo, D. J., "Corrosion Resistant Materials Handbook," Noyes Data Corp., Park Ridge, N.J., 1985.
4. "Definition of Terms Relating to Corrosion and Corrosion Testing," ASTM G15.
5. "Terminology Relating to Erosion and Wear," ASTM G40.
6. "Recommended Practice for Conducting Plant Corrosion Tests," ASTM G4.

7. "Industrial Cleaning Manual," NACE (National Association of Corrosion Engineers), TPC Publication 8, Houston, Tex., 1992.
8. "Recommended Practice for Cleaning and Descaling Stainless Steel Parts, Equipment, and Systems," ASTM A380.
9. Rausch, W., "The Phosphating of Metals," ASM International, Metals Park, Ohio, 1990.
10. Freeman, D. B., "Phosphating and Metal Pretreatment," Industrial Press, Inc., New York, 1986.
11. "Specification for Anodic Oxide Coatings on Aluminum," ASTM B580.
12. Nathan, C. C., ed., "Corrosion Inhibitors," NACE, Houston, Tex., 1979.

CHAPTER 2
INDUSTRIAL CHEMICAL CLEANING METHODS

Robert Haydu
President and Chief Chemist, Bronz-Glow Coatings Corp., Jacksonville, Fla.

W. Emerson Brantley III
Marketing Director, Bronz-Glow Coatings Corp., Jacksonville, Fla.

Jerry Casenhiser
Senior Chemist, Bronz-Glow Coatings Corp., Jacksonville, Fla.

CHEMICAL CLEANING

Each year corporations spend millions of dollars to clean plants and process equipment to ensure product quality and useful equipment life, conserve energy, and maintain a safe work environment. This expense is minimal, however, compared with the high costs of equipment repair or replacement, substandard production, and the constant upward increase in energy consumption that results from fouled, poorly maintained equipment. In this age of new technology, where process equipment and process soils (unwanted deposits or films in a system) are becoming more complex, the need for industrial chemical cleaning service companies with special knowledge of chemistry and specialized equipment has become more prevalent. Before selecting a chemical cleaning process, however, special consideration must be given to the metallurgical makeup of a system and the corresponding deposits (contaminants formed). Accurate procedures should be utilized to pinpoint the most efficient and economical chemical cleaning process to use. Because deposits often consist of multiple contaminants, a complete analysis of their chemical composition is of primary importance in selecting the most appropriate solvents to dissolve and/or remove these deposits.

Today, the high cost of energy and downtime (nonproduction hours) is an important factor in considering chemical cleaning of systems or equipment. To minimize operating costs while maintaining high operating efficiency, many companies try to correlate their chemical cleaning requirements with scheduled plant or department shutdowns.

Removal and control of deposit buildup is essential to equipment operating efficiency, as well as equipment life and product quality. Consequently, chemical cleaning solvents or mechanical cleaning tools are regularly employed to remove unwanted soils or deposits from the process-side and water-side segments of industrial process equipment. Neglect in removing organic or inorganic deposits on the process or water sides of industrial equipment may result in irrevocable damage to the equipment, leading to costly repairs and premature replacement of portions or the entire unit.

Process-side deposits are normally of a carbonaceous or organic structure and can be difficult to classify because modern technology encompasses a broad spectrum of processes in the manufacturing of products. Mineral or inorganic deposits generally found on the water side of process equipment are easier to classify since their components (minerals) can be more readily identified.

The structuring of complex molecules (polymerization) during processing of a given product can form unwanted by-products (deposits) on the process side. These by-products or deposits may develop at any given point in the system since time, temperature, and reactants are all elements that contribute to the formation of these organic deposits. On the cooling process side of a heat exchanger, fouling may be caused from organic polymeric compounds which lead or precipitate out from the process stream during the cooling phase. Unlike utility boilers where most of the scale-forming components of the feed water have been removed by pretreatment, cooling water systems are not generally provided with pretreated water because of the large volumes required and cost of treatment. Thus scale buildup formed by the dissolved or undissolved solids in the water is usually more rapid and the scales more varied in composition. The presence of calcium salts, one of the more common deposits found in cooling water systems, is a contributing factor to scale buildup since some calcium salts have low solubility at high temperatures.

Industrial boilers are treated with sludge conditioners to help hold sludge in suspension and keep it soft and nonadherent, allowing for these solids to settle out in the mud drum or be removed from the system during blowdowns. When solids become concentrated and build up on the boiler tube wall, they can cause blistering or bursting of tubes due to the overheating of the underlying metal.

Open cooling water systems present fouling problems in that algae, slime, molds, and other types of plant and animal life, which tend to develop in the system, adversely affect operating efficiency. As the water cascades over the tower, it becomes highly aerated and thus can become corrosive. In an effort to control corrosive attack, corrosion inhibitors may be added (while maintaining proper pH values) to prevent scale formation. Open water cooling systems are difficult to maintain, since generally a large volume of makeup water is required to replace the water lost through evaporation, leaks, or spillage. The algae and slime organisms are often transmitted from the tower directly into the process equipment as the cooling water passes through the system, where it may precipitate, causing plugging or fouling of the equipment.

DEGRADATION

Formation of deposits can occur through the decomposition of certain compounds during operations. For example, a decomposition product results from the use of sulfonated lignin-type compounds in low- or medium-pressure boilers for water treatment. Lignin, a substance derived from wood, is the cement that binds the woody cells together. The lignin binder is dissolved from the wood as the first step in the manufacture of paper. The sulfonated lignin compounds are effective in providing protection to boilers in operation. However, a decomposition product resembling small sootlike particles can be formed by agglomerating and baking out on the metal surfaces. Removing this organic matter requires special consideration. A similar organic deposit can be found in some boilers when small quantities of oil leak from pumps, and the like, into the boiler and become emulsified and degraded. Such deposits may mix and be baked on the metal surfaces along with normal boiler deposits. A common occurrence is where sulfides or oxides of iron are bound by an organic matrix.* Such deposits may require a multistep chemical cleaning procedure to guarantee the removal of the organic and inorganic deposits.

Table 2.1 shows common water-formed deposits and where they might be expected to occur.

---

*"Industrial Cleaning Manual," NACE TPC Publication 8, 1982, pp. 5–6.

TABLE 2.1 Common Water-Formed Deposits and Their Expected Locations

Deposit	Steam generators		Heat exchangers
	Low pressure	High pressure	
Hydroxylapatite	x	x	
Anhydrite	x	x	
Calcium carbonate	x	x	
Dolomite	x	x	
Silica	x	x	x
Serpentine	x		x
Magnetite	x	x	x
Copper		x	
Cuprite		x	
Nickel oxide*		x	
Zinc oxide		x	
Hermatite			x

*The actual forms of nickel and zinc deposits on steel equipment are not usually the oxides. They may be present as:
   $NiFe_2O_4$—nickel ferrite or Trevorite
   $ZnFe_2O_4$—zinc ferrite or Franklinite

Source: "Industrial Cleaning Manual," NACE TPC Publication 8, 1982, p. 4.

PREOPERATIONAL DEPOSITS

When new equipment is fabricated, erected, or installed, engineers and plant management must concern themselves with soils of deposits which have accumulated on the metal surfaces during construction or with deposits that have been applied in the manner of protective coatings at the mill. These soils may appear in the form of dirt, weld splatter, iron oxide, paint, lacquer, or oil coatings. The removal of these deposits from a newly fabricated process system is referred to as preoperational cleaning, because thorough surface cleaning should be performed prior to initial equipment start-up.

Preoperational cleaning is desirable for several reasons. The most notable is that on the process side deposits can contaminate the product, reduce efficiency through restricting heat transfer, damage delicate or sensitive instrumentation and valves, and aid in accelerating corrosion. Since the unwanted deposits may be of an organic or inorganic nature, the application of a multistep cleaning procedure may be required to ensure removal of all unwanted deposits. Sometimes preoperational cleaning is performed on exterior surfaces strictly for cosmetic purposes. It may be as complex (as in food handling, pharmaceutical, or other clean room types of industries) as requiring the complete removal of all detectable residual chemical films and contaminants, even those invisible to ordinary inspection methods.

CHEMICAL CLEANING SOLVENTS

The composition of the unwanted deposits, and the overall objective of the cleaning process, are factors to be considered in choosing the chemical cleaning products for the job. In addition, other important factors are metallurgy and mechanical design of equipment (to be cleaned), time and temperature restraints, and disposal requirements of spent chemical solutions. As discussed, special care in selection of cleaning solvents must be taken when cleaning food-grade surfaces. Some of the more commonly used cleaning solvent systems are listed below.

Alkaline cleaners—primarily for degreasing of metal surfaces, and used prior to acid cleaning or etching

Organic (*carbon-based*) *acids*—in various forms remove oxides, mill scale, and other impurities from the metal surfaces

Inorganic and mineral acids—to remove water-side deposits

Organic (*carbon-based*) *solvents*—useful to remove aromatic and aliphatic organic deposits

Complexing, chelating, or sequestering agents ("wetting agents")—react with hardness ions, forming water-soluble complexes

Alkaline Cleaners

Caustic Soda-Surfactant Alkalies. These are used to remove oil and grease deposits from equipment. They may be used in the first phase of a two-phase cleaning process for the removal of oil and grease prior to acid cleaning. Alkalies are also used after acid cleaning to ensure neutralization of the acid. They will also provide a certain degree of passivation of newly pickled steel surfaces. Some of the more common alkalies used in chemical cleaning are sodium silicates, trisodium phosphate, potassium and sodium hydroxide, and soda ash.

Caustic Soda plus Potassium Permanganate. An excellent oxidizing solution, this is widely used in treating sulfide deposits where sour crude is being processed. The solution converts the sulfide to iron oxide and sulfur, which can then be removed with ammoniated citric acid or another acid. Little or no hydrogen sulfide is generated with this technique. An insoluble reaction product of manganese is formed when the permanganate reacts. Care should be taken if hydrochloric acid is to be used as the next stage of the cleaning process, because chlorine gas will be liberated when they react. Oxalic acid is usually added to the hydrochloric acid to react with any chlorine which is formed.*

Organic Acids

Monoammoniated Citric Acid. One of the more versatile acids, monoammoniated citric acid can be used to dissolve iron oxide as well as clean and brighten stainless steel. One of its special properties is the citrate anion, which is a chelating agent for iron ions. Removal of iron oxides is normally performed at a pH range of 3.5 with citric acid. The pH of the solution can then be slowly adjusted to a pH value of 9 and an oxidant such as sodium nitrate added. A passive film is thus formed to prevent rerusting of steel surfaces while the process equipment is out of use.

Inorganic Acids

Inhibited Muriatic Acid (HCl). An excellent material for dissolving iron oxides and calcium-containing scales. When silica is present in the deposits to be removed, ammonium bifluoride or hydrofluoric acid may be blended with the muriatic acid to dissolve the silicates.

Inhibited Sulfuric Acid. This is another material used for removing iron deposits. Sulfuric acid is used in some special cases for cleaning stainless steel. The advantage is that it is less expensive than

---

*"Industrial Cleaning Manual," NACE TPC Publication 8, 1982, pp. 5–6.

nitric acid and does not contain chlorides as does muriatic acid. Sulfuric acid is generally used at concentrations of 5 to 15 percent by volume. It is one of the most aggressive acids and requires extreme care in blending and handling.

Nitric Acid. Nitric acid is commonly used to passivate austenitic stainless steel. A strong mineral acid and oxidizer, nitric acid removes foreign particles or impurities from the stainless-steel surface. Nitric acid solutions may also be used to brighten aluminum surfaces.

Sulfamic Acid. This material is best used to remove calcium and other carbonate scales. Sulfamic acid, with the addition of a proprietary inhibitor, is the only acid which is recommended to be used on galvanized metals.

Phosphoric Acid. While used for selective chemical cleaning procedures, phosphoric acid is probably most commonly used in the cleaning and brightening of aluminum piping or equipment. As the acid becomes partly spent, there is a danger of insoluble reaction products being formed. Phosphoric acid may also be blended with surfactants where removal of light oil deposits is required in the process of cleaning and brightening aluminum.

ORGANIC SOLVENTS

The petroleum fractions such as kerosene and Stoddard solvents are flammable. Chlorinated solvents are not flammable, but the vapors are toxic to varying degrees. Organic solvents are excellent for removing grease and oil deposits but are usually costly to purchase and quite expensive when it comes to disposal. However, the spent solutions are recyclable (through distillation) and may be sold to reclaiming companies for eventual resale. Solvents like Freon K or PCA may be used for test solutions to determine cleanliness of critical systems or equipment in the field of nuclear energy. For example, this type of testing (normally performed during preoperational cleaning) ensures that no contaminants are present which could have an adverse effect on the equipment or the process. An infrared spectrophotometer "fingerprints" the solvent, using specified wavelengths and sensitivity. After being flushed through the system, the effluent solvent is then tested and compared with the initial test sample of the solvent to determine if the level of organic contaminants is within acceptable limits. This testing is generally performed in a special room with carefully controlled atmospheric conditions. Solvents such as M-Pyrol (manufactured by GAF Corp.) may be used in the cleaning of PVC reactors. Other organic solvents include carbon tetrachloride, trichloroethylene, acetic acid, and acetone.

Complexing, Chelating, or Sequestering Agents

These are expensive cleaning solvents that may be used singularly or in blends with other solvents for more effective removal of various deposits. Chelates selectively combine with other metal ions which are in a solution, and are so precise that they can be chosen so as to combine (complex) only with the metal ions which form the unwanted deposits, while having no effect on the metal alloys of the system itself. This action allows the primary solvent(s) to attack the deposits more effectively.

These agents are often used to enhance the actions of other solvents in this manner, often in the flush stage of cleaning, where they specifically enhance the removal of the loosened deposits from the system (see Fig. 2.1).

These advantages, along with the relative ease and safety with which they may be used, often offset any cost disadvantages. EDTA, gluconates, and polyphosphonates are some such agents.

Copper Complexers. Materials such as thiourea may be added to hydrochloric acid in proper concentration to effectively dissolve copper and hold it in solution. Without complexers the copper will separate from the solution and replate as the magnetite dissolves. A nonacidic method of copper

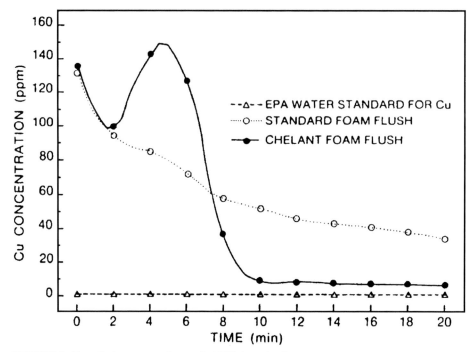

FIGURE 2.1 Foam flush stage tests after standard HCl foam cleaning stage.

removal is sodium bromate, an oxidizing agent, with ammonium hydroxide. The copper is oxidized and goes into solution as the tetraamine copper ion. The copper will remain in solution as long as proper pH values are maintained.

Inhibitors, Passivators, and Oxidizing Agents

These agents, when added to a solvent, retard the corrosive action of solvent upon the structural metal of the system being cleaned. They also create a passive (protective) oxide layer which further acts as a corrosion inhibitor. The application of chromates, phosphates, and silicates is employed to substantially decrease the corrosion rate of iron and other metals by the repair and formation of passive oxide films. These are also referred to as conversion coatings and are sometimes used for the purpose of preventing rusting while equipment is out of service.

Often process equipment, after being acid-cleaned, is not immediately placed back into service. In such cases it is sometimes desirable to improve the corrosion resistance of the metal surface by means of chemical passivation. These solutions are often alkaline and are anodic (oxidizing) inhibitors, the most commonly used being the nitrites and nitrates.

Note: It is assumed that the reader recognizes that the cleaning materials discussed are properly inhibited where appropriate.

Caution: It is recommended that complete data regarding the use, handling, storage, and disposal of all chemicals be obtained from the manufacturer or your chemical cleaning engineer prior to their use.

The first category of solvent or chemical cleaning can be further classified by about eight general application methods.

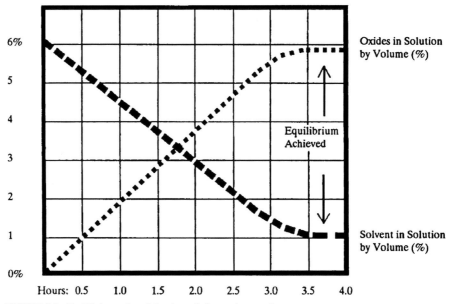

FIGURE 2.2 Equilibrium point of cleaning solutions and contaminants.

CIRCULATION

Of chemical cleaning solutions, circulation is probably the most widely used method of cleaning (removing contaminants from) a process system or unit. This practice is used extensively in the cleaning of industrial boilers. The system is filled with the solution and then recirculated by a properly sized pump to maintain an adequate solution flow rate through the system. The pump is normally fed on the suction side from a holding tank which also acts as a return reservoir for the cleaning solvent as it circulates and recirculates through the equipment, allowing periodic sampling of the solution. Monitoring of the solution allows personnel to determine the progression of the cleaning process, as well as the point where optimum system cleaning is accomplished (see Fig. 2.2). Also, precise temperature control and concentration of cleaning material can be maintained using this method. Undissolved or insoluble deposits from the equipment are deposited in the reservoir tank, where they can easily be removed for disposal.

CASCADE METHOD

This technique is used mostly in chemical plants and refineries to remove carbonaceous deposits from the interior surfaces of a tower. The chemicals are generally pumped up through a reflux line and allowed to cascade down over the surfaces of the trays and interior tower. As the chemical passes over the surfaces, it removes the deposits. Obviously, since towers may be as high as 100 ft or more, the chemical cleaning service company must have equipment with sufficient pumping capacity to generate the high-volume flow rates and necessary high head pressures required to ensure that chemicals are in contact with all the interior surfaces continuously during the cleaning. A disadvantage of this is that soils that have accumulated on the bottom side of the trays or bubble caps are not always completely dissolved or removed since the chemical solution has contact with those areas only intermittently. Occasionally an inert gas may be injected at the base of the tower in an attempt to improve contact time with the cleaning solution in these inaccessible areas.

FILL AND SOAK

This method is probably the simplest of all forms of chemical cleaning. Small items may be immersed in tanks for a given period to remove or loosen deposits and passivate metal surfaces where desired. The surface is then rinsed to flush loosened deposits from the equipment. Sometimes vessels or boilers may be cleaned in a similar manner by filling the vessel or boiler with the chemical cleaning solution and letting it soak to loosen deposits. This method is also useful in cleaning heat exchangers, piping systems, and tanks. In some cases, where such large equipment is filled and soaked, periodic circulation, such as 15 min each hour, may be employed to prevent stratification of chemical solutions and to displace loosened layers of deposits. Proper venting should be maintained as the chemical reactions between the cleaning solvent and the soil deposits often release gases which may be toxic and/or explosive. When utilizing the fill-and-soak method, care should be taken during the rinse stage to ensure that flushing is as thorough as possible, so that dissolved and loosened deposits are removed from the surface.

ONSTREAM CLEANING

This method is used to remove contaminants without disrupting the production mode of the plant or equipment, thus saving costs by eliminating downtime. It must be determined in advance that damage or contamination to the finished product or equipment will not occur. Proper concentration of the cleaning solution is introduced upstream from the process equipment and allowed to pass through heat exchangers, coolers, and jacketed vessels to remove the unwanted deposits which reduce heat-transfer efficiency (process temperature) in the system. Care must be taken not to remove an excessive amount of deposits too quickly, as this may cause plugging downstream in smaller lines. Identifying points in the stream where periodic bleed-off of the cleaning solution can occur will prevent a high concentration of solids buildup in the system during the cleaning process. These bleed-offs can be used for sampling purposes to determine the progression of the cleaning process (concentrations of contaminants). Once the desired level of cleaning is achieved, the cleaning solution is flushed via the normal blowdown procedures, and the spent residues neutralized and disposed of, or recycled for future use.

Steam Vapor Phase Cleaning

In this method, chemical cleaning solvents are introduced into a high-pressure (usually 100 psi or greater) steam flow that is passed through the process equipment at a controlled rate. Soils, greases, and oils are dissolved and carried out in the steam vapors. Passivating solutions may also be applied in the same manner.

Foam Cleaning

By foaming cleaning solvents onto a surface, the contact time of the chemical is increased, especially on vertical surfaces. Using a foam generator (process developed by Dowell Schlumberger), a continuous stream of foamed solvent solution is applied to the surfaces to be cleaned. Some fluids require a foam-enhancing additive for optimum results. Where heavy grain dust or powdered metals dust is being removed from a vertical surface such as a silo or column, foam also has the characteristic property of reducing static electricity charges which may cause combustion. Foam cleaning is usually more cost-efficient than filling a system with liquid solvent (circulation or fill-and-soak), as it requires a lower volume of chemical solution. Aerating (foaming) of a solution also reduces total volume weight, important when the weight of a solution in a system creates a structural integrity concern. Also, the effectiveness of foaming is largely unaffected by system leaks.

Gel Cleaning

A method similar to foam cleaning is simply to add a thickening agent to a cleaning solvent to produce a gel-type cleaning reagent, which may be sprayed or brushed onto a vertical surface to remove corrosion deposits. The gel cleaner is then removed by a pressure rinse or spray. This method is frequently used to remove iron oxides from metal surfaces prior to painting. A commonly used material of this type is "naval gel."

Pickling and Passivating Pipelines and Vessels

The term "pickling" refers to specialized industrial cleaning processes which utilize chemical cleaners that also act as corrosion inhibitors and passivators. There are numerous procedures used, depending on deposits to be removed, metallurgy of the surfaces, and corrosive elements to be resisted. For carbon steel, the use of hydrochloric, sulfuric, phosphoric, or citric acids will perform this dual function. On various alloys, nitric or ammoniated citric acids are most commonly used (Citri-Solv process). George Radney, recognized leading authority in this highly specialized field, notes that in all cases, a pretreatment of sodium hydroxide, trisodium phosphate, and sodium metasilicate should be circulated at 160 to 170°F for 4 to 6 hr, followed by a water rinse, prior to the procedure being performed. This removes any pipe varnishes or hydrocarbon soils, which would adversely affect the pickling and passivating process.

It is not unusual for procedures such as this to require several weeks or even months to complete properly, as in a recent project for a major oil company, involving a special module containing 15,000 ft of stainless-steel piping for undersea application. This project alone, cleaned to NASA specifications, required over 10 months of cleaning and passivating procedures.*

Special adherence to desired pH guidelines must be observed to achieve proper passivation of the metal surfaces. Proper pH monitoring can determine whether a protective corrosion-inhibitive barrier is produced, while inadequate controls can contribute to further corrosion of the surface.

MECHANICAL AND COMBINATION CLEANING METHODS

In addition to chemical cleaning, which relies on cleaning solvents to remove unwanted deposits or corrosion, a second, equally important method of industrial cleaning is the use of nonchemical media to remove deposits or corrosion—mechanical cleaning.

In some cases, a combination of both chemical and mechanical means are employed in the cleaning process.

High-Pressure Water Jetting

In this method, also known as hydroblasting, water is used at pressures of 1000 to 10,000 psi or greater to remove corrosion deposits from equipment such as boiler tubes and exchanger tube bundles.

---

*George Radney, letter, Dec. 14, 1992.

9.26 CHEMICAL CORROSION CONTROL AND CLEANING

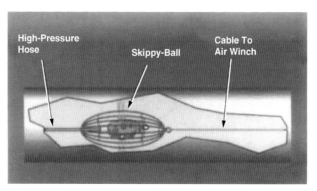

FIGURE 2.3 Schematic drawing of Roto-Jet.

The sheer force of the water stream, often exceeding the speed of sound, is the sole cleaning agent. This force is achieved only by use of special high-pressure pumps and rigid or flexible lances with specially designed nozzles. These lances fit inside the tube, and as they are passed through the tube, the water pressure knocks the deposits from the tube wall. One such system (see Fig. 2.3) employs an automated ball, which carries the sprayhead through piping, tanks, or vessels. This method is superior to manually applied pressure washing, which can only safely deliver 50 to 60 lb of force in such confined areas.*

It should be noted that, while high-pressure water jetting is highly efficient, it can also be extremely dangerous if performed without proper safeguards. Because highly pressurized water can cut through even harder substances with relative ease, the potential effect on the human body can be deadly. For this reason, safety "deadman" switches on all spray equipment, as well as proper training in the handling and control of the spray wand, should be considered mandatory for operator safety.

Hydrodrilling

Specially designed drills are used in conjunction with water to cut through heavy deposits on tube walls. This technique is generally employed when tubes are blocked or plugged with deposits. The water acts as a lubricant and aids in flushing out loosened deposits during the drilling process.

Pigs, Plugs, and Crawlers

Mechanical devices such as these may be used alone or with chemical solvents. These cleaning instruments are generally made to travel through a pipeline by the force of fluid or chemical cleaning solution.

The same flow forces that create the layering of deposits on the inner wall of a pipeline also assist these unique devices in their cleaning. This can be clearly seen in the progressive diagrams presented in Fig. 2.4.

As the lower velocity of the fluid makes the wall susceptible to deposit formation, the employment of a pig causes a shift of velocity, around the pig, aiding the mechanical abrasion action of the pig as it cuts loose the harder deposits. The unwanted matter is flushed forward by this fluid flow, which also propels the pig itself downline.

---

*Roto-Jet Service, Dowell Schlumberger, Inc., 1992.

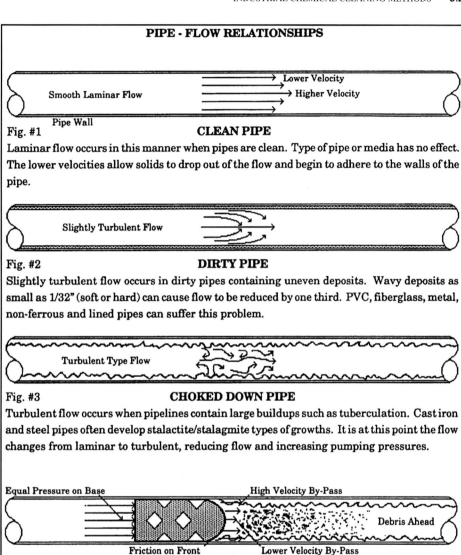

FIGURE 2.4 Schematic, pipe-flow relationships. (*Courtesy of Girard Industries, Inc., Houston, Tex.*)

9.28 CHEMICAL CORROSION CONTROL AND CLEANING

FIGURE 2.5 Pipeline plugs showing deterioration.

"Plugs" (Fig. 2.5) are specifically used early in the cleaning process, and provide information as to the degree of deposits existing in the pipeline's interior. They are usually made of a low-density material which will shear off as the plug is propelled through the line. The resulting plug indicates the current interior diameter of the pipeline, and the appropriate-sized pig can be chosen. This is important for a number of reasons: if the pig is too small, it will have little or no abrasive effect; too large, and it may become lodged in the line.

Consequently, pipeline pigs are often used in stages, beginning with the smallest effective size or abrasiveness, and increasing accordingly.

Proper chemical cleaning sequences (acid-rinsing, passivating, and neutralizing phases) can be achieved by using plugs or stops between the individual solutions. This reduces the volume of chemicals, as well as the time required to clean long, larger-diameter lines.

The "pipe crawlers" consist of inflatable rubber balls of varying sizes, fitted with chain mesh (Fig. 2.6), which tumbles through the pipeline, propelled by the fluid flow pressure. As it moves through, this abrasive tumbling action progressively removes deposits from the inner wall of the pipe.

FIGURE 2.6 Pipe crawlers are available in sizes $1^1/_2$ to 36 in. They are made of reinforced rubber and have an air valve for easy inflation. The scraper chain cleans pipe walls mechanically as the crawler moves with the fluid flow. (*Courtesy of Rotary Pipe Crawlers, Corona, Calif.*)

FIGURE 2.7 Typical pipeline pig. It is equipped with disks to propel it with the fluid flow and with stiff wire brushes for mechanical cleaning. Pigs sometimes are used in connection with chemical cleaning solutions for pipeline cleaning. (*T. D. Williamson Co., Tulsa, Okla.*)

Foam pigs, made of flexible open cell foam of varying densities, have special cleaning qualities when the line has obstacles present such as 90° "ells," "tees," and alternating changes in diameter. Patented rings (Fig. 2.7) also utilize a bullet shape for decreased forward resistance, and a concave base for improved propulsion characteristics. The spiral and crisscross rows of abrasive material also cause the pig to spin as it travels downline, increasing its effectiveness. The increased use of such pigs, and the high degree of specialized cleaning functions they can perform (batching, scraping, special cleaning, wiping, and drying), has caused the development of numerous variations of size, density, and abrasiveness (see Fig. 2.8).

FIGURE 2.8 Specialized foam pigs are available for almost every pipeline cleaning requirement, from light cleanings, wiping, and drying (squeegeeing), to heavy buildup removal and scraping. Pigs can be used in lines of less than 2 in. diameter to 144 in. (12 ft) diameter. These various pigs have a concave base and conical head design to increase speed and effectiveness, and the spiral/crisscross bands of abrasive material create a patented rotating motion, further enhancing cleaning. (*Courtesy of Girard Industries, Inc., Houston, Tex.*)

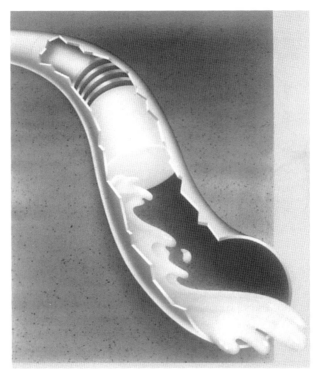

FIGURE 2.9 The "GellyPIG" is a patented injectable gel pig that is flexible and completely conformable. It provides a multitude of cleaning functions, including "batching" (separating of cleaning fluids in multistep cleaning procedures) and condensate removal, and enhances drying and dewatering procedures. (*Courtesy of Dowell Schlumberger, Inc.*)

More recently, highly viscoelastic gels have been utilized which can be injected into a pipeline, becoming customized pigs, completely conforming to the surface that requires cleaning. This increases overall surface contact and surface tension, enhancing the cleaning process (see Fig. 2.9).

Abrasion Method

Abrasion is considered a mechanical method for removing a contaminated layer from the surface. Besides the removal of oxides, abrasion deforms the surface into an amorphous metal layer with a high dislocation density and increased roughness.* This can be a desirable feature when a protective coating or paint is to be applied after cleaning, because the new surface demonstrates enhanced adhesive qualities.

Some abrasive techniques include grinding, sanding, wire brushing, scrubbing, scraping, and the use of a nail gun. Abrasive blasting is also a widely used method to clean metal surfaces. Some of the media used for this type of cleaning are sand, glass bead, steel shot, and grit. Increased environmental concerns regarding spent media have decreased the use of abrasive blasting in recent years.

---

*"Surface Contamination, Genesis, Detection, and Control," vol. 2, Plenum Press, New York, 1979, pp. 726–727.

SELECTION OF CLEANING PROCESSES

The preceding elements should all be considered in determining the most effective, efficient, and economical method(s) for specific industrial needs. Overall, the cleaning process may further be categorized as specific or general.*

An example of a general cleaning process is the removal of iron oxide from a carbon steel surface with an inhibited acid solution known to be acceptable for this purpose.

A specific cleaning process would be the selection of specialized cleaning solvents to remove only a specific type of deposit without affecting or changing the metal surface.

Chemical cleaning engineers and specialists are usually an excellent consultation source and can assist plant maintenance personnel in determining the proper chemical, mechanical, or specific cleaning combinations to best meet their particular requirements. The selection of an industrial service company specializing in chemical cleaning is a major factor in the planning of a successful cleaning project. Though there are many such industrial cleaning companies, some are equipped with research and testing laboratories that lend support to the field service group. The complexity of an industrial chemical cleaning project is not necessarily determined by size, number of units, or fluid volume of the system. Some of the most difficult or technical projects can be of minute volume or size. The complexity is often determined by metallurgies, equipment design, type of deposit, and so on. It is important that the desired end result of the cleaning be specifically defined in advance by plant maintenance personnel to aid the industrial cleaning specialist in selecting the proper methods, solvents, and procedures.

Corrosion itself is often defined as the deterioration of a material because of a reaction with its environment.† The most desirable situation would be for equipment used in industry to be constructed of materials which will not corrode (react) in their particular industrial environment. However, a multitude of factors make this rarely feasible or practical.

For example, standard heat-transfer fin-tube coils, like those found on heating, ventilation, air conditioning, and refrigeration systems, are most often constructed of a base aluminum finstock with copper tubes (or coils). In most corrosive environments, the wafer-thin fins sacrifice almost immediately to the more noble copper. The use of copper fins and copper tubes might be more resistant, but structural weight and added cost may make this option unacceptable. Exotic alloys, such as stainless steel, may be extremely corrosion-resistant but are usually even more expensive, more difficult to work with (i.e., repair, handling, welding), and do not exhibit desirable thermal-transfer properties.

Regular and periodic cleaning of the aluminum fins removes contaminants but also has a deteriorating effect on the fins themselves. Therefore, industries often look for other protective methods beyond basic industrial maintenance cleaning. In many cases, specialized protective coatings provide a solution, extending equipment life and reducing the need for continuous cleaning and removal of corrosive deposits. As with the cleaning processes available, there are also highly specialized protective processes. A secondary problem, however, is that many industrial operating environments often contain a multitude of corrosive elements, often spanning the pH scale (1.0 to 14.0) even in localized areas of production.

Only in recent years have high-tech multistage coatings processes been developed to address this increasingly perplexing problem, using complex, chain-linked polyelastomers, along with highly developed substrates. These complex chains, able to exceed 3000 to 4000 and more hours of salt spray exposure (ASTM B117.85) with *no* corrosion present, offer industry the capability of radically extending equipment life, decreasing energy costs and maintenance expense, and in many cases stopping corrosion's onslaught. By comparison, earlier protective coatings, such as phenolics and epoxies, usually withstand less than 1500 hr in this stringent and comprehensive test (see Fig. 2.10).

---

*"Surface Contamination, Genesis, Detection, and Control," vol. 1, Plenum Press, New York, 1979, p. 394.
†"Technical Manual and Reference Guide," Bronze-Glow Coatings Corp., 1991, p. 3.1.

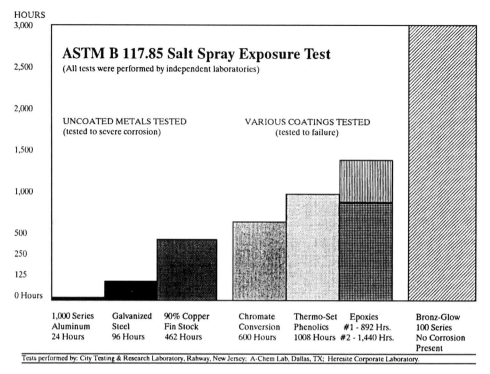

FIGURE 2.10 Corrosion resistance of various metals and coatings.

Such complex processes were pioneered, developed, and patented only as recently as the early 1980s and have only more recently been expanded to wider fields of industrial applications with corresponding positive results. Products such as Bronz-Glow Coatings' 100 series, a patented 18-step process developed originally for unprotected fin-tube coils, have been developed and modified so as to be applicable to structural steel, piping, and concrete. Future applications could alter significantly what has historically been considered an unattainable goal: protecting equipment to the point that the effects of corrosion are substantially reduced, extending equipment life three to five times longer than previous methods.

Environmental Concerns

Proper disposal of waste deposits has become an increasingly important consideration, especially when spent chemical solvents are also involved. In today's economy this is often an extremely costly part of the project. It is necessary to ensure that all material removed is disposed of in compliance with federal and state regulations, even when outside chemical cleaning service companies perform this service. Disposal permit numbers should be recorded on the hauler's manifest and kept on file for future reference.

Much research is being performed to determine satisfactory methods of recycling, reclamation, and possible reuse of contaminated chemicals, blast media, waste sludge, and other industrial cleaning by-products. Effective removal of debris from the solvent fluids reduces the volume of material which must be disposed of. In cases where various filtration methods are successful, the solvent or media may be reused as well, further reducing overall expense in many instances.

These and other environmental concerns are important considerations and will become major deciding factors in a growing number of cleaning applications in the foreseeable future. In all cases

TABLE 2.2 Professional Organizations That May Be of Assistance to the Reader in Gathering Data on Cleaning Problems

National Association of Corrosion Engineers, 1440 South Creek Dr., Houston, TX 77084
American Society for Testing and Materials, 1916 Race St., Philadelphia, PA 19104
American Society of Mechanical Engineers, 345 East 47th St., New York, NY 10017
Edison Electrical Institute, 1111 19th Street, N.W., Washington, DC 20036
American Institute of Chemical Engineers, 345 East 47th St., New York, NY 10017

Source: ASHRAE, The American Society of Heating, Refrigeration, and Air Conditioning Engineers, Inc., 1791 Tullie Circle N.E., Atlanta, GA 30329.

they should be addressed, along with the other important factors regarding the choice of methods, equipment, and solvents.

Only by properly addressing all of the pertinent factors of a particular application, in the preplanning phase of a project, can successful results be expected. Chance is not a friend for those who attempt to shortcut any of the necessary steps in an industrial cleaning application. Fortunately, there are many professional organizations whose members and technical expertise can be of great benefit in almost any project, no matter how difficult or complex.

While the importance of professional industrial cleaning is measured today in millions of dollars, the economics of industrial efficiency dictate that the importance of proper, consistent cleaning methods, and of individuals with the knowledge and expertise necessary to perform such applications, will only increase in the foreseeable future (Table 2.2).

BIBLIOGRAPHY

Atkinson, J. T. N., and H. VanDroffelar, "Corrosion and Its Control," National Association of Corrosion Engineers, Texas, 1982.
Banov, Abel, "Paints and Coatings Handbook," 2d ed., Structures Publishing Company, Michigan, 1978.
"Concurrence Sheet: SSC Facility Cleanliness Requirements for Propellant, Gas, and Hydraulic Systems," NASA Standards, John C. Stennis Space Center, SSC 79-001, Rev. 10/19/1984.
Hoy, Edgar F., "Enhanced Foam Cleaning of Copper-Alloy Surface Condensers," 47th Annual Meeting, International Water Conference, Pittsburgh, Pa., Oct. 27–29, 1986, Dowell Schlumberger, Inc., Tulsa, Okla.
"Industrial Cleaning Manual," NACE TPC Publication 8, National Association of Corrosion Engineers, Texas, 1982.
Mittal, K. L., "Surface Contamination Genesis, Detection, and Control," vols. 1, 2, Plenum Press, New York, London, 1979.
"Polly-Pigs (TM) Industrial Pipeline Cleaners," Girard Industries, Inc., Houston, Tex.
"PreOperational Cleaning," Dowell Schlumberger, Inc., publication TSL-3058, 1992, Tulsa, Okla.
Silman, H., G. Isserlis, and A. F. Averill, "Protective and Decorative Coatings for Metals," Finishing Publications, Ltd., England, 1978.
"Technical Manual and Reference Guide," Bronz-Glow Coatings Corp., Jacksonville, Fla., 1991.
Tomashov, Nikon D., and Galina P. Chernova, "Passivity and Protection of Metals against Corrosion," Plenum Press, New York, 1967.

ACKNOWLEDGMENTS

The senior author wishes to thank the NACE T-3M Committee (chemical cleaning) for their continued efforts to keep industry apprised of modern techniques, and George W. Radney of Retsco, a specialist in servicing offshore rigs in preoperational and maintenance chemical cleaning.

This chapter is dedicated to the memory of the author's mentor, R. J. Becker, P.E., corrosion engineer.

CHAPTER 3
PAINTING AND PROTECTIVE COATINGS

Bryant (Web) Chandler
Cannon Sline, Philadelphia, Pa.

MAINTENANCE PAINTING: WHY?

Maintenance painting is performed for five basic reasons:

1. Beautification, aesthetics, or awareness
2. Corrosion protection or plant life extension
3. Prevention of product contamination
4. Removal or encapsulation of lead-based paints
5. Safety

BENEFITS TO MAINTENANCE PAINTING

The five basic categories of maintenance painting all have individual and mutual benefits. It might be arguable among some that the health and safety reasons outweigh any monetary considerations for maintenance painting, to which this author agrees; however, I will endeavor to treat all the benefits as having equal importance.

Corrosion costs industry billions of dollars each year in replacement costs of capital equipment and structures, most of which could have been prevented by good maintenance painting practices. Unfortunately, maintenance painting is a low-priority item in most facility budgets because painting doesn't help generate revenue or make product.

If the benefits of maintenance painting might help you in selling the concept or need, here are several items to consider:

- Will painting ease the eyestrain and raise the light reflectance level, thus achieving better productivity, less absenteeism, or other positive human factor reasons?
- Will painting make housekeeping easier?
- Will corrosion of the structure or equipment be arrested and thus provide some life extension? The services of a knowledgeable contractor or consultant will be helpful in determining what might be required in this area.

- Will product contamination be eliminated?
- Will energy be saved by better heat transfer, reduction of friction, or other efficiency improvement in materials or fluid handling?
- Will color coding minimize or eliminate human errors that might cause unnecessary equipment or plant shutdowns?
- Will monetary fines or negative publicity be reduced when all plant equipment is operating properly and not taken out of service because of corrosion-related problems?
- Will accidents be reduced by better highlighting, awareness, encapsulation, or elimination?

There may be many other benefits you can think of, and they all may be evaluated in terms of dollars, which is a universal language that approval authorities recognize. It may take some digging or research to determine the benefits, and other colleagues may be your best resource. Accounting should know operating costs, energy costs, equipment value or replacement costs, maybe even lost time due to illness, and the cost of fines when operating out of regulatory limits. Safety records will certainly show root causes of injuries which might get corrected by proper maintenance painting.

After approval, implementation, and completion of your maintenance painting project, don't fail to continue tracking your benefits. The next project will be much easier to sell if the approval authorities know that the benefits were realized and the monetary expenditure is valid.

MAINTENANCE PAINTING: SURVEY

A good maintenance painting program starts with a complete plant survey to prioritize the painting requirements. The survey will provide the road map for the painting program so as to maximize the investment, especially if there are budget constraints, which is normally the case.

The type of plant will determine the elements of the survey. Corrosion of steel structures and equipment is primary in such industries as chemicals, pulp and paper, refineries, fossil-fuel generating stations, waste-water treatment, and other facilities where moisture and chemicals are prevalent. The food, beverage, and pharmaceutical industries may have other primary concerns due to the need to maintain cleanliness, and therefore, a smooth, tight paint or coating which won't harbor contamination is important. Nuclear plants are concerned about the need to maintain good housekeeping, which reduces airborne radioactive contamination. Therefore, sealed concrete walls and good floor coatings have a high priority in addition to corrosion-resistant steel inside containment vessels and steam suppression pools.

A survey of a large industrial complex should be broken down into manageable units or areas using column lines or other easily distinguishable boundaries. By keeping the areas small, the budgetary problems can be better managed, and subcontract maintenance painting contracts can be adequately specified. In defining areas, use nomenclature that is familiar to plant personnel, and try to group similar items, such as a complete pipe support bridge, a small chemical processing unit, and so on.

Once the areas have been defined, they can be broken down further into like items. A pipe support bridge might be broken down into four categories: structural steel, piping, hanger/saddles, and grating/handrail. These categories will be fairly standard for buildings and structures.

The survey should include a critical assessment by an experienced contractor, inspector, or consultant using some or all of the following elements:

1. Critical evaluation of the degree of rusting in accordance with ASTM D610, "Evaluating the Degree of Rusting."
2. Adhesion test of existing paint system per ASTM D3359-83, "Measuring Adhesion by Tape Test," or ASTM D4541, "Pull-off Strength of Coatings Using Portable Adhesion Testers."
3. An evaluation of the chalking of the existing paint per ASTM D4214, "Evaluating the Degree of Chalking of Exterior Paint Films."

4. Determination of the thickness of existing coating per ASTM D1186, "Non-Destructive Measurement of Dry-Film Thickness of Non-Magnetic Coatings Applied to a Ferrous Base," ASTM D1400, "Non-Destructive Measurement of Dry-Film Thickness of Non-Conductive Coatings Applied to a Nonferrous Metal Base," or ASTM D4138, "Measurement of Dry-Film Thickness of Protective Coating Systems by Destructive Means" (for metallic and nonmetallic substrates).

The rusting and adhesions evaluations will largely determine the severity of damage to or deterioration of the existing paint system and the need for recoating. The degree of chalking, paint film thickness, and other local factors will help in determining if the existing paint system can be topcoated or must be removed.

A sufficient number of inspections should be done and recorded so as to provide a good sampling, and the inspection locations should be varied (i.e., vertical, overhead, underside, top surface, windward side, leeward side, and so on).

Once all the values have been recorded from the inspection results, a weighing system should be applied to the individual values. The degree of rusting should be weighted heavier than the adhesion value of the existing paint and more than chalking to determine when to repaint. A numerical rating will result which should reflect the priority of maintenance painting.

The use of videotapes and photographs will help support the survey data, help sell the need for maintenance painting, and provide excellent historical value of the worth of the maintenance painting program.

Once the priorities of maintenance painting have been established via the survey, it is then important to schedule the maintenance painting work. Ideally, if the painting contractor has a small crew working year round at the plant or facility, the permanent crew can be the foremen or leaders for additional painters hired to perform maintenance painting work.

Plant outages or shutdowns in the wintertime do not provide the ideal conditions for repaint work, since most paint systems require 50°F (10°C) and greater to properly cure. Enclosures, heating, flame-sprayed coatings, and other viable options are available at some additional cost, but they do provide solutions. Your industrial painting contractor can provide the cost variables associated with these options, for example, cost to erect enclosures, cost to heat, cost to perform repaint on a 24-hr around-the-clock basis to reduce heating time or eliminate other interferences, cost to flame spray with metals or thermoplastics. All the costs can be presented to management for their decision, where other options may be forthcoming, since they know the bigger picture.

The most cost-effective maintenance repaint program is one that provides a scheduled survey or inspection and repaint procedure. In this manner, the life of the paint or coating is prolonged through preventative maintenance, which includes spot cleaning and spot painting when necessary.

MAINTENANCE PAINTING FOR BEAUTIFICATION, AESTHETICS, OR AWARENESS

Maintenance painting for beautification, aesthetics, or awareness includes such things as office painting, machinery painting, wall graphics, color coding, floor coating, pipe or utility color coding, or company logo display. The finish coat color of paint plays a major role in this category, more so than any other, since color affects light reflectance, affects sizes of rooms or areas, highlights hazardous areas, allows conformance to standards or specifications such as in pipe color coding, compliments furnishings and landscaping, or for the simple reason that the boss likes the color. Much research has been done and numerous articles and books published as to the effect color has on people's moods and attitudes. A consultant may be a wise investment when planning a major repaint of the office, factory, or building exterior to assist you in coordinating and planning the overall color scheme.

In general, a much wider range of material choices and colors is available when painting for beautification, aesthetics, or awareness. Gloss, semigloss, flat, stippled, and smooth are some of the finishes available. Most of the finish paint systems can be applied over drywall, brick, steel, wood, plaster, concrete block, and masonry, as well as over existing paint systems. When applied over existing paint, test patches should be applied to check adhesion, color hiding, or bleed-through over stains. A bigger test area may be prudent if there are special cleaning or environmental concerns. A special graffiti-washable paint may be a wise selection for exterior walls or buildings. Consult your contractor or paint store before making your final paint selection, and especially read the remaining section of this chapter dealing with surface preparation, paint application, inspection, and safety.

MAINTENANCE PAINTING FOR CORROSION PROTECTION OR PLANT LIFE EXTENSION

Maintenance painting for corrosion protection or plant life extension is often considered to be an industrial or heavy-duty painting project. These projects are exposed to the industrial environment of fumes, chemical splashes and spills, immersion, heat, ultraviolet light, or other effects that accelerate corrosion.

The corrosion that most people are familiar with is rusting, which occurs in carbon steel, cast iron, or other metallic materials that contain a high percentage of iron. Other forms of corrosion exist, such as stress corrosion cracking (SCC) associated with the nickel-chromium alloys or stainless steels and microbiologic-induced corrosion (MIC) associated with many metallic materials, carbon steel included. Stress corrosion cracking is usually found adjacent to welds in stainless steels and may be arrested by the use of a coating or special paint. Seek the services of a consultant knowledgeable in metallurgy and painting to assist you in determining the proper course of action. Stress corrosion cracking, if left unattended, can cause catastrophic failure of the equipment or component. Microbiologic-induced corrosion is found in equipment or systems where water may be present even in small amounts, such as in fuel storage tanks. Microbiologic-induced corrosion can be arrested with maintenance painting or special coatings; however, special precleaning surface preparation techniques are required to remove the biologic element before application of the selected paint system. This type of corrosion may be present throughout an entire water system, and extensive treatment may be required to eliminate it. Proper cleaning and coating may be short-lived in a tank interior if MIC will return due to external causes.

Substrates that we try to protect by maintenance painting to obtain plant life extension can involve materials other than carbon steel and cast iron. These materials, although they may not corrode, include concrete, wood, aluminum, copper and its alloys, stainless steel, and any other substrate where painting will add to the life expectancy. Concrete, unless it is the acid-resistant type, requires protection from acid and caustic substances to prevent severe degradation. Waterproofing or sealing of concrete is required to prevent chemicals such as salt from corroding the reinforcing steel inside the concrete structures. This causes severe spalling or breaking of concrete due to the considerable volume expansion from rust formation.

Painting of concrete will require some caution, especially painting of concrete floors, decks, and slabs. A thin horizontal layer of weak concrete may exist at the surface called *laitance,* and this must be removed to achieve good bonding of the paint. Acid etching, mechanical abrading, or abrasive blasting will remove the laitance and expose a strong concrete to bond to. Moisture in concrete will want to escape, causing pinholes, especially if the painting is done when the ambient temperature is rising, so try to paint or seal the concrete in the late afternoon when temperatures normally decline.

Wood may require painting to reduce the effects of weathering, erosion, or both. Use caution in painting wood, because it may have excess moisture, contain pitch, or be overcoated with an old dried, flaking paint. A constantly wet or moist wood may be mossy or have a green algae growth, and this needs a special surface treatment and drying before painting.

Aluminum, which is widely used for its atmospheric corrosion resistance, can corrode. When it is in contact with concrete, the alkalinity can affect bare aluminum, and certain environmental conditions may cause deterioration. Special pretreatments may be necessary before painting aluminum. Consult your contractor or an expert.

If a piece of equipment fabricated from carbon steel has been modified and a more noble or corrosion-resistant metal such as titanium has been added, then accelerated corrosion may be experienced. This will occur when the two different metals are in contact in an immersion or damp environment such as condensers or heat exchangers. Their tubes may be changed from carbon steel to stainless steel or admiralty brass to titanium, in which case the original tubesheet and shell, if left uncoated, will corrode adjacent to the more corrosion-resistant tubes as a result of galvanic corrosion. In this case, careful coating of the shell and tubesheet by a specialty contractor will arrest the condition.

MAINTENANCE PAINTING FOR THE PREVENTION OF PRODUCT CONTAMINATION

Maintenance painting is performed for purposes of preventing product contamination. The purity of wet or dry chemicals or materials may be rendered nonusable if contaminated by the container or vessel they are stored or conveyed in. Painting or coating the contact surfaces with acceptable materials is frequently done, thus allowing less expensive materials of construction.

There are several associations, laboratories, and foundations that test paint and coating systems for contact with materials. The best known standard is the American National Standards Institute/National Sanitation Foundation ANSI/NSF Standard 61 for potable water tanks.

When painting the insides of tanks or vessels for product storage, it is suggested that heat and continuous ventilation be used during and after painting to make sure that the paint is completely cured and that there are no traces of uncatalyzed paint, from which taste or odor may be absorbed by the product. One taste test that is frequently used is to place a pat of creamery butter on the newly painted and cured surface, leave it for several hours, and then taste it. The taste of uncatalyzed paint will be absorbed into the butter.

Industrial paint companies perform extensive testing of their paints and coatings that come in contact with many different products. Discuss your product exposure with the technical personnel at a reputable industrial paint company. Inquire as to whether they have performed direct exposure tests and where the recommended paint system has been used elsewhere; then check it out. Make absolutely sure you advise the technical personnel at the paint company about *all* the ingredients in your product(s), since even trace amounts of some substances can affect the paint. If your product formula is proprietary, then have the paint company provide you with prepainted panels for testing with your product. The paint company should be willing to discuss a testing program with you and assist you in evaluation of the test panels after testing.

MAINTENANCE PAINTING FOR ENCAPSULATION OR AFTER REMOVAL OF LEAD-BASED PAINTS

Maintenance painting articles or books prior to 1990 were not concerned about removal or encapsulation of lead-based paints. Now the subject generates a great deal of interest, including complete books, symposiums, and federal regulations and standards. Because of the health hazard to painters and other personnel exposed to lead-based paints, special mention is required in this chapter.

Any repaint or maintenance painting of older equipment, facilities, structures, and so on may encounter a previously applied lead-based paint system, and one should approach the work with this assumption in mind. Testing kits for lead-based paint are available. Consult Steel Structures Painting Council, in Pittsburgh, Pa., for suppliers or a listing of certified contractors.[1] OSHA (Occupation Safety and Health Administration), as of September 1992, is in the process of developing a regulation for worker exposure to lead, but until the document is complete, there are NIOSH (National Institute for Occupational Safety and Health) alerts and NIOSH/OSHA guidelines available. Your contractor or consultant should be aware of these alerts and guidelines and advise you accordingly, or you should consult the Steel Structures Painting Council in Pittsburgh, Pa., for reprints and other appropriate literature.

Surface preparation work for repainting will create the greatest concentration of airborne lead. Such operations may include grinding, sanding, abrasive blasting, water blasting, burning, chipping, chemical stripping, washing, or other similar operations. Vacuum-shrouded tools, negative-ventilation containment systems, water runoff retention, or other combination of positive captive devices or systems should be used.

Complete worker physical examinations should be performed with blood lead level testing before and after the maintenance painting operation and periodically during work on long-duration projects. Industrial hygienists should be consulted to set up a proper worker/employee health program. Other considerations will include showering and dressing facilities for the lead paint abatement workers, air monitoring, hazardous waste disposal, and proper worker training and certification. It is suggested that owners utilize contractors certified for lead-based paint removal by the Steel Structures Painting Council (SSPC) in Pittsburgh, Pa. A certification program by the SSPC requires that contractors submit written proof that they have the equipment, financial stability, technical expertise, safety record, and personnel qualifications to properly perform lead-based paint abatement work.

MAINTENANCE PAINTING FOR SAFETY AWARENESS

Maintenance painting performed for safety awareness would include such things as

- Handrail identification with yellow paint
- Stair tread highlighting
- Special color highlights on floors and walls for safety showers, shutoff valves or switches, or other critical function identification
- Red painting of areas around fire extinguishers, hose stations, stand pipes, and sprinkler system piping
- Lane marking for equipment, vehicular traffic, or pedestrians inside and outside of plants and warehouses
- Nonskid application to slippery or special traffic areas
- Application of a fire-retardant paint to permanent structures or to lumber or plywood that is used for temporary purposes such as scaffold plank or enclosure materials
- Color coding of piping per ANSI (American National Standards Institute) standards or another rational color system that standardizes plant piping systems
- Application of reflecting paint for proper highlighting or identification at night

There are some special concerns when considering the safety painting in some of the above categories. Yellow paint can contain pigments that are hazardous, such as chromates, so consult your supplier or manufacturer before purchasing to avoid health and environmental problems during paint application. Fire-retardant intumescent paint should be applied at the thickness specified by the manufacturer to perform properly.

MATERIAL SELECTION

The paint and coatings available today are considerably different from those used 5 years ago. The major changes have been brought about by regulatory requirements in 1991 for reduced volatile organic compounds (VOCs). The elimination of lead pigments, chromates, asbestos fillers, and other similar hazardous materials brought about further change. The benefit to the industrial community is the availability of high-tech, high-performance paint and coating systems that provide superior corrosion protection with reduced health hazards and volatile organic compounds in the atmosphere.

In selecting a paint or coating system, there are many key factors that should be considered:

1. Project requirements:
   - Time available for application.
   - What is the life expectancy desired?
   - Will the primer be applied in the shop or in the field?
   - Accessibility for future maintenance painting?
   - Is special approval required, such as in food contact?
2. Exposure conditions:
   - What is the climate?
   - Is the exposure atmospheric or immersion?
   - Is cathodic protection also required?
   - Is there contact with chemicals?
   - What is the surface temperature?
   - Is there impact and abrasion?
3. Surface preparation conditions:
   - What techniques and manpower are available?
   - Can surface contamination be completely removed?
   - Can the desired cleanliness level be achieved?
   - What is the time available for surface preparation?
   - Can the proper profile be achieved?
   - Are there controlling environmental requirements?
   - Is access available for proper preparation?
4. Application and safety:
   - What techniques and manpower are available?
   - What is the expected climatic conditions?
   - What are the access limitations?
   - What is the time availability and the number of coats required?
   - What are the environmental and safety requirements or restrictions?
5. Economics:
   - What is the available budget?
   - Can economies be made when evaluating all the factors?

There may be other pertinent factors that will affect your paint-selection decisions, and they all need to be considered in order for the paint system to function properly. For example, a coating system requiring the best surface preparation possible will fail early if conditions make it impossible to achieve or if a 20-year paint life is desired when only a hand clean and primer can be budgeted.

SURFACE-TOLERANT COATINGS

Surface-tolerant coatings are available that can perform well in industrial maintenance painting. Generic surface-tolerant paint materials include some coal tars, oil-based coatings, the newer moisture-cured urethanes, various epoxy mastics, and calcium sulfonate-based paints and coatings.

The surface-tolerant coatings represent a trade-off between cost and performance. Steel covered with rust, millscale, oil, grease, and moisture is not the ideal substrate on which to apply coatings. The surface-tolerant coating will partially overcome the deficiencies by providing a combination of good wetting, barrier protection, and in some cases other means of suppressing the natural tendency of oxides to absorb moisture and expand. A surface-tolerant coating will have a better chance of success where forces promoting corrosion are not severe. Circumstances to avoid include:

- Highly aggressive exposure environments (e.g., heavy chemical fumes, high temperatures, frequent immersion)
- Highly contaminated substrates (e.g., containing chlorides, surfates, other soluble salts, or high quantities of grease or other contaminants)

When there is doubt about the performance of a surface-tolerant paint or coating, make sure to consult your paint manufacturer or try test patches in several locations. Where surface-tolerant coatings or any paint is applied over existing painted surfaces, adhesion will only be as good as that of the original paint.

GENERIC TYPES OF COATINGS OR PAINTS

Paints and coatings can vary widely within the same generic class by varying the curing agents, solvents, wetting agents, and other additives. As a result, the chemical and physical properties, durability, and performance under different exposure conditions can vary enormously among products. Therefore, research your requirements, and specify paint materials by product name rather than generic type.

Waterborne

Waterborne coatings such as latex paints have been used extensively as interior and exterior architectural coatings and in industrial situations on concrete and wood. They have been used as topcoats over compatible solvent-borne primers.

Waterborne primers for steel, however, are a relatively new development. Complete waterborne systems are currently available for protection of steel, and their use is increasing.

Different types of coatings go by the name *waterborne.* There are coatings in which water is the solvent, emulsion coatings such as latex paints, and water-reducible paints such as paints containing organic solvents that are compatible with water. Of these, emulsion paints and water-reducible paints are favored for industrial use because they can protect the substrate adequately. Latex waterborne paints contain synthetic resins dispersed in water. The resin is dispersed in the water as very tiny particles instead of being dissolved in the water.

Other organic solvents in addition to water are included in high-performance waterborne coatings for aid in coalescence or fusing of the tiny resin particles into a solid film. The organic solvents present in latex paint are sufficiently low in VOCs to meet the current regulations. Latex paints are actually complex mixtures of hiding pigments, extenders, wetting and dispersion aids, thickeners, defoamers, fungicides, stabilizers, and other ingredients. Some latex paints are *thixotropic,* meaning they are fluid when stressed by mixing, brushing, rolling, or spraying and form a gel when relaxed.

Curing of latex paint begins with water evaporation, which brings the tiny dispersed resin particles closer together. These particles will eventually fuse together to form the film. When viewed under magnification, the cured film will resemble a honeycomb.

The safe handling of waterborne paints is just as important as that of solvent-borne paints. The major difference with waterborne paints is the higher flash point than solvent-borne paints; therefore, the hazard of fire or explosion is lower. Consult the manufacturer's data sheet for ventilation recommendations, personnel protection, respirators, and hygiene, as well as storage, mixing, and application equipment, since the characteristics may be different from those you may be familiar with. The Material Safety Data Sheets are to be consulted for further recommendations.

The use of waterborne paints will increase with time. Primers and complete waterborne paint systems are available to provide adequate protection for steel in moderate environments. Cleanup is easy with water and soap, thus avoiding the harsh, flammable chemicals needed with solvent-borne paint systems.

Alkyd

The alkyd coatings were probably those most generally used in the past for general maintenance purposes. Like every type of paint, the alkyd has positive and negative qualities. On the plus side, it has exterior durability, and it is the most economical in cost per square foot per year in average conditions, although it may soon be surpassed by waterborne paint. It contributes good film thickness and mild chemical resistance and, with longer oil, good flexibility. It has good drying characteristics and good color retention for the type of vehicles that can be used in home enamels and general industry (manufacturing).

On the negative side, alkyd coatings are caustic, and acid (liquid) resistance is quite low, mainly owing to poor moisture resistance. Therefore, when not used in chemical exposure or wet or damp areas, performance should be satisfactory.

Zinc Silicate Coatings or Inorganic Zinc Coatings

Zinc-filled inorganic coatings are unique in that they are completely inorganic in composition. Since they contain no resins, plasticizers, or drying oils, they represent a completely different approach to the art of formulating coatings. The films are outstanding in their resistance to severe weathering conditions, alcohols and solvents, and services involving extreme abrasion. They are also good for low or high temperatures and under thermal insulation.

Their principal limitation is their inability to withstand exposure to strong mineral acids and alkalies.

Zinc-filled inorganic coatings may be used as primers for epoxies, urethane, and similar high-performance topcoats. Surface preparation requirements are more critical than those for conventional coatings; therefore, sandblasting to white or near-white metal is mandatory.

Silicones

Silicones may be used alone or in combination with alkyds, epoxy-esters, and other synthetics according to their formulation, and they provide better weather resistance and color and gloss retention, as well as increased water resistance. The silicones as coating resins find great use in heat-resistant coatings. Formulations that will withstand temperatures up to 1000°F (537°C) are available, and typical applications include stacks, mufflers, furnaces, high-temperature processing equipment, stoves, space heaters, and jet engine parts.

The use of silicone for water-repellent masonry finishes is also popular. Silicones are valuable here because they prevent water from penetrating the masonry. Water penetration normally results in the leaching out of salts, leaving unsightly white or yellow deposits on the surface. Also, spalling and cracking are retarded, so the life of concrete and masonry is increased.

Epoxy Coatings

The resins used in these coatings were developed in the United States during an investigation to obtain a film-forming substance that would have the properties of chemical resistance, flexibility, solvent resistance, and a high degree of adhesion to the majority of surfaces. There are five types of epoxy resin-based coatings:

Air-drying

Catalyzed and modified single-component

Coal-tar epoxide coatings

Baking finishes

High-build epoxide coatings—water-emulsion epoxies

There are several types of high-build epoxy coatings:

- *Epoxy mastics.* By formulating with high pigmentation, it is possible to produce high-build catalyzed systems. The solids are in the range of 60 to 85 percent by volume. The chemical resistance is usually superior to the normal-solution catalyzed epoxides owing no doubt to better solvent release and hence a greater resistance to blistering. The maximum dry-film thickness obtainable is about 8 mil per coat. Solvent resistance is very good.
- *Liquid-resin type.* Here, liquid resins are employed instead of resins that need to be dissolved in solvents. Consequently, coatings can be formulated that have high solids content by volume—usually in the range of 80 to 90 percent. Dry-film thicknesses of 8 to 10 mil per coat can be obtained. Chemical and solvent resistance may be rated as good to excellent.
- *Coal-tar epoxy coatings.* These are blends of the epoxide resin, coal tar or bitumen, extender pigments, and solvents. Again, a catalyst is used to cross-link the resin. The main advantages are improved water resistance and improved temperature sensitivity. The conventional solvent-type catalyzed epoxide coatings have only fair water resistance because of their tendency to blister. This blistering is an osmotic phenomenon connected with the film's retaining appreciable amounts of hydrophobic material. A film-former such as coal tar overcomes this defect, so water resistance can be rated as excellent. Coal-tar epoxy coatings cure by a chemical reaction, whereby the catalyst causes the resin to cross-link into large three-dimensional molecules. This curing reaction is temperature-dependent, and at temperatures lower than 50°F (10°C), the reaction is very slow. These coatings cannot be used in contact with gasolines or white-petroleum cargoes.

Urethanes

Urethanes are a series of coatings exhibiting good chemical resistance, high-gloss retention, and cost-effective protection. Abrasion resistance is a distinct factor along with excellent impact properties. The advent of aliphatic isocyanide-based urethane coatings has greatly enhanced the older formulations as to gloss, color retention, and weatherability, equal or superior to acrylic coatings. Aliphatic urethane coatings, although more costly, have physical properties and chemical resistances of both the older version one and two package aromatic isocyanide systems.

Solvents for these systems are generally aromatics, esters, and ketones. Therefore, proper ventilation is necessary, along with special precautions, recognizing the potential toxicity and dermatology problems. Consult the manufacturer's Material Safety Data Sheets (MSDS).

SURFACE PREPARATION

A surface is not "painted" or protected unless there is a bond between it and the coating material. When that union is lacking, the surface is merely "covered" or "bridged," affording little protection.

A painted surface will provide protection against deterioration, and the film will gradually wear; that is, while the paint film may become thinner, it remains fixed to the surface. This gradual wearing gives maximum return for the money (labor, materials, shutdowns, and so on) invested. There is also an added dividend. If repainting is done before the film is too far gone, surface preparation costs are held to a minimum. Although surfaces receptive to paint are essential to a good job, good preparation is ignored so frequently that paint manufacturers annually market a great deal more of their product than is actually needed for efficient industrial maintenance. With poor surface preparation, the repaint cycle is shortened.

Qualified industrial painting contractors are fully aware that proper surface preparation is important. At times, 70 percent of their bid on the job may be for "cleaning." Where paint failure or corrosion exists, 50 to 75 percent of the total cost is justified for the cleaning and priming. Dirt and dust removal, in the average plant, will constitute from 5 to 10 percent, and preparing structural steel may be up to 40 percent more. These are only a few examples to serve as guides and to counteract the tendency to underestimate the importance of surface preparation.

Ordinary dirt, dust, and microscopic grease film are the most prevalent obstacles to paint adhesion in industrial manufacturing maintenance. Of these, grease is the worst villain. Grease films travel with air currents, settling on exterior surfaces.

Traces of grease must be scrubbed and washed off. Where a detectable amount of grease is present, steam detergent cleaning will remove the film and dirt. In production areas, the need for complete removal of dirt before painting is obvious. Compressed-air and vacuum cleaning will take care of the dust but will not remove a thin, greasy film. There is an air-blasting process called *Sponge Jet*[2] that propels small bits of sponge at the substrate that absorb and remove grease. A paint system sold by a national manufacturer[3] is available that encompasses surface oils in the curing process, thus negating the need for removal of oils; however, consult the manufacturer.

Straight solvent cleaning, if required, involves washing the surface with environmentally acceptable solvents or detergent to remove grease and dirt, provided the former is not too thick. It is important to wash off the rinse residue afterwards.

Abrasive blasting is necessary when heavy accumulations of rust, chemical scale, old coatings, or other deleterious matter are present on the surface. Where abrasive blasting is not feasible, power-driven wire brush, sanders, or pneumatic or electric hammers may do an acceptable job. The dry dustless or vacuum sandblast method is valuable to industry. Abrasive blasting can be achieved with no spent abrasive released to the plant atmosphere, but the method is four to five times as expensive as conventional blasting. Blasting with dry ice or liquid nitrogen pellets is also commercially available and works well where abrasives cannot be tolerated or thick layers of paint may exist. If the cost of abrasive or chemical cleaning agent disposal is not feasible or too expensive, ice pellet blasting may be an alternative, since only the material removed is left after the ice evaporates.

Hand wire brushing is used where the surface does not show a heavy scale or rust buildup. This procedure will take care of light rusting and plain dirt. Scrapers are satisfactory when only lightly adhering scale or dirt and mud may be caked in spots on the surface.

Electrically powered tools should never be used where a spark could ignite combustible gases or materials. Such conditions call for pneumatic power, provided that the tools themselves, such as wire brushes, cannot cause trouble by sparking. For the same reasons—fire hazards—combustible solvents should not be used in confined places where flames or sparks could ignite them.

New steel exposed to the weather should not be primed until all mill scale is removed. The most economical way is to permit the mill scale to rust off. It requires 1 to 2 years for this to happen. If a coating is put over new steel, the mill scale will chip off in time and take the paint, prime, or finish coat as well.

Soluble salts such as chlorides, sulfates, and acids and caustics can collect on surfaces downstream or down wind of cooling water systems, towers, or processing units. The soluble salts must be removed; otherwise, paint and coating failure is guaranteed, regardless of whether the substrate is steel, concrete, wood, and so on. Testing for soluble salts, acids, and caustics can be done using various techniques that your industrial painting contractor can perform. Chloride ion contamination can be detected using a potassium ferrocyanide test or a Skat Kit made specifically for detecting chlorides; other tests also may be available. Acids and caustics are detected by using litmus paper moistened with demineralized water and the paper placed on the surface. A color change of the litmus paper indicates acid or caustic as described in the instructions and color chart with the litmus paper.

Removal of the soluble-salt contamination can be accomplished by pressure washing, steam cleaning, or scrubbing with or without chemical additives to assist the removal. Make sure that the water used for the contamination removal is clean and not contributing to the problem. A thorough rinsing is required at the completion of the washing/scrubbing. Retest the surface to make sure that all the salts are removed. Microbiologic-induced corrosion (MIC) may be present in water systems which are at ambient temperature and normally not active, such as in pipe ends, storage vessels, or at the fuel-to-water interface in storage tanks. If you suspect MIC, acquire the services of an industrial painting contractor experienced with MIC or a laboratory or consultant with prior experience. If MIC is not identified and eliminated, then aggressive corrosion can occur. If proper sterilization techniques are not used before application of a paint or coating system, then the applied system will surely fail within a short period of time, as evidenced by blistering.

Surface preparation of lead-based paint must be performed by contractors with experience in that work or personnel supervised by industrial hygienists or consultants with prior training. Sanding, washing, power tool cleaning, and abrasive blasting must all be performed with adequate precautions in place for personnel and environmental protection. The Steel Structures Painting Council in Pittsburgh, Pa., can provide a listing of contractors certified as possessing the capabilities to perform lead-based paint removal or encapsulation work.[1]

A listing of surface preparation specifications appears in Appendix 3A and should be used for defining the type of preparation desired from your painting contractor. The sponsoring society or association will provide copies for a fee on request.

APPLYING PAINT

To brush, roll, or spray is the leading question in paint application. One is usually more efficient. Seldom do the choices have equal merit.

Potentially, spraying has the edge where risks of product contamination, damage to equipment, and air carries are at a minimum. The method is also necessary in using the high-performance or exotic coatings to get the required film thickness for proper protection. Rough surfaces such as concrete, cinder and concrete block, brickwork, and masonry and places inaccessible to a person with a brush or surfaces where brushing would be difficult are generally suited to spraying. So are clusters of beams or intricate mazes of piping, even though extra costs for protection of nearby areas are necessary.

Paint spraying equipment has had to keep up with the development of new coatings, especially the high or 100 percent solids paint systems. These paint systems are comprised of two components that require mixing together. The pot life of some can be measured in seconds, thereby requiring special plural-component spray systems in which the two components are mixed at the spray gun nozzle. The individual components may be heated in their respective containers and/or the spray hose heated to further reduce the pot life and speed up the curing process. This type of equipment should only be used by experienced industrial painting contractors who have personnel schooled and trained in its safe and efficient operation.

Conventional spray equipment using both air and fluid hoses; high-volume, low-pressure spray equipment; and airless spray equipment, where paint under high pressure at the spray gun tip is atomized, is equipment that has been available for some time. Proper operation for effective and safe paint application also requires training.

Air currents, even mild ones, can play peculiar tricks with atomized spray particles. The mist can whisk through corridors and swirl up elevator shafts, leaving no trace of passage, and then settle on machinery, products, or walls several floors or many feet away. Outside wind carries are just as eccentric. They may skip an automobile within 100 yards and mottle another a mile away.

Until one of these remote spottings is actually experienced, it is hard to believe the damage or trouble they can cause. It is these quirks that make it imperative that all spray-work areas be completely sealed off or carefully controlled.

The operator also must be well versed in spraying's limitations. To cite one, the *nozzle* must be as close as possible to right angles with the surface to do a good job. If it isn't, the coatings will vary in thickness, requiring maintenance and rework that might have been avoided. Manufacturers of spray equipment will be glad to provide you with the safe and proper instructions on videotapes and in written text.

INSPECTIONS

Periodic inspection of all surfaces is the surest and best method of determining the need for painting. The key phrase is *periodic inspection of all surfaces.*

To be of value, such inspections cannot be made casually or by an amateur. A good paint inspector should have these qualities:

1. Practical (and commensurate technical) knowledge of the effects of time, wear, and corrosion on paint film and underlying substrates. Preferably, the person should come from the engineering department. An inspector should know the cause of all defects (Table 3.1) that are found so that he can assist in prescribing the cure through proper surface preparation and choice of coatings. Consultation with a technical expert or consultant is frequently advisable and recommended.

TABLE 3.1 Symptoms and Causes of Deterioration

Symptom	How to recognize it	Cause
Alligatoring	Looks like an alligator hide (cracks may or may not reach down to base surface)	Covering a relatively soft coat with a relatively hard one.
Blistering	Unsightly blisters that usually break open and fall off	Water seeping from base surface pushes off the paint film. Sometimes skin drying of the paint is too rapid. Vapor pressure is a factor.
Bubbling	Bubbles on the surface	Moisture or sap in unseasoned wood rises to surface and collects in bubbles. Solvent release in exotic coatings.
Chalking	Premature dull chalky appearance (all paint chalks mildly with age)	Paint may have been poorly formulated, or paint was applied over a badly weathered and porous base surface which absorbed oil out of new paint. Typical with epoxy paints.
Checking	Paint film is cracked open in spots	Shrinkage of film and breaking at weaker portions. This is due to uneven coating and poor bonding qualities (between coats or finish coating and primer); also due to improper compounding where certain ingredients did not dry in proper sequence; sometimes brought about by poor surface preparation.
Chipping (also known as flaking)	Paint film completely broken away	Paint film lacks adhesion. Usually base surface expands or shrinks at a different rate than film and film pulls free. Sudden changes in temperature may cause poor surface preparation.
Crazing	Interlacing "checking" in a gridlike pattern	Same cause as checking.
Discoloration	Off-shade or dark spots	Many reasons, some involving impurities under paint film, such as rust, too much active lime in plaster or cement, resin or sap in woodwork, or exterior attack from fumes or gas.
Spalling	Areas of brick or concrete crack, split, and break off	Moisture seeps into fine hair cracks and expands when it freezes, or salts are absorbed into concrete and corrodes reinforcing steel, which expands.
Wrinkling	Looks like a grained leather surface	Improper drying of paint; coat was too thick or failed to dry properly because weather changed or job was done in unseasonable weather.
Hidden rusting	Streaked surfaces	Failure to clean properly and flood with paint the hard-to-reach corners, angles, trusses, especially those with angles back to back, and narrow intervening spaces between steel members.
Peeling	Paint film peels off cleanly; chipping of large pieces	Moisture getting behind paint film through exposed ends; condensation in wall space or break of putty in sash; failure of caulking compound or cracks in the sash-frame-wall joints, excessive heat (hot pipes, etc.).

2. An inquisitive mind, never satisfied with superficial appearance, that readily absorbs detail and is imbued with a passion for thoroughness.
3. Physical ability to crawl around in places not easily accessible. Such places include beams, roof structures, tank grillage, mazes of pipes, and all other places difficult to examine. The inspector must not be afraid of getting dirty. Like a mouse, the good inspector must run around the foundations and up to the rafters, poking into remote corners and hidden places.
4. Authority and, above all, the confidence of management to the point that when the inspector says that painting is needed, management is willing to accept the diagnosis.

Tools of the inspector's trade (instruments) include destructive and nondestructive types (to measure wet- and dry-film thickness), pull-test modules to gauge adhesion to substrates (especially concrete), and a range of instruments, all available to the trade. A partial list includes:

Dry-film thickness gauge
Wet-film thickness gauge
Surface profile comparator
Surface thermometers
Sling psychrometer
Temperature recorder
Pocket thermometers

Note: Keeping these gadgets properly calibrated is very important.

A good inspector must be reconciled to paperwork. Running notes must be transferred to records that clearly show the status of every surface in the plant. These fall into three categories:

1. Bad areas that need painting at once
2. Areas that show signs of trouble but not to the point that would justify immediate painting
3. Areas in good shape

All three classifications are important, even the last one, although it is frequently not made a matter of record. However, without all the information, the overall picture is incomplete and continues to leave much to guesswork or memory, thus depreciating the value of inspections. Possibly a fourth item might be added to the basic three. It would include painted places that should have been left unpainted and protected with penetrating oil, light grease film, or both. These would include sprinkler heads, fire-escape joints and hinges, valve stems, and so on. Paint on floor columns that are constantly being bumped by mobile equipment which breaks the paint film should be encased in concrete. Other obvious notations also can be recorded.

When assembling data for an inspecting schedule, several things should be kept in mind: Interior inspection can be made at any time of the year but should not be attempted in hot weather. Heat makes the task so disagreeable that even the conscientious inspector will skip the tough spots. Exterior inspections should be made only in good weather for the same reasons. Excessive heat or cold discourages thoroughness. Springtime is best if it works in with the correct time interval.

The best way to tackle the problem is set up an inspection routine based on the worst prevailing conditions. Then, by working backward to surfaces with the greatest durability, the complete plant inspection interval can be ascertained. Meanwhile, critical areas are kept under surveillance. At the same time, a schedule is being developed for surfaces that fail most quickly. And finally, most managers don't realize the value of a follow-up inspection after a painting job. Postpainting inspections forestall false reliance on inadequate surface protection from poor or careless painting jobs. Again, those hard-to-get-at spots should receive the most careful scrutiny on checkup inspections.

A complete listing of painting inspection standards and the respective professional associations or societies is included in Appendix 3A. An understanding of the applicable specifications and standards will greatly assist the inspection work. If inspections are performed by different personnel, then consistent recording of conditions is essential; such standards are listed in the appendix.

SAFETY

On painting jobs, insurance statistics show the common ladder leads to more injuries than any other type of equipment. It is debatable whether this is due to frequency of use or carelessness because of a ladder's uncomplicated construction.

Unless someone is around to stop them, workers will persist in mounting ladders that are not held by another person or anchored or tied in at the base. To reach greater heights, they will place ladders too straight. To cover an extra few inches, they will lean out to the side. And no one should ever stand on any of the top three rungs of a ladder, regardless of its type. Keep aluminum ladders away from electrical wiring.

No repaired ladder should ever be used. The same applies to those showing signs of deterioration, such as minute cracks and splits. Ladders used or stored in the presence of intense heat or excessive dampness or where there are contact chemical or acid fumes should be examined very carefully.

Rigging and scaffolding are not for novices. Experts should be hired. Most industrial painting contractors keep several people on their permanent payroll with the proper training and expertise.

Rope that has been used around acid, caustic, or chemical fumes should be discarded. Rope should never be stored where rats can get to the strands and should be kept so that air can circulate freely around it. Rope that has seen service should be discarded after 1 year regardless of its condition.

There are times and places when safety nets should be strung to protect workers above, personnel below, and materials. Safety standards and regulations exist, and specialty contractors are available for consultation.

Swing falls, blocks, and all metal used to support painters should be watched carefully for weaknesses, and metal should be tested periodically for fatigue, rust, or deterioration. Testing of cable climbers is required by law and should be done only by competent personnel. Sufficient warning and danger signs should be placed when workers are working overhead, and areas should be roped off.

Workers wearing respirators or goggles in a spraying operation—and they should be wearing them—must have life lines or safety lines strung when working high. A painter with fogged goggles and no guide rope can easily slip off into space. A life line, properly secured, is the law.

Workers painting from a scaffold board should never be allowed to reach beyond its edge. And the planks should be kept clean. Accumulated paint should be burned or scraped off. Planks and scaffolds are safer when kept clean and in good shape.

Danger also lurks in spots that are not so obvious. In a plant where the floor temperature may be 70 to 80°F (21 to 26°C), it might be 100 to 110°F (37 to 43°C) directly under a 50-ft ceiling. Ceiling ventilation must be provided in those conditions or heat stress considered. Heat and paint fumes, when mixed, create a condition that puts not only painters but also persons below in jeopardy.

At times, ventilating fans aren't sufficient, and respirators must be worn. This is especially true of some paints with the newer vehicles and thinners in them. In confined areas, the concentration of toxic or hazardous fumes must be analyzed and monitored with calibrated instruments. Numerous respirators are available that will provide different protection factors. A fresh-air full-face respirator is the best if it fits the worker properly.

Always consult the MSDS for the paint being applied. The proper protection criteria are fully described, including other very important information, such as LELs (lower explosion limits), disposal and cleanup methods, firefighting techniques, and so on.

Thick accumulations of dust atop ducts and pipes high in the ceiling require care from sparking that would set off a flash. In such places, wood-handled dust brooms or vacuum cleaning equipment (nonsparking) should be used for cleaning. Likewise, nonsparking tools should always be used in a chemical plant or where there might be illuminating- or natural-gas leakage, gasoline fumes, or similar hazards. In addition, the workers must not have steel nails in their shoes. Sewed rubber or leather soles should be required.

Compressors and other motor-driven equipment be serviced continuously to make sure that they are not fire risks.

Soiled paint rags can easily cause spontaneous combustion. They should be put outside of buildings and in sealed containers at the close of every working period. For the same reason, drop sheets should not be folded or stacked unless the paint on them is thoroughly dry and all loose dirt is removed.

Thinners and paints should be kept in sealed containers and never exposed to flames or where fumes can drift toward an open fire.

Your insurance company can provide valuable service in working with you to reduce accidents and set up an effective program. Safety consultants, when used properly, will save their clients money through reduced insurance premiums resulting from reduced accident claims.

THE PAINTERS

The best returns to management from industrial painting are derived when the proper ingredients are applied by skilled workers under competent administration and supervision. Giving a person a can of paint and brush, roller, or spray gun doesn't make them a good painter, and if they aren't, management loses.

No large plant should be without its skilled maintenance crew. There are always retouch, cleanup, and emergency jobs to be done. This is good maintenance and of value to management.

It is the major periodic jobs that raise a question as to who is to do them. Basically, there are three answers—all with gimmicks. There is the choice of (1) keeping a large force of painters on the payroll, (2) augmenting a small regular painting crew temporarily, or (3) calling in an outside contractor or applicator. Consider also that the average employer today has a *burden* over and above actual wages averaging 42 to 47 percent. This burden includes fringes, social security, insurance, unemployment taxes, and so on. These costs are continuously escalating.

To keep a large force busy, management frequently, and in fact with few exceptions, must create work, thereby increasing maintenance costs.

To augment a small key crew temporarily, management must assume considerable responsibility. It must provide the technical and proficient supervisory personnel; handle labor relations, including the unions; plan safety precautions and insurance; secure, maintain, and store extra equipment; make all purchases; and finally, assume all responsibility for the quality of the finished job.

If outside contractors are decided on, management must still tread circumspectly. Going into the competitive market and obtaining a "firm" bid may appear a sure way of knowing the cost. But, and this is a big *but,* straight competitive bidding is likely to put management into the hands of a contractor willing to skimp to show a profit.

By adequate inspection throughout the job, management, of course, can insist that proper standards be met. Sooner or later this means wrangling and more corner cutting than may be detected easily. Further, while the hassle goes on, the job stops or goes forward at a snail's pace. We are in the "age of litigation," and protracted legal matters are costly. Moreover, the dilution of management time is incalculable.

On a cost-plus form contract, there is no incentive for the contractor to get the job done as economically as possible. The more it costs, the more the contractor makes. Again, management must depend on the integrity of the contractor.

A third approach lies between these two: It is the guaranteed-cost contract. Here the contractor tells management the job will not exceed a certain figure. If the contractor makes a mistake and goes over that amount, the contractor takes the licking. If the cost is under, management gains. Since contractors do not like to bid themselves out of a reasonably profitable business, they keep the guaranteed-cost figure competitive, at the same time making provisions for a quality job. This means that the contractor must be highly experienced. The contractor's organization must include technical experts and top supervisory personnel. Otherwise, the guaranteed cost bid would be out of line, one way or the other. There are variations to this approach involving fixed fees plus incentives, extended coating warranties, and so on. All tend to assure the purchaser of quality services and on-time completion, coupled with fair and reasonable costs. The old adage about getting what you pay for is readily applicable to protective coatings contracting.

Still, management is in the hands of the contractor, but not to the same extent as in the other two bid approaches. Management identifies its cost exposure at the outset. It is a settled issue. Next, the quality of the work can be checked, and if it is below par, the owner can insist on corrections without encountering the same resistance as on a low-bid contract.

Suffice it to say that we are in the age of specialization. The sophistication of the protective coatings field demands the input of professional contractors or applicators. Hence management is urged to consult its own in-house maintenance supervisor, its selected and qualified contractor, and the coatings manufacturer to assume the best investment for each "painting" dollar. Third-party consulting specialists are also readily available. All the foregoing are resources available to management. Utilizing all these resources is an intelligent approach in today's sophisticated industrial world.

REFERENCES

1. Steel Structures Painting Council (SSPC), 4516 Henry Street, Suite 301, Pittsburgh, PA 15213-3728 (412-687-1113).
2. Sponge-Jet, Inc., Post Office Box 206, Dover, NH 03820.
3. Valspar Corp., 1401 Severn Street, Baltimore, MD 21230.

APPENDIX 3A

American Society for Testing and Materials (ASTM)
1916 Race Street
Philadelphia, PA 19103
215-299-5400(FAX: 215-977-9679)

Preparation of Surfaces for Painting

Method no.

D2092	Preparation of Zinc-Coated (Galvanized) Steel Surfaces for Paint
D4258	Surface Cleaning Concrete for Coating
D4259	Abrading Concrete
D4260	Acid Etching Concrete
D4261	Surface Cleaning Concrete Unit Masonry for Coatings
D4262	pH of Chemically Cleaned or Etched Concrete Surfaces
D4263	Indicating Moisture in Concrete by the Plastic Sheet Method
D4285	Indicating Oil or Water in Compressed Air
D4417	Field Measurement of Surface Profile of Blast Cleaned Steel
D4940	Conductimetric Analysis of Water Soluble Ionic Contamination of Blasting Abrasives
E337	Measuring Humidity with a Psychrometer (Measurement of Wet- and Dry-Bulb Temperature)

Physical Properties—Wet Films

Method no.

D3925	Sampling Liquid Paints and Related Pigmented Coatings
D4212	Viscosity by Dip-Type Viscosity Cups

Physical Properties—Cured Films

Method no.

D2240	Rubber Property—Durometer Hardness
D2583	Indentation Hardness of Rigid Plastics by Means of a Barcol Impressor
D3359	Measuring Adhesion by Tape Test
D3363	Film Hardness by Pencil Test
D4541	Pull-Off Strength of Coatings Using Portable Adhesion Testers
D4752	Measuring MEK Resistance of Ethyl Silicate (Inorganic) Zinc-Rich Primers by Solvent Rub
D5064	Conducting a Patch Test to Assess Coating Compatibility

Thickness Measurement

Method no.

D1005	Measurement of Dry-Film Thickness of Organic Coatings Using Micrometers
D1186	Non-Destructive Measurement of Dry-Film Thickness of Non-Magnetic Coatings Applied to a Ferrous Base
D1400	Non-Destructive Measurement of Dry-Film Thickness of Non-Conductive Coatings Applied to a Non-Ferrous Metal Base
D4138	Measurement of Dry-Film Thickness of Protective Coating Systems by Destructive Means
D4414	Measurement of Wet Film Thickness by Notch Gauges

Holiday Detection

Method no.

D4787	Continuity Verification of Liquid or Sheet Linings Applied to Concrete Substrates
D5162	Discontinuity (Holiday) Testing on Non-Conductive Protective Coating on Metallic Substrates

Visual Examination and Appearance

Method no.

D4214	Evaluating the Degree of Chalking of Exterior Paint Films
D4610	Determining the Presence of and Removing Microbial (Fungal or Algal) Growth on Paint and Related Coatings

General Topics

Method no.

D16	Paint, Varnish, Lacquer, and Related Products, Terminology
D3276	Painting Inspectors (Metal Substrates) Standard Guide
D4227	Qualification of Journeyman Painters for Application of Coatings to Concrete Surfaces of Safety-Related Areas in Nuclear Facilities
D4228	Qualification of Journeyman Painters for Application of Coatings to Steel Surfaces of Safety-Related Areas in Nuclear Facilities
D4537	Establishing Procedures to Qualify and Certify Inspection Personnel for Coating Work in Nuclear Facilities
D4538	Protection Coating and Lining Work for Power Generation Facilities, Terminology

Steel Structures Painting Council (SSPC)
4516 Henry Street, Suite 301
Pittsburgh, PA 15213-3728
412-687-1113 (FAX: 412-687-1153)

SSPC-VIS-1-89	Visual Standard for Abrasive Blast Cleaned Steel
SSPC-VIS-2-82	Standard Method of Evaluating Degree of Rusting on Painted Steel Surfaces
SSPC-SP1	Solvent Cleaning
SSPC-SP2	Hand Tool Cleaning
SSPC-SP3	Power Tool Cleaning
SSPC-SP5	White Metal Blast Cleaning
SSPC-SP6	Commercial Blast Cleaning
SSPC-SP7	Brush-off Blast Cleaning
SSPC-SP8	Pickling
SSPC-SP10	Near-White Blast Cleaning
SSPC-SP11	Power Tool Cleaning to White Metal
SSPC-PA1	Shop, Field and Maintenance Painting
SSPC-PA2	Measurement of Dry Paint Thickness with Magnetic Gauges
SSPC-AB1	Mineral and Slag Abrasives
SSPC-PA Guide 3	A Guide to Safety in Paint Application

National Association of Corrosion Engineers (NACE)
P.O. Box 218
Houston, TX 77218-8340
713-492-0535

Recommended Practices

RP-01-72	Surface Preparation of Steel and other Hard Materials by Water Washing
RP-0178-89	Fabrication Details, Surface Finish Requirements, and Proper Design Considerations for Tanks and Vessels to be Lined for Immersion Service
RP-0184-84	Repair of Lining Systems

RP-0287-87	Field Measurement of Surface Profile on Abrasive Blast Cleaned Steel Surfaces Using a Replica Tape
RP-0288-88	Inspection of Linings on Steel and Concrete

Test Methods

TM-01-70	Visual Standard for Surfaces of New Steel Airblast Cleaned with Sand Abrasive
TM-0170	Visual Comparator for Surfaces of New Steel Airblast Cleaned with Slag Abrasive
TM-0175-75	Visual Standard for Surfaces of New Steel Centrifugally Blast Cleaned with Steel Grit and Shot
TM-0186-89	Holiday Detection of Internal Tubular Coatings of 10 to 130 mils Dry Film Thickness
TM-0384-89	Holiday Detection of Internal Tubular Coatings of Less than 10 mils Dry Film Thickness

CHAPTER 4
PIPING*

Tyler G. Hicks
Mechanical Engineer, Rockville Centre, N.Y.

The first step in any piping maintenance program involves eliminating, as far as possible, basic conditions that make excessive maintenance necessary. These may include severe corrosion, water hammer, or poor piping layout. In the case of both corrosion and water hammer, little can be done until one understands the cause.

Corrosion. Probably the biggest single piping-maintenance problem is corrosion. Books have been written on its theory, but the important point is that internal corrosion of piping is generally caused by atmospheric oxygen dissolved in water, and it stops when oxygen is removed or used up by its attack on the metal. Water coming into a system from the outside is always saturated with oxygen, and will continue to corrode the piping until the oxygen is consumed in the process. That is why service-water lines (always supplied with new water and new oxygen) rust faster than hot-water heating lines, which constantly recirculate the same water.

In the steam-water circuit of power plants, dissolved air (oxygen) enters through the makeup water, and through leaks, into parts of the system under vacuum. The accepted cure is to minimize all such leaks by joint and packing maintenance, and then deaerate the feedwater in a suitably designed heater. Sometimes sodium sulfite is used to remove the last traces of oxygen. Corrosion of condensate lines of heating systems is usually caused by air getting in (through vents, reliefs, and joints) at points where the system is under vacuum.

External corrosion may be rapid where a pipe is frequently wet from "sweating" or other moisture—and particularly if the wet surface is repeatedly exposed to air containing sulfurous or acid fumes. To cure it, remove the cause of the sweating or waterproof the pipe. Pipe buried in cinders or soil will often corrode, particularly if the soil is damp and acidic. A practical protection is a watertight covering, generally of asphaltic or similar material, applied directly to the pipe or a spiral wrapping of strong fabric. Normally pipe with perforations or cracks from corrosion or other causes is replaced at once. Where this is not possible because of operating conditions, emergency patches, like those shown in Fig. 4.1, may save a shutdown. These are used on iron and steel pipes.

United States Navy practice for a substantial brazed repair of leaking copper or brass pipe is as follows: Shape the copper patch to fit. Clean the mating surfaces with a file, emery cloth, and hydrochloric acid. Wire patch securely in place. Brick in an enclosure to confine the heat. Heat with an acetylene torch, but do not burn patch or piping. Run in spelter solder (with borax flux) between surfaces. Keep turning pipe back and forth so spelter will run between all parts of braze. When patch is cool, test with water pressure.

---

*For much of the information in this chapter, the author is indebted to the magazine *Power*, New York, and Crane Company, Chicago: also to other manufacturers who participated from time to time in the preparation of the *Power* material.

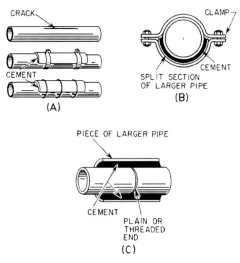

FIGURE 4.1 Emergency repairs for piping leaks. (*A*) To seal pipe crack, apply iron cement and bind it tight with a metal sheet. (*B*) Clamp on a half-shell cut from next larger pipe size and seal with cement or soft gasket. (*C*) For an emergency pipe joint, slide ends into larger pipe and caulk with iron cement.

A small hole in brass or copper pipe may be closed with a rivet or screw plug. Small weak areas may be temporarily reinforced by tightly wrapping with wire thoroughly soldered together in layers to make a solid band.

Water Hammer. This occurs when a moving column of water in a pipe is suddenly stopped or retarded. If the cause is too sudden closing of a valve, the cure is either a mechanical speed limit on the valve or a tag urging cautious handling. Where a pipe is being constantly hammered by connected reciprocating equipment, anchor the pipe firmly and try relieving the shock by air chambers, surge tanks, or similar devices.

Drainage. Failure to remove condensate from steam lines is a major cause of water hammer. Drain all condensate pockets. Make sure that the traps are operating and that no pipe sags so far as to create a pocket. Watch out for condensate caught above closed valves in vertical lines or in back of globe valves in horizontal lines. If water hammer occurs only when steam is admitted to a cold system, it indicates that the system is not adequately pitched or trapped to take care of the large initial condensation. Gradual preheating may ease the situation. Where water hammer has continued for some time, inspect the pipe guides, anchors, and adjacent walls for serious cracks.

When the more obvious ills of a piping system have been remedied, it is time to set up an orderly maintenance procedure to forestall future trouble. In piping, as with other equipment, proper maintenance means preventive maintenance—fixing things before they break. That, in turn, implies organization, records, and definite inspection schedules. The starting point should be a complete set of drawings of the piping system, on which changes and repairs can be noted and dated as made.

Organization. It is not enough to have good piping mechanics, important as this is. Failure to organize maintenance can cause much trouble and unnecessary expense. Complete drawings of the system, with all changes and corresponding dates indicated, will repeatedly save ripping out this or that piece to rediscover facts that should be in the office files. Moreover, recorded installation dates on piping elements will warn the experienced maintenance engineer when trouble may be expected from the everyday causes that lead to piping failure.

When a leak occurs, the adjacent piping should be studied to locate anything else that needs repair, so that the whole job may be done at one time. If the hookup is deficient in unions, or otherwise difficult to maintain, the situation should be remedied while making necessary repairs. General routine inspection crews should check leaks, look for signs of corrosion or weakness, make sure anchors are holding and expansion joints working freely, and check hangers for alignment and distribution of load.

Temperatures. Inspection of an old system should consider any increased pressure or temperature since installation, because these may exceed the safe limits of the materials installed. The ANSI code "Pressure Piping," B31, sets the upper temperature limits for all common piping materials used in industrial plants. Consult the latest edition of the code to determine the proper operational temperature for various materials. Note that compliance with code requirements is purely voluntary.

Pipe Threads. Any detailed discussion of piping maintenance must start with the practical art of pipe threading for two reasons: (1) Badly made threaded joints increase maintenance and endanger plant operation, and (2) the pipe maintenance man is always a piping erector to some degree.

Good threaded joints are mainly a matter of making good threads. These, in turn, require a clean understanding of the proper shape and dimensions of the desired thread and of the die or cutting head that forms it. A brief review of the general scheme of standard American pipe and thread dimensions may bring out certain points often overlooked. Pipe 14 in or larger is named by its actual outside diameter. Usual OD sizes are 14, 16, 18, 20, 24, 30 in, and larger.

Pipe Data. For 12 in and smaller, the nominal pipe size is very roughly the inside diameter of so-called "Standard" pipe. The two heavier series called "Extra Strong" and "Double Extra Strong" have the same external diameters but smaller internal diameters. These traditional names will eventually be displaced by the more logical nomenclature of the American National Standards Institute, which sets up a series of pipe schedule numbers of progressively increasing thickness to cover the great variety of modern conditions. For pipe sizes up to 8 or 10 in, Schedule 40 is identical with "Standard," "Schedule 80" with "Extra Strong," "Schedule 160" with "Double Extra Strong." The complete size schedules and other specifications for piping are contained in the ANSI code for "Pressure Piping." Regardless of "weight" or "schedule," pipe of a given nominal size follows the outside and thread dimensions given by the American Standard pipe threads.

Most threads are cut with a set of four or six chasers mounted in a hand die or a machine-operated head. The lip angles are important, and Fig. 4.2 shows lip rake and cutting angles. Note that the cutting angle equals the lip angle only when the face of the chaser lies along a diameter of the pipe. If the

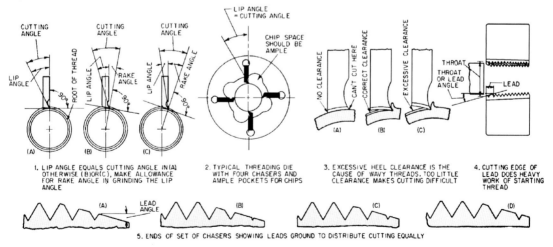

FIGURE 4.2 Lip, rake, and cutting angles for thread chasers.

chaser is not in this position, allow for it when grinding the lip angle. Since the chaser acts much like a lathe tool, there are similar variations in cutting angle. The cutting angle should be very small for brass, not over 16° for wrought iron, from 15° to 20° for bessemer pipe, and at least 25° for open-hearth pipe.

Heel clearance (Fig. 4.2) is formed in the chaser threads at the factory and cannot be altered or restored with maintenance tools. Clearance reduces as the front part of the chaser wears away with long use. The only cure is a new set. To get the chasers well started on the pipe, they are beveled at the entrance end with a heavy "lead" angle (Fig. 4.2) to a diameter larger than the end of the pipe. Figure 4.2 also shows the lead ends of four successive chasers from the same head. The lead angle is the same in all four, but the cutting edge of each is advanced a little over the next one so that the cutting load is distributed evenly. Maintaining this equal cutting by proper grinding is a matter of first importance.

Figure 4.3 shows how to grind the lip angle on a narrow chaser such as is used in a hand die. This method (using the side of the wheel) is not suitable for the wide, flat chasers used in machine-threading heads. These can be ground on a surface grinder or equivalent rig. Proper grinding of the lead requires the fixture shown (or equivalent), in which the table can be turned about either horizontal or vertical axes to give universal adjustments. Before grinding chaser leads, arrange them in serial order 1, 2, 3, 4. Test the angles of the fixture to make sure the wheel bears squarely on the existing face of metal to be ground. Then take chasers in rotation and remove the same amount of metal from each. If any in the set is then found to be insufficiently ground, start back with number one and give each an equal additional grinding. This even treatment should be watched and tested to make sure each chaser carries its share of the cutting load.

Figure 4.4 gives a number of useful pointers in the care of broken and damaged threaded joints of many different types. Figure 4.5 shows ways to salvage the flanges of pipes in which the threads rust, preventing easy removal of the flange. The normal engagement to make tight joints is given in Fig. 4.6.

Flanges. These are designed to ensure a tight joint that can conveniently be broken for piping changes and repairs. Yet poor selection of flanges, bolts, and gaskets, plus careless joint makeup, can cause endless trouble. Once the desired face has been selected (see Fig. 4.7A), flange dimensions and materials are established by the ANSI code for "Pressure Piping." The customary flange bolt for low-pressure piping is square-headed with a hex nut and American Standard Coarse Thread Series threads. Above 160 psi and 450°F, alloy-steel studs with hex nuts on both ends are often used. These nuts are generally of carbon steel or at best a less strong alloy than the studs. Extreme strength is not needed and the difference in metals in contact reduces chances of thread "freezing."

For high-pressure-flange studs the 8-pitch thread series, which provides eight threads per inch for all bolts 1 in or larger, is often used. Standard flange-drilling templates come in multiples of four holes—four, eight, twelve, sixteen, and so on. The flange holes in valves and fittings are set to straddle the centerline. Of the flange faces shown in Fig. 4.7A some engineers prefer the raised face or the pipe-lap, with a ring gasket, because these joints (unlike the male-female and tongue-and-groove) do not have to be sprung apart to break a joint or remove a gasket. Grooved joints with thick, soft gaskets are often used for low-temperature, high-pressure hydraulic joints. With either male-female or tongue-and-groove joints, the gasket should be thinner than the depression to avoid "mushrooming."

Gaskets. Widely used gasket materials range from soft rubber for cold water to narrow, solid iron rings for high-pressure steam joints. Table 4.1 shows typical uses and limiting temperatures. Hand-cut gaskets for raised-face flanges should fit nearly inside the bolts and extend to, but not beyond, the edge of the pipe opening. If the joint has to be broken frequently, coat one side of the gasket with graphite to prevent sticking. When a joint has been newly made with soft packing, take up on the bolts again after the line has been hot for some time.

Thin gaskets are less likely to blow than thick. If flanges fail to meet, it may be risky to fill the gap with a thick, soft gasket. It is better to use a metal filler, gasketed on both sides. Raised faces of flanges are often serrated. These serrations, like corrugations in gaskets and the use of narrow gaskets, are means for increasing the gasket pressure per square inch by reducing contact area. Most leaky joints are the result of insufficient bolt tension and gasket pressure. According to Crocker,* the

---

*"Piping Handbook," 5th ed., McGraw-Hill Book Company, New York, 1967.

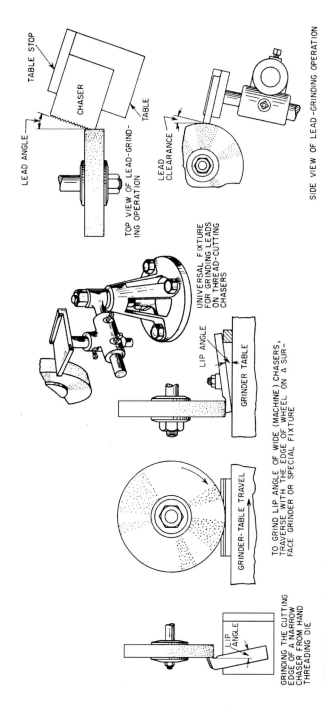
FIGURE 4.3 Steps in grinding chasers.

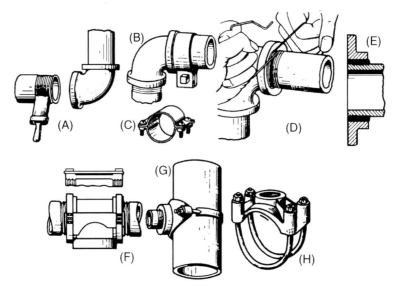

FIGURE 4.4 Repairs for threaded joints. Iron cements specially compounded for the service may be applied as paste or putty to seal pipe joints and cracks. Cement used thin makes a good new-thread dope. (*A*) For best results, leading joints should be remade with cement; otherwise cement can be caulked under pipe clamp (*B*) alongside leaky thread. (*C*) Another thread-leak repair involves winding soft wire (*D*) around cement putty at leak. Allow cement to harden before using. In emergency, a threaded pipe may be cemented into a flange of larger pipe size (*E*). Tamp cement solidly in place. Crack in fitting (*F*) sealed with cement backed by metal plate. Outlet saddles for water (*G*) and steam (*H*) are clamped to pipe over drilled holes to make emergency branch connections. Saddles may be packed with soft gaskets or cement.

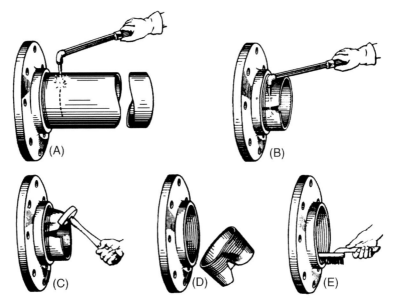

FIGURE 4.5 Salvaging flanges when the threads are frozen. (*A*) Server pipe near flange. (*B*) Cut V notch in pipe stub. (*C*) Collapse pipe with hammer. (*D*) Pipe falls out, leaving flange threads unharmed. (*E*) Clean threads thoroughly.

DIMENSIONS, IN INCHES
Dimensions given do not allow for variations in tapping or threading.

Size	1/8	1/4	3/8	1/2	3/4	1	1¼	1½	2	2½	3	3½	4	5	6	8	10	12	14
A	1/4	3/8	3/8	1/2	9/16	11/16	11/16	11/16	3/4	15/16	1	11/16	1⅛	1¼	15/16	17/16	15/8	1¾	
B*						9/16	5/8	5/8	5/8	7/8	15/16	1	11/16	13/16	1¼	1⅜	19/16	1 11/16	1⅞

FIGURE 4.6 Normal engagement between male and female threads to make tight joints.

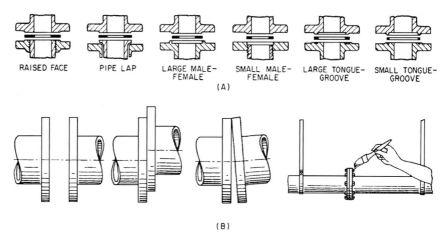

FIGURE 4.7 (*A*) Standard cast-steel flanges. (*B*) Align flanges to avoid trouble; careful aligning before bolting prevents excessive bolting stresses in valves, fittings, and pipe flanges. Use of thread lubricant cuts friction, protects threads, and makes joints easier to break for necessary repairs.

TABLE 4.1 Gasket-Material Selection

Gasket material	Fluid	Usual maximum temp, °F
Red rubber	Steam, air, water	250
Asbestos composition	Steam, water, oil	750
Fiber and paper	Oil	200
Synthetic rubber	Oil	200
Copper, corrugated or plain	Steam or water	600
Steel, corrugated or plain	Steam or water	1000
Stainless steel, 12–14% chromium, corrugated	Steam or water	1000
Hydrogen-annealed furniture iron	Steam or water	1000
Monel, corrugated or plain	Steam or water	1000
Ingot iron, special gasket for ring-type joint	Steam, water, oil	1000

From Crocker, "Piping Handbook," 5th ed., McGraw-Hill Book Company, New York, 1967.

initial gasket pressure should be at least 4000 psi for rubber, 12,000 psi for laminated asbestos in serrated joints, and 30,000 to 60,000 psi for solid-metal gaskets.

Ordinarily a tensile stress of about 7000 psi in the bolt would balance the actual steam pressure, yet the alloy-steel bolts of high-pressure joints are often stressed to 30,000 psi and sometimes to 60,000 psi to flow the gasket into the uneven surface of the flange and thereby ensure tightness. Tests have been made and tabulated showing how much wrench pull will create a given bolt tension in well-lubricated threads of a given pitch. The practical value of such tables is limited because of the great variation of pull with thread lubrication and also because it is rarely convenient to measure wrench pull, particularly where the wrench must be sledged. From the practical angle, in the case of all medium- and low-pressure work, the fitter may as well continue to tighten by "feel" and experience. When this will not do, as with 1400-psi steam lines, the only reliable determination of bolt tension is by micrometer measurement of bolt elongation, using studs with machined micrometer pads and taking care that before-and-after measurements are taken cold and at the same temperature. The measured elongation must be referred to the grip distance between the nuts and not to the full length of the stud.

For steel of any composition a stress of 30,000 psi corresponds closely to a stretch of 0.001 in/in, with other stresses and stretches in proportion up to the elastic limit. Thus, for a bolt with 3-in grip, the total stretch should be 0.002 in to create a unit tension of 20,000 psi in the bolt.

Leaks. Like other piping leaks, those in flange joints grow rapidly. Fix them at the start. A frequent cause is poor alignment of piping. It should not be necessary to spring flanges into line (Fig. 4.7B). All steamfitters agree that nut tightening should follow a definite sequence. Two are shown in Fig. 4.8A—one "round and round," the other "crisscross." Either should be satisfactory.

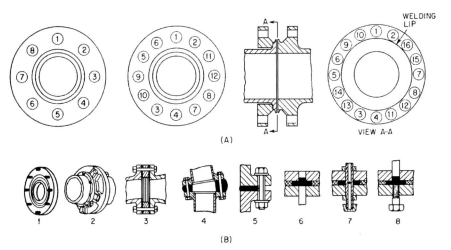

FIGURE 4.8 (A) Two sequences of bolt tightening. In either case, first set nuts finger-tight. For method at left, tighten nuts moderately in order 1, 2, 3, etc. Then make another round to set bolts a little tighter—and so on until bolts are equally and sufficiently tight and feelers show equal flange separation all around. Many prefer crossover method. Bolt tightening for seal-welded joint: First put in temporary bolts all around, hammer-tight. Remove 1 and 2, seal weld there, and replace temporary with alloy bolts set up hammer-tight. Repeat this operation at 3 and 4, 5 and 6, etc., crisscrossing as shown. Follow some general procedure for a different number of bolts, crisscrossing as before. (B) Iron cement aids in making and repairing flanged joints. Corrugated-iron gasket coated both sides with iron cement will make tight joints (1) despite irregular flange faces. To plug leaky thread at flange, set band clamp with flat edge against flange (2) and tamp groove full of cement. If flanges cannot be aligned, turn a metal "Dutchman" (3,4) and seal it on both sides with iron cement or cement-plastered gaskets. To make a tight joint with rough-cast flanges (5), separate flanges by rope of soft spacers; then fill joint with cement. Localized leaks in flanges force cement through other side by hammering on a close-fitting rod. Or (7) inject cement with grease gun through pipe nipple with locknuts and a central side-outlet hole. If stud screws into flange, reverse stud (8) making a backstop for cement.

An improvement on the second might be to tighten a single bolt on one side, then on the opposite—then move 90° and repeat. Figure 4.8*B* gives a number of pointers on the maintenance of flanged joints.

Valves. Damage in storage or handling and poor installation handicap a valve from the start. Complete wrapping, wooden crating, or thin metal caps over the ends protect valves as shipped from the factory. Keep this protection in place and store the valve under cover until it is installed. Penetration of sand into working parts often follows storage on the ground, exposed to the weather. Rough handling easily damages valves; place them where they cannot fall and where other material cannot fall on them.

Installation starts with removal of the valve's protective covering; then clean all grit and dirt from the inside. Pipe cleaning is just as necessary if damage to seats and disks is to be avoided. Blow out the valve with compressed air or flush it with water; clean the pipe in the same manner, or pull a swab through, to remove dirt and metal chips left from threading operations or storage.

Future troubles can be minimized by mounting valves properly, protecting them against outside damage, and locating them at the most suitable point in the line. Except for split-wedge and double-disk gates, most valves can be mounted at any angle, although it is always better, from the valve standpoint, to mount it with stem pointing upward. Any position with the stem pointing downward brings the bonnet under the line of flow, forming a pocket to catch pipe scale and other foreign matter. This soon cuts and destroys inside stem threads. On lines exposed to freezing temperatures, moisture trapped at this point may cause frozen and burst bonnets. Even when the valve is mounted with its stem upright, take the precaution of installing a drain plug in the bottom of the body. Figures 4.9 and 4.10 show a number of installation pointers.

Flow Direction. Direction of flow through globe valves depends on the nature of service and can usually be determined by asking the question, "Should the valve open or shut if the disk and stem part company?" The ASME Boiler Code requires pressure under the disk for globe valves in boiler-feed lines so that the loose disk will not act as a check and stop water flow. If the valve controls equipment which might overspeed, applying pressure over the disk forms a check valve which will shut down the unit if the disk comes loose. Drain valves with pressure under the disk will vibrate open if not tightly closed. If a valve is persistently left cracked open, reversing it to put pressure over the disk will ensure tighter closing.

To summarize: Unless the service clearly requires pressure under the disk, install the valve to put pressure on top of the disk. Pressure over the disk aids in keeping the valve closed, tending to compensate for any stem contraction caused by temperature changes. Valves of large size, 12 in and over, present generous disk areas to line pressure. In unbalanced service, such as discharging from high to low pressure, this pressure makes valve operation increasingly difficult as exposed disk area increases. This is less pronounced in gates than in globes, because disk movement in the former crosses the line of flow (Fig. 4.11), whereas in the latter it opposes the flow. To minimize pressure difficulties, equip all gate valves 12 in and over and all globe valves over 6 in with throttling-globe bypass valves.

In pipelines handling sludge or other suspended matter, keep valves out of vertical lines whenever possible; stoppage of flow allows suspended matter to settle and choke a closed valve. This is especially troublesome when it interferes with check-valve operation. Never install a valve without thought as to access. To be sure valves are opened and closed correctly, make it convenient and safe to do so. Do not expect a man standing on a ladder reaching for an overhead valve to exert any great amount of force. Install the valve horizontally, and fit it with a chain operator which can be reached from floor level. Overhead valves on large lines, requiring two-man operation, should be vertical, to permit operators to stand on the pipe. If possible install a working platform, extend the valve stem through the floor above, and provide an impactor handwheel or a power operator.

Beware of valves just within fingertip reach of a normal man; rather than hunt for something to stand on, he'll stretch just enough to reach the handwheel. The valve cannot be closed tight, and leakage will develop. Always allow sufficient clearance for rising stems and for removing the bonnet and stem. It is much easier to leave the valve in the line and remove only internal parts for inspection and cleaning. Providing easy access to all valves represents the first step to correct operation, regular

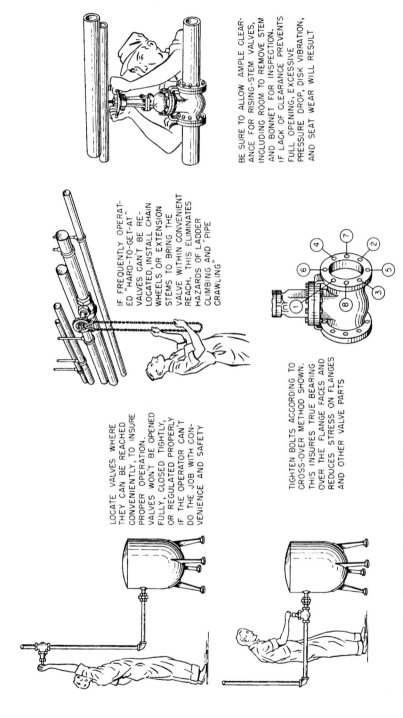

FIGURE 4.9 Long valve life begins with proper installation.

Be sure to ream out burrs that impede flow and sometimes damage equipment

Blow dirt and sand from pipe before making up joints; foreign matter may score valve seats

Apply dope on male threads only to keep it from getting into pipe and equipment

  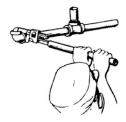

Use right-size wrench; too much leverage may twist valve bodies or crack cast fittings

Never use a hickey except to *break* a stubborn joint in taking down a line

FIGURE 4.10 Suggestions for pipe fitters.

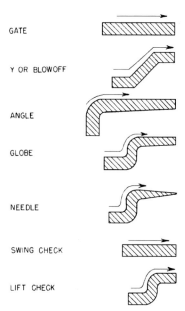

FIGURE 4.11 Valve flow chart.

inspection, and careful maintenance. Good check-valve service depends to a large extent on meeting special needs. Correct position is most important; install swing and tilt checks with the pin horizontal and so that gravity will close the disk; place lift checks so that the lift is vertical.

Diaphragm-operated control valves present problems connected with the diaphragm as well as with the inner valve. Inverted installation on oil or chemical lines exposes the diaphragm to leakage through the stem packing. Heat from adjacent steam lines may cause early deterioration of the rubber. On steam-operated valves, protect the diaphragm against steam contact by installing a water leg or accumulator and filling it with water before admitting steam. When using the regulated pressure as the actuating medium, connect the pilot line, controlled by a lock-shield globe valve, on the downstream side of the valve. Do not make the connection in an elbow or pipe bend, because erratic pressures occur at these points. Select as a connection point a straight section of pipe at least 10 ft from the valve. The globe valve permits throttling the pilot supply to smooth out pressure pulsations, and the lock shield protects against tampering.

Regulating-valve bodies, marked as to flow direction, must be inserted in the line to conform with the marking (Fig. 4.12). A correctly chosen valve will not necessarily match the pipe size. For the connections, reducing flanges or bushings save space, but bell or venturi reducers give better flow conditions. Install a strainer ahead of the valve to remove foreign matter and protect the seats. Impingement noises can be minimized by installing the valve in a straight pipe run. Do not install an elbow or bend immediately downstream of the valve. To facilitate the regular inspection regulating valves deserve, connect a throttling globe valve as a bypass.

Check every valve installation for ways to simplify maintenance. When applying pipe insulation, end the permanent covering at bonnet bolts. Use a removable section to cover the bonnet; otherwise internal inspection may be neglected because of hesitation in breaking a perfect-looking insulated joint. After the system has been heated to operating temperature, tighten all body and bonnet bolts and adjust the stem packing.

Valve Maintenance. Neglecting a valve until it must be replaced or fitted with new parts wastes expensive materials; frequent and regular inspections uncover leaks and reveal other conditions, such as corrosion, incrustation, and wiredrawing, that mean future trouble unless corrected. Caught early, these defects can be repaired without major difficulty or expense; allowed to go unattended, they may require expensive parts and materials and can cause production shutdowns. A good maintenance program includes correct operation, regular and systematic inspection, proper lubrication of all rotating or sliding parts, replacement of stem packing when leakage or excessive friction develops, and refacing leaking seats and disks. Valve parts such as stem threads, thrust washers, and disk-spacing wedges or cams must be kept free of corrosion, incrustation, or foreign materials and must be adequately lubricated as recommended by the manufacturers. Plug cocks, with their large metal surfaces, require frequent lubrication to prevent galling and seizing of the sliding parts.

Packing. Proper maintenance of packing and correct adjustment of packing nuts are essential to satisfactory stem life and good valve performance. New packing, impregnated with graphite, lubricates the stem; after this lubricant disappears, friction between the stem and packing increases. If the packing nut must be tightened to a point where it is difficult to turn the stem, the packing has become dry and hard or is otherwise unsuitable for the service. In either case it should be discarded; it imposes an additional burden that will rapidly shorten stem-thread life. Excessive packing compression causes uncertainty as to whether or not the valve is fully seated. As a result, operators frequently seat valves tighter than they need be. This excessive closing effort, often applied with a wrench, injures the stem threads.

Stem packing that has been subjected to high-temperature steam and then allowed to cool often leaks a small amount when the line goes into operation again. Expansion and contraction of bonnet, stem, packing, gland, and packing nut cause this condition. It does not necessarily call for adjustment of the packing nut or gland; as soon as the valve becomes hot it will, in most cases, stop leaking. Packing maintenance is particularly important on automatically operated valves; excessive friction causes erratic movement. When lubricators are provided, keep the stem and packing adequately lubricated. Use stem packing that fits; preformed rings enter easily and draw up evenly. When winding coil packing spirally around the valve stem, force it to the outer edge of the stuffing box instead of wrapping it tightly to the stem. After adding the maximum number of rings, draw up

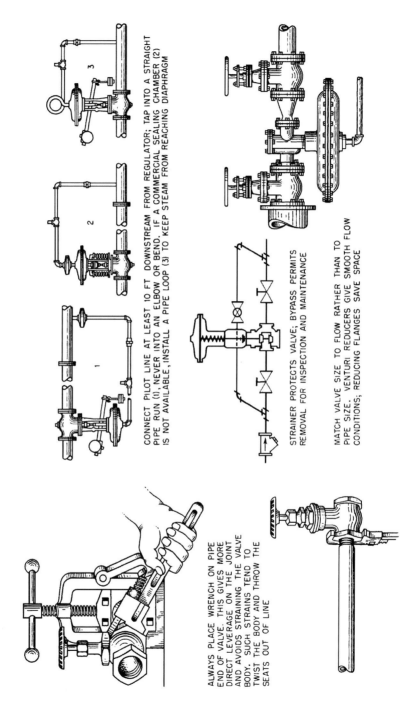

FIGURE 4.12 Good regulator hookups.

the gland evenly, with a wrench, until the packing is forced into a snug position. Then slack off on the gland and make the nut finger tight. A good valve stem, sufficient packing of the right kind, and a finger-tight gland will hold all moderate pressures. Higher pressures require a tighter gland to prevent the applied pressure from getting in under the packing.

Many valves contain a back seat closing off the packing gland against pressure, when the valve is open. This arrangement permits repacking the stem with the valve under pressure. Be certain a valve has this feature before attempting to repack under pressure. Maintenance and inspection involve frequent dismantling and reassembling of valves. Knowing the best way to do these operations simplifies the job and avoids damage to valves from improper handling. Before starting to remove a bonnet, open the valve so that no bending stress is placed on the stem during removal. Likewise, put the stem in the "open" position before replacing a bonnet. U-clamp gate-valve bodies have been split by tightening the bonnet joint with the wedge in its extreme closed position; union-bonnet gate-valve seats have been sprung apart by wedging action in the same manner.

Bolt Tension. On valves designed for high-temperature high-pressure service, bonnet bolts are usually tightened until a known tension is imposed on the bolt. Before loosening nuts, clean the bolt ends and measure the bolt length with a micrometer. Keep a record of individual bolt lengths, and elongate them the same amount when reassembling the valve, making sure the valve and bolt temperatures are the same as before. Always draw up body bolts evenly until the bonnet tip is true and square with the body. Most actual maintenance and repair operations are concerned with keeping seats and disks in leakproof condition. The specific methods used depend on the valve construction, condition of seats and disks, and equipment available. Modern methods of building up metallic surfaces (hard soldering, brazing, or welding) now offer means of salvaging valves with badly eroded seats and disks. Such building up makes repair possible without removal of the parent metal, greatly extending valve life.

Build up bronze seats with hard solder or bronze rod; alloy-steel seats for temperatures below 750°F can be repaired by brazing, which, although not as resistant as the original metal, will give good service. Building up alloy trim with supposedly identical metals can easily lead to trouble unless complete information is available as to the composition and hardness of the parent metal. Before building up disks on automatic valves for service on high pressures and temperatures, consult the manufacturer, because any major change in seat or disk contour may cause serious operating difficulties. If facilities for building up special trim metal are not available, the valve can be returned to the factory, although this practice may interfere with plant production. When it is absolutely necessary to buy new parts, buy seats and disks in pairs so that only a minimum amount of grinding is needed. Salvage all good parts of damaged valves for use as spare parts to rebuild other valves when they become damaged.

Seats and disks can be ground or refaced in many ways. The procedure for grinding seats and disks in union-bonnet globe valves is, perhaps, the simplest (Fig. 4.13). It is necessary only to pin the stem and disk together and use the bonnet as a guide for lapping the disk against the seat. Screwed bonnets cannot be used as guides for grinding; this job requires a grinding kit or a drill press. If a drill press is available, remove the stem from the bonnet and insert it in the drill chuck. Clamp the valve body in the drill-press vise; level on the top edge of the body-bonnet joint which parallels the seat surfaces. Pin the disk and stem, apply compound, and grind with the drill press at low speed and light spindle pressure.

Check Valves. Check-valve disks can be lapped against the seats; the disk usually contains a slot for a screwdriver to apply the turning movement. When lapping stainless iron disks and seats against each other, mix white lead and oil with the grinding compound to provide lubrication; otherwise the metal will drag and ruin the surfaces. The use of a grinding compound with small grain size reduces the tendency to gall. Grind with light strokes, lifting the disk frequently to a new position and cleaning the surfaces often. Stellited seats and disks can be ground in the same manner as that recommended for stainless iron, except that in some cases it may be necessary to use silicon-carbide grit.

Balanced-pressure double-seat regulating valves require particular care in grinding. Both seats must be established at the same time. Watch the disks to see which seat touches first; put compound on this disk with just a trace on the other. Grind until both seats are established uniformly. To produce a true seat under operating conditions, provide a steam connection to heat the body and stem to operating temperature, before grinding.

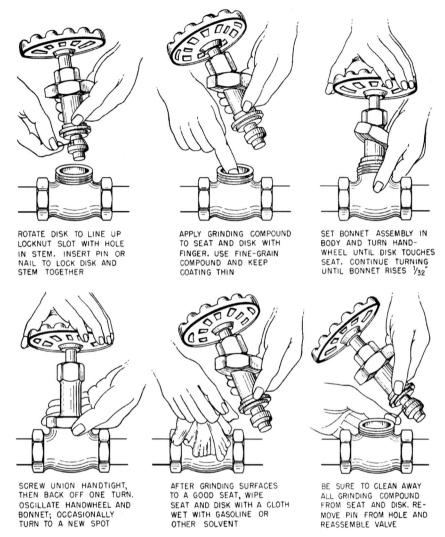

FIGURE 4.13 Regrinding globe-valve seats.

Machining. On globe valves which are badly worn, or where eroded areas have been built up, machining eliminates excessive lapping. Seats can be machined in a lathe, or ground with an emery-cloth-covered metal disk mounted in a drill press. A similar grinding disk, mounted on a motor-driven flexible shaft, can also be used on small valves. Repeated machining or grinding may reduce seat thickness dangerously. On integral-seat globe valves too small to insert soldering or brazing equipment, the seat opening can be reamed and threaded to take a renewable seat ring, which can be made up from material available in the plant shop. While it is desirable to replace the seat with the same metal, a different one can be used in an emergency. It is not advisable to use stainless iron of the same hardness for both seat and disk unless the valve is to handle oil.

Seat Rings. Repeated refacing of shoulder-design seat rings reduces the shoulder thickness to a point where metal contact pressure against the underside causes concaving of the outer or seating surface. The remedy lies in replacing the ring or increasing its thickness by welded overlays. Leakage past seat-ring threads must be repaired immediately; it can be done by welding and rethreading or by reaming and threading to take an oversize ring. If the damage is too great for these remedies, weld the ring solidly in the valve body.

Good service from composition globe-valve disks depends on care. When one side of such a disk is eroded, either machine it to a new face or reverse it and use the opposite side. Disk stocks can be stretched by using substitutes such as leather, scrap rubber belting, or lead machined smooth. Although not ideal, these materials will give satisfactory service until new disks arrive.

Refacing. Refacing of gate-valve seats and disks usually requires machining in a lathe or grinding in a drill press, although disks with surfaces not too badly worn can be refaced with a sanding wheel or by hand grinding, and seats can be ground with a hand brace. Lathe machining offers no major difficulties for parallel-seat gate bodies, but wedge-gate bodies require cumbersome holders. The drill-press grinding method is convenient.

Parallel-seat disks may be chucked readily for lathe refacing; a taper block to fit the faceplate helps with wedge disks. The drill-press grinding method can be used on wedges as well as seats. Special holding jigs, milled out to hold the wedge snug and level and keep it from turning, offer maximum convenience, but a jig is required for each valve size. One flat tapered plate, with clamps, will serve for many sizes and can be used on either a drill press or a lathe. Wedge disks can also be refaced by holding the surface against a motor-driven sanding or grinding disk. Keep the wedge centered on the disk and hold it with uniform pressure, or uneven grinding will ruin the taper. Hand grinding on an emery-cloth-covered flat surface proves satisfactory where only light scratches or machine marks need be removed.

Reseating Kits. Most of the methods described so far require that the valve be removed from the line. Valve reseating kits are available for refacing seat rings without removing the body from the line. This eliminates the need for breaking pipe joints, reduces possibilities of leaks, and saves time and labor.

Valves subjected to corrosive conditions soon become covered with "barnacles" which build up around seat and disk rings. If allowed to increase to any great extent, the barnacles soon creep over the seat edges and prevent the valve from closing tight. Cleaning the valve with a sandblast and applying a good paint or metal-spray coating on the areas around the seats greatly retards this growth. Timely repairs make valves last longer and save metal.

Hard-Facing. This is a useful maintenance tool for protecting steel valve parts against severe abrasive wear and wiredrawing action. Cobalt-chromium-tungsten alloys (stellite) retain their hardness at red heat, making them particularly suited to surfaces in friction and to parts exposed to high temperature. Before applying hard-facing alloy, prepare the part by grooving the surface to a depth of from $3/32$ to $1/8$ in (Fig. 4.14), leaving a ridge on each side. Round off all sharp corners, because sharp edges melt easily and interalloying between the base metal and the hard-facing alloy might occur. Such "dilution" results in decreased wear resistance and frequently causes blowholes. If the seat or part is too narrow for grooving, machine it flat and round off the edges. After grooving, clean the surface thoroughly, removing all dirt, scale, and grease.

If the part is small, say under 3 in. in diameter, and the welding flame is large enough to keep it red hot during welding, no furnace preheating is necessary; preliminary heating can be done with the

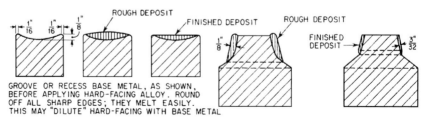

FIGURE 4.14 Correct preparation means good hard-facing.

torch. Larger parts require preheating with several torches or in a temporary furnace. Raise temperature slowly to 800 to 1200°F, or just under the point where steel begins to scale. This is a faint red heat, just visible in a dark room.

Maintain Heat. If possible deposit the alloy while the valve part is in the furnace; when this is impractical, reheat whenever the part cools to 600°F. Fixtures can be purchased or made that will rotate the work during the hard-facing. Such a device is not absolutely necessary; a helper can turn the part.

Hard-facing requires an excess acetylene flame (Fig. 4.15). Adjust the acetylene feather length, measured from the blowpipe tip, to three times the inner-cone length, measured from the welding-tip end. This flame prepares steel by melting an extremely thin surface layer, giving the steel a watery, glazed appearance called "sweating." This sweating, produced only by an excess acetylene flame, is necessary for successful hard-facing on steel.

Torch Angle. Hold the torch to direct the flame at a 30° to 60° angle to the surface, with the inner-cone tip about $1/8$ in from the steel. Keep this position until the steel under the flame suddenly glazes. The extent of sweating area varies with size of welding tip, but for a medium tip, steel will sweat about $1/4$ in around flame. Withdraw the torch slightly, and bring the welding rod between the inner cone of flame and the steel surface. The inner-cone tip should almost touch the rod, and the rod should lightly touch the sweating area. The melting rod forms a puddle on the steel. If the first few drops foam or bubble, or do not spread evenly, the steel is too cold and should be brought to the recommended temperature.

Some steels foam slightly when brought to sweating heat. When this occurs, do not deposit metal until foaming stops. If deposit is started before foaming is noticed, direct the torch at the foaming spot and agitate the molten metal with flame until foaming stops. Depositing metal during foaming causes blowholes and poor results. To spread molten alloy over the area, remove the rod from the flame and direct the flame into the puddle. Return the rod and melt off more alloy as required. Now direct the flame so that it plays partly on the edge of the puddle and partly on the adjoining steel surface. As steel approaches sweating heat, a puddle of hard-facing alloys spreads. As it spreads, bring the rod quickly into the flame again to add more metal as needed. If any dirt or scale appears on the steel or in the puddle, float it to the surface with the flame or dislodge it with the end of the welding rod.

With a little practice the right amount of alloy can be added to make the desired thickness. It is better to do this in one operation than to go back over the entire job to add another layer. During the operation, move the flame back to melt a thin surface layer of the deposit, to smooth out high spots as the work progresses. Do this quickly, without letting the front edge of the puddle solidify and without interrupting the steady forward travel of the work. After completing the deposit, use the flame to smooth out remaining rough surfaces. On this second pass, take care to melt only the hard-facing surface and not the base metal. This avoids bringing iron from the base metal to the hard-facing deposit.

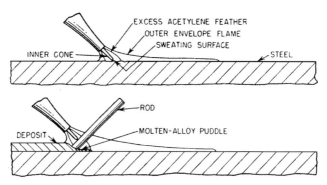

FIGURE 4.15 Sweating work surface with a torch.

Prevent Cracks. When the deposit reaches desired size and thickness, remove the flame slowly to prevent formation of shrinkage cracks and blowholes. If these occur, remelt the deposit and remove particles of scale from the pool. If holes still show, grind the alloy deposit down to steel, heat the area with flame, gradually, and deposit additional metal. Make sure that no slag, dirt, or scale is covered or embedded in the deposit to cause pinholes.

Slow cooling is absolutely essential to produce a deposit free from cracks and internal stresses. Parts showing a strong tendency to crack, such as large gate-valve wedges and seat rings, or parts on which the deposit is circular or large in area, should be returned to the preheating furnace while still hot from welding. Bring them slowly to a low red heat; then let them cool in the furnace. If a furnace is not available, place the part in dry powdered lime, ashes, or other insulating material, so that at least 2 in of material covers and protects every point of the part.

Some alloy steels used for valve trim require heat treatment in order to maintain corrosion resistance. When this is necessary, follow the steel manufacturer's instructions, with but one exception: never cool hard-faced parts by quenching in water or in an air blast. This will set up strains and cause cracks in hard-facing. If quenching is considered necessary, use only oil. After the hard-faced part has cooled, excess metal must be removed. This can be done by grinding or by machining with a tungsten-carbide tool.

Traps. With these, too, good operation and low-cost maintenance starts with proper installation. For example, hard-to-get-at traps will be neglected; easy access encourages regular inspection. If a trap is exposed to low temperatures, protect it against freezing. Install impulse or thermostatic drains in the trap or piping inlet to release all water when pressure is shut off and the condensate temperature falls.

Trap location directly affects operation. Wherever physically possible, install traps below equipment to be drained, so condensate can flow by gravity. Avoid U bends or water seals; they obstruct free flow and cause "steam binding." A slug of water flowing toward the trap immediately after discharge lies in the pocket until steam, remaining beyond the pocket and in the trap chamber, condenses. Where traps must be installed above equipment to be drained, install a check valve and water seal or U bend in the connective piping. The check prevents backflow and loss of prime and also prevents back drainage from a common-return header, or entrance of air when the unit is shut off. The water seal, although acting as an obstruction, serves as a sump for condensate collection, allowing it to be carried to the trap in slugs. Without the seal it would be possible for the bottom coil to remain partly filled with condensate.

Where traps discharge into common lines and connections do not require a check value in the inlet line, a check in the discharge prevents backflow from other units or drainage back to an idle unit when the trap has an individual overhead discharge line. Even when condensate flows vertically downward to a trap, a heavy rush of condensate chokes the line, prevents backward escape of trapped steam, and requires time for steam to escape or condense before water enters the trap. Although obstructing bends installed on steam lines only delay trap action, more serious trouble occurs on air lines unless the trap vents back to the vessel being drained. Such vent lines can be used only when the trap stands below the drained equipment.

If allowed to enter traps, pipe scale and sediment prevent tight seating and cause blow-through. Cleaning the pipe before installation fails to protect fully because temperature changes and flow loosen other particles. Install a strainer ahead of the trap, or if this is not possible, fabricate a "dirt leg" from pipe, to act as a catch pocket (Fig. 4.16). Uniform piping connections help in removing or exchanging traps for inspection and repair. Test valves and a tee in the discharge from each trap facilitate checking trap action. Installing a bypass around the trap permits adequate drainage and removal for repair when no spares are available.

Many traps include a valve, an operating device (bucket, float, bellows, and so on) and, if necessary, a linkage and bearings between valve and operator. Trap maintenance usually involves cleaning to remove foreign matter that might interfere with valve or linkage action, reseating valves when necessary, removing lost motion from linkage, and renewing the body gaskets. Moving parts located inside the body, in contact with moisture, offer lubrication difficulties and may show considerable wear. Excessive wear and lost motion, if allowed to continue, prevent positive operation and may reach the stage where the valve no longer seats.

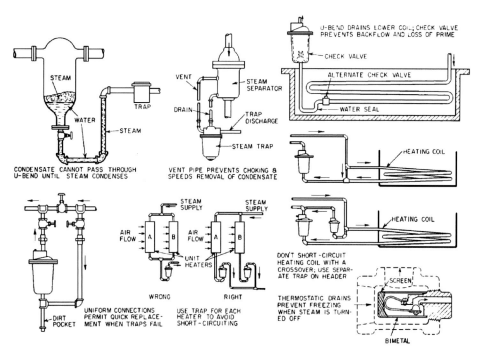

FIGURE 4.16 Pointers for trap installation.

Normal wear of some valves gradually enlarges the seat area at the point of disk contact. This increases the surface acted on by differential pressure across the valve, making it more difficult to operate. The area may increase to such an extent that the operating device is no longer powerful enough to move the valve. When this happens, the valve must be refaced to regain the original area. Whenever a trap fails to operate, and the reason is not readily apparent, observe the trap discharge by opening a test cock and breaking the discharge connection. Live steam usually indicates a leaking valve; it may be caused by the trap losing prime. Failure to discharge can be caused by a leaking bypass valve, inlet piping or trap obstructed by sediment and pipe scale, return line too small, or an obstructed outlet. Figure 4.17 shows a number of steps in trap care.

Valve leakage represents a common cause of trouble; worn seats are not so much to blame as are particles that prevent tight closure. Other trap troubles include rusting and sticking of the mechanism, lost motion in linkage, gasket and connection leaks, float leaks, and bent lever arms. Wear in levers and pins of a continuous-discharge trap causes intermittent operation. For most of these difficulties, simple mechanical remedies suffice, once the cause of the trouble is spotted. When valves do not seat properly yet appear to be in good condition, check linkage and stem length. Repeated regrinding may have shortened the stem. Make adjustments of stem length at or near full operating temperature.

There are numerous ways of checking trap operation. A slight temperature difference between inlet and outlet indicates a working trap; no difference indicates a leaking trap; a large change in temperature indicates no condensate passing. Intermittent-discharge traps produce a light clicking sound at each operation; constant-discharge traps can be checked with a listening rod or a stethoscope applied to the trap body. When visible discharge is not satisfactory evidence, passing the discharge into a vessel of water forms a positive test. Weigh the original quantity of water and check its temperature. After discharge, weigh the water again and check its temperature. Heat given up by trap

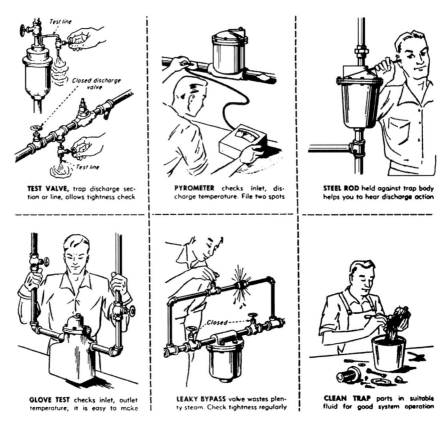

FIGURE 4.17 Trap maintenance tests.

discharge in falling to final temperature equals heat gained by original water quantity rising to same temperature. The chart in Fig. 4.18 simplifies computations, eliminating the use of steam tables. Table 4.2 gives trap troubleshooting data.

Pipe Supports. An ideal piping system would float like a layer of logs on a smooth pond, each part self-supported and imposing no stress on any other part. All elements in the system would hold their correct relative positions and alignment despite thermal expansion and contraction. Actual well-built systems approach this theoretical ideal by the intelligent use of anchors to fix certain points of the system, and expansion joints, supports, and guides to combine support with free movement for all the rest of the piping in the system. Without being an expert in the mathematical design of piping systems, the maintenance engineer should understand the duty to be expected of each part of the system under his care so he can check whether that job is actually being performed as it should.

First take the system as it stands, cold or at some fixed temperature. Anchors should securely lock the anchored points in the piping to heavy steelwork or other dependable footing. Between each pair of anchors should be an expansion joint or bend designed to absorb all possible movement from temperature differences. Pipe supports should be spaced closely enough to prevent undue sag in the span. All steam and air lines should be properly pitched for condensate drainage and checked to make certain that sag does not bring the center point of any span below its lower support, thus forming a condensation pocket leading to water hammer and other troubles.

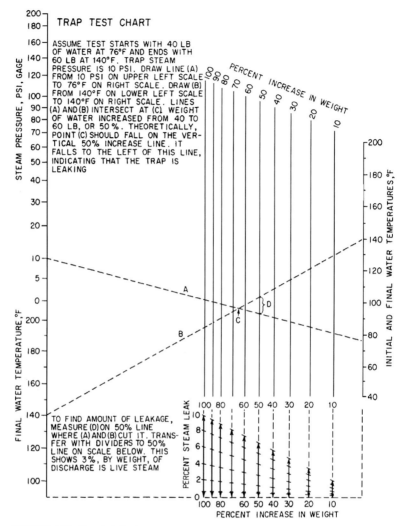

FIGURE 4.18 Trap test chart.

Check each hanger or other support to make sure that it carries its share of the load; that pipes track truly on fitted rollers or other guides; and that supports and their attachments are amply strong for the load and set to carry the weight, yet permit free pipe movement in the direction of expansion. To avoid trouble these conditions should be met whether the line is hot or cold. To make a single set of adjustments serve for both extremes of temperature is a most difficult job for both the designer and the maintenance engineer. One requirement is an understanding of thermal expansion, and here are the main points:

When an unrestrained steel body is uniformly heated, no forces whatever are set up, either internal or external. The body expands gently and proportionately in all directions. Thus, if a piping system is completely unrestrained except that the weight is carried at all points by fully "floating" supports, heating to a higher temperature will produce no forces whatever, nor any change in the

TABLE 4.2 Troubleshooting Chart for Steam Traps

Trouble	Possible cause and cure
Trap doesn't discharge	1. Steam pressure too high, pressure regulating out of order, boiler pressure gage reads low, steam pressure raised without altering or adjusting trap. On the last item consult trap maker. He can supply parts for higher pressure or tell you how to adjust trap. 2. Plugged strainer, valve, or fitting ahead of trap; clean. 3. Internal parts of trap plugged with dirt or scale; take trap apart and clean. Fit strainer ahead of trap. 4. Bypass open or leaking; close or repair. 5. Internal parts damaged or broken; dismantle trap, repair.
Trap won't shut off	1. Trap too small for load; figure condensate quantity to be handled and put in correct-size trap. 2. Defective mechanism holds trap open; repair. 3. Larger condensate load from (*a*) boiler foaming or priming, leaky steam coils, kettles or other units, or (*b*) greater process load; find cause of increased condensate flow and cure, or install larger trap. *Note:* Traps made to discharge continuously won't show these symptoms. Instead, the condensate line to trap overloads; water backs up.
Trap blows steam	1. Open or leaky bypass valve; close or repair. 2. Trap has lost prime; check for sudden or frequent drops in steam pressure. 3. Dirt or scale in trap; take apart and clean. 4. Inverted bucket trap too large, blows out seal; use smaller orifice or replace with smaller trap.
Trap capacity suddenly falls	1. Inlet pressure too low; raise to trap rating, fit larger trap, change pressure parts or setting. 2. Back pressure too high; look for plugged return line, traps blowing steam into return, open bypass or plugged vent in return line. 3. Back pressure too low; raise.
Condensate won't drain from system	1. System is air-bound; fit suitable vent or trap with larger air capacity to get rid of the air. 2. Steam pressure low; raise to the right value. 3. Condensate short-circuits; use a trap for each unit.
Not enough steam heat	1. Defective thermostatic elements in radiator traps; remove, test, and replace damaged elements. 2. Boiler priming; reduce boiler-water level. If boiler foams, check fires and feed with fresh water while blowing down boiler at quarter-minute intervals. 3. Scored or out-of-round valve seat in trap; grind seat or replace old trap body with new one. 4. Vacuum pump runs continuously; look for a cracked radiator, split return main, cracked pipe fitting, or a loose union connection. Or pump shaft's packing may leak. 5. Too much water hammer in system; check drip-trap size. Undersized drip traps can't handle all condensate formed during warm-up so hammering results. Fit larger trap if drip lines are clean and scale-free. Size for warm-up load, not for load with mains hot. 6. System run-down; older heating plants are sometimes troublesome because a large number of trap elements are defective. Easiest cure is replacement of all thermostatic elements in the radiators. This is low-cost, sure.

TABLE 4.2 Troubleshooting Chart for Steam Traps (*Continued*)

Trouble	Possible cause and cure
Traps freeze in winter	1. Discharge line has long horizontal run where water collects; make discharge line as short as possible and pitch away from trap. 2. Trap and piping not insulated; fit insulation to outdoor traps and piping connected to them.
Back flow in return line	1. Trap below return main doesn't have right fittings; use check valve and a water seal, or both, depending on what the trap maker recommends. 2. High-pressure traps discharge into a low-pressure return; flashing may cause high back pressure. Change piping to prevent return pressure from exceeding trap rating. 3. No cooling leg ahead of a thermostatic trap that drips a main; condensate may be too hot to allow trap to open right. Use a 4- to 6-ft cooling leg ahead of thermostatic traps on this service. Fit strainer in cooling leg to keep solids out of trap.

Courtesy of *Power* magazine.

proportions of the layout. All dimensions will be slightly increased, as if the new layout were a slightly enlarged photograph of the original. The coefficient of expansion for any grade of iron or steel is approximately 0.000007/°F. This simply means that the expansion is 7 parts per million per degree. For example, if the temperature of a steel pipe is raised by 400°F, its length will increase by about 400 × 7 = 2800 parts per million, or about 0.28 in per 100 in. The coefficient of expansion varies somewhat with the actual temperature; so expansions should be taken directly from Table 4.3 or similar data if accurate results are desired.

With either corrugated-metal or well-lubricated sliding expansion joints, the forces to be handled by the anchors are substantial, but much less than with the expansion pipe bends commonly used for the higher pressures. Figuring such bends is a rather mathematical branch of design engineering, but the maintenance engineer may be able to learn from the designer how much force the bend exerts for each inch of compression as the connected piping expands.

Expansion. When piping is heated, it moves straight out from the anchor by an amount proportional to the temperature rise and to the distance from the anchor. In an elaborate piping system

TABLE 4.3 Thermal Expansion of Steam Pipe

Temp, °F	Cast iron pipe	Steel pipe	Wrought iron pipe	Copper pipe
−20	0	0	0	0
0	0.127	0.145	0.152	0.204
100	0.787	0.898	0.939	1.338
200	1.495	1.691	1.778	2.500
300	2.233	2.519	2.630	3.665
400	3.008	3.375	3.521	4.870
500	3.847	4.296	4.477	6.110
600	4.725	5.247	5.455	7.388
700	5.629	6.229	6.481	8.676
800	6.587	7.250	7.508	9.992
900	7.579	8.313	8.639	11.360
1000	8.617	9.421	9.776	12.741

Condensed from Crocker, "Piping Handbook," 5th ed., McGraw-Hill Book Company, New York, 1967.

many complications can arise. If a long horizontal run without an expansion joint ends in a riser, expansion will push the riser out of plumb unless the top end is so suspended that it can move out too. If the weight of a riser, when cold, is equally distributed over several rigid hangers, one above the other, heating the line will expand the pipe and thereby unload all upper hangers, shifting the entire load to the bottom hanger. Many other similar effects will be observed, and generally can be sized up on the spot by the exercise of commonsense mechanics.

Here are some practical points to remember: Supports for a horizontal pipe (Fig. 4.19) far from an anchor must allow ample roll, slide, or swing in the direction of the pipe expansion. If temperature changes cause any piping (whether horizontal or vertical) to rise and fall, such sections must be carried on properly designed spring supports. For large movements in a vertical direction, the springs must be "soft." Such springs permit substantial up and down movement with only a moderate change in the support delivered to the pipe.

Maintenance of a piping system will naturally start with obvious ills—leaks, water hammer, swaying, and vibration. Leaks may be caused by joints improperly designed or made up, by expansion forces, or by improper support. Piping should always be supported on both sides of every large valve. Where leaks persist or recur despite good joint technique, check the alignment and condition of neighboring anchors, supports, and expansion joints to make sure the leak is not caused by external forces. Other common causes of leaks are pipe swaying and water hammer. Water hammer may result from reciprocating pumps or the too quick closing of valves. A common cause, traceable to poor maintenance, is undrained condensate caught in low points of the piping, back of globe valves, and so forth. Check the system to make sure that low points are raised and all other pockets drained. Sags caused by misalignment can usually be cured by simple hanger adjustments.

When all such obvious ills have been cured, the job shifts to preventive maintenance—regular routine inspection to make sure that anchors are holding and showing no signs of breaking or slipping; that walls and footings near anchors are not showing distress cracks; that sliding expansion joints are not leaking or sticking; that supports everywhere are in line with pipe and tracking true; that supporting rolls turn freely and support their share of the load whether the line is hot or cold; and that bolts, turnbuckles, and other stressed members give no sign of distress or possible early failure.

Pipe Insulation. A good maintenance program keeps insulation in perfect condition because necessary inspections and repairs save many times their cost in fuel. Here are practical pointers on maintaining and repairing covering of piping, valves, and fittings—chiefly those containing steam or hot water:

A good job of hot-pipe insulation will have the following characteristics: (1) efficient insulating material applied to economic thickness, (2) material able to stand ordinary handling, (3) inner layer able to stand pipe temperature, (4) insulation bound securely to pipe, (5) joints closely fitted and staggered (if double layer), (6) insulation well covered and painted, if necessary, and (7) complete waterproofing for outdoor or underground lines.

Materials. Widely used heat-insulating materials for plant piping include calcium silicate, laminated asbestos-felt, and various forms of mineral wool (both molded and felted), including glass wool, as well as other materials. The calcium covering is molded in sections or blocks. Both calcium and laminated-asbestos coverings may be used safely up to 600°F. Mineral-wool insulation can withstand temperatures higher than 1000°F.

For pipes above the temperature limit of calcium and asbestos, double-layer insulation is common. An outer layer of calcium or asbestos is protected from overheating by an inner layer of molded covering composed of calcined diatomaceous silica, asbestos fiber, and cementing materials. This high-temperature covering looks like calcium and has lower insulating efficiency and the ability to resist temperatures well above 1000°F. Both mineral wool and laminated asbestos are particularly suited for points subjected to heavy vibration or shock. Table 4.4 gives the insulation thickness commonly recommended for laminated-asbestos and calcium coverings. Note that the correct thickness increases with both pipe size and temperature.

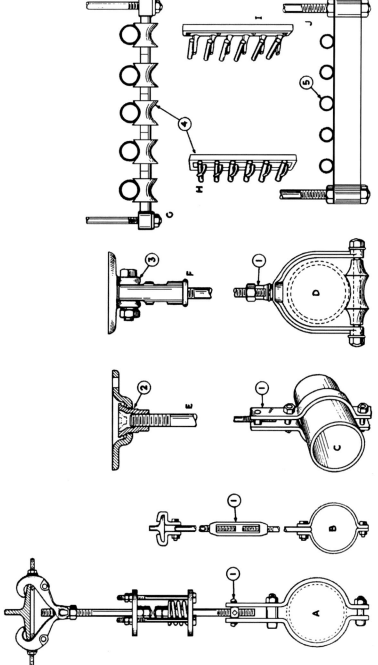

FIGURE 4.19 Typical pipe supports. Hangers generally permit vertical adjustment (1) to permit piping alignment to be maintained and to allow for proper division of load among supports. Pivoted hangers may permit universal movement (2) or may permit one-way movement as shown at (3). Multiple supports for banks of small pipe may be grooved for axial movement only (4) or may have a flat surface (5) to allow a certain amount of sidewise movement.

TABLE 4.4 Recommended Thicknesses of Pipe Covering

Plus inner layer of HT for high temperature

Pipe size, in.	Temperature of hot surface, °F														
	170	270	370	470	570	670		770		870		970		1070	
	Temperature difference, °F														
	100	200	300	400	500	600		700		800		900		1000	
	Calcium					HT	Cal	HT	Cal	HT	Cal	HT	Cal	HT	Cal
1	S	S	1½	2	2	2		2		2		2½		2½	
2	S	S	1½	2	2	1½	1½	1½	2	2	2	2½	2	2½	2½
3	S	S	1½	2	2	1½	1½	1½	2	2	2	2½	2	2½	2½
4	S	S	1½	2	2	1½	1½	1½	2	2	2	2½	2	2½	2½
5	S	S	2	DS	DS	1½	2	1½	2	2	2	2½	2	2½	2½
6	S	S	2	DS	DS	1½	2	1½	2	2	2	2½	2	2½	2½
8	S	S	2	DS	DS	1½	2	2	2	2	2½	2½	2½	3	2
10	S	S	2	DS	DS	1½	2	2	2	2	2½	2½	2½	3	2
12	S	S	2	DS	DS	1½	2	2	2	2	2½	2½	2½	3	2
14	S	S	2	DS	DS	1½	2	2	2	2	2½	2½	2½	3	2
16	S	S	2	DS	DS	1½	2	2	2	2	2½	2½	2½	3	2
18	S	S	2	DS	DS	1½	2	2	2	2	2½	2½	2½	3	2
20	S	S	2	DS	DS	1½	2	2	2	2	2½	2½	2½	3	2
Flat	1½	1½	2	2½	3	1½	2	2	2	2	2½	2½	2½	3	2

Note: HT = high-temperature covering, S = standard thick, DS = double standard.

	Laminated Asbestos		
Temperature of heated surface, °F	Pipe size, in.		
	Under 2	2 to 4	4½ and up
Up to 300	1	1½	1½
301 to 400	1½	1½	2
401 to 500	2	2	2½
501 to 600	2	2½	3

Figure 4.20 shows standard methods of covering piping and fittings. Such applications are fairly simple for any mechanic. Calcium and other molded coverings can be easily sawed to trim length and beveled with a knife at the ends that face the flanges. Coverings on bends, fittings, and other irregular surfaces can be built up by wiring on odds and ends of calcium blocks or pipe coverings and filling the remaining spaces with calcium cement. Since this cement is of exactly the same composition as the solid pieces, the mass sets as a homogeneous whole.

Insulation for cold lines may range all the way from a simple antisweat jacket for cold water to elaborate built-up, thick insulation for low-temperature refrigerating lines. Materials used include hair felt, cork, and mineral-wool felt. An essential characteristic of low-temperature insulation is complete sealing against the penetration of moisture that would otherwise destroy the insulation. Check periodically to make certain that such coverings remain hermetically sealed and completely free of internal water or ice. The application of refrigeration insulation is a specialty for experts, as is most heat insulation. Routine maintenance of warm-pipe insulation should include prompt repair of damaged surfaces, repainting and waterproofing, tightening bands and wires, and repairing torn

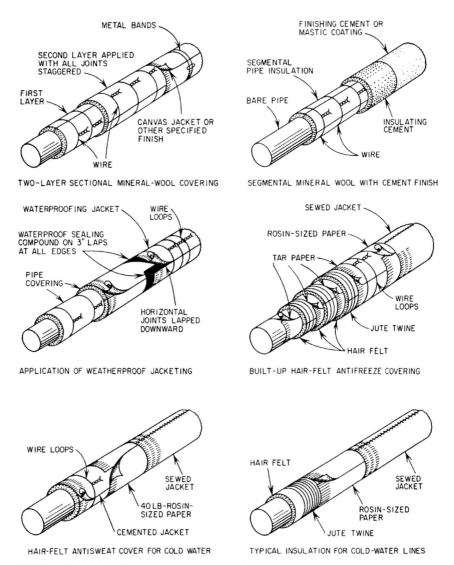

FIGURE 4.20 Methods for applying some typical pipe insulations.

canvas jackets. Look out for shrinkage, loosening, and the effect of moisture, fumes, and vibration. Make sure that steam temperatures have not been raised above the safe limit for the material used. Check carefully for steam and water leaks concealed by insulation.

Often a casual inspection will reveal bare flanges—either originally bare or left so by a recent replacement of the gasket. A single large bare flange can waste a ton of coal per year; so all such should be covered—preferably with replaceable covers.

Flanges. The preservation of flange covering is a major maintenance problem. Even supposedly removable covers are often broken in removal, particularly when an emergency requires quick access to the flange. Replacement of the insulation is a nuisance, too often delayed or omitted because the

9.82 CHEMICAL CORROSION CONTROL AND CLEANING

plant can run without it. It should be a standard rule to protect flange covers as far as possible and replace them promptly.

Silicate of soda (water glass) is a convenient and powerful adhesive for cementing tears in asbestos laminations or canvas jackets. Where calcium covering is broken, a monolithic repair can be made by wiring in a calcium "Dutchman" and filling voids with cement, which has the same composition as the block. See Fig. 4.21 for pointers on insulation maintenance.

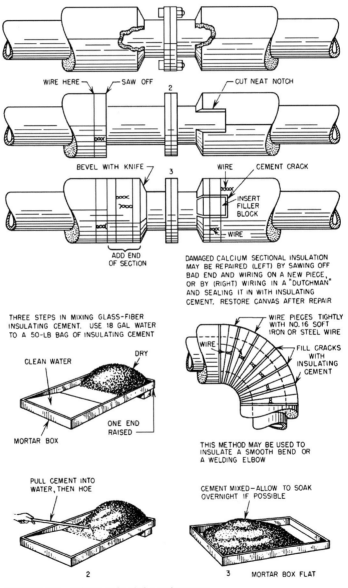

FIGURE 4.21 Pointers on insulation maintenance.

Maintenance Welding. Each element of piping maintenance has two aspects: (1) how to select and install to reduce maintenance and (2) how to maintain. In the case of welded joints the second is practically eliminated. If the joint is rightly selected and made, it should need no maintenance for the life of the plant. In general, maximum use of welding means minimum maintenance except where joints must be broken from time to time. Flange connections should be used at such points.

Most engineers are familiar with the standard lines of welding fittings—ells, tees, crosses, flange necks, and the like—also values with welding necks. No attempt will be made here to show how to make pipe welds. That information is available in concise booklets issued by the manufacturers and in standard codes for the making and testing of welds and for the qualification of welders. The ANSI code "Pressure Piping" covers welded joints from the specification aspect.

The elaborate sleeves, patches, and reinforcements of the early days of welding reflected the user's lack of confidence in the process. Modern practice favors the plain butt weld, made by officially qualified welders following standard welding and testing procedures. Results have been so good that welding is preferred today for the highest pressures and temperatures. In such lines flanges are used only where joints must be breakable.

Joints. In low-pressure work, welding has the disadvantage that cast-iron valves cannot be welded in, so one must either use the more expensive steel valves or install the cast-iron valves with flanged joints. For pipe wall up to $3/4$-in thick, the sides of the standard butt joint are beveled $37\frac{1}{2}°$, with a $1/16$-in land. For thicker pipe the bevel is U-shaped to avoid the need for excessive welding. Chill rings ensure full penetration of the weld without the formation of dangerous "icicles" in the pipe. Except for high-velocity steam, the extra cost of a flush chill (Fig. 4.22) is rarely warranted.

Flanges can be installed in a welded system in the three ways shown in Fig. 4.22. The lap-joint stub end, welded to the pipe and backed by a slip-on flange, gives the highest type of lap joint. A quick field connection can be improvised by joining a slip-on flange to the pipe end by two fillet welds as shown. Welding may be used also to seal the thread where a flange is screwed onto a pipe.

In high-pressure work it is best to build up welds in $1/8$-in layers, cleaning and inspecting after each layer. This catches defects before they are buried. Moreover, the heat of each layer improves the grain structure of the underlying layer.

It is considered unsafe to weld "carbon-moly" steel without preheating; so induction and resistance electric heaters have been designed to keep the pipe at 300 to 600°F throughout the welding. Similar heaters are used to stress-relieve high-strength welds at 1200°F.

Corrosion-Resistant Piping. In recent years a number of nonmetallic materials have been introduced for corrosion-resistant service—plastic, glass, and so on. Older corrosion-resistant materials include transite, stainless steel, cast iron, and coated piping. The coating may be plastic, bitumastic, rubber, and the like. The general maintenance procedures for piping and fittings made of special materials resemble those for iron and steel piping, except for differences resulting from the materials involved. For specific procedures in maintaining special piping materials, consult the manufacturer. There are many variations in procedures and methods which must be carefully observed. The corrosion questionnaire, Table 4.5, can be helpful as a reminder of factors to be considered.

Pipe Materials. For a comprehensive list of piping materials see the ANSI code "Pressure Piping." For maximum safety it is advisable to design all industrial piping in accordance with code requirements, even though local regulations may not make this mandatory. Correct design, with adequate provisions for maintenance, is the key to long trouble-free operation with minimum attention. Routine inspection of piping systems and their associated equipment ensures finding minor defects before they become major ones.

Flange Bolting. Table 4.6 gives data determined by the Crane Company in a number of extensive surveys of field-erected flanged joints. Experience shows these stresses are satisfactory for American Standard steel flanges. It is recommended that the initial bolt stress be about 45,000 psi. Figure 4.23 shows a micrometer for measuring bolt-stud elongation.

This company also recommends that the bolting in all flanged joints operating at temperatures over 500°F be pulled up after the first shutdown. At high temperatures, where creep may be expected

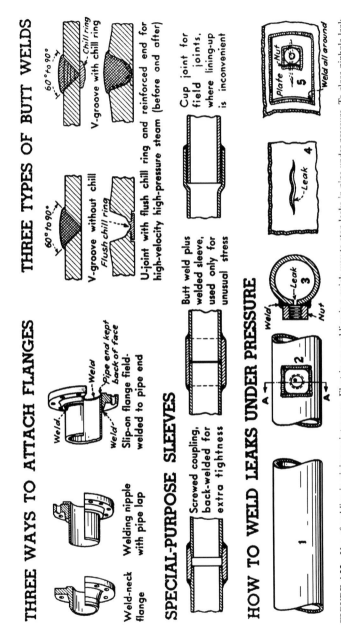

FIGURE 4.22 Use of welding in piping maintenance. Electric-arc welding is a quick way to stop leaks in pipe under pressure. To close pinhole leak (1), weld an ordinary square-head machine nut around leak (2, 3). This permits water, etc. to escape through the hole in the nut during welding, making it possible to complete the weld. To stop the leak, screw a bolt into the nut with sealing compound. A cracked or split pipe (4) can be closed in the same way. This kind of leak usually is too large to be covered by a nut; so a piece of plate stock is applied as a patch. Drill a small hole in the plate and shape the plate to the pipe. Next, weld a nut over the hole and weld the plate to the pipe (5). Then screw in a bolt or plug, using sealing compound.

TABLE 4.5 Corrosion Questionnaire

1. What is the name or composition of fluid to be handled?
2. What is the concentration (percentage strength, specific gravity, pH value, etc.)?
3. What is the operating temperature and pressure in pipelines?
4. If the fluid is not a water solution but a gas, organic fluid, etc., is water or water vapor apt to be present at any time or at any particular location? If so, explain.
5. If the fluid is a water solution, are substances other than water (abrasive solids, oil, etc.) present at any time or at any particular location? If so, explain.
6. Is there much opportunity for air-in leakage?
7. Is the flow through the system intermittent or continuous?
8. If intermittent, is the system ever drained entirely of fluid and allowed to dry?
9. Is the piping system washed or rinsed at regular intervals, and if so, with what materials?
10. Is a slight amount of corrosion objectionable from the standpoint of contamination or discoloration of product?
11. If so, what metals are particularly objectionable?
12. What materials are being used or proposed for piping, tanks, etc.?
13. Has any specific trouble been experienced with these materials?
14. What materials have been used for valves and fittings?
15. In general, what is the comparative life of the different materials which have been used?
16. What packings give the best service?

Courtesy of Crane Company.

to occur, it is recommended that bolting be pulled up at least once during the first 200 hr of service, regardless of whether the line has been shut down or not. Check bolt stress periodically during the life of the installation, as part of the routine maintenance program.

As a general rule, it is only necessary to check the elongation in two or three diametrically opposite bolt studs, using the average of the values obtained as the elongation for the remainder. If the

TABLE 4.6 Flange-bolting Data

Size of alloy-steel bolt-stud[*]	Average stress applied manually, psi[†]	Approximate torque to obtain stress, ft-lb[‡]	Elongation, in. per in. of effective length[§]
3/4	52,000	175	0.00173
7/8	48,000	255	0.00160
1	45,000	370	0.00150
1 1/8	42,500	500	0.00142
1 1/4	40,000	665	0.00133
1 3/8	38,000	860	0.00127
1 1/2	36,500	975	0.00122
1 5/8	35,000	1,285	0.00117
1 3/4	34,000	1,700	0.00113
1 7/8	33,000	2,200	0.00110
2	32,000	2,350	0.00107

[*]Coarse thread series, 1 in., and smaller; 8-pitch thread series, 1 1/8 in., and larger.
[†]Average stress applied by maintenance men in assembly, using a lever and wrench or by sledging.
[‡]Based upon well-lubricated threads.
[§]Based on a modulus of elasticity of 30,000,000. The effective length of bolt-stud equals the distance from center of one nut to center of the other.

Courtesy of Crane Company.

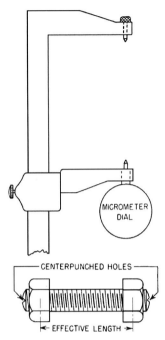

FIGURE 4.23 Micrometer for measuring bolt elongation.

bolt elongation is to be determined while the joint is in service, one bolt should be checked and pulled up before another is loosened. Measurements should be taken immediately after releasing the load before the bolt temperature decreases. The following procedure should be used in checking bolt elongation: (1) Determine the length of the bolt in the assembled joint. (2) Release the load on the bolt by loosening the nut, and remeasure the length. (3) Subtract the second reading from the first. (4) Divide this value by the effective length of the bolt. (Effective length equals the distance from the center of one nut to the center of the other.) If the residual elongation is less than 70 percent of the values given in Table 4.7, the bolts should be pulled up so that the final elongation approximates the figures shown.

Liquid Velocity. Table 4.7 gives condensed data from the experience of two hydraulic organizations with liquid velocities for pipes and ditches of many different types. It is useful in choosing or changing velocities to reduce pipe maintenance and deterioration.

Leak Detection. Where leaks are difficult or impossible to detect visually, radioactive isotopes are widely used to determine the general location of the opening in the pipe. The isotope is injected into the liquid stream and its flow traced by a Geiger counter. A leak is detected by a change in the radioactivity level. Since this method exposes personnel to some radioactivity and requires special equipment, it is usual practice to have a firm specializing in this type of testing do the work.

Acid Cleaning. Piping, valves, heat exchangers, process vessels, and so forth, are often cleaned today by means of acid. This is pumped through the piping and vessels under controlled conditions of concentration, velocity, temperature, and time. After the acid is removed from the system, a neutralizing agent is generally applied, followed by a flushing with clean water. Since a rather specialized knowledge and equipment setup is required for acid cleaning, it is usual practice to have this

TABLE 4.7 Water Velocities to Reduce Piping Maintenance

Conditions to be prevented	Type of flow	Pipe or ditch material	Velocity limits, ft per sec
Deposits of silt and mud	Vertical upward	All types of pipes and ditches	24 min
	45° upward		13 min
	9° upward		5 min
	3° upward		4 min
	Horizontal		3.3 min
	3° downward		2.6 min
	9° downward		Almost zero
Rust formation	All types	All corrosive pipe materials	26 min
Deterioration of pipe walls	All types	Concrete pipe, carrying pure water	20 max
		Concrete pipe, carrying sand-laden water	10 max
		Steel and cast-iron pipe	50 max
		Wood-stave pipe	40 max
Deterioration of ditch walls	All types	Fine-grained sand	0.6 max
		Coarse sand	1.2 max
		Small stones	2.4 max
		Coarse stones	4.0 max
		Rock	25 max
		Concrete carrying sandy water	10 max
		Concrete ditch, carrying pure water	20 max
		Sandy loam, 40% clay	1.8 max
		Loamy soil, 65% clay	3.0 max
		Clay loam, 85% clay	4.8 max
		Soil, 95% clay	6.2 max
		Clay	7.3 max
Formation of ice in ditch or race	All types	All types of ditches or races	5.0 min

These velocities realize special conditions.

work done by a firm having the required experience and machinery. Since portable equipment is suitable, acid cleaning can be done in the plant during a routine shutdown. Correctly applied acid cleaning can often reduce piping-system maintenance costs considerably.

Pipe Identification. Colored bands in accordance with the ANSI "Scheme for Identification of Piping Systems" are valuable in plant maintenance operations. They permit more ready identification of the piping, eliminating errors.

Relief Devices. Safety and relief valves must be kept in good working order at all times to ensure safe operation and prevent the loss of valuable liquids. Routine tests of relieving capacity should be part of the piping maintenance program. Safety heads (Fig. 4.24A) must be replaced immediately after function (Fig. 4.24B).

Plastic Piping. Metal piping is being replaced in many applications by plastic piping of various types. Plastic piping has internal, external, and electrolyte corrosion resistance. It is easy to install, is not subject to caking of the fluid on the walls, and cannot be pitted by tuberculation. Since plastic pipe does not corrode, it is ideal for high-purity systems.

Plastic pipe, however, has both temperature and pressure limitations. The usual maximum allowable operating temperature is in the 200°F range and depends on the material of which the pipe is made. Certain glass-fiber-reinforced pipe can withstand temperatures up to about 250°F; asbestos-reinforced pipe can operate at temperatures in the 400°F range. New plastic materials are rapidly

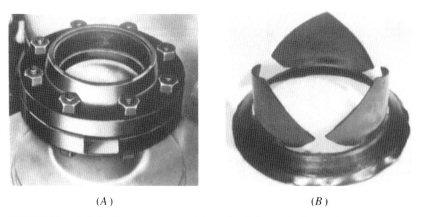

FIGURE 4.24 *A.* Safety heads on compressed-air bottles. *B.* Ruptured safety head.

being introduced—some allow operation at over 500°F. In general, though, plastic piping is usually confined to operating temperatures less than 200°F.

Allowable working pressure of plastic pipe decreases as operating temperature increases. Most plastic-pipe manufacturers recommended that the working pressure not exceed 20 percent of the bursting pressure. Typical bursting pressures range from a high of about 1500 psi at 70°F for smaller-diameter pipe to a low of about 100 psi for 6-in pipe at 70°F. Plastic pipe is more expensive than galvanized metal pipe, but the many advantages of plastic often make the extra investment worthwhile.

Valves, elbows, tees, flanges, and other pipe fittings are also made of plastic. Figure 4.25 shows a typical molded plastic valve. These parts have the same advantages as plastic piping and are subject to the same general pressure and temperature limitations.

Maintenance of Plastic Piping.   Relatively little maintenance is required for plastic piping. Since the external surface of plastic piping resists corrosion, painting is never required for protection, though the surface may be painted for purposes of appearance.

FIGURE 4.25 Polyvinyl chloride molded plastic Y-type globe valve for use in plastic piping systems.

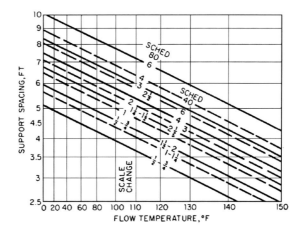

FIGURE 4.26 Recommended support spacing for uninsulated polyvinyl chloride piping. Chart is for plastic pipe carrying fluids of up to 1.35 gravity. For insulated piping reduce spans by 30 percent.

Inspect plastic piping regularly for leaks, sagging, or out-of-roundness. Any of these conditions can be caused by excessive operating temperatures or pressures. Repair leaks using the solvent cement recommended by the pipe manufacturer. Be certain to drain pipe and dry it thoroughly before applying cement. Brush the cement carefully over the entire surface being repaired. Apply liberal amounts of the cement to ensure complete repair of the leaking joint. At least a 10-hr drying time is required before the pipe can be subjected to operating pressure and temperature. Where possible, allow a 48-hr drying time for development of full strength in the pipe.

Leaks caused by sagging or out-of-round pipes cannot be repaired without replacing the affected section of pipe, unless the damage is only minor. Sagging is caused by too few supports. Figure 4.26 shows the usual support spacing for polyvinyl chloride plastic piping recommended by one pipe manufacturer. Note that the recommended spacing is a function of flow temperature, pipe thickness (schedule number), and pipe diameter. Where pipe sagging causes leaks, check the distances between supports, the flow temperature, and the ambient temperature at the pipe. Fit more support if the distance between the existing supports exceeds that recommended in Fig. 4.26. Continuous supports are sometimes used for short spans in place of spaced supports. All supports should allow the pipe to expand axially without damage to the exterior surface of the pipe. Therefore, the support must not clamp the pipe tightly.

Where sagging is caused by excessive flow or ambient temperature, reduce the fluid or air temperature. Excessive temperature can cause sagging even when the pipe has properly spaced supports.

Out-of-roundness usually results from excessive operating temperature. Reduce the fluid or air temperature before installing replacement sections for the out-of-round pipe.

Since plastic piping is particularly sensitive to shock resulting from being struck by a hard object, protect the piping in areas of heavy traffic. Use a guard railing or low wall to protect the piping from fork-lift trucks and other vehicles used in industry. Plastic pipe used in outdoor service is often buried under ground to protect it from moving vehicles and the ultraviolet rays of sunlight. Direct sunlight will shorten the life of plastic piping made of certain materials.

CHAPTER 5
SCAFFOLDS AND LADDERS

Colin P. Bennett
Scaffolding Consultant

Major strides have been made in recent years in the availability of different types and wider choices of scaffolds and ladders for above-the-ground work. Today's plant manager has more varied and versatile equipment available than ever before; consequently, without realistic guidance, it is easier to make the wrong choice of equipment.

In general, the wise plant maintenance manager should insist upon having a variety of access equipment at his disposal; the days have passed when a plant could "make do" with the old weather-beaten stepladder and one extension ladder in poor condition. Certain types of ladders and scaffolds are especially suited to certain specific tasks, but they can often be most unsuitable for work of another nature that requires more specifically suited scaffolding. Consideration must always be given to the type of equipment which will produce the highest specific work output and at the same time accomplish this in a safe manner with safe equipment. Sturdy work platforms of high stability and ample size result in freely moving, confident workers; the solidity of the work platform is a very important psychological factor and one which is commonly overlooked. Many people are unable to climb more than a few feet above the ground without feeling the necessity to hang onto something tightly with at least one hand; consequently, a person assigned to work on a ladder or at a precarious perch may produce only a small fraction of work output as when doing the same work at ground level.

In considering the type of equipment for specific tasks, the following typical questions should be analyzed:

1. How high is the work?
2. Is it spot work or is it continuous in a horizontal- or vertical-pass direction?
3. Is it work which can be most easily done from top to bottom over a length of 10, 20, or 30 ft?
4. Is it a large area which will require a combination of frequent horizontal and vertical passes?
5. Is the area below the work suitable for support from the ground?
6. Is there a wall or an unobstructed space to support a ladder or scaffold?
7. Must people or conveyances pass unobstructed beneath the work?
8. Depending on question 2, how frequently is it necessary to move the work support?
9. How many men and how much equipment must be supported on the scaffold?

The following is a list of scaffolds and ladders followed by basic descriptions to enable you to determine the best equipment to use for individual circumstances:

Stepladders and extension ladders—wood, aluminum, and fiberglass

Welded aluminum folding or sectional scaffolding (rolling towers)

Welded sectional steel scaffolds and rolling towers

Safety swinging scaffolds, one- and two-point suspension

Tube and coupler scaffolds (steel and aluminum)

Modular systems scaffolds

Special-design scaffolds

STEPLADDERS AND EXTENSION LADDERS

Stepladders and extension ladders are preferably used for spot work at relatively low heights with no obstructions. The size and weight of a ladder are important; there is room for only one man on a ladder, but if two men are required to carry and lift the ladder into position, then one man is working only a small part of the time.

Choice of Materials for Ladders—Wood, Metal, Fiberglass. **Wood** is preferred for regular use in relatively low-height operations and is the workhorse of the industry. However, a wood ladder (or any other type for that matter) can be abused through overfamiliarity. Frequent inspection, properly documented, is essential. Wood is often abused because of its ability to withstand high excessive loads for short periods of time, such as walking or working on it while it is in a *flat* position. Wood is organic and therefore biodegradable; therefore, wood-ladder life can be extended by water-repellent and anti-fungicide treatments such as pentachlorophenol. Ladders should never be painted or coated with an opaque substance; this would cover over any defects which may develop in use. Clear varnish or shellac are sometimes used but are of little practical benefit; in fact they may make a ladder more slippery when wet and, under some circumstances, can enhance the growth of bacterial fungi (decay).

In recent years long-length, ladder-grade lumber such as hemlock has become difficult, if not impossible, to obtain with any assurance of regularity and cost control. Far-sighted manufacturers looked for alternative materials. One developed a glue-laminated construction in which siderails are constructed using four or five full-width strips glued together to make the required rail size. They are manufactured, glued, and cured using identical quality control techniques as for glue-laminated structural beams; in fact they meet all test requirements of the American Institute of Timber Construction (AITC).

In general, laminated woods equal the strength of the solid woods from which the strips are cut and surpass the performance of solid wood with respect to durability, longevity, and consistency of strength. They meet all requirements of Underwriters' Laboratories, ANSI, and OSHA as well as the private specifications of many larger users, such as the power generating and telephone industries for transmission line maintenance.

Minor damage or imperfections which develop in solid wood can be propagated into long cracks across or through the piece, whereas in a laminated ladder any such local damage is retained within the laminate in which the damage occurs and a minor crack does not continue past the adjacent gluelines; this means that the piece retains most of its structural integrity and has a longer useful life.

In the foreseeable future, ladder-grade wood may be available in relatively short lengths; the manufacturing technology is already developed so that shorter individual strips can be finger-jointed together. Laminated members are used for siderails, steps, and stepladder tops.

Some experiments have also been made using plywood-type construction with very thin plies. While adequate for strength, the industrial techniques are such that currently only various fir species are used which have a weight disadvantage over the lighter hemlock or other softwoods traditionally used for ladder construction.

Aluminum ladders are preferred by many for their lighter weight. However, Underwriters' Laboratories, Inc., issue the following "CAUTION" printed on their "listing" label:

> ELECTRICAL SHOCK HAZARD—METAL LADDERS SHOULD NOT BE USED WHERE CONTACT MAY BE MADE WITH ELECTRICAL CIRCUITS. REFER TO INSTRUCTION LABEL.

Fiberglass ladders are preferred for use in workplaces involving electrical hazards, are about the same weight as wood ladders, and have the additional advantage of insulation properties. Fiberglass is less subject to the effects of abuse (resulting from being dropped or otherwise mishandled) than either wood or metal ladders, since the fiberglass siderails have superb recovery from bending or distortion. For proper maintenance, the fiberglass ladder should be coated with a hard floor wax or car wax to reduce the tendency of the glass fibers to "bloom" at areas of friction and scraping.

Some things are so simple, both in the way they are made and in the ways they are used, that the scientific principles upon which they are based are barely visible. With all the simplicity it has constantly retained, the ladder as we know it in its best form today represents a high degree of engineering skill, scientific accuracy, and most important, dependable safety.

Vital differences in ladders may not be detected except by the expert; hence industry finds it profitable to seek expert ladder advice. Proper weight, exact balance, scientific proportions, dependable quality of materials, character of workmanship, and of utmost importance, adaptability of a certain type of ladder to the particular kind of service for which it is intended—these factors are essential in the modern ladder even though they may not always be visible to the uninitiated.

A program of ladder upkeep and care should be as much a part of any company's safety program as is its maintenance of other plant and machinery.

Regulations and Standards. The design, manufacture, labeling, and use of ladders are nowadays quite rigidly controlled by both governmental regulations and national consensus and laboratory standards. Please refer to the following sources for specific details of use, design, manufacturing, and testing.

OSHA—Federal Regulations. Some states have other reference identification.

1. Occupational Safety and Health Standards for Industry
    CFR Title 29 Part 1910.25 "Portable Wood Ladders"
                  Part 1910.26 "Portable Metal Ladders"
                  Part 1910.27 "Fixed Ladders"
    These regulations do not cross-reference other standards.

2. Occupational Safety and Health Regulations for Construction
    CFR Title 29 Part 1926.450 "Ladders"
    These regulations cross-reference ANSI Standards listed below.

American National Standards Institute (ANSI).

1. ANSI A14.1—1982, "Safety Requirements for Portable Wood Ladders"
2. ANSI A14.2—1982, "Safety Requirements for Portable Metal Ladders"
3. ANSI A14.3—1982, "Safety Requirements for Fixed Ladders"
4. ANSI A14.5—1982, "Safety Requirement for Portable Reinforced Plastic Ladders"

These standards include highly instructive selection and care, and they use information plus various appendices which can be useful in the documentation of industrial accidents.

Underwriters' Laboratories, Inc.

1. UL 112-1980 Standard for Safety, "Portable Wood Ladders"
2. UL 184-1980 Standard for Safety, "Portable Metal Ladders"

The ANSI and UL standards are essentially similar in terms of design, manufacture, testing, and labeling, but their functions are quite different. Except as specified in OSHA 1926.450, ANSI standards are voluntary and in fact give manufacturers the option of stating on ladder products their compliance with ANSI standards. ANSI conducts no tests nor issues any certification of compliance.

Underwriters' Laboratories, Inc., serves industry in more specific ways. First they function as testing laboratories. Products are submitted to them, tested, and certified to the manufacturer as to test performance. Products are then listed under a certification number as being in compliance therewith. UL also has a follow-up service which involves periodic visits by them to the manufacturing location to inspect for continuing quality compliance and the application of serially numbered seals placed on the products.

Consumer Product Safety Commission. In 1971 the Consumer Product Safety Commission required all extension ladders to be marked with a statement of maximum extended length. Until that time ladder manufacturers had traditionally referred to their sizes as the sum of the nominal lengths of the sections, omitting the amount of overlap between sections. This requirement is now included in the above standards.

Ladder Duty Ratings. The sizes, raw materials, and hardware specifications for ladder components are keyed to some extent to the severity of intended use, but retain some historical references to load rating in terms of capacity. These capacities, stated below, form the basis for laboratory testing of ladders and are not intended to influence *purchase* of ladders in terms of users' weight.

Type IA is an extra-heavy-duty industrial ladder, 300-lb duty rating

Type I is a heavy-duty industrial ladder, 250-lb duty rating

Type II is a medium-duty commercial ladder, 225-lb duty rating

Type III is a light-duty household ladder, 200-lb duty rating

Type IA and I are the obvious choice for general industrial use by utilities, contractors, and the like. Type II are for light maintenance such as occasional office use. Type III are as stated and are the type generally sold in supermarkets and department stores; they are not recommended for industrial or commercial use.

LADDER GROUPS

A. Rung Ladders

Single Ladders. These are for work at constant heights and are limited in length to 30 ft for Types IA and I, 20 ft for Type II, and 14 ft for Type III. For rough usage at construction workplaces the Type IA single ladder is preferred by masons and builders.

Extension Ladders. These are a maximum of 60 ft (nominal length) and are comprised of two overlapping, sliding sections extendable for height adjustment. The amount of overlap depends on the total nominal ladder length. They must have a positive stop to prevent overextension. Extension ladder locks should allow automatic ladder height adjustment by one man from the ground in either direction of extension or retraction by means of a pulley rope, without need for additional manipulation. Siderail size, width, and other dimensions are all variable based upon the quality of wood used and extended length: Minimum sizes based on lumber types are specified in the standards. In earlier days, three-section extension ladders were available because of their ability to retract shorter than two-section ladders. However, three-section ladders must have two section-overlaps which drastically decreases the available working length. A 40-ft, two-section ladder will extend to approximately 36 ft. A 40-ft, three-section ladder will extend only to 32 ft; economy dictates that a 36-ft, two-section ladder would reach the same height at a lower cost. If required, three-sectional ladders are now only available as "special-purpose" ladders. A typical, good-quality extension ladder is shown in Fig. 5.1.

Sectional Ladders. These are often known as window cleaners' ladders, but in fact are used widely by utilities for manhole access. They consist of separate sections which are overlapped in the same plane and joined together at the protruding ends of the top and bottom rungs of each section by means of slots in the siderails. They are generally 6 ft long, although some utilities use 8- and 10-ft lengths. They are of two basic types: continuous taper (each section progressively tapering toward the top) and interchangeable (each section of constant taper). The maximum assembled length of sectional ladders is restricted to 31 ft.

FIGURE 5.1 Extension ladder meeting ANSI Safety Standard A14.1 and listed by UL, Inc.

In general all rung ladders must be blind-bored to allow a minimum bearing of the end tenons in the siderails of $7/8$ in, or must be through-bored. Rungs must be prevented from rotating, generally by nailing through the tenons. The quantity of pressed steel rung braces present in a ladder are indicative of quality since these assist the ladder to resist "racking" from its original rectangular configuration due to loosening of the tenons. All separate ladders should be equipped with safety feet suitable for the intended use.

B. Stepladders. These are available in a wide variety of styles, weights, and features for use by various trades having different ladder needs. The ladders have steps instead of rungs which should be level and horizontal with the ladder in the open position. When open, they must be stable, self-supporting, and free from racking, have safety spreaders which hold the ladder firmly in an open position and with a shield over the riveted spreader joint which will prevent injury to hands when opening or closing the ladder. Trussed, threaded rods are placed below the steps to add strength and assist stability; top-quality stepladders also incorporate knee braces at the sides of the steps. Stepladders with built-in, folding shelves for pails and other utensils are the favorite of many trades. Many styles are available to satisfy differing needs for various types of work.

Top-quality stepladders have rungs between the two rear legs; lower-quality ones use flat slats. Two styles are worthy of special mention. The platform stepladder gives a top platform on which a person can stand, with a "guard" on three sides as shown in Fig. 5.2. This gives a firm footing and allows both hands to be free. The top "back rest" generally is made with holes in it to insert tools to prevent them from falling. The other is a double-front stepladder which has steps on both front and rear sides. As a general rule the high strength and sturdy construction of the Type IA ladders far offset their slightly higher cost by extended service life.

FIGURE 5.2 Platform stepladder. Preferred because it permits worker to have both hands free.

C. Special-Purpose Ladders. These are most commonly made for the specific needs of an industry or plant but in general are a variation of either a rung ladder or a stepladder. In all cases they should comply with the appropriate ANSI and UL Standards for the type and duty rating.

Precautionary Measures. Where special groups are using ladders, such as plumbers, electricians, and millwrights, the ladders should be properly identified, with the members of each craft held responsible for their particular ladders. The use of just any ladder the worker comes across is likely to lead to costly disaster. Instructing workers as to ladder usage is extremely important.

As mentioned previously fiberglass (FRP) ladders are preferred by many industries having a need for their insulation properties. However, the user should be aware of two attendant hazards in their use which may not casually be apparent. First, a soaking-wet fiberglass ladder will conduct electricity through a continuous moisture path. Second, steps and rear rungs are generally made of aluminum riveted to the FRP siderails; electricity can "jump" along the plastic surface between the rivets of adjacent steps or rungs since this distance may be only 8 or 9 in. Therefore, fiberglass ladders do not give *absolute* assurance of insulation under all conditions, and possibly a well-made, dry wood ladder may be equally as effective for many users.

LADDER SAFETY RULES

LADDER SAFETY RULES #402W FOR WOOD
SINGLE & EXTENSION LADDERS

FOLLOW THESE INSTRUCTIONS FOR YOUR SAFETY
AND THAT OF OTHERS

1. Inspect ladder carefully on receipt and before EACH use. Test all working parts for proper attachment and operation. Ladders found to be damaged, defective or with missing parts should be withdrawn from use and marked "DO NOT USE." Never use a ladder known to have been dropped until it has been carefully reinspected for damage of any nature.
2. Install and use this ladder in compliance with the Regulations of the Occupational Safety and Health Act—1970, and with all other applicable governmental regulations, codes and ordinances. Ladder usage must be restricted to the purpose for which the ladder is designed.
3. Keep nuts, bolts, and other fastenings tight. Oil moving metal parts frequently. Obtain replacement parts from the manufacturer. Do not allow makeshift repairs. Replace frayed or badly worn rope promptly. Keep rungs free of grease, oil, paint, snow, ice, or other slippery substances.
4. Ladders must stand on a firm level surface. Always use safety feet and other suitable precautions if ladder is to be used on a slippery surface. Never use an unstable ladder. Ladders should not be placed on temporary supports to increase the working length or adjust for uneven surfaces.
5. Face ladder when ascending or descending. Always place ladder close enough to work to avoid dangerous over-reaching. Keep work centered between side rails. Side loading should be avoided.
6. Sectional (Window Cleaners) Ladders must be assembled in proper sequence with a base section at the bottom, and equipped with safety feet where slippery conditions exist. Maximum assembled length must not exceed 21 ft for Standard Sectional Ladders or 31 ft for Heavy Duty. Do not intermingle Sectional Ladders of different types or strength. All Safety Rules printed herein apply to Sectional Ladders except nos. 10 and 11.
7. Never place ladders in front of doors or openings unless appropriate precautions are taken.
8. Before installing an extension or single ladder always insure that working length of ladder will reach support height required. It should be lashed or otherwise secured at top to prevent slipping and should extend at least 3 ft above a roof or other elevated platform. Never stand on top three rungs of an extension or single ladder.
9. Install a single or extension ladder so that the horizontal distance of that ladder foot from the top support is $1/4$ of the effective extended length of the ladder ($75 1/2°$ angle). Always insure that both siderails are fully supported top and bottom. Never support ladder by top rung.
10. Overlap extension ladder sections by at least: 3' each overlap for total nominal lengths up to and including 36'; 4' each overlap for total nominal lengths over 36', up to and including 48'; 5' each overlap for total nominal lengths over 48', up to and including 60'. At overlaps, fly or upper sections must always be outermost so as to rest on lower section(s).
11. Be sure all locks on extension ladders are securely hooked over rungs before climbing. Make adjustments of extension ladder heights only when standing at the base of the ladder. Never extend a ladder while standing on it. For three-section ladders, always fully extend top section first. Ladders must not be tied or fastened together to provide longer sections other than manufactured for.
12. Water conducts electricity. Do not use wet ladders where direct contact with a live power source is possible. Use extreme caution around electrical wires, services, and equipment. Provide for temporary insulation of any exposed electrical conductors near place of work.
13. A ladder is intended to carry only one person at a time. Do not overload. For support of two persons special ladders are available. NEVER use a ladder in a horizontal position, never sit on a ladder when it is on edge and never use a ladder in a flat position as a scaffold plank.
14. Store ladders on edge in such a manner to provide easy access for inspection. Provide sufficient supports to prevent sagging. Never use ladders after prolonged immersion in water, or exposure to fire, chemicals, fumes, or other conditions that could affect their strength.
15. Only premium grade extension ladders should be used in conjunction with ladder jacks and stages or planks.

16. For further instruction on the care of Wood Single and Extension Ladders refer to the American National Standard, Safety Code for Portable Wood Ladders, ANSI A14.1—1981.

LADDER SAFETY RULES #402A FOR ALUMINUM
SINGLE & EXTENSION LADDERS

FOLLOW THESE INSTRUCTIONS FOR YOUR SAFETY
AND THAT OF OTHERS

1. Inspect ladder carefully on receipt and before EACH use. Test all working parts for proper attachment and operation. Ladders found to be damaged, defective or with missing parts should be withdrawn from use and marked "DO NOT USE." Never use a ladder which has been dropped until it has been carefully reinspected for damage of any nature.
2. Install and use this ladder in compliance with the Regulations of the Occupational Safety and Health Act—1970, and with all other applicable governmental regulations, codes, and ordinances. Ladder usage must be restricted to the purpose for which the ladder is designed.
3. Keep nuts, bolts, and other fastenings tight. Oil moving metal parts frequently. Obtain replacement parts from manufacturer. Do not allow makeshift repairs. Never straighten or use a bent ladder. Replace frayed or badly worn rope promptly.
4. Ladders must stand on a firm, level surface. Always use safety feet. If ladder is to be used on a slippery surface take additional precautions. Never use an unstable ladder. Ladders should not be placed on temporary supports to increase the working length or adjust for uneven surfaces.
5. Face ladder when ascending or descending. Always place ladder close enough to work to avoid dangerous over-reaching. Keep work centered between siderails. Side loading should be avoided.
6. Keep rungs free of grease, oil, paint, snow, ice, or other slippery substances.
7. Never place ladders in front of doors or openings unless appropriate precautions are taken.
8. Before installing an extension or single ladder always insure that working length of ladder will reach support height required. It should be lashed or otherwise secured at top to prevent slipping and should extend at least 3 feet above a roof or other elevated platform. Never stand on top 3 rungs of an extension or single ladder.
9. Install a single or extension ladder so that the horizontal distance of that ladder foot from the top support is $1/4$ of the effective working length of the ladder ($75 1/2 °$ angle). Always insure that both siderails are fully supported top and bottom. Never support ladder by top rung.
10. Overlap extension ladder sections by at least:

    3' each overlap for total nominal lengths up to and including 36'.

    4' each overlap for total nominal lengths over 36' up to and including 48'.

    5' each overlap for total nominal lengths over 48', up to and including 60'.

    At overlaps, fly or upper sections must always be outermost so as to rest on lower section(s).
11. Be sure all locks on extension ladders are securely hooked over rungs before climbing. Make adjustments of extension ladder heights only when standing at the base of the ladder. Never extend a ladder while standing on it. For three-section ladders, always fully extend top section first. Ladders must not be tied or fastened together to provide longer sections than manufactured for.
12. Metal and water conduct electricity. Do not use metal, metal reinforced or wet ladders where direct contact with a live power source is possible. Use extreme caution around electrical wires, services, and equipment. Provide for temporary insulation of any exposed electrical conductors near place of work.
13. A ladder is intended to carry only one person at a time. Do not overload. For support of two persons special ladders are available. NEVER use a ladder in a horizontal position, never sit on a ladder when it is on edge and never use a ladder in a flat position as a scaffold plank.

14. Store ladders on edge in such a manner to provide easy access for inspection. Provide sufficient supports to prevent sagging. Never use ladders after exposure to fire, chemicals, fumes, or other conditions which could affect their strength.
15. Portable ladders are designed as one-man working ladders, including any material supported by the ladder. There are three classifications:

    Type I—Heavy Duty for users requiring not more than a 250-lb load capacity for maintenance, construction or heavy duty work.

    Type II—Medium Duty for users requiring not more than a 225-lb load capacity for painting, or other medium duty work.

    Type III—Light Duty for users requiring not more than a 200-lb load capacity for service requirements such as general household use. Not for use with stages or planks.

16. Only Type I and Type II extension ladders should be used in conjunction with ladder jacks and stages or planks.
17. For further instructions on the care of Aluminum Single and Extension Ladders refer to the American National Standard, Safety Code for Portable Metal Ladders, ANSI A14.2—1982.

LADDER SAFETY RULES #403W
FOR WOOD STEPLADDERS

FOLLOW THESE INSTRUCTIONS FOR YOUR SAFETY
AND THAT OF OTHERS

1. Inspect ladder carefully on receipt and before EACH use. Test all working parts for proper attachment and operation. Ladders found to be damaged, defective, or with missing parts should be withdrawn from use and marked "DO NOT USE." Never use a ladder that has been dropped or tipped over until it has been reinspected for damage of any nature.
2. Install and use this ladder in compliance with the Regulations of the Occupational Safety and Health Act—1970, and with all other applicable governmental regulations, codes, and ordinances. Ladder usage must be restricted to the purpose for which the ladder is designed.
3. Keep nuts, bolts, and other fastenings tight. Oil moving metal parts frequently. Obtain replacement parts from the manufacturer. Do not allow makeshift repairs.
4. Ladders must stand on a firm level surface. Never use an unstable ladder. Never "walk" a stepladder while on it. Ladders should not be placed on temporary supports to increase the working length or to adjust for uneven surfaces.
5. Face ladder when ascending or descending. Always place ladder close enough to work to avoid dangerous over-reaching. Keep work centered between siderails. Side loading should be avoided.
6. Keep steps free of grease, oil, paint, snow, ice, or other slippery substances.
7. Insure that stepladders are fully opened with spreaders locked. Do not stand on top, pail rest, or rear rungs of stepladders.
8. Never place ladders in front of doors or openings unless appropriate precautions are taken.
9. Water conducts electricity. Do not use wet ladders where direct contact with a live power source is possible. Use extreme caution around electrical wires, services, and equipment. Provide for temporary insulation of any exposed electrical conductors near place of work.
10. A ladder is intended to carry only one person at a time. Do not overload. For support of two persons special ladders are available. NEVER use a stepladder in a closed or horizontal position, never sit on a ladder when it is on edge and never use a ladder in a flat position as a scaffold plank.
11. Store ladders on edge in such a manner as to provide easy access for inspection. Provide sufficient supports to prevent sagging. Never use ladders after exposure to fire, chemicals, fumes, or other conditions which could affect their strength.

9.100 CHEMICAL CORROSION CONTROL AND CLEANING

12. Portable ladders are designed as one-man working ladders, including any material supported by the ladder. There are three classifications:

    Type I—Industrial—for Heavy Duty and Industrial use.

    Type II—Commercial—for Medium Duty and Light Industrial use.

    Type III—Household—for Light Duty such as light household use.

13. For further instructions on the use and care of Wood Stepladders refer to the American National Standard, Safety Code for Portable Wood Ladders, A14.1—1982.

LADDER SAFETY RULES #403A
FOR ALUMINUM STEPLADDERS

FOLLOW THESE INSTRUCTIONS FOR YOUR SAFETY
AND THAT OF OTHERS

1. Inspect ladder carefully on receipt and before EACH use. Test all working parts for proper attachment and operation. Ladders found to be damaged, deformed, defective or with missing parts should be withdrawn from use and marked "DO NOT USE." Never use a ladder that has been dropped until it has been carefully reinspected for damage of any nature.

2. Install and use this ladder in compliance with the Regulations of the Occupational Safety and Health Act—1970, and with all other applicable governmental regulations, codes, and ordinances. Ladder usage must be restricted to the purpose for which the ladder is designed.

3. Keep nuts, bolts, and other fastenings tight. Oil moving metal parts frequently. Obtain replacement parts from manufacturer. Do not allow makeshift repairs. Never straighten or use a bent ladder.

4. Ladders must stand on a firm, level surface. Always use safety feet. If ladder is to be used on a slippery surface, take additional precautions. Never use an unstable ladder. Never "walk" a stepladder while on it. Ladders should not be placed on temporary supports to increase the working length or to adjust for uneven surfaces.

5. Face ladder when ascending or descending. Always place ladder close enough to work to avoid dangerous over-reaching. Keep work centered between siderails. Side loading should be avoided.

6. Keep steps free of grease, oil, paint, snow, ice, or other slippery substances.

7. Insure that stepladders are fully opened with spreaders locked. Do not stand on top, pail rest, or rear rungs of stepladders.

8. Never place ladders in front of doors or openings unless appropriate precautions are taken.

9. Metal and water conduct electricity. Do not use metal, metal reinforced or wet ladders where direct contact with a live power source is possible. Use extreme caution around electrical wires, services, and equipment. Provide for temporary insulation of any exposed electrical conductors near place of work.

10. A ladder is intended to carry only one person at a time. Do Not overload. For support of two persons special ladders are available. NEVER use a stepladder in a closed or horizontal position, never sit on a ladder when it is on edge and never use a ladder in a flat position as a scaffold plank.

11. Store ladders on edge in such a manner as to provide easy access for inspection. Provide sufficient supports to prevent sagging. Never use ladders after exposure to fire, chemicals, fumes, or other conditions which could affect their strength.

12. Portable ladders are designed as one-man-working ladders, including any material supported by the ladder. There are three duty classifications:

    Type I—Heavy Duty—for users requiring not more than 250-lb load capacity for maintenance, construction or heavy duty work.

Type II—Medium Duty—for users requiring not more than a 225-lb load capacity for painting, or other medium duty work.

Type III—Light Duty—for users requiring not more than a 200-lb load capacity or service requirements such as general household use. Light duty ladders should not be used with scaffold planks.

13. For further instructions on the care of Aluminum Stepladders refer to the American National Standard, Safety Code for Portable Metal Ladders, ANSI A14.2—1982.

SCAFFOLDING

Regulations and Standards. The design, manufacture, and use of scaffolds are nowadays quite rigidly controlled by both governmental regulations and national consensus standards. Please refer to the following sources for specific details of design, manufacturing, and use.

OSHA—Federal Regulations. Some states have other reference identification.

1. Occupational Safety and Health Standards for Industry

    CFR Title 29 Part 1910.28 "Safety Requirements for Scaffolding"
    Part 1910.29 "Manually Propelled Mobile Ladder Stands and Scaffolds (Towers)"
2. Occupational Safety and Health Regulations for Construction

    CFR Title 29 Part 1926.451 "Scaffolding"

    These regulations do not cross-reference other standards.

American National Standards Institute (ANSI)

1. ANSI A10.8 "Scaffolding—Safety Requirements"

    This standard establishes safety requirements for the construction, operation, maintenance, and use of scaffolds used in the demolition and maintenance of buildings and structures.

WELDED ALUMINUM SCAFFOLDS

This type of scaffold affords firm, solid work platforms for use by one or more men. Because of their lightness, they are fast and easy to erect and are therefore preferred where a number of off-the-floor jobs are required to be done in a large number of positions. Their lightness, mobility, and ease of erection make them most suitable for light-duty work, especially where the equipment requires frequent erection and dismantling.

These scaffolds are prefabricated from high-strength aluminum-alloy tubing and are equipped with casters as necessary for easy mobility of the erected scaffold. The types usually most practical for maintenance work are aluminum rolling scaffolds with internal stairways, and aluminum ladder scaffolds.

Folding ladder scaffolds (Fig. 5.3) are built in one-piece base sections which speed the erection and dismantling process. The ladder-type base sections are 29 in or 4 ft 6 in wide, with spans of 6, 8, or 10 ft between frames. In this type of unit, the two diagonal braces and one horizontal brace are integral parts of the folding unit. Intermediate, extension, and guardrail sections can be placed atop the folding unit, using individual end frames and braces.

A larger folding-type scaffold has base dimensions of 4 ft 6 in by 6 ft. This unit has an internal stairway, and the upper sections as well as the base section are one-piece folding units. When the scaffold must be erected higher than recommended for a base of this size, outriggers can be used. They clamp to the legs of the base section. Means for leveling, to compensate for uneven ground, are

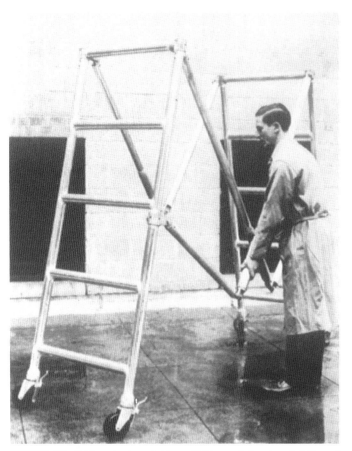

FIGURE 5.3 Folding aluminum ladder scaffolds are designed so that the end frames will not fall over at any point during the erection or dismantling process. It is a completely free-standing unit at all times.

FIGURE 5.4 Aluminum sectional scaffold with internal steps. For free-standing scaffolds OSHA requires that the maximum platform height must not exceed a ratio of 4 to 1 compared with the smallest dimension: This ratio is 3 to 1 in California. The scaffold shown is 36 ft high; outriggers were used to increase the base dimensions to 10 ft. Alternatively, bases can be widened using additional frames and other components.

part of the leg equipment. The casters on the legs are locked at both wheel and swivel. Folding scaffold sections are, of necessity, heavier than individual components of demountable sectional scaffolding.

Sectional aluminum stairway scaffolds are designed with end frames of various heights to provide different working levels, adjustable bottom sections with casters but without the folding feature, intermediate sections, half sections, and guardrail sections. All components are demountable so as to be light and easy to handle for erection and dismantling. Outriggers may be used to increase the base area (see Fig. 5.4).

The folding, sectional-stairway, and ladder scaffolds are used for outdoor cleaning and maintenance work—ladder scaffolds for low to medium height and one-man jobs, and folding or sectional-stairway types for higher or heavier work. They are especially suitable when the work is horizontal. Indoors, they simplify work on walls and ceilings, and often are suitable for group lamp replacement.

All aluminum scaffolds can be equipped with similar internal stairways. The narrower ladder scaffolds can be additionally equipped with platforms having sliding closeable hatches at each level for intermediate working levels and for safety while climbing the ladders.

WELDED SECTIONAL STEEL SCAFFOLDS

Used in situations similar to those of aluminum scaffolds, welded sectional steel scaffolds are heavier and therefore more suitable for heavy-duty work requiring relatively infrequent erections and dismantling. They can be assembled as a rolling scaffold and have mobility similar to aluminum scaffolds but are heavier and more cumbersome to handle. Some end frames have integral exterior ladders. Adjustable extension legs may be used for leveling. Casters lock at the wheel and swivel.

Steel ladder scaffolds, similar to the aluminum ones, are used for heavy-duty work in restricted spaces. Steel pivoted diagonal cross bracing is used as with larger steel frames. An often useful accessory for the steel ladder scaffold is the bridging trestle. This replaces the diagonal cross braces at the bottom level and permits the scaffold to clear obstructions or permits the passage of traffic beneath the scaffold without interference.

For access to high work areas where relatively heavy work has to be done, such as the replacement of a crane motor or large heating unit, the steel scaffold is unsurpassed in strength and versatility. It is now available with OSHA-complying external stand-off access ladders (Fig. 5.5), or even internal stair-type systems called "step units" (Figs. 5.6 and 5.7).

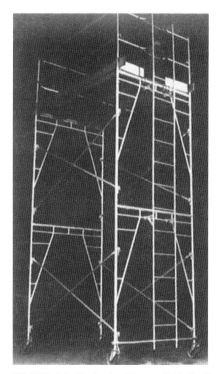

FIGURE 5.5 Sectional rolling scaffold. Steel frames are 5 ft wide and joined by pivoted diagonal braces of lengths from 4 to 10 ft. A "hook-on" type of access ladder is used incorporating a 3-ft-high grabrail at top, guarded by steel wire rope with snap hooks. The complete lift of the ladder is set back $7\frac{1}{2}$ in. from the scaffold frames.

FIGURE 5.6 "Tubelox" tube and coupler scaffolding with integral step units for access to 330-ft-high curved stack. The step units are available with guardrail and midrail panels, plus the opportunity of positioning landing or rest platforms at any desired level.

FIGURE 5.7 Trouble-saver sectional scaffolding with built-in step units for silo access during construction. Similar installations are quickly erected for many industrial maintenance operations when the safe, speedy vertical movement of persons is needed. (*The Patent Scaffolding Co., Inc. Photo by Tracy O'Neal.*)

SAFETY SWINGING SCAFFOLDS

These are generally two-men work platforms, either wood or aluminum, suspended from roof supports either inside or outside buildings (see Fig. 5.8). They are most suited for successive work operations vertically above each other. Swing-stage platforms are generally available in lengths from 8 to 32 ft, and the industry norm is 28 in. in width. They can be used in conjunction with hoisting apparatus consisting of rope blocks and falls, manually operated steel-wire-rope hoisting mechanisms, and air or electrically operated machines at the platform level. All hanging equipment requires most extreme care in safe rigging procedures by experienced personnel. In general, they are seldom economical for work at heights of less than 30 ft, *unless* access to such work is impractical by other types of ladders or scaffolds.

Swinging scaffolds are particularly suitable for cleaning and painting, tuck pointing, window washing, and similar jobs on exterior walls or tanks where large vertical range and quick up-down mobility are required. They are used also where the surface below the work is crowded, or where conditions are unsuitable for support of ground-based scaffolding. They are recommended for light and medium loads.

Safety belts and separately attached lifelines are essential for proper worker safety as well as insistence on the installation and use of guardrails, midrails, and toeboards. Wire mesh between guardrail and toeboard is required by OSHA *only* when "employees" are "required" to work under the scaffold.

As well as the conventional two-point suspension swing-stage platforms, additional items for specialized work are available in the form of one-man work cages and bosun's chairs. The cages are often available with extensions attached each side of the cage for use by two men. Generally, the cages are used with power-operated winches and the bosun's chairs with powered winches or blocks and falls.

FIGURE 5.8 Safety swinging scaffold (two-point suspension). Utilizes steel-wire rope with ratchet-action raising and crack-handle lowering. Note use of guardrails, midrails, and toeboards. Similar scaffolds are also available with power machines and aluminum platforms. Safety belts, lifelines, and lanyards must always be available and used on all swinging scaffolds.

TUBE AND COUPLER SCAFFOLDS (STEEL AND ALUMINUM)

These scaffolds provide the greatest versatility in scaffolding odd shapes such as processing works and refineries and in erecting to extreme heights. They are erected from four basic components: baseplates, interlocking tubing or pipe, bolt-activated couplers for making right-angle connections, and adjustable couplers for making connections at other than right angles.

Horizontal runners can be placed at any point on the vertical posts and, in turn, bearers at any point on the runner, thereby obtaining maximum versatility. This type of scaffolding is unsurpassed in providing work platforms for spheres, cylinders, and other odd-shaped vessels such as those in refineries. It also can be used to build storage racks of virtually any size and capacity, and is even the most suitable type of equipment to scaffold certain buildings having uneven exteriors and projections (Fig. 5.9).

FIGURE 5.9 This shows Tubelox scaffolding used to scaffold building surfaces with a difficult roof-overhang condition. The versatility of Tubelox is similarly utilized for cooling towers, spheres, and refinery structures.

FIGURE 5.10 Aluminum ladder scaffold. Basic unit consists of 6-ft ladder frames, diagonal braces, platform, and adjustable casters. Note use of guardrails, midrails, and toeboards.

This scaffold also is available in all-aluminum components (Fig. 5.10), making it particularly useful in corrosive atmospheres. Both types can be made into rolling scaffolds with the addition of special casters at the base.

MODULAR SYSTEMS SCAFFOLDS

During the 1970s decade, a new scaffolding design principle was first introduced into Europe and then into the United States. While a number of manufacturers and distributors use various descriptive trade names for their proprietary brands, a term gaining in popularity is "modular systems scaffolds." The systems combine the dimensional flexibility of tube and coupler scaffolds with fixed component sizes and the ease of assembly of sectional steel scaffolding.

All systems use fixed-length posts or uprights having proprietary connective modules affixed to them at regular vertical spacings of $19\frac{1}{2}$ to 21 in depending on the manufacturer. These modules generally embody a ring, cup, or slot principle which is used to receive, locate, and fasten the horizontal bearers and runners of a scaffold which in turn have mating connective devices attached at their ends to engage the post modules. Such connections generally use a loose or captive wedge principle to achieve varying degrees of rigidity at connections between posts and horizontals.

In all systems the modular connections are effected without the use of wrenches, threaded bolts, and the like; all one needs to construct the basic scaffold is a mallet and a spirit level and the requisite fixed-length components. For rigidity, diagonal bracing is attached using similar connecting methods as for the horizontal members.

Many industrial operations use structures, vessels, and containers which require periodic shutdown for maintenance and repair, such as the chemical, petrochemical, power plant, and pulp and paper industries, just to name a few. Such operations must be carried out with maximum speed and efficiency; when scaffolding is needed to do this work, such as for internal boiler repair, vessel cleaning, and insulation, most industries have traditionally used tube and coupler scaffolding because of its ability to cope with irregular dimensions and complicated geometric shapes. However, to effect a simple, right-angle joint, normal couplers require two threaded nuts to be loosened, the coupler positioned to an exact dimension, partially tightened, accurately leveled, and finally fully tightened; each horizontal tube member requires at least two, separate couplers (four threading operations) to attach it to two others. These are extremely labor-intensive operations normally requiring knowledgeable and experienced personnel for maximum productivity.

In contrast, modular systems scaffolds achieve great economy of erection labor; a horizontal is connected to two vertical members with only two wedges (or other devices) instead of four bolts and nuts. Horizontal dimensional positioning is automatically done by the fixed-length components; vertical positioning requires no measuring because of the regular, known spacing of the connection modules on the posts. Once the scaffold base is installed to be plumb and level, the scaffold can continue to be built up similarly plumb and level in the same manner as sectional frames.

While labor savings cannot be quantified for a number of reasons, many industrial users have indicated that modular systems can effectively lower scaffolding labor costs by very substantial percentages. Naturally, each application has its own peculiarity of dimension, size, ease (or otherwise) of access to the work site, and other factors. Manufacturers have produced new designs of access ladders, stairs, fabricated planking, and landing modules, to name but a few, plus special-use components geared to special industry requirements.

The type shown in this chapter is representative of a number of available systems and it is not appropriate to compare the features and claims of the individual manufacturers here. The system shown is the QES (Quick Erect Scaffold) wholly manufactured in the United States by Patent Scaffolding Co., A Division of Harsco Corporation. The posts use slotted rings as modules; captive wedges in the slotted ends of horizontals and diagonals give fast and accurate connections as shown in Figs. 5.11, 5.12, 5.13, and 5.14. Figure 5.15 shows a simple modular scaffold for work on a round storage vessel. Scaffolds for interior boiler maintenance, for instance, are infinitely more complex in scope and execution; the more complex the work, the more it can benefit from prescheduling and preplanning. Virtually all manufacturers and distribution centers for these scaffolds offer in-depth technical support services from which the industrial user may expect to receive planning and design assistance.

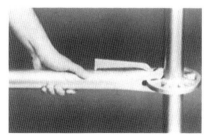

FIGURE 5.11 Preliminary step in connecting a QES™ horizontal to a prespaced ring module of a post. The horizontal's slots are merely aligned with any one of the ring slots. (*Photo by Bill Mitchell.*)

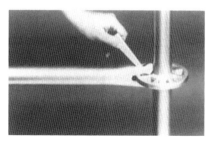

FIGURE 5.12 The wedge is inserted by hand into the ring slot while holding the end of the horizontal in line with the other hand. (*Photo by Bill Mitchell.*)

9.108 CHEMICAL CORROSION CONTROL AND CLEANING

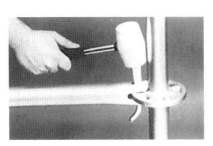

FIGURE 5.13 The wedge is lightly tapped into firm engagement with a hammer or mallet. Note the rivet in the thin end of the wedge which captivates it in the horizontal tube. (*Photo by Bill Mitchell.*)

FIGURE 5.14 A completed joint of two in-line horizontals connected to the modular ring of a post. Other horizontal members such as bearers are positioned at right angles to these. Diagonal bracing members are connected to the other angularly positioned slots. Eight identically sized and spaced slots are positioned in each modular ring. (*Photo by Bill Mitchell.*)

FIGURE 5.15 Circular access scaffold for a storage vessel using QES™ modular system scaffold. Note the ease with which ducts, piping, and other obstructions are bypassed.

SPECIAL-DESIGN SCAFFOLDS

Special-purpose scaffolds are available specifically designed for access purposes where standard scaffolding components cannot be used easily or are even impossible to use. Such scaffolds, stationary and mobile, are frequently used in the aircraft and aerospace industries, as well as for special requirements in processing plants. Such specials can be designed using any metals and other materials on an individual basis.

Safety Requirements. Almost all present-day steel and aluminum scaffolds are listed under the Reexamination Service of Underwriters' Laboratories, Inc. The UL seal on maintenance scaffolds means not only that the product is properly designed to sustain the loads for which it was intended but also that certain manufacturing standards are included in the design to assure maximum strength. One of the first things a maintenance department should look for is a UL listing sticker or label. After purchase of the equipment, it becomes the maintenance department's responsibility to make sure not only that the scaffold is used properly, but also that it is maintained properly. Of course, all equipment must comply with the appropriate OSHA regulations concerning its manufacture, installation, and use. Figure 5.16 is an OSHA scaffolding checklist for scaffolds of various types which can be used to assist in assuring that OSHA compliance is being maintained.

OSHA SCAFFOLDING CHECKLIST

Project _____

Inspection area _____ Area supervisor _____

Inspected by _____ Date _____

Ladders and scaffolding	Yes	No	Action/comments
Manually propelled mobile scaffolds			
1. When using free-standing mobile scaffold towers, do you restrict tower heights to four times the minimum base dimension? (3 and 3½ in some states)			
2. Do casters have a positive locking device that will hold the scaffold?			
3. Are mobile scaffolds properly braced by cross and horizontal bracing?			
4. Is the cross bracing of such length as will automatically square and align vertical members so that the erected scaffold is plumb, square, and rigid?			
5. Are all brace connections secured?			
6. Are platforms tightly planked for the full width of such scaffolds, except for the necessary entrance opening?			
7. Do you provide a ladder or stairway for proper access and exit?			
8. Is such a ladder or stairway affixed to or built into the scaffold?			
9. Is it so located that when in use it will tend to tip the scaffold?			
10. Are landing platforms provided at intervals not greater than 30 ft?			
11. Is the force necessary to move your mobile scaffold applied near or as close to base as practicable?			
12. Do you make adequate provision to stabilize the tower during movement from one location to another?			
13. Do you permit scaffolds to be moved only on level floors, free of obstructions and openings?			
General scaffolds			
14. When scaffolds are in use, do they rest upon a suitable footing?			
15. Are they plumb?			
16. Have you installed guardrails at all open sides and ends of the scaffolds?			
17. Are such guardrails made of not less than 2- by 4-in. lumber (or other material providing equivalent protection)?			
18. Are they approximately 42 in. high, with a midrail of 1- by 6-in. lumber (or material providing equivalent protection), and toeboards 4 in. high?			
19. Is wire mesh installed between the toeboard and the guardrail, extending along the entire opening?°			
20. Does such mesh consist of No. 19 gage U.S. Standard wire ½-in. mesh, or the equivalent?			
21. Is the planking used of scaffold-plank quality, even if not officially "graded" as scaffold plank?			

FIGURE 5.16 OSHA scaffolding checklist.

22. Does the span of the planks exceed the maximum allowable depending on the designation light duty, medium duty, or heavy duty? (1926.451(a)(10))
23. Are your men instructed always to replace guardrails, midrails, and bracing if they have had to be temporarily removed for passing materials and equipment?
24. Do your sectional scaffolds over 125 ft in height have drawings designed by a registered professional engineer? They must.
25. Do you obtain safety rules and instructions on use of scaffolds from your scaffolding supplier? You should—they are free.

Ladders

26. Are both legs of rung ladders supported at top and bottom?
27. Is the footing slippery, and if so are safety feet installed?
28. Can the work height be reached without standing on the top of stepladders or the top three rungs of rung ladders?
29. Does the extension ladder "fly" section rest on top of the base section and not hang underneath?
30. Does the wood look crumbly? If so, check by inserting a sharp knife end under a sliver of wood and pry up. If it results in a long splinter, it is O.K.; if it results in "crumbling" of the wood, it is decayed.
31. Never drop ladders or allow them to fall unless they are thoroughly and minutely inspected for damage before reuse.
32. Never use a bent aluminum ladder or one which has been bent and restraightened; the material has been overstressed on both occasions and is no longer reliable.

°Wire mesh required only when employees are required to work or pass underneath the scaffold.

FIGURE 5.16 (*Continued*)   OSHA scaffolding checklist.

Safe Use and Safety Rules. Any reputable manufacturer of ladders and scaffolds will furnish (with or attached to their equipment) information on specific items and include a set of safety rules which, if followed, can drastically reduce many industrial accidents associated with the use of this equipment. Copies of safety rules should be freely available from the manufacturer of the product—not only at the time of first delivery, but cheerfully and freely in later years when requested. Printed safety rules are *not* intended to be retained with the delivery slip (to which they are frequently attached) and sent into the office for billing purposes and thence forgotten. Neither are they for the purpose of padding out a foreman's hip pocket while the men involved in doing the work have never seen or read them. Safety rules must be read and clearly understood by the men *doing* the work; it is up to the judgment of the workman's immediate superior as to whether best results are obtained by having the workman read the safety rules or by having them read to him along with an explanation of the reasons why certain safety precautions are vital to freedom from injury.

The need for maximum employee safety in the erection and use of scaffolds is strongly emphasized. The OSHA regulations should be thoroughly read and observed, especially where the use of certain safety components has traditionally not been customary. Briefly, the OSHA regulations merely describe in detail the safe use of scaffolds, and therefore it is to the benefit of employer and employee to be completely familiar with such requirements pertinent to the work to be done. There is no shortcut to OSHA compliance.

In general, all scaffold platforms, walkways, mezzanines, and the like must be equipped with guardrails, midrails, and toeboards at all openings, open sides, and ends. Although no precise definition of "open side" currently exists in OSHA regulations, the following serves as a commonsense guide.

Open Side or End.   The side or end of a scaffold platform or walkway used for persons to work or walk thereupon is considered "open" if it is at a distance of more than 14 in from a solid structural surface or if the solid structural surface has openings or recesses adjacent to the platform into which a person could fall a distance of more than 4 ft.

Where persons must work or pass under a working platform, an 18 gage, $1/2$-in wire mesh or its equivalent should be installed between the guardrail and toeboard. Guardrails should be minimum 234 in, midrails 136 in, and toeboards 134 in nominal lumber sizes. Guardrail supports should be spaced at a maximum of 8 ft apart. Other materials of equivalent strength or properly designed and used components of a proprietary scaffolding system are acceptable.

OSHA SCAFFOLDING CHECKLIST

The checklist shown in Fig. 5.16 highlights certain basic safety precautions concerning scaffolds and can be used as a base for expansion to cover your own particular maintenance requirements. This list is basic and does not purport to be all-inclusive or to encompass all circumstances. Such determinations must be made by the employer in his own individual circumstances.

APPLICATION

Let us examine a typical example, and first assume that the exterior windows of a 400-ft-long by 24-ft-high building are to be cleaned. The building front is brick up to 6 ft, with standard glass panes from 8 to 24 ft. One man is assigned to do the job, with occasional additional help if he needs it. Factors are:

1. The work is within ladder range.
2. The window frames afford support.
3. The area below is asphalted, with some grass sections.
4. No traffic need pass below—wheeled or pedestrian.
5. The work ranges both horizontally and vertically.
6. There is no chance of electrical contact.

At first consideration, most factors indicate this work could be done by one man with a 28-ft aluminum extension ladder, which has a maximum extended height of 25 ft; if the man does not use the upper three rungs (safe practice), he will stand on a rung at 22 ft and be able to reach 28 ft high. He can move the ladder easily by himself. He can reach 2 ft either side of the ladder and can cover a 5-ft-wide strip of windows in a vertical pass. This is obviously the correct choice, yes? No!

The man will spend excessive time climbing up and down and relocating the ladder. His *productive* working time will be not more than 50 percent. Remember—a ladder must be installed one-fourth of its extended length away from the building at base; therefore, although he can reach the upper-level windows easily, he will have to lower and reposition the ladder frequently to reach the lower levels. Resorting to over-reaching sideways and behind the ladder is unsafe and is the initiating factor in many ladder accidents. A platform stepladder with the platform 18 ft high is a better possibility. The best choice, however, is a 29-in-wide by 10-ft-long aluminum ladder scaffold, about 19 ft high to top working level, with outboard safety supports and a "climb-through" wood platform which has a hatch which can be slid out of the way to climb through and thence replaced to make a full platform cover. The reasons for the choice of this equipment are:

1. The man can achieve all heights by varying the levels of the plywood platform for a full working width of 10 ft; from one work level plus his own reach he can clean say 60 ft^2, compared with a typical 6 ft^2 from one extension-ladder position.
2. He can reach all heights with one initial placement plus only *two* platform repositionings.
3. He can cover 10 ft by 24 ft, less 6 ft brick, that is, 180 ft^2, with minor downtime for repositioning.

4. When a 10-ft vertical pass is completed, he can *roll* the scaffold himself to the next position. With certain youth and agility he could even erect it himself; the aluminum components are extremely light; otherwise, he would require help for initial installation.
5. For the grass areas (or even at the rear of ornamental-shrub beds), the scaffold can be rolled on leveled planks. Four-inch steel channels can also be used as wheel guides and bridges over minor humps and deviations from level. Level such earth or grass with sand so that there are no "holes" *under* the planks.

Note: Aluminum scaffolds are standardly equipped with 5-in-diameter casters but are optionally available with 8-in ones; both have a 24-in range of screw-leg adjustment for support from different levels. The 8-in wheels provide easier rolling over rough surfaces such as old asphalt.

Maintenance. Aluminum scaffolds of all types require minimum maintenance. Stairways, ladders, and platforms should be inspected frequently, and any grease or oil should be removed immediately. Make sure the plywood platforms or platform planks are solid, with no splits. Do not store platforms near excessive heat, to avoid drying and warping. Casters should be cleaned and lubricated, and brakes should be checked for satisfactory operation. Threads on extension legs should be cleaned and lubricated periodically for smooth operation. Coupling pins used to join frames vertically should be kept clean so that upper frames slip over the pins easily and freely.

With aluminum scaffolds, slight bends in the tubing due to severe impact or mishandling should be straightened. The spring-lock devices used to fasten the braces should be kept free from dirt to ensure proper operation of the lock. On the more popular types of aluminum scaffolding, this mechanism is exposed and can be cleaned easily with a wire brush.

The steel types of scaffolding should be kept clean by scraping or wire brushing; any rusted spots on frames or braces should be scraped and touched up with quick-drying enamel. Stud threads should be lubricated, and wing nuts run off and on to ensure fast, secure fastening during erection. All frames should be checked frequently for missing vertical-coupling sprockets and the pins that lock them in place, as well as wing nuts. Cross braces should be straightened if bent, and the alignment of the tops of the frames should be checked and braces realigned if necessary.

For maximum safety, swinging scaffolds must be properly maintained. The operating mechanism, or winch, should be kept free of dirt and grit at all times. A wire brush usually is satisfactory for this work. All operating parts should be properly lubricated as outlined in the manufacturer's instructions. The safety devices in the winches should be inspected frequently. Pawls and pawl springs should be checked for proper working condition. Teeth on the drum casting should be inspected, and if broken or worn, the manufacturer should be consulted about replacement. The steel cable should be run off and checked for excessive damage and kinks and then rewound through an oily rag to clean the cable and give it a thin coat of oil. Worm and gear mechanisms should be checked for excessive play, cleaned, and repacked with fiber grease. The stirrup should be checked for alignment and straightened, and all painted surfaces of the machine should be recoated where necessary for rust prevention.

Wooden platforms require careful inspection and maintenance. Grease or oil spilled on the platform should be removed immediately. After a job, the platform should be placed across horses for inspection, overhaul, and repair. Mortar, concrete, and paint should be removed with a wire brush and scraper. Broken rungs, slats, and damaged or missing toeboards should be replaced, as should missing hinges and hooks and eyes. After necessary repairs have been made, the platform should be given a thick coat of quick-drying paint. Finally, clean and examine the S and L hooks from which the scaffold is hung, clean and inspect center stanchions, and replace missing or defective wing nuts and bolts.

Tube and coupler scaffolding should undergo a systematic inspection. Bent tubes should be straightened and, in case of seriously damaged tubes, discarded or cut into shorter lengths for short bearers. Very dirty or rusty tubes should be cleaned with a wire brush. During this operation, inspect the male and female fittings for damage which would affect their safety and then clean with a wire brush. Remove damaged couplers from stock. Studs should be kept covered with a light film of oil, and catch bolts should be checked for stripped threads.

Ladders and scaffolds are vital accessories for all "off-the-ground" work. It behooves all persons, employers and employees alike, to keep their equipment in first-class operating condition and always use it with safety as the maximum prerequisite.

S·E·C·T·I·O·N · 10

MAINTENANCE WELDING

CHAPTER 1
ARC WELDING IN MAINTENANCE

J. E. Hinkel
The Lincoln Electric Company, Cleveland, Ohio

THE ROLE OF WELDING IN A MAINTENANCE DEPARTMENT

An important use of arc welding is the repair of plant machinery and equipment. In this respect, welding is an indispensable tool without which production operations would soon shut down. Fortunately, welding machines and electrodes have been developed to the point where reliable welding can be accomplished under the most adverse circumstances. Frequently, welding must be done under something less than ideal conditions, and therefore, equipment and operators for maintenance welding should be the best.

Besides making quick, on-the-spot repairs of broken machinery parts, welding offers the maintenance department a means of making many items needed to meet a particular demand promptly. Broken castings, when new ones are no longer available, can be replaced with steel weldments fashioned out of standard shapes and plates (see Fig. 1.1). Special machine tools required by production for specific operations often can be designed and made for a fraction of the cost of purchasing a standard machine and adapting it to the job. Material-handling devices can be made to fit the plant's physical dimensions. Individual jib cranes can be installed. Conveyors, either roll-down or pallet-type, can be tailor-made for specific applications. Tubs and containers can be made to fit products. Grabs, hooks, and other handling equipment can be made for shipping and receiving. Jigs and fixtures, as well as other simple tooling, can be fabricated in the maintenance department as either permanent tooling or as temporary tooling for a trial lot.

The almost infinite variety of this type of welding makes it impossible to do more than suggest what can be done. Figures 1.2 through 1.5 provide just a few examples of the imaginative applications of welding technology achieved by some maintenance technicians. The welding involved should present no particular problems if the operators have the necessary training and background to provide them with a knowledge of the many welding techniques that can be used.

A maintenance crew proficient in welding can fabricate and erect many of the structures required by a plant, even to the extent of making structural steel for a major plant expansion. Welding can be done either in the plant maintenance department or on the erection site. Structures must, of course, be adequately designed to withstand the loads to which they will be subjected. Such loads will vary from those of wind and snow in simple sheds to dynamic loads of several tons where a crane is involved. Materials and joint designs must be selected with a knowledge of what each can do. Then the design must be executed by properly trained and qualified welders. Structural welding involves out-of-position work, so a welder must be able to make good welds

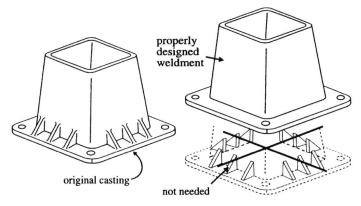

FIGURE 1.1 Replacing casting (*left*) with weldment (*right*).

FIGURE 1.2 Long delivery time prompted welding of this cast iron punch-press frame.

FIGURE 1.3 Plant-made racks for holding steel.

FIGURE 1.4 Typical welding jig and positioner that can be readily fabricated.

FIGURE 1.5 Maintenance department fabricating trusses for plant expansion. Trusses made from channels and angles. A jig was laid out on plate in the plant yard.

under all conditions. Typical joints that are used in welded structures are shown in Figs. 1.6 through 1.9.

Standard structural shapes can be used, including pipe, which makes an excellent structural shape. Electrodes such as the E6010-11 types are often the welder's first choice for this kind of fabrication welding because of their all-position characteristics. These electrodes, which are not low-hydrogen types, may be used providing the weldability of the steel is such that neither weld cracks nor severe porosity is likely to occur.

Scrap materials often can be put to good use. When using scrap, however, it is best to weld with a low-hydrogen E7016-18 type of electrode, since the analysis of the steel is unlikely to be known, and some high-carbon steels may be encountered. Low-hydrogen electrodes minimize cracking tendencies. Structural scrap frequently comes from dismantled structures such as elevated railroads, which used rivet-quality steel that takes little or no account of the carbon content.

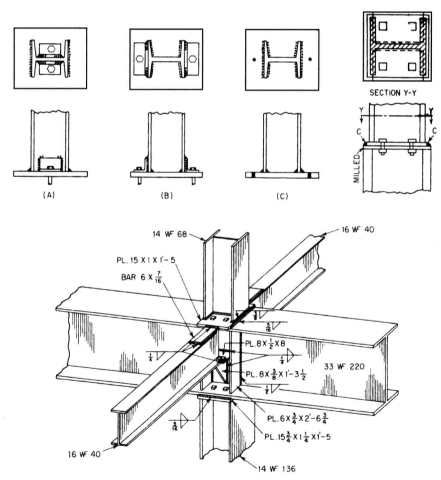

FIGURE 1.6 Typical column bases, column splices, and beam-to-column connections that can be used in structural welding.

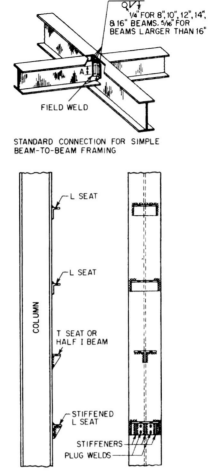

FIGURE 1.7 Beam-to-beam framing and methods for seating beams on columns.

SHIELDED METAL ARC WELDING (SMAW)

Shielded metal arc welding is the most widely used method of arc welding. With SMAW, often called "stick welding," an electric arc is formed between a consumable metal electrode and the work. The intense heat of the arc, which has been measured at temperatures as high as 13,000°F, melts the electrode and the surface of the work adjacent to the arc. Tiny globules of molten metal rapidly form on the tip of the electrode and transfer through the arc, in the "arc stream," and into the molten "weld pool" or "weld puddle" on the work's surface (see Fig. 1.10).

Within the shielded metal arc welding process, electrodes are readily available in tensile strength ranges of 60,000 to 120,000 psi (see Table 1.1). In addition, if specific alloys are required to match the base metal, these too are readily available (see Table 1.2).

ARC WELDING IN MAINTENANCE **10.7**

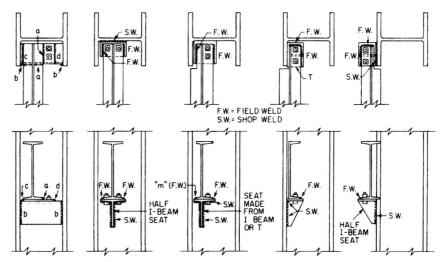

FIGURE 1.8 Different ways of connecting beams to columns when an offset is required.

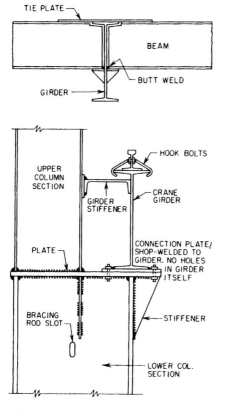

FIGURE 1.9 A beam-and-girder connection and a column detail showing craneway.

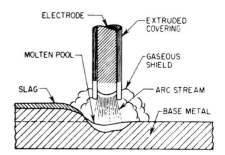

FIGURE 1.10 Shielded metal arc welding process.

TABLE 1.1 Electrodes Commonly Used for Maintenance Welding

AWS classification	Tensile strength, min. psi	Yield strength, min. psi
E6010-11	62,000	50,000
E7010-11	70,000	57,000
E7016-18	70,000	57,000
E8016-18	80,000	67,000
E9016-18	90,000	77,000
E10016-18	100,000	87,000
E11016-18	110,000	97,000
E12016-18	120,000	107,000

Note: E6010-11 and E7010-11 are cellulosic electrodes. All others are low-hydrogen electrodes and are better suited to welding higher-strength steels.

TABLE 1.2 Chemical Requirements

AWS Classification[a]	Chemical composition, percent[b]								
	C	Mn	P	S	Si	Ni	Cr	Mo	V
Chemical composition, percent[b]									
E7010-A1	0.12	0.60	0.03	0.04	0.40	—	—	0.40–0.65	—
E7011-A1		0.60			0.40				
E7015-A1		0.90			0.60				
E7016-A1		0.90			0.60				
E7018-A1		0.90			0.80				
E7020-A1		0.60			0.40				
E7027-A1		1.00			0.40				
Chromium-molybdenum steel electrodes									
E8016-B1 / E7017-B1	0.05 to 0.12	0.90	0.03	0.04	0.60 / 0.80	—	0.40–0.65	0.40–0.65	—
E8015-B2L	0.05	0.90	0.03	0.04	1.00	—	1.00–1.50	0.40–0.65	—
E8016-B2 / E8018-B2	0.05 to 0.12	0.90	0.03	0.04	0.60	—	1.00–1.50	0.40–0.65	—
E8018-B2L	0.05	0.90	0.03	0.04	0.80	—	1.00–1.50	0.40–0.65	—
E9015-B3L	0.05	0.90	0.03	0.04	1.00	—	2.00–2.50	0.90–1.20	—
E9015-B3 / E9016-B3 / E9018-B3	0.05 to 0.12	0.90	0.03	0.04	0.60 / 0.60 / 0.80	—	2.00–2.50	0.90–1.20	—
E9018-B3L	0.05	0.90	0.03	0.04	0.80	—	2.00–2.50	0.90–1.20	—
E8015-B4L	0.05	0.90	0.03	0.04	1.00	—	1.75–2.25	0.40–0.65	—
B8016-B5	0.07 to 0.15	0.40 to 0.70	0.03	0.04	0.30–0.60	—	0.40–0.60	1.00–1.25	0.05
Nickel steel electrodes									
E8016-C1 / E8018-C1	0.12	1.25	0.03	0.04	0.06 / 0.08	2.00–2.75	—	—	—
E7015-C1L / E7016-C1L / E7018-C1L	0.05	1.25	0.03	0.04	0.50	2.00–2.75	—	—	—

(*Continued*)

TABLE 1.2 Chemical Requirements (*Continued*)

AWS Classification[a]	Chemical composition, percent[b]								
	C	Mn	P	S	Si	Ni	Cr	Mo	V
Nickel steel electrodes									
E8016-C2 E8018-C2	0.12	1.25	0.03	0.04	0.60 0.80	3.00–3.75	—	—	—
E7015-C2L E7016-C2L E7018-C2L	0.05	1.25	0.03	0.04	0.50	3.00–3.75	—	—	—
E8016-C3[c] E8018-C3[c]	0.12	0.40–1.25	0.03	0.03	0.80	0.80–1.10	0.15	0.35	0.05
Nickel-molybdenum steel electrodes									
E8018-NM[d]	0.10	0.80–1.25	0.02	0.03	0.60	0.80–1.10	0.05	0.40–0.65	0.02
Manganese-molybdenum steel electrodes									
E9015-D1 E9018-D1	0.12	1.25–1.75	0.03	0.04	0.60 0.80	—	—	0.25–0.45	—
E8016-D3 E8018-D3	0.12	1.00–1.75	0.03	0.04	0.60 0.80	—	—	0.40–0.65	—
E10015-D2 E10016-D2 E10018-D2	0.15	1.65–2.00	0.03	0.04	0.60 0.60 0.80	—	—	0.25–0.45	—
All other low-alloy steel electrodes[e]									
EXX10-G[e] EXX11-G EXX13-G EXX15-G EXX16-G EXX18-G E7020-G	—	1.00 min[f]	—	—	0.80 min[f]	0.50 min[f]	0.30 min[f]	0.20 min[f]	0.10 min[f]
E9018-M[c] E10018-M[c] E11018-M[c] E12018-M[c] E12018-M1[c] E7018-W[g] E8018-W	0.10 0.10 0.10 0.10 0.10 0.12 0.12	0.60–1.25 0.75–1.70 1.30–1.80 1.30–2.25 0.80–1.60 0.40–0.70 0.50–1.30	0.030 0.030 0.030 0.030 0.015 0.025 0.03	0.030 0.030 0.030 0.030 0.012 0.025 0.04	0.80 0.60 0.60 0.60 0.65 0.40–0.70 0.35–0.80	1.40–1.80 1.40–2.10 1.25–2.50 1.75–2.50 3.00–3.80 0.20–0.40 0.04–0.80	0.15 0.35 0.40 0.30–1.50 0.65 0.15–0.30 0.45–0.70	0.35 0.25–0.50 0.25–0.50 0.30–0.55 0.20–0.30 — —	0.05 0.05 0.05 0.05 0.05 0.08 —

Note: Single values shown are *maximum* percentages, except where otherwise specified.

[a] The suffixes A1, B3, C2, etc., designate the chemical composition of the electrode classification.

[b] For determining the chemical composition, DECN (electrode negative) may be used where DC, both polarities, is specified.

[c] These classifications are intended to conform to classifications covered by the military specifications for similar compositions.

[d] Copper shall be 0.10% max and aluminum shall be 0.05% max for E8018-NM electrodes.

[e] The letters "XX" used in the classification designations in this table stand for the various strength levels (70, 80, 90, 100, 110, and 120) of electrodes.

[f] In order to meet the alloy requirements of the G group, the weld deposit need have the minimum, as specified in the table, of only one of the elements listed. Additional chemical requirements may be agreed between supplier and purchaser.

[g] Copper shall be 0.30 to 0.60% for E7018-W electrodes.

FLUX-CORED ARC WELDING (FCAW)

Flux-cored arc welding is generally applied as a semiautomatic process. It may be used with or without external shielding gas depending on the electrode selected. Either method utilizes a fabricated flux-cored electrode containing elements within the core that perform a scavenging and deoxidizing action on the weld metal to improve the properties of the weld.

FCAW with Gas

If gas is required with a flux-cored electrode, it is usually CO_2 or a mixture of CO_2 and another gas. These electrodes are best suited to welding relatively thick plate (not sheet metal) and for fabricating and repairing heavy weldments.

FCAW Self-Shielded

Self-shielded flux-cored electrodes, better known as Innershield,* are also available. In effect, these are stick electrodes turned inside out and made into a continuous coil of tubular wire. All shielding, slagging, and deoxidizing materials are in the core of the tubular wire. No external gas or flux is required.

Innershield electrodes offer much of the simplicity, adaptability, and uniform weld quality that account for the continuing popularity of manual welding with stick electrodes, but as a semiautomatic process, they get the job done faster. This is an open-arc process that allows the operator to place the weld metal accurately and to visually control the weld puddle. These electrodes operate in all positions: flat, vertical, horizontal, and overhead.

The electrode used for semiautomatic and fully automatic flux-cored arc welding is mechanically fed through a welding gun or welding jaws into the arc from a continuously wound coil that weighs approximately 50 lb. Only the fabricated flux-cored electrodes are suited to this method of welding, since coiling extruded flux-coated electrodes would damage the coating. In addition, metal-to-metal contact at the electrode's surface is necessary to transfer the welding current from the welding gun into the electrode. This is impossible if the electrode is covered.

A typical application of semiautomatic and fully automatic equipment for FCAW is shown in Fig. 1.11. For a given cross section of electrode wire, much higher welding amperage can be applied with semiautomatic and fully automatic processes. This is because the current travels only a very short distance along the bare metal electrode, since contact between the current-carrying gun and the bare metal electrode occurs close to the arc. In manual welding, the welding current must travel the entire length of the electrode, and the amount of current is limited to the current-carrying capacity of the wire. The higher currents used with automatic welding result in a high weld metal deposition. This increases travel speed and reduces welding time, thereby lowering costs.

SUBMERGED ARC WELDING (SAW)

With submerged arc welding, the arc is completely hidden under a small mound of granular flux which is automatically deposited around the electrode as it is fed to the work (see Fig. 1.12). Because the arc is hidden, it is impractical for many maintenance welding applications. However, the process has some inherent advantages. With ingenuity and the help of some form of mechanization, it can be used to very good advantage in certain maintenance situations, especially when welding on heavy members.

While SAW entails the use of considerable amounts of flux, the unfused flux may be collected and reused. Precautions should be taken to keep the flux and the work clean in order to prevent weld contamination and to maintain weld quality.

---

*Innershield is a registered trademark of the Lincoln Electric Company.

FIGURE 1.11 Semiautomatic flux-cored arc welding (FCAW).

The high currents required for SAW also contribute a deep penetrating arc characteristic. Consequently, either no groove or a smaller than normal groove may be used depending on the thickness of the base metal. This results in reductions in both welding time and the amount of filler metal required. For example, no chamfering is necessary for two-pass butt joints in steel up $5/8$ to in thick. Complete penetration also can be obtained in fillet welds for material up to $3/8$ in thick without chamfering. For joints in thicker material, a double V-groove weld is used.

With submerged arc welding, distortion is minimized because of the high welding speeds, a minimal number of passes, and efficient application of heat. This means that less heat is applied to the weld area and that the heat is applied more uniformly than would be the case with manual welding. Distortion due to an unbalanced heat condition, as in single-groove multiple-pass welded joints, can be corrected by presetting the base metal parts to offset angular movement. The other methods of controlling distortion, discussed later in this chapter, also can be applied.

Although SAW is used primarily for production welding, its potential for maintenance use remains only partially exploited. The process is particularly suited to rebuilding worn surfaces and developing abrasion-resistant surfaces for parts that are exposed to severe wear and metal-to-metal contact.

GAS-SHIELDED METAL ARC WELDING (GMAW)

The GMAW process, sometimes called *metal inert gas* (MIG) welding, incorporates the automatic feeding of a continuous consumable electrode that is shielded by an externally supplied gas. Since the equipment provides for automatic control of the arc, only the travel speed, gun positioning, and guidance are controlled manually. Process control and function are achieved through the basic elements of equipment shown in Fig. 1.13. The gun guides the consumable electrode and conducts the electric current and shielding gas to the workpiece. The electrode feed unit and power source are used in a system that provides automatic regulation of the arc length. The basic combination used to produce this regulation consists of a constant-voltage power source (characteristically providing an essentially flat volt-ampere curve) in conjunction with a constant-speed electrode feed unit.

10.12 MAINTENANCE WELDING

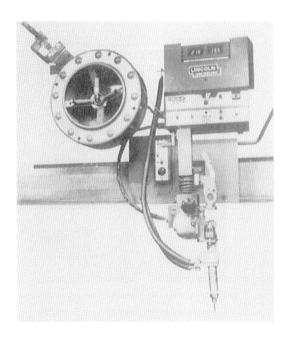

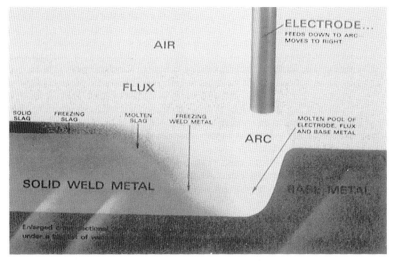

FIGURE 1.12 Elements of the submerged arc welding process.

GMAW for Maintenance Welding

In terms of maintenance welding applications, GMAW has the following advantages over SMAW:

1. Can be used in all positions with the low-energy modes.
2. Produces virtually no slag to remove or be trapped in weld.

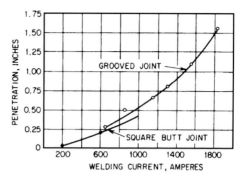

FIGURE 1.13 Penetration versus applied current for submerged arc welding.

3. Requires less operator training time than SMAW.
4. Adaptable to semiautomatic or machine welding.
5. Low-hydrogen process.
6. Faster welding speeds than SMAW.
7. Suitable for welding carbon steels, alloy steels, stainless steels, aluminum, and other non-ferrous metals. Table 1.3 lists recommended filler metals for GMAW.

Gas Selection for GMAW

There are many different gases and combinations of gases that can be used with the GMAW process. These choices vary with the base metal, whether a spray arc or short-circuiting arc is desired, or sometimes just according to operator preference. Recommended gas choices are given in Tables 1.4 and 1.5.

GAS TUNGSTEN ARC WELDING (GTAW)

Process Description

The GTAW process, also referred to as the *tungsten inert gas* (TIG) process, derives the heat for welding from an electric arc established between a tungsten electrode and the part to be welded (Fig. 1.14). The arc zone must be filled with an inert gas to protect the tungsten electrode and molten metal from oxidation and to provide a conducting path for the arc current. The process was developed in 1941 primarily to provide a suitable means for welding magnesium and aluminum, where it was necessary to have a process superior to the shielded metal arc (stick electrode) process. Since that time, GTAW has been refined and has been used to weld almost all metals and alloys.

The GTAW process requires a gas- or water-cooled torch to hold the tungsten electrode; the torch is connected to the weld power supply by a power cable. In the lower-current gas-cooled torches (Fig. 1.15), the power cable is inside the gas hose, which also provides insulation for the conductor. Water-cooled torches (Fig. 1.16) require three hoses: one for the water supply, one for the water return, and one for the gas supply. The power cable is usually located in the water-return hose. Water cooling of the power cable allows use of a smaller conductor than that used in a gas-cooled torch of the same current rating.

TABLE 1.3 Recommended Filler Metals for GMAW

Base metal type	Recommended electrode		AWS filler metal specification (use latest edition)	Electrode diameter		Current range Amperes
	Material type	Electrode classification		in	mm	
Aluminum and aluminum alloys	1100	ER1100 or ER4043		0.030	0.8	50–175
	3003, 3004	ER1100 or ER5356		1/64	1.2	90–250
	5052, 5454	ER5554, ER5356 or ER5183	A5.10	1/16	1.6	160–350
	5083, 5086, 5456	ER5556 or ER5356		3/32	2.4	225–400
	6061, 6063	ER4043 or ER5356		1/8	3.2	350–475
Magnesium alloys	AZ10A	ERAZ61A, ERAZ92A				
	AZ31B, AZ61A, AZ80A	ERAZ61A, ERAZ92A		0.040	1.0	150–300*
	ZE10A	ERAZ61A, ERAZ92A		3/64	1.2	160–320*
	ZK21A	ERAZ61A, ERAZ92A		1/16	1.6	210–400*
	AZ63A, AZ81A, AZ91C		A5.19	3/32	2.4	320–510*
	AZ92A, AM100A	ERAZ92A		1/8	3.2	400–600*
	HK31A, HM21A, HM31A	ERAZ92A				
	LA141A	EREZ33A				
		EREZ33A				
Copper and copper alloys	Silicon Bronze	ERCuSi-A				
	Deoxidized copper	ERCu		0.035	0.9	150–300
	Cu-Ni alloys	ERCuNi	A5.7	0.045	1.2	200–400
	Aluminum bronze	ERCuAl-A1, A2 or A3		1/16	1.6	250–450
	Phosphor bronze	ERCuSn-A		3/32	2.4	350–550
Nickel and nickel alloys				0.020	0.5	—
	Monel* Alloy 400	ERNiCu-7		0.030	0.8	—
	Inconel† Alloy 600	ERNiCrFe-5	A5.14	0.035	0.9	100–160
				0.045	1.2	150–260
				1/16	1.6	100–400
Titanium and titanium alloys	Commercially pure	Use a filler metal one or two grades lower		0.030	0.8	—
	Ti-0.15 Pd	ERTi-0.2 Pd	A5.16	0.035	0.9	—
	Ti-5Al-2.5Sn	ERTi-5A1-2.5Sn or commercially pure		0.045	1.2	—
Austenitic stainless steels	Type 201	ER308		0.020	0.5	—
	Types 301, 302, 304 & 308	ER308		0.025	0.6	—
				0.030	0.8	75–150
	Type 304	ER308L		0.035	0.9	100–160
	Type 310L	ER310	A5.9	0.045	1.2	140–310
	Type 316	ER316		1/16	1.6	280–450
	Type 321	ER321		5/64	2.0	—
	Type 347	ER347		3/32	2.4	—
				7/64	2.8	—
				1/8	3.2	—
Steel	Hot rolled or cold-drawn plain carbon steels	ER70S-3 or ER70S-1		0.020	0.5	—
		ER70S-2, ER70S-4		0.025	0.6	—
		ER70S-5, ER70S-6		0.030	0.8	40–220
				0.35	0.9	60–280
			A5.18	0.045	1.2	125–380
				0.052	1.3	260–460
				1/16	1.6	275–450
				5/64	2.0	—
				3/32	2.4	—
				1/8	3.2	—
Steel	Higher strength carbon steels and some low alloy steels	ER80S-D2		0.035	0.9	60–280
		ER80S-Ni1		0.045	1.2	125–380
		ER100S-G		1/16	1.6	275–450
			A5.28	5/64	2.0	—
				3/32	2.4	—
				1/8	3.2	—
				5/32	4.0	—

*Spray transfer mode.
†Trademark-International Nickel Co.

TABLE 1.4 Selection of Gases for GMAW with Spray Transfer

Metal	Shielding gas	Advantages
Aluminum	Argon	0 to 1 in (0 to 25 mm) thick: best metal transfer and arc stability; least spatter.
	35% argon + 65% helium	1 to 3 in (25 to 76 mm) thick: higher heat input than straight argon; improved fusion characteristics with 5XXX series Al-Mg alloys.
	25% argon + 75% helium	Over 3 in (76 mm) thick: highest heat input; minimizes porosity.
Magnesium	Argon	Excellent cleaning action.
Carbon steel	Argon + 1.5% oxygen	Improves arc stability; produces a more fluid and controllable weld puddle; good coalescence and bead contour; minimizes undercutting; permits higher speeds than pure argon.
	Argon 1 3–10% CO_2	Good bead shape; minimizes spatter; reduces chance of cold lapping; can not weld out of position.
Low-alloy steel	Argon + 2% oxygen	Minimizes undercutting; provides good toughness.
Stainless steel	Argon + 1% oxygen	Improves arc stability; produces a more fluid and controllable weld puddle, good coalescence and bead contour; minimizes undercutting on heavier stainless steels.
	Argon + 2% oxygen	Provides better arc stability, coalescence, and welding speed than 1 percent oxygen mixture for thinner stainless steel materials.
Copper, nickel, and their alloys	Argon	Provides good wetting; decreases fluidity of weld metal for thickness up to $1/8$ in (3.2 mm).
	Argon + helium	Higher heat inputs of 50 & 75 percent helium mixtures offset high heat dissipation of heavier gages.
Titanium	Argon	Good arc stability; minimum weld contamination; inert gas backing is required to prevent air contamination on back of weld area.

Applicability of GTAW

The GTAW process is capable of producing very high quality welds in almost all metals and alloys. However, it produces the lowest metal deposition rate of all the arc welding processes. Therefore, it normally would not be used on steel, where a high deposition rate is required and very high quality usually is not necessary. The GTAW process can be used for making root passes on carbon and low-alloy steel piping with consumable insert rings or with added filler metal. The remainder of the groove would be filled using the coated-electrode process or one of semiautomatic processes such as GMAW (with solid wire) or FCAW (with flux-cored wire).

TABLE 1.5 Selection of Gases for GMAW with Short Circuiting Transfer

Metal	Shielding gas	Advantages
Carbon steel	75% argon + 25% CO_2	Less than $1/8$ in (3.2 mm) thick: high welding speeds without burn-thru; minimum distortion and spatter.
	75% argon + 25% CO_2	More than $1/8$ in (3.2 mm) thick: minimum spatter; clean weld appearance; good puddle control in vertical and overhead positions.
	CO_2	Deeper penetration; faster welding speeds.
Stainless steel	90% helium + 7.5% argon + 2.5% CO_2	No effect on corrosion resistance; small heat-affected zone; no undercutting; minimum distortion.
Low alloy steel	60–70% helium + 25–35% argon + 4–5% CO_2	Minimum reactivity; excellent toughness; excellent arc stability, wetting characteristics, and bead contour, little spatter.
	75% argon + 25% CO_2	Fair toughness; excellent arc stability, wetting characteristics, and bead contour; little spatter.
Aluminum, copper, magnesium, nickel, and their alloys	Argon & argon + helium	Argon satisfactory on sheet metal; argon-helium preferred on thicker sheet material [over $1/8$ in (3.2 mm)].

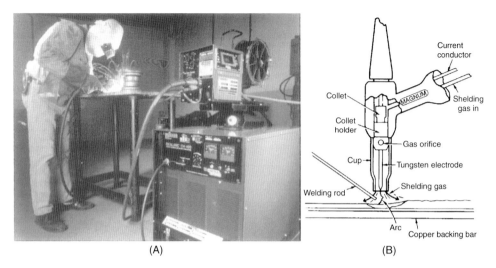

FIGURE 1.14 (*A*) Gas-shielded metal arc welding. (*B*) Gas tungsten arc welding.

A constant-current or drooping-characteristic power supply is required for GTAW, either dc or ac and with or without pulsing capabilities. For water-cooled torches, a water cooler circulator is preferred over the use of tap water.

For automatic or machine welding, additional equipment is required to provide a means of moving the part in relation to the torch and feeding the wire into the weld pool. A fully automatic system may require a programmer consisting of a microprocessor to control weld current, travel speed, and filler wire feed rate. An inert-gas supply (argon, helium, or a mixture of these), including pressure regulators, flowmeters, and hoses, is required for this process. The gases may be supplied from cylinders or liquid containers. A schematic diagram of a complete gas tungsten arc welding arrangement is shown in Fig. 1.17.

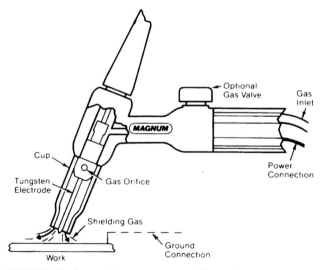

FIGURE 1.15 Gas cooled gas tungsten arc-welding torch.

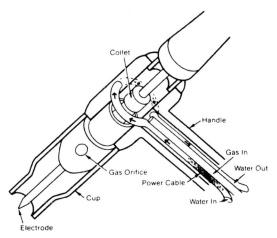

FIGURE 1.16 Section of typical gas tungsten arc-welding electrode holder (water cooled).

GTAW would be used for those alloys where high-quality welds and freedom from atmospheric contamination are critical. Examples of these are the reactive and refractory metals such as titanium, zirconium, and columbium, where very small amounts of oxygen, nitrogen, and hydrogen can cause loss of ductility and corrosion resistance. It can be used on stainless steels and nickel-base superalloys, where welds exhibiting high quality with respect to porosity and fissuring are required. The GTAW process is well suited for welding thin sheet and foil of all weldable metals because it can be controlled at the very low amperages (2 to 5 A) required for these thicknesses. GTAW would not be used for welding the very low melting metals such as tin-lead solders and zinc-base alloys because the high temperature of the arc would be difficult to control.

Advantages and Disadvantages of GTAW

The main advantage of GTAW is that high-quality welds can be made in all weldable metals and alloys except the very low melting alloys. This is because the inert gas surrounding the arc and weld zone protects the hot metal from contamination. Another major advantage is that filler metal can be

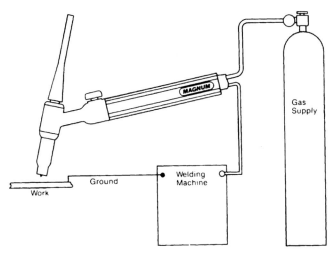

FIGURE 1.17 Gas tungsten arc-welding equipment arrangement.

added to the weld pool independently of the arc current. With other arc welding processes, the rate of filler metal addition controls the arc current. Additional advantages are very low spatter, portability in the manual mode, and adaptability to a variety of automatic and semiautomatic applications.

The main disadvantage of GTAW is the low filler metal deposition rate. Further disadvantages are that it requires greater operator skill and is generally more costly than other arc welding processes.

Principles of Operating GTAW

In the GTAW process, an electric arc is established in an inert-gas atmosphere between a tungsten electrode and the metal to be welded. The arc is surrounded by the inert gas, which may be argon, helium, or a mixture of these two. The heat developed in the arc is the product of the arc current times the arc voltage, where approximately 70 percent of the heat is generated at the positive terminal of the arc. Arc current is carried primarily by electrons (Fig. 1.18) which are emitted by the heated negative terminal (cathode) and obtained by ionization of the gas atoms. These electrons are attracted to the positive terminal (anode), where they generate approximately 70 percent of the arc heat. A smaller portion of the arc current is carried by positive gas ions which are attracted to the negative terminal (cathode), where they generate approximately 30 percent of the arc heat. The cathode loses heat by the emission of electrons, and this energy is transferred as heat when the electrons deposit or condense on the anode. This is one reason why a significantly greater amount of heat is developed at the anode than at the cathode.

The voltage across an arc is made up of three components: the cathode voltage, the arc column voltage, and the anode voltage. In general, the total voltage of the gas tungsten arc will increase with arc length (Fig. 1.19), although current and shielding gas have effects on voltage, which will be discussed later. The total arc voltage can be measured readily, but attempts to measure the cathode and anode voltages accurately have been unsuccessful. However, if the total arc voltage is plotted against arc length and extrapolated to zero arc length, a voltage that approximates the sum of cathode voltage plus anode voltage is obtained. The total cathode plus anode voltage determined in this manner is between 7 and 10 V for a tungsten cathode in argon. Since the greater amount of heat is generated at the anode, the GTAW process is normally operated with the tungsten electrode or cathode negative (negative polarity) and the work or anode positive. This puts the heat where it is needed, at the work.

Polarity and GTAW

The GTAW process can be operated in three different modes: electrode-negative (straight) polarity, electrode-positive (reverse) polarity, or ac (Fig. 1.20). In the electrode-negative mode, the greatest amount of heat is developed at the work. For this reason, electrode-negative (straight) polarity is used

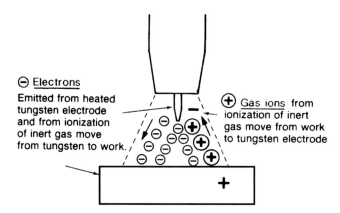

FIGURE 1.18 Schematic of gas tungsten arc, direct-current electrode negative.

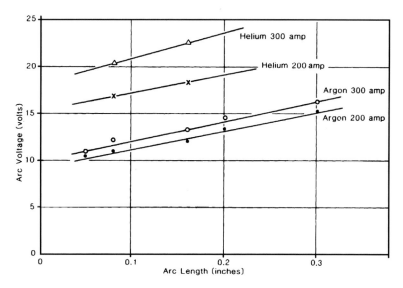

FIGURE 1.19 Effect of arc length and current on arc voltage. (Metric conversion: 1 in. = 2.54 mm; 2 in. = 5.08 mm; 3 in. = 7.62 mm.)

with GTAW for welding most metals. Electrode-negative (straight) polarity has one disadvantage—it does not provide cleaning action on the work surface. This is of little consequence for most metals because their oxides decompose or melt under the heat of the arc so that molten metal deposits will wet the joint surfaces. However, the oxides of aluminum and magnesium are very stable and have melting points well above that of the metal. They would not be removed by the arc heat and would remain on the metal surface and restrict wetting.

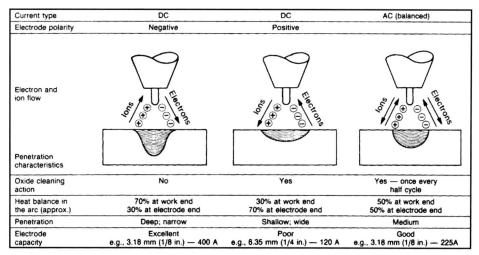

FIGURE 1.20 Characteristics of current types for gas tungsten arc welding. (*From AWS Welding Handbook, 7th ed., vol. 2.*)

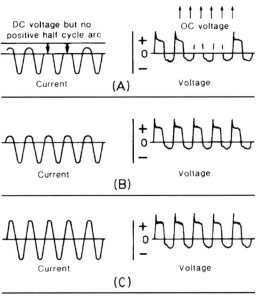

FIGURE 1.21 Voltage and current waveforms for ac welding: (*A*) partial and complete rectification; (*B*) with arc stabilization; (*C*) with current balancing. (*From AWS Welding Handbook, 7th ed., vol. 2.*)

In the electrode-positive (reverse) polarity mode, cleaning action takes place on the work surface by the impact of gas ions. This removes a thin oxide layer while the surface is under the cover of an inert gas, allowing molten metal to wet the surface before more oxide can form.

When ac gas tungsten arc welding aluminum, rectification occurs, and more current will flow when the electrode is negative (Fig. 1.21). This condition exists because the clean aluminum surface does not emit electrons as readily as the hot tungsten electrode. It will occur with standard ac welding power supplies. More advanced GTA welders incorporate circuits which can balance the negative- and positive-polarity half-cycles. Generally, this balanced condition is desirable for welding aluminum. The newest GTA power supplies include solid-state control boards which allow adjusting the ac current so as to favor either the positive- or negative-polarity half-cycle. These power supplies also chop the tip of the positive and negative half-cycles to produce a squarewave ac rather than a sinusoidal ac. When maximum cleaning is desired, the electrode-positive mode is favored; when maximum heat is desired, the electrode-negative mode is favored.

GTAW Shielding Gases and Flow Rates

Any of the inert gases could be used for GTAW. However, only helium (atomic weight 4) and argon (atomic weight 40) are used commercially because they are much more plentiful and much less costly than the other inert gases. Typical flow rates are 15 to 40 ft^3/hr.

Argon is used more extensively than helium for GTAW because

1. It produces a smoother, quieter arc action.
2. It operates at a lower arc voltage for any given current and arc length.
3. There is greater cleaning action in the welding of materials such as aluminum and magnesium in the ac mode.

4. Argon is more available and lower in cost than helium.
5. Good shielding can be obtained with lower flow rates.
6. Argon is more resistant to arc zone contamination by cross drafts.
7. The arc is easier to start in argon.

The density of argon is approximately 1.3 times that of air and 10 times that of helium. For this reason, argon will blanket a weld area and be more resistant than helium to cross drafts. Helium, being much lighter than air, tends to rise rapidly and cause turbulence, which will bring air into the arc atmosphere. Since helium costs about 3 times as much as argon and its required flow rate is 2 to 3 times that for argon, the cost of helium used as a shielding gas can be as much as 9 times that of argon.

Although either helium or argon can be used successfully for most GTAW applications, argon is selected most frequently because of the smoother arc operation and lower overall cost. Argon is preferred for welding thin sheet to prevent melt-through. Helium is preferred for welding thick materials and materials of high thermal conductivity such as copper and aluminum.

Electrode Material for GTAW

In selecting electrodes for GTAW, five factors must be considered: material, size, tip shape, electrode holder, and nozzle. Electrodes for GTAW are classified as pure tungsten, tungsten containing 1 or 2 percent thoria, tungsten containing 0.15 to 0.4 percent zirconia, and tungsten which contains an internal lateral segment of thoriated tungsten. The internal segment runs the full length of the electrode and contains 1 or 2 percent thoria. Overall, these electrodes contain 0.35 to 0.55 percent thoria. All tungsten electrodes are normally available in diameters from 0.010 to 0.250 in and lengths from 3 to 24 in. Chemical composition requirements for these electrodes are given in AWS A5.12, "Specification for Tungsten Arc Welding Electrodes."

Pure tungsten electrodes, which are 99.5 percent pure, are the least expensive but also have the lowest current-carrying capacity on ac power and a low resistance to contamination. Tungsten electrodes containing 1 or 2 percent thoria have greater electron emissivity than pure tungsten and, therefore, greater current-carrying capacity and longer life. Arc starting is easier, and the arc is more stable, which helps make the electrodes more resistant to contamination from the base metal. These electrodes maintain a well-sharpened point for welding steel.

Tungsten electrodes containing zirconia have properties in between those of pure tungsten and thoriated tungsten electrodes with regard to arc starting and current-carrying capacity. These electrodes are recommended for ac welding of aluminum over pure tungsten or thoriated tungsten electrodes because they retain a balled end during welding and have a high resistance to contamination. Another advantage of the tungsten-zirconia electrodes is their freedom from the radioactive element thorium, which, although not harmful in the levels used in electrodes, is of concern to some welders.

GTAW Electrode Size and Tip Shape

The electrode material, size, and tip shape (Fig. 1.22) will depend on the welding application, material, thickness, type of joint, and quantity. Electrodes used for ac or electrode-positive polarity will be of larger diameter than those used for electrode-negative polarity.

The total length of an electrode will be limited by the length that can be accommodated by the GTAW torch. Longer lengths allow for more redressing of the tip than short lengths and are therefore more economical. The extension of the electrode from the collet or holder determines the heating and voltage drop in the electrode. Since this heat is of no value to the weld, the electrode extension should be kept as short as necessary to provide access to the joint.

It is recommended that electrodes to be used for dc negative-polarity welding be of the 2 percent thoria type and be ground to a truncated conical tip. Excessive current will cause the electrode to overheat and melt. Too low a current will permit cathode bombardment and erosion caused by the low operating temperature and resulting arc instability. Although a sharp point on the tip promotes easy arc starting, it is not recommended because it will melt and form a small ball on the end.

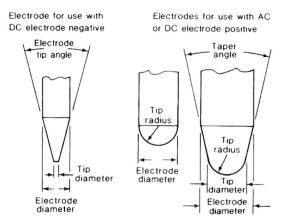

FIGURE 1.22 Shapes of tungsten electrodes.

For ac and dc electrode-positive welding, the desirable electrode tip shape is a hemisphere of the same diameter as the electrode. This shape tip on the larger electrodes required for ac and dc electrode-positive welding provides a stable surface within the operating current range. Zirconia-type electrodes are preferred for ac and dc electrode-positive operation because they have a higher current-carrying capacity than the pure tungsten electrodes, yet they will readily form a molten ball under standard operating conditions. Thoriated electrodes do not ball readily and, therefore, are not recommended for ac or dc electrode-positive welding.

The degree of taper on the electrode tip affects weld penetration, where the smaller taper angles tend to reduce the width of the weld bead and thus increase penetration. When preparing the tip angle on an electrode, grinding should be done parallel to the length of the electrode. Special machines are available for grinding electrodes. These can be set to accurately grind any angle required.

GTAW Electrode Holders and Gas Nozzles

Electrode holders usually consist of a two-piece collet made to fit each standard-sized tungsten electrode. These holders and the part of the GTAW torch into which they fit must be capable of handling the required welding current without overheating. These holders are made of a hardenable copper alloy.

The function of the gas nozzle is to direct the flow of inert gas around the holder and electrode and then to the weld area. The nozzles are made of a hard, heat-resistant material such as ceramic and are available in various sizes and shapes. Large sizes give a more complete inert gas coverage of the weld area but may be too big to fit into restricted areas. Small nozzles can provide adequate gas coverage in restricted areas where features of the component help keep the inert gas at the joint. Most nozzles have internal threads which screw over threads on the electrode holder. Some nozzles are fitted with a washer-like device that consists of several layers of fine-wire screen or porous powder metal. These units provide a nonturbulent or lamellar gas flow from the torch which results in improved inert gas coverage at a greater distance from the nozzle. In machine or automatic welding, more complete gas coverage may be provided by backup gas shielding from the fixture and a trailer shield attached to the torch.

Characteristics of GTAW Power Supplies

Power supplies for use with GTAW should be of the constant-current, drooping-voltage type (Fig. 1.23). They may have other optional features such as up slope, down slope, pulsing, and current programming capabilities. Constant-voltage power supplies should not be used for GTAW.

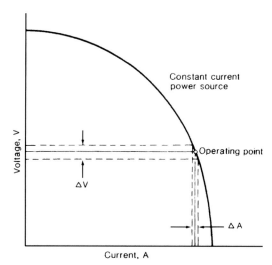

FIGURE 1.23 Static volt-ampere characteristics. (*From Lincoln GS-100.*)

The power supply may be a single-phase transformer-rectifier which also can supply ac for welding aluminum. Engine generator-type power supplies are usually driven by a gasoline or diesel engine and will produce dc with either a constant-current, drooping-voltage or constant-voltage characteristics. Engine alternator power supplies will produce ac for GTAW. A power supply capable of operating on either constant current or constant voltage should be set for the constant-current mode for GTAW.

Power supplies made specifically for GTAW normally will include a high-frequency source for arc starting and valves that control the flow of inert gas and cooling water for the torch. Timers allow the valves to be opened a short time before the arc is initiated and closed a short time after the arc is extinguished. The high frequency is necessary for arc starting instead of torch starting, where tungsten contamination of the weld is likely. It should be possible to set the high frequency for arc starting only or for continuous operation in the ac mode.

Power supplies should include a secondary contactor and a means of controlling arc current remotely. For manual welding, a foot pedal would perform these functions of operating the contactor and controlling weld current. A power supply with a single current range is desirable because it allows the welder to vary the arc current between minimum and maximum without changing a range switch.

The more advanced power supplies incorporate features that permit pulsing the current in the dc mode with essentially square pulses. Both background and pulse peak current can be adjusted, as well as pulse duration and pulsing frequency (Fig. 1.24). In the ac mode, the basic 60-Hz sine wave can be modified to produce a rectangular wave. Other controls permit the ac wave to be balanced or varied to favor the positive or negative half-cycles. This feature is particularly useful when welding aluminum and magnesium, where the control can be set to favor the positive half-cycle for maximum cleaning. In the dc mode, the pulsing capability allows welds to be made in thin material, root passes, and overhead with less chance of melt-through or droop.

GTAW Torches

A torch for GTAW must perform the following functions:

1. Hold the tungsten electrode so that it can be manipulated along the weld path.
2. Provide an electrical connection to the electrode.
3. Provide inert-gas coverage of the electrode tip, arc, and hot weld zone.
4. Insulate the electrode and electrical connections from the operator or mounting bracket.

10.24 MAINTENANCE WELDING

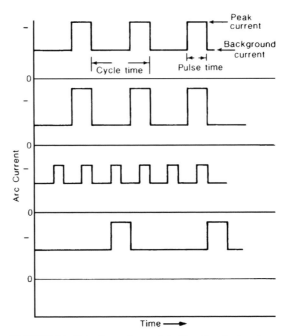

FIGURE 1.24 Pulsed waveshapes attainable with pulsed gas tungsten arc power supply.

Typical GTAW torches were shown in Figs. 1.14, 1.15, and 1.16. The torch consists of a metallic body, a collet holder, a collet, and a tightening cap to hold the tungsten electrode. The electrical cable is connected to the torch body, which is enclosed in a plastic insulating outer sheath. For manual torches, a handle is connected to the sheath. Power, gas, and water connections pass through the handle or, in the case of automatic operation, through the top of the torch. In the smaller, low-current torches, the electrode, collet, and internal components are cooled by the inert-gas flow. Larger, high-current torches are water-cooled and require connections to tap water and a drain or to a water cooler circulator. A cooler circulator with distilled or deionized water is preferred to prevent buildup of mineral deposits from tap water inside the torch.

Inert gas flows through the torch body and through holes in the collet holder to the arc end of the torch. A cup or nozzle is fitted over the arc end of the torch to direct inert gas over the electrode and the weld pool. The nozzles normally screw onto the torch and are made of a hard, heat-resistant ceramic. Some are made of high-temperature glass such as Vicor and are pressed on over a compressible plastic taper. Some nozzles can be fitted with an insert washer made up of several layers of fine-wire screen sometimes called a *gas lens*. This produces a lamellar rather than turbulent flow of inert gas to increase the efficiency of shielding.

On most manual GTAW torches, the handle is fixed at an angle of approximately 70° to the torch body. Some makes of torches have a flexible neck between the handle and torch body which allows the angle between the handle and the torch body to be adjusted over a range from about 50° to 90°.

Manual GTAW Techniques

To become proficient in manual gas tungsten arc welding, the welder must develop skills in manipulating the torch with one hand while controlling weld current with a foot pedal or thumb control and feeding filler metal with the other hand (see Fig. 1.25). Before welding is started on any job, a rough idea of the welding conditions is needed, such as filler material, current, shielding gas, and so forth.

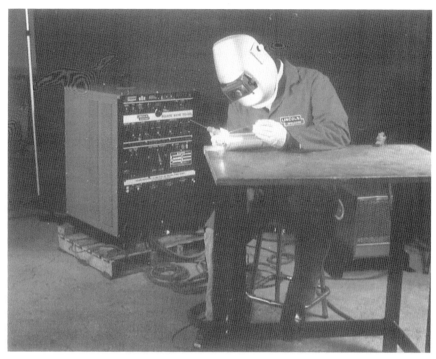

FIGURE 1.25 Fast pedal control.

Establishing Welding Parameters for GTAW

The material, thickness, joint design, and service requirements will determine the weld current, inert gas, voltage, and travel speed. This information may be available in a "welding procedure specification" (WPS) or from handbook data on the material and thickness. If welding parameters are not provided in a WPS, the information given in Tables 1.6, 1.7, and 1.8 can be used as starting-point parameters for carbon and low-alloy steels, stainless steels, and aluminum. These should be considered starting values; final values should be established by running a number of test parts.

Gas Tungsten Arc Starting Methods

The gas tungsten arc may be started by touching the work with the electrode, by a superimposed high frequency pulse, or by a high-voltage pulse. The touch method is not recommended for critical work because there is a strong possibility of tungsten contamination with this technique. Most weld power supplies intended for GTAW contain a high-frequency generator (usually a spark-gap oscillator) which superimposes the high-frequency pulse on the main weld power circuit. When welding with dc electrode-negative or electrode-positive, the high-frequency switch should be set in the HT start position. When the welder presses the foot pedal to start welding, a timer is activated which starts the high-frequency pulse and stops it when the arc initiates. Once started, the arc will continue after the high-frequency pulse stops as long as the power and proper arc gap are maintained. When welding with ac, the switch should be set in the HF continuous position to ensure that the arc restarts after voltage reversal on each half-cycle. High-frequency generators on welders produce frequencies in the radio communications range. Therefore, manufacturers of power supplies must certify that the

TABLE 1.6 Recommended Wire Sizes for Input Power Cable for Typical Motor-Generator-Type Welder (*Based on National Electrical Code®*)

Welder size	60-Hz input voltage	Ampere rating	3 wires in conduit or 3-conductor cable, Type R	Grounding conductor
200	230	44	8	8
	460	22	12	10
	575	18	12	14
300	230	62	6	8
	460	31	10	10
	575	25	10	12
400	230	78	6	6
	460	39	8	8
	575	31	10	10
600	230	124	2	6
	460	62	6	8
	575	50	8	8
900	230	158	1	3
	460	79	6	6
	575	63	4	8

TABLE 1.7 Typical Input Cable Sizes for A-C/D-C Welder

Welder	Volts input	Amp input		Wire size (3 in conduit)			Wire size (3 in free air)		
		With condsr.	Without condsr.	With condsr.	Without condsr.	Ground conduct.	With condsr.	Without condsr.	Ground conduct.
300	200	84	104	2	1	1	4	4	4
	440	42	52	6	6	6	8	8	8
	550	38	42	8	6	6	10	8	8
400	220	115	143	0	00	00	3	1	1
	440	57.5	71.5	4	3	3	6	6	6
	550	46	57.2	6	4	4	8	6	6
500	220	148	180	000	0000	0000	1	0	0
	440	74	90	3	2	2	6	4	4
	550	61	72	4	3	3	6	6	6

TABLE 1.8 Welding Cable Sizes, Motor-Generator Welder

Machine size, amp	Cable sizes for lengths (electrode plus ground)		
	Up to 50 ft	50–100 ft	100–250 ft
200	2	2	1/0
300	1/0	1/0	3/0
400	2/0	2/0	4/0*
600	3/0	3/0	4/0*
900	Automatic application only		

*Recommended longest length of 4/0 cable for 400-A welder, 150 ft; for 600-A welder, 100 ft. For greater distances, cable size should be increased; this may be a question of cost—consider ease of handling versus moving of welder closer to work.

radiofrequency radiation from the power supply does not exceed limitations established by the Federal Communications Commission (FCC). The allowable radiation may be harmful to some computer and microprocessor systems and to communications systems. These possibilities for interference should be investigated before high-frequency starting is used. Installation instructions provided with the power supply should be studied and followed carefully.

OXYACETYLENE CUTTING

Steel can be cut with great accuracy using an oxyacetylene torch (see Fig. 1.26). However, not all metals cut as readily as steel. Cast iron, stainless steel, manganese steels, and nonferrous materials cannot be cut and shaped satisfactorily with the oxyacetylene process because of their reluctance to oxidize. In these cases, plasma arc cutting is recommended.

The cutting of steel is a chemical action. The oxygen combines readily with the iron to form iron oxide. In cast iron, this action is hindered by the presence of carbon in graphite form, so cast iron cannot be cut as readily as steel. Higher temperatures are necessary, and cutting is slower. In steel, the action starts at bright-red heat, whereas in cast iron, the temperature must be nearer the melting point in order to obtain a sufficient reaction.

Because of the very high temperature, the speed of cutting is usually fairly high. However, since the process is essentially one of melting without any great action, tending to force the molten metal out of the cut, some provision must be made for permitting the metal to flow readily away from the cut. This is usually done by starting at a point from which the molten metal can flow readily. This method is followed until the desired amount of metal has been melted away.

AIR-CARBON ARC CUTTING AND GOUGING

Air-carbon arc cutting (CAC-A) is a physical means of removing base metal or weld metal using a carbon electrode, an electric arc, and compressed air (see Fig. 1.27). In the air-carbon arc process, the intense heat of the arc between the carbon electrode and the workpiece melts a portion of the base

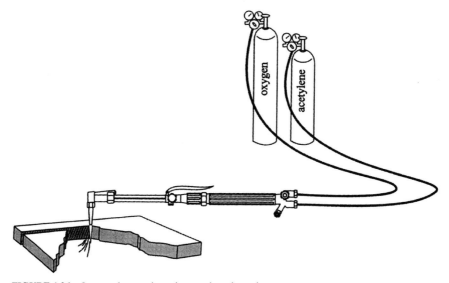

FIGURE 1.26 Oxyacetylene cutting using a carbon electrode.

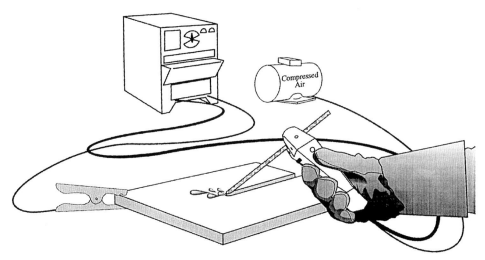

FIGURE 1.27 Oxyacetylene cutting using compressed air.

metal or weld. Simultaneously, a jet of air is passed through the arc of sufficient volume and velocity to blow away the molten material. This sequence can be repeated until the required groove or cut has been obtained. Since CAC-A does not depend on oxidation to maintain the cut, it is capable of cutting metals that oxyacetylene cutting will not cut. It is used to cut carbon steel, stainless steel, many copper alloys, and cast iron.

Arc gouging can be used to remove material approximately five times as fast as chipping. Depth of cut can be closely controlled, and welding slag does not deflect or hamper the cutting action, as it would with cutting tools. Gouging equipment generally costs less to operate than chipping hammers or gas cutting torches. An arc-gouged surface is clean and smooth and usually can be welded without further preparation. Drawbacks of the process include the fact that it requires large volumes of compressed air and it is not as good as other processes for through-cutting.

Electrode type and air supply specifications for arc gouging are outlined in Tables 1.9 and 1.10.

TABLE 1.9 Electrode Type and Polarity Recommended for Arc Gouging

Material	Electrode	Power
Steel	DC	DCEP
	AC	AC
Stainless steel	DC	DCEP
	AC	AC
Iron (cast iron, ductile iron, malleable iron)	AC	AC or DCEN
	DC	DCEP (high-amperage)
Copper alloys	AC	AC or DCEN
	DC	DCEP
Nickel alloys	AC	AC or DCEN

Source: AWS Handbook, 6th ed., Section 3A.
Note: AC is not the preferred method.

TABLE 1.10 Air Consumption for Arc Gouging

Maximum electrode size (in)	Application	Pressure (psi)	Consumption (cfm)
1/4	Intermittent-duty, manual torch	40	3
1/4	Intermittent-duty, manual torch	80	9
3/8	General-purpose	80	16
3/4	Heavy-duty	80	29
5/8	Semiautomatic mechanized torch	80	25

Applications

The CAC-A process may be used to provide a suitable bevel or groove when preparing plates for welding (see Fig. 1.28). It also may be used to back-gouge a seam prior to welding the side. CAC-A provides an excellent means of removing defective welds or misplaced welds and has many applications in metal fabrication, casting finishing, construction, mining, and general repair and maintenance. When using CAC-A, normal safety precautions must be taken and, in addition, ear plugs must be worn.

Power Sources

While it is possible to arc air gouge with ac, this is not the preferred method. A dc power source of sufficient capacity and a minimum of 60 open circuit volts, either rectifier or motor generator, will give the best results. With dc, it is operated with DCEP (electrode-positive). Arc voltages normally range from about 35 to 56 V. Table 1.11 lists recommended power sources, and Table 1.12 lists suggested current ranges. It is recommended that the power source have overload protection in the output circuit. High current surges of short duration occur with arc gouging, and these surges can overload the power source.

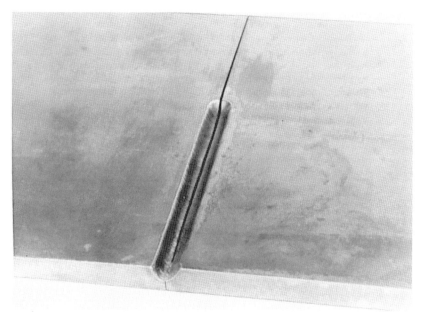

FIGURE 1.28 Preparing plate for welding.

TABLE 1.11 Power Sources for Arc Gouging

Type of current	Type of power source	Remarks
DC	Variable-voltage motor-generator, rectifier or resistor-grid equipment	Recommended for all electrode sizes
DC	Constant-voltage motor-generator, or rectifier	Recommended only for electrodes above $1/4$-in diameter
AC	Transformer	Should be used only with AC electrodes
AC-DC	Rectifier	DC supplied by three-phase transformer-rectifier is satisfactory. DC from single-phase source not recommended. AC from AC-DC power source is satisfactory if AC electrodes are used

PLASMA ARC CUTTING

Plasma arc cutting has become an essential requirement for any properly equipped maintenance department. It provides the best, fastest, and cheapest method of cutting carbon or alloy steel, stainless steel, aluminum, nonferrous metals, and cast iron. In fact, it will cut any conductive material. The cuts are clean and precise, with very little dross or slag to remove. The heat is so concentrated within the immediate area of the cut that very little distortion takes place (Fig. 1.29). On gage thickness material, the speed is limited only by the skill of the operator.

Plasma arc cutting operates on the principle of passing an electric arc through a quantity of gas through a restricted outlet. The electric arc heats the gas as it travels through the arc. This turns the gas into a plasma that is extremely hot. It is the heat in the plasma that heats the metal. A typical power source for plasma arc cutting is about the size of a small transformer welder and comes equipped with a special torch, as shown in Fig. 1.30.

Plasma arc torch consumables include the electrode and the orifice. The electrode is copper, with a small hafnium insert at the center tip. The arc emanates from the hafnium, which gradually erodes with use, and this requires electrode replacement periodically. The torch tip contains the orifice which constricts the plasma arc. Various sized orifices are available, and the smaller diameters operate at lower amperages and produce a more constricted narrow arc than the larger diameters. The orifices also gradually wear with use and must be replaced when the arc becomes too wide.

The plasma arc cutting torch can be used for arc gouging. The main changes are that the tip orifice is larger than for cutting, and the torch is held at an angle of about 30° from horizontal rather than at 90°, as in cutting. Plasma gouging can be used on all metals and is particularly suitable for aluminum and stainless steels, where oxyacetylene cutting is ineffective, and carbon arc gouging tends to cause carbon contamination.

TABLE 1.12 Recommended Range of Current for Arc Gouging

Type of electrode and power	Maximum and minimum current (amp)					
	Electrode size (in)					
	$5/32$	$3/16$	$1/4$	$5/16$	$3/8$	$1/2$
DC Electrodes, DECP Power	90–150	150–200	200–400	250–450	350–600	600–1000
AC Electrodes, AC Power	—	150–200	200–300	—	300–500	400–600
AC Electrodes, DCEN Power	—	150–180	200–250	—	300–400	400–500

ARC WELDING IN MAINTENANCE **10.31**

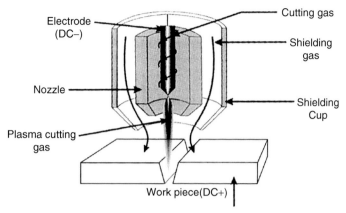

FIGURE 1.29 Plasma arc cutting.

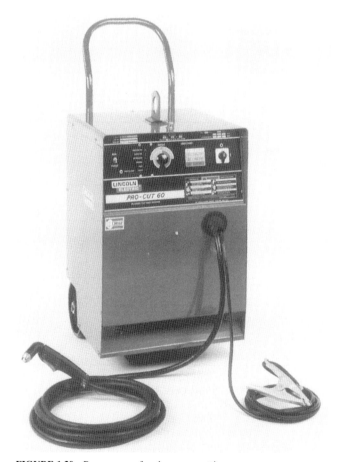

FIGURE 1.30 Power source for plasma arc cutting.

WELDING PROCEDURES

Much of the welding done by maintenance welders does not normally require detailed written welding procedures. The judgment and skill of the welders generally are sufficient to get the job done properly. However, there are some maintenance applications which demand the attention of a welding engineer or a supervisor, someone who knows more than the typical welder about the service conditions of the weldment or perhaps the weldability of it. Or there may be metallurgical factors or high-stress service requirements which must be given special consideration. A distinction should be made between "casual" and "critical" welding. When an application is considered critical, a welding engineer or a properly qualified supervisor should provide the welding operator with a detailed written procedure specification (WPS) providing all the information needed to make the weld properly. Suggested forms for selecting the proper process and writing a WPS are shown in Fig. 1.31A and B.

CONDITIONS	SMAW (Stick)	GMAW (MIG)	FCAW Gas	FCAW Self Shielded	GTAW (TIG)	SAW (Sub Arc)
CARBON STEEL						
a. Sheet Metal	**	****		**	****	
b. Plate	***	***	****	*****		****
STAINLESS						
a. Sheet Metal	*	****			****	
b. Plate	**	***			**	***
ALUMINUM						
a. Sheet Metal					****	
b. Plate		***			**	
NON-FERROUS Copper, Bronze, etc.					****	
HIGH DEP. RATE	*	**	***	***		****
PORTABILITY OF EQUIPMENT	***	***	***	***	**	
USER FRIENDLY	*	**	**	*	****	*
MIN. WELD CLEANING	*	***	**	*	****	*
ALL POSITION WELDING	****	****	**	**	****	

Rating:
 ****   Excellent
 ***   Very Good
 **   Good
 *   Fair
 (none)   Not Recomended

FIGURE 1.31 (A) Process selection based on typical maintenance applications.

Weldment: _____   Quality Category:

Drawing No. _____   Critical ☐

Functional ☐

Process	Electrode Size/Type	Polarity	Volts	WFS/AMP (either)	ESO	Shield. Gas/Flux
SMAW						
GMAW						
SAW						
TIG						
FCAW						

Sketch of the Joint

Special Instructions: _____
_____
_____
_____
_____
_____

Welder: _____   Proced. Apprvd by: _____

Welder: _____   Date: _____

FIGURE 1.31 (*Continued*) (*B*) Welding procedure sheet.

QUALIFICATION OF WELDERS

If the nature of the welding is critical and requires a written procedure, then equal consideration should be given to making sure the operator is qualified to do the kind of welding called for in the procedure. This can be done by having the operator make test welds which simulate the real thing. These welds can then be examined either destructively or nondestructively to see if the welder has demonstrated the required skills.

Note: Some welding may fall under local, state, or federal code requirements which do not permit welding to be done by in-house maintenance welders unless they have been certified.

OTHER WELDING PROCESSES

The welding processes and equipment described up to this point have potential for use in the typical maintenance welding department. Many other welding processes are used in industrial or production applications which, admittedly, have limited maintenance use. These will be summarized briefly. Additional information about specific processes can be obtained from the American Welding Society, 550 N.W. LeJeune Road, Miami, FL 33126.

Atomic Hydrogen Welding

Atomic hydrogen welding differs from other arc welding methods in that the arc is formed between the two tungsten electrodes and the work is not part of the welding circuit. A stream of hydrogen gas is passed through the arc and, in the heat of the arc, changes from molecular to atomic form, giving off an intense heat. The hydrogen acts as an effective heat-transfer medium, resulting in high heat being applied close to the work. A filler rod is used to supply additional metal to the joint. The process has some advantages in welding thin sheet where a high finish is needed.

Electroslag Welding

Electroslag welding is the metal arc welding process employing the principles of submerged arc welding. This process involves fusion of the base metal and continuously fed filler metal under a substantial layer of high-temperature, electrically conductive molten flux. By feeding one or a combination of two or three electrodes simultaneously into the arc, the operator can join plates ranging from 1 to 14 in thick in a single pass. Application of the process is generally limited to very heavy weldments. Welds are made with the joint vertical and with welding progressing from bottom to top.

Electrogas Welding

Electrogas welding applies a procedure similar to electroslag welding, utilizing retaining shoes or dams to keep the molten metal in the joint. Usually, electrogas is used for welding thick plates which are positioned vertically. Consumable electrodes are flux-cored (Innershield) and solid wires, and heating of the weld joint is applied uniformly by the arc, which is oscillated mechanically. The external gas is directed through small ports of the "gas box" mounted on the retaining shoes. Gas composition is usually 80 percent argon and 20 percent CO_2. Economic feasibility depends mainly on the length and width of the joint involved. The longer the welding joint, the smaller the cost of fixturing setup per unit length of the weld. The process is suitable for carbon steels, pressure-vessel steels, and structural steels.

Gas-Shielded Spot Welding

This method of welding combines either gas-shielded tungsten arc welding or gas-shielded metal arc welding with an electrical timing control system that automatically starts and maintains the arc for a controlled time period. Two lapped pieces of metal are spot welded together by applying heat from an electric arc to the top surface of the joint. Welding action is controlled by the current input to the arc and the time the arc dwells on the material being welded. Shielding of the arc, electrode (consumable metal or nonconsumable tungsten), and fluid weld puddle are similar to those of conventional gas-shielded arc welding. The resulting spot weld parallels that produced by resistance welding techniques; however, no electrode pressure is required, and the welding is done from one of the plates with no weld backup required. Industrial use of both inert-gas and CO_2 spot welding is expanding.

Plasma Arc Welding

Plasma arc welding exists in several forms. The basic principle is that of an arc or jet created by the electrical heating of a plasma-forming gas (such as argon with additions of helium or hydrogen) to such a high temperature that its molecules become ionized atoms possessing extremely high energy.

When properly controlled, this process results in very high melting temperatures. Plasma arc holds the potential solution to the easier joining of many hard-to-weld materials. Another application is the depositing of materials having high melting temperatures to produce surfaces of high resistance to extreme wear, corrosion, or temperature. As discussed earlier, when plasma arc technology is applied to metal cutting, it achieves unusually high speeds and has become an essential tool for a variety of maintenance applications.

Resistance Welding

Resistance welding processes are designed primarily for production welding. In resistance welding, the joining of the parts is accomplished by the heat obtained from the resistance of the work to the flow of electric current in a circuit of which the work is a part and by the application of pressure. Spot welding, the most common resistance welding process, can be used effectively in maintenance applications. It is usually employed in the welding of thin metal sheets and is accomplished by placing the sheets between movable copper-alloy electrodes. The electrodes carry the welding current and can be actuated to apply the proper pressure during the welding cycle. Resistance welding of aluminum and copper parts presents special problems, since these metals exhibit high electrical conductivity. Although production shops usually have large, floor-mounted resistance welding equipment, maintenance departments are more likely to use small, handheld spot welding guns to fabricate sheet metal.

Stud Welding

Stud welding is the end welding of a stud, ordinarily a machine screw, at a particular spot on the work by fusion. An electric arc, struck between the stud serving as the electrode and the baseplate, brings the tip of the stud and the surface of the work adjacent to the stud to a molten state. Light pressure is applied, forcing the stud into the molten weld puddle. Current flow is discontinued, and the stud fuses to the work surface as it cools. A compact unit, called a *stud welder,* supplies the welding current. The arc may be shielded or unshielded.

BASE METALS

The Carbon Steels

Carbon steels are widely used in all types of manufacturing. The weldability of the different types (low, medium, and high) varies considerably. The preferred analysis range of the common elements found in carbon steels is shown in Table 1.13. Welding metals whose elements vary above or below the range usually calls for special welding procedures.

Low-Carbon Steels (0.10 to 0.30 Percent Carbon). Steels of low-carbon content represent the bulk of the carbon steel tonnage used by industry. These steels are usually more ductile and easier to form than higher-carbon steels. For this reason, low-carbon steels are used in most applications requiring considerable cold forming, such as stampings and rolled or bent shapes in bar stock, structural shapes, or sheet. Steels with less than 0.13 percent carbon and 0.30 percent manganese have a slightly greater tendency to internal porosity than steels of higher carbon and manganese content.

TABLE 1.13 Preferred Analysis Range of Carbon Steels

	Low, %	Preferred, %	High, %
Carbon	0.06	0.10–0.25	0.35
Manganese	0.30	0.35–0.80	1.40
Silicon		0.10 or under	0.30 max
Sulfur		0.035 or under	0.05 max
Phosphorus		0.03 or under	0.04 max

Medium-Carbon Steels (0.31 to 0.45 Percent). The increased carbon content in medium-carbon steel usually raises the tensile strength, hardness, and wear resistance of the material. These steels are selectively used by manufacturers of railroad equipment, farm machinery, construction machinery, material-handling equipment, and other similar products. The medium-carbon steels can be welded successfully with the E60XX electrode if certain simple precautions are taken and the cooling rate is controlled to prevent excessive hardness.

High-Carbon Steels (0.46 Percent and Higher). The high-carbon steels are generally used in a hardened condition. This group includes most of the steels used in tools for forming, shaping, and cutting. Tools used in metalworking, woodworking, mining, and farming, such as lathe tools, drills, dies, knives, scraper blades, and plowshares, are typical examples. The high-carbon steels are often described as being "difficult to weld" and are not suited to mild steel welding procedures. Usually, low-hydrogen electrodes or processes are required, and controlled welding procedures, including preheating and postheating, are needed to produce crack-free welds.

The higher the carbon content of the steel, the harder the material becomes when it is quenched from above the critical temperature. Welding raises steel above the critical temperature, and the cold mass of metal surrounding the weld area creates a quench effect. Hardness and the absence of ductility result in cracking as the weld cools and contracts. Preheating from 300 to 600°F and slow cooling will usually prevent cracking. Figure 1.32 shows a calculator for determining preheat and interpass temperatures.

For steels in the higher carbon ranges (over 0.30 percent), special electrodes are recommended. The lime ferritic low-hydrogen electrodes (E7016 or E7018) can be used to good advantage in overcoming the cracking tendencies of high-carbon steels. A 308 stainless steel electrode also can be used to give good physical properties to a weld in high-carbon steel.

Cast Iron. Cast iron is a complex alloy with a very high carbon content. Quickly cooled cast iron is harder and more brittle than slowly cooled cast iron. The metal also naturally exhibits low ductility, which results in considerable strain on parts of a casting when one local area is heated. The brittleness and the uneven contraction and expansion of cast iron are the principal concerns when welding it.

Each job must be analyzed to predetermine the effect of welding heat so that corresponding procedures can be adopted. Welds can be deposited in short lengths, allowing each to cool. Peening of the weld metal while it is red hot may be used to stretch the weld deposit. Steel, cast iron, carbon, or nonferrous electrodes may be used. All oil, dirt, and foreign matter must be removed from the joint before welding. With steel electrodes, intermittent welds no longer than 3 in should be used with light peening. To reduce contraction, the work should never be allowed to get too hot in one spot. Preheating will help to reduce hardening of the deposit to make it more machinable.

For the most machinable welds, a nonferrous-alloy rod should be used. A two-layer deposit will have a softer fusion zone than a single-layer deposit. When it is practical, heating of the entire casting to a dull red heat is recommended in order to further soften the fusion zone and burn out dirt and foreign matter. When the weld is in a deep groove, it is general practice to use a steel electrode for welding cast iron to fill up the joint to within approximately $1/8$ in of the surface and then finish the weld with the more machinable nonferrous deposit, usually a 95 to 98 percent nickel electrode.

The Alloy Steels

High-Tensile Low-Alloy Steels. These steels are finding increasing use in metal fabricating because their higher strength levels permit the use of thinner sections, thereby saving metal and reducing weight. They are made with a number of different alloys and can be readily welded with specially designed electrodes that produce excellent welds of the same mechanical properties as the base metal. However, it is not necessary to have a core wire of exactly the same composition as the steel.

Stainless Steels. Electrodes are made to match various types of stainless steels so that corrosion-resistance properties are not destroyed in welding. The most commonly used types of stainless steels for welded structures are the 304, 308, 309, and 310 groups. Group 304 stainless, with a maximum carbon content of 0.8 percent, is commonly specified for weldments.

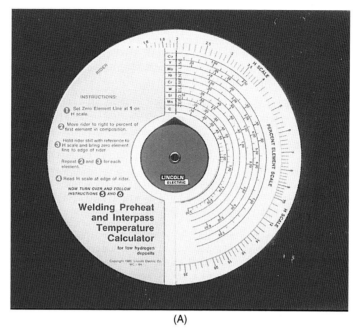

FIGURE 1.32 (*A*) Preheat calculator (observe). (*B*) Preheat calculator (reverse).

Welding procedures are much the same as for welding mild steel, except that one must take into account the higher electrical resistance of stainless and reduce the current accordingly. It is important to work carefully, cleaning all edges of foreign material. Light-gage work must be clamped firmly to prevent distortion and buckling. Small-diameter and short electrodes should be used to prevent loss of chromium and undue overheating of the electrode. The weld deposit should be approximately the same analysis as the plate (see Tables 1.14 and 1.15).

Stainless Clad Steel. The significant precautions in welding this material are in joint design, including edge preparation, procedure, and choice of electrode. The electrode should be of the correct analysis for the cladding being welded. The joint must be prepared and welded to prevent dilution of the clad surface by the steel backing material. The backing material is welded with a mild steel electrode but in multiple passes to prevent excessive penetration into the cladding. The clad side is also welded in small passes to prevent penetration into the backing material and resulting dilution of the stainless joint. When welding thin-gage material and it is necessary to make the weld in one pass, a 309 stainless electrode should be used for the steel side as well as the stainless side. The design and preparation of the joint can do much to prevent iron pickup, as well as reduce the labor costs of making the joint.

Straight Chromium Steels. The intense air-hardening property of these steels, which is proportional to the carbon and chromium content, is the chief consideration in establishing welding procedures. Considerable care must be taken to keep the work warm during welding, and it must be annealed afterward; otherwise, the welds and the areas adjacent to the welds will be brittle. It is a good idea to consult steel suppliers for specific details of proper heat treatment.

TABLE 1.14 Typical Compositions of Austenitic Stainless Steels

AISI type	Composition* (%)			
	Carbon	Chromium	Nickel	Other†
201	0.15	16.0–18.0	3.5–5.5	0.25 N, 5.5–7.5 Mn, 0.060 P
202	0.15	17.0–19.0	4.0–6.0	0.25 N, 7.5–10.0 Mn, 0.060 P
301	0.15	16.0–18.0	6.0–8.0	—
302	0.15	17.0–19.0	8.0–10.0	—
302B	0.15	17.0–19.0	8.0–10.0	2.0–3.0 Si
303	0.15	17.0–19.0	8.0–10.0	0.20 P, 0.15 S (min), 0.60 Mo (opt)
303Se	0.15	17.0–19.0	8.0–10.0	0.20 P, 0.06 S, 0.15 Se (min)
304	0.08	18.0–20.0	8.0–12.0	—
304L	0.03	18.0–20.0	8.0–12.0	—
305	0.12	17.0–19.0	10.0–13.0	—
308	0.08	19.0–21.0	10.0–12.0	—
309	0.20	22.0–24.0	12.0–15.0	—
309S	0.08	22.0–24.0	12.0–15.0	—
310	0.25	24.0–26.0	19.0–22.0	1.5 Si
310S	0.08	24.0–26.0	19.0–22.0	1.5 Si
314	0.25	23.0–26.0	19.0–22.0	1.5–3.0 Si
316	0.08	16.0–18.0	10.0–14.0	2.0–3.0 Mo
316L	0.03	16.0–18.0	10.0–14.0	2.0–3.0 Mo
317	0.08	18.0–20.0	11.0–15.0	3.0–4.0 Mo
321	0.08	17.0–19.0	9.0–12.0	Ti (5 × %C min)
347	0.08	17.0–19.0	9.0–13.0	Cb + Ta (10 × %C min)
348	0.08	17.0–19.0	9.0–13.0	Cb + Ta (10 × %C min but 0.10 Ta max), 0.20 Co

*Single values denote maximum percentage unless otherwise noted.

†Unless otherwise noted, other elements of all alloys listed include maximum contents of 2.0% Mn, 1.0% Si, 0.045% P, and 0.030% S. Balance is Fe.

TABLE 1.15 Typical Compositions of Ferritic Stainless Steels

AISI type	Composition* (%)			
	Carbon	Chromium	Manganese	Other†
405	0.08	11.5–14.5	1.0	0.1–0.3 Al
430	0.12	14.0–18.0	1.0	—
430F	0.12	14.0–18.0	1.25	0.060 P, 0.15 S (min), 0.60 Mo (opt)
430FSe	0.12	14.0–18.0	1.25	0.060 P, 0.060 S, 0.15 Se (min)
442	0.20	18.0–23.0	1.0	—
446	0.20	23.0–27.0	1.5	0.25 N

*Single values denote maximum percentage unless otherwise noted.
†Unless otherwise noted, other elements of all alloys listed include maximum contents of 1.0% Si, 0.040% P, and 0.030% S. Balance is Fe.

High-Manganese Steels. High-manganese steels (11 to 14 percent Mn) are very tough and are work-hardening, which makes them ideally suited for surfaces which must resist abrasion or wear, as well as shock. When building up parts made of high-manganese steel, an electrode of similar analysis should be used.

The Nonferrous Metals

Aluminum. Most fusion welding of aluminum alloys is done with either the gas metal arc (GMAW) process or the gas tungsten arc (GTAW) process. In either case, inert-gas shielding is used.

With GMAW, the electrode is aluminum filler fed continuously from a reel into the weld pool. This action propels the filler metal across the arc to the workpiece in line with the axis of the electrode, regardless of the orientation of the electrode. Because of this, and because of aluminum's qualities of density, surface tension, and cooling rate, horizontal, vertical, and overhead welds can be made with relative ease. High deposition rates are practical, producing less distortion, greater weld strength, and lower welding costs than can be attained with other fusion welding processes.

GTAW uses a nonconsumable tungsten electrode, with aluminum alloy filler material added separately, either from a handheld rod or from a reel. Alternating current (ac) is preferred by many users for both manual and automatic gas tungsten arc welding of aluminum because ac GTAW achieves an efficient balance between penetration and cleaning.

Copper and Copper Alloys. Copper and its alloys can be welded with shielded metal arc, gas-shielded carbon arc, or gas tungsten arc welding. Of all these, gas-shielded arc welding with an inert gas is preferred. Decrease in tensile strength as temperature rises and a high coefficient of contraction may make welding of copper complicated. Preheating usually is necessary on thicker sections because of the high heat conductivity of the metal. Keeping the work hot and pointing the electrode at an angle so that the flame is directed back over the work will aid in permitting gases to escape. It is also advisable to put as much metal down per bead as is practical.

CONTROL OF DISTORTION

The heat of welding can distort the base metal; this sometimes becomes a problem in welding sheet metal or unrestrained large sections. The following suggestions will help in overcoming problems of distortion:

1. Reduce the effective shrinkage force.
    a. Avoid overwelding. Use as little weld metal as possible by taking advantage of the penetrating effect of the arc force.

b. Use correct edge preparation and fit-up to obtain required fusion at the root of the weld.
  c. Use fewer passes.
  d. Place welds near a neutral axis.
  e. Use intermittent welds.
  f. Use back-step welding method.
2. Make shrinkage forces work to minimize distortion.
  a. Preset the parts so that when the weld shrinks, they will be in the correct position after cooling.
  b. Space parts to allow for shrinkage.
  c. Prebend parts so that contraction will pull the parts into alignment.
3. Balance shrinkage forces with other forces (where natural rigidity of parts is insufficient to resist contraction).
  a. Balance one force with another by correct welding sequence so that contraction caused by the weld counteracts the forces of welds previously made.
  b. Peen beads to stretch weld metal. Care must be taken so that weld metal is not damaged.
  c. Use jigs and fixtures to hold the work in a rigid position with sufficient strength to prevent parts from distorting. Fixtures can actually cause weld metal to stretch, preventing distortion.

SUBMERGED ARC WELDING (SAW): EQUIPMENT, ELECTRODES, AND FLUX

Welding Equipment

The welding heads normally used for fully automatic submerged arc welding perform the triple function of progressively depositing flux along the joint, feeding the electrode, and transmitting welding current to the electrode. The flux is usually supplied from a hopper either mounted directly on the head or connected to the head by tubing. The bare electrode or wire is fed into the welding head from a coil mounted on a reel. The distance between the end of the electrode and the base metal is held constant by special controls which automatically regulate the electrode feed motor speed or welding current.

Automatic submerged arc welding is increasingly used as a maintenance process. Both fully automatic and semiautomatic equipment can be used, and the chief maintenance application is for hard surfacing. Automatic SAW permits the rapid deposition of large amounts of uniformly excellent weld metal. Semiautomatic equipment is relatively inexpensive and can be adapted to existing welding equipment of larger amperage outputs. Fully automatic equipment costs more, and only a large volume of work is likely to justify its installation for maintenance applications. Equipment manufactured for semiautomatic SAW performs the same functions as that for fully automatic welding. The welding head, however, now consists of a welding gun and wire feeder unit. The flux for semiautomatic welding is supplied by a canister mounted on the welding gun or a "continuous-flow" flux feed from a pressurized tank. With semiautomatic welding equipment, the electrode wire is mechanically fed to the work, but the welding gun is manually moved along the joint being welded. This procedure gives added flexibility to SAW by permitting its use on irregular shapes and contours, thereby promoting expanded use of the process.

Direct current is used with both semiautomatic and fully automatic SAW, whereas alternating current is usually limited to the fully automatic application of the process. Welding voltages will range from 28 to 55 V. Currents used for submerged arc welding are generally higher than those used for other processes and range from a low of 200 A to as high as 1200 to 1500 A, single wire.

FIGURE 1.33 Semiautomatic submerged arc welding.

Alternating current may be supplied from one or more heavy-duty welding transformers. Direct current may be supplied by one or more motor-generator or rectifier welding machines with capacity suitable for the application. Direct-current power supplies can be constant-potential or variable-voltage types depending on the application and the manufacturer's recommendations. Installations of semiautomatic and fully automatic welding equipment are illustrated in Figs. 1.33 and 1.34.

Electrodes and Fluxes

The ferrous and nonferrous electrodes commonly used for submerged arc welding are bare rods or wires with clean, bright surfaces to facilitate the introduction of relatively high currents. Electrodes are normally used in the form of coils ranging in weight from 25 to 200 lb. On very high production welding installations, the electrode is frequently fed from a coil in a drum. These drums range up to 1000 lb in weight. Ferrous wire of a composition that might readily rust is copper coated to retard rusting and improve the contact surfaces.

The fluxes used with submerged arc welding are granulated fusible mineral materials essentially free from substances that would create large amounts of gases during welding. These fluxes are made to a variety of chemical specifications which develop particular performance characteristics. The flux has a number of functions to perform, including prevention of atmospheric contamination and performing a scavenging-deoxidizing action on the molten metal in the weld crater. Some special fluxes perform the additional function of contributing alloying elements to the weld deposit, thereby developing specific weld-metal characteristics of higher strength or even abrasion resistance. The choice of flux depends on the welding procedure to be employed, the type of joint, and the composition of the electrode to be used with it.

FIGURE 1.34 Fully automatic welding equipment.

SPECIAL APPLICATIONS

Sheet Metal Welding

Plant maintenance frequently calls for sheet metal welding. The principles of good welding practice apply in welding sheet metal as elsewhere, but welding thin-gage metals poses the specific challenges of potential distortion and/or burn-through. Special attention should therefore be given to all the factors involved in controlling distortion: the speed of welding, the choice of proper joints, good fit-up, position, proper current selection, use of clamping devices and fixtures, number of passes, and sequence of beads.

Good welding practice normally calls for the highest arc speeds and the highest currents within the limits of good weld appearance. In sheet metal work, however, there is always the limitation imposed by the threat of burn-through. As the gap in the work increases in size, the current must be decreased to prevent burn-through; this, of course, will reduce welding speeds. A clamping fixture will improve the fit-up of joints, making higher speeds possible. If equipped with a copper backing strip, the clamping fixture will make welding easier by decreasing the tendency to burn-through and also removing some of the heat that can cause warpage. Where possible, sheet metal joints should be welded downhill at about a 45° angle with the same currents that are used in the flat position or slightly higher. Tables 1.16 and 1.17 offer guides to the selection of proper current, voltage, and electrodes for the various types of joints used with 20- to 8-gage sheet metal.

TABLE 1.16 Welding Currents for Sheet Metal (SMAW)

Type of welded joint	20 ga			18 ga			16 ga			14 ga			12 ga			10 ga			8 ga		
	F*	V*	O*	F	V	O	F	V	O	F	V	O	F	V	O	F	V	O	F	V	O
Plain butt	30†	30†	30†	40†	40†	40†	70†	70†	70†	85†	80	85†	115	110	110	135	120	115	190	130	120
Lap	40†	40†	40†	60†	60†	60†	100	100	100	130	130	130	135	120	120	155	130	120	165	140	120
Fillet				40†	40†	40†	70†	70†	70†	100	90	85	150	140	120	160	150	130	160	160	130
Corner	40†	40†	40†	60†	60†	60†	90†	90†	90†	90	80	75	125	110	110	140	130	125	175	130	125
Edge	40†	40†	40†	60†	60†	60†	80†	80†	80†	110	80	80	145	110	110	150	120	120	160	120	120

*F—flat position; V—vertical; O—overhead.
†Electrode negative, work positive.

Hard Surfacing

The building up of a layer of metal or a metal surface by electric arc welding, commonly known as *hard surfacing*, has important and useful applications in equipment maintenance. These may include restoring worn cutting edges and teeth on excavators, building up worn shafts with low- or medium-carbon deposits, lining a carbon-steel bin or chute with a stainless steel corrosion-resistant alloy deposit, putting a tool-steel cutting edge on a medium-carbon steel base, and applying wear-resistant surfaces to metal machine parts of all kinds. The dragline bucket shown in Fig. 1.35 is being returned to "new" condition by rebuilding and hard surfacing. Arc weld surfacing techniques include, but are not limited to, hard surfacing. There are many buildup applications which do not require hard surfacing. Excluding the effects of corrosion, wear of machinery parts results from various combinations of abrasion and impact. Abrasive wear results from one material scratching another, and impact wear results from one material hitting another.

Resisting Abrasive Wear

Abrasive wear is resisted by materials with a high scratch hardness. Sand quickly wears metals with a low scratch hardness, but under the same conditions, it will wear a metal of high scratch hardness very slowly. Scratch hardness, however, is not necessarily measured by standard hardness tests. Brinell and Rockwell hardness tests are not reliable measures for determining the abrasive wear resistance of a metal. A hard-surfacing material of the chromium carbide type may have a hardness of 50 Rockwell C. Sand will wear this material at a slower rate than it will a steel hardened to 60 Rockwell C. The sand will scratch all the way across the surface of the steel. On the surfacing alloy,

TABLE 1.17 Sizes of Electrodes for Sheet Metal (SMAW)

Type of welded joint	20 ga			18 ga			16 ga			14 ga			12 ga			10 ga			8 ga		
	F*	V*	O*	F	V	O	F	V	O	F	V	O	F	V	O	F	V	O	F	V	O
Plain butt	3/32	3/32	3/32	3/32	3/32	3/32	1/8	1/8	1/8	1/8	1/8	1/8	5/32	5/32	5/32	5/32	5/32	5/32	3/16	5/32	5/32
Lap	3/32	3/32	3/32	3/32	3/32	3/32	1/8	1/8	1/8	5/32	5/32	5/32	5/32	5/32	5/32	3/16	3/16	5/32	3/16	3/16	5/32
Fillet				3/32	3/32	3/32	1/8	1/8	1/8	1/8	1/8	1/8	5/32	5/32	5/32	3/16	5/32	5/32	3/16	5/32	5/32
Corner	3/32	3/32	3/32	3/32	3/32	3/32	1/8	1/8	1/8	1/8	1/8	1/8	3/16	5/32	5/32	3/16	5/32	5/32	3/16	5/32	5/32
Edge	3/32	3/32	3/32	3/32	3/32	3/32	1/8	1/8	1/8	1/8	1/8	1/8	3/16	5/32	5/32	3/16	5/32	5/32	3/16	5/32	5/32

*F—flat position; V—vertical; O—overhead.

FIGURE 1.35 Shielded metal arc welding is used to rebuild and to hard-surface worn areas of a dragline bucket.

the scratch will progress through the matrix material and then stop when the sand grain comes up against one of the microscopic crystals of chromium carbide, which has a higher scratch hardness than sand. If two metals of the same type have the same kind of microscopic constituents, however, the metal having the higher Rockwell hardness will be more resistant to abrasive wear.

Resisting Impact Wear

Whereas abrasive wear is resisted by the surface properties of a metal, impact wear is resisted by the properties of the metal beneath the surface. To resist impact, a tough material is used, one that does not readily bend, break, chip, or crack. It yields so as to distribute or absorb the load created by impact, and the ultimate strength of the metal is not exceeded. Included in impact wear is that caused by bending or compression at low velocity without impact, resulting in loss of metal by cracking, chipping, upsetting, flowing, or crushing.

Types of Surfacing Electrodes

Many different kinds of surfacing electrodes are available. The problem is to find the best one to do a given job. Yet because service conditions vary so widely, no universal standard can be established for determining the ability of the surfacing to resist impact or abrasion. Furthermore, there is no ideal surfacing material that resists impact and abrasion equally well. In manufacturing the surfacing electrodes, it is necessary to sacrifice one quality somewhat to gain the other. High impact resistance is gained by sacrificing abrasion resistance, and vice versa.

Price is no index to electrode quality. An expensive electrode ingredient does not necessarily impart wear resistance. Therefore, the user of surfacing materials must rely on a combination of the manufacturer's recommendations and the user's own tests to select the best surfacing material for a particular purpose.

Choosing Hard-Facing Material

The chart shown in Fig. 1.36 lists the relative characteristics of manual hard-facing materials. This chart is a guide to selecting the two items in the following list:

1. The hard-facing electrode best suited for a job not hard-faced before.
2. A more suitable hard-facing electrode for a job where the present material has not produced the desired results.

Example 1

Application: Dragline bucket tooth, as shown in Fig. 1.37.
Service: Sandy gravel with some good-sized rocks.

Maximum wear that can be economically obtained is the goal of most hard-facing applications. The material chosen should rate as highly as possible in the resistance-to-abrasion column, unless some other characteristics shown in the other columns make it unsuited for this particular application.

First, consider the tungsten carbide types. Notice that they are composed of very hard particles in a softer and less abrasion-resistant matrix. Although such material is the best for resisting sliding abrasion on hard material, in sand the matrix is apt to scour out slightly, and then the brittle particles are exposed. These particles are rated poor in impact resistance, and they may break and spall off when they encounter the rocks.

Next best in abrasion, as listed in the chart, is the high-chromium carbide type shown in the electrode size column to be a powder. It can be applied only in a thin layer and also is not rated high in impact resistance. This makes it of dubious use in this rocky soil.

The rod-type high-chromium carbides also rate very high in abrasion resistance but do not rate high in impact resistance. However, the second does show sufficient impact rating to be considered if two or three different materials are to be tested in a field test. Given the possibility that it has enough impact resistance to do this job, there may be reluctance to pass up its very good wearing properties.

Nevertheless, the semiaustenitic type is balanced in both abrasion and impact resistance. It is much better in resistance to impact than the materials that rate higher in abrasion resistance. Therefore, the semiaustenitic is the first choice for this job, considering that the added impact resistance of the austenitic type is not necessary, since the impact in this application is not extreme.

Example 2

Application: Same dragline tooth used in Example 1.
Service: Soil changed to clay and shale.

The semiaustenitic type selected in the first example stands up well, but the teeth wear only half as long as the bucket lip. With double the wear on the teeth, only half the downtime periods would be needed for resurfacing, and both teeth and bucket could be done together. Since impact wear is now negligible with the new soil conditions, a material higher in the abrasion column should be considered. A good selection would be the first high-chromium carbide rod, which could give twice the wear by controlling the size bead applied while still staying within a reasonable cost range.

Example 3

Application: Same dragline tooth used as in Examples 1 and 2.
Service: Soil changed to obtain large rocks.

If the earth contains many hard and large rocks, and the teeth are failing because of spalling under impact, one should move down the abrasion-resisting column to a more impact-resistant material, such as the semiaustenitic type.

These examples demonstrate that where a dragline operates in all kinds of soils, a material that is resistant to both impact and abrasion, such as a semiaustenitic type, is the best choice. When the same type of reasoning is used to check the important characteristics, an appropriate material can be

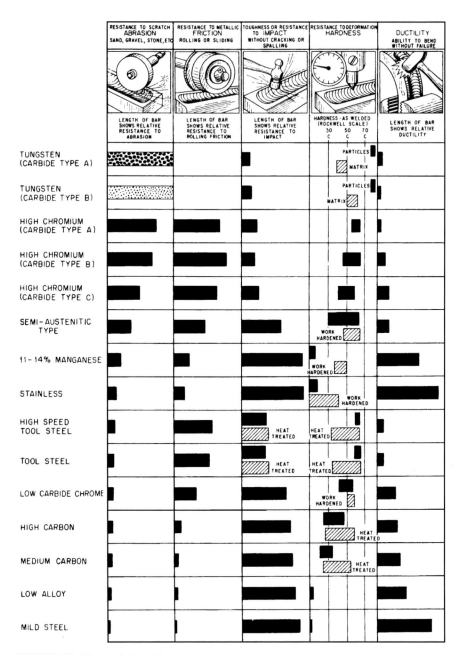

FIGURE 1.36 Hard-surfacing guide.

ARC WELDING IN MAINTENANCE   10.47

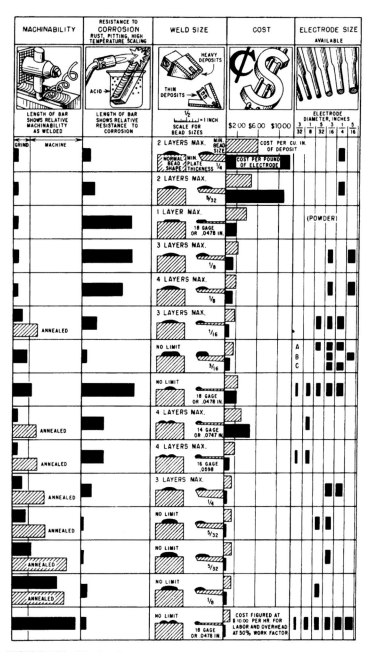

FIGURE 1.36 (*Continued*)

FIGURE 1.37 Bucket teeth have been rebuilt and hard-surfaced.

chosen for any application. If, for any reason, the first choice does not prove satisfactory, it is usually easy to improve the next application by choosing a material that is rated higher in the characteristic that was lacking. Where failures occur because of cracking or spalling, it usually indicates that a material higher in impact or ductility rating should be used. Where normal wear alone seems too rapid, a material with a higher abrasion rating is indicated.

Check Welding Procedure

Often, hard-facing failures due to cracking or spalling may be caused by improper welding procedures. Before changing the hard-surfacing material, consider whether or not the material has been properly applied. For almost any hard-facing application, very good results can be obtained by following these precautions:

1. Do not apply hard-surfacing material over cracked or porous areas. Remove any defective areas down to sound base metal.
2. Preheat. Preheating to 400 to 500°F improves the resistance to cracking and spalling. This minimum temperature should be maintained until welding is completed. The exception to the rule is 11 to 14 percent manganese steel, which should be kept cool.
3. Cool slowly. If possible, allow the finished weldment to cool under an insulating material such as lime or sand.
4. Do not apply more than the recommended number of layers.

When more than normal buildup is required, apply intermediate layers of either medium-carbon or stainless steel. This will provide a good bond to the base metal and will eliminate excessively thick layers of hard-surfacing material which might otherwise spall off. Stainless steel is also an excellent choice for intermediate layers on manganese steels or for hard-to-weld steels where preheating is not practical.

Check before the Part Is Completely Worn

Whenever possible, examine a surfaced part when it is only partly worn. Examination of a part after it is completely worn is unsatisfactory. Did the surface crumble off, or was it scratched off? Is a tougher surface needed, or is additional abrasion resistance required? Should a heavier layer of surfacing be used? Should surfacing be reduced? All these questions can be answered by examination of a partially worn part and with a knowledge of the surfacing costs and service requirements.

FIGURE 1.38 Mild-steel die on the edge of which tool steel has been deposited by means of tool-steel electrode.

When it is impossible to analyze the service conditions thoroughly in advance, it is always on the safe side to choose a material tougher than is thought to be required. A tough material will not spall or chip off and will offer some resistance to abrasion. A hard, abrasion-resistant material is more susceptible to chipping, and surfacing material does no good if it falls off.

After some experience with surfacing materials, various combinations of materials can be tried to improve product performance. For example, on a part which is normally surfaced with a tough, semiaustenitic electrode, it may be possible to get additional abrasion resistance without sacrificing resistance to cracking. A little of the powdered chromium carbide material can be fused to critical areas where additional protection is needed.

Many badly worn parts are first built up to almost finished size with a high-carbon electrode, then surfaced with an austenitic rod, and finally a few beads of chromium carbide deposit are placed in spots requiring maximum protection against abrasion. Regardless of the circumstances, a careful analysis of the surfacing problem will be well worthwhile. Examples of jobs are shown in Figs. 1.38 to 1.40.

Hard Surfacing with SAW

The submerged arc process offers several advantages for hard surfacing. The greater uniformity of the surface makes for better wearing qualities. The speed of SAW creates major economies in hard-surfacing areas which require the deposition of large amounts of metal. These areas may be either flat or curved surfaces. Mixer bottom plates, scraper blades, fan blades, chutes, and refinery vessels are examples of the flat plate to be surfaced. Shafts, blooming mill spindles, skelp rolls, crane wheels, tractor idlers and rollers, and rams are examples of cylindrical surfaces (Figs. 1.41 to 1.44).

FIGURE 1.39 Cone used for uncoiling steel. Hardfacing material has been deposited on mild-steel base. Surface is ready for grinding.

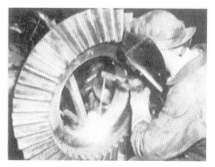

FIGURE 1.40 Using a mild-steel electrode to build up the inside diameter and all the teeth of a 25 year-old cast-steel gear that could not be replaced.

FIGURE 1.41 Steel-mill coke pushers being hard-faced by submerged arc process. Mild-steel wire and alloy flux. Fully automatic equipment in foreground, semiautomatic in background.

FIGURE 1.42 Hard-surfacing wire mill roll by submerged arc process. Mild-steel wire and alloy flux. Gas torch keeps roll up to temperature.

FIGURE 1.43 Automatic head adapted for oscillating and for two electrodes, being used to deposit 3-in. beads on a flat mixer bottom plate.

FIGURE 1.44 Submerged arc welding being used to hard-surface a cylindrical surface.

The process can be used with either fully automatic or semiautomatic equipment depending on the economics of the application. Fully automatic equipment can be quickly fitted with auxiliary accessories, resulting in more economical metal deposition. An oscillating device can be added to an automatic head to create a bead up to 3 in wide in a single pass. Another attachment permits the feeding of two electrode wires through a single head and a single contact jaw. Both these attachments are useful in hard surfacing.

Hard surfacing with a submerged arc can be done with several different types of materials. The hard-surfacing deposit can be created by using solid alloy wires and a neutral granular flux. It also can be created by using a solid mild-steel wire and an agglomerated-alloy flux, the alloys being added to the deposit through the flux rather than through the wire. Also available are tubular wires which contain alloying material in the hollow portion of the mild-steel tube. All the methods have specific advantages. With SAW, considerable variation in the hard-surfacing deposit can be made by changing the welding procedure to control admixture and the heat-treatment effect of the welding cycle. Methods and procedures should be established with the help of qualified engineers.

SELECTION AND MAINTENANCE OF EQUIPMENT

Machines

Satisfactory welding can be accomplished with either alternating or direct welding current. Each type of current, however, has particular advantages which make it best suited for certain types of welding and welding conditions. The chief advantage of alternating current is its elimination of arc blow, which may be encountered when welding on heavy plate or into a corner. The magnetic fields set up in the plate deflect the path of the arc. Alternating current tends to minimize this deflection and also will increase the speed of welding with larger electrodes, over $^3/_{16}$ in diameter, and with the iron powder type of electrodes.

The chief advantages of direct current are the stability of the arc and the fact that the current output of the motor-generator type of welder will remain constant in spite of variations in the input voltage which affect a transformer-type welder. Direct current, therefore, is a more versatile welding current. Certain electrodes, such as stainless, require a very stable arc; these electrodes operate much better with

FIGURE 1.45 A 300-amp dc motor-generator set for work in a power plant.

direct current. Direct current, because of its stability, is also better for sheet metal welding, where the danger of burn-through is present. The dc arc also can be more readily varied to meet different welding conditions. A wider range of control over both voltage and current permits closer adjustment of the arc for difficult welding conditions, such as might be encountered in vertical or overhead welding. Because of its versatility, direct current should be available for maintenance welding.

Direct-current welders (Figs. 1.45 and 1.46) are made either as motor-generator sets or as transformer-rectifier sets. Motor-generator sets are powered by ac or dc motors. Generators are also powered by small air-cooled gasoline engines (Fig. 1.47). The advantage of this type of set is that for on-the-spot maintenance welding, it is not necessary to string electric power lines to the job site. Engine-driven welders powered by gasoline engines are also available and come in larger sizes than the air-cooled engine sets (Fig. 1.48). These are suitable where the size of the plant maintenance operation warrants a larger welder.

For most general maintenance welding, a 250-A output capacity is ample. Several manufacturers make compact, portable machines especially for this type of welding. Higher amperages may be required in particular applications; for these, heavy-duty machines should be used.

Another type of welding machine is one that produces both alternating and direct welding current, either of which is available at the flip of a switch (Fig. 1.49). This is ideal for maintenance welding, since it makes any kind of welding arc available, offering complete flexibility.

Accessory Equipment

The varied and severe service demands made on maintenance welding equipment require that the best in accessories be used. Most maintenance welders make racks or other storage conveniences which they attach directly to the welding machine to facilitate storing and transporting electrodes and accessories. While these arrangements will vary to suit individual tastes and needs, the end result is to have everything immediately available for use.

A fire extinguisher is an essential accessory. Many electrode holders are available, but only a few combine all the desirable features. The operator holds the electrode clamped in a holder, and the current from the welding set passes through the holder to the electrode. The clamping device should be designed to hold the electrode securely in position yet permit the quick and easy exchange of electrodes. It should be light in weight, properly balanced and easy to handle, yet sturdy enough to withstand rough use. It should be designed to remain cool enough to be handled comfortably (see Fig. 1.50).

FIGURE 1.46 Motor-generator with rectifier.

FIGURE 1.47 Direct-current welder.

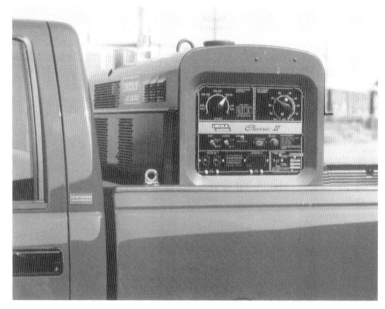

FIGURE 1.48 Air-cooled engine set.

FIGURE 1.49 An ac or dc unit.

FIGURE 1.50 Fully insulated electrode holder designed especially for cool operation.

Face or head shields are generally constructed of some kind of pressed fiber insulating material, usually black to reduce glare. The shield should be light in weight and comfortable to wear. The glass windows in the shield should be of a material that absorbs infrared rays, ultraviolet rays, and most visible rays emanating from the arc. The welding lens should be protected from molten metal spatter and breakage by a chemically treated clear "nonspatter" glass covering the exposed side of the lens. The operator should always wear a protective shield when welding and should never look at the arc with the naked eye. When a new lens is put into the shield, care should be taken to make sure no light leaks in around the glass. If practical, the welding room should be painted a dead black or some other dark color to prevent the reflection of light and glare. Others working around the welding area can be easily shielded from light and sparks by the use of portable screens.

Special goggles are used by welders' helpers, foremen, supervisors, inspectors, and others working close to a welding arc to protect their eyes from occasional flashes. A good set of goggles has an adjustable elastic head band and is light weight, cool, well-ventilated, and comfortable. Clear cover glasses and tinted lenses in various shades are available for this type of goggle.

During the arc welding process, some sparks and globules of molten metal are thrown out from the arc. For protection from possible burns, the operator is advised to wear an apron of leather or other protective material. Some operators also wear spats or leggings and sleevelets of leather or other fire-resistant material. Some sort of protection should be provided for the operator's ankles and feet, since a globule of molten metal can cause a painful burn before it can be extracted from the shoe. A gauntlet type of glove, preferably made of leather, is generally used by operators to protect their hands from the arc rays, spatters of molten metal, sparks, and the like. Gloves also provide protection when the operator is handling the work.

Other tools of value in any shop where welding is done include wire brushes for cleaning the welds, cold chisels for chipping, clamps for holding work in position for welding, wedges, and, where work is large or heavy, a crane or chain block. A drill, air hammer, and grinder are also valuable accessories.

INSTALLATION OF EQUIPMENT

Good welding begins with the proper installation of equipment. Installations should be made in locations that are as clean as possible, and there should be provisions for a continuous supply of clean air for ventilation. It is important to provide separate enclosures if the atmosphere is excessively moist or contains corrosive vapors. If welding must be done where the ambient temperature is high, place the equipment in a different location. Sets operated outdoors should be equipped with protection against inclement weather.

When installing welding equipment, consider the following:

1. Contact the local power company to ensure an adequate supply of electric power.
2. Provide an adequate and level support for the equipment.
3. Protect adequately against mechanical abuse and atmospheric conditions.
4. Provide fresh air for ventilation and cooling.
5. Electrically ground the frame of the welder.
6. Check electrical connections to make sure they are clean and mechanically tight.
7. The fuses for a motor-generator welder should be of the "high lag" type and be rated two or three times the input-current rating of the welder.
8. Provide welding leads of sufficient capacity to handle the required current.
9. Check the set before operating it to make sure that no parts are visibly loose or in poor condition.

EQUIPMENT OPERATION AND MAINTENANCE

The following precautions will do much to ensure maximum service and performance from arc welding equipment.

Keep the Machine Clean and Cool

Because of the large volume of air pulled through welders by the fans in order to keep the machines cool, the greatest enemies of continuous efficient performance are airborne dust and abrasive materials. Machines which are exposed to ordinary dust should be blown out at least once a week with dry, clean compressed air at a pressure not exceeding 30 psi. Higher pressures may damage windings.

In foundries or machine shops, where cast iron or steel dust is present, vacuum cleaning should be substituted for compressed air. Compressed air under high pressure tends to drive the abrasive dust into the windings.

Abrasive material in the atmosphere grooves and pits the commutator and wears out brushes. Greasy dirt or lint-laden dust quickly clogs air passages between coils and causes them to overheat. Since resistance of the coils is raised and the conductivity lowered by heat, it reduces efficiency and can result in burned-out coils if the machine is not protected against overload. Overheating makes the insulation between coils dry and brittle. Neither the air intake nor the exhaust vents should be blocked, because this will interrupt the flow of air through the machine. The welder covers should be kept on; removing them destroys the proper path of ventilation.

Do Not Abuse the Machine

Never leave the electrode grounded to the work. This can create a "dead" short circuit. The machine is forced to generate much higher current than it was designed for, which can result in a burned-out machine.

Do Not Work the Machine over Its Rated Capacity

A 200-A machine will not do the work of a 400-A machine. Operating above capacity causes overheating, which can destroy the insulation or melt the solder in the commutator connections.

Use extreme care in operating a machine on a steady load other than arc welding, such as thawing water pipes, supplying current for lighting, running motors, charging batteries, or operating heating equipment. For example, a dc machine, NEMA-rated 300 A to 40 V or 12 kW, should not be used for any continuous load greater than 9.6 kW and not more than 240 A. This precaution applies to machines with a duty cycle of at least 60 percent. Machines with lower load-factor ratings must be operated at still lower percentages of the rated load.

Do Not Handle Roughly

A welder is a precisely aligned and balanced machine. Mechanical abuse, rough handling, or severe shock may disturb the alignment and balance of the machine, resulting in serious trouble. Misalignment can cause bearing failure, bracket failure, unbalanced air gap, or unbalance in the armature.

Never pry on the ventilating fan or commutator to try to move the armature. To do so will damage the fan or commutator. If the armature is jammed, inspect the unit for the cause of the trouble. Check for dirt or foreign particles between the armature and frames. Inspect the banding wire on the armature. Look for a frozen bearing.

Do not neglect the engine if the welder is an engine-driven unit. It deteriorates rapidly if not properly cared for. Follow the engine manufacturer's recommendations. Change the oil regularly. Keep air filters and oil strainers clean. Do not allow grease and oil from the engine to leak back into the generator. Grease quickly accumulates dirt and dust, clogging the air passages between the coils.

Maintain the Machine Regularly

Bearings. The ball bearings in modern welders have sufficient grease to last the life of the machine under normal conditions. Under severe conditions—heavy use or a dirty location—the bearings should be greased about once a year. An ounce of grease a year is sufficient for each bearing. A pad of grease approximately 1 in^3 in volume weighs close to 1 oz. Dirt is responsible for more bearing failures than any other cause. This dirt may get into the grease cup when it is removed to refill, or it may get into the grease in its original container. Before the grease cup or pipe plug is removed, it is important to wipe it absolutely clean. A piece of dirt no larger than the period at the end of this sentence may cause a bearing to fail in a short time. Even small particles of grit that float around in the factory atmosphere are dangerous.

If too little grease is applied, bearings fail. If the grease is too light, it will run out. Grease containing solid materials may ruin antifriction bearings. Rancid grease will not lubricate. Dirty grease or dirty fittings or pipes can cause bearing failures.

Generally, bearings do not need inspection. They are sealed against dirt and should not be opened. If bearings must be pulled, it should be done using a special puller designed to act against the inner race.

Never clean new bearings before installing them. Handle them with care. Put them in place by driving against the inner race. Make sure that they fit squarely against the shoulders.

Brackets or End Bolts. If it becomes necessary to remove a bracket, to replace a bearing, or to disassemble the machine, do so by removing the bolts and tapping lightly and evenly with a babbitt hammer all around the outside diameter of the bracket ring. Do not drive off with a heavy steel hammer. The bearing may become worn over size, caused by the pounding of the bearing when the armature is out of balance. The bearing should slide into the housing with a light drive fit. Replace the bracket if the housing is over size.

Brushes and Brush Holders. Set brush holders approximately $1/32$ to $3/32$ in above the surface of the commutator. If brush holders have been removed, be certain that they are set squarely in the rocker slot when replaced. Do not force the brush holder into the slot by driving on the insulation. Check to ensure that the brush holder insulation is squarely set. Tighten brush holders firmly. When properly set, they are parallel to the mica segments between commutator bars. Use the grade of brushes recommended by the manufacturer of the welding set. Brushes that are too hard or too soft may damage the commutator. Brushes will be damaged by excessive clearance in the brush holder or uneven brush spring pressure. High commutator bars, high mica segments, excessive brush spring pressure, and abrasive dust also will wear out brushes rapidly.

Inspect brushes and holders regularly. A brush may wear down and lose spring tension. It will then start to arc, with damage to the commutator and other brushes. Keep the brush contact surface of the holder clean and free from pit marks. Brushes must be able to move freely in the holder. Replace them when the pigtails are within $1/8$ in of the commutator or when the limit of spring travel is reached.

New brushes must be sanded in to conform to the shape of the commutator. This may be done by stoning the commutator with a stone or by using fine sandpaper (not emery cloth or paper). Place the sandpaper under the brush, and move it back and forth while holding the brush down in the normal position under slight pressure with the fingers. See that the brush holders and springs seat squarely and firmly against the brushes and that the pigtails are fastened securely.

Commutators. Commutators normally need little care. They will build up a surface film of brown copper oxide, which is highly conductive, hard, and smooth. This surface helps to protect the commutator. Do not try to keep a commutator bright and shiny by constant stoning. The brown copper oxide film prevents the buildup of a black abrasive oxide film that has high resistance and causes excessive brush and commutator wear. Wipe clean occasionally with a rag or canvas to remove grease discoloration from fumes or other unnatural film. If brushes are chattering because of high bars, high mica, or grooves, stone by hand or remove and turn in a lathe, if necessary.

Most commutator trouble starts because the wrong grade of brushes is used. Brushes that contain too much abrasive material or have too high a copper content usually scratch the commutator and prevent the desired surface film from building up. A brush that is too soft may smudge the surface with the same result as far as surface film is concerned. In general, brushes that have a low voltage drop will give poor commutation. Conversely, a brush with high voltage drop commutates better but may cause overheating of the commutator surface.

If the commutator is burned, it may be dressed down by pressing a commutator stone against the surface with the brushes raised. If the surface is badly pitted or out of round, the armature must be removed from the machine and the commutator turned in a lathe. It is good practice for the commutator to run within a radial tolerance of 0.003 in. The mica separating the bars of the commutator is undercut to a depth of $1/32$ to $1/16$ in. Mica exposed at the commutator surface causes brush and commutator wear and poor commutation. If the mica is even with the surface, undercut it. When the commutator is operating properly, there is very little visible sparking. The brush surface is shiny and smooth, with no evidence of scratches.

Generator Frame. The generator frame and coils need no attention other than inspection to ensure tight connections and cleanliness. Blow out dust and dirt with compressed air. Grease may be cleaned off with naphtha. Keep air gaps between armature and pole pieces clean and even.

Armature. The armature must be kept clean to ensure proper balance. Unbalance in the set will pound out the bearings and wear the bearing housing oversize. Blow out the armature regularly with clean, dry compressed air. Clean out the inside of the armature thoroughly by attaching a long pipe to the compressed air line and reaching into the armature coils.

Motor Stator. Keep the stator clean and free from grease. When reconnecting it for use on another voltage, solder all connections. If the set is to be used frequently on different voltages, it may save time to place lugs on the ends of all the stator leads. This eliminates the necessity for loosening and resoldering to make connections, since the lugs may be safely joined with a screw, nut, and lock washer.

Exciter Generator. If the machine has a separate exciter generator, its armature, coils, brushes, and brush holders will need the same general care recommended for the welder set. Keep the covers over the exciter armature, since the commutator can be damaged easily.

Controls. Inspect the controls frequently to ensure that the ground and electrode cables are connected tightly to the output terminals. Loose connections cause arcing that destroys the insulation around the terminals and burns them. Do not bump or hit the control handles—it damages the controls, resulting in poor electrical contacts. If the handles are tight or jammed, inspect them for the cause. Check the contact fingers of the magnetic starting switch regularly. Keep the fingers free from deep pits or other defects that will interfere with a smooth, sliding contact. Copper fingers may be filed lightly. All fingers should make contact simultaneously. Keep the switch clean and free from dust. Blow out the entire control box with low-pressure compressed air.

Connections of the leads from the motor stator to the switch must be tight. Keep the lugs in a vertical position. The line voltage is high enough to jump between the lugs on the stator leads if they are allowed to become loose and cocked to one side or the other. Keep the cover on the control box at all times.

Condensers. Condensers may be placed in an ac welder to raise the power factor. When condensers fail, it is not readily apparent from the appearance of the condenser. Consequently, to check a condenser, one should see if the input current reading corresponds to the nameplate amperes at the rated input voltage and with the welder drawing the rated output load current. If the reading is 10 to 20 percent more, at least one condenser has failed. *Caution:* Never touch the condenser terminals without first disconnecting the welder from the input power source; then discharge the condenser by touching the two terminals with an insulated screwdriver.

TABLE 1.18 Typical Arc Welding Troubleshooting Chart

Trouble	Cause	Remedy
Welder will not start (Starter not operating)	Power circuit dead	Check voltage
	Broken power lead	Repair
	Wrong supply voltage	Check name plate against supply
	Open power switches	Close
	Blown fuses	Replace
	Overload relay tripped	Let set cool. Remove cause of overloading
	Open circuit to starter button	Repair
	Defective operating coil	Replace
	Mechanical obstruction in contractor	Remove
Welder will not start (Starter operating)	Wrong motor connections	Check connection diagram
	Wrong supply voltage	Check name plate against supply
	Rotor stuck	Try turning by hand
	Power circuit single-phased	Replace fuse; repair open line
	Starter single-phased	Check contact of starter tips
	Poor motor connection	Tighten
	Open circuit in windings	Repair
Starter operates and blows fuse	Fuse too small	Should be two to three times rated motor current
	Short circuit in motor connections	Check starter and motor leads for insulation from ground and from each other
Welder starts but will not deliver welding current	Wrong direction of rotation	Check connection diagram
	Brushes worn or missing	Check that all brushes bear on commutator with sufficient tension
	Brush connections loose	Tighten
	Open field circuit	Check connection to rheostat, resistor, and auxiliary brush studs
	Series field and armature circuit open	Check with test lamp or bell ringer

Delay Relays. The delay relay contacts may be cleaned by passing a cloth soaked in naphtha between them. Do not force the contact arms or use any abrasives to clean the points. Do not file the silver contacts. The pilot relay is enclosed in a dustproof box and should need no attention. Relays are usually adjusted at the factory and should not be tampered with unless faulty operation is obvious. Table 1.18, a troubleshooting chart, may prove to be a great timesaver.

SAFETY

Arc welding can be done safely provided sufficient measures are taken to protect the operator from the potential hazards. If the proper measures are ignored or overlooked, welding operators can be exposed to such dangers as electrical shock and overexposure to radiation, fumes and gases, and fire and explosion, any of which could cause severe injury or even death. With the diversification of the welding which may be done by maintenance departments, it is vitally important that the appropriate safety measures be evaluated on a job-by-job basis and that they be rigidly enforced.

A quick guide to welding safety is provided in Fig. 1.51. All the potential hazards, as well as the proper safety measures, may be found in ANSI Z-49.1, published by the American National Standards Institute and the American Welding Society. A similar publication, "Arc Welding Safety," is available from The Lincoln Electric Company.

Hazard	Factors to Consider	Precaution Summary
Electric shock can kill	• Wetness • Welder in or on workpiece • Confined space • Electrode holder and cable insulation	• Insulate welder from workpiece and ground using *dry* insulation. Rubber mat or dry wood. • Wear *dry, hole-free* gloves. (Change as necessary to keep dry.) • Do not touch electrically "hot" parts or electrode with bare skin or wet clothing. • If wet area and welder cannot be insulated from workpiece with dry insulation, use a semiautomatic, constant-voltage welder or stick welder with voltage reducing device. • Keep electrode holder and cable insulation in good condition. Do not use if insulation damaged or missing.
Fumes and gases can be dangerous	• Confined area • Positioning of welder's head • Lack of general ventilation • Electrode types, i.e., manganese, chromium, etc. See MSDS • Base metal coatings, galvanize, paint	• Use ventilation or exhaust to keep air breathing zone clear, comfortable. • Use helmet and positioning of head to minimize fume in breathing zone. • Read warnings on electrode container and material safety data sheet (MSDS) for electrode. • Provide additional ventilation/exhaust where special ventilation requirements exist. • Use special care when welding in a confined area. • Do not weld unless ventilation is adequate.
Welding sparks can cause fire or explosion	• Containers which have held combustibles • Flammable materials	• Do not weld on containers which have held combustible materials (unless strict AWS F4.1 procedures are followed). Check before welding. • Remove flammable materials from welding area or shield from sparks, heat. • Keep a fire watch in area during and after welding. • Keep a fire extinguisher in the welding area. • Wear fire retardant clothing and hat. Use earplugs when welding overhead.
Arc rays can burn eyes and skin	• Process: gas-shielded arc most severe	• Select a filter lens which is comfortable for you while welding. • Always use helmet when welding. • Provide non-flammable shielding to protect others. • Wear clothing which protects skin while welding.
Confined space	• Metal enclosure • Wetness • Restricted entry • Heavier than air gas • Welder inside or on workpiece	• Carefully evaluate adequacy of ventilation especially where electrode requires special ventilation or where gas may displace breathing air. • If basic electric shock precautions cannot be followed to insulate welder from work and electrode, use semiautomatic, constant-voltage equipment with cold electrode or stick welder with voltage reducing device. • Provide welder helper and method of welder retrieval from outside enclosure.
General work area hazards	• Cluttered area	• Keep cables, materials, tools neatly organized.
	• Indirect work (welding ground) connection	• Connect work cable as close as possible to area where welding is being performed. Do *not* allow alternate circuits through scaffold cables, hoist chains, ground leads.
	• Electrical equipment	• Use only double insulated or properly grounded equipment. • Always disconnect power to equipment before servicing.
	• Engine-driven equipment	• Use in only open, well ventilated areas. • Keep enclosure complete and guards in place. • See Lincoln service shop if guards are missing. • Refuel with engine off. • If using auxiliary power, OSHA may require GFI protection or assured grounding program (or isolated windings if less than 5KW).
	• Gas cylinders	• Never touch cylinder with the electrode. • Never lift a machine with cylinder attached. • Keep cylinder upright and chained to support.

FIGURE 1.51 Welding safety checklist.

BIBLIOGRAPHY

ANSI/ASC Standard Z49.1, "Safety in Welding and Cutting," the American Welding Society, P. O. Box 351040, Miami, FL 33135.

ANSI Z87.1, "Practice for Occupational and Educational Eye and Face Protection," American National Standards Institute, 1430 Broadway, New York, NY 10018.

AWS F1.1, "Methods for Sampling Airborne Particulates Generated by Welding and Allied Processes," the American Welding Society.

AWS F4.1, "Recommended Safe Practices for the Preparation for Welding and Cutting of Containers and Piping that Have Held Hazardous Substances," the American Welding Society.

Jefferson, T. B., and Woods, Gorham, *Metals and How to Weld Them,* 2nd Ed., 1962, The James F. Lincoln Arc Welding Foundation, P. O. Box 17035, Cleveland, OH 44177-0035.

The Procedure Handbook of Arc Welding, 12th Ed., 1973, The Lincoln Electric Company, 22801 St. Clair Avenue, Cleveland, OH 44117.

In addition, the following booklets and pocket guides are available from The Lincoln Electric Company, Book Division, P. O. Box 17035, Cleveland, OH 44177-0035.

"Arc Welding Safety"

"Flux Cored Arc Welding Guide"

"Gas Metal Arc Welding Guide"

"Gas Tungsten Arc Welding Guide"

"Handyman's Guide to Semiautomatic Welding"

"Prevention and Control of Distortion"

"Welding Safely—The Way the Pros Do"

CHAPTER 2
GAS WELDING IN MAINTENANCE

Engineers of L-TEC Welding and Cutting Systems
Florence, S.C.

AIR-ACETYLENE SOLDERING, HEATING, AND BRAZING

An air-acetylene appliance produces a flame with a temperature of approximately 4000°F by mixing acetylene with atmospheric air in much the same way that air is mixed with city gas in a kitchen range. The correct mixture produces a pale blue flame with a bright, sharp inner cone that is hot enough for light silver soldering (brazing), for most soft soldering, and for hundreds of heating jobs. Air-acetylene appliances are used throughout industry as companion equipment to the oxyacetylene torch for applications requiring clean, ready-to-use heat but not the extremely high temperatures of the oxyacetylene flame.

An air-acetylene outfit consists of torch handle, a torch stem or tip, a pressure-reducing regulator, a cylinder of acetylene, and a hose for connecting the torch to the regulator and tank. Interchangeable stems or tips that give various sizes and types of flames are available (see Fig. 2.1). The acetylene cylinders themselves come in all sizes, including small portable units (see Fig. 2.2).

Precautions and Safe Practices

1. *Do not* let acetylene escape near any possible source of ignition. Accumulations of acetylene in certain proportions may explode if ignited.
2. *Never* store acetylene tanks in a closed or confined space, such as a closet.
3. *Never* solder a container that contains or has contained flammable liquids or vapors (including gasoline, benzene, solvents, and other similar or dissimilar materials) unless the container has been thoroughly purged of all traces of flammable material and vapors. Be sure that any container you work on is vented. We urge that before you do work of this kind, you get Booklet A-6.0.40 from the American Welding Society, 2501 Northwest 7th St., Miami, FL 33125.
4. *Never* use a tank with a leaking valve.
5. *Do not* make any repairs to an acetylene tank, except to tighten the packing-gland nut on the valve.
6. *Do not* abuse or drop tanks or handle them roughly.
7. *Never* use a tank as a roller. Never use a wrench or pliers on the tank valve. Always use a valve key.
8. *Never* allow full tank pressure to enter a stopped hose. Always use a regulator when there is a needle valve on the torch handle.
9. Examine your hose for leaks frequently. Dipping it into a bucket of clean water, with the pressure in the hose, is the quickest and easiest way.
10. *Do not* use hose that is worn or any equipment that is in need of repair.

10.64 MAINTENANCE WELDING

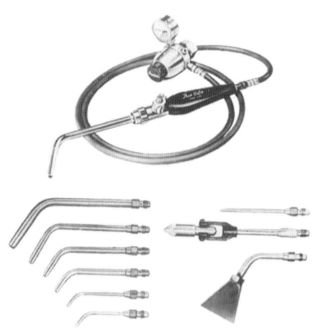

FIGURE 2.1 A typical air-acetylene outfit consisting of a regulator, torch handle, and attached torch stem with interconnecting hose. Also shown are some of the typical interchangeable torch stems available. Notice the special-purpose stems—the hatchet-shaped paint-burner stem and the soldering-iron stem.

FIGURE 2.2 Connecting a typical air-acetylene outfit. The standard type of portable outfit is shown. Smaller tanks also are available.

> *Always read and understand the manufacturers' operating instructions and your employer's safety practices before operating and maintaining gas welding and cutting equipment. Always read and understand all precautionary labels and instructions before using equipment and materials.*

Soldering

The air-acetylene torch is used extensively for all kinds of soldering with both soft and silver (hard) solder. Although soft soldering is more widely used, silver soldering (also referred to as *brazing*) is sometimes used for soldering sweat-type fittings in addition to the more precise soldering associated with jewelry and instrument manufacturing. With an air-acetylene torch, the silver solder used must have a melting point lower than 1500°F. If sweat-type fittings being silver-soldered are larger than $1\frac{1}{2}$ in. diameter, or if a great number of joints are being made, an oxyacetylene torch is recommended, since its greater flame temperature speeds the work. Silver-soldering commercial metals over $\frac{1}{32}$ in thick is also best done with an oxyacetylene torch. In contrast to silver soldering, practically all soft soldering can be done with an air-acetylene torch.

> *Warning: Silver soldering requires a special rod and a special flux, usually in paste form. Care should be taken to follow the manufacturer's directions. The fumes from fluxes and some silver solders are toxic; therefore, special ventilating precautions are necessary.*

When using air-acetylene appliances, you have a choice of two soldering methods:

1. The open (direct) flame method. The flame heats the workpiece, and the workpiece melts the solder in conjunction with the flame. The advantages of the open (direct) flame method include:

a. Speed (no copper intermediary to be heated).
   b. Greater diversity in the uses to which the flame can be put.
   c. Greater efficiency in the use of fuel (the gas goes further because it is applied directly to the workpiece).
   d. More heat because of direct application of the flame.
2. The enclosed (indirect) flame method. The flame is applied to the soldering copper. The copper in turn heats the workpiece. The workpiece, in conjunction with the soldering copper, melts the solder where it is needed. The advantages of the enclosed (indirect) flame method include:
   a. Heat is better controlled.
   b. Less experience is needed on the operator's part.
   c. More delicate work is possible, especially where damage to the adjacent materials might result from the use of an open flame.

   See Table 2.1 for commonly used soft solders and Table 2.2 for soldering fluxes.

Sheet Metal Working

Sheet metal soldering can be done with either the enclosed (indirect) flame method or the open (direct) flame method depending on the choice of the operator. Many types of joints can be made in sheet metal. Joints described on the following pages are most widely used.

1. The lap joint (see Fig. 2.3):
   a. Thoroughly clean the edges to be joined.
   b. Flux the edges by dipping them in a bath of hydrochloric (muriatic) acid or, using a brush, paint them with it.
   c. If you are using a soldering iron, tin the iron first and then tin the edges. If you are using a soldering torch, tin the edges. The edges should be tinned along their entire length and then placed so that the tinned edges overlap. Use C clamps to hold them together if you have them.

TABLE 2.1 Commonly Used Soft Solders

Composition, %	Melting range, °F	Gives best results when used for
Tin, 63; lead, 37	361–361	Critical electronic work, coatings for printed-circuit boards
Tin, 60; lead, 40	361–374	
Tin, 50; lead, 50	361–420	General purposes
Tin, 40; lead, 60	361–460	Automobile radiators, roofing seams, wiped joints in plumbing, dip coatings
Tin, 35; lead, 65	361–478	
Tin, 30; lead, 70	361–496	Filling dents in automobile bodies
Tin, 20; lead, 80	361–534	Apply by wiping; some dip coating
Tin, 15; lead, 85	438–553	
Tin, 10; lead, 90	514–574	Where higher-melting-point solders are necessary
Tin, 5; lead, 95	574–596	
Tin, 96; silver, 4	430–430	Food-handling equipment, plumbing, heating, refrigeration tube joints where higher-temperature or higher-strength solders are necessary
Tin 95; antimony, 5	450–464	Some electrical and copper tubing joints. Do not use on zinc or galvanized sheet

TABLE 2.2 Soldering Fluxes

Metal	Flux to use*
Aluminum	Aluminum application flux
Block tin	Rosin or zinc chloride
Brass	Rosin, zinc chloride, or muriatic acid†
Cast iron	Zinc chloride or muriatic acid
Chromium	Muriatic acid
Copper	Rosin, zinc chloride, or muriatic acid
Gun metal	Rosin, zinc chloride, or sal ammoniac
Inconel	Strong zinc chloride
Iron (galvanized)	Muriatic acid
Iron (tin-coated)	Rosin or zinc chloride
Lead	Rosin, zinc chloride, or muriatic acid
Monel	Zinc chloride or muriatic acid
Nickel	Zinc chloride or muriatic acid
Pewter	Rosin, pewter application flux
Stainless steel	Strong zinc chloride
Steel (plain)	Zinc chloride
Steel (galvanized)	Muriatic acid
Steel (tin-coated)	Rosin or zinc chloride
Terne plate	Rosin or zinc chloride
Tin	Rosin or zinc chloride
Zinc	Strong zinc chloride or muriatic acid

*Nearly all these fluxes are available commercially in paste form. Pastes are usually preferred because they give excellent results on most jobs and are easy to use.
†Muriatic acid is a mild form of hydrochloric acid.

d. Next, pressing down on the soldering iron, run it up and down over the seam until a fillet of solder is visible. If you are using a soldering torch, move it back and forth with the flame touching the work until the fillet appears. In both cases, where no fillet appears, add more solder.

e. When making a long seam with a plain lap joint, it is best to "tack" the seam first. Tacking means applying drops or spots of solder at intervals along a seam to hold it in place. Clean, flux, and tin the entire job. Heat the seam, and apply solder spot by spot. Then do the regular soldering job on the whole seam. If the "tacks" tend to melt or the seams to pull apart when you're near them with the torch, proceed as follows:

(**1**) Press the pieces of metal together at the "trouble spot" with a stick.

(**2**) Reheat the "tacks" and the solder that has been previously applied as tinning. Keep pressing the heated area together with the stick until the solder has cooled and formed a bond. Proceed with the soldering job.

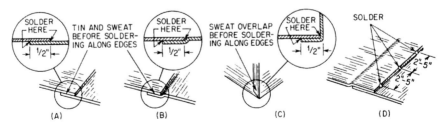

FIGURE 2.3 Three variations of the lap joint (A, B, and C); at the right (D), how to tack long seam by applying spots of solder at intervals.

f. When the joint is finished, wipe off all excess solder with a stiff bristle brush and wash off the excess flux with hot water.
2. Lock joint (see Fig. 2.4):
   a. Thoroughly clean surfaces that will form the joint.
   b. Form the lock joint between the two sheets.
   c. Pound the joint tight with a composition mallet, or use a block of wood between the sheets and a steel hammer. Try to get the joint as flat and tight as possible.
   d. Apply acid flux along the seam, and heat the seam.
   e. Apply just enough solder to seal the seam. (You have already made the seam mechanically strong by hammering and forming the lock joint.)
   f. If the seam is fairly long, you can run the flame a few inches ahead of the solder instead of heating and soldering a section at a time.
   g. Remove all excess solder with a stiff bristle brush, and wash off excess flux with hot water.
3. Flange joint:
   a. A flange joint is generally used in conjunction with rivets or spot welds. The solder is used to make the seam tight to air, gas, or water.
   b. Before the joint is formed, the area to which the solder will be applied must be thoroughly cleaned and must remain clean until the seam is finished.
   c. A tinning coat of solder can be applied to the seam before it is riveted or spot-welded.
   d. Either use acid core solder, or flux the joint with hydrochloric (muriatic) acid.
   e. Heat the joint with either a torch or soldering iron. Capillary attraction will draw the solder into the seam. Fill the joint with the desired amount of solder.
   f. Remove all excess solder with a stiff bristle brush; wash off excess flux with hot water.

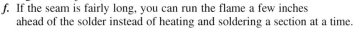
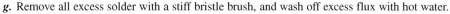

FIGURE 2.4 A lock joint where mechanical strength is provided by the joint rather than the solder bond.

Automobile Body Soldering

Automobile body soldering is done to fill in dents that cannot be hammered out completely, rough spots, and welded seams. Either soldering method can be used, direct (open) flame or indirect (closed) flame. Where the deposits of solder to be made are considerable or in places where an open flame would not damage chrome finishes or glass, we recommend the open flame method because of its speed and the rapidity with which the solder can be deposited. For the places adjacent to glass or chrome finishes, use the enclosed flame method. When you have decided which method to use, proceed as follows:

1. Grind away the paint from the dented area, and polish with steel wool or emery cloth.
2. Flux thoroughly, and after heating, apply enough solder to tin the dent.
3. Fill in the dent by adding solder from a bar and smoothing with a maple paddle. Take care not to melt the solder until it runs. Melt it just enough to make it pasty; then smooth with the paddle.
4. When the dent is filled in, heat the solder slightly and smooth it again before letting it cool.
5. Finish the job with rasps, body files, and emery cloth. Clean, prime, and paint.

Electrical Connections

For soldering electrical connections (Fig. 2.5), the enclosed (indirect) flame method is preferred. Prepare the electrical connections the way you usually do, and proceed as follows:

1. Thoroughly clean the connections.
2. Apply a noncorrosive flux paste.

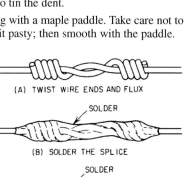

FIGURE 2.5 Soldering electrical connections.

3. Tin the soldering iron with a thin coat of solder.
4. Tin the wires, and melt enough solder onto them to be sure you have a good electrical connection.

Note: Where very large connections are to be made, an open-flame stem can be used.

Installing Sweat-Type Fittings

The following is the most efficient method for making sweat-type joints as recommended by two of the leading copper-tube manufacturers. The air-acetylene torch with a direct (open) flame is used by literally thousands of plumbers and is universally recognized as the best means of making these joints. The torch saves time and money, and a relatively inexperienced workman can do a good job with very little training and practice.

There are two basic types of sweat-type fitting: the *plain type* and *cast type*. With the plain type, the solder is fed at the point where the fitting and the tube join. With the cast type, the solder is fed in through precast holes in the fitting itself. The instructions below will work equally well with both types (see Figs. 2.6 and 2.7).

1. Cut the tube to the length required with a hacksaw (32 teeth to the inch) or a disk cutter. Make certain that the tube ends are cut square. Special vises which hold the tube securely and guide the saw blade are furnished by a number of manufacturers.
2. Ream the tube, and remove burrs on the outside. Using a sizing tool if necessary to correct any possible distortion of the tube from handling. The point of a sizing tool is inserted in the end of the tube and is hammered until the tube is again round.
3. Clean the outside surface of the tube and the inside surface of the fitting until the metal is bright. All traces of discoloration must be removed. This must be done even though the tube may appear to be perfectly clean, and it is particularly important when soldering larger-size joints. No. 00 steel wool is very satisfactory for cleaning tubes and fittings. Do not use files or rough sandpaper, and the likes, as they score the surface and may result in a poor joint.
4. Apply a thin, uniform, and complete coating of a reliable brand of soldering flux or paste to the cleaned portion of both tube and fitting. Do not apply the flux too thickly, as excess flux may form bubbles when heated and prevent the solder from creeping into the joint. After the tube has been inserted into the fitting as far as it will go, revolve the fitting once or twice to spread the flux evenly.

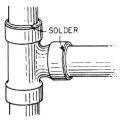

FIGURE 2.6 Where to solder plain-type sweat fittings.

FIGURE 2.7 Soldering 1-in. copper tubing and fitting with precast holes.

5. Apply the flame evenly all around the circumference of the fitting, and as it becomes heated, move the flame back and forth to prevent overheating. Occasionally test the heat by touching the fitting with solder where the tube and fitting join. Do not let the flame touch the solder while testing the temperature of the joint.

    It is important not to overheat the joint. If the connection is heated too much, the flux may be burned out from inside the joint and the solder will not spread properly. An overheated joint causes the solder to seep through the joint and run away.

    During the heating operation, adjacent wood surfaces should be protected from the heat by heat-resistant shielding or wet clothes. Because of its narrow, concentrated flame, the air-acetylene torch can be used very close to wood surfaces without scorching them.

6. Remove the flame, and apply solder to the edge of the fitting where it comes in contact with the tube as soon as the fitting has reached the correct temperature to melt the solder. Be sure that enough solder is used. Enough solder to make an efficient joint will be automatically sucked in by capillary attraction. When a line of solder shows completely around the fitting, that is, a fillet of solder appears in the chamfer at the end of the fitting, the joint has all the solder it will take. Wipe off any excess solder or flux.

7. Slightly reheat the connection in order to help the solder permeate the metal. Remove the flame, and continue to feed solder to make certain the joint is filled.

8. Permit the connection to cool for a fraction of a minute. A rag or wad of waste saturated with water will hasten the cooling. Remove all surplus solder from around the edges with a brush. This operation will show whether or not the solder has filled the joint.

9. When disconnecting a soldered tube from a fitting on which other soldered connections are to be left intact, the application of wet cloths to the parts which are not to be disconnected will prevent melting of the solder at such connections.

10. More than ordinary care should be exercised in soldering fittings $2\frac{1}{2}$ in. in diameter and larger. It is essential that the heat be uniformly distributed around the entire circumference of the fitting and not concentrated in one spot.

    When making large-diameter joints, a tip producing a large flame should be used. The flame should be directed on the fitting to avoid any unnecessary annealing of the tube. For assembling lines 3 in. in diameter or over, it may be advisable to use two or three torches. Solder should then be applied simultaneously at two or more points.

11. In applying solder to a tee, feed solder from both ends of the fitting.

12. Solder when confined between two surfaces will run uphill (by capillary attraction), and joints can be made in almost any position.

13. In sweating male and female adapters, care should be taken to allow more time for the solder to set, as these heavier fittings do not cool so quickly.

Paint Burning

An air-acetylene torch with a paint-burning stem is a quick, easy, and economical means for removing old, cracked, and checked paint from a surface that can stand a moderate amount of heat and where there is no chance of creating a fire in hidden crevices, recesses, and welts. Only use when there is no fire risk to other materials. The number of coats of paint is not important; it just takes a little more time to remove them. Paint can be removed from wood, canvas, brick, stone, or metal.

Warning: Avoid inhaling any dust or fumes that may be given off in the paint-burning operation. Such dust and fumes may be toxic, particularly if the paint being removed contains lead or cadmium compounds.

There are two methods of removing paint. They are listed below as *Method A* and *Method B*. We suggest you try both methods. You can then use the one that suits your particular type of work. Once the old paint is removed and rough spots smoothed, the surface is ready for a new coat of paint.

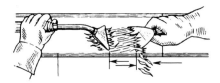

FIGURE 2.8 Paint burning (method *A*).   **FIGURE 2.9** Paint burning (method *B*).

Method A (Fig. 2.8)

1. Hold the paint burner in your left hand. Hold the putty knife (with a stiff blade about 3 in wide) in your right hand.
2. Move the torch backward and forward 1 in from the painted surface about 6 in at a stroke. Follow the movements of the torch with a steady forward movement of the putty knife, keeping the putty knife hot with the flame.
   Note: You will find it advisable to wear asbestos or other heavy flame-resistant gloves when burning paint. The putty knife gets very hot after a while; so you should protect your right hand. Cloth (cotton) gloves are not satisfactory.
3. Moving the torch back and forth changes the paint to a plastic state and keeps the putty-knife blade hot. A hot blade reduces the tendency of the paint to stick to it.

Method B (Fig. 2.9)

1. Move the torch more or less steadily from right to left over the painted surface. Bring the paint to a bubbly plastic state. Scrape off the paint as soon as it bubbles. Do not let the flame touch the blade of the putty knife.
2. The putty knife should have a back-and-forth motion which will intermittently expose the scraped area to the flame. This method is recommended for particularly heavy or stubborn paint.

Miscellaneous Air-Acetylene Applications

There are many repair and maintenance jobs aside from soldering and brazing that can be done efficiently by skilled operators with an air-acetylene torch. A few of these applications are given below.

Loosening Nuts and Bolts. Frequently you come across a bolt that resists all attempts to loosen it with a wrench. Heat the nut for several minutes and let it cool; then try the wrench again. Generally, you will now find the nut ready to turn.

Freeing Frozen Shafts. A frozen shaft of small diameter can be freed by heating the collar that holds it. Heat the collar, not the shaft. You will find that you can separate the parts quite quickly no matter how tightly they are frozen together.

Lead Working. The air-acetylene torch can be used to build up lead battery terminals. Any of the standard stems can be used depending on the amount of work to be done and the speed with which you want to do it.

It is recommended that you use a form, where possible, to keep the lead in the shape of a battery terminal and to prevent it from running on the battery. Put the form over the old terminal and keep adding melted lead until the desired height and shape of the terminal are attained.

The air-acetylene torch can be used to repair lead-lined vats, wipe joints in lead pipe and lead-covered cable, and solder battery-cable lugs.

Warning: When working with lead in a confined space, be very sure of your ventilation and, if possible, use a suitable air-line mask.

Anchoring Bolts in Concrete or Stone.   Firmly anchoring a large bolt in concrete or stone can be solved as follows:

1. Drill a hole in the concrete or stone with a star or other type drill. It is best to dish or widen the bottom of the hole slightly to increase the stability of the bolt after the solder sets. Make certain all free moisture or water is removed from the cavity.
2. Heat the solder (bar solder is best) in a ladle with an air-acetylene torch until the solder is molten.
3. Place the bolt in the hole thread-end up, and pour molten solder around it until the solder is level with the floor.
4. This type of mounting will give years of satisfactory service; if it should become loose, just reheat the solder with the torch and it will be as tight as ever.

Cutting Asphalt Tile.   The air-acetylene torch has been used with good success by asphalt-tile contractors for heating tiles that have to be bent, formed, or cut. After a few seconds of heating, the tile can be shaped or cut with great ease.

Cutting Safety Glass.   Using an air-acetylene torch with a medium-sized stem, the following procedure can be used when cutting safety glass.

1. Score both sides of the glass with a glass cutter and break the glass.
2. Soften the plastic filler by running the torch back and forth along the lines of the cut.
3. Wobble the glass from side to side several times. Then hold the glass to one side while you cut the heat-softened plastic filler with a razor blade.

OXYACETYLENE WELDING, CUTTING, GOUGING, AND HARD FACING

Metal production, fabrication, and repair, as they are known today, would be impossible without the oxyacetylene process and its flame of approximately 6000°F. The oxyacetylene process is built on two principles: (1) acetylene burned with an equal amount of oxygen produces an intensely hot flame that will melt and fuse most metals, and (2) a jet of oxygen striking a piece of ferrous metal that has been heated to its kindling temperature will rapidly burn the metal away.

Precautions and Safe Practices

Gas welding has several potential hazards. These are the fumes and gases released by the heat of the flame, the fire hazard due to the flame, and the toxic hazard of many of the fluxes required. A summary of some of the potential hazards is discussed here. For a complete discussion of all the hazards, refer to American National Standards Institute publication ANSI/ASC-Z49.1, "Safe Practices for Welding and Cutting," available from the American Welding Society, P.O. Box 351040, Miami, FL 33135.

In addition, refer to Precautions and Safe Practices in the first section of this chapter.

Welding and Brazing

Welding with an oxyacetylene torch is simple. You put two pieces of metal together, then melt the edges with an oxyacetylene flame. The molten metal flows together and forms a single, solid piece of metal. Welding rod similar to the base metal is usually added to strengthen the joint. This is known as *fusion welding.* If you use a steel rod, the process is sometimes called *steel welding.*

Braze welding is another method. In braze welding, the two pieces being joined are heated to a dull red. They are not melted. A flux is added to clean the metal and protect it from air. When the pieces are dull red, molten-bronze welding rod is added to form a strong bond. This bronze weld is generally as strong as the base metal.

10.72 MAINTENANCE WELDING

Building up worn parts with bronze or steel welding rod, heating and forming of metals, gouging, hard facing, and soldering are other jobs done by the oxyacetylene flame.

Cutting

Oxygen cutting is similar to the eating away of steel by ordinary rusting, only it is very much faster. In rusting, the oxygen—in the air or in water—affects the metal slowly. Directing a jet of pure oxygen at metal heated almost to the melting point actually speeds up chemical reaction of rusting to such an extent that the metal ignites and burns away. The iron oxide melts and runs off as molten slag to expose more iron to the action of the oxygen jet. This makes it possible to cut iron and steel leaving a smooth, narrow cut. An oxyacetylene outfit requires some basic equipment (see Fig. 2.10).

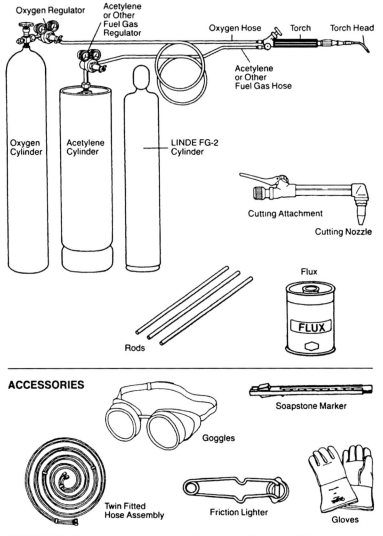

FIGURE 2.10 The basic equipment essential in any oxyfuel gas welding process must include the pictured tools and accessories.

Setting Up an Outfit

Suggestions and recommendations for safe handling of oxyacetylene equipment have been set forth by the American National Standards Institute (*Form: ANSI/ASC-Z49.1 "Safety in Welding and Cutting"*). They are summarized here in brief form.

Note: *Be sure to obtain and read the complete document.*

1. Fasten the cylinders to be used in an upright position.
2. Be sure to keep a clear space between cylinders and the work so that cylinders and pressure-reducing regulators can always be reached quickly.
3. "Crack" the cylinder valve.
   a. Stand at one side or to the rear of the cylinder outlet.
   b. Open the cylinder valve slightly for an instant, and then close it. This will clean the valve of dust or dirt which may have accumulated during storage.

   Note: Valves on flammable-gas cylinders should *not* be "cracked" near welding or cutting work in progress, or near sparks, flame, or other possible sources of ignition.
4. Always attach a regulator to the gas supply.
   a. Make certain that the regulator is proper for the particular gas and service pressure.
5. Do not use adaptors unless you follow the approved instructions of the manufacturer.
6. Never force connections that do not fit properly.
7. Only use hoses fitted with connections made especially for oxygen-fuel gas equipment.

   Note: New hoses may contain some loose talc resulting from the manufacturing process. The talc should be blown out of both oxygen and fuel gas hoses using about 5 psi of oxygen for a few seconds. Do not use compressed air or fuel gas for this purpose.
8. Before opening a cylinder valve, make sure the regulator pressure-adjusting screw is released by rotating it counterclockwise until it turns freely.
9. While opening a cylinder valve:
   a. Stand to one side of the regulator.
   b. *Slowly* open the cylinder valve. *Never* open a cylinder valve suddenly.
   c. Do not open the acetylene cylinder valve more than $1\frac{1}{2}$ turns. Leave the T-wrench in place.
10. Always make sure there is enough oxygen and fuel gas available in the cylinders to complete your welding, cutting, or heating work.
11. Check all connections and joints for leaks.
12. Never tighten a leaky connection between a regulator and cylinder when under pressure.
13. Attach the proper welding tip or cutting nozzle and use correct pressures.
14. Do not use matches for lighting torches.
15. If the flame goes out and burns back within the torch, producing a pronounced hissing or squealing noise, *immediately* shut off the torch; otherwise, the flame will burn through the torch and may cause injury or death.
16. Do not relight flames on hot work in a pocket or small confined space.
17. As soon as you have finished working, or if you are going to disconnect the regulator, do the following:
    a. Close the cylinder valve.
    b. Open the torch valve to release all pressure from the hose and regulator.
    c. When the gage pressure drops to zero, turn out (counterclockwise) the pressure-adjusting screw.
    d. Close the torch valve once more.

FLAME ADJUSTMENT

The three basic types of flames for an oxyacetylene torch are shown in Fig. 2.11.

1. To adjust to an excess acetylene (carburizing) flame, start with both torch valves closed. Then:
   a. "Crack" the torch oxygen valve; open the acetylene valve about a full turn. Light the torch.
   b. Increase the oxygen supply until you see three distinct parts to the flame: a brilliant inner cone, a whitish acetylene feather, and a bluish outer envelope. This is a carburizing flame.
   c. The amount of excess acetylene in the flame is expressed as a ratio of the total length of the feather to the total length of the inner cone. Thus, in a $2\times$ flame, the acetylene feather is twice as long as the inner cone.
2. To adjust to a neutral flame:
   a. Proceed as above, but keep adding oxygen until the acetylene feather just disappears.
   b. This leaves two parts to the flame: a brilliant white inner cone and a bluish outer envelope.
   c. At this point, the torch is burning equal amounts of oxygen and acetylene. This is a neutral flame.
3. To adjust to an oxidizing flame:
   a. Proceed as above until a neutral flame is obtained.
   b. Keep adding oxygen beyond the point where the acetylene feather disappears until the inner cone shortens (about 20 percent shorter than a neutral inner cone) and becomes "necked-in."
   c. A harsh sound also characterizes this oxidizing flame unless a very low flow is used.
4. Carburizing, neutral, and oxidizing flames can be harsh or soft. You get a harsh flame when using almost the maximum flow through a tip; you get a soft flame when using less than normal flow.
   a. In a harsh flame, the pressures approach "blowoff"; that is, a slight increase in pressure causes a gap to appear between the flame and the tip.
   b. In a soft flame, the gas flow is reduced with the torch valves. The inner cone is about half as long as that in a harsh flame.

Braze Welding (See Fig. 2.12)

Braze welding is a process which enables you to weld various metals and alloys without melting the base metal. Using a bronze rod (which melts between 1500 and 1650°F) as a filler metal, you can make strong joints in many metals and alloys. The process is similar to soldering, the difference being that solder melts at a much lower temperature than the bronze rod does and is of much lower strength.

In braze welding, a slightly oxidizing flame is generally used, since a carburizing flame gives off certain gases that dissolve in the molten puddle and leave weak, porous spots in the weld.

You can braze-weld cast, malleable, wrought, and galvanized iron; carbon steels and alloy steels; copper; brass and bronze; nickel; Monel; Inconel; and other metals. Here are some of the features and advantages of braze welding:

1. Braze welding can be used for many repair and fabrication jobs.
2. It is faster than fusion welding because less heat is required to melt the filler metal.

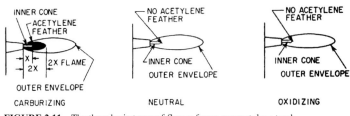

FIGURE 2.11 The three basic types of flames for an oxyacetylene torch.

FIGURE 2.12 A braze weld.

a. This means that you use less gas, so costs are lower.
   b. Less heat means less distortion in the piece being braze-welded.
   c. More work can be done in less time with this fast process.
3. Braze welding produces good strong joints.
   a. Bronze rod, properly deposited, can have a tensile strength up to 56,000 psi.
   b. Tensile strength of plain low-carbon steel is about 52,000 psi.
4. Braze welding can be used for joining dissimilar metals: cast iron to steel, iron to copper, etc.
5. Braze welding can be used to join malleable iron parts and to repair large castings.

Fusion Welding (See Fig. 2.13)

Fusion welding is the joining of metal by melting and fusing the edges together. The joint is a thorough mixture of the base metal and the welding or filler rod used to build up the seam. There is no sharp line of demarcation as with a braze weld. The filler rod is used in all cases, except when you are welding sheet metal, and should be about the same composition as the base metal. For example, you use steel rods when welding various plain-carbon or alloy steels while cast iron rod is used for welding iron castings.

FIGURE 2.13 A fusion weld.

Fusion welding is used mainly where you cannot use braze welding. It has a wide appeal to small users for light-gage, mostly nonproduction work and for maintenance jobs, although it has largely been displaced for large-scale production work by electric arc welding. Fusion welding, rather than braze welding, is necessary for parts that will be in use at high temperatures. As a braze-welded joint becomes heated, it loses its strength rapidly, since the filler metal will melt at about 1650°F.

Parts subjected to great tensile stresses, that is, great pulling loads, should be fusion welded. For example, some steels have tensile strengths up to 90,000 psi and more. These exceed the tensile strength of a braze weld (up to 56,000 psi) and must be fusion welded with a special steel rod if the joint strength is to equal or exceed the strength of the base metal.

Fusion welding also can be used where an approximate color match between welded parts is necessary.

Fusion welding uses a neutral or slightly oxidizing flame, since a carburizing flame can cause entrapments in the filler metal.

Weld Preparation

1. As a part of your preparation for welding, you should select and prepare the proper joint design for your work (see Fig. 2.14).
   a. Square-edge butt and flange welds are commonly used in sheet-metal work. In the latter, the edges are turned up and melted to form the joint and no filler rod is needed.
   b. Lap joints are rarely used except where one cylindrical section fits inside another.
   c. The butt joint with beveled edges (the V) is most widely used. For plate over $3/16$ in, bevel the edges by oxygen cutting, grinding, or machining to an included angle of 90°.
   d. The corner fillet and double-V joints also can be used, depending on the demands of the job.
   e. In the double-V joint, the plates are welded from both sides. It is generally used for work thicker than $1/2$ in.

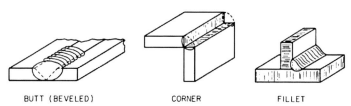

FIGURE 2.14 Butt, corner, and fillet joints.

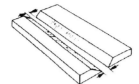

FIGURE 2.15 When making a butt weld of pieces over $1/8$ in thick, "progressive spacing" will counteract distortion caused by expansion.

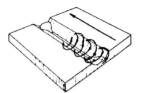

FIGURE 2.16 Torch motion for fusion and braze welding.

2. The second step in preparation is to clean the edges of any oil or grease, dirt, scale, or rust with steel wool, a wire brush, or some other means.

3. Before welding, the pieces must be properly spaced, since they will tend to expand during welding. Two types of spacing are used to counteract expansion. On material $1/8$ in thick or less, the edges are generally placed parallel to each other about $1/8$ in apart. They should then be tack-welded every 6 in or so to prevent undue distortion. This space between the plates allows the molten filler metal to flow to the bottom of the joint. Good penetration is thus ensured. For material over $1/8$ in thick, a progressive method of spacing is used (see Fig. 2.15). For every foot of weld length, the pieces should be spread apart about $1/8$ in. For example, welding pieces 2 ft long, you would leave the pieces approximately $1/4$ in apart at the finishing end of the weld. At the starting end, the pieces should be spaced about $1/16$ in. Slightly more spacing is required for unbeveled edges.

4. Tack weld the pieces at start and finish. In sheet metal, tack weld about every 6 in. This keeps the pieces in alignment and prevents them from drawing too close together as they are heated during welding. There is a great deal of strain on a tack weld as a result of these internal expansion and contraction stresses. So make your tack welds carefully, and make them strong.

Torch Motion

There is no hard and fast rule which will tell anyone exactly how to move a torch when welding. Each welder develops a natural motion after a little practice. Figure 2.16 gives a suggested pattern for moving the torch while welding. The motion is effectively a series of semicircles, wide enough to ensure heating beyond the limits of beveling and moving forward to a slight extent in each torch swing. For fusion welding, it is important to try to move the flame around in front of the rod at the end of each sweep so that complete melting of the edges is obtained. With braze welding, it is not necessary to melt the edges of the joint; so the torch motion is generally faster.

Making a Weld

The procedures for making a braze or fusion weld are essentially the same except that for fusion welding, the edges of pieces being joined are actually melted, while for braze welding, the edges of the pieces being joined are heated to a dull red. In both fusion and braze welding, the filler rod is melted to furnish the filler metal for the seam. The following brief discussion applies to both fusion and braze welding. Nevertheless, the difference between the two welding methods should always be remembered.

1. Steel thicker than $3/16$ in can be welded in two or more passes, while material over $1/2$ in should always be multipass welded (see Fig. 2.17).

2. Lay in the first pass or root weld from 2 to 3 in long. After making this beginning section of the root weld, go back and build up the finishing weld to the desired reinforcement.

3. There are two reasons for making a root weld for about $2\frac{1}{2}$ in and then returning for the second pass or finishing weld before continuing with the root weld (see Fig. 2.18).
   a. You take advantage of the heat left in the plate when you made the beginning section of the root weld. If you continued all the way across the plate, this heat would be lost when you returned for a second pass.

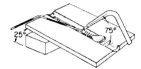

FIGURE 2.17 Braze or fusion welding of two pieces thicker than $3/16$ in. is best done with two passes. Notice that the weld proceeds in stages.

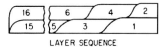

FIGURE 2.18 Sequence of root and finishing welds for a weld made in two passes (layers).

    b. Experience shows that best results in strength, uniformity, and appearance are achieved when this system is used.
4. The following points are important in making a good weld.
    a. Do not add filler rod until you have formed a molten puddle (fusion welding) or heated to a dull red (braze welding).
    b. Keep the rod in the puddle.
    c. Keep your eye on the leading edge of the puddle to ensure that you always have thorough fusion or heating.
    d. Remember that the rod is deposited evenly by constantly melting it into the molten puddle, not by applying the flame directly to the rod.
5. The torch should be tilted slightly, to an angle of 75° with the plate surface, to ensure a certain amount of preheat as the weld proceeds. The plate may be tilted upward to an angle of perhaps 25° to aid an even buildup.
6. The torch should be directed squarely into the V between the plates so that both sides will be heated evenly.
7. Proper weld sequence is shown below. The first root weld is made for about $2\frac{1}{2}$ in.; the first finishing weld is about half this length. Each successive pass, both root and finishing, is about the same length as the first section. The final section of finishing weld will be a bit longer than any other part to make up for the shortness of the first finishing weld.
8. Never make a flush weld if maximum strength is desired. Always provide reinforcement; that is, make sure that the lowest ripple is $1/16$ in above the surface of the plate.

Heavy Braze Welding

The following points apply to such heavy jobs as repairing heavy steel or iron castings.

1. Preparation of the work is important. First, vee out the crack with your gouging nozzle on steel parts or by chipping, grinding, or machining if the piece is cast iron. Be sure to clean thoroughly a generous space on each side of the V to permit the crown of the weld to lap over and give additional strength.
2. On cast-iron pieces, it is fairly certain that graphite (pure carbon) flakes are embedded in the surface and have been smeared by machining or grinding. In the presence of an oxidizing flame, this carbon will unite with oxygen and burn off as a gas. Use steel wool or a wire brush on the surface to complete the cleaning job.
3. Choose a location where it will be possible to set up a temporary preheat furnace. The reasons for preheating are:
    a. It is easier to braze weld if heat is stored up in heavy pieces. A fairly small and convenient welding flame can be used if the pieces are at about 500°F.
    b. If the pieces are cold, heat from the welding flame would be rapidly drawn away.
    c. Preheating will help prevent excessive internal stresses from occurring as the piece cools.
4. Depending on the shape of the piece, it may be possible to make a double V and have a welder work on each side of the joint.

5. Since it is easier to build up a weld in successive horizontal layers, position the work if possible so that your weld line is flat.
    a. If possible, it is desirable to support your starting weld on a carbon block or piece of firebrick. If this cannot be done, use a piece of 10-gage sheet or carbon plate fitted to the bottom of the abutting pieces. When the weld is finished and the casting is cool, you can remove the sheet if necessary by chipping.
    b. Use plenty of flux or flux-coated rod so the tinning action will take place automatically and stay well ahead of the weld itself.
6. When completed, the weld can be cleaned by starting at the top and working downward with a large oxidizing flame, melting the runovers.
7. In cases where the castings are spread out, be sure that they are well supported, since cast iron is weakened when heated to a high temperature.

Fusion Welding Cast Iron

1. Cast iron does not have the strength and ductility that steel has. Without careful cleaning before and proper cooling after welding, a casting may become hard and brittle and possibly crack.
    a. Clean off any dirt, scale, and grease that might weaken the final weld with a wire brush, a grinder, or a file.
    b. In order to equalize internal expansion and contraction stresses introduced during welding, preheat small castings locally with your blowpipe. Large castings should be placed entirely in a preheat furnace and raised to a temperature of approximately 500°F. The stresses of concentrated welding heat might crack the casting without this preheating.
2. Molten cast iron is very fluid and may tend to fall through. It is also a good idea to weld "in the flat" with some sort of backup where possible. Carbon blocks may be removed after the weld has cooled.
3. Bevel the edges, by chipping or grinding, to an included angle of about 90°.
4. To help further in cleaning the edges so that a clean, sound weld will be obtained, use a flux that will chemically float out dirt, slag, and oxide inclusions.
5. Add just enough flux so that all the impurities are cleaned and fluxed out of the weld zone.
6. Use only one pass. It is not necessary to fill in a root weld and then a finishing weld as was the case with steel.
7. Cast iron must be cooled slowly after welding. Sudden chilling of a recently welded cast-iron part can cause it to crack. Fast cooling also makes a casting hard, brittle, and subject to being cracked easily; slow cooling imparts softness and ease of machinability. Small parts can be placed into a can of lime or cement or some similar material so that they will cool properly. Larger castings can be left in the preheat furnace for slow, even cooling.

For recommended welding methods see Tables 2.3 and 2.4.

Oxygen Cutting

Iron burns (all burning is an oxidation process) as readily as wood or paper if it is heated to the right temperature and is exposed to a large amount of pure oxygen. Metals like aluminum, stainless steel, and magnesium also oxidize, but it takes even more heat to melt their oxides than it does for iron oxide. Other means must be used to cut them. Oxygen cutting is primarily intended for cutting ferrous metals (iron or steel).

The first step in oxygen-cutting ferrous metals is to preheat the metal until it is red hot. At this point, the metal is said to be at its kindling or ignition temperature—it is ready to burn away. The actual cut in the metal is started by directing the pure oxygen stream from a cutting torch at the preheated metal. The hot iron and the oxygen react instantly, producing so much heat that the oxide formed melts and

TABLE 2.3 Recommended Welding Methods (Ferrous)

Metal	Welding method	Flame adjustment	Recommended welding rod	Flux
Steel, cast	Fusion weld	Neutral	High-test steel	None
Steel pipe	Fusion weld Steel welding	Neutral Carburizing	High-test steel CMS steel	None
Steel plate	Fusion weld Steel welding	Neutral Carburizing	Drawn iron High-test steel CMS steel	None
Steel sheet	Fusion weld Bronze weld	Neutral Slightly oxidizing	Drawn iron High-test steel Bronze Cupro	None Brazing None
High-carbon steel	Fusion weld	Carburizing	High-test steel CMS steel	None
Manganese steel	Fusion weld	Slightly oxidizing	Same composition as base metal	None
Cromansil steel	Fusion weld	Neutral	High-test steel CMS steel	None
Wrought iron	Fusion weld	Neutral	High-test steel	None
Galvanized iron	Fusion weld Fusion weld Bronze weld	Neutral Neutral Slightly oxidizing	Drawn iron High-test steel Bronze	None None Brazing
Cast iron, gray	Fusion weld Bronze weld	Neutral Slightly oxidizing	Cast iron Bronze	Ferrous Brazing
Cast iron, malleable	Bronze weld	Slightly oxidizing	Bronze	Brazing
Cast iron pipe, gray	Fusion weld Bronze weld	Neutral Slightly oxidizing	Cast iron Bronze	Oxweld ferrous Brazing
Cast iron pipe	Fusion weld	Neutral	Cast iron Same composition as base metal	Ferrous
Chromium-nickel	Bronze weld	Slightly oxidizing	Bronze	Brazing
Chromium-nickel steel castings	Fusion weld	Neutral	Same composition as base metal 25-12 chromium-nickel steel Columbium-bearing 18-8 Stainless steel	Stainless steel
Chromium-nickel steel (18-8)	Fusion weld	Neutral	Columbium-bearing 18-8 Stainless steel	Stainless steel
Chromium-nickel steel (25-12)	Fusion weld	Neutral	Same composition as base metal	Stainless steel
Chromium steel	Fusion weld	Neutral	25-12 chromium-nickel steel Columbium-bearing 18-8 Stainless steel	Stainless steel
Chromium steel (4–6 percent)	Fusion weld	Neutral	Columbium-bearing 18-8 Stainless steel	Stainless steel
Chromium iron	Fusion weld	Neutral	25-12 chromium-nickel steel Columbium-bearing 18-8 Stainless steel Same composition as base metal	Stainless steel

TABLE 2.4 Recommended Welding Methods (Nonferrous)

Metal	Welding method	Flame adjustment	Recommended welding rod	Flux
Aluminum	Fusion weld	Slightly carburizing	Aluminum	Aluminum
Brass	Fusion weld	Oxidizing	Bronze	Brazing
	Bronze weld	Slightly oxidizing	Bronze	
Bronze	Fusion weld	Neutral	Bronze	Brazing
	Bronze weld	Slightly oxidizing	Bronze	
Copper (deoxidized)	Fusion weld	Neutral	Deoxidized copper	None
	Bronze weld	Slightly oxidizing	Bronze	Brazing
Copper (electrolytic)	Fusion weld	Neutral	Cupro	None
	Bronze weld	Slightly oxidizing	Bronze	Brazing
Everdur bronze	Fusion weld	Slightly oxidizing	Everdur bronze	Silicon
Nickel	Fusion weld	Slightly carburizing	Same composition as base metal	None
Monel metal	Fusion weld	Slightly carburizing	Same composition as base metal	Monel
Inconel	Fusion weld	Slightly carburizing	Same composition as base metal	Iconel
Lead	Fusion weld	Slightly carburizing	Same composition as base metal	None

flows or is blown away. As the oxide flows away, the cut progresses through the metal as the next layer of metal is exposed to the oxygen. When the torch is moved along the line of cut, the heat of the reaction between the iron and oxygen raises the temperature of these successive layers of metal.

Oxygen cutting is used almost everywhere—for cutting straight lines and circles in plate, for cutting shapes to accurate dimensions in single pieces and in stacks, for trimming plate to size and beveling it for welding, for piercing holes, for cutting I beams and other structural members to size, and for many other uses. The oxygen-cutting torch is also a prime fabricating tool in industry for preparing plates and cutting structural members in the shipbuilding, heavy machinery, and building construction industries. Oxygen cutting is also extensively used for demolishing and scrapping of machinery, obsolete equipment, unsafe or unwanted structures; for cutting heavy scrap to smaller size; for removing bolts and rivets; and for similar work (see Fig. 2.19).

Oxygen cutting is very versatile in that steel, wrought iron, and cast iron can be cut in almost any form, of almost any thickness. Hand cutting is restricted to thicknesses of about 1 ft. Machine cuts have been made, however, in material of about 6 ft in thickness.

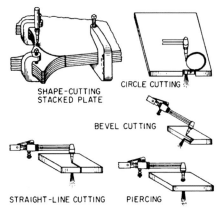

FIGURE 2.19 Some of the jobs done by oxygen cutting.

The process is inexpensive. Initial equipment cost and subsequent upkeep costs are very low compared with other means of doing the same job. The gas costs are almost negligible when you consider the variety and quality of the work done. The equipment needed for oxygen cutting is easily portable and can be taken almost anywhere for "on-the-job" use. The process is very fast. Depending on the thickness of the material, speeds up to 500 fph can be attained. The process is easily learned. The correct techniques can be studied and picked up in a few minutes.

Oxygen-Cutting Equipment (See Fig. 2.20)

Oxygen cutting requires the same equipment needed for welding, including a welding torch fitted with a cutting attachment and a special nozzle. Where you are going to do oxygen cutting for long periods, a cutting torch is more desirable than a cutting attachment. The cutting nozzles come in various sizes. The thickness of the metal and its surface condition determine the size of the nozzle needed. For example, five different-sized nozzles handle all thicknesses up to 12 in.

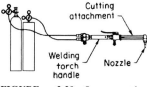

FIGURE 2.20 Oxygen-cutting equipment.

Various accessories, which supplement basic equipment, are available for making special types of cuts. In freehand guided cutting, the torch head can be drawn along a bar or straightedge. This will assure an accurate square or beveled straight-line cut. Circles or disks with 2 in or greater diameters can be accurately made with circle-cutting attachments. Where high accuracy is required in cutting straight lines, circles, or shapes, special machines are available which mechanically hold, guide, and advance the torch over the work. Little or no finishing is required on these high-quality machine cuts.

Preparation for Cutting

1. First, select a suitable place for working—make sure there is no combustible material at hand. Use heat-resistant or sheet metal shields to protect wood floors, etc., where necessary. Protect your legs and feet from sparks and slag.
2. A clean metal surface means lower gas consumption and a good-quality cut. So remove all the dirt, paint, etc., you can by scraping or wire brushing.
3. Look at the instruction sheet for your cutting attachment or torch to find out what size nozzle to use for the thickness of metal you are cutting.
4. The adjustment of the flame for a cutting attachment or torch is different from that for a welding torch because the latter has no cutting oxygen stream.
    a. If the cutting oxygen valve is opened after the preheat flames are adjusted to neutral, the preheat flames will lack oxygen. This is because both preheat and cutting oxygen come from the same source and part of the preheat supply has been diverted to form the cutting-oxygen stream.
    b. To correct this, the preheat flames should be adjusted with the cutting oxygen lever down.
    c. Also, oxygen-flow adjustments must always be made with the needle valve on the cutting attachment. Open the torch oxygen valve wide, leave it that way while cutting, and adjust the flame with the other valve.

Making the Cut

1. During cutting, hold the torch in one hand and guide the torch by resting it on your other hand.
    a. A piece of firebrick on the plate will provide a rest for your hand as well as indicate the proper spacing of the torch from the work.
    b. Make sure nothing will prevent you from finishing the cut without interruption.
2. Hold the torch so that the preheat cones just lick the work surface. Preheat the starting point on the edge to a bright red (see Fig. 2.21).

FIGURE 2.21 Preheating.

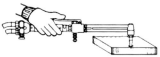
FIGURE 2.22 Beginning a cut.

3. Start the cut by slowly pressing down the cutting-oxygen lever (see Fig. 2.22).
   a. Keep the tip vertical and always the same height above the work.
   b. Do not advance the torch until the cut is completely through the metal.
   c. Continue the cutting action by moving the torch along the line of cut at a uniform rate (see Fig. 2.23).
4. If you move the torch too slowly, you will melt over the edges of the cut and give it a ragged appearance.
5. If you move the torch too fast, the cutting jet will not penetrate the metal completely and you will "lose the cut." In this case, release the cutting oxygen lever, go back to where you "lost the cut," and start over again.
   a. Experience is the only way to learn exactly how fast to move the torch.
   b. When the cut is finished, release the cutting-oxygen lever and turn off the preheat flames.

Gouging

Gouging or grooving is merely a special type of oxygen cutting. It is a means of removing a narrow strip of metal from the surface of a plate. You use the same equipment for gouging that you use for cutting, except that you must have a large-bore, low-velocity nozzle. As in cutting, the operation centers around three main steps: preheating, starting the groove, and progressing. Other things to be watched during gouging include:

1. Pulling the nozzle back along the plate surface after preheating, then opening the cutting oxygen lever. This ensures that the stream will fall on hot metal, not on relatively cold metal ahead of the preheated spot.
2. Keeping the flames low. If the inner cones of the preheat flames on the lower side of the nozzle are just barely touching the work, you will get maximum efficiency from the preheat flames.
3. Keeping the torch moving in a straight line. When making a long groove, there is a tendency to move the torch toward you as the groove proceeds and describe a long arc instead of a straight line in the plate.

With the "step-back" method of gouging, you will have less tendency to lose the cut or swing out of line than if you gouge in one continuous pass.

1. The groove is carried progressively across the plate in a series of short gouges.
2. Start the groove, then continue it for about 3 in. Lift up the torch, bring it back about $1/2$ in, and restart the groove.

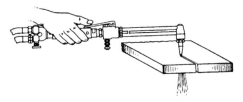
FIGURE 2.23 Cutting.

3. As each short pass is completed, the nozzle is drawn back slightly to restart the groove.
4. Repeat these steps until you have reached the full length of the desired groove.

Gouging is used in three main applications:

1. Removing defective welds. When a weld does not have a good appearance or is not as strong as it should be, it can be removed by gouging and replaced. You can also remove the old weld and have the piece ready to be rewelded all in one operation by gouging.
2. Opening up cracks in castings so that sound repairs can be made by welding.
3. Dismantling welded structures to permit reuse of most of the parts, thus obtaining maximum salvage.

Using a special gouging nozzle, you can cut grooves from $3/8$ to $1/2$ in wide by $1/8$ to $7/16$ in deep. These variations in groove dimensions are controlled by three factors:

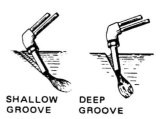

FIGURE 2.24 The angle of the nozzle with respect to the work controls the depth of the groove.

1. By the angle of the nozzle with respect to the work (see Fig. 2.24). A flat angle gives a shallow groove, and a steeper angle a deep groove.
2. By the speed of travel of the torch. The faster you move, the shallower the gouge becomes.
3. By the oxygen pressure. High pressures wash a bit more metal out of the groove than lower pressures.

HARD FACING

Hard facing is the process of applying a layer of special alloy on a metal part or surface to protect it from wear. The big difference between hard facing and the fusion welding is that the hard-facing alloy does not mix with the base metal to any extent. In fusion welding, complete penetration is necessary, but in hard facing, it should be avoided. This is important because mixing of the base metal with the hard-facing alloy would dilute and soften the deposit. In hard facing, the surface of the steel picks up carbon from an excess acetylene flame. The carbon lowers the melting point of the steel and causes it to melt quickly to a depth of only a few thousandths of an inch. This very thin film of melted steel fuses with the hard-face deposit to make a strong bond between the deposit and the steel.

Metals that can be hard-faced include carbon and low-alloy steels (covering 95 percent of the wear problems you will normally encounter), all forms of cast iron (except chilled), and many other special alloys.

With the longer life of hard-faced parts (2 to 25 times longer), the reduction of maintenance labor and of replacement parts used is dollars saved. Table 2.5 shows a few typical examples of how hard facing increases the life of parts.

TABLE 2.5 Relationship of Increased Part Life to Hard Facing

Part hard-faced	Times longer life
Pump shaft	3
Clutch plate	7
Valves, valve seats	7
Valve-seat inserts	15
Hand shovels	3
Spray nozzle disks	12
Cams	6
Shear blades	10
Mill hammers	5
Punches	13

Hard-Facing Rods

There are a number of hard-facing rods available to help you solve particular wear problems resulting from such factors as abrasion, impact, corrosion, and heat. Very often more than one cause of wear is present. Your problem then is to choose the hard-facing alloy best suited to combat the combination of factors. You should consider every job as a special problem. The same rod used for one job will not necessarily work on the same or similar part in another instance. If you are in doubt about which rod to select, test several under actual conditions. Manufacturer's data will usually help you select the proper rod, but often you must make the final decision in light of what you can find out about the wear conditions involved.

How to Hard-Face Steel

1. Clean the surface to be hard-faced by filing, wire brushing, or grinding. Edges or corners that might become overheated during hard facing should be grooved out as shown and filled with hard-facing deposit. Use your cutting torch or attachment and grooving nozzle for this (see Fig. 2.25). If an edge or corner of the part takes a lot of pounding or impact in use, machine the corner or edge as shown in Fig. 2.26. The dotted lines in the illustration show how the hard-face deposit should be built up to the original contour of the part.

2. Parts more than 2 in. in thickness should be preheated throughout to prevent the deposit or the part itself from cracking when it cools. You can preheat medium-sized parts with your torch. Use a neutral flame. Move the flame in a wide circle over the part. Gradually make the circles smaller and smaller until the part turns a dull red color. Large surfaces or bulky parts should be preheated in a furnace. Heat the part until it turns a dull red color.

3. Deposits up to $1/8$ in. in thickness can be made in one pass. Best impact resistance is obtained from deposits $1/16$ in. in thickness, never over $1/8$ in. If you want to build up a badly worn surface with hard facing to a depth greater than $1/8$ in, you should use more than one deposit.

4. Hard-facing rods are applied with a carburizing flame—a flame using more acetylene than oxygen. The extra acetylene shows up as a whitish "feather" around the inner cone. Use the amount of excess acetylene specified by the rod manufacturer.

5. Low- and medium-carbon steels are the most widely used metals and are the easiest to hard-face. The following instructions are for the hard facing of these steels.

6. If you have selected a rod, prepared the part, and set up your welding outfit, you are ready to start hard facing. Begin by heating the part (see Fig. 2.27).

7. Now adjust to a carburizing flame. Reduce the amount of oxygen until you have the proper flame, depending on the rod you are using.

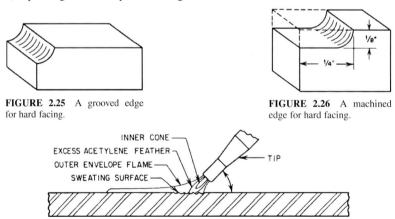

FIGURE 2.25 A grooved edge for hard facing.

FIGURE 2.26 A machined edge for hard facing.

FIGURE 2.27 Heating the surface before depositing the hard-facing rod.

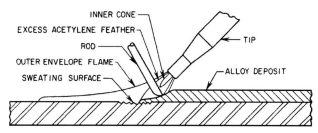

FIGURE 2.28 Depositing the hard-facing rod.

8. Hold the carburizing flame over the heated area. The tip of the inner cone should be just off the steel surface—about $1/8$ in as shown. Hold the flame there until the metal under the flame starts to "sweat."
9. Next, lift the welding torch a little and put the rod into the flame so that it just touches the "sweating" surface. Lower the torch until the inner cone of the flame just touches the rod and is about $1/8$ in from the steel surface, as shown in Fig. 2.28. A small puddle of melted rod will form on the sweating surface. If the first few drips of the melted rod foam or bubble or do not spread evenly, the surface is too cold. Take the rod away and start over again.
10. Next, take the rod out of the puddle. Spread the puddle over the "sweating" surface by pointing the flame into it—do not use the rod to spread it. If there is not enough hard-facing deposit to cover the wearing surface, continue the process.
11. Point the flame so that it touches the forward end of the puddle and the steel surface.
12. When the surface sweats, add more metal to the puddle from the rod. Then, as you did before, remove the rod and spread the puddle with the flame. Repeat until the entire surface is covered.
13. Allow the part to cool slowly to prevent cracks and stresses in the hard face. Small and medium-size parts can be cooled in air. Large or bulky parts should be wrapped in asbestos paper or buried in asbestos, slaked lime, wood ashes, or some other insulating material until they cool. Parts that are liable to crack should be put in the preheating furnace while they are still hot from hard facing. Then they should be brought to an even red heat and, with the heat turned off, allowed to cool overnight in the furnace with the door closed.

Hard-Facing Cast Iron

1. Cast iron does not sweat like steel, and it melts at about the same temperature as the rod. *So be careful*—do not melt the base metal too deeply.
2. Use a little less acetylene in the flame than you would for steel.
3. Use a cast iron brazing flux when you apply the rod.
4. A crust will form over the surface of the cast iron when it is heated. To get a good bond, you will have to break the crust with the end of your rod.
5. Very thin cast-iron parts should be backed up with wet asbestos or carbon paste to keep them from melting.

Finishing the Hard-Facing Deposit

1. *Heat treating* of the hard-faced part is *usually not necessary*. The only time you will heat—treat a part after hard facing it is when you want to toughen the base metal. To do that, heat the whole part to a dull red heat. Then dip it in oil. Do not use water for the quench because it may crack the deposit and base metal.
2. *Surface cracks* are usually caused by insufficient preheat or by cooling the part too quickly. You will find, however, that a surface crack will not harm the properties of the hard face or the strength of the part. If you want to repair a cracked surface:

a. Preheat the piece as for hard facing.
b. Heat the metal around the crack to a dull red.
c. Then melt the edges down into the crack.
d. Add a little metal from the rod.
e. Now slowly move the flame away from the hot spot to prevent quick cooling.

3. You can grind a hard-faced part to exact size or remove high spots on the surface. Use a grinding wheel not coarser than 46 or finer than 60 in Grade I or J of the Norton system. The speed of the wheel should be between 2800 and 4200 sfpm. Higher speeds might crack the hard-face surface.

See Fig. 2.29 for melting points of metals and alloys.

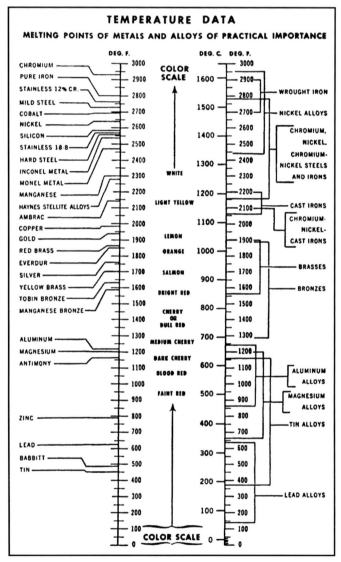

FIGURE 2.29 Melting points of metals and alloys.

TABLE 2.6 Codes, Specifications, and Welding Standards*

Title	Published by	Field of Application	Source
General			
AWSV Welding terms and Definitions, A3.0	ASW,‡ 1980, 80 pp.	Welding cutting, brazing, soldering, and thermal spraying	AWS‡
Standard for fire prevention in use of Cutting and Welding Processes ANSI/NFPA 51B	NFPA, 1984, 11 pp.	General welding	National Fire Prevention Assoc., Batterymarch Park, Quincy, Mass 02269
General Industry Standards 29CFR 1910 Subpart Q	OSHA, 1983, 838 pp.	General welding	Dept. of Labor, OSHA, U.S. Govt. Printing Office, Washington, DC 20402
AWS Symbols for Welding and Nondestructive Testing (including brazing) AWS A2.4	AWS,‡ 1970, 80 pp.	Engineering-shop drawings	AWS‡
Electric Arc-Welding Power Sources EW1, 1983	National Electrical Manufacturers Association, 1983	General welding	National Electrical Mfg. Assoc., 2101 L Street, N.W., Suite 300, Washington, DC 20037
Safety in Welding and Cutting, AWS/ASC Z49.1	AWS, 1983	General welding	AWS
Safety Standard for Transformer-Type Arc-Welding Machines, ANSI/UL 591-1980	Underwriters' Laboratories, 1981, 28 pp.	General welding	ANSI¶
Welding Symbols Chart AWS A2.1 (WC and DC)	AWS,‡ 1979	Wall size 22 by 28 in. Desk size 11 by 17.	AWS‡ AWS‡
Boilers and pressure vessels			
ASME Boiler and Pressure Vessel Code Sec. I, Power Boilers	ASME,§ 1977	Power boilers in stationary service	ASME§
Sec. III, Nuclear Power Plant Components	ASME,§ 1977	Nuclear power plants	ASME§
Sec. IV, Heating Boilers	ASME,§ 1977	Boilers in operation at less than 15 psig and for hot-water heating and supply	ASME§
Sec. VIII, Pressure Vessels	ASME,§ 1977	Pressure vessels	ASME§
Piping			
Power Piping, ANSI B31.1	ANSI,¶ 1977	Pressure piping systems	ANSI¶
Standard for Welding Pipelines and Related Facilities (Std 1104)	API,† 1983	Cross-country petroleum and natural-gas pipelines	API‡
Structural Welding Code, Aluminum D 1.2	AWS, 1983, 152 pp.	Primary code governing structural steel welded fabrication	AWS

(Continued)

TABLE 2.6 Codes, Specifications, and Welding Standards* (*Continued*)

Title	Published by	Field of Application	Source
Piping			
Specifications for Qualification of Welding Procedures and Welders for Piping and Tubing D 10.9	AWS, 1980, 60 pp.	All piping systems	AWS
Structural (building)			
Rules for Arc and Gas Welding and Oxygen Cutting of Steel Covering the Specifications for Design, Fabrication, and Inspection of Arc and Gas Welded Steel Structures and Qualifications of Welders and Supervisors. Cal. 1-38-SR.	Board of Standards and Appeals, New York, 1968, V. 53, Bull. 51, p. 1338; Amendments: 1974, V. 59, Bull. 4, p. 34; V. 59, Bull. 13, p. 259; 1977, V. 62, Bull. 22, p. 500	Buildings in New York	Board of Standards and Appeals, City of New York, 80 Lafayette St., New York, N.Y. 10013
Structural Welding Code, Steel D1.1	AWS,‡ 1983, 315 pp.	Highway, railway, bridges, buildings, and tubular structures	AWS‡
Recommended Safe Practices for Welding and Cutting Containers and Piping that have held hazardous substances F 4.1	AWS,‡ 1965, 16 pp.	Shops engaged in welding or cutting operations on containers of combustible solids, liquids, or gases	AWS‡
Specifications for Field-Welded Tanks for Storage of Production Liquids, 8th ed. (Spec. 12D and supplement)	API,† 1983	Oil-field service—capacities over 500 bbl	API†
Specifications for Shop-Welded Tanks for Storage of Production Liquids, 7th ed. (Spec. 12F and supplement)	API,† 1983	Oil-field service—capacities to 440 bbl	API†
Recommended Rules for Design and Construction of Large, Welded, Low-Pressure Storage Tanks, 7th ed. (Std. 620)	API,† 1982	Petroleum products storage—for internal pressures of 15 psig or less	API†
Welded Steel Tanks for Oil Storage, 7th ed. (Std. 650)	API,† 1981	Oil storage at atmospheric pressure	API†
Standard for Welded Steel Elevated Tanks, Standpipes, and Reservoirs for Water Storage, D5.2	American Water Works Association and AWS,‡ 1979, 63 pp.	Elevated steel water tanks, standpipes, and reservoirs	AWS‡

TABLE 2.6 Codes, Specifications, and Welding Standards* (*Continued*)

Title	Published by	Field of Application	Source
Structural (building)			
Water Tanks for Private Fire Protection, NFPA Std. No. 22	National Fire Protection Association, 1976	Field-welded tanks, gravity and pressure towers, etc.	National Fire Protection Association, Batterymarch Park, Quincy, MA 02269
Inspection and Qualification			
A Guide to AWS Welding Inspector Qualification, including AWS QC1-84, Standard for Qualification and Certification of Welding Inspectors, WIQC	AWS, 1984, 32 pp.	Explanation of the AWS welding inspector certification program	AWS‡
Welding Inspection, WI	AWS, 1980, 222 pp.	Reference material for the welding inspector	AWS
Welding Procedure and Performance Qualification, B3.0	AWS,‡ 1977, 97 pp.	Industry, welding instructors, and codewriting bodies wishing to prescribe methods	AWS‡
Standard Methods for Mechanical testing for Welds	AWS,‡ 1977, 60 pp.	Mechanical testing	AWS‡
Welding and Brazing Qualifications, Boiler and Pressure Vessel Code, Sec. IX	ASME,§ 1977	Boilers and pressure vessels	ASME§

*All are available from sponsoring organization. For convenience, AWS is given as source when possible.
†American Petroleum Institute, 1220 L St., N.W., Washington, DC 20005.
††American Petroleum Institute, 211 N. Ervay, Suite 1700, Dallas, TX 79201 (alternate source).
‡American Welding Society, PO Box 351040, Miami, FLA. 33135.
§The American Society of Mechanical Engineers, Order Department, United Engineering Center, 345 East 47th St., New York, NY 10017.
¶American Natural Standards Institute, 1430 Broadway, New York, NY 10018.

BIBLIOGRAPHY

"Brazing Alcoa Aluminum," Aluminum Company of America, 1967.

Jefferson, T. B.: *The Welding Encyclopedia,* Jefferson Publications, 17th ed., 1974.

Linnert, G.E.: *Welding Metallurgy,* American Welding Society, 3d ed., vol. 1, *Fundamentals,* 1965; vol. 2, *Technology,* 1967.

Precautions and Safe Practices for Gas Welding, Cutting and Heating, L-TEC Welding and Cutting Systems, Form 52-529.

"Stellite Hard-Facing Products," Stellite Division, Cabot Corporation, 1975.

The Oxy-Acetylene Handbook, L-TEC Welding and Cutting Systems, 5th ed., 1955.

"Welding Alcoa Aluminum," Aluminum Company of America, 1972.

Welding Handbook, American Welding Society, 7th ed., vol. 1, *Fundamentals of Welding,* 373 pp.; vol. 2, *Welding Processes: Arc and Gas Welding and Cutting, Brazing, and Soldering,* 592 pp.; vol. 3, *Welding Processes: Resistance and Solid State Welding and Other Joining Processes,* 459 pp.; vol. 4, *Metals and Their Weldability,* 582 pp.

INDEX

A3 process, 3.155–3.159
    form for, 3.158 3.159
    steps of, 3.156–3.158
Ability, in law of intelligent action, 3.91
Abrasion, with ventilating fans, 4.99–4.100
Abrasion method, 9.30
Abrasion resistance, 4.21
Abrasive wear, 10.43–10.44
ABS, for corrosion control, 9.10
Absolute-pressure elements, 7.18
Absorbers, gas, 4.113, 4.114
Absorption, atmospheric, 7.113
Absorption machines:
    leak testing for, 4.66
    leaktightness, 4.65
    maintenance of, 4.64–4.65
    pumps, 4.66
    purge units, 4.66
    safeties, 4.66
    service valves, 4.66
Absorptivity, 7.113
AC (alternating current), 7.115
AC induction motors, 6.3–6.9
    construction of, 6.6
    design of, 6.7
    insulation for, 6.7–6.8
    speed-torque characteristics of, 6.6–6.7
    temperature of, 6.8–6.9
    theory for, 6.6
    troubleshooting, 6.20–6.25
AC instruments, 7.56
AC motors, 6.3–6.6
AC volts, 7.53–7.55
Accelerating voltages, 7.66
Acceleration, of vibration, 7.75, 7.77
Accelerometers, 7.97
Acceptable operating envelope, 3.171
Acceptance testing, 7.69
Accessory equipment, 10.52–10.55
Accounting function, 2.106–2.107
Acetal homopolymers, 9.10
Acid(s):
    citric, 9.20
    cleaning with, 9.20–9.21
    inhibited muriatic acid (HCl), 9.20
    inhibited sulfuric acid, 9.20–9.21
    inorganic, 9.20–9.21
        cleaning with, 9.20–9.21
        inhibited muriatic acid (HCl), 9.20
        inhibited sulfuric acid, 9.20–9.21
        nitric acid, 9.21

Acid(s): inorganic (*Cont.*)
        phosphoric acids, 9.21
        sulfuric acid, 9.21
    monoammoniated citric acids, 9.20
    nitric, 9.21
    organic, 9.20
    phosphoric, 9.21
    sulfuric, 9.21
Acid cleaning, 9.86, 9.87
Acid number, 8.31
Across-the-flex element:
    of flexible couplings, 5.67–5.71
    horizontal alignment solution for, 5.70–5.71
    vertical alignment solution for, 5.67–5.70
Action item matrix (AIM), 1.58–1.60
Active descriptions, 3.140
A/D conversion, 7.113
Adjustable pulleys, 4.74
Adjustable while in motion (nonenclosed) friction drives, 5.145–5.151
Adjustable-type relays, 6.68
Administrative skills, 1.35
Aesthetics, painting for, 9.37–9.38
Aftercoolers, 5.193
AIM (action item matrix), 1.58–1.60
Air, entrained, 5.217–5.218
Air compressors, 5.185–5.196
    and aftercoolers, 5.193
    and air filters, 5.186, 5.187
    and air receivers, 5.187
    bearings in, 5.192–5.193
    capacity of, 5.187
    cleaning, 5.193
    foundation of, 5.185–5.186
    and intercoolers, 5.193
    location of, 5.185, 5.187
    lubrication of, 5.188–5.189
    nonlubricated cylinders in, 5.189
    packing, 5.193–5.196
    piston rings in, 5.191–5.192
    starting new, 5.188
    and suction lines, 5.186, 5.187
    unloading, 5.193
    valves of, 5.189–5.191
Air conditioners (*see* Air-conditioning equipment)
Air filters, 4.76–4.77, 5.186, 5.187
Air motors, 4.84
Air receivers, 5.187
Air washers, 4.66–4.67
    cleaning, 4.66
    fans in, 4.66

Air washers (*Cont.*)
    lubrication of, 4.66–4.67
    maintenance of, 4.66–4.67
Air-acetylene appliances, 10.63–10.71
    applications of, 10.70–10.71
    automobile body soldering, 10.67
    electrical connections, 10.67–10.68
    paint burning, 10.69–10.70
    precautions with, 10.63–10.64
    safety considerations with, 10.63–10.64
    sheet metal working, 10.65–10.67
    soldering, 10.64–10.65
    sweat-type fittings, 10.68–10.69
Air-carbon arc cutting (CAC-A), 10.27–10.30
Air-cleaning (dust control) equipment, 4.111–4.124
    electrostatic precipitators, 4.117–4.118
        fans in, 4.121–4.122
        maintenance of, 4.120–4.122
        sludge settling tanks, 4.121
    exhaust ducts
        maintenance of, 4.122
        preventative maintenance of, 4.123
    exhaust hoods
        maintenance of, 4.122
        preventative maintenance of, 4.123
    fabric filtration, 4.116–4.117, 4.120
    fans in, 4.121–4.122
    gas absorbers, 4.113, 4.114
    inertial (dry centrifugal), 4.112, 4.119
    maintenance of, 4.118–4.124
        electrostatic precipitators, 4.120–4.122
        exhaust ducts, 4.122
        exhaust hoods, 4.122
        fabric filtration, 4.120
        inertial equipment, 4.119
        wet collectors, 4.119–4.120
    orifice type scrubbers, 4.120
    packed scrubbers, 4.119
    particulate scrubbers, 4.114, 4.115
    and sludge settling tanks, 4.121
    types of, 4.112–4.118
    wet centrifugal scrubbers, 4.119–4.120
    wet collectors (scrubbers), 4.112, 4.113
        maintenance of, 4.119–4.120
        orifice type, 4.120
        packed scrubbers, 4.119
        wet centrifugal, 4.119–4.120
        wet dynamic, 4.120
    wet dynamic scrubbers, 4.120
Air-compressor check valve, 5.201–5.202

I.1

Air-conditioning equipment, 4.53–4.85
  absorption machines, 4.64–4.65
  and air motors, 4.84
  air washers, 4.66–4.67
  air-cooled condensers, 4.69
  belt drives, 4.73–4.74
  centrifugal compressors, 4.67–4.68
  coils, 4.68–4.69, 4.84
  components of, 4.57–4.58
  and compressed-air lines, 4.84
  and condensers, 4.70, 4.84
  coolers, 4.71
  cooling loads, 4.53–4.55
    latent loads, 4.54
    from lights, 4.54
    occupancy loads, 4.54
    outside-air loads, 4.54
    removal of, 4.54
    size of, 4.54–4.55
    sun loads, 4.53
    transmission loads, 4.53–4.54
  cooling towers, 4.71–4.73
  dampers, 4.73
  dehumidifiers, 4.80
  economizers, 4.74–4.75
  equipment for, 4.57
  evaporative condensers, 4.69–4.70, 4.84
  fans, 4.75–4.76
  filters, 4.76–4.78
  freezing issues with
    air motors, 4.84
    coils, 4.68–4.69, 4.84
    compressed-air lines, 4.84
    condensers, 4.70, 4.71, 4.84
    damage from, 4.83
    evaporative condensers, 4.70, 4.84
    prevention of, 4.83–4.84
    susceptible areas, 4.84
    water coils, 4.84
    water lines, 4.84
    water-cooled condensers, 4.71
    well pumps, 4.84
  heat pumps, 4.79–4.80
  heaters, 4.78–4.79
  and heating loads, 4.55
  humidifiers, 4.80
  installation of, 4.61
  latent loads, 4.54
  loads on, 4.53–4.55
    cooling, 4.53–4.55
    heating, 4.55
  maintenance of, 4.61–4.85
    absorption machines, 4.64–4.65
    air washers, 4.66–4.67
    air-cooled condensers, 4.69
    belt drives, 4.73–4.74
    centrifugal compressors, 4.67–4.68
    coils, 4.68–4.69
    coolers, 4.71
    cooling towers, 4.71–4.73
    dampers, 4.73
    dehumidifiers, 4.80
    economizers, 4.74–4.75
    evaporative condensers, 4.69–4.70
    fans, 4.75–4.76
    filters, 4.76–4.78
    with freeze protection, 4.83–4.84
    heat pumps, 4.79–4.80
    heaters, 4.78–4.79

Air-conditioning equipment, maintenance of (*Cont.*)
    humidifiers, 4.80
    operating logs for, 4.63
    pumps, 4.80–4.81
    requirements for, 4.63, 4.64
    responsibility for, 4.61
    rooftop units, 4.81–4.82
    room air conditioners, 4.82–4.83
    self-contained units, 4.83
    spare parts, 4.63
    training for, 4.63
    and water conditioning, 4.84–4.85
    water-cooled condensers, 4.70–4.71
  occupancy loads, 4.54
  and operating logs, 4.63
  outside-air loads, 4.54
  preventative maintenance for, 4.61
  pumps, 4.80–4.81
  refrigeration, 4.55–4.56
    equipment for, 4.58–4.61
    measuring, 4.55–4.56
    and oil, 4.58
    and refrigerant, 4.58
  responsibility for, 4.61
  rooftop units, 4.81–4.82
  room air conditioners, 4.82–4.83
  self-contained, 4.83
  spare parts for, 4.63
  sun loads, 4.53
  training for, 4.63
  transmission loads, 4.53–4.54
  and water coils, 4.84
  water conditioning, 4.84–4.85
    corrective measures for, 4.85
    definitions of, 4.85
    problems with, 4.84–4.85
  and water lines, 4.84
  water-cooled condensers, 4.70–4.71
  and well pumps, 4.84
Air-cooled condensers:
  cleaning of, 4.69
  inspection of, 4.69
  lubrication of, 4.69
  maintenance of, 4.69
Aircraft engines, 7.71
Air-receivers, 5.187
Alarm (alert) limits, 7.94–7.95
Algae, in cooling towers, 4.72
Alignment:
  of belt drives, 4.74
  of cooling towers, 4.72
  of face/rim, 5.63, 5.66, 5.67
  of fans, 4.76
  of flexible couplings, 5.54–5.56
  horizontal, 5.61–5.63, 5.66, 5.67, 5.70–5.71
  vertical, 5.59–5.61, 5.63, 5.66–5.70
Alkaline cleaners, 9.20
Alkyd coatings, 9.43
Alligatoring, 9.47
Allowance determination, 3.59, 3.60
Alloy steels, 10.36–10.39
  high-manganese steels, 10.39
  high-tensile low-alloy steels, 10.36
  stainless clad steels, 10.38
  stainless steels, 10.36, 10.38
  straight chromium steels, 10.38
  and welding, 10.36–10.39
Alterations, maintenance for, 1.10

Alternate displays, 7.66
Alternating current (ac), 7.115 (*See also* AC)
Alternating current-instruments, 7.56
Alternators:
  defined, 7.115
  motor, 7.116
Aluminum:
  for corrosion control, 9.9
  and welding, 10.39
Aluminum complex soaps, 8.10
Aluminum ladders, 9.92
Aluminum scaffolding, 9.101–9.103, 9.105, 9.106
Aluminum soaps, 8.9
Ambient measurements, 7.110
Ambient temperature, 7.113
American National Standards Institute (ANSI), 9.93
Ammeters, 7.51–7.52
Ampacity, 7.115
Amperage, 7.115
Amperes, 7.53, 7.115
Amplitude, 7.77
Analog infrared equipment, 7.110
Analog multimeters, 7.44–7.48
Analogue data, 7.113
Analysis(-es):
  break-even, 3.14
  broadband data for, 7.104
  cause-and-effect, 3.138–3.139
  computerized systems for, 2.88–2.90
  cost-benefit, 7.135
    for lubricating oil analysis, 7.135
    for tribology, 7.135
  cost-effectiveness, 3.14
  electric motor analysis, 2.29–2.30
    insulation resistance, 2.29–2.30
    testing, 2.30
  of equipment, 2.90
  failure analysis
    for FSIs, 2.11–2.12
    root-cause, 7.104
  first level logic tree analysis, 2.12–2.13
  fuel, 8.31–8.32
  infrared, 8.32
  life cycle cost analyses, 3.136
  logic tree analysis, 2.12–2.14
    first level, 2.12–2.13
    second level, 2.14
  lubricant, 8.30–8.32
    and microorganisms, 8.32
    and sediment, 8.32
    and sugar, 8.32
    and sulfur, 8.32
    and viscosity, 8.32
    and viscosity index, 8.32
    and water, 8.32
  lubricating oil analysis, 2.25–2.26, 7.128–7.129
    contamination, 2.25, 7.128
    cost-benefit analysis for, 7.135
    fuel dilution, 2.25, 7.128
    fuel soot, 2.26, 7.128
    nitration, 2.26, 7.129
    oxidation, 2.26, 7.128
    particle count, 2.26, 7.129
    solids content, 2.25, 7.128
    total acid number, 2.26, 7.129
    total base number, 2.26, 7.129
    viscosity, 7.128

Analysis(-es) (Cont.)
  in lubrication programs, 8.29–8.30
  maintainability, 3.136–3.137
  narrowband data for, 7.104
  oil, 7.129
  operating dynamics analysis, 2.31
  Pareto, 3.141–3.142
  of parts inventory, 2.89
  root cause analysis (RCA), 153–3.176
    A3 process, 3.155–3.159
    for acceptable operating envelope, 3.171
    analysis for, 3.159
    application review, 3.172–3.174
    capacity restrictions, 3.176
    for clarifying problems, 3.166–3.168
    concept of, 3.154–3.155
    damage, 3.175–3.176
    data for, 3.168–3.169
    design review, 3.169–3.172
    for economic performance, 3.176
    for equipment damage/failure, 3.175–3.176
    event-reporting format for, 3.165–3.166
    and facts, 3.168–3.169
    failure, 3.175–3.176
    and five whys (5W), 3.161
    formal process for, 3.161–3.163
    identifying events with, 3.163–3.166
    incident classification, 3.174–3.176
    for incoming-project specifications, 3.171
    methodology for, 3.154–3.155, 3.163
    objectives of, 3.171
    for operating performance, 3.176
    for output-product specifications, 3.171
    personnel requirements for, 3.163
    physical evidence for, 3.168–3.169
    problem solving with, 3.174
    process for, 3.159–3.163
    for product quality, 3.176
    and regulatory compliance, 3.176
    reporting for, 3.165–3.166
    requirement of, 3.169
    for safety, 3.176
    for system damage/failure, 3.175–3.176
    and visual inspection, 3.159–3.161
    for work performed, 3.171
  root-cause failure analysis, 7.104
  second level logic tree analysis, 2.14
  sequence-of-events analysis, 3.139–3.141
    active descriptions in, 3.140
    in logical order, 3.139
    precision of, 3.140
    qualifiers in, 3.141
  SFMEA, 3.133–3.137
    and life cycle cost analyses, 3.136
    and maintainability analysis, 3.136–3.137
    and mean time between failures (MTBF), 3.136
    and reliability block diagram (RBD), 3.136
    and reliability predictions, 3.136
    and Simplified Failure Mode, Effect, and Criticality Analysis (SFMECA), 3.135
  SFMECA, 3.135
  signature, 2.23
  spectrographic, 7.129
  statistical analyses tools, 3.141–3.152
  techniques for, 7.104
  for tribology, 7.135

Analysis(-es) (Cont.)
  of vibration, 7.69–7.71
  wear particle analysis, 2.26–2.28, 7.129–7.134
    cost-benefit analysis for, 7.135
    cutting wear, 2.27, 7.131, 7.132
    monitoring and trending, 7.130
    for particulate matter, 7.130–7.134
    rolling fatigue, 2.27
    rolling wear, 2.27
    rubbing wear, 2.27
    sliding wear, 2.27
  of work controls, 2.89
  of work orders, 2.88
Analytical cost estimate model, 3.15
Analytical ferrography, 8.31
Analytical models, 3.15
Angular (face) solution, 5.63, 5.66
ANSI (American National Standards Institute), 9.93
Application review, 3.172–3.174
  installation, 3.172
  maintenance history, 3.173
  maintenance procedures, 3.174
  operating envelope, 3.172
  operating procedures, 3.173
Aqueous solutions, 9.14–9.15
Arc chambers, 6.45–6.46
Arc welding, 10.3–10.60
  flux-cored (FCAW), 10.10
  gas-shielded metal (GMAW), 10.11–10.13
  role of, 10.3–10.6
  shielded metal (SMAW), 10.6–10.9
  submerged (SAW), 10.10–10.11
Armature, 7.115
Arresters, 7.115, 7.116
Ash, 8.31
Asset care, 1.19
Asset management, life cycle, 1.18
Assignment maintenance mechanics, 3.58–3.59
Astigmatism, 7.66
ASTM standards, 9.51–9.54
Atmospheric absorption, 7.113
Atmospheric attenuation, 7.113
Atmospheric emission, 7.113
Atmospheric pressure, 7.8–7.10
Atmospheric radiance, 7.113
Atmospheric reflectance, 7.113
Atomic hydrogen welding, 10.34
Attenuation, atmospheric, 7.113
Attenuators, 7.66, 7.115
Authority, 1.48–1.49
Automatic costs allocation, 2.106–2.107
Automatic dampers, 4.73
Automatic lubrication system, 8.13–8.26
  centralized, 8.14–8.16
  duoline systems, 8.20, 8.21
  ejection systems, 8.23–8.24
  injection systems, 8.24
  injector systems, 8.17
  oil mis systems, 8.16
  orifice-control systems, 8.16–8.17
  pump-to-point systems, 8.20–8.21
  selection of, 8.25–8.26
  series-progressive systems, 8.17–8.19
  twin-line systems, 8.19–8.20
  types of, 8.16
  zone-control systems, 8.21, 8.22
Automatic triggering, 7.66

Automobile body soldering, 10.67
Autoranging, 7.50
Averages/ratios, 3.111
Averaging bulbs, 7.23
Awareness, painting for, 9.37–9.38
Axial fans, 4.95

Backlash, 7.83
Backlogs, 2.79
Backward-curved-blade centrifugal fans, 4.96
Balanced scorecard (BSC), 3.124–3.127
Ball check valve, 5.201–5.202
Ball valves, 5.207–5.208
Banded (joint) belts, 5.120–5.121
Bands, 7.113, 7.114
Bandwidth:
  defined, 7.113
  of oscilloscopes, 7.66–7.67
Bar coding, 2.71–2.73
Base metals, 10.35–10.39
  alloy steels, 10.36–10.39
  carbon steels, 10.35–10.36
  nonferrous metals, 10.39
  and welding, 10.35–10.39
Base numbers, 8.31
Basic gear drives, 5.164
Batteries, 6.79–6.103
  charging
    lead-acid batteries, 6.81–6.85
    nickel-cadmium batteries, 6.93–6.94
    nickel-iron batteries, 6.100–6.101
    of standby-power batteries, 6.100–6.101
  cleaning, 6.101–6.102
  constant-current method for, 6.84
  cycled, 6.100
  and electrolytes, 6.102–6.103
  equalizing charge for, 6.95
  float charging, 6.94
  for infrared equipment, 7.110
  installation of
    lead-acid batteries, 6.80–6.81
    nickel-cadmium batteries, 6.92–6.93
  internal inspection for, 6.86–6.87
  lead-acid, 6.80–6.89
    charging, 6.81–6.85
    condition of, 6.86–6.87
    constant-current method for, 6.84
    installation of, 6.80–6.81
    internal inspection for, 6.86–6.87
    maintenance of, 6.85–6.86
    modified constant-voltage method for, 6.82–6.83
    operation of, 6.80–6.81
    overdischarging, 6.86
    placement of, 6.81
    records for, 6.86
    repairs to, 6.88–6.89
    taper method for, 6.84
    test discharge for, 6.86
    troubleshooting, 6.87–6.88
    two-rate method for, 6.84
  life expectancy for, 6.91–6.92
  maintenance of
    lead-acid, 6.85–6.86
    nickel-cadmium, 6.95–6.99
    nickel-iron batteries, 6.101–6.103
  modified constant-voltage method for, 6.82–6.83
  nickel-cadmium, 6.89–6.99

I.4   INDEX

Batteries, nickel-cadmium (*Cont.*)
  charging, 6.93–6.94
  construction of, 6.90
  equalizing charge for, 6.95
  float charging, 6.94
  installation of, 6.92–6.93
  life expectancy for, 6.91–6.92
  maintenance of, 6.95–6.99
  performance of, 6.91
  plate processing for, 6.90
  and pocket plates, 6.90–6.91
  selection of, 6.92
  and sintered plates, 6.91
  trickle charging, 6.94–6.95
  voltage of, 6.91
 nickel-iron, 6.99–6.103
  charging, 6.100–6.101
  cleaning, 6.101–6.102
  cycling, 6.100, 6.102
  and electrolytes, 6.102–6.103
  laying up, 6.102
  maintenance for, 6.101–6.103
  operation of, 6.99–6.100
  watering, 6.101
 and overdischarging, 6.86
 performance of, 6.91
 plate processing for, 6.90
 and pocket plates, 6.90–6.91
 and sintered plates, 6.91
 standby-power, 6.100–6.101
 taper method for, 6.84
 test discharge for, 6.86
 trickle charging, 6.94–6.95
 troubleshooting, 6.87–6.88
 two-rate method for, 6.84
 voltage of, 6.91
 watering, 6.101
Beam finders, 7.67
Beams, dual, 7.67
Bearing mountings, 5.28–5.32
Bearings:
 in air compressors, 5.192–5.193
 antifriction, 6.12
 of centrifugal compressors, 4.68
 dismounting, 5.38–5.39
 dynamics of, 7.80–7.82
 fan, 4.75–4.76
 grease-lubricated antifriction bearings, 6.12
 hot, 5.118, 5.123, 5.137, 5.142
 oil-lubricated sleeve bearings, 6.12–6.13
 overheated, 4.110
 plain, 5.3–5.18
  bimetals for, 5.5
  cleaning, 5.3–5.4
  conformability of, 5.4
  corrosion resistance of, 5.4
  design of, 5.6–5.9
  embedability of, 5.4
  fatigue strength of, 5.4
  inspection of, 5.9–5.15
  location of, 5.7–5.8
  and lubrication, 5.8–5.9
  materials for, 5.4–5.5
  monometals for, 5.5
  preventative maintenance for, 5.3–5.4
  reconditioning, 5.9, 5.15–5.18
 rolling-element, 5.19–5.44, 7.80–7.81
  bearing mountings, 5.28–5.32
  boundary dimensions for, 5.20, 5.23

Bearings: rolling-elemen (*Cont.*)
  design for, 5.20
  dismounting of, 5.33, 5.38–5.40
  failure of, 5.19
  fit of, 5.25–5.26
  housing of, 5.25–5.26
  load ratings for, 5.24–5.25
  lubrication for, 5.41–5.44
  mounting of, 5.33–5.38
  nomenclature for, 5.20
  series, 5.23, 5.24
  shafts of, 5.25–5.26
  sleeve, 6.12–6.13, 7.81–7.82
  used, 5.9–5.14
  ventilating fans, 4.107, 4.110
  vibration of, 7.80–7.82
Beautification, painting for, 9.37–9.38
Behavioral expectations, 2.111
Bellows-actuated flowmeters, 7.5–7.7, 7.16–7.18
  design of, 7.5–7.6
  operation of, 7.6–7.7
Belt drives, 4.73–4.74
  adjustable pulleys in, 4.74
  alignment of, 4.74
  and conventional mount synchronous sprockets, 5.130
  couplings in, 4.74
  curvilinear, 5.124
  direct-connected, 4.74
  dynamics of, 7.83–7.84
  friction-type, 5.145–5.147
  maintenance of, 4.73–4.74
  modified curvilinear, 5.124
  packaged, 5.155–5.156
  replacement of, 4.74
  and reverse mount synchronous sprockets, 5.130
  service life of, 4.73, 4.74
  synchronous, 5.123–5.127, 5.129–5.142
   application guidelines for, 5.129
   curvilinear belts, 5.124
   guidelines for, 5.129
   installation of, 5.129–5.131
   maintenance of, 5.131–5.135
   modified curvilinear belts, 5.124
   nomenclature for, 5.126–5.127
   and QD bushings, 5.130–5.131
   tensioning, 5.135–5.136
   timing belts, 5.124
   tooth pitch, 5.124
   tooth profiles, 5.124–5.125
   troubleshooting, 5.136–5.142
  synchronous sprockets, 5.127–5.129
   conventional mount, 5.130
   diameter of, 5.127
   guidelines for, 5.129
   reverse mount, 5.130
  and tensioning, 4.74
   in synchronous belts, 5.135–5.136
   in V-belt drives, 5.115–5.116
  timing belts, 5.124
  troubleshooting
   synchronous belts, 5.136–5.142
   V-belt drives, 5.117–5.123
  V-, 5.99–5.143
   application guidelines for, 5.103–5.108
   datum length of, 5.101
   dimensions of, 5.99–5.103
   effective length of, 5.101, 5.102

Belt drives, V- (*Cont.*)
  length of, 5.101–5.102
  maintenance of, 5.112–5.115
  QD bushings, 5.111
  tension of, 5.115–5.116
  tensioning, 5.115–5.116
  and TL bushings, 5.111
  troubleshooting, 5.117–5.123
  and V-belt sheaves, 5.102–5.103, 5.108–5.112
  and ventilating fans, 4.106, 4.108
  vibration of, 7.83–7.84
Belt guards, 4.106
Belt transmissions, 5.153–5.155
Belts (*see* Belt drives)
Benefit-cost method, 3.14
Bent shafts, 7.89
Best practices:
  inventory management, 2.61
  for key performance indicators (KPIs), 3.129
BHP (brake horsepower), 5.220
Bills of material, 2.94
Bimetals, for plain bearings, 5.5
Bi-modal histograms, 3.149
Bipolar input, 7.49
Bitumen roofing membranes, modified, 4.4
Blades, 7.83
Blanking, 7.67
Blisters, 4.12, 9.47
Block-and-jaw (Oldham) couplings, 5.49–5.50
Blowers, 7.91
Bolt tension (*see* Tensioning)
Bolting, 9.83–9.86
Bourdon tubes, 7.15–7.18
Braided shunts, 6.46
Brake horsepower (BHP), 5.220
Braze welding, 10.74–10.75
Brazing, 10.71–10.72
Breakdown maintenance, 1.41–1.42, 2.3
Break-even analysis, 3.14
Brick masonry structures, 4.27–4.40
  bronze stains on, 4.38
  brown (manganese) stains on, 4.36–4.37
  cleaning, 4.31–4.40
   failures of, 4.31–4.32
   hand washing, 4.35
   procedures for, 4.32–4.36
   of stains, 4.36–4.40
  copper stains on, 4.38
  cracked walls, 4.30–4.31
   replacing units for, 4.30–4.31
   and vertical expansion joints, 4.31
  dirt stains on, 4.38
  and efflorescence, 4.36
  egg splatter on, 4.39
  face grouting for, 4.29
  failures of, 4.31–4.32
  flashings, 4.29
  green stain (vanadium salts) on, 4.36
  hand washing for, 4.35
  and high-pressure cold water, 4.33
  and high-pressure steam, 4.33
  historic structures, 4.40
  inspection of, 4.27
  iron stains on, 4.37–4.38
  ivy stains on, 4.39
  leaky walls, 4.27–4.30
   face grouting, 4.29
   flashings, 4.29

Brick masonry structures, leaky walls (*Cont.*)
  tuckpointing, 4.28–4.29
    water repellents for, 4.30
  oil stains on, 4.38
  paint stains on, 4.37
  paper stains on, 4.39
  plant growth on, 4.39
  and poultice, 4.37
  routine inspection of, 4.27
  sandblasting, 4.33, 4.35
  smoke stains on, 4.38
  stains on, 4.36–4.40
    from bronze, 4.38
    brown (manganese) stain, 4.36–4.37
    cleaning of, 4.36–4.40
    from copper, 4.38
    from dirt, 4.38
    from egg splatter, 4.39
    green stain (vanadium salts), 4.36
    from iron, 4.37–4.38
    from ivy, 4.39
    from oil, 4.38
    from paint, 4.37
    from paper, 4.39
    from plant growth, 4.39
    from poultice, 4.37
    from smoke, 4.38
    from straw, 4.39
    from tar, 4.38
    of unknown origin, 4.39–4.40
    from welding splatter, 4.38
    from white scum, 4.39
  straw stains on, 4.39
  tar stains on, 4.38
  tuckpointing, 4.28–4.29
  and vertical expansion joints, 4.31
  water repellents, 4.30
  welding splatter on, 4.38
  white scum on, 4.39
Brightness, 7.67
Broadband data, 7.104
Broadband trending, 2.22–2.23
Broken strands, 7.112
Bronze check valves, 5.201–5.203
  air-compressor check valve, 5.201–5.202
  ball check valve, 5.201–5.202
  nonmetallic-disk lift-check valve, 5.201
  regrinding lift-check valve, 5.201
  Renewo lift-check valve, 5.201
  swing-check valve, 5.201
Bronze gate valves, 5.203–5.204
  double-wedge rising-stem, 5.204
  renewable-wedge-and-seat bronze, 5.204
  screw-and-yoke rising-stem, 5.204
  single-wedge nourishing-stem, 5.204
  solid-wedge rising-stem, 5.204
Bronze globe valves, 5.200–5.201
  600-Brinnel flat-seat, 5.201
  nonmetallic-disk, 5.201
  plug-type Renewo, 5.201
  threaded-bonnet, 5.200
  union-bonnet regrinding valve, 5.200
Bronze stains, 4.38
Bronze valves, 5.199–5.204
  check valves, 5.201–5.203
  gate valves, 5.203–5.204
  globe valves, 5.200–5.201
Brown (manganese) stains, 4.36–4.37
Brushes, 7.115

BSC (balanced scorecard), 3.124–3.127
Bubbling, 9.47
Budget:
  CMMS for, 2.102
  preparation of, 2.102
  tracking, 2.102, 2.107
Building maintenance, 1.10
Built-up roofing membranes, 4.4, 4.11
Bulbs, 7.23
Bulk storage, 8.7
Burning, of gear-tooth, 5.182
Bus support insulators, 6.75
Bushings:
  QD, 5.111, 5.130–5.131
  TL, 5.111, 5.131
  of transformers, 7.112
Business plan, 1.36
Busy ducts, 7.111
Butterfly valves, 5.208
Bypass operation, 5.222

°C (Celsius degree), 7.113
CAC-A (air-carbon arc cutting), 10.27–10.30
Calcium complex soaps, 8.11
Calcium soaps, 8.10
Calibration:
  field calibration, 7.38–7.39
  of flowmeters, 7.33–7.35
  of pressure gages, 7.38–7.40
  of relative-humidity-measuring instruments, 7.40
  shop calibration, 7.39–7.40
  of thermometers, 7.36
Capacitors:
  defined, 7.115
  and infrared technology, 7.112
Capacity:
  of air compressors, 5.187
  of air-receivers, 5.187
  of fans, 4.75
  in law of intelligent action, 3.91–3.92
  and predictive maintenance, 2.33
  restrictions, 3.176
  of ventilating fans, 4.109
Capital cost, 7.134
Carbon steels, 10.35–10.36
  cast iron, 10.36
  and welding, 10.35–10.36
Cascade method, 9.23–9.24
Cast iron:
  carbon steels, 10.36
  fusion welding, 10.78
  hard facing for, 10.85
  welding, 10.36
Cathodic protection, 9.13
Cause-and-effect analysis, 3.138–3.139
Cause-&-effect diagrams, 1.64–1.67
Caustic soda plus potassium permanganate, 9.20
Caustic soda-surfactant alkalies, 9.20
Cavitation:
  corrosion, 9.6
  and plain bearings, 5.14
Celsius degrees (°C), 7.113
Centralization, 1.45–1.46
Centralized automatic lubrication system, 8.14–8.16
  efficiency with, 8.15
  and housekeeping, 8.16
  and operating costs, 8.16

Centralized automatic lubrication system (*Cont.*)
  productivity with, 8.16
  safety with, 8.14
Centrifugal compressors, 4.67–4.68
  bearings of, 4.68
  leak testing for, 4.68
  lubrication of, 4.67–4.68
  maintenance of, 4.67–4.68
  oil heater of, 4.68
  purge unit, 4.68
  and refrigerant, 4.68
  safeties of, 4.68
  shutdown of, 4.68
Centrifugal fans, 4.97
Centrifugal pumps, 5.213–5.222
  brake horsepower (BHP) for, 5.220
  bypass operation of, 5.222
  configuration of, 5.213–5.215
    impeller orientation in, 5.214
    impeller types in, 5.213
    in-line, 5.214
    with opposed impellers, 5.215
  and discharge-piping configuration, 5.221–5.222
  and entrained air, 5.217–5.218
  foundation of, 5.221
  hydraulic curve of, 5.218–5.220
  impeller orientation in, 5.214
  impeller types in, 5.213
  and inlet-piping configuration, 5.221
  in-line configuration of, 5.214
  installation of, 5.220–5.222
  net positive suction head, 5.215–5.216
  operating conditions for, 5.222
  operating methods for, 5.222
  with opposed impellers, 5.215
  performance of, 5.215
  and piping, 5.221
  startup procedures for, 5.222
  suction conditions of, 5.215–5.218
    and entrained air, 5.217–5.218
    net positive suction head, 5.215–5.216
    volume, 5.216–5.217
  total dynamic head (TDH) requirement for, 5.218
  total system head (TSH) requirement of, 5.218
  and volume, 5.216–5.217
Ceramics, 9.9
Cetane number, 8.31
Chain couplings, 5.45–5.46
Chain drives, 5.78–5.80
  and contamination, 5.79
  and heat, 5.79
  installation of, 5.78–5.79
  lubrication of, 5.79–5.80
  and oil viscosity, 5.79, 5.80
  precision, 5.79–5.82
  and windage, 5.79
Chain hoists, 5.91–5.98
  design of, 5.98
  hand-chain manually operated, 5.91–5.92
  manually lever-operated, 5.91
  performance of, 5.98
  powered, 5.93–5.96
  preventative maintenance for, 5.98
  selection of, 5.96–5.98
  types of, 5.91–5.96

Chains, for power transmissions:
  bath lubrication for, 5.80
  and chain drives, 5.78–5.82
  and contamination, 5.79
  disk lubrication for, 5.80
  drip lubrication for, 5.80
  lubrication for, 5.80
  manual lubrication for, 5.80
  oil-stream lubrication for, 5.80
  service factors for, 5.75–5.76
  special inverted-tooth chains, 5.76–5.78
Chalking, 9.47
Change(s):
  to density, 4.91
  documentation of, 1.25
  KPIs for, 3.124
  to plant engineering, 1.25
  resistance to, 1.35
Charging:
  with constant current, 6.94
  of cycled batteries, 6.100
  equalizing charge, 6.95
  float charging, 6.94
  for lead-acid batteries, 6.81–6.85
  of nickel-cadmium batteries, 6.93–6.95
  nickel-iron batteries, 6.100–6.101
  of standby-power batteries, 6.100–6.101
  and trickle charges, 6.94–6.95
  trickle charging, 6.94–6.95
Chart(s):
  control, 3.145–3.147
    application of, 3.146
    example of, 3.146, 3.147
    key points of, 3.146
  flow, 3.144–3.145
  flow process, 3.95–3.98
  multiple-activity process, 3.99–3.102
  for work simplification, 3.95–3.102
Check valves, 9.68
  air-compressor, 5.201–5.202
  ball, 5.201–5.202
  bronze, 5.201–5.203
  iron, 5.204
  nonmetallic-disk lift-, 5.201
  regrinding lift-, 5.201
  Renewo lift-, 5.201
  swing-, 5.201
Check welding procedure, 10.48–10.49
Checking (coating failure), 9.47
Chemical cleaning, 9.17–9.33
  abrasion method, 9.30
  cascade method, 9.23–9.24
  and circulation, 9.23
  and copper complexers, 9.21–9.22
  with crawlers, 9.26–9.30
  and degradation, 9.18–9.19
  environmental concerns with, 9.32–9.33
  fill and soak method, 9.24
  foam cleaning, 9.24–9.25
  gel cleaning, 9.25
  high-pressure water jetting, 9.25, 9.26
  hydrodrilling, 9.26
  and mechanical cleaning, 9.25–9.30
    abrasion method, 9.30
    with crawlers, 9.26–9.30
    high-pressure water jetting, 9.25, 9.26
    hydrodrilling, 9.26
    with pigs, 9.26–9.30
    with plugs, 9.26–9.30

Chemical cleaning (Cont.)
  onstream cleaning, 9.24–9.25
    foam cleaning, 9.24–9.25
    gel cleaning, 9.25
    passivating, 9.25
    pickling, 9.25
    steam vapor phase cleaning, 9.24
  organic solvents, 9.21–9.23
    copper complexers, 9.21–9.22
    oxidizing agents, 9.22
  and oxidizing agents, 9.22
  passivating, 9.25
  pickling, 9.25
  with pigs, 9.26–9.30
  with plugs, 9.26–9.30
  for preoperational deposits, 9.19
  process selection for, 9.31–9.33
  solvents for, 9.19–9.23
    alkaline cleaners, 9.20
    copper complexers, 9.21–9.22
    inorganic acids, 9.20–9.21
    organic acids, 9.20
    organic solvents, 9.21–9.23
    oxidizing agents, 9.22
  steam vapor phase cleaning, 9.24
Chipping (flaking), 9.47
Chloride ion concentration, 8.31
Choice, theory of, 3.3
Chopped display, 7.67
Chopping transient blanking, 7.67
Circuit breakers, 6.57
  defined, 7.115
  distribution panels, 7.111
  and infrared technology, 7.111, 7.112
Circuit interrupters, 6.56
Circuit protectors, 7.115
Circuits, 7.115
Circulation, 9.23
Clamp-on instruments, 7.51–7.52
  ammeter, 7.51–7.52
  power-factor meter, 7.52
  wattmeter, 7.52
Clamp-on wattmeters, 7.52
Cleaning:
  abrasion method, 9.30
  after plant growth, 4.39
  of air compressors, 5.193
  of air washers, 4.66
  of air-cooled condensers, 4.69
  brick masonry structures, 4.31–4.40
    failures of, 4.31–4.32
    procedures for, 4.32–4.36
    stains in, 4.36–4.40
  bronze stains, 4.38
  of brown (manganese) stain, 4.36–4.37
  cascade method for, 9.23–9.24
  chemical, 9.17–9.33
    cascade method for, 9.23–9.24
    and circulation, 9.23
    and degradation, 9.18–9.19
    environmental concerns with, 9.32–9.33
    fill and soak method for, 9.24
    and mechanical cleaning, 9.25–9.30
    onstream cleaning, 9.24–9.25
    organic solvents, 9.21–9.23
    for preoperational deposits, 9.19
    process selection for, 9.31–9.33
    solvents for, 9.19–9.23
  coils, 4.69

Cleaning (Cont.)
  concrete floor surfaces, 4.24–4.25
  control components, 6.42
  cooling towers, 4.72
  copper stains, 4.38
  for corrosion control, 9.12
  dehumidifiers, 4.80
  dirt stains, 4.38
  of efflorescence, 4.36
  egg splatter, 4.39
  environmental concerns with, 9.32–9.33
  of evaporative condensers, 4.70
  fans, 4.75
  fill and soak method for, 9.24
  filters, 4.77
  foam cleaning, 9.24–9.25
  gel cleaning, 9.25
  of green stain (vanadium salts), 4.36
  hand washing, 4.35
  high-pressure cold water, 4.33
  high-pressure steam, 4.33
  high-pressure water jetting, 9.25, 9.26
  of historic structures, 4.40
  humidifiers, 4.80
  hydrodrilling, 9.26
  iron stains, 4.37–4.38
  ivy stains, 4.39
  of low-sloped roofs, 4.14–4.15
  mechanical cleaning, 9.25–9.30
  nickel-iron batteries, 6.101–6.102
  oil stains, 4.38
  onstream cleaning, 9.24–9.25
  paint stains, 4.37
  paper stains, 4.39
  passivating, 9.25
  pickling, 9.25
  for piping, 9.86, 9.87
  plain bearings, 5.3–5.4
  of poultice, 4.37
  for preoperational deposits, 9.19
  procedures for, 4.32–4.36
    hand washing, 4.35
    high-pressure cold water, 4.33
    high-pressure steam, 4.33
    sandblasting, 4.33, 4.35
  reasons for, 4.40
  rooftop air-conditioning units, 4.82
  sandblasting, 4.33, 4.35
  smoke stains, 4.38
  of stains, 4.36–4.40
    from bronze, 4.38
    from copper, 4.38
    from dirt, 4.38
    from egg splatter, 4.39
    from iron, 4.37–4.38
    from ivy, 4.39
    from oil, 4.38
    from paint, 4.37
    from paper, 4.39
    from plant growth, 4.39
    from poultice, 4.37
    from smoke, 4.38
    from straw, 4.39
    from tar, 4.38
    of unknown origin, 4.39–4.40
    from welding splatter, 4.38
    from white scum, 4.39
  steam vapor phase cleaning, 9.24
  straw stains, 4.39

Cleaning (*Cont.*)
  tar stains, 4.38
  water-cooled condensers, 4.70–4.71
  welding splatter, 4.38
  white scum, 4.39
Closed circuits, 7.115
Closed tanks, 7.10–7.11
Cloud points, 8.31
CMMS (*see* Computer-based maintenance management system)
Coal-tar epoxy coatings, 9.44
Coated smooth-surfaced roofs, 4.11
Coatings, 9.42–9.44
  alkyd, 9.43
  application of, 9.46
  ASTM standards for, 9.51–9.54
  coal-tar epoxy, 9.44
  conversion, 9.12–9.13
  corrosion control with, 9.12–9.13
  discoloration of, 9.47
  epoxy, 9.43–9.44
  epoxy mastics, 9.44
  inorganic zinc, 9.43
  inspections of, 9.46–9.49
  lead-based, 9.39–9.40
  liquid-resin type epoxy, 9.44
  organic, 9.13
  and personnel, 9.49–9.50
  removal of, 9.39–9.40
  safety considerations with, 9.49–9.50
  silicones, 9.43
  surface preparation for, 9.44–9.46
  surface-tolerant, 9.41–9.42
  urethanes, 9.44
  waterborne, 9.42
  zinc, 9.43
  zinc silicate, 9.43
Coiled bulbs, 7.23
Coils, 4.68–4.69
  cleaning, 4.69
  defined, 7.115
  freeze protection for, 4.68–4.69
  freezing of, 4.68–4.69, 4.84
  heating, 4.79
  maintenance of, 4.68–4.69
  preheating, 4.84
  tightness of, 4.68
  water, 4.84
Cold mounting, 5.33
Collaboration, 2.112
Color test, 8.31
Comb histograms, 3.149
Common-mode rejection, 7.49
Common-mode signals, 7.67
Communications:
  for control, 1.50–1.51
  KPIs for, 3.124
  operating policies for, 1.50–1.51
Comparative job standards, 3.111, 3.114
Comparative time estimating, 3.130–3.131
Compensation program, 3.65–3.66
Compound-wound motors, 6.9
Compressed-air lines, 4.84
Compressors:
  air, 5.185–5.196
    and aftercoolers, 5.193
    and air filters, 5.186, 5.187
    and air receivers, 5.187
    bearings in, 5.192–5.193

Compressors: air (*Cont.*)
  capacity of, 5.187
  cleaning, 5.193
  foundation of, 5.185–5.186
  and intercoolers, 5.193
  location of, 5.185, 5.187
  lubrication of, 5.188–5.189
  nonlubricated cylinders in, 5.189
  packing, 5.193–5.196
  piston rings in, 5.191–5.192
  starting new, 5.188
  and suction lines, 5.186, 5.187
  unloading, 5.193
  valves of, 5.189–5.191
  centrifugal, 4.67–4.68
    bearings of, 4.68
    leak testing for, 4.68
    lubrication of, 4.67–4.68
    maintenance of, 4.67–4.68
    oil heater of, 4.68
    purge unit, 4.68
    and refrigerant, 4.68
    safeties of, 4.68
    shutdown of, 4.68
  vibration of, 7.91–7.92
Computer-based maintenance management system (CMMS), 2.91–2.113
  for accounting function, 2.106–2.107
  for automatic costs allocation, 2.106–2.107
  and behavioral expectations, 2.111
  and collaboration, 2.112
  communications affecting, 2.112
  and confrontation, 2.112
  and consultants, 2.112
  cost accounting data, 2.96–2.97
  for cost history evaluation, 2.107
  craft/skill data, 2.97
  for data management, 2.107
  and deliverables, 2.111
  downtime scheduling, 2.103–2.104
  effort affecting, 2.111
  for engineering function, 2.102–2.103
  equipment/asset bills of material, 2.94
  equipment/asset hierarchies, 2.92–2.94
  equipment/asset identification and specifications, 2.92
  equipment/asset repair history, 2.104
  and expectations, 2.111
  and expertise, 2.112
  failure of, 2.109–2.113
    and behavioral expectations, 2.111
    and collaboration, 2.112
    communications affecting, 2.112
    and confrontation, 2.112
    and consultants, 2.112
    and deliverables, 2.111
    effort affecting, 2.111
    and expectations, 2.111
    and expertise, 2.112
    from lack of resources, 2.110–2.111
    from partial implementation, 2.110
    from staffing issues, 2.111
    and work culture, 2.113
  for files, 2.92–2.99
    cost accounting data, 2.96–2.97
    craft/skill data, 2.97
    equipment/asset bills of material, 2.94
    equipment/asset hierarchies, 2.92–2.94

Computer-based maintenance management system (CMMS), for files (*Cont.*)
  equipment/asset identification and specifications, 2.92
  preventative maintenance plans, 2.96
  purchase orders, 2.98–2.99
  purchase requisitions, 2.98
  repetitive maintenance plans, 2.96
  spare parts, 2.94–2.95
  stores inventory, 2.94–2.95
  work order history, 2.97
  work orders, 2.95–2.96
  for finance function, 2.106–2.107
  functionality of, 2.91
  functions of, 2.107–2.109
  for inventory function, 2.104–2.105
    for direct-buy items, 2.105
    material safety data sheets (MSDS), 2.105
    parts requirements for planned work, 2.105
    parts to equipment/asset, 2.104–2.105
    parts usage history, 2.104
    purchase orders, 2.105
    reorder/stocking requirements, 2.105
    repositioning parts, 2.105
    work orders, 2.105
  for inventory replenishment program, 2.108
  for inventory to equipment/asset where-used, 2.94–2.95
  for job scheduling, 2.108
  for maintenance function, 2.99–2.102
    budget preparation and tracking, 2.102
    craft utilization, 2.102
    equipment/asset repair history, 2.101–2.102
    preventative maintenance planning, 2.100–2.101
    requisition of parts and services, 2.101
    resource scheduling, 2.101
    work order initiation, 2.99–2.100
    work order planning, 2.100
    work order scheduling, 2.101
  and maintenance managers, 2.108
  for management function, 2.107
  modification of, 2.112–2.113
  and planners, 2.108–2.109
  preventative maintenance plans, 2.96
  for preventative maintenance programs, 2.107–2.108
  and product quality, 2.109
  for production function, 2.103–2.104
    downtime scheduling, 2.103–2.104
    equipment/asset repair history, 2.104
    repair request backlog, 2.104
  purchase orders, 2.98–2.99
  purchase requisitions, 2.98
  for purchasing function, 2.105–2.106
  and reliability, 2.109
  repair request backlog, 2.104
  repetitive maintenance plans, 2.96
  and resources, 2.110–2.111
  spare parts, 2.94–2.95
  staffing issues with, 2.111
  stores inventory, 2.94–2.95
  usage for, 2.99–2.107
  and work assignments, 2.109
  and work culture, 2.113
  work order history, 2.97
  work orders, 2.95–2.96

Computerized systems, 2.79–2.90
  for analysis, 2.88–2.90
    of equipment, 2.90
    of parts inventory, 2.89
    of work controls, 2.89
    of work orders, 2.88
  for equipment, 2.90
  implementation of, 2.81–2.82
  for on-line inquiries, 2.82–2.83
  operating characteristics of, 2.82–2.84
    on-line inquiries, 2.82–2.83
    performance reports, 2.84
    report generation, 2.83–2.84
    work orders, 2.82–2.83
  organizing for, 2.81
  for parts inventory, 2.89
  for performance reports, 2.84
  and planning, 2.86–2.87
  priority system, 2.85
  purpose for, 2.80–2.81
  for report generation, 2.83–2.84
  for reports, 2.88–2.90
  and scheduling, 2.87
  for work controls, 2.89
  work order system, 2.84–2.85
  for work orders, 2.82–2.83, 2.88
Concrete floor surfaces, 4.17–4.26
  abrasion resistance, 4.21
  cleaning, 4.24–4.25
  corrosion-resistant, 4.26
  cost of, 4.18
  creak control for, 4.20
  design for, 4.19
  dusting floors, 4.25
  finishes for, 4.25–4.26
  joints of, 4.19–4.20
  maintenance of, 4.24–4.25
  monolithically finished, 4.21
  new, 4.18–4.23
  reinforcement of, 4.20
  repair of, 4.25
  resurfacing, 4.24
  sealers for, 4.25–4.26
  separate, 4.21–4.23
  shrinkage of, 4.19–4.20
  slab design for, 4.19
  subgrade for, 4.19
  superflat floors, 4.23
  toppings for, 4.21–4.23, 4.26
    corrosion-resistant, 4.26
    separate, 4.21–4.23
  water-cement ratio of, 4.18–4.19
Condensers:
  air-cooled, 4.69
  evaporative, 4.69–4.70
    cleaning of, 4.70
    fan section of, 4.70
    and freezing, 4.70
    freezing of, 4.70, 4.84
    lubrication of, 4.70
    refrigerant joints, 4.70
  freezing of, 4.70, 4.84
  water-cooled, 4.70–4.71
    cleaning, 4.70–4.71
    freeze protection for, 4.71
    freezing of, 4.71
    maintenance of, 4.70–4.71
    and refrigerated joints, 4.71
    and water treatment, 4.71

Conduction, 7.109
  defined, 7.113
  and infrared technology, 7.109
Conductivity, thermal, 5.4
Conductors:
  ground, 7.113
  and infrared technology, 7.112, 7.113
Configuration management, 1.18–1.19
Confrontation, 2.112
Connecting rods, 5.17
Constant current, 6.94
Constant voltage, 6.93–6.94
Constant-current method, 6.84
Consultants, 2.112
Consumer Product Safety Commission, 9.94
Contact(s), 6.47–6.51
  bounce, 6.47
  dressing for, 6.49–6.51
  and motor controllers, 6.47–6.51
  overtravel measurement, 6.49
  pressure of, 6.48–6.49
  vacuum contactors, 6.51
  wear of, 6.49
Contactors, 6.51–6.56
  inspection of, 6.51–6.55
  troubleshooting, 6.52–6.55
  vacuum, 6.51
Contactor-type control relays, 6.65
Contamination:
  and chains, 5.79
  of fuel, 8.31
  lubricating oil analysis for, 7.128
  of lubricating oils, 2.25
  and painting, 9.39
  in power transmissions, 5.79
  and precision chain drives, 5.79
  product, 9.39
  of solids, 8.32
Continuous improvement:
  reliability for, 1.7
  and self-directed work teams, 1.35
  Six Sigma for, 1.58
  Six Sigma safety for, 1.58
Continuously wound rheostats, 6.60–6.61
Continuous-process systems, 7.79, 7.93
Contract maintenance, 4.49
Contractors, outside, 1.43–1.44
Control(s):
  communications for, 1.50–1.51
  of coolers, 4.71
  KPIs for, 3.124
  and operating policies, 1.50–1.51
  (See also related topics, e.g.: Cost control)
Control charts, 3.145–3.147
  application of, 3.146
  example of, 3.146, 3.147
  key points of, 3.146
Control components, 6.39–6.77
  adjustable-type relays, 6.68
  cleaning, 6.42
  contactors, 6.51–6.55
  contactor-type control relays, 6.65
  and contacts, 6.47–6.51
  control circuit components, 6.64–6.72
  dashpots, 6.66
  DC relay maintenance, 6.71
  disconnect devices, 6.56–6.57
  drum controllers, 6.62–6.63
  enclosures for, 6.40–6.41

Control components (*Cont.*)
  equipment grounding, 6.63–6.64
  lubrication for, 6.42
  magnetic circuit modification, 6.70–6.71
  and magnetic hysteresis, 6.70
  magnetic overload relays, 6.66
  maintenance of, 6.72–6.73
  manual/machine-operated, 6.64
  mechanical checks, 6.42
  motor controllers, 6.43–6.51
  and personnel, 6.39–6.40
  power circuit components, 6.43
  preventative maintenance for, 6.41–6.42
  rheostats, 6.57–6.61
  solid-state overload relays, 6.66–6.68
  thermal overload relays, 6.65–6.66
  time-tactors, 6.68–6.70
Control reports, 2.79
Controllers, motor, 6.43–6.51
Convection, 7.109
  defined, 7.113
  and infrared technology, 7.109
Conventional built-up roofing membranes, 4.4
Conventional modes, 7.67
Conventional mount, 5.110
Conventional mount synchronous sprockets, 5.130
Conversion coatings, 9.12–9.13
Coolers, 4.71
Cooling, for motors, 6.3–6.4
Cooling loads, 4.53–4.55
  and air-conditioning equipment, 4.53–4.55
  from lights, 4.54
  occupancy loads, 4.54
  outside-air loads, 4.54
  removal of, 4.54
  size of, 4.54–4.55
  sun loads, 4.53
  transmission loads, 4.53–4.54
Cooling towers, 4.71–4.73
  algae in, 4.72
  alignment of, 4.72
  cleaning, 4.72
  eliminators, 4.72
  exposure of, 4.72
  fans, 4.72
  lubrication of, 4.72
  maintenance of, 4.71–4.73
  water distribution in, 4.72
  water level of, 4.72
  winter protection for, 4.73
Cooperative partnerships, 1.23–1.30
  with engineering function, 1.29
  interdependency, 1.26–1.27
  with maintenance function, 1.24–1.25, 1.28
  with management function, 1.25
  with operations, 1.27–1.28
  with plant engineering, 1.25
  with procurement function, 1.25, 1.29–1.30
  with production function, 1.24
  and responsibilities, 1.23–1.25, 1.27–1.30
  with sales function, 1.23–1.24
Copper, 10.39
Copper alloys, 10.39
Copper complexers, 9.21–9.22
Copper corrosion, 8.31
Copper stains, 4.38
Corona, 7.113

Corrective maintenance, 2.3–2.6
  maintenance history database for, 2.5
  planning for, 2.4–2.5
  prerequisites of, 2.4
  role of, 2.6
  trained maintenance planners for, 2.5
Corrosion, 9.3–9.8
  cavitation corrosion, 9.6
  copper, 8.31
  crevice corrosion, 9.7
  and defects, 9.6–9.8
  defined, 9.3
  dezincification, 9.7–9.8
  dissolution, 9.7–9.8
  erosion corrosion, 9.6
  forms of, 9.6–9.8
  galvanic corrosion, 9.6
  interfilm corrosion, 9.8
  intergranular corrosion, 9.7
  localized, 9.5–9.6
  in low-sloped roofs, 4.15
  mechanisms of, 9.4–9.6
  of metals, 4.15
  microbiological corrosion, 9.7
  of piping, 9.55–9.56
  pitting corrosion, 9.6–9.7
  of polymeric materials, 9.8
  selective leaching, 9.7
  stress-cracking corrosion, 9.7
  types of, 9.4–9.6
  uniform, 9.5, 9.6
  and ventilating fans, 4.99
Corrosion control, 9.3–9.4, 9.8–9.15
  and ABS, 9.10
  and acetal homopolymers, 9.10
  and aluminum, 9.9
  in aqueous solutions, 9.14–9.15
  cathodic protection, 9.13
  and ceramics, 9.9
  cleaning practices for, 9.12
  with conversion coatings, 9.12–9.13
  defined, 9.3–9.4
  design for, 9.11–9.12
  and galvanized steel, 9.9
  and graphite, 9.9
  and inhibitors, 9.13–9.15
  material selection for, 9.8–9.11
    ABS, 9.10
    acetal homopolymers, 9.10
    aluminum, 9.9
    ceramics, 9.9
    galvanized steel, 9.9
    graphite, 9.9
    polyethersulfone, 9.10
    polyethylene (PE), 9.10
    polymers, 9.9–9.10
    polystyrene, 9.10
    polyurethane, 9.9–9.10
    polyvinylchloride (PVC), 9.9
    rubber elastomers, 9.10
    stainless steel, 9.9
    steel, 9.9
    synthetic elastomers, 9.10
    zinc, 9.9
  for new surfaces, 9.12–9.13
  with organic coatings, 9.13
  painting for, 9.38–9.39
  of plain bearings, 5.4
  with plating, 9.12

Corrosion control (Cont.)
  and polyethersulfone, 9.10
  and polyethylene (PE), 9.10
  and polymers, 9.9–9.10
  and polystyrene, 9.10
  and polyurethane, 9.9–9.10
  and polyvinylchloride (PVC), 9.9
  and rubber elastomers, 9.10
  with rust-preventive oils, 9.14
  and stainless steel, 9.9
  and steel, 9.9
  and synthetic elastomers, 9.10
  and zinc, 9.9
Corrosion-resistant floors, 4.26
Corrosion-resistant piping, 9.83
Corrosive wear, 5.182
Cost(s):
  capital, 7.134
  of concrete floor surfaces, 4.18
  for elevators, 4.51
  of maintenance, 2.57
  for predictive maintenance, 2.31
  production unit costs, 1.5
  and reliability, 1.5
  for repair, 3.107
  of tribology, 7.134
Cost accounting, for reliability, 3.16–3.17
Cost accounting data, 2.96–2.97
Cost control:
  indexes for, 1.52–1.53
  standard-practice sheets for, 1.52–1.53
  systems for, 1.53
Cost estimate model, 3.15
Cost estimates, 3.107–3.110
  for emergency services, 3.108
  from engineers, 3.109–3.110
  from foreman, 3.109
  and job classifications, 3.107–3.108
  labor, 3.110–3.117
    and deferring maintenance, 3.116
    with factor development, 3.115
    judgment for, 3.110–3.111
    PERT approach to, 3.116
    quickread factors for, 3.115–3.116
    with rule of seven, 3.115
    and slotting, 3.111, 3.115
    standards per unit, 3.111, 3.114–3.116
    technique selection for, 3.116–3.117
  for materials, 3.117–3.118
  for overhead, 3.118–3.119
  for planned maintenance, 3.108
  from planners, 3.110
  preparation of, 3.109–3.110
  prerequisites for, 3.107
  from rate setters, 3.110
  for repetitive maintenance, 3.108
  usage of, 3.108–3.109
Cost estimation methods, 3.14–3.16
Cost history evaluation, 2.107
Cost-benefit analysis, 7.135
Cost-effectiveness analysis, 3.14
Counterflashings, 4.13
Counters, 7.63–7.65
Counter-timers, 7.64
Coupler scaffolds, 9.105
Coupling(s):
  in belt drives, 4.74
  chain, 5.45–5.46
  diaphragm, 5.51, 5.52

Coupling(s) (Cont.)
  in direct-connected drives, 4.74
  elastomeric, 5.52–5.53
  flexible (see Flexible couplings)
  floating-shaft gear, 5.46
  gear, 5.46
  laminated disk-ring, 5.50–5.51
  metallic-grid, 5.48–5.49
  Oldham (block-and-jaw), 5.49–5.50
  for power transmissions (see Flexible couplings)
  spindle, 5.46, 5.48
  and ventilating fans, 4.107–4.108
Coupling guards, 4.106
Coverage, for work force, 1.44–1.45
Cracked walls, 4.31
Cracks, grinding, 5.182, 5.183
Craft personnel:
  formal instruction for, 1.15
  informal instruction for, 1.15
  on-the-job training for, 1.15
  selection of, 1.14–1.15
  training of, 1.15
  utilization of, 2.102
Craft/skill data, 2.97
Cranes, 5.83–5.89
  governmental regulations for, 5.89
  inspection of, 5.85
  installation of, 5.83–5.85
  maintenance of, 5.85–5.89
Crankcase:
  distortion of, 5.14
  reconditioning, 5.17–5.18
Crankshaft distortion, 5.14
Crawlers, cleaning with, 9.26–9.30
Crazing, 9.47
Creak control, 4.20
Crevice corrosion, 9.7
Critical speeds, 7.84–7.85
Criticality ranking, 2.11
Cross-craft supervision, 1.14
Cross-training, 1.35
Crush, 5.14
Cryogenic-liquid-level measurement, 7.12–7.13
Culture:
  and CMMS failure, 2.113
  for Six Sigma, 1.55–1.56
Current transformers, 7.57
Currents:
  AC volts, 7.53–7.55
  alternating current (ac), 7.115
  alternating current-instruments, 7.56
  and amperage, 7.53
  constant, 6.94
  DC volts, 7.53
  defined, 7.115
  direct current (dc), 7.115
  direct current-instruments, 7.53, 7.54
  and single-function instruments, 7.53–7.56
  and wattage, 7.53
Curvilinear belts, 5.124
Cutting:
  oxyacetylene, 10.27
  and oxyacetylene appliances, 10.72
  oxygen, 10.78–10.82
  oxygen cutting, 10.78–10.82
  plasma arc cutting, 10.30–10.31
Cutting wear, 2.27, 7.131, 7.132
Cycled batteries, 6.100

I.10 INDEX

Cycling nickel-iron batteries, 6.100, 6.102
Cylinders, nonlubricated, 5.189

Dampers, 4.73
　automatic, 4.73
　in fans, 4.76
　fire, 4.73
　hand, 4.73
　maintenance of, 4.73
　outlet, 4.102
　relief, 4.73
　splitter, 4.73
　and ventilating fans, 4.106
　vortex, 4.76
Dashpots, 6.66
Data:
　analogue, 7.113
　broadband, 7.104
　cost accounting data, 2.96–2.97
　craft/skill data, 2.97
　elemental basic data, 3.114
　methods-time measurement (MTM) data, 3.41–3.52
　nameplate data, 3.170
　narrowband data, 7.104
　for piping, 9.57, 9.58
Data management, 2.107
Data-acquisition frequency, 7.93–7.94
Databases:
　for predictive maintenance program, 7.93–7.95, 7.101–7.102
　for vibration, 7.93–7.95
DC (direct current), 7.115
DC motors, 6.3–6.6, 6.9–6.11
　AC motors vs., 6.3–6.6
　application data for, 6.9–6.11
　design of, 6.9
　insulation for, 6.9–6.11
　theory for, 6.9
　troubleshooting, 6.26–6.36
　types of, 6.9
DC relay maintenance, 6.71
DC volts, 7.53
Decentralization, 1.45–1.46
Decision logic tree analysis (*see* Logic tree analysis)
Decision-making, 3.105
Default logic, 2.14–2.17
Deflection blanking, 7.67
Deflection factors, 7.67
Degradation, 9.18–9.19
Dehumidifiers, 4.80
Delay pickoff, 7.67
Delayed sweep, 7.67
Delays, signal, 7.68
Deliverables, 2.111
Deliveries:
　late, 1.25
　and procurement function, 1.25
　and sales function, 1.24
　unrealistic, 1.24
Delta temperature (ΔT), 7.113
Density, of supervisory personnel, 1.14
Design:
　of AC induction motors, 6.7
　of Bellows-actuated flowmeters, 7.5–7.6
　of chain hoists, 5.98
　of concrete floor surfaces, 4.19
　for corrosion control, 9.11–9.12

Design (*Cont.*)
　of DC motors, 6.9
　for lubrication, 5.8–5.9
　of machines, 7.71
　of motors, 6.7, 6.9
　of plain bearings, 5.6–5.9
　plant engineering, 1.25
　for rolling-element bearings, 5.20
　slab, 4.19
　thermal-system bulb designs, 7.22–7.23
　and vibration, 7.71
Desire, in law of intelligent action, 3.90–3.91
Dew-point measurement, 7.25
Dezincification, 9.7–9.8
Diagrams, flow, 3.98
Diaphragm couplings, 5.51, 5.52
Diaphragm-actuated flowmeter transmitters, 7.26
Digital infrared equipment, 7.110
Digital instruments:
　long-term stability of, 7.49
　resolution of, 7.49
　response time of, 7.49
　temperature coefficient of, 7.50
Digital multimeters, 7.48–7.50
Digits, number of, 7.49
Diodes, 7.115
Direct current (dc), 7.115
Direct current-instruments, 7.53, 7.54
Direct reading ferrography, 8.31
Direct-buy items, 2.105
Direct-connected drives, 4.74
Dirt:
　on brick masonry structures, 4.38
　and plain bearings, 5.9–5.10
Discharge-piping configuration, 5.221–5.222
Discoloration, of coatings, 9.47
Disconnect devices, 6.56–6.57
　circuit breakers, 6.57
　circuit interrupters, 6.56
　disconnect switches, 6.56
　motor circuit protectors, 6.57
Disconnect switches, 4.106, 6.56
Dismounting, of rolling-element bearings, 5.33, 5.38–5.40
Dispatching, 1.41
Dispension, of lubricants, 8.33
Displacement, of vibration, 7.75, 7.77
Displacement (eddy-current) probes, 7.96–7.97
Displays, alternate, 7.66
Disposal:
　of lubricants, 8.6
　as maintenance, 1.11
　of waste, 1.11
Dissolution, 9.7–9.8
Distillation, 8.31
Distortion:
　harmonic, 7.116
　and welding, 10.39–10.40
Distribution, 7.115
Documentation, 1.25
Do-it-yourself maintenance, 4.49
Double-wedge rising-stem gate valve, 5.204
Downtime scheduling, 2.103–2.104
Drainage, 9.56
Drivers, 4.110
Drives:
　basic gear, 5.164
　belt, 4.73–4.74
　　adjustable pulleys in, 4.74

Drives: belt (*Cont.*)
　　alignment of, 4.74
　　and conventional mount synchronous sprockets, 5.130
　　couplings in, 4.74
　　curvilinear belts, 5.124
　　direct-connected drives, 4.74
　　maintenance of, 4.73–4.74
　　modified curvilinear belts, 5.124
　　replacement of, 4.74
　　and reverse mount synchronous sprockets, 5.130
　　service life of, 4.73, 4.74
　　synchronous belts, 5.123–5.127
　　synchronous sprockets, 5.127–5.129
　　and tensioning, 4.74
　　timing belts, 5.124
　　troubleshooting, 5.117–5.123, 5.136–5.142
　　V-belt drives, 5.99–5.143
　　and V-belt sheaves, 5.102–5.103
　couplings in, 4.74
　direct-connected, 4.74
　epicyclic gear, 5.164, 5.165, 5.167
　flat-belt, 5.160
　friction-disk-type, 5.156–5.157
　gear, 5.161
　　basic, 5.164
　　epicyclic, 5.164, 5.165, 5.167
　　gear-tooth wear and failure, 5.177–5.184
　　installation of, 5.169, 5.172
　　lubrication of, 5.174–5.177
　　maintenance of, 5.177
　　selection of, 5.167–5.169
　　shutdown of, 5.177
　geared differential, 5.157–5.159
　mechanical variable-speed, 5.145–5.160
　　friction drives, 5.145–5.160
　　friction-type belts, 5.145–5.147
　packaged belt, 5.155–5.156
　selection of, 5.167–5.169
　shutdown of, 5.177
　start-up, 5.174
　troubleshooting, 5.177
　types of, 5.161–5.164
　start-up, 5.174
　static-adjustment type, 5.151–5.153
　traction-type, 5.157
Drum controllers, 6.62–6.63
Dry centrifugal air-cleaning equipment (*see* Inertial air-cleaning equipment)
Dry dynamic air-cleaning equipment, 4.112, 4.119
Dry type transformers, 7.112
Dry-bulb thermometers, 7.24, 7.25
Dry-water column procedure, 7.33–7.35
Dual beams, 7.67
Dual traces, 7.67
Ducts:
　busy, 7.111
　exhaust, 4.122, 4.123
　and infrared technology, 7.111
Duoline systems, 8.20, 8.21
Dust control equipment (*see* Air-cleaning equipment)
Dusting floors, 4.25

Economizers, 4.74–4.75
Eddy-current (displacement) probes, 7.96–7.97
Edge-peaked histograms, 3.150

INDEX **I.11**

Efficiency:
    with centralized automatic lubrication system, 8.15
    estimation of, 3.60–3.62
    with lubrication programs, 8.15
    of personnel, 1.44–1.45
    of work force, 1.44–1.45
    and work measurement, 3.60–3.62
    and work simplification, 3.104–3.105
Efflorescence, 4.36
Effort, 2.111
Egg splatter, 4.39
80/20 rule, 3.142–3.144
    application of, 3.143
    example of, 3.143–3.144
    interpretation of, 3.143
    key points for, 3.142–3.143
    options in, 3.142
    procedure for, 3.142
Ejection systems, 8.23–8.24
Elastomeric couplings, 5.52–5.53
Elastomers:
    for corrosion control, 9.10
    low-sloped roofs with, 4.11
    rubber, 9.10
    synthetic, 9.10
    vulcanized, 4.11
Electric motor analysis, 2.29–2.30
    insulation resistance, 2.29–2.30
    testing, 2.30
Electric motors, 7.90
Electrical instruments, 7.43–7.68
    and AC volts, 7.53–7.55
    alternating current-instruments, 7.56
    ammeter, 7.51–7.52
    and amperes, 7.53
    analog multimeters, 7.44–7.48
    clamp-on instruments, 7.51–7.52
        ammeter, 7.51–7.52
        power-factor meter, 7.52
        wattmeter, 7.52
    clamp-on wattmeters, 7.52
    counters, 7.64
    counter-timers, 7.64
    current transformers, 7.57
    and currents, 7.53–7.56
        AC volts, 7.53–7.55
        alternating current-instruments, 7.56
        amperes, 7.53
        DC volts, 7.53
        direct current-instruments, 7.53, 7.54
        watts, 7.53
    and DC volts, 7.53
    digital multimeters, 7.48–7.50
    direct current-instruments, 7.53, 7.54
    frequency counters, 7.63–7.65
        counters, 7.64
        counter-timers, 7.64
        oscilloscopes, 7.64, 7.65
    high-resistance determination with, 7.63–7.64
    low-resistance determination with, 7.62–7.63
    meggers, 7.59–761
        high-resistance determination with, 7.64
        precautions when using, 7.61
        resistance measurements with, 7.59–761
    multimeters, 7.44–7.50
        analog multimeters, 7.44–7.48
        digital multimeters, 7.48–7.50
        special-purpose industrial multimeters, 7.50

Electrical instruments (*Cont.*)
    ohmmeters, 7.59, 7.64
    oscilloscopes, 7.64, 7.65
    potential transformers, 7.58
    for power measurements, 7.57–7.58
        current transformers, 7.57
        potential transformers, 7.58
        single-phase connections, 7.57
    power-factor meter, 7.52
    for resistance measurements, 7.58–7.63
        high-resistance determination, 7.63–7.64
        low-resistance determination, 7.62–7.63
        megger, 7.59–761
        ohmmeter, 7.59
        Wheatstone bridge, 7.61–7.62
    selection of, 7.43–7.44
    single-function instruments, 7.50–7.66
        clamp-on instruments, 7.51–7.52
        and currents, 7.53–7.56
        frequency counters, 7.63–7.65
        for power measurements, 7.57–7.58
        for resistance measurements, 7.58–7.63
        solid-state circuit testers, 7.65–7.66
        tube checkers, 7.65
    single-phase connections, 7.57
    solid-state circuit testers, 7.65–7.66
    special-purpose industrial multimeters, 7.50
    three-phase wattmeters, 7.58
    transformers
        current, 7.57
        potential, 7.58
    tube checkers, 7.65
    wattmeters, 7.52
        clamp-on, 7.52
        three-phase, 7.58
    and watts, 7.53
    Wheatstone bridge, 7.61–7.62
Electrical interlocks, 6.76
Electrical motors (*see* Motor(s))
Electrodes:
    and submerged arc welding, 10.41–10.42
    surfacing, 10.44
    for welding, 10.21–10.22
Electrogas welding, 10.34
Electrolytes:
    and nickel-iron batteries, 6.102–6.103
    renewing, 6.102–6.103
    spilled, 6.103
Electromagnetic interference (EMI), 7.115
Electromagnetic spectrum:
    defined, 7.113
    and infrared technology, 7.108–7.109
Electrostatic precipitators, 4.117–4.118
    fans in, 4.121–4.122
    maintenance of, 4.120–4.122
    sludge settling tanks, 4.121
Elemental basic data, 3.114
Elevators, 4.43–4.52
    and age adjustment, 4.52
    contract maintenance for, 4.49
    cost to maintain, 4.51
    dependability of, 4.44
    do-it-yourself maintenance for, 4.49
    economic factors of, 4.44
    emergency service for, 4.48–4.49

Elevators (*Cont.*)
    inspections for, 4.44–4.46, 4.50–4.51
    maintenance objectives for, 4.43–4.44
    maintenance organization for, 4.46–4.49
        do-it-yourself maintenance, 4.49
        emergency service, 4.48–4.49
        personnel, 4.46–4.48
    manufacturer of
        and age adjustment, 4.52
        experience of, 4.51
        integrity of, 4.51
        maintenance by, 4.50
        premaintenance repairs from, 4.52
        preventative maintenance from, 4.51
        safety inspections by, 4.50–4.51
        services provided by, 4.51–4.52
        warranties from, 4.51–4.52
    parts for, 4.50
    performance of, 4.44
    and personnel, 4.46–4.48
    premaintenance repairs for, 4.52
    preventative maintenance for, 4.51
    repairs for, 4.52
    safety considerations with, 4.44, 4.50–4.51
    safety inspections for, 4.44–4.46, 4.50–4.51
    services for, 4.51–4.52
    warranties for, 4.51–4.52
Eliminators, 4.72
Embedability, 5.4
Emergencies, 1.36
Emergency leak repair, 4.9–4.10
Emergency services:
    cost estimates for, 3.108
    for elevators, 4.48–4.49
EMI (electromagnetic interference), 7.115
Emission, atmospheric, 7.113
Emissivity, 7.109
    defined, 7.114
    and infrared technology, 7.109
    of materials, 7.118–7.125
Emittance, 7.114
Employee ownership, 1.6
Employees:
    and organized plant lubrication program, 8.4
    and reliability, 1.6–1.7
    safety of, 1.6–1.7
Encapsulation, 9.39–9.40
Enclosures:
    for control components, 6.40–6.41
    of and motor control centers, 6.74
Engineering, 1.17–1.20, 2.102–2.103
    CMMS for, 2.102–2.103
    cooperative partnerships with, 1.29
    equipment/asset modification history, 2.103
    equipment/asset specifications, 2.103
    and maintenance, 1.29
    maintenance function from, 1.28–1.29
    plant, 1.25
    project planning and tracking, 2.103
    reliability, 1.32–1.33
    responsibilities of, 1.19–1.20
    and vibration, 7.71
    (*See also* Reliability engineering)
Engineers:
    cost estimates from, 3.109–3.110
    and work measurement, 3.19–3.21
Engines, aircraft, 7.71
Entrained air, 5.217–5.218

Environmental issues, 1.7
  of chemical cleaning, 9.32–9.33
  risk of, 1.7
Epicyclic gear drives, 5.164, 5.165, 5.167
Epoxy coatings, 9.43–9.44
Epoxy mastics, 9.44
Equalizing charge, 6.95
Equipment:
  accessory, 10.52–10.55
  air-cleaning, 4.111–4.124
    electrostatic precipitators, 4.117–4.118
    exhaust ducts, 4.122, 4.123
    exhaust hoods, 4.122, 4.123
    fabric filtration, 4.116–4.117, 4.120
    fans in, 4.121–4.122
    gas absorbers, 4.113, 4.114
    inertial (dry centrifugal), 4.112, 4.119
    maintenance, 4.118–4.124
    orifice type scrubbers, 4.120
    packed scrubbers, 4.119
    particulate scrubbers, 4.114, 4.115
    preventative maintenance, 4.123
    and sludge settling tanks, 4.121
    types of, 4.112–4.118
    wet centrifugal scrubbers, 4.119–4.120
    wet collectors (scrubbers), 4.112, 4.113
    wet dynamic scrubbers, 4.120
  air-conditioning, 4.53–4.85
    absorption machines, 4.64–4.65
    and air motors, 4.84
    air washers, 4.66–4.67
    air-cooled condensers, 4.69
    belt drives, 4.73–4.74
    centrifugal compressors, 4.67–4.68
    coils, 4.68–4.69
    components of, 4.57–4.58
    and compressed-air lines, 4.84
    and condensers, 4.70, 4.84
    coolers, 4.71
    cooling loads, 4.53–4.55
    cooling towers, 4.71–4.73
    dampers, 4.73
    dehumidifiers, 4.80
    economizers, 4.74–4.75
    evaporative condensers, 4.69–4.70, 4.84
    fans, 4.75–4.76
    filters, 4.76–4.78
    and freeze protection, 4.83–4.84
    and freezing, 4.68–4.69, 4.84
    heat pumps, 4.79–4.80
    heaters, 4.78–4.79
    and heating loads, 4.55
    humidifiers, 4.80
    installation of, 4.61
    latent loads, 4.54
    and loads, 4.53–4.55
    maintenance of, 4.61–4.85
    occupancy loads, 4.54
    and operating logs, 4.63
    outside-air loads, 4.54
    preventative maintenance for, 4.61
    pumps, 4.80–4.81
    refrigeration, 4.55–4.56
    responsibility for, 4.61
    rooftop units, 4.81–4.82
    room air conditioners, 4.82–4.83
    self-contained units, 4.83
    spare parts for, 4.63
    sun loads, 4.53

Equipment: air-conditioning (*Cont.*)
    training for, 4.63
    transmission loads, 4.53–4.54
    and water coils, 4.84
    water conditioning, 4.85
    and water conditioning, 4.84–4.85
    and water lines, 4.84
    water-cooled condensers, 4.70–4.71
    and well pumps, 4.84
  analog infrared, 7.110
  analysis of, 2.90
  availability of, 2.44–2.45
  digital infrared, 7.110
  dry dynamic air-cleaning, 4.112, 4.119
  effectiveness of, 2.44, 2.46–2.47
  for flowmeters, 7.33–7.34
  inertial (dry centrifugal) air-cleaning, 4.112, 4.119
  infrared
    digital, 7.110
    lenses for, 7.110
    portability of, 7.110
    qualitative measurements of, 7.110
    quantitative measurements of, 7.110
    software for, 7.110
    wave lengths of, 7.110
  for infrared technology, 7.107–7.110
    analysis of, 7.109–7.110
    infrared imaging, 7.107–7.108
    infrared thermometers, 7.107
    line scanners, 7.107
  inspection of, 1.10
  life of, 1.5–1.6
  for low-sloped roofs, 4.15
  maintenance of, 1.48–1.49
  multiple centrifugal air-cleaning, 4.112, 4.119
  packaged ventilating, 4.98
  and personnel, 1.47–1.48
  plant, 1.10
  production, 1.47–1.48
  records of, 3.31
  for refrigeration, 4.58–4.61
  shutting down, 1.48–1.49
  simple cyclone air-cleaning, 4.112, 4.119
  storage of, 4.15
  for welding, 10.52–10.55
  and work measurement, 3.26–3.27
  (*See also specific equipment, e.g.:* Line scanners)
Equipment grounding, 6.63–6.64
Equipment/asset:
  bills of material, 2.94
  hierarchies, 2.92–2.94
  identification, 2.92
  modification history, 2.103
  parts for, 2.104–2.105
  repair history, 2.101–2.102, 2.104
  specifications, 2.92, 2.103
  specifications review for, 2.103
Erosion corrosion, 9.6
Errors, operator, 1.24
Estimators, training for, 3.119–3.120
Evaluation:
  KPIs for, 3.124
  of personnel, 3.65
Evaporative condensers, 4.69–4.70
  cleaning of, 4.70
  fan section of, 4.70
  and freezing, 4.70

Evaporative condensers (*Cont.*)
  freezing of, 4.70, 4.84
  lubrication of, 4.70
  maintenance of, 4.69–4.70
  refrigerant joints, 4.70
Event-reporting format, 3.165–3.166
Exhaust ducts:
  maintenance of, 4.122
  preventative maintenance of, 4.123
Exhaust hoods:
  maintenance of, 4.122
  preventative maintenance of, 4.123
Existing, 1.10
Expectations, and CMMS failure, 2.111
Expertise, and CMMS failure, 2.112
Exposure, of cooling towers, 4.72
Extension ladders, 9.92–9.94
Extension necks, 7.23

°F (Fahrenheit degree), 7.113
Fabric filtration, 4.116–4.117, 4.120
Face grouting, 4.29
Face solution (*see* Angular solution)
Face/rim alignment:
  angular (face) solution for, 5.63, 5.66
  flexible couplings on, 5.63–5.67
  horizontal alignment solution for, 5.66, 5.67
  parallel-offset (rim) solution for, 5.66
  vertical alignment for, 5.63, 5.66
Facilities:
  for lubricants, 8.6
  maintenance for, 1.10
  painting for, 9.38–9.39
  plant, 9.38–9.39
  (*See also* Plant facilities)
Fahrenheit degrees (°F), 7.113
Failure(s):
  of bearings, 5.19
  and behavioral expectations, 2.111
  from bent shafts, 7.89
  of blowers, 7.91
  of brick masonry structures, 4.31–4.32
  of CMMS, 2.109–2.113
  and collaboration, 2.112
  of compressors, 7.91–7.92
  and confrontation, 2.112
  consequences of, 2.37
  and consultants, 2.112
  of continuous-process lines, 7.93
  of couplings, 5.53–5.55
  and culture, 2.113
  and deliverables, 2.111
  effects of, 2.36
  and effort, 2.111
  of electric motors, 7.90
  of equipment, 3.175–3.176
  and expectations, 2.111
  and expertise, 2.112
  of fans, 7.91
  of flexible couplings, 5.53–5.55
  functional, 2.36
  in gear drives, 5.177–5.184
  of gearboxes, 7.90–7.91
  of gear-tooth, 5.177–5.184
  of generators, 7.92
  from imbalance, 7.87–7.88
  and infrared technology, 7.111
  from looseness, 7.89
  mean time between, 3.136

INDEX **I.13**

Failure(s) (*Cont.*)
  from mechanical rub, 7.89
  from misalignment, 7.88–7.89
  modes of, 2.36
  and personnel, 2.111
  premature
    of synchronous belts, 5.136–5.138
    of V-belt drives, 5.117, 5.118
  of pumps, 7.92–7.93
  redesign for, 2.14
  and reliability centered maintenance, 2.36–2.37
  of rolling-element bearings, 5.19
  and root cause analysis, 3.175–3.176
  run to, 2.39
  of synchronous belts, 5.136–5.138
  of systems, 3.175–3.176
  of V-belt drives, 5.117, 5.118
  and vibration, 7.87–7.93
  and work culture, 2.113
Failure analysis:
  for FSIs, 2.11–2.12
  root-cause, 7.104
Failure-finding tasks, 2.38–2.39
Fan(s), 4.75–4.76
  in air washers, 4.66
  alignment of, 4.76
  axial, 4.95
  backward-curved-blade centrifugal, 4.96
  bearings, 4.75–4.76, 4.110
  capacity of, 4.75
  cleaning, 4.75
  in cooling towers, 4.72
  and corrosion, 4.99
  in electrostatic precipitators, 4.121–4.122
  in evaporative condensers, 4.70
  forward-curved-blade centrifugal, 4.96
  in-line-flow centrifugal, 4.97
  lubrication for, 4.76
  maintenance of, 4.75–4.76
  propeller, 4.95
  radial-blade centrifugal, 4.96, 4.97
  rotation of, 4.76
  straight-blade centrifugal, 4.96, 4.97
  ventilating, 4.87–4.110
    and abrasion, 4.99–4.100
    and bearings, 4.110
    capacity of, 4.109
    and corrosion, 4.99
    drivers, 4.110
    fan laws, 4.91, 4.94
    and flammable gases, 4.100
    and flow, 4.87
    forward-curved-blade centrifugal fans, 4.96
    handling of, 4.105
    in-line-flow centrifugal fans, 4.97
    inspection of, 4.104, 4.107–4.109
    installation of, 4.104–4.106
    jet pumps, 4.98
    noise of, 4.101–4.102
    packaged ventilating equipment, 4.98
    performance data for, 4.88–4.91
    power roof exhausters, 4.98
    pressure of, 4.109
    propeller fans, 4.95
    protective devices, 4.107
    radial-blade centrifugal fans, 4.96, 4.97
    safety devices, 4.106

Fan(s), ventilating (*Cont.*)
    selection of, 4.98–4.100
    storage of, 4.105
    straight-blade centrifugal fans, 4.96, 4.97
    and temperatures, 4.98–4.99
    terminology, 4.87–4.88
    types of, 4.94–4.98
    uses of, 4.94–4.98
    vibration, 4.109–4.110
    volume-control devices for, 4.102–4.104
  vibration of, 7.91
  vortex dampers in, 4.76
  wheel of, 4.76
Fan laws, 4.91, 4.94
Fan wheels, 4.105–4.106
Far infrared, 7.114
Fasteners, 4.12
Fatigue:
  of gear-tooth, 5.180–5.181
  and plain bearings, 5.10–5.11
  surface, 5.180–5.181
Fatigue spallwear, 7.131, 7.132
Fatigue strength, 5.4
FCAW (*see* Flux-cored arc welding)
FCAW (flux-cored arc welding), 10.10
Federal regulations (*see* Regulations)
Feeders, parallel, 7.113
Ferrography, 2.27–2.28
  analytical, 8.31
  direct reading, 8.31
  and tribology, 7.129
Fiberglass ladders, 9.93
Field calibration, 7.38–7.39
Field maintenance:
  of flowmeters, 7.33
  for relative-humidity-measuring instruments, 7.40
Fill and soak method, 9.24
Filled-system thermometers, 7.19–7.24
  classifications of, 7.19–7.23
  fully compensated liquid-filled, 7.20
  gas-filled thermal systems, 7.22
  liquid-filled, 7.19
  mercury-filled thermal systems, 7.22–7.23
  and thermal-system bulb designs, 7.22–7.23
  vapor-pressure thermal systems, 7.20–7.22
Filters, 4.76–4.78
  air, 5.186, 5.187
  air filters, 4.76–4.77
  cleaning, 4.77
  defined, 7.115–7.116
  maintenance of, 4.76–4.78
  oil filters, 4.77–4.78
  purpose of, 4.76
  types of, 4.76–4.77
  water filters, 4.77–4.78
Filtration, fabric, 4.116–4.117, 4.120
Finance function, 2.106–2.107
Finishes, for concrete floor surfaces, 4.25–4.26
Fire dampers, 4.73
Fire point, 8.31
Fire protection, 8.7
First level logic tree analysis, 2.12–2.13
Fit, of rolling-element bearings, 5.25–5.26
Five whys (5W), 3.141, 3.161
"5S" program, 2.67–2.68
Five-sigma control, 1.57
Flaking (chipping), 9.47

Flame adjustment, 10.74–10.83
Flammable gases, 4.100
Flanges:
  bolting, 9.83–9.86
  in low-sloped roofs, 4.13
  for piping, 9.58, 9.81–9.86
Flash point, 8.31
Flashings:
  and counterflashings, 4.13
  defects with, 4.7, 4.12
  for leaky walls, 4.29
  metal-base, 4.13
  open, 4.12
  permanent repair of, 4.12, 4.13
  replacement of, 4.29
  sagging, 4.12
Flashovers, 7.116
Flat-belt drives, 5.160
Flexible couplings, 5.45–5.71
  across-the-flex element of, 5.67–5.71
    horizontal alignment solution for, 5.70–5.71
    vertical alignment solution for, 5.67–5.70
  alignment of, 5.54–5.56
  angular (face) solution for, 5.63, 5.66
  chain couplings, 5.45–5.46
  diaphragm couplings, 5.51, 5.52
  elastomeric couplings, 5.52–5.53
  on face/rim, 5.63–5.67
  failure of, 5.53–5.55
  floating-shaft gear couplings, 5.46
  gear couplings, 5.46
  horizontal alignment solution for, 5.61–5.63, 5.66, 5.67, 5.70–5.71
  and indicator sag, 5.57
  installation of, 5.54
  laminated disk-ring couplings, 5.50–5.51
  metallic-grid couplings, 5.48–5.49
  Oldham (block-and-jaw) couplings, 5.49–5.50
  parallel-offset (rim) solution for, 5.66
  readings for, 5.57–5.58
  reverse indicator for, 5.58–5.63
  selection of, 5.53–5.54
  and shaft center line, 5.56
  and shaft misalignment, 5.56–5.67
    angular (face) solution for, 5.63, 5.66
    on face/rim, 5.63–5.67
    horizontal alignment solution for, 5.61–5.63, 5.66, 5.67
    indicator sag, 5.57
    parallel-offset (rim) solution for, 5.66
    readings for, 5.57–5.58
    reverse indicator for, 5.58–5.63
    soft foot, 5.56, 5.57
    and thermal growth, 5.58
    vertical alignment on, 5.59–5.61, 5.63, 5.66
  and soft foot, 5.56, 5.57
  spindle couplings, 5.46, 5.48
  and thermal growth, 5.58
  types of, 5.45–5.53
  vertical alignment on, 5.59–5.61, 5.63, 5.66
  vertical alignment solution for, 5.67–5.70
Float charging, 6.94
Floating-shaft gear couplings, 5.46
Floors:
  concrete, 4.17–4.26
  joints in, 4.19–4.20

I.14 INDEX

Flow:
  direction of, 9.63, 9.66
  of flow transmitters, 7.34
  of gear-tooth, 5.179–5.180
  plastic, 5.179–5.180
  of valves, 9.63, 9.66
  and ventilating fans, 4.87
Flow charts, 3.144–3.145
  key points in, 3.145
  procedure for, 3.144
Flow diagrams, 3.98
Flow process chart, 3.95–3.98
  creation of, 3.95, 3.98
  and flow diagrams, 3.98
  points in, 3.98
  work simplification with, 3.95–3.98
Flow rates, 10.20–10.21
Flow transmitters, flow of, 7.34
Flowmeter transmitters, diaphragm-actuated, 7.26
Flowmeters, 7.5–7.7, 7.32–7.34
  Bellows-actuated, 7.5–7.7
  calibration of, 7.33–7.35
    dry-water column procedure for, 7.33–7.35
    equipment for, 7.33–7.34
  dry-water column procedure for, 7.33–7.35
  equipment for, 7.33–7.34
  field maintenance of, 7.33
  installation of, 7.32–7.33
  maintenance of, 7.32–7.34
  mechanical, 7.5–7.7
  at zero, 7.33
Flow-of-work requests, 1.40–1.41
Fluid management programs, 8.4–8.6
Fluids management, 8.30
Flux-cored arc welding (FCAW), 10.10
Fluxes, and submerged arc welding, 10.41–10.42
Foam tests, 8.31
Focus, 7.67
Force-deflection method:
  for tensioning, 5.135
  for V-belt drives, 5.115–5.116
Foreman, maintenance of, 3.109
Formal instruction, 1.15
Forward-curved-blade centrifugal fans, 4.96
Four-sigma control, 1.57
Freeze protection:
  for coils, 4.68–4.69
  maintenance with, 4.83–4.84
  for water coils, 4.84
  for water-cooled condensers, 4.71
Freezing:
  of air motors, 4.84
  and air-conditioning equipment, 4.84
  areas prone to, 4.84
  of coils, 4.68–4.69, 4.84
  of compressed-air lines, 4.84
  of condensers, 4.70, 4.84
  damage from, 4.83
  and evaporative condensers, 4.70
  of evaporative condensers, 4.70, 4.84
  prevention of, 4.83–4.84
  of water coils, 4.84
  of water lines, 4.84
  of water-cooled condensers, 4.71
  of well pumps, 4.84
Frequency:
  data-acquisition, 7.93–7.94
  of vibration, 7.76

Frequency counters, 7.63–7.65
  counters, 7.64
  counter-timers, 7.64
  oscilloscopes, 7.64, 7.65
Frequency domain, 7.76
Frequency method:
  for tensioning, 5.135–5.136
  for V-belt drives, 5.115–5.116
Friction drives, 5.145–5.160
  adjustable while in motion (nonenclosed), 5.145–5.151
  belt transmissions, 5.153–5.155
  flat-belt drives, 5.160
  friction-disk-type drives, 5.156–5.157
  geared differential drives, 5.157–5.159
  packaged belt drives, 5.155–5.156
  static-adjustment type, 5.151–5.153
  traction-type drives, 5.157
Friction-disk-type drives, 5.156–5.157
Friction-type belts, 5.145–5.147
Front line supervision, 1.32
FSIs (see Functionally significant items)
Fuel analysis, 8.31–8.32
Fuel contamination, 8.31
Fuel soot, 2.26, 7.128
Fuels:
  dilution of, 2.25, 7.128
  lubricating oil analysis for, 7.128
  testing for, 8.31–8.32
Fully compensated liquid-filled system thermometers, 7.20
Functionally significant items (FSIs), 2.11–2.12
Fuses:
  defined, 7.116
  and infrared technology, 7.112
  and motor control centers, 6.75
Fusion welding, 10.75
Fusion welding cast iron, 10.78

Gages, pressure (see Pressure gages)
Galvanic corrosion, 9.6
Galvanized steel, 9.9
Gas:
  flux-cored arc welding with, 10.10
  for GMAW, 10.13
Gas absorbers, 4.113, 4.114
Gas tungsten arc welding (GTAW), 10.13–10.27
  advantages of, 10.17–10.18
  applicability of, 10.15–10.17
  disadvantages of, 10.17–10.18
  electrode material for, 10.21–10.22
  flow rates, 10.20–10.21
  manual techniques, 10.24
  mechanical welding parameters for, 10.25
  methods for, 10.25–10.27
  and polarity, 10.18–10.20
  power supplies, 10.22–10.23
  principles of, 10.18
  process for, 10.13–10.15
  shielding gases, 10.20–10.21
  torches, 10.23–10.24
Gas welding, 10.63–10.89
  air-acetylene appliances, 10.63–10.71
  flame adjustment for, 10.74–10.83
  and hard facing, 10.83–10.89
  oxyacetylene appliances, 10.71–10.74

Gases:
  flammable, 4.100
  shielding, 10.20–10.21
Gas-filled thermal systems, 7.22
Gaskets, for piping, 9.58, 9.61–9.62
Gas-shielded metal arc welding (GMAW), 10.11–10.13
Gas-shielded spot welding, 10.34
Gate valves:
  bronze, 5.203–5.204
  double-wedge rising-stem gate valve, 5.204
  iron valves, 5.204–5.205
  renewable-wedge-and-seat bronze gate valve, 5.204
  screw-and-yoke rising-stem gate valve, 5.204
  single-wedge nourishing-stem gate valve, 5.204
  solid-wedge rising-stem gate valve, 5.204
Gaussian response, 7.67
Gear(s):
  and backlash, 7.83
  dynamics of, 7.82–7.83
  in positive-displacement pumps, 5.224
  vibration of, 7.82–7.83
Gear couplings, 5.46
Gear drives, 5.161
  basic, 5.164
  epicyclic, 5.164, 5.165, 5.167
  gear-tooth wear and failure, 5.177–5.184
    surface deterioration, 5.177–5.183
    tooth breakage, 5.177, 5.183–5.184
  installation of, 5.169, 5.172
  lubrication of, 5.174–5.177
  maintenance of, 5.177
  selection of, 5.167–5.169
  shutdown of, 5.177
  start-up, 5.174
  and surface deterioration, 5.177–5.183
  and tooth breakage, 5.177, 5.183–5.184
  troubleshooting, 5.177
  types of, 5.161–5.164
Gear mesh, 7.82–7.83
Gearboxes, 7.90–7.91
Geared differential drives, 5.157–5.159
Gear-tooth wear and failure:
  burning, 5.182
  corrosive wear, 5.182
  grinding cracks, 5.182, 5.183
  interference wear, 5.182, 5.183
  plastic flow, 5.179–5.180
  scoring, 5.180
  surface deterioration, 5.177–5.183
    burning, 5.182
    corrosive wear, 5.182
    grinding cracks, 5.182, 5.183
    interference wear, 5.182, 5.183
    plastic flow, 5.179–5.180
    scoring, 5.180
    surface fatigue, 5.180–5.181
    wear, 5.178–5.179
  surface fatigue, 5.180–5.181
  tooth breakage, 5.177, 5.183–5.184
  wear, 5.178–5.179
Generalizing, in scientific method, 3.92–3.93
Generators:
  motor, 7.116
  vibration of, 7.92
Geometry (CRT), 7.67

Globe valves, 5.200–5.201
  bronze valves, 5.200–5.201
  iron valves, 5.204
  nonmetallic-disk globe valve, 5.201
  plug-type Renewo globe valve, 5.201
  600-Brinnel flat-seat globe valve, 5.201
  threaded-bonnet globe valve, 5.200
  union-bonnet regrinding valve, 5.200
Glycol, 8.32
GMAW (gas-shielded metal arc welding), 10.11–10.13
Gouging, 10.82–10.83
Government regulations (*see* Regulations)
Gradients, temperature, 7.114
Graphite, for corrosion control, 9.9
Graticules, 7.67, 7.68
Gravimetric solids, 8.32
Gravity, and lubricant analysis, 8.32
Grease(s), 8.7–8.12
  compatibility of, 8.11
  components of, 8.8
  and lubrication programs, 8.7–8.12
  manufacture of, 8.11–8.12
  properties of, 8.8–8.9
  testing for, 8.8, 8.9
  thickeners for, 8.9–8.11
Grease lubrication, 5.41–5.43
Grease thickeners, 8.9–8.11
  aluminum complex soaps, 8.10
  aluminum soaps, 8.9
  calcium complex soaps, 8.11
  calcium soaps, 8.10
  lithium complex soaps, 8.11
  lithium soaps, 8.10
  nonsoap organoclay, 8.11
  nonsoap polyurea, 8.11
  sodium soaps, 8.9, 8.10
Grease-lubricated antifriction bearings, 6.12
Green stain (vanadium salts), 4.36
Grinding cracks, of gear-tooth, 5.182, 5.183
Ground(s):
  defined, 7.116
  maintenance for, 1.10
  plant, 1.10
Ground conductors, 7.113
GTAW (*see* Gas tungsten arc welding)
Gutter joints, 4.13

Hand dampers, 4.73
Hand washing, 4.35
Hand-chain manually operated chain hoists, 5.91–5.92
Hard facing, 10.83–10.89
  for cast iron, 10.85
  finishing for, 10.85–10.89
  and gas welding, 10.83–10.89
  material for, 10.45–10.48
  rods, 10.84
  for steel, 10.84–10.85
Hard facing rods, 10.84
Hard surfacing:
  with submerged arc welding, 10.49–10.51
  and welding, 10.43
Hard-facing piping, 9.70–9.72
Harmonic (order), 7.76–7.77
Harmonic distortion, 7.116
Harmonics, 7.116
HCl (inhibited muriatic acid), 9.20

Heat:
  and chain drives, 5.79
  and power transmissions, 5.79
  and precision chain drives, 5.79
  spiral, 7.113
Heat pumps, 4.79–4.80
Heat transfer, 7.109
  conduction, 7.109
  convection, 7.109
  emissivity, 7.109
  and infrared technology, 7.109
  radiation, 7.109
Heaters, 4.78–4.79
  heating coils, 4.79
  maintenance of, 4.78–4.79
  water heaters, 4.79–4.80
Heating coils, 4.79
Heavy braze welding, 10.76–10.77
Helix tubes, 7.16
Hertz (Hz), 7.116
Hidden rusting, 9.47
High voltage switchgear, 7.112
High-carbon steels, 10.36
High-manganese steels, 10.39
High-pressure cold water, 4.33
High-pressure steam, 4.33
High-pressure water jetting, 9.25, 9.26
High-resistance determination, 7.63–7.64
High-tensile low-alloy steels, 10.36
Histograms, 3.147–3.150
  bi-modal, 3.149
  comb, 3.149
  edge-peaked, 3.150
  interpretations of, 3.148–3.150
  isolated peaked, 3.150
  key points for, 3.149
  normal, 3.149
  plateau, 3.149
  procedure for, 3.148
  skewed, 3.150
  truncated, 3.150
Historic structures, 4.40
History, equipment/asset, 2.104
Hoists, chain, 5.91–5.98
Holes, in low-sloped roofs, 4.12, 4.15
Homopolymers, acetal, 9.10
Hoods, exhaust, 4.122, 4.123
Horizontal alignment:
  for face/rim alignment, 5.66, 5.67
  for flexible couplings, 5.70–5.71
  for shaft misalignment, 5.61–5.63, 5.66, 5.67
Hot bearings:
  troubleshooting, 5.118, 5.123, 5.137, 5.142
  and V-belt drives, 5.118, 5.123
Hot spots, 7.114
Hot-short phenomenon, 5.13
Hourly personnel, 1.13–1.14
  amount of, 1.13–1.14
  crafts, 1.14
  and operation personnel, 1.13–1.14
Housekeeping:
  and centralized automatic lubrication system, 8.16
  for low-sloped roofs, 4.6, 4.14
Humidifiers, 4.80
Humidity measurements, 7.24–7.25
  dew-point measurement, 7.25
  nylon-ribbon hygrometer method, 7.24–7.25

Hydraulic curve, 5.218–5.220
Hydraulic removal, 5.39, 5.40
Hydrodrilling, 9.26
Hygrometer method, nylon-ribbon, 7.24–7.25
Hysteresis, magnetic, 6.70
Hz (hertz), 7.116

Imaging, infrared, 7.107–7.108
Imbalance, vibration from, 7.87–7.88
Impact wear, 10.44
Impedance (Z):
  defined, 7.116
  input, 7.50
Impellers:
  opposed, 5.215
  orientation of, 5.214
  types of, 5.213
Improvement:
  KPIs for, 3.124
  methods improvement, 3.93–3.95
  proactive, 1.35
  proactive improvement, 1.35
  quality, 2.21
  total productive maintenance for, 2.51–2.52
  (*See also* Continuous improvement)
Incentives, for self-directed work teams, 1.35
Incoming inspections, 3.12–3.13
Incoming-project specifications, 3.171
Indexes, for cost control, 1.52–1.53
Indicator sag, 5.57
Induced loads, 7.86–7.87
Inductance, 7.116
Inductors, 7.116
Industrial chemical cleaning (*see* Chemical cleaning)
Industrial multimeters, 7.50
Industrywide averages/ratios, 3.111
Inertial (dry centrifugal) air-cleaning equipment, 4.112
  dry dynamic, 4.112, 4.119
  maintenance of, 4.119
  multiple centrifugal, 4.112, 4.119
  simple cyclone, 4.112, 4.119
Informal instruction, 1.15
Infrared analysis, 8.32
Infrared bands, 7.114
Infrared equipment:
  digital, 7.110
  lenses for, 7.110
  portability of, 7.110
  qualitative measurements of, 7.110
  quantitative measurements of, 7.110
  software for, 7.110
  wave lengths of, 7.110
Infrared imaging, 2.24, 7.107–7.108
Infrared radiation, 7.114
Infrared technology, 7.105–7.125
  and broken strands, 7.112
  and busy duct, 7.111
  and capacitors, 7.112
  and circuit breakers, 7.111, 7.112
  and conduction, 7.109
  and conductors, 7.112
  and convection, 7.109
  and dry type transformers, 7.112
  and electrical terminology, 7.114–7.117
  and electromagnetic spectrum, 7.108–7.109
  and emissivity, 7.109

Infrared technology (*Cont.*)
  equipment for, 7.107–7.110
    analog, 7.110
    analysis of, 7.109–7.110
    infrared imaging, 7.107–7.108
    infrared thermometers, 7.107
    line scanners, 7.107
  and failure, 7.111
  far, 7.114
  and fuses, 7.112
  and ground conductors, 7.113
  and heat transfer concepts, 7.109
  infrared imaging for, 7.107–7.108
  infrared thermometers for, 7.107
  line scanners for, 7.107
  and load break switches, 7.112
  and looseness, 7.111
  and materials, 7.118–7.125
  and motors, 7.111
  and oil-filled transformers, 7.111–7.112
  and parallel feeders, 7.113
  problems with, 7.111–7.113
  quantitative, 7.114
  and radiation, 7.109
  safety with, 7.111
  scanning procedures for, 7.110–7.113
  and spiral heating, 7.113
  and switches, 7.112
  and switchgear, 7.111, 7.112
  terms for, 7.113–7.114
  theory for, 7.108
  and transformers, 7.111–7.112
Infrared thermometers, 2.24, 7.107
Inhibited muriatic acid (HCl), 9.20
Inhibited sulfuric acid, 9.20–9.21
Inhibitors:
  for aqueous solutions, 9.14–9.15
  and corrosion control, 9.13–9.15
In-house work force, 1.43–1.44
Injection systems, 8.24
Injector systems, 8.17
Inlet-piping configuration, 5.221
In-line configuration, 5.214
In-line-flow centrifugal fans, 4.97
Inorganic acids, 9.20–9.21
  cleaning with, 9.20–9.21
  inhibited muriatic acid (HCl), 9.20
  inhibited sulfuric acid, 9.20–9.21
  nitric acid, 9.21
  phosphoric acids, 9.21
  sulfuric acid, 9.21
Inorganic zinc coatings, 9.43
Input, 7.49
Input impedance, 7.50
Input RC characteristics, 7.67
Inquiries, on-line, 2.82–2.83
Insolubles, 8.32
Inspection:
  of air-cooled condensers, 4.69
  for appearance, 5.9
  of brick masonry structures, 4.27
  for cavitation, 5.14
  of contactors, 6.51–6.55
  of cranes, 5.85
  for crankcase distortion, 5.14
  for crankshaft distortion, 5.14
  for crush, 5.14
  for dirt, 5.9–5.10
  for fatigue, 5.10–5.11

Inspection (*Cont.*)
  for foreign materials, 5.11, 5.13
  for hot-short phenomenon, 5.13
  internal, 6.86–6.87
  for lead-acid batteries, 6.86–6.87
  of low-sloped roofs, 4.5–4.7
  of plain bearings, 5.9–5.15
  preliminaries for, 5.9
  of rooftop air-conditioning units, 4.81–4.82
  for synchronous belts, 5.132
  of used bearings, 5.9–5.14
  of V-belt drives, 5.112
  of ventilating fans, 4.104, 4.107–4.109
  visual, 2.30
    and root cause analysis, 3.159–3.161
    for synchronous belts, 5.132
    of V-belt drives, 5.112
  for wear, 5.9, 5.11
Inspections:
  of coatings, 9.46–9.49
  for elevators, 4.44–4.46, 4.50–4.51
  incoming, 3.12–3.13
Installation:
  of air-conditioning equipment, 4.61
  application review, 3.172
  of batteries, 6.80–6.81, 6.92–6.93
  of centrifugal pumps, 5.220–5.222
  of chain drives, 5.78–5.79
  of cranes, 5.83–5.85
  of flexible couplings, 5.54
  of flowmeters, 7.32–7.33
  of gear drives, 5.169, 5.172
  of lead-acid batteries, 6.80–6.81
  maintenance for, 1.10
  of nickel-cadmium batteries, 6.92–6.93
  of positive-displacement pumps, 5.227
  of pressure gages, 7.37–7.38
  of QD-type sheaves, 5.110
  of refrigeration equipment, 4.61
  of relative-humidity-measuring instruments, 7.40
  root cause analysis (RCA), 3.172
  of synchronous belts, 5.129–5.131
  of thermometers, 7.35–7.36
  of TL-type sheaves, 5.110–5.111
  of V-belt drives, 5.112
  for V-belt sheaves, 5.108–5.112
  of ventilating fans, 4.104–4.106
  of welding equipment, 10.55
Installed systems, 7.100–7.101
Instruction:
  for craft personnel, 1.15
  formal, 1.15
  informal, 1.15
Instruction manuals, 1.51–1.53
Instrumentation, 1.49–1.50
Instruments (*see* Electrical instruments; Mechanical instruments)
Insulation:
  for AC induction motors, 6.7–6.8
  bus support insulators, 6.75
  for coolers, 4.71
  for DC motors, 6.9–6.11
  for piping, 9.78–9.81
Insulation resistance, 2.29–2.30
Insulators, 7.116
Insurance administration, 1.11
Intelligent action, law of (*see* Law of intelligent action)

Intensity modulation, 7.68
Interchangeable lenses, 7.110
Intercoolers, 5.193
Interdependency, 1.26–1.27
Interference, 7.115
Interference wear, of gear-tooth, 5.182, 5.183
Interfilm corrosion, 9.8
Intergranular corrosion, 9.7
Internal graticules, 7.68
Internal inspection, 6.86–6.87
Intervals, 2.17–2.18
Intraplant relations, 1.47–1.50
  authority, 1.48–1.49
  and instrumentation, 1.49–1.50
  maintenance personnel, 1.47–1.48
  operating policies for, 1.47–1.50
  and safety, 1.49
Inventory, 2.59–2.78
  analysis of, 2.89
  best practices inventory management for, 2.60–2.61
  carrying cost for, 2.64–2.65
  computerized maintenance management systems (CMMS) for, 2.75–2.78
    enterprise asset management (EAM), 2.75–2.76
    stock keeping units (SKU), 2.76–2.78
  control of, 2.61
  criticality of, 2.63–2.64
  and loss elimination, 2.62–2.63
  MRO carrying cost for, 2.64–2.65
  objectives of, 2.59
  obsolete, 2.78
  parts, 2.89
  and replacement asset value (RAV), 2.61
  and risk management, 2.62
  stock keeping units (SKU), 2.76–2.78
  stocking levels for, 2.64
  and storeroom management, 2.65–2.75
  types of, 2.60
Inventory function, 2.104–2.105
  CMMS for, 2.104–2.105
  for direct-buy items, 2.105
  material safety data sheets (MSDS), 2.105
  parts requirements for planned work, 2.105
  parts to equipment/asset, 2.104–2.105
  parts usage history, 2.104
  purchase orders, 2.105
  reorder/stocking requirements, 2.105
  repositioning parts, 2.105
  work orders, 2.105
Inventory replenishment program, 2.108
Inventory to equipment/asset where-used, 2.94–2.95
Inverters, 7.116
Iron stains, 4.37–4.38
Iron valves, 5.204–5.205
  check valves, 5.204
  gate valves, 5.204–5.205
  globe valves, 5.204
ISO 9000 compliance, 2.107
Isolated peaked histograms, 3.150
Isolation, 7.116
Isothermal mapping, 7.114
Ivy stains, 4.39

Jet pumps, 4.98
Jitter, 7.68

INDEX **I.17**

Job(s):
  size of, 1.40
  wage rate for, 3.70
Job classifications, 3.69–3.89
  and cost estimates, 3.107–3.108
  wage rate for, 3.70
Job evaluation, 3.66–3.69
  classifications, 3.69–3.89
  factors in, 3.66–3.68
  and job analysis, 3.68–3.69
  principles of, 3.66
Job scheduling, 2.108
Job sequence, 2.79
Job standards, comparative, 3.111, 3.114
Joint(s):
  of concrete floor surfaces, 4.19–4.20
  and coolers, 4.71
  and cracked walls, 4.31
  and evaporative condensers, 4.70
  refrigerant, 4.71
  refrigerated, 4.71
  saw cutting of, 4.31
  shrinkage of, 4.19–4.20
  vertical expansion, 4.31
  and water-cooled condensers, 4.71
Joint (banded) belts, 5.120–5.121

Key performance indicators (KPIs), 3.121–3.132
  and balanced scorecard (BSC), 3.124–3.127
  best practices for, 3.129
  for change, 3.124
  for communication, 3.124
  and comparative time estimating, 3.130–3.131
  for control, 3.124
  for evaluation, 3.124
  for improvement, 3.124
  in knowledge-based organizations, 3.121–3.124
  for long-term focus, 3.124
  for measurement, 3.124
  for motivation, 3.124
  need for, 3.124
  for planning, 3.124
  PMS models for, 3.124
  requirements of, 3.128–3.129
  for resource allocation, 3.124
  and work measurement, 3.129–3.132
    application of, 3.131
    comparative time estimating, 3.130–3.131
    familiarity for, 3.130
    levels of, 3.132
kHz (kilohertz), 7.116
Kilivolt-ampere (kVA), 7.116
Kilohertz (kHz), 7.116
Knowledge-based organizations, 3.121–3.124
KPIs (*see* Key performance indicators)
kVA (kilivolt-ampere), 7.116

Labor cost estimates, 3.110–3.117
  and deferring maintenance, 3.116
  with factor development, 3.115
  judgment for, 3.110–3.111
  PERT approach to, 3.116
  quickread factors for, 3.115–3.116
  with rule of seven, 3.115
  and slotting, 3.111, 3.115
  standards per unit, 3.111, 3.114–3.116
  technique selection for, 3.116–3.117

Ladders, 9.91–9.101
  aluminum, 9.92
  American National Standards Institute (ANSI) on, 9.93
  Consumer Product Safety Commission on, 9.94
  extension ladders, 9.92–9.94
  federal regulations for, 9.93
  fiberglass, 9.93
  materials for, 9.92–9.93
    aluminum, 9.92
    fiberglass, 9.93
  OSHA regulations on, 9.93
  ratings for, 9.94
  regulations for, 9.93–9.94
  rung ladders, 9.94–9.95
    extension ladders, 9.94
    sectional ladders, 9.94, 9.95
    single ladders, 9.94
  safety considerations with, 9.96–9.101
  sectional ladders, 9.94, 9.95
  single ladders, 9.94
  special-purpose ladders, 9.96
  standards for, 9.93–9.94
  stepladders, 9.92–9.96
  Underwriters' Laboratories, Inc. on, 9.93–9.94
Laminated disk-ring couplings, 5.50–5.51
Latent loads, 4.54
Law of intelligent action, 3.90–3.92
  ability, 3.91
  capacity, 3.91–3.92
  desire, 3.90–3.91
Leaching, selective, 9.7
Lead time, for scheduling, 1.40
Lead-acid batteries, 6.80–6.89
  charging, 6.81–6.85
  condition of, 6.86–6.87
  constant-current method for, 6.84
  installation of, 6.80–6.81
  internal inspection for, 6.86–6.87
  maintenance of, 6.85–6.86
  modified constant-voltage method for, 6.82–6.83
  operation of, 6.80–6.81
  overdischarging, 6.86
  placement of, 6.81
  records for, 6.86
  repairs to, 6.88–6.89
  taper method for, 6.84
  test discharge for, 6.86
  troubleshooting, 6.87–6.88
  two-rate method for, 6.84
Lead-based paints, 9.39–9.40
Leaks:
  in absorption machines, 4.66
  in centrifugal compressors, 4.68
  detection of, 7.70, 9.86
  in piping, 9.62–9.63, 9.86
  testing for
    for absorption machines, 4.66
    for centrifugal compressors, 4.68
  and vibration, 7.70
Leaktightness, 4.65
Lenses, 7.110
Life cycle asset management, 1.18
Life cycle cost analyses, 3.136
Life expectancy, 6.91–6.92
Lifts (*see* Elevators)

Lighting, 8.6
Lightning arrester, 7.116
Lights, 4.54
Line scanners, 2.24, 7.107
Linear-motion machinery, 7.78
Liquid velocity, 9.86
Liquid-filled system thermometers, 7.19
Liquid-level instruments, 7.35
Liquid-level measurement, 7.8–7.14
  under atmospheric pressure, 7.8–7.10
  in closed tanks, 7.10–7.11
  cryogenic-liquid-level measurement, 7.12–7.13
  purge-type liquid-level measurement, 7.13–7.14
  of sealing liquids, 7.11–7.12
Liquid-resin type epoxy, 9.44
Liquids, sealing, 7.11–7.12
Lithium complex soaps, 8.11
Lithium soaps, 8.10
Load(s), 4.53–4.55
  and air-conditioning equipment, 4.53–4.55
  cooling, 4.53–4.55
    and air-conditioning equipment, 4.53–4.55
    from lights, 4.54
    occupancy loads, 4.54
    outside-air loads, 4.54
    removal of, 4.54
    size of, 4.54–4.55
    sun loads, 4.53
  heating, 4.55
  induced, 7.86–7.87
  latent, 4.54
  occupancy loads, 4.54
  outside-air loads, 4.54
  sun loads, 4.53
  total work load, 1.40
  transmission, 4.53–4.54
  transmission loads, 4.53–4.54
  work load, 1.40
Load break switches, 7.112
Load ratings, 5.24–5.25
Lobe-type pumps, 5.226
Localized corrosion, 9.5–9.6
Logic, default, 2.14–2.17
Logic tree analysis, 2.12–2.14
  first level, 2.12–2.13
  second level, 2.14
Logical order, 3.139
Long-term focus, 3.124
Loose fasteners, 4.12
Loose parts, 7.70
Looseness:
  and infrared technology, 7.111
  vibration from, 7.89
Loss elimination, 1.19
Low-carbon steels, 10.35
Low-resistance determination, 7.62–7.63
Low-sloped roofs, 4.3–4.16
  bitumen membranes on, 4.11
  blisters on, 4.12
  built-up bitumen membranes on, 4.11
  cleaning, 4.14–4.15
  coated smooth-surfaced roofs, 4.11
  conventional built-up roofing membranes, 4.4
  corrosion of, 4.15
  counterflashings, 4.13
  defects with, 4.5–4.7
    flashings, 4.7, 4.12
    housekeeping, 4.6
    masonry, 4.7

I.18   INDEX

Low-sloped roofs, defects with (*Cont.*)
  membrane damage, 4.6–4.7
  roof accessory components, 4.7
  surfacing, 4.6
 emergency leak repair, 4.9–4.10
 and equipment, 4.15
 equipment for, 4.15
 fasteners, 4.12
 flanges, 4.13
 flashings, 4.7, 4.12, 4.13
 gutter joints, 4.13
 and holes, 4.12, 4.15
 holes in, 4.15
 housekeeping, 4.14
 inspection of, 4.5–4.7
 maintenance prevention for, 4.14–4.15
 masonry, 4.7
 membrane damage, 4.6–4.7
 metals, 4.15
 mineral surfaced roofs, 4.11
 modified bitumen membranes on, 4.11
 modified bitumen roofing membranes, 4.4
 with nonvulcanized elastomers, 4.11
 permanent repair of, 4.10–4.14
  for blisters, 4.12
  of built-up bitumen membranes, 4.11
  of coated smooth-surfaced roofs, 4.11
  for counterflashings, 4.13
  for fasteners, 4.12
  for flanges, 4.13
  for flashings, 4.12, 4.13
  for gutter joints, 4.13
  for holes, 4.12
  of mineral surfaced roofs, 4.11
  of modified bitumen membranes, 4.11
  with nonvulcanized elastomers, 4.11
  for punctures, 4.12
  with thermoplastic membranes, 4.11–4.12
  with vulcanized elastomers, 4.11
 preventative maintenance for, 4.14–4.15
  corrosion, 4.15
  and holes, 4.15
  housekeeping, 4.14
  metals, 4.15
  roof drains, 4.15
  rusting, 4.15
  and storage, 4.15
  surface cleaning, 4.14–4.15
  and traffic, 4.15
 protected roof membrane assembly
  ("upside-down" roof system), 4.5
 punctures, 4.12
 record keeping for, 4.3
 repair materials for, 4.8–4.9
 repair of, 4.10–4.14
 replacement of, 4.16
 reroofing of, 4.16
 roof accessory components, 4.7
 roof drains, 4.15
 rusting, 4.15
 single-ply roofing membranes, 4.5
 and storage, 4.15
 storage on, 4.15
 superimposition of, 4.16
 surface cleaning, 4.14–4.15
 surfacing, 4.6
 temporary repair of, 4.9–4.10
 with thermoplastic membranes, 4.11–4.12

Low-sloped roofs (*Cont.*)
  and traffic, 4.15
  traffic on, 4.15
  with vulcanized elastomers, 4.11
Low-voltage release devices, 6.59
Lubricant(s):
 application of, 8.29–8.30
 bulk storage of, 8.7
 dispensing, 8.33
 disposal of, 8.6
 facilities for, 8.6
 fire protection for, 8.7
 and glycol, 8.32
 and gravity, 8.32
 handling of, 8.6, 8.33–8.34
 and lighting, 8.6
 location of, 8.6
 and lubrication, 8.4–8.7
 and lubrication programs, 8.4–8.7
 maintenance of, 8.5, 8.6
 monitoring programs for, 8.5
 and personnel, 8.6
 protection of, 8.6–8.7
 purchasing, 8.4–8.5
 safety considerations with, 8.33
 selection of, 8.4–8.5, 8.27–8.30
 source of, 8.35
 storage of, 8.7, 8.33
 testing for, 8.31–8.32
Lubricant analysis, 8.30–8.32
 and glycol, 8.32
 and gravity, 8.32
 and microorganisms, 8.32
 and sediment, 8.32
 and sugar, 8.32
 and sulfur, 8.32
 and viscosity, 8.32
 and viscosity index, 8.32
 and water, 8.32
Lubricating oil analysis, 2.25–2.26, 7.128–7.129
 contamination, 2.25, 7.128
 cost-benefit analysis for, 7.135
 fuel dilution, 2.25, 7.128
 fuel soot, 2.26, 7.128
 nitration, 2.26, 7.129
 oxidation, 2.26, 7.128
 particle count, 2.26, 7.129
 solids content, 2.25, 7.128
 total acid number, 2.26, 7.129
 total base number, 2.26, 7.129
 viscosity, 2.25, 7.128
Lubrication:
 of air compressors, 5.188–5.189
 of air washers, 4.66–4.67
 of air-cooled condensers, 4.69
 automatic lubrication system, 8.13–8.26
 centralized automatic lubrication system,
   8.14–8.16
 of centrifugal compressors, 4.67–4.68
 of chains, for power transmissions, 5.79–5.80
 for control components, 6.42
 of cooling towers, 4.72
 of dehumidifiers, 4.80
 design considerations with, 5.8–5.9
 of evaporative condensers, 4.70
 for fans, 4.76
 of gear drives, 5.174–5.177
 grease lubrication, 5.41–5.43

Lubrication (*Cont.*)
  and greases, 8.7–8.12
   compatibility of, 8.11
   components of, 8.8
   manufacture of, 8.11–8.12
   properties of, 8.8–8.9
   testing for, 8.8, 8.9
   thickeners for, 8.9–8.11
  of humidifiers, 4.80
  and lubricants, 8.4–8.7
  maintenance for, 1.10
  oil lubrication, 5.41
  and plain bearings, 5.8–5.9
  for power transmissions, 5.79–5.80
  of precision chain drives, 5.79–5.80
  for rolling-element bearings, 5.41–5.44
  for rooftop air-conditioning units, 4.82
  selection of, 5.43–5.44
  supply of, 5.3
Lubrication programs, 8.13–8.35
 analysis in, 8.29–8.30
 automatic lubrication system, 8.13–8.26
  centralized, 8.14–8.16
  duoline systems, 8.20, 8.21
  ejection systems, 8.23–8.24
  injection systems, 8.24
  injector systems, 8.17
  oil mis systems, 8.16
  orifice-control systems, 8.16–8.17
  pump-to-point systems, 8.20–8.21
  selection of, 8.25–8.26
  series-progressive systems, 8.17–8.19
  twin-line systems, 8.19–8.20
  types of, 8.16
  zone-control systems, 8.21, 8.22
 centralized automatic lubrication system,
   8.14–8.16
   efficiency with, 8.15
   and housekeeping, 8.16
   and operating costs, 8.16
   productivity with, 8.16
   safety with, 8.14
 devices for, 8.13
 duoline systems, 8.20, 8.21
 efficiency with, 8.15
 ejection systems, 8.23–8.24
 employee responsibility for, 8.4
 fluid management programs for, 8.4–8.6
 and fluids management, 8.30
 and greases, 8.7–8.12
 and housekeeping, 8.16
 implementation of, 8.28, 8.34
 injector systems, 8.17
 lubricant selection for, 8.27–8.30
 and lubricants, 8.4–8.7
 and management, 8.27
 management of, 8.3–8.12
  fluid management programs for, 8.4–8.6
  greases, 8.7–8.12
  lubricants, 8.4–8.7
  organized plant lubrication program, 8.3–8.4
 management responsibility for, 8.3–8.4
 oil mis systems, 8.16
 and operating costs, 8.16
 organized plant lubrication program, 8.3–8.4
  employee responsibility for, 8.4
  management responsibility for, 8.3–8.4
  safety considerations with, 8.4

Lubrication programs (*Cont.*)
  orifice-control systems, 8.16–8.17
  planning for, 8.28
  and plant lubrication surveys, 8.29
  productivity with, 8.16
  pump-to-point systems, 8.20–8.21
  and quality assurance, 8.30
  responsibility for, 8.3–8.4
  safety considerations with, 8.4, 8.14
  scheduling in, 8.29–8.30
  selection of, 8.25–8.26
  series-progressive systems, 8.17–8.19
  training for, 8.27–8.28
  twin-line systems, 8.19–8.20
  zone-control systems, 8.21, 8.22

Machine design, 7.71
Machine vibration, 7.74
Machine/manual-operated control circuits, 6.64
Machinery (*see* Machine(s))
Machines and machinery:
  absorption machines
    leak testing for, 4.66
    leaktightness, 4.65
    maintenance of, 4.64–4.65
    pumps, 4.66
    purge units, 4.66
    safeties, 4.66
    service valves, 4.66
  linear-motion machinery, 7.78
  plant machinery
    bearings, 7.80–7.82
    belt drives, 7.83–7.84
    blades, 7.83
    critical speeds, 7.84–7.85
    dynamics of, 7.79–7.87
    gears, 7.82–7.83
    induced loads, 7.86–7.87
    mode shape, 7.85–7.86
    preloads, 7.86–7.87
    process variables, 7.87
    resonance, 7.86
    rolling-element bearings, 7.80–7.82
    running speeds, 7.84
    sleeve bearings, 7.81–7.82
    vanes, 7.83
    vibration of, 7.77–7.87
  reciprocating machinery, 7.78
  rotating machinery, 7.77–7.78
  and vibration, 7.77–7.78
  for welding, 10.51–10.52
Magnet (operating) coils, 6.44–6.45
Magnet constructions, 6.43–6.44
Magnetic circuit modification, 6.70–6.71
Magnetic hysteresis, 6.70
Magnetic motor controllers, 6.43
Magnetic overload relays, 6.66
Magnified sweeps, 7.68
Maintainability analysis, 3.136–3.137
Maintenance, 1.9–1.16
  of absorption machines, 4.64–4.65
  of air washers, 4.66–4.67
  of air-cleaning equipment, 4.118–4.124
  of air-conditioning equipment, 4.61–4.85
  of air-cooled condensers, 4.69
  for alterations, 1.10
  of belt drives, 4.73–4.74
  breakdown, 1.41–1.42, 2.3

Maintenance (*Cont.*)
  of centrifugal compressors, 4.67–4.68
  of coils, 4.68–4.69
  of concrete floor surfaces, 4.24–4.25
  contract, 4.49
  of control circuit components, 6.72–6.73
  of coolers, 4.71
  of cooling towers, 4.71–4.73
  cost estimates for (*see* Cost estimates)
  of cranes, 5.85–5.89
  of dampers, 4.73
  of dehumidifiers, 4.80
  do-it-yourself, 4.49
  of dry dynamic air-cleaning equipment, 4.119
  of economizers, 4.74–4.75
  of electrostatic precipitators, 4.120–4.122
  for elevators, 4.43–4.44, 4.46–4.49
  and engineering function, 1.29
  of equipment, 1.48–1.49
  equipment inspection, 1.10
  of evaporative condensers, 4.69–4.70
  of exhaust ducts, 4.122
  of exhaust hoods, 4.122
  of fabric filtration, 4.120
  of fans, 4.75–4.76
  field, 7.33, 7.40
  of filters, 4.76–4.78
  of flowmeters, 7.32–7.34
  with freeze protection, 4.83–4.84
  of gear drives, 5.177
  of heat pumps, 4.79–4.80
  of heaters, 4.78–4.79
  of humidifiers, 4.80
  of inertial air-cleaning equipment, 4.119
  for installations, 1.10
  insurance administration as, 1.11
  of lead-acid batteries, 6.85–6.86
  of liquid-level instruments, 7.35
  of lubricants, 8.5, 8.6
  lubrication, 1.10
  manpower requirements for, 1.13–1.14
  of mechanical instruments, 7.32–7.35
  of multiple centrifugal air-cleaning equipment, 4.119
  of nickel-cadmium batteries, 6.95–6.99
  for nickel-iron batteries, 6.101–6.103
  operating logs for, 4.63
  and operations, 1.27–1.28
  organization for, 1.11–1.13
  planned, 3.108
  and planning, 1.24–1.25
  of plant buildings and grounds, 1.10
  of plant equipment, 1.10
  plant protection as, 1.11
  of plastic piping, 9.88–9.89
  of precision chain drives, 5.81–5.82
  primary functions of, 1.10
  procedures for, 2.5
  and procurement function, 1.29–1.30
  program maintenance, 7.102
  of pumps, 4.80–4.81
  for relative-humidity-measuring instruments, 7.40
  repetitive, 3.108
  responsibilities of, 1.9–1.11
    primary functions, 1.10
    secondary functions, 1.11

Maintenance (*Cont.*)
  responsibility for, 4.61
  of rooftop A/C units, 4.81–4.82
  of room air conditioners, 4.82–4.83
  salvage as, 1.11
  for scaffolding, 9.112–9.113
  secondary functions of, 1.11
  of self-contained A/C units, 4.83
  of simple cyclone air-cleaning equipment, 4.119
  of spare parts, 4.63
  storeskeeping as, 1.11
  of synchronous belts, 5.131–5.135
  of thermometers, 7.36
  training for, 1.15, 1.16, 4.63
  and utilities, 1.10
  of valves, 9.66
  of V-belt drives, 5.112–5.115
  for vibration, 7.102
  waste disposal as, 1.11
  and water conditioning, 4.84–4.85
  of water-cooled condensers, 4.70–4.71
  of welding equipment, 10.57–10.59
  of wet collectors, 4.119–4.120
  (*See also* Breakdown maintenance)
  (*See also* Corrective maintenance)
  (*See also* Preventative maintenance)
Maintenance engineering (*see* Engineering)
Maintenance function:
  budget preparation and tracking, 2.102
  CMMS for, 2.99–2.102
  cooperative partnerships with, 1.24–1.25, 1.28
  craft utilization, 2.102
  and engineering, 1.28–1.29
  equipment/asset repair history, 2.101–2.102
  and operations, 1.28
  preventative maintenance planning, 2.100–2.101
  and procurement, 1.29
  for reliability, 1.24–1.25
  requisition of parts and services, 2.101
  resource scheduling, 2.101
  responsibilities of, 1.28–1.29
  and work orders, 2.99–2.101
Maintenance history database, 2.5
Maintenance organization, 1.31–1.38
  business plan of, 1.36
  considerations with, 1.33, 1.36–1.37
  for emergencies, 1.36
  for overhauling, 1.36
  for preventative maintenance, 1.36
  rebuilding, 1.36
  responsibilities of, 1.32–1.33
  with self-directed work teams, 1.33–1.35
  span of control ratios for, 1.37–1.38
  structure of, 1.33, 1.36–1.37
  supervision of, 1.37–1.38
  support for, 1.37–1.38
  work types of, 1.36–1.37
Maintenance painting (*See* Painting)
Maintenance personnel (*see* Personnel)
Maintenance planning (*see* Planning)
Maintenance programs, 1.19–1.20
Maintenance scheduling (*see* Scheduling)
Maintenance supervision (*see* Supervision)
Maintenance workers (*see* Personnel)

Management:
  best practices inventory management, 2.61
  computer-based maintenance management system (CMMS), 2.91–2.113
  configuration management, 1.18–1.19
  data management, 2.107
  fluid management programs, 8.4–8.6
  fluid management programs for, 8.4–8.6
  fluids management, 8.30
  of greases, 8.7–8.12
  life cycle asset management, 1.18
  of lubricants, 8.4–8.7
  of lubrication, 8.3–8.12
    fluid management programs for, 8.4–8.6
    greases, 8.7–8.12
    lubricants, 8.4–8.7
    organized plant lubrication program, 8.3–8.4
  lubrication programs, 8.27
  organized plant lubrication program, 8.3–8.4
  and organized plant lubrication program, 8.3–8.4
  production management, 2.21
  risk management, 1.18
  storeroom management, 2.65–2.75
  total plant management, 2.20
Management function, 2.107
  budget tracking, 2.107
  CMMS for, 2.107
  cooperative partnerships with, 1.25
  ISO 9000 compliance, 2.107
  for reliability, 1.25
Management theory, 1.11–1.12
Manganese (brown) stains, 4.36–4.37
Manpower (*see* Personnel)
Manual motor controllers, 6.46–6.47
Manual techniques, 10.24
Manually lever-operated chain hoists (pullers), 5.91
Manual/machine-operated control circuits, 6.64
Manuals, instruction, 1.51–1.53
Mapping, isothermal, 7.114
Maps, process, 1.64
Masonry, 4.7
Material(s):
  for corrosion control, 9.8–9.11
  cost estimates for, 3.117–3.118
  emissivity of, 7.118–7.125
  foreign, 5.11, 5.13
  infrared technology, 7.118–7.125
  and infrared technology, 7.118–7.125
  inspection of, 5.11, 5.13
  for ladders, 9.92–9.93
  for low-sloped roofs, 4.8–4.9
  for painting, 9.40–9.41
  for piping, 9.83
  for plain bearings, 5.4–5.5
  for repair, 4.8–4.9
  storage of, 4.15
  for thermography, 7.118–7.125
Material cost estimates, 3.117–3.118
Material safety data sheets (MSDS), 2.105
Matrix form (tabular) models, 3.15
Matrix models, 3.15–3.16
Mean time between failures (MTBF), 3.136, 7.116

Measurement(s):
  ambient, 7.110
  cryogenic-liquid-level measurement, 7.12–7.13
  dew-point measurement, 7.25
  humidity measurements, 7.24–7.25
  of infrared equipment, 7.110
  KPIs for, 3.124
  liquid-level measurement, 7.8–7.14
  overtravel measurement, 6.49
  power measurements, 7.57–7.58
  purge-type liquid-level measurement, 7.13–7.14
  qualitative measurements, 7.110
  quantitative measurements, 7.110
  resistance measurements, 7.58–7.63
  tangential noise measurement, 7.68
  total flow measurement, 7.7–7.8
  work measurement, 3.19–3.63
Mechanical checks, 6.42
Mechanical cleaning:
  abrasion method, 9.30
  and chemical cleaning, 9.25–9.30
  with crawlers, 9.26–9.30
  high-pressure water jetting, 9.25, 9.26
  hydrodrilling, 9.26
  with pigs, 9.26–9.30
  with plugs, 9.26–9.30
Mechanical flowmeters, 7.5–7.7
Mechanical instruments, 7.3–7.71
  Bellows-actuated flowmeters, 7.5–7.7
    design of, 7.5–7.6
    operation of, 7.6–7.7
  cryogenic-liquid-level measurement, 7.12–7.13
  dew-point measurements, 7.25
  diaphragm-actuated flowmeter transmitter, 7.26
  filled-system thermometers, 7.19–7.24
    classifications of, 7.19–7.23
    fully compensated liquid-filled, 7.20
    gas-filled thermal systems, 7.22
    liquid-filled, 7.19
    mercury-filled thermal systems, 7.22–7.23
    and thermal-system bulb designs, 7.22–7.23
    vapor-pressure thermal systems, 7.20–7.22
  flowmeters, 7.5–7.7, 7.32–7.34
    calibration of, 7.33–7.35
    field maintenance of, 7.33
    installation of, 7.32–7.33
    maintenance of, 7.32–7.34
    at zero, 7.33
  fully compensated liquid-filled thermometers, 7.20
  gas-filled thermal systems, 7.22
  for humidity measurements, 7.24–7.25
    dew-point measurement, 7.25
    nylon-ribbon hygrometer method, 7.24–7.25
  liquid-filled system thermometers, 7.19
  liquid-level instruments, 7.35
  for liquid-level measurement, 7.8–7.14
    under atmospheric pressure, 7.8–7.10
    in closed tanks, 7.10–7.11
    cryogenic-liquid-level measurement, 7.12–7.13
    purge-type liquid-level measurement, 7.13–7.14
    of sealing liquids, 7.11–7.12
  maintenance of, 7.32–7.35

Mechanical instruments, maintenance of (*Cont.*)
  flowmeters, 7.32–7.34
  liquid-level instruments, 7.35
  and mechanical-pressure instruments, 7.14–7.19
  mercury-filled thermal systems, 7.22–7.23
  meter-body operation, 7.27
  miniature, 7.4–7.5, 7.29–7.32
  nylon-ribbon hygrometer method, 7.24–7.25
  pneumatic pressure transmitters, 7.28–7.29
  pneumatic temperature transmitters, 7.28–7.29
  pneumatic transmission systems, 7.25–7.28
    diaphragm-actuated flowmeter transmitter, 7.26
    meter-body operation, 7.27
  pressure gages, 7.37–7.40
    calibration of, 7.38–7.40
    installation of, 7.37–7.38
  pressure transmitters, pneumatic, 7.28–7.29
  for process monitoring/control, 7.3
  purge-type liquid-level measurement, 7.13–7.14
  recording instruments, 7.4–7.5
  recording type, 7.29–7.32
  relative-humidity-measuring instruments, 7.40–7.41
  temperature transmitters, pneumatic, 7.28–7.29
  and thermal-system bulb designs, 7.22–7.23
  thermometers, 7.35–7.36
    calibration of, 7.36
    filled-system, 7.19–7.24
    installation of, 7.35–7.36
    maintenance of, 7.36
  for total flow measurement, 7.7–7.8
  transmitters
    pneumatic pressure transmitters, 7.28–7.29
    pneumatic temperature transmitters, 7.28–7.29
  types of, 7.4
  vapor-pressure thermal systems, 7.20–7.22
  (*See also specific types, e.g.:* Plain bearings)
Mechanical interlocks, 6.76–6.77
Mechanical rub, 7.89
Mechanical variable-speed drives, 5.145–5.160
  adjustable while in motion (nonenclosed) friction drives, 5.145–5.151
  friction drives, 5.145–5.160
    adjustable while in motion (nonenclosed), 5.145–5.151
    belt transmissions, 5.153–5.155
    flat-belt drives, 5.160
    friction-disk-type drives, 5.156–5.157
    geared differential drives, 5.157–5.159
    packaged belt drives, 5.155–5.156
    static-adjustment type, 5.151–5.153
    traction-type drives, 5.157
  friction-type belts, 5.145–5.147
Mechanical welding, 10.25
Mechanical-pressure instruments, 7.14–7.19
  absolute-pressure elements for, 7.18
  bellows, 7.16–7.18
  Bourdon tubes, 7.15–7.18
  helix tubes, 7.16
  pressure gages, 7.15–7.18
    absolute-pressure elements for, 7.18
    bellows, 7.16–7.18

Mechanical-pressure instruments, pressure gages (*Cont.*)
    Bourdon tubes, 7.15–7.18
    helix tubes, 7.16
    spiral actuating elements for, 7.16
    springs, 7.16–7.18
  spiral actuating elements for, 7.16
  springs, 7.16–7.18
  types of, 7.14–7.15
Medium-carbon steels, 10.36
Megahertz (MHz), 7.116
Meggers:
  high-resistance determination with, 7.64
  precautions when using, 7.61
  and resistance measurements, 7.59–761
  resistance measurements with, 7.59–761
  using, 7.61
Mercury-filled thermal systems, 7.22–7.23
Metal(s):
  alloy steels, 10.36–10.39
  base, 10.35–10.39
  bimetals, 5.5
  carbon steels, 10.35–10.36
  corrosion of, 4.15
  and low-sloped roofs, 4.15
  monometals, 5.5
  nonferrous metals, 10.39
  rusting of, 4.15
  spectrometals, 8.32
  spectrometric, 8.32
  spectrometric metals, 8.32
  trimetals, 5.5
  and welding, 10.35–10.39
Metal-base flashings, 4.13
Metallic-grid couplings, 5.48–5.49
Methods-time measurement (MTM) data, 3.41–3.52
MHz (megahertz), 7.116
Microbiological corrosion, 9.7
Microorganisms, 8.32
Micropatch study, 8.32
Microwave bands, 7.114
Mid-infrared, 7.114
Mineral surfaced roofs, 4.11
  modified bitumen roofing membranes, 4.4
  power roof exhausters, 4.98
Miniature mechanical instruments, 7.4–7.5, 7.29–7.32
Misalignment:
  failure from, 7.88–7.89
  and flexible couplings, 5.56–5.67
  shaft, 5.63, 5.66
  of synchronous belts, 5.132
  V-belt drives, 5.112, 5.113
  vibration from, 7.88–7.89
Mixed sweeps, 7.68
Mode shape, 7.85–7.86
Modified bitumen roofing membranes, 4.4, 4.11
Modified constant-voltage method, 6.82–6.83
Modified curvilinear belts, 5.124
Modular systems scaffolds, 9.106–9.107
Modulation, intensity, 7.68
Monitoring programs, 8.5
Monoammoniated citric acids, 9.20
Monolithically finished concrete floor surfaces, 4.21
Monometals, for plain bearings, 5.5
Monopolar input, 7.49

Motion economy, 3.102–3.104
Motivation:
  KPIs for, 3.124
  of self-directed work teams, 1.35
Motor(s), 6.3–6.37
  AC induction motors, 6.3–6.9
    construction of, 6.6
    DC motors vs., 6.3–6.6
    design of, 6.7
    insulation for, 6.7–6.8
    speed-torque characteristics of, 6.6–6.7
    temperature of, 6.8–6.9
    theory for, 6.6
    troubleshooting, 6.20–6.25
  adjustable-type relays, 6.68
  arc chambers, 6.45–6.46
  braided shunts, 6.46
  brushes for, 6.14–6.19
  and bus support insulators, 6.75
  circuit breakers, 6.57
  circuit interrupters, 6.56
  circuit protectors, motor`, 6.57
  common characteristics of, 6.3–6.6
  compound-wound, 6.9
  and contact bounce, 6.47
  contactors, 6.51–6.56
    inspection of, 6.51–6.55
    and motor control centers, 6.75
    troubleshooting, 6.52–6.55
    vacuum, 6.51
  contactor-type control relays, 6.65
  contacts, 6.47–6.51
    bounce, 6.47
    dressing for, 6.49–6.51
    overtravel measurement, 6.49
    pressure of, 6.48–6.49
    vacuum contactors, 6.51
    wear of, 6.49
  continuously wound rheostats, 6.60–6.61
  control centers, motor, 6.73–6.77
    and bus support insulators, 6.75
    and contactors, 6.75
    control devices, 6.76
    and disconnect devices, 6.75
    electrical interlocks, 6.76
    enclosures of, 6.74
    and fuses, 6.75
    mechanical interlocks, 6.76–6.77
    and terminal connections, 6.74
    thermal overload relays, 6.75–6.76
  control circuit components, 6.64–6.72
    adjustable-type relays, 6.68
    contactor-type control relays, 6.65
    dashpots, 6.66
    DC relay maintenance, 6.71
    magnetic circuit modification, 6.70–6.71
    and magnetic hysteresis, 6.70
    magnetic overload relays, 6.66
    maintenance of, 6.72–6.73
    manual/machine-operated, 6.64
    solid-state overload relays, 6.66–6.68
    thermal overload relays, 6.65–6.66
    time-tactors, 6.68–6.70
  control components, 6.39–6.77
    cleaning, 6.42
    contactors, 6.51–6.55
    and contacts, 6.47–6.51
    control circuit components, 6.64–6.72

Motor(s), control components (*Cont.*)
    disconnect devices, 6.56–6.57
    drum controllers, 6.62–6.63
    enclosures for, 6.40–6.41
    equipment grounding, 6.63–6.64
    lubrication for, 6.42
    mechanical checks, 6.42
    motor controllers, 6.43–6.51
    and personnel, 6.39–6.40
    power circuit components, 6.43
    preventative maintenance for, 6.41–6.42
    rheostats, 6.57–6.61
  control devices, 6.76
  controllers, motor, 6.43–6.51
    arc chambers, 6.45–6.46
    braided shunts, 6.46
    and contacts, 6.47–6.51
    and magnet constructions, 6.43–6.44
    magnetic, 6.43
    manual, 6.46–6.47
    operating (magnet) coils, 6.44–6.45
  cooling methods for, 6.3–6.4
  dashpots, 6.66
  DC motors, 6.3–6.6, 6.9–6.11
    AC motors vs., 6.3–6.6
    application data for, 6.9–6.11
    design of, 6.9
    insulation for, 6.9–6.11
    theory for, 6.9
    troubleshooting, 6.26–6.36
    types of, 6.9
  DC relay maintenance, 6.71
  design of, 6.7, 6.9
  disconnect devices, 6.56–6.57
    circuit breakers, 6.57
    circuit interrupters, 6.56
    disconnect switches, 6.56
    motor circuit protectors, 6.57
    and motor control centers, 6.75
  disconnect switches, 6.56
  drum controllers, 6.62–6.63
  electrical interlocks, 6.76
  enclosures, 6.74
  equipment grounding, 6.63–6.64
  and fuses, 6.75
  and infrared technology, 7.111
  insulation for, 6.7–6.8
  low-voltage release devices, 6.59
  lubrication for, 6.42
  and magnet constructions, 6.43–6.44
  magnetic circuit modification, 6.70–6.71
  and magnetic hysteresis, 6.70
  magnetic motor controllers, 6.43
  magnetic overload relays, 6.66
  manual motor controllers, 6.46–6.47
  manual/machine-operated control circuits, 6.64
  mechanical checks, 6.42
  mechanical interlocks, 6.76–6.77
  operating (magnet) coils, 6.44–6.45
  overtravel measurement for, 6.49
  permanent magnet motors, 6.9
  and personnel, 6.39–6.40
  and potentiometers, 6.60–6.61
  power circuit components, 6.43
  predictive maintenance program for, 6.19
  preventative maintenance for, 6.41–6.42
  ratings of, 6.3

Motor(s) (*Cont.*)
  relays
    adjustable-type relays, 6.68
    contactor-type control relays, 6.65
    DC relay maintenance, 6.71
    magnetic overload relays, 6.66
    solid-state overload relays, 6.66–6.68
    thermal overload relays, 6.65–6.66, 6.75–6.76
  and resistor replacement, 6.59
  rheostats, 6.57–6.61
    contact pressure settings, 6.59
    continuously wound, 6.60–6.61
    locating, 6.59–6.60
    with loose connection, 6.60
    low-voltage release devices, 6.59
    and potentiometers, 6.60–6.61
    and resistor replacement, 6.59
    servicing, 6.58–6.59
    and wound-rotor motors, 6.61
  scheduled maintenance for, 6.11–6.19
  series-wound, 6.9
  shunt-wound motors, 6.9
  solid-state overload relays, 6.66–6.68
  and temperature, 6.13–6.14
  and terminal connections, 6.74
  thermal overload relays, 6.65–6.66, 6.75–6.76
  time-tactors, 6.68–6.70
  and vacuum contactors, 6.51
  and vibration, 6.13–6.14
  vibration of, 7.90
  wear of, 6.49
  and wound-rotor motors, 6.61
Motor alternators, 7.116
Motor circuit protectors, 6.57
Motor control centers, 6.73–6.77
  and bus support insulators, 6.75
  and contactors, 6.75
  control devices, 6.76
  and disconnect devices, 6.75
  electrical interlocks, 6.76
  enclosures of, 6.74
  and fuses, 6.75
  mechanical interlocks, 6.76–6.77
  and terminal connections, 6.74
  thermal overload relays, 6.75–6.76
Motor controllers, 6.43–6.51
  arc chambers, 6.45–6.46
  braided shunts, 6.46
  and contacts, 6.47–6.51
  and magnet constructions, 6.43–6.44
  magnetic, 6.43
  manual, 6.46–6.47
  operating (magnet) coils, 6.44–6.45
Motor generators, 7.116
Motor ratings, 6.3
Mounting:
  reverse, 5.110
  of rolling-element bearings, 5.33–5.38
  of V-belt sheaves, 5.110
  for vibration, 7.98
Mountings:
  bearing, 5.28–5.32
  cold mounting, 5.33
  tapered-bore bearings, 5.35–5.38
  temperature mountings, 5.33, 5.35
  for ventilating fans, 4.105

MRO parts, 1.25
MSDS (material safety data sheets), 2.105
MTBF (*see* Mean time between failures)
MTM (methods-time measurement) data, 3.41–3.52
μ (unit ohms), 7.113
Multimeters, 7.44–7.50
  analog multimeters, 7.44–7.48
  digital multimeters, 7.48–7.50
  special-purpose industrial multimeters, 7.50
Multiple centrifugal air-cleaning equipment, 4.112, 4.119
Multiple-activity process chart, 3.99–3.102
  creation of, 3.99
  usage of, 3.99–3.100
  work simplification with, 3.99–3.102
Multirating tables, 4.88, 4.91
Multitrace, 7.68

Nameplate data, 3.170
Narrowband data, 7.104
Narrowband trending, 2.23
Near-infrared, 7.114
Net positive suction head, 5.215–5.216
Neutral, 7.117
Nickel-cadmium batteries, 6.89–6.99
  charging, 6.93–6.94
    with constant current, 6.94
    with constant voltage, 6.93–6.94
    and trickle charges, 6.94–6.95
  construction of, 6.90
  equalizing charge for, 6.95
  float charging, 6.94
  installation of, 6.92–6.93
  life expectancy for, 6.91–6.92
  maintenance of, 6.95–6.99
  performance of, 6.91
  plate processing for, 6.90
  and pocket plates, 6.90–6.91
  selection of, 6.92
  and sintered plates, 6.91
  trickle charging, 6.94–6.95
  voltage of, 6.91
Nickel-iron batteries, 6.99–6.103
  charging, 6.100–6.101
    of cycled batteries, 6.100
    of standby-power batteries, 6.100–6.101
  cleaning, 6.101–6.102
  cycling, 6.100, 6.102
  and electrolytes, 6.102–6.103
    renewing, 6.102–6.103
    spilled, 6.103
  laying up, 6.102
  maintenance for, 6.101–6.103
  operation of, 6.99–6.100
  watering, 6.101
Nitration:
  and lubricating oil analysis, 2.26
  lubricating oil analysis for, 7.129
Nitric acid, 9.21
No I.I., 3.12
Noise:
  from electrical motors, 6.13–6.14
  from synchronous belts, 5.137, 5.141
  tangential measurement, 7.68
  troubleshooting, 5.117, 5.122
  from V-belt drives, 5.117, 5.122

Noise (*Cont.*)
  from ventilating fans, 4.101–4.102
    elimination of, 4.101–4.102
    as problem, 4.101–4.102
    volume-control devices for, 4.102–4.104
Noise control, 7.70
Nonenclosed (adjustable while in motion) friction drives, 5.145–5.151
Nonferrous metals, 10.39
Nonlubricated cylinders, 5.189
Nonmetallic-disk globe valve, 5.201
Nonmetallic-disk lift-check valve, 5.201
Nonsoap organoclay, 8.11
Nonsoap polyurea, 8.11
Nonvulcanized elastomers, 4.11
Normal histograms, 3.149
Normal-mode rejection, 7.49
Number of digits, 7.49
Nylon-ribbon hygrometer method, 7.24–7.25

Occupancy loads, 4.54
Occupational Safety and Heatlh Administration (OSHA):
  on ladders, 9.93
  on scaffolding, 9.111
Ohm (Ω), 7.117
Ohmmeters:
  high-resistance determination with, 7.64
  for resistance measurements, 7.59
Ohm's law, 7.117
Oil:
  in chains, 5.79, 5.80
  and coolers, 4.71
  corrosion control with, 9.14
  for power transmissions, 5.79, 5.80
  quality of, 8.34
  and refrigeration, 4.58
  rust-preventive, 9.14
  testing for, 8.34
  viscosity of, 5.79, 5.80
Oil analysis, 7.129
Oil filters, 4.77–4.78
Oil heaters, 4.68
Oil lubrication, 5.41
Oil mis systems, 8.16
Oil stains, 4.38
Oil-filled transformers, 7.111–7.112
Oil-lubricated sleeve bearings, 6.12–6.13
Oldham (block-and-jaw) couplings, 5.49–5.50
O&M (operating and maintenance) manuals, 3.171
Ω (Ohm), 7.117
One sigma control, 1.56
100% test only, 3.12
100% test-stress-retest, 3.12
On-line inquiries, 2.82–2.83
Onstream cleaning, 9.24–9.25
  foam cleaning, 9.24–9.25
  gel cleaning, 9.25
  passivating, 9.25
  pickling, 9.25
  steam vapor phase cleaning, 9.24
On-the-job training, 1.15
Open circuits, 7.115
Operating and maintenance (O&M) manuals, 3.171

Operating (magnet) coils, 6.44–6.45
Operating costs, 8.16
Operating dynamics analysis, 2.31
Operating logs, 4.63
Operating performance, 3.176
Operating policies, 1.39–1.53
    for communications, 1.50–1.51
    and control, 1.50–1.51
    and instruction manuals, 1.51–1.53
    for intraplant relations, 1.47–1.50
    and standard-practice sheets, 1.51–1.53
    for work allocation, 1.39–1.43
    for work force, 1.43–1.47
Operational basic data, 3.114
Operations:
    cooperative partnerships with, 1.27–1.28
    and maintenance, 1.27–1.28
    maintenance function from, 1.28
    responsibilities of, 1.27–1.28
Operator errors, 1.24
Order(s):
    purchase orders, 2.105, 2.106
    and sales function, 1.24
    size of, 1.24
    (*See also* Work orders)
Order (harmonic), 7.76–7.77
Organic acids, 9.20
Organic coatings, 9.13
Organic solvents, 9.21–9.23
    chemical cleaning, 9.21–9.23
        copper complexers, 9.21–9.22
        oxidizing agents, 9.22
    cleaning with, 9.21–9.23
Organization, for maintenance, 1.11–1.13
    for computerized systems, 2.81
    factors affecting, 1.12
    management theory for, 1.11–1.12
    for piping, 9.56–9.57
    reporting structure, 1.12
    and specialized personnel, 1.13
Organized plant lubrication program, 8.3–8.4
    employee responsibility for, 8.4
    management responsibility for, 8.3–8.4
    safety considerations with, 8.4
Organoclay, nonsoap, 8.11
Orifice type wet collectors, 4.120
Orifice-control systems, 8.16–8.17
Oscilloscopes, bandwidth of, 7.64–7.67
OSHA (*see* Occupational Safety and Heatlh Administration)
Outlet dampers, 4.102
Output-product specifications, 3.171
Outside contractors, 1.43–1.44
Outside-air loads, 4.54
Overdischarging, 6.86
Overhauling, 1.36
Overhead cost estimates, 3.118–3.119
Overload, on drivers, 4.110
Overrange, 7.49
Overtime, 1.6
Overtravel measurement, 6.49
Ownership, employee, 1.6
Oxidation:
    and lubricating oil analysis, 2.26
    lubricating oil analysis for, 7.128
Oxidizing agents, 9.22

Oxyacetylene appliances, 10.71–10.74
    and brazing, 10.71–10.72
    and cutting, 10.72
    precautions for, 10.71
    safety considerations with, 10.71
    setup for, 10.73
Oxyacetylene cutting, 10.27
Oxygen cutting, 10.78–10.82
    equipment for, 10.81
    preparation for, 10.81
    process for, 10.81–10.82

Packaged belt drives, 5.155–5.156
Packaged ventilating equipment, 4.98
Packaging:
    for air compressors, 5.193–5.196
    for valves, 9.66, 9.68
Packed scrubbers, 4.119
Paint burning, 10.69–10.70
Paint stains, 4.37
Painters, 9.50–9.51
Painting, 9.35–9.41
    for aesthetics, 9.37–9.38
    for awareness, 9.37–9.38
    for beautification, 9.37–9.38
    benefits of, 9.35–9.36
    for corrosion protection, 9.38–9.39
    and encapsulation, 9.39–9.40
    and lead-based paints, 9.39–9.40
    material selection for, 9.40–9.41
    need for, 9.35
    for plant facilities, 9.38–9.39
    and product contamination, 9.39
    safety considerations with, 9.40
    survey for, 9.36–9.37
Paints (*see* Coatings)
Paper stains, 4.39
Parallel circuits, 7.115
Parallel feeders, 7.113
Paralleling, 2.14–2.17
Parallel-offset (rim) solution, 5.66
Pareto analysis, 3.141–3.142
Pareto charts, 1.61–1.63
Particle count:
    and lubricating oil analysis, 2.26
    lubricating oil analysis for, 7.129
Particle counts, 8.32
Particulate matter, 7.130–7.134
Particulate scrubbers, 4.114, 4.115
Partnerships (*see* Cooperative partnerships)
Parts:
    CMMS for, 2.94–2.95
    for elevators, 4.50
    to equipment/asset, 2.104–2.105
    MRO, 1.25
    for planned work, 2.105
    pre-burned, 3.12
    repositioning, 2.105
    requisition of, 2.101
    spare, 2.94–2.95, 5.208, 5.210–5.211
    spare parts inventory, 1.6
    tested, 3.12
    usage history, 2.104
Parts inventory analysis, 2.89
Passivating, 9.25
Past-performance method, 3.52–3.56
PE (polyethylene), 9.10
Peeling, 9.47

Pentane, 8.32
Performance:
    of batteries, 6.91
    of centrifugal pumps, 5.215
    of chain hoists, 5.98
    economic, 3.176
    of elevators, 4.44
    of nickel-cadmium batteries, 6.91
    operating, 3.176
    of positive-displacement pumps, 5.226
    of pumps, 5.226
    and RCA, 3.176
    reports for, 2.84
    of synchronous belts, 5.137, 5.142
    and total productive maintenance, 2.45
    of V-belt drives, 5.118, 5.123
    of ventilating fans, 4.88–4.91
Performance reports, 2.84
Periodic inspection, 4.107–4.109
Permanent magnet motors, 6.9
Permanent repair, 4.10–4.14
Personnel, 1.13–1.14, 1.43–1.47
    centralization of, 1.45–1.46
    and CMMS failure, 2.111
    compensation program for, 3.65–3.66
    and control components, 6.39–6.40
    coverage for, 1.44–1.45
    craft, 1.14–1.15
    decentralization of, 1.45–1.46
    efficiency of, 1.44–1.45
    and elevators, 4.46–4.48
    hourly
        amount of, 1.13–1.14
        crafts, 1.14
        and operation personnel, 1.13–1.14
    hourly personnel, 1.13–1.14
    in-house, 1.43–1.44
    with intraplant relations, 1.47–1.48
    job evaluation for, 3.66–3.69
        classifications, 3.69–3.89
        factors in, 3.66–3.68
        and job analysis, 3.68–3.69
        principles of, 3.66
    and lubricants, 8.6
    for maintenance, 1.13–1.14
    operating policies for, 1.43–1.47
    operation, 1.13–1.14
    outside contractors, 1.43–1.44
    painters, 9.49–9.50
    and production equipment, 1.47–1.48
    rating and evaluating, 3.65
    recruitment of, 1.46
    for root cause analysis, 3.163
    selection of, 1.14–1.16
    specialized, 1.13
    standardization of, 3.27–3.31
    supervision, 1.14
    supervisory, 1.15–1.16
        cross-craft supervision, 1.14
        density of, 1.14
    training of, 1.15, 1.16, 1.46–1.47
    wage rate for, 3.70
    and work measurement, 3.27–3.31
    and work simplification, 3.105
Phosphoric acids, 9.21
Pickling, 9.25
Pickoff, delay, 7.67
Pigs, cleaning with, 9.26–9.30

Pipe threading, 9.57
Piping, 9.55–9.89
    acid cleaning for, 9.86, 9.87
    bolt tension for, 9.68
    and centrifugal pumps, 5.221
    check valves for, 9.68
    cleaning, 9.25
    cleaning for, 9.86, 9.87
    corrosion of, 9.55–9.56
    corrosion-resistant piping, 9.83
    data for, 9.57, 9.58
    and drainage, 9.56
    flanges for, 9.58, 9.81–9.86
    flow direction of, 9.63, 9.66
    gaskets for, 9.58, 9.61–9.62
    hard-facing, 9.70–9.72
    identification of, 9.87
    insulation for, 9.78–9.81
    leaks in, 9.62–9.63, 9.86
    and liquid velocity, 9.86
    machining, 9.69
    maintenance of, 9.66
    materials for, 9.83
    organization for, 9.56–9.57
    packing for, 9.66, 9.68
    pickling, 9.25
    and pipe threading, 9.57
    plastic, 9.88–9.89
    plastic piping, 9.87–9.89
    relief devices for, 9.87
    reseating kits, 9.70
    seat rings, 9.69–9.70
    and stem traps, 9.76–9.77
    support for, 9.74–9.78
    temperatures for, 9.57
    and traps, 9.72–9.74
    and valves, 5.211, 9.63, 9.66, 9.68–9.70
        bolt tension for, 9.68
        check valves, 9.68
        flow direction of, 9.63, 9.66
        machining, 9.69
        maintenance of, 9.66
        packing for, 9.66, 9.68
        refacing, 9.70
        reseating kits, 9.70
        seat rings, 9.69–9.70
    and water hammer, 9.56
    welding for, 9.83
Piston rings, 5.191–5.192
Pitting corrosion, 9.6–9.7
Plain bearings, 5.3–5.18
    appearance of, 5.9
    bimetals for, 5.5
    cavitation of, 5.14
    cleaning, 5.3–5.4
    conformability of, 5.4
    construction of, 5.5
    corrosion resistance of, 5.4
    crankcase distortion, 5.14
    crankshaft distortion, 5.14
    and crush, 5.14
    design of, 5.6–5.9
    and dirt, 5.9–5.10
    embedability of, 5.4
    fatigue, 5.10–5.11
    fatigue strength of, 5.4
    and foreign materials, 5.11, 5.13
    and hot-short phenomenon, 5.13

Plain bearings (*Cont.*)
    inspection of, 5.9–5.15
        for appearance, 5.9
        for cavitation, 5.14
        for crankcase distortion, 5.14
        for crankshaft distortion, 5.14
        for crush, 5.14
        for dirt, 5.9–5.10
        for fatigue, 5.10–5.11
        for foreign materials, 5.11, 5.13
        for hot-short phenomenon, 5.13
        preliminaries for, 5.9
        used bearings, 5.9–5.14
        for wear, 5.9, 5.11
    location of, 5.7–5.8
    and lubrication
        design considerations with, 5.8–5.9
        supply of, 5.3
        types of, 5.4
    materials for, 5.4–5.5
        construction of, 5.5
        requirements for, 5.4
    monometals for, 5.5
    preventative maintenance for, 5.3–5.4
    reconditioning, 5.9, 5.15–5.18
        of connecting rods, 5.17
        of crankcase, 5.17–5.18
        for reassembly, 5.18
    requirements for, 5.4
    and retention, 5.7–5.8
    surface action of, 5.4
    temperature strength of, 5.4
    thermal conductivity of, 5.4
    trimetals for, 5.5
    used, 5.9–5.14
    wear, 5.9, 5.11
Planned maintenance, 3.108
Planners:
    and CMMS, 2.108–2.109
    cost estimates from, 3.110
Planning:
    and computerized systems, 2.86–2.87
    for corrective maintenance, 2.4–2.5
    follow-up for, 2.87–2.88
    KPIs for, 3.124
    for lubrication programs, 8.28
    and maintenance, 1.24–1.25
    maintenance history database for, 2.5
    in maintenance organizations, 1.36
    for production function, 1.24
    project, 2.103
    responsibilities of, 1.32–1.33
    statistical, 3.52–3.56
    supervision, 1.32–1.33
    trained maintenance planners for, 2.5
    and work measurement, 3.31–3.32
    work order, 2.100
    work plans, 2.79
Plant buildings, 1.10
Plant engineering, 1.25
    changes to, 1.25
    cooperative partnerships with, 1.25
    design, 1.25
    and documentation, 1.25
    modifications, 1.25
    and reliability, 1.25
Plant equipment, 1.10

Plant facilities:
    existing, 1.10
    low-sloped roofs, 4.3–4.16
        conventional built-up roofing
            membranes, 4.4
        defects with, 4.5–4.7
        emergency leak repair, 4.9–4.10
        inspection of, 4.5–4.7
        maintenance prevention for, 4.14–4.15
        modified bitumen roofing membranes, 4.4
        permanent repair of, 4.10–4.14
        protected roof membrane assembly
            ("upside-down" roof system), 4.5
        record keeping for, 4.3
        repair materials for, 4.8–4.9
        single-ply roofing membranes, 4.5
        temporary repair of, 4.9–4.10
    maintenance for, 1.10
    painting for, 9.38–9.39
    (*See also specific parts of, e.g.:* Low-sloped
        roofs)
Plant growth, 4.39
Plant lubrication surveys, 8.29
Plant machinery:
    bearings, 7.80–7.82
        rolling-element bearings, 7.80–7.81
        sleeve bearings, 7.81–7.82
        vibration of, 7.80–7.82
    belt drives, 7.83–7.84
    blades, 7.83
    critical speeds, 7.84–7.85
    dynamics of, 7.79–7.87
        bearings, 7.80–7.82
        belt drives, 7.83–7.84
        blades, 7.83
        critical speeds, 7.84–7.85
        gears, 7.82–7.83
        induced loads, 7.86–7.87
        mode shape, 7.85–7.86
        preloads, 7.86–7.87
        process variables, 7.87
        resonance, 7.86
        running speeds, 7.84
        vanes, 7.83
    gears, 7.82–7.83
    induced loads, 7.86–7.87
    mode shape, 7.85–7.86
    preloads, 7.86–7.87
    process variables, 7.87
    resonance, 7.86
    rolling-element bearings, 7.80–7.82
    running speeds, 7.84
    sleeve bearings, 7.81–7.82
    vanes, 7.83
    vibration of, 7.77–7.87
Plant protection, 1.11
Plantwide averages/ratios, 3.111
Plasma arc cutting, 10.30–10.31
Plasma arc welding, 10.34–10.35
Plastic flow, of gear-tooth, 5.179–5.180
Plastic piping, 9.87–9.89
Plate processing, 6.90
Plateau histograms, 3.149
Plating, corrosion control with, 9.12
Plots, scatter, 3.150–3.152
Plugs, cleaning with, 9.26–9.30
Plug-type Renewo globe valve, 5.201
PMS models, 3.124

Pneumatic pressure transmitters, 7.28–7.29
Pneumatic temperature transmitters, 7.28–7.29
Pneumatic transmission systems, 7.25–7.28
   diaphragm-actuated flowmeter transmitter, 7.26
   meter-body operation, 7.27
   operation of, 7.27–7.28
Pocket plates, 6.90–6.91
Polarity, and welding, 10.18–10.20
Polyethersulfone, 9.10
Polyethylene (PE), 9.10
Polymeric materials, corrosion of, 9.8
Polymers, for corrosion control, 9.9–9.10
Polystyrene, for corrosion control, 9.10
Polyurea, nonsoap, 8.11
Polyurethane, for corrosion control, 9.9–9.10
Polyvinylchloride (PVC), 9.9
Portability, of infrared equipment, 7.110
Positive-displacement pumps, 5.222–5.227
   configuration of, 5.223–5.226
   gears in, 5.224
   installation of, 5.227
   lobe-type pump, 5.226
   operating methods for, 5.227
   performance of, 5.226
   rotating speed of, 5.226
   screws in, 5.224–5.225
   suction of, 5.226
   vane pump, 5.225–5.226
   viscosity of, 5.226
Potassium permanganate, 9.20
Potential transformers, 7.58
Potentiometers, and rheostats, 6.60–6.61
Poultice, 4.37
Pour points, 8.32
Power circuit components, 6.43
Power measurements, 7.57–7.58
   current transformers, 7.57
   potential transformers, 7.58
   single-function instruments for, 7.57–7.58
   single-phase connections, 7.57
Power roof exhausters, 4.98
Power transmissions:
   chains for, 5.73–5.82
      bath lubrication for, 5.80
      and chain drives, 5.78–5.82
      disk lubrication for, 5.80
      drip lubrication for, 5.80
      lubrication for, 5.80
      manual lubrication for, 5.80
      oil-stream lubrication for, 5.80
      service factors for, 5.75–5.76
      special inverted-tooth chains, 5.76–5.78
   flexible couplings for (*see* Flexible couplings)
Powered chain hoists, 5.93–5.96
Power-factor meters, 7.52
Pre-burned parts, 3.12
Precipitators, electrostatic, 4.117–4.118
Precision, of sequence-of-events analysis, 3.140
Precision chain drives, 5.79–5.82
   and contamination, 5.79
   and heat, 5.79
   lubrication of, 5.79–5.80
   maintenance of, 5.81–5.82
   and oil viscosity, 5.79, 5.80
   and windage, 5.79

Predictive maintenance, 2.19–2.34
   benefits of, 2.32–2.34
      and breakdown losses, 2.32, 2.33
      and capacity factor, 2.33
      maintenance costs of, 2.33–2.34
      and quality defects, 2.33
   definition of, 2.19
   electric motor analysis, 2.29–2.30
      insulation resistance, 2.29–2.30
      testing, 2.30
   maintenance management, 2.20–2.21
   operating dynamics analysis, 2.31
   process parameters, 2.28–2.29
   production management, 2.21
   program costs for, 2.31
   quality improvement, 2.21
   for synchronous belts, 5.133
   techniques for, 2.21–2.22
   thermography, 2.23–2.24, 2.32
   total plant management, 2.20
   tribology, 2.24–2.28, 2.32
      ferrography, 2.27–2.28
      lubricating oil analysis, 2.25–2.26
      wear particle analysis, 2.26–2.28
   ultrasonics, 2.30–2.32
   of V-belt drives, 5.115
   and vibration, 7.69
   vibration monitoring, 2.22–2.23, 2.31
      broadband trending, 2.22–2.23
      narrowband trending, 2.23
      signature analysis, 2.23
   visual inspection, 2.30
Predictive maintenance program, 7.101–7.104
   analysis techniques for, 7.104
   with databases, 7.93–7.95, 7.101–7.102
   for motors, 6.19
   program maintenance, 7.102
   trending techniques for, 7.102
   for vibration, 7.101–7.104
Preheating, of coils, 4.84
Preloads, 7.86–7.88
Premaintenance repairs, 4.52
Premature failure:
   of synchronous belts, 5.136–5.138
   of V-belt drives, 5.117, 5.118
Preoperational deposits, 9.19
Preparation:
   of budget, 2.102
   for coatings, 9.44–9.46
   of cost estimates, 3.109–3.110
   for oxygen cutting, 10.81
   for welding, 10.75–10.76
Present worth method, 3.14
Pressure:
   atmospheric, 7.8–7.10
   of contacts, 6.48–6.49
   of ventilating fans, 4.109
Pressure gages, 7.15–7.18, 7.37–7.40
   absolute-pressure elements for, 7.18
   bellows, 7.16–7.18
   Bourdon tubes, 7.15–7.18
   calibration of, 7.38–7.40
      field calibration of, 7.38–7.39
      shop calibration of, 7.39–7.40
   field calibration of, 7.38–7.39
   helix tubes, 7.16
   installation of, 7.37–7.38
   shop calibration of, 7.39–7.40

Pressure gages (*Cont.*):
   spiral actuating elements for, 7.16
   springs, 7.16–7.18
Pressure transmitters, pneumatic, 7.28–7.29
Prestart check, 4.105
Preventative maintenance, 1.25, 2.3–2.4, 2.7–2.18
   for air-conditioning equipment, 4.61
   breakdown maintenance vs., 1.41–1.42
   for chain hoists, 5.98
   CMMS for, 2.96, 2.100–2.101, 2.107–2.108
   for control components, 6.41–6.42
   for corrosion, 4.15
   for elevators, 4.51
   for exhaust ducts, 4.123
   for exhaust hoods, 4.123
   for holes, 4.15
   for housekeeping, 4.14
   information collection for, 2.10
   for low-sloped roofs, 4.14–4.15
   for metals, 4.15
   for plain bearings, 5.3–5.4
   planning for, 2.100–2.101
   for roof drains, 4.15
   for rusting, 4.15
   for and storage, 4.15
   with surface cleaning, 4.14–4.15
   system analysis for, 2.10
   and system functions, 2.10
   and traffic, 4.15
   work allocation for, 1.41–1.42
Priority:
   computerized, 2.79
   and scheduling, 1.41
Priority system, 2.85
Proactive improvement, 1.35
Probes, displacement (eddy-current), 7.96–7.97
Process control, 7.3
Process maps, for Six Sigma safety, 1.64
Process monitoring, 7.3
Process stability, 1.5
Process variables, 7.87
   mechanical instruments for (*see* Mechanical instruments)
   and vibration, 7.87
Procurement function, 1.25
   cooperative partnerships with, 1.25, 1.29–1.30
   deliveries, 1.25
   and maintenance, 1.29–1.30
   and maintenance function, 1.29
   MRO parts, 1.25
   for reliability, 1.25
   specifications for, 3.170
   of vendors, 1.25
Product contamination, 9.39
Product mix, 1.23–1.24
Product quality, 3.176
   and CMMS, 2.109
   and root cause analysis (RCA), 3.176
Product safety, 3.11–3.12
Production, 7.71
Production equipment, 1.47–1.48
Production function, 1.24
Production management, 2.21
Production run, 1.24
Production unit costs, 1.5
Productivity, 8.16
Program maintenance, 7.102
Project planning, 2.103

Propeller fans, 4.95
Protected roof membrane assembly ("upside-down" roof system), 4.5
Protection:
    cathodic, 9.13
    for cooling towers, 4.73
    corrosion, 9.38–9.39
    fire, 8.7
    freeze, 4.68–4.69
        for coils, 4.68–4.69
        maintenance with, 4.83–4.84
        for water coils, 4.84
        for water-cooled condensers, 4.71
    for lubricants, 8.7
    of lubricants, 8.6–8.7
    painting for, 9.38–9.39
    winter, 4.73
Protective screens, 4.106
Pullers (manually lever-operated chain hoists), 5.91
Pullet wear, 5.112
Pulleys, adjustable, 4.74
Pumps, 4.66, 4.80–4.81, 5.213–5.227
    centrifugal, 5.213–5.222
        brake horsepower (BHP) for, 5.220
        bypass operation of, 5.222
        configuration of, 5.213–5.215
        and discharge-piping configuration, 5.221–5.222
        and entrained air, 5.217–5.218
        foundation of, 5.221
        hydraulic curve of, 5.218–5.220
        impeller orientation in, 5.214
        impeller types in, 5.213
        and inlet-piping configuration, 5.221
        in-line configuration of, 5.214
        installation of, 5.220–5.222
        net positive suction head, 5.215–5.216
        operating conditions for, 5.222
        operating methods for, 5.222
        with opposed impellers, 5.215
        performance of, 5.215
        and piping, 5.221
        startup procedures for, 5.222
        suction conditions of, 5.215–5.218
        total dynamic head (TDH) requirement for, 5.218
        total system head (TSH) requirement of, 5.218
        and volume, 5.216–5.217
    freezing of, 4.84
    heat, 4.79–4.80
    jet pumps, 4.98
    lobe-type pumps, 5.226
    maintenance of, 4.80–4.81
    positive-displacement, 5.222–5.227
        configuration of, 5.223–5.226
        gears in, 5.224
        installation of, 5.227
        lobe-type pump, 5.226
        operating methods for, 5.227
        performance of, 5.226
        rotating speed of, 5.225
        screws in, 5.224–5.225
        suction of, 5.226
        vane pump, 5.225–5.226
        viscosity of, 5.226
    problems with, 4.81

Pumps (Cont.)
    vane pumps, 5.225–5.226
    vibration of, 7.92–7.93
    well freezing, 4.84
    well pumps, 4.84
Pump-to-point systems, 8.20–8.21
Punctures, in low-sloped roofs, 4.12
Purchase orders, 2.106
    CMMS for, 2.98–2.99, 2.105
    receipts against, 2.106
Purchase requisitions, 2.98
Purchasing function, 2.105–2.106
    for purchase orders, 2.106
    for requisitions, 2.106
    for stores stock inventory, 2.106
Purge units, 4.66, 4.68
Purge-type liquid-level measurement, 7.13–7.14
PVC (polyvinylchloride), 9.9

QD bushings:
    and synchronous belt drives, 5.130–5.131
    and V-belt drives, 5.111
    and V-belt sheaves, 5.111
QD-type sheaves, 5.110
QD-type sprockets, 5.130
Qualifiers, in sequence-of-events analysis, 3.141
Qualitative infrared thermography, 7.114
Qualitative measurements, 7.110
Quality:
    of oil, 8.34
    and total productive maintenance, 2.45–2.46
    (See also Product quality)
Quality assurance, 8.30
Quality control, 7.70
Quality improvement, 2.21
Quantitative infrared thermography, 7.114
Quantitative measurements, 7.110
Questioning, in scientific method, 3.92

R (see Resistance)
Radar bands, 7.114
Radial-blade centrifugal fans, 4.96, 4.97
Radiance, atmospheric, 7.113
Radiation, 7.109
    defined, 7.114, 7.117
    infrared, 7.114
    and infrared technology, 7.109
Radio bands, 7.114
Random sampling, 3.56–3.58, 3.60
Ranking, criticality, 2.11
Rate of return method, 3.14
Rate setters, cost estimates from, 3.110
Rating, personnel, 3.65
Ratings:
    for ladders, 9.94
    motor, 6.3
    for rolling-element bearings, 5.24–5.25
Ratio delay, 3.56–3.62
    accuracy of, 3.60
    for allowance determination, 3.59, 3.60
    and work simplification, 3.59
RBD (reliability block diagram), 3.136
RC characteristics, input, 7.67
RCA (see Root cause analysis)
Ready backlogs, 2.79
Reassembly, reconditioning for, 5.18
Rebuilding, of maintenance organization, 1.36

Reciprocating machinery, 7.78
Reconditioning:
    of connecting rods, 5.17
    of crankcase, 5.17–5.18
    for plain bearings, 5.9, 5.15–5.18
    for reassembly, 5.18
Record keeping, 4.3
Recording instruments, 7.4–7.5
Recordkeeping, 4.107
Records, for lead-acid batteries, 6.86
Recruitment, of work force, 1.46
Rectifiers, 7.117
Redesign, for failure, 2.14
Refacing valves, 9.70
Reflectance, 7.113
Reflectivity, 7.114
Refrigerant(s), 4.58
    and centrifugal compressors, 4.68
    and refrigeration, 4.58
Refrigerant float chambers, 4.71
Refrigerant joints:
    and coolers, 4.71
    and evaporative condensers, 4.70
    and water-cooled condensers, 4.71
Refrigeration, 4.55–4.56
    and air-conditioning equipment, 4.55–4.56
    equipment for, 4.58–4.61
        installation of, 4.61
        oil, 4.58
        refrigerant, 4.58
    installation for, 4.61
    measuring, 4.55–4.56
    and oil, 4.58
    and refrigerant, 4.58
Regrinding lift-check valve, 5.201
Regulations:
    compliance to, 3.176
    for cranes, 5.89
    federal, 9.93
    for ladders, 9.93–9.94
    and root cause analysis (RCA), 3.176
    for ventilating fans, 4.88
    (See also specific organizations, e.g.: Occupational Safety and Health Administration)
Reinforcement, of concrete floor surfaces, 4.20
Rejection, 7.49
Relative-humidity-measuring instruments, 7.40–7.41
Relays:
    adjustable-type relays, 6.68
    contactor-type control relays, 6.65
    DC relay maintenance, 6.71
    defined, 7.117
    magnetic overload relays, 6.66
    solid-state overload relays, 6.66–6.68
    thermal overload relays, 6.65–6.66
Reliability, 1.3–1.8
    benefits of, 1.5–1.7
    for continuous improvement, 1.7
    cost accounting for, 3.16–3.17
    cost estimation methods for, 3.14–3.16
    economic evaluation for, 3.13–3.14
    and employee ownership, 1.6
    and employee safety, 1.6–1.7
    and environmental issues, 1.7
    and equipment life, 1.5–1.6
    incoming inspections for, 3.12–3.13

Reliability (*Cont.*)
  issues with, 3.11–3.13
  and maintenance costs, 1.5
  maintenance function for, 1.24–1.25
  management function for, 1.25
  and overtime, 1.6
  and plant engineering, 1.25
  problems with, 1.23–1.25
  and process stability, 1.5
  procurement function for, 1.25
  and product safety, 3.11–3.12
  production function for, 1.24
  and production unit costs, 1.5
  responsibility for, 1.23–1.25
  sales function for, 1.23–1.24
  and self-directed work teams, 1.7–1.8
  and spare parts inventory, 1.6
Reliability block diagram (RBD), 3.136
Reliability centered maintenance (RCM), 2.35–2.39
  and failure, 2.36–2.37
  failure-finding tasks, 2.38–2.39
  functions of, 2.35–2.36
  proactive tasks, 2.37–2.38
Reliability engineering, 1.32–1.33
  and asset care, 1.19
  and configuration management, 1.18–1.19
  and life cycle asset management, 1.18
  and loss elimination, 1.19
  responsibilities of, 1.17–1.19, 1.32–1.33
  and risk management, 1.18
  and work allocation, 1.42–1.43
Reliability predictions, 3.136
Relief dampers, 4.73
Relief devices, 9.87
Renewable-wedge-and-seat bronze gate valve, 5.204
Renewo lift-check valve, 5.201
Reorder requirements, 2.105
Repair(s):
  for elevators, 4.52
  of emergency leaks, 4.9–4.10
  of equipment/asset, 2.101–2.102, 2.104
  history of, 2.101–2.102, 2.104
  to lead-acid batteries, 6.88–6.89
  for low-sloped roofs, 4.9–4.10
  of low-sloped roofs, 4.9–4.14
  materials for, 4.8–4.9
  permanent, 4.10–4.14
  premaintenance, 4.52
  procedures for, 2.5
  request backlog, 2.104
  temporary, 4.9–4.10
  time for, 2.6
  verification for, 2.6
Repair costs, 3.107
Repetitive maintenance:
  CMMS for, 2.96
  cost estimates for, 3.108
Replacement:
  of belt drives, 4.74
  of flashings, 4.29
  of low-sloped roofs, 4.16
Replacement studies, 3.14
Reports and reporting:
  computerized systems for, 2.83–2.84, 2.88–2.90
  control reports, 2.79

Reports and reporting (*Cont.*)
  generation of, 2.83–2.84
  organization for, 1.12
  performance, 2.84
  for performance, 2.84
  performance reports, 2.84
  for root cause analysis, 3.165–3.166
  in storerooms, 2.66–2.67
  structure for, 1.12
Requisitions, 2.106
Reroofing, 4.5, 4.16
Reseating kits, 9.70
Resistance (R):
  to change, 1.35
  defined, 7.117
Resistance measurements, 7.58–7.63
  high-resistance determination, 7.63–7.64
  low-resistance determination, 7.62–7.63
  megger, 7.59–761
  with meggers, 7.59–761
  ohmmeter, 7.59
  single-function instruments for, 7.58–7.63
  Wheatstone bridge, 7.61–7.62
Resistance welding, 10.35
Resistors:
  defined, 7.117
  replacement of, 6.59
Resolution, 7.49, 7.68
Resonance, 7.86
Resource allocation, 3.124
Resource scheduling, 2.101
Resources, lack of, 2.110–2.111
Response time, 7.49
Responsibilities:
  for air-conditioning equipment, 4.61
  of cooperative partnerships, 1.23–1.25, 1.27–1.30
  of employees, 8.4
  of engineering, 1.19–1.20
  of front line supervision, 1.32
  for lubrication programs, 8.3–8.4
  for maintenance, 1.9–1.11, 4.61
  of maintenance organization, 1.32–1.33
  of management, 8.3–8.4
  of operations, 1.27–1.28
  for organized plant lubrication program, 8.3–8.4
  of planning, 1.32–1.33
  for reliability, 1.17–1.19, 1.23–1.25, 1.32–1.33
  of reliability engineering, 1.17–1.19, 1.32–1.33
  of scheduling, 1.32–1.33
  of supervision, 1.32
Resurfacing, for concrete floor surfaces, 4.24
Retention, and plain bearings, 5.7–5.8
Reverse indicators, 5.58–5.63
Reverse mount, V-belt sheaves, 5.110
Reverse mount synchronous sprockets, 5.130
Review, of equipment/asset specifications, 2.103
Rheostats, 6.57–6.61
  contact pressure settings, 6.59
  continuously wound, 6.60–6.61
  locating, 6.59–6.60
  with loose connection, 6.60
  low-voltage release devices, 6.59
  and potentiometers, 6.60–6.61
  and resistor replacement, 6.59

Rheostats (*Cont.*)
  servicing, 6.58–6.59
  and wound-rotor motors, 6.61
Rim (parallel-offset) solution, 5.66
Risetime, 7.68
Risk management, 1.18
Rolling fatigue wear, 2.27, 7.131–7.133
  defined, 7.131
  and sliding wear, 2.27, 7.132–7.133
Rolling wear, 2.27
Rolling-element bearings, 5.19–5.44, 7.80–7.82
  bearing mountings, 5.28–5.32
  boundary dimensions for, 5.20, 5.23
  cold mounting, 5.33
  design for, 5.20
  dismounting of, 5.33, 5.38–5.40
    bearings, 5.38–5.39
    hydraulic removal, 5.39, 5.40
  failure of, 5.19
  fit of, 5.25–5.26
  housing of, 5.25–5.26
  load ratings for, 5.24–5.25
  lubrication for, 5.41–5.44
    grease lubrication, 5.41–5.43
    oil lubrication, 5.41
    selection of, 5.43–5.44
  mounting of, 5.33–5.38
    cold mounting, 5.33
    tapered-bore bearings, 5.35–5.38
    temperature mountings, 5.33, 5.35
  nomenclature for, 5.20
  series, 5.23, 5.24
  shafts of, 5.25–5.26
  tapered-bore bearings, 5.35–5.38
  temperature mountings, 5.33, 5.35
  vibration of, 7.80–7.81
Roll-off, 7.68
Roof(s):
  bitumen roofing membranes, 4.4
  built-up roofing membranes, 4.4
  coated smooth-surfaced roofs, 4.11
  conventional built-up roofing membranes, 4.4
  low-sloped, 4.3–4.16
  low-sloped roofs, 4.3–4.16
  mineral surfaced, 4.11
  mineral surfaced roofs, 4.11
  permanent repair of, 4.11
  storage on, 4.15
Roof accessory components, 4.7
Roof drains, 4.15
Roofing membranes, 4.4, 4.5
Rooftop air-conditioning units, 4.81–4.82
  cleaning, 4.82
  inspection of, 4.81–4.82
  lubrication for, 4.82
  maintenance of, 4.81–4.82
Room air conditioners, 4.82–4.83
  applications for, 4.82–4.83
  instructions for, 4.83
  maintenance of, 4.82–4.83
  types of, 4.82
Root cause analysis (RCA), 3.153–3.176
  A3 process, 3.155–3.159
    form for, 3.158–3.159
    steps of, 3.156–3.158
  for acceptable operating envelope, 3.171
  analysis for, 3.159
  application review, 3.172–3.174

Root cause analysis (RCA), application review (*Cont.*)
  installation, 3.172
  maintenance history, 3.173
  maintenance procedures, 3.174
  operating envelope, 3.172
  operating procedures, 3.173
 capacity restrictions, 3.176
 for clarifying problems, 3.166–3.168
 concept of, 3.154–3.155
 damage, 3.175–3.176
 data for, 3.168–3.169
 design review, 3.169–3.172
 for economic performance, 3.176
 for equipment damage/failure, 3.175–3.176
 event-reporting format for, 3.165–3.166
 and facts, 3.168–3.169
 failure, 3.175–3.176
 and five whys (5W), 3.161
 formal process for, 3.161–3.163
 identifying events with, 3.163–3.166
 incident classification, 3.174–3.176
 for incoming-project specifications, 3.171
 methodology for, 3.154–3.155, 3.163
 objectives of, 3.171
 for operating performance, 3.176
 for output-product specifications, 3.171
 personnel requirements for, 3.163
 physical evidence for, 3.168–3.169
 problem solving with, 3.174
 process for, 3.159–3.163
  analysis for, 3.159
  and five whys (5W), 3.161
  formal, 3.161–3.163
  personnel requirements for, 3.163
  purpose of, 3.162
  and visual inspection, 3.159–3.161
 for product quality, 3.176
 and regulatory compliance, 3.176
 reporting for, 3.165–3.166
 requirement of, 3.169
 for safety, 3.176
 for system damage/failure, 3.175–3.176
 and visual inspection, 3.159–3.161
 for work performed, 3.171
Root-cause failure analysis, 7.104
Rotating machinery, 7.77–7.78
Rotating speed, 5.226
Rotation, of fans, 4.76
Rotors, 7.117
Rub, mechanical, 7.89
Rubber elastomers, 9.10
Rubbing wear, 2.27, 7.131
Rung ladders, 9.94–9.95
 extension ladders, 9.94
 sectional ladders, 9.94, 9.95
 single ladders, 9.94
Running speeds, 7.84
Rusting:
 hidden, 9.47
 in low-sloped roofs, 4.15
 of metals, 4.15
Rust-preventive oils, 9.14

Safeties, 4.66, 4.68
Safety considerations:
 for air-acetylene appliances, 10.63–10.64

Safety considerations (*Cont.*)
 for centralized automatic lubrication system, 8.14
 for coatings, 9.49–9.50
 for elevators, 4.44–4.46, 4.50–4.51
 for employees, 1.6–1.7
 in infrared technology, 7.110
 and intraplant relations, 1.49
 for ladders, 9.96–9.101
 for lubricants, 8.33
 for meggers, 7.61
 for organized plant lubrication program, 8.4
 for painting, 9.40
 for product, 3.11–3.12
 and reliability, 3.11–3.12
 and root cause analysis, 3.176
 for scaffolding, 9.108–9.111
 for welding, 10.59–10.60
 (*See also* Six Sigma safety)
Safety swinging scaffolds, 9.104
Safety taskforces, 1.60–1.61
Sagging:
 in flashings, 4.12
 indicator sag, 5.57
Sales function:
 cooperative partnerships with, 1.23–1.24
 and delivery, 1.24
 and orders, 1.24
 and product mix, 1.23–1.24
 and production run, 1.24
 for reliability, 1.23–1.24
Salvage, as maintenance, 1.11
Sampling:
 random, 3.56–3.58, 3.60
 work, 3.56
Sandblasting, 4.33, 4.35
SAW (*see* Submerged arc welding)
Saw cutting, 4.31
Scaffolding, 9.91–9.92, 9.101–9.113
 aluminum, 9.101–9.103, 9.105, 9.106
 application of, 9.111–9.113
 coupler scaffolds, 9.105
 maintenance for, 9.112–9.113
 modular systems scaffolds, 9.106–9.107
 OSHA on, 9.111
 safety considerations with, 9.108–9.111
 safety swinging scaffolds, 9.104
 special-design scaffolds, 9.108–9.111
 steel, 9.103, 9.105, 9.106
 tube scaffolds, 9.105, 9.106
 welded aluminum scaffolds, 9.101–9.102
 welded steel scaffolds, 9.103
Scanners, line, 2.24, 7.107
Scanning procedures, 7.110–7.113
Scatter plots, 3.150–3.152
Scheduled maintenance, 6.11–6.19
Scheduling, 1.39–1.41
 amount of, 1.39–1.40
 CMMS for, 2.108
 and computerized systems, 2.87
 coordinating, 1.41
 and dispatching, 1.41
 downtime, 2.103–2.104
 with flow-of-work requests, 1.40–1.41
 job, 2.108
 lead time for, 1.40
 in lubrication programs, 8.29–8.30
 and priority, 1.41

Scheduling (*Cont.*)
 for production function, 1.24
 resource, 2.101
 responsibilities of, 1.32–1.33
 by size of job, 1.40
 supervision, 1.32–1.33
 systems for, 1.40–1.41
 by total work load, 1.40
 for work allocation, 1.39–1.41
 work measurement for, 3.31–3.32
 work order, 2.101
 by work unit, 1.40
Scientific method:
 and application, 3.93
 generalizing in, 3.92–3.93
 for methods improvement, 3.93–3.95
 questioning in, 3.92
 wondering in, 3.92
 work simplification with, 3.92–3.95
Scoring, of gear-tooth, 5.180
SCR (semiconductor-/silicon- controlled rectifier), 7.117
Screw-and-yoke rising-stem gate valve, 5.204
Screws, in positive-displacement pumps, 5.224–5.225
Scrubbers (*see* Wet collectors)
Scum, 4.39
Sealers, for concrete floor surfaces, 4.25–4.26
Sealing liquids, 7.11–7.12
Seat rings, 9.69–9.70
Second level logic tree analysis, 2.14
Secondary switchgear, 7.111
Sectional ladders, 9.94, 9.95
Sediment, 8.32
Selective leaching, 9.7
Self-contained air conditioners, 4.83
Self-directed work teams, 1.33–1.35
 administrative skills for, 1.35
 and continuous improvement, 1.35
 cross-training for, 1.35
 implementation of, 1.35
 incentives for, 1.35
 motivation of, 1.35
 and proactive improvement, 1.35
 and reliability, 1.7–1.8
 and resistance to change, 1.35
 teamwork in, 1.35
Self-shielded flux-cored arc welding, 10.10
Semiconductor- (silicon-) controlled rectifier (SCR), 7.117
Sequence-of-events analysis, 3.139–3.141
 active descriptions in, 3.140
 in logical order, 3.139
 precision of, 3.140
 qualifiers in, 3.141
Series circuits, 7.115
Series rolling-element bearings, 5.23, 5.24
Series-progressive systems, 8.17–8.19
Series-wound motors, 6.9
Service life, of belt drives, 4.73, 4.74
Service valves, 4.66
Severe wear, 7.133–7.134
SFMEA (*see* Simplified Failure Modes and Effects Analysis)
SFMECA (Simplified Failure Mode, Effect, and Criticality Analysis), 3.135
Shaft center line, 5.56

INDEX

Shafts:
  bent, 7.89
  misalignment of, 5.56–5.67
  of rolling-element bearings, 5.25–5.26
  and vibration, 7.89
Sheaves:
  QD-type sheaves, 5.110
  TL-type sheaves, 5.110–5.111
  V-belt drives, 5.117, 5.121, 5.123
  V-belt sheaves, 5.102–5.103
    conventional mount, 5.110
    diameter of, 5.103
    installation procedure for, 5.108–5.112
    and QD bushings, 5.111
    QD-type sheaves, 5.110
    reverse mount, 5.110
    and TL bushings, 5.111
    TL-type sheaves, 5.110–5.111
    vibration, 5.117, 5.122
Sheet metal working, 10.42–10.43, 10.65–10.67
Shielded metal arc welding (SMAW), 10.6–10.9
Shielding, 7.117
Shielding gases, 10.20–10.21
Shop calibration, 7.39–7.40
Shrinkage, of concrete floor surfaces, 4.19–4.20
Shunts, 7.117
Shunt-wound motors, 6.9
Shutdown:
  of centrifugal compressors, 4.68
  of gear drives, 5.177
Signal delays, 7.68
Signals, common-mode, 7.67
Signature analysis, 2.23, 7.94
Silicon- (semiconductor-) controlled rectifier (SCR), 7.117
Silicone coatings, 9.43
Simple cyclone air-cleaning equipment, 4.112, 4.119
Simplification (*see* Work simplification)
Simplified Failure Mode, Effect, and Criticality Analysis (SFMECA), 3.135
Simplified Failure Modes and Effects Analysis (SFMEA), 3.133–3.137
  and life cycle cost analyses, 3.136
  and maintainability analysis, 3.136–3.137
  and mean time between failures (MTBF), 3.136
  and reliability block diagram (RBD), 3.136
  and reliability predictions, 3.136
  and Simplified Failure Mode, Effect, and Criticality Analysis (SFMECA), 3.135
Sine waves, 7.117
Single ladders, 9.94
Single phases, 7.117
Single-function instruments, 7.50–7.66
  clamp-on instruments, 7.51–7.52
  and currents, 7.53–7.56
  frequency counters, 7.63–7.65
  for power measurements, 7.57–7.58
  for resistance measurements, 7.58–7.63
  solid-state circuit testers, 7.65–7.66
  tube checkers, 7.65
Single-phase connections, 7.57
Single-ply roofing membranes, 4.5
Single-wedge nourishing-stem gate valve, 5.204
Sintered plates, 6.91

Six Sigma safety, 1.55–1.67
  action item matrix (AIM) for, 1.58–1.60
  cause-&-effect diagrams for, 1.64–1.67
  for continuous improvement, 1.58
  control levels for, 1.56–1.57
  culture for, 1.55–1.56
  five-sigma control, 1.57
  four-sigma control, 1.57
  one sigma control, 1.56
  outcomes of, 1.58
  Pareto charts for, 1.61–1.63
  process for, 1.58
  process maps for, 1.64
  purpose of, 1.58
  safety taskforces for, 1.60–1.61
  six-sigma control, 1.57
  three-sigma control, 1.56
  tools for, 1.57–1.58, 1.61–1.67
    cause-&-effect diagrams, 1.64–1.67
    Pareto charts, 1.61–1.63
    process maps, 1.64
  two-sigma control, 1.56
600-Brinnel flat-seat globe valve, 5.201
Six-sigma control, 1.57
Skewed histograms, 3.150
Slabs, design for, 4.19
Sleeve bearings, 7.81–7.82
  oil-lubricated sleeve bearings, 6.12–6.13
  rolling-element bearings, 7.81–7.82
Sliding wear, 2.27, 7.132–7.134
  and rolling fatigue wear, 2.27, 7.132, 7.133
  severe, 2.27, 7.133–7.134
Sludge settling tanks, 4.121
SMAW (shielded metal arc welding), 10.6–10.9
Smoke stains, 4.38
Soaps:
  aluminum complex soaps, 8.10
  aluminum soaps, 8.9
  calcium complex soaps, 8.11
  calcium soaps, 8.10
  lithium complex soaps, 8.11
  lithium soaps, 8.10
  sodium soaps, 8.9, 8.10
Sodium soaps, 8.9, 8.10
Soft foot, 5.56, 5.57
Software, for infrared equipment, 7.110
Soldering, 10.64–10.65, 10.67
Solenoids, 7.117
Solids, gravimetric, 8.32
Solids contamination, 8.32
Solids content:
  and lubricating oil analysis, 2.25
  lubricating oil analysis for, 7.128
Solid-state circuit testers, 7.65–7.66
Solid-state overload relays, 6.66–6.68
Solvents:
  alkaline cleaners, 9.20
  caustic soda plus potassium permanganate, 9.20
  caustic soda-surfactant alkalies, 9.20
  for chemical cleaning, 9.19–9.23
  copper complexers, 9.21–9.22
  inorganic acids, 9.20–9.21
    inhibited muriatic acid (HCl), 9.20
    inhibited sulfuric acid, 9.20–9.21
    nitric acid, 9.21
    phosphoric acids, 9.21
    sulfuric acid, 9.21

Solvents (*Cont.*):
  monoammoniated citric acids, 9.20
  organic, 9.21–9.23
    copper complexers, 9.21–9.22
    oxidizing agents, 9.22
  organic acids, 9.20
  organic solvents, 9.21–9.23
  oxidizing agents, 9.22
Spalling, 9.47
Span of control ratios, 1.37–1.38
Spare parts:
  for air conditioning units, 4.63
  CMMS for, 2.94–2.95
  identification of, 5.208, 5.210
  ordering, 5.208, 5.210–5.211
  valves, 5.208, 5.210–5.211
Spare parts inventory, 1.6
Special lifts (*see* Elevators)
Special-design scaffolds, 9.108–9.111
Specialized personnel, 1.13
Special-purpose industrial multimeters, 7.50
Special-purpose ladders, 9.96
Specific job standards, 3.114
Spectral bands, 7.114
Spectrographic analysis, 7.129
Spectrometals, 8.32
Spectrometric metals, 8.32
Speed:
  critical, 7.84–7.85
  critical speeds, 7.84–7.85
  of plant machinery, 7.84–7.85
  of positive-displacement pumps, 5.226
  rotating, 5.226
  running, 7.84
  running speeds, 7.84
  and vibration, 7.84–7.85
Speed reducers, 5.161
Spindle couplings, 5.46, 5.48
Spiral actuating elements, 7.16
Spiral heating, 7.113
Splitter, dampers, 4.73
Spot welding, gas-shielded, 10.34
Springs, 7.16–7.18
Sprockets:
  conventional mount synchronous, 5.130
  QD-type, 5.130
  reverse mount, 5.130
  reverse mount synchronous, 5.130
  synchronous, 5.127–5.129
    conventional mount, 5.130
    diameter of, 5.127
    guidelines for, 5.129
    reverse mount, 5.130
    troubleshooting, 5.137, 5.141
    wear, 5.132
  TL-type, 5.130
Stability:
  of digital instruments, 7.49
  of process, 1.5
Stainless clad steels, 10.38
Stainless steels:
  for corrosion control, 9.9
  and welding, 10.36, 10.38
Stains:
  on brick masonry structures, 4.36–4.40
  from bronze, 4.38
  brown (manganese) stain, 4.36–4.37
  from copper, 4.38

Stains (*Cont.*)
    from dirt, 4.38
    from egg splatter, 4.39
    green stain (vanadium salts), 4.36
    from iron, 4.37–4.38
    from ivy, 4.39
    from oil, 4.38
    from paint, 4.37
    from paper, 4.39
    from plant growth, 4.39
    from poultice, 4.37
    from smoke, 4.38
    from straw, 4.39
    from tar, 4.38
    of unknown origin, 4.39–4.40
    from welding splatter, 4.38
    from white scum, 4.39
Standard-practice sheets, 1.51–1.53
    for cost control, 1.52–1.53
    and operating policies, 1.51–1.53
Standards (*see* Regulations)
Standby-power batteries, 6.100–6.101
Start-up:
    for centrifugal pumps, 5.222
    for gear drives, 5.174
    and ventilating fans, 4.106
Static-adjustment type drives, 5.151–5.153
Statistical analyses tools, 3.141–3.152
    control charts, 3.145–3.147
    80/20 rule, 3.142–3.144
    flow charts, 3.144–3.145
    histograms, 3.147–3.150
    Pareto analysis, 3.141–3.142
    scatter plots, 3.150–3.152
Statistical plans, 3.52–3.56
Stators, 7.117
Steam, 4.33
Steel scaffolding, 9.103, 9.105, 9.106
Steel valves, 5.205–5.207
Steels:
    alloy, 10.36–10.39
    carbon, 10.35–10.36
    for corrosion control, 9.9
    galvanized, 9.9
    and hard facing, 10.84–10.85
    high-carbon, 10.36
    high-manganese, 10.39
    high-tensile low-alloy, 10.36
    low-carbon, 10.35
    medium-carbon, 10.36
    stainless, 9.9, 10.36, 10.38
    stainless clad, 10.38
    straight chromium, 10.38
    and welding, 10.36–10.39
Stem traps, 9.76–9.77
Stepladders, 9.92–9.96
Stocking requirements, 2.105
Storage:
    bulk, 8.7
    of equipment, 4.15
    on low-sloped roofs, 4.15
    of lubricants, 8.7, 8.33
    of materials, 4.15
    on roofs, 4.15
    of ventilating fans, 4.105
Storeroom management, 2.65–2.75
Storerooms:
    bar coding for, 2.71–2.73
    environment of, 2.68–2.70

Storerooms (*Cont.*)
    functions of, 2.66
    housekeeping for, 2.67–2.68
    KPIs for, 2.68
    locator systems for, 2.70–2.71
    operation of, 2.74–2.75
    reporting in, 2.66–2.67
    storage equipment for, 2.73–2.74
    supervisor for, 2.65–2.66
Stores inventory, 2.94–2.95
Stores stock inventory, 2.106
Storeskeeping, as maintenance, 1.11
Straight chromium steels, 10.38
Straight-blade centrifugal fans, 4.96, 4.97
Straw stains, 4.39
Stress-cracking corrosion, 9.7
Stud welding, 10.35
Subgrade, for concrete floor surfaces, 4.19
Submerged arc welding (SAW), 10.10–10.11,
        10.40–10.42
    and electrodes, 10.41–10.42
    equipment for, 10.40–10.41
    and fluxes, 10.41–10.42
    hard surfacing with, 10.49–10.51
Substitution, of MRO parts, 1.25
Suction:
    in centrifugal pumps, 5.215–5.218
    and entrained air, 5.217–5.218
    net positive suction head, 5.215–5.216
    of positive-displacement pumps, 5.226
    and volume, 5.216–5.217
Suction conditions, 5.215–5.218
Suction lines, 5.186, 5.187
Sugar, 8.32
Sulfated ash, 8.31
Sulfur, 8.32
Sulfuric acid, 9.21
Sun loads, 4.53
Superflat floors, 4.23
Superimposition, 4.16
Supervision, 1.14
    cross-craft, 1.14
    front line, 1.32
    front line supervision, 1.32
    of maintenance organization, 1.37–1.38
    planning, 1.32–1.33
    reliability engineering, 1.32–1.33
    responsibilities of, 1.32
    scheduling, 1.32–1.33
Supervisory personnel, 1.14
    cross-craft supervision, 1.14
    density of, 1.14
    selection of, 1.15–1.16
    training of, 1.16
Supply, of lubrication, 5.3
Support:
    for maintenance organization, 1.37–1.38
    for piping, 9.74–9.78
Surface cleaning, 4.14–4.15
Surface deterioration, 5.177–5.183
    burning, 5.182
    corrosive wear, 5.182
    in gear-tooths, 5.177–5.183
    grinding cracks, 5.182, 5.183
    interference wear, 5.182, 5.183
    plastic flow, 5.179–5.180
    scoring, 5.180
    surface fatigue, 5.180–5.181
    wear, 5.178–5.179

Surface fatigue, of gear-tooth,
        5.180–5.181
Surface-tolerant coatings, 9.41–9.42
Surfacing:
    electrodes, 10.44
    low-sloped roofs, 4.6
Surveys:
    for painting, 9.36–9.37
    plant lubrication surveys, 8.29
Sweat-type fittings, 10.68–10.69
Sweeps, 7.67, 7.68
Swing-check valve, 5.201
Swinging scaffolds, 9.104
Switches:
    defined, 7.117
    disconnect, 4.106
    and infrared technology, 7.112
    load break switches, 7.112
Switchgear:
    high voltage, 7.112
    and infrared technology, 7.111, 7.112
    secondary, 7.111
Synchronous belt drives:
    and QD bushings, 5.130–5.131
    and TL bushings, 5.131
Synchronous belts, 5.123–5.127,
        5.129–5.142
    application guidelines for, 5.129
    curvilinear belts, 5.124
    guidelines for, 5.129
    and hot bearings, 5.137, 5.142
    installation of, 5.129–5.131
    maintenance of, 5.131–5.135
        during installation, 5.133, 5.134
        misalignment, 5.132
        predictive maintenance, 5.133
        and takeup allowance, 5.133, 5.134
        visual inspection, 5.132
        wear, 5.132
    misalignment of, 5.132
    modified curvilinear belts, 5.124
    and noise, 5.137, 5.141
    nomenclature for, 5.126–5.127
    performance of, 5.137, 5.142
    predictive maintenance for, 5.133
    premature failure of, 5.136–5.138
    and takeup, 5.137, 5.140
    and takeup allowance, 5.133, 5.134
    tensioning, 5.135–5.137, 5.140
        force-deflection method for, 5.135
        frequency method for, 5.135–5.136
    timing belts, 5.124
    tooth pitch, 5.124
    tooth profiles, 5.124–5.125
    tracking behavior, 5.137, 5.139–5.140
    troubleshooting, 5.136–5.142
        hot bearings, 5.137, 5.142
        noise, 5.137, 5.141
        performance, 5.137, 5.142
        premature failure, 5.136–5.138
        sprockets, 5.137, 5.141
        and takeup, 5.137, 5.140
        tensioning, 5.137, 5.140
        tracking behavior, 5.137, 5.139–5.140
        vibration, 5.137, 5.140
        wear, 5.136, 5.138–5.139
    vibration, 5.137, 5.140
    visual inspection for, 5.132
    wear, 5.132, 5.136, 5.138–5.139

Synchronous sprockets, 5.127–5.129
  conventional mount, 5.130
  diameter of, 5.127
  guidelines for, 5.129
  reverse mount, 5.130
  troubleshooting, 5.137, 5.141
  wear, 5.132
Synthetic elastomers, 9.10
Systems:
  analysis of, 2.10
  functions of, 2.10
  identification of, 2.10
  selection of, 2.11
  (See also specific types, e.g.: Automatic lubrication system)

Tabular (matrix form) models, 3.15
Takeup:
  and synchronous belts, 5.137, 5.140
  and V-belt drives, 5.117, 5.122
Takeup allowance:
  and synchronous belts, 5.133, 5.134
  and V-belt drives, 5.113
Tangential noise measurement, 7.68
Tanks, liquid-level measurement in, 7.10–7.11
Taper method, 6.84
Tapered-bore bearings, 5.35–5.38
Tar stains, 4.38
Tasks, 2.17
  categories of, 2.38
  determination of, 2.14
  failure-finding tasks, 2.38–2.39
  frequencies of, 2.17–2.18
  of maintenance programs, 1.19–1.20
  proactive tasks, 2.37–2.38
  selection of, 2.12–2.14
TDH (total dynamic head) requirement, 5.218
Teams (see Self-directed work teams)
Teamwork, 1.35
Temperature:
  of AC induction motors, 6.8–6.9
  ambient, 7.113
  Celsius degree, 7.113
  delta temperature (ΔT), 7.113
  of digital instruments, 7.50
  and electrical motors, 6.13–6.14
  Fahrenheit degree, 7.113
  for piping, 9.57
  of plain bearings, 5.4
  and ventilating fans, 4.98–4.99
Temperature gradients, 7.114
Temperature mountings, 5.33, 5.35
Temperature transmitters, pneumatic, 7.28–7.29
Temporary repair, 4.9–4.10
Tensioning, 9.68
  in belt drives, 4.74
  force-deflection method for, 5.135
  frequency method for, 5.135–5.136
  synchronous belts, 5.135–5.137, 5.140
    force-deflection method for, 5.135
    frequency method for, 5.135–5.136
  in V-belt drives, 5.115–5.116
Terminal connections, 6.74
Test discharge, 6.86
Tested parts, 3.12
Testing:
  acceptance, 7.69
  color test, 8.31

Testing (Cont.)
  electric motor analysis, 2.30
  with foam tests, 8.31
  for fuels, 8.31–8.32
  for greases, 8.8, 8.9
  for lubricants, 8.31–8.32
  for oil, 8.34
  100% test only, 3.12
  100% test-stress-retest, 3.12
Theory of choice, 3.3
Thermal conductivity, 5.4
Thermal emittance, 7.114
Thermal growth, 5.58
Thermal overload relays, 6.65–6.66, 6.75–6.76
Thermal-system bulb designs, 7.22–7.23
  averaging bulbs, 7.23
  coiled bulbs, 7.23
  extension necks, 7.23
  union-connected bulbs, 7.23
Thermography, 2.23–2.24, 2.32, 7.105
  defined, 7.114
  with infrared imaging, 2.24
  infrared thermometers, 2.24
  line scanners for, 2.24
  and materials, 7.118–7.125
  qualitative infrared, 7.114
  quantitative infrared, 7.114
  (See also Infrared technology)
Thermometers, 7.35–7.36
  calibration of, 7.36
  dry-bulb, 7.24, 7.25
  filled-system, 7.19–7.24
    classifications of, 7.19–7.23
    fully compensated liquid-filled, 7.20
    gas-filled thermal systems, 7.22
    liquid-filled, 7.19
    mercury-filled thermal systems, 7.22–7.23
    and thermal-system bulb designs, 7.22–7.23
    vapor-pressure thermal systems, 7.20–7.22
  fully compensated liquid-filled, 7.20
  and gas-filled thermal systems, 7.22
  infrared, 7.107
  infrared thermometers, 2.24
  installation of, 7.35–7.36
  liquid-filled, 7.19
  maintenance of, 7.36
  and mercury-filled thermal systems, 7.22–7.23
  and thermal-system bulb designs, 7.22–7.23
  and vapor-pressure thermal systems, 7.20–7.22
  wet-bulb, 7.24, 7.25
Thermoplastic membranes, 4.11–4.12
Thickeners, grease, 8.9–8.11
Threaded-bonnet globe valve, 5.200
Threading, 9.57
Three phase, 7.117
Three-phase wattmeters, 7.58
Three-sigma control, 1.56
Tightness, of coils, 4.68
Time domain, 7.76
Time Study:
  advantages/disadvantages of, 3.39–3.41
  and job standards, 3.32–3.34
  and standard data, 3.34–3.39
  work measurement with, 3.32–3.41
Time-tactors, 6.68–6.70
Timing belts, 5.124

TL bushings:
  an V-belt drives, 5.111
  and synchronous belt drives, 5.131
  and V-belt sheaves, 5.111
TL-type sheaves, 5.110–5.111
TL-type sprockets, 5.130
Toluene, 8.32
Tooth pitch, 5.124
Tooth profiles, 5.124–5.125
Torch motion, 10.76
Total acid number, 2.26, 7.129
Total backlogs, 2.79
Total base number, 2.26
Total dynamic head (TDH) requirement, 5.218
Total flow measurement, 7.7–7.8
Total plant management, 2.20
Total productive maintenance (TPM), 2.41–2.58
  autonomous maintenance, 2.41–2.42, 2.54
  and autonomous maintenance, 2.47
  barriers to, 2.48–2.49, 2.52
  benefits of, 2.55–2.58
  components of, 2.54–2.55
  conflict in, 2.49
  data availability for, 2.49
  education for, 2.54
  and equipment, 2.44–2.45
  focus of, 2.50
  implementation of, 2.47–2.48
  for improvement, 2.51–2.52
  and inventory, 2.57–2.58
  with just-in-time training, 2.50
  master plan for, 2.47
  metrics for, 2.46–2.47
  and morale, 2.58
  objectives of, 2.53
  and operator ownership, 2.50
  and performance, 2.45
  planned maintenance, 2.42–2.44
  and planned maintenance, 2.48
  preventative maintenance, 2.54–2.55
  priority for, 2.49
  and process losses, 2.53–2.54
  and productivity, 2.57
  quality rate for, 2.45–2.46
  for reduction of maintenance, 2.44, 2.48
  and reliability engineering, 2.55
  and safety considerations, 2.58
  and start-up management, 2.55
  strategic direction for, 2.49
  support for, 2.50
  training for, 2.54
  and union buy-in, 2.51
Total system head (TSH) requirement, 5.218
Total work load, 1.40
Towers, cooling, 4.71–4.73
Traces, dual, 7.67
Tracking behavior, 5.137, 5.139–5.140
Traction-type drives, 5.157
Traffic, on low-sloped roofs, 4.15
Trained maintenance planners, 2.5
Training:
  for air-conditioning equipment, 4.63
  of craft personnel, 1.15
  cross-training, 1.35
  for estimators, 3.119–3.120
  formal instruction, 1.15
  informal instruction, 1.15
  for lubrication programs, 8.27–8.28

I.32 INDEX

Training (*Cont.*)
 for maintenance, 1.15, 1.16, 4.63
 on-the-job training, 1.15
 of personnel, 1.15, 1.16
 for self-directed work teams, 1.35
 of supervisory personnel, 1.16
 for work force, 1.46–1.47
Transducers, velocity, 7.97
Transformers:
 bushings of, 7.112
 current, 7.57
 defined, 7.117
 and infrared technology, 7.111–7.112
 oil-filled, 7.111–7.112
 potential, 7.58
Transmission loads, 4.53–4.54
Transmissions, belt, 5.153–5.155
Transmittance, 7.114
Transmitters:
 diaphragm-actuated flowmeter, 7.26
 flow, 7.34
 pneumatic, 7.27–7.28
 pneumatic pressure, 7.28–7.29
 pneumatic temperature, 7.28–7.29
Traps:
 and piping, 9.72–9.74
 stem traps, 9.76–9.77
Trending:
 broadband, 2.22–2.23
 narrowband, 2.23
 tribology, 7.130
 for vibration, 7.102
 for wear particle analysis, 7.130
Tribology, 2.24–2.28, 2.32, 7.127–7.136
 accuracy of samples for, 7.135
 capital cost of, 7.134
 cost of, 7.134
 cost-benefit analysis for, 7.135
 ferrography, 2.27–2.28, 7.129
 limitations of, 7.134–7.136
 lubricating oil analysis, 2.25–2.26, 7.128–7.129
  for contamination, 7.128
  cost-benefit analysis for, 7.135
  for fuel dilution, 7.128
  for fuel soot, 7.128
  for nitration, 7.129
  for oxidation, 7.128
  for particle count, 7.129
  for solids content, 7.128
  for total acid number, 7.129
  for total base number, 7.129
  for viscosity, 7.128
 monitoring, 7.130
 and particulate matter, 7.130–7.134
 recurring cost of, 7.134
 results of, 7.135
 samples for, 7.135
 spectrographic analysis, 7.129
 trending, 7.130
 wear particle analysis, 2.26–2.28, 7.129–7.134
  cost-benefit analysis for, 7.135
  monitoring and trending, 7.130
  for particulate matter, 7.130–7.134
Trickle charges, 6.94–6.95
Triggering, automatic, 7.66
Triggers, 7.68
Trimetals, 5.5
Truncated histograms, 3.150

TSH (total system head) requirement, 5.218
Tube checkers, 7.65
Tube scaffolds, 9.105, 9.106
Tuckpointing, 4.28–4.29
Twin-line systems, 8.19–8.20
Two-rate method, 6.84
Two-sigma control, 1.56

Ultraviolet bands, 7.114
Unblanking, 7.68
Underwriters' Laboratories, Inc., 9.93–9.94
Uniform cash flow method, 3.14
Uniform corrosion, 9.5, 9.6
Union-bonnet regrinding valve, 5.200
Union-connected bulbs, 7.23
Unit ohms (μ), 7.113
Unloading, of air compressors, 5.193
"Upside-down" roof system (protected roof membrane assembly), 4.5
Urethanes, 9.44
Utilities, 1.10

V (*see* Volts)
Vacuum contactors, 6.51
Valves, 5.197–5.211
 of air compressors, 5.189–5.191
 air-compressor check valve, 5.201–5.202
 ball, 5.207–5.208
 ball check valve, 5.201–5.202
 bolt tension for, 9.68
 bronze, 5.199–5.204
  check valves, 5.201–5.203
  gate valves, 5.203–5.204
  globe valves, 5.200–5.201
 bronze check valves, 5.201–5.203
 bronze gate valves, 5.203–5.204
 bronze globe valves, 5.200–5.201
 butterfly, 5.208
 check valves, 9.68
  air-compressor check valve, 5.201–5.202
  ball check valve, 5.201–5.202
  bronze, 5.201–5.203
  nonmetallic-disk lift-check valve, 5.201
  regrinding lift-check valve, 5.201
  Renewo lift-check valve, 5.201
  swing-check valve, 5.201
 description of, 5.197–5.199
 double-wedge rising-stem gate valve, 5.204
 flow direction of, 9.63, 9.66
 gate valves, 5.203–5.204
  bronze, 5.203–5.204
  double-wedge rising-stem, 5.204
  renewable-wedge-and-seat bronze, 5.204
  screw-and-yoke rising-stem, 5.204
  single-wedge nourishing-stem, 5.204
  solid-wedge rising-stem, 5.204
 globe valves
  bronze, 5.200–5.201
  nonmetallic-disk, 5.201
  plug-type Renewo, 5.201
  600-Brinnel flat-seat, 5.201
  threaded-bonnet, 5.200
  union-bonnet regrinding valve, 5.200
 identification of, 5.210–5.211
 iron, 5.204–5.205
  check valves, 5.204
  gate valves, 5.204–5.205
  globe valves, 5.204

Valves (*Cont.*)
 machining, 9.69
 maintenance of, 9.66
 nonmetallic-disk globe valve, 5.201
 nonmetallic-disk lift-check valve, 5.201
 packing for, 9.66, 9.68
 and piping, 5.211, 9.63, 9.66, 9.68–9.70
 plug-type Renewo globe valve, 5.201
 refacing, 9.70
 regrinding lift-check valve, 5.201
 renewable-wedge-and-seat bronze gate valve, 5.204
 Renewo lift-check valve, 5.201
 reseating kits, 9.70
 screw-and-yoke rising-stem gate valve, 5.204
 seat rings, 9.69–9.70
 service, 4.66
 single-wedge nourishing-stem gate valve, 5.204
 600-Brinnel flat-seat globe valve, 5.201
 solid-wedge rising-stem gate valve, 5.204
 spare parts, 5.208, 5.210–5.211
 steel, 5.205–5.207
 swing-check valve, 5.201
 threaded-bonnet globe valve, 5.200
 union-bonnet regrinding valve, 5.200
Vanadium salts (green stain), 4.36
Vane(s):
 dynamics of, 7.83
 variable-inlet, 4.103, 4.106
 and vibration, 7.83
Vane pumps, 5.225–5.226
Vapor-pressure thermal systems, 7.20–7.22
Variable-inlet vanes, 4.103, 4.106
Variable-speed controls, 4.104
V-belt drives, 5.99–5.143
 application guidelines for, 5.103–5.108
 and banded belts, 5.120–5.121
 datum length of, 5.101
 dimensions of, 5.99–5.103
 effective length of, 5.101, 5.102
 force-deflection method for, 5.115–5.116
 frequency method for, 5.115–5.116
 and hot bearings, 5.118, 5.123
 installation of, 5.113
 length of, 5.101–5.102
 maintenance of, 5.112–5.115
  for installation, 5.113
  misalignment, 5.112, 5.113
  predictive, 5.115
  pullet wear, 5.112
  takeup allowance, 5.113
  visual inspection, 5.112
 misalignment, 5.112, 5.113
 and noise, 5.117, 5.122
 and performance, 5.118, 5.123
 predictive maintenance of, 5.115
 and premature failure, 5.117, 5.118
 and pullet wear, 5.112
 QD bushings, 5.111
 and sheaves, 5.117, 5.121, 5.123
 and takeup, 5.117, 5.122
 and takeup allowance, 5.113
 tension of, 5.115–5.116
 tensioning, 5.115–5.116
  force-deflection method for, 5.115–5.116
  frequency method for, 5.115–5.116
 and TL bushings, 5.111
 troubleshooting, 5.117–5.123

INDEX **I.33**

V-belt drives, troubleshooting (*Cont.*)
  banded (joint) belts, 5.120–5.121
  banded (joint) problems, 5.119–5.120
  hot bearings, 5.118, 5.123
  noise, 5.117, 5.122
  performance, 5.118, 5.123
  premature failure, 5.117, 5.118
  sheaves, 5.117, 5.121, 5.123
  and takeup, 5.117, 5.122
  vibration, 5.117, 5.122
  wear, 5.117, 5.119–5.120
 and V-belt sheaves, 5.102–5.103, 5.108–5.112
 and ventilating fans, 4.106, 4.108
 and vibration, 5.117, 5.122
 visual inspection of, 5.112
 and wear, 5.117, 5.119–5.120
V-belt sheaves, 5.102–5.103
  conventional mount, 5.110
  diameter of, 5.103
  installation procedure for, 5.108–5.112
  and QD bushings, 5.111
  QD-type sheaves, 5.110
  reverse mount, 5.110
  and TL bushings, 5.111
  TL-type sheaves, 5.110–5.111
  vibration of, 5.117, 5.122
Velocity:
  liquid, 9.86
  and piping, 9.86
  of vibration, 7.77
Velocity transducers, 7.97
Vendors:
  procurement function of, 1.25
  selection of, 1.25
  specifications for, 3.170
Ventilating fans, 4.87–4.110
 and abrasion, 4.99–4.100
 axial fans, 4.95
 backward-curved-blade centrifugal fans, 4.96
 balancing of, 4.108–4.109
 bearings, 4.107
 and bearings, 4.107
 bearings, overheated, 4.110
 and belt guards, 4.106
 capacity of, 4.109
 and corrosion, 4.99
 and coupling guards, 4.106
 couplings, 4.107–4.108
 and couplings, 4.107–4.108
 and dampers, 4.106
 and disconnect switches, 4.106
 drivers, 4.110
 fan laws, 4.91, 4.94
 and fan wheel, 4.105–4.106
 and fan wheels, 4.105–4.106
 and flammable gases, 4.100
 and flow, 4.87
 forward-curved-blade centrifugal fans, 4.96
 handling of, 4.105
 in-line-flow centrifugal fans, 4.97
 inspection of, 4.104, 4.107–4.109
  balancing, 4.108–4.109
  bearings, 4.107
  couplings, 4.107–4.108
  for installation, 4.104
  periodic, 4.107–4.109
  problems with, 4.109–4.110

Ventilating fans, inspection of (*Cont.*)
  recordkeeping for, 4.107
  V-belt drives, 4.108
 installation of, 4.104–4.106
  considerations with, 4.104–4.105
  dampers, 4.106
  fan wheel, 4.105–4.106
  foundations for, 4.105
  mountings for, 4.105
  prestart check for, 4.105
  and startup, 4.106
  variable-inlet vanes, 4.106
  V-belt drive, 4.106
 jet pumps, 4.98
 mountings for, 4.105
 multirating tables for, 4.88, 4.91
 noise of, 4.101–4.102
  elimination of, 4.101–4.102
  as problem, 4.109–4.110
  volume-control devices for, 4.102–4.104
 packaged ventilating equipment, 4.98
 performance data for, 4.88–4.91
  on density changes, 4.91
  multirating tables for, 4.88, 4.91
  standards for, 4.88
 periodic inspection of, 4.107–4.109
 power roof exhausters, 4.98
 pressure of, 4.109
 prestart check for, 4.105
 propeller fans, 4.95
 protective devices, 4.107
 and protective screens, 4.106
 radial-blade centrifugal fans, 4.96, 4.97
 recordkeeping for, 4.107
 safety devices, 4.106
  belt guards, 4.106
  coupling guards, 4.106
  disconnect switches, 4.106
  protective screens, 4.106
 selection of, 4.98–4.100
 standards for, 4.88
 and startup, 4.106
 storage of, 4.105
 straight-blade centrifugal fans, 4.96, 4.97
 and temperatures, 4.98–4.99
 terminology, 4.87–4.88
 types of, 4.94–4.98
 uses of, 4.94–4.98
 and variable-inlet vanes, 4.106
 and V-belt drive, 4.106
 and V-belt drives, 4.108
 vibration, 4.109–4.110
 volume-control devices for, 4.102–4.104
  outlet damper, 4.102
  variable-inlet vanes, 4.103
  variable-speed control, 4.104
Vertical alignment:
 for face/rim alignment, 5.63, 5.66
 for flexible couplings, 5.67–5.70
 and shaft misalignment, 5.59–5.61, 5.63, 5.66
Vertical expansion joints, 4.31
Vessels, 9.25
Vibration, 7.69–7.104
 acceleration of, 7.75, 7.77
 and accelerometers, 7.97
 and acceptance testing, 7.69
 and aircraft engines, 7.71
 alert (alarm) limits for, 7.94–7.95

Vibration (*Cont.*)
 amplitude of, 7.77
 analysis of, 7.69–7.71
  for acceptance testing, 7.69
  for aircraft engines, 7.71
  for engineering, 7.71
  for leak detection, 7.70
  for loose parts detection, 7.70
  for machine design, 7.71
  for noise control, 7.70
  parameters for, 7.94
  for predictive maintenance, 7.69
  for production optimization, 7.71
  for quality control, 7.70
  techniques for, 7.104
 of bearings, 7.80–7.82
 of belt drives, 7.83–7.84
 and bent shafts, 7.89
 of blades, 7.83
 of blowers, 7.91
 broadband data for, 7.104
 causes of, 7.74
 characteristics of, 7.74–7.77
  acceleration, 7.75, 7.77
  amplitude, 7.77
  displacement, 7.75, 7.77
  frequency, 7.76
  frequency domain, 7.76
  order (harmonic), 7.76–7.77
  phase, 7.75–7.76
  time domain, 7.76
  velocity, 7.77
 of compressors, 7.91–7.92
 of continuous-process systems, 7.79, 7.93
 and critical speeds, 7.84–7.85
 with data-acquisition frequency, 7.93–7.94
 and databases, 7.93–7.95
  alert (alarm) limits for, 7.94–7.95
  analysis parameters for, 7.94
  with data-acquisition frequency, 7.93–7.94
  signature analysis boundaries for, 7.94
 defining, 7.72–7.73
 displacement of, 7.75, 7.77
 and displacement (eddy-current) probes, 7.96–7.97
 and electric motors, 7.90
 and electrical motors, 6.13–6.14
 and engineering, 7.71
 and failure, 7.87–7.93
  from bent shafts, 7.89
  of blowers, 7.91
  of compressors, 7.91–7.92
  of continuous-process lines, 7.93
  of electric motors, 7.90
  of fans, 7.91
  of gearboxes, 7.90–7.91
  of generators, 7.92
  from imbalance, 7.87–7.88
  from looseness, 7.89
  from mechanical rub, 7.89
  from misalignment, 7.88–7.89
  of pumps, 7.92–7.93
 of fans, 7.91
 frequency domain of, 7.76
 frequency of, 7.76
 and gear mesh, 7.82–7.83
 of gearboxes, 7.90–7.91
 of gears, 7.82–7.83

Vibration (*Cont.*)
  of generators, 7.92
  from imbalance, 7.87–7.88
  and induced loads, 7.86–7.87
  and installed systems, 7.100–7.101
  and leak detection, 7.70
  and linear-motion machinery, 7.78
  and loose parts detection, 7.70
  from looseness, 7.89
  and machine design, 7.71
  machine vibration, 7.74
  measurement of, 7.96–7.101
    with accelerometers, 7.97
    with displacement (eddy-current) probes, 7.96–7.97
    with installed systems, 7.100–7.101
    mounting techniques for, 7.98
    with velocity transducers, 7.97
    with vibration analyzers, 7.98–7.100
    with vibration meters, 7.98
  from mechanical rub, 7.89
  from misalignment, 7.88–7.89
  and mode shape, 7.85–7.86
  mounting techniques for, 7.98
  and narrowband data, 7.104
  and noise control, 7.70
  order (harmonic) of, 7.76–7.77
  phase of, 7.75–7.76
  of plant machinery, 7.77–7.87
    bearings, 7.80–7.82
    belt drives, 7.83–7.84
    blades, 7.83
    continuous-process systems, 7.79
    critical speeds, 7.84–7.85
    dynamics of, 7.79–7.87
    gears, 7.82–7.83
    induced loads, 7.86–7.87
    linear-motion machinery, 7.78
    mode shape, 7.85–7.86
    preloads, 7.86–7.87
    process variables, 7.87
    reciprocating machinery, 7.78
    resonance, 7.86
    rotating machinery, 7.77–7.78
    running speeds, 7.84
    vanes, 7.83
  and predictive maintenance, 7.69
  predictive maintenance program for, 7.101–7.104
    analysis techniques for, 7.104
    with databases, 7.93–7.95, 7.101–7.102
    program maintenance, 7.102
    trending techniques for, 7.102
  and preloads, 7.86–7.87
  and process variables, 7.87
  and production optimization, 7.71
  program maintenance for, 7.102
  of pumps, 7.92–7.93
  and quality control, 7.70
  and reciprocating machinery, 7.78
  and resonance, 7.86
  of rolling-element bearings, 7.80–7.81
  and root-cause failure analysis, 7.104
  and rotating machinery, 7.77–7.78
  and running speeds, 7.84
  signature analysis boundaries for, 7.94
  of synchronous belts, 5.137, 5.140
  time domain of, 7.76

Vibration (*Cont.*)
  trending techniques for, 7.102
  and vanes, 7.83
  of V-belt drives, 5.117, 5.122
  of V-belt sheaves, 5.117, 5.122
  velocity of, 7.77
  and velocity transducers, 7.97
  of ventilating fans, 4.109–4.110
  and vibration analyzers, 7.98–7.100
  and vibration meters, 7.98
Vibration analyzers, 7.98–7.100
Vibration meters, 7.98
Vibration monitoring, 2.22–2.23, 2.31
  broadband trending, 2.22–2.23
  narrowband trending, 2.23
  signature analysis, 2.23
Viscosity:
  chain drives, 5.79, 5.80
  and lubricant analysis, 8.32
  and lubricating oil analysis, 2.25, 7.128
  of oil, 5.79, 5.80
  of positive-displacement pumps, 5.226
Viscosity index, 8.32
Visible bands, 7.114
Visual inspection, 2.30
  and root cause analysis, 3.159–3.161
  for synchronous belts, 5.132
  of V-belt drives, 5.112
Volts (V):
  AC volts, 7.53–7.55
  accelerating, 7.66
  accelerating voltages, 7.66
  constant
    charging with, 6.93–6.94
  constant voltage, 6.93–6.94
  DC volts, 7.53
  defined, 7.117
Volume, and suction, 5.216–5.217
Volume-control devices:
  outlet damper, 4.102
  variable-inlet vanes, 4.103
  for variable-speed control, 4.104
  for ventilating fans, 4.102–4.104
Vortex dampers, 4.76
Vulcanized elastomers, 4.11

W (watts), 7.117
Wage rate, 3.70
Walls:
  cracked, 4.30–4.31
  leaky, 4.27–4.30
Warranties, for elevators, 4.51–4.52
Waste disposal, 1.11
Water:
  and lubricant analysis, 8.32
  treatment of, 4.71
Water coils, 4.84
Water conditioning:
  and air-conditioning equipment, 4.84–4.85
  corrective measures for, 4.85
  definitions of, 4.85
  problems with, 4.84–4.85
Water distribution, 4.72
Water filters, 4.77–4.78
Water hammers, 9.56
Water heaters, 4.79–4.80
Water jetting, high-pressure, 9.25, 9.26

Water level, 4.72
Water lines, 4.84
Water repellents, 4.30
Waterborne coatings, 9.42
Water-cooled condensers, 4.70–4.71
  cleaning, 4.70–4.71
  freeze protection for, 4.71
  freezing of, 4.71
  maintenance of, 4.70–4.71
  and refrigerated joints, 4.71
  and water treatment, 4.71
Watering, of nickel-iron batteries, 6.101
Wattmeters, 7.52
  clamp-on, 7.52
  three-phase, 7.58
Watts, 7.53
Watts (W), 7.117
Wave lengths, 7.110
Wear:
  abrasive, 10.43–10.44
  of contacts, 6.49
  corrosive, 5.182
  cutting, 2.27, 7.131, 7.132
  fatigue spallwear, 7.131, 7.132
  in gear drives, 5.177–5.184
  of gear-tooth, 5.177–5.184
  impact, 10.44
  of plain bearings, 5.9, 5.11
  pullet, 5.112
  rolling, 2.27
  rolling fatigue, 2.27, 7.131–7.133
  rubbing, 2.27, 7.131
  severe, 2.27, 7.133–7.134
  sliding, 2.27, 7.132–7.134
    and rolling fatigue wear, 7.132, 7.133
    severe, 7.133–7.134
  of synchronous belts, 5.132, 5.136, 5.138–5.139
  of synchronous sprockets, 5.132
  and V-belt drives, 5.112, 5.117, 5.119–5.120
  and welding, 10.44
Wear particle analysis, 2.26–2.28
  cutting wear, 2.27, 7.131, 7.132
  monitoring and trending, 7.130
  for particulate matter, 7.130–7.134
    cutting wear, 7.131, 7.132
    fatigue spallwear, 7.131, 7.132
    rolling fatigue wear, 7.131–7.133
    rubbing wear, 7.131
    sliding wear, 7.132–7.134
  rolling fatigue, 2.27
  rolling wear, 2.27
  rubbing wear, 2.27
  sliding wear, 2.27
Welded aluminum scaffolds, 9.101–9.102
Welded steel scaffolds, 9.103
Welding, 10.3–10.89
  and abrasive wear, 10.43–10.44
  accessory equipment for, 10.52–10.55
  air-acetylene appliances, 10.63–10.71
    applications of, 10.70–10.71
    automobile body soldering, 10.67
    electrical connections, 10.67–10.68
    paint burning, 10.69–10.70
    precautions with, 10.63–10.64
    safety considerations with, 10.63–10.64
    sheet metal working, 10.65–10.67

INDEX **I.35**

Welding, air-acetylene appliances (*Cont.*)
   soldering, 10.64–10.65
   sweat-type fittings, 10.68–10.69
air-carbon arc cutting (CAC-A), 10.27–10.30
and alloy steels, 10.36–10.39
   high-manganese steels, 10.39
   high-tensile low-alloy steels, 10.36
   stainless clad steels, 10.38
   stainless steels, 10.36, 10.38
   straight chromium steels, 10.38
and aluminum, 10.39
arc welding, 10.3–10.60
   flux-cored arc welding (FCAW), 10.10
   gas-shielded metal arc welding (GMAW), 10.11–10.13
   role of, 10.3–10.6
   shielded metal arc welding (SMAW), 10.6–10.9
   submerged arc welding (SAW), 10.10–10.11
atomic hydrogen welding, 10.34
and automobile body soldering, 10.67
and base metals, 10.35–10.39
   alloy steels, 10.36–10.39
   carbon steels, 10.35–10.36
   nonferrous metals, 10.39
braze welding, 10.74–10.75
and carbon steels, 10.35–10.36
   cast iron, 10.36
   high-carbon steels, 10.36
   low-carbon steels, 10.35
   medium-carbon steels, 10.36
check welding procedure, 10.48–10.49
and copper, 10.39
and copper alloys, 10.39
and distortion, 10.39–10.40
and electrical connections, 10.67–10.68
electrode material for, 10.21–10.22
electrogas welding, 10.34
equipment for
   accessory equipment for, 10.52–10.55
   installation of, 10.55
   machines for, 10.51–10.52
   maintenance of, 10.57–10.59
   operation of, 10.56–10.57
flame adjustment
   in braze welding, 10.74–10.75
   in fusion welding, 10.75
   fusion welding cast iron, 10.78
   and gouging, 10.82–10.83
   heavy braze welding, 10.76–10.77
   oxygen cutting, 10.78–10.82
   torch motion, 10.76
   weld preparation, 10.75–10.76
flame adjustment for, 10.74–10.83
flux-cored arc welding (FCAW), 10.10
   with gas, 10.10
   self-shielded, 10.10
fusion welding, 10.75
fusion welding cast iron, 10.78
gas tungsten arc welding (GTAW), 10.13–10.27
   advantages of, 10.17–10.18
   applicability of, 10.15–10.17
   disadvantages of, 10.17–10.18
   electrode material for, 10.21–10.22
   flow rates, 10.20–10.21
   manual techniques, 10.24
   mechanical welding parameters for, 10.25

Welding, gas tungsten arc welding (GTAW) (*Cont.*)
   methods for, 10.25–10.27
   and polarity, 10.18–10.20
   power supplies, 10.22–10.23
   principles of, 10.18
   process for, 10.13–10.15
   shielding gases, 10.20–10.21
   torches, 10.23–10.24
gas welding, 10.63–10.89
   air-acetylene appliances, 10.63–10.71
   flame adjustment for, 10.74–10.83
   and hard facing, 10.83–10.89
   oxyacetylene appliances, 10.71–10.74
gas-shielded metal arc welding (GMAW), 10.11–10.13
gas-shielded spot welding, 10.34
and gouging, 10.82–10.83
and hard facing, 10.83–10.89
   for cast iron, 10.85
   finishing for, 10.85–10.89
   material for, 10.45–10.48
   rods, 10.84
   for steel, 10.84–10.85
and hard surfacing, 10.43
heavy braze welding, 10.76–10.77
and high-carbon steels, 10.36
and high-manganese steels, 10.39
and high-tensile low-alloy steels, 10.36
and impact wear, 10.44
and low-carbon steels, 10.35
machines for, 10.51–10.52
mechanical welding parameters for, 10.25
and medium-carbon steels, 10.36
and nonferrous metals, 10.39
oxyacetylene appliances, 10.71–10.74
   and brazing, 10.71–10.72
   and cutting, 10.72
   precautions for, 10.71
   safety considerations with, 10.71
   setup for, 10.73
oxyacetylene cutting, 10.27
oxygen cutting, 10.78–10.82
   equipment for, 10.81
   preparation for, 10.81
   process for, 10.81–10.82
and paint burning, 10.69–10.70
for piping, 9.83
plasma arc cutting, 10.30–10.31
plasma arc welding, 10.34–10.35
and polarity, 10.18–10.20
and power supplies, 10.22–10.23
procedures for, 10.32–10.33
qualifications for, 10.33
resistance welding, 10.35
safety considerations with, 10.59–10.60
sheet metal welding, 10.42–10.43
and sheet metal working, 10.65–10.67
shielded metal arc welding (SMAW), 10.6–10.9
and shielding gases, 10.20–10.21
soldering, 10.64–10.65, 10.67
spot welding, 10.34
and stainless clad steels, 10.38
and stainless steels, 10.36, 10.38
for steel, 10.84–10.85
and straight chromium steels, 10.38
stud welding, 10.35

Welding (*Cont.*)
   submerged arc welding (SAW), 10.10–10.11, 10.40–10.42
   and electrodes, 10.41–10.42
   equipment for, 10.40–10.41
   and fluxes, 10.41–10.42
   hard surfacing with, 10.49–10.51
   and surfacing electrodes, 10.44
   and sweat-type fittings, 10.68–10.69
   torch motion, 10.76
Welding splatter, 4.38
Welding torches, 10.23–10.24
Well pumps, 4.84
Wet centrifugal scrubbers, 4.119–4.120
Wet collectors (scrubbers), 4.112, 4.113
   maintenance of, 4.119–4.120
   orifice type, 4.120
   packed scrubbers, 4.119
   particulate, 4.114, 4.115
   wet centrifugal, 4.119–4.120
   wet dynamic, 4.120
Wet dynamic scrubbers, 4.120
Wet-bulb thermometers, 7.24, 7.25
Wheatstone bridge, 7.61–7.62
Wheels, fan, 4.76
White scum, 4.39
Windage, 5.79
Winter protection, for cooling towers, 4.73
Wondering, in scientific method, 3.92
Work allocation:
   for breakdown maintenance, 1.41–1.42
   operating policies for, 1.39–1.43
   for preventative maintenance, 1.41–1.42
   and reliability engineering, 1.42–1.43
   scheduling, 1.39–1.41
Work controls, 2.89
Work culture, 2.113
Work force (*see* Personnel)
Work load, 1.40
Work measurement, 3.19–3.63
   application of, 3.131
   and assignment maintenance mechanics, 3.58–3.59
   comparative time estimating, 3.130–3.131
   for efficiency, 3.60–3.62
   and efficiency estimation, 3.60–3.62
   and engineers, 3.19–3.21
   and equipment, 3.26–3.27
   equipment records, 3.31
   for estimations
      of efficiency, 3.60–3.62
      standards, 3.62
   familiarity for, 3.130
   and KPIs, 3.129–3.132
   levels of, 3.132
   methods for, 3.21–3.26
   methods-time measurement (MTM) data for, 3.41–3.52
   with past-performance method, 3.52–3.56
   and personnel, 3.27–3.31
   for planning, 3.31–3.32
   programs for, 3.63
   random sampling with, 3.56–3.58, 3.60
   with ratio delay, 3.56–3.62
      accuracy of, 3.60
      for allowance determination, 3.59, 3.60
      and assignment maintenance mechanics, 3.58–3.59

Work measurement, with ratio delay (*Cont.*)
    and efficiency estimation, 3.60–3.62
    random sampling with, 3.56–3.58, 3.60
    studies on, 3.60
    and work simplification, 3.59
  for scheduling, 3.31–3.32
  and standards, 3.62
  with statistical plans, 3.52–3.56
  with Time Study, 3.32–3.41
  with work sampling, 3.56
Work order system, 2.84–2.85
Work orders:
  analysis of, 2.88
  and CMMS, 2.95–2.97, 2.105
  computerized, 2.79
  computerized systems for, 2.82–2.83
  history of, 2.97
  initiation, 2.99–2.100
  planning, 2.100
  scheduling, 2.101
Work plans, 2.79
Work requests, 2.79

Work sampling, 3.56
Work simplification, 3.89–3.105
  applications of, 3.104–3.105
  charting techniques for, 3.95
  and decision-making, 3.105
  and efficiency, 3.104–3.105
  and flow diagrams, 3.98
  with flow process chart, 3.95–3.98
    creation of, 3.95, 3.98
    and flow diagrams, 3.98
    points in, 3.98
  with law of intelligent action, 3.90–3.92
    ability, 3.91
    capacity, 3.91–3.92
    desire, 3.90–3.91
  and manpower, 3.105
  with motion economy, 3.102–3.104
    checklist for, 3.103–3.104
    principles of, 3.102–3.103
  with multiple-activity process chart, 3.99–3.102
    creation of, 3.99
    usage of, 3.99–3.100

Work simplification (*Cont.*)
  and ratio delay, 3.59
  with scientific method, 3.92–3.95
    and application, 3.93
    generalizing in, 3.92–3.93
    for methods improvement, 3.93–3.95
    questioning in, 3.92
    wondering in, 3.92
  and technical decisions, 3.105
Work teams (*see* Self-directed work teams)
Work unit, scheduling by, 1.40
Workers (*see* Personnel)
Wound-rotor motors, 6.61
Wrinkling, 9.47

X-ray, 7.114

Z (*see* Impedance)
Zinc, for corrosion control, 9.9
Zinc coatings, 9.43
Zone-control systems, 8.21, 8.22